FINITE ELEMENT ANALYSIS

FROM CONCEPTS TO APPLICATIONS

DAVID S. BURNETT

AT&T Bell Laboratories   *Whippany, New Jersey*

ADDISON-WESLEY PUBLISHING COMPANY

Reading, Massachusetts • Menlo Park, California
Don Mills, Ontario • Wokingham, England • Amsterdam
Sydney • Singapore • Tokyo • Madrid • Bogotà
Santiago • San Juan

The programs and applications presented in this book have been included for their instructional value. They have been tested with care but are not guaranteed for any particular purpose. The publisher does not offer any warranties or representations, nor does it accept any liabilities with respect to the programs or applications.

Library of Congress Cataloging in Publication Data

Burnett, David S.
    Finite element analysis.

    Bibliography: p.
    Includes index.
    1. Finite element method.   I. Title.
TA347.F5B87   1987   620'.001'515353   85-20107
ISBN 0-201-10806-2

ABCDEFGHIJ–HA–8987

To my children,

Diana and **James**

Reason,

when used for discovery

rather than defense,

is an intellectually honest enterprise

and hence one of life's greatest pleasures.

May your lives be filled with the joy of discovery!

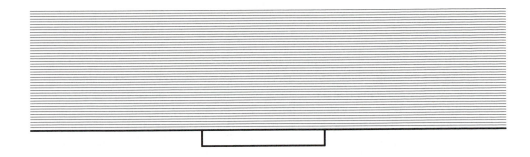

Preface

For many people, their first impression of this book is its rather formidable size. This is not the accidental result of unconstrained verbosity, but the natural concomitant of a deliberate pedagogical approach embodying a variety of special features, as explained below. Foremost amongst them are: a careful, graduated introduction of basic concepts in the early chapters; very thorough and detailed explanations; and supplementary commentary that continually interrelates material in different parts of the book. As a result, you may find, as many have already, that the text can be read, and understood, at a faster than normal rate, thereby requiring less time to complete the book than might be expected. Indeed, it is my intention and hope that your total time invested in learning the subject (including studies and experience beyond this book) will be significantly shortened. In other words, I have invested thousands of hours so that you won't have to!

Why I Wrote This Book

This book and the accompanying computer program were developed at AT&T Bell Laboratories to provide introductory level instruction and training in the finite element method (FEM) for members of the technical staff. The project has spanned ten years, including several teachings at five of the company locations and culminating in a final intensive development phase.

The primary objective was to provide a solid *foundation* in the basic concepts and methods of FE analysis. I have therefore tried to present a thorough, detailed treatment of the basics, rather than a cursory, encyclopedic overview of the entire field. The FEM seems to encompass an endless variety of topics and modeling techniques, with new methods and capabilities constantly evolving and finding their way into the popular commercial programs. This requires analysts to continually update and expand their knowledge. Hopefully the type of foundation provided by this book will greatly *facilitate this ongoing educational process*.

Philosophy

The approach used in this book was motivated by two primary considerations: audience and pedagogy. Regarding audience, the purpose was to make the method understandable and usable to people in all fields of engineering and the physical sciences, not just in mechanical engineering, where it originated. Regarding pedagogy, the purpose was to make the technical derivations and explanations as lucid and as simple as possible, without being simplistic. Let's take a closer look at each of these two guidelines.

The audience consideration was inspired by the observation that even though Bell Laboratories has an extremely diversified technical environment, usage of the FEM during the mid-1970's (when this project began) was limited almost exclusively to mechanical engineers. This was unfortunate since the FEM is a general and powerful *numerical analysis* technique, applicable to a broad variety of mathematical problems that arise in almost all areas of science and engineering. Of course, the reason for this imbalance was due primarily to the historical "accident" that the practical implementation of the FEM began and matured in the field of solid mechanics, e.g., elasticity, plasticity, and dynamics. Because of these origins, most of the FE literature is permeated with mechanical concepts such as forces, moments, displacements, rotations, masses, dampers, springs, rods, beams, plates, and shells. Understandably, many physicists, applied mathematicians, and nonmechanical engineers have concluded that this is not a field for them.

To help solve this problem, the present book intentionally avoids reference to mechanical concepts throughout all the general theoretical derivations, allowing only for conventional terminology (such as "stiffness" and "mass"). Solid mechanics is treated as only one of many different applications of the FEM. Consequently, mechanical concepts are used only where they are needed — in the mechanical illustrative problems as well as the entirety of Chapters 12 and 16, which are devoted to dynamics and elasticity, respectively.

Of course, the only technical language understandable to all fields is mathematics, so that is the approach used here. This, however, presented a challenge: how to avoid a lot of the sophisticated mathematical concepts and specialized mathematical jargon prevalent in much of the mathematically-oriented FE literature, while preserving only those concepts necessary for intelligent application of the FEM. This challenge was met by the second of the above considerations: pedagogy.

My basic thinking/learning/teaching principle is best expressed by the oft-quoted statement attributed to A. Einstein, "Everything should be made as simple as possible, *but not simpler*" (author's emphasis). This principle has always seemed to be good epistemology, good pedagogy, and good psychology. An additional and closely related teaching principle which I have developed is the following: "When a new concept is first introduced, use the simplest possible context and elaborate all details." Applying these principles to the above "mathematics challenge" has resulted in the following features in this book.

Special Features

- Approximately the first two-thirds of the text deals with one-dimensional (1-D) problems. Although most practical finite element problems are 2-D and 3-D, most of the fundamental and commonly used concepts can be explained and illustrated in a 1-D setting. This includes not only the elementary concepts but also more advanced topics such as accuracy, mesh refinement, isoparametric elements, optimal flux-sampling ("stress"-sampling) points, infinite domains, eigenproblems, initial-value problems, and others. The outstanding advantage of working in a 1-D rather than 2-D setting is that ideas are not obscured by long and tedious mathematical manipulations.

The last third of the book deals with 2-D problems. If the reader has learned the 1-D material well, then the analogous but more complicated derivations in 2-D will be better appreciated and more easily grasped. Indeed, many engineers who studied the manuscript for this book found the transition from 1-D to 2-D to be anticlimactical.

Because the parallels between 2-D and 3-D are even greater than between 1-D and 2-D, 3-D problems are not treated in the text. (See further discussion in Chapter 2, Section 2.1.)

- The conversational writing style reflects the author's desire to *teach* the subject rather than to merely describe it. Unusually thorough explanations provide reasons and purpose for every mathematical step. Explanations and derivations are sometimes repeated, though in different contexts, since repetition is a key to reinforcing learning. In the same spirit, the mathematical derivations are very thorough in detail. The reader is always told in advance the purpose or goal of the ensuing material, including sometimes a brief outline of the path that will be followed. This occurs at the "macro" level (Chapter 2 for the entire book), the "mini" level (Introduction sections for each chapter), and the "micro" level (short discussions preceding each development).

- In Chapters 3 and 4 — Foundations I and Foundations II — the basic structure of the FEM is first developed *without using the element concept*, since the latter fragments the problem, obscuring other concepts. In many respects, this is equivalent to dealing with *one*-element models.

- A logically ordered 12-step procedure, introduced in Chapter 4, is used repeatedly throughout the remainder of the book for each new class of problems. Steps 1 through 6 compose the theoretical development that yields the element equations. Steps 7 through 12 compose the numerical computation, forming the flowchart for a computer program.

- Chapter 5 — The Element Concept — employs a "one, two, many" principle to progress from one element to two elements to many elements. The rule for assembly is objectively derived from a continuity condition (using two elements), rather than by heuristically appealing to physical analogies.

- The principal developmental chapters in Part III (Chapters 7, 8, 10, and 11) are each structured into three phases: theory, programming, and application. The latter two phases provide reinforcement of the theoretical ideas as well as immediate practical experience.

- Part IV (2-D problems) follows a pattern similar to that of Part III (1-D problems), although the programming is de-emphasized. Classroom experience has shown that interest in programming is generally high at the beginning of the course (to see how the theoretical equations are implemented into code), but it tapers off thereafter. This is perhaps because it is seldom cost-effective for engineers to do applications programming in traditional FE areas since outside organizations specializing in such endeavors can more efficiently develop, maintain, and support large software systems.

- A computer program is developed as the book progresses, growing from chapter to chapter. The reader therefore participates in the development and use of a complete 8-element "general purpose" FE program (UNAFEM), similar in structure to much larger commercial programs. Parts of the code are reproduced in the text as the development proceeds, and a complete listing is in the appendix. The program is quite readable; e.g., one-fourth of the code is comment statements. It has been found to be an excellent research and educational tool.

- There are 8 illustrative problems which are solved in their entirety. Solutions include the physical and mathematical description, mesh and accuracy considerations, data input, evaluation of output, mesh refinement techniques, and error estimation. The problems are accompanied by comments, suggestions, advice, etc., on the practical aspects of problem solving. Topics include: steady-state heat conduction, static deflection and vibration of a stretched cable, wave functions and energy levels of a quantum mechanical oscillator, transient heat conduction (two problems, involving shock and postshock responses), acoustic modes in a vehicular/pedestrian tunnel, and stresses in an I-beam subjected to torsion.

- There are 171 homework exercises. Considerable effort was expended to create exercises that illustrate an important point, as my publisher can attest. In fact, many of the exercises extend the material in the text by means of additional observations and explanations. Almost all of the exercises are pencil and paper type, requiring at most a hand calculator. This was done to make the exercises independent of UNAFEM so that the instructor would not have to spend time trying to get UNAFEM and its graphics running on his or her particular computer system. For those willing to make the effort, though, UNAFEM would provide an easy source of program-type exercises that would complement the written exercises and be compatible with the material in the text.

Topical Summary

The principal developmental chapters, 3 through 11 and 13 through 15, deal with a general class of boundary-value (equilibrium) problems, eigenproblems (resonance), and first-order initial-boundary-value (diffusion) problems. They work with the 2-D quasiharmonic equation and its 1-D analogue, as well as their associated eigenvalue and initial-boundary-value equations. These general equations include the classic Laplace, Poisson, Sturm-Liouville, Helmholtz and time-independent Schroedinger equations. Hence they encompass a broad spectrum of important physical applications, e.g., heat conduction (steady state and transient), electromagnetism, acoustics, hydrodynamics, quantum mechanics, deflection of cables and membranes, torsion of elastic cylinders, and many others. In addition, Chapter 12 treats second-order initial-boundary-value problems (dynamics), and Chapter 16 deals with elasticity.

Essentially the entire book is devoted to linear analysis. However, a brief introduction to nonlinear analysis is provided by a carefully structured set of homework exercises at the end of Chapter 15.

The reader is referred to Chapter 2 for further elaboration. Chapter 2 serves as a *technical* preface for the book in that it provides a technical outline of Chapters 3 through 16, including mathematical details, so that the reader can begin with an overall perspective and a "road map."

Prerequisites

The mathematical background required for almost the entire book (Chapters 1 through 15) is intermediate-level calculus, i.e., about 3 to 5 semesters of calculus. For example, differentiation and integration (including integration by parts), convergence of sequences, and elementary notions of vectors and matrices are adequate to understand almost all of the 1-D material. The 2-D material requires the 2-D extensions of some of these concepts, e.g., divergence theorem, gradient, and line and surface integrals. In addition to the mathematics, it is recommended that one have had the equivalent of 2 or 3 years of college-level education in science or engineering.

The last chapter (Chapter 16) deals with the theory of elasticity which requires a bit more from the reader than the preceding chapters. This subject (as well as any other field theory with multiple unknown functions — e.g., electromagnetism or fluid dynamics) definitely benefits from a liberal use of matrix operations, including an occasional reference to basic tensor-transformation rules. This chapter therefore provides an introduction to the more advanced areas of the FEM and appropriately belongs at the end of this beginning-level book.

Judging from my own experience and that of my colleagues, topics such as functional analysis and variational calculus are definitely not needed to understand and use the FEM. Nor is a prior course in ordinary or partial differential equations needed (although some very elementary notions might be helpful), since the FEM is itself a technique for numerically solving differential equations.

In summary, the book would be suitable as early as the junior or senior year in engineering, mathematics, or physics.

Recommended Syllabus for a One-Semester Course

The book contains sufficient material for one and one-half to two semesters. For a one-semester course it is recommended that the instructor concentrate on only boundary-value problems, i.e., Chapters 1 through 9, 13, and 16. These chapters include all the 1-D and 2-D material as well as application to elasticity.

This is an excellent introductory sequence since boundary-value problems are fundamental to the FEM; i.e., they are the type of problem most often analyzed by the FEM and most of the techniques used for boundary-value problems are also used for eigenproblems and initial-boundary-value problems. Since the sequence ends with elasticity, instructors in mechanical and civil engineering will have an opportunity to introduce their own FE stress-analysis programs. Instructors in other fields of engineering, physics, or mathematics could substitute a different topic for Chapter 16 or else continue with a study of eigenproblems or initial-boundary-value problems in the other chapters.

Acknowledgments

I am most grateful to AT&T Bell Laboratories for two major sources of support: the In-Hours Continuing Education Program for providing me the opportunity to develop and teach a course on FE analysis — and the Book Review Board for authorizing the use of company resources to transform the course notes into a publishable book. Without this support, this book would not have been written.

My deep appreciation is extended to my former department head Tim Delchamps for his support, enthusiasm, and encouragement throughout the entire project. His unwavering confidence in the value of the final product gave me energy when the road ahead seemed long. I am also grateful to my past and present managers Dan Bevins, Les Kleinberg, John Mariano, and Frank Labianca for their consideration and efforts in accommodating this project, along with my other responsibilities, to the needs of the organization.

I am indebted to several individuals who gave generously of their time: Terrence Lenahan for his good counsel and clear thinking, from abstract concepts to homework exercises; Richard Holford for his rigorous and clear explanations of some of the finer mathematical points; Paul Franklin for his assistance in generating data, particularly the error plots in Chapter 9; and Bill Brodsky for helpful ideas during the early planning stages. For reviewing parts of the manuscript my appreciation goes to Harry Scholz, Franz Geyling, Phyllis Fox, and the many reviewers for the Book Review Board. Too numerous to mention here are the names of my In-Hours students from over the years; their criticism and praise provided valuable guidance that altered or confirmed my pedagogical techniques.

A very warm and special thank-you goes to Lorraine Shallop for her extraordinary effort in typing the entire manuscript as well as numerous earlier versions dating back to the original course notes in 1976. Her meticulous and conscientious craftsmanship and cheerful attitude certainly lightened my load over the years. Excellent support was also provided by the AT&T Bell Laboratories technical library staff; in particular, I am grateful to Sue Liebman for being so helpful throughout the entire project.

It has truly been a pleasure working with everyone at Addison-Wesley. I want to especially thank Tom Robbins and Mary Coffey for their patience, when at times it must have been difficult, and their gentle but firm guidance. My thanks also go to Susanah Michener and Joseph Vetere.

A project of this magnitude has a profound impact on one's personal life. To my children, Diana and James: Your love and affection has been comforting. I can only hope that our playful camaraderie and the intellectual excitement we've shared have partially compensated for the fewer times we had together. To Alice: Thank you for encouraging me to undertake this project and, once begun, for patiently accommodating the needs of the family to my long hours. Your support was deeply appreciated; your contribution was substantial. To my friend Ted Peters: I extend my deep appreciation for your truly understanding the human side of this project. Our innumerable dinner discussions on the pedagogical and philosophical issues and your letting me share with you both the joys and frustrations of this creative endeavor were emotionally energizing. And to all my friends who saw less of me and kept asking when it would be finished: It's finished!

David S. Burnett
Whippany, New Jersey

Special Acknowledgment

One person in particular deserves special recognition for an exceptional contribution. W. John Denkmann wrote the entire UNAFEM program. It would have required a herculean effort on my part to write both the text and the program. By taking over the entire programming effort, John considerably eased my burden and enabled the timely publication of this book.

John is a member of the technical staff at AT&T Consumer Product Laboratories in Indianapolis, Indiana, responsible for providing analytical and FE support for the design of telephone components and systems. He was with AT&T Bell Laboratories from 1963 to 1984. He has had over 20 years experience as an FE analyst and teacher. I was one of John's students in 1971 in the first FE course to be offered at Bell Laboratories.

Because of John's FE expertise, his contribution was also felt in other ways. Since the program was developed in parallel with the writing of the text, there was, of necessity, an ongoing collaboration that ranged from coding details to theoretical debates. Surely this collaboration enhanced the quality of the text itself and it certainly provided a much needed source of technical companionship in what otherwise would have been a very lonely project. Thanks, John, for your friendship, your counsel, and your excellent technical support.

D. S. B.

Contents

Part I Introduction

Part II Basic Concepts

Part III   One-Dimensional Problems:
Theory, Programming and Applications

Part IV Two-Dimensional Problems:
Theory, Programming, and Applications

P A R T

I

Introduction

1

What Is the
Finite Element Method?

Introduction

This chapter presents a cursory overview of the subject. It serves three purposes:

1. For the FE[1] novice it is an advertisement that identifies the wide spectrum of applications and the benefits to be realized, so that he or she may judge the relevancy to his or her own field of application.

2. For all readers it offers a broad picture of the terrain the FEM covers, thereby providing a context for particular developments in subsequent chapters.

3. As an introduction to the FEM, it is meant to provide a "feel" for the subject, to give an overall general impression of what it's all about. Mathematical details are reserved for subsequent chapters.

1.1 Definition and Description of the FEM

Let's begin with an introductory definition of the FEM that identifies the broad context of the subject:

> The FEM is a computer-aided mathematical technique for obtaining approximate numerical solutions to the abstract equations of calculus that predict the response of physical systems subjected to external influences.

Such problems arise in many areas of engineering, science, and applied mathematics. Applications to date have occurred principally in the areas of solid mechanics (e.g., elasticity, plasticity, statics, and dynamics), heat transfer (conduction, convection, and radiation),

---

[1] Throughout the book "finite element(s)" and "finite element method" will be abbreviated as FE and FEM, respectively.

3

fluid mechanics (inviscid or viscous), acoustics, and electromagnetism, as well as in coupled interaction of these phenomena (e.g., fluid-solid interaction and electrothermoelasticity). New areas of application are continually being discovered, recent ones being solid state physics and quantum mechanics.

To be more specific, and using somewhat more mathematical language, the FEM can handle problems possessing any or all of the following characteristics:

- Any mathematical or physical problem described by the equations of calculus, e.g., differential, integral, integrodifferential, or variational equations. This would include all the examples cited above.
- Boundary-value problems (also called equilibrium or steady-state problems); eigen-problems (resonance and stability phenomena); and initial-value problems[2] (diffusion, vibration, and wave propagation).
- The domain of the problem (e.g., the region of space occupied by the system — see Section 1.3) may be any geometric shape, in any number of dimensions. Complicated geometries are as straightforward to handle as simple geometries, the only difference being that the former may require a bit more time and expense. For example, a quite simple geometry would be the shape of a circular cylindrical waveguide for acoustic or electromagnetic waves (fiber optics). A more complicated geometry would be the shape of an automobile chassis, perhaps being analyzed for the dynamic stresses in-duced by a rough road surface.
- Physical properties (e.g., density, stiffness, permeability, conductivity) may vary throughout the system.
- The external influences, generally referred to as *loads* or *loading conditions*, may be in any physically meaningful form. The loads are typically applied to the bound-ary of the system (boundary conditions), to the interior of the system (interior loads), or at the beginning of time (initial conditions).
- Problems may be linear or nonlinear.

Problems typically solved by the FEM usually possess several of these characteristics. There now exist dozens of so-called general-purpose FE programs, available commer-cially and publicly, whose capabilities include all of the above list. Most are limited to one or two fields of application, such as solid mechanics and heat conduction, or fluid mechanics or electromagnetism. Within these fields they can solve wide classes of prob-lems of extraordinary complexity, in most cases the only limitation being the size of the user's budget. There also exist hundreds, perhaps thousands, of smaller special-purpose programs whose capabilities are more limited; many are available commercially, publicly, or privately.

---

[2] Problems that are purely initial value (time is the only independent variable) are usually handled with the finite *difference* method (FDM). With problems that are both initial value and boundary value (independent variables include both space and time, which is quite common), the FEM is used for the spatial part and the FDM is used for the time part (see Chapters 11, 12, and 15). However, FD algorithms can be derived using FE principles (see Section 11.3.3), thereby justifying the rather common practice of including initial-value problems within the scope of the FEM. It is the author's opinion, though, that most FE practitioners still think in terms of finite differences, not finite elements, when dealing with the time dimension; and, in the same spirit, many people prefer to identify the FEM only with boundary-value problems (and eigenproblems, which are closely related).

Most people's contact with the FEM will be as users of one of these programs, rather than as developers of new ones. As a user only, one does not become directly involved with the underlying mathematics. However, experience has shown that it is difficult to be an effective user of an FE program without understanding some of the basic concepts and mathematical techniques employed by the method.

1.2 How the FEM Works

Before outlining the central steps in the FE problem-solving procedure, we need to first describe some concepts and related terminology that are germane to every problem that the FEM will analyze. The following brief digression is intended to be a review for the reader, not an introduction.

The problem begins with the engineer or analyst who wants to describe or predict the response of a system that is subjected to external influences (forces, voltages, temperatures, etc.) that change the state of the system. He or she is basically looking for a numerical solution to the governing equations and loading conditions that characterize and determine the behavior of that system. The problem thus becomes a mathematical one.

Let's define four concepts: system, domain, governing equations, and loading conditions. The definitions are descriptive rather than mathematical. Figure 1.1 shows these concepts symbolically and illustrates them with two simple examples.

The *system* is typically, but not necessarily, a physical object composed of various materials: solids, liquids, gases, plasmas, or combinations thereof. For example, a structural system would involve solid materials. A simple example would be a mechanical linkage; a complicated example would be an entire airplane. An acoustic system would involve fluids, such as a body of water or air in which pressure (acoustic) waves are propagating. An electromagnetic system might involve materials and/or "empty" space.

The *domain* of the problem is typically the region of space occupied by the system. It may also be the interval of time during which the changes in the system take place. In mathematical parlance, it is the region or interval of the independent variables, which are usually (but not necessarily) space and time. However, in many FE applications the domain will refer to only some of the independent variables (usually space, but not time); the intended meaning should be clear from context.

The *governing equations* may be differential equations expressing a conservation or balance of some physical property such as mass, momentum, or energy. They may also be integral equations expressing a variational principle, such as the minimization of potential energy for conservative mechanical systems. They also include constitutive equations, which describe particular types of material behavior; these equations contain experimentally determined physical properties (physical constants) of the materials that constitute the system.

Loading conditions are externally originating forces, temperatures, currents, fields, etc., that interact with the system, causing the state of the system to change. Loads acting in the interior of the domain, which we will call *interior loads* (some people prefer *volume loads*), appear as part of the governing equations. Loads acting on the boundary of the domain, which we will call *boundary loads*, appear in separate equations called *boundary conditions*.

As an additional example of the use of this terminology, consider the problem of predicting the steady-state temperature distribution in an automobile engine block (Fig. 1.2).

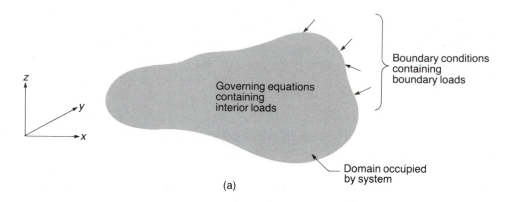

(a)

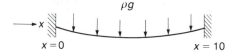

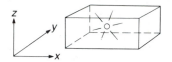

System: A flexible cable with tension T, hanging under its own weight ρg

System: A sound source at frequency ω and intensity f radiating inside a closed room with (ideally) rigid ceiling, walls, and floor

Domain:	Interval $0 < x < 10$ along x-axis
Boundary of domain:	The two end points $x = 0$ and $x = 10$
Governing equation:	$\dfrac{d}{dx}\left(T\dfrac{dW}{dx}\right) = \rho g$ where W is the vertical displacement
Interior load:	ρg
Boundary conditions:	$W = 0$ at $x = 0$ $W = 0$ at $x = 10$

Domain:	Volume occupied by room
Boundary of domain:	The ceiling, walls, and floor
Governing equation:	$\dfrac{\partial^2 p}{\partial x^2} + \dfrac{\partial^2 p}{\partial y^2} + \dfrac{\partial^2 p}{\partial z^2} - \dfrac{\omega^2}{c^2}p = f$ where p is the acoustic pressure
Interior load:	f
Boundary conditions:	$\dfrac{\partial p}{\partial n} = 0$ on ceiling, walls, and floor (n is normal to surfaces)

(b)   (c)

FIGURE 1.1 ● Mathematical description of a physical problem: (a) symbolic diagram of a general problem; (b) example of a one-dimensional mechanical system; (c) example of a three-dimensional acoustic system.

The system is the block itself and the domain is the region of space it occupies. There are two governing equations:

Conservation of energy ●

$$\nabla \cdot \mathbf{q} = Q \qquad \left[\text{in one dimension, } \frac{dq}{dx} = Q\right] \tag{1.1}$$

Fourier's law of heat conduction ●

$$q = -k\nabla T \qquad \left[\text{in one dimension, } q = -k\frac{dT}{dx}\right] \qquad (1.2)$$

Here q is a heat flux vector, ∇ is the gradient symbol $[(\partial/\partial x)\boldsymbol{i} + (\partial/\partial y)\boldsymbol{j} + (\partial/\partial z)\boldsymbol{k}$ in rectangular coordinates], Q is the rate of internal heat generation, T is temperature, and k is the experimentally measured thermal conductivity (see Chapters 6 and 13). Conservation of energy expresses a balance between energy lost by flowing across the boundaries, $\nabla \cdot q$, and energy generated internally, Q. Fourier's law states that a temperature gradient ∇T will cause heat to flow, q, the rate of flow being proportional to k. These two equations are usually combined into a single partial differential equation known as the *quasiharmonic equation* (Chapter 13):

$$-\nabla \cdot (k\nabla T) = Q \qquad \left[\text{in one dimension, } -\frac{d}{dx}\left(k\frac{dT}{dx}\right) = Q\right] \qquad (1.3)$$

The loading conditions consist only of boundary loads: the temperature T_0 imposed on the cylinder walls and convection from the exterior surfaces. The latter requires another experimentally measured physical property — h, the convective heat transfer coefficient.

The reader is referred to other textbooks for a detailed development of the governing equations and loading conditions pertinent to each field of application. From here on it will be assumed that the reader is familiar with the equations that characterize his or her problem. The problems treated in this book will be preceded by a discussion of the relevant governing equations and loading conditions.

With this understanding, we can now provide a thumbnail sketch of how the FEM works. The following operations, accompanied by an illustrative problem in Section 1.3,

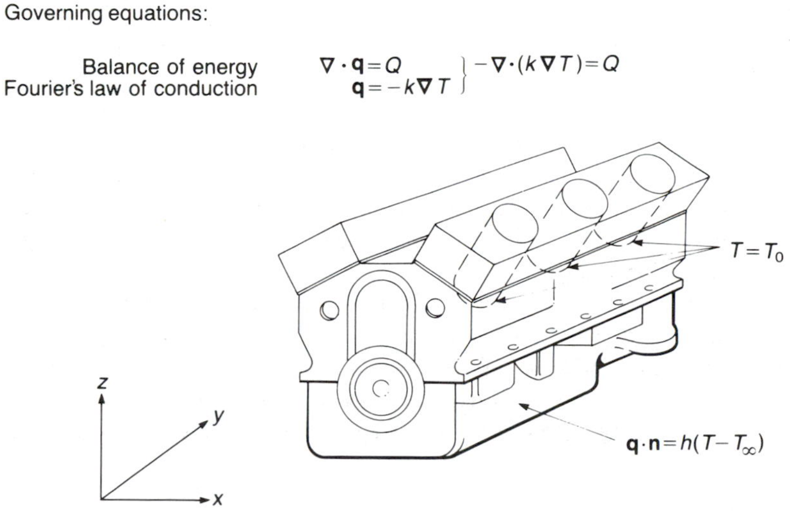

Governing equations:

Balance of energy $\nabla \cdot q = Q$ $\left.\right\}$ $-\nabla \cdot (k\nabla T) = Q$
Fourier's law of conduction $q = -k\nabla T$

$T = T_0$

z

y

$q \cdot n = h(T - T_\infty)$

x

FIGURE 1.2 ● Mathematical description of heat conduction in an automobile engine block.

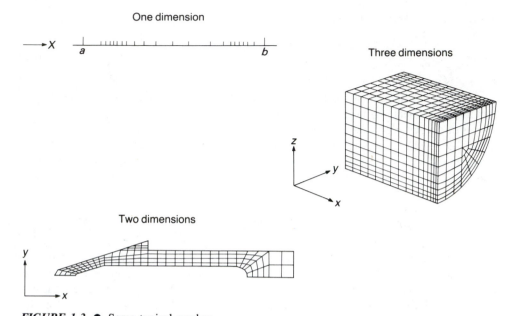

FIGURE 1.3 ● Some typical meshes.

identify the salient features present in every FE analysis.

● The domain of the problem is divided (partitioned) into smaller regions (subdomains) called *elements* (Fig. 1.3). Adjacent elements touch without overlapping, and there are no gaps between the elements. (It might be helpful to visualize a sharp knife slicing up the domain into smaller pieces.) The shapes of the elements are intentionally made as simple as possible, such as triangles and quadrilaterals in two-dimensional domains, and tetrahedra, pentahedra ("wedges" or "pyramids"), and hexahedra ("bricks") in three dimensions. The entire mosaic-like pattern of elements is called a *mesh*.[3]

Figure 1.3 illustrates some typical meshes in domains of one, two, and three dimensions. The one-dimensional domain is the interval from x_a to x_b along the x-axis, say; it is shown partitioned into 21 elements, which are merely shorter intervals between x_a and x_b. The two-dimensional mesh contains mostly quadrilateral elements and a few triangular elements. The three-dimensional mesh contains tetrahedra and wedges (and possibly pyramids and bricks).

Mesh generation, the process of partitioning a domain into a mesh of elements, was performed manually during the early years of the FEM, usually on large pieces of grid paper; it was a laborious, time-consuming, error-prone procedure. However, present-day computer programs called *preprocessors* (see Chapter 7) have automated the procedure to a high degree so that it is now a much simpler and less time-consuming task.

---

[3] See Chapter 5 for a more complete definition of a mesh.

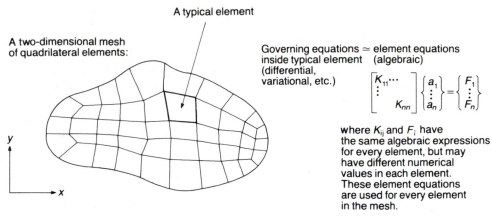

A typical element

A two-dimensional mesh
of quadrilateral elements:

Governing equations $\simeq$ element equations
inside typical element   (algebraic)
(differential,
variational, etc.)

$$\begin{bmatrix} K_{11} \cdots \\ \vdots \\ \phantom{K_{11}} K_{nn} \end{bmatrix} \begin{Bmatrix} a_1 \\ \vdots \\ a_n \end{Bmatrix} = \begin{Bmatrix} F_1 \\ \vdots \\ F_n \end{Bmatrix}$$

where K_{ij} and F_i have
the same algebraic expressions
for every element, but may
have different numerical
values in each element.
These element equations
are used for every element
in the mesh.

FIGURE 1.4 ● Element equations usually need to be derived for only one or two elements in the mesh.

- In each element the governing equations, usually in differential or variational (integral) form, are transformed into algebraic equations, called the *element equations*, which are an approximation of the governing equations.[4] Algebraic equations are much easier to work with than calculus equations for both people and computers; they are relatively easy to solve, either by hand or by computer.

  The derivation of the element equations is a theoretical procedure performed by the analyst (or program developer); it will be explained in careful detail in this text. Two features are especially noteworthy. First, the element equations are algebraically identical for all elements of the same type: those possessing similar geometry (all triangles, for instance) and other characteristics to be described later. Consequently, element equations usually need to be derived for only one or two typical elements, not every element in the mesh. Second, since the element is of simple geometric shape (generally much simpler than the entire domain), the derivation is usually straightforward; and since the element is small — i.e., covers only a small part of the entire domain — a good approximation may be obtained with only a few algebraic equations, typically 2 to 20, occasionally more (Fig. 1.4). In brief, the analytical effort for the entire problem has been reduced to deriving a few algebraic equations for usually only one or two small elements.

- The terms in the element equations are numerically evaluated for each element in the mesh, a process best performed on a computer. The resulting numbers are *assembled* (combined) into a much larger set of algebraic equations called the *system equations*. The latter characterize the response of the entire system so, not surprisingly, usually comprise a very large number of equations, typically hundreds or thousands, occasionally tens or hundreds of thousands. Such huge systems of equations can be solved

---

[4] Sometimes this may involve two steps (see Chapters 11 and 12) in which the governing equations are first approximated by a simpler, but nonalgebraic form (e.g., partial differential equations simplified to ordinary differential equations), and the latter are then further approximated by algebraic equations.

economically because the matrix of coefficients is ''sparse'' [see Section 1.3, Eq. (1.10)].

- At this point we have transformed the governing equations, which include the interior loads, but we have not dealt with the boundary conditions, which contain the boundary loads. These are now imposed by modifying the system equations. This involves adding values to existing terms and/or shifting terms from one side of the equations to the other. Both are relatively simple operations.
- The system equations are then solved on a computer using conventional numerical analysis techniques that have been popular for many years, having evolved prior to

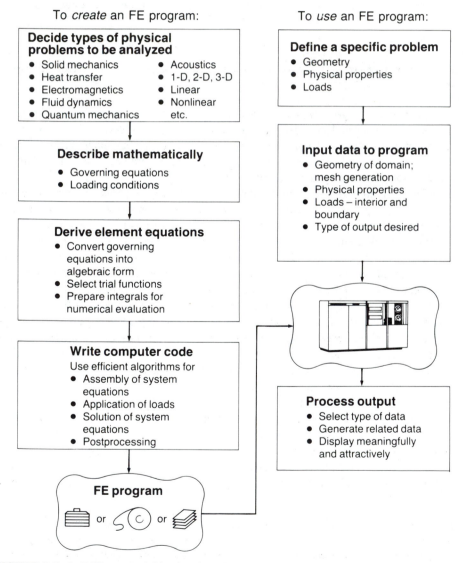

To *create* an FE program:

Decide types of physical problems to be analyzed
- Solid mechanics
- Heat transfer
- Electromagnetics
- Fluid dynamics
- Quantum mechanics
- Acoustics
- 1-D, 2-D, 3-D
- Linear
- Nonlinear
  etc.

Describe mathematically
- Governing equations
- Loading conditions

Derive element equations
- Convert governing equations into algebraic form
- Select trial functions
- Prepare integrals for numerical evaluation

Write computer code
Use efficient algorithms for
- Assembly of system equations
- Application of loads
- Solution of system equations
- Postprocessing

FE program
or   or

To *use* an FE program:

Define a specific problem
- Geometry
- Physical properties
- Loads

Input data to program
- Geometry of domain; mesh generation
- Physical properties
- Loads – interior and boundary
- Type of output desired

Process output
- Select type of data
- Generate related data
- Display meaningfully and attractively

FIGURE 1.5 ● Different activities involved in creating versus using an FE program.

the FEM. These techniques are implemented on the computer in special ways that take full advantage of characteristics peculiar to how FE system equations are formed.

• The final operation, called *postprocessing*, displays the solution to the system equations in tabular, graphical, or pictorial form. Other physically meaningful quantities might be derived from the solution and also displayed.

Subsequent chapters will fill in the details of the above procedure. It will be seen that the order of these operations and the relative effort spent on each depends on whether one is developing a new program or using an already existing one, the latter being the most common situation. Figure 1.5 illustrates the different activities involved in creating versus using an FE program.

Much of the work described above is done only once, during the creation of a new program (the left column in Fig. 1.5). However, when using an existing program (the right column in Fig. 1.5) one needs only to supply to the program the data for a specific problem, a few commands to guide the mesh-generation procedure, and instructions to control the data output. This book will follow both paths; a program called UNAFEM will be developed, step by step, from the mathematical description to the computer coding, and then the program will be used to solve a variety of specific problems. Because of the intrinsic modularity of the FEM, both development and use will alternate back and forth as the book progresses. Chapter 2 provides a detailed outline of how this development will proceed.

1.3 An Illustrative Problem

Let's illustrate the preceding discussion with a specific problem. Figure 1.6 shows a flat metal bar cut from a steel plate and containing a hole in the center. The bar will be required to support a tensile load of 10,000 lb applied to each end, as indicated by the arrows. This load will change the shape of the bar ever so slightly (elastic deformation) and produce stresses throughout the bar that vary from one position to another. We want to know how high these stresses will be and, in particular, whether they will be high enough to exceed the yield strength at any point, causing the bar to permanently deform and perhaps even to break. This is the type of problem that might be dealt with by a mechanical engineer.

How would this problem be described mathematically? Here the "system" is the bar and the "loading condition" is the applied tension. To predict the response of the system (stresses induced in the bar) to the external influences (applied tension), we will find a

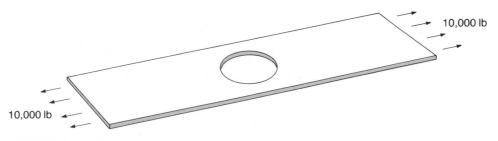

FIGURE 1.6 ● A flat steel bar with a hole, under tension.

numerical solution to the governing equations and loading conditions. For this problem, the governing equations are derived from the theory of elasticity (see Chapter 16). They are shown below, reproduced from Chapter 16. It is not important here that the reader understand these equations, since mechanical engineering may be of no interest to him or her. What is important at the moment is merely to appreciate that in order to answer the physical questions posed about the bar, *we represent the physical problem as a mathematical problem.*

Equations of stress equilibrium. •

$$\frac{\partial \sigma_x}{\partial x} + \frac{\partial \tau_{xy}}{\partial y} = -f_x$$

$$\frac{\partial \tau_{xy}}{\partial x} + \frac{\partial \sigma_y}{\partial y} = -f_y \tag{1.4}$$

where σ_x, σ_y, and τ_{xy} are components of stress, and f_x and f_y are interior loads that act on every material point inside the bar (such as gravity); they happen to be zero in this problem.

Constitutive relations (stress-strain equations) that describe the elastic response of steel •

$$\sigma_x = \frac{E}{1-\nu^2} \epsilon_x + \frac{E\nu}{1-\nu^2} \epsilon_y$$

$$\sigma_y = \frac{E\nu}{1-\nu^2} \epsilon_x + \frac{E\nu}{1-\nu^2} \epsilon_y$$

$$\tau_{xy} = \frac{E}{2(1+\nu)} \gamma_{xy} \tag{1.5}$$

where ϵ_x, ϵ_y, and γ_{xy} are components of strain, and E and ν are experimentally measured physical properties of steel (Young's modulus and Poisson's ratio).

Strain-displacement relations that describe the purely geometric aspects of the deformation •

$$\epsilon_x = \frac{\partial u}{\partial x}$$

$$\epsilon_y = \frac{\partial v}{\partial y}$$

$$\gamma_{xy} = \frac{\partial u}{\partial y} + \frac{\partial v}{\partial x} \tag{1.6}$$

where u and v are the components of displacement (movement of a material point in the bar) in the x- and y-directions, respectively.

These equations can all be combined [by substituting Eqs. (1.6) into Eqs. (1.5), and then Eqs. (1.5) into Eqs. (1.4)] to yield two governing equations:

$$\frac{E}{1-\nu^2} \frac{\partial^2 u}{\partial x^2} + \frac{E}{2(1-\nu)} \frac{\partial^2 v}{\partial x \partial y} + \frac{E}{2(1+\nu)} \frac{\partial^2 u}{\partial y^2} = -f_x$$

$$\frac{E}{1-\nu^2} \frac{\partial^2 v}{\partial y^2} + \frac{E}{2(1-\nu)} \frac{\partial^2 u}{\partial x \partial y} + \frac{E}{2(1+\nu)} \frac{\partial^2 v}{\partial x^2} = -f_y \tag{1.7}$$

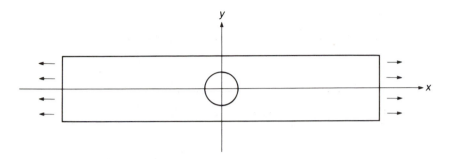

FIGURE 1.7 ● Domain in the x,y-plane for the bar problem.

Equations (1.7) are partial differential equations[5] containing two unknown functions, $u(x,y)$ and $v(x,y)$, the two components of displacement.

The domain of the problem is a region in the x,y-plane in the shape of the bar, as shown in Fig. 1.7. The tensile loads of 10,000 lb are applied to the ends of the bar, which form part of the boundary of the domain. Hence these loads are represented as boundary conditions,

$$\sigma_x = \frac{10,000}{A} \tag{1.8}$$

applied to both ends where A is the area of the end surface of the bar.

We seek numerical solutions for the unknown functions $u(x,y)$ and $v(x,y)$ which satisfy (approximately) both the governing equations [Eqs. (1.7)] and the boundary conditions [Eq. (1.8)].[6] The solutions for $u(x,y)$ and $v(x,y)$ may then be substituted into Eqs. (1.6) to determine strains, and the strains into Eqs. (1.5) to determine stresses, which are the quantities our mechanical engineer is seeking.

Turning now to the solution of these equations, the first thing the alert engineer will notice is that the problem has two lines of *symmetry*: the x- and y-axes, as drawn in Fig.

---

[5] Partial differential equations are so-named because they are equations that contain partial derivatives. In contrast, ordinary differential equations contain total (i.e., ordinary) derivatives.

[6] For the sake of technical accuracy, we note in passing that the boundary conditions [Eq. (1.8)] are incomplete in two respects. First, boundary conditions must be specified for every point on the boundary of the domain. In this problem the sides of the bar and the circumference of the hole have no loads applied to them (and they are free to move anywhere), so zero loads must be specified along these parts of the boundary.

Second, even these boundary conditions are insufficient for a well-posed problem because the bar remains free to translate freely anywhere in the x,y-plane. This implies an infinity of different positions and therefore solutions. To ensure a unique solution, the engineer must also constrain the displacement of enough points along the boundary to prevent such arbitrary translations. This requirement, which will be discussed in Section 16.2.3 (see also Sections 6.6 and 13.1) is implied by the form of the governing equations [Eqs. (1.7)]. Although the requirement is perhaps subtle to the nonmechanical engineer, it does point to a much wider problem: It is important that the engineer have a sufficient physical/mathematical understanding of the problem so that he or she can pose a meaningful problem before even attempting a solution, whether by the FEM or any other method.

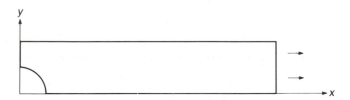

FIGURE 1.8 ● Reduction of the domain prior to the FE analysis.

1.7. These axes divide the bar into four quadrants. Within each quadrant the geometry, physical properties, and boundary conditions are identical (or mirror images of each other), and hence the solutions will be identical (or mirror images of each other) within each quadrant. This simplifies the problem because we will only need to find a solution within a single quadrant, as shown in Fig. 1.8.[7] Real problems frequently possess symmetries that can be exploited in this manner to reduce the size of the problem before the FE analysis begins.

    If our mechanical engineer is using an existing FE computer program, he or she will begin by inputting data to the program that defines the geometric shape of the domain (shown in Fig. 1.8), the physical properties of steel, and the tensile loads. In addition, he or she will partition the domain into a mesh of elements (mesh generation). The choice of shape and size of the elements and how many to use — that is, what type of mesh to construct — must be made by the engineer.[8]

    In general, there is a wide latitude in the type of mesh that may be constructed; i.e., there is no one best mesh. Ten skilled FE analysts doing the same problem will likely construct ten different meshes, and all might well yield solutions of acceptable, though different, accuracy. It might be said that one mesh is better than another if it provides better accuracy at less cost in computer fees and the analyst's time. Constructing a "good" mesh rather than a "bad" mesh is a skill that comes with experience and by acquiring an understanding of the FEM. This book will help develop some of that understanding and will offer practical guidelines.

    Figure 1.9 shows one possible mesh for the bar. There are a lot of smaller elements along the inner circular boundary where the analyst is expecting the solution to be more complicated; this will improve the accuracy of the solution in this region (one of the aforementioned guidelines that we will meet later on). Remember, the FEM yields *approx-imate* solutions, and therefore accuracy is a central concern of the FE user. An experienced engineer, before beginning the analysis, can usually make an intelligent estimate of what the general nature of the solution will be and can use such an estimate as a guide to choosing an appropriate mesh. Lacking any such prior knowledge, one can still con-

---

[7] Although three quadrants are removed from the analysis, their effect on the remaining quadrant must still be maintained. This is done by imposing additional boundary conditions along the edges touching the *x*- and *y*-axes. This detail is not important in the present context but will be explained in Section 13.7.

[8] Some current research is exploring ways to automate this selection process.

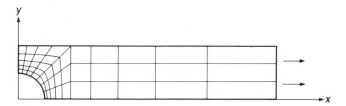

FIGURE 1.9 ● A mesh for analysis of the bar [1.1].

struct a good mesh although it will usually involve more trial and error. The mesh in Fig. 1.9 was constructed by a mesh-generation program that requires the user to define only a few elements; the program generates all the rest.

The next step is to evaluate numerically the element equations for each of the elements in the mesh. Recall that the element equations are algebraic equations derived from, and approximately equivalent to, the governing equations [Eqs. (1.7)]. For the particular type of element used here,[9] they consist of a set of 16 linear equations in 16 unknowns, which may be written in matrix form as follows:

$$
\begin{bmatrix}
K_{11} & K_{12} & K_{13} & \cdots & K_{1,16} \\
K_{21} & K_{22} & K_{23} & \cdots & K_{2,16} \\
K_{31} & K_{32} & K_{33} & \cdots & K_{3,16} \\
\cdot & \cdot & \cdot & & \cdot \\
\cdot & \cdot & \cdot & & \cdot \\
\cdot & \cdot & \cdot & & \cdot \\
K_{16,1} & K_{16,2} & K_{16,3} & \cdots & K_{16,16}
\end{bmatrix}
\begin{Bmatrix}
a_1 \\ a_2 \\ a_3 \\ \cdot \\ \cdot \\ \cdot \\ a_{16}
\end{Bmatrix}
=
\begin{Bmatrix}
F_1 \\ F_2 \\ F_3 \\ \cdot \\ \cdot \\ \cdot \\ F_{16}
\end{Bmatrix}
\tag{1.9}
$$

The coefficients $K_{ij}(i,j=1,2,\ldots,16)$ are related to the partial derivative terms on the left-hand side of Eqs. (1.7), and the forcing terms $F_i(i=1,2,\ldots,16)$ are related to the interior loading terms on the right-hand side of Eqs. (1.7). All the K_{ij} and F_i have definite expressions that the program can numerically evaluate; the resulting numerical values generally vary from one element to the next.

The $a_i(i=1,2,\ldots,16)$ are the unknowns. They are related to the unknown functions $u(x,y)$ and $v(x,y)$. Each element uses a different set of a_i, but with the following important exception: Neighboring elements have some of their a_i in common. This fact will be seen (Chapter 5) to play a crucial role because it is what connects all the element equations together to form the system equations.

As the element equations are evaluated for each element, the resulting values for the K_{ij} and F_i from each element are *assembled*, resulting in a rather large system of linear

---

[9] This problem used what are called "C^0-quadratic isoparametric quadrilateral" elements, each element containing eight "nodes" and two "degrees of freedom" at each node, hence 16 unknowns. All of these concepts will be carefully explained in subsequent chapters. (See especially Chapter 13, Section 13.4.2; also Chapter 16.)

algebraic equations, called the *system equations*. The present problem, using the mesh in Fig. 1.9, yields a system of 410 equations in 410 unknowns, shown symbolically below:

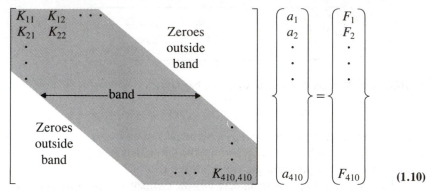

$$(1.10)$$

It should be noted that the matrix of coefficients is sparse; that is, most of the coefficients are zero. In addition, the nonzero terms are clustered in a narrow band along the main diagonal. This *bandedness* is characteristic of FE system equations; it results in a tremendous saving in both computer time and storage, and means a reduction in cost, typically by a factor of 100 to 1000, from the cost of solving a full matrix. Because of these savings, it becomes practical to solve very complicated problems whose system equations contain tens and even hundreds of thousands of unknowns.

After assembly of the system equations, the boundary conditions are imposed. For example, the 10,000-lb tensile force is added to some of the terms in the forcing vector on the right-hand side. The system equations are now solved, using any of the conventional algebraic equation-solving techniques, modified slightly to take advantage of the bandedness.

In the final step, called *postprocessing*, the solution to the system equations (numerical values for all the a_i's) may be listed in tabular fashion and/or displayed graphically. In our bar problem it would then be necessary to use this solution in conjunction with Eqs. (1.5) and (1.6) to calculate the stresses in the bar. The stresses may be listed tabularly, element by element say, or they may be plotted as stress contours as shown in Fig. 1.10.

Rapid advances in computer graphics technology have made it possible to display FE solutions on color graphics terminals in a variety of informative and attractive formats. For example, the stress contours shown in Fig. 1.10 could also have been displayed by the same program in a gradation of colors from the maximum stress value to the minimum.

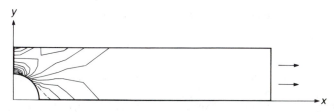

FIGURE 1.10 ● Stress contours in the bar [1.1].

Impressive graphical displays can sometimes instill a false confidence that one has *the* solution to the problem. However, an FE solution is an *approximation*, so it generally has some error relative to the exact solution of the governing equations. Therefore, for a solution to be meaningful, the engineer must be able to estimate the accuracy and decide whether the solution is acceptable or should be improved. Both estimation and improvement are usually accomplished by repeating the entire analysis using a different mesh of the same types of elements (see Chapter 7) or by keeping the same mesh but changing the type of element (see Chapter 8), and then observing the differences between the two (or more) solutions.

It is hoped that this cursory outline of the salient features of a typical FE problem has provided the reader with a flavor of what the FEM is all about. It is also intended to provide a "preview of coming attractions," since the remaining chapters will delve into all the details that have been omitted here.

1.4 Some Applications to Real Problems

The following examples illustrate a variety of applications in which the FEM has been successful in providing answers to engineering problems. Each example will state the field of application and will include three items: a brief description of the physical system and the question to be answered or the information sought, a diagram of the FE mesh, and a comparison of the FE solution with experimental data or another independently derived solution.

■ *EXAMPLE 1.1* Solid mechanics, elasticity (static equilibrium)
Figure 1.11 shows a cross-sectional view of a thick-walled cylinder that provides a protective housing for delicate electronics used in oceanographic work at great depths. There was concern whether the high hydrostatic pressure on the outside of the housing would create stresses in the wall of the housing that would exceed the yield strength, resulting in a permanent (plastic) deformation. The housing was idealized to have rotational symmetry, which simplified the problem to a two-dimensional analysis in the plane of the cross section. The good agreement between the FE-predicted strains (which are proportional to stresses) and the experimentally measured strains provided confidence in using the FE model to optimize the design. ■

■ *EXAMPLE 1.2* Solid mechanics, elasticity (transient dynamic, nonlinear)
Figure 1.12 shows a cable-laying ship paying out a communications cable for installation on the ocean floor. Rough seas cause heaving of the ship, which produces time-varying stresses in the cable. Problem: What is the roughest sea in which cables can be installed without seriously risking cable damage from these dynamic stresses? The mesh of cable elements is shown symbolically on the drawing. The results for a particular cable system, ship, and sea state are plotted, showing dynamic tension (total tension minus static tension) at the bow as a function of time, for a period of 5 min. This shows excellent agreement with experimentally measured tensions from a full-scale test at sea. This model subsequently provided valuable warnings that high stresses from resonances could occur when a junction between dissimilar cables is present in the suspended portion. ■

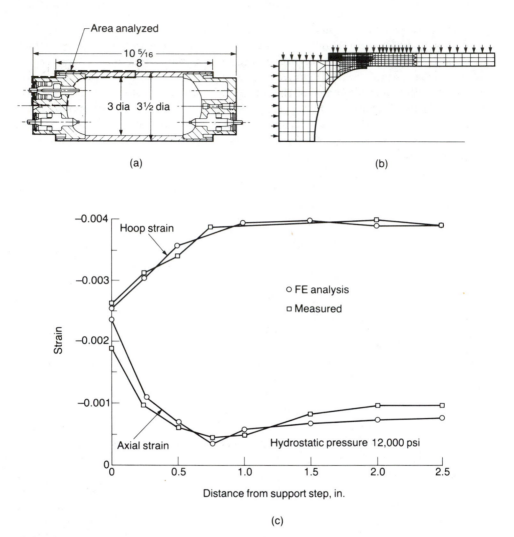

FIGURE 1.11 ● High-pressure housing for oceanographic application [1.2]: (a) cross section of housing; (b) mesh; (c) comparison of FE solution and experimental data.

■ *EXAMPLE 1.3* Solid mechanics, elasticity (resonance)

Related to the previous example is the problem of characterizing the types of stress waves that can propagate along cables. Figure 1.13 shows the cross sections of two typical ocean cables composed of several layers of helically wound fibers and/or extruded polymers. An FE model for this problem was derived from the dynamic equations of elasticity, and a FORTRAN program was written, requiring altogether about 10 days of effort. A typical mesh consisted of concentric ring-shaped elements. The program calculated the natural modes of propagation (a resonance phenomenon) and plotted their ''dispersion'' relations, which are curves of frequency versus reciprocal wavelength. For the special case of a

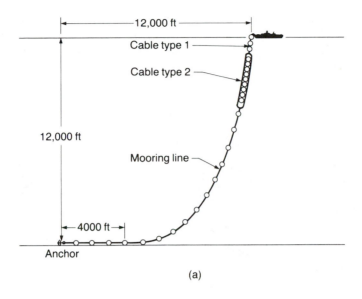

(a)

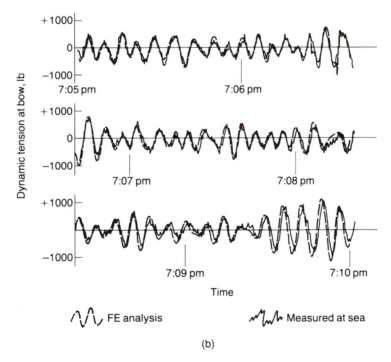

(b)

FIGURE 1.12 ● Installation of communications cable on the ocean floor [1.3]: (a) suspended cable and its FE mesh; (b) comparison of FE solution and experimental data.

homogeneous, isotropic cylinder, an exact theoretical solution exists [1.4], which is shown in Fig. 1.13, along with the FE solution. Independent verification like this, by comparison with solutions from other sources or experimental data, is most valuable in providing confidence in an FE model. ∎

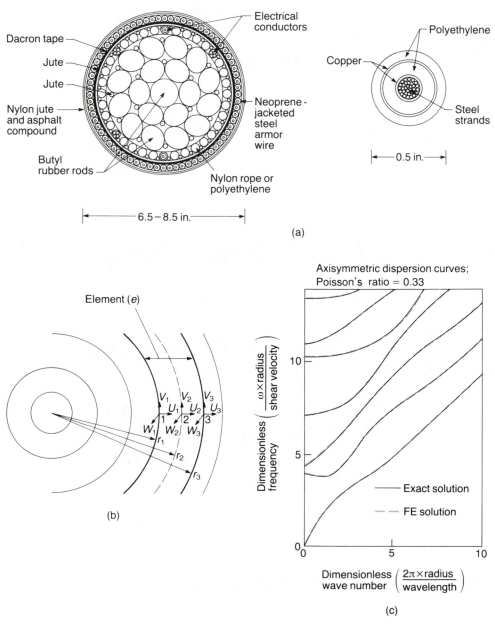

FIGURE 1.13 ● Stress waves in multilayered ocean cables [1.5]: (a) cross section of two types of cables; (b) FE mesh; (c) comparison of FE and exact theoretical solutions for isotropic, homogeneous cable.

■ *EXAMPLE 1.4* Solid mechanics, plasticity (transient dynamic, nonlinear)
Various parts of a modern strategic bomb are designed to mitigate considerable kinetic energy when impacting a hard surface in order to ensure the survivability of the internal components. Figure 1.14 shows the mesh used for a three-dimensional analysis of a steel nose cone during the few milliseconds immediately following impact. The crushing of the cone predicted by the analysis is revealed by the plot of the deformed mesh, which agrees extremely well with the photograph of an actual test specimen. ■

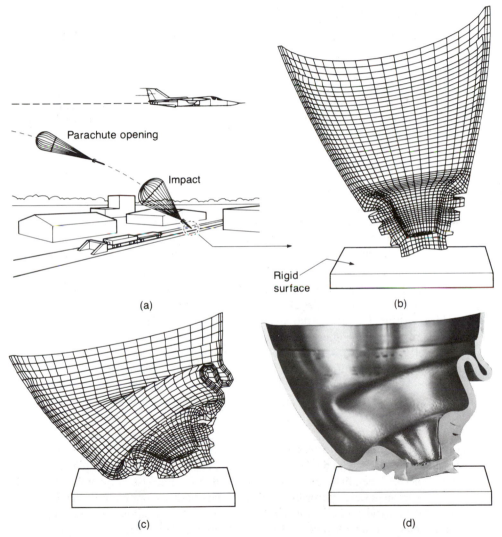

FIGURE 1.14 ● Impact crushing of a steel nose cone [1.6]: (a) delivery and impact of bomb; (b) initial mesh of nose cone (before impact); (c) deformed mesh (15 msec after impact); (d) photograph of steel test specimen subjected to impact. (After M. L. Chiesa and M. L. Callabresi, "Nonlinear Analysis of a Mitigating Steel Nose Cone" in A. K. Noor and H. G. McComb, Jr. (eds.), *Computational Methods in Nonlinear Structural and Solid Mechanics*, Pergamon, New York, 1981, pp. 295-300.)

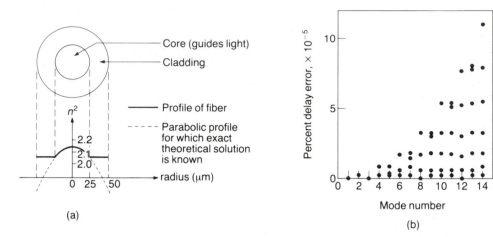

FIGURE 1.15 ● Modes of light propagation in optical fiber [1.7]: (a) cross section of fiber and profile of index of refraction, n; (b) errors in FE-computed mode delays for infinite parabolic profile. (FE mesh uses concentric ring elements, as in Fig. 1.13.)

■ *EXAMPLE 1.5* Electromagnetism

This example is very similar to Example 1.3 in that it involves waves propagating in a cylindrical waveguide. Here, though, we are dealing with electromagnetic waves in an optical fiber. Figure 1.15 shows the cross section of an optical fiber. There are typically two layers: an inner core that guides the light and in which the index of refraction varies radially according to some profile, and an outer cladding in which the index of refraction is usually uniform. In the design of optical fibers for telecommunications it is important to determine the propagation modes and mode delays (reciprocals of group wave velocities) associated with different index-of-refraction profiles in order to choose an optimum profile.

For this problem the engineer derived a relatively simple one-dimensional FE model from Maxwell's equations. A mesh of circular ring elements was used, just like in Fig. 1.13(b). Initial verification of the model was accomplished by comparison of the FE calculations with the known exact solution for an infinitely extending parabolic profile (shown in Fig. 1.15a), as well as by comparison with experimental data. ■

■ *EXAMPLE 1.6* Magnetostatics

Power conversion circuits sometimes employ inductive elements called "swinging chokes," whose inductance varies with the DC current applied to it. Such elements greatly enhance the performance between light- and heavy-load conditions. Because of the nonlinear properties of the core materials from which they are fabricated, an optimum design of choke geometry and material properties generally requires many iterations. Performing these iterative designs with an FE model, prior to building a prototype, can save substantial time and costs.

In this problem the engineer utilized a commercial FE program that solves Maxwell's equations for two-dimensional static magnetic fields. Figure 1.16 shows a cross-sectional view of a swinging choke with a stepped gap in the center leg. The mesh of triangular

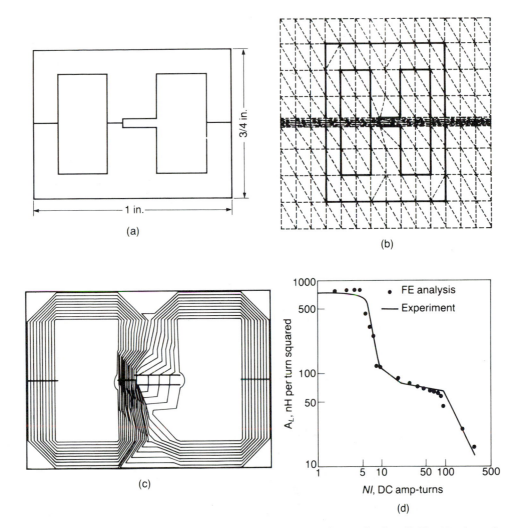

FIGURE 1.16 ● Variable-inductance choke for power conversion applications [1.8]: (a) schematic of the choke; (b) FE mesh (vertical dimension magnified ×2 to enhance view of mesh in gap); (c) computed flux lines at 55 ampere-turns; (d) comparison of FE solution and experimental data.

elements is refined (having a greater number of smaller elements) in the gap region where the solution is expected to show considerable variation. The calculated magnetic flux lines are shown, along with a plot of inductance versus current that compares both the FE results and experimental data. ■

■ *EXAMPLE 1.7* Heat conduction
The thermal performance of integrated-circuit (IC) packages used in telecommunication systems is of critical concern to both the circuit- and physical-design engineer. Heat generated in some of the components (such as transistors and resistors), if not dissipated

rapidly enough to the surrounding coolant air, can increase the IC temperature sufficiently to impair its performance. The designer must therefore choose packaging materials and geometry that will maximize the rate of heat dissipation and hence minimize the operating temperature.

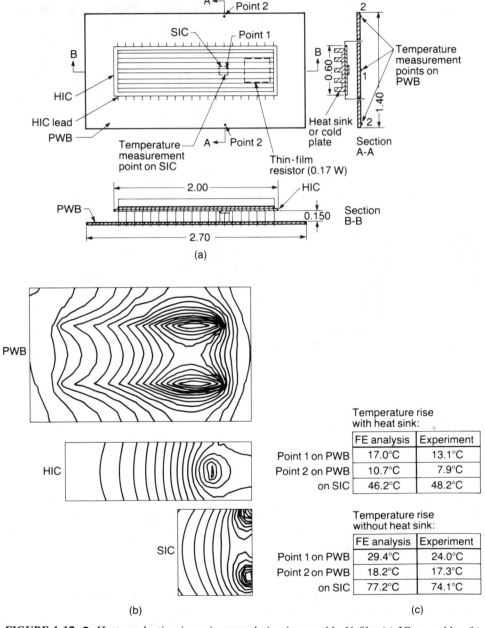

FIGURE 1.17 ● Heat conduction in an integrated-circuit assembly [1.9]: (a) IC assembly; (b) temperature contours (dimensions not to scale); (c) verification of FE model.

Figure 1.17(a) shows an orthographic drawing of an IC assembly for which an FE heat conduction study was performed. The assembly consists of an epoxy-glass printed wiring board (PWB), only part of which is shown here and included in the FE model. A 40-pin hybrid integrated circuit (HIC) is through-hole mounted onto the PWB. A smaller, square-shaped silicon integrated circuit (SIC) with 44 beam leads is in turn bonded to the HIC on the surface facing the PWB. On the other side of the HIC is an aluminum heat sink.

The FE mesh for the complete IC assembly is more easily described in words than depicted graphically. It consists of three separate layers of 2-D quadrilateral elements, stacked on top of each other, representing the PWB, HIC, and SIC. The meshes for the PWB and SIC each have a regular pattern of square elements, similar to the mesh in Fig. 1.20(c). The mesh for the HIC has an irregular pattern of rectangular and nonrectangular elements. In between the layers and perpendicular to them are 1-D "rod" elements representing each of the interconnecting pins and beam leads.

Figure 1.17(b) shows computed temperature contours, and Fig. 1.17(c) shows a comparison between experimentally measured temperatures and analytically predicted values. The good agreement provided increased confidence in the accuracy of the FE model. The engineer then performed many analyses with the model, varying several parameters to determine the effects of various proposed design modifications. This parametric study revealed several ways to improve the original design. ∎

■ *EXAMPLE 1.8* Acoustics
In designing automobiles it is important for passenger comfort and safety to provide a quiet passenger compartment. To achieve this goal, the engineer must determine the acoustic modes associated with a proposed body design. Here again there are likely to be many design iterations that an FE model can perform quickly and relatively cheaply. Figure 1.18 shows a two-dimensional mesh of triangular elements for a typical irregularly-shaped automobile interior. The predicted frequencies and mode shapes show excellent agreement with experimental values. A finite difference model (not shown) was also analyzed and the results were in very close agreement with those of the FE model, thereby providing further confidence in the analytical models. ∎

■ *EXAMPLE 1.9* Coupled phenomena: piezoelectricity (elasticity and electricity) and acoustics
Figure 1.19 shows a schematic drawing of a piezoelectric ceramic underwater sound projector. In designing such a projector, one of the important performance parameters is the radiation pattern and how it varies with the frequency of the projected sound. To optimize the design, an FE model was developed. The ceramic cylinder was modeled with three-dimensional elements for which the element equations approximate the equations of piezoelectricity (coupling between electric fields and elastic deformation). The surrounding fluid employed acoustic elements (not shown here) out to some finite radius sphere, depicted by a dashed circle. From the sphere out to infinity, classical expressions for acoustic radiation (which don't employ finite elements) are appropriately discretized and "matched" to the discrete acoustic elements touching the circle from the inside. The figure shows a radiation pattern predicted by the model, along with an experimentally measured pattern. This example demonstrates that the FEM is flexible enough to be effectively combined with other analysis methods. (See Zienkiewicz et al. [1.11].) ∎

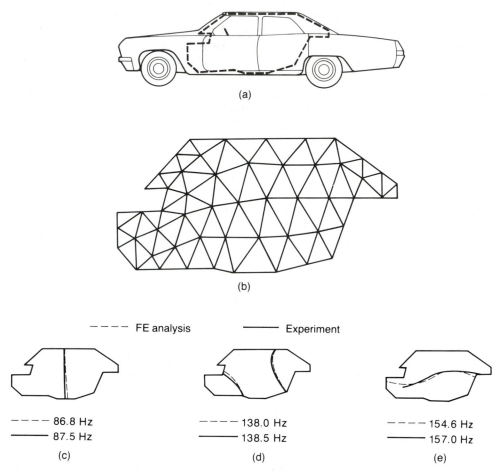

FIGURE 1.18 ● Acoustic modes in an automobile passenger compartment [1.10]: (a) passenger compartment (outlined by dashes); (b) FE mesh; (c) through (e) mode shapes and frequencies.

■ *EXAMPLE 1.10* Coupled phenomena: fluid dynamics and heat transport
Crystals of gallium arsenide, a semiconducting material, are used for substrates in integrated circuits. The crystals are often grown in horizontal boats by the moving-furnace (horizontal Bridgman) method. In this process, the boat (shown in Fig. 1.20a) is surrounded by a furnace that produces a temperature gradient along the length of the boat. The furnace moves slowly lengthwise, while the boat is held stationary, creating a moving liquid/solid interface in the molten semiconductor. This is a problem of thermal convection and conduction in a liquid, involving the coupling of thermal and fluid dynamic effects.

As the demand increases for crystals with lower defect densities, it is important to understand melt-flow phenomena and how they relate to crystal yield and quality. It is known from previous research that the flow patterns in the melt can be quite intricate, including many swirling eddies. The flow is 3-D because the heat from the furnace not only varies along the length of the melt but it also enters the melt by flowing transversely through the walls of the boat and the melt's free surface. In addition, no-slip conditions

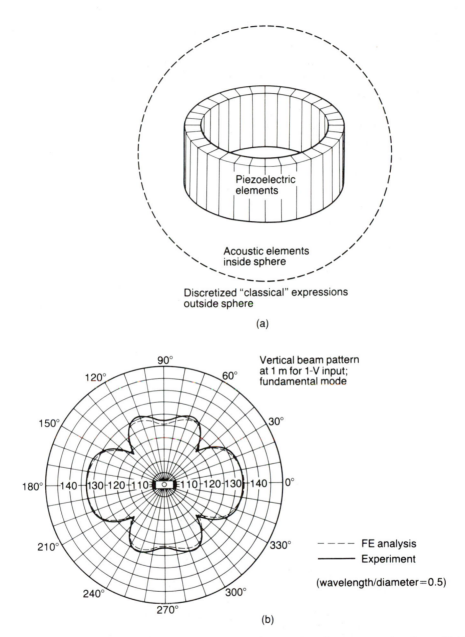

FIGURE 1.19 ● Acoustic radiation from a piezoelectric ceramic underwater sound projector [1.12]: (a) projector and its FE mesh; (b) comparison of FE solution and experimental data.

on the walls cause the fluid velocity to vary transversely. Since experimental data is extremely difficult to obtain, the engineers on this problem turned to analytical methods to obtain a solution to the coupled 3-D Navier-Stokes and heat transport equations.

Before analyzing the very complicated 3-D problem, the engineers gained valuable experience and insight by first considering an idealized 2-D problem. This was achieved by simplifying the boat geometry (shown in Fig. 1.20b) and considering only the flow in the vertical plane of symmetry midway between the side walls. Such a simplification greatly reduced computational costs yet still yielded important information. The domain is therefore simply a rectangle, for which the mesh was an 8-by-32 array of square elements (shown in Fig. 1.20c).

To help verify the FE model, the problem was also analyzed using the finite difference method. By using two independent modeling techniques and achieving good agreement with both (as shown in Fig. 1.20d and e), the engineers' confidence was significantly increased. In addition, they had acquired a model which would, in turn, help to verify the subsequent 3-D model. ■

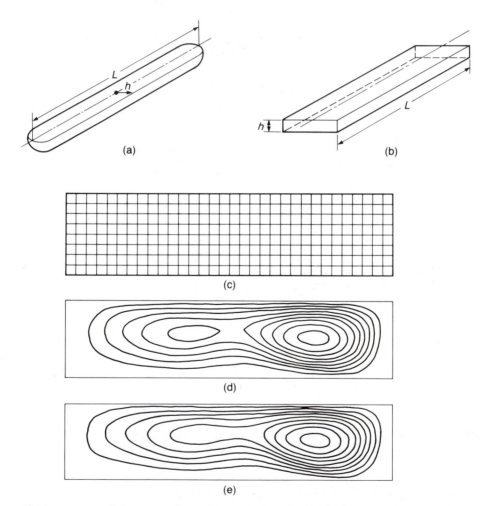

FIGURE 1.20 ● Fluid flow in semiconductor crystal-growth melts [1.13]: (a) actual boat geometry; (b) idealized boat geometry; (c) FE mesh; (d) streamlines computed by FEM; (e) streamlines computed by finite difference method.

These 10 examples have illustrated some of the more usual fields of application. Examples can also be cited from other fields: quantum mechanics [1.14, 1.15]; solid state physics [1.16, 1.17]; random processes [1.18]; and physical, chemical, and biological aspects of marine and estuarial systems [1.19–1.21].

Since the practical implementation of the FEM originated and matured in the field of structural analysis (see Section 1.6), there have been many notable engineering accomplishments in this field in which the FEM has played a significant role — for example, the Boeing 747 [1.22] (and subsequent designs), the GM Cadillac Seville [1.23] and Ford Escort [1.24], and the NASA Columbia space shuttle, [1.25], to name a few. One of the largest FE models to date was the structural analysis of an offshore oil-drilling platform for use in the North Sea [1.26]. The model used approximately 50,000 elements, which involved approximately 720,000 unknowns (and therefore 720,000 simultaneous algebraic equations).

It would be irresponsible to leave the reader with the impression that the history of the FEM is marked only by successes. There have been some dramatic and tragic failures of large engineering structures, presumably designed with the aid of the FEM; for example, the collapse of the roof on the Hartford (Connecticut) Civic Center Coliseum and the collapse of an offshore oil platform in the North Sea (both about 1980). Documentation of these failures is not easily available, so it is difficult to judge to what extent the use (or abuse) of the FEM was responsible. I leave it to the reader to guess how many other less dramatic — and unpublished — failures have occurred with FE applications.

There is a lesson to be learned from both the successes and failures. One cannot simply pick up a user's manual to an FE program and, lacking an understanding of basic FE concepts, expect to generate a solution to an engineering problem that is meaningful, reliable, and whose accuracy has been estimated and judged to be acceptable. Someone working on the problem (the engineer at the computer terminal, the project leader, a helpful colleague providing consultation) must have this understanding, or else the project is likely to be a failure rather than a success. It is hoped that this text will help to provide such an understanding of basic FE concepts and their practical application.

1.5 Benefits from Using the FEM

The most obvious benefit is that the FEM can provide solutions to many complicated problems that would be intractable by other techniques. This has been most apparent in the fields of mechanical and structural engineering; in fact, it is generally recognized that over the last quarter century the FEM has produced a revolution in these fields. It now appears that similar revolutions are on the horizon as the FEM continues to spread to other disciplines.

The FEM is a very modular technique. The element equations can be used repeatedly, not only for all the elements in a particular mesh but also for other problems and in other programs. New types of elements (and hence new sets of element equations) can be added to programs as the need arises, gradually building up an *element library*, that can be moved from program to program. A user can select from the element library to create a mesh of different element types, much like a child using different blocks to build a structure. This has a definite impact on human resources, because when a person develops an FE computer program, it can be used to solve not just one specific problem but a whole class of problems that differ substantially in geometry, boundary conditions, and other properties. This is a capital investment for the future.

There is a more subtle long-range benefit. All FE applications use basically the same types of mathematical techniques — most of which can be learned at the undergraduate college level and some of which are introduced at the high school level. Thus considerable proficiency can be gained by the average user from a few years of formal schooling plus some practical experience, and the repeated exposure to the same techniques during subsequent years maintains a sharpness of skill that yields efficient use of one's time. In short, a few techniques will solve a broad spectrum of problems. To borrow a familiar phrase, "a little knowledge goes a long way."

1.6 Historical Development

The ideas that gave birth to the FEM evolved gradually from the independent contributions of many people in the fields of engineering, applied mathematics, and physics. The essential ideas, though, began to appear in publications principally during the 1940s (see Fig. 1.21).

Courant's 1943 paper [1.27] is a classic. Although one of the earliest and soon forgotten, it detailed an approach similar to that which evolved again in the middle of the 1960s. Thus, to solve the torsion problem in elasticity, he defined piecewise linear polynomials over a triangularized region. Schoenberg's paper in 1946 [1.28] gave birth to the theory of splines, recommending the use of piecewise polynomials for approximation and interpolation. In the early and mid-1950s Polya [1.29, 1.30], Hersch [1.31], and Weinberger [1.32, 1.33] used ideas similar to Courant's to estimate bounds for eigenvalues. In 1959 Greenstadt [1.34] divided a domain into "cells," assigned a different function to each cell, and applied a variational principle. White [1.35] and Friedrichs [1.36] used triangular elements to develop difference equations from variational principles.

In the physics community, the work of Prager and Synge [1.37] led to the development of the hypercircle method that provided a geometric interpretation for the minimum principles of classical elasticity theory. Synge [1.38, 1.39] used piecewise linear functions defined over a triangularized region with a Ritz variational procedure. McMahon [1.40] solved a three-dimensional electrostatic problem using tetrahedral elements and linear trial functions.

In the engineering community [1.41], Hrenikoff proposed the idea in 1941 [1.42] that the elastic behavior of a physically continuous plate would be similar, under certain loading conditions, to a framework of *physically separate* one-dimensional rods and beams, connected together at discrete points. The problem could then be handled by familiar computational methods for trusses and frameworks. McHenry [1.43] and Newmark [1.44] further refined this idea.

With the commercial introduction of high-speed stored-program digital computers in the early 1950s, several people, notably Langefors [1.45] in 1952 and Argyris [1.46] in a series of papers from 1954 to 1955, took the well-established framework-analysis procedures and reformulated them into a matrix format ideally suited for efficient automatic computation.

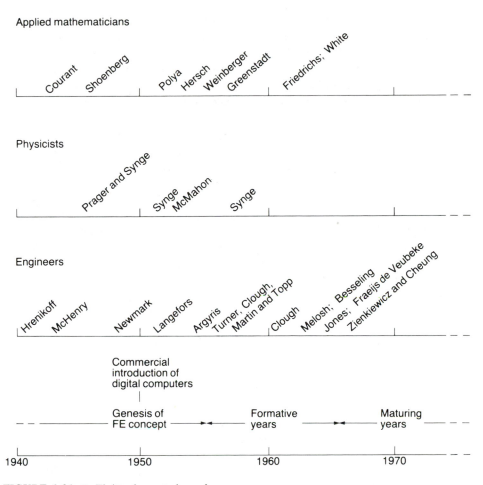

FIGURE 1.21 ● Finite element chronology.

In 1956 Turner, Clough, Martin, and Topp [1.47] advanced the framework concept a step further by modeling the odd-shaped wing panels of high-speed aircraft as an assemblage of smaller panels of simple triangular shape. This was a conceptual breakthrough because it made it possible to more realistically model two- or three-dimensional structures as assemblages of similar two- or three-dimensional pieces rather than of one-dimensional bars — e.g., shells made of smaller shells or solids made of smaller solids. The abstract of their paper contained the very prophetic observation, ''Considerable extension of the material presented in this paper is possible.''

All the ingredients had now been assembled: computers, matrix methods, and the element concept. Engineers quickly recognized that they had a new, powerful, and practical tool. This perhaps marked the beginning of the steady growth of the FEM. The next five to ten years saw a profusion of papers dealing with practical engineering applications to aerospace and civil engineering structures.

The name of the method, "finite elements," first appeared in Clough's paper in 1960 [1.48]. Work by Melosh [1.49] and Besseling [1.50] in 1963, and by Jones [1.51], and Fraeijs de Veubeke [1.52] in 1964 showed that the FEM could be identified as a form of the Ritz variational method using piecewise-defined trial functions. In 1965 Zienkiewicz and Cheung [1.53] broadened the scope of the FEM tremendously by demonstrating that it was applicable to all field problems that could be placed in variational form.

The mid-1960s might be identified as the transition from the early formative years to the beginning of the "modern era" — or, as one author suggested, the "golden era" — of the FEM. The first decade (mid-1950s to mid-1960s) had been dominated strongly by large-scale structural applications, and the approach was characterized by the physical concept of imagining the structure to be composed of physically separate pieces.

From the mid-1960s to the end of the 1970s, the FEM spread beyond the original confines of structural analysis to many other fields of application, and the trend became more mathematical. These were the "maturing" years that saw the appearance of over 7000 publications by the year 1976 [1.54, 1.55]. Mathematicians began placing the method on firmer foundations with rigorous proofs for convergence, error bounds, and stability. Techniques were extended beyond variational methods to include weighted residuals and global energy-balance principles. Numerical algorithms were further refined to fully exploit the bandedness of FE equations. Automatic mesh-generation codes, interactive graphics, and other pre- and postprocessing capabilities were developed. By 1981 it could be said that the practical implementation of the FEM had spanned its first quarter century [1.56].

References

[1.1]   ANSYS, a large-scale, general-purpose finite element program, developed and maintained by Swanson Analysis Systems, Inc., Houston, Pa.

[1.2]   R. D. Smith, unpublished work, AT&T Bell Laboratories, Whippany, N. J., 1977.

[1.3]   D. S. Burnett, unpublished work, AT&T Bell Laboratories, Whippany, N. J., 1979.

[1.4]   J. Zemanek, Jr., "An Experimental and Theoretical Investigation of Elastic Wave Propagation in a Cylinder," *J Acous Soc Amer*, **51**(1): 265-283 (1972).

[1.5]   D. S. Burnett, unpublished work, AT&T Bell Laboratories, Whippany, N. J., 1974.

[1.6]   M. L. Chiesa and M. L. Callabresi, "Nonlinear Analysis of a Mitigating Steel Nose Cone," *Computers and Structures*, **13**: 295-301 (1981).

[1.7]   T. A. Lenahan, "Calculation of Modes in an Optical Fiber Using the Finite Element Method and EISPACK," *Bell Sys Tech J*, **62**(9): 2663-2694 (Nov. 1983).

[1.8]   J. M. Dishman, D. R. Kressler, and R. Rodriguez, "Characterization, Modeling and Design of Swinging Inductors for Power Conversion Applications," *Proc Powercon 8*, San Diego, April 1981, paper B-3 (AT&T Bell Laboratories).

[1.9]   C. C. Mandrone, unpublished work, AT&T Bell Laboratories, North Andover, MA, 1982.

[1.10] T. Shuku and K. Ishihara, "The Analysis of the Acoustic Field in Irregularly Shaped Rooms by the Finite Element Method," *J Sound Vibration*, **29**: 67-76 (1973).

[1.11] O. C. Zienkiewicz, D. W. Kelly, and P. Bettess, "Marriage a la mode — The Best of Both Worlds (Finite Elements and Boundary Integrals)" in R. Glowinski, E. Y. Rodin, and O. C. Zienkiewicz (eds.), *Energy Methods in Finite Element Analysis*, Wiley, 1979, pp. 81-107.

[1.12] J. T. Hunt, M. R. Knittel, and D. Barach, "Finite Element Approach to Acoustic Radiation from Elastic Structures," *J Acoust Soc Am*, **55**(2): 269-280 (1974).

[1.13] M. J. Crochet, F. T. Geyling, and J. J. VanSchaftingen, "Numerical Simulation of the Horizontal Bridgman Growth of a Gallium Arsenide Crystal," *J Crystal Growth*, **65**: 166-172 (1983).

[1.14] M. Friedman, A. Rabinovitch, and R. Thieberger, "Application of the Finite Element Method to the Hydrogen Atom in a Box in an Electric Field," *J Comp Phys*, **33**: 359-368 (1979).

[1.15] S. W. Schoombie and J. F. Botha, "Error Estimates for the Solution of the Radial Schrödinger Equation by the Rayleigh-Ritz Finite Element Method," *IMA J Num Anal*, **1**: 47-63 (1981).

[1.16] P. E. Cottrell and E. M. Buturla, "Steady State Analysis of Field Effect Transistors via the Finite Element Method," *IEEE Int Electron Devices Meeting Tech Digest*, Washington, D.C., Dec. 1975, pp. 51-54.

[1.17] J. J. Barnes and R. J. Lomax, "Finite-Element Methods in Semiconductor Device Simulation," *IEEE Trans Elec Devices, ED-***24**(8): 1082-1089 (1977).

[1.18] Tze-Chien Sun, "A Finite Element Method for Random Differential Equations with Random Coefficients," *SIAM J Numer Anal*, **16**(6): 1019-1035 (1979).

[1.19] G. F. Pinder, "A Galerkin Finite Element Simulation of Groundwater Contamination on Long Island, New York," *Water Resources Research*, **9**(6): 1657-1669 (1973).

[1.20] C. A. Brebbia and P. Partridge, "Finite Element Models for Circulation Studies," *Proc Int Conf Math Models Environ Probs* (University of Southampton, England, Sept. 1975), Halsted, 1976, pp. 141-159.

[1.21] R. Willis, "A Planning Model for the Management of Groundwater Quality," *Water Resources Research,* **15**(6): 1305-1312 (1979).

[1.22] S. D. Hansen, G. L. Anderton, N. E. Connacher, and C. S. Dougherty, "Analysis of the 747 Aircraft Wing-Body Intersection," *Proc 2d Conf Matrix Meth Struct Mech*, Wright-Patterson Air Force Base, Dayton, Ohio, 1968, AFFDL-TR-68-150, pp. 743-788.

[1.23] C. E. W., "Cadillac's Small Car," *Machine Design*, April 17, 1975, pp. 10-12.

[1.24] D. J. Hughes, "Ford — Structural Analysis in Europe," *Finite Element News*, Feb. 1982, pp. 26-27.

[1.25] J. Parmater, "Space Shuttle Launch Climaxes Years of Effort," *SDRC Newsletter*, Struct. Dyn. Res. Corp., Milford, Ohio, June 1981, pp. 3-8. (Other organizations also contributed to structural design of the shuttle.)

[1.26] T. Harwiss, "Large Analyses in A. S. Computas," *Finite Element News*, Sept. 1979, pp. 6-9.

[1.27]   R. Courant, "Variational Methods for the Solution of Problems of Equilibrium and Vibrations," *Bull Am Math Soc*, **49**: 1-23 (1943).

[1.28]   I. J. Schoenberg, "Contributions to the Problem of Approximation of Equidistant Data by Analytic Functions. Part A. On the Problem of Smoothing or Graduation. A First Class of Analytic Approximation Formulae," *Quart Appl Math*, **4**: 45-99 (1946).

[1.29]   G. Polya, "Sur Une Interpretation de la Methode des Differences Finies qui peut Fournir des Bornes Superieures ou Inferieures," *C R Acad Sci Paris*, **235**: 995-997 (1952).

[1.30]   G. Polya, "Estimates for Eigenvalues," *Studies in Mathematics and Mechanics Presented to Richard von Mises*, Academic Press, New York: 1954, pp. 200-207.

[1.31]   J. Hersch, "Equations Differentielles et Functions de Cellules," *C R Acad Sci Paris*, **240**: 1602-1604 (1955).

[1.32]   H. F. Weinberger, "Upper and Lower Bounds for Eigenvalues by Finite Difference Methods," *Commun Pure Appl Math*, **9**: 613-623 (1956).

[1.33]   H. F. Weinberger, "Lower Bounds for Higher Eigenvalues by Finite Difference Methods," *Pacific J Math*, **8**: 339-368 (1958).

[1.34]   J. Greenstadt, "On the Reduction of Continuous Problems to Discrete Form," *IBM J Res Div*, **3**: 355-363 (1959).

[1.35]   G. N. White, "Difference Equations for Plane Thermal Elasticity," LAMS-2745, Los Alamos Scientific Laboratories, Los Alamos, N. M., 1962.

[1.36]   K. O. Friedrichs, "A Finite Difference Scheme for the Neumann and the Dirichlet Problem," NYO-9760, Courant Institute of Mathematical Science, New York University, New York, 1962.

[1.37]   W. Prager and J. L. Synge, "Approximations in Elasticity Based on the Concept of Function Space," *Quart Appl Math*, **5**: 241-269 (Oct. 1947).

[1.38]   J. L. Synge, "Triangulation in the Hypercircle Method for Plane Problems," *Proc Roy Irish Acad*, **54**(A): 341-367 (1952).

[1.39]   J. L. Synge, *The Hypercircle in Mathematical Physics*, Cambridge University Press, New York, 1957.

[1.40]   J. McMahon, "Lower Bounds for the Electrostatic Capacity of a Cube," *Proc Roy Irish Acad*, **55**(A): 133-167 (1953).

[1.41]   J. B. Spooner, "A History of the Finite Element Method" in *State-of-the-Art of Finite Element Methods in Structural Mechanics*, Proc First World Congress on Finite Element Methods, Bournemouth, England, Oct. 13-17, 1975, pp. A1-A22.

[1.42]   A. Hrenikoff, "Solution of Problems in Elasticity by the Framework Method," *J Appl Mech*, **8**: 169-175 (1941).

[1.43]   D. McHenry, "A Lattice Analogy for the Solution of Plane Stress Problems," *J Inst Civil Eng*, **21**: 59-82 (1943).

[1.44]   N. M. Newmark, "Numerical Methods of Analysis of Bars, Plates and Elastic Bodies" in L. E. Grinter (ed.), *Numerical Methods of Analysis in Engineering*, Macmillan, New York, 1949.

[1.45]   B. Langefors, "Analysis of Elastic Structures by Matrix Transformation, with Special Regard to Semimonocoque Structures," *J Aeron Sci*, **19**(7): 451-458 (1952).

[1.46] J. H. Argyris, "Energy Theorems and Structural Analysis," *Aircraft Eng*, **26** (Oct.-Nov. 1954); **27** (Feb.-May 1955). Reprinted in J. H. Argyris and S. Kelsey, *Energy Theorems and Structural Analysis*, Butterworth, London, 1960.

[1.47] M. J. Turner, R. W. Clough, H. C. Martin, and L. J. Topp, "Stiffness and Deflection Analysis of Complex Structures," *J Aeron Sci*, **23**(9): 805-823,854 (1956).

[1.48] R. W. Clough, "The Finite Element Method in Plane Stress Analysis," *Proc 2d ASCE Conf. Electronic Computation*, Pittsburgh, PA., Sept. 1960, pp. 345-378.

[1.49] R. J. Melosh, "Basis for the Derivation of Matrices for the Direct Stiffness Method," *AIAA J*, **1**: 1631-1637 (1963).

[1.50] J. F. Besseling, "The Complete Analogy Between the Matrix Equations and the Continuous Field Equations of Structural Analysis," International Symposium on Analogue and Digital Techniques Applied to Aeronautics, Liege, Belgium, 1963.

[1.51] R. E. Jones, "A Generalization of the Direct-Stiffness Method of Structural Analysis," *AIAA J*, **2**: 821-826 (1964).

[1.52] B. Fraeijs de Veubeke, "Upper and Lower Bounds in Matrix Structural Analysis," B. F. de Veubeke (ed.), *Matrix Methods of Structural Analysis*, AGARD, **72**, Pergamon, New York, 1964, pp. 165-201.

[1.53] O. C. Zienkiewicz and Y. K. Cheung, "Finite Elements in the Solution of Field Problems," *The Engineer*, **220**: 507-510 (1965).

[1.54] J. E. Akin, D. L. Fenton, and W. C. T. Stoddart, "The Finite Element Method — A Bibliography of Its Theory and Applications," Rept. EM 72-1, Department of Engineering Mechanics, University of Tennessee, 1971.

[1.55] J. R. Whiteman, *A Bibliography for Finite Elements*, Academic Press, New York, 1975.

[1.56] R. W. Clough, "The Finite Element Method after Twenty-Five Years: A Personal View," *Computers and Structures*, **12**: 361-370 (1980).

2

Outline of
Problems Treated in This Book

Introduction

The problems treated in this book cover many fields of application in the physical sciences. It is a happy fact of nature that only a few governing equations are necessary to describe many of these different applications. This book presents a very detailed and thorough explanation of the FE analysis of these few equations, beginning with the FE theory, then the coding into a computer program, and finally the application of the program to specific problems in the various fields.

The present chapter identifies these governing equations and their fields of application and provides a brief outline of the book.

2.1 One-, Two-, and Three-Dimensional Considerations

As explained in the preface, there are a variety of reasons that favor introducing new concepts in their simplest possible context. For the FEM this means working initially with problems in only *one* dimension i.e., with only one independent variable, which hereafter is represented by the letter x. In this book the governing equations are differential equations. For a one-dimensional (1-D) problem the governing equations will therefore be *ordinary* differential equations (sometimes abbreviated ODEs), which are equations containing an unknown function of x, say $U(x)$, and one or more of its total (or *ordinary*) derivatives — dU/dx, d^2U/dx^2, etc. [See, for example, Eqs. (2.1), (2.3), (2.5), (2.7), and (2.10).] The mathematical concepts involved in the FE analysis of an ordinary differential equation presuppose a knowledge of two or three semesters of basic calculus, a familiarity with matrix notation (such as the representation of a system of linear algebraic equations as the product of a matrix and a vector), two years of high school algebra that includes techniques for solving simultaneous linear algebraic equations, and perhaps a brief exposure to elementary numerical integration techniques.

It is fortunate that so many of the basic FE ideas can be taught in this 1-D context, and still more fortunate that these ideas are just as relevant to problems in two or more dimensions. Although the practical benefits of the FEM are usually greatest with problems in two or more dimensions, it should by no means be inferred that the 1-D problem is worthy only as a teaching tool. Many industrial problems, especially those in the early research and development phases, can be meaningfully investigated by studying 1-D models (recall Examples 1.3 and 1.5 in Section 1.4). It is, in fact, the mature and skillful engineer who can reduce an apparent 2-D or 3-D problem to an approximately equivalent 1-D problem. Also, the reasonableness of a complicated 2-D or 3-D analysis can frequently be tested by modeling parts of the problem as approximately equivalent 1-D models. We can thus see that there are both pedagogical and practical benefits to mastering the 1-D problem first. Consequently, Part II (Chapters 3 through 5) and Part III (Chapters 6 through 12) concentrate entirely on 1-D problems.[1]

Part IV (Chapters 13 through 16) works with problems in *two* dimensions, i.e., with two independent variables, which hereafter are represented by the letters x and y. For a two-dimensional (2-D) problem the governing equations will therefore be *partial* differential equations (sometimes abbreviated *PDEs*), which are equations containing an unknown function of x and y — say $U(x,y)$ — and one or more of its partial derivatives, $\partial U/\partial x$, $\partial U/\partial y$, $\partial^2 U/\partial x^2$, etc. [See, for example, Eqs. (2.14), (2.15), (2.17), (2.19), and (2.21).] The mathematical concepts involved in the FE analysis of a partial differential equation presuppose a knowledge of one or two additional semesters of calculus, what might be termed "intermediate calculus." The student will discover striking parallels between the 1-D and 2-D problems, with many of the differences involving only the underlying mathematical details.

Three-dimensional problems will not be treated here because the parallels between 2-D and 3-D, as regards both concepts and practical techniques, are even greater than those between 1-D and 2-D. Perhaps the principal challenge in 3-D problems is the efficient management of computer resources, since the amount of data processing can be extremely large.

The sections that follow in this chapter look at each major division of the book.

2.2 Part II: Basic Concepts

As the title indicates, Part II, which comprises Chapters 3, 4, and 5, develops the basic concepts of the FEM, those which are present in all FE analyses. More specialized concepts are developed throughout the remainder of the book.

The most important and fundamental of all the concepts is the element; it pervades almost every aspect of the practical application of the FEM. Nevertheless, in what may seem to be a paradoxical statement, a considerable amount of the mathematical structure of the FEM is independent of the element concept. This fact explains the structure of Chapters 3 through 5.

Chapters 3 and 4 are entitled Foundations I and II, respectively, because they develop as much of the structure as possible without using the element concept, which is not

---

[1] Chapters 11 and 12 involve two dimensions, space and time, but only the space dimension is treated by FE concepts, so the problems may be classified as 1-D as regards the FE analysis. See further remarks in Section 2.3.

introduced until Chapter 5. There are compelling logical and pedagogical reasons for this approach.

Breaking a problem into elements (recall Section 1.2), ipso facto, fragments the problem into separate pieces, a process that spawns additional computational details that easily obscure many important relationships — a case of not being able to see the forest for the trees. These details are distracting, especially when they are not logically necessary. Conversely, eliminating these details (by not using elements) permits a clearer and more smoothly flowing presentation of all the other concepts.

In addition, there are properties of the system as a whole that are mimicked by each separate element. If this can be seen and appreciated, it provides additional insight.

If the element concept were introduced at the outset, then the ensuing development might leave the reader with the false impression that virtually the entire structure is to be associated only with the FEM. On the contrary, most of the FEM consists of ideas borrowed from non-FEM applications. In a sense, the FEM is not the result of a discovery of a single revolutionary concept; rather, it is the result of a clever and effective integration of many traditional concepts. Consequently, many readers will be pleasantly surprised to discover that they already know quite a bit about the FEM.

The above comments can perhaps be better appreciated by the following observation: If a problem is not broken into elements, one can consider the entire problem to be *one* element. In this sense, Chapters 3 and 4 can be viewed as treating a one-element problem, and Chapter 5 as treating two or more elements.

By the end of Chapter 4, most of the basic mathematical structure will have been developed. The introduction of the element concept in Chapter 5 will then require only a few additional details to be added to this structure. For example, Section 5.3 will break the problem into two halves (i.e., two elements), and then each half will be analyzed by the method used in Chapter 4. Sections 5.4 and 5.5 will break the problem into still more elements, yet each element will again be analyzed by the method introduced in Chapter 4.

As the concepts in Chapters 3, 4, and 5 are developed, they will be applied to a specific illustration problem, in order to provide an immediate concrete example. The same problem will be used for all three chapters. This is because the focus in these chapters is the *method* of solution; by keeping the problem constant, any changes that occur from one chapter to the next must be due solely to the method itself, not to changes in the problem. (In the remainder of the book we will then apply the method to many different problems, and, at the same time, develop additional concepts as we go along.)

The specific problem used for illustration is a 1-D *boundary-value* problem, also called an *equilibrium problem*. The governing equation is the following ordinary differential equation:[2]

$$\frac{d}{dx}\left(x \, \frac{dU(x)}{dx} \right) = \frac{2}{x^2} \tag{2.1}$$

Here x is the independent variable. The domain is the interval $1 < x < 2$ along the x-axis (Fig. 2.1).

---

[2] The reason for choosing a differential equation in the particular form of Eq. (2.1) is explained in Section 3.2.

FIGURE 2.1 ● Domain for the illustrative 1-D boundary-value problem in Part II.

The boundary of the domain is the two end points $x=1$ and $x=2$. The boundary conditions at these points are

$$U(1) = 2$$

$$\left(-x\frac{dU}{dx}\right)_{x=2} = \frac{1}{2} \tag{2.2}$$

$U(x)$ is the dependent variable, sometimes referred to as the *state* variable because U and its derivatives completely describe the state of the system. The symbol U reminds us that $U(x)$ is an *un*known function at the beginning of the problem.

Equations (2.1) and (2.2) together are called a boundary-value problem because numerical values (data) must be specified at every point of the boundary. The data appears in equations, called *boundary conditions* (BCs), which specify numerical values for $U(x)$ and/or some of its derivatives on the boundary.[3]

The number of boundary conditions that must be specified at each boundary point depends on the order of the governing differential equation, which is the order of the highest derivative in the differential equation. For example, Eq. (2.1) is of second-order because the highest derivative (after differentiating the terms in parentheses) is d^2U/dx^2. In general, if the differential equation is of order $2m$, then m boundary conditions must be specified at each boundary point (see Section 4.5). Therefore Eq. (2.1) must have one boundary condition specified at each of its two boundary points — as in Eqs. (2.2).

The purpose of the FE analysis is to find an explicit expression for $U(x)$, in terms of known functions, which *approximately* satisfies Eq. (2.1) in the domain and Eqs. (2.2) on the boundary. Such an expression constitutes an approximate solution to the problem, and will be denoted by a tilde over the U: $\tilde{U}(x)$. Indeed, Chapters 3, 4, and 5 will find several different approximate solutions, all slightly different from each other yet close to the exact solution. Figure 2.2 demonstrates symbolically that there exists only one exact solution, $U(x)$, but an unlimited number of approximate solutions, $\tilde{U}_1(x)$, $\tilde{U}_2(x)$, . . . , that may be close to the exact solution.

This illustration problem has several physical applications. Equation (2.1) is a special case of Eq. (2.3), described in the next section, which is the more general 1-D ordinary differential equation treated in Part III. Therefore this problem may represent any of the physical applications listed in Table 2.1 below and described more fully in Part III, Chapter 6.

---

[3] Mathematicians frequently refer to 1-D boundary-value problems as *two-point* boundary-value problems, since the boundary of a 1-D problem consists of two points, i.e., one at either end of a line. (It also provides a distinction from 1-D *initial-value* problems which require boundary data at only one point.) Such terminology, however, has no meaningful extension to 2-D problems, for which the boundary is one or more curves, or to 3-D problems, for which the boundary is one or more surfaces. Therefore the author prefers the consistent terminology: 1-D, 2-D, and 3-D.

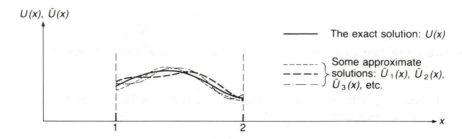

$U(x), \bar{U}(x)$

———— The exact solution: $U(x)$

$\left.\begin{array}{l} \text{------} \\ \text{-- -- --} \\ \text{-- · --} \end{array}\right\}$ Some approximate solutions: $\bar{U}_1(x), \bar{U}_2(x), \bar{U}_3(x)$, etc.

x

1   2

FIGURE 2.2 ● The FE analysis seeks to find one (or more) of the unlimited number of approximate solutions that are close to the exact solution.

TABLE 2.1 ● Some physical applications for the 1-D boundary-value problem, Eq. (2.3), analyzed in Part III.

Mechanics: Transverse deflection of a flexible cable
Elasticity: Longitudinal deformation of an elastic rod
Heat conduction: 1-D temperature distribution in a solid structure
Electrostatics: 1-D electric potential distribution
Magnetostatics: 1-D magnetic potential distribution
Hydrodynamics: 1-D flow pattern in an inviscid, incompressible fluid

---

Part II contains no computer programming because the details of programming are reserved principally for Part III.

2.3 Part III: One-Dimensional Problems

The purpose of Part III is to carry through a *complete* FE analysis of several types of 1-D problems, as flowcharted in Fig. 1.5. This means performing the FE analysis of the governing equations, programming the resulting FE equations, and finally using the program to solve some specific problems — both in the text and in end-of-chapter exercises.

Chapters 6 through 8 work with a general 1-D boundary-value (equilibrium) problem. The governing equation is the following ordinary differential equation:

$$-\frac{d}{dx}\left[\alpha(x)\frac{dU(x)}{dx}\right] + \beta(x)U(x) = f(x) \tag{2.3}$$

Here x is the independent variable, $U(x)$ is the unknown function (state variable), and $\alpha(x)$, $\beta(x)$, and $f(x)$ are known functions. The domain is any finite or infinite interval $x_a < x < x_b$ along the x-axis (Fig. 2.3). The functions $\alpha(x)$ and $\beta(x)$ usually represent material or physical

Domain

x_a x_b x

FIGURE 2.3 ● Domain for the 1-D boundary-value problem in Part III.

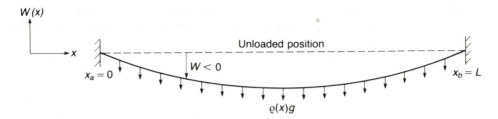

FIGURE 2.4 ● A flexible cable hanging under its own weight.

properties of the system. We will call $f(x)$ an *interior load* (or, sometimes, *volume load*) since it represents loading applied to the interior (i.e., the domain) of the system.

The boundary of the domain is the two end points $x=x_a$ and $x=x_b$. The boundary conditions at x_a and x_b may be either of two conditions:

At $x = x_a$,

$$U(x_a) = U_a \quad \text{or} \quad \left(-\alpha(x) \, \frac{dU(x)}{dx} \right)_{x_a} = \tau_a$$

At $x = x_b$,

$$U(x_b) = U_b \quad \text{or} \quad \left(-\alpha(x) \, \frac{dU(x)}{dx} \right)_{x_b} = \tau_b \tag{2.4}$$

where U_a, U_b, τ_a, and τ_b have known values. We will call U_a, U_b, τ_a, and τ_β *boundary loads* since they represent loading on the boundary of the system.[4] The quantity $-\alpha(dU/dx)$, represented by the symbol τ, has a definite physical meaning in each different physical application of Eq. (2.3). Lacking a general name appropriate to all applications, we will call it a *flux* since it frequently represents a flow of some quantity, although in some problems (e.g., the cable deflection problem described next) it represents a force.

Equation (2.3) occurs in several different physical applications, some of which are listed in Table 2.1.

To illustrate one of these applications, consider the transverse deflection of a flexible cable. Figure 2.4 shows such a cable of length L, stretched horizontally with a tension

---

[4] The author uses the word "load" to mean any influence or agent that causes the state of the system to change — for example, a force, displacement, temperature, voltage, heat flux, or pressure, to mention only a few. Loads may be applied to the interior and/or the boundary of a system; they are part of the *known* data of the problem. This is a mathematically motivated definition of the concept "load" since it is independent of any particular physical application.

It should be noted, though, that many FE authors prefer a more physically motivated definition, in which only some of the influences or agents are called "loads" and others are called "constraints." For example, in conventional FE formulations of mechanical problems, forces would still be called loads but displacements would be called "constraints." This terminology arose from certain historically popular mathematical techniques; however, in other unconventional techniques, displacements may be either constrained or unconstrained (see Section 4.5). Because of this potential ambiguity the author favors the broader mathematical definition.

T, and fastened at its ends $x_a=0$ and $x_b=L$. It is hanging under its own weight, $\varrho(x)g$, where $\varrho(x)$ is the mass per unit length (which may vary with position x) and g is the acceleration of gravity.

The governing equation is

$$-\frac{d}{dx}\left(T\frac{dW(x)}{dx}\right) = -\varrho(x)g \qquad (2.5)$$

where $W(x)$ is the vertical (transverse) deflection of the cable from the unloaded horizontal position, and $\varrho(x)g$ is an interior load acting upon all points of the cable along its length (i.e., interior to the domain $0<x<L$). The boundary conditions are

$$W(0) = 0$$

$$W(L) = 0 \qquad (2.6)$$

We see that Eq. (2.5) has the same form as Eq. (2.3), where $\alpha(x)=T$, $\beta(x)=0$, and $f(x)=-\varrho(x)g$. In this application the flux is $\tau=-T(dW/dx)$, which, for small deflections, is the vertical component of tension (see Chapter 6).

The complete derivation of Eq. (2.5) from physical principles is reserved for Chapter 6. Indeed, the governing equations for several of the applications in Table 2.1, represented abstractly by Eq. (2.3), are derived in Chapter 6, starting from the appropriate physical principles in each case. Chapter 6 provides an introduction to Part III by explaining the physical problems whose behavior is described (ideally) by the governing equation analyzed in Chapters 7 and 8.[5]

Chapter 7 is, in a sense, what the previous six chapters have been leading up to. Here the reader is carried step by step through a complete FE analysis, from theory to programming to application, utilizing the simple element developed in Chapter 5. The FE equations derived from Eq. (2.3) have been coded into a computer program called UNAFEM.[6] Parts of the program listing (in the appendix) are reproduced in enlarged format in Chapter 7 to illustrate how the equations are translated into code. UNAFEM is then applied to two specific problems: heat conduction in a multilayered material and a flexible cable supporting concentrated weights. Complete details of input data and numerical results are shown.

The appendix also contains the data input instructions for UNAFEM. If the reader is able to install UNAFEM on a computer, it will provide excellent practical experience in using an FE program to solve specific application problems. A few such problems are included as end-of-chapter exercises.

Chapter 8 continues to work with Eq. (2.3) by improving the approximation capability developed in Chapter 7. This is done by developing three different "higher-order" elements, which provide (usually, but not always) increasingly greater accuracy than provided by the simple element used in Chapter 7. The approximating FE equations are derived

---

[5] Some readers may wish to peruse Chapter 6 even before proceeding with Part II.

[6] This program implements most of the analysis presented in Parts III and IV. The appendix contains a complete, photo-reduced listing.

in complete detail for each of the three elements, and their implementation into the UNAFEM code is explained. Finally, the program is applied to the same two problems in Chapter 7. End-of-chapter exercises provide additional opportunity for working with the different types of elements.

Chapter 9 assesses the relative merits of the previous four elements. A considerable amount of numerical data demonstrates their accuracies in different situations. This in turn provides insight into the effective use of the different elements.

Chapter 10 adds an extra term to Eq. (2.3), which converts it into a related *eigenproblem*:

$$-\frac{d}{dx}\left(\alpha(x)\frac{dU(x)}{dx}\right) + \beta(x)U(x) - \lambda\gamma(x)U(x) = 0 \tag{2.7}$$

where $\gamma(x)$ is a known function and λ is an unknown scalar-valued parameter called an *eigenvalue*. The domain is the same as for the boundary-value problem shown in Fig. 2.3. The boundary conditions at x_a and x_b may be either of two conditions:

At $x = x_a$,

$$U(x_a) = 0 \quad \text{or} \quad \left(-\alpha(x)\frac{dU(x)}{dx}\right)_{x_a} = 0$$

At $x = x_b$,

$$U(x_b) = 0 \quad \text{or} \quad \left(-\alpha(x)\frac{dU(x)}{dx}\right)_{x_b} = 0 \tag{2.8}$$

It should be noted that the interior load — the right-hand side of Eq. (2.7) — and the boundary loads — the right-hand side of Eqs. (2.8) — are both zero; i.e., no loads are driving the system, a condition that characterizes the eigenproblem. A solution to this problem requires finding both a function for $U(x)$, called an *eigenfunction* (or *mode*), and a corresponding value for the eigenvalue λ. In fact, there are usually an infinite number of eigenfunction/eigenvalue pairs in any specific application, although fortunately only a few are generally of practical interest. Equation (2.7) has several physical applications, some of which are listed in Table 2.2.

To illustrate the eigenproblem, consider again the flexible cable shown in Fig. 2.4, but without the load $\varrho(x)g$, so that there are neither interior nor boundary loads on the cable. We ask whether the cable can vibrate "freely" (no loads), which means each point

TABLE 2.2 ● Some physical applications for the 1-D eigenproblem, Eq. (2.7), analyzed in Part III.

Mechanics: Transverse modes of vibration of a flexible cable

Elasticity: Longitudinal modes of vibration of an elastic rod

Acoustics: 1-D modes of vibration in an inviscid, compressible fluid

Electromagnetism: 1-D modes of vibration of electric and magnetic field components

Quantum mechanics: Probability amplitude functions for stationary states of a particle in a 1-D potential field

---

of the cable is moving back and forth sinusoidally in time t. Such a "free vibration" is of the form

$$w(x,t) = W(x) \sin \omega t \qquad (2.9)$$

where $W(x)$ is the amplitude of vibration at each location x, and $\omega = 2\pi f$ is the circular frequency. As shown in Section 10.1, $W(x)$ is the solution to the eigenproblem

$$-\frac{d}{dx}\left(T\frac{dW(x)}{dx}\right) - \omega^2 \varrho(x)W(x) = 0 \qquad (2.10)$$

with the boundary conditions

$$W(0) = 0$$

$$W(L) = 0 \qquad (2.11)$$

We see that Eq. (2.10) has the same form as Eq. (2.7), where $\alpha(x) = T$, $\beta(x) = 0$, $\lambda = \omega^2$, and $\gamma(x) = \varrho(x)$. Here the eigenvalue is ω^2, so that each calculated eigenvalue represents a critical (or "natural") frequency corresponding to a particular mode of vibration.

The complete derivation of Eq. (2.10) from physical principles is presented in Section 10.1, along with a physical interpretation of Eq. (2.7) for some of the other applications listed in Table 2.2. The FE equations for Eq. (2.7) are then derived, the resulting expressions are added to the UNAFEM program, and the coding is explained. UNAFEM is then applied to two specific problems: a vibrating cable and a quantum mechanical oscillator. Complete details of input data and numerical results are shown. End-of-chapter exercises provide additional experience for the reader in applying UNAFEM to other types of eigenproblems.

Chapter 11 adds an extra term (a first-order time derivative) to Eq. (2.3), converting it into the related *initial-boundary-value* problem,

$$\mu(x)\frac{\partial U(x,t)}{\partial t} - \frac{\partial}{\partial x}\left(\alpha(x)\frac{\partial U(x,t)}{\partial x}\right) + \beta(x)U(x,t) = f(x,t) \qquad (2.12)$$

There are now two independent variables, x and t, where t usually represents time. The derivatives must therefore be partial rather than total, and the problem is now two-dimensional. However, as will be shown in Chapter 11, only the space dimension x is dealt with by the FEM. The time dimension is dealt with separately, using an algorithm that is usually derived by the finite-difference method. (But, as will be shown, it can also be derived by the FEM.) Consequently the problem is one-dimensional as regards the FE analysis, and the four 1-D elements developed in Chapters 7 and 8 will be usable here.

The boundary conditions are the same as in Eqs. (2.4), except that they may be a function of time. In addition, there must also be specified *initial conditions*. Both are explained in Section 11.1.

Equation (2.12) describes *diffusion* problems, examples of which are listed in Table 2.3. A solution is therefore changing in time, i.e., transient, beginning from an initial instant and proceeding indefinitely into the future.

Chapter 10 derives the FE equations for Eq. (2.12), and the resulting expressions are added to the UNAFEM program. UNAFEM is then applied to a transient-heat-conduction problem.

TABLE 2.3 ● Some physical applications for the 1-D initial-boundary-value problem, Eq. (2.12), analyzed in Part III.

Transient heat conduction: Diffusion of heat in a solid material
Fluid/solid percolation: Flow of liquid through a porous solid
Fluid/fluid mixing: Flow of one liquid (solute) through another liquid (solvent)
Atomic physics: Diffusion of electrons in a gas
Nuclear physics: Diffusion of neutrons in matter

Chapter 12 considers another class of initial-boundary-value problems, those dealing with wave propagation and vibration, frequently referred to as *dynamics* problems. The governing differential equation is obtained by adding another time-derivative term (second order) to the diffusion equation:

$$\varrho(x) \frac{\partial^2 U(x,t)}{\partial t^2} + \mu(x) \frac{\partial U(x,t)}{\partial t} - \frac{\partial}{\partial x}\left(\alpha(x) \frac{\partial U(x,t)}{\partial x}\right) + \beta(x)U(x,t) = f(x,t) \tag{2.13}$$

As with the diffusion problem, the space and time dimensions are treated separately, and the four 1-D elements from Chapters 7 and 8 will once again be usable. Chapter 12 derives the FE equations for Eq. (2.13). The resulting expressions are not added to the UNAFEM program, although their implementation would be similar to that in the diffusion problem.

This completes the one-dimensional work. In this context we have covered the basic concepts and acquired practical training in problems occurring in the three principal types of engineering problems: boundary-value, eigenvalue, and initial-boundary-value.

2.4 Part IV: Two-Dimensional Problems

The purpose of Part IV is to extend the concepts learned in Part III to two dimensions. The presentation is shorter than the 1-D treatment, relying on frequent analogies with the 1-D material.

The essential difference between the 1-D work in Part III and the 2-D work in Part IV is, of course, the dimensionality, and this feature increases the complexity of the calculus. The domain of 2-D problems will be the x,y-plane; i.e., there will be two independent variables, x and y.[7] Therefore derivatives will be partial derivatives and governing equations will be partial differential equations.

Chapter 13 works with a general 2-D boundary-value (equilibrium) problem; it is the 2-D counterpart of the 1-D problem in Eq. (2.3). The governing equation is the following partial differential equation:

$$-\frac{\partial}{\partial x}\left(\alpha_x(x,y) \frac{\partial U(x,y)}{\partial x}\right) - \frac{\partial}{\partial y}\left(\alpha_y(x,y) \frac{\partial U(x,y)}{\partial y}\right) + \beta(x,y)U(x,y) = f(x,y) \tag{2.14}$$

---

[7] The initial-boundary-value problem will have a third independent variable, time, but it will be treated by finite differences, and the variables x and y will be treated by finite elements — in complete analogy with the 1-D problems in Part III.

Here x and y are the independent variables, $U(x,y)$ is the unknown function (state variable), and $\alpha_x(x,y)$, $\alpha_y(x,y)$, $\beta(x,y)$, and $f(x,y)$ are known functions with meanings similar to those in the 1-D problem. The domain is any finite region in the x,y-plane (Fig. 2.5). The boundary is the curve enclosing the domain; it may be any arbitrary shape. Boundary conditions similar to those in Eqs. (2.4), but involving partial derivatives, must be applied along the entire boundary curve, i.e., at every point of the boundary.

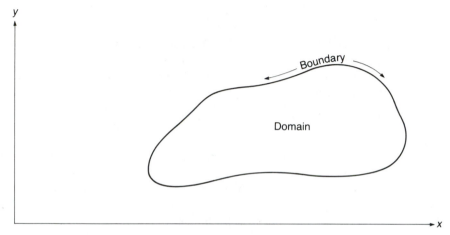

FIGURE 2.5 ● Domain for the 2-D boundary-value problem in Part IV.

TABLE 2.4 ● Some physical applications for the 2-D boundary-value problem, Eq. (2.14), analyzed in Part IV.

Mechanics: Transverse deflection of a flexible membrane
Heat conduction: 2-D temperature distribution in a solid structure
Electrostatics: 2-D electric potential distribution
Magnetostatics: 2-D magnetic potential distribution
Hydrodynamics: 2-D flow pattern in an inviscid, incompressible fluid

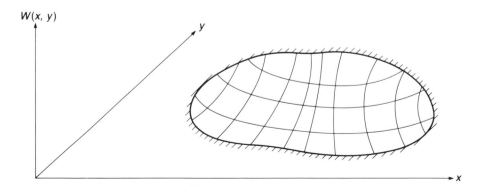

FIGURE 2.6 ● A flexible membrane hanging under its own weight.

Table 2.4 lists some of the physical applications of Eq. (2.14). These are the direct extensions of the 1-D problems in Table 2.1 to two dimensions. An exception is the 1-D elastic rod, which becomes the plane stress problem in 2-D (see Chapter 16).

To illustrate, consider the transverse deflection of a flexible membrane (the 2-D extension of the 1-D flexible cable in Fig. 2.4). Figure 2.6 shows such a membrane, stretched horizontally with a tension per unit length, T, and fastened along the entire boundary. It is hanging under its own weight, $\varrho(x,y)g$, where $\varrho(x,y)$ is the mass per unit area. The governing equation is

$$-\frac{\partial}{\partial x}\left(T\frac{\partial W(x,y)}{\partial x}\right) - \frac{\partial}{\partial y}\left(T\frac{\partial W(x,y)}{\partial y}\right) = -\varrho(x,y)g \qquad (2.15)$$

where $W(x,y)$ is the vertical deflection from the unloaded horizontal position. [Note that Eq. (2.5) is the 1-D analogue of Eq. (2.15).] The boundary condition is

$$W(B) = 0 \qquad (2.16)$$

where B represents the boundary, i.e., every point along the boundary. We see that Eq. (2.15) has the same form as Eq. (2.14), where $\alpha_x(x,y)=\alpha_y(x,y)=T$; $\beta(x,y)=0$; and $f(x,y)=-\varrho(x,y)g$.

Chapter 13 derives the FE equations for four different 2-D elements. These equations have been coded into UNAFEM, and parts of the program listing (in the appendix) are reproduced in Chapter 13 to illustrate the implementation of the theory into code. UNAFEM is then applied to the problem of torsion of an I-beam, showing complete details from the input data to the numerical results. End-of-chapter exercises provide additional problem-solving experience.

Chapter 14 adds an extra term to Eq. (2.14), converting it into the related eigenproblem:

$$-\frac{\partial}{\partial x}\left(\alpha_x(x,y)\frac{\partial U(x,y)}{\partial x}\right) - \frac{\partial}{\partial y}\left(\alpha_y(x,y)\frac{\partial U(x,y)}{\partial y}\right)$$
$$+ \beta(x,y)U(x,y) - \lambda\gamma(x,y)U(x,y) = 0 \qquad (2.17)$$

where $\gamma(x,y)$ is a known function and λ is the unknown eigenvalue. The domain is of general form (see Fig. 2.5). Boundary conditions similar to those in Eqs. (2.8), but involving partial derivatives, must be applied at every point along the boundary.

Table 2.5 lists some of the physical applications of Eq. (2.17). These are the direct extensions of the 1-D problems in Table 2.2 to two dimensions, with the exception of the elastic rod for the reason given above.

TABLE 2.5 ● Some physical applications for the 2-D eigenproblem, Eq. (2.17), analyzed in Part IV.

Mechanics: Transverse modes of vibration of a flexible membrane

Acoustics: 2-D modes of vibration in an inviscid, compressible fluid

Hydrodynamics: 2-D modes of vibration of shallow surface waves in an inviscid, incompressible fluid

Electromagnetism: 2-D modes of vibration of electric and magnetic field components (e.g., an electromagnetic waveguide)

Quantum mechanics: Probability amplitude functions for stationary states of a particle in a 2-D potential field

---

If we use the flexible membrane again for illustration, the transverse motion for free vibration is described by

$$w(x,y,t) = W(x,y) \sin \omega t \qquad (2.18)$$

where $W(x,y)$ is the amplitude of vibration. $W(x,y)$ is the solution to the eigenproblem,

$$-\frac{\partial}{\partial x}\left(T \frac{\partial W(x,y)}{\partial x}\right) - \frac{\partial}{\partial y}\left(T \frac{\partial W(x,y)}{\partial y}\right) - \omega^2 \varrho(x,y) W(x,y) = 0 \qquad (2.19)$$

with the boundary condition

$$W(B) = 0 \qquad (2.20)$$

where B represents the boundary. Clearly Eq. (2.19) has the same form as Eq. (2.17), and, as in the 1-D vibrating cable example, the eigenvalue is ω^2.

Chapter 14 derives the FE expressions for the extra term and adds them to the UNAFEM program. UNAFEM is then applied to an acoustics problem.

Chapter 15 adds an extra term to Eq. (2.14), converting it into the related initial-boundary-value problem:

$$\mu(x,y) \frac{\partial U(x,y,t)}{\partial t} - \frac{\partial}{\partial x}\left(\alpha_x(x,y) \frac{\partial U(x,y,t)}{\partial x}\right) - \frac{\partial}{\partial y}\left(\alpha_y(x,y) \frac{\partial U(x,y,t)}{\partial y}\right)$$
$$+ \beta(x,y)U(x,y,t) = f(x,y,t) \qquad (2.21)$$

Equation (2.21) is the 2-D analogue of the 1-D initial-boundary-value problem in Eq. (2.12). The physical applications of Eq. (2.21) are the same as in the 1-D case (Table 2.3). Chapter 15 derives the FE expressions for the extra term and adds them to UNAFEM, which is then applied to a transient heat conduction problem.

After having completed Chapters 7 through 15, the reader will have participated in the detailed development of a small, "general-purpose" FE computer program, capable of solving an interesting assortment of one- and two-dimensional problems, and similar in structure to many of the much larger, commercially available, general-purpose FE programs. This development is summarized in Fig. 2.7, which is a specialization of the general flowchart shown in Fig. 1.5.

Chapter 16, the final chapter, deals with a special topic, elasticity, and is included here for both technical and historical reasons. Technically, the governing equations for all the problems in the previous chapters have contained only one unknown function, symbolized by $U(x)$ in 1-D and $U(x,y)$ in 2-D. In Chapter 16 we want to look at how to deal with problems for which the governing equations contain two or more unknown functions. Many, perhaps most, important industrial problems fall into this category, e.g., solid mechanics (elasticity, plasticity, and other topics), fluid mechanics, and electromagnetism. In these applications there are typically two or three unknown functions, representing the components of a vector quantity. For example, in elasticity the displacement of a material point is usually the unknown. Displacement is a vector quantity with two components in 2-D and three components in 3-D. In fluid mechanics the unknown functions may be the components of a velocity vector, and electromagnetism may deal with the components of a vector potential. These ideas will be explained in the context of several types of 2-D boundary-value problems in linear elasticity: plane stress, plane strain, and axisymmetric bodies of revolution.

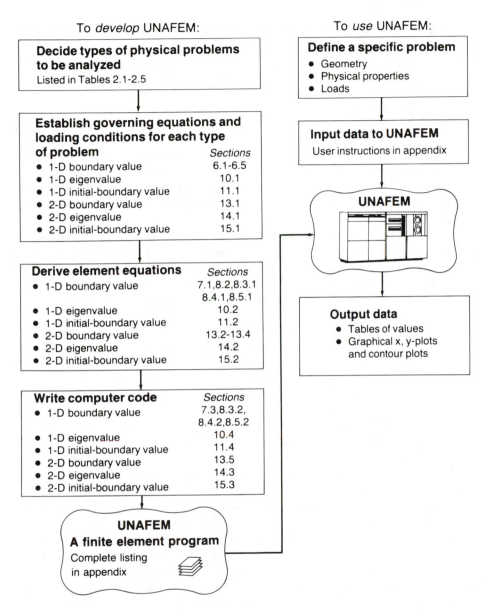

To *develop* UNAFEM:

Decide types of physical problems to be analyzed
Listed in Tables 2.1-2.5

Establish governing equations and loading conditions for each type of problem

	Sections
• 1-D boundary value	6.1-6.5
• 1-D eigenvalue	10.1
• 1-D initial-boundary value	11.1
• 2-D boundary value	13.1
• 2-D eigenvalue	14.1
• 2-D initial-boundary value	15.1

Derive element equations

	Sections
• 1-D boundary value	7.1,8.2,8.3.1 8.4.1,8.5.1
• 1-D eigenvalue	10.2
• 1-D initial-boundary value	11.2
• 2-D boundary value	13.2-13.4
• 2-D eigenvalue	14.2
• 2-D initial-boundary value	15.2

Write computer code

	Sections
• 1-D boundary value	7.3,8.3.2, 8.4.2,8.5.2
• 1-D eigenvalue	10.4
• 1-D initial-boundary value	11.4
• 2-D boundary value	13.5
• 2-D eigenvalue	14.3
• 2-D initial-boundary value	15.3

UNAFEM A finite element program
Complete listing in appendix

To *use* UNAFEM:

Define a specific problem
• Geometry
• Physical properties
• Loads

Input data to UNAFEM
User instructions in appendix

UNAFEM

Output data
• Tables of values
• Graphical x, y-plots and contour plots

FIGURE 2.7 ● Development and use of UNAFEM.

The topic of elasticity is chosen because it is historically the oldest and most thoroughly developed area of application of the FEM. Indeed, most other FE textbooks deal exclusively or predominantly with this topic. Consequently, a brief treatment of elasticity here can provide a bridge, or introduction, to those other books that deal with more advanced or more specialized topics.

The plane stress problem was already illustrated in the example problem in Section 1.3. For convenience we repeat the governing equations here:

$$\frac{E}{1-v^2}\frac{\partial^2 u(x,y)}{\partial x^2} + \frac{E}{2(1-v)}\frac{\partial^2 v(x,y)}{\partial x \partial y} + \frac{E}{2(1+v)}\frac{\partial^2 u(x,y)}{\partial y^2} = -f_x(x,y)$$

$$\frac{E}{1-v^2}\frac{\partial^2 v(x,y)}{\partial y^2} + \frac{E}{2(1-v)}\frac{\partial^2 u(x,y)}{\partial x \partial y} + \frac{E}{2(1+v)}\frac{\partial^2 v(x,y)}{\partial x^2} = -f_y(x,y) \qquad (2.22)$$

The two unknown functions are $u(x,y)$ and $v(x,y)$, representing the components in the x and y directions, respectively, of the displacement of a material point, located initially (before applying loads) at the position with coordinates x,y. A more complete mathematical and physical explanation of Eqs. (2.22) and associated boundary conditions is presented in Chapter 16.

There are two governing equations rather than one because there are two unknown functions. In general, the number of governing equations must equal the number of unknown functions. Each governing equation has a structure similar to the 2-D boundary-value problem in Chapter 13: a second-order partial differential equation (that is, the highest derivative is of second order). But the equations cannot be solved separately, as was done in Chapter 13, because they are *coupled* together; i.e., both $u(x,y)$ and $v(x,y)$ appear in each equation. This coupling is the source of the few and relatively minor differences between problems with one unknown and problems with several unknowns.

The FE analysis of Eqs. (2.22) follows the same pattern used in the previous chapters. For example, the 2-D elements used in Chapter 13 for the unknown function $U(x,y)$ can be used in Chapter 16 for each of the functions $u(x,y)$ and $v(x,y)$. It will be seen, in fact, that dealing with two (or more) unknown functions instead of one introduces no new FE concepts; it merely increases the amount of calculus and algebra and creates more terms to keep track of. The new problems are therefore more of a bookkeeping than of an analytical nature.

Chapter 16 derives the FE equations corresponding to a more generalized version of Eqs. (2.22) and indicates how they might be programmed (essentially the same as UNAFEM, in fact). No program is included in this book because so many 2-D elasticity FE programs are already in existence that most readers of this book will have access to one.

A final observation: Essentially the entire book is devoted to linear analysis. Nonlinear problems are not treated, apart from occasional passing comments and a series of tutorial exercises at the end of Chapter 15, because, in the author's opinion, one should first achieve some degree of familiarity and proficiency with linear analysis (the goal of this book) before attempting a nonlinear analysis. Nonlinearities add to the mathematical complexity of the theory and programming, and this in turn adds to the difficulty in performing a reliable nonlinear analysis.[8] However, after completing this textbook, the reader will have the tools to delve profitably into nonlinear analysis.

---

[8] Despite these challenges, the FEM is admirably well suited for treating virtually all kinds of nonlinear problems. Indeed, many of today's very large general-purpose FE programs contain extensive nonlinear capabilities, and there exist many smaller special-purpose programs devoted primarily to nonlinear analysis.

P A R T

II

Basic Concepts

3

Foundations I:
Trial-Solution Methods

Introduction

This chapter provides a skeletal framework for the mathematical structure of the FEM. It identifies three principal operations that are present in every FE analysis: construction of a trial solution, application of an optimizing criterion, and estimation of accuracy. These operations are briefly described, then illustrated by application to a specific problem.

The emphasis here is more on ideas than on mathematics. These ideas will provide a conceptual structure that will help tie together the many details in subsequent chapters.

3.1 The Trial-Solution Procedure: Three Principal Operations

Chapter 2 presented several types of physical problems and their corresponding mathematical formulations in terms of differential equations and boundary conditions. These formulations contain an unknown function, denoted by $U(x)$ for 1-D problems and $U(x,y)$ for 2-D problems. For most practical problems it is impossible to determine the *exact* solution to these equations, — i.e., to find an explicit expression for U, in terms of known functions, which exactly satisfies the governing equation(s) and boundary conditions. As an alternative, the FEM seeks an *approximate* solution: an explicit expression for U, in terms of known functions, which only approximately satisfies the governing equation(s) and boundary conditions.[1] We will denote such an approximate solution by the letter U with a tilde over it. Thus, $\tilde{U}$ will denote an approximate solution, whereas U will denote the exact solution.

The FEM obtains an approximate solution by using the classical *trial-solution* procedure. This procedure forms the basic structure of every FE analysis, irrespective of whether the problem is formulated in terms of differential equations (equations containing

---

[1] To be precise, the approximate solution may satisfy some of the boundary conditions exactly.

derivatives of U), or integral or variational equations (which contain integrals of U and/or its derivatives), or integrodifferential equations.

The trial-solution procedure is characterized by three principal operations. Listed in their order of application, they are

1. Construction of a *trial solution* for $\tilde{U}$

2. Application of an *optimizing criterion* to $\tilde{U}$

3. Estimation of the accuracy of $\tilde{U}$

Before illustrating the procedure with a specific problem (Sections 3.2 through 3.6), we want to briefly describe each of the three operations and show how they are interrelated.

3.1.1 Construction of a Trial Solution

The first operation involves the construction of a trial solution $\tilde{U}(x;a)$ in the form of a *finite* sum of functions:

$$\tilde{U}(x;a) = \phi_0(x) + a_1\phi_1(x) + a_2\phi_2(x) + \ldots + a_N\phi_N(x) \tag{3.1}$$

Here x represents all the independent variables in the problem.[2] The functions $\phi_0(x)$, $\phi_1(x),\ldots,\phi_N(x)$ are known functions called *trial functions* (or sometimes *basis* or *coordinate functions*). The coefficients $a_1, a_2,\ldots, a_N$ are undetermined parameters,[3] frequently called *degrees of freedom (DOF)* or sometimes *generalized coordinates*. We would say that $\tilde{U}(x;a)$ in Eq. (3.1) has N DOF. The symbol a in $\tilde{U}(x;a)$ represents all the parameters $a_1, a_2,\ldots ,a_N$ on the right-hand side; it reminds us that $\tilde{U}$ is a function of x as well as of $a_1, a_2,\ldots, a_N$.

The trial function $\phi_0(x)$ is not multiplied by any parameter. Its purpose is to satisfy some or all of the boundary conditions. This will be illustrated in Section 3.3.

The construction of a trial solution consists of constructing expressions for each of the trial functions in terms of specific, *known* functions. From a practical standpoint it is important to use functions that are algebraically as simple as possible and also easy to work with — for example, powers of x (polynomials) or trigonometric sines and cosines.[4] One of the principal attractions of the FEM is that it provides a systematic procedure for constructing trial functions, and the procedure can be automated on a computer. Indeed, the very essence of the FEM lies in the special manner in which the trial functions are constructed.

Let's assume for the moment that we have somehow (to be explained later) constructed a trial solution in the form of Eq. (3.1); i.e., we have established specific expressions for each of the trial functions $\phi_i(x)$. Only the parameters a_i remain undetermined.

---

[2] For example, in a 2-D problem with x and y the independent variables, Eq. (3.1) would be written

$$\tilde{U}(x,y;a) = \phi_0(x,y) + a_1\phi_1(x,y) + a_2\phi_2(x,y) + \ldots + a_N\phi_N(x,y).$$

[3] The coefficients a_i are parameters in the conventional FE formulation of boundary-value problems and eigenproblems. However, in mixed initial-value/boundary-value problems (see Chapters 11, 12, and 15) they may be functions of time.

[4] Approximating a function by a finite sum of known, simple functions [e.g., approximating $U(x)$ by $\tilde{U}(x;a)$] is one of the oldest approximation techniques, dating back at least to the eighteenth century with the work of Euler and Lagrange, and the nineteenth century with the work of Fourier.

3.1.2 Application of an Optimizing Criterion

The purpose of the optimizing criterion is to determine specific numerical values for each of the parameters a_1, a_2, ..., a_N. Observe that a particular set of values for all the a_i uniquely defines a particular solution, because then all the a_i and $\phi_i(x)$ in Eq. (3.1) are uniquely determined. Since each a_i can assume an infinity of possible values ($-\infty < a_i < +\infty$), there is an N-fold infinity of possible solutions. It is the job of the optimizing criterion to select from all these possibilities a best (or optimum) solution[5] — i.e., a best set of values for the a_i. By "best," we mean that the solution is *as close as possible, in some sense, to the exact solution.*

There are two types of optimizing criteria that have played a dominant role historically (pre-FEM) as well as in the FEM:

1. *Methods of weighted residuals* (MWR), which are applicable when the governing equations are differential equations
2. *Ritz variational method* (RVM), which is applicable when the governing equations are variational (integral) equations.

MWR criteria seek to minimize an expression of error in the differential equation (not the unknown function itself). There are many different MWR criteria [3.1, 3.2]. Section 3.4 will demonstrate four of the most popular MWR criteria:

1. The collocation method
2. The subdomain method
3. The least-squares method
4. The Galerkin method

Variational principles, sometimes referred to as *extremum* or *minimum* principles, seek to minimize, or find an extremum in, some physical quantity, such as energy. Examples can be cited from many fields of engineering and physics [3.1–3.8] — e.g., the principle of minimum potential energy in solid mechanics. Section 3.5 will demonstrate the Ritz variational method.

Now let's assume we have applied one of these criteria to our trial solution $\tilde{U}(x;a)$, thereby determining a best set of values for all the a_i; i.e., we have a best solution. We will call this best solution an *approximate solution* since hopefully, and usually, it is a reasonable approximation to the exact solution. And we will designate it by $\tilde{U}(x)$, with no a in the argument, because it is now only a function of the variable x (the a_i having been numerically determined). Since $\tilde{U}(x)$ is only approximate, there are questions of accuracy that naturally arise.

3.1.3 Estimation of Accuracy

We would like to get some indication of the closeness of the approximate solution $\tilde{U}(x)$ to the exact solution $U(x)$. Without *some* indication, the solution is effectively worthless.

---

[5] We can only speak of *a* best, not *the* best, solution because different optimizing criteria generally (though not always) produce different optimum solutions, as will be demonstrated in Sections 3.2 through 3.6.

The closeness may be expressed by the *error E(x)*, which is simply the difference between $U(x)$ and $\tilde{U}(x)$:

$$E(x) = U(x) - \tilde{U}(x) \qquad (3.2)$$

Equation (3.2) is also referred to as the *pointwise error* because it expresses the error at each point x ("pointwise") in the domain.

There are other expressions for error — i.e., other ways to measure the closeness of two functions — which are used more in theoretical studies (see Section 3.6.2). In practical applications one is usually interested in a solution, and hence its error, at specific locations (or small regions) in the domain, and therefore the pointwise error is a natural and meaningful measure of accuracy.

As a practical matter, Eq. (3.2) cannot be used to calculate $E(x)$ since it contains the exact solution, which is generally unknown. In fact, we can never expect to calculate $E(x)$ exactly in an actual numerical problem; if we could, then we would merely add $E(x)$ to $\tilde{U}(x)$ to get $U(x)$, in which case the error would be zero. We must therefore look for ways to *estimate* $E(x)$. There are two practical ways for estimating $E(x)$. Section 3.6 will demonstrate one of these ways; Chapter 4 and subsequent chapters will demonstrate both ways.

Now let's assume we have made an error estimate, and we find that the magnitude of our estimated error is too large to be acceptable. Is there any way to decrease the error? Yes. We return to the first operation and construct a different trial solution, one that contains more DOF than the first one. An easy way to do this is to add a few more trial functions (and hence DOF) to the previous trial solution. Repeating the second and third operations will then generate a second approximate solution, which hopefully (and usually) will yield a lower error estimate. If the new estimate is still unacceptable, then the cycle can be repeated again and again with successively improved trial solutions (more and more DOF) until an acceptable error estimate is obtained.

The terminology "trial solution" seems quite apropos because we keep trying different trial solutions until we find one that yields an acceptable error estimate. In the same spirit, we might describe the FEM as a "trial-and-error" procedure. Figure 3.1 shows

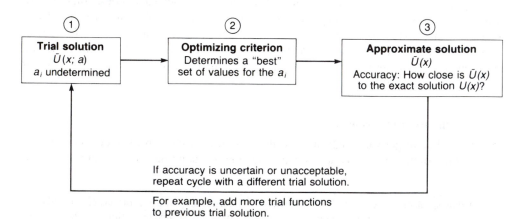

FIGURE 3.1 ● The trial-solution procedure.

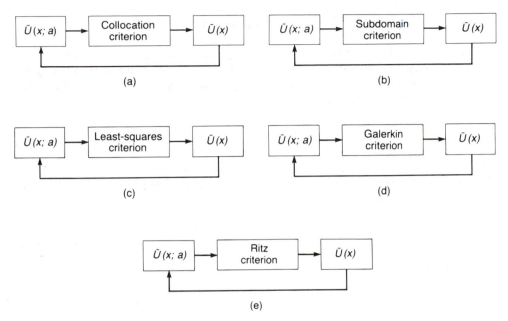

FIGURE 3.2 ● Five trial-solution methods: (a) the collocation method; (b) the subdomain method; (c) the least-squares method; (d) the Galerkin method; (e) the Ritz method.

symbolically the sequence of operations in the trial-solution procedure.

The popular terminology "trial-solution method" in the title of this chapter refers to the complete trial-solution procedure (all three operations), employing a *particular* optimizing criterion. Thus we could say there is a different trial-solution method for each different optimizing criterion, although the difference is only in the optimizing criterion, because the trial solution and accuracy estimation may be the same in all cases. In other words, all trial-solution methods use the procedure shown in Fig. 3.1, the only difference between the methods being the optimizing criterion.[6] This is illustrated in Fig. 3.2 for the five criteria mentioned above.

It was also mentioned above that the essence of the FEM is in the special form of the trial solution (to be introduced in Chapter 5). An FE trial solution may be used with any of the trial-solution methods, thereby creating a *finite element method*. For example, if each of the $\tilde{U}(x;a)$ in Fig. 3.2 were an FE trial solution, then the captions could be changed to "the collocation FEM," "the subdomain FEM," and so on. Since the essential characteristic of the FE concept is the special form of the trial solution, not the particular optimizing criterion, this author will use the phrase "the FEM" to refer generically or collectively to any or all trial-solution methods that use an FE trial solution.

This completes a brief description of the trial-solution procedure. Sections 3.2 through 3.6 will illustrate these ideas by applying them to a specific problem.

---

[6] The phrase "optimizing criterion" was coined by this author in order to have a generic concept that subsumes all residual, variational, and yet-to-be-discovered criteria. In addition, it refers only to the second of the three principal operations, so it is a narrower concept than the trial-solution method.

3.2 An Illustrative Problem: Description

We consider the 1-D boundary-value (equilibrium) problem described in Section 2.2. The governing equation is an ordinary differential equation. It is of second order (the highest-order derivative is second order) and is linear (the unknown function $U(x)$ and its derivatives appear only to the first power):

$$\frac{d}{dx}\left(x\,\frac{dU(x)}{dx}\right) = \frac{2}{x^2}$$

(3.3)

The special structure of this equation deserves an explanation, for which we digress briefly.

We note that the left-hand side of Eq. (3.3) could be expanded by using the chain rule of differentiation, yielding

$$x\,\frac{d^2U(x)}{dx^2} + \frac{dU(x)}{dx} = \frac{2}{x^2}$$

(3.4)

This is a special case of a general second-order differential equation,

$$c_2(x)\,\frac{d^2U(x)}{dx^2} + c_1(x)\,\frac{dU(x)}{dx} + c_0(x)U(x) = f(x)$$

(3.5)

where $c_2(x)=x$, $c_1(x)=1$, $c_0(x)=0$, and $f(x)=2/x^2$. When the coefficients $c_2(x)$ and $c_1(x)$ satisfy the condition

$$c_1(x) = \frac{dc_2(x)}{dx}$$

(3.6)

then Eq. (3.5) may be written as

$$\frac{d}{dx}\left(c_2(x)\,\frac{dU(x)}{dx}\right) + c_0(x)U(x) = f(x)$$

(3.7)

which is the form of Eq. (3.3).

The differential equation for this illustration problem was intentionally constructed so that the coefficients would satisfy Eq. (3.6), thereby enabling us to write the differential equation in the form of Eq. (3.7) rather than Eq. (3.5). There is a compelling reason for limiting ourselves to an equation of this special form. Equation (3.6) expresses a type of *symmetry* that is probably more apparent to a mathematician than to an engineer. However, this symmetry property is fundamentally an aspect of the physical world, and it manifests itself in various ways in our differential equations and numerical analyses. We are speaking here, not of a geometric symmetry, but of a cause-and-effect symmetry, i.e., an "interchangeability" between cause and effect at any two points in the physical system.[7]

This type of symmetry is exhibited by a broad spectrum of applications — in fact, most of the applications dealt with in industrial design problems. The governing equations for all these applications satisfy a symmetry condition analogous to that expressed

---

[7] A mathematician would describe this symmetry by saying the differential equation is *self-adjoint*. The physicist or engineer might describe the same symmetry by saying the physical phenomenon has *reciprocity*.

FIGURE 3.3 ● Domain for the illustrative problem.

by Eq. (3.6) (see Chapters 6 and 13). Therefore, rather than being "limiting," this illustration problem is actually representative of a broad spectrum of typical applications. In subsequent chapters there will be frequent talk of symmetric equations or symmetric matrices; their symmetry is a direct consequence of this basic symmetry in the governing differential equation. This completes the digression.

The domain of our illustrative problem will be the interval $1 < x < 2$, as shown in Fig. 3.3. The boundary of the domain is the two end points $x = 1$ and $x = 2$. The boundary condition (BC) at each point is

$$U(1) = 2$$

$$\left(-x \frac{dU}{dx} \right)_{x=2} = \frac{1}{2} \tag{3.8}$$

We note that the BC at $x = 2$ could be written more simply as $(dU/dx)_{x=2} = -1/4$ by substituting the value 2 for x in the term $-xdU/dx$. However, it was noted in Section 2.3 that when the governing differential equation has the form of Eq. (2.3), the quantity $-\alpha dU/dx$ usually has a physical interpretation, and we gave it the name "flux" and represented it by the symbol τ.[8] Comparing Eq. (3.3) with Eq. (2.3) indicates that for our illustrative problem the expression for the flux is

$$\tau(x) = -x \frac{dU}{dx} \tag{3.9}$$

In physically motivated problems, boundary data is usually specified for the flux rather than for just the derivative itself. Thus, the BC at $x = 2$ specifies data for the flux. We can rewrite Eqs. (3.8) more meaningfully as

$$U(1) = 2$$

$$\tau(2) = \frac{1}{2} \tag{3.10}$$

Equation (3.3) has several physical applications, such as those listed in Table 2.1. Since in many applications one is interested in determining a solution for the flux as well as for the function, we will find approximate solutions for both $U(x)$ and $\tau(x)$ for the illustrative problem.

3.3 Construction of a
Trial Solution for the Illustrative Problem

In choosing expressions for the trial functions $\phi_i(x)$, an important practical consideration is to use functions that are easy to work with because, as will be seen shortly, we fre-

---

[8] See further comments at end of Section 6.5.

quently must calculate derivatives and integrals of the $\phi_i(x)$.[9] Powers of x are certainly the easiest for these operations, so a logical choice for a trial solution would be the first few terms of a power series, i.e., a polynomial:

$$\tilde{U}(x;a) = a_1 + a_2x + a_3x^2 + \ldots + a_Nx^{N-1} \qquad (3.11)$$

For the present problem we will find it convenient (not necessary) to write this polynomial in a different form because of considerations involving the BCs. This deserves another short digression for some explanation.

A solution to any boundary-value problem must always satisfy (exactly or approximately) two separate conditions: (1) the governing equation (differential or variational, say) in the *interior* of the domain and (2) the BCs on the *boundary* of the domain. The optimizing criterion pertains to the first condition; that is, it tries to make $\tilde{U}(x;a)$ satisfy the differential equation or variational equation in the interior.[10] It requires separate operations, though, to make $\tilde{U}(x;a)$ satisfy the BCs. We will speak of these separate operations as "applying the boundary conditions."

There are two general methods (or methodologies or approaches, if you will) for applying the BCs. Because there seems to be no readily available terminology to distinguish between the two methods, this book will refer to them as the *theoretical method* and the *numerical method*.

The theoretical method • The BCs (some or all of them — see Chapter 4) are applied directly to the trial solution at the *beginning* of the analysis by constructing the *theoretical expressions* for the trial functions in such a form that they will satisfy "appropriate" conditions at the boundaries.

The numerical method • The BCs are applied to the weighted residual (or variational) equations at a *later stage* in the analysis by manipulating *numbers* in these equations.

Both methods yield the same solution (see Exercise 4.1) because they both perform basically the same operations but in a different order. The choice of which to use may be determined by the following considerations.

Prior to the advent of computers the theoretical method was probably employed universally because it significantly reduces the amount of numerical computation required. However, in exchange for this benefit, it usually requires considerable ingenuity and analysis to construct trial functions in the required form, a task that quickly becomes impractical or impossible for even moderately complex problems. Consequently, this method is

---

[9] We note in passing that the $\phi_i(x)$ must all be "linearly independent." In mathematical parlance, this means it is impossible to find values for the a_i such that $a_1\phi_1(x)+a_2\phi_2(x)+\ldots+a_N\phi_N(x)=0$ for all values of x. In other words, it is impossible for any of the $\phi_i(x)$, $i=1,2,\ldots N$, to be expressed as a linear combination of the others. If this were possible, then one of the $\phi_i(x)$ would be redundant, which would cause computational problems. The property of linear independence is easily established in the FEM (in fact, virtually assured by the nature of FE trial functions), so this point need not concern the application-oriented reader any further.

[10] There exist methods for using weighted residual criteria on the boundaries ("boundary residual" and "boundary element" methods), but they have not yet been implemented in most commercially available FE computer programs.

generally limited to only very simple problems (such as the present illustrative problem), although occasionally it can be used to advantage in special circumstances (see Section 16.4.2, step 5).

The numerical method has the advantages and disadvantages reversed: The trial functions are easy to construct, but the subsequent application of the BCs can be numerically very "messy." This situation, though, is eminently well suited for a computer, since a computer can do all the numerical work. Because of the power of present-day computers, there is almost no limit to the complexity of problems that can be solved with this approach. Not surprisingly, the numerical method is the standard approach used in computer applications of the FEM.

For the present illustration problem, however, we will use the theoretical method for applying BCs because (1) we want to focus principally on the basic structure of the trial-solution procedure without being distracted by a lot of arithmetic, (2) the problem is simple enough for the theoretical method to be relatively straightforward, and (3) understanding the theoretical method will aid in understanding the numerical method.[11] In Chapter 4 we will then introduce the numerical method as the preferred approach for practical problem solving with computers, and continue with it for the remainder of the book. This ends the digression.

The *theoretical method* may be described by the following rule:

Construct the expressions for the trial functions in such a form that the trial solution $\tilde{U}(x;a)$ will be forced to satisfy the BCs *for all values of the a_i.*

Using this rule, the required form for the trial functions could be derived as follows.

Consider the first BC in Eqs. (3.8), $U(1)=2$, and apply it directly to the general trial-solution expansion in Eq. (3.1):

$$\tilde{U}(1;a) = \phi_0(1) + a_1\phi_1(1) + a_2\phi_2(1) + \ldots + a_N\phi_N(1) = 2 \tag{3.12}$$

Our rule would then be satisfied by requiring that

$$\phi_0(1) = 2$$

$$\phi_i(1) = 0 \qquad i = 1,2,\ldots,N \tag{3.13}$$

The strategy here is to make all the trial functions that are multiplied by a parameter vanish at $x=1$. $\tilde{U}$ at $x=1$ would then be independent of all the a_i; in fact, it would be equal only to $\phi_0(1)$, so we must then require $\phi_0(1)$ to satisfy the BC.

In a similar manner, the second BC in Eqs. (3.8) may be applied to $\tilde{U}(x;a)$:

$$\left(-x\frac{d\tilde{U}}{dx}\right)_{x=2} = \left(-x\frac{d\phi_0}{dx}\right)_{x=2} + a_1\left(-x\frac{d\phi_1}{dx}\right)_{x=2}$$

$$+ a_2\left(-x\frac{d\phi_2}{dx}\right)_{x=2} + \ldots + a_N\left(-x\frac{d\phi_N}{dx}\right)_{x=2} = \frac{1}{2} \tag{3.14}$$

---

[11] The theoretical method can, on occasion, be very effective in special situations, so it is valuable as more than just a convenient introductory teaching aid.

and again the rule would be satisfied by requiring

$$\left(-x\frac{d\phi_0}{dx}\right)_{x=2} = \frac{1}{2}$$

$$\left(-x\frac{d\phi_i}{dx}\right)_{x=2} = 0 \qquad i = 1,2,\ldots,N \tag{3.15}$$

If we could find trial functions that satisfy Eqs. (3.13) and (3.15), then clearly $\tilde{U}(x;a)$ would satisfy both BCs *for all values of the* a_i. The significance of this is that no matter what numerical values are subsequently determined for the a_i by the optimizing criterion, the resulting approximate solution would necessarily satisfy both BCs. This property of "locking in" or *constraining* the trial solution to satisfy BCs is a prominent feature in trial-solution methods.

In general, then, we require the trial function $\phi_0(x)$, the only one not multiplied by a parameter, to satisfy the BCs; and we require all the other trial functions, those multiplied by parameters, to satisfy the same form of BCs but with the values set equal to zero.[12]

Let's now construct some actual trial functions that satisfy Eqs. (3.13) and (3.15). We begin with the polynomial that we originally proposed in Eq. (3.11), and we will somewhat arbitrarily set $N=4$ (a cubic polynomial) since this will keep the analysis simple yet interesting:

$$\tilde{U}(x;a) = a_1 + a_2x + a_3x^2 + a_4x^3 \tag{3.16}$$

Applying each BC directly to $\tilde{U}(x;a)$ in Eq. (3.16),

$$\tilde{U}(1;a) = a_1 + a_2 + a_3 + a_4 = 2$$

$$\left(-x\frac{d\tilde{U}}{dx}\right)_{x=2} = -2a_2 - 8a_3 - 24a_4 = \frac{1}{2} \tag{3.17}$$

or, simplifying the algebra,

$$a_1 + a_2 + a_3 + a_4 = 2$$

$$a_2 + 4a_3 + 12a_4 = -\frac{1}{4} \tag{3.18}$$

Equations (3.18) are called *constraint equations* (or *linear constraint equations*, to emphasize that all the a_i appear to the first power) because they constrain the a_i's so that they are no longer independent from each other. Each constraint equation means one less independent a_i in Eq. (3.16). Since there are four a_i in Eq. (3.16) and there are two constraint equations, then there can only be two independent a_i left ($4-2=2$). We must therefore eliminate two of the a_i in Eq. (3.16).

---

[12] Chapter 4 will show how to transform the original differential equation so that it will only be necessary to constrain $\tilde{U}(x;a)$ to satisfy some, not all, of the BCs. Such a transformation is the conventional approach to the FEM and will be used throughout the remainder of the book. [An exercise in Chapter 4 illustrates an unconventional transformation that makes it unnecessary to constrain $\tilde{U}(x;a)$ to satisfy any of the BCs — they are satisfied approximately instead.]

One way to eliminate the dependent a_i is to substitute Eqs. (3.18) into Eq. (3.16). In general, we can choose to eliminate any two of the a_i (see Exercise 3.2). The easiest choice is any parameter that appears in only one of the constraint equations, such as a_1 in Eqs. (3.18). Thus, solving the first of Eqs. (3.18) for a_1 yields

$$a_1 = 2 - a_2 - a_3 - a_4 \tag{3.19}$$

Substituting Eq. (3.19) into Eq. (3.16) yields

$$\tilde{U}(x;a) = (2-a_2-a_3-a_4) + a_2 x + a_3 x^2 + a_4 x^3$$
$$= 2 + a_2(x-1) + a_3(x^2-1) + a_4(x^3-1) \tag{3.20}$$

Now we can solve the second of Eqs. (3.18) for any of the remaining a_i, say a_2, yielding

$$a_2 = -\frac{1}{4} - 4a_3 - 12a_4 \tag{3.21}$$

Substituting Eq. (3.21) into Eq. (3.20) yields

$$\tilde{U}(x;a) = 2 + \left(-\frac{1}{4} - 4a_3 - 12a_4\right)(x-1) + a_3(x^2-1) + a_4(x^3-1)$$
$$= 2 - \frac{1}{4}(x-1) + a_3(x-1)(x-3) + a_4(x-1)(x^2+x-11) \tag{3.22}$$

This is the desired form of the trial solution.[13]

To simplify the subscripting, let's write Eq. (3.22) in the general form,

$$\tilde{U}(x;a) = \phi_0(x) + a_1\phi_1(x) + a_2\phi_2(x) \tag{3.23}$$

where

$$\phi_0(x) = 2 - \frac{1}{4}(x-1)$$
$$\phi_1(x) = (x-1)(x-3)$$
$$\phi_2(x) = (x-1)(x^2+x-11)$$

The notation a_3 and a_4 has been changed to a_1 and a_2, respectively, because there are only two parameters left, so it is more convenient to call them numbers 1 and 2.

The trial solution expression for the flux follows from applying Eq. (3.9) to Eq. (3.23):

$$\tilde{\tau}(x;a) = -x\frac{d\tilde{U}(x;a)}{dx}$$
$$= \left(-x\frac{d\phi_0(x)}{dx}\right) + a_1\left(-x\frac{d\phi_1(x)}{dx}\right) + a_2\left(-x\frac{d\phi_2(x)}{dx}\right) \tag{3.24}$$

---

[13] If Eqs. (3.18) were to contain all the a_i in both equations, then a_1 would have to be eliminated from the other constraint equation as well as from the trial solution. (Otherwise a_1 would reappear in the trial solution when the second constraint equation was applied.) See Exercise 3.2.

where

$$-x\,\frac{d\phi_0(x)}{dx} = \frac{1}{2} + \frac{1}{4}(x-2)$$

$$-x\,\frac{d\phi_1(x)}{dx} = -2x(x-2)$$

$$-x\,\frac{d\phi_2(x)}{dx} = -3x(x-2)(x+2)$$

Figure 3.4 shows plots of the trial functions and their fluxes. It is clear from both the figure and Eqs. (3.23) and (3.24) that Eqs. (3.13) and (3.15) are satisfied. This completes the first operation, the construction of a trial solution.

The reader can perhaps discern from this example that the theoretical method for applying BCs can quickly become very complicated with more than a few trial functions. Also, this example was only a one-dimensional problem, for which the boundary consists of only two points and hence there were only two BCs. For a two-dimensional problem the boundary is a curve, so there would be BCs along the entire curve, resulting in many constraint equations similar to Eqs. (3.18). It is obvious that applying BCs by the theoretical method would seriously limit the practicality of trial-solution methods. The reader will therefore appreciate the practicality of the numerical method in Chapter 4 since it obviates all this messy manipulation of the trial functions (making the computer do the work instead). And in Chapter 5, introduction of the element concept will greatly simplify even the numerical method.

We now move on to the second operation, that of applying an optimizing criterion to determine the ''best'' numerical values for a_1 and a_2. In Sections 3.4 and 3.5 we will apply five of the most popular criteria, each time using the same trial solution, Eq. (3.23). This will yield five different approximate solutions, all fairly close to the exact solution.

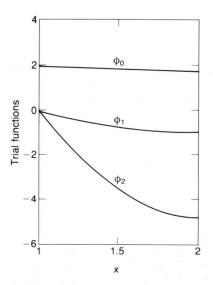
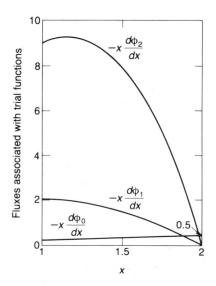

FIGURE 3.4 ● (a) Trial functions and (b) associated fluxes used in the illustrative problem.

3.4 Four Approximate Solutions
Using Methods of Weighted Residuals

We first must define a quantity called the residual. Transfer all terms to the left-hand side (LHS) of Eq. (3.3) so that zero is on the right-hand side (RHS):

$$\frac{d}{dx}\left(x\,\frac{dU(x)}{dx}\right) - \frac{2}{x^2} = 0 \tag{3.25}$$

Equation (3.25) says that if the exact solution were substituted for $U(x)$ on the LHS, then the RHS would be identically zero over the entire domain. If any other function, such as an approximate trial solution $\tilde{U}(x;a)$, were substituted for $U(x)$, the result would be a nonzero function, called the *residual* of the equation and denoted by $R(x;a)$:

$$R(x;a) = \frac{d}{dx}\left(x\,\frac{d\tilde{U}(x;a)}{dx}\right) - \frac{2}{x^2} \neq 0 \tag{3.26}$$

Thus, substituting Eq. (3.23) into Eq. (3.26) yields

$$R(x;a) = -\frac{1}{4} + 4(x-1)a_1 + 3(3x^2-4)a_2 - \frac{2}{x^2} \tag{3.27}$$

The central idea of all the MWR criteria is this: We try to find numerical values for a_1 and a_2 in Eq. (3.27) which will make $R(x;a)$ as close to zero as possible for all values of x throughout the entire domain. The logic is as follows. The exact solution, by definition, is the function that satisfies the differential equation over the entire domain and the BCs on the boundary. Any function that satisfies the differential equation over the entire domain must also make the residual zero over the entire domain, and vice versa. If we can therefore find a $\tilde{U}$ that makes $R(x;a)=0$ everywhere in the domain, and if the boundary conditions are also satisfied exactly, then $\tilde{U}$ must be the exact solution U. This conclusion is valid for any reasonably well-posed problem for which there exists only one exact solution.

It would also seem reasonable that if a particular $\tilde{U}$ makes $R(x;a)$ deviate only "slightly" from zero, then $\tilde{U}$ is probably "very close" to U. This is of course only an intuitive argument, but it explains the basic rationale behind the more rigorous mathematical arguments. [3.1, 3.3]

Note that we are dealing with two quite distinct concepts of closeness here. On the one hand we are concerned with how close the residual is to zero; this measures an error in satisfying the governing differential equation. On the other hand, we are also concerned with how close $\tilde{U}$ is to U — i.e., with the error in the solution itself, $E(x)=U(x)-\tilde{U}(x)$. Of course, it is the latter error that is of primary interest. We employ the technique of minimizing the residual error because, by the above rationale, it tends to simultaneously minimize the solution error.

Application of an MWR criterion (to any boundary-value problem) produces a set of algebraic equations, as will be shown below; their solution is the "best" set of numerical values for the a_i. Each different MWR criterion will determine a different set of values (sometimes two sets may be identical), resulting in many different approximate solutions. Depending on the trial functions chosen, the different solutions may all be close to each other and to the exact solution, in which case they would probably all be acceptable to the user. In practice, one needs to use only one criterion.

We now apply four MWR criteria to our illustration problem, using the residual in Eq. (3.27).

3.4.1 The Collocation Method

For each undetermined parameter a_i, choose a point x_i in the domain. At each x_i, force the residual to be exactly zero:

$$R(x_1;a) = 0$$

$$R(x_2;a) = 0$$

$$\vdots$$

$$R(x_N;a) = 0 \tag{3.28}$$

For a trial solution with N parameters, we therefore produce a system of N residual equations. The points x_i are called *collocation points*; they may be located anywhere in the domain and on the boundary, not necessarily in any particular pattern. [See references in Chapter 9 for guidelines for choosing good locations.]

The present problem has two parameters so we must select two collocation points. It might be reasonable to distribute them uniformly, e.g., at $x_1=4/3$ and $x_2=5/3$, as shown in Fig. 3.5.

Substituting these points into Eqs. (3.27) yields a system of two algebraic equations:

$$\frac{4}{3}a_1 + 4a_2 = \frac{11}{8}$$

$$\frac{8}{3}a_1 + 13a_2 = \frac{97}{100} \tag{3.29}$$

whose solution is

$$a_1 = 2.0993$$

$$a_2 = -0.3560 \tag{3.30}$$

The approximate solution for our problem is obtained by substituting Eq. (3.30) into Eqs (3.23) and (3.24):

$$\tilde{U}_C(x) = 2 - \frac{1}{4}(x-1) + 2.0993(x-1)(x-3) - 0.3560(x-1)(x^2+x-11)$$

$$\tilde{\tau}_C(x) = \frac{1}{2} + \frac{1}{4}(x-2) - 4.1986x(x-2) + 1.0680x(x-2)(x+2) \tag{3.31}$$

The subscript C identifies this as the collocation solution; subsequent solutions will use other subscripts.

FIGURE 3.5 ● Collocation points at $x_1=4/3$ and $x_2=5/3$.

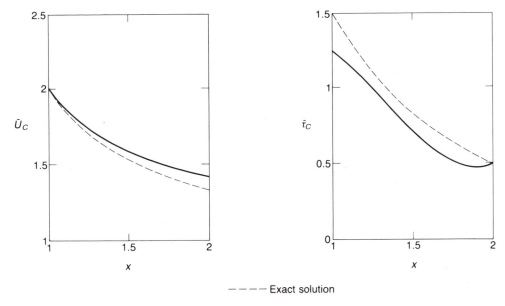

FIGURE 3.6 ● Collocation solution.

$\bar{U}_C(x)$ and $\tilde{\tau}_C(x)$ are plotted in Fig. 3.6, along with the exact solution. Observe that the BCs are satisfied exactly, which is not surprising, since our trial solution was constrained to satisfy them exactly. It perhaps also should be noted from the figure that the error in $\bar{U}_C$ and $\tilde{\tau}_C$ is not zero at the collocation points. This is usually the case; at the collocation points it is the error in the residual that is zero, not the error in the solution. (Recall the above discussion on the two concepts of closeness, or error.)

3.4.2 The Subdomain Method

For each undetermined parameter a_i choose an interval Δx_i, in the domain. Then force the *average* of the residual in each interval to be zero:

$$\frac{1}{\Delta x_1} \int_{\Delta x_1} R(x;a) \, dx = 0$$

$$\frac{1}{\Delta x_2} \int_{\Delta x_2} R(x;a) \, dx = 0$$

$$\vdots$$

$$\frac{1}{\Delta x_N} \int_{\Delta x_N} R(x;a) \, dx = 0 \tag{3.32}$$

Again, for a trial solution with N parameters, we have a system of N residual equations. The intervals Δx_i are called *subdomains*. They may be chosen in any fashion, even overlapping or with gaps in between.

FIGURE 3.7 ● Subdomains $\Delta x_1 = 1 < x < 1.5$ and $\Delta x_2 = 1.5 < x < 2$.

For the present problem we must choose two subdomains. A reasonable choice would be to simply divide the domain into two equal halves — that is, $\Delta x_1 = 1 < x < 1.5$ and $\Delta x_2 = 1.5 < x < 2$, as shown in Fig. 3.7.

Substituting the residual, Eq. (3.27), into Eqs. (3.32), and using these subdomains, the residual equations become

$$\int_1^{1.5} \left(-\frac{1}{4} + 4(x-1)a_1 + 3(3x^2-4)a_2 - \frac{2}{x^2} \right) dx = 0$$

$$\int_{1.5}^2 \left(-\frac{1}{4} + 4(x-1)a_1 + 3(3x^2-4)a_2 - \frac{2}{x^2} \right) dx = 0 \tag{3.33}$$

where the constant $1/\Delta x_i$ in front of each integral has been dropped. Carrying out the integration yields two algebraic equations

$$\frac{1}{2}a_1 + \frac{9}{8}a_2 = \frac{19}{24}$$

$$\frac{3}{2}a_1 + \frac{63}{8}a_2 = \frac{11}{24} \tag{3.34}$$

whose solution is

$$a_1 = 2.5417$$

$$a_2 = -0.4259 \tag{3.35}$$

Substituting Eqs. (3.35) into Eqs. (3.23) and (3.24) yields the approximate subdomain solution:

$$\tilde{U}_S(x) = 2 - \frac{1}{4}(x-1) + 2.5417(x-1)(x-3) - 0.4259(x-1)(x^2+x-11)$$

$$\tilde{\tau}_S(x) = \frac{1}{2} + \frac{1}{4}(x-2) - 5.0834x(x-2) + 1.2777x(x-2)(x+2) \tag{3.36}$$

$\tilde{U}_S(x)$ and $\tilde{\tau}_S(x)$ are plotted in Fig. 3.8, along with the exact solution. Observe again that the BCs are satisfied exactly.

3.4.3 The Least-Squares Method

With this criterion we *minimize*, with respect to each a_i, the integral over the entire domain of the square of the residual — i.e., it is a least-mean-square criterion. The integral of the square of the residual is a function of the a_i's, so minimization requires setting the

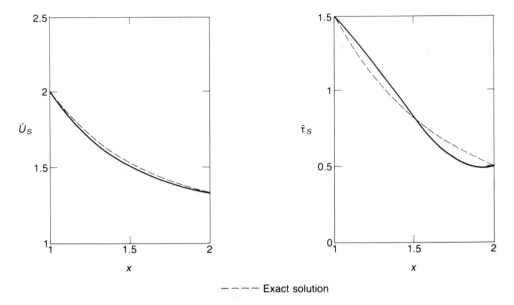

FIGURE 3.8 ● Subdomain solution.

partial derivatives with respect to each a_i equal to zero:

$$\frac{\partial}{\partial a_1} \int_1^2 R^2(x;a)\,dx = 0$$

$$\frac{\partial}{\partial a_2} \int_1^2 R^2(x;a)\,dx = 0$$

$$\vdots$$

$$\frac{\partial}{\partial a_N} \int_1^2 R^2(x;a)\,dx = 0 \tag{3.37}$$

Carrying out the derivative on the integrand, this becomes

$$\int_1^2 R(x;a)\frac{\partial R(x;a)}{\partial a_1}\,dx = 0$$

$$\int_1^2 R(x;a)\frac{\partial R(x;a)}{\partial a_2}\,dx = 0$$

$$\vdots$$

$$\int_1^2 R(x;a)\frac{\partial R(x;a)}{\partial a_N}\,dx = 0 \tag{3.38}$$

where the constant 2 in front of each integral has been dropped. Once again, a trial solution with N parameters yields a system of N residual equations.

From Eq. (3.27),

$$\frac{\partial R(x;a)}{\partial a_1} = 4(x-1)$$

$$\frac{\partial R(x;a)}{\partial a_2} = 3(3x^2-4) \tag{3.39}$$

Hence our two residual equations become

$$\int_1^2 \left(-\frac{1}{4} + 4(x-1)a_1 + 3(3x^2-4)a_2 - \frac{2}{x^2}\right) 4(x-1) \, dx = 0$$

$$\int_1^2 \left(-\frac{1}{4} + 4(x-1)a_1 + 3(3x^2-4)a_2 - \frac{2}{x^2}\right) 3(3x^2-4) \, dx = 0 \tag{3.40}$$

Carrying out the integration yields two algebraic equations:

$$\frac{16}{3}a_1 + 27a_2 = 8 \ln 2 - \frac{7}{2}$$

$$27a_1 + \frac{711}{5}a_2 = \frac{33}{4} \tag{3.41}$$

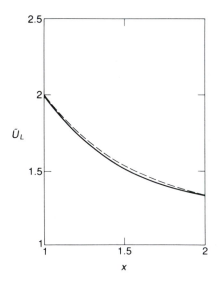 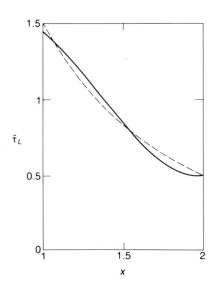

---- Exact solution

FIGURE 3.9 ● Least-squares solution.

whose solution is

$$a_1 = 2.3155$$

$$a_2 = -0.3816 \tag{3.42}$$

Substituting Eqs. (3.42) into Eqs. (3.23) and (3.24) yields the approximate least-squares solution:

$$\tilde{U}_L(x) = 2 - \frac{1}{4}(x-1) + 2.3155(x-1)(x-3) - 0.3816(x-1)(x^2+x-11)$$

$$\tilde{\tau}_L(x) = \frac{1}{2} + \frac{1}{4}(x-2) - 4.6310x(x-2) + 1.1448x(x-2)(x+2) \tag{3.43}$$

$\tilde{U}_L(x)$ and $\tilde{\tau}_L(x)$ are plotted in Fig. 3.9, along with the exact solution.

3.4.4 The Galerkin Method[14]

For each parameter a_i we require that a *weighted* average of $R(x;a)$ over the entire domain be zero. *The weighting functions are the trial functions* $\phi_i(x)$ *associated with each* a_i:

$$\int_1^2 R(x;a)\phi_1(x) \, dx = 0$$

$$\int_1^2 R(x;a)\phi_2(x) \, dx = 0$$

$$\vdots$$

$$\int_1^2 R(x;a)\phi_N(x) \, dx = 0 \tag{3.44}$$

And again a trial solution with N parameters yields a system of N residual equations. Using Eqs. (3.23) and (3.27), our two residual equations become

$$\int_1^2 \left(-\frac{1}{4} + 4(x-1)a_1 + 3(3x^2-4)a_2 - \frac{2}{x^2} \right) (x-1)(x-3) \, dx = 0$$

$$\int_1^2 \left(-\frac{1}{4} + 4(x-1)a_1 + 3(3x^2-4)a_2 - \frac{2}{x^2} \right) (x-1)(x^2+x-11) \, dx = 0 \tag{3.45}$$

---

[14] This is also known as the Bubnov-Galerkin method. See "Introduction: Historical Review" in Mikhlin [3.3].

Carrying out the integration yields two algebraic equations:

$$-\frac{5}{3}a_1 - \frac{41}{5}a_2 = \frac{29}{6} - 8 \ln 2$$

$$-\frac{41}{5}a_1 - \frac{81}{2}a_2 = \frac{211}{16} - 24 \ln 2 \tag{3.46}$$

whose solution is

$$a_1 = 2.1378$$

$$a_2 = -0.3477 \tag{3.47}$$

Substituting Eqs. (3.47) into Eqs. (3.23) and (3.24) yields the approximate Galerkin solution:

$$\tilde{U}_G(x) = 2 - \frac{1}{4}(x-1) + 2.1378(x-1)(x-3) - 0.3477(x-1)(x^2+x-11)$$

$$\tilde{\tau}_G(x) = \frac{1}{2} + \frac{1}{4}(x-2) - 4.2756x(x-2) + 1.0431x(x-2)(x+2) \tag{3.48}$$

$\tilde{U}_G(x)$ and $\tilde{\tau}_G(x)$ are plotted in Fig. 3.10, along with the exact solution.

It was observed by Crandall [3.2] that all of the above criteria (plus many others) could be put in the general form,

$$\int_D R(x;a)W_1(x) \, dx = 0$$

$$\int_D R(x;a)W_2(x) \, dx = 0$$

$$\vdots$$

$$\int_D R(x;a)W_N(x) \, dx = 0 \tag{3.49}$$

where $R(x;a)$ is the residual; $W_1(x)$, $W_2(x),\ldots,$ $W_N(x)$ are N different weighting functions; and the integration is over the domain D of the problem. There must be N equations, each using a different $W_i(x)$, in order to uniquely determine values for the N parameters a_i in the trial solution. Figure 3.11 shows the type of weighting functions used by each of the above four criteria. Since Eqs. (3.49) require different weighted averages of the residual to vanish, Crandall suggested the term *weighted residuals* to characterize these methods.

Looking back at these four trial-solution methods, we see that eac'ı has transformed the original differential equation into an approximately equivalent system of *algebraic* equations. We might say, therefore, that the *raison d'être* of the trial-solution procedure is that it transfers the "unknown-ness" of the solution from an unknown *function U(x)* to unknown *parameters a_i*, which is equivalent to converting the problem from its calculus formulation (e.g., differential and/or integral equations) to an algebraic formulation, the latter being much easier to solve than the former. In this sense, then, the FEM can be

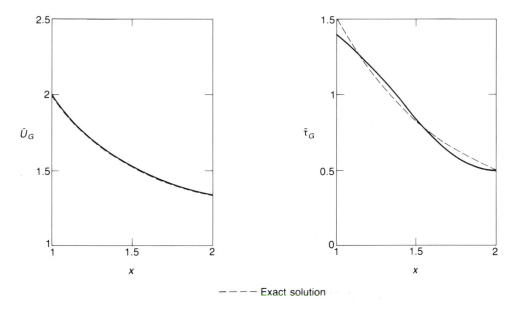

---- Exact solution

FIGURE 3.10 ● Galerkin solution.

viewed as a procedure that transforms unsolvable calculus problems into approximately equivalent but solvable algebra problems.

No mention was made in the above problems as to how the resulting algebraic equations were actually solved. This is because in each case there were only two (linear) equations in two unknowns, and it is expected that anyone studying this text learned the proper technique in a previous algebra course. If we had chosen a trial solution with more DOF, we would merely have had a larger system of algebraic equations to solve. In the ensuing chapters we will see that the boundary-value and initial-boundary-value problems[15] always result in a *system of linear algebraic equations*.

There are many well-developed, classical methods for solving systems of linear algebraic equations. They fall into two categories, direct and iterative. Common to all direct methods is the technique of successively eliminating parameters from each of the equations. Perhaps the most fundamental of all the methods is *Gaussian elimination*. (This is the method generally taught in algebra courses, where it is sometimes referred to as the method of substitution.) FE computer programs almost universally use Gaussian elimination.

A detailed knowledge of the linear-equation-solver algorithms used by FE programs is not necessary for efficacious use of the programs (although it would be necessary in order to develop such programs). Students of this text may nevertheless wish to review some of the basic concepts, assumptions, and computer algorithms associated with Gaussian

---

[15] Eigenproblems also yield a system of linear algebraic equations but with an extra feature; hence they require a different solution technique (see Chapter 10).

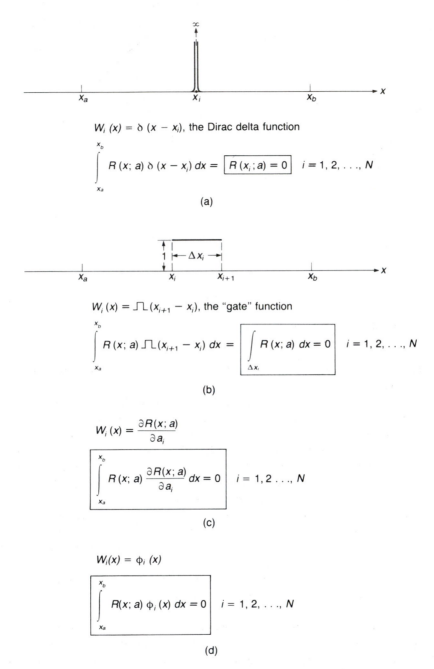

FIGURE 3.11 ● Weighting functions $W_i(x)$ associated with four different weighted-residual methods: (a) collocation; (b) subdomain; (c) least squares; (d) Galerkin.

elimination. The subject is treated in any of the hundreds of books on basic algebra, linear algebra, or numerical analysis. The references cited here [3.9–3.12] are only a sampling of the vast body of literature.

For those who are less ambitious, Exercise 3.1 provides a brief reminder of the Gaussian elimination technique, sufficient for solving any of the other exercises that involve hand calculations. Of course, many programmable pocket calculators contain linear equation solvers that could also be used for small problems. Later chapters will touch on some of the strategies used for the efficient computer implementation of Gaussian elimination for large FE problems, namely, data-management techniques that take advantage of special properties possessed by algebraic systems generated from FE analyses.

3.5 A Fifth Approximate Solution
Using the Ritz Variational Method[16]

The name "variational" refers to the calculus of variations, a topic about which many dozens of textbooks have been written. References [3.1–3.8] are only a representative sample. It is a subject involving many subtleties and sometimes long, intricate mathematical proofs. Most of the subject is not relevant to understanding or practicing the FEM. The name "Ritz" refers to a much narrower concept: a specific, well-defined technique for obtaining approximate numerical solutions to problems in a variational form.

To use the Ritz variational method in the FEM, one need only grasp a few elementary concepts from the variational calculus before proceeding directly with the Ritz technique. The curious reader is referred to the listed references for an introductory treatment of the variational calculus.

For the present we need only observe that the original differential equation and BCs in Eqs. (3.3) and (3.8), respectively, are completely equivalent to the following *variational* formulation:

$$\delta I(U) = 0 \tag{3.50}$$

where

$$I(U) = \int_1^2 \left[\frac{1}{2} x \left(\frac{dU}{dx} \right)^2 + \frac{2}{x^2} U \right] dx + \left[\left(-x \frac{dU}{dx} \right) U \right]_1^2$$

By "equivalent" we mean that the exact solution to Eqs. (3.3) and (3.8) is identical to the exact solution to Eq. (3.50). The integral $I(U)$ is a scalar-valued function*al* of the function $U(x)$. We may think of the function $U(x)$ as an independent variable, and of I as a function (called "functional") of the "variable" $U(x)$. The variational operator δ acting on $I(U)$ involves a process of varying the function $U(x)$ within a small neighborhood about the exact solution (somewhat analogous to the way a differential operator d/dx involves a small variation in the independent variable x).

---

[16] This section may be omitted without any loss of continuity in the surrounding material or the remainder of the book.

If this variational operation were performed theoretically (not numerically), in the context of the calculus of variations, the solution would be the differential equation (3.3)! Unfortunately, this does not bring us any closer to a numerical solution. To resolve this dilemma, we turn to the Ritz method.

The Ritz method[17] is a practical technique to obtain an approximate numerical solution directly from the variational formulation. We substitute for U in Eq. (3.50) a trial solution, such as Eq. (3.23). Since the $\phi_i(x)$ all have specific expressions, the integral with respect to the variable x can be evaluated, and I therefore becomes an ordinary function (not a functional) of parameters,

$$I\left(\tilde{U}(x;a)\right) = I(a) \tag{3.51}$$

This transforms the problem from the realm of functional analysis (in which functions are the independent variables) to the realm of conventional calculus (in which parameters are the independent variables). We can now make $I(a)$ stationary by applying the usual condition from the differential calculus that $dI=0$. Thus

$$dI = \frac{\partial I}{\partial a_1}da_1 + \frac{\partial I}{\partial a_2}da_2 + \ldots + \frac{\partial I}{\partial a_N}da_N = 0 \tag{3.52}$$

Because $a_1, a_2,\ldots, a_N$ can each be varied independently, then the coefficient of each da_i in Eq. (3.52) must vanish separately:

$$\frac{\partial I}{\partial a_1} = 0$$

$$\frac{\partial I}{\partial a_2} = 0$$

$$\vdots$$

$$\frac{\partial I}{\partial a_N} = 0 \tag{3.53}$$

Just as with the residual methods, we once again have a system of N equations.

Let's now apply the Ritz method to our illustration problem. We will use the same trial solution that was used for the weighted residual methods, namely, Eq. (3.23). Substituting Eq. (3.23) into Eq. (3.50) yields

$$I(\tilde{U}) = \int_1^2 \left[\frac{1}{2}x\left(\frac{d\phi_0}{dx} + a_1\frac{d\phi_1}{dx} + a_2\frac{d\phi_2}{dx}\right)^2 + \frac{2}{x^2}(\phi_0+a_1\phi_1+a_2\phi_2)\right] dx$$

$$+ \left[\left(-x\frac{d\tilde{U}}{dx}\right)(\phi_0+a_1\phi_1+a_2\phi_2)\right]_1^2 \tag{3.54}$$

---

[17] Developed by W. Ritz in 1909 for the solution of equilibrium problems, and by Lord Rayleigh in 1870 for the solution of vibration problems. Hence the Ritz method is sometimes called the Rayleigh-Ritz method.

The trial solution was not substituted into the boundary term $-xd\tilde{U}/dx$ because the BC at $x=2$ will be applied by assigning the value 1/2 to this term.[18]

Next apply the stationarity conditions

$$\frac{\partial I}{\partial a_1} = 0$$

$$\frac{\partial I}{\partial a_2} = 0 \tag{3.55}$$

to Eq. (3.54), which yields, after rearranging terms,

$$\left(\int_1^2 \frac{d\phi_1}{dx} x \frac{d\phi_1}{dx} dx \right) a_1 + \left(\int_1^2 \frac{d\phi_1}{dx} x \frac{d\phi_2}{dx} dx \right) a_2$$

$$= - \int_1^2 \frac{2}{x^2} \phi_1 dx - \int_1^2 \frac{d\phi_1}{dx} x \frac{d\phi_0}{dx} dx - \left[\left(-x \frac{d\tilde{U}}{dx} \right) \phi_1 \right]_1^2$$

$$\left(\int_1^2 \frac{d\phi_2}{dx} x \frac{d\phi_1}{dx} dx \right) a_1 + \left(\int_1^2 \frac{d\phi_2}{dx} x \frac{d\phi_2}{dx} dx \right) a_2$$

$$= - \int_1^2 \frac{2}{x^2} \phi_2 dx - \int_1^2 \frac{d\phi_2}{dx} x \frac{d\phi_0}{dx} dx - \left[\left(-x \frac{d\tilde{U}}{dx} \right) \phi_2 \right]_1^2 \tag{3.56}$$

Substituting the expressions for ϕ_i from Eq. (3.23) yields

$$\left(\int_1^2 4x(x-2)^2 dx \right) a_1 + \left(\int_1^2 6x(x-2)^2(x+2) dx \right) a_2$$

$$= - \int_1^2 \frac{2}{x^2}(x-1)(x-3) dx + \int_1^2 \frac{x}{2}(x-2) dx + \frac{1}{2}$$

$$\left(\int_1^2 6x(x-2)^2(x+2) dx \right) a_1 + \left(\int_1^2 9x(x-2)^2(x+2)^2 dx \right) a_2$$

$$= - \int_1^2 \frac{2}{x^2}(x-1)(x^2+x-11) dx + \int_1^2 \frac{3x}{4}(x-2)(x+2) dx + \frac{5}{2} \tag{3.57}$$

---

[18] We could have used a trial solution that satisfied only the BC $U(1)=2$ since the boundary term in Eq. (3.54) provides a means for applying the other BC, $(-xdU/dx)_{x=2}=1/2$. By using the trial solution in Eq. (3.23), which already satisfies the BC at $x=2$, we are wasting a bit of effort here. No matter, since our purpose is to use the same trial solution for all five optimizing criteria. (See also comments in Section 4.5.)

where the boundary terms were evaluated using Eqs. (3.8) and (3.23)[19] — that is,

$$\left[\left(-x\frac{d\tilde{U}}{dx}\right)\phi_1\right]_1^2 = \left(-x\frac{d\tilde{U}}{dx}\right)_{x=2}\phi_1(2) - \left(-x\frac{d\tilde{U}}{dx}\right)_{x=1}\phi_1(1)^{\,0} = \frac{1}{2}(-1)$$

$$\left[\left(-x\frac{d\tilde{U}}{dx}\right)\phi_2\right]_1^2 = \left(-x\frac{d\tilde{U}}{dx}\right)_{x=2}\phi_2(2) - \left(-x\frac{d\tilde{U}}{dx}\right)_{x=1}\phi_2(1)^{\,0} = \frac{1}{2}(-5) \qquad (3.58)$$

Carrying out the integration in Eqs. (3.57) yields two algebraic equations:

$$\frac{5}{3}a_1 + \frac{41}{5}a_2 = 8 \ln 2 - \frac{29}{6}$$

$$\frac{41}{5}a_1 + \frac{81}{2}a_2 = 24 \ln 2 - \frac{211}{16} \qquad (3.59)$$

These equations are identical to Eqs. (3.46) for the Galerkin residual method, so of course the resulting approximate solutions, $\tilde{U}_R(x)$ and $\tilde{T}_R(x)$, are also identical. This is not a coincidence for this particular problem but a manifestation of a general principle that the Ritz variational method and the Galerkin residual method always produce identical solutions when using the same trial solution. (Some authors emphasize this identity by combining the two names as Ritz-Galerkin or Galerkin-Ritz.) More will be said about this in Section 3.7.

3.6 Estimation of Accuracy of the Solutions

At this point we have five approximate solutions to our illustrative problem; only four are different because the Galerkin and Ritz solutions are identical. Figures 3.6, 3.8, 3.9, and 3.10 plotted each of these solutions, along with the exact solution. In real problems we generally don't know the exact solution, so such simple visual comparisons as these figures afford are not really possible. We must therefore learn how to examine approximate solutions and estimate their accuracy.

3.6.1 Comparison of the Five Solutions

Figure 3.12 shows all five approximate solutions plotted on the same graph. Let's assume for the moment we have no knowledge of the exact solution, so that these curves are all we have to work with. The most important observation is the *relative closeness* of all the curves. The curves for $\tilde{U}(x)$ are within $\pm 5\%$ of each other at every point x in the domain;

---

[19] Note that the trial functions ϕ_1 and ϕ_2 both vanish at $x=1$, as indeed they were constructed to do back in Section 3.3. Consequently the terms $(-xd\tilde{U}/dx)_{x=1}$, for which there is no prescribed boundary value, conveniently drop out of the expressions. This was not a coincidence but a direct result of (and reason for) constructing the trial functions as in Section 3.3. Since the Ritz variational method is not being used in this text, nor is it being advocated for general use (see Section 3.7), further discussion of this point is not necessary, although several relevant remarks may be found in Section 4.5.

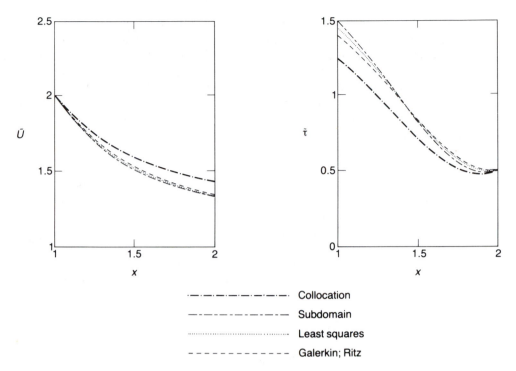

FIGURE 3.12 ● All five approximate solutions.

the curves for $\tilde{\tau}(x)$ are within $\pm 10\%$ of each other. (The derivative of a function is usually less accurate than the function itself.)

There is only one exact solution to the problem, so it is encouraging to see that the different approximate solutions are all close to each other. After all, if the curves were all widely separated, only one could possibly be close to the exact solution (or, of course, none of them might be close), and even then we would have no way of telling which one is the close one. So a minimum (i.e., a necessary) requirement, when we have several different approximate solutions, is that they all be close to each other.

But is this sufficient to conclude that all the curves must also be close to the exact solution? The answer is clearly no, since it is quite possible that they are all close to the wrong function, which could be far removed from the exact solution. Therefore, relative closeness of several different approximate solutions is not sufficient, per se, to ensure that any one of them is near to the exact solution. However, by introducing the additional concept of *convergence of a sequence of approximate solutions*, we can effectively reverse this conclusion, so that relative closeness will imply (though not guarantee), in almost all practical applications, a high likelihood of closeness to the exact solution.

3.6.2 Convergence of a Sequence of Approximate Solutions

The subject of convergence of approximate solutions to differential and/or variational equations involves several abstract mathematical concepts that require sophisticated mathematical treatment. Much of the rigorous treatment is in the realm of functional analysis and therefore outside the scope of this book. The following discussion will intentionally simplify the subject by attempting to distill the essential ideas while avoiding as much mathematical formalism as possible. The purpose here is to give the reader an intuitive appreciation for these *ideas* since they have important ramifications in practical applications, as will be demonstrated by the illustrative problems. (Conversely, an understanding of the mathematical underpinnings is not necessary in order to skillfully use FE programs.) Those who wish to "dig deeper" should consult the literature cited in this chapter and in Section 5.6.

To employ the concept of convergence we must work with only one optimizing criterion, not several different ones. And we must derive two or more different approximate solutions by using two or more different trial solutions. The object is to generate a *sequence of approximate solutions*:

$$\tilde{U}_1(x), \ \tilde{U}_2(x), \ \tilde{U}_3(x), \ \ldots \tag{3.60}$$

To generate such a sequence, we simply keep repeating the cycle shown in Fig. 3.1 with progressively "better" trial solutions. Thus we would create a first trial solution,

$$\tilde{U}_1(x;a) = \phi_0(x) + \sum_{j=1}^{l} a_j \phi_j(x) \tag{3.61}$$

which would generate a corresponding approximate solution $\tilde{U}_1(x)$.

Then we would create a second trial solution, $\tilde{U}_2(x;a)$, which would contain more DOF than the first one. An easy way (but not the only way) to do this is to add one or more trial functions to $\tilde{U}_1(x;a)$. For example,

$$\tilde{U}_2(x;a) = \tilde{U}_1(x;a) + \sum_{j=l+1}^{m} a_j \phi_j(x) \tag{3.62}$$

which would yield a second approximate solution, $\tilde{U}_2(x)$. [In general, the numerical values computed for the a_j, $j=1,2,\ldots,l$, in $\tilde{U}_2(x)$ may be different from the values computed for $\tilde{U}_1(x)$.]

We could then create a third trial solution, $\tilde{U}_3(x;a)$, by adding still more trial functions to $\tilde{U}_2(x;a)$. For example,

$$\tilde{U}_3(x;a) = \tilde{U}_2(x;a) + \sum_{j=m+1}^{n} a_j \phi_j(x) \tag{3.63}$$

which would yield a third approximate solution, $\tilde{U}_3(x)$. The cycle can be repeated as often as desired, and as long as one's computing resources last. In practice, two or three solutions are usually adequate.

The rationale for this procedure is as follows: As we continue to add more and more DOF to the trial solution, we would expect the resulting approximate solutions to get closer and closer to the exact solution. That is, we would hope that the sequence of approximate solutions will *converge* to the exact solution as the number of DOF increases indefinitely:

$$|U(x) - \tilde{U}_n(x)| \to 0 \quad \text{as } n \to \infty \quad 1 \le x \le 2 \tag{3.64}$$

where $\tilde{U}_n(x)$ has more DOF than $\tilde{U}_{n-1}(x)$.

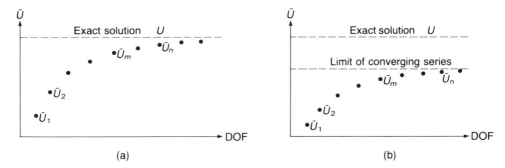

FIGURE 3.13 ● Symbolic representation of convergence to correct and incorrect limits: (a) sequence of approximate solutions converging to the exact solution (a desirable situation); (b) sequence of approximate solutions also converging, but to the wrong function (a generally undesirable situation).

Equation (3.64) is a statement of *pointwise convergence*; it describes the approximate solution as approaching arbitrarily (i.e., infinitesimally) close to the exact solution, at each point in the domain, as the number of DOF increases indefinitely. There are other ways to measure convergence, depending on how one measures the closeness of two functions. For example, *uniform convergence* requires that the *maximum* value of $|U(x) - \tilde{U}_n(x)|$ in the domain vanish as $n \to \infty$. This is a bit stronger than pointwise convergence since it requires a uniform rate of convergence at every point in the domain [3.13]. Two other types of convergence — namely, *convergence in energy* and *convergence in the mean* — involve an *average* of a function of the pointwise error over the domain. Convergence in energy is the most natural type for theoretical proofs and it will be discussed below. It will be seen there that mathematical theorems can predict convergence in energy but not pointwise convergence, so we must rely on numerical experiments as well as theory to make (weaker) predictions about pointwise convergence.

We emphasize pointwise convergence here [Eq. (3.64)] because of its practical utility. An engineer usually seeks a solution for $U(x)$ or $\tau(x)$ [rather than energy], either at particular points or averaged over small regions. In addition, in an FE analysis the most accurate values of $\tilde{U}(x)$ and $\tilde{\tau}(x)$ occur at specific points in the domain (namely, nodes and other superconvergent points, as explained in Chapters 7, 8, and 9). For these reasons we are especially interested in the pointwise convergence of $\tilde{U}(x)$ and $\tilde{\tau}(x)$.

Let's assume for the moment that we have a sequence of approximate solutions that is converging pointwise, i.e., according to Eq. (3.64). Then two different approximate solutions in the sequence must approach arbitrarily close *to each other*;[20] that is,

$$|\tilde{U}_m(x) - \tilde{U}_n(x)| \to 0 \quad \text{as } m,n \to \infty \quad 1 \le x \le 2 \quad (3.65)$$

It is important to realize that even though Eq. (3.64) implies Eq. (3.65), the reverse is not true; that is, Eq. (3.65) does *not* imply Eq. (3.64). Thus, as shown in Fig. 3.13(b) (and demonstrated in Chapter 9), we might observe a sequence converging according to

---

[20] Equation (3.65) follows directly from Eq. (3.64) by the following inequality:

$$|U - \tilde{U}_n| + |U - \tilde{U}_m| \ge |(U - \tilde{U}_n) - (U - \tilde{U}_m)| = |\tilde{U}_m - \tilde{U}_n|$$

Mathematicians refer to Eq.(3.65) as *Cauchy convergence*.

Eq. (3.65), but it may be converging to the wrong limit, in which case the sequence is not satisfying Eq. (3.64). Despite this important caveat, we will see in a moment that if we construct trial functions so that they satisfy certain conditions, then there will be a reasonably high likelihood that a sequence converging according to Eq. (3.65) is converging to the exact solution. In addition, Eq. (3.65) is independent of the exact solution, whereas Eq. (3.64) is not. For these reasons, we will find Eq. (3.65) to be a practical way to estimate accuracy. In other words, we will estimate accuracy by generating a sequence of two or more approximate solutions and examining the differences between successive solutions.

Let's demonstrate these ideas by applying them to our illustrative problem. Since the concept of convergence requires working with only one optimizing criterion for all solutions, we can use only one of our previously computed solutions. We'll use the Galerkin solution since that is the method that will be used throughout the rest of the book (see Section 3.7). We therefore have available only the solid curves in Fig. 3.10. These are repeated in Fig. 3.14 but without the dotted curves for the exact solution in order to simulate what is experienced in practice.

The solution in Fig. 3.14 employed the 2-DOF trial solution in Eq. (3.23). We could generate another trial solution by dropping the second DOF from Eq. (3.23), leaving a 1-DOF trial solution:

$$\tilde{U}_1(x;a) = 2 - \frac{1}{4}(x-1) + a_1(x-1)(x-3) \tag{3.66}$$

The resulting approximate solution, $\tilde{U}_1(x)$, will probably be less accurate than $\tilde{U}_2(x)$. The order in which we generate the sequence is clearly immaterial; it only matters that we ultimately have two or more solutions to compare with each other. After making the estimate we would then actually use the solution with the greatest number of DOF, since it is usually the most accurate.

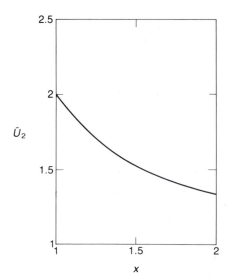
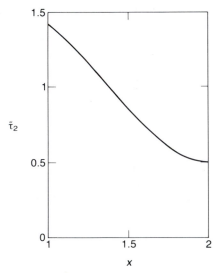

FIGURE 3.14 ● The 2-DOF Galerkin solution $\tilde{U}_G$ and $\tilde{\tau}_G$ from Fig. 3.10.

Let's also generate a third solution, by adding a third DOF to Eq. (3.23):

$$\tilde{U}_3(x; a) = 2 - \frac{1}{4}(x-1) + a_1(x-1)(x-3) + a_2(x-1)(x^2+x-11)$$

$$+ a_3(x-1)(x^3+x^2+x-31) \qquad (3.67)$$

Following the same procedure as in Section 3.3 for $\phi_1(x)$ and $\phi_2(x)$, the new trial function and its corresponding flux are

$$\phi_3(x) = (x-1)(x^3+x^2+x-31)$$

$$-x\frac{d\phi_3(x)}{dx} = -4x(x-2)(x^2+2x+4) \qquad (3.68)$$

Equations (3.68) clearly satisfy the derived boundary conditions in Eqs. (3.13) and (3.15):

$$\phi_3(1) = 0 \qquad \left(-x\frac{d\phi_3}{dx}\right)_{x=2} = 0 \qquad (3.69)$$

Repeating the same type of calculations as shown in Section 3.4.4, using first $\tilde{U}_1(x; a)$ and then $\tilde{U}_3(x; a)$, we generate two new solutions, $\tilde{U}_1(x)$ and $\tilde{U}_3(x)$. The complete sequence of three solutions is shown in Eqs. (3.70) and is plotted in Fig. 3.15.

$$\tilde{U}_1(x) = 2 - \frac{1}{4}(x-1) + 0.4271(x-1)(x-3)$$

$$\tilde{U}_2(x) = 2 - \frac{1}{4}(x-1) + 2.1378(x-1)(x-3) - 0.3477(x-1)(x^2+x-11)$$

$$\tilde{U}_3(x) = 2 - \frac{1}{4}(x-1) + 3.3725(x-1)(x-3) - 0.8881(x-1)(x^2+x-11)$$

$$+ 0.0864(x-1)(x^3+x^2+x-31) \qquad (3.70)$$

$$\tilde{\tau}_1(x) = \frac{x}{4} - 0.8542x(x-2)$$

$$\tilde{\tau}_2(x) = \frac{x}{4} - 4.2756x(x-2) + 1.0431x(x-2)(x+2)$$

$$\tilde{\tau}_3(x) = \frac{x}{4} - 6.7450x(x-2) + 2.6643x(x-2)(x+2) - 0.3456x(x-2)(x^2+2x+4)$$

The curves in Fig. 3.15 are getting closer to each other as the number of DOF increases; i.e., they appear to be converging pointwise. As discussed above, we must be cautious about jumping to the conclusion that they are converging to the exact solution rather than to some other limiting function. However, there exist convergence theorems that ensure that a sequence of approximate solutions must converge to the exact solution (assuming no computational errors) if the trial functions satisfy certain conditions.[21] The theorems

---

[21] For the present illustrative problem, which uses the Galerkin method without finite elements, see Chapter IX in Mikhlin [3.3].

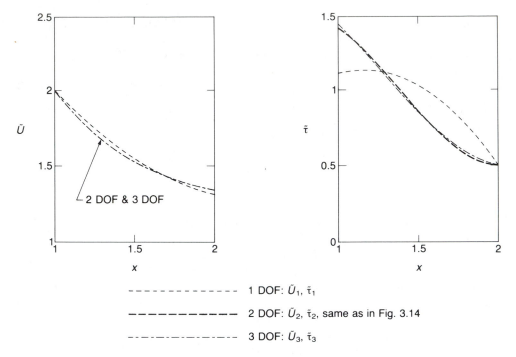

FIGURE 3.15 ● A sequence of three approximate solutions using the Galerkin method.

ensure convergence only in an average sense over the entire domain (e.g., with respect to energy); they cannot ensure pointwise convergence (except in very special circumstances noted below).

The convergence theorems require the trial functions to satisfy two conditions, called the *completeness* and *continuity* conditions. The completeness condition might be viewed as the fundamental, or broader, concept, with continuity being a property that is needed in order to define completeness.

The completeness condition ensures that it is possible for a sequence of approximate solutions to get *arbitrarily close* to the exact solution. The condition requires that the trial functions belong to an infinite sequence of functions such that a linear combination of an arbitrarily large but *finite* number of the functions is capable of approximating the exact solution arbitrarily closely. The closeness is measured by the *energy error*, which is an integral over the entire domain (i.e., a global average) of a function of the pointwise error:

$$\text{Energy error} = \left(\int_{\text{domain}} E(x)A[E(x)] \; dx \right)^{1/2}$$

$$(3.71)$$

where

$$E(x) = U(x) - \tilde{U}(x)$$

and A is the differential operator for the governing differential equation; i.e., it contains the derivatives operating on U (and any functions multiplying U) in the governing differential equation. [Thus, in Eq. (3.3), $A=d/dx(xd/dx)$.] In physical applications the expression $E(x)A[E(x)]$, or similarly $U(x)A[U(x)]$, generally corresponds to an energy density.

An infinite sequence of functions with the above properties is said to be *complete in energy*. If the trial functions belong to such a sequence, the trial solutions can then converge to the exact solution with respect to the energy; i.e., the energy error can be made as close to zero as desired.

Eq. (3.71) can be meaningful only if the derivatives in the operator A exist, which in turn requires continuity of the trial functions and perhaps one or more of their derivatives, depending on the order of the derivatives in A. This is the source of the continuity condition.[22]

Section 5.6 will present the specific requirements of the completeness and continuity conditions appropriate for an FE context. The requirements will not be presented here because the context of this chapter (and Chapter 4) does not involve the use of elements and therefore the requirements for completeness would be different.

For the present illustrative problem it will simply be observed that our polynomial trial solutions [Eq. (3.11)] will converge in energy as $N \to \infty$ because the powers of x (x^N, $N=0,1,2,...$) have sufficient continuity and they form an infinite sequence of functions which is complete in energy for Eq. (3.3) and its boundary conditions. We can therefore conclude that the sequence of approximate solutions in Fig. 3.15 is converging in energy to the exact solution, provided we made no mistakes in the calculations.

It has been emphasized that the convergence theorems are only able to ensure convergence in energy, i.e., in an average sense over the entire domain,[23] and that they cannot ensure pointwise convergence, i.e., at each and every point in the domain.[24] Nevertheless, experience shows that if a solution converges with respect to the energy error, then in practically all physically well-posed problems it will also converge pointwise. Combining this observation with the preceding remarks, we conclude from Fig. 3.15 that there is a reasonable likelihood that the pointwise error in $\tilde{U}_3$ and $\tilde{\tau}_3$ is about 1 and 5%, respectively.

All of the error discussion has so far focused on only one kind of error, one that is at the very heart of the FEM and, in a sense, is the most important one. This is the error introduced by using a trial solution with a finite number of terms to approximate the exact solution. It is often referred to as the *discretization error* because the "unknown-ness" of the exact solution is transferred from a function $U(x)$ to a discrete set of parameters

---

[22] It will be seen in Chapter 4 that integration by parts will lower the order of derivatives in the integrals in the residual equations [and the integral in Eq. (3.71)] and therefore lower the continuity requirements.

[23] The convergence theorems require that the operator A be *positive-definite*, a property that characterizes most problems of practical importance. The operator also usually satisfies a slightly stronger condition, that of being *positive-bounded-below*, in which case convergence in energy implies *convergence in the mean*. The latter measures closeness by the *mean error* (also known as the *root-mean-square error*), which is Eq. (3.71) without the A operator; i.e., the integrand is simply $E^2(x)$.

[24] In the special circumstance where the entire loading for a problem is a point load, the energy error can be directly related to a pointwise error, and then the energy convergence theorems can ensure pointwise convergence [3.14].

$a_1, a_2, \ldots, a_N$. We will have a lot more to say about discretization error in subsequent chapters.

The convergence theorems discussed above deal only with the discretization error; they assume all other errors are absent. However, there are many other sources of error in an FE analysis. We could classify them as either intentional or unintentional.

Intentional errors are those due to approximations that are purposely included in the FE mathematical model. These will be referred to as *approximation errors*. [They are sometimes referred to as *modeling errors*; however, we will use this term to refer to only certain types of approximation errors (see Section 7.2).] Obviously the discretization error falls in this class. We will meet several other types of approximation errors in both the theoretical and computational phases of the analysis.

Unintentional errors include essentially uncontrollable errors (e.g., truncation and round-off errors due to representing numbers in a computer by a finite number of bits) and just plain mistakes (e.g., programming bugs and wrong input data). Being human, we are likely to inject careless mistakes into the analysis at any of a thousand different places. Accuracy, therefore, cannot be guaranteed by convergence theorems. Even a nicely converging sequence of approximate solutions should not be totally trusted. Our approximate solutions ought therefore to be *verified* by independent analysis or experimental data. Such independent verification can go a long way to mitigate most of the theoretical and numerical uncertainties inherent in an FE analysis. This principle of independent verification is demonstrated by all 10 of the examples from industry shown in Chapter 1. (See also comments in Section 4.4, step 12.)

Let's summarize the important points from the above discussion:

- It is possible and desirable to estimate (rather than to determine precisely) the error in an approximate solution.
- To do so, two or more trial solutions are constructed, each containing progressively more trial functions (and more DOF).
- The trial functions should satisfy the completeness and continuity conditions of any available and appropriate convergence theorems. (See Section 5.6 for the conditions appropriate for a broad class of problems using FE trial functions.)
- The resulting sequence of approximate solutions "must" then converge with respect to the energy error (i.e., in an average sense) to the exact solution. (This assumes there are no significant unintentional errors of any kind — often not a very good assumption.)
- In almost all physically well-posed problems, convergence in energy also entails pointwise convergence.
- Therefore pointwise error may be estimated (usually reliably, but not always) by comparing the differences between two or more approximate solutions generated in the above manner.
- Because of the many sources of error, both intentional and unintentional — as well as the occasional exceptions to the usually observed pointwise convergence behavior — it may be desirable to provide greater confidence in such pointwise error estimates. This may be done by comparing the FE results with an independent analysis or experimental data or any other simple checks and tests to help establish the reasonableness of the results.

3.7 Choice of the Galerkin Method for This Book

There are many different types of optimizing criteria to choose from. The number of possible residual methods is virtually unlimited, and there are one or more variational principles available for many of the typical physical applications. Sections 3.4 and 3.5 illustrated the methods used most frequently in practice. Of these, the Galerkin residual and Ritz variational methods have been used in almost all FE work.

The Ritz variational method was first historically, being used almost exclusively throughout the 1960s for applications in solid mechanics. This was probably due in part to the availability of an extensive body of classical literature on the calculus of variations, providing convergence theorems for variational formulations. Indeed, most fundamental theoretical work in the FEM is still done with a variational formulation. In addition, variational principles were (and are) most abundant in the field of solid mechanics where the early FE work was concentrated.

However, as the FEM spread to other disciplines during the 1970s the Galerkin method gained in popularity, probably because its applicability was wider than that of the Ritz method. Thus, quoting Schryer,

> Since Rayleigh-Ritz requires an ''energy'' functional [i.e., a variational principle], its use is limited to differential equations which have such functionals. Galerkin's method does not need or use an energy functional and can thus be applied to equations where Rayleigh-Ritz cannot [3.15].

And

> Galerkin's method can be applied to literally any differential equation, but when applied to a differential equation with an ''energy'' functional, it agrees exactly with the Rayleigh-Ritz solution [3.15].

A similar observation is made by Finlayson and Scriven:

> ...whereas MWR can be applied to any problem, whether or not it is linear and self-adjoint, the variational method is only applicable to those problems for which a variational principle exists and has been found, a common situation only when the system of equations is linear and self-adjoint [3.16].

Thus Galerkin's method is the broader principle, being applicable to a wider class of problems than the Ritz method. And where Galerkin and Ritz are both applicable, they produce *identical solutions* when the same trial solution is used (as was illustrated in Section 3.5).

Another consideration is that many people find the MWR concept easier to understand than the variational concept with its notion of virtual operators. This author agrees with the following sentiments by Finlayson and Scriven:

> From a pragmatic standpoint the main shortcoming of variational formulations is that the variational methods which they support provide no approximation scheme that cannot be set up more simply and more quickly as one of the standard direct approximating methods, in particular, one or another version of MWR [3.16].

And later in the same paper:

> What practical utility the variational formulations have results from the corresponding variational methods for approximate solutions....But in fact these approximation schemes are far more readily set up as the straightforward Galerkin method or another closely related version of the method of weighted residuals. That these direct approximation procedures avoid completely the effort and mathematical embellishment of a variational formulation has not been emphasized adequately in the literature....Apart from self-adjoint, linear systems, which are comparatively rare,[25] there is no practical need for variational formalism. When approximate solutions are in order the applied scientist and engineer are better advised to turn immediately to direct approximation methods for their problems, rather than search for or try to understand quasi-variational formulations and restricted variational principles [3.16].

One advantage of the Ritz over the Galerkin formulation occurs in the case of *positive-definite* problems (which includes most FE applications). An example would be a physical system in *stable* equilibrium, corresponding to a *minimum* in its potential energy; i.e., the governing variational principle would be a minimum principle. In these problems the sequence of values of the integral $I(\tilde{U})$, corresponding to a sequence of approximate solutions, will converge *monotonely* if the sequence of trial solutions is constructed so that each trial solution includes as a subset all solutions representable by the previous trial solution. (This may be accomplished by adding trial functions without disturbing the previous ones, as was done in Section 3.6.2.) Because of the monotone convergence, an upper bound can be established for $I(\tilde{U})$. If a complementary variational principle also exists, both upper and lower bounds can be established. Placing bounds on $I(\tilde{U})$ can be of value when $I(\tilde{U})$ represents a quantity of interest, e.g., the potential energy of an equilibrium configuration or the resonant frequency of a vibrating system.[26] However, in FE practice this bounding property of monotone convergence is not often exploited.

Based on the above observations and the author's own experience, both in teaching and in practical applications, it would appear that the Galerkin method is probably the most generally useful and easily understandable optimizing criterion for FE work. Few engineers have the time or inclination to develop a proficiency with several different optimizing criteria; and it is not necessary, because for most problems analyzed by the FEM, any of several optimizing criteria could produce a solution of acceptable accuracy, but with varying costs in computer time and the engineer's time. It is the latter that is always most expensive, so an engineer looks for a versatile and "robust" criterion that is easy to understand and will do the job well in practically all problems. Since this appears

---

[25] This author takes exception to this statement. Most of the classical field theories, which form the hard core of most commercial FE programs, are linear and self-adjoint.

[26] When the loads on a system are of a particularly simple nature, the integral $I(\tilde{U})$ can sometimes be directly related to certain pointwise properties [3.14]. Such simple loadings tend to occur more in research problems than in real design problems.

to be the Galerkin method, all problems in the remainder of the book will employ that technique.[27]

In addition, the Galerkin method is fully compatible with all commercial FE programs that use variational formulations since, as observed above, it produces identical results (using the same trial solution).

Exercises

3.1 This exercise provides a review of the Gaussian elimination procedure for solving systems of linear algebraic equations [3.9–3.12]. The procedure consists of two phases: *forward elimination* (also called *forward reduction* or *triangularization*), followed by *backward* (or *back*) *substitution*. The coefficient matrix is symmetric and positive-definite, since these are typical properties in FE applications. (Positive-definiteness obviates pivoting, i.e., interchanging the order of equations and DOF to avoid zero divisors.)

Solve the following system:

$$5a_1 + 2a_2 + a_3 + a_4 = 3$$
$$2a_1 + 10a_2 + 4a_3 + 3a_4 = 2$$
$$a_1 + 4a_2 + 8a_3 + 2a_4 = 6$$
$$a_1 + 3a_2 + 2a_3 + 12a_4 = 4$$

a) Show that after forward elimination the *triangularized* system of equations is as follows:

$$5a_1 + 2a_2 + a_3 + a_4 = 3$$
$$\frac{46}{5}a_2 + \frac{18}{5}a_3 + \frac{13}{5}a_4 = \frac{4}{5}$$
$$\frac{147}{23}a_3 + \frac{18}{23}a_4 = \frac{117}{23}$$
$$\frac{1075}{98}a_4 = \frac{125}{49}$$

Hint: Eliminate all the coefficients of a_1 below the first equation by subtracting appropriate multiples of the first equation from the second, third, and fourth equations. Similarly, eliminate a_2 by subtracting multiples of the (now-reduced) second equation from the third and fourth. Finally eliminate a_3.

b) Complete the solution by backward substituting. Thus, solve the last equation for a_4, then substitute a_4 back into the third equation and solve for a_3, etc., ending with the first equation for a_1.

c) Verify your solution by seeing if it satisfies the original system of equations.

---

[27] Although the Galerkin method is being recommended over the variational method for general FE applications, those readers who are mathematically and philosophically inclined may nevertheless wish to peruse some of the literature on the variational calculus. It is a mathematically elegant subject with a rich philosophical history. Lanczos [3.4] is most enjoyable reading.

3.2 This exercise illustrates how to apply multiple constraint equations when a DOF to be eliminated appears in more than one constraint equation. Repeat the problem in Section 3.3, this time eliminating the DOF a_3 and a_4.

a) Modify the trial solution in Eq. (3.16). Proceed as follows: Rewrite the two constraint equations [Eqs. (3.18)] as a system of linear algebraic equations with a_3 and a_4 as the unknown terms on the LHS:

$$a_3 + a_4 = 2 - a_1 - a_2$$

$$4a_3 + 12a_4 = -\frac{1}{4} - a_2$$

Triangularize the LHS by performing forward elimination as described in Exercise 3.1; i.e., eliminate a_3 from the second equation. Now proceed as in the text, using the first equation to eliminate a_3 from the trial solution, then the second (reduced) equation to eliminate a_4.

b) Complete the analysis using any of the five optimizing criteria. Compare your solution with the one in the text using the same criterion. [Recall that a_3 and a_4 in Eq. (3.16) were relabeled a_1 and a_2 in Eq. (3.23).] Are the two solutions identical? Should they be identical?

3.3 Express the following differential equations in the standard form of Eq. (2.3) or (3.7) — namely, the "symmetric" (or "self-adjoint") form.

a) $$x^2 \frac{d^2U}{dx^2} + 2x \frac{dU}{dx} + U = f$$

b) $$\frac{d^2U}{dx^2} - \frac{s}{x} \frac{dU}{dx} + x^sU = x^sf \qquad s \text{ is a constant}$$

Hint: In (b) multiply the equation by an appropriate term. What is the expression for the flux in each case?

3.4 Consider the hanging cable described in Section 2.3. The governing differential equation is

$$-\frac{d}{dx}\left(T\frac{dW}{dx}\right) = -\varrho g \qquad 0 < x < L$$

and the BCs are $W(0)=0$ and $W(L)=0$. $W(x)$ is the vertical deflection of the cable. Use the values $L=10$ m, $\varrho=1$ kg/m, $g=9.8$ m/sec², and $T=98$ N. What is the deflection of the cable at the center and at $L/4$ from either end?

a) Derive three approximate solutions using the Galerkin method. Employ the trial solutions,

$$\tilde{W}_1(x;a) = a_1 \sin \frac{\pi x}{10}$$

$$\tilde{W}_2(x;a) = a_1 \sin \frac{\pi x}{10} + a_2 \sin \frac{3\pi x}{10}$$

$$\tilde{W}_3(x;a) = a_1 \sin \frac{\pi x}{10} + a_2 \sin \frac{3\pi x}{10} + a_3 \sin \frac{5\pi x}{10}$$

Note that these trial solutions satisfy the BCs.

b) Evaluate your sequence of approximate solutions at the points $x=2.5, 5.0,$ and

7.5. Does the sequence appear to be converging? Estimate the accuracy (point-wise error) of $\tilde{W}_3(x)$. Considering the exact solution, $W(x)=x(x-10)/20$, was your error estimate comparable to the actual error?

3.5 If the trial solution in the hanging-cable problem in Exercise 3.4 were a quadratic polynomial, $\tilde{W}(x;a)=a_1+a_2x+a_3x^2$, do you think the optimizing criteria would yield the exact solution (which is also quadratic)? Try it, using any of the five criteria. Remember to modify $\tilde{W}$ to satisfy the BCs for all a_i.

3.6 In the hanging-cable problem in Exercise 3.4 there is a bilateral symmetry about the center, $x=5$, because the properties, loads, and geometry are all uniform and hence symmetrical about the center. In other words, the solution will be identical (a mirror reflection) on either side of center. A more efficient model would therefore analyze only half the cable. Derive a solution using only the section of cable from $x=0$ to $x=5$ m.

a) What BC should be applied to the "end" at $x=5$? (Consider both the required symmetry and continuity at $x=5$.)

b) Construct a 2-DOF trial solution using trigonometric functions that satisfy the appropriate BCs.

c) Complete the analysis and compare the accuracy with the previous 2-DOF solution for the whole cable.

3.7 Consider again the hanging cable in Exercise 3.4, but include an *elastic foundation* underneath the cable, distributed uniformly along its length (see Section 6.3 and Fig. 6.5). The governing differential equation is

$$-\frac{d}{dx}\left(T\frac{dW}{dx}\right) + kW = -\varrho g$$

where k is the stiffness of the foundation (24.5 N/m-m). The other constants have the values given in Exercise 3.4. What is the deflection of the cable at the center and at $L/4$ from either end?

a) Derive two approximate solutions using the Galerkin method. For trial solutions utilize a three-term power series (quadratic polynomial) and a four-term power series (cubic polynomial). Modify them to satisfy the BCs, $W(0)=0$ and $W(10)=0$, for all values of the DOF. (Recall the technique illustrated in Exercise 3.2.)

b) Evaluate your solutions at $x=2.5$, 5.0, and 7.5. Estimate the accuracy of the cubic solution. Compare with the exact solution,

$$W(x) = 0.4\left(\frac{\sin\,[5-(x/2)] + \sinh\,(x/2)}{\sinh 5}\right)$$

How good was your estimate?

3.8 Consider the steady-state flow of heat in a cylindrical rod with circular cross section, with convection heat loss from the surface (see Section 6.1.2). The temperature in the rod, $T(x)$, is described by the differential equation

$$-\frac{d}{dx}\left(k\frac{dT(x)}{dx}\right) + \frac{2h}{r}T(x) = \frac{2h}{r}T_\infty \qquad 0 < x < 10$$

subject to the BCs $T(0)=100\,°F$ and $q(10)=0.2$ BTU/sec-in^2, where q is the heat flux $-kdT/dx$. The other constants in the equation are as follows: k is the thermal conductivity (22.0 BTU/sec-in.- °F), h is the convective heat transfer coefficient (0.193×10^{-5} BTU/sec-in²- °F), r is the radius of the rod (0.5 in.), and T_∞ is the temperature of the fluid in which the rod is immersed (72 °F).

Use the Galerkin method to obtain two approximate solutions for both the temperature and the heat flux, using a three-term power series (quadratic polynomial) and a four-term power series (cubic polynomial) for trial solutions. Compare your solutions and estimate accuracy.

3.9 Steady-state heat conduction in a spherical nuclear-fission fuel element may be described by the differential equation

$$-\frac{d}{dr}\left(r^2 k \frac{dT}{dr}\right) = Cr^2\left(1 + \frac{r^2}{R^2}\right) \qquad 0 < r < R$$

where $T(r)$ is the temperature, r is a radial coordinate, R is the radius of the sphere (0.04 ft), k is the thermal conductivity (1 BTU/hr-ft- °F), and C is the amplitude of the interior heat source (3×10^6 BTU/hr-ft³). The BCs are

$$\frac{dT}{dr} = 0 \text{ at } r = 0 \qquad \text{symmetry and continuity at center}$$

$$k\frac{dT}{dr} + hT = 0 \text{ at } r = R \qquad \text{convection at outer surface}$$

where h is a convective heat transfer coefficient (400 BTU/hr-ft²- °F).

a) Beginning with a three-term power series (quadratic polynomial) for the trial solution, show that the modified trial solution that satisfies the BCs for all DOF values is as follows:

$$\tilde{T}(r;a) = a\left[r^2 - R^2\left(1 + \frac{4}{N}\right)\right]$$

where N, the Nusselt number, is equal to $2Rh/k$.

b) Using the Galerkin method, show that the approximate solution for the temperature distribution in the sphere is as follows:

$$\tilde{T}(r) = \frac{CR^2}{60k}\frac{20(5N+56)}{7(N+10)}\left[\left(1+\frac{4}{N}\right) - \left(\frac{r}{R}\right)^2\right]$$

The exact solution is

$$T(r) = \frac{CR^2}{60k}\left[\left(13+\frac{64}{N}\right) - 10\left(\frac{r}{R}\right)^2 - 3\left(\frac{r}{R}\right)^4\right]$$

Graphically compare both the approximate and exact solutions.

3.10 Consider the differential equation

$$\frac{d^2U}{dx^2} + \frac{1}{4}U = 0 \qquad 0 < x < \pi$$

with the BCs $U(0)=1$ and $U(\pi)=0$. Using a cubic polynomial for the trial solution

$\tilde{U}(x;a)=a_1+a_2x+a_3x^2+a_4x^3$, apply the collocation method with $x=\pi/3$ and $x=2\pi/3$ as collocation points to find an approximation to both the function and the flux. Remember to modify $\tilde{U}(x;a)$ to satisfy the BCs for all values of the a_i. The exact solution is $U(x)=\cos x/2$.

3.11 Repeat Exercise 3.10 using the Galerkin method.

3.12 Repeat Exercise 3.10 using the least-squares method.

3.13 Consider the differential equation

$$\frac{d}{dx}\left[(x+1)\frac{dU(x)}{dx}\right] = 0 \qquad 1 < x < 2$$

with the BCs $U(1)=1$ and $\tau(2)=1$. The flux τ is $-(x+1)dU/dx$.

a) Using the Galerkin method with a quadratic polynomial for the trial solution $\tilde{U}(x;a)=a_1+a_2x+a_3x^2$, obtain an approximate solution for both the function and the flux.

b) Obtain a second approximate solution using a cubic polynomial for the trial solution.

c) Compare the two solutions and estimate their accuracy. [Recall Eq. (3.65) and accompanying discussion.]   Also compare with the exact solution $U(x)=1-\ln[(x+1)/2]$ and $\tau(x)=1$. How good was your estimate? Do your two solutions appear to be converging to the exact solution?

3.14 Consider the differential equation in Exercise 3.13:

$$\frac{d}{dx}\left[(x+1)\frac{dU(x)}{dx}\right] = 0 \qquad 1 < x < 2$$

but with the BCs $U(1)=1$ and $U(2)=2$. Perform the same analysis as before; i.e., obtain two solutions and estimate accuracy. (When modifying the trial solution to satisfy the BCs, you may wish to review the technique illustrated in Exercise 3.2.) The exact solution is

$$U(x) = \frac{\ln (x+1) + \ln (3/4)}{\ln (3/2)}$$

$$\tau(x) = \frac{-1}{\ln (3/2)}$$

3.15 Consider the differential equation

$$\frac{d}{dx}\left[x^2\frac{dU(x)}{dx}\right] = \frac{1}{12}(-30x^4+204x^3-351x^2+110x) \qquad 0 < x < 4$$

with the BCs $U(0)=1$ and $U(4)=0$. Use the Galerkin method to obtain two approximate solutions, using a three-term power series (quadratic polynomial) and a four-term power series (cubic polynomial) for trial solutions. (When modifying the trial solution to satisfy the BCs, you may wish to review the technique illustrated in Exercise 3.2.) Compare your solutions and estimate their accuracy. Also compare with the exact solution

$$U(x) = \frac{1}{24}(-3x^4+34x^3-117x^2+110x+24)$$

How good was your estimate? Do your two solutions appear to be converging to the exact solution?

3.16 The variation of temperature with time of a body being heated internally and losing heat to its surroundings is described by the following differential equation [see Eq. (11.1)]:

$$c\frac{dT}{dt} + kT = f \qquad t > 0$$

where $T(t)$ is the temperature, t is time, c is the heat capacity (1 cal/cm^3-°C), k is the convection loss coefficient (0.1 cal/sec-cm^3-°C) and f is the interior heat source (4 cal/sec-cm^3). This is a *first*-order differential equation describing a transient change in a system for all times after an initial time; hence it only requires one "boundary" condition, called an *initial condition*, which for this problem is

$$T = 40°C \qquad \text{at} \qquad t = 0$$

Although the exact solution involves an exponential decay that extends until $t=\infty$, we are interested in the solution only during the first 10 sec.

a) Use a three-term power series (quadratic polynomial) for the trial solution and modify it to satisfy the initial condition for all values of the DOF.

b) Use the Galerkin method to derive an approximate solution $\tilde{T}(t)$. Compare the values at $t=5$ sec and $t=10$ sec with the exact solution [Eq. (11.9)].

c) Also compare your solution and the exact solution at $t=100$ sec. How could you revise your analysis to improve the accuracy at $t=100$ sec?

3.17 Analyze the transient heat problem in Exercise 3.16 using the same trial solution, but employ the collocation method. Compare the accuracy with that of the previous Galerkin solution.

References

[3.1]   B. A. Finlayson, *The Method of Weighted Residuals and Variational Principles*, Academic Press, New York, 1972.

[3.2]   S. H. Crandall, *Engineering Analysis*, McGraw-Hill, New York, 1956.

[3.3]   S. G. Mikhlin, *Variational Methods in Mathematical Physics*, Pergamon, New York, 1964.

[3.4]   C. Lanczos, *The Variational Principles of Mechanics*, University of Toronto Press, Toronto, 1966.

[3.5]   K. Rektorys, *Variational Methods in Mathematics, Science, and Engineering*, Reidel, Dordrecht, Holland, 1980.

[3.6]   K. Washizu, *Variational Methods in Elasticity and Plasticity*, Pergamon, New York, 1975.

[3.7]   O. Bolza, *Lectures on the Calculus of Variations*, Dover, New York, 1961.

[3.8]   P. M. Morse and H. Feshbach, *Methods of Theoretical Physics*, vols. I and II, McGraw-Hill, New York, 1953.

[3.9]   F. B. Hildebrand, *Methods of Applied Mathematics*, Prentice-Hall, Englewood Cliffs, N. J., 1952.

[3.10] H. Anton, *Elementary Linear Algebra*, 3d ed., Wiley, New York, 1981.

[3.11] L. W. Johnson and R. D. Riess, *Introduction to Linear Algebra*, Addison-Wesley, Reading, Mass., 1981.

[3.12] G. E. Forsythe and C. B. Moler, *Computer Solution of Linear Algebraic Systems*, Prentice-Hall, Englewood Cliffs, N. J., 1967.

[3.13] W. Kaplan, *Advanced Calculus*, Addison-Wesley, Reading, Mass., 1984.

[3.14] R. D. Cook, *Concepts and Applications of Finite Element Analysis*, Wiley, New York, 1974, p. 68.

[3.15] N. L. Schryer, unpublished work, AT&T Bell Laboratories, Murray Hill, N. J. (Document focuses primarily on linear 1-D self-adjoint boundary-value problems.)

[3.16] B. A. Finlayson and L. E. Scriven, ''On the Search for Variational Principles,'' *Int J Heat and Mass Transfer*, **10**: 799-821 (1967).

4

Foundations II:
A General 12-Step
Trial-Solution Procedure

Introduction

This chapter will present a 12-step procedure for solving boundary-value problems, eigenproblems, and mixed initial-value/boundary-value problems by trial-solution methods, with or without the element concept. Whereas the focus in Chapter 3 was on broad, underlying concepts (namely, the three principal operations), the focus here will be on procedural details. As each step is explained it will be applied to the illustrative problem used in Chapter 3.

As a point of contrast, Chapter 3 took several shortcuts that would not be practical for solving large problems on a computer; that was done in order to simplify the initial demonstration of the trial-solution procedure. In this chapter there will be no shortcuts. Here we want to explain the actual procedure that will be used throughout the remainder of the book, a procedure that is better suited for numerical work on computers. This will require two principal changes to the approach used in Chapter 3: (1) integrating by parts the residual equations and (2) using the numerical method, rather than the theoretical method, for applying boundary conditions.

Only one ingredient will be missing: the element concept. However, much of the basic mathematical structure of the FEM will have been developed by the end of this chapter, without any reference to the concept of an element. Chapter 5 will then concentrate on the few additional details needed to introduce and apply this concept.

4.1 Outline of the 12-Step Procedure

The following 12-step procedure will be used to solve all the problems in this book. The procedure carries the analysis from the original governing equations to the final numerical solution. It may be used with or without the element concept. This chapter will explain the procedure as it would be used *without* the element concept (or, what is the same thing, as it would be used with only one element). Chapter 5 will then modify the procedure slightly in order to incorporate the element concept, so that it may be used with more than one element.

The 12-step procedure is shown in Fig. 4.1. There is a definite overall pattern and logical order to the steps that will become apparent as the chapter proceeds. More readily apparent is the division into two halves, six steps each, which are logically and functionally distinct. The first six steps involve a *theoretical* pencil-and-paper manipulation of equations by the analyst/engineer; the second six steps involve *numerical* computation usually performed on a computer (or a hand calculator for problems with only a few DOF like the present one). Comparing Fig. 4.1 with the flowchart shown in Fig. 2.7, it can be seen that steps 1 through 6 correspond to the left column and steps 7 through 12 correspond to the right column.

Steps 1 through 3 are short, formal steps involving only a few manipulations of the residual equations. The form of the equations at the end of step 2 determines certain mathematical properties that the trial solution must possess, namely, the completeness and continuity conditions, which were discussed in general terms in Section 3.6.2 and which will be explained in detail in Section 5.6. Step 3 substitutes into the equations the general

---

Theoretical Development

Step 1: Write the Galerkin residual equations.

   2: Integrate by parts.

   3: Substitute the general form of the trial solution into interior integrals in residual equations.

The resulting formal expressions
are the system equations.

   4: Develop specific expressions for the trial functions.

   5: Substitute the trial functions into the system equations, and transform the integrals into a form appropriate for numerical evaluation.

   6: Prepare expressions for the flux, using the trial functions.

Numerical Computation

Step 7: Specify numerical data for a particular problem.

   8: Evaluate the interior terms in the system equations.

   9: Apply the boundary conditions to the system equations.

  10: Solve the system equations.

  11: Evaluate the flux.

  12: Display the solution and estimate its accuracy.

FIGURE 4.1 ● Twelve-step trial-solution procedure, not using the element concept. (See Section 5.9 for procedure using the element concept.)

form of the trial solution, as expressed in Eq. (3.1);[1] that is,

$$\tilde{U}(x;a) = \phi_0(x) + a_1\phi_1(x) + a_2\phi_2(x) + \ldots + a_N\phi_N(x) \qquad (4.1)$$

The form of the equations at the end of step 3 will be referred to as the *system equations*. They consist of various integrals that contain the trial functions $\phi_j(x)$. Step 4 then proceeds with the actual development of the $\phi_j(x)$ — that is, the determination of the number to use, N, and the derivation of specific expressions for each one. Most of the detailed theoretical work is done in steps 4 through 6.

The numerical computations in steps 7 through 12 will be simple enough in this and the next chapter to be carried out by hand. In Chapter 7, however, these steps will be coded into a computer program. In fact, steps 7 through 12 describe the basic structure of a typical FE program.

4.2 An Illustrative Problem

For continuity with Chapter 3, we will analyze the same illustrative problem described in Section 3.2, repeated here for convenience.

Differential equation and domain •

$$\frac{d}{dx}\left(x\,\frac{dU(x)}{dx}\right) = \frac{2}{x^2} \qquad 1 < x < 2 \qquad (4.2a)$$

Boundary conditions •

$$U(1) = 2$$

$$\left(-x\,\frac{dU}{dx}\right)_{x=2} = \frac{1}{2} \qquad (4.2b)$$

Before applying the 12 steps to this problem, let's make one minor modification. The domain is defined to be the interval $1 < x < 2$. We will indeed obtain an approximate numerical solution for this specific problem. However, in steps 1 through 6 we will let the interval be completely general, namely, $x_a < x < x_b$. Then in step 7, which is normally used to define data for a specific problem, including the location of the boundary, we will set $x_a = 1$ and $x_b = 2$. This approach is more instructive since it won't cover up important general relationships by introducing numbers in steps 1 through 6. In addition, it simulates a real-life design problem in which the desired location for the boundary is one of the uncertainties. In such cases we would want our theoretical development (steps 1 through 6) to be general enough to include any boundary location; then during the numerical computations (steps 7 through 12) we could generate solutions for several different boundaries until we found the "best" location.

So let's begin.

4.3 Steps 1 through 6: Theoretical Development

Step 1: Write the Galerkin residual equations, one for each unknown parameter a_i.

---

[1] It will be seen later (Section 4.3, step 4) that the $\phi_0(x)$ term can be ignored during construction of the trial solution whenever the BCs are applied by the numerical method rather than by the theoretical method. This is the situation normally encountered in the FEM.

In principle, we could use any of the optimizing criteria discussed in Chapter 3; however, we will use the Galerkin criterion for the reasons stated in Section 3.7.[2]

The residual for the governing equation [Eq. (4.2a)] is

$$R(x;a) = \frac{d}{dx}\left(x\frac{d\tilde{U}}{dx}\right) - \frac{2}{x^2} \tag{4.3}$$

Recalling Eqs. (3.44), the N Galerkin residual equations are

$$\int_{x_a}^{x_b} R(x;a)\phi_i(x)\,dx = 0 \qquad i = 1,2,...,N \tag{4.4}$$

Substituting Eq. (4.3),

$$\int_{x_a}^{x_b}\left[\frac{d}{dx}\left(x\frac{d\tilde{U}}{dx}\right) - \frac{2}{x^2}\right]\phi_i(x)\,dx = 0 \qquad i = 1,2,...,N \tag{4.5}$$

Step 2: Integrate by parts the highest derivative term in the residual equations.

In this step we want to transform the integral in Eq. (4.5) that contains the second-order derivative[3]; the reasons for doing this will be given following step 3.

The rule for integration by parts can be found in almost any elementary calculus book, but we will derive it here. Recall the "chain rule" for differentiating the product of two functions, say $f(x)$ and $g(x)$:

$$\frac{d}{dx}(fg) = \frac{df}{dx}g + f\frac{dg}{dx} \tag{4.6}$$

Integrate both sides over some interval, say x_a to x_b:

$$\int_{x_a}^{x_b}\frac{d}{dx}(fg)\,dx = \int_{x_a}^{x_b}\frac{df}{dx}g\,dx + \int_{x_a}^{x_b}f\frac{dg}{dx}\,dx \tag{4.7}$$

The LHS is a perfect differential:

$$LHS = \int_{x_a}^{x_b}\frac{d}{dx}(fg)\,dx$$

$$= \int_{x_a}^{x_b}d(fg)$$

$$= [fg]_{x_a}^{x_b} \tag{4.8}$$

---

[2] If a Ritz variational method were used, the governing equations would be in the form of a variational principle. Steps 1 and 2 would therefore be unnecessary and one would begin with step 3.

[3] If the collocation method were used instead of the Galerkin, this step would be omitted since there would be no integral to integrate by parts.

where

$$[fg]_{x_a}^{x_b} = [fg]_{x=x_b} - [fg]_{x=x_a}$$

Thus Eq. (4.7) may be written, after rearranging the terms,

$$\int_{x_a}^{x_b} \frac{df}{dx} g \, dx = [fg]_{x_a}^{x_b} - \int_{x_a}^{x_b} f \frac{dg}{dx} \, dx \tag{4.9}$$

which is the desired formula for integration by parts.[4] Note that integration by parts transforms an integral into two new terms: a *boundary* term and a different integral.

Now apply Eq. (4.9) to the term in Eqs. (4.5) containing the highest (second-order) derivative, namely,

$$\int_{x_a}^{x_b} \frac{d}{dx}\left(x \frac{d\tilde{U}}{dx} \right) \phi_i \, dx \tag{4.10}$$

This has the same form as the LHS of Eq. (4.9), where f corresponds to $x(d\tilde{U}/dx)$ and g to ϕ_i. Therefore Eq. (4.10) becomes

$$\int_{x_a}^{x_b} \frac{d}{dx}\left(x \frac{d\tilde{U}}{dx} \right) \phi_i \, dx = -\left[\left(-x \frac{d\tilde{U}}{dx} \right) \phi_i \right]_{x_a}^{x_b} - \int_{x_a}^{x_b} x \frac{d\tilde{U}}{dx} \frac{d\phi_i}{dx} \, dx \tag{4.11}$$

A minus sign has been added directly in front of the boundary term $x(d\tilde{U}/dx)$ so that it may be identified as flux.

Substituting Eq. (4.11) into Eqs. (4.5), the residual equations become

$$\int_{x_a}^{x_b} x \frac{d\tilde{U}}{dx} \frac{d\phi_i}{dx} \, dx = -\int_{x_a}^{x_b} \frac{2}{x^2} \phi_i \, dx - \left[\left(-x \frac{d\tilde{U}}{dx} \right) \phi_i \right]_{x_a}^{x_b} \qquad i = 1,2,\dots,N \tag{4.12}$$

Note that terms are arranged in Eqs. (4.12) so that all *loading* terms, in the interior and on the boundary,[5] are on the RHS. More on this in a moment.

Step 3: Substitute the general form of the trial solution into interior integral on the LHS of the residual equations.

The derivative of $\tilde{U}$ appears in the integral in Eqs. (4.12). From Eq. (4.1),

$$\frac{d\tilde{U}}{dx} = \frac{d\phi_0}{dx} + \sum_{j=1}^{N} a_j \frac{d\phi_j}{dx} \tag{4.13}$$

---

[4] We note in passing that Eq. (4.6) requires that $f(x)$ and $g(x)$ both be continuous functions, and therefore Eq. (4.9) presumes and is valid only when $f(x)$ and $g(x)$ are continuous in the interval $x_a < x < x_b$. This presents no difficulties in the present problem because the trial functions and their first derivatives (which f and g will represent) are continuous over the entire domain. However, it will require care when dealing with FE trial functions (see Chapter 5), since the latter inherently contain discontinuities.

[5] The interior loading terms are all those terms in the original differential equation not containing the unknown $U(x)$ or its derivatives. The boundary loading terms are all those terms evaluated on the boundary.

Substitute Eq. (4.13) into the integral on the LHS of Eqs. (4.12):

$$\text{LHS of Eqs. (4.12)} = \int_{x_a}^{x_b} x \frac{d\tilde{U}}{dx} \frac{d\phi_i}{dx} \, dx$$

$$= \int_{x_a}^{x_b} x \left(\frac{d\phi_0}{dx} + \sum_{j=1}^{N} a_j \frac{d\phi_j}{dx} \right) \frac{d\phi_i}{dx} \, dx$$

$$= \sum_{j=1}^{N} \left(\int_{x_a}^{x_b} \frac{d\phi_i}{dx} x \frac{d\phi_j}{dx} \, dx \right) a_j + \int_{x_a}^{x_b} \frac{d\phi_i}{dx} x \frac{d\phi_0}{dx} \, dx \tag{4.14}$$

Substituting Eq. (4.14) back into Eqs. (4.12), the residual equations become

$$\sum_{j=1}^{N} \left(\int_{x_a}^{x_b} \frac{d\phi_i}{dx} x \frac{d\phi_j}{dx} \, dx \right) a_j = - \int_{x_a}^{x_b} \frac{2}{x^2} \phi_i \, dx - \left[\left(-x \frac{d\tilde{U}}{dx} \right) \phi_i \right]_{x_a}^{x_b}$$

$$- \int_{x_a}^{x_b} \frac{d\phi_i}{dx} x \frac{d\phi_0}{dx} \, dx \qquad i = 1,2,\ldots,N \tag{4.15}$$

This completes the first three steps. Let's examine what we have at this point. Equations (4.15) may be written out in full:

$$\left(\int_{x_a}^{x_b} \frac{d\phi_1}{dx} x \frac{d\phi_1}{dx} \, dx \right) a_1 + \left(\int_{x_a}^{x_b} \frac{d\phi_1}{dx} x \frac{d\phi_2}{dx} \, dx \right) a_2 + \ldots + \left(\int_{x_a}^{x_b} \frac{d\phi_1}{dx} x \frac{d\phi_N}{dx} \, dx \right) a_N$$

$$= - \int_{x_a}^{x_b} \frac{2}{x^2} \phi_1 \, dx - \left[\left(-x \frac{d\tilde{U}}{dx} \right) \phi_1 \right]_{x_a}^{x_b} - \int_{x_a}^{x_b} \frac{d\phi_1}{dx} x \frac{d\phi_0}{dx} \, dx$$

$$\left(\int_{x_a}^{x_b} \frac{d\phi_2}{dx} x \frac{d\phi_1}{dx} \, dx \right) a_1 + \left(\int_{x_a}^{x_b} \frac{d\phi_2}{dx} x \frac{d\phi_2}{dx} \, dx \right) a_2 + \ldots + \left(\int_{x_a}^{x_b} \frac{d\phi_2}{dx} x \frac{d\phi_N}{dx} \, dx \right) a_N$$

$$= - \int_{x_a}^{x_b} \frac{2}{x^2} \phi_2 \, dx - \left[\left(-x \frac{d\tilde{U}}{dx} \right) \phi_2 \right]_{x_a}^{x_b} - \int_{x_a}^{x_b} \frac{d\phi_2}{dx} x \frac{d\phi_0}{dx} \, dx$$

$$\vdots \qquad \qquad \vdots \qquad \qquad \vdots$$

$$\left(\int_{x_a}^{x_b} \frac{d\phi_N}{dx} x \frac{d\phi_1}{dx} \, dx \right) a_1 + \left(\int_{x_a}^{x_b} \frac{d\phi_N}{dx} x \frac{d\phi_2}{dx} \, dx \right) a_2 + \ldots + \left(\int_{x_a}^{x_b} \frac{d\phi_N}{dx} x \frac{d\phi_N}{dx} \, dx \right) a_N$$

$$= - \int_{x_a}^{x_b} \frac{2}{x^2} \phi_N \, dx - \left[\left(-x \frac{d\tilde{U}}{dx} \right) \phi_N \right]_{x_a}^{x_b} - \int_{x_a}^{x_b} \frac{d\phi_N}{dx} x \frac{d\phi_0}{dx} \, dx \tag{4.16}$$

or, equivalently, in matrix form:

$$
\begin{bmatrix}
\int_{x_a}^{x_b} \dfrac{d\phi_1}{dx} x \dfrac{d\phi_1}{dx}\,dx & \int_{x_a}^{x_b} \dfrac{d\phi_1}{dx} x \dfrac{d\phi_2}{dx}\,dx & \cdots & \int_{x_a}^{x_b} \dfrac{d\phi_1}{dx} x \dfrac{d\phi_N}{dx}\,dx \\[4mm]
\int_{x_a}^{x_b} \dfrac{d\phi_2}{dx} x \dfrac{d\phi_1}{dx}\,dx & \int_{x_a}^{x_b} \dfrac{d\phi_2}{dx} x \dfrac{d\phi_2}{dx}\,dx & \cdots & \int_{x_a}^{x_b} \dfrac{d\phi_2}{dx} x \dfrac{d\phi_N}{dx}\,dx \\[4mm]
\vdots & \vdots & & \vdots \\[4mm]
\int_{x_a}^{x_b} \dfrac{d\phi_N}{dx} x \dfrac{d\phi_1}{dx}\,dx & \int_{x_a}^{x_b} \dfrac{d\phi_N}{dx} x \dfrac{d\phi_2}{dx}\,dx & \cdots & \int_{x_a}^{x_b} \dfrac{d\phi_N}{dx} x \dfrac{d\phi_N}{dx}\,dx
\end{bmatrix}
\begin{Bmatrix}
a_1 \\[4mm] a_2 \\[4mm] \vdots \\[4mm] a_N
\end{Bmatrix}
$$

$$
=
\begin{Bmatrix}
-\displaystyle\int_{x_a}^{x_b} \dfrac{2}{x^2}\phi_2\,dx - \left[\left(-x\dfrac{d\tilde{U}}{dx}\right)\phi_1\right]_{x_a}^{x_b} - \int_{x_a}^{x_b} \dfrac{d\phi_1}{dx} x \dfrac{d\phi_0}{dx}\,dx \\[6mm]
-\displaystyle\int_{x_a}^{x_b} \dfrac{2}{x^2}\phi_2\,dx - \left[\left(-x\dfrac{d\tilde{U}}{dx}\right)\phi_2\right]_{x_a}^{x_b} - \int_{x_a}^{x_b} \dfrac{d\phi_2}{dx} x \dfrac{d\phi_0}{dx}\,dx \\[6mm]
\vdots \\[6mm]
-\displaystyle\int_{x_a}^{x_b} \dfrac{2}{x^2}\phi_N\,dx - \left[\left(-x\dfrac{d\tilde{U}}{dx}\right)\phi_N\right]_{x_a}^{x_b} - \int_{x_a}^{x_b} \dfrac{d\phi_N}{dx} x \dfrac{d\phi_0}{dx}\,dx
\end{Bmatrix}
\tag{4.17}
$$

Defining

$$
K_{ij} = \int_{x_a}^{x_b} \dfrac{d\phi_i}{dx} x \dfrac{d\phi_j}{dx}\,dx
$$

$$
F_i = -\int_{x_a}^{x_b} \dfrac{2}{x^2}\phi_i\,dx - \left[\left(-x\dfrac{d\tilde{U}}{dx}\right)\phi_i\right]_{x_a}^{x_b} - \int_{x_a}^{x_b} \dfrac{d\phi_i}{dx} x \dfrac{d\phi_0}{dx}\,dx
\tag{4.18}
$$

Equations (4.17) may be written more compactly as

$$
\begin{bmatrix}
K_{11} & K_{12} & \cdots & K_{1N} \\
K_{21} & K_{22} & \cdots & K_{2N} \\
\vdots & \vdots & & \vdots \\
\vdots & \vdots & & \vdots \\
K_{N1} & K_{N2} & \cdots & K_{NN}
\end{bmatrix}
\begin{Bmatrix}
a_1 \\ a_2 \\ \vdots \\ \vdots \\ a_N
\end{Bmatrix}
=
\begin{Bmatrix}
F_1 \\ F_2 \\ \vdots \\ \vdots \\ F_N
\end{Bmatrix}
\tag{4.19}
$$

or, in abbreviated matrix notation:

$$[K]\{a\} = \{F\} \tag{4.20}$$

Equations (4.15), (4.16), (4.17), (4.19), and (4.20) are, of course, merely five different representations of the same set of N equations.

It is conventional in FE analysis to use the uppercase letter K to represent the matrix of coefficients that multiply the vector of unknown parameters. This matrix is usually referred to as the *stiffness matrix*. The vector of loading terms on the RHS is usually referred to as the *load vector* (or *force vector*). This terminology originated in the field of structural engineering where $Kx=F$ describes the displacement x of a simple linear spring of stiffness K being pulled by a force F. In a multiple-degree-of-freedom elastic structure the FE equations for static equilibrium can always be written in the matrix form $[K]\{x\} = \{F\}$, where the elements of $[K]$, $\{x\}$, and $\{F\}$ have the same physical meaning of stiffness, displacement, and force. In other fields of application (electromagnetism, heat conduction, fluid dynamics, etc.) the residual equations are also in the form of Eq. (4.20), and $[K]$ is still called the stiffness matrix and $\{F\}$ the load vector (or force vector). We will refer to the set of N equations as the *system equations*.

We can now identify three benefits that were achieved as a result of integrating by parts.

1. The highest order of the derivatives of the trial solution or trial functions appearing in the system equations has been *lowered* from two to one. If we had not integrated by parts, one of the trial functions would appear as a second-order derivative, and the other would not be differentiated at all; after integration by parts, though, both trial functions appear as a first-order derivative.

   Lowering the order of the derivatives is generally not important when using trial functions that are defined over the entire domain by a single function (as in Chapter 3 and here), because such functions are usually continuous and have continuous derivatives to any order. However, it will be seen in Chapter 5 that it is of considerable importance when using FE trial functions, because such functions inherently possess discontinuities, so their derivatives above a certain order do not exist at the discontinuities.

2. The stiffness matrix was made *symmetric*; that is, $K_{ij}=K_{ji}$, as can be seen from Eqs. (4.18), where both trial functions appear as a first derivative (instead of one as a second derivative and the other not differentiated at all). Symmetry yields computational advantages because only half as many stiffness integrals need to be evaluated and stored.

   This symmetry was possible to achieve because the governing differential equation [Eq. (4.2a)] has symmetry, as discussed in Section 3.2. When such symmetry exists it is generally wise to take full advantage of it, not only for computational economy but also for the rich physical meaning that is generally present.[6] This type of sym-

---

[6] There are preliminary indications, in work not yet published, that if the governing differential equation has symmetry, then the system equations in their *final numerical form* — after BCs have been imposed (i.e., at the end of step 9 in Section 4.4) — can always be made symmetric, irrespective of whether or not integration by parts has been performed, or indeed, how many times it has been performed. For example, see Eqs. (3.46), where no integration by parts was performed, and also see Exercises 4.11 and 4.12, where integration by parts is performed twice. If this hypothesis is true, then symmetry of the stiffness matrix would no longer be a compelling argument for integration by parts.

metry is present in a wide spectrum of physical phenomena; it will be present in all the equations in this book.

3. A *boundary term was created*:

$$\left[\left(-x\,\frac{d\tilde{U}}{dx}\right)\phi_i\right]_{x_a}^{x_b}$$

This contains the flux $\tau = -xd\tilde{U}/dx$, which is contained in the second BC in Eq. (4.2b). This will enable us to apply the second BC very easily; we merely substitute for this term the value from Eq. (4.2b). On the other hand, the first BC will not be so simple to apply. (We will have more to say in Section 4.5 on the interrelationship of integration by parts and boundary conditions.)

These first three steps are quite general and relatively brief; they involve only the governing differential equation and the formal expression for the trial solution in Eq. (4.1). The next three steps are more specific, involving a particular choice of expressions for the trial functions and the subsequent manipulation of all the integrals in the stiffness matrix and load vector.

Step 4: Develop specific expressions for the trial functions $\phi_i(x)$.

Recall that in Chapter 3 we considered a polynomial for the trial solution, namely,

$$\tilde{U}(x;a) = a_1 + a_2x + a_3x^2 + a_4x^3 + \ldots + a_Nx^{N-1} \tag{4.21}$$

because powers of x are the easiest functions to differentiate and integrate, as will be required in the expressions for K_{ij} and F_i in Eqs. (4.18). In Chapter 3 we then modified the trial functions (powers of x) so that the resulting trial solution would satisfy the BCs for all values of the a_i; this was the theoretical method for applying the BCs. However, here and in the remainder of the book, we will totally ignore the BCs at this stage because they will be applied later, in step 9, using the numerical method.

It remains only to choose the number of terms N in Eq. (4.21). We will choose $N=3$ because it includes enough terms to make the problem interesting without complicating the analysis. Thus our trial solution will be

$$\tilde{U}(x;a) = a_1 + a_2x + a_3x^2$$

$$= \sum_{j=1}^{3} a_j\phi_j(x) \tag{4.22}$$

The trial functions are therefore[7]

$$\phi_1(x) = 1 \qquad \phi_2(x) = x \qquad \phi_3(x) = x^2 \tag{4.23}$$

---

[7] The completeness and continuity conditions for the trial functions were discussed in Section 3.6.2. It was observed there that the powers of x (x^N, $N=0,1,2,\ldots$), defined over the entire domain, satisfy both conditions for the present illustrative problem. Section 5.6 will present the requirements for the completeness and continuity conditions in the context of elements, i.e., when the trial functions are defined only over parts of the domain.

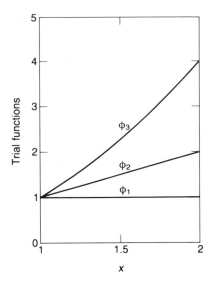

FIGURE 4.2 ● Trial functions for the illustrative problem.

These are shown in Fig. 4.2 for the interval $1 \leq x \leq 2$.

Note that no $\phi_0(x)$ term is necessary in this step because the BCs, which are the usual source that generates a $\phi_0(x)$, won't be applied until step 9. In principle, one could still include a $\phi_0(x)$ term, for whatever reason desired,[8] but in practice it is rarely done when the BCs are applied by the numerical method.

Step 5: Substitute the trial functions into stiffness and load terms in the system equations, and transform the integrals into a form appropriate for later numerical evaluation in step 8.

Consider first the K_{ij} terms from Eqs. (4.18):

$$K_{ij} = \int_{x_a}^{x_b} \frac{d\phi_i}{dx} \, x \, \frac{d\phi_j}{dx} \, dx$$

The derivatives of the trial functions follow from Eqs. (4.23):

$$\frac{d\phi_1}{dx} = 0$$

$$\frac{d\phi_2}{dx} = 1$$

$$\frac{d\phi_3}{dx} = 2x \tag{4.24}$$

---

[8] In special circumstances one might know a function that is close to the exact solution. By using ϕ_0 to represent that function, the remaining trial functions would need only to approximate the small difference between the known function and the exact solution.

Since K_{ij} is symmetric ($K_{ij}=K_{ji}$), we need only work on the terms for which $j \geq i$.

$$i = 1 \qquad K_{11} = 0 \qquad \text{(because } d\phi_1/dx = 0)$$

$$K_{12} = 0$$

$$K_{13} = 0 \qquad\qquad \textbf{(4.25a)}$$

$$i = 2 \qquad K_{22} = \int_{x_a}^{x_b} 1 \cdot x \cdot 1 \; dx = \frac{1}{2}\left(x_b^2 - x_a^2\right)$$

$$K_{23} = \int_{x_a}^{x_b} 1 \cdot x \cdot 2x \; dx = \frac{2}{3}\left(x_b^3 - x_a^3\right) \qquad\qquad \textbf{(4.25b)}$$

$$i = 3 \qquad K_{33} = \int_{x_a}^{x_b} 2x \cdot x \cdot 2x \; dx = \left(x_b^4 - x_a^4\right) \qquad\qquad \textbf{(4.25c)}$$

Consider next the F_i terms from Eqs. (4.18):

$$F_i = -\int_{x_a}^{x_b} \frac{2}{x^2}\phi_i \; dx - \left[\left(-x\frac{d\tilde{U}}{dx}\right)\phi_i\right]_{x_a}^{x_b}$$

This is the sum of an integral over the interior load FI_i and a boundary term FB_i:

$$FI_i = -\int_{x_a}^{x_b} \frac{2}{x^2}\phi_i \; dx$$

$$FB_i = -\left[\left(-x\frac{d\tilde{U}}{dx}\right)\phi_i\right]_{x_a}^{x_b} \qquad\qquad \textbf{(4.26)}$$

First the interior integral:

$$i = 1 \qquad FI_1 = -\int_{x_a}^{x_b} \frac{2}{x^2} 1 \; dx = 2\left(\frac{1}{x_b} - \frac{1}{x_a}\right)$$

$$i = 2 \qquad FI_2 = -\int_{x_a}^{x_b} \frac{2}{x^2} x \; dx = -2 \ln\frac{x_b}{x_a}$$

$$i = 3 \qquad FI_3 = -\int_{x_a}^{x_b} \frac{2}{x^2} x^2 \; dx = -2(x_b - x_a) \qquad\qquad \textbf{(4.27)}$$

Next the boundary term:

$$FB_i = \left(-x \frac{d\tilde{U}}{dx}\right)_{x_a} \phi_i(x_a) - \left(-x \frac{d\tilde{U}}{dx}\right)_{x_b} \phi_i(x_b) \qquad (4.28a)$$

Equation (4.28a) contains flux terms, $-x d\tilde{U}/dx$, evaluated at each boundary point. We will defer treatment of the flux until step 9, which deals with the numerical application of the BCs. However, the terms $\phi_i(x_a)$ and $\phi_i(x_b)$ may be replaced, using Eqs. (4.23):

$$i = 1 \qquad FB_1 = \left(-x \frac{d\tilde{U}}{dx}\right)_{x_a} - \left(-x \frac{d\tilde{U}}{dx}\right)_{x_b}$$

$$i = 2 \qquad FB_2 = \left(-x \frac{d\tilde{U}}{dx}\right)_{x_a} x_a - \left(-x \frac{d\tilde{U}}{dx}\right)_{x_b} x_b$$

$$i = 3 \qquad FB_3 = \left(-x \frac{d\tilde{U}}{dx}\right)_{x_a} x_a^2 - \left(-x \frac{d\tilde{U}}{dx}\right)_{x_b} x_b^2 \qquad (4.28b)$$

Step 6: Prepare expressions for the flux using the trial functions.

$$\tilde{\tau} = \text{flux} = -x \frac{d\tilde{U}}{dx}$$

$$= -x \sum_{j=1}^{3} a_j \frac{d\phi_j}{dx}$$

$$= -a_2 x - 2a_3 x^2 \qquad (4.29)$$

In many problems the flux is as important as the function $U(x)$ itself. And sometimes there are other physical quantities of interest. In general, then, this step carries out whatever algebra or calculus operations are necessary to express the flux (and/or other quantities) in terms of the parameters a_i. Then, when the system equations are solved in step 10 for the a_i, the resulting numerical values can be substituted into Eq. (4.29).

This completes the first phase, steps 1 through 6, which involves the theoretical preparation of all the necessary equations. We now have a system of three linear algebraic equations. Let's write them out in full, by substituting Eqs. (4.25), (4.27), and (4.28b) into Eq. (4.19):

$$\begin{bmatrix} 0 & 0 & 0 \\ 0 & \frac{1}{2}\left(x_b^2 - x_a^2\right) & \frac{2}{3}\left(x_b^3 - x_a^3\right) \\ 0 & \frac{2}{3}\left(x_b^3 - x_a^3\right) & \left(x_b^4 - x_a^4\right) \end{bmatrix} \begin{Bmatrix} a_1 \\ a_2 \\ a_3 \end{Bmatrix}$$

$$= \begin{Bmatrix} 2\left(\frac{1}{x_b} - \frac{1}{x_a}\right) \\ -2 \ln \frac{x_b}{x_a} \\ -2(x_b - x_a) \end{Bmatrix} + \begin{Bmatrix} \left(-x \frac{d\tilde{U}}{dx}\right)_{x_a} - \left(-x \frac{d\tilde{U}}{dx}\right)_{x_b} \\ \left(-x \frac{d\tilde{U}}{dx}\right)_{x_a} x_a - \left(-x \frac{d\tilde{U}}{dx}\right)_{x_b} x_b \\ \left(-x \frac{d\tilde{U}}{dx}\right)_{x_a} x_a^2 - \left(-x \frac{d\tilde{U}}{dx}\right)_{x_b} x_b^2 \end{Bmatrix} \qquad (4.30)$$

4.4 Steps 7 through 12: Numerical Computation

Step 7: Specify all numerical data for the problem.

1. *Geometric data.* Specify the coordinates that define the boundary. For a 1-D problem this is a trivial step, since only two points need to be specified. From the problem statement in Eq. (4.1a),

$$x_a = 1 \qquad x_b = 2 \tag{4.31}$$

We note that for a 2-D problem the boundary is a curve, for a 3-D problem it is a surface, etc. In these cases we would specify all the coordinates necessary to define a curve, a surface, etc.

2. *Physical properties and applied loads.* In our present problem these have already been defined. The coefficient x on the LHS of the governing differential equation [Eq. (4.1a)] corresponds to a physical property; the term $2/x^2$ on the RHS corresponds to an interior load [recall Eq. (2.3)]. The boundary conditions in Eqs. (4.1b) specified the boundary loads. (In Chapter 7 the terms corresponding to physical properties and applied loads in the governing differential equation and BCs will all be in a general form so that a computer program can be written for a general class of problems. This step will then correspond to inputting to the program numerical data for these terms.)

Step 8: Numerically evaluate all the interior terms in the system equations (that is, all terms except the BC terms).

The interior terms include the stiffness integrals (K_{ij} terms) and the interior load integrals (FI_i terms). In addition, we can also evaluate the coefficients of the BC terms.[9] Thus, using the values from Eqs. (4.31), Eqs. (4.30) become

$$\begin{bmatrix} 0 & 0 & 0 \\ 0 & \dfrac{3}{2} & \dfrac{14}{3} \\ 0 & \dfrac{14}{3} & 15 \end{bmatrix} \begin{Bmatrix} a_1 \\ a_2 \\ a_3 \end{Bmatrix} = \begin{Bmatrix} -1 \\ -2\ln 2 \\ -2 \end{Bmatrix} + \begin{Bmatrix} \left(-x\dfrac{d\tilde{U}}{dx}\right)_{x=1} - \left(-x\dfrac{d\tilde{U}}{dx}\right)_{x=2} \\ \left(-x\dfrac{d\tilde{U}}{dx}\right)_{x=1} - 2\left(-x\dfrac{d\tilde{U}}{dx}\right)_{x=2} \\ \left(-x\dfrac{d\tilde{U}}{dx}\right)_{x=1} - 4\left(-x\dfrac{d\tilde{U}}{dx}\right)_{x=2} \end{Bmatrix} \tag{4.32}$$

The system equations are now almost ready to solve; only the BCs remain unspecified.

Step 9: Apply the BCs to the system equations.

Equations (4.2b) specified one BC at each of the two boundary points:

$$U(1) = 2$$

$$\text{Flux} = \left(-x\frac{dU}{dx}\right)_{x=2} = \frac{1}{2}$$

---

[9] It will be seen in Chapter 5 that the BC terms have no coefficients when the element concept is used. Therefore, in a normal FE analysis, step 8 would involve only the interior terms — i.e., the stiffness and interior load integrals.

These two BCs will be applied to the system equations in distinctly different ways: One will be "constrained" and the other will be "unconstrained." This terminology will be explained in Section 4.5.

Application of the BC $(-xdU/dx)_{x=2} = 1/2$ • We'll do this BC first because it is very simple to apply. The term $(-xd\tilde{U}/dx)_{x=2}$ *appears explicitly* in the boundary term vector on the RHS of Eqs. (4.32), so we only need to substitute the value 1/2 everywhere this term appears:

$$
\begin{bmatrix}
0 & 0 & 0 \\
0 & \dfrac{3}{2} & \dfrac{14}{3} \\
0 & \dfrac{14}{3} & 15
\end{bmatrix}
\begin{Bmatrix} a_1 \\ a_2 \\ a_3 \end{Bmatrix}
=
\begin{Bmatrix} -1 \\ -2\ \ln 2 \\ -2 \end{Bmatrix}
+
\begin{Bmatrix}
\left(-x\dfrac{d\tilde{U}}{dx}\right)_{x=1} - \dfrac{1}{2} \\[2mm]
\left(-x\dfrac{d\tilde{U}}{dx}\right)_{x=1} - 1 \\[2mm]
\left(-x\dfrac{d\tilde{U}}{dx}\right)_{x=1} - 2
\end{Bmatrix}
\tag{4.33}
$$

The fact that the BC in Eqs. (4.2b) involves the exact solution U, whereas the terms in Eqs. (4.32) involve the approximate trial solution $\tilde{U}$, is of no consequence. We are merely assigning a value (the exact value) that we would like the approximate solution to have.

Note that Eqs. (4.33) still contain the term $(-xd\tilde{U}/dx)_{x=1}$. This corresponds to a flux BC at $x=1$. However, we don't have any numerical value to specify for this term because the other BC specifies U at $x=1$, not the flux. This is no problem, though, because we'll see shortly that this term will disappear when the other BC is applied.

It will be seen later, in step 12, that the solution to this problem will only *approximately* satisfy the flux BC $(-xdU/dx)_{x=2}=1/2$, despite the fact that we substituted the exact value into the system equations. The trial solution, and resulting approximate solution, are not constrained to satisfy the flux BC, and we will speak of this BC as being *unconstrained*. More on this point in Section 4.5.

Application of the BC $U(1) = 2$ • Application of this BC is more complicated than the previous flux BC and hence will require more explanation. It is important to note, though, that the complication is due primarily to the fact that the trial functions are defined over the *entire* domain of the problem. When using element trial functions (which are defined only over small parts of the domain — see Chapter 5 and rest of book), the procedure for applying this BC degenerates to a very simple case of the procedure we are about to explain here.

We begin by observing that there are no terms in the system equations [Eqs. (4.33)] for which we can substitute the value 2 specified by this BC. So we resort to the only other alternative: *Apply the BC directly to the trial solution itself*. Thus, applying $U(1)=2$ to Eq. (4.22) yields

$$
a_1 + a_2 + a_3 = 2
\tag{4.34}
$$

This is a constraint equation. It constrains the a_i's so that they are no longer independent. Thus, only two of the a_i's in Eq. (4.34) can be independent, the third being determined by Eq. (4.34).

It will be recalled that a similar approach was used in Section 3.3; i.e., the BCs were applied directly to the trial solution to generate constraint equations [Eqs. (3.18)]. There the constraint equations were immediately substituted back into the trial solution to *modify the trial functions*. The procedure was performed at the beginning of the problem, in what we would now call step 4. That approach was referred to as the *theoretical method*.

Here we want to *modify the system equations* by applying the constraint equation [Eq. (4.34)] to Eqs. (4.33), a procedure that will involve manipulation of the numbers in Eqs. (4.33). Hence this approach is referred to as the *numerical method*. It has the decided practical advantage that it can be done on a computer. (We will not implement the method on a computer until after the element concept is developed in Chapter 5 because that concept will greatly simplify the implementation.)

The basic idea of the numerical method is as follows. The system equations [Eqs. (4.33)] assume that a_1, a_2, and a_3 are independent, but the constraint equation [Eq. (4.34)] allows only two of these three parameters to be independent. We must therefore eliminate one of the a_i from Eqs. (4.33) so that there will be only two equations involving two independent a_i. [When the two equations are subsequently solved, the resulting values for the two a_i may be substituted back into Eq. (4.34) to solve for the third a_i.]

The modification of Eqs. (4.33) to eliminate one of the a_i involves a three-step procedure. This will require a short digression to explain. The three steps will be demonstrated by applying them to a general system of three equations in three unknowns:

$$K_{11}a_1 + K_{12}a_2 + K_{13}a_3 = F_1$$
$$K_{21}a_1 + K_{22}a_2 + K_{23}a_3 = F_2$$
$$K_{31}a_1 + K_{32}a_2 + K_{33}a_3 = F_3 \tag{4.35a}$$

or, equivalently, in matrix notation,

$$\begin{bmatrix} K_{11} & K_{12} & K_{13} \\ K_{21} & K_{22} & K_{23} \\ K_{31} & K_{32} & K_{33} \end{bmatrix} \begin{Bmatrix} a_1 \\ a_2 \\ a_3 \end{Bmatrix} = \begin{Bmatrix} F_1 \\ F_2 \\ F_3 \end{Bmatrix} \tag{4.35b}$$

These three equations will be modified by a general linear constraint equation:

$$c_1a_1 + c_2a_2 + c_3a_3 = d \tag{4.36}$$

First step: Solve the constraint equation [Eq. (4.36)] for one of the a_i in terms of all the others.

In general it makes no difference which a_i is used, so long as its coefficient c_i is not zero. For numerical stability, a good choice is the a_i with the largest coefficient. Let's choose a_3. Then

$$a_3 = \frac{d}{c_3} - \frac{c_1}{c_3}a_1 - \frac{c_2}{c_3}a_2 \tag{4.37}$$

It is the third DOF, a_3, that will now be eliminated from the system equations [Eqs. (4.35)].

Second step: Eliminate the chosen DOF (in this case, a_3) from the system equations.

Substituting Eq. (4.37) for a_3 in each equation in Eqs. (4.35) yields

$$K_{11}a_1 + K_{12}a_2 + K_{13}\left(\frac{d}{c_3} - \frac{c_1}{c_3}a_1 - \frac{c_2}{c_3}a_2\right) = F_1$$

$$K_{21}a_1 + K_{22}a_2 + K_{23}\left(\frac{d}{c_3} - \frac{c_1}{c_3}a_1 - \frac{c_2}{c_3}a_2\right) = F_2$$

$$K_{31}a_1 + K_{32}a_2 + K_{33}\left(\frac{d}{c_3} - \frac{c_1}{c_3}a_1 - \frac{c_2}{c_3}a_2\right) = F_3 \qquad \textbf{(4.38)}$$

Combining coefficients of a_1 and a_2, and moving the constant terms to the RHS, we have

$$\left(K_{11} - \frac{c_1}{c_3}K_{13}\right)a_1 + \left(K_{12} - \frac{c_2}{c_3}K_{13}\right)a_2 = F_1 - \frac{d}{c_3}K_{13}$$

$$\left(K_{21} - \frac{c_1}{c_3}K_{23}\right)a_1 + \left(K_{22} - \frac{c_2}{c_3}K_{23}\right)a_2 = F_2 - \frac{d}{c_3}K_{23}$$

$$\left(K_{31} - \frac{c_1}{c_3}K_{33}\right)a_1 + \left(K_{32} - \frac{c_2}{c_3}K_{33}\right)a_2 = F_3 - \frac{d}{c_3}K_{33} \qquad \textbf{(4.39a)}$$

or, in matrix notation,

$$\begin{bmatrix} K_{11} - \frac{c_1}{c_3}K_{13} & K_{12} - \frac{c_2}{c_3}K_{13} \\[2mm] K_{21} - \frac{c_1}{c_3}K_{23} & K_{22} - \frac{c_2}{c_3}K_{23} \\[2mm] K_{31} - \frac{c_1}{c_3}K_{33} & K_{32} - \frac{c_2}{c_3}K_{33} \end{bmatrix} \begin{Bmatrix} a_1 \\[2mm] a_2 \end{Bmatrix} = \begin{Bmatrix} F_1 - \frac{d}{c_3}K_{13} \\[2mm] F_2 - \frac{d}{c_3}K_{23} \\[2mm] F_3 - \frac{d}{c_3}K_{33} \end{Bmatrix} \qquad \textbf{(4.39b)}$$

We may think of this second step as a series of *column operations* in which multiples of the third column of stiffness terms K_{i3} in Eqs. (4.35b) are added to (or subtracted from) the other columns; the multipliers are the coefficients of the a_i and the constant term in the constraint equation [Eq. (4.37)]. Thus, Eqs. (4.39b) could also have been derived by performing the following column operations on Eqs. (4.35b):

- Multiply the third column of stiffness terms K_{i3} by $-c_1/c_3$ and add the product to the first column of stiffness terms.
- Multiply the third column of stiffness terms K_{i3} by $-c_2/c_3$ and add the product to the second column of stiffness terms.
- Multiply the third column of stiffness terms K_{i3} by d/c_3 and subtract the product from the RHS terms.
- Finally, delete the third column of stiffness terms K_{i3} (and therefore delete a_3 too).

Third step: Eliminate the equation corresponding to the chosen DOF (in this case, the third equation).

This step may be thought of as a series of *row operations*, directly parallel to the column operations in the second step. These operations are performed on Eqs. (4.39b) as follows:

- Multiply the third row in Eqs. (4.39b) [or, equivalently, the third equation in Eqs. (4.39a)] by $-c_1/c_3$ and add the product to the first row.
- Multiply the third row by $-c_2/c_3$ and add the product to the second row.
- Finally, delete the third row.

This leaves the following two equations:

$$
\left[
\begin{array}{cc}
\left(K_{11} - \dfrac{c_1}{c_3} K_{13} \right) - \dfrac{c_1}{c_3} \left(K_{31} - \dfrac{c_1}{c_3} K_{33} \right) & \left(K_{12} - \dfrac{c_2}{c_3} K_{13} \right) - \dfrac{c_1}{c_3} \left(K_{32} - \dfrac{c_2}{c_3} K_{33} \right) \\[2ex]
\left(K_{21} - \dfrac{c_1}{c_3} K_{23} \right) - \dfrac{c_2}{c_3} \left(K_{31} - \dfrac{c_1}{c_3} K_{33} \right) & \left(K_{22} - \dfrac{c_2}{c_3} K_{23} \right) - \dfrac{c_2}{c_3} \left(K_{32} - \dfrac{c_2}{c_3} K_{33} \right)
\end{array}
\right]
\left\{
\begin{array}{c}
a_1 \\[2ex]
a_2
\end{array}
\right\}
$$

$$
= \left\{
\begin{array}{c}
\left(F_1 - \dfrac{d}{c_3} K_{13} \right) - \dfrac{c_1}{c_3} \left(F_3 - \dfrac{d}{c_3} K_{33} \right) \\[2ex]
\left(F_2 - \dfrac{d}{c_3} K_{23} \right) - \dfrac{c_2}{c_3} \left(F_3 - \dfrac{d}{c_3} K_{33} \right)
\end{array}
\right\}
\tag{4.40}
$$

which is the desired form of the system equations — two equations in two unknowns. Note that if the stiffness matrix is symmetric in Eqs. (4.35), then it is still symmetric in Eqs. (4.40).

The reasons for the row operations are not as obvious as for the column operations. However, it is straightforward to show (see Exercise 4.1) that

- The column operations correspond to *modifying the trial functions* so that the trial solution will satisfy the constrained BC exactly for all values of the a_i (just as in the theoretical method).
- The row operations correspond to *modifying the weighting functions* in the residual equations so that they will be the same as the modified trial functions, as required by the Galerkin method.

Thus, the theoretical and numerical methods of applying BCs both yield identical results because both perform the same modification on the trial functions and weighting functions. The difference is that the theoretical method does the modification *explicitly* at the beginning of the analysis, whereas the numerical method does it *implicitly* later on in the analysis. This completes the digression.

We can now apply the constraint equation [Eq. (4.34)] to the system equations [Eqs. (4.33)] using the above three-step procedure.

First step: Solve the constraint equation [Eq. (4.34)] for one of the a_i in terms of all the others.

We could choose a_1, a_2, or a_3. a_1 would be the easiest to use in this particular problem because all the zeroes in row 1 and column 1 of the stiffness matrix in Eqs. (4.33) would eliminate most of the computation. However, it is instructive to see this computation, so let's choose another one, say a_3:

$$
a_3 = 2 - a_1 - a_2
\tag{4.41}
$$

Second step: Eliminate a_3 from Eqs. (4.33).

A comparison of Eqs. (4.41) and (4.37) shows that $c_1/c_3=1$, $c_2/c_3=1$, and $d/c_3=2$. Therefore the column operations on Eqs. (4.33) correspond to subtracting the third column of stiffness terms from the first and second columns, then subtracting two times the third column from the RHS, and finally deleting the third column (and a_3). The result is

$$
\begin{bmatrix}
0 & 0 \\
-\dfrac{14}{3} & \dfrac{3}{2}-\dfrac{14}{3} \\
-15 & \dfrac{14}{3}-15
\end{bmatrix}
\begin{Bmatrix} a_1 \\ a_2 \end{Bmatrix}
=
\begin{Bmatrix}
\left(-x\dfrac{d\tilde{U}}{dx}\right)_{x=1} - \dfrac{3}{2} \\
\left(-x\dfrac{d\tilde{U}}{dx}\right)_{x=1} - 2\ln 2 - 1 - \dfrac{14}{3}(2) \\
\left(-x\dfrac{d\tilde{U}}{dx}\right)_{x=1} - 4 - 15(2)
\end{Bmatrix}
\tag{4.42}
$$

or simplifying,

$$
\begin{bmatrix}
0 & 0 \\
-\dfrac{14}{3} & -\dfrac{19}{6} \\
-15 & -\dfrac{31}{3}
\end{bmatrix}
\begin{Bmatrix} a_1 \\ a_2 \end{Bmatrix}
=
\begin{Bmatrix}
\left(-x\dfrac{d\tilde{U}}{dx}\right)_{x=1} - \dfrac{3}{2} \\
\left(-x\dfrac{d\tilde{U}}{dx}\right)_{x=1} - 2\ln 2 - \dfrac{31}{3} \\
\left(-x\dfrac{d\tilde{U}}{dx}\right)_{x=1} - 34
\end{Bmatrix}
\tag{4.43}
$$

Third step: Eliminate the third equation from Eqs. (4.43).

Again using $c_1/c_3=1$ and $c_2/c_3=1$, the row operations on Eqs. (4.43) reduce to simply subtracting the third row from the first and second rows, then deleting the third row. The result is

$$
\begin{bmatrix}
15 & \dfrac{31}{3} \\
15-\dfrac{14}{3} & \dfrac{31}{3}-\dfrac{19}{6}
\end{bmatrix}
\begin{Bmatrix} a_1 \\ a_2 \end{Bmatrix}
=
\begin{Bmatrix}
34-\dfrac{3}{2} \\
34-2\ln 2 - \dfrac{31}{3}
\end{Bmatrix}
\tag{4.44}
$$

or simplifying,

$$
\begin{bmatrix}
15 & \dfrac{31}{3} \\
\dfrac{31}{3} & \dfrac{43}{6}
\end{bmatrix}
\begin{Bmatrix} a_1 \\ a_2 \end{Bmatrix}
=
\begin{Bmatrix}
\dfrac{65}{2} \\
\dfrac{71}{3}-2\ln 2
\end{Bmatrix}
\tag{4.45}
$$

Equations (4.45) are the final form of the system equations after applying both BCs. Note that the stiffness matrix is still symmetric, just as it was in Eqs. (4.33).

Perhaps the most conspicuous feature in the third step is that the boundary terms $(-xd\tilde{U}/dx)_{x=1}$ in Eqs. (4.43), for which there is no prescribed boundary value, canceled each other out during the row operations. [Recall Eqs. (3.56) and (3.57) in Section 3.5.] This cancellation was not just a coincidence for this particular problem; it was a natural concomitant to the method (theoretical or numerical) of applying BCs by constraint equations. The reason for the cancellation is related to the previous observation that the effect of the row and column operations is to modify the trial functions (and weighting functions). As shown in Exercise 4.1, the modified trial functions, which multiply the unknown boundary terms, always vanish at those boundary points where a BC is constrained [as exemplified by Eqs. (3.13)]. These relationships are not at all obvious, so the reader is encouraged to work through Exercise 4.1 for further insight.

As a final comment, it will be seen later, in step 12, that the solution to this problem will satisfy the BC $U(1)=2$ *exactly*. This is to be expected since this BC was applied by a constraint equation which, as explained above, effectively constrains the trial solution to satisfy the BC for all values of the a_i (just as in the theoretical method in Section 3.3). We will therefore speak of this BC as being *constrained*. More on this point in Section 4.5.

Step 10: Solve the system equations.

The solution to Eqs. (4.45) can easily be done by hand, yielding

$$a_1 = 3.719$$

$$a_2 = -2.254 \tag{4.46a}$$

These values for a_1 and a_2 must now be substituted back into the constraint equation [Eq. (4.34)]. (They must not be substituted into the deleted system equation since that has been replaced by the constraint equation.) Thus

$$a_3 = 0.535 \tag{4.46b}$$

The resulting approximate solution is

$$\tilde{U}(x) = 3.719 - 2.254x + 0.535x^2 \tag{4.47}$$

Step 11: Evaluate the flux (and, in general, any other quantities related to the trial solution) using the values determined for the a_i.

Substituting the values for a_2 and a_3 into Eq. (4.29) yields

$$\tilde{\tau}(x) = 2.254x - 1.070x^2 \tag{4.48}$$

Step 12: Plot the solution and estimate its accuracy.

Figure 4.3 is a plot of Eq. (4.47) and Eq. (4.48). Our principal concern is the accuracy of $\tilde{U}(x)$ and $\tilde{\tau}(x)$. How close are these curves to the exact solution?

An immediate observation is that $\tilde{U}(x)$ satisfies the constrained BC at $x=1$ exactly, but the flux $\tilde{\tau}(x)$ satisfies the unconstrained BC at $x=2$ only approximately. This difference in the behavior of the two types of boundary conditions — one being satisfied exactly and the other approximately — was alluded to in step 9 and will be explained in Section 4.5.

We can take advantage of this approximate satisfaction of the flux BC to provide us with an error indication. We know the exact error in the flux at the boundary point, not just an estimate, because we already know the exact solution at this point. Thus we specified an exact value of 1/2, but the system equations yielded a value of 0.228, an error of 54%. This is not very helpful in determining the error in the flux at other locations in the domain; however, it would seem reasonable to estimate that the error is comparable to 54% within a small region "close" to this boundary point.

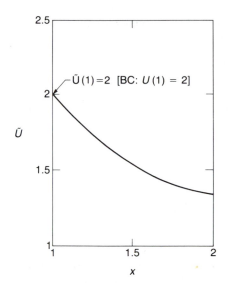
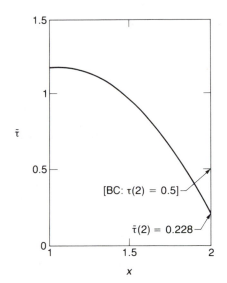

FIGURE 4.3 ● An approximate solution using three DOF.

For the present problem this error calculation is not as useful as the sequence of approximate solutions we will examine next. However, it will be seen in Chapter 5 and the subsequent chapters that when the domain is subdivided into elements, a similar effect involving unconstrained "interelement" conditions will occur at all the boundaries between the elements, thereby providing error estimates for the flux at all interelement boundaries.

In order to estimate the accuracy of the function $\tilde{U}(x)$ (as well as $\tilde{\tau}(x)$ away from the boundary), we apply the procedure described in Section 3.6.2: Generate a sequence of two or more approximate solutions, each using successively more (or fewer) trial functions, and then compare the differences between them. So let's calculate two more solutions: One will use a trial solution that removes one term from our previous trial solution; the other will use a trial solution that adds an extra term to our previous trial solution. These extra solutions will presumably be less accurate and more accurate, respectively. The fact that the one solution may be less accurate is not important. Our purpose is to estimate accuracy and for this we only need a *sequence* of two or more solutions; the order in which we generate the sequence is immaterial.

Our present solution is therefore the second member of a sequence, generated from the following sequence of trial solutions:[10]

$$\tilde{U}_1(x;a) = a_1 + a_2x$$

$$\tilde{U}_2(x;a) = a_1 + a_2x + a_3x^2$$

$$\tilde{U}_3(x;a) = a_1 + a_2x + a_3x^2 + a_4x^3 \tag{4.49}$$

---

[10] A linear polynomial is the lowest-degree polynomial that can be used. If only a constant term were used, e.g., $\tilde{U}_0(x;a)=a_1$, then all the K_{ij} integrals in the stiffness matrix, Eqs. (4.18), would vanish because each integrand contains the first derivative of the trial functions. In this case there would be no system equations.

Repeating the analysis (steps 1 through 11) with $\tilde{U}_1(x;a)$ and $\tilde{U}_3(x;a)$ yields the following sequence of approximate solutions:

$$\tilde{U}_1(x) = 2.591 - 0.591x$$

$$\tilde{U}_2(x) = 3.719 - 2.254x + 0.535x^2$$

$$\tilde{U}_3(x) = 4.963 - 4.908x + 2.340x^2 - 0.395x^3 \qquad \textbf{(4.50a)}$$

and their corresponding expressions for flux:

$$\tilde{\tau}_1(x) = 0.591x$$

$$\tilde{\tau}_2(x) = 2.254x - 1.070x^2$$

$$\tilde{\tau}_3(x) = 4.908x - 4.680x^2 + 1.185x^3 \qquad \textbf{(4.50b)}$$

which are plotted in Fig. 4.4.

The convergence theorems referred to in Section 3.6.2 assure us that the solution must converge (assuming no computational mistakes) in an average sense, i.e., with respect to the energy error. And, as observed there, experience suggests that the solution is highly likely to also converge pointwise. Both $\tilde{U}(x)$ and $\tilde{\tau}(x)$ in Fig. 4.4 appear to be converging at all values of x (pointwise), although $\tilde{\tau}(x)$ is converging more slowly. A visual estimate would suggest that the error for $\tilde{U}_3(x)$ is less than 5% at all values of x, but about 30% for $\tilde{\tau}_3(x)$.

Feeling a bit uncertain about these estimates, especially for $\tilde{\tau}_3(x)$, we will try one more trial solution, this one containing an additional term and therefore, we hope, more accuracy:

$$\tilde{U}_4(x;a) = a_1 + a_2x + a_3x^2 + a_4x^3 + a_5x^4 \qquad \textbf{(4.51)}$$

This yields the approximate solution,

$$\tilde{U}_4(x) = 6.250 - 8.557x + 6.123x^2 - 2.097x^3 + 0.281x^4$$

$$\tilde{\tau}_4(x) = 8.557x - 12.246x^2 + 6.290x^3 - 1.123x^4 \qquad \textbf{(4.52)}$$

All four approximate solutions are plotted in Fig. 4.5. They show a consistent converging trend; i.e., successive solutions are closer and closer together. Our estimates of 5% error for $\tilde{U}_3$ and 30% error for $\tilde{\tau}_3$ seem like good ones, perhaps a bit conservative. In fact it is tempting to now estimate 1% error for $\tilde{U}_4$ and 10% error for $\tilde{\tau}_4$.

The results in Fig. 4.5, buttressed by the convergence theorems, are certainly reassuring. Can we be certain, though, of those 1 and 10% error estimates? Of course not. The probability of computational mistakes is always present in every problem, so, despite any mathematical proofs "guaranteeing" convergence, it is perhaps prudent to view Fig. 4.5 (supported by convergence theorems) as providing us with a *level of confidence* (not certainty) that (1) our approximate solutions are probably converging pointwise to the exact solution, and (2) the pointwise error is probably about 1% for $\tilde{U}_4(x)$ and about 10% for $\tilde{\tau}_4(x)$.

In practical applications where there may be considerable human value at stake, we might demand a much greater level of confidence than that afforded by the above arguments, which essentially employ a deductive approach (appealing to broad principles, namely, convergence theorems, to conclude that our particular problem ought to converge).

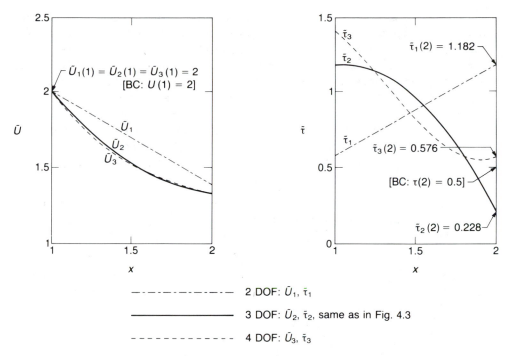

FIGURE 4.4 ● A sequence of three approximate solutions.

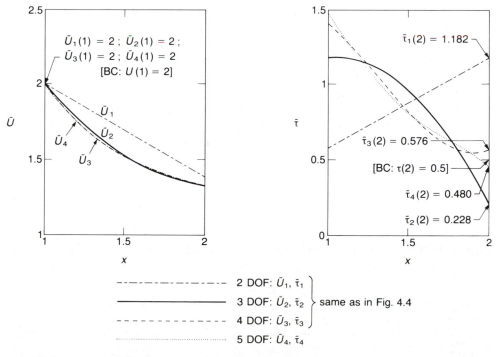

FIGURE 4.5 ● A sequence of four approximate solutions.

An inductive approach to increasing confidence is to apply the same analysis procedure to other, similar problems with known solutions, i.e., test problems. By "similar" we mean that the test problems use the same, or about the same, algorithmic steps as the real problem; in the context of a computer program, this means that essentially the same parts of the code are used. If the results of these test problems are correct, we make the "inductive leap" that the solution to our real problem is likely to be correct too.[11]

Another approach to increasing confidence is to acquire additional *independent* evidence, such as experimental data or another analysis of the same problem using a different (non-FEM) method, as was illustrated by each of the 10 industrial examples in Section 1.4. This approach can provide a relatively high level of confidence because it is based on the following logic:

> Consider two *independently* derived solutions (analyses, experimental data, etc.) that agree with each other. If they are wrong, then both have the same error. But the likelihood of two *independent* solutions containing identical (or nearly identical) errors is small. Consequently, the likelihood is high that both are correct.

For example, in the present problem, if we had only a single experimental data point within a few percent of the $\tilde{U}_4(x)$ curve, we could have much greater confidence (though still not certainty) in our error estimate.

This completes the 12-step procedure. It will be modified slightly in Chapter 5 to incorporate the element concept. But before proceeding to Chapter 5, let's take a closer look at some of the concepts we have dealt with thus far.

4.5 Boundary Conditions: Terminology and Concepts

In this and the previous chapter we encountered several different ways of applying BCs, as well as a variety of terminology — for example, the theoretical method, the numerical method, and constrained and unconstrained BCs. This section will bring these ideas together into a more coherent picture.

We present first a traditional method for classifying BCs, one which is still widely used by engineers and mathematicians. Consider a boundary-value problem in which the governing differential equation is of order $2m$, where m is an integer:

$$a_{2m}(x) \frac{d^{2m}U(x)}{dx^{2m}} + a_{2m-1} \frac{d^{2m-1}U(x)}{dx^{2m-1}} + \ldots + a_1(x) \frac{dU(x)}{dx}$$

$$+ a_0(x)U(x) = f(x) \qquad (4.53)$$

The highest-order derivative is even — 2, 4, 6, 8, etc. We consider only even-order derivatives (for the highest derivative) because this is representative of the classical field theories, which are the "bread and butter" of virtually all engineering applications —

---

[11] Developers of commercial FE programs are applying this inductive approach when they qualify their programs by applying them to a large collection of test problems. Nevertheless, it is advisable for users of such programs to still perform one or more tests before analyzing a particular problem, as will be demonstrated by the illustrative problems in subsequent chapters.

solid and fluid mechanics, acoustics, heat transfer, and electromagnetism, to name a few. Equation (4.53) illustrates a 1-D differential equation, but the remarks in this section apply also to problems in any number of dimensions.

A boundary-value problem of order $2m$ requires m BCs to be specified at every point of the boundary [4.1]. A BC is an equation relating the values of U and/or some of its derivatives from order 1 up to order $2m-1$, at points on the boundary.[12] It is conventional to classify the BCs into two principal types:[13]

1. *Essential BC.* An equation relating the values of U and/or any of its derivatives up to order $m-1$, at points on the boundary

2. *Natural BC.* An equation relating the values of any of the derivatives of U from order m to $2m-1$, at points on the boundary.[14]

This classification divides the hierarchy of derivatives into two halves: the highest m derivatives (from m to $2m-1$) and the lowest m derivatives (from 0 to $m-1$, where 0 implies the function itself).

Essential BCs involve the lower-order derivatives. They are sometimes also referred to as Dirichlet, kinematic, displacement, or geometric BCs. In solid mechanics they usually correspond to displacements and sometimes slopes, in heat transfer to temperature, and in other applications to a potential.

Natural BCs involve the higher-order derivatives. They are sometimes also referred to as Neumann, dynamic, force, or stress BCs. In solid mechanics they correspond to forces, stresses, and moments; in heat transfer to heat flux; and in potential problems to the gradient of the potential.

The two cases $m=1$ and $m=2$ cover almost all practical applications, so we summarize them here.

---

[12] For problems in two or more dimensions, the only derivatives involved are those normal (perpendicular) to the boundary. See Chapter 13.

[13] There is also a third type, called a *mixed BC.* As its name implies, it is a mix of the other two, containing U and/or any of its derivatives up to order $2m-1$. In the FEM it is treated operationally like a natural BC (see Exercise 4.6).

[14] The words "essential" and "natural" appear to have originated in the context of the calculus of variations during the nineteenth century, but are now applied to any boundary-value problem, irrespective of whether it is in variational form or differential equation form. There appears to be some rationale to this choice of words. Thus, in the most common formulations of the classical field theories (those which were principally studied during the nineteenth century and which still govern most applications today) it is necessary — i.e., essential — that at least some of the BCs be essential BCs (or mixed BCs, which contain some essential terms) in order for the problem to have a unique solution. The mathematical reason is that in these classical formulations the unknown function U appears only in derivatives in the governing differential equation(s). Therefore, if the BCs were only natural BCs, i.e., only derivatives of U, then the entire problem would be formulated only in terms of derivatives of U, resulting in an infinity of solutions differing by at least an arbitrary constant. The word "natural" may allude to the fact that in the variational formulation, the natural BC terms are the only ones present (i.e., they appear "naturally") in the governing variational equation.

TABLE 4.1 ● Second-order differential equations ($m=1$).

Type of BC	Specifies boundary values for
Essential	U
Natural	$\dfrac{dU}{dx}$

Examples: elasticity (plane stress, plane strain, axisymmetric, three-dimensional), heat conduction, acoustics, fluid dynamics, and electromagnetism

TABLE 4.2 ● Fourth-order differential equations ($m=2$).

Type of BC	Specifies boundary values for
Essential	$U, \dfrac{dU}{dx}$
Natural	$\dfrac{d^2U}{dx^2}, \dfrac{d^3U}{dx^3}$

Examples: beam theory, plate and shell theories, and special formulations of items in Table 4.1

The illustrative problem in this and the preceding chapter (see Sections 3.2 and 4.2) is a second-order problem — that is, $m=1$. Therefore its BCs are classified as follows:

$$U(1) = 2 \qquad \text{essential BC}$$

$$\left(-x\,\frac{dU}{dx} \right)_{x=2} = \frac{1}{2} \qquad \text{natural BC} \tag{4.54}$$

There is a "symmetry" in the physical meaning of the various orders of derivatives: To each essential type of derivative there corresponds a natural type of derivative, such that the product of the two derivatives (and perhaps other physical parameters) represents *energy*. Thus the orders of derivatives may be grouped naturally into the following *pairs*: orders m and $m-1$, $m+1$ and $m-2,\ldots,2m-1$, and 0. To illustrate, beam theory in solid mechanics is governed by a fourth-order differential equation ($m=2$). The second and first derivatives (m and $m-1$) correspond to a moment and a slope, respectively, and their product represents the energy expended by a moment acting through an angular rotation. The third and zero-th derivatives ($m+1$ and $m-2$) correspond to a shear stress and a displacement, respectively, and their product represents the energy expended by a force (stress times area) acting through a distance. These remarks may help to explain why so many physical phenomena are governed by *even*-order differential equations.

The classification of BCs as essential or natural is determined simply by the order of the governing differential equations. It is completely independent of the method used to solve the differential equation.

This author has found it very useful, when dealing with the many different trial-solution methods, to introduce a different classification of BCs, one which reflects *how they are applied* — i.e., an operational classification. For this purpose we may classify BCs as *constrained* or *unconstrained*, a terminology that was introduced in step 9. Before defining these terms, let's take a closer look at how the concepts "constrained" and "unconstrained" arise in trial-solution applications.

Recall from Fig. 3.1 that in the trial-solution procedure the resulting approximate solution (item 3) is derived from just two ingredients: a trial solution (item 1) and an optimizing criterion (item 2), the latter producing a system of equations. Therefore, in order for the approximate solution to satisfy a BC, there exist only two avenues for introducing the BC into the analysis: by applying the BC to the trial solution itself (directly via the theoretical method in step 4 or indirectly via the numerical method in step 9) or by substituting the value of the BC for a term that appears in the system equations.

For example, in the problem treated in this chapter, the system equations contained a boundary term for the flux so the flux BC was applied by simply substituting a number for the term. This did not constrain the trial solution itself, so the resulting solution only satisfied the flux BC approximately. On the other hand, there was no boundary term for the function, so the BC that specified a value for the function had to be applied instead to the trial solution (using the numerical method). This created a constraint equation, and the resulting solution was constrained to satisfy the BC exactly.

The mechanism that created the boundary term for the flux was integration by parts. It is integration by parts that determines whether a BC will be constrained or unconstrained. Thus, recall Eq. (4.11) in which the integral containing the highest (second) derivative was integrated by parts:

$$\int_{x_a}^{x_b} \frac{d}{dx}\left(x\frac{d\tilde{U}}{dx}\right)\phi_i\,dx = -\left[\left(-x\frac{d\tilde{U}}{dx}\right)\phi_i\right]_{x_a}^{x_b} - \int_{x_a}^{x_b} x\frac{d\tilde{U}}{dx}\frac{d\phi_i}{dx}\,dx$$

We note that this created a boundary term for the flux, as well as a new integral. The RHS of Eq. (4.11) was the form used in the analysis in Sections 4.3 and 4.4.

If desired, however, the new integral on the RHS of Eq. (4.11) could also be integrated by parts:

$$\int_{x_a}^{x_b} x\frac{d\tilde{U}}{dx}\frac{d\phi_i}{dx}\,dx = \left[x\tilde{U}\frac{d\phi_i}{dx}\right]_{x_a}^{x_b} - \int_{x_a}^{x_b} \tilde{U}\frac{d}{dx}\left(x\frac{d\phi_i}{dx}\right)dx \tag{4.55}$$

Substituting Eq. (4.55) into Eq. (4.11) yields

$$\int_{x_a}^{x_b} \frac{d}{dx}\left(x\frac{d\tilde{U}}{dx}\right)\phi_i\,dx = -\left[\left(-x\frac{d\tilde{U}}{dx}\right)\phi_i\right]_{x_a}^{x_b} - \left[x\tilde{U}\frac{d\phi_i}{dx}\right]_{x_a}^{x_b}$$

$$+ \int_{x_a}^{x_b} \tilde{U}\frac{d}{dx}\left(x\frac{d\phi_i}{dx}\right)dx \tag{4.56}$$

Observe that each successive integration by parts accomplishes two things:

1. Another boundary term is created, in which the order of the derivative of $\tilde{U}$ is *lowered* by one.

2. The integral is replaced by a new integral in which the order of the derivative of $\tilde{U}$ is *lowered* by one but the order of the derivative of ϕ_i is *raised* by one.

Since the original integral on the LHS of Eq. (4.11) contains $\tilde{U}$ differentiated twice and ϕ_i not at all, integrating by parts twice has transferred the second derivative from $\tilde{U}$ to ϕ_i in the integral. This is as far as we can go because two more integrations by parts of the integral in Eq. (4.56) would only transfer the derivative back again from ϕ_i to $\tilde{U}$, reproducing the original integral on the LHS of Eq. (4.11).

For a general boundary-value problem in which the governing differential equation is of order $2m$, the integral containing the derivative of order $2m$ in the residual equations has the form

$$\int_{x_a}^{x_b} \frac{d^{2m}\tilde{U}}{dx^{2m}} \phi_i \, dx$$

$$(4.57)$$

This could be integrated by parts any number of times from 1 to $2m$. For example, after integrating by parts m times we would have

$$\int_{x_a}^{x_b} \frac{d^{2m}\tilde{U}}{dx^{2m}} \phi_i \, dx = (-1)^m \int_{x_a}^{x_b} \frac{d^m\tilde{U}}{dx^m} \frac{d^m\phi_i}{dx^m} \, dx$$

$$(4.58)$$

$+ \ m$ boundary terms containing derivatives of
$\tilde{U}$ of order $2m-1$, $2m-2$, ..., m (corresponding
to the natural BCs)

After integrating by parts $2m$ times we would have

$$\int_{x_a}^{x_b} \frac{d^{2m}\tilde{U}}{dx^{2m}} \phi_i \, dx = \int_{x_a}^{x_b} \tilde{U} \frac{d^{2m}\phi_i}{dx^{2m}} \, dx$$

$$(4.59)$$

$+ \ 2m$ boundary terms containing derivatives of
$\tilde{U}$ of order $2m-1$, $2m-2$, ..., 1, 0 (corresponding
to both the essential and natural BCs)

The presence or absence of boundary terms in the system equations, which determines whether or not the corresponding BCs are constrained or unconstrained, is clearly a direct consequence of how many times one integrates by parts.

These observations are summarized in the following definitions:

1. *Unconstrained BC.* A BC for which a corresponding boundary term appears in the system equations. The term appears as the result of integrating by parts the residual equations.[15] The BC is applied simply by substituting a number for the term. The resulting approximate solution satisfies the unconstrained BC only approximately.

---

[15] In a variational formulation the term appears "naturally" without integrating by parts.

2. *Constrained BC.* A BC for which there is no corresponding boundary term in the system equations. The BC is applied by a constraint equation, using either the theoretical method (Chapter 3) or the numerical method (Chapter 4). The resulting approximate solution satisfies the constrained BC exactly.

In principle, essential BCs and natural BCs may be either constrained or unconstrained, depending on how many times one integrates by parts. For example, if one didn't integrate by parts at all, then there would be no boundary terms in the system equations, so both the essential and natural BCs would have to be constrained. This was the case in Chapter 3, where we chose to use the theoretical method to constrain both BCs.

Another choice is to integrate by parts exactly m times, as in Eq. (4.58), making all the natural BCs unconstrained and all the essential BCs constrained. This was the approach used in this chapter. Thus we integrated by parts only once, thereby permitting[16] the natural BC to be unconstrained but still requiring the essential BC to be constrained. We chose to use the numerical method to apply the essential BC.

Yet another choice is to integrate by parts $2m$ times, as in Eq. (4.59), making the natural and essential BCs all unconstrained. For the illustrative problem in Chapters 3 and 4, this would correspond to integrating by parts twice [Eq. (4.56)], permitting both the essential and natural BCs to be unconstrained. This approach is demonstrated by Exercises 4.11 and 4.12.

There are obviously many different ways to obtain a variety of approximate solutions to a given problem, depending on how many times one chooses to integrate by parts. Is there an optimum number of times to integrate by parts? From purely technical considerations, there is not (yet) a clear answer. More work needs to be done in this area.

However, at this time (mid-1980s), it appears that all commercial FE programs have followed only one path: integrating by parts m times, making all essential BCs constrained and all natural BCs unconstrained. This focus on only the one approach is probably due principally to historical "accident": Most (or all) of the convergence theorems used in the FEM were originally developed for the classical variational principles. These principles are all in a form equivalent to using the Galerkin method and integrating by parts m times. Therefore these principles require essential BCs to be constrained and permit natural BCs to be unconstrained. Thus there has been a strong historical legacy that has driven the field almost exclusively in one direction, resulting in little or no attempt to develop convergence theorems for these other possible approaches. Until and unless such theorems are developed, it would be prudent to continue using the conventional approach for all serious FE applications. Therefore, in the remainder of this book all second-order differential equations will be integrated by parts just once, thereby requiring the essential BCs to be constrained and permitting the natural BCs to be unconstrained.

Exercises

4.1 The purpose of this exercise is to demonstrate the equivalence of the theoretical and numerical methods for applying BCs, and to provide meaning to the row/column

---

[16] The presence of a boundary term in the system equations *permits*, but does not require, the BC to be unconstrained. If desired, one could also constrain it, as was illustrated in the Ritz method in Section 3.5.

operations in the numerical method (see Section 4.4, step 9). The reader will be prompted through several steps.

a) Consider the general trial solution in Eq. (4.1) and an essential BC, $\tilde{U}(x_a)=d$. Apply the latter to the former to produce a constraint equation. We must eliminate one of the DOF (from the system equations), say a_N. Show that the constraint equation has the form

$$a_N = \frac{d-c_0}{c_N} - \sum_{j=1}^{N-1} \frac{c_j}{c_N} a_j, \qquad \text{where} \qquad c_j = \phi_j(x_a) \tag{1}$$

b) *Theoretical method*: Apply Eq. (1) to Eq. (4.1). Show that the modified trial solution is

$$\tilde{U}(x;a) = \psi_0(x) + \sum_{j=1}^{N-1} a_j \psi_j(x) \tag{2}$$

where

$$\psi_0(x) = \phi_0(x) + \frac{d-c_0}{c_N} \phi_N(x)$$

$$\psi_j(x) = \phi_j(x) - \frac{c_j}{c_N} \phi_N(x) \qquad j = 1,2,\ldots,N-1$$

What BCs are satisfied by $\psi_0(x)$ and $\psi_j(x)$ at x_a? Write the system equations corresponding to the modified trial solution. *Hint*: Simply alter Eqs. (4.15) in an appropriate fashion.

c) *Numerical method*: Apply Eq. (1) to the system equations [Eqs. (4.15)] using the row/column operations described in Section 4.4. Substitute expressions for ψ_0 and ψ_j from Eq. (2). Observe that the resulting system equations are identical to those in (b) above. Q.E.D.

d) Review the discussion in Section 4.4, step 9, while examining the details in the above development. Note especially: (1) The modified trial functions $\psi_j(x)$ vanish at x_a, which eliminates from the system equations all natural BCs at x_a; (2) the row/column operations modify the weighting/trial functions, respectively.

4.2 a) Apply steps 1 through 3 of the 12-step procedure (see Fig. 4.1) to derive the formal expressions for the system equations for the following general problem:

$$-\frac{d^2U}{dx^2} + \beta U = f \qquad x_a < x < x_b$$

where β and f are constants. Use the general expression for $\tilde{U}(x;a)$ in Eq. (4.1), but omit the $\phi_0(x)$ term.

b) Finish the development of the system equations by applying steps 4 through 6. For a trial solution use a cubic polynomial

$$\tilde{U}(x;a) = a_1 + a_2 x + a_3 x^2 + a_4 x^3$$

Evaluate all integrals with "closed-form" expressions. How would you alter the resulting system equations for a quadratic polynomial trial solution?

4.3 **a)** Using the system equations in Exercise 4.2 corresponding to a quadratic polynomial trial solution, apply steps 7 through 12 of the 12-step procedure to obtain an approximate solution to the following specific problem:

$$\frac{d^2U}{dx^2} + U = 0 \qquad 0 < x < \frac{\pi}{2}$$

with the BCs $\tau(0) = -1$ and $\tau(\pi/2) = 0$. (Note that the expression for the flux τ is $-dU/dx$.) Graphically compare your approximate solution with the exact solution, $U(x) = \sin x$.

b) Consider the same differential equation and domain as in (*a*) above but change the BCs to $U(0) = 0$ and $U(\pi/2) = 1$. Thus the BCs have merely been changed from natural to essential, although the exact solution remains the same. Again use the system equations in Exercise 4.2 corresponding to a quadratic polynomial trial solution and apply steps 7 through 12 to obtain an approximate solution. Apply the essential BCs using the row/column operations in Section 4.4, step 9. (Recall also Exercise 3.2.) Graphically compare this solution with the one in (*a*) above. Why would you expect them to be different?

4.4 Use the system equations developed in Exercise 4.2 to analyze the hanging-cable problem in Exercise 3.4. Obtain two approximate solutions corresponding to the trial solutions:

$$\tilde{W}_1(x;a) = a_1 + a_2x + a_3x^2$$
$$\tilde{W}_2(x;a) = a_1 + a_2x + a_3x^2 + a_4x^3$$

Evaluate your solutions at $x = 2.5$, 5.0, and 7.5 and compare with the exact solution. (See Exercise 3.4.) Do your solutions appear to be converging?

4.5 Use the system equations developed in Exercise 4.2 to analyze the hanging cable on an elastic foundation in Exercise 3.7. Obtain two approximate solutions corresponding to the trial solutions:

$$\tilde{W}_1(x;a) = a_1 + a_2x + a_3x^2$$
$$\tilde{W}_2(x;a) = a_1 + a_2x + a_3x^2 + a_4x^3$$

Evaluate your solutions at $x = 2.5$, 5.0, and 7.5 and compare with the exact solution. (See Exercise 3.7.) Do your solutions appear to be converging?

4.6 Analyze the nuclear-fuel-element problem in Exercise 3.9 using again a three-term power series (quadratic polynomial) for the trial solution, but this time follow the 12-step procedure. This will entail integrating by parts, as well as applying both BCs by the numerical method in step 9. (Hence do not modify the trial solution, as was done in Exercise 3.9.)

This problem illustrates the application of mixed BCs — namely, the convection BC at $r = R$. Thus, the term $d\tilde{T}(R)/dr$ will appear on the RHS of the system equations. Merely substitute the BC, $-(h/k)\tilde{T}(R)$, using the trial solution for $\tilde{T}(R)$. This will produce a_i terms which are then transferred back to the LHS, resulting in a modification of the stiffness matrix. Note that the latter remains symmetric. Show that the approximate solution for the temperature distribution in the sphere is as follows:

$$\tilde{T}(r) = \frac{CR^2}{60k}\left[\left(\frac{82}{7}+\frac{64}{N}\right) + \frac{48}{7}\frac{r}{R} - \frac{130}{7}\left(\frac{r}{R}\right)^2\right]$$

Graphically compare this solution with the ones in Exercise 3.9.

4.7 Consider the steady-state heat flow problem described in Exercise 3.8:

$$-\frac{d}{dx}\left(k\frac{dT(x)}{dx}\right) + \frac{2h}{r}T(x) = \frac{2h}{r}T_\infty \qquad 0 < x < 10$$

Use the same data for the BCs and the constants in the differential equation. For a trial solution use the quadratic polynomial $\tilde{U}(x;a)=a_1+a_2x+a_3x^2$. Apply the 12-step procedure to obtain an approximate solution for the temperature and the heat flux. Compare your solution to the solution obtained in Exercise 3.8. Should they be the same or different?

4.8 Consider the differential equation in Exercise 3.13:

$$\frac{d}{dx}\left[(x+1)\frac{dU(x)}{dx}\right] = 0 \qquad 1 < x < 2$$

with the BCs $U(1)=1$ and $\tau(2)=1$.

a) For a trial solution use the linear polynomial $\tilde{U}_1(x;a)=a_1+a_2x$. Apply the 12-step procedure to obtain an approximate solution for both the function and the flux.

b) Repeat the analysis using the quadratic polynomial $\tilde{U}_2(x;a)=a_1+a_2x+a_3x^2$. Compare the solution $\tilde{U}_2(x)$ with the linear solution $\tilde{U}_1(x)$ and the exact solution (Exercise 3.13). Is $\tilde{U}_2(x)$ different from the solution obtained in Exercise 3.13? Why or why not?

4.9 Consider the differential equation in Exercise 3.14:

$$\frac{d}{dx}\left[(x+1)\frac{dU(x)}{dx}\right] = 0 \qquad 1 < x < 2$$

with the BCs $U(1)=1$ and $U(2)=2$. For a trial solution use the quadratic polynomial: $\tilde{U}(x;a)=a_1+a_2x+a_3x^2$. Apply the 12-step procedure to obtain an approximate solution for both the function and the flux. Is this solution different from the solution obtained in Exercise 3.14? Why or why not? (Recall the discussion in Section 4.4, step 9, and in Exercise 4.1.)

4.10 Consider the differential equation in Exercise 3.15:

$$\frac{d}{dx}\left(x^2\frac{dU(x)}{dx}\right) = \frac{1}{12}(-30x^4+204x^3-351x^2+110x) \qquad 0 < x < 4$$

with the boundary conditions $U(0)=1$ and $U(4)=0$. For a trial solution use the quadratic polynomial $\tilde{U}(x;a)=a_1+a_2x+a_3x^2$. Apply the 12-step procedure to obtain an approximate solution. Compare your solution to the solution obtained in Exercise 3.15. Are they the same or different? Why?

4.11 Analyze the problem in Section 4.2 using the 12-step procedure, but in step 2 integrate by parts *twice* so that both essential and natural BC terms appear in the system equations [see Eq. (4.56)]. Try both quadratic and cubic polynomials for trial solutions. This exercise illustrates how to formulate a problem so that all types of BCs are treated in the same manner of approximation; i.e., they are all unconstrained (see Section 4.5).

There will be boundary terms on the RHS for which no BCs are available — namely, a natural BC at $x=1$ and an essential BC at $x=2$. For these terms merely substitute the trial solution, then transfer all a_i terms to the LHS. Note that the stiffness matrix becomes symmetric after the boundary terms have all been evaluated.

4.12 Use the 12-step procedure and a quadratic polynomial trial solution to derive two approximate solutions to the following problem:

$$\frac{d^2U}{dx^2} = -\pi^2 \sin \pi x \qquad 0 < x < 1$$

with the BCs $U(0)=2$ and $U(1)=5$.

a) *First solution*: Employ the conventional technique of integrating by parts *once*, thereby yielding constrained essential BCs.

b) *Second solution*: Employ the unconventional technique of integrating by parts *twice*, thereby yielding unconstrained essential BCs. (See Exercise 4.11 for how to treat the extra BC terms.)

c) Graphically compare the two solutions as well as the exact solution, $U(x)=2+3x+\sin \pi x$. Note how the approximation of the essential BCs in the second solution affects the accuracy away from the boundaries.

4.13 Consider the transient heating problem in Exercise 3.16. In that analysis there was no integration by parts so the initial condition was *constrained* by modifying the trial solution. This time apply the 12-step procedure (and a quadratic polynomial trial solution) and integrate by parts *once* over the interval 0 to 10 sec. The initial condition will now appear as a term on the RHS, so it can be applied in the same manner as a natural BC. However, there will also be a "boundary" term at $t=10$. This is treated as explained in some of the previous exercises: Insert the trial solution and transfer all a_i terms to the LHS. Compare this solution with the one in Exercise 3.16. Note that in the present solution the initial condition is *unconstrained*. How does this affect the accuracy away from $t=0$?

Reference

[4.1]   S. H. Crandall, *Engineering Analysis*, McGraw-Hill, New York, 1956.

5

The Element Concept

Introduction

This chapter derives several more approximate solutions to the illustrative problem in Chapters 3 and 4. The principal difference, though, is that here the domain of the problem will be partitioned into several pieces, called *elements*, and trial functions will be defined separately over each element. The 12-step procedure introduced in Chapter 4, modified slightly, can then be used to analyze each element.

The sections are logically organized according to a "one, two, many" principle: Section 5.2 derives a one-element solution, Section 5.3 a two-element solution, and Sections 5.4 and 5.5 a four- and an eight-element solution, respectively, i.e., many elements.

In the one-element solution in Section 5.2 the domain is not partitioned. The entire domain is therefore one element. The reason for examining a one-element solution is that it enables us to focus on two important features that characterize an element, irrespective of whether the element occurs by itself (as in Section 5.2) or with others (as in Sections 5.3, 5.4, and 5.5). The first feature involves the construction of the trial functions. We will once again use a polynomial trial solution, but the terms will be rearranged so that the resulting trial functions will have special properties on the boundary of the element. These special properties will yield significant computational advantages when dealing with two or more elements. The second feature is that the resulting system equations for the one element, called the *element equations*, will be able to be used later in the multiple-element solutions. The concept of element equations plays a central role in the FEM.

Section 5.3 will begin by partitioning the domain into two elements, then deriving an exact formulation for the partitioned problem that is equivalent in all respects to the original formulation for the nonpartitioned problem; i.e., the exact solution to both problems must be identical. Each element can then be analyzed separately, as though each were a separate problem. The resulting element equations for each element will be algebraically identical and, in fact, identical to the element equations for the one-element problem in Section 5.2, thereby illustrating that the properties of each piece of the whole system are similar to the properties of the system as a whole. The focus will then shift to the important concept of *assembly*, or how to join together the two sets of element equations to yield the assembled system equations for the complete problem.

The analysis for both a four- and an eight-element solution in Sections 5.4 and 5.5, respectively, will generalize from the two-element experience by writing the element equations for a typical element using *local node numbers*. These equations can then be used for each of the other elements simply by replacing the local node numbers with the actual global node numbers.

In Section 5.6 we present the completeness and continuity conditions for ensuring *convergence*. They are then applied to the sequence of one-, two-, four-, and eight-element solutions and it is observed that this process of *mesh refinement* has produced a converging sequence.

Sections 5.7 and 5.8 provide more insight into the concepts of assembly and element trial functions. Section 5.9, which concludes the chapter and Part II, summarizes the 12-step procedure as it will be used in the remainder of the book.

5.1 An Illustrative Problem

For continuity with Chapters 3 and 4, we will analyze the same illustrative problem described in Sections 3.2 and 4.2, repeated here for convenience.

Differential equation and domain •

$$\frac{d}{dx}\left(x\,\frac{dU(x)}{dx} \right) = \frac{2}{x^2} \qquad 1 < x < 2 \tag{5.1a}$$

Boundary conditions •

$$U(1) = 2 \qquad \text{essential BC (constrained)}$$

$$\left(-x\,\frac{dU}{dx} \right)_{x=2} = \frac{1}{2} \qquad \text{natural BC (unconstrained)} \tag{5.1b}$$

5.2 A One-Element Solution: The Element Equations

In this section we let the entire domain be one element. The reader will see later in this chapter that when the domain is divided into elements, we will construct trial functions only within each element, not over the entire domain as was done in Chapters 3 and 4. Therefore, when we use only one element, as in the present section, that one element becomes identical with the entire domain and the corresponding trial functions also cover the entire domain. Consequently, all the solutions in Chapters 3 and 4 can be viewed as one-element solutions.

In the present one-element solution we will again use a polynomial trial solution, as was used in Chapters 3 and 4, but the terms will be rearranged into a special form. We will follow the 12-step procedure introduced in Chapter 4.

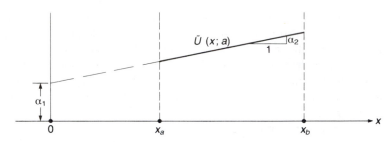

FIGURE 5.1 ● An inappropriate form for an FE trial solution — a power series expansion.

Steps 1 through 3: The first three steps remain the same as in Chapter 4. The reader may wish to review those steps, which culminated in the system equations, represented by any of the expressions given in Eqs. (4.15), (4.16), (4.17), (4.19), and (4.20). We will use the general notation x_a and x_b for the coordinates of the two boundary points, and then in step 7 substitute the particular numerical values $x_a=1$ and $x_b=2$.

Step 4: Develop specific expressions for the trial functions $\phi_i(x)$.

Let's consider a linear polynomial, written in the form of a power series:

$$\tilde{U}(x;\alpha) = \alpha_1 + \alpha_2 x \tag{5.2}$$

A linear polynomial is the lowest degree that can be used to satisfy the completeness condition, as will be explained in Section 5.6. Figure 5.1 shows a plot of $\tilde{U}(x,\alpha)$ and indicates the meaning of α_1 and α_2.

Parameters that have meanings such as those shown in Fig. 5.1 are not appropriate for FE analysis, so it will be necessary to transform the power series in Eq. (5.2) into a different form. We note that there are no technical reasons to prohibit the use of power series (recall examples in Chapters 3 and 4); however, they do not yield any of the computational advantages that make the FEM a practical analysis tool.

There are ways to write polynomials other than in the form of a power series. In particular, the FEM requires that the polynomial be written in the form of an *interpolation polynomial*.[1] The principle of interpolation is actually quite simple:

Each parameter a_i must represent the value of the trial solution at a specific point in the element.[2] Each such point is called a *node*.

Figure 5.2 shows two nodes, one at each end of the element, numbered 1 and 2, and a parameter associated with each node, labeled a_1 and a_2, respectively. It is evident from

---

[1] Here again the FEM makes use of classical concepts. The reader will find an expanded discussion of interpolation polynomials in any elementary numerical-analysis text.

[2] There are extensions to this principle in which the parameters may also represent values of the *derivatives* of the trial solution at specified points (see Section 8.6). There are also exceptions to this principle (in which the trial functions are called "bubble" functions or "nodeless" trial functions), but such exceptions are used in addition to the basic type of trial functions described here.

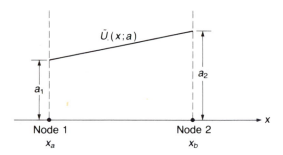

FIGURE 5.2 ● The recommended form for an FE trial solution — an interpolation polynomial.

the figure that the parameters a_1 and a_2 are defined so as to satisfy the principle defined above:

$$\tilde{U}(x_a;a) = a_1$$

$$\tilde{U}(x_b;a) = a_2 \qquad (5.3)$$

Thus a_1 is the value of $\tilde{U}$ at node 1 (at $x=x_a$), and a_2 is the value of $\tilde{U}$ at node 2 (at $x=x_b$).

We now want to express the trial solution in Eq. (5.2) in terms of a_1 and a_2 rather than α_1 and α_2. This may be done by imposing Eqs. (5.3) on Eq. (5.2):[3]

$$\alpha_1 + \alpha_2 x_a = a_1$$

$$\alpha_1 + \alpha_2 x_b = a_2 \qquad (5.4)$$

Solving for α_1 and α_2 in terms of a_1 and a_2 yields

$$\alpha_1 = \frac{x_b a_1 - x_a a_2}{x_b - x_a}$$

$$\alpha_2 = \frac{a_2 - a_1}{x_b - x_a} \qquad (5.5)$$

Substituting Eqs. (5.5) into Eq. (5.2) yields

$$\tilde{U}(x;a) = \left(\frac{x_b a_1 - x_a a_2}{x_b - x_a}\right) + \left(\frac{a_2 - a_1}{x_b - x_a}\right)x \qquad (5.6)$$

---

[3] Writing Eqs. (5.4) in matrix form,

$$\begin{bmatrix} 1 & x_a \\ 1 & x_b \end{bmatrix} \begin{Bmatrix} \alpha_1 \\ \alpha_2 \end{Bmatrix} = \begin{Bmatrix} a_1 \\ a_2 \end{Bmatrix}$$

the determinant of the matrix of coefficients is

$$\begin{vmatrix} 1 & x_a \\ 1 & x_b \end{vmatrix}$$

This determinant (as well as all its extensions to polynomials of higher degree) is called a *Vandermonde* determinant.

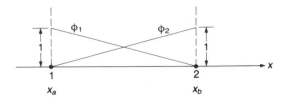

FIGURE 5.3 ● Trial functions in Eqs. (5.7b).

Combining coefficients of a_1 and a_2, Eq. (5.6) becomes

$$\tilde{U}(x;a) = a_1\phi_1(x) + a_2\phi_2(x) \qquad (5.7a)$$

where the *trial functions* $\phi_1(x)$ and $\phi_2(x)$ have the form,

$$\phi_1(x) = \frac{x_b - x}{x_b - x_a}$$

$$\phi_2(x) = \frac{x - x_a}{x_b - x_a} \qquad (5.7b)$$

$\tilde{U}(x;a)$ in Eqs. (5.7) is the desired form for our linear-polynomial trial solution. It is still a linear polynomial because both $\phi_1(x)$ and $\phi_2(x)$ are linear polynomials. $\phi_1(x)$ and $\phi_2(x)$ satisfy the following important properties at the boundary points of the element:[4]

$$\phi_1(x_a) = 1$$
$$\phi_1(x_b) = 0 \qquad (5.8a)$$

and

$$\phi_2(x_a) = 0$$
$$\phi_2(x_b) = 1 \qquad (5.8b)$$

The trial functions are shown in Fig. 5.3. The relationship between $\tilde{U}(x;a)$, $\phi_1(x)$, and $\phi_2(x)$, as described by Eq. (5.7a), is shown in Fig. 5.4.

When written in the form of Eqs. (5.7), $\tilde{U}$ is called a *Lagrange interpolation polynomial*, and it is said that $\tilde{U}$ interpolates the points (x_a, a_1) and (x_b, a_2). More generally, a function (such as a polynomial) is said to be *interpolatory* (or *interpolating* or an *interpolation function*) if it is defined to be equal to particular values at a number of specific and separate points (nodes). (The values may be temporarily represented by parameters that will later be assigned particular numerical values.)

---

[4] Note the striking parallel between the present development and that in Chapter 4, step 4. There the trial functions were constructed in order to satisfy conditions on the boundary of the *domain*. Here the trial functions are constructed in order to satisfy somewhat similar conditions on the boundary of the *element*. (In the present circumstance, of course, the boundary of the element also happens to be the boundary of the domain; nevertheless, it will be seen in Sections 5.3 through 5.5 that Eqs. (5.8) will also apply to each element when there is more than one element in the domain.)

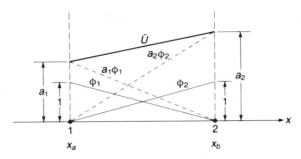

FIGURE 5.4 ● Trial solution and trial functions in Eqs. (5.7).

The trial functions themselves, Eqs. (5.7b), are also interpolatory because of Eqs. (5.8). That is, each one is defined to be equal to *unity* at one node and *zero* at the other node. This special property can be exhibited better if we make a minor notational change. Substituting x_1 for x_a and x_2 for x_b, Eqs. (5.7b) appear as

$$\phi_1(x) = \frac{x_2 - x}{x_2 - x_1}$$

$$\phi_2(x) = \frac{x - x_1}{x_2 - x_1} \tag{5.9}$$

and Eqs. (5.8) can then be written compactly as

$$\phi_j(x_i) = \delta_{ji} \tag{5.10}$$

where δ_{ji} is called a *Kronecker delta* and has the property

$$\delta_{ji} = \begin{cases} 1 & \text{if } j = i \\ 0 & \text{if } j \neq i \end{cases} \tag{5.11}$$

Equation (5.10), or equivalently Eqs. (5.8), will be referred to as the *interpolation property*. It is a general property that characterizes almost all FE trial functions used for second-order differential equations.[5] It will be used in this and all remaining chapters.

We could also have derived the trial functions in Eqs. (5.7b) by working directly from the interpolation property — an approach that is frequently used in practice. It is instructive to see this alternative approach. Returning to the beginning of step 4 we would argue as follows. We want a trial solution that is a linear polynomial so we write the trial solution in the general form,[6]

$$\tilde{U}(x;a) = a_1\phi_1(x) + a_2\phi_2(x) \tag{5.12}$$

---

[5] For higher-order differential equations, a similar but more generalized interpolation property is used, involving derivatives of the trial functions (see Section 8.6).

[6] No $\phi_0(x)$ term is needed because we will use the numerical method in step 9 to apply the constrained BC.

and require that each trial function be a linear polynomial:

$$\phi_1(x) = \beta_1 + \beta_2 x$$

$$\phi_2(x) = \gamma_1 + \gamma_2 x \tag{5.13}$$

The coefficients β_1, β_2, γ_1, γ_2 are determined from the interpolation property. For example, applying Eqs. (5.8a) to $\phi_1(x)$,

$$\beta_1 + \beta_2 x_a = 1$$

$$\beta_1 + \beta_2 x_b = 0 \tag{5.14}$$

which yields

$$\beta_1 = \frac{x_b}{x_b - x_a}$$

$$\beta_2 = -\frac{1}{x_b - x_a} \tag{5.15}$$

Substituting Eqs. (5.15) into Eqs. (5.13) yields

$$\phi_1(x) = \frac{x_b - x}{x_b - x_a} \tag{5.16}$$

which agrees with the first of Eqs. (5.7b). $\phi_2(x)$ is treated similarly.

Note that Eq. (5.12) was written with two unknown parameters (and two corresponding trial functions) because a linear polynomial is uniquely defined by two parameters. This is also why we chose two nodes, and hence two parameters, in Fig. 5.2. Similarly, a quadratic polynomial, $\alpha_1 + \alpha_2 x + \alpha_3 x^2$, is determined uniquely by three parameters, so three nodes and corresponding parameters would be required; and Eq. (5.12) would require three terms, each containing a quadratic polynomial. Chapter 8 will develop the appropriate trial functions for quadratic, cubic, and quartic polynomials.

There is an important requirement on the location of the nodes within the element that deserves some preliminary comments (details will follow later). In the above development, the two nodes were placed on the *boundary* of the element — i.e., one node at each of the two boundary points. If our only objective were to define a linear polynomial, then in principle the nodes could be located at any two separate points. But this is not our only objective. In addition, as will be seen in the next three sections, we must also be concerned with connecting (assembling) the equations derived for each element. Since adjacent elements touch each other only at their boundaries, it should not be surprising that locating nodes on element boundaries will significantly affect the assembly process. In fact, the many practical computational advantages of the FEM can be realized only if a certain (minimum) number of nodes are placed on the element boundaries; additional nodes might also be placed in the interior of the element, but only for the purpose of defining higher-degree polynomials in order to achieve greater accuracy. Rather than state a general principle here (which would be too general to be meaningful at this point), the placement of nodes can better be explained by specific examples. Therefore subsequent chapters will explain and demonstrate the proper placement of nodes for a variety of different types of elements.

Step 5: Substitute the trial functions into stiffness and load terms, and transform the integrals into a form appropriate for numerical evaluation.

Since the trial solution contains two trial functions, then the system equations in step 3 are Eqs. (4.18) and (4.19) with $N=2$:

$$\begin{bmatrix} K_{11} & K_{12} \\ K_{21} & K_{22} \end{bmatrix} \begin{Bmatrix} a_1 \\ a_2 \end{Bmatrix} = \begin{Bmatrix} F_1 \\ F_2 \end{Bmatrix} \tag{5.17a}$$

where

$$K_{ij} = \int_{x_a}^{x_b} \frac{d\phi_i}{dx} \, x \, \frac{d\phi_j}{dx} \, dx \qquad \begin{array}{l} i = 1,2 \text{ and} \\ j = 1,2 \end{array}$$

$$F_i = FI_i + FB_i$$

$$= -\int_{x_a}^{x_b} \frac{2}{x^2} \, \phi_i \, dx - \left[\left(-x \frac{d\tilde{U}}{dx} \right) \phi_i \right]_{x_a}^{x_b} \tag{5.17b}$$

The derivatives of the trial functions follow from Eqs. (5.7b):

$$\frac{d\phi_1}{dx} = -\frac{1}{x_b - x_a}$$

$$\frac{d\phi_2}{dx} = \frac{1}{x_b - x_a} \tag{5.18}$$

Hence the stiffness terms become

$$K_{11} = \int_{x_a}^{x_b} \left(-\frac{1}{x_b - x_a} \right) \, x \, \left(-\frac{1}{x_b - x_a} \right) \, dx = \frac{1}{2} \frac{x_b + x_a}{x_b - x_a}$$

$$K_{12} = K_{21} = -K_{11}$$

$$K_{22} = K_{11} \tag{5.19}$$

The interior load terms become

$$FI_1 = -\int_{x_a}^{x_b} \frac{2}{x^2} \frac{x_b - x}{x_b - x_a} \, dx = -\frac{2}{x_a} + \frac{2}{x_b - x_a} \ln \frac{x_b}{x_a}$$

$$FI_2 = \frac{2}{x_b} - \frac{2}{x_b - x_a} \ln \frac{x_b}{x_a} \tag{5.20}$$

And the boundary load terms become

$$FB_1 = -\left(-x \frac{d\tilde{U}}{dx} \right)_{x_b} \phi_1(x_b) + \left(-x \frac{d\tilde{U}}{dx} \right)_{x_a} \phi_1(x_a) = \left(-x \frac{d\tilde{U}}{dx} \right)_{x_a}$$

$$FB_2 = -\left(-x \frac{d\tilde{U}}{dx} \right)_{x_b} \tag{5.21}$$

Note that the interpolation property of the trial functions has eliminated their appearance in the boundary terms, leaving only the natural BC expression (the flux) at each boundary.

Step 6: Prepare expressions for flux using the trial functions.

$$\tau(x) = \text{flux} = -x \frac{d\tilde{U}}{dx}$$

$$= -x \left(a_1 \frac{d\phi_1}{dx} + a_2 \frac{d\phi_2}{dx} \right)$$

$$= \frac{x}{x_b - x_a} (a_1 - a_2) \tag{5.22}$$

This completes steps 1 through 6, which involve the theoretical analysis to produce the system equations and corresponding flux expression. The final form of the system equations may be seen by substituting Eqs. (5.19) through (5.21) into Eqs. (5.17a):

$$
\begin{bmatrix}
\dfrac{1}{2}\dfrac{x_b+x_a}{x_b-x_a} & -\dfrac{1}{2}\dfrac{x_b+x_a}{x_b-x_a} \\[2mm]
-\dfrac{1}{2}\dfrac{x_b+x_a}{x_b-x_a} & \dfrac{1}{2}\dfrac{x_b+x_a}{x_b-x_a}
\end{bmatrix}
\begin{Bmatrix} a_1 \\[2mm] a_2 \end{Bmatrix}
$$

$$
= \begin{Bmatrix}
-\dfrac{2}{x_a} + \dfrac{2}{x_b-x_a}\ln\dfrac{x_b}{x_a} \\[3mm]
\dfrac{2}{x_b} - \dfrac{2}{x_b-x_a}\ln\dfrac{x_b}{x_a}
\end{Bmatrix}
+ \begin{Bmatrix}
\left(-x\dfrac{d\tilde{U}}{dx}\right)_{x_a} \\[3mm]
-\left(-x\dfrac{d\tilde{U}}{dx}\right)_{x_b}
\end{Bmatrix} \tag{5.23}
$$

Equations (5.23) may also be called the *element equations* because they are the residual equations for a single element. In fact, it would have been quite appropriate to refer to the system equations at the end of step 3 in Chapter 4 [e.g., Eqs. (4.18) and (4.19)] as the element equations, since the entire domain could have been considered to be a single element. This nomenclature will become more meaningful when we work with domains containing more than one element in Sections 5.3 through 5.5.

We now continue with the second part of the analysis, steps 7 through 12, which involves all the numerical computation.

Step 7: Specify the numerical data for the problem.

1. *Geometric data.* For the present one-element analysis, the only geometric data is the coordinates of the two boundary points:

$$x_a = 1$$

$$x_b = 2 \tag{5.24}$$

2. *Physical properties and applied loads.* These have already been defined for this specific illustration problem. See comments under step 7 in Section 4.4.

Step 8: Numerically evaluate all the interior (nonboundary) terms in the element equations (which are the system equations for this one-element analysis).

Substituting Eqs. (5.24) into Eqs. (5.23) yields

$$
\begin{bmatrix} \dfrac{3}{2} & -\dfrac{3}{2} \\[2mm] -\dfrac{3}{2} & \dfrac{3}{2} \end{bmatrix}
\begin{Bmatrix} a_1 \\[2mm] a_2 \end{Bmatrix}
=
\begin{Bmatrix} -2 + 2\ln 2 \\[2mm] 1 - 2\ln 2 \end{Bmatrix}
+
\begin{Bmatrix} \left(-x\,\dfrac{d\tilde{U}}{dx}\right)_{x=1} \\[3mm] -\left(-x\,\dfrac{d\tilde{U}}{dx}\right)_{x=2} \end{Bmatrix}
\tag{5.25}
$$

Step 9: Apply the boundary conditions.

First let's do the unconstrained (natural) BC, which is always the easier of the two types. From Eqs. (5.1b), we simply substitute the value 1/2 for the term $(-x d\tilde{U}/dx)_{x=2}$ on the RHS of Eqs. (5.25), yielding

$$
\begin{bmatrix} \dfrac{3}{2} & -\dfrac{3}{2} \\[2mm] -\dfrac{3}{2} & \dfrac{3}{2} \end{bmatrix}
\begin{Bmatrix} a_1 \\[2mm] a_2 \end{Bmatrix}
=
\begin{Bmatrix} -2 + 2\ln 2 \\[2mm] 1 - 2\ln 2 \end{Bmatrix}
+
\begin{Bmatrix} \left(-x\,\dfrac{d\tilde{U}}{dx}\right)_{x=1} \\[3mm] -\dfrac{1}{2} \end{Bmatrix}
\tag{5.26}
$$

The essential BC, $U(1)=2$, is a constrained BC because there are no terms appearing explicitly on the RHS for which we can substitute the value 2 for U. So we must constrain the trial solution itself to satisfy the essential BC (recall Section 4.4, step 9). Therefore, applying the essential BC in Eqs. (5.1b) to Eq. (5.7a) yields

$$
\tilde{U}(1;a) = a_1\phi_1(1) + a_2\phi_2(1) = 2
\tag{5.27}
$$

Because of the interpolation property, $\phi_1(1)=1$ and $\phi_2(1)=0$, so Eq. (5.27) reduces to

$$
a_1 = 2
\tag{5.28}
$$

Equation (5.28) is a *constraint equation* that may be applied to Eqs. (5.26) by the row/column operations described in Section 4.4, step 9. This constraint equation is in the simplest possible form, so the row/column operations reduce to two simple operations:

1. Multiply column 1 in the $[K]$ matrix by 2 and subtract the product from the RHS load vector; then delete column 1 (and therefore delete a_1 too).

2. Delete row 1 in the $[K]$ matrix and load vector.

Here we see how the use of *interpolatory trial functions with boundary nodes* (which are the essential properties of element trial functions — see Section 5.8) has simplified the application of constrained (essential) BCs. It will be seen in all the subsequent analyses that the use of element trial functions always reduces essential BCs to the general form,

$$
a_i = d
\tag{5.29}
$$

That is, *only one* parameter is assigned the value of the essential BC. Application of the constraint Eq. (5.29) then consists of the following two operations:

1. Multiply column i in the $[K]$ matrix by d and subtract the product from the RHS load vector; then delete column i (and therefore delete a_i too).

2. Delete row i in the $[K]$ matrix and load vector.

Applying the constraint Eq. (5.28) to Eqs. (5.26) in the above manner, the first of the two equations is eliminated and we are left with only the second equation in the following form (which the reader should verify):

$$\frac{3}{2} a_2 = \frac{7}{2} - 2 \ln 2 \tag{5.30}$$

Equation (5.30) is our system equation — the only one left.

Step 10: Solve the system equations.

From Eq. (5.30),

$$a_2 = \frac{7}{3} - \frac{4}{3} \ln 2 = 1.409 \tag{5.31}$$

The parameter a_1 has already been determined by the essential BC, Eq. (5.28). Substituting Eqs. (5.28) and (5.31) into Eqs. (5.7) yields the approximate solution:

$$\tilde{U}(x) = 2(2-x) + 1.409(x-1)$$

$$= 2.591 - 0.591x \tag{5.32}$$

Step 11: Evaluate the flux.

From Eq. (5.22),

$$\tilde{\tau}(x) = x(a_1 - a_2) = 0.591x \tag{5.33}$$

Step 12: Plot the solution and estimate its accuracy.

Figure 5.5 is a plot of $\tilde{U}(x)$ and $\tilde{\tau}(x)$ in Eqs. (5.32) and (5.33). It should be noted that this solution is identical to the solution $\tilde{U}_1(x)$ and $\tilde{\tau}_1(x)$ in Eqs. (4.50) in Section 4.4. This

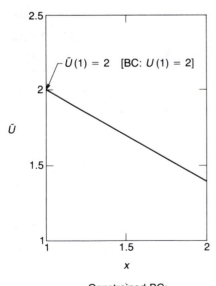

Constrained BC:
$\tilde{U}(1) = U(1)$

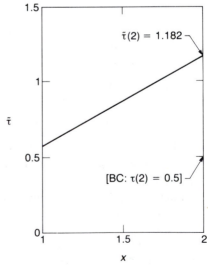

Unconstrained BC:
$\tilde{\tau}(2) \neq \tau(2)$

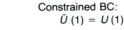

FIGURE 5.5 ● One-element solution.

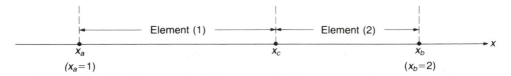

FIGURE 5.6 ● Partitioning the domain into two elements.

is to be expected because both used the same trial solution, a linear polynomial, although in a different algebraic form. Note also that the constrained (essential) BC at $x=1$ is satisfied exactly, whereas the unconstrained (natural) BC at $x=2$ is only satisfied approximately.

The only clue we have about the accuracy is the error in $\tilde{\tau}(x)$ at the unconstrained (natural) BC at $x=2$. The calculated value is $\tilde{\tau}(2)=1.182$, whereas the exact value is $\tau(2)=1/2$, an error of 136%! To estimate the accuracy of $\tilde{\tau}$ elsewhere in the domain, and the accuracy in $\tilde{U}$, we will need to generate another approximate solution. We will do this in the next section by partitioning the domain into two elements.

5.3 A Two-Element Solution: Assembly of Element Equations

Figure 5.6 shows the domain of the problem, $x_a<x<x_b$, partitioned into two pieces, or subdomains: $x_a<x<x_c$ and $x_c<x<x_b$. Each subdomain is called an *element*. There is no gap between the elements, nor is there any overlap. In other words, the domain of the problem is identical to the sum (union) of the two elements plus the interelement boundary point x_c. The elements may be of unequal length (as suggested by the figure); that is, x_c may be located anywhere between x_a and x_b.[7]

In order to keep track of the analysis and computations associated with each element, it will be convenient to label each element. This is normally done using the numbers (1), (2), (3), etc., placed in parentheses and attached as a superscript to all other notation. For example, the trial solutions for the two elements would be labeled $\tilde{U}^{(1)}(x;a)$ and $\tilde{U}^{(2)}(x;a)$, respectively.

5.3.1 Equivalent Formulation of the Problem in Terms of Elements

The original problem statement in Section 5.1, Eqs. (5.1), may be written in a different, but *equivalent*, form that is appropriate for the partitioned domain in Fig. 5.6. By "equivalent" we mean that the *exact* solution to the alternative formulation is identical to the *exact* solution to the previous formulation; neither formulation contains more information than the other. The following discussion pertains only to the exact solution, not to any approximate solutions.

Since the problem is being broken into two elements, questions naturally arise concerning the *continuity* of the solution at the interelement boundary point x_c. We will need

---

[7] This description of an element refers only to its geometric properties. Additional properties will be added as they are developed. The concept of an element is so fundamental and central to the FEM that the word "element" has come to mean a variety of things, but all are related to the basic geometric entity.

the following notation:

A function is said to be of class C^n (in a specified domain) if the function and its first n derivatives are continuous (in the domain).

For example, a continuous function is of class C^0. If its first derivative is also continuous, then it is of class C^1; if its second derivative is also continuous, then it is of class C^2, etc. A function of class C^n must also be of class C^{n-1}, C^{n-2},..., C^0, but it may or may not be of class C^{n+1}, C^{n+2}, C^{n+3}, etc.[8]

From the basic definition of a derivative (as a limit of the ratio of differences), it follows that if a function of class C^n is differentiated, then the derivative is of class C^{n-1}, the second derivative is of class C^{n-2}, etc. Conversely, from the basic definition of an integral (as a limit of a sum), the integral of a function of class C^{n-2} yields a function of class C^{n-1}, a second integration yields a function of class C^n, etc. In words, differentiation lowers the order of continuity, whereas integration raises it. This property will be used below.

Let's now reformulate the problem for the partitioned domain. Let $U^{(1)}(x)$ be the exact solution within element (1), and $U^{(2)}(x)$ be the exact solution within element (2). The governing differential equation within each element may then be written as follows:

Element (1)

$$\frac{d}{dx}\left(x \frac{dU^{(1)}}{dx}\right) = \frac{2}{x^2} \qquad x_a < x < x_c \tag{5.34a}$$

Element (2)

$$\frac{d}{dx}\left(x \frac{dU^{(2)}}{dx}\right) = \frac{2}{x^2} \qquad x_c < x < x_b \tag{5.34b}$$

Equations (5.34) are exactly equivalent to Eq. (5.1a) in the domain $x_a < x < x_b$, excepting only the point x_c.

All we need now is a statement about the continuity between $U^{(1)}(x)$ and $U^{(2)}(x)$ at x_c. For the present illustration problem this is quite easy because the terms in the original differential equation (5.1a) are "very smooth." We can, in fact, make the following observations about any point in the domain $x_a < x < x_b$ (not just x_c). The RHS of Eq. (5.1a) is the term $2/x^2$. This term is continuous over $x_a < x < x_b$ and so are its derivatives of all orders; i.e., it is of class C^∞. Since the RHS is equal to the LHS, then the LHS must also be C^∞. Integrating the LHS repeatedly will therefore (recall above discussion) produce continuous functions, ad infinitum.

For example, let's integrate both sides of Eq. (5.1a) over an infinitesimally small interval about any point x_p in the domain — i.e., from $x_p - \epsilon$ to $x_p + \epsilon$:

$$\int_{x_p-\epsilon}^{x_p+\epsilon} \frac{d}{dx}\left(x \frac{dU}{dx}\right) dx = \int_{x_p-\epsilon}^{x_p+\epsilon} \frac{2}{x^2} dx \tag{5.35}$$

---

[8] Saying that a function is of class C^n carries with it an implication about the $(n+1)^{st}$ and higher derivatives: Either they are discontinuous, or else we don't know or don't care! The label "C^n" is generally used to describe only the highest order of differentiation one is interested in, ignoring any higher orders that might also happen to be continuous. In other words, a function is of class C^n if *at least* its first n derivatives are continuous.

Carrying out the integration yields

$$\left[\left(x\frac{dU}{dx}\right)\right]_{x_p-\epsilon}^{x_p+\epsilon} = \left[-\frac{2}{x}\right]_{x_p-\epsilon}^{x_p+\epsilon} \tag{5.36}$$

or

$$\left(x\frac{dU}{dx}\right)_{x_p+\epsilon} - \left(x\frac{dU}{dx}\right)_{x_p-\epsilon} = -\frac{2}{x_p+\epsilon} + \frac{2}{x_p-\epsilon} \tag{5.37}$$

Now let $\epsilon \to 0$. The *RHS* $\to 0$, leaving

$$\left(x\frac{dU}{dx}\right)_{x_p^+} = \left(x\frac{dU}{dx}\right)_{x_p^-} \tag{5.38}$$

where x_p^+ means evaluate xdU/dx arbitrarily close to the point x_p, approaching from the right (the $+$ sign). Similarly, x_p^- means evaluate xdU/dx arbitrarily close to x_p, but approaching from the left (the $-$ sign).

Equation (5.38) is a statement of continuity of xdU/dx at the point x_p. Since it applies to any point in the domain, then xdU/dx is continuous throughout the domain. And since x is continuous everywhere, then dU/dx is continuous. A similar integration would then show that U is continuous at any point x_p:

$$U(x_p^+) = U(x_p^-) \tag{5.39}$$

For the particular point x_c, we can write $U^{(1)}$ for any point to the left and $U^{(2)}$ for any point to the right. Therefore the continuity conditions [Eqs. (5.38) and (5.39)] can be written for just this point in the form,[9]

$$\left(-x\frac{dU^{(1)}}{dx}\right)_{x_c} = \left(-x\frac{dU^{(2)}}{dx}\right)_{x_c} \tag{5.40a}$$

$$U^{(1)}(x_c) = U^{(2)}(x_c) \tag{5.40b}$$

This completes the reformulation.[10] Equations (5.1a and b) are equivalent to Eqs. (5.34a and b), (5.40a and b), and (5.1b); the former imply the latter and vice versa.[11] Figure 5.7 summarizes the reformulated problem.

Equations (5.40) will be called *interelement boundary conditions* (IBCs) because in many respects they bear a strong resemblance to our familiar BCs (the latter might also appropriately be labeled ''domain BCs''). Thus, IBCs are conditions that must be satisfied by the exact solution on the *boundaries between elements*, whereas BCs are conditions

---

[9] It is convenient to include a minus sign on both sides of Eq. (5.40a) so that the terms will represent flux (rather than negative flux).

[10] Chapter 7 will consider more general situations involving discontinuities on both the LHS and RHS — situations that occur frequently in practice. When the differential equation is integrated over some types of discontinuities, Eq. (5.40a) may be altered.

[11] Continuity of the second derivative is implied by the differential equation itself because the RHS is continuous. Repeated differentiation of the differential equation yields relations that imply continuity of all higher derivatives as well since the RHS is of class C^∞.

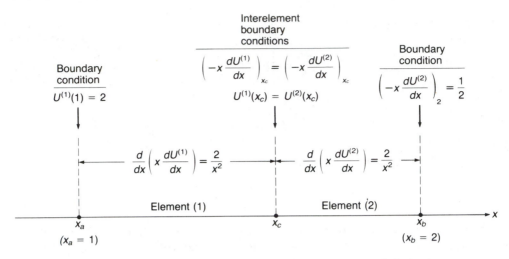

FIGURE 5.7 ● Equivalent formulation of the problem (for the exact solution).

that must be satisfied by the exact solution on the *boundaries of the domain*. Both are necessary properties of the exact solution.

Equation (5.40a) is analogous to a natural BC because it places a condition on the flux; however, instead of setting the flux equal to a specified value, it is set equal to the flux in the adjacent element. Similarly, Eq. (5.40b) is analogous to an essential BC because it places a condition on the function; however, instead of setting the function equal to a specified value, it is set equal to the function in the adjacent element. In addition, it will be seen in the next section that when the IBCs are applied, Eq. (5.40a) will be unconstrained and Eq. (5.40b) will be constrained, again in complete analogy with natural and essential BCs. For these reasons we will refer to Eq. (5.40a) as a *natural IBC* and to Eq. (5.40b) as an *essential IBC*.

We can now treat the original problem as though it were two separate problems, applying the 12-step procedure to éach element as we did in Section 5.2.[12] In addition, though, the essential IBC [continuity condition on $U(x)$] will be used to connect (assemble) the equations derived for each element.

5.3.2 Steps 1 through 6: Theoretical Development for Each Element

We will derive the equations for element (1) first, using the results from Section 5.2. Then element (2) will be treated in a similar manner.

---

[12] The author used to teach the traditional approach wherein the governing differential equation is first written over the entire domain and then integrated by parts. Alert students discovered subtle difficulties in that approach, involving integrating by parts over discontinuous functions (which violates the assumptions behind integration by parts — recall the footnote in step 2, Section 4.3). A rigorous resolution requires arguments too mathematically sophisticated for most tastes. Gradually the entire approach was abandoned and replaced by the present approach which avoids those difficulties, yet at the same time is rigorous, conceptually simpler, and generally preferred by colleagues. For a similar approach see the articles by J. Greenstadt [5.1] and E. R. de Arantes e Oliveira [5.2].

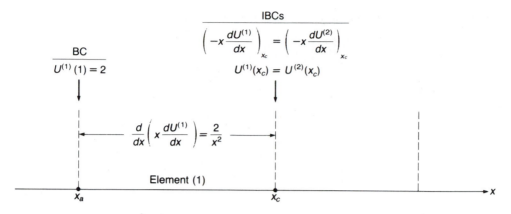

FIGURE 5.8 ● Statement of the problem (for the exact solution) for element (1).

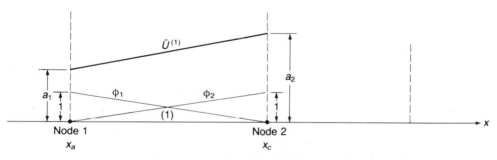

FIGURE 5.9 ● Trial solution and trial functions for element (1).

Element (1). Figure 5.8 shows only that portion of Fig. 5.7 that is relevant to element (1). We will use the same trial solution used in Section 5.2 and shown in Fig. 5.9. The trial functions follow directly from Eqs. (5.7) and Fig. 5.4. Thus

$$\tilde{U}^{(1)}(x;a) = a_1\phi_1(x) + a_2\phi_2(x) \qquad (5.41a)$$

where

$$\phi_1(x) = \frac{x_c - x}{x_c - x_a}$$

$$\phi_2(x) = \frac{x - x_a}{x_c - x_a} \qquad (5.41b)$$

$\tilde{U}^{(1)}(x;a)$ is the *element trial solution* for element (1).

It should be clear that the problem shown in Fig. 5.8, using the trial solution in Fig. 5.9 and Eqs. (5.41), is essentially the same as the one-element problem in Section 5.2, with only the following differences:

1. *Notation.* U and $\tilde{U}$ changed to $U^{(1)}$ and $\tilde{U}^{(1)}$, respectively; x_b changed to x_c.
2. *Conditions at the interelement boundary.* BC at x_b (now x_c) changed to two IBCs.

The IBCs do not enter the analysis until the numerical computation phase (steps 7 through 12). Therefore the element equations at the end of step 6 in Section 5.2 can be copied from Eqs. (5.23), making only the notational changes $\tilde{U}$ to $\tilde{U}^{(1)}$ and x_b to x_c:

$$
\begin{bmatrix}
\dfrac{1}{2}\dfrac{x_c+x_a}{x_c-x_a} & -\dfrac{1}{2}\dfrac{x_c+x_a}{x_c-x_a} \\[3mm]
-\dfrac{1}{2}\dfrac{x_c+x_a}{x_c-x_a} & \dfrac{1}{2}\dfrac{x_c+x_a}{x_c-x_a}
\end{bmatrix}
\begin{Bmatrix} a_1 \\[3mm] a_2 \end{Bmatrix}
$$

$$
= \begin{Bmatrix}
-\dfrac{2}{x_a} + \dfrac{2}{x_c-x_a}\ln\dfrac{x_c}{x_a} \\[3mm]
\dfrac{2}{x_c} - \dfrac{2}{x_c-x_a}\ln\dfrac{x_c}{x_a}
\end{Bmatrix}
+ \begin{Bmatrix}
\left(-x\,\dfrac{d\tilde{U}^{(1)}}{dx}\right)_{x_a} \\[3mm]
-\left(-x\,\dfrac{d\tilde{U}^{(1)}}{dx}\right)_{x_c}
\end{Bmatrix}
\tag{5.42}
$$

Equations (5.42) are the element equations for element (1).

It will be useful in subsequent developments to be able to refer back to these element equations in their general matrix symbolism — i.e., Eqs. (5.17a) applied to element (1). Thus Eqs. (5.42) may be written symbolically as

$$
\begin{bmatrix}
K_{11}^{(1)} & K_{12}^{(1)} \\[2mm]
K_{21}^{(1)} & K_{22}^{(1)}
\end{bmatrix}
\begin{Bmatrix} a_1 \\[2mm] a_2 \end{Bmatrix}
=
\begin{Bmatrix} F_1^{(1)} \\[2mm] F_2^{(1)} \end{Bmatrix}
\tag{5.43}
$$

The superscript (1) identifies the stiffness and load terms as belonging to element (1). The superscripts will be useful in distinguishing the terms in element (1) from similar terms in element (2). The parameters a_i don't need an element superscript, for reasons that will become clear by the end of Section 5.3.

As a general rule, superscripts will refer to elements and subscripts to nodes. For example, $K_{12}^{(1)}$ belongs to element (1), and it relates, or "couples," the parameters associated with nodes 1 and 2, namely, a_1 and a_2. Similarly, $F_1^{(1)}$ belongs to element (1) and is a load related to, or "applied at," node 1. In physical applications (see subsequent chapters) these statements will become more meaningful.

Equations (5.42) complete steps 1 through 5. For step 6 we want to write the general expression for the flux in element (1). Using Eq. (5.41a),

$$
\tilde{\tau}^{(1)}(x;a) = -x\,\frac{d\tilde{U}^{(1)}(x;a)}{dx} = -x\left(a_1\,\frac{d\phi_1(x)}{dx} + a_2\,\frac{d\phi_2(x)}{dx}\right)
$$

Hence, from Eq. (5.41b),

$$
\tilde{\tau}^{(1)}(x;a) = \frac{a_1 - a_2}{x_c - x_a}\,x
\tag{5.44}
$$

Element (2). Figures 5.10 and 5.11 are directly analogous to Figs. 5.8 and 5.9, respectively. Note that the node at x_c is the same node used in element (1). It would therefore seem natural to use the label already assigned to it, namely, node 2. (In fact, that is the normal FE procedure — to use a single node number for each node.) However, for instructional purposes only, we are giving this node a different label in element (2), namely, node 3, because at this point in the development the formulations for elements (1) and (2) are totally independent of each other. This is only a temporary situation because in Section 5.3.3 the two elements will be linked together; that will enable us (in Section 5.3.4)

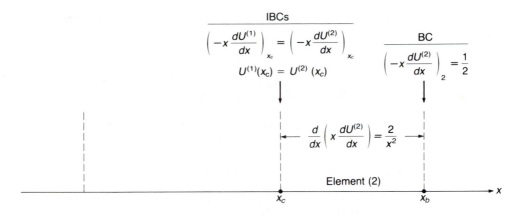

FIGURE 5.10 ● Statement of the problem (for the exact solution) for element (2).

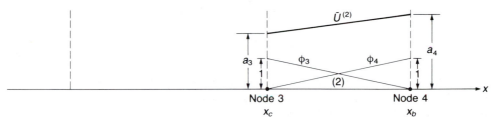

FIGURE 5.11 ● Trial solution and trial functions for element (2).

to drop this redundant notation and adopt the conventional FE notation, which will be used in the remainder of the book.

Again the trial functions may be copied from Eqs. (5.7), using the notation in Fig. 5.11:

$$\tilde{U}^{(2)}(x;a) = a_3\phi_3(x) + a_4\phi_4(x) \qquad (5.45a)$$

where

$$\phi_3(x) = \frac{x_b - x}{x_b - x_c} \qquad \phi_4(x) = \frac{x - x_c}{x_b - x_c} \qquad (5.45b)$$

$\tilde{U}^{(2)}(x;a)$ is the element trial solution for element (2).

And again the element equations can be copied from Eqs. (5.23), making the notational changes $\tilde{U}$ to $\tilde{U}^{(2)}$ and x_a to x_c:

$$
\begin{bmatrix}
\dfrac{1}{2}\dfrac{x_b+x_c}{x_b-x_c} & -\dfrac{1}{2}\dfrac{x_b+x_c}{x_b-x_c} \\[3mm]
-\dfrac{1}{2}\dfrac{x_b+x_c}{x_b-x_c} & \dfrac{1}{2}\dfrac{x_b+x_c}{x_b-x_c}
\end{bmatrix}
\begin{Bmatrix}
a_3 \\[3mm]
a_4
\end{Bmatrix}
$$

$$
= \begin{Bmatrix}
-\dfrac{2}{x_c} + \dfrac{2}{x_b-x_c}\ln\dfrac{x_b}{x_c} \\[3mm]
\dfrac{2}{x_b} - \dfrac{2}{x_b-x_c}\ln\dfrac{x_b}{x_c}
\end{Bmatrix}
+ \begin{Bmatrix}
\left(-x\,\dfrac{d\tilde{U}^{(2)}}{dx}\right)_{x_c} \\[3mm]
-\left(-x\,\dfrac{d\tilde{U}^{(2)}}{dx}\right)_{x_b}
\end{Bmatrix} \qquad (5.46)
$$

Equations (5.46) are the element equations for element (2). They may be written symbolically as

$$
\begin{bmatrix} K_{33}^{(2)} & K_{34}^{(2)} \\ K_{43}^{(2)} & K_{44}^{(2)} \end{bmatrix} \begin{Bmatrix} a_3 \\ a_4 \end{Bmatrix} = \begin{Bmatrix} F_3^{(2)} \\ F_4^{(2)} \end{Bmatrix}
\tag{5.47}
$$

where the subscript and superscript notation has the same meaning described previously for element (1). For example, $K_{34}^{(2)}$ belongs to element (2), and it relates, or "couples," the parameters associated with nodes 3 and 4, namely, a_3 and a_4.

The expression for the flux in element (2) follows from Eqs. (5.45):

$$
\begin{aligned}
\tilde{\tau}^{(2)}(x;a) &= -x\,\frac{d\tilde{U}^{(2)}(x;a)}{dx} \\
&= -x\left(a_3\,\frac{d\phi_3}{dx} + a_4\,\frac{d\phi_4}{dx} \right) \\
&= \frac{a_3 - a_4}{x_b - x_c}\,x
\end{aligned}
\tag{5.48}
$$

At this point in the analysis we have two completely separated problems, i.e., two pairs of element equations, each containing its own parameters a_i. The set of all four equations, written together in matrix form, is

$$
\begin{bmatrix}
\dfrac{1}{2}\dfrac{x_c+x_a}{x_c-x_a} & -\dfrac{1}{2}\dfrac{x_c+x_a}{x_c-x_a} & 0 & 0 \\[2ex]
-\dfrac{1}{2}\dfrac{x_c+x_a}{x_c-x_a} & \dfrac{1}{2}\dfrac{x_c+x_a}{x_c-x_a} & 0 & 0 \\[2ex]
0 & 0 & \dfrac{1}{2}\dfrac{x_b+x_c}{x_b-x_c} & -\dfrac{1}{2}\dfrac{x_b+x_c}{x_b-x_c} \\[2ex]
0 & 0 & -\dfrac{1}{2}\dfrac{x_b+x_c}{x_b-x_c} & \dfrac{1}{2}\dfrac{x_b+x_c}{x_b-x_c}
\end{bmatrix}
\begin{Bmatrix} a_1 \\[2ex] a_2 \\[2ex] a_3 \\[2ex] a_4 \end{Bmatrix}
$$

$$
= \begin{Bmatrix}
-\dfrac{2}{x_a} + \dfrac{2}{x_c-x_a}\ln\dfrac{x_c}{x_a} \\[2ex]
\dfrac{2}{x_c} - \dfrac{2}{x_c-x_a}\ln\dfrac{x_c}{x_a} \\[2ex]
-\dfrac{2}{x_c} + \dfrac{2}{x_b-x_c}\ln\dfrac{x_b}{x_c} \\[2ex]
\dfrac{2}{x_b} - \dfrac{2}{x_b-x_c}\ln\dfrac{x_b}{x_c}
\end{Bmatrix}
+ \begin{Bmatrix}
\left(-x\,\dfrac{d\tilde{U}^{(1)}}{dx}\right)_{x_a} \\[2ex]
-\left(-x\,\dfrac{d\tilde{U}^{(1)}}{dx}\right)_{x_c} \\[2ex]
\left(-x\,\dfrac{d\tilde{U}^{(2)}}{dx}\right)_{x_c} \\[2ex]
-\left(-x\,\dfrac{d\tilde{U}^{(2)}}{dx}\right)_{x_b}
\end{Bmatrix}
\tag{5.49}
$$

and in symbolic form,

$$
\begin{bmatrix}
K_{11}^{(1)} & K_{12}^{(1)} & 0 & 0 \\
K_{21}^{(1)} & K_{22}^{(1)} & 0 & 0 \\
0 & 0 & K_{33}^{(2)} & K_{34}^{(2)} \\
0 & 0 & K_{43}^{(2)} & K_{44}^{(2)}
\end{bmatrix}
\begin{Bmatrix}
a_1 \\ a_2 \\ a_3 \\ a_4
\end{Bmatrix}
=
\begin{Bmatrix}
F_1^{(1)} \\ F_2^{(1)} \\ F_3^{(2)} \\ F_4^{(2)}
\end{Bmatrix}
\tag{5.50}
$$

Note that the element (1) stiffness terms do not overlap (i.e., are not added to) any of the element (2) stiffness terms. This is a consequence of the fact that the two pairs of element equations are uncoupled. (They will become coupled in step 8 when we impose the essential IBCs.)

Equations (5.50) are the *system equations* for the entire problem. As a point of terminology, the system equations are the collection of all the element equations. For the special case of a one-element problem (see, for example, Section 5.2 and Chapters 3 and 4), the element equations *are* the system equations.

5.3.3 Steps 7 through 12: Numerical Computation, Including Assembly of Separate Element Equations

Step 7: Specify numerical data for the problem.

1. *Geometric data.*

$$x_a = 1$$

$$x_b = 2 \tag{5.51}$$

We will define the elements to be of equal length since there is no reason in this problem to favor one end of the domain over the other. Hence

$$x_c = \frac{3}{2} \tag{5.52}$$

The location of the nodes are therefore

$$\text{Node 1 at } x_a = 1$$

$$\text{Node 2 and 3 at } x_c = \frac{3}{2}$$

$$\text{Node 4 at } x_b = 2 \tag{5.53}$$

The nodes and elements are collectively referred to as a *mesh*. The process of defining the lengths (or sizes and shapes in 2-D and 3-D) and locations of the elements, defining the locations of nodes, and assigning numbers to each node and element is called *mesh generation*. Figure 5.12 shows the mesh for this problem.

2. *Physical properties and applied loads.* These have already been defined for this specific illustration problem. See comments under step 7 in Section 4.4.

FIGURE 5.12 ● Mesh for the two-element problem.

Step 8: Numerically evaluate all the interior (nonboundary) terms in both sets of element equations; then assemble the two sets of element equations.

Substituting Eqs. (5.51) and (5.52) into Eqs. (5.49) yields

$$
\begin{bmatrix}
\dfrac{5}{2} & -\dfrac{5}{2} & 0 & 0 \\[2mm]
-\dfrac{5}{2} & \dfrac{5}{2} & 0 & 0 \\[2mm]
0 & 0 & \dfrac{7}{2} & -\dfrac{7}{2} \\[2mm]
0 & 0 & -\dfrac{7}{2} & \dfrac{7}{2}
\end{bmatrix}
\begin{Bmatrix}
a_1 \\[2mm] a_2 \\[2mm] a_3 \\[2mm] a_4
\end{Bmatrix}
=
\begin{Bmatrix}
-2 + 4\ln\dfrac{3}{2} \\[2mm]
\dfrac{4}{3} - 4\ln\dfrac{3}{2} \\[2mm]
-\dfrac{4}{3} + 4\ln\dfrac{4}{3} \\[2mm]
1 - 4\ln\dfrac{4}{3}
\end{Bmatrix}
+
\begin{Bmatrix}
\left(-x\,\dfrac{d\tilde{U}^{(1)}}{dx}\right)_{x=1} \\[2mm]
-\left(-x\,\dfrac{d\tilde{U}^{(1)}}{dx}\right)_{x=\frac{3}{2}} \\[2mm]
\left(-x\,\dfrac{d\tilde{U}^{(2)}}{dx}\right)_{x=\frac{3}{2}} \\[2mm]
-\left(-x\,\dfrac{d\tilde{U}^{(2)}}{dx}\right)_{x=2}
\end{Bmatrix}
$$

(5.54)

We next want to apply one of the two interelement boundary conditions (IBCs), namely, the essential IBC, Eq. (5.40b). To the beginner it might seem more logical to consider this a part of step 9 (with the BCs), or as a separate step by itself. It will be seen in a moment, though, that applying the essential IBC involves a very simple operation. From a computer programming standpoint (see Chapter 7), it is most efficient to combine this operation directly with the numerical evaluations in step 8.

Recall Eq. (5.40b), evaluated at the point $x_c = 3/2$:

$$
U^{(1)}\left(\frac{3}{2}\right) = U^{(2)}\left(\frac{3}{2}\right)
$$

(5.55)

which expresses continuity of $U(x)$ across the interelement boundary. The procedure for applying Eq. (5.55) is identical to that used for the essential BC in Chapter 4. Thus we observe that there are no terms in Eqs. (5.54) for which we can substitute Eq. (5.55). The only other way to apply this condition is to constrain the trial functions themselves. We must therefore require

$$
\tilde{U}^{(1)}\left(\frac{3}{2}; a\right) = \tilde{U}^{(2)}\left(\frac{3}{2}; a\right)
$$

(5.56)

Substituting Eqs. (5.41) and (5.45) into Eq. (5.56) yields

$$
a_2 = a_3
$$

(5.57)

which we recognize as a constraint equation. It may be applied to Eqs. (5.54) by the now-familiar row/column operations described in Section 4.4, step 9. They simplify considerably for Eq. (5.57): Combine (add) columns 2 and 3 of the stiffness matrix, and combine (add) equations 2 and 3. Therefore Eqs. (5.54) become

$$
\begin{bmatrix}
\dfrac{5}{2} & -\dfrac{5}{2} & 0 \\[2mm]
-\dfrac{5}{2} & \dfrac{5}{2}+\dfrac{7}{2} & -\dfrac{7}{2} \\[2mm]
0 & -\dfrac{7}{2} & \dfrac{7}{2}
\end{bmatrix}
\begin{Bmatrix} a_1 \\ a_2 \\ a_4 \end{Bmatrix}
=
\begin{Bmatrix}
-2 + 4\ln\dfrac{3}{2} \\[2mm]
\dfrac{4}{3} - 4\ln\dfrac{3}{2} - \dfrac{4}{3} + 4\ln\dfrac{4}{3} \\[2mm]
1 - 4\ln\dfrac{4}{3}
\end{Bmatrix}
$$

$$
+
\begin{Bmatrix}
\left(-x\,\dfrac{d\tilde{U}^{(1)}}{dx}\right)_{x=1} \\[3mm]
\left(-x\,\dfrac{d\tilde{U}^{(2)}}{dx}\right)_{x=\frac{3}{2}} - \left(-x\,\dfrac{d\tilde{U}^{(1)}}{dx}\right)_{x=\frac{3}{2}} \\[3mm]
-\left(-x\,\dfrac{d\tilde{U}^{(2)}}{dx}\right)_{x=2}
\end{Bmatrix}
\tag{5.58}
$$

or, simplifying the numbers,

$$
\begin{bmatrix}
\dfrac{5}{2} & -\dfrac{5}{2} & 0 \\[2mm]
-\dfrac{5}{2} & 6 & -\dfrac{7}{2} \\[2mm]
0 & -\dfrac{7}{2} & \dfrac{7}{2}
\end{bmatrix}
\begin{Bmatrix} a_1 \\ a_2 \\ a_4 \end{Bmatrix}
$$

$$
=
\begin{Bmatrix}
-2 + 4\ln\dfrac{3}{2} \\[2mm]
4\ln\dfrac{8}{9} \\[2mm]
1 - 4\ln\dfrac{4}{3}
\end{Bmatrix}
+
\begin{Bmatrix}
\left(-x\,\dfrac{d\tilde{U}^{(1)}}{dx}\right)_{x=1} \\[3mm]
\left(-x\,\dfrac{d\tilde{U}^{(2)}}{dx}\right)_{x=\frac{3}{2}} - \left(-x\,\dfrac{d\tilde{U}^{(1)}}{dx}\right)_{x=\frac{3}{2}} \\[3mm]
-\left(-x\,\dfrac{d\tilde{U}^{(2)}}{dx}\right)_{x=2}
\end{Bmatrix}
\tag{5.59}
$$

Equations (5.59) are the assembled system equations.

This procedure of applying the essential IBC [continuity of $U(x)$] to the element equations is called *assembly*. Stated more formally:

Assembly is the enforcement of continuity between the element trial solutions.

When using element trial functions (interpolatory trial functions with boundary nodes), the essential IBCs always reduce to a constraint equation of the general form,

$$a_i = a_j \tag{5.60}$$

Assembly then consists simply of combining (adding) rows i and j and columns i and j in the stiffness matrix and rows i and j in the load vector.

It will be seen in step 12 that the resulting solution will be continuous at the interelement boundary because, as we learned in Chapter 4, constraint equations are constraints placed directly on the trial solution and hence on the resulting approximate solution as well. Thus the essential IBC is a constrained IBC, in direct analogy to the essential BC.

It is helpful to visualize the assembly procedure in terms of the general matrix symbolism. Thus, performing the above assembly operation on Eqs. (5.50) yields

$$\begin{bmatrix} K_{11}^{(1)} & K_{12}^{(1)} & 0 \\ K_{21}^{(1)} & K_{22}^{(1)}+K_{33}^{(2)} & K_{34}^{(2)} \\ 0 & K_{43}^{(2)} & K_{44}^{(2)} \end{bmatrix} \begin{Bmatrix} a_1 \\ a_2 \\ a_4 \end{Bmatrix} = \begin{Bmatrix} F_1^{(1)} \\ F_2^{(1)}+F_3^{(2)} \\ F_4^{(2)} \end{Bmatrix} \tag{5.61}$$

It can be seen that assembly consists merely of *adding* some of the terms from one pair of element equations to some of the terms in the other pair of element equations. Thus one does not need to perform (as was done above) the intermediate step of writing both pairs of element equations separately, with a 4-by-4 stiffness matrix [Eqs. (5.54)], and then doing the row/column operations to reduce it to a 3-by-3 stiffness matrix [Eqs. (5.59)]. Instead, we can go directly from the two separate pairs of equations to a 3-by-3 system by simply adding the second pair of equations to the first pair, as shown in Eqs. (5.61).

Let's illustrate this procedure. We would begin with the first element, evaluating the terms in its equations, resulting in Eqs. (5.43):

$$\begin{bmatrix} K_{11}^{(1)} & K_{12}^{(1)} \\ K_{21}^{(1)} & K_{22}^{(1)} \end{bmatrix} \begin{Bmatrix} a_1 \\ a_2 \end{Bmatrix} = \begin{Bmatrix} F_1^{(1)} \\ F_2^{(1)} \end{Bmatrix}$$

Then we would evaluate the terms for the second element, resulting in Eqs. (5.47):

$$\begin{bmatrix} K_{33}^{(2)} & K_{34}^{(2)} \\ K_{43}^{(2)} & K_{44}^{(2)} \end{bmatrix} \begin{Bmatrix} a_3 \\ a_4 \end{Bmatrix} = \begin{Bmatrix} F_3^{(2)} \\ F_4^{(2)} \end{Bmatrix}$$

However, instead of writing Eqs. (5.43) and (5.47) together as in Eqs. (5.50), we write them together as in Eqs. (5.61), noting that because of the constraint equation [Eq. (5.57)], a_3 can be changed to a_2 in Eqs. (5.47).

The present notation is a bit awkward for computer implementation, but was necessary to explain the concept of assembly and to mathematically derive the rule for its implementation. In Section 5.3.4 we will introduce a preferred method for labeling the nodes and the parameters a_i. Then Section 5.7 will consolidate all these assembly explanations into a simple, easy-to-use general assembly rule.

Step 9: Apply the BCs and the natural IBC to the (assembled) system equations.

The natural IBC • Recall Eq. (5.40a), evaluated at the point $x_c = 3/2$:

$$\left(-x\frac{dU^{(1)}}{dx}\right)_{x=\frac{3}{2}} = \left(-x\frac{dU^{(2)}}{dx}\right)_{x=\frac{3}{2}} \tag{5.62}$$

which expresses continuity of flux across the interelement boundary. The procedure for applying Eq. (5.62) is identical to that used for a natural BC. Thus, rewriting Eq. (5.62) in the form,

$$\left(-x\frac{dU^{(2)}}{dx}\right)_{x=\frac{3}{2}} - \left(-x\frac{dU^{(1)}}{dx}\right)_{x=\frac{3}{2}} = 0 \tag{5.63}$$

the expression in Eq. (5.63) appears as a term on the RHS of Eqs. (5.59), so we merely substitute the value zero from Eq. (5.63). Equations (5.59) become

$$\begin{bmatrix} \frac{5}{2} & -\frac{5}{2} & 0 \\[2mm] -\frac{5}{2} & 6 & -\frac{7}{2} \\[2mm] 0 & -\frac{7}{2} & \frac{7}{2} \end{bmatrix} \begin{Bmatrix} a_1 \\[2mm] a_2 \\[2mm] a_4 \end{Bmatrix} = \begin{Bmatrix} -2 + 4\ln\frac{3}{2} \\[2mm] 4\ln\frac{8}{9} \\[2mm] 1 - 4\ln\frac{4}{3} \end{Bmatrix} + \begin{Bmatrix} \left(-x\frac{d\tilde{U}^{(1)}}{dx}\right)_{x=1} \\[2mm] 0 \\[2mm] -\left(-x\frac{d\tilde{U}^{(2)}}{dx}\right)_{x=2} \end{Bmatrix} \tag{5.64}$$

The natural IBC is clearly an unconstrained IBC, in direct analogy with the natural BC, and we will see in step 12 that the resulting solution will satisfy Eq. (5.63) only approximately; i.e., the flux will be discontinuous at $x=3/2$.

From a computer programming standpoint (see Chapter 7), the application of the natural IBC (continuity of flux) usually involves no operation at all because in the computer there are no theoretical expressions like Eq. (5.63), only empty storage locations waiting to be assigned numerical values for each of the terms in Eqs. (5.64). It is customary to assign initial zero values to all locations at the start of an analysis, and during the analysis to add only the nonzero values from each of the element equations. Clearly, then, there is no need to add a zero.[13]

---

[13] It will be shown in Chapter 7 that it is possible for the RHS of Eq. (5.63) to be nonzero, a situation that occurs when there is a concentrated load applied at the node.

The natural BC • The natural BC in Eq. (5.1b) may be written in the form,

$$\left(-x\,\frac{dU^{(2)}}{dx}\right)_{x=2} = \frac{1}{2} \tag{5.65}$$

where U in Eq. (5.1b) may be replaced by $U^{(2)}$, since $U^{(2)}$ is that part of U that touches the node at $x=2$. Substituting the value $1/2$ from Eq. (5.65) into the term on the RHS of Eqs. (5.64) yields

$$\begin{bmatrix} \dfrac{5}{2} & -\dfrac{5}{2} & 0 \\[2mm] -\dfrac{5}{2} & 6 & -\dfrac{7}{2} \\[2mm] 0 & -\dfrac{7}{2} & \dfrac{7}{2} \end{bmatrix} \begin{Bmatrix} a_1 \\[2mm] a_2 \\[2mm] a_4 \end{Bmatrix} = \begin{Bmatrix} -2 + 4\ln\dfrac{3}{2} \\[2mm] 4\ln\dfrac{8}{9} \\[2mm] 1 - 4\ln\dfrac{4}{3} \end{Bmatrix} + \begin{Bmatrix} \left(-x\,\dfrac{d\tilde{U}^{(1)}}{dx}\right)_{x=1} \\[2mm] 0 \\[2mm] -\dfrac{1}{2} \end{Bmatrix} \tag{5.66}$$

The essential BC • Since the essential BC is applied to the point $x=1$, which is in element (1), then we apply the BC to the trial solution $U^{(1)}(x;a)$ in Eq. (5.41a):

$$\tilde{U}^{(1)}(1;a) = a_1\phi_1(1) + a_2\phi_2(1) = 2 \tag{5.67}$$

Because of the interpolation property, $\phi_1(1)=1$ and $\phi_2(1)=0$, so Eq. (5.67) reduces to

$$a_1 = 2 \tag{5.68}$$

This type of constraint equation [see Eq. (5.29)] is applied by the following row/column operations:

1. Multiply column 1 in the stiffness matrix by 2 and subtract from the RHS load vector; then delete column 1.
2. Delete row 1 in the stiffness matrix and load vector.

Performing these operations on Eqs. (5.66) yields[14]

$$\begin{bmatrix} 6 & -\dfrac{7}{2} \\[2mm] -\dfrac{7}{2} & \dfrac{7}{2} \end{bmatrix} \begin{Bmatrix} a_2 \\[2mm] a_4 \end{Bmatrix} = \begin{Bmatrix} 4\ln\dfrac{8}{9} + 5 \\[2mm] 1 - 4\ln\dfrac{4}{3} - \dfrac{1}{2} \end{Bmatrix} \tag{5.69}$$

Equations (5.69) are the final form of the system equations, ready to be solved.

Step 10: Solve the system equations.

---

[14] We note once again that the unknown boundary term $(-xd\tilde{U}^{(1)}/dx)_{x=1}$ automatically drops out of the system equations when the essential BC is applied. (See Section 4.4, step 9.)

The solution to Eqs. (5.69) is

$$a_2 = 1.551$$

$$a_4 = 1.365 \tag{5.70a}$$

From Eqs. (5.57) and (5.68),

$$a_1 = 2.000$$

$$a_3 = 1.551 \tag{5.70b}$$

Substituting Eqs. (5.70) into Eqs. (5.41) and (5.45) yields

$$\tilde{U}^{(1)}(x) = 2.000\phi_1(x) + 1.551\phi_2(x)$$

$$= 2.000 \left(\frac{1.5-x}{0.5}\right) + 1.551 \left(\frac{x-1.0}{0.5}\right) \tag{5.71a}$$

and

$$\tilde{U}^{(2)}(x) = 1.551\phi_3(x) + 1.365\phi_4(x)$$

$$= 1.551 \left(\frac{2.0-x}{0.5}\right) + 1.365 \left(\frac{x-1.5}{0.5}\right) \tag{5.71b}$$

Step 11: Evaluate the flux.
Substituting Eqs. (5.70) into Eqs. (5.44) and (5.48) yields

$$\tilde{\tau}^{(1)}(x) = 0.898x$$

$$\tilde{\tau}^{(2)}(x) = 0.372x \tag{5.72}$$

Step 12: Plot the solution and estimate its accuracy.
Figure 5.13 is a plot of $\tilde{U}(x)$ and $\tilde{\tau}(x)$ in Eqs. (5.71) and (5.72). It should be noted that the essential BC at $x=1$ was constrained so $\tilde{U}(x)$ satisfies the condition exactly. Likewise, the essential IBC at $x=3/2$ (continuity of $\tilde{U}$) was also constrained, so $\tilde{U}(x)$ also satisfies this condition exactly. The natural BC at $x=2$, however, was unconstrained, so $\tilde{\tau}(x)$ satisfies this condition only approximately. And the natural IBC at $x=3/2$ (continuity of flux) was also unconstrained, so $\tilde{\tau}(x)$ also satisfies this condition only approximately.[15]

   To estimate error, we can get a first clue regarding the flux error by examining the points where the unconstrained BC and IBC were applied. At $x=2$, the value $\tilde{\tau}(2)=0.744$ is in error by 49% from the exact value. At $x=3/2$, the *spread* in values at the discontinuity (1.347 versus 0.558) gives a range of $\pm41\%$ about the mean (0.953).

---

[15] The mismatch in $\tilde{\tau}$ at $x=3/2$ seems pretty terrible, yet this is not unusual for such an extremely coarse mesh, namely, one having only two elements. We will see in a moment that as more elements are added (which is easy to do on a computer), this mismatch can be made as small as desired. We will also see in Chapters 7 and 8 that there are special ways to evaluate the flux so that more accurate results can be obtained.

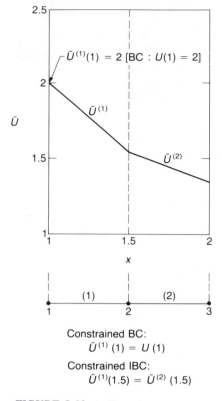

Constrained BC:
$$\tilde{U}^{(1)}(1) = U(1)$$
Constrained IBC:
$$\tilde{U}^{(1)}(1.5) = \tilde{U}^{(2)}(1.5)$$

Unconstrained BC:
$$\tilde{\tau}^{(2)}(2) \neq \tau(2)$$
Unconstrained IBC:
$$\tilde{\tau}^{(1)}(1.5) \neq \tilde{\tau}^{(2)}(1.5)$$

FIGURE 5.13 ● Two-element solution.

To estimate the error in both $\tilde{U}(x)$ and $\tilde{\tau}(x)$ at all points throughout the domain, we need to compare the two-element solution with the one-element solution. (Recall the discussion in Section 3.6.2 on the convergence of a sequence of approximate solutions.) Figure 5.14 shows both the one- and two-element solutions from Figs. 5.5 and 5.13, labeled as $\tilde{U}_1(x)$, $\tilde{\tau}_1(x)$ and $\tilde{U}_2(x)$, $\tilde{\tau}_2(x)$, respectively, to indicate the first and second approximate solutions we have obtained.

The pointwise error between $\tilde{U}_1(x)$ and $\tilde{U}_2(x)$ is less than 10% over the entire domain; that is,

$$\left| \frac{\tilde{U}_2(x) - \tilde{U}_1(x)}{\tilde{U}_2(x)} \right| < 0.1 \qquad \text{for } 1 < x < 2 \tag{5.73}$$

The pointwise error between $\tilde{\tau}_1(x)$ and $\tilde{\tau}_2(x)$ is less than 60% over the entire domain; that is,

$$\left| \frac{\tilde{\tau}_2(x) - \tilde{\tau}_1(x)}{\tilde{\tau}_2(x)} \right| < 0.6 \qquad \text{for } 1 < x < 2 \tag{5.74}$$

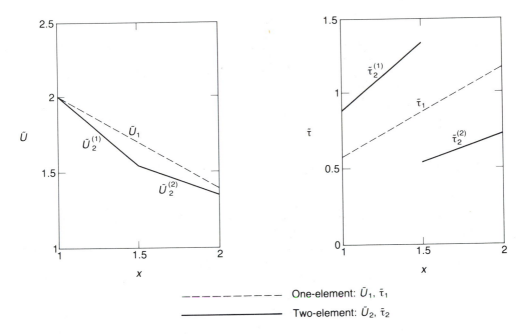

FIGURE 5.14 ● One- and two-element solutions.

Based on Figs. 5.13 and 5.14, we estimate that the error in $\tilde{U}_2(x)$ is in the neighborhood of 10% and the error in $\tilde{\tau}_2(x)$ is in the neighborhood of 50 to 60%. Errors of this magnitude would seem quite reasonable for such a coarse mesh — i.e., only two elements and a straight-line approximation in each element. Our intuition might tell us that in order to achieve accuracy on the order of 1%, we'll probably need a much finer mesh — i.e., one which partitions the domain into much smaller elements. Section 5.4 will explain a more organized way to deal with many elements. But before leaving Section 5.3, a few comments are in order regarding the two-element analysis.

5.3.4 Comments on Assembly — and a Rational Notation

Assembly ● It was explained in Section 4.4 that application of constraint equations has the effect of modifying the original trial functions. To show this graphically, consider again the constraint equation [Eq. (5.57)] resulting from the essential IBC:

$$a_2 = a_3$$

Equation (5.57) enforces continuity between $\tilde{U}^{(1)}$ and $\tilde{U}^{(2)}$ at the node at x_c. Substituting Eq. (5.57) directly into the trial solution $\tilde{U}^{(2)}(x;a)$ in Eq. (5.45a),

$$\tilde{U}^{(2)}(x;a) = a_2\phi_3(x) + a_4\phi_4(x) \tag{5.75a}$$

And let's rewrite $\tilde{U}^{(1)}(x;a)$ from Eq. (5.41a):

$$\tilde{U}^{(1)}(x;a) = a_1\phi_1(x) + a_2\phi_2(x) \tag{5.75b}$$

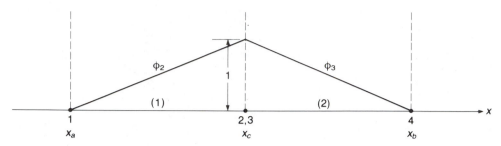

FIGURE 5.15 ● Assembled trial functions, providing continuity in U at interelement boundary.

We originally defined two parameters, a_2 and a_3, at the node at x_c (recall Figs. 5.9 and 5.11). However, as a result of Eq. (5.57), we only need one of these parameters, say a_2. Since $\phi_2(x)$ and $\phi_3(x)$ both have the value 1 at x_c and both share the same parameter in their trial solutions [a_2 in Eqs. (5.75)], then $\phi_2(x)$ and $\phi_3(x)$ are effectively *joined together as a new trial function* that is continuous at x_c, as shown in Fig. 5.15.[16] This shows graphically how the original trial functions were modified by Eq. (5.57). It also illustrates a wider principle: Assembly (i.e., enforcement of essential IBCs) always has the effect of joining together element trial functions in adjacent elements to form a single new trial function that is continuous at the common nodes.[17]

A rational notation ● At this point it is meaningful to introduce a conventional and more rational notation for labeling the nodes and parameters. We have just seen that the continuity condition on $\tilde{U}$ (the essential IBC) forces trial solutions in adjacent elements to share parameters at the node on their common boundary, resulting in a single parameter at each node.[18] Therefore we could have drawn Fig. 5.16 as an alternative to Figs. 5.9 and 5.11.

The trial solutions could then have been written as follows:

$$\tilde{U}^{(1)}(x;a) = a_1\phi_1^{(1)}(x) + a_2\phi_2^{(1)}(x)$$

$$\tilde{U}^{(2)}(x;a) = a_2\phi_2^{(2)}(x) + a_3\phi_3^{(2)}(x) \tag{5.76}$$

where superscripts (1) and (2) have been added to the trial functions to indicate the element to which they belong. The subscript indicates the node at which the trial function has the value 1. For example, $\phi_3^{(2)}(x)$ is a trial function in element (2) and equals 1 at node 3. The nodes are conveniently labeled with integers, starting with the number 1. There are computational advantages to numbering the nodes sequentially in one direction, but it is not necessary (see Section 5.7).

---

[16] The upside-down-V-shaped function illustrated in Fig. 5.15 is sometimes referred to as a *hat* function.

[17] Assembly of 2-D and 3-D elements establishes continuity not only at the common nodes but usually also along the common lines and surfaces (see Chapter 13).

[18] This "sharing" principle for enforcing continuity conditions, may lead to more than one parameter at each node for higher-order differential equations (Section 8.6) or problems with more than one unknown function (Chapter 16). In all cases, though, only one node number would be assigned to each node.

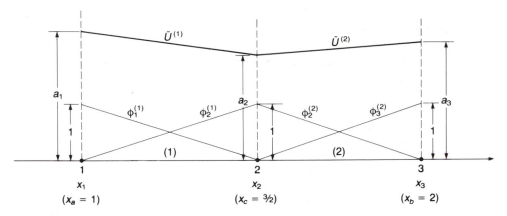

FIGURE 5.16 ● Preferred notation for trial functions and nodes.

It is important to note that Fig. 5.16 and Eqs. (5.76) represent only notational changes; the algebraic form of all the equations in this section remain the same. To see an immediate advantage in this notation, consider the form of the element equations for each element in Eqs. (5.43) and (5.47). They would now appear as follows:

Element (1)

$$\begin{bmatrix} K_{11}^{(1)} & K_{12}^{(1)} \\ K_{21}^{(1)} & K_{22}^{(1)} \end{bmatrix} \begin{Bmatrix} a_1 \\ a_2 \end{Bmatrix} = \begin{Bmatrix} F_1^{(1)} \\ F_2^{(1)} \end{Bmatrix} \qquad \text{(the same as before)}$$

(5.77a)

Element (2)

$$\begin{bmatrix} K_{22}^{(2)} & K_{23}^{(2)} \\ K_{32}^{(2)} & K_{33}^{(2)} \end{bmatrix} \begin{Bmatrix} a_2 \\ a_3 \end{Bmatrix} = \begin{Bmatrix} F_2^{(2)} \\ F_3^{(2)} \end{Bmatrix} \qquad \text{(different from before)}$$

(5.77b)

and the assembled equations would have the appearance,

$$\begin{bmatrix} K_{11}^{(1)} & K_{12}^{(1)} & 0 \\ K_{21}^{(1)} & K_{22}^{(1)}+K_{22}^{(2)} & K_{23}^{(2)} \\ 0 & K_{32}^{(2)} & K_{33}^{(2)} \end{bmatrix} \begin{Bmatrix} a_1 \\ a_2 \\ a_3 \end{Bmatrix} = \begin{Bmatrix} F_1^{(1)} \\ F_2^{(1)}+F_2^{(2)} \\ F_3^{(2)} \end{Bmatrix}$$

(5.78)

This is clearly a more consistent and meaningful notation than that used in Eqs. (5.61).

Perhaps the most important advantage of this new notation is that the element equations for each element will employ identical notation, excepting only the numerical value of the subscripts and superscripts. This lends itself very efficiently to an automated (i.e., computer) treatment, as will be seen in the next section.

5.4 A Four-Element Solution: The Typical Element

5.4.1 Steps 1 through 6: Theoretical Development for a Typical Element

Now let's derive a four-element solution, using the improved notation shown in the preceding section in Fig. 5.16 and Eqs. (5.76). Figure 5.17 shows the domain partitioned into four elements. The nodes and corresponding parameters are numbered sequentially from left to right, starting with the number 1; the elements are numbered in a similar manner. (Section 5.7 demonstrates random numbering of nodes and elements.) Also shown are the trial solutions and trial functions within each element.

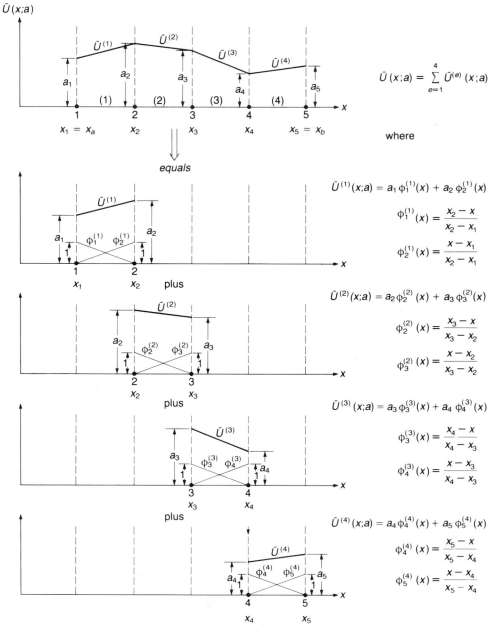

$$\bar{U}(x;a) = \sum_{e=1}^{4} \bar{U}^{(e)}(x;a)$$

where

$$\bar{U}^{(1)}(x;a) = a_1 \phi_1^{(1)}(x) + a_2 \phi_2^{(1)}(x)$$

$$\phi_1^{(1)}(x) = \frac{x_2 - x}{x_2 - x_1}$$

$$\phi_2^{(1)}(x) = \frac{x - x_1}{x_2 - x_1}$$

$$\bar{U}^{(2)}(x;a) = a_2 \phi_2^{(2)}(x) + a_3 \phi_3^{(2)}(x)$$

$$\phi_2^{(2)}(x) = \frac{x_3 - x}{x_3 - x_2}$$

$$\phi_3^{(2)}(x) = \frac{x - x_2}{x_3 - x_2}$$

$$\bar{U}^{(3)}(x;a) = a_3 \phi_3^{(3)}(x) + a_4 \phi_4^{(3)}(x)$$

$$\phi_3^{(3)}(x) = \frac{x_4 - x}{x_4 - x_3}$$

$$\phi_4^{(3)}(x) = \frac{x - x_3}{x_4 - x_3}$$

$$\bar{U}^{(4)}(x;a) = a_4 \phi_4^{(4)}(x) + a_5 \phi_5^{(4)}(x)$$

$$\phi_4^{(4)}(x) = \frac{x_5 - x}{x_5 - x_4}$$

$$\phi_5^{(4)}(x) = \frac{x - x_4}{x_5 - x_4}$$

FIGURE 5.17 ● Four-element trial solution.

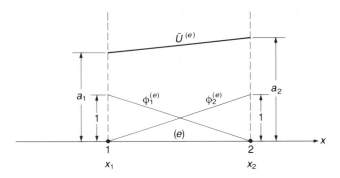

FIGURE 5.18 ● Typical element, using local node numbers.

A perusal of Fig. 5.17 shows that the notation is essentially the same for each element, *the only difference being the numerical values of the subscripts and superscripts.* Because of this similarity, there is no need to derive the element equations (steps 1 through 6) for each element. In fact, they only need to be derived for one of the elements; the resulting equations can then be used for all the other elements, simply by substituting the correct numerical values of the subscripts and superscripts for each element in turn.

To implement this procedure, we introduce the idea of a *typical* (or "generic" or "representative") element, as shown in Fig. 5.18. It uses a trial solution and trial functions like each of the four elements in Fig. 5.17, namely,

$$\tilde{U}^{(e)}(x;a) = a_1\phi_1^{(e)}(x) + a_2\phi_2^{(e)}(x) \tag{5.79a}$$

where

$$\phi_1^{(e)}(x) = \frac{x_2 - x}{x_2 - x_1}$$

$$\phi_2^{(e)}(x) = \frac{x - x_1}{x_2 - x_1} \tag{5.79b}$$

The typical element uses a standard notational convention that is well suited for computer implementation. The nodes are numbered 1, 2, 3,..., n, where n is the number of nodes in the element ($n=2$ in Fig. 5.18). These are popularly referred to as *local node numbers*, to distinguish them from the regular (or *global*) numbers used in the complete mesh, as in Fig. 5.17. Similarly, the coordinates are labeled $x_1, x_2, x_3,..., x_n$, and the parameters are labeled $a_1, a_2, a_3,..., a_n$.[19] There is no need, as far as the computer is concerned, to assign a special element number for the typical element; however, we will find it useful in our theoretical expressions to use the label (e) to signify the (e)th element.

Expressions (5.79) can be made identical to each of the four sets of expressions in Fig. 5.17 if the subscripts (node numbers) and superscripts (element numbers) are replaced

---

[19] If there is more than one parameter at each node, then the parameters are labeled in a slightly different fashion, as explained in Chapter 16.

by the numbers in the following *connectivity table:*

TABLE 5.1 ● Connectivity table for mesh, Fig. 5.17.

Element	Local node 1 becomes node	Local node 2 becomes node
(1)	1	2
(2)	2	3
(3)	3	4
(4)	4	5

The table is constructed simply from a visual examination of the node and element numbers in the mesh at the top of Fig. 5.17. (The procedure is easily automated on a computer, as demonstrated by the UNAFEM program in Chapter 7.) The purpose of the table is to show which nodes are connected to which elements and thus how the node and element numbers for each element in the mesh are related to the typical element.

Now that we have a "mapping" scheme from the typical element to each of the four elements, all we need are the element equations for the typical element. These were already derived; they may be copied from either the one-element problem, Eqs. (5.23), or the two-element problem, Eqs. (5.42) or Eqs. (5.46), but using instead the notation of the typical element:

$$
\begin{bmatrix}
\dfrac{1}{2}\dfrac{x_2+x_1}{x_2-x_1} & -\dfrac{1}{2}\dfrac{x_2+x_1}{x_2-x_1} \\[2mm]
-\dfrac{1}{2}\dfrac{x_2+x_1}{x_2-x_1} & \dfrac{1}{2}\dfrac{x_2+x_1}{x_2-x_1}
\end{bmatrix}
\begin{Bmatrix} a_1 \\ a_2 \end{Bmatrix}
$$

$$
= \begin{Bmatrix}
-\dfrac{2}{x_1} + \dfrac{2}{x_2-x_1}\ln\dfrac{x_2}{x_1} \\[3mm]
\dfrac{2}{x_2} - \dfrac{2}{x_2-x_1}\ln\dfrac{x_2}{x_1}
\end{Bmatrix}
+ \begin{Bmatrix}
\left(-x\,\dfrac{d\tilde{U}^{(e)}}{dx}\right)_{x_1} \\[3mm]
-\left(-x\,\dfrac{d\tilde{U}^{(e)}}{dx}\right)_{x_2}
\end{Bmatrix}
\qquad \text{(5.80)}
$$

or in general matrix notation,

$$
\begin{bmatrix}
K^{(e)}_{11} & K^{(e)}_{12} \\
K^{(e)}_{21} & K^{(e)}_{22}
\end{bmatrix}
\begin{Bmatrix} a_1 \\ a_2 \end{Bmatrix}
= \begin{Bmatrix} F^{(e)}_1 \\ F^{(e)}_2 \end{Bmatrix}
\qquad \text{(5.81)}
$$

Similarly we can copy the expression for the element flux from either the one-element problem, Eq. (5.22), or the two-element problem, Eq. (5.44) or Eq. (5.48), using instead the typical element notation:

$$
\tilde{\tau}^{(e)}(x;a) = \frac{a_1 - a_2}{x_2 - x_1}\, x
\qquad \text{(5.82)}
$$

This completes the theoretical development, steps 1 through 6, because we have the element equations and element flux for the typical element.[20] The numerical computation phase will use these equations, along with the connectivity table, to evaluate the terms in the element equations for each element.

5.4.2 Steps 7 through 12: Numerical Computation

Step 7: Specify numerical data for the problem.

1. *Mesh generation* (geometric data).

   a) Node definition: Specify the coordinates of each node and assign a number to each node. As in the two-element problem we will define the elements to be of equal length.[21] Therefore the nodes shown in Fig. 5.17 should be placed as follows:

$$x_1 = 1.00$$

$$x_2 = 1.25$$

$$x_3 = 1.50$$

$$x_4 = 1.75$$

$$x_5 = 2.00 \tag{5.83}$$

   b) Element definition: Specify the node numbers associated with each element. The node and element numbers were defined in Fig. 5.17. In a typical FE program (such as the UNAFEM program in this book) these numbers would be input to the computer in this step and stored in a *connectivity array*, similar to the connectivity table (Table 5.1).

   The resulting mesh is shown in Fig. 5.19.

2. *Physical properties and applied loads*. These have already been defined for this specific illustration problem. See comments under step 7 in Section 4.4.

Step 8: Numerically evaluate all the interior (non-boundary) terms in the element equations for each element, and assemble the element equations to form system equations.

We will treat each element one at a time, in a manner quite similar to the way a computer program might proceed (see Chapter 7).

Element (1). Using the entries in the connectivity table (Table 5.1) for element (1), we can substitute into the element equations (5.80) node number 1 for local node 1 and node number 2 for local node 2. Since both sets of numbers are identical for this first element, Eqs. (5.80) are unaltered. Next, substitute from Eq. (5.83) the values $x_1 = 1$ and

---

[20] There was no need to explicitly perform each of the six steps for this four-element problem since that would have repeated most of the material in Section 5.2 for the one-element problem. The emphasis here was only on the new feature — the typical element and its notation convention. Subsequent chapters will explicitly perform all six steps only when new types of elements are developed.

[21] We could make some elements shorter in those regions of the domain where we might want more accuracy. This will be demonstrated in subsequent chapters.

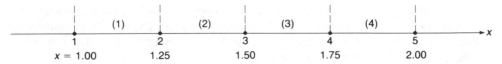

FIGURE 5.19 ● Mesh for the four-element problem.

$x_2=1.25$ into Eqs. (5.80). This yields

$$\begin{bmatrix} \dfrac{9}{2} & -\dfrac{9}{2} \\[2mm] -\dfrac{9}{2} & \dfrac{9}{2} \end{bmatrix} \begin{Bmatrix} a_1 \\[2mm] a_2 \end{Bmatrix} = \begin{Bmatrix} -2 + 8\ln\dfrac{5}{4} \\[2mm] \dfrac{8}{5} - 8\ln\dfrac{5}{4} \end{Bmatrix} + \begin{Bmatrix} \left(-x\,\dfrac{d\tilde{U}^{(1)}}{dx}\right)_{x=1} \\[3mm] -\left(-x\,\dfrac{d\tilde{U}^{(1)}}{dx}\right)_{x=1.25} \end{Bmatrix}$$

(5.84)

or in general matrix notation,

$$\begin{bmatrix} K_{11}^{(1)} & K_{12}^{(1)} \\[2mm] K_{21}^{(1)} & K_{22}^{(1)} \end{bmatrix} \begin{Bmatrix} a_1 \\[2mm] a_2 \end{Bmatrix} = \begin{Bmatrix} F_1^{(1)} \\[2mm] F_2^{(1)} \end{Bmatrix}$$

(5.85)

Of course, no assembly is possible with only one element so we proceed to the second element.

Element (2). Using the entries in the connectivity table for element (2), substitute into Eqs. (5.80) node number 2 for local node 1 and node number 3 for local node 2:

$$\begin{bmatrix} \dfrac{1}{2}\dfrac{x_3+x_2}{x_3-x_2} & -\dfrac{1}{2}\dfrac{x_3+x_2}{x_3-x_2} \\[4mm] -\dfrac{1}{2}\dfrac{x_3+x_2}{x_3-x_2} & \dfrac{1}{2}\dfrac{x_3+x_2}{x_3-x_2} \end{bmatrix} \begin{Bmatrix} a_2 \\[4mm] a_3 \end{Bmatrix}$$

$$= \begin{Bmatrix} -\dfrac{2}{x_2} + \dfrac{2}{x_3-x_2}\ln\dfrac{x_3}{x_2} \\[3mm] \dfrac{2}{x_3} - \dfrac{2}{x_3-x_2}\ln\dfrac{x_3}{x_2} \end{Bmatrix} + \begin{Bmatrix} \left(-x\,\dfrac{d\tilde{U}^{(2)}}{dx}\right)_{x_2} \\[3mm] -\left(-x\,\dfrac{d\tilde{U}^{(2)}}{dx}\right)_{x_3} \end{Bmatrix}$$

(5.86)

Substitute from Eqs. (5.83) the values $x_2=1.25$ and $x_3=1.5$ into Eqs. (5.86). This yields

$$\begin{bmatrix} \dfrac{11}{2} & -\dfrac{11}{2} \\[2mm] -\dfrac{11}{2} & \dfrac{11}{2} \end{bmatrix} \begin{Bmatrix} a_2 \\[2mm] a_3 \end{Bmatrix} = \begin{Bmatrix} -\dfrac{8}{5} + 8\ln\dfrac{6}{5} \\[2mm] \dfrac{4}{3} - 8\ln\dfrac{6}{5} \end{Bmatrix} + \begin{Bmatrix} \left(-x\,\dfrac{d\tilde{U}^{(2)}}{dx}\right)_{x=1.25} \\[3mm] -\left(-x\,\dfrac{d\tilde{U}^{(2)}}{dx}\right)_{x=1.5} \end{Bmatrix}$$

(5.87)

or in general matrix notation,

$$\begin{bmatrix} K_{22}^{(2)} & K_{23}^{(2)} \\[2mm] K_{32}^{(2)} & K_{33}^{(2)} \end{bmatrix} \begin{Bmatrix} a_2 \\[2mm] a_3 \end{Bmatrix} = \begin{Bmatrix} F_2^{(2)} \\[2mm] F_3^{(2)} \end{Bmatrix}$$

(5.88)

Assembly of the element equations (5.84) and (5.87) follows the pattern explained for the two-element problem in step 8, Section 5.3.3 [Eqs. (5.58) and (5.61)], and Section 5.3.4 [Eqs. (5.78)]. Thus

$$
\begin{bmatrix}
\dfrac{9}{2} & -\dfrac{9}{2} & 0 \\[2mm]
-\dfrac{9}{2} & \dfrac{9}{2}+\dfrac{11}{2} & -\dfrac{11}{2} \\[2mm]
0 & -\dfrac{11}{2} & \dfrac{11}{2}
\end{bmatrix}
\begin{Bmatrix} a_1 \\ a_2 \\ a_3 \end{Bmatrix}
=
\begin{Bmatrix}
-2 + 8\ln\dfrac{5}{4} \\[2mm]
\dfrac{8}{5} - 8\ln\dfrac{5}{4} - \dfrac{8}{5} + 8\ln\dfrac{6}{5} \\[2mm]
\dfrac{4}{3} - 8\ln\dfrac{6}{5}
\end{Bmatrix}
$$

$$
+
\begin{Bmatrix}
\left(-x\dfrac{d\tilde{U}^{(1)}}{dx}\right)_{x=1} \\[4mm]
\left(-x\dfrac{d\tilde{U}^{(2)}}{dx}\right)_{x=1.25} - \left(-x\dfrac{d\tilde{U}^{(1)}}{dx}\right)_{x=1.25} \\[4mm]
-\left(-x\dfrac{d\tilde{U}^{(2)}}{dx}\right)_{x=1.5}
\end{Bmatrix}
\tag{5.89}
$$

or in general matrix notation,

$$
\begin{bmatrix}
K_{11}^{(1)} & K_{12}^{(1)} & 0 \\[2mm]
K_{21}^{(1)} & K_{22}^{(1)}+K_{22}^{(2)} & K_{23}^{(2)} \\[2mm]
0 & K_{32}^{(2)} & K_{33}^{(2)}
\end{bmatrix}
\begin{Bmatrix} a_1 \\ a_2 \\ a_3 \end{Bmatrix}
=
\begin{Bmatrix}
F_1^{(1)} \\[2mm]
F_2^{(1)} + F_2^{(2)} \\[2mm]
F_3^{(2)}
\end{Bmatrix}
\tag{5.90}
$$

where the contributions from each element have been indicated.

Element (3). Using the entries in the connectivity table (Table 5.1) for element (3), substitute into Eqs. (5.80) node number 3 for local node 1 and node number 4 for local node 2:

$$
\begin{bmatrix}
\dfrac{1}{2}\dfrac{x_4+x_3}{x_4-x_3} & -\dfrac{1}{2}\dfrac{x_4+x_3}{x_4-x_3} \\[2mm]
-\dfrac{1}{2}\dfrac{x_4+x_3}{x_4-x_3} & \dfrac{1}{2}\dfrac{x_4+x_3}{x_4-x_3}
\end{bmatrix}
\begin{Bmatrix} a_3 \\ a_4 \end{Bmatrix}
$$

$$
=
\begin{Bmatrix}
-\dfrac{2}{x_3} + \dfrac{2}{x_4-x_3}\ln\dfrac{x_4}{x_3} \\[3mm]
\dfrac{2}{x_4} - \dfrac{2}{x_4-x_3}\ln\dfrac{x_4}{x_3}
\end{Bmatrix}
+
\begin{Bmatrix}
\left(-x\dfrac{d\tilde{U}^{(3)}}{dx}\right)_{x_3} \\[3mm]
-\left(-x\dfrac{d\tilde{U}^{(3)}}{dx}\right)_{x_4}
\end{Bmatrix}
\tag{5.91}
$$

Substitute from Eqs. (5.83) the values $x_3 = 1.5$ and $x_4 = 1.75$ into Eq. (5.91). This yields

$$\begin{bmatrix} \dfrac{13}{2} & -\dfrac{13}{2} \\[2mm] -\dfrac{13}{2} & \dfrac{13}{2} \end{bmatrix} \begin{Bmatrix} a_3 \\[2mm] a_4 \end{Bmatrix} = \begin{Bmatrix} -\dfrac{4}{3} + 8 \ln \dfrac{7}{6} \\[2mm] \dfrac{8}{7} - 8 \ln \dfrac{7}{6} \end{Bmatrix}$$

$$+ \begin{Bmatrix} \left(-x \dfrac{d\tilde{U}^{(3)}}{dx} \right)_{x=1.5} \\[4mm] -\left(-x \dfrac{d\tilde{U}^{(3)}}{dx} \right)_{x=1.75} \end{Bmatrix} \qquad \textbf{(5.92)}$$

or in general matrix notation,

$$\begin{bmatrix} K_{33}^{(3)} & K_{34}^{(3)} \\[2mm] K_{43}^{(3)} & K_{44}^{(3)} \end{bmatrix} \begin{Bmatrix} a_3 \\[2mm] a_4 \end{Bmatrix} = \begin{Bmatrix} F_3^{(3)} \\[2mm] F_4^{(3)} \end{Bmatrix} \qquad \textbf{(5.93)}$$

Element (3) is assembled to element (2) in exactly the same manner that element (2) was assembled to element (1). Thus Eqs. (5.92) are added to Eqs. (5.89):

$$\begin{bmatrix} \dfrac{9}{2} & -\dfrac{9}{2} & 0 & 0 \\[2mm] -\dfrac{9}{2} & \dfrac{9}{2}+\dfrac{11}{2} & -\dfrac{11}{2} & 0 \\[2mm] 0 & -\dfrac{11}{2} & \dfrac{11}{2}+\dfrac{13}{2} & -\dfrac{13}{2} \\[2mm] 0 & 0 & -\dfrac{13}{2} & \dfrac{13}{2} \end{bmatrix} \begin{Bmatrix} a_1 \\[2mm] a_2 \\[2mm] a_3 \\[2mm] a_4 \end{Bmatrix} = \begin{Bmatrix} -2 + 8 \ln \dfrac{5}{4} \\[2mm] \dfrac{8}{5} - 8 \ln \dfrac{5}{4} - \dfrac{8}{5} + 8 \ln \dfrac{6}{5} \\[2mm] \dfrac{4}{3} - 8 \ln \dfrac{6}{5} - \dfrac{4}{3} + 8 \ln \dfrac{7}{6} \\[2mm] \dfrac{8}{7} - 8 \ln \dfrac{7}{6} \end{Bmatrix}$$

$$+ \begin{Bmatrix} \left(-x \dfrac{d\tilde{U}^{(1)}}{dx} \right)_{x=1} \\[4mm] \left(-x \dfrac{d\tilde{U}^{(2)}}{dx} \right)_{x=1.25} - \left(-x \dfrac{d\tilde{U}^{(1)}}{dx} \right)_{x=1.25} \\[4mm] \left(-x \dfrac{d\tilde{U}^{(3)}}{dx} \right)_{x=1.5} - \left(-x \dfrac{d\tilde{U}^{(2)}}{dx} \right)_{x=1.5} \\[4mm] -\left(-x \dfrac{d\tilde{U}^{(3)}}{dx} \right)_{x=1.75} \end{Bmatrix} \qquad \textbf{(5.94)}$$

or, in general matrix notation, adding Eqs. (5.98) to Eqs. (5.95):

$$
\begin{bmatrix}
K_{11}^{(1)} & K_{12}^{(1)} & 0 & 0 \\
K_{21}^{(1)} & K_{22}^{(1)} + K_{22}^{(2)} & K_{23}^{(2)} & 0 \\
0 & K_{32}^{(2)} & K_{33}^{(2)} + K_{33}^{(3)} & K_{34}^{(3)} \\
0 & 0 & K_{43}^{(3)} & K_{44}^{(3)}
\end{bmatrix}
\begin{Bmatrix}
a_1 \\ a_2 \\ a_3 \\ a_4
\end{Bmatrix}
=
\begin{Bmatrix}
F_1^{(1)} \\
F_2^{(1)} + F_2^{(2)} \\
F_3^{(2)} + F_3^{(3)} \\
F_4^{(3)}
\end{Bmatrix}
\tag{5.95}
$$

where the contributions from each element have been indicated.

Element (4). Using the entries in the connectivity table (Table 5.1) for element (4), substitute into Eqs. (5.80) node number 4 for local node 1 and node number 5 for local node 2:

$$
\begin{bmatrix}
\dfrac{1}{2}\dfrac{x_5+x_4}{x_5-x_4} & \dfrac{1}{2}\dfrac{x_5+x_4}{x_5-x_4} \\[2ex]
-\dfrac{1}{2}\dfrac{x_5+x_4}{x_5-x_4} & \dfrac{1}{2}\dfrac{x_5+x_4}{x_5-x_4}
\end{bmatrix}
\begin{Bmatrix}
a_4 \\[2ex] a_5
\end{Bmatrix}
$$

$$
=
\begin{Bmatrix}
-\dfrac{2}{x_4} + \dfrac{2}{x_5-x_4}\ln\dfrac{x_5}{x_4} \\[2ex]
\dfrac{2}{x_5} - \dfrac{2}{x_5-x_4}\ln\dfrac{x_5}{x_4}
\end{Bmatrix}
+
\begin{Bmatrix}
\left(-x\,\dfrac{d\tilde{U}^{(4)}}{dx}\right)_{x_4} \\[2ex]
-\left(-x\,\dfrac{d\tilde{U}^{(4)}}{dx}\right)_{x_5}
\end{Bmatrix}
\tag{5.96}
$$

Substitute from Eqs. (5.83) the values $x_4 = 1.75$ and $x_5 = 2$ into Eqs. (5.95). This yields

$$
\begin{bmatrix}
\dfrac{15}{2} & -\dfrac{15}{2} \\[2ex]
-\dfrac{15}{2} & \dfrac{15}{2}
\end{bmatrix}
\begin{Bmatrix}
a_4 \\[2ex] a_5
\end{Bmatrix}
=
\begin{Bmatrix}
-\dfrac{8}{7} + 8\ln\dfrac{8}{7} \\[2ex]
1 - 8\ln\dfrac{8}{7}
\end{Bmatrix}
$$

$$
+
\begin{Bmatrix}
\left(-x\,\dfrac{d\tilde{U}^{(4)}}{dx}\right)_{x=1.75} \\[2ex]
-\left(-x\,\dfrac{d\tilde{U}^{(4)}}{dx}\right)_{x=2}
\end{Bmatrix}
\tag{5.97}
$$

or, in general matrix notation,

$$
\begin{bmatrix}
K_{44}^{(4)} & K_{45}^{(4)} \\[1ex]
K_{54}^{(4)} & K_{55}^{(4)}
\end{bmatrix}
\begin{Bmatrix}
a_4 \\[1ex] a_5
\end{Bmatrix}
=
\begin{Bmatrix}
F_4^{(4)} \\[1ex] F_5^{(4)}
\end{Bmatrix}
\tag{5.98}
$$

Finally, assemble Eqs. (5.97) to Eqs. (5.94):

$$
\begin{bmatrix}
\dfrac{9}{2} & -\dfrac{9}{2} & 0 & 0 & 0 \\[2mm]
-\dfrac{9}{2} & \dfrac{9}{2}+\dfrac{11}{2} & -\dfrac{11}{2} & 0 & 0 \\[2mm]
0 & -\dfrac{11}{2} & \dfrac{11}{2}+\dfrac{13}{2} & -\dfrac{13}{2} & 0 \\[2mm]
0 & 0 & -\dfrac{13}{2} & \dfrac{13}{2}+\dfrac{15}{2} & -\dfrac{15}{2} \\[2mm]
0 & 0 & 0 & -\dfrac{15}{2} & \dfrac{15}{2}
\end{bmatrix}
\begin{Bmatrix}
a_1 \\[2mm] a_2 \\[2mm] a_3 \\[2mm] a_4 \\[2mm] a_5
\end{Bmatrix}
$$

$$
=\left\{
\begin{array}{c}
-2+8\ln\dfrac{5}{4} \\[3mm]
\dfrac{8}{5}-8\ln\dfrac{5}{4}-\dfrac{8}{5}+8\ln\dfrac{6}{5} \\[3mm]
\dfrac{4}{3}-8\ln\dfrac{6}{5}-\dfrac{4}{3}+8\ln\dfrac{7}{6} \\[3mm]
\dfrac{8}{7}-8\ln\dfrac{7}{6}-\dfrac{8}{7}+8\ln\dfrac{8}{7} \\[3mm]
1-8\ln\dfrac{8}{7}
\end{array}
\right\}
+\left\{
\begin{array}{c}
\left(-x\dfrac{d\tilde{U}^{(1)}}{dx}\right)_{x=1} \\[3mm]
\left(-x\dfrac{d\tilde{U}^{(2)}}{dx}\right)_{x=1.25}-\left(-x\dfrac{d\tilde{U}^{(1)}}{dx}\right)_{x=1.25} \\[3mm]
\left(-x\dfrac{d\tilde{U}^{(3)}}{dx}\right)_{x=1.5}-\left(-x\dfrac{d\tilde{U}^{(2)}}{dx}\right)_{x=1.5} \\[3mm]
\left(-x\dfrac{d\tilde{U}^{(4)}}{dx}\right)_{x=1.75}-\left(-x\dfrac{d\tilde{U}^{(3)}}{dx}\right)_{x=1.75} \\[3mm]
-\left(-x\dfrac{d\tilde{U}^{(4)}}{dx}\right)_{x=2}
\end{array}
\right\}
$$

$$\text{(5.99)}$$

In general matrix notation, adding Eqs. (5.98) to (5.95) yields

$$
\begin{bmatrix}
K_{11}^{(1)} & K_{12}^{(1)} & 0 & 0 & 0 \\[2mm]
K_{21}^{(1)} & K_{22}^{(1)}+K_{22}^{(2)} & K_{23}^{(2)} & 0 & 0 \\[2mm]
0 & K_{32}^{(2)} & K_{33}^{(2)}+K_{33}^{(3)} & K_{34}^{(3)} & 0 \\[2mm]
0 & 0 & K_{43}^{(3)} & K_{44}^{(3)}+K_{44}^{(4)} & K_{45}^{(4)} \\[2mm]
0 & 0 & 0 & K_{54}^{(4)} & K_{55}^{(4)}
\end{bmatrix}
\begin{Bmatrix}
a_1 \\[2mm] a_2 \\[2mm] a_3 \\[2mm] a_4 \\[2mm] a_5
\end{Bmatrix}
=
\begin{Bmatrix}
F_1^{(1)} \\[2mm]
F_2^{(1)}+F_2^{(2)} \\[2mm]
F_3^{(2)}+F_3^{(3)} \\[2mm]
F_4^{(3)}+F_4^{(4)} \\[2mm]
F_5^{(4)}
\end{Bmatrix}
$$

$$\text{(5.100)}$$

where the contributions from each element have been indicated.

This completes the numerical evaluation of all four sets of element equations and their assembly into the system equations (5.99). The addition operations in Eqs. (5.99) were purposely not summed in order to illustrate the pattern of assembly. Carrying out the addition would simplify Eqs. (5.99) to the following:

$$
\begin{bmatrix}
\dfrac{9}{2} & -\dfrac{9}{2} & 0 & 0 & 0 \\[2mm]
-\dfrac{9}{2} & 10 & -\dfrac{11}{2} & 0 & 0 \\[2mm]
0 & -\dfrac{11}{2} & 12 & -\dfrac{13}{2} & 0 \\[2mm]
0 & 0 & -\dfrac{13}{2} & 14 & -\dfrac{15}{2} \\[2mm]
0 & 0 & 0 & -\dfrac{15}{2} & \dfrac{15}{2}
\end{bmatrix}
\begin{Bmatrix}
a_1 \\[2mm] a_2 \\[2mm] a_3 \\[2mm] a_4 \\[2mm] a_5
\end{Bmatrix}
=
\begin{Bmatrix}
-2 + 8 \ln \dfrac{5}{4} \\[2mm]
8 \ln \dfrac{24}{25} \\[2mm]
8 \ln \dfrac{35}{36} \\[2mm]
8 \ln \dfrac{48}{49} \\[2mm]
1 - 8 \ln \dfrac{8}{7}
\end{Bmatrix}
$$

$$
+
\begin{Bmatrix}
\left(-x \dfrac{d\tilde{U}^{(1)}}{dx} \right)_{x=1} \\[4mm]
\left(-x \dfrac{d\tilde{U}^{(2)}}{dx} \right)_{x=1.25} - \left(-x \dfrac{d\tilde{U}^{(1)}}{dx} \right)_{x=1.25} \\[4mm]
\left(-x \dfrac{d\tilde{U}^{(3)}}{dx} \right)_{x=1.5} - \left(-x \dfrac{d\tilde{U}^{(2)}}{dx} \right)_{x=1.5} \\[4mm]
\left(-x \dfrac{d\tilde{U}^{(4)}}{dx} \right)_{x=1.75} - \left(-x \dfrac{d\tilde{U}^{(3)}}{dx} \right)_{x=1.75} \\[4mm]
- \left(-x \dfrac{d\tilde{U}^{(4)}}{dx} \right)_{x=2}
\end{Bmatrix}
\qquad \textbf{(5.101)}
$$

Figure 5.20 shows the complete trial solution corresponding to Eqs. (5.101). Also shown are the assembled trial functions, multiplied by a parameter a_i, which are now continuous at their interelement boundaries (recall Fig. 5.15).

In the above development the algebraic expressions for the element equations were written out in full for each element because it is instructive to actually see (at least once during one's lifetime) their *repetitive* nature; i.e., only the numerical values of the subscripts and superscripts are altered. However, in a computer program (see Chapter 7) the element equations [Eqs. (5.80)] need to be coded only once, with the subscripts and superscripts

$\bar{U}(x;a)$

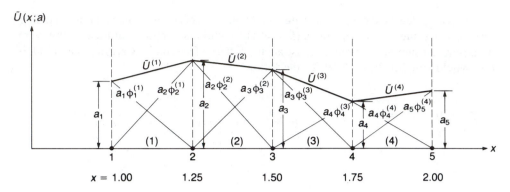

FIGURE 5.20 ● Complete trial solution and assembled trial functions.

represented by array indices that change numerically for each element according to the connectivity table (Table 5.1).[22]

Step 9: Apply the BCs and the natural IBCs.

The natural BC and natural IBCs ● The natural BC in Eqs. (5.1b) may be written in the form,

$$\left(-x\frac{dU^{(4)}}{dx}\right)_{x=2} = \frac{1}{2} \tag{5.102}$$

where U in Eqs. (5.1b) may be replaced by $U^{(4)}$, since $U^{(4)}$ is that part of the exact solution that touches the node at $x=2$.

The natural IBCs are obtained by applying Eq. (5.40a) to each of the interelement nodes:

$$\left(-x\frac{dU^{(2)}}{dx}\right)_{x=1.25} - \left(-x\frac{dU^{(1)}}{dx}\right)_{x=1.25} = 0$$

$$\left(-x\frac{dU^{(3)}}{dx}\right)_{x=1.50} - \left(-x\frac{dU^{(2)}}{dx}\right)_{x=1.50} = 0$$

$$\left(-x\frac{dU^{(4)}}{dx}\right)_{x=1.75} - \left(-x\frac{dU^{(3)}}{dx}\right)_{x=1.75} = 0 \tag{5.103}$$

where $U^{(1)}$, $U^{(2)}$, $U^{(3)}$, and $U^{(4)}$ represent the exact solution in each element. Each of the three equations in Eqs. (5.103) are statements of continuity of flux between elements. Such continuity is a property of the exact solution, and relations such as Eqs. (5.103) may be derived for each interelement boundary point in precisely the same manner as was done for Eq. (5.40a) in Section 5.3.1. However, in practice such derivations aren't necessary because there is a physical interpretation for the natural IBCs, to be explained in Section 7.6, which will make it obvious from the physical statement of a problem what values to specify for the natural IBCs.

Substituting the values from Eqs. (5.102) and (5.103) into the terms on the RHS of

---

[22] In Chapter 7 we will deal with a more general governing differential equation, so the algebraic form of the element equations will be different from Eqs. (5.80); the latter are applicable only to the governing differential equation (5.1a).

Eqs. (5.101) yields

$$
\begin{bmatrix}
\dfrac{9}{2} & -\dfrac{9}{2} & 0 & 0 & 0 \\[2mm]
-\dfrac{9}{2} & 10 & -\dfrac{11}{2} & 0 & 0 \\[2mm]
0 & -\dfrac{11}{2} & 12 & -\dfrac{13}{2} & 0 \\[2mm]
0 & 0 & -\dfrac{13}{2} & 14 & -\dfrac{15}{2} \\[2mm]
0 & 0 & 0 & -\dfrac{15}{2} & \dfrac{15}{2}
\end{bmatrix}
\begin{Bmatrix}
a_1 \\ a_2 \\ a_3 \\ a_4 \\ a_5
\end{Bmatrix}
$$

$$
= \begin{Bmatrix}
-2 + 8\ln\dfrac{5}{4} \\[2mm]
8\ln\dfrac{24}{25} \\[2mm]
8\ln\dfrac{35}{36} \\[2mm]
8\ln\dfrac{48}{49} \\[2mm]
1 - 8\ln\dfrac{8}{7}
\end{Bmatrix}
+
\begin{Bmatrix}
\left(-x\,\dfrac{d\tilde{U}^{(1)}}{dx}\right)_{x=1} \\[2mm]
0 \\[2mm]
0 \\[2mm]
0 \\[2mm]
-\dfrac{1}{2}
\end{Bmatrix}
\tag{5.104}
$$

The essential BC • The condition $U(1)=2$ is applied as a constrained BC, in exactly the same manner as for the one- and two-element problems (see step 9 in Sections 5.2 and 5.3, respectively). Thus using the trial solution $\tilde{U}^{(1)}$ in Fig. 5.17 yields the constraint equation $a_1=2$. This is applied to the system equations [Eqs. (5.104)] by multiplying column 1 in the stiffness matrix by 2 [the result of substituting $a_1=2$ into Eqs. (5.104)] and then subtracting the first column from the RHS load vector. Finally, the first column and first equation (row) are deleted. The result is

$$
\begin{bmatrix}
10 & -\dfrac{11}{2} & 0 & 0 \\[2mm]
-\dfrac{11}{2} & 12 & -\dfrac{13}{2} & 0 \\[2mm]
0 & -\dfrac{13}{2} & 14 & -\dfrac{15}{2} \\[2mm]
0 & 0 & -\dfrac{15}{2} & \dfrac{15}{2}
\end{bmatrix}
\begin{Bmatrix}
a_2 \\ a_3 \\ a_4 \\ a_5
\end{Bmatrix}
=
\begin{Bmatrix}
9 + 8\ln\dfrac{24}{25} \\[2mm]
8\ln\dfrac{35}{36} \\[2mm]
8\ln\dfrac{48}{49} \\[2mm]
\dfrac{1}{2} - 8\ln\dfrac{8}{7}
\end{Bmatrix}
\tag{5.105}
$$

This is the final form of the system equations, after all the BCs and IBCs have been applied.

Step 10: Solve the system equations.

Equations (5.105) could be solved by hand, using the Gaussian elimination procedure (see Exercise 3.1); the zeroes in the stiffness matrix would simplify the calculations. Or one could use any algebraic equation-solving program available on most computers and many hand calculators, including the subroutine GAUSEL in the UNAFEM program in the appendix. In any case, the solution to Eqs. (5.105) is

$$a_2 = 1.714$$
$$a_3 = 1.540$$
$$a_4 = 1.427$$
$$a_5 = 1.352 \tag{5.106a}$$

and from the constraint equation for the essential BC,

$$a_1 = 2.000 \tag{5.106b}$$

Substituting Eqs. (5.106) into each of the trial-solution expressions in Fig. 5.17 and using the nodal coordinates from Eqs. (5.83) yields the following expressions for the approximate solution in each element:

$$\tilde{U}^{(1)}(x) = 2.000 \left(\frac{1.25 - x}{0.25} \right) + 1.714 \left(\frac{x - 1.00}{0.25} \right)$$

$$\tilde{U}^{(2)}(x) = 1.714 \left(\frac{1.50 - x}{0.25} \right) + 1.540 \left(\frac{x - 1.25}{0.25} \right)$$

$$\tilde{U}^{(3)}(x) = 1.540 \left(\frac{1.75 - x}{0.25} \right) + 1.427 \left(\frac{x - 1.50}{0.25} \right)$$

$$\tilde{U}^{(4)}(x) = 1.427 \left(\frac{2.00 - x}{0.25} \right) + 1.352 \left(\frac{x - 1.75}{0.25} \right) \tag{5.107}$$

Step 11: Evaluate the flux.

The expressions for the flux in each element follow directly from Eq. (5.82), in which the subscripts are permuted for each element according to the connectivity table (Table 5.1):

$$\tilde{\tau}^{(1)}(x) = \frac{a_1 - a_2}{x_2 - x_1} x$$

$$\tilde{\tau}^{(2)}(x) = \frac{a_2 - a_3}{x_3 - x_2} x$$

$$\tilde{\tau}^{(3)}(x) = \frac{a_3 - a_4}{x_4 - x_3} x$$

$$\tilde{\tau}^{(4)}(x) = \frac{a_4 - a_5}{x_5 - x_4} x \tag{5.108}$$

Substituting Eqs. (5.106) and (5.83) into Eqs. (5.108) yields

$$\tilde{\tau}^{(1)}(x) = 1.144x$$

$$\tilde{\tau}^{(2)}(x) = 0.696x$$

$$\tilde{\tau}^{(3)}(x) = 0.452x$$

$$\tilde{\tau}^{(4)}(x) = 0.300x \qquad \textbf{(5.109)}$$

Step 12: Plot the solution and estimate its accuracy.

Figure 5.21 is a plot of $\tilde{U}(x)$ and $\tilde{\tau}(x)$ in Eqs. (5.107) and (5.109). Once again we see that the essential BC and essential IBCs, which were constrained, are satisfied exactly. Thus $\tilde{U}(x)$ is continuous at all the interelement boundaries. The natural BC and natural IBCs were unconstrained, so they are satisfied only approximately. Thus $\tilde{\tau}(x)$ is discontinuous at all the interelement boundaries.

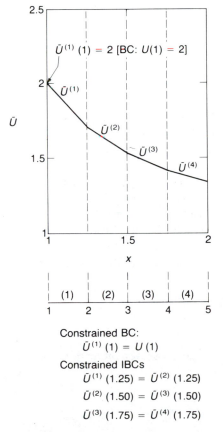

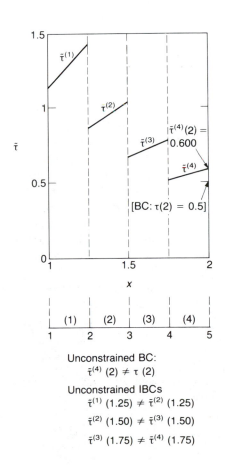

FIGURE 5.21 ● Four-element solution.

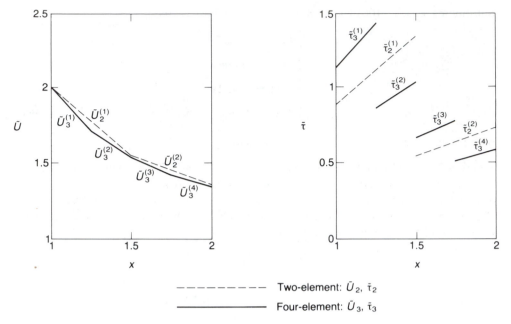

FIGURE 5.22 ● Two- and four-element solutions.

As before, the error in the flux can be estimated in the "immediate vicinity" of all points where an unconstrained BC or IBC is applied. Thus, at $x=2$, the calculated value $\tilde{\tau}(2)=0.600$ is in error by 20% from the exact value. At the interelement points x_2, x_3, and x_4 the spread in values of the discontinuities, relative to the mean, yields errors of ±24, ±21, and $\pm20\%$, respectively.

To estimate the error in both $\tilde{U}(x)$ and $\tilde{\tau}(x)$ at all points throughout the domain, we need to compare the four-element solution with the two-element solution. Figure 5.22 shows both the two- and four-element solutions from Figs. 5.13 and 5.21, labeled as $\tilde{U}_2(x)$, $\tilde{\tau}_2(x)$ and $\tilde{U}_3(x)$, $\tilde{\tau}_3(x)$, respectively, to indicate the second and third approximate solutions we have obtained.

The pointwise error between $\tilde{U}_2(x)$ and $\tilde{U}_3(x)$ is less than 3.6% over the entire domain; that is,

$$\left| \frac{\tilde{U}_3(x) - \tilde{U}_2(x)}{\tilde{U}_3(x)} \right| < 0.036 \qquad \text{for } 1 < x < 2$$

$$\text{(5.110)}$$

The pointwise error between $\tilde{\tau}_2(x)$ and $\tilde{\tau}_3(x)$ is less than 29% over the entire domain; that is,

$$\left| \frac{\tilde{\tau}_3(x) - \tilde{\tau}_2(x)}{\tilde{\tau}_3(x)} \right| < 0.29 \qquad \text{for } 1 < x < 2$$

$$\text{(5.111)}$$

5.5 An Eight-Element Solution

Having derived the element equations for a typical element [Eqs. (5.80)], we can now easily generate solutions for meshes with any number of elements. Indeed, we may simply follow the procedure just demonstrated with the four-element solution. Thus we would begin by constructing a mesh of numbered nodes and elements. Each node is defined by specifying its coordinates and assigning a node number, and each element is defined by specifying the node numbers that are associated with it (in a connectivity table). Using the nodal coordinates and connectivity table, the terms in the element equations can be numerically evaluated for each element in the mesh. As the terms for each element are evaluated, they are assembled (added) to the previous element terms, resulting in the system equations after all elements have been assembled. Finally, BCs and natural IBCs are applied to the system equations and the latter solved.

This brief outline is actually a description of steps 7 through 12, the numerical computation phase. In other words, the present eight-element solution does not need to repeat steps 1 through 6 once the element equations have been derived for a typical element; the same element equations can be used over and over again for as many elements as desired. This is a central feature of the FEM and a principal reason for its being a practical and efficient analysis tool. The burden on the analyst/engineer for working out theoretical expressions for the residual (or variational) equations is greatly reduced, involving only one element rather than the entire domain. The remainder of the work is numerical and can be given to a computer to perform. In the following steps 7 through 12 it will be noted that the procedure is essentially the same as for the four-element problem. And it will be essentially the procedure followed in Chapter 7 for programming steps 7 through 12.

Step 7: Specify numerical data for the problem.

1. *Mesh generation* (geometric data). We will define a mesh of eight elements, as shown in Fig. 5.23.

   a) Node definition: Specify the coordinates of each node and assign a number to each node. Again, let's make all the elements the same length. (In subsequent chapters we will construct meshes with different-size elements.)

$$x_1 = 1.000$$
$$x_2 = 1.125$$
$$x_3 = 1.250$$
$$x_4 = 1.375$$
$$x_5 = 1.500$$
$$x_6 = 1.625$$
$$x_7 = 1.750$$
$$x_8 = 1.875$$
$$x_9 = 2.000 \qquad \text{(5.112)}$$

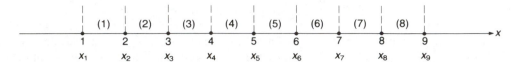

FIGURE 5.23 ● Mesh for the eight-element problem.

b) Element definition: Construct a connectivity table listing node numbers associated with each statement.

TABLE 5.2. ● Connectivity table for mesh, Fig. 5.23.

Element	Local node 1 becomes node	Local node 2 becomes node
(1)	1	2
(2)	2	3
(3)	3	4
(4)	4	5
(5)	5	6
(6)	6	7
(7)	7	8
(8)	8	9

2. *Physical properties and applied loads.* As noted before these have already been defined for this specific illustration problem. See comments under step 7 in Section 4.4.

Step 8: Numerically evaluate all the interior (nonboundary) terms in the element equations for each element and assemble the element equations to form the system equations.

The procedure here is identical to step 8 for the four-element problem, except that we have eight elements to do rather than four. (We could clearly benefit from a computer at this point.) For each element, substitute the actual node numbers from the connectivity table (Table 5.2) into the element equations [Eqs. (5.80)]; then substitute the coordinate values from Eqs. (5.112). The final result, after assembling all eight elements, is given in Eqs. (5.113).

It is helpful to have a clear picture of how each set of element equations contributes to the overall system equations. Therefore, Eqs. (5.113) are written in terms of general matrix notation in Eqs. (5.114), where the contributions from each element have been indicated.

After performing the addition operations, the system equations [Eqs. (5.113)] appear as shown in Eqs. (5.115).

$$
\begin{bmatrix}
\dfrac{17}{2} & -\dfrac{17}{2} & 0 & 0 & 0 & 0 & 0 & 0 & 0 \\[2mm]
-\dfrac{17}{2} & \dfrac{17}{2}+\dfrac{19}{2} & -\dfrac{19}{2} & 0 & 0 & 0 & 0 & 0 & 0 \\[2mm]
0 & -\dfrac{19}{2} & \dfrac{19}{2}+\dfrac{21}{2} & -\dfrac{21}{2} & 0 & 0 & 0 & 0 & 0 \\[2mm]
0 & 0 & -\dfrac{21}{2} & \dfrac{21}{2}+\dfrac{23}{2} & -\dfrac{23}{2} & 0 & 0 & 0 & 0 \\[2mm]
0 & 0 & 0 & -\dfrac{23}{2} & \dfrac{23}{2}+\dfrac{25}{2} & -\dfrac{25}{2} & 0 & 0 & 0 \\[2mm]
0 & 0 & 0 & 0 & -\dfrac{25}{2} & \dfrac{25}{2}+\dfrac{27}{2} & -\dfrac{27}{2} & 0 & 0 \\[2mm]
0 & 0 & 0 & 0 & 0 & -\dfrac{27}{2} & \dfrac{27}{2}+\dfrac{29}{2} & -\dfrac{29}{2} & 0 \\[2mm]
0 & 0 & 0 & 0 & 0 & 0 & -\dfrac{29}{2} & \dfrac{29}{2}+\dfrac{31}{2} & -\dfrac{31}{2} \\[2mm]
0 & 0 & 0 & 0 & 0 & 0 & 0 & -\dfrac{31}{2} & \dfrac{31}{2}
\end{bmatrix}
\begin{Bmatrix}
a_1 \\ a_2 \\ a_3 \\ a_4 \\ a_5 \\ a_6 \\ a_7 \\ a_8 \\ a_9
\end{Bmatrix}
$$

$$
=\begin{Bmatrix}
-2+16\ln\dfrac{9}{8} \\[2mm]
\dfrac{16}{9}-16\ln\dfrac{9}{8}-\dfrac{16}{9}+16\ln\dfrac{10}{9} \\[2mm]
\dfrac{16}{10}-16\ln\dfrac{10}{9}-\dfrac{16}{10}+16\ln\dfrac{11}{10} \\[2mm]
\dfrac{16}{11}-16\ln\dfrac{11}{10}-\dfrac{16}{11}+16\ln\dfrac{12}{11} \\[2mm]
\dfrac{16}{12}-16\ln\dfrac{12}{11}-\dfrac{16}{12}+16\ln\dfrac{13}{12} \\[2mm]
\dfrac{16}{13}-16\ln\dfrac{13}{12}-\dfrac{16}{13}+16\ln\dfrac{14}{13} \\[2mm]
\dfrac{16}{14}-16\ln\dfrac{14}{13}-\dfrac{16}{14}+16\ln\dfrac{15}{14} \\[2mm]
\dfrac{16}{15}-16\ln\dfrac{15}{14}-\dfrac{16}{15}+16\ln\dfrac{16}{15} \\[2mm]
1-16\ln\dfrac{16}{15}
\end{Bmatrix}
+\begin{Bmatrix}
\left(-x\,\dfrac{d\tilde{U}^{(1)}}{dx}\right)_{x=1} \\[2mm]
\left(-x\,\dfrac{d\tilde{U}^{(2)}}{dx}\right)_{x=1.125}-\left(-x\,\dfrac{d\tilde{U}^{(1)}}{dx}\right)_{x=1.125} \\[2mm]
\left(-x\,\dfrac{d\tilde{U}^{(3)}}{dx}\right)_{x=1.250}-\left(-x\,\dfrac{d\tilde{U}^{(2)}}{dx}\right)_{x=1.250} \\[2mm]
\left(-x\,\dfrac{d\tilde{U}^{(4)}}{dx}\right)_{x=1.375}-\left(-x\,\dfrac{d\tilde{U}^{(3)}}{dx}\right)_{x=1.375} \\[2mm]
\left(-x\,\dfrac{d\tilde{U}^{(5)}}{dx}\right)_{x=1.500}-\left(-x\,\dfrac{d\tilde{U}^{(4)}}{dx}\right)_{x=1.500} \\[2mm]
\left(-x\,\dfrac{d\tilde{U}^{(6)}}{dx}\right)_{x=1.625}-\left(-x\,\dfrac{d\tilde{U}^{(5)}}{dx}\right)_{x=1.625} \\[2mm]
\left(-x\,\dfrac{d\tilde{U}^{(7)}}{dx}\right)_{x=1.750}-\left(-x\,\dfrac{d\tilde{U}^{(6)}}{dx}\right)_{x=1.750} \\[2mm]
\left(-x\,\dfrac{d\tilde{U}^{(8)}}{dx}\right)_{x=1.875}-\left(-x\,\dfrac{d\tilde{U}^{(7)}}{dx}\right)_{x=1.875} \\[2mm]
-\left(-x\,\dfrac{d\tilde{U}^{(8)}}{dx}\right)_{x=2}
\end{Bmatrix}
$$

(5.113)

$$
\begin{bmatrix}
K_{11}^{(1)} & K_{12}^{(1)} & 0 & 0 & 0 & 0 & 0 & 0 & 0 \\
K_{21}^{(1)} & K_{22}^{(1)} + K_{22}^{(2)} & K_{23}^{(2)} & 0 & 0 & 0 & 0 & 0 & 0 \\
0 & K_{32}^{(2)} & K_{33}^{(2)} + K_{33}^{(3)} & K_{34}^{(3)} & 0 & 0 & 0 & 0 & 0 \\
0 & 0 & K_{43}^{(3)} & K_{44}^{(3)} + K_{44}^{(4)} & K_{45}^{(4)} & 0 & 0 & 0 & 0 \\
0 & 0 & 0 & K_{54}^{(4)} & K_{55}^{(4)} + K_{55}^{(5)} & K_{56}^{(5)} & 0 & 0 & 0 \\
0 & 0 & 0 & 0 & K_{65}^{(5)} & K_{66}^{(5)} + K_{66}^{(6)} & K_{67}^{(6)} & 0 & 0 \\
0 & 0 & 0 & 0 & 0 & K_{76}^{(6)} & K_{77}^{(6)} + K_{77}^{(7)} & K_{78}^{(7)} & 0 \\
0 & 0 & 0 & 0 & 0 & 0 & K_{87}^{(7)} & K_{88}^{(7)} + K_{88}^{(8)} & K_{89}^{(8)} \\
0 & 0 & 0 & 0 & 0 & 0 & 0 & K_{98}^{(8)} & K_{99}^{(8)}
\end{bmatrix}
\begin{Bmatrix}
a_1 \\ a_2 \\ a_3 \\ a_4 \\ a_5 \\ a_6 \\ a_7 \\ a_8 \\ a_9
\end{Bmatrix}
$$

$$
=
\begin{Bmatrix}
F_1^{(1)} \\
F_2^{(1)} + F_2^{(2)} \\
F_3^{(2)} + F_3^{(3)} \\
F_4^{(3)} + F_4^{(4)} \\
F_5^{(4)} + F_5^{(5)} \\
F_6^{(5)} + F_6^{(6)} \\
F_7^{(6)} + F_7^{(7)} \\
F_8^{(7)} + F_8^{(8)} \\
F_9^{(8)}
\end{Bmatrix}
\tag{5.114}
$$

$$
\begin{bmatrix}
\dfrac{17}{2} & -\dfrac{17}{2} & 0 & 0 & 0 & 0 & 0 & 0 & 0 \\[2mm]
-\dfrac{17}{2} & 18 & -\dfrac{19}{2} & 0 & 0 & 0 & 0 & 0 & 0 \\[2mm]
0 & -\dfrac{19}{2} & 20 & -\dfrac{21}{2} & 0 & 0 & 0 & 0 & 0 \\[2mm]
0 & 0 & -\dfrac{21}{2} & 22 & -\dfrac{23}{2} & 0 & 0 & 0 & 0 \\[2mm]
0 & 0 & 0 & -\dfrac{23}{2} & 24 & -\dfrac{25}{2} & 0 & 0 & 0 \\[2mm]
0 & 0 & 0 & 0 & -\dfrac{25}{2} & 26 & -\dfrac{27}{2} & 0 & 0 \\[2mm]
0 & 0 & 0 & 0 & 0 & -\dfrac{27}{2} & 28 & -\dfrac{29}{2} & 0 \\[2mm]
0 & 0 & 0 & 0 & 0 & 0 & -\dfrac{29}{2} & 30 & -\dfrac{31}{2} \\[2mm]
0 & 0 & 0 & 0 & 0 & 0 & 0 & -\dfrac{31}{2} & +\dfrac{31}{2}
\end{bmatrix}
\begin{Bmatrix}
a_1 \\ a_2 \\ a_3 \\ a_4 \\ a_5 \\ a_6 \\ a_7 \\ a_8 \\ a_9
\end{Bmatrix}
$$

$$
= \left\{ \begin{array}{c}
-2 + 16\ln\dfrac{9}{8} \\[3mm]
16\ln\dfrac{80}{81} \\[3mm]
16\ln\dfrac{99}{100} \\[3mm]
16\ln\dfrac{120}{121} \\[3mm]
16\ln\dfrac{143}{144} \\[3mm]
16\ln\dfrac{168}{169} \\[3mm]
16\ln\dfrac{195}{196} \\[3mm]
16\ln\dfrac{224}{225} \\[3mm]
1 - 16\ln\dfrac{16}{15}
\end{array} \right\}
+ \left\{ \begin{array}{c}
\left(-x\,\dfrac{d\tilde{U}^{(1)}}{dx}\right)_{x=1} \\[3mm]
\left(-x\,\dfrac{d\tilde{U}^{(2)}}{dx}\right)_{x=1.125} - \left(-x\,\dfrac{d\tilde{U}^{(1)}}{dx}\right)_{x=1.125} \\[3mm]
\left(-x\,\dfrac{d\tilde{U}^{(3)}}{dx}\right)_{x=1.250} - \left(-x\,\dfrac{d\tilde{U}^{(2)}}{dx}\right)_{x=1.250} \\[3mm]
\left(-x\,\dfrac{d\tilde{U}^{(4)}}{dx}\right)_{x=1.375} - \left(-x\,\dfrac{d\tilde{U}^{(3)}}{dx}\right)_{x=1.375} \\[3mm]
\left(-x\,\dfrac{d\tilde{U}^{(5)}}{dx}\right)_{x=1.500} - \left(-x\,\dfrac{d\tilde{U}^{(4)}}{dx}\right)_{x=1.500} \\[3mm]
\left(-x\,\dfrac{d\tilde{U}^{(6)}}{dx}\right)_{x=1.625} - \left(-x\,\dfrac{d\tilde{U}^{(5)}}{dx}\right)_{x=1.625} \\[3mm]
\left(-x\,\dfrac{d\tilde{U}^{(7)}}{dx}\right)_{x=1.750} - \left(-x\,\dfrac{d\tilde{U}^{(6)}}{dx}\right)_{x=1.750} \\[3mm]
\left(-x\,\dfrac{d\tilde{U}^{(8)}}{dx}\right)_{x=1.875} - \left(-x\,\dfrac{d\tilde{U}^{(7)}}{dx}\right)_{x=1.875} \\[3mm]
-\left(-x\,\dfrac{d\tilde{U}^{(8)}}{dx}\right)_{x=2}
\end{array} \right\}
$$

(5.115)

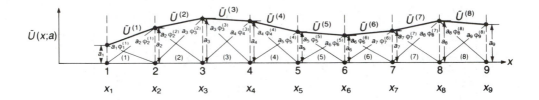

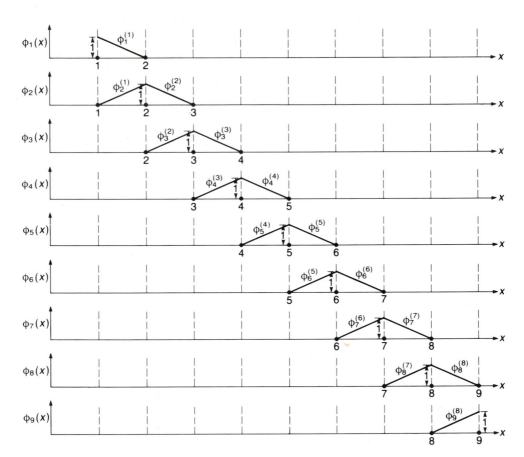

FIGURE 5.24 ● Trial solution and assembled trial functions for the eight-element problem: (a) Complete trial solution; each element trial solution $\tilde{U}^{(e)}$ is a linear sum of two element trial functions; (b) Assembled trial functions; assembly enforces interelement continuity.

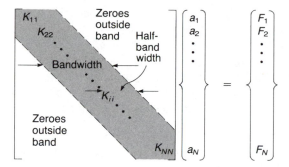

FIGURE 5.25 ● A banded stiffness matrix.

A striking feature of Eqs. (5.115) is that most of the terms in the stiffness matrix are zero. Such a matrix is said to be *sparse*. This is a direct consequence of the fact that each of the assembled trial functions is nonzero over only one or two elements and zero over all the remaining elements, as shown in Fig. 5.24.[23] This "zeroness" is what produces all the zeroes in the stiffness matrix. The sparseness of the stiffness matrix is an important characteristic of the FEM, making it feasible and economical to solve very large problems involving tens of thousands and even hundreds of thousands of equations.

Another noticeable feature of the above stiffness matrix is that it is *banded*; that is, all the nonzero terms are clustered within a narrow band about the main diagonal,[24] as shown symbolically in Fig. 5.25. The band is defined by two lines, one on either side of the main diagonal and parallel to it, which just enclose all the nonzero terms. For symmetric matrices these two lines are always equidistant from the main diagonal. Some terms inside the band may also be zero.

The *bandwidth* is the (maximum) number of terms along a row from one side of the band to the other. (Note that the rows in the upper-left and lower-right corners of the band contain fewer terms than the bandwidth.) The *half-bandwidth* is the number of terms in a row to one side (or the other) of the main diagonal, plus the main diagonal. Designating the half-bandwidth by b, the bandwidth is $2b-1$. The stiffness matrix in Eq. (5.115) has a half-bandwidth of 2.

It might be noted that if the nonzero terms were only on the main diagonal (that is, a *diagonal* stiffness matrix), then all the equations would be uncoupled, since each would be in the form

$$K_{ii}a_i = F_i \qquad i = 1,2,...N$$

and the solution would be trivially easy:

$$a_i = \frac{F_i}{K_{ii}} \qquad i = 1,2,...N$$

---

[23] The element trial functions were actually never defined outside their own elements. However, they could be trivially extended by defining them to be identically zero in all other elements.

[24] The main diagonal is an imaginary line drawn from the upper left to the lower right of the matrix. A stiffness term is on the main diagonal if its row and column subscripts are equal: $K_{11}, K_{22},..., K_{NN}$.

It should not be surprising, therefore, that equations containing a banded stiffness matrix can be solved much faster than if the matrix were full.

Bandedness is a direct consequence of the order — i.e., the pattern — in which we number the nodes in the mesh. In Fig. 5.23 we numbered the nodes sequentially from one end of the domain to the other end; this produced the very narrow band shown above. If we had not numbered sequentially, but in a random pattern, say, the bandwidth would be greater. This will be explained and demonstrated in the next section.[25]

Step 9: Apply the BCs and the natural IBCs.

The natural IBCs express continuity of the flux at each of the interelement boundaries [recall Eqs. (5.40) and (5.103)]:

$$\left(-x\frac{dU^{(2)}}{dx}\right)_{x=1.125} - \left(-x\frac{dU^{(1)}}{dx}\right)_{x=1.125} = 0$$

$$\left(-x\frac{dU^{(3)}}{dx}\right)_{x=1.250} - \left(-x\frac{dU^{(2)}}{dx}\right)_{x=1.250} = 0$$

$$\vdots \qquad \vdots \qquad \vdots$$

$$\left(-x\frac{dU^{(8)}}{dx}\right)_{x=1.875} - \left(-x\frac{dU^{(7)}}{dx}\right)_{x=1.875} = 0 \qquad \textbf{(5.116)}$$

The natural BC at $x=2$, which touches element (8), takes the form,

$$\left(-x\frac{dU^{(8)}}{dx}\right)_{x=2} = \frac{1}{2} \qquad \textbf{(5.117)}$$

Substituting the values from Eqs. (5.116) and (5.117) into the terms on the RHS of Eqs. (5.115) yields Eqs. (5.118).

The essential BC, $U(1)=2$, yields, as before, the constraint equation $a_1=2$. This is applied in exactly the same manner as for the one-, two- and four-element problems: Multiply column 1 of the stiffness matrix by 2 and then subtract the column from the RHS load vector. Finally, delete the first column and first equation. The result is Eqs. (5.119).

Step 10: Solve the system equations.

The solution to Eqs. (5.119), including the constraint equation $a_1=2$, is

$$a_1 = 2.000 \qquad a_6 = 1.475$$
$$a_2 = 1.837 \qquad a_7 = 1.424$$
$$a_3 = 1.712 \qquad a_8 = 1.382$$
$$a_4 = 1.615 \qquad a_9 = 1.348$$
$$a_5 = 1.537 \qquad \textbf{(5.120)}$$

---

[25] Section 13.6.5 will explain how to number nodes in 2-D meshes in order to help minimize the bandwidth.

$$\begin{bmatrix} \dfrac{17}{2} & -\dfrac{17}{2} & 0 & 0 & 0 & 0 & 0 & 0 & 0 \\[2mm] -\dfrac{17}{2} & 18 & -\dfrac{19}{2} & 0 & 0 & 0 & 0 & 0 & 0 \\[2mm] 0 & -\dfrac{19}{2} & 20 & -\dfrac{21}{2} & 0 & 0 & 0 & 0 & 0 \\[2mm] 0 & 0 & -\dfrac{21}{2} & 22 & -\dfrac{23}{2} & 0 & 0 & 0 & 0 \\[2mm] 0 & 0 & 0 & -\dfrac{23}{2} & 24 & -\dfrac{25}{2} & 0 & 0 & 0 \\[2mm] 0 & 0 & 0 & 0 & -\dfrac{25}{2} & 26 & -\dfrac{27}{2} & 0 & 0 \\[2mm] 0 & 0 & 0 & 0 & 0 & -\dfrac{27}{2} & 28 & -\dfrac{29}{2} & 0 \\[2mm] 0 & 0 & 0 & 0 & 0 & 0 & -\dfrac{29}{2} & 30 & -\dfrac{31}{2} \\[2mm] 0 & 0 & 0 & 0 & 0 & 0 & 0 & -\dfrac{31}{2} & \dfrac{31}{2} \end{bmatrix} \begin{Bmatrix} a_1 \\ a_2 \\ a_3 \\ a_4 \\ a_5 \\ a_6 \\ a_7 \\ a_8 \\ a_9 \end{Bmatrix}$$

$$= \begin{Bmatrix} -2 + 16 \ln \dfrac{9}{8} \\[2mm] 16 \ln \dfrac{80}{81} \\[2mm] 16 \ln \dfrac{99}{100} \\[2mm] 16 \ln \dfrac{120}{121} \\[2mm] 16 \ln \dfrac{143}{144} \\[2mm] 16 \ln \dfrac{168}{169} \\[2mm] 16 \ln \dfrac{195}{196} \\[2mm] 16 \ln \dfrac{224}{225} \\[2mm] 1 - 16 \ln \dfrac{16}{15} \end{Bmatrix} + \begin{Bmatrix} \left(-x \dfrac{d\tilde{U}^{(1)}}{dx}\right)_{x=1} \\[2mm] 0 \\ 0 \\ 0 \\ 0 \\ 0 \\ 0 \\ 0 \\ -\dfrac{1}{2} \end{Bmatrix}$$

$$\text{(5.118)}$$

$$
\begin{bmatrix}
18 & -\dfrac{19}{2} & 0 & 0 & 0 & 0 & 0 & 0 \\[2mm]
-\dfrac{19}{2} & 20 & -\dfrac{21}{2} & 0 & 0 & 0 & 0 & 0 \\[2mm]
0 & -\dfrac{21}{2} & 22 & -\dfrac{23}{2} & 0 & 0 & 0 & 0 \\[2mm]
0 & 0 & -\dfrac{23}{2} & 24 & -\dfrac{25}{2} & 0 & 0 & 0 \\[2mm]
0 & 0 & 0 & -\dfrac{25}{2} & 26 & -\dfrac{27}{2} & 0 & 0 \\[2mm]
0 & 0 & 0 & 0 & -\dfrac{27}{2} & 28 & -\dfrac{29}{2} & 0 \\[2mm]
0 & 0 & 0 & 0 & 0 & -\dfrac{29}{2} & 30 & -\dfrac{31}{2} \\[2mm]
0 & 0 & 0 & 0 & 0 & 0 & -\dfrac{31}{2} & \dfrac{31}{2}
\end{bmatrix}
\begin{Bmatrix}
a_2 \\ a_3 \\ a_4 \\ a_5 \\ a_6 \\ a_7 \\ a_8 \\ a_9
\end{Bmatrix}
=
\begin{Bmatrix}
17 + 16\ln\dfrac{80}{81} \\[2mm]
16\ln\dfrac{99}{100} \\[2mm]
16\ln\dfrac{120}{121} \\[2mm]
16\ln\dfrac{143}{144} \\[2mm]
16\ln\dfrac{168}{169} \\[2mm]
16\ln\dfrac{195}{196} \\[2mm]
16\ln\dfrac{224}{225} \\[2mm]
\dfrac{1}{2} - 16\ln\dfrac{16}{15}
\end{Bmatrix}
$$

(5.119)

The resulting expressions for the approximate solution in each element are as follows:

$$\tilde{U}^{(1)}(x) = 2.000 \left(1.125 - \frac{x}{0.125} \right) + 1.837 \left(x - \frac{1.000}{0.125} \right)$$

$$\tilde{U}^{(2)}(x) = 1.837 \left(1.250 - \frac{x}{0.125} \right) + 1.712 \left(x - \frac{1.125}{0.125} \right)$$

$$\vdots \qquad \vdots \qquad \vdots$$

$$\tilde{U}^{(8)}(x) = 1.382 \left(2.000 - \frac{x}{0.125} \right) + 1.348 \left(x - \frac{1.875}{0.125} \right) \qquad \textbf{(5.121)}$$

Step 11: Evaluate the flux.

The expressions for the flux in each element follow directly from Eq. (5.82) for the typical element; the correct subscripts for each element are determined by the connectivity table

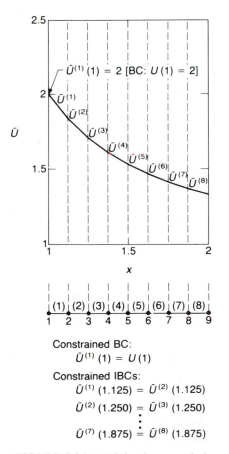

Constrained BC:
$$\tilde{U}^{(1)}(1) = U(1)$$

Constrained IBCs:
$$\tilde{U}^{(1)}(1.125) = \tilde{U}^{(2)}(1.125)$$
$$\tilde{U}^{(2)}(1.250) = \tilde{U}^{(3)}(1.250)$$
$$\vdots$$
$$\tilde{U}^{(7)}(1.875) = \tilde{U}^{(8)}(1.875)$$

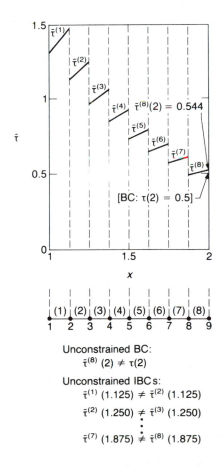

Unconstrained BC:
$$\tilde{\tau}^{(8)}(2) \neq \tau(2)$$

Unconstrained IBCs:
$$\tilde{\tau}^{(1)}(1.125) \neq \tilde{\tau}^{(2)}(1.125)$$
$$\tilde{\tau}^{(2)}(1.250) \neq \tilde{\tau}^{(3)}(1.250)$$
$$\vdots$$
$$\tilde{\tau}^{(7)}(1.875) \neq \tilde{\tau}^{(8)}(1.875)$$

FIGURE 5.26 ● Eight-element solution.

(Table 5.2). The resulting expressions are as follows:

$$\tilde{\tau}^{(1)}(x) = 1.304x$$

$$\tilde{\tau}^{(2)}(x) = 1.000x$$

$$\vdots \qquad \vdots$$

$$\tilde{\tau}^{(8)}(x) = 0.272x \tag{5.122}$$

Step 12: Plot the solution and estimate its accuracy.

Figure 5.26 is a plot of $\tilde{U}(x)$ and $\tilde{\tau}(x)$. As before, note especially the interelement continuity in $\tilde{U}(x)$, the interelement discontinuities in $\tilde{\tau}(x)$, the exact satisfaction of the essential BC at $x=1$ and the approximate satisfaction of the natural BC at $x=2$.

At $x=2$ the calculated value $\tilde{\tau}(2)=0.544$ is in error by 8.8% from the exact value. At the interelement points $x_2, x_3,\ldots, x_8$, the spread in values of the discontinuities, relative to the mean, yields errors of $\pm 13.2, \pm 12.6, \pm 10.9, \pm 11.4, \pm 9.7, \pm 9.7,$ and $\pm 10.5\%$, respectively.

Figure 5.27 shows both the four- and eight-element solutions from Figs. 5.21 and 5.26, labeled as $\tilde{U}_3(x), \tilde{\tau}_3(x)$ and $\tilde{U}_4(x), \tilde{\tau}_4(x)$, respectively, to indicate the third and fourth approximate solutions we have obtained. The pointwise error between $\tilde{U}_3(x)$ and $\tilde{U}_4(x)$ is less than 1.1% over the entire domain; that is,

$$\left| \frac{\tilde{U}_4(x) - \tilde{U}_3(x)}{\tilde{U}_4(x)} \right| < 0.011 \qquad \text{for } 1 < x < 2$$

$$\tag{5.123}$$

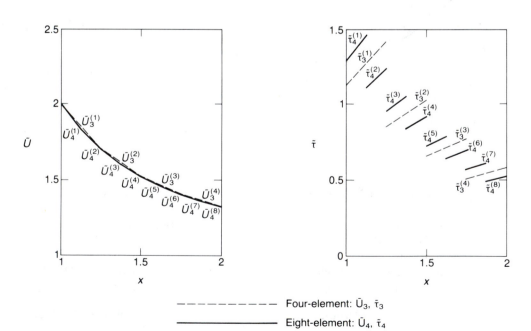

Four-element: $\tilde{U}_3, \tilde{\tau}_3$

Eight-element: $\tilde{U}_4, \tilde{\tau}_4$

FIGURE 5.27 ● Four- and eight-element solutions.

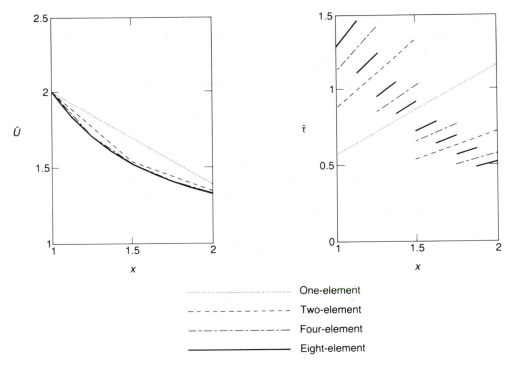

$$Ũ$$

$$\tilde{\tau}$$

$$x$$

$$x$$

.... One-element

— — — — — — — Two-element

—·—·—·—·—·— Four-element

———————— Eight-element

FIGURE 5.28 ● A comparison of the one-, two-, four-, and eight-element solutions.

The pointwise error between $\tilde{\tau}_3(x)$ and $\tilde{\tau}_4(x)$ is less than 14.4% over the entire domain; that is,

$$\left| \frac{\tilde{\tau}_4(x) - \tilde{\tau}_3(x)}{\tilde{\tau}_4(x)} \right| < 0.144 \qquad \text{for } 1 < x < 2$$

$$(5.124)$$

5.6 Mesh Refinement and Conditions for Convergence

Figure 5.28 shows the one-, two-, four-, and eight-element solutions together. There is clearly a converging trend. It also appears that $Ũ$ is converging pointwise "faster" than is $\tilde{\tau}$; i.e., as the elements are reduced in size, there is a greater improvement in accuracy for $Ũ$ than for $\tilde{\tau}$. This is a general property, related to the fact that the derivative of $Ũ$ (which the flux is proportional to) is a lower-degree polynomial.[26]

The one-, two-, four-, and eight-element analyses have therefore yielded an apparently converging sequence of approximate solutions. As discussed in Section 3.6.2, in order

---

[26] Chapters 7, 8, and 9 will discuss flux convergence in more detail; it will be seen that there usually exist special locations within each element (called superconvergent points) where the rate of convergence of the flux is faster than anywhere else in the element.

to ensure that a sequence is converging to the exact solution, rather than to some other limiting function, the trial functions should be made to satisfy appropriate completeness and continuity conditions. This, indeed, was done in the preceding sections, although the actual conditions were not explained there.

We will now present the completeness and continuity conditions that are applicable to a broad class of problems, namely, those characterized by governing equations in which the differential operator is linear, self-adjoint (symmetric), and positive-bounded-below, which includes all those discussed in this book as well as many more [5.2-5.11]. It is assumed that the governing differential equation is of order $2m$, where $m=1,2,\ldots$, and that the residual equations have been integrated by parts m times. (Recall Section 4.5, especially Tables 4.1 and 4.2.) Therefore m is the highest-order derivative of the element trial solution $\tilde{U}^{(e)}$ [and trial functions $\phi_i^{(e)}$] that appears in the integrals in the element equations at the end of step 2, and it is these integrals that are referred to in the following conditions.[27] The completeness condition pertains to the interior of an element; the continuity condition pertains to the boundaries between elements.

Completeness condition. The element trial solution $\tilde{U}^{(e)}$ and any of its derivatives up to order m appearing in the integrals should be able to assume any constant value within an element when, in the limit, the size of the element decreases to zero.

Continuity condition. At interelement boundaries the element trial solutions should be C^{m-1}-continuous; that is, $\tilde{U}^{(e)}$ and its derivatives up to order $m-1$ should be continuous.

If the element trial functions $\phi_i^{(e)}$ are constructed so that $\tilde{U}^{(e)}$ satisfies both conditions, then a sequence of approximate solutions, corresponding to a sequence of successively refined meshes, will converge energywise to the exact solution (assuming no other errors) in the limit as $h_{max}\to0$. Here h represents the size of an element (i.e., the length of a 1-D element or a representative dimension of a higher-dimensional element). Hence h_{max} is the size of the largest element, and the stipulation $h_{max}\to0$ states that all elements must shrink to zero for convergence to the exact solution (see further comments below). This type of convergence is referred to as *h-convergence* since it is achieved by letting the *size of the elements become progressively smaller.*

The completeness condition is necessary for convergence to the exact solution. An incomplete element will generally converge to some other limiting function which will be as close to the exact solution as the trial solution will permit. Hence it is possible in some circumstances for an incomplete element to still yield an acceptable error [5.12]. However, it is generally difficult to predict the error performance of an incomplete element so the practical guideline should be to satisfy the completeness condition (which is not difficult to satisfy).

The continuity condition is not necessary for convergence. The purpose of the condition is to ensure that discontinuities at the interelement boundaries are not severe enough

---

[27] If a Ritz variational method is used instead of the Galerkin method, one would begin with step 3 (steps 1 and 2 being unnecessary), and m would again be the highest-order derivative of $\tilde{U}^{(e)}$ appearing in the variational integrals. The completeness and continuity conditions would then refer to the variational integrals.

to introduce errors in addition to the discretization error. The condition can sometimes be relaxed, though, if the errors that are introduced decrease to zero fast enough as the mesh is refined, in which case convergence to the exact solution would still occur as $h_{max} \to 0$.

If the continuity condition is satisfied, along with the completeness condition, then convergence is ensured; i.e., the two conditions together are sufficient for convergence. An element that satisfies both conditions is called a *conforming*, or *compatible*, element. These names emphasize the satisfaction of the continuity condition. An element that satisfies the completeness condition but not the continuity condition, yet still exhibits convergence, is called a *nonconforming*, or *incompatible*, element (see Section 8.6). Such elements often exhibit superior performance relative to conforming elements.

Since polynomials are used almost universally for trial solutions, let's reexamine the completeness condition in the context of polynomial trial solutions. First we need a definition: A polynomial is *complete to degree p* if all the powers from 0 (the constant term) to p inclusive are present.[28] If a particular power is omitted, any higher powers that are present do not contribute to completeness. For example, the polynomials $a_1 + a_2x + a_3x^2$ and $a_1 + a_2x + a_3x^2 + a_4x^4$ are both complete to degree 2. The second polynomial has a higher power than 2 but because the cubic term is missing, it is only complete to degree 2; if the cubic term were present, then it would be complete to degree 4.

With the preceding definition, we have the following alternative completeness condition:

> *Completeness condition using polynomials.* If a polynomial is used for the element trial solution $\tilde{U}^{(e)}$, the polynomial should be complete at least to degree m;[29] that is, $p \geq m$.

As in the more general completeness condition stated previously, m is the highest-order derivative of $\tilde{U}^{(e)}$ that appears in the element equations at the end of step 2.

A polynomial complete to degree m is continuous and has continuous derivatives up to order m, each controlled by a different parameter as the element shrinks to zero. Therefore $\tilde{U}^{(e)}$ and each of its derivatives up to order m can clearly assume any constant value as the element shrinks to zero, as required by the more general completeness condition. Hence this alternative completeness condition and the continuity condition are together sufficient to ensure convergence.

We can now return to our sequence of one-, two-, four-, and eight-element solutions shown in Fig. 5.28 and verify that the element trial solutions do indeed satisfy the conditions for convergence. Thus the element is clearly a conforming element because (1) the element trial solution $\tilde{U}^{(e)}$ is a complete linear polynomial and hence is complete to degree

---

[28] When the word "complete" is used here to describe a polynomial, it means simply that *all* terms up to some degree are present. It should not be confused with the broader concept of completeness used in the convergence theorems. Despite this distinction, both uses of the word are closely related. In particular, the completeness of an infinite power series is an important property when dealing with *p-convergence*, which is the classical type of convergence, appropriate to Chapters 3 and 4, and discussed elsewhere in the text.

[29] The completeness condition applies to the element trial solution $\tilde{U}^{(e)}$, not to each of its trial functions separately. Therefore some of the trial functions may be (and often are) incomplete polynomials.

m ($m=1$ for this problem), and (2) $\tilde{U}^{(e)}$ is C^0-continuous ($m-1=0$) at the interelement boundaries. In each successive mesh, h_{max} was made smaller (by progressively halving). We can therefore conclude that this sequence is converging energywise to the exact solution (assuming no other errors). And recalling the discussion in Section 3.6.2, experience indicates that there is a high likelihood that the sequence is converging pointwise too.

We could continue this process of *mesh refinement* as long as we like, partitioning the domain into more and more elements of progressively smaller size until our error estimates are as small as desired. In practice, one may refine the mesh in almost any manner one wishes. There is no need to follow such an orderly procedure of doubling the number of elements and halving their size, as was done in these examples.[30] Thus we could have added one element at a time, or jumped from four to a hundred elements. Or we could have held some elements constant while refining the others, a process called *local mesh refinement*. Since the complexity of a solution usually varies considerably from one area of the domain to another, local refinement is an efficient way to provide more DOF only where the solution is more complicated or where more local accuracy is desired. It is a technique used routinely in practice and will be illustrated in subsequent chapters.

One must be careful, however, not to locally refine too much. The reason for this is contained in the above convergence theorem, which states that convergence to the exact solution will occur as $h_{max} \to 0$, not merely as $h \to 0$. The "max" subscript is important because in order to converge to the *exact* solution, *every* element must become arbitrarily small. (Clearly, if the largest element becomes arbitrarily small, then so must every other element.) Even though we never seek the exact solution (only the theoretical proofs do), this nevertheless places a restraint on excessive local refinement. For example, if we should let a few elements remain fixed in size while the others become progressively smaller, a finite error will remain in those few elements which will put a limit on how small an error can be achieved in the others, no matter how small they become. Thus, even though successive solutions will exhibit convergence, they will be converging to the wrong limit (recall Fig. 3.13). This effect will be demonstrated in Chapter 9.

In this chapter we have discussed only *h-refinement*, which is the process of improving accuracy by increasing the number of elements (while continuing to use the same type of element). We will see in subsequent chapters that we can also refine by keeping the elements the same size but increasing the degree p of the polynomial trial solution in each element, a procedure appropriately referred to as *p-refinement*. Convergence to the exact solution will then occur as $p_{min} \to \infty$; that is, the degree of the lowest-degree element trial solution becomes arbitrarily large. This type of convergence is correspondingly referred to as *p-convergence*.[31] This was the approach used in Chapters 3 and 4 because those chapters did not employ elements (or, what amounts to the same thing, the entire domain

---

[30] Such an orderly refinement can affect the *manner* in which convergence occurs. Thus, if a mesh is refined by adding more DOF in such a way that all possible solutions contained in the previous trial solution are still contained in the new trial solution (e.g., by halving elements), then the energy error will converge *monotonely*; otherwise, convergence will still occur, but not necessarily monotonely. In addition, the *rate* of convergence is affected by the procedure used to add DOF (see Section 9.3).

[31] The names *p*-refinement and *p*-convergence (and, similarly, *h*-refinement and *h*-convergence) refer to different aspects of the same process. The distinction is that refinement describes what the analyst does in modeling the problem, whereas convergence describes what happens to the solution as a result of the refinement. Loosely speaking, refinement pertains to input and convergence to output.

was one element); hence the only way to achieve convergence was to progressively increase the degree of the single polynomial trial solution.

This brief discussion of mesh-refinement techniques is intended to be only an introduction to some important practical ideas that will be discussed in more detail throughout Parts III and IV and demonstrated in all of the illustrative problems.

5.7 Random Node and Element Numbering: A General Assembly Rule

Node and element numbers are arbitrary labels to distinguish one node (or element) from another. Any numbers may be used (within a finite range permitted by the particular computer being used) and in any pattern desired. The choice will not affect the resulting approximate solution, but it may significantly affect the efficiency of the computations performed · in arriving at the solution. The difference between a well-numbered mesh and a poorly numbered mesh could easily be one or two orders of magnitude in computational costs. For 1-D meshes the optimal rule is simple: Number sequentially from one end of the domain to the other, as will be explained below. For 2-D and 3-D meshes optimal numbering strategies are more involved, but there are a few simple guidelines (see Section 13.5). For now we only want to learn how to deal with randomly numbered meshes and, in the process, we will discover a simple and general rule for assembly that is ideally suited for computer implementation.

Let's consider the two-element mesh shown in Fig. 5.29. The trial solution and DOF are also shown.

Following our previous pattern of writing and assembling the element equations for each element in turn, going from left to right across the domain, the assembled system equations for this mesh would have the following form,

$$\begin{bmatrix} K^{(2)}_{22} & K^{(2)}_{23} & \\ K^{(2)}_{32} & K^{(2)}_{33} + K^{(1)}_{33} & K^{(1)}_{31} \\ & K^{(1)}_{13} & K^{(1)}_{11} \end{bmatrix} \begin{Bmatrix} a_2 \\ a_3 \\ a_1 \end{Bmatrix} = \begin{Bmatrix} F^{(2)}_2 \\ F^{(2)}_3 + F^{(1)}_3 \\ F^{(1)}_1 \end{Bmatrix} \qquad (5.125)$$

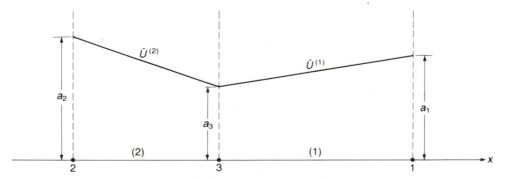

FIGURE 5.29 ● A randomly (nonsequentially) numbered two-element mesh, and trial solution.

where the contributions from each element have been indicated. We could solve these three equations by hand in exactly the same manner we would solve any three linear algebraic equations; the fact that the parameters are labeled in the order 2, 3, 1 instead of 1, 2, 3 would not affect the procedure or the solution.

However, in a computer, matrices and vectors are stored in arrays in which the subscripts (indices) are arranged in sequential order 1,2,3,.... . Therefore, because of programming considerations, it is important that we impart *some* order to the subscripts or superscripts in our matrices and vectors. That is, we want to rearrange the system equations [Eqs. (5.125)] so that either the node numbers or element numbers are in sequential order. In a randomly numbered mesh we can't, in general, rearrange these equations so that both the node and element numbers are in sequence. So we have just two choices: Place either the node numbers or the element numbers in sequential order (in the system equations, not the mesh). The choice will significantly alter the programming of the Gaussian elimination equation-solving algorithm.

The approach that orders the node numbers is called the *bandwidth method*; the approach that orders the element numbers is called the *wave-front method* (or *frontal-solution method*). Both methods are used in commercial FE programs, and each has advantages and disadvantages. Since the programming for the bandwidth method is considerably less complicated, we will choose the bandwidth method as being more suitable for the present textbook. (The choice mainly affects program developers; it has only a minor effect on program users.)

Having chosen the bandwidth method, we now want to rearrange Eqs. (5.125) so that the DOF vector will be in the form,

$$\begin{Bmatrix} a_1 \\ a_2 \\ a_3 \end{Bmatrix}$$

(5.126)

It should be noted that the very simple rearrangement of terms performed in Eqs. (5.127–5.130) is not an operation one would normally go through in practice. The purpose of this rearrangement is to reveal an assembly rule that will enable us to assemble element equations from a randomly numbered mesh directly into the desired form of the system equations — i.e., a form in which the DOF vector is in sequential order, as in Eq. (5.126).

First write out Eqs. (5.125) in algebraic (nonmatrix) form:

$$K_{22}^{(2)}a_2 + \qquad\qquad K_{23}^{(2)}a_3 \qquad\qquad = F_2^{(2)}$$

$$K_{32}^{(2)}a_2 + \left[K_{33}^{(2)} + K_{33}^{(1)} \right] a_3 + K_{31}^{(1)}a_1 = F_3^{(2)} + F_3^{(1)}$$

$$K_{13}^{(1)}a_3 + K_{11}^{(1)}a_1 = F_1^{(1)}$$

(5.127)

In order to make the DOF vector appear as it does in Eq. (5.126), reorder the terms in each equation so that a_1 is first, a_2 second, and a_3 third:

$$K_{22}^{(2)}a_2 + \qquad\qquad K_{23}^{(2)}a_3 = F_2^{(2)}$$

$$K_{31}^{(1)}a_1 + K_{32}^{(2)}a_2 + \left[K_{33}^{(2)} + K_{33}^{(1)} \right] a_3 = F_3^{(2)} + F_3^{(1)}$$

$$K_{11}^{(1)}a_1 \qquad\qquad + \qquad\qquad K_{13}^{(1)}a_3 = F_1^{(1)}$$

(5.128)

Now reorder the equations so that the first equation is the one in which the first subscript on all the K_{ij} terms (the *row* subscript) is 1, the second equation is the one in which the row subscript is 2, etc. This means that in Eqs. (5.128) the third equation should be placed first, the first equation placed second, and the second equation placed third:

$$K_{11}^{(1)}a_1 \qquad\qquad + \qquad\qquad K_{13}^{(1)}a_3 = F_1^{(1)}$$

$$K_{22}^{(2)}a_2 + \qquad\qquad K_{23}^{(2)}a_3 = F_2^{(2)}$$

$$K_{31}^{(1)}a_1 + K_{32}^{(2)}a_2 + \left[K_{33}^{(2)} + K_{33}^{(1)} \right] a_3 = F_3^{(2)} + F_3^{(1)} \qquad\qquad \textbf{(5.129)}$$

Finally, rewriting in matrix form again,

$$
\begin{bmatrix}
K_{11}^{(1)} & 0 & K_{13}^{(1)} \\
0 & K_{22}^{(2)} & K_{23}^{(2)} \\
K_{31}^{(1)} & K_{32}^{(2)} & K_{33}^{(2)} + K_{33}^{(1)}
\end{bmatrix}
\begin{Bmatrix}
a_1 \\
a_2 \\
a_3
\end{Bmatrix}
=
\begin{Bmatrix}
F_1^{(1)} \\
F_2^{(2)} \\
F_3^{(2)} + F_3^{(1)}
\end{Bmatrix}
\qquad\qquad \textbf{(5.130)}
$$

which is the desired arrangement, since the DOF vector is now arranged sequentially.

The most conspicuous feature of Eqs. (5.130) is that each K_{ij} term is in the row and column corresponding to its subscripts [which was not true in Eqs. (5.125)]; similarly, each F_i term is in the row corresponding to its subscript. This is certainly not surprising, since the above operations were defined to do just that. However, the point of all this is to show that we could have assembled the two sets of element equations directly into the form of Eqs. (5.130) simply by *adding each K_{ij} term to the ith row and jth column and each F_i term to the ith row.*

More generally, if the (e)th element in a mesh has node numbers s and r, as shown in Fig. 5.30, then the element equations would have the form,

$$
\begin{bmatrix}
K_{ss}^{(e)} & K_{sr}^{(e)} \\
K_{rs}^{(e)} & K_{rr}^{(e)}
\end{bmatrix}
\begin{Bmatrix}
a_s \\
a_r
\end{Bmatrix}
=
\begin{Bmatrix}
F_s^{(e)} \\
F_r^{(e)}
\end{Bmatrix}
\qquad\qquad \textbf{(5.131)}
$$

and these equations would be assembled by adding the terms to the sth and rth rows and columns of the system matrix equations:

Column s Column r

$$\qquad\qquad\qquad \textbf{(5.132)}$$

The general rule for assembly is therefore as follows:

Add the element stiffness term $K_{ij}^{(e)}$ to the ith row and jth column of the system stiffness matrix, and add the element load term $F_i^{(e)}$ to the ith row of the system load vector.

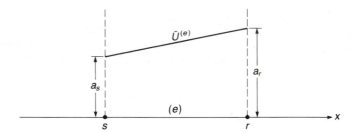

FIGURE 5.30 ● A typical element in a randomly numbered mesh.

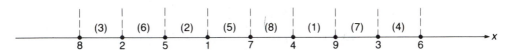

FIGURE 5.31 ● A randomly numbered eight-element mesh.

TABLE 5.3 ● Connectivity table for mesh, Fig. 5.31.

Element	Local node 1 becomes node	Local node 2 becomes node
(1)	4	9
(2)	5	1
(3)	8	2
(4)	3	6
(5)	1	7
(6)	2	5
(7)	9	3
(8)	7	4

Chapter 7 will show how this rule is coded into the UNAFEM program.

Let's illustrate the rule by reconsidering the eight-element problem analyzed in the previous section. We will renumber the mesh as shown in Fig. 5.31; the corresponding connectivity table is shown in Table 5.3.

Applying the above assembly rule, the assembled system equations would then have the form given in Eqs. (5.133).

The reader should carefully study Eqs. (5.133) to be satisfied that the location of each term, using the connectivity table (Table 5.3), follows the assembly rule given above. For example, according to the connectivity table, the element equations for element (3) have the form,

$$\begin{bmatrix} K_{88}^{(3)} & K_{82}^{(3)} \\ K_{28}^{(3)} & K_{22}^{(3)} \end{bmatrix} \begin{Bmatrix} a_8 \\ a_2 \end{Bmatrix} = \begin{Bmatrix} F_8^{(3)} \\ F_2^{(3)} \end{Bmatrix}$$

(5.134)

$$
\begin{bmatrix}
K_{11}^{(2)}+K_{11}^{(5)} & 0 & 0 & 0 & K_{15}^{(2)} & 0 & K_{17}^{(5)} & 0 & 0 \\
0 & K_{22}^{(3)}+K_{22}^{(6)} & 0 & 0 & K_{25}^{(6)} & 0 & 0 & K_{28}^{(3)} & 0 \\
0 & 0 & K_{33}^{(4)}+K_{33}^{(7)} & 0 & 0 & K_{36}^{(4)} & 0 & 0 & K_{39}^{(7)} \\
0 & 0 & 0 & K_{44}^{(1)}+K_{44}^{(8)} & 0 & 0 & K_{47}^{(8)} & 0 & K_{49}^{(1)} \\
K_{51}^{(2)} & K_{52}^{(6)} & 0 & 0 & K_{55}^{(2)}+K_{55}^{(6)} & 0 & 0 & 0 & 0 \\
0 & 0 & K_{63}^{(4)} & 0 & 0 & K_{66}^{(4)} & 0 & 0 & 0 \\
K_{71}^{(5)} & 0 & 0 & K_{74}^{(8)} & 0 & 0 & K_{77}^{(5)}+K_{77}^{(8)} & 0 & 0 \\
0 & K_{82}^{(3)} & 0 & 0 & 0 & 0 & 0 & K_{88}^{(3)} & 0 \\
0 & 0 & K_{93}^{(7)} & K_{94}^{(1)} & 0 & 0 & 0 & 0 & K_{99}^{(1)}+K_{99}^{(7)}
\end{bmatrix}
\begin{Bmatrix}
a_1 \\ a_2 \\ a_3 \\ a_4 \\ a_5 \\ a_6 \\ a_7 \\ a_8 \\ a_9
\end{Bmatrix}
=
\begin{Bmatrix}
F_1^{(2)}+F_1^{(5)} \\
F_2^{(3)}+F_2^{(6)} \\
F_3^{(4)}+F_3^{(7)} \\
F_4^{(1)}+F_4^{(8)} \\
F_5^{(2)}+F_5^{(6)} \\
F_6^{(4)} \\
F_7^{(5)}+F_7^{(8)} \\
F_8^{(3)} \\
F_9^{(1)}+F_9^{(7)}
\end{Bmatrix}
\qquad (5.133)
$$

The assembly rule therefore places these terms in rows and columns 2 and 8 in Eqs. (5.133).

If numerical values are substituted for the terms in Eqs. (5.133), using the same nodal coordinates as for the eight-element problem in Section 5.5 [i.e., using Eqs. (5.99), with the subscripts appropriately relabeled], then Eqs. (5.133) would appear as given in Eqs. (5.135). A comparison of the system equations [Eqs. (5.135) and (5.118)] reveals that all the same numbers are present; the only difference is a reordering of the equations and the parameters a_i. Thus each equation in Eqs. (5.135) is identical to a corresponding equation in Eqs. (5.118).

Numbering a mesh in a random pattern clearly does not affect the sparsity of the stiffness matrix (the number of zero terms); it only rearranges the pattern of zero and nonzero

$$
\begin{bmatrix}
22 & 0 & 0 & 0 & -\dfrac{21}{2} & 0 & -\dfrac{23}{2} & 0 & 0 \\[2mm]
0 & 18 & 0 & 0 & -\dfrac{19}{2} & 0 & 0 & -\dfrac{17}{2} & 0 \\[2mm]
0 & 0 & 30 & 0 & 0 & -\dfrac{31}{2} & 0 & 0 & -\dfrac{29}{2} \\[2mm]
0 & 0 & 0 & 26 & 0 & 0 & -\dfrac{25}{2} & 0 & -\dfrac{27}{2} \\[2mm]
-\dfrac{21}{2} & -\dfrac{19}{2} & 0 & 0 & 20 & 0 & 0 & 0 & 0 \\[2mm]
0 & 0 & -\dfrac{31}{2} & 0 & 0 & \dfrac{31}{2} & 0 & 0 & 0 \\[2mm]
-\dfrac{23}{2} & 0 & 0 & -\dfrac{25}{2} & 0 & 0 & 24 & 0 & 0 \\[2mm]
0 & -\dfrac{17}{2} & 0 & 0 & 0 & 0 & 0 & \dfrac{17}{2} & 0 \\[2mm]
0 & 0 & -\dfrac{29}{2} & -\dfrac{27}{2} & 0 & 0 & 0 & 0 & 28 \\
\end{bmatrix}
\begin{Bmatrix}
a_1 \\ a_2 \\ a_3 \\ a_4 \\ a_5 \\ a_6 \\ a_7 \\ a_8 \\ a_9
\end{Bmatrix}
$$

$$
= \begin{Bmatrix}
16 \ln \dfrac{120}{121} \\[3mm]
16 \ln \dfrac{80}{81} \\[3mm]
16 \ln \dfrac{224}{225} \\[3mm]
16 \ln \dfrac{168}{169} \\[3mm]
16 \ln \dfrac{99}{100} \\[3mm]
1 - 16 \ln \dfrac{16}{15} \\[3mm]
16 \ln \dfrac{143}{144} \\[3mm]
-2 + 16 \ln \dfrac{9}{8} \\[3mm]
16 \ln \dfrac{195}{196}
\end{Bmatrix}
+ \begin{Bmatrix}
0 \\[3mm]
0 \\[3mm]
0 \\[3mm]
0 \\[3mm]
0 \\[3mm]
-\dfrac{1}{2} \\[3mm]
0 \\[3mm]
\left(-x\,\dfrac{d\tilde{U}^{(3)}}{dx}\right)_{x=1} \\[3mm]
0
\end{Bmatrix}
\tag{5.135}
$$

terms. However, by rearranging the pattern, it does affect the bandwidth. Thus the half-bandwidth in Eqs. (5.135) is 7, in contrast to only 2 for the sequentially numbered mesh.

This can significantly affect the efficiency of a bandwidth-type equation-solving program. It will be seen in Chapter 7 that the bandwidth-type equation-solving program in UNAFEM works on all the terms inside the band and ignores the zeroes outside the band. If there are also a lot of zeroes inside the band because of a randomly ("badly") numbered mesh, then there is a lot of wasted computation due to adding zeroes and multiplying by zeroes.[32] Therefore, in order to lower the cost of computation, it is advisable (when using a bandwidth-type equation solver) to number the nodes in a 1-D mesh sequentially from one end of the domain to the other. Chapters 13 through 16 will illustrate effective ways to number 2-D meshes (see especially Section 13.6.5).

A final comment regarding assembly: Whenever in doubt about the proper rule for assembly, such as when using unusual types of element trial functions (see note below), one can always return to fundamentals and *derive* the assembly rule, as was done in this chapter. Thus one would first write the trial solutions for two completely separated elements — i.e., solutions having no a_i parameters in common (recall Section 5.3). Then apply the essential IBCs (continuity of U, and any other conditions that may be applicable) at the interelement boundary to generate constraint equations. Finally, apply the constraint equations using the row/column operations [see Eq. (5.60) or Section 4.4, step 9]. From this two-element example a general pattern can then be discerned.[33]

5.8 Element Trial Functions: Terminology and Concepts

It is generally recognized that the essence of the FEM lies in the special nature of the trial functions it uses; in particular, the trial functions are *local* rather than *global*:

A global trial function is nonzero (excepting isolated points) over the *entire* (global) domain of the problem.

and

A local trial function is nonzero (excepting isolated points) in only a *small part* of the domain, being *identically zero everywhere else*.

Element trial functions are local because they are nonzero (excepting isolated points) in only a single element; they may be considered to be identically zero in all the other elements. As observed before, it is all the "zeroness" inherent in local trial functions that produces so many zero terms in the stiffness matrix, greatly reducing the amount of computation.

---

[32] A variant of the bandwidth method, called the *skyline method* [5.13], deals more efficiently with zeroes inside the band.

[33] A traditional and popular approach to explaining assembly is to simply state the rule following Eq. (5.132) and then appeal to physical analogies with electrical networks or simple structural systems of bars. Although such heuristic and analogic arguments can sometimes provide additional insight, they provide no objective means to derive assembly rules under all circumstances. Thus the assembly rule following Eq. (5.132) is only applicable if (as is usually the case) the element trial functions are interpolatory with some of the nodes on the boundaries (recall comments in Section 5.2). If these conditions are not met, then the technique described above can be used to derive the proper assembly rule.

However, even though we initially define trial functions over single elements, we have seen that the process of assembly has the effect of joining together the trial functions in adjacent elements (recall Figs. 5.15 and 5.24), making the combined trial functions "less local" than before. It was shown in Section 5.3.3 that assembly is the enforcement of continuity between element trial solutions, and consists of application of a constraint equation derived from the element trial solutions. What is to prevent such constraint equations from progressively linking together trial functions from all the elements into global trial functions (leading to a full stiffness matrix and thereby destroying the practical advantage of the FEM)?

The answer is the special way we construct the element trial functions. Thus it was explained in Section 5.2 that (1) the element trial functions must be interpolatory, and (2) some of the interpolation points (nodes) must be located on the element boundaries. It is these two properties that preserve the localness of the assembled trial functions and hence all the practical advantages of the FEM.

The principal practical advantages of the FEM (i.e., of local rather than global trial functions) are

1. The stiffness matrix is sparse (usually very sparse), making it possible to solve problems involving even hundreds of thousands of equations.

2. The expressions for the remaining nonzero terms in the system equations may be derived from only a single typical element, which can then be used repeatedly for all the other elements.

3. The equations for each element are easily assembled by merely adding each term to the appropriate location in the system equations.

4. Boundary conditions, no matter how complicated, are easily applied.

There are also conceptual advantages, some of which were mentioned in Section 1.6. On the other hand, there are occasional circumstances where the combining of global and local trial functions can be very effective [5.14].

We have seen in this chapter that the basic building block of an FE analysis is the element. That is, the total trial solution is a sum (union) of all the element trial solutions. And each element trial solution is itself a linear sum of element trial functions, the latter designated by the notation $\phi_i^{(e)}(x)$. Therefore, the most fundamental component of an FE trial solution, as well as the resulting approximate solution, is the element trial function.[34]

The $\phi_i^{(e)}(x)$ functions play a central role in the theoretical development (steps 1 through 6) for the typical element. The resulting element equations are then used to generate the equations for all the other elements. Following the solution of the system equations, the $\phi_i^{(e)}(x)$ are needed once again to calculate flux and any other physical properties in each element. In short, element trial functions tend to pervade almost all aspects of an FE analysis. Perhaps because of this they have acquired their own special name:

A trial function defined over one element — i.e., an element trial function — is also called a *shape function*.

---

[34] One could think of elements as the "molecules" of the FE model and the element trial functions as the "atoms."

Shape functions and element trial functions are therefore synonymous terms.[35] In the remainder of the book we will use the more conventional terminology, "shape function," but once in a while we'll revert to "element trial function" because it is a bit more descriptive.

Shape functions completely determine the quality of approximation within each element. From a practical standpoint, it would be advisable for an FE user to be aware of and to understand the shape functions that are used in the particular type of element he or she is working with, in order to know what kind of performance (such as accuracy) to expect from the element. Subsequent chapters will have plenty to say about shape functions. In particular, the influence of different types of shape functions on the quality of solutions will be amply demonstrated.

5.9 Summary of the 12-Step Procedure Using the Element Concept

Figure 5.32 summarizes the 12 steps that were used to solve the two-, four- and eight-element problems. They should be compared with the steps outlined in Fig. 4.1 to see how they were modified to accommodate the element concept. It will be seen that most of the theoretical and computational details are virtually identical, with or without the element concept. In one case the details pertain to the entire domain; in the other, to only a single element.

The only significant differences pertain to the problem of connecting the separate element trial solutions. In step 4 the shape functions are required to be interpolatory with (some) nodes placed on the element boundary. In step 8 the essential interelement BCs (continuity of the function between elements) are imposed by the operation of assembly. In step 9 the natural interelement BCs (continuity of the flux between elements) are imposed, in addition to and in the same manner as the natural BCs. Thus, where the flux (for the exact solution) should be continuous, the natural interelement BCs are zero, so nothing needs to be added to the (initially zeroed) load vector; where the flux should be discontinuous, the natural interelement BCs represent a concentrated flux, so a nonzero value must be added to the load vector.

As was noted previously, in the theoretical development the bulk of the calculus and algebra manipulations occur in Steps 4 through 6. Steps 1 through 3 are merely short, formal operations. The general form of the element trial solution in step 3 is simply the formal representation of the unknown function as a finite sum of shape functions [see Eq. (4.1)]; the $\phi_0(x)$ term is generally excluded, but could be included for special situations (e.g., as a global function). If a Ritz variational method is used instead of the Galerkin method, then the governing equations are in the form of a variational principle and steps 1 and 2 would be unnecessary; one would begin with step 3. If a collocation method is used, there would be no integral to integrate by parts, so step 2 would be omitted.

The 12-step procedure outlined in Fig. 5.32 will be used throughout the remainder of the book.

---

[35] Many authors use the notation $N_i^{(e)}(x)$ rather than $\phi_i^{(e)}(x)$ for shape functions.

Theoretical Development

Step 1: Write the Galerkin residual equations for a typical element.

2: Integrate by parts.

3: Substitute the general form of the element trial solution into interior integrals in residual equations.

The resulting formal expressions
are the element equations.

4: Develop specific expressions for the shape functions (element trial functions).

5: Substitute the shape functions into the element equations, and transform the integrals into a form appropriate for numerical evaluation.

6: Prepare expressions for the flux, using the shape functions.

Numerical Computation

Step 7: Specify numerical data for a particular problem.

8: Evaluate the interior terms in the element equations for each element, and assemble the terms into system equations.

9: Apply the boundary conditions, including the natural interelement boundary conditions, to the system equations.

10: Solve the system equations.

11: Evaluate the flux.

12: Display the solution and estimate its accuracy.

FIGURE 5.32 ● Twelve-step procedure for finite element analysis

---

Exercises

5.1 This exercise applies the element concept to Exercise 4.2. Apply steps 1 through 6 of the 12-step procedure (Fig. 5.32) to develop the element equations for the following general problem:

$$-\frac{d^2U}{dx^2} + \beta U = f \qquad x_a < x < x_b$$

where β and f are constants. Use the linear element in Fig. 5.18. *Hint:* Note the similarities with Exercise 4.2 in steps 1 through 3.

5.2 Using the element equations in Exercise 5.1, apply steps 7 through 12 to obtain an approximate solution to the following specific problem:

$$\frac{d^2U}{dx^2} + U = 0 \qquad 0 < x < \frac{\pi}{2}$$

with the BCs $\tau(0) = -1$ and $\tau(\pi/2) = 0$. (Note that the expression for the flux τ is $-dU/dx$.) Use a mesh of two elements of equal length. Graphically compare your two-element solution with the one-element solution in Exercise 4.3(a), as well as with the exact solution, $U(x) = \sin x$.

5.3   Use the element equations developed in Exercise 5.1 to analyze the hanging-cable problem in Exercise 3.4. Obtain three approximate solutions, using uniform meshes of two, three, and four elements. (A one-element solution is easily available "by inspection," but it is not very interesting. What is it?) Plot the three solutions. Do they appear to be converging? Estimate the accuracy (pointwise error) of the four-element solution. Considering the exact solution (see Exercise 3.4), how good is your estimate?

5.4   Use the element equations developed in Exercise 5.1 to analyze the hanging cable on an elastic foundation in Exercise 3.7. Obtain three approximate solutions, using uniform meshes of two, three, and four elements. (A one-element solution is easily available "by inspection," but it is not very interesting. What is it?) Plot the three solutions. Do they appear to be converging? Estimate the accuracy (pointwise error) of the four-element solution. Considering the exact solution (see Exercise 3.7), how good is your estimate?

5.5   Using the element equations from Exercise 5.1, apply steps 7 through 12 to derive an approximate solution to the following specific problem:

$$\frac{d^2U}{dx^2} + U = 0 \qquad 0 < x < \frac{3\pi}{2}$$

with the BCs $\tau(0) = -1$ and $\tau(3\pi/2) = 0$. This is the same as Exercise 5.2 except for the larger domain. However, in the present case use a mesh of at least two elements, and make the length of element (1) equal to $\sqrt{3}$.

Note that $K_{11} = 0$ and hence the conventional Gaussian elimination procedure *without pivoting*, used universally in FE programs, will fail, since the first few elimination operations would divide by K_{11}. The matrix has *ceased to be positive-definite* (the assumption that obviates pivoting) at a critical value of the element length. In the more general equation [Eq. (2.3)], which is used in Part III of this book and of which the present problem is a special case, this can occur whenever $\alpha(x)$ and $\beta(x)$ are of opposite sign, yielding two stiffness terms that cancel each other. Nevertheless, there is still a unique solution to the system equations, since the original problem is well posed with a unique solution. The remedy is simply to solve the system equations using Gaussian elimination *with* pivoting. This means you must solve for the DOF in some order other than sequentially from 1 to N. Try it, and compare your solution with the exact solution, $U(x) = \sin x$.

5.6   Consider the problem of heat conduction in the spherical nuclear-fission element described in Exercise 3.9.

a)   Apply steps 1 through 6 to develop the element equations. Use the linear element in Fig. 5.18. Evaluate all integrals with closed-form expressions.

b)   Apply steps 7 through 12 to obtain an approximate solution using a two-element mesh. The technique for applying the mixed (convection) BC at $r = R$ is described in Exercise 4.6.

c)   Obtain another solution by refining the mesh to three or four elements. Estimate your accuracy by comparing the two solutions. Considering the exact solution in Exercise 3.9, is your estimate about right?

5.7 Consider the steady-state heat flow problem described in Exercise 3.8:

$$-\frac{d}{dx}\left[k\,\frac{dT(x)}{dx}\right] + \frac{2h}{r}\,T(x) = \frac{2h}{r}\,T_\infty \qquad 0 < x < 10$$

using the same data for the BCs and the constants in the differential equation.

a) Apply steps 1 through 6 of the 12-step procedure (Fig. 5.32) to develop the element equations.

b) Apply steps 7 through 12 to obtain a one-element solution for the temperature and the heat flux.

c) Refine the mesh, obtaining two- and three-element solutions.

d) Plot the three solutions. Does the sequence of approximate solutions appear to be converging? Estimate the pointwise error of the three-element solution.

5.8 Consider the differential equation in Exercise 3.13:

$$\frac{d}{dx}\left[(x+1)\,\frac{dU(x)}{dx}\right] = 0 \qquad 1 < x < 2$$

with the BCs $U(1)=1$ and $\tau(2)=1$.

a) Apply steps 1 through 6 of the 12-step procedure (Fig. 5.32) to develop the element equations.

b) Apply steps 7 through 12 to obtain a one-element solution for the function and the flux.

c) Refine the mesh, obtaining two- and three-element solutions.

d) Plot the three solutions. Does the sequence of approximate solutions appear to be converging? Estimate the pointwise error of the three-element solution.

5.9 Consider the differential equation in Exercise 3.14:

$$\frac{d}{dx}\left[(x+1)\,\frac{dU(x)}{dx}\right] = 0 \qquad 1 < x < 2$$

with the BCs $U(1)=1$ and $U(2)=2$.

a) Apply steps 1 through 6 of the 12-step procedure (Fig. 5.32) to develop the element equations.

b) Apply steps 7 through 12 to obtain a two-element solution for the function and the flux. A one-element analysis would be a valid, well-posed problem; however, it would also be a waste of time because the solution is trivially easy. Why? What is the one-element solution?

c) Refine the mesh, obtaining three- and four-element solutions.

d) Plot the three solutions. (You might also include the one-element solution, which is easily available "by inspection.") Does the sequence of approximate solutions appear to be converging? Estimate the pointwise error of the four-element solution.

5.10 Consider the differential equation in Exercise 3.15:

$$\frac{d}{dx}\left[x^2\,\frac{dU(x)}{dx}\right] = -\frac{1}{12}\,(-30x^4 + 204x^3 - 351x^2 + 110x) \qquad 0 < x < 4$$

with the BCs $U(0)=1$ and $U(4)=0$.

a) Apply steps 1 through 6 of the 12-step procedure (Fig. 5.32) to develop the element equations.

b) Apply steps 7 through 12 to obtain a two-element solution for $U(x)$. (As in Exercise 5.9, the one-element solution is easily available "by inspection.")

c) Refine the mesh, obtaining three- and four-element solutions.

d) Plot all the solutions. Does the sequence of approximate solutions appear to be converging? Estimate the pointwise error of the four-element solution.

5.11 a) Using sine and cosine functions, develop two interpolatory shape functions for a two-node element. They should provide C^0-continuity; i.e., they must satisfy the interpolation property $\phi_i(x_j)=\delta_{ij}$.

b) Using these trigonometric elements and the 12-step procedure, obtain an approximate solution to the hanging-cable problem in Exercise 3.4. Employ a two-element mesh.

5.12 Consider the following nonsequentially numbered mesh of four elements:

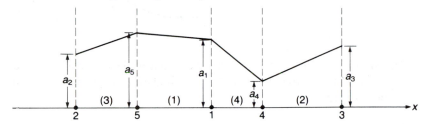

Using K and F notation, assemble the element stiffnesses $K_{ij}^{(e)}$ and loads $F_i^{(e)}$ into the system stiffness matrix and load vector using the appropriate values for the subscripts i,j and superscript (e). What is the half-bandwidth of the system stiffness matrix? Note that it is determined by the element with the maximum difference between its node numbers [see Eq. (13.198)].

Note: For the next four exercises, it may be helpful to study Section 13.6.5. The discussion in that section does not rely on any 2-D concepts, and it is pertinent to problems in all dimensions.

5.13 Consider the following 1-D domain, which is in the form of a circle. On such a domain the independent variable would be the arc length, s, along the circle, measured from some arbitrary origin. The interesting feature of this domain is that there are no ends.

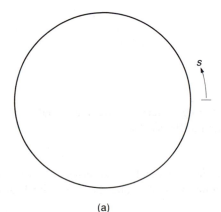

(a)

Consider a mesh of 10 elements (using linear shape functions so there is a node at the end of each element):

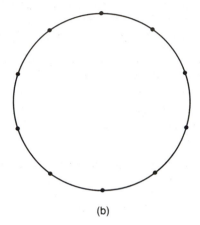

(b)

Devise a numbering pattern for the nodes that achieves a half-bandwidth of only 3 for the system stiffness matrix. Why is numbering sequentially around the circle a poor strategy?

5.14 Consider the star-shaped domain shown below, with the indicated mesh. This could be a 1-D problem if we imagined the rays to be, say, heat-conducting rods, such as the spokes of an umbrella. The 2-D geometry does not necessarily make it a 2-D analysis; the spokes could all be folded together along the same direction (e.g., as in the closed umbrella) without affecting the heat conduction within each spoke.

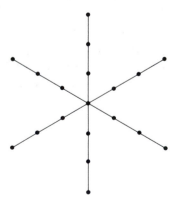

Devise a numbering pattern for the nodes that achieves a half-bandwidth of only 4 for the system stiffness matrix.

5.15 Let's combine the previous two exercises and create a spoked wheel. If you like, you may imagine the spokes and the rim to be heat-conducting rods.

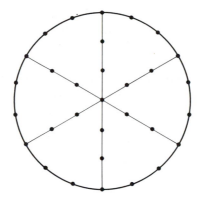

Devise a numbering pattern for the nodes that achieves a half-bandwidth of only 7 for the system stiffness matrix.

5.16 Consider the pattern of interconnected rods shown below. The indicated mesh assigns one element per rod. As in the previous exercises, you can imagine each of the rods to be conducting heat, with a 1-D element modeling the conduction along each rod.

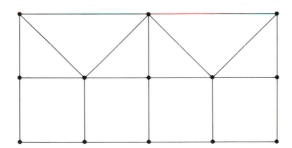

Devise a numbering pattern for the nodes that achieves a half-bandwidth of only 5 for the system stiffness matrix.

References

[5.1]   J. Greenstadt, ''On the Reduction of Continuous Problems to Discrete Form,'' *IBM J Res Div*, **3**: 355-363 (1959).

[5.2]   E. R. de Arantes e Oliveira, ''Theoretical Foundations of the Finite Element Method,'' *Int. J. Solids Structures*, **4**: 929-952 (1968).

[5.3]   S. G. Mikhlin, *Variational Methods in Mathematical Physics*, Pergamon Press, New York, 1964.

[5.4]   G. Strang and G. F. Fix, *An Analysis of the Finite Element Method*, Prentice-Hall, Englewood Cliffs, N. J., 1973.

[5.5]   J. T. Oden and J. N. Reddy, *An Introduction to the Mathematical Theory of Finite Elements*, Wiley, New York, 1976.

[5.6]   R. J. Melosh, "Basis for the Derivation of Matrices for the Direct Stiffness Method," *AIAA*, **1**(7): 1631-1637 (1963).

[5.7]   P. Tong and T. H. H. Pian, "The Convergence of Finite Element Method in Solving Linear Elastic Problems," *Int J Solids Structures*, **3**: 865-879 (1967).

[5.8]   C. Patterson, "Sufficient Conditions for Convergence in the Finite Element Method for Any Solution of Finite Energy," in J. R. Whiteman (ed)., *The Mathematics of Finite Elements and Applications*, Academic Press, New York, 1973.

[5.9]   E. R. de Arantes e Oliveira, "The Patch Test and the General Convergence Criteria of the Finite Element Method," *Int J Solids Structures*, **13**: 159-178 (1977).

[5.10]   O. C. Zienkiewicz, *The Finite Element Method*, McGraw-Hill, New York, 1977.

[5.11]   D. H. Norrie and G. deVries, *An Introduction to Finite Element Analysis*, Academic Press, New York, 1978.

[5.12]   M. D. Olson and T. W. Bearden, "A Simple Flat Triangular Shell Element Revisited," *Int J Num Meth Eng,* **14**: 51-68 (1979).

[5.13]   K. J. Bathe and E. L. Wilson, *Numerical Methods in Finite Element Analysis*, Prentice-Hall, Englewood Cliffs, N. J., 1976.

[5.14]   C. D. Mote, Jr., "Global-Local Finite Element," *Int J Num Meth Eng*, **3**: 565-574 (1971).

One-Dimensional Problems: Theory, Programming, and Applications

6

Physical Applications for
1-D Boundary-Value Problems

Introduction

In Part III we will work with a general 1-D second-order ordinary differential equation with general boundary conditions. This equation describes a variety of different physical phenomena: thermal, mechanical, electrical, magnetic, and fluid flow, to name a few. In the following sections we derive this governing differential equation for four types of physical applications. The equation is then written in a general notation that will be used in the theoretical analysis and computer-program development in Chapters 7 and 8. We conclude with a few comments on the type of boundary condition data that should be specified for a well-posed problem.

6.1 Heat Conduction

6.1.1 1-D Heat Flow

Figure 6.1 shows two situations in which heat flows predominantly in only one direction. In Fig. 6.1(a) we have a long, thin rod of length L and variable cross-sectional area A. The cross-sectional area and shape are permitted to vary "gradually" along the length; too rapid a variation would make the heat flow essentially 2-D or 3-D. The lateral surface is perfectly insulated; therefore heat flowing just inside the surface must flow parallel to the surface. And since the rod is long relative to its width, then the heat flowing across the entire cross section will flow predominantly parallel to the axis of the rod. Only near the ends of the rod might the flow begin to deviate appreciably from 1-D. Of course a cylindrical rod with constant area (hence parallel sides) and uniform loading conditions on the end faces would yield exactly 1-D heat flow.

In Fig. 6.1(b) we have a thin wall or panel in which the thickness w of the panel is much less than the lateral dimensions — geometrically the converse of the case in Fig. 6.1(a). Throughout the central region of the panel, heat flows essentially perpendicular to the faces; near the edges of the panel the flow might deviate appreciably from 1-D. For purposes of analysis, the dashed lines depict an imaginary disc of cross-sectional area A and thickness w, cut from the central region, its axis perpendicular to the panel faces. The heat flow inside this imaginary disc would be the same as the flow inside an insulated cylindrical rod with constant area.

Both cases represent mathematical idealizations that approximate to one degree or another events in the real world. After all, insulation is not perfect and dimensions are finite (introducing end or edge effects). It is the job of the engineer to decide when such 1-D models are sufficiently accurate in real situations.

The governing differential equation for both cases is derived from the following two principles.

Balance (conservation) of energy •

Rate of heat energy   rate of heat energy
added to   = lost from
differential element   differential element

From Fig. 6.1,

$$qA + QA\,dx = (q+dq)(A+dA) \tag{6.1}$$

or, neglecting second-order terms,

$$\frac{d}{dx}\Big(q(x)A(x)\Big) = Q(x)A(x) \tag{6.2}$$

where

$q(x) = $ heat flux,[1] E/tL^2

q is the heat energy flowing per second in the x-direction, per cross-sectional area.

$Q(x) = $ interior (volume) heat source, E/tL^3

Q is the heat energy produced in, or injected into, the interior of the differential element per second, per volume of material. It comes from some external source and hence is a known quantity. ["External" means that the source itself, i.e., the heat-producing mechanism, is excluded from the analytical model. Only the effect of the source, $Q(x)$, is included, and this effect is located interior to the differential element.] If $Q<0$, it is a heat sink, i.e., heat is removed from the system.

---

[1] The symbols E/tL^2 represent the dimensions of heat flux, namely, thermal energy per time per length squared. These and other symbols are defined in Table 7 in the appendix. In particular applications, such as the illustrative problems in subsequent chapters, one would substitute for these symbols any consistent system of physical units, for example, foot-pound-second (British Engineering), centimeter-gram-second (cgs), or meter-kilogram-second (mks) units. The latter are the base of the SI units, which are listed in Table 7.

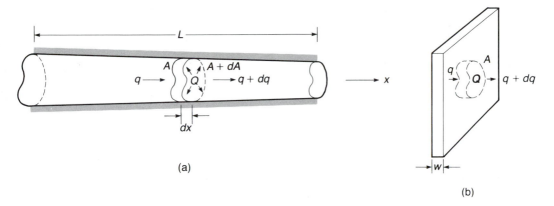

FIGURE 6.1 ● Two models for 1-D heat flow: (a) a heat-conducting rod, with its lateral surface insulated; (b) a wall or panel.

Constitutive equation (**or physical law**) ● Fourier's law of heat conduction states that

$$q(x) = -k(x)\frac{dT(x)}{dx}$$

(6.3)

where

$T(x)$ = temperature, T

$k(x)$ = thermal conductivity, E/tLT

k is an empirically measured material property that indicates the ease with which heat flows in the material.[2] From Eq. (6.3), it is the heat flux per temperature gradient.

Fourier's law is often described by the statement that "heat flows downhill" on a plot of temperature versus distance, as shown in Fig. 6.2. Thus a negative temperature gradient, $dT/dx < 0$, produces a positive heat flux, $q > 0$, and vice versa.

The governing differential equation is obtained by substituting Eq. (6.3) into Eq. (6.2):

$$-\frac{d}{dx}\left[k(x)A(x)\frac{dT(x)}{dx}\right] = Q(x)A(x)$$

(6.4a)

---

[2] Some materials are "directional," meaning that the value of k varies in different directions; i.e., heat flows more easily in one direction than another. Such a material is said to be *anisotropic*. If k is the same in all directions, then the material is *isotropic*. The concept of anisotropy will be explained in Section 13.1.2, where it is more meaningful in the context of 2-D problems. In order for a problem to be modeled as 1-D (rather than two or more dimensions), we must assume that we are dealing either with isotropic materials or else with anisotropic materials in which a principal material axis (see Section 13.1.2) is parallel to the x-axis. Since anisotropy can be an attribute of many other physical properties besides thermal conductivity, this same assumption will also apply to all the material properties used in the 1-D problems in this book.

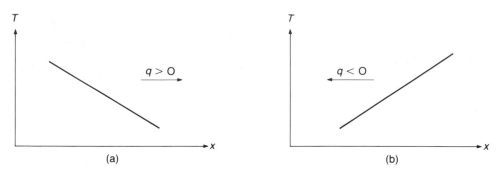

FIGURE 6.2 ● Illustration of Fourier's law [Eq. (6.3)]: (a) $dT/dx < 0$; (b) $dT/dx > 0$.

or, in the usual case of a cylindrical rod with constant area,

$$-\frac{d}{dx}\left[k(x)\,\frac{dT(x)}{dx}\right] = Q(x)$$

(6.4b)

Here $T(x)$ is the unknown function, $k(x)$ is a material (or physical) property, and $Q(x)$ is the interior load term. Since Eqs. (6.4a and b) are second-order differential equations, we must specify one boundary condition at each of the two boundary points (recall Section 4.5). In general this will be either an essential or a natural BC[3]:

Essential BC: Specify temperature T
Natural BC: Specify heat flux q

6.1.2 1-D Model of 2-D Heat Flow

Figure 6.3 shows a long, thin rod, the same as seen in Fig. 6.1(a), but here the lateral surface is not insulated. It may lose heat to the surrounding medium. If the medium is a gas, then the principal mechanism of heat loss will be convection, as suggested in the figure.

The flow is no longer 1-D since it has components both parallel and perpendicular to the axis of the rod. Although the temperature would vary over the cross section, we might be interested only in the average temperature at each cross section, i.e., a single temperature at each "station" along the rod. The heat loss from the surface could also be treated as an average loss over the cross section and hence represented as an average interior heat loss, i.e., a single value at each station along the rod. The equations in Section 6.1.1 would be modified as follows.

Balance of energy ● From Fig. 6.3,

$$qA + QA\,dx = (q+dq)(A+dA) + hl\,dx(T-T_\infty)$$

(6.5)

---

[3] A third possibility is a convective BC, $q=h(T-T_\infty)$, where the terms are as explained in Section 6.1.2. This is a linear sum of the other two types of BCs, and is treated in a manner similar to a natural BC. (See Exercise 4.6.)

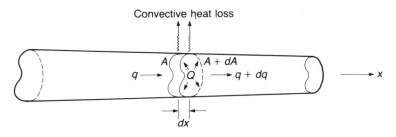

Convective heat loss

T_∞ (Ambient temp.)

FIGURE 6.3 ● A heat-conducting rod, with convection from its lateral surface.

or, neglecting second-order terms,

$$\frac{d}{dx}\left(q(x)A(x)\right) + hlT(x) = Q(x)A(x) + hlT_\infty$$

(6.6)

where

$hl\,dx(T-T_\infty)$ = rate of convective heat loss across the lateral surface, E/t

h = convective heat transfer coefficient, E/tL^2T

h is an empirically measured material/physical property that depends on the properties of both the cylinder and the gas, as well as the geometry of the cylinder surface.

l = circumference of the cylinder, L

T_∞ = ambient temperature, T

The infinity subscript signifies a location far enough away for the temperature to be undisturbed by the presence of the rod.

Constitutive equation ● Fourier's law remains the same, namely, Eq. (6.3).
Substituting Eq. (6.3) into Eq. (6.6) yields the governing differential equation,

$$-\frac{d}{dx}\left(k(x)A(x)\frac{dT(x)}{dx}\right) + hlT(x) = Q(x)A(x) + hlT_\infty$$

(6.7a)

or, in the case of a cylindrical rod with constant area,

$$-\frac{d}{dx}\left(k(x)\frac{dT(x)}{dx}\right) + \frac{hl}{A}T(x) = Q(x) + \frac{hlT_\infty}{A}$$

(6.7b)

The interior load is the sum of an applied load $Q(x)$, plus part of the convective heat loss hlT_∞. Equations (6.7a and b) differ from Eqs. (6.4a and b) in the addition of an extra convective term on both sides of the equations. The extra term on the LHS involves the unknown temperature $T(x)$; the extra term on the RHS acts like a known interior load.

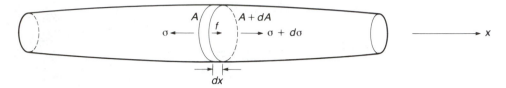

FIGURE 6.4 ● An elastic rod, in approximately a uniaxial stress state.

6.2 Elastic Rod

The derivations in this and the next two sections follow a pattern similar to that of the previous heat-conduction problem, so only the equations and terminology will be summarized. Figure 6.4 shows an elastic rod parallel to the x-axis, with a cross-sectional area and shape that may vary "gradually" along the length. The significant forces and stresses act only in the x-direction.

Balance of forces in x-direction ●

$$(\sigma+d\sigma)(A+dA) + fA\,dx - \sigma A = 0 \tag{6.8}$$

or, neglecting second-order terms,

$$-\frac{d}{dx}\left[\sigma(x)A(x)\right] = f(x)A(x) \tag{6.9}$$

where

$\sigma(x) = $ *stress*, F/L^2

$f(x) = $ interior (volume) force, F/L^3

f is a force that acts on points interior to the rod. It is produced by an external source, e.g., gravitational or centrifugal force.

Constitutive equation ● Hooke's linear stress-strain law states that

$$\sigma(x) = E(x)\frac{du(x)}{dx} \tag{6.10}$$

where

$E(x) = $ Young's modulus, F/L^2

E is a material property that measures the stiffness of the rod, i.e., its resistance to stretching when subjected to stress.

$u(x) = $ displacement of a material point in the x-direction, L

du/dx is the strain in the rod.

Substituting Eq. (6.10) into Eq. (6.9) yields the governing differential equation,

$$-\frac{d}{dx}\left[E(x)A(x)\frac{du(x)}{dx}\right] = f(x)A(x) \tag{6.11a}$$

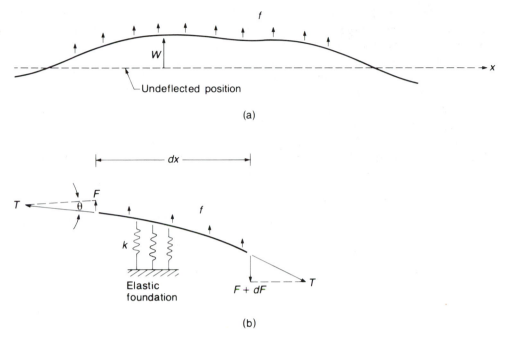

FIGURE 6.5 ● Transverse deflection of a cable: (a) the cable; (b) forces on the differential length of cable.

or, in the case of a cylindrical rod with constant area,

$$-\frac{d}{dx}\left[E(x)\,\frac{du(x)}{dx}\right] = f(x)$$

(6.11b)

The displacement $u(x)$ is the unknown function. Stress plays the role of flux. The BCs have the following interpretation:

Essential BC: Specify displacement u
Natural BC: Specify stress σ

6.3 Cable Deflection

Figure 6.5(a) shows a length of cable (or rope or string). When unloaded, the cable is oriented in a straight line parallel to the x-axis, stretched under a tension $T(x)$.

The tension may vary along the length (for example, if the cable were hanging vertically in a gravitational field). The only loads on the cable will be forces or displacements acting transverse (perpendicular) to the x-axis, resulting in deflections that are also transverse to the x-axis.[4] In addition, the model will include an elastic foundation, i.e., a transverse stiffness $k(x)$, due to an external agent, distributed along the length, as shown in Fig. 6.5(b).

---

[4] The terms "transverse" and "parallel" are preferred to "vertical" and "horizontal" since the latter are appropriate only in the presence of a gravitational field.

In the following derivation only "small" deflections will be permitted. This restriction has several implications. First, it assumes that the tension $T(x)$ remains the same in both the loaded and unloaded configurations. (A more precise theory would show that the tension may change as the cable deflects, but it is a second-order effect.) Second, it assumes that the slope θ of any section of the cable relative to the x-axis (Fig. 6.5b), is small enough that $\sin \theta \approx \tan \theta$, or, equivalently, $\cos \theta \approx 1$ [see Eq. (6.15)]. Finally, it assumes that bending stresses generated by the change in curvature are small relative to the axial stresses [$T(x)$ divided by the cross-sectional area]. In such cases the cable is sometimes said to be "flexible," or it is referred to as a "string" rather than a cable. However, the term "string" is not very appropriate when dealing with high-strength, heavy-duty industrial cables, which can sometimes be modeled quite effectively by this theory.

Balance of transverse forces • From Fig. 6.5(b),

$$F + f \, dx - kW \, dx - (F + dF) = 0 \tag{6.12}$$

or

$$\frac{dF(x)}{dx} + k(x)W(x) = f(x) \tag{6.13}$$

where

$F(x)$ = transverse component of tension, F

$f(x)$ = transverse force, distributed along the cable, F/L

f is a known load, applied by an external agent.

$W(x)$ = transverse displacement, or deflection from an unloaded straight line, L

$k(x)$ = elastic modulus of foundation, F/L^2

k is a material property of the foundation. It is a measure of the transverse stiffness of the foundation, F/L, per length of foundation in the x-direction, L.

Constitutive equation • From Fig. 6.5(b), the geometric relation of $T(x)$ to its transverse component is

$$F(x) = T(x) \sin \theta \tag{6.14}$$

For our small deflection model we must assume θ is small enough such that

$$\sin \theta \approx \tan \theta = -\frac{dW}{dx} \tag{6.15}$$

Then Eq. (6.14) becomes

$$F(x) = -T(x) \frac{dW(x)}{dx} \tag{6.16}$$

Substituting Eq. (6.16) into (6.13) yields the governing differential equation,

$$-\frac{d}{dx}\left(T(x) \frac{dW(x)}{dx}\right) + k(x)W(x) = f(x) \tag{6.17}$$

The transverse displacement $W(x)$ is the unknown function. The transverse component of tension $F(x)$ plays the role of flux. The BCs have the following interpretation:

> Essential BC: Specify transverse displacement W
> Natural BC: Specify transverse component of tension F

A particular application of this cable problem was discussed briefly in Section 2.3.

6.4 Electrostatic Field

Figure 6.6 shows two situations in which an electrostatic field is effectively one-dimensional. The panel might represent a parallel plate capacitor. Both situations are analogous to the 1-D heat flow models of Fig. 6.1.

Conservation of electric charge (**Gauss' law**) ●

$$DA + \varrho A\, dx = (D+dD)(A+dA) \tag{6.18}$$

or, neglecting second-order terms,

$$\frac{d}{dx}\left(D(x)A(x)\right) = \varrho(x)A(x) \tag{6.19}$$

where

$\quad D(x)$ = displacement of electric charge, Q/L^2

$\quad \varrho(x)$ = charge density, Q/L^3

Constitutive equation ●

$$D(x) = \epsilon(x)E(x) \tag{6.20}$$

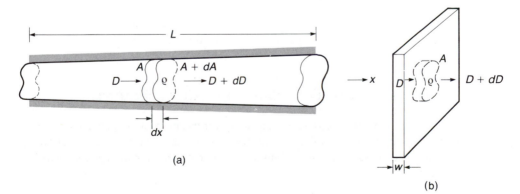

(a)

(b)

FIGURE 6.6 ● Two models for a 1-D electrostatic field: (a) a dielectric rod, with its lateral surface electrically insulated; (b) a wall or panel of dielectric material.

where

$E(x)$ = electrostatic field, V/L

$\epsilon(x)$ = permittivity, C/L

ϵ is a material property that measures how easily charged particles are displaced in a dielectric. It is related to another property, the dielectric constant K, by the relation $K = \epsilon/\epsilon_0$, where ϵ_0 is the permittivity of a vacuum.

In addition we also need the relation

$$E(x) = -\frac{d\Phi(x)}{dx} \tag{6.21}$$

where

$\Phi(x)$ = electrostatic potential, V

Equation (6.21) is a corollary of Coulomb's law [6.1], which is a physical principle describing the nature of electrostatic forces between charged particles.

Combining Eqs. (6.21) and (6.20) yields the constitutive equation,

$$D(x) = -\epsilon(x)\frac{d\Phi(x)}{dx} \tag{6.22}$$

Substituting Eq. (6.22) into (6.19) yields the governing differential equation,

$$-\frac{d}{dx}\left(\epsilon(x)A(x)\frac{d\Phi}{dx}\right) = \varrho(x)A(x) \tag{6.23a}$$

or, in the case of a cylindrical rod with constant area,

$$-\frac{d}{dx}\left(\epsilon(x)\frac{d\Phi(x)}{dx}\right) = \varrho(x) \tag{6.23b}$$

The potential $\Phi(x)$ is the unknown function. The displacement $D(x)$ plays the role of flux. The BCs have the following interpretation:

Essential BC: Specify electrostatic potential Φ
Natural BC: Specify electric displacement D

6.5 A General Formulation for All Applications

It is helpful to see the governing differential equations for the above four applications written all together. For simplicity, we'll list the equations for the usual case of a constant cross-sectional area. (The subsequent conclusions would be the same for the equations with a variable area.)

Heat conduction:

$$-\frac{d}{dx}\left(k(x)\frac{dT(x)}{dx}\right) + \frac{hl}{A}T(x) = Q(x) + \frac{hlT_\infty}{A}$$

Elastic rod:

$$-\frac{d}{dx}\left(E(x)\,\frac{du(x)}{dx}\right) = f(x)$$

Cable deflection:

$$-\frac{d}{dx}\left(T(x)\,\frac{dW(x)}{dx}\right) + k(x)W(x) = f(x)$$

Electrostatics:

$$-\frac{d}{dx}\left(\epsilon(x)\,\frac{d\Phi(x)}{dx}\right) = \varrho(x) \tag{6.24}$$

They all have the following general form:

$$-\frac{d}{dx}\left(\alpha(x)\,\frac{dU(x)}{dx}\right) + \beta(x)U(x) = f(x) \tag{6.25}$$

where $\beta(x)$ is zero for the elastic rod and electrostatics problems. In short, we really only have one governing differential equation.[5] If we write an FE computer program to solve Eq. (6.25), as indeed we will do in Chapters 7 and 8, then the program can be used to solve problems in each of the physical applications represented by Eqs. (6.24).

Table 6.1 summarizes Eqs. (6.24) and (6.25) and notes the physical meaning of all the terms.[6] The governing differential equation in each case is the result of combining two separate equations: a balance equation and a constitutive equation.

A *balance equation* (or conservation principle)

$$\frac{d\tau(x)}{dx} + \beta(x)U(x) = f(x) \tag{6.26}$$

The quantity $d\tau/dx$ represents a loss of energy from the system due to a ''flow'' across the boundaries; $\beta(x)U(x)$ represents a loss of energy from the interior of the system, where $\beta(x)$ is a known proportionality constant, representing a physical or material property that is experimentally measured; and $f(x)$, which is a known quantity, represents a gain of energy to the interior of the system. Equation (6.26) states that the energy lost from the system equals the energy gained.

A *constitutive equation* (or physical law)

$$\tau(x) = -\alpha(x)\,\frac{dU(x)}{dx} \tag{6.27}$$

---

[5] Equation (6.25) is occasionally referred to as a *Sturm-Liouville* boundary-value problem, probably because of its close relationship to the classical Sturm-Liouville eigenproblem [see Eq. (2.7) and Chapter 10]. However, the name Sturm-Liouville has traditionally been applied only to the eigenproblem.

[6] Other physical applications could be added to Table 6.1, such as magnetostatics, incompressible fluid flow, fluid seepage through porous media, etc.

is an empirical relationship which describes the materials that constitute the system. The quantity $\alpha(x)$ is an experimentally measured physical or material property; it is part of the prescribed physical data of the problem. We call $\tau(x)$ the *flux*, a conventional but not always appropriate terminology. In some applications it represents a flow of some quantity, in others a force or stress. Equation (6.27) states that the flux is proportional to the gradient (derivative) of $U(x)$.

Substituting Eq. (6.27) into Eq. (6.26) yields Eq. (6.25).

TABLE 6.1 ● One-dimensional boundary-value problems.

Application	Unknown	Physical/material properties	Interior load	Flux	Balance equation	Constitutive equation	Governing equation (balance + constitutive eqs.)
General mathematical formulation	U	α	β	f	τ	$\dfrac{d\tau}{dx}+\beta U=f$	$\tau=-\alpha\dfrac{dU}{dx}$	$-\dfrac{d}{dx}\left(\alpha\dfrac{dU}{dx}\right)+\beta U=f$
Heat conduction	T (temperature)	k (thermal conductivity)	$\dfrac{hl}{A}$ (convection loss coefficient) $\dfrac{hl}{A}T_\infty=$ part of ambient convection	$Q+\dfrac{hl}{A}T_\infty$ Q = heat source	q (heat flux)	$\dfrac{dq}{dx}+\dfrac{hl}{A}T$ $=Q+\dfrac{hl}{A}T_\infty$	$q=-k\dfrac{dT}{dx}$	$-\dfrac{d}{dx}\left(k\dfrac{dT}{dx}\right)+\dfrac{hl}{A}T$ $=Q+\dfrac{hl}{A}T_\infty$
Elasticity	u (longitudinal displacement)	E (Young's modulus)	f (body force per unit volume)	σ (stress)	$-\dfrac{d\sigma}{dx}=f$	$\sigma=E\dfrac{du}{dx}$	$-\dfrac{d}{dx}\left(E\dfrac{du}{dx}\right)=f$
Cable deflection	W (transverse displacement)	T (tension in cable)	k (elastic modulus of foundation)	f (distributed transverse force)	F (transverse component of tension)	$\dfrac{dF}{dx}+kW=f$	$F=-T\dfrac{dW}{dx}$	$-\dfrac{d}{dx}\left(T\dfrac{dW}{dx}\right)+kW=f$
Electrostatics	Φ (electrostatic potential)	ϵ (permittivity)		ϱ (charge density)	D (electric displacement)	$\dfrac{dD}{dx}=\varrho$	$D=-\epsilon\dfrac{d\Phi}{dx}$	$-\dfrac{d}{dx}\left(\epsilon\dfrac{d\Phi}{dx}\right)=\varrho$

The two principal types of BCs are as follows:

Essential BC: Specify the function U
Natural BC: Specify the flux τ

Designating the boundary of the domain by the two end points x_a and x_b, there must be one BC specified at each end point. It may be either an essential BC or a natural BC.

At $x=x_a$, specify

$$U(x_a) = U_a \quad \text{or} \quad \left(-\alpha\frac{dU}{dx}\right)_{x_a} = \tau_a$$

(6.28a)

At $x=x_b$, specify

$$U(x_b) = U_b \quad \text{or} \quad \left(-\alpha\, \frac{dU}{dx}\right)_{x_b} = \tau_b \qquad \text{(6.28b)}$$

where U_a, τ_a, U_b, and τ_b are known constants.[7]

It should be remarked that in the flux expressions for all four applications the quantities $k(x)$, $E(x)$, $T(x)$, and $\epsilon(x)$ are always positive, which is the source of the positive-definiteness of their stiffness matrices. However, only three of the applications, namely, heat conduction, cable deflection, and electrostatics, include a minus sign in the flux definition; the elastic rod does not. The explanation requires that we look at the 2-D and 3-D extensions of these problems. Thus in 2-D and 3-D the above three applications (and many others as well) are all governed by the *quasiharmonic* equation [see Eq. (2.14), as well as Chapter 13]. Eq. (6.25) is the 1-D version of the quasiharmonic equation. In any application that is governed by the quasiharmonic equation in all dimensions, the definition of flux includes a minus sign and $\alpha(x)$ is a positive quantity, as in Eq. (6.27).

In elasticity theory a minus sign appears in the balance equation(s) rather than the constitutive equation(s) because of a sign convention that is used in defining stress (which plays the role of flux). In addition, the governing equations in 2-D and 3-D are a system of coupled equations [see Eqs. (2.22), as well as Chapter 16], which are obviously different from the quasiharmonic equation. However, in the special case of a 1-D elastic rod, the coupled equations reduce to the single 1-D quasiharmonic equation (with the aforementioned shift in the minus sign from the constitutive to the balance equation).

In summary, since Chapters 3 through 15 deal essentially with 1-D and 2-D applications of the quasiharmonic equation, as well as the related eigenproblems and initial-boundary-value problems, flux is defined with a minus sign throughout these chapters. The reader will be aware that the 1-D elastic rod uses a different sign convention that, nevertheless, still produces the same governing equation as for the 1-D quasiharmonic applications.

6.6 A Well-Posed Problem

Chapters 7 and 8 will treat the general problem described by Eqs. (6.25) and (6.28), over a domain $x_a < x < x_b$. It was stated above that we must choose one BC from Eq. (6.28a) and one BC from (6.28b). In some cases we cannot choose any combination of these BCs,

---

[7] These two conditions occur most frequently in physical applications. There is an occasional need for a third condition, sometimes called a *mixed BC*, which is a linear combination of the essential and natural BCs:

$$\left(-\alpha\, \frac{dU}{dx}\right)_{x_a} + hU(x_a) = c_a$$

where h and c_a are known constants. A similar condition could exist at x_b. Such a BC would be treated in the FE analysis in a manner similar to the treatment of a natural BC. (Recall the footnote at the very end of Section 6.1.1 illustrating a mixed BC for the heat conduction problem. Also see Exercise 4.6.)

because certain combinations would correspond to a physical situation that is incompatible with the equations.

Let's be more specific. The problem arises when $\beta(x)=0$. In this case Eq. (6.25) becomes

$$-\frac{d}{dx}\left(\alpha(x)\,\frac{dU(x)}{dx}\right) = f(x) \tag{6.29}$$

For this problem we cannot specify a natural BC at both end points. Or, stated equivalently, at least one of the BCs must be an essential BC.[8]

The mathematical explanation is as follows. Let U_1 be a solution to Eq. (6.29), with natural BCs at both end points. Then $U_2=U_1+c$ (where c is an arbitrary constant) is also a solution. Why? Because everywhere U appears in Eq. (6.29) or the BCs, it appears as a derivative. Since $dc/dx=0$, then $dU_2/dx=dU_1/dx$, so U_2 is also a solution. Therefore the problem has an infinity of solutions, all differing from each other by an arbitrary constant. Not surprisingly, this makes a mess of the numerical analysis.

A physical explanation may be obtained by looking at, say, the elastic rod, which has no $\beta(x)$ term. Its natural BC corresponds to an axial force. If we were to specify a force at both ends (the same force, so that the rod is in equilibrium) the rod could be located anywhere along the x-axis [e.g., by adding an arbitrary constant to the displacement $u(x)$], and the differential equation and force BCs would be satisfied at any location. To preclude this ambiguity in the problem, it is therefore necessary to specify the displacement (an essential BC) at at least one of the end points.

In the case where $\beta(x)\neq0$, $U(x)$ appears as a function in the differential equation, so one cannot add a constant to a solution and create another solution. In this case there is a unique solution and the problem is well posed for any combination of the BCs.[9] An example would be the cable with an elastic foundation, Eq. (6.17); essential BCs are not necessary because the foundation prevents a uniform transverse displacement of the cable [adding a constant to $W(x)$] without a corresponding change in the transverse forces.

We will meet this problem of how to specify BCs for a well-posed problem again in Part IV for 2-D problems. It should be noted that the question of what constitutes proper BC data for a problem is not an FE question but rather a mathematical physics question. Thus the engineer should know how to pose a mathematically and physically meaningful problem even before attempting an FE analysis.

Reference

[6.1]   J. D. Jackson, *Classical Electrodynamics*, Wiley, New York, 1962, p. 7.

---

[8] Recall the footnote to the explanation of essential BCs in Section 4.5.

[9] The exception to this statement occurs when the BCs both specify a zero value and the interior load is zero; i.e., all the loads on the system are zero. In this case, if $\beta(x)$ is given any of certain critical values, called "eigenvalues" (see Chapter 10), there would again be an infinity of solutions.

7

1-D Boundary-Value
Problems: Linear Elements

Introduction

The purpose of this chapter is to perform a complete FE analysis of a problem, including computer implementation. This includes the theoretical development (steps 1 through 6) for a typical element, the numerical computation (steps 7 through 12) by means of a computer program, and then use of the program to obtain numerical results for two illustration problems. The linear element developed in Chapter 5, which is the simplest possible type of element for the present 1-D problem, will be used in the analysis.

7.1 Statement of the Problem

We will analyze the general 1-D boundary-value problem described in Chapter 6. For convenience, we summarize here the complete description.

Differential equation and domain •

$$-\frac{d}{dx}\left(\alpha(x)\,\frac{dU(x)}{dx}\right) + \beta(x)U(x) = f(x) \qquad x_a < x < x_b \tag{7.1a}$$

Boundary conditions •

At $x = x_a$,

$$U(x_a) = U_a \qquad \text{essential BC}$$

or

$$\left(-\alpha\,\frac{dU}{dx}\right)_{x_a} = \tau_a \qquad \text{natural BC} \tag{7.1b}$$

221

At $x=x_b$,

$$U(x_b) = U_b \qquad \text{essential BC}$$

or

$$\left(-\alpha \frac{dU}{dx}\right)_{x_b} = \tau_b \qquad \text{natural BC} \tag{7.1c}$$

where U_a or τ_a and U_b or τ_b are specified numbers.

7.2 Derivation of Element Equations:
Theoretical Development (Steps 1 through 6)

We learned in Chapter 5 that it is necessary to derive element equations for only one element, namely, a typical element (e). These element equations can then be used for each of the elements in any mesh constructed over the domain (Fig. 7.1).

The typical element trial solution can always be written in the general form,

$$\tilde{U}^{(e)}(x;a) = \sum_{j=1}^{n} a_j \phi_j^{(e)}(x) \tag{7.2}$$

We use the local node number convention of numbering the unknown parameters (DOF) $1,2,\ldots,n$, where n is the number of DOF in the element. In the following theoretical development (steps 1 through 6), the first three steps are short, formal operations using only the general form [Eq. (7.2)]. It is not until step 4 that we decide on the value of n and the specific form of each of the shape functions $\phi_j^{(e)}(x)$.

Step 1: Write the Galerkin residual equations for a typical element.

The residual for Eq. (7.1a) is

$$R(x;a) = -\frac{d}{dx}\left[\alpha(x) \frac{d\tilde{U}^{(e)}(x;a)}{dx}\right] + \beta(x)\tilde{U}^{(e)}(x;a) - f(x) \tag{7.3}$$

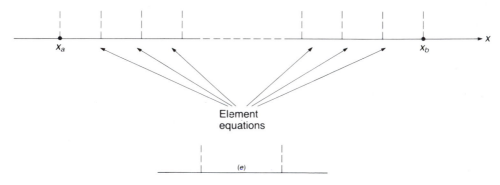

FIGURE 7.1 ● Element equations from a typical element (e) are used for each element in the mesh.

We need one weighted residual equation for each DOF in Eq. (7.2):

$$\overset{(e)}{\int} R(x;a)\phi_i^{(e)}(x)\,dx = 0 \qquad i = 1,2,\ldots,n \qquad (7.4)$$

where $\overset{(e)}{\int}$ denotes integration over element (e). Substituting Eq. (7.3) into Eq. (7.4) yields

$$\overset{(e)}{\int} \left[-\frac{d}{dx}\left(\alpha(x)\,\frac{d\tilde{U}^{(e)}(x;a)}{dx}\right) + \beta(x)\tilde{U}^{(e)}(x;a) - f(x) \right] \phi_i^{(e)}(x)\,dx = 0$$

$$i = 1,2,\ldots,n \qquad (7.5)$$

Step 2: Integrate by parts.

Using the general formula for integration by parts, Eq. (4.9), the first term in Eqs. (7.5) becomes

$$\overset{(e)}{\int} \left[-\frac{d}{dx}\left(\alpha(x)\,\frac{d\tilde{U}^{(e)}(x;a)}{dx}\right) \right] \phi_i^{(e)}(x)\,dx$$

$$= \left[\left(-\alpha(x)\,\frac{d\tilde{U}^{(e)}}{dx}\right)\phi_i^{(e)}(x) \right]^{(e)} + \overset{(e)}{\int}\alpha(x)\,\frac{d\tilde{U}^{(e)}(x;a)}{dx}\,\frac{d\phi_i^{(e)}(x)}{dx}\,dx \qquad (7.6)$$

where the notation $[\ ]^{(e)}$ for the boundary term has the meaning,

$$\left[\left(-\alpha(x)\,\frac{d\tilde{U}^{(e)}}{dx}\right)\phi_i^{(e)}(x) \right]^{(e)}$$

$$= \left[\left(-\alpha(x)\,\frac{d\tilde{U}^{(e)}}{dx}\right)\phi_i^{(e)}(x) \right]_{\substack{\text{right end} \\ \text{of element}}} - \left[\left(-\alpha(x)\,\frac{d\tilde{U}^{(e)}}{dx}\right)\phi_i^{(e)}(x) \right]_{\substack{\text{left end} \\ \text{of element}}} \qquad (7.7)$$

Substituting Eqs. (7.6) into Eqs. (7.5), the residual equations become

$$\overset{(e)}{\int} \left[\alpha(x)\,\frac{d\tilde{U}^{(e)}(x;a)}{dx}\,\frac{d\phi_i^{(e)}(x)}{dx} + \beta(x)\tilde{U}^{(e)}(x;a)\phi_i^{(e)}(x) \right] dx$$

$$= \overset{(e)}{\int} f(x)\phi_i^{(e)}(x)\,dx - \left[\left(-\alpha(x)\,\frac{d\tilde{U}^{(e)}}{dx}\right)\phi_i^{(e)}(x) \right]^{(e)} \qquad i = 1,2,\ldots,n \qquad (7.8)$$

All load terms are placed on the RHS; this includes the boundary term as well as the integral containing the interior load $f(x)$. The terms containing integrals of the trial solution over the element are placed on the LHS.

Step 3: Substitute the element trial solution into integrals on the LHS.

The derivative of Eq. (7.2) is

$$\frac{d\tilde{U}^{(e)}(x;a)}{dx} = \sum_{j=1}^{n} a_j\,\frac{d\phi_j^{(e)}(x)}{dx} \qquad (7.9)$$

Substituting Eqs. (7.2) and (7.9) into Eqs. (7.8) yields

$$
\sum_{j=1}^{n} \left[\int^{(e)} \frac{d\phi_i^{(e)}(x)}{dx} \, \alpha(x) \, \frac{d\phi_j^{(e)}(x)}{dx} \, dx \; + \; \int^{(e)} \phi_i^{(e)}(x)\beta(x)\phi_j^{(e)}(x) \, dx \right] a_j
$$

$$
= \int^{(e)} f(x)\phi_i^{(e)}(x) \, dx \; - \; \left[\left(-\alpha(x) \, \frac{d\tilde{U}^{(e)}}{dx} \right) \phi_i^{(e)}(x) \right]^{(e)} \qquad i = 1,2,\ldots,n
$$

$$\tag{7.10}$$

These are the element equations for the typical element. They may be written in the conventional matrix form,

$$
\begin{bmatrix}
K_{11}^{(e)} & K_{12}^{(e)} & \cdots & K_{1n}^{(e)} \\
K_{21}^{(e)} & K_{22}^{(e)} & \cdots & K_{2n}^{(e)} \\
\vdots & \vdots & & \vdots \\
K_{n1}^{(e)} & K_{n2}^{(e)} & \cdots & K_{nn}^{(e)}
\end{bmatrix}
\begin{Bmatrix}
a_1 \\ a_2 \\ \vdots \\ a_n
\end{Bmatrix}
=
\begin{Bmatrix}
F_1^{(e)} \\ F_2^{(e)} \\ \vdots \\ F_n^{(e)}
\end{Bmatrix}
$$

$$\tag{7.11a}$$

where the expressions for the stiffness and load terms are, respectively,

$$
K_{ij}^{(e)} = \int^{(e)} \frac{d\phi_i^{(e)}(x)}{dx} \, \alpha(x) \, \frac{d\phi_j^{(e)}(x)}{dx} \, dx \; + \; \int^{(e)} \phi_i^{(e)}(x)\beta(x)\phi_j^{(e)}(x) \, dx
$$

$$
F_i^{(e)} = \int^{(e)} f(x)\phi_i^{(e)}(x) \, dx \; - \; \left[\left(-\alpha(x) \, \frac{d\tilde{U}^{(e)}}{dx} \right) \phi_i^{(e)}(x) \right]^{(e)}
$$

$$\tag{7.11b}$$

Step 4: Develop specific expressions for the shape functions $\phi_i^{(e)}(x)$ (that is, the element trial functions).

The governing equation [Eq. (7.1a)] is a second-order differential equation, just like the illustrative problem in Chapters 3 through 5. Therefore the same completeness and continuity convergence requirements apply to the present problem. Recalling the discussion in Section 5.6, let's review these requirements. The highest derivative of the element trial solution $\tilde{U}^{(e)}$ in Eqs. (7.8) is of order $m=1$. Therefore the completeness condition using polynomials requires that the polynomial for $\tilde{U}^{(e)}$ be complete at least to degree 1 — that is, a linear polynomial. The continuity condition requires that $\tilde{U}^{(e)}$ be C^0-continuous at interelement boundaries.

In other words, we can use the same two-node linear element that was used in Chapter 5 [recall Fig. 5.18 and Eqs. (5.79)]. For convenience, we repeat those results here:

$$
\tilde{U}^{(e)}(x;a) = \sum_{j=1}^{2} a_j \phi_j^{(e)}(x)
$$

$$\tag{7.12a}$$

where

$$\phi_1^{(e)}(x) = \frac{x_2 - x}{x_2 - x_1}$$

$$\phi_2^{(e)}(x) = \frac{x - x_1}{x_2 - x_1} \tag{7.12b}$$

This element is frequently referred to as a C^0-linear element (Fig. 7.2), in obvious reference to the continuity and completeness properties it possesses.

Step 5: Substitute the shape functions into the element equations, and transform the integrals into a form appropriate for numerical evaluation.

The element contains two DOF, so Eqs. (7.11) become

$$\begin{bmatrix} K_{11}^{(e)} & K_{12}^{(e)} \\ K_{21}^{(e)} & K_{22}^{(e)} \end{bmatrix} \begin{Bmatrix} a_1 \\ a_2 \end{Bmatrix} = \begin{Bmatrix} F_1^{(e)} \\ F_2^{(e)} \end{Bmatrix} \tag{7.13a}$$

where

$$K_{ij}^{(e)} = \int_{x_1}^{x_2} \frac{d\phi_i^{(e)}(x)}{dx} \alpha(x) \frac{d\phi_j^{(e)}(x)}{dx} \, dx + \int_{x_1}^{x_2} \phi_i^{(e)}(x)\beta(x)\phi_j^{(e)}(x) \, dx$$

$$F_i^{(e)} = \int_{x_1}^{x_2} f(x)\phi_i^{(e)}(x) \, dx - \left[\left(-\alpha(x) \frac{d\tilde{U}^{(e)}}{dx} \right) \phi_i^{(e)}(x) \right]_{x_1}^{x_2} \tag{7.13b}$$

Consider first the $K_{ij}^{(e)}$ terms. These require the derivatives of the shape functions,

$$\frac{d\phi_1^{(e)}(x)}{dx} = -\frac{1}{x_2 - x_1}$$

$$\frac{d\phi_2^{(e)}(x)}{dx} = \frac{1}{x_2 - x_1} \tag{7.14}$$

Substituting Eqs. (7.12b) and (7.14) into Eqs. (7.13b), the $K_{11}^{(e)}$ term becomes

$$K_{11}^{(e)} = \int_{x_1}^{x_2} \left(-\frac{1}{x_2 - x_1} \right) \alpha(x) \left(-\frac{1}{x_2 - x_1} \right) dx$$

$$+ \int_{x_1}^{x_2} \left(\frac{x_2 - x}{x_2 - x_1} \right) \beta(x) \left(\frac{x_2 - x}{x_2 - x_1} \right) dx \tag{7.15}$$

The integrals cannot be evaluated until the material properties $\alpha(x)$ and $\beta(x)$ are defined, i.e., represented mathematically by specific functions. However, $\alpha(x)$ and $\beta(x)$ will, in general, be different functions in different problems. If we want to write a computer program to handle a wide variety of problems, then we will have to provide some way to represent $\alpha(x)$ and $\beta(x)$ by a variety of different functions. Let's digress briefly to discuss how this can be done.

There are two common methods for defining $\alpha(x)$ and $\beta(x)$ in a computer program. One method is to provide for user-written subroutines that allow the user to specify any

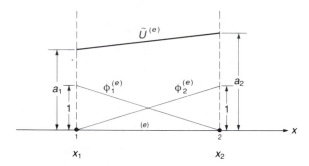

FIGURE 7.2 ● A C^0-linear element.

functions he or she desires. This has the advantage that $\alpha(x)$ and $\beta(x)$ may be defined exactly, i.e., as accurately as they are known in the original problem statement in Section 7.1. Since the form of $\alpha(x)$ and $\beta(x)$ for each problem cannot be known in advance by the program developer, the K_{ij} integrals cannot be evaluated using exact, closed-form expressions. They would therefore have to be evaluated approximately using numerical integration (quadrature) formulas (see Chapters 8 and 13). It would also require the user, for each analysis, to compile and link the subroutines with the rest of the program, making the program less portable unless additional machine-dependent code is provided. This method is recommended for special applications in which $\alpha(x)$ or $\beta(x)$ tend to have complicated expressions. It will not be used here because the method explained next is simpler and is adequate for most applications.

The second method is for the program developer to select a few functions, and then represent $\alpha(x)$ and $\beta(x)$ in terms of these. For example, we might include a few powers of x, an exponential, and a sinusoidal term:

$$\alpha(x) = \alpha_0 + \alpha_1 x + \alpha_2 x^2 + \ldots + \alpha_m x^m + \alpha_{m+1} e^{\alpha_{m+2} x} + \alpha_{m+3} \sin \alpha_{m+4} x$$

$$\beta(x) = \beta_0 + \beta_1 x + \beta_2 x^2 + \ldots + \beta_m x^m + \beta_{m+1} e^{\beta_{m+2} x} + \beta_{m+3} \sin \beta_{m+4} x$$

$$(7.16)$$

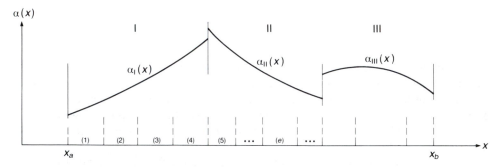

FIGURE 7.3 ● An example of using several quadratic polynomials (in this case, three) to define the physical property $\alpha(x)$.

These expressions (or approximations to them, as explained below) could be substituted into the integrals in Eq. (7.15) and the latter integrated by exact integral formulas or approximate quadrature formulas. A program user would then only need to specify numerical values for the α_i and β_i parameters. We will employ this method.

In the interest of keeping both the theoretical development and program coding brief and uncomplicated, we will limit Eqs. (7.16) to quadratic polynomials. However, to accommodate properties more complex than a quadratic, we will let the user define $\alpha(x)$ and $\beta(x)$ as *piecewise* quadratic polynomials. Thus the user may define up to 20 different pairs of quadratic polynomials:[1]

$$\alpha(x) = \alpha_0 + \alpha_1 x + \alpha_2 x^2$$

$$\beta(x) = \beta_0 + \beta_1 x + \beta_2 x^2 \tag{7.17}$$

each pair being applicable to a different region of the domain. In many of the typical applications of Eqs. (7.1) one pair of these polynomials will be adequate to define $\alpha(x)$ and $\beta(x)$ over the entire domain. In other applications it may require several pairs. For example, Fig. 7.3 illustrates a problem in which the property $\alpha(x)$ is defined by three separate polynomials, in regions I, II, and III. (It is noted in passing that there may be any number of elements within each different region, but elements should not overlap regions with different properties — more on this in Section 7.4.3.)

For some problems $\alpha(x)$ and $\beta(x)$ cannot be represented exactly by Eqs. (7.17), even if defined piecewise. In these cases we will be introducing a *modeling error*, meaning that we have altered the data in our original problem statement (Section 7.1). More specifically, the original problem statement includes (besides the general form of the governing differential equation and BCs) three categories of data: physical properties [for example, $\alpha(x)$ and $\beta(x)$], loads [$f(x)$ and values of BCs], and geometry (location of boundary). This data *defines the mathematical model*. If any of this data is changed in the course of the FE analysis, then the model is changed, and we are therefore analyzing *a problem that is different from the original one*. This concept will prove useful in discussions of convergence (see step 6 below), since we will be able to identify the limit of the convergence process as the (exact) solution of the altered model rather than of the original model.

If $\alpha(x)$ or $\beta(x)$ is so complicated that the altered model appears to the analyst to be unacceptably poor, then the user-written subroutine approach should be implemented instead. This is an issue that requires judgment on the part of the analyst, which in turn usually requires some experience. Modeling errors are one part of FE analysis that still require judgment and experience; such decisions cannot yet be automated.

Modeling errors occur quite frequently in FE analysis. They may be classified as approximation errors, intentionally included in the analysis (see Section 3.6.2). In 1-D problems the most common modeling errors will involve physical properties and loads, the latter being illustrated in Section 7.5. Geometry errors need not arise in 1-D because the boundary consists of only two points whose locations can always be represented exactly. However, in 2-D and 3-D problems most modeling errors involve geometry because curved boundaries frequently cannot be represented exactly by the limited element shapes that are available (see Section 13.6.3). This ends the digression.

---

[1] The number 20 is the limit shown in the program listing in the appendix; if desired, it may be changed to another number.

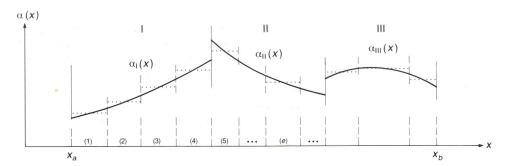

FIGURE 7.4 ● Using the example from Fig. 7.3, the horizontal dotted lines show how the polynomials $\alpha_I(x)$, $\alpha_{II}(x)$, and $\alpha_{III}(x)$ are approximated by a constant within each element.

Having defined $\alpha(x)$ and $\beta(x)$ for the program, perhaps approximately, we can now address the problem of evaluating the K_{ij} integrals. If the quadratic polynomials in Eqs. (7.17) were substituted for $\alpha(x)$ and $\beta(x)$ in Eq. (7.15), the resulting integrands would be simple enough to be integrated exactly with closed-form expressions that, in turn, could be easily programmed. This would be a very acceptable procedure. However, for reasons to be given in a moment, instead of substituting Eqs. (7.17) into Eq. (7.15), we will further simplify the analysis by requiring the program to approximate the polynomials by a *constant within each element*, as illustrated in Fig. 7.4. Thus the polynomials will be replaced by a stepped (piecewise-constant) approximation that yields, in general, a different constant value within each element. Here we are introducing yet another type of approximation error.[2] This is essentially a *computational error* since it is due to simplifying the numerical evaluation of an integral.

How do we choose the "best" constant value for each element, i.e., the one that will give a *sufficiently* accurate value (not necessarily the most accurate value) for the K_{ij} integrals? The answer requires a theoretical argument that is better suited to Chapter 8. But that argument can be distilled as follows. Recall from Section 3.6 that there is an inherent error in the FEM, called the discretization error, that is due solely to representing the exact solution by a finite number of trial functions and a corresponding discrete set of parameters. The error exists even if all numerical computation were to be done exactly. It is therefore computationally wasteful to evaluate integrals to a much greater accuracy (e.g., exactly) than that which is present due to the discretization error, since the latter will limit the best obtainable accuracy anyway. For the present type of element (C^0-linear), it will be seen in Section 8.3.1 that the best constant value for $\alpha(x)$ and $\beta(x)$ is obtained by evaluating $\alpha(x)$ and $\beta(x)$ at the *center* of each element, as illustrated in Fig. 7.5 for $\alpha(x)$ for the typical element.

---

[2] By the end of the book the reader will have discovered that a typical FE analysis contains quite a few different approximation errors, and these are all in addition to any unintentional errors such as programming bugs and input data mistakes.

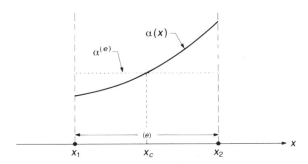

FIGURE 7.5 ● Approximating $\alpha(x)$ within a typical element by the value at the center, $\alpha^{(e)}$.

The coordinate of the center of the (e)th element, x_c, is

$$x_c = \frac{1}{2}(x_1 + x_2)$$

(7.18)

Substituting Eq. (7.18) into Eqs. (7.17) yields the desired constant values $\alpha^{(e)}$ and $\beta^{(e)}$ for the (e)th element:

$$\alpha^{(e)} = \alpha(x_c) = \alpha_0 + \alpha_1 x_c + \alpha_2 x_c^2$$
$$\beta^{(e)} = \beta(x_c) = \beta_0 + \beta_1 x_c + \beta_2 x_c^2$$

(7.19)

The constants $\alpha^{(e)}$ and $\beta^{(e)}$ may now be substituted for $\alpha(x)$ and $\beta(x)$ in Eq. (7.15) and the integrals evaluated using closed-form expressions. Before doing this, though, some further observations may be helpful.

It should be intuitively appealing from Fig. 7.5 that as elements become smaller by refining the mesh, the segment of $\alpha(x)$ within each element approaches a straight (generally sloped) line, in which case the center values approach arbitrarily close to the exact mean of $\alpha(x)$ within each element. In other words, the center value seems to be a reasonable choice if the element is "small enough." This might help to explain the fact that the theory that will be encountered in Section 8.3.1 is only valid in the limit of "small-enough" elements. (Such a theory is often referred to as an *asymptotic* theory.) The point here is that if $\alpha(x)$ or $\beta(x)$ vary appreciably over the domain, then a coarse mesh of a few large elements will probably yield poor approximations for the K_{ij} integrals and a correspondingly poor FE solution. This is not cause for alarm, however, because (1) a large error would immediately become apparent when a second solution is generated using a refined mesh, and (2) a variable $\alpha(x)$ and $\beta(x)$ will usually produce a correspondingly variable solution, necessitating a finer mesh anyway.

Let's summarize the above developments:

1. As a user convenience, in order to minimize the amount of input data, the physical/material properties $\alpha(x)$ and $\beta(x)$ may be defined to the program by specifying the coefficients for one or more (up to 20) quadratic polynomials over the domain.

2. To simplify the theoretical analysis and the subsequent programming effort, the K_{ij} integrals for each element will approximate $\alpha(x)$ and $\beta(x)$ by their (constant) values at the element centers. The justification for this procedure is that as a mesh is refined,

any computational error introduced by this approximation should be comparable to[3] the inherent discretization error.

Continuing now with the analysis, we substitute $\alpha^{(e)}$ for $\alpha(x)$ and $\beta^{(e)}$ for $\beta(x)$ in the integrals in Eq. (7.15) and integrate the latter using closed-form expressions:

$$
K_{11}^{(e)} = \int_{x_1}^{x_2} \left(-\frac{1}{x_2-x_1} \right) \alpha^{(e)} \left(-\frac{1}{x_2-x_1} \right) dx
$$

$$
+ \int_{x_1}^{x_2} \left(\frac{x_2-x}{x_2-x_1} \right) \beta^{(e)} \left(\frac{x_2-x}{x_2-x_1} \right) dx = \frac{\alpha^{(e)}}{L} + \frac{\beta^{(e)}L}{3} \tag{7.20a}
$$

where $L = x_2 - x_1$. Similarly,

$$
K_{12}^{(e)} = -\frac{\alpha^{(e)}}{L} + \frac{\beta^{(e)}L}{6}
$$

$$
K_{21}^{(e)} = K_{12}^{(e)} \qquad \text{(symmetry)}
$$

$$
K_{22}^{(e)} = K_{11}^{(e)} \tag{7.20b}
$$

Now consider the $F_i^{(e)}$ terms in Eqs. (7.13b). Substituting Eqs. (7.12b) into Eqs. (7.13b), $F_1^{(e)}$ and $F_2^{(e)}$ become

$$
F_1^{(e)} = \int_{x_1}^{x_2} f(x) \frac{x_2 - x}{x_2 - x_1} dx + \left(-\alpha^{(e)} \frac{d\tilde{U}^{(e)}}{dx} \right)_{x_1}
$$

$$
F_2^{(e)} = \int_{x_1}^{x_2} f(x) \frac{x - x_1}{x_2 - x_1} dx - \left(-\alpha^{(e)} \frac{d\tilde{U}^{(e)}}{dx} \right)_{x_2} \tag{7.21}
$$

where the interpolation property $\phi_i^{(e)}(x_j) = \delta_{ij}$ and Eqs. (7.19) were used to simplify the boundary terms. The latter are recognized as the expressions for flux,

$$
\tilde{\tau}^{(e)}(x_1) = \left(-\alpha^{(e)} \frac{d\tilde{U}^{(e)}}{dx} \right)_{x_1}
$$

$$
\tilde{\tau}^{(e)}(x_2) = \left(-\alpha^{(e)} \frac{d\tilde{U}^{(e)}}{dx} \right)_{x_2} \tag{7.22}
$$

So Eqs. (7.21) can be written,

$$
F_1^{(e)} = \int_{x_1}^{x_2} f(x) \frac{x_2 - x}{x_2 - x_1} dx + \tilde{\tau}^{(e)}(x_1)
$$

$$
F_2^{(e)} = \int_{x_1}^{x_2} f(x) \frac{x - x_1}{x_2 - x_1} dx - \tilde{\tau}^{(e)}(x_2) \tag{7.23}
$$

---

[3] By "comparable to" we mean the same (or better) *rate* of convergence, a concept discussed in step 6 below, as well as in Chapters 8 and 9.

The interior load $f(x)$ may be handled in the same manner that $\alpha(x)$ and $\beta(x)$ were. Thus, to make the integration error in Eqs. (7.23) comparable to (i.e., converge at the same rate as) the discretization error, we only need to approximate $f(x)$ by its value at the center of the element. This is equivalent to representing $f(x)$ as a constant in each element, equal to its value at the center:

$$f(x) \simeq f(x_c) = f^{(e)} \qquad \text{in } (e)\text{th element} \tag{7.24}$$

The manner in which the values of $f^{(e)}$ are calculated is basically a programming consideration. They can either be calculated by hand and then input directly to the program, or else a general function for $f(x)$ can be specified [like the piecewise quadratic for $\alpha(x)$ and $\beta(x)$] which the program would then evaluate at each of the element centers. Since many of the interesting applications of Eqs. (7.1) involve only a constant $f(x)$, we opted for the former approach. If the constant is the same over many elements, then the user need only specify the constant for one element, and the program will duplicate the value for all the other elements.

Substituting Eq. (7.24) into Eqs. (7.23), moving $f^{(e)}$ outside the integrals, and using exact integration formulas for the remaining integration yields

$$F_1^{(e)} = \frac{f^{(e)}L}{2} + \tilde{\tau}^{(e)}(x_1)$$

$$F_2^{(e)} = \frac{f^{(e)}L}{2} - \tilde{\tau}^{(e)}(x_2) \tag{7.25}$$

This is an intuitively appealing result. The interior load $f(x)$ is a distributed load, having the units of some physical quantity (e.g., force, heat, charge) *per unit length*. Therefore the product $f^{(e)}L$ is the total load acting on the element [since $f^{(e)}$ is constant over the element], and Eqs. (7.25) say that half the total load acts at each node.

The final expression for the element equations is as follows:

$$
\begin{bmatrix}
\dfrac{\alpha^{(e)}}{L} + \dfrac{\beta^{(e)}L}{3} & -\dfrac{\alpha^{(e)}}{L} + \dfrac{\beta^{(e)}L}{6} \\[2ex]
\text{Symmetric} & \dfrac{\alpha^{(e)}}{L} + \dfrac{\beta^{(e)}L}{3}
\end{bmatrix}
\begin{Bmatrix} a_1 \\[2ex] a_2 \end{Bmatrix}
=
\begin{Bmatrix} \dfrac{f^{(e)}L}{2} \\[2ex] \dfrac{f^{(e)}L}{2} \end{Bmatrix}
+
\begin{Bmatrix} \tilde{\tau}^{(e)}(x_1) \\[2ex] -\tilde{\tau}^{(e)}(x_2) \end{Bmatrix}
\tag{7.26}
$$

Step 6: Prepare expression for the flux.
From Eqs. (7.12), (7.14), and (7.19),

$$\tilde{\tau}^{(e)}(x) = -\alpha(x)\frac{d\tilde{U}^{(e)}(x;a)}{dx}$$

$$= -\alpha(x)\left[a_1\left(\frac{-1}{x_2 - x_1}\right) + a_2\left(\frac{1}{x_2 - x_1}\right)\right]$$

$$= -\alpha(x)\frac{a_2 - a_1}{x_2 - x_1} \tag{7.27}$$

Equation (7.27) is the approximate solution for $\tilde{\tau}^{(e)}(x)$ within the (e)th element, where $\alpha(x)$ may be evaluated using the quadratic polynomial of Eqs. (7.17).

The accuracy of Eq. (7.27) can vary appreciably over the element. However, there is one point in the element where the accuracy is better than anywhere else, *if* the element is "small enough." This special point is at the *center* of the element.[4] The following discussion will attempt to explain the reasons for the existence of such a point.

We want to first establish the general accuracy of Eq. (7.27) over the (*e*)th element. We consider the pointwise error in $\tilde{\tau}^{(e)}(x)$ due solely to the polynomial approximation for $U(x)$ (ignoring any computational errors), i.e., the discretization error. Thus,

$$|\tau(x) - \tilde{\tau}^{(e)}(x)| = \left| \left(-\alpha(x) \frac{dU(x)}{dx} \right) - \left(-\hat{\alpha}(x) \frac{d\tilde{U}^{(e)}(x)}{dx} \right) \right| \tag{7.28}$$

where $\hat{\alpha}(x)$ represents any modeling approximation of $\alpha(x)$, such as might be introduced by Eqs. (7.17). For the moment let's assume there is no approximation to $\alpha(x)$; that is, $\hat{\alpha}(x) = \alpha(x)$. (We'll comment on this important assumption later.) Therefore Eq. (7.28) may be written

$$|\tau(x) - \tilde{\tau}^{(e)}(x)| = |\alpha(x)| \left| \frac{dU(x)}{dx} - \frac{d\tilde{U}^{(e)}(x)}{dx} \right| \tag{7.29}$$

so that the pointwise error is proportional to the difference in the two derivatives. Let's examine each derivative separately.

The element trial solution, $\tilde{U}^{(e)}(x)$, is a complete linear polynomial; therefore $d\tilde{U}^{(e)}/dx$ is a constant, where the constant is arbitrary. The exact solution $U(x)$ is unknown, but we may represent it within any element by a Taylor series about some point x_a inside the element:

$$U(x) = c_0 + c_1(x - x_a) + c_2(x - x_a)^2 + c_3(x - x_a)^3 + \ldots \tag{7.30}$$

where $|x - xa| \leq h$, the size (length) of the element. The derivative is therefore

$$\frac{dU(x)}{dx} = c_1 + 2c_2(x - x_a) + 3c_3(x - x_a)^2 + \ldots \tag{7.31}$$

Since $d\tilde{U}^{(e)}/dx$ is an arbitrary constant, it is capable of approaching arbitrarily close to the constant c_1 in dU/dx as $h \to 0$, but it can't approach arbitrarily close (over the entire element) to the linear polynomial $c_1 + 2c_2(x - x_a)$ or any higher-degree polynomial. Therefore the difference between dU/dx and $d\tilde{U}^{(e)}/dx$ will be dominated by the $2c_2(x - x_a)$ term; that is,

$$\left| \frac{dU(x)}{dx} - \frac{d\tilde{U}^{(e)}(x)}{dx} \right| \to 2c_2 h + 3c_3 h^2 + \ldots \qquad \text{as } h \to 0 \tag{7.32a}$$

This is usually written as follows:

$$\left| \frac{dU(X)}{dx} - \frac{d\tilde{U}^{(e)}(x)}{dx} \right| = O(h) \qquad \text{as } h \to 0 \tag{7.32b}$$

---

[4] The center location is characteristic of the present 1-D C^0-linear element. It will be seen in Chapters 8 and 13 that other types of elements also have special points, but they are not always located at the center.

where "$= O(h)$" is read "is of order h." This terminology means that if h is small enough — that is, if $h < h^*$, then

$$\left| \frac{dU(x)}{dx} - \frac{d\tilde{U}^{(e)}(x)}{dx} \right| \leq K(h^*)h \qquad \text{as } h \to 0 \qquad (h < h^*)$$

$$(7.32c)$$

where $K(h^*)$ is a constant (for a given h^*) that can be made arbitrarily close to $2c_2$ as $h^* \to 0$.

Expressions (7.32a, b, and c) are all equivalent. They tell us that the *rate* of convergence (if convergence occurs at all) is $O(h)$, meaning that if the element size is halved ($h \to h/2$), then the pointwise error is halved (for "small-enough" elements). Note that if c_2 happens to be zero in a given element as $h \to 0$, then the dominant term would be $3c_3(x - x_a)^2$ and the rate of convergence would be $O(h^2)$. Therefore when we say that the rate of convergence is $O(h)$, it should be understood that we mean *at least* $O(h)$; i.e., it might be even faster at random points.

Equation (7.29) may now be written

$$|\tau(x) - \tilde{\tau}^{(e)}(x)| = O(h) \qquad \text{as } h \to 0 \qquad (7.33)$$

Therefore the approximation error (or accuracy) in the flux at any general location in each element is $O(h)$.

Next we want to show that the accuracy in the flux is *one order better*, namely, $O(h^2)$, at the center of the element. The theoretical argument [7.1] is quite abstract, certainly beyond the scope of this book. Nevertheless, the essential ideas will be presented here, albeit in somewhat heuristic[5] fashion, with the hope that the reader will gain some insight into the special way that the FEM deals with the flux solution.[6]

The argument in Moan [7.1] involves the following three central ideas:

1. If the governing equations are linear and self-adjoint (as they are in this book — recall Sections 3.2 and 3.7), then the FE procedure provides a *weighted least-squares fit* of $d^m \tilde{U}^{(e)}/dx^m$ to $d^m U/dx^m$, in the limit as $h \to 0$. The weighting is the factor $\alpha(x)$. Here m is the order of the highest derivative of $\tilde{U}^{(e)}$ in the expression for flux [which is also the highest derivative of the shape functions in the stiffness integrals].

2. Because the trial functions are local, the least-squares fit is approximately achieved within each element separately.

3. From polynomial approximation theory [7.2] it is known that if a complete 1-D polynomial of degree q forms a weighted least-squares fit to a 1-D polynomial of degree

---

[5] Heuristic 1: serving to indicate or point out; stimulating interest as a means of furthering investigation 2: encouraging the student to discover for himself (Random House, 1966).

[6] This situation with the flux stands in marked contrast to that of the function, $\tilde{U}^{(e)}(x)$. The approximation error of $\tilde{U}^{(e)}(x)$ can be shown by the same method to be $O(h^2)$, one order better than the general error for the flux. This is to be expected since the flux is proportional to the derivative, and differentiation generally decreases accuracy. What is interesting is that this loss of accuracy in the flux can be recovered — as explained above.

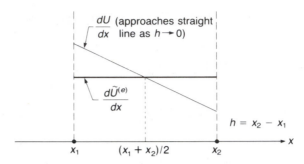

FIGURE 7.6 ● Least-squares fit of an approximate solution, $d\tilde{U}^{(e)}/dx$, to the exact solution, dU/dx, in a 1-D C^0-linear element. The intersection approaches the center, $(x_1+x_2)/2$, as $h \to 0$.

$q+1$, then the two polynomials will equal each other at $q+1$ points.[7] These points will be the *zeroes* of the $(q+1)$st-degree polynomial in a family of *orthogonal polynomials* [7.3] *with weight* $\alpha(x)$.

To provide some meaning to all this abstraction, let's apply it to our 1-D C^0-linear element. For this element, $m=1$. Therefore items 1 and 2 in the list above state that the FE analysis essentially produces a weighted least-squares fit of $d\tilde{U}^{(e)}/dx$ to dU/dx within each element, as the elements become "small enough."

Next consider item 3 and recall that $d\tilde{U}^{(e)}/dx$ is a constant, so $q=0$. Item 3 states that if a constant ($q=0$) forms a weighted least-squares fit to a *linear* polynomial ($q+1=1$), then the former will intersect the latter at a special point whose location is known (see below). From Eq. (7.31),

$$\frac{dU(x)}{dx} = c_1 + 2c_2(x-x_a) + O(h^2) \qquad \text{as } h \to 0 \tag{7.34}$$

Thus, for h small enough, dU/dx is a linear polynomial to within an error $O(h^2)$. Putting items 1 and 3 together, we see that the constant $d\tilde{U}^{(e)}/dx$ will form a weighted least-squares fit with dU/dx, and $d\tilde{U}^{(e)}/dx$ will intersect dU/dx at the location of the (one) zero of a particular first-degree ($q+1=1$) orthogonal polynomial.

The type of orthogonal polynomial is determined by the weighting factor $\alpha(x)$. We will be primarily interested in the case where $\alpha(x)$ is a constant (the reasons are given below). The orthogonal polynomials corresponding to a constant weighting factor are the *Legendre polynomials* [7.3]. The zero of the first-degree ($q+1=1$) Legendre polynomial is at the *center* of the element. Therefore the special point where $d\tilde{U}^{(e)}/dx$ intersects dU/dx, as $h \to 0$, is the center of the element.

Figure 7.6 illustrates all these relationships. In summary, as the element size gets smaller, the exact flux dU/dx approaches a linear polynomial with accuracy $O(h^2)$; and $d\tilde{U}^{(e)}/dx$, which is a constant, intersects this exact solution very close to the center of the element. (The intersection is not exactly at the center because dU/dx is not exactly a straight

---

[7] In 2-D and 3-D there are similar least-squares fits, but they involve incomplete polynomials.

line.) Most important, though, is the fact that the accuracy of $d\tilde{U}^{(e)}/dx$ at the center is $O(h^2)$, *one order better* than given by Eq. (7.33) for any general location. Since the flux is the product of $\alpha(x)$ and the derivative, then the accuracy of the flux is also $O(h^2)$; that is,

$$|\tau(x_c) - \tilde{\tau}^{(e)}(x_c)| = O(h^2) \qquad \text{as } h \to 0 \tag{7.35}$$

where x_c is the center of the element.

We conclude that the center of the element is the best point in the element to evaluate, or "sample," the flux in Eq. (7.27). This special location is called an *optimal flux-sampling point*,[8] and it is said that the flux is *superconvergent* at this point [7.4] (see Exercise 7.1).

In the argument above we considered the case where $\alpha(x)$ is a constant. If $\alpha(x)$ varies over the element, then the orthogonal polynomials corresponding to $\alpha(x)$ as a weighting factor would not be Legendre polynomials but some other kind, and therefore the zero (optimal point) would likely shift from the center to some other point. In practice it would necessitate very complicated programming to calculate the optimal points for any arbitrary $\alpha(x)$. However, this is not really necessary because the entire theory is *asymptotic*, i.e., valid only in the limit of "small-enough" elements. Therefore any variable $\alpha(x)$ can be approximated by a constant within each element, and the error in such an approximation will become arbitrarily small as $h \to 0$. This is compatible with the asymptotic nature of the theory.

We had assumed in the argument above that there is no modeling error for $\alpha(x)$. Now let's drop that assumption and consider an approximate (i.e., different) representation, $\hat{\alpha}(x)$. Corresponding to $\hat{\alpha}(x)$ is an exact solution for the function $\hat{U}(x)$ and the flux $\hat{\tau}(x)$. Equations (7.28) and (7.29) would then be written as follows:

$$|\hat{\tau}(x) - \tilde{\tau}^{(e)}(x)| = \left| \left(-\hat{\alpha}(x) \frac{d\hat{U}(x)}{dx} \right) - \left(-\hat{\alpha}(x) \frac{d\tilde{U}^{(e)}(x)}{dx} \right) \right|$$

$$= |\hat{\alpha}(x)| \left| \frac{d\hat{U}(x)}{dx} - \frac{d\tilde{U}^{(e)}(x)}{dx} \right| \tag{7.36}$$

On the RHS we can apply the same arguments to the exact solution $\hat{U}(x)$ as we did before to $U(x)$, so that we can substitute $\hat{U}(x)$ for $U(x)$ in Eqs. (7.30) through (7.33). And Eq. (7.35) would become instead,

$$|\hat{\tau}(x_c) - \tilde{\tau}^{(e)}(x_c)| = O(h^2) \qquad \text{as } h \to 0 \tag{7.37}$$

The important point here is that convergence still occurs, and at the same rate, but *to a different limit* — namely, the exact solution corresponding to $\hat{\alpha}(x)$. This conclusion applies to any modeling approximations. It is therefore recommended that the analyst always be aware of the modeling approximations being made so that he or she will then look for convergence to the exact solution of the approximate model rather than of the original model in the problem statement.

---

[8] Usually called an *optimal stress-sampling point* by people in solid mechanics, since flux corresponds to stress in elasticity theory (see Chapter 16).

It should be noted that the optimal flux-sampling point occurs at the element center only for the present type of element, namely, C^0-linear. Chapter 8 will analyze three higher-order elements in which the element trial solutions are C^0-quadratic, C^0-cubic, and C^0-quartic. It will be shown there, by extension of the above argument, that there are two, three, and four optimal flux-sampling points, respectively, in those elements, and their locations are at points in addition to, or besides, the center point. In short, there is an infinite hierarchy of optimal flux-sampling points, corresponding to the zeroes of the infinite family of Legendre polynomials.

As a point of terminology, these zero points of the Legendre polynomials happen to also be the sampling points in *Gauss-Legendre quadrature*, a type of approximate numerical integration used extensively in FE analysis and explained later in Chapters 8 and 13. It has therefore been conventional in FE practice to refer to these zero points as *Gauss points* (tabulated in Fig. 8.12). We will adhere to this convention and from now on say that the optimal flux-sampling point for the C^0-linear element is at the Gauss point (corresponding to this type of element.) And later, in Chapters 8 and 13, we will refer to the optimal flux-sampling points in those elements as the Gauss points (corresponding to those types of elements).[9] This Gauss-point terminology will become more meaningful after reading Section 8.3.1.

It bears emphasizing that the above optimal flux-sampling point theory is an asymptotic theory; i.e., the elements must be "small enough" in order for the most accurate flux values to occur at the Gauss point(s). Conversely, if an element is too large, the theory is invalid, meaning that the values at the Gauss points are not necessarily the most accurate.

How small is "small enough"? This is generally not easy to determine in practice. For example, in a sequence of refined meshes, the Gauss-point values corresponding to the coarsest meshes may vary erratically from one mesh to the next. As the mesh is refined further, a steady convergence will probably be observed, but it may initially be at a rate that only approaches that predicted by the asymptotic theory (see Chapter 9). Also, with more complicated elements, such as those studied in Chapters 8 and 13, it is difficult to keep a Gauss point at the same coordinate location as a mesh is refined because the Gauss points are at "odd" locations in those elements (see Fig. 8.12). Therefore the rate of convergence of Gauss-point values at a given coordinate location usually won't be observable

---

[9] This convention of using the term "Gauss points" (rather than "zeroes of Legendre polynomials") to refer to the locations of the optimal flux-sampling points, has apparently been a source of some confusion. The term "Gauss points" tends to suggest that the existence, per se, of the optimal points is related to the use of approximate numerical quadrature — namely, the Gauss-Legendre quadrature commonly used to evaluate integrals in the element equations (see Chapter 8). On the contrary, the existence of the optimal points has nothing to do with numerical quadrature. The theory described above is quite abstract, involving functional analysis and polynomial approximation theory; as such, it implicitly assumes that any and all numerical computations, were they to be performed, would be exact. If, in fact, the integrals should be evaluated approximately, that wouldn't alter the existence of the optimal points, although their locations might be shifted slightly because of the errors in the approximation. Because of these observations, it would be *conceptually* more accurate (though verbally awkward, perhaps) to refer to the optimal points as the "zeroes of Legendre polynomials" rather than as "Gauss points," even though both sets of points are numerically identical since both are derived from properties of the Legendre polynomials.

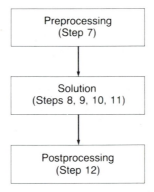

FIGURE 7.7 ● The three principal phases of an FE computer program.

in practical problems. Despite these difficulties, it would seem that the best bet is always to evaluate the flux at the Gauss points because the accuracy there will usually be better than anywhere else.

In summary, the program will calculate the flux within each (C^0-linear) element by evaluating Eq. (7.27) only at the center of each element. This completes steps 1 through 6, the theoretical development.

7.3 Implementation into a Computer Program: Numerical Computation (Steps 7 through 12)

7.3.1 Preprocessing, Solution, and Postprocessing

For purposes of computer programming, steps 7 through 12 may be logically grouped into three distinct phases: preprocessing, solution,[10] and postprocessing (Fig. 7.7). In large programs these phases may each comprise a wide variety of tasks. In fact, preprocessor and postprocessor programs have become so complicated that some on the market today are developed by companies separate from those that develop the solution programs. We will briefly describe each of the three phases, and at the same time provide a more programming-oriented description of steps 7 through 12 (recall Fig. 5.32).

Preprocessing refers to the preparation of all data necessary for the solution phase (Fig. 7.8). For most problems the bulk of the data is mesh definition and load definition. A good preprocessor program will automatically generate most of this bulk data and also check, alter, or optimize it.

The procedure usually goes something like this. A user begins an analysis with only a handful of data that describes the physical problem: some physical properties obtained from tables or experiments, some coordinates sufficient to define the shape of a body (i.e., the boundary of a domain), and perhaps a few simple loads or a few parameters describing complicated loads. A mesh-generator program then takes this initial coordinate data

---

[10] The solution phase is also frequently called the analysis phase.

Step 7: Specify numerical data.

1. Control parameters

   - Select the type of problem; other options.

2. Geometry

   - (Optional) Define the shape of the real physical object, or region of space, that is being modeled. This includes all external boundaries and any internal boundaries, e.g., holes, material interfaces, etc. This database could be used as input to a mesh-generation program, and it would interface with other computer-aided design and manufacturing processes.

   - Mesh generation
      - Nodes — define coordinates and assign node numbers.

      - Elements — define the pattern of node numbers associated with each element; calculate the bandwidth (or wavefront) of mesh.

      - Mesh plot — graphically display the mesh to verify its accuracy.

3. Material and physical properties

4. Loads

   - Define all interior and boundary loads.

5. Control of output data

   - Specify type of data desired and how it should be displayed. (Some or all of this may be done during postprocessing.)

FIGURE 7.8 ● Typical data input by a user or generated by a program during preprocessing.

Step 8: Evaluate the interior terms in the element equations for each element, and assemble the terms into system equations.

Form the system stiffness matrix $[K]$ and load vector $\{F\}$ by performing two steps for every element in the mesh:

1. Form the element stiffness matrix $[K]^{(e)}$ and load vector $\{F\}^{(e)}$ by evaluating terms in the element equations.

2. Assemble $[K]^{(e)}$ into $[K]$ and $\{F\}^{(e)}$ into $\{F\}$.

Step 9: Apply the boundary conditions, including the natural interelement boundary conditions, to the system equations.

1. Natural BCs and natural IBCs are unconstrained. Add values to $\{F\}$.

2. Essential BCs are constrained. Impose constraints by modifying $[K]$ and $\{F\}$.

Step 10: Solve the system equations.

Determine values for the parameters a_i by solving the set of linear algebraic equations $[K]\{a\} = \{F\}$. This yields the approximate solution $\bar{U}(x)$.

Step 11: Evaluate the flux.

Use the shape functions and the computed values of the parameters to compute the flux $\tilde{\tau}(x)$ [and any other quantities related to the flux] in some or all of the elements.

FIGURE 7.9 ● Principal activities during the solution phase.

and generates a mesh, creating coordinate values for all the nodes and node numbers for all the elements. A load-generator program may take the load parameters and create loads to be applied to various nodes or element boundaries or interiors. The user must supply commands to guide the mesh- and load-generation procedure. The result of the preprocessing is a greatly expanded volume of data, which is then input to the solution phase.

The *solution* phase includes steps 8, 9, 10, and 11 (Fig. 7.9). This entails extensive computation with the data input from the preprocessing phase, a process that is often graphically referred to as "number-crunching." The computer costs are usually greatest for this phase. The details shown in Fig. 7.9 are strictly for the boundary-value type of problem. The eigenproblem in Chapter 10 and the initial-boundary-value problem in Chapters 11 and 12 also use steps 8 through 11, but with a variation or two in the details that will be explained in those chapters.

Postprocessing involves the selection and display of data, and finally, an estimation of accuracy (Fig. 7.10). In some of the more sophisticated programs it may also include

---

Step 12: Display the solution and estimate its accuracy.

1. Select which data should be displayed. The data may be chosen according to many different criteria, such as

   - Values associated with a set of element numbers
   - Values associated with a set of node numbers
   - Values associated with a region of the domain defined by a range of coordinates
   - Values associated with a region occupied by a particular material
   - Values within a tolerance range

2. Print tabular lists of data

   - Node by node
   - Element by element
   - Perhaps sort the values from minimum to maximum

3. Plot graphs that meaningfully and attractively summarize large amounts of data
   - x,y-plots or x,y,z-plots of any variable(s) versus any other variable(s)
   - Bar graphs
   - Contour plots
       - Over all or part of the domain in 2-D problems
       - Over cross-sectional slices or surfaces in 3-D problems
   - Isometric views, with or without hidden line removal
   - Animated displays on CRTs, for time-dependent problems

4. (Optional) Generate new data from the FE solution by performing non-FE operations on the FE data. For example,

   - Summing (and differencing) or multiplying (and dividing)
   - Differentiating or integrating
   - Fourier transforms
   - Statistical measures (means, variances, correlations, etc.)

5. Estimate accuracy

   - Compare two or more solutions from a sequence of refined meshes
   - Compare solution(s) with experimental data or with a solution from another independent analysis

---

FIGURE 7.10 ● Typical activities during postprocessing

the generation of new data from the FE solution using non-FE calculations. In many problems the user is interested in obtaining information about only a few small regions of the
domain, yet the solution phase generates information over the entire domain.[11] For large
problems the amount of data produced can be so great that it becomes impractical for
a user to attempt to survey it all. One of the important functions of a postprocessor program is therefore to filter out only the important information and present it to the user
in a concise, meaningful manner.

7.3.2 The UNAFEM Program

An FE program called UNAFEM[12] implements most of the theory developed in this book.
The program follows the general structure outlined above. It is a complete, working program that is meant to provide training experience for beginners and to be an experimental
tool for more experienced users. The complete listing and data input instructions are in
the appendix.

Very few readers are likely to be interested in studying every line of the listing. (There
are about 4500 lines.) However, there are a few sections in the code that are the direct
implementation of the equations derived in the text. Most students have found it helpful
to peruse these sections since they reinforce the theory. Therefore these sections of the
listing will be reproduced in the appropriate chapters, along with a few comments. Lines
of code that are not pertinent to the immediate discussion (e.g., specification statements)
will be omitted from the chapter reproductions. A vertical column of dots will indicate
omitted lines of code.

Since the program is intended primarily for instruction, rather than for real applications, highest priority has been placed on readability and lower priority on computational
efficiency.[13] Since UNAFEM is meant to provide training for the typical commercial
and publicly available programs, we employ a generally accepted program structure. We
also employ good programming style, including many comment statements. As an aid
to beginners, there are almost three dozen error diagnostics, principally in the preprocessing
phase, to help find errors in the input data.

A user who wants to develop his or her own FE program for application to small-
scale problems will find UNAFEM a good model to emulate. For large-scale problems
and extensive applications, one might follow a similar logical flow, but it would be advisable
to incorporate more efficient storage schemes and data-management techniques.

---

[11] Step 10 always generates the solution for all the a_i parameters in the problem (although, in principle,
this is not absolutely necessary). The data generated in step 11 (e.g., flux) is optional, though almost
always wanted at least somewhere in the domain. For 2-D and especially 3-D problems, step 11 could
involve extensive computation, so some programs allow the user to control the amount of data generated
(by setting control parameters during preprocessing), thereby providing a significant reduction in the
volume of data from the solution phase, and a corresponding cost reduction.

[12] *Un*derstanding and *a*pplying the *f*inite *e*lement *m*ethod (pronounced YOO-na-fem).

[13] For small problems (on the order of 100 DOF), computational costs are almost negligible relative to
the cost of the user's time. Hence, small instructional programs that minimize user's time rather than
computer time may be close to optimal efficiency after all.

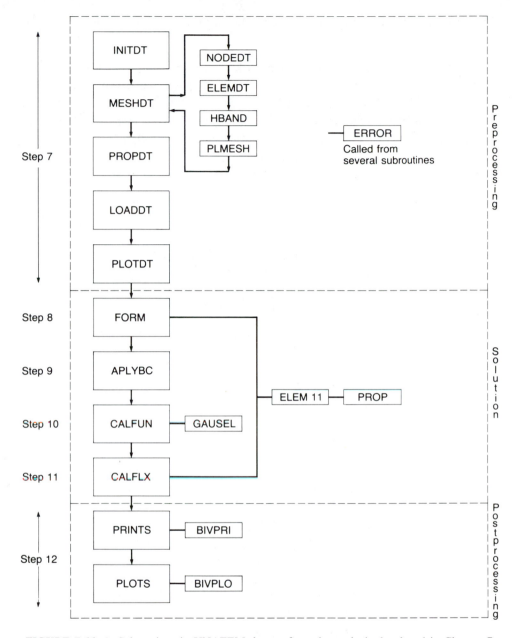

FIGURE 7.11 ● Subroutines in UNAFEM that perform the analysis developed in Chapter 7.

Figure 7.11 is a partial flowchart of UNAFEM, showing only those subroutines in UNAFEM that are used to perform the analysis developed in this chapter.[14] Figure 7.12 is the listing for the main routine, showing the calls to each of the principal subroutines in the flowchart.

---

[14] In subsequent chapters, as each new theoretical topic is developed, the flowchart will be reproduced, indicating the new additions to the program as a result of implementing the theory into code. In this way the reader can observe the steady growth of the program's capabilities.

Now let's take a more detailed look at each of the subroutines listed in Fig. 7.11. In particular we will point out those lines in the code corresponding to the equations derived in the theoretical development (steps 1 through 6) in Section 7.2.

The preprocessing phase • Subroutine INITDT calculates values for various constants that are used throughout the program, mostly in conjunction with the three types of elements

```
C                        ***************
C                        * U N A F E M *
C                        ***************
C
C            (UN)DERSTANDING AND (A)PPLYING THE (F)INITE (E)LEMENT (M)ETHOD
C
C
C            THIS PROGRAM WAS WRITTEN BY W. JOHN DENKMANN, OF AT&T
C            CONSUMER PRODUCT LABORATORIES, IN COLLABORATION WITH
C            DAVID S. BURNETT, OF AT&T BELL LABORATORIES, FOR THE
C            TEXTBOOK "FINITE ELEMENT ANALYSIS: FROM CONCEPTS TO
C            APPLICATIONS."
C
C            THE PROGRAM SOLVES THREE TYPES OF ONE- AND TWO-DIMENSIONAL
C            PROBLEMS USING THE FINITE ELEMENT METHOD:
C
C               1.  BOUNDARY-VALUE PROBLEMS
C                   (INCLUDES LAPLACE AND POISSON EQUATIONS)
C
C                  1-D
C                   -D/DX(ALPHA(X)*DU/DX) + BETA(X)*U = F(X)
C                  2-D
C                   -D/DX(ALPHAX(X,Y)*DU/DX) - D/DY(ALPHAY(X,Y)*DU/DY)
C                        + BETA(X,Y)*U = F(X,Y)
C
C               2.  EIGENPROBLEMS
C                   (INCLUDES STURM-LIOUVILLE, HELMHOLTZ, AND
C                    TIME-INDEPENDENT SCHROEDINGER EQUATIONS)
C
C                  1-D
C                   -D/DX(ALPHA(X)*DU/DX) + BETA(X)*U - LAMBDA*GAMMA(X)*U = 0
C                  2-D
C                   -D/DX(ALPHAX(X,Y)*DU/DX) - D/DY(ALPHAY(X,Y)*DU/DY)
C                        + BETA(X,Y)*U - LAMBDA*GAMMA(X,Y)*U = 0
C
C               3.  INITIAL-BOUNDARY-VALUE PROBLEMS
C                   (INCLUDES DIFFUSION EQUATION)
C
C                  1-D
C                   MU(X)*DU/DT - D/DX(ALPHA(X)*DU/DX) + BETA(X)*U = F(X,T)
C                  2-D
C                   MU(X,Y)*DU/DT - D/DX(ALPHAX(X,Y)*DU/DX)
C                    - D/DY(ALPHAY(X,Y)*DU/DY) + BETA(X,Y)*U = F(X,Y,T).
C
C            ALL LIMITS ARE SET IN SUBROUTINE INITDT
C
      COMMON /PROBLM/ DIMEN,PROB,TITLE(18)
      DATA XIBVP/'IBVP'/
      LOGICAL MORDAT
C----------------------------------------------------------------------------
C
C*************** PREPROCESSING (INPUT) *******************************
C
   10 CONTINUE
C--------------- INITIALIZATION DATA
      CALL INITDT
C--------------- MESH DATA
      CALL MESHDT
C--------------- PHYSICAL PROPERTY DATA
      CALL PROPDT
C--------------- LOAD DATA
   20 CONTINUE
```
(Continued)

FIGURE 7.12 • The main routine for UNAFEM.

FIGURE 7.12 ● (Continued)

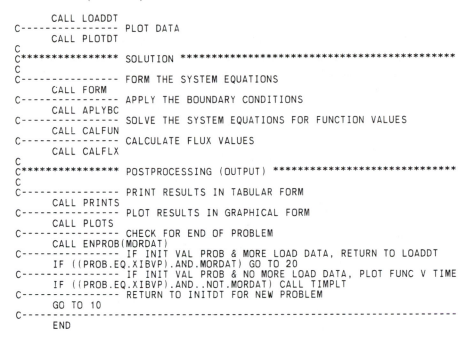

```
          CALL LOADDT
C---------------- PLOT DATA
          CALL PLOTDT
C
C*************** SOLUTION ***********************************
C
C---------------- FORM THE SYSTEM EQUATIONS
          CALL FORM
C---------------- APPLY THE BOUNDARY CONDITIONS
          CALL APLYBC
C---------------- SOLVE THE SYSTEM EQUATIONS FOR FUNCTION VALUES
          CALL CALFUN
C---------------- CALCULATE FLUX VALUES
          CALL CALFLX
C
C*************** POSTPROCESSING (OUTPUT) *********************
C
C---------------- PRINT RESULTS IN TABULAR FORM
          CALL PRINTS
C---------------- PLOT RESULTS IN GRAPHICAL FORM
          CALL PLOTS
C---------------- CHECK FOR END OF PROBLEM
          CALL ENPROB(MORDAT)
C---------------- IF INIT VAL PROB & MORE LOAD DATA, RETURN TO LOADDT
          IF ((PROB.EQ.XIBVP).AND.MORDAT) GO TO 20
C---------------- IF INIT VAL PROB & NO MORE LOAD DATA, PLOT FUNC V TIME
          IF ((PROB.EQ.XIBVP).AND..NOT.MORDAT) CALL TIMPLT
C---------------- RETURN TO INITDT FOR NEW PROBLEM
          GO TO 10
C----------------------------------------------------------------
          END
```

developed in Chapter 8. It also defines the limits that are currently set on the sizes of several parameters and associated arrays.

Subroutine MESHDT is a control program that sequentially calls four other subroutines that generate the mesh, calculate the half-bandwidth, and plot the mesh. Their functions are given below.

1. NODEDT (*node data*) defines each node with a unique node number and a location (*x*-coordinate). The nodes may be defined in any order — i.e., starting with any node in the mesh, continuing to any other node, etc., until all nodes have been defined. If the nodes in the mesh are numbered in some regular pattern (for example, sequentially), then the user can take advantage of NODEDT's modest node-generation capability: If there is a set of uniformly spaced nodes in the mesh, with successive node numbers incremented by the same value, then the user needs to define only the first and last nodes in the set and NODEDT will define the ones in between. For example, if node number 8 is defined at $x=6$ and node number 16 at $x=30$, with a node number increment of 2 for the nodes in between, then NODEDT will generate nodes 10, 12, and 14 at $x=12$, 18, and 24, respectively. (The data-input instructions in the appendix contain additional explanation for this and all the other input subroutines.)

2. ELEMDT defines each element with a unique element number and the numbers of the nodes attached to that element, as well as other data. The elements must be defined in numerical order, starting with the number 1; i.e., element number 1 is defined first, then element number 2, etc. ELEMDT has a modest element-generation capability: If there is a set of elements in which the element numbers form an uninterrupted sequence, and the node numbers in each element are incremented from those in the preceding element by the same amount, then the user needs to define only the first element in the set and ELEMDT will define all the others. For example, if ele-

ment number 12 with nodes 20 and 21 is specified to be the first of a set of five elements, with a node number increment of 3, then ELEMDT will generate the other four elements — namely, 13, 14, 15, and 16, with nodes 23 and 24, 26 and 27, 29 and 30, and 32 and 33, respectively.

3. HBAND calculates the half-bandwidth that the assembled system stiffness matrix will have (recall Fig. 5.25). This is determined solely by the configuration of elements and the node-numbering pattern. The value of the half-bandwidth is used later during solution of the assembled system equations.

4. PLMESH plots the mesh. This provides an immediate check on the mesh generation, usually an error-prone procedure, especially for 2-D and 3-D problems that involve a large amount of data handling.

Subroutine PROPDT defines the physical/material properties $\alpha(x)$ and $\beta(x)$. Values are read in for the coefficients of the quadratic polynomials in Eqs. (7.17) — namely, α_0, α_1, α_2, β_0, β_1, and β_2. The polynomials may be defined for as many as 20 different regions in the domain.

Subroutine LOADDT defines all the applied loads, i.e., in the interior and on the boundary. For the interior loads, values are read in for the constants $f^{(e)}$ in Eq. (7.24). A different value may be specified for each element, for as many elements as desired. There is a modest load-generation capability that will generate the same load $f^{(e)}$ over a set of elements. For example, if a load of 10 units is specified for elements 30 and 50, with an element number increment of 5, then LOADDT will generate a load of 10 units for elements 35, 40, and 45.

For the boundary loads, LOADDT reads in values for essential and natural boundary conditions, i.e., function and flux values. In addition, a discontinuity in the flux at inter-element boundaries (the physical meaning of such a discontinuity is explained in Sections 7.5.2 and 7.6) may be specified at any interelement nodes desired.

Subroutine PLOTDT reads in data that scales the x,y-axes for the various output plots (item 5 in Fig. 7.8). Subroutine ERROR contains the error diagnostic messages. All errors are considered fatal, terminating the execution of the program.

The solution phase • Figure 7.13 is a partial listing of subroutine FORM. This subroutine forms the system stiffness matrix $[K]$ and load vector $\{F\}$ by evaluating the terms in the element equations for each element in the mesh, and adding (i.e., assembling) these terms to the appropriate locations in $[K]$ and $\{F\}$. Most of the routine, therefore, is enclosed in a single loop (DO 500) over all the elements. Within the loop there are two steps: The element routine, ELEM11,[15] is called to evaluate the element stiffness matrix $[K]^{(e)}$ and the element load vector $\{F\}^{(e)}$, and then $[K]^{(e)}$ is added to $[K]$ and $\{F\}^{(e)}$ to $\{F\}$.

Before beginning the loop over all the elements, $[K]$ and $\{F\}$ must be "initialized"; i.e., all the locations that will be used in the K(I,J) and F(I) arrays must be set equal to zero. This initialization is necessary because, during assembly, terms from the element equations are added to the terms *already present* in K(I,J) and F(I) from other elements.

---

[15] UNAFEM has eight element routines, each for a different type of element. The names of the routines are ELEM11, ELEM12, ELEM13, ELEM14, ELEM21, ELEM22, ELEM23, and ELEM24. The first digit in the name refers to the dimension of the problem, either 1-D or 2-D. The second digit reflects the order in which the element is developed in the text.

```
      SUBROUTINE FORM
C
C          THIS ROUTINE FORMS THE SYSTEM MATRICES BY PERFORMING
C              TWO TASKS FOR EACH ELEMENT IN THE MESH:
C                 1.  FORMING THE ELEMENT MATRICES BY CALLING THE
C                     APPROPRIATE ELEMENT ROUTINES, AND
C                 2.  ASSEMBLING THE SYSTEM MATRICES FROM THE
C                     ELEMENT MATRICES.
                 .
                 .
                 .
C---------- INITIALIZE ARRAYS FOR THE SYSTEM MATRICES
                 .
                 .
                 .
C*********** BEGIN LOOP OVER ALL ELEMENTS *****************************
C
      DO 500 IELNO=1,NUMEL
          ITYPE = NTYPE(IELNO)
          NND = NNODES(ITYPE)
C---------- FORM ELEMENT MATRICES
          IF (ITYPE.EQ.11) CALL ELEM11(ELFORM)
                 .
                 .
                 .
C---------- ASSEMBLE THE SYSTEM MATRICES FROM THE ELEMENT MATRICES
          IF (PROB.EQ.BDVP) GO TO 100
          IF (PROB.EQ.EIGP) GO TO 200
          IF (PROB.EQ.XIBVP.AND.THETA.EQ.0.) GO TO 300
          IF (PROB.EQ.XIBVP.AND.THETA.NE.0.) GO TO 400
C
C************** ASSEMBLE MATRICES FOR BOUNDARY-VALUE PROBLEM ***********
C
  100     CONTINUE
          DO 140 I=1,NND
              II = ICON(I,IELNO)
              F(II) = F(II)+FE(I)
              DO 120 J=1,NND
                  JJ = ICON(J,IELNO)
                  JSHIFT = JJ-(II-1)
                  IF (JJ.GE.II) K(II,JSHIFT) = K(II,JSHIFT)+KE(I,J)
  120         CONTINUE
  140     CONTINUE
          GO TO 500
C
C************** ASSEMBLE MATRICES FOR EIGENPROBLEM ********************
C
  200     CONTINUE
                 .
                 .
                 .
          GO TO 500
C
C************** ASSEMBLE MATRICES FOR
C                  EXPLICIT INITIAL-BOUNDARY-VALUE PROBLEM *************
C
  300     CONTINUE
                 .
                 .
                 .
          GO TO 500
C
C************** ASSEMBLE MATRICES FOR
C                  IMPLICIT INITIAL-BOUNDARY-VALUE PROBLEM *************
C
  400     CONTINUE
                 .
                 .
                 .
  500 CONTINUE
      RETURN
C--------------------------------------------------------------------
      END
```

FIGURE 7.13 ● Subroutine FORM, a partial listing.

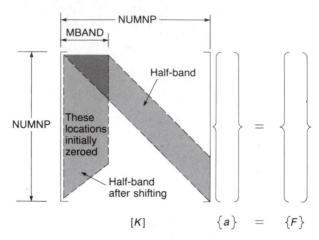

FIGURE 7.14 ● Shifting of the half-band in the stiffness matrix.

Therefore the *first* time a location in K(I,J) or F(I) has a term added to it, there should be a zero present.

UNAFEM stores the system stiffness matrix $[K]$ in a special way in order to utilize space in main memory more efficiently (Fig. 7.14). We recall from Fig. 5.25 that $[K]$ is typically banded, with a half-bandwidth of MBAND (calculated in subroutine HBAND). Since the equation-solver routine (GAUSEL — see the appendix) ignores all the zeroes outside the band, it would be very wasteful of main memory to store all these zeroes. In other words, there is no need to reserve a square array of NUMNP rows by NUMNP columns, where NUMNP (calculated in subroutine NODEDT) is the number of nodal points (DOF) in the problem.[16] Since $[K]$ is symmetric, locations need to be reserved only for terms above (or below), and including, the main diagonal. A popular way to do this is to shift each row over to the left until its main diagonal term is in column 1. This results in a rectangular array with NUMNP rows by MBAND columns, as shown in Fig. 7.14.

This shifting may be observed in two places in the listing in Fig. 7.13. During initialization a rectangular array of dimension NUMNP by MBAND is zeroed. During assembly, the stiffness terms are all shifted II-1 columns to the left, where II is the row number and therefore is the number of columns between column 1 and the main diagonal.

As regards the overall structure of the routine, note that there are four branches to four different assembly procedures. UNAFEM solves four different kinds of problems: boundary-value problems, eigenproblems, and two types of initial-boundary-value problems. Each involves some differences in the way data is stored and handled, resulting in differences in coding. This four-way branching is used in three of the principal subroutines in the solution phase: FORM, APLYBC, and CALFUN (recall Fig. 7.11).

Figure 7.15 is a partial listing of subroutine ELEM11 showing clearly the terms from the element equations [Eqs. (7.26)] in step 5. The load vector containing the boundary

---

[16] It is the number of DOF in a problem that determines the size of $[K]$. Since all the problems treated by UNAFEM have one DOF at each node, the number of nodes and the number of DOF can be used interchangeably in this program.

```
      SUBROUTINE ELEM11(OPT)
C
C          THIS SUBROUTINE IS THE MASTER ROUTINE FOR ELEMENT 11, A
C          2-NODE 1-D C0-LINEAR ELEMENT.
            .
            .
            .
C*************FORM ELEMENT MATRICES AND VECTORS*************************
C
      I = ICON(1,IELNO)
      J = ICON(2,IELNO)
      IPHY = NPHYS(IELNO)
C-----------EVALUATE TERMS IN ELEMENT EQUATIONS
      XCENT = (X(I)+X(J))/2.
      YCENT = 0.
      CALL PROP(IPHY,XCENT,YCENT,ALPHA,ALPHAY,BETA,GAMMA,MU)
      L = ABS(X(J)-X(I))
      KE(1,1) =   ALPHA/L+BETA*L/3.
      KE(1,2) = -ALPHA/L+BETA*L/6.
      KE(2,1) = KE(1,2)
      KE(2,2) = KE(1,1)
      FE(1) = FINT(IELNO)*L/2.
      FE(2) = FE(1)
            .
            .
            .
```

Eq. (7.18) → XCENT line

Eq. (7.26) → KE and FE block

FIGURE 7.15 ● Subroutine ELEM11, a partial listing showing only the evaluation of terms in the element equations.

flux terms is not included since values for these terms are applied later to the assembled system equations. Note the call to subroutine PROP, which evaluates the properties $\alpha(x)$ and $\beta(x)$ at the center of the element, according to Eqs. (7.19) in step 4. Figure 7.16 is a partial listing of subroutine PROP, showing only the terms used in 1-D boundary-value problems. [The subroutine also evaluates properties for eigenproblems (Chapter 10), initial-boundary-value problems (Chapter 11), and the 2-D counterparts of all the 1-D problems (Chapters 13-15). In particular, note that in 2-D the property α has x- and y-components $\alpha_x(x,y)$ and $\alpha_y(x,y)$; hence, α for the 1-D problem is stored in the array used for the α_x, which is called ALPHAX.]

Figure 7.17 is a partial listing of subroutine APLYBC, which applies both the essential and natural boundary conditions to the system stiffness matrix and load vector. The natural boundary loads, stored in array TAUNBC(N), are simply added to the system load vector F(I). The application of an essential BC follows the technique for applying a constraint equation explained in Sections 4.4 and 5.2. Recall that the technique requires deleting

```
      SUBROUTINE PROP(IPHY,X,Y,AX,AY,B,G,M)
C
C          THIS ROUTINE CALCULATES THE PHYSICAL PROPERTIES
            .
            .
            .
      AX =   ALPHAX(1,IPHY)   + ALPHAX(2,IPHY)*X   + ALPHAX(3,IPHY)*X*X
            .
            .
      B  =   BETA(1,IPHY)   +   BETA(2,IPHY)*X   +   BETA(3,IPHY)*X*X
            .
            .
            .
```

Eqs. (7.17)

FIGURE 7.16 ● Subroutine PROP, a partial listing showing only 1-D boundary-value calculations.

```
      SUBROUTINE APLYBC
C
C        THIS ROUTINE APPLIES THE NATURAL AND ESSENTIAL BOUNDARY
C             CONDITIONS TO THE SYSTEM MATRICES.
                .
                .
                .
C******** APPLY THE NONZERO NATURAL BOUNDARY CONDITIONS, IF ANY, *******
C                BY ADDING TERMS TO LOAD VECTOR
C
      IF (NUMNBC.EQ.0) GO TO 90
C
      DO 80 N=1,NUMNBC
          II = INP(N)
          JJ = JNP(N)
          KK = KNP(N)
C-------- IF JJ EQ 0 WE HAVE A CONCENTRATED FLUX AT NODE II
          IF (JJ.NE.0) GO TO 20
          F(II) = F(II)+TAUNBC(N)
          GO TO 80
  20      CONTINUE
                .
                .
  80 CONTINUE
  90 CONTINUE
C******** APPLY THE ESSENTIAL BOUNDARY CONDITIONS, IF ANY, ************
C                BY IMPOSING CONSTRAINT EQUATIONS
C
      IF (NUMEBC.LE.0) RETURN
      IF (PROB.EQ.BDVP) GO TO 100
      IF (PROB.EQ.EIGP) GO TO 200
      IF (PROB.EQ.XIBVP.AND.THETA.EQ.0.) GO TO 300
      IF (PROB.EQ.XIBVP.AND.THETA.NE.0.) GO TO 400
C
C********** MODIFY BOUNDARY-VALUE PROBLEM
C
 100 CONTINUE
      DO 140 IEB=1,NUMEBC
          I = IUEBC(IEB)
          DO 120 J=2,MBAND
              IR1 = I+(J-1)
              IR2 = I-(J-1)
              IF (IR1.LE.NUMNP) F(IR1)=F(IR1)-K(I,J)*UEBC(IEB)
              IF (IR1.LE.NUMNP) K(I,J)=0.
              IF (IR2.GT.0) F(IR2)=F(IR2)-K(IR2,J)*UEBC(IEB)
              IF (IR2.GT.0) K(IR2,J)=0.
  120     CONTINUE
          F(I) = UEBC(IEB)
          K(I,1) = 1.
 140 CONTINUE
      RETURN
C
C********** MODIFY EIGENPROBLEM
C
 200 CONTINUE
                .
                .
C********** MODIFY EXPLICIT INITIAL-BOUNDARY-VALUE PROBLEM
C
 300 CONTINUE
                .
                .
C********** MODIFY IMPLICIT INITIAL-BOUNDARY-VALUE PROBLEM
C
 400 CONTINUE
                .
                .
                .
```

FIGURE 7.17 ● Subroutine APLYBC, a partial listing.

a row and column. This is awkward to do on a computer because the missing row and column would cause problems in the equation-solving algorithm. One resolution would be to shift all the remaining rows and columns over (or up) one, to fill in the gap. But this could involve a great deal of data manipulation. Fortunately, there is a much simpler solution: Replace the deleted row and column by zeroes, put a 1 on the main diagonal at the intersection of the row and column, and add the value of the essential BC to the load vector. This is equivalent to replacing the deleted row by the constraint equation.

To illustrate, consider the following 4-by-4 system equations:

$$\begin{bmatrix} K_{11} & K_{12} & K_{13} & K_{14} \\ K_{21} & K_{22} & K_{23} & K_{24} \\ K_{31} & K_{32} & K_{33} & K_{34} \\ K_{41} & K_{42} & K_{43} & K_{44} \end{bmatrix} \begin{Bmatrix} a_1 \\ a_2 \\ a_3 \\ a_4 \end{Bmatrix} = \begin{Bmatrix} F_1 \\ F_2 \\ F_3 \\ F_4 \end{Bmatrix} \tag{7.38}$$

and the essential BC,

$$a_2 = d \tag{7.39}$$

Applying Eq. (7.39) to Eq. (7.38) in the manner described in step 9 in Section 5.2 would yield

$$\begin{bmatrix} K_{11} & & K_{13} & K_{14} \\ & & & \\ K_{31} & & K_{33} & K_{34} \\ K_{41} & & K_{43} & K_{44} \end{bmatrix} \begin{Bmatrix} a_1 \\ a_3 \\ a_4 \end{Bmatrix} = \begin{Bmatrix} F_1 - K_{12}d \\ F_3 - K_{32}d \\ F_4 - K_{42}d \end{Bmatrix} \tag{7.40}$$

which shows the deleted second row and column. Performing the above operations on Eq. (7.40) would then yield

$$\begin{bmatrix} K_{11} & 0 & K_{13} & K_{14} \\ 0 & 1 & 0 & 0 \\ K_{31} & 0 & K_{33} & K_{34} \\ K_{41} & 0 & K_{43} & K_{44} \end{bmatrix} \begin{Bmatrix} a_1 \\ a_2 \\ a_3 \\ a_4 \end{Bmatrix} = \begin{Bmatrix} F_1 - K_{12}d \\ d \\ F_3 - K_{32}d \\ F_4 - K_{42}d \end{Bmatrix} \tag{7.41}$$

Equation (7.41) preserves the 4-by-4 structure of Eq. (7.38) as well as the locations of all terms not in the second row or column. The second row in Eq. (7.41) is clearly Eq. (7.39). Note that the coding in APLYBC takes into account the shifting of the half-band in $[K]$.

Figure 7.18 is a partial listing of subroutine CALFUN (calculate function) which controls the solution of the system equations for each of the four types of problems in UNAFEM. For the boundary-value problem, CALFUN merely calls another subroutine, GAUSEL, the algebraic equation solver that implements the Gaussian elimination method. The listing for GAUSEL is in the appendix. As noted previously, GAUSEL avoids operating

```
      SUBROUTINE CALFUN
C
C          THIS ROUTINE SOLVES THE SYSTEM EQUATIONS FOR THE FUNCTION VALUES
             .
             .
             .
C--------------- BRANCH TO PROPER EQUATION SOLVING METHOD
        IF (PROB.EQ.BDVP) GO TO 100
        IF (PROB.EQ.EIGP) GO TO 200
        IF (PROB.EQ.XIBVP.AND.THETA.EQ.0.) GO TO 300
        IF (PROB.EQ.XIBVP.AND.THETA.NE.0.) GO TO 400
C
C*************** BOUNDARY-VALUE PROBLEM
C
   100 CONTINUE
        CALL GAUSEL(NUMNP,MBAND,K,F,A,0)
        RETURN
C
C**************** EIGENPROBLEM
C
   200 CONTINUE
             .
             .
             .
C*************** EXPLICIT INITIAL-BOUNDARY-VALUE PROBLEM
C
   300 CONTINUE
             .
             .
             .
C************** IMPLICIT INITIAL-BOUNDARY-VALUE PROBLEM
C
   400 CONTINUE
             .
             .
             .
```

FIGURE 7.18 ● Subroutine CALFUN, a partial listing.

on all the zeroes outside the band. This provides a reasonably efficient algorithm for small-to-medium-size problems (say up to one or two thousand DOF), and hence is suitable for a program like UNAFEM. For much larger problems, additional efficiencies can be gained by more sophisticated data-management techniques [7.5 through 7.7].

Figure 7.19 is a partial listing of subroutine CALFLX. This subroutine controls the calculation of fluxes in each element by calling the element subroutine ELEM11 for every element in the mesh. Figure 7.20 shows the portion of subroutine ELEM11 that does the flux calculations.

We recall from the discussion in step 6 above that the optimal flux-sampling point for the present 1-D C^0-linear element is at the center (the Gauss point). Figure 7.20 shows the flux being evaluated using Eq. (7.27), in which the property $\alpha(x)$ has been evaluated at the center (XCENT) by a call to subroutine PROP. In addition to the center value, ELEM11 also calculates the fluxes at the two boundary nodes, but *not* by using Eq. (7.27). This requires an explanation.

It was argued in step 6 that, for a "small-enough" element, the Gauss point is the most accurate location in the element to evaluate the flux expression based on the trial solution, Eq. (7.27). And it was therefore recommended that Eq. (7.27) be evaluated *only* at the Gauss point and that it be ignored everywhere else. This is all quite reasonable but it leaves us with another problem: We have only a single point value of flux inside each element.

```
        SUBROUTINE CALFLX
C
C          THIS ROUTINE CONTROLS THE CALCULATION OF FLUXES IN EACH
C             ELEMENT, AND CALCULATES THE NODAL FLUXES.
                .
                .
                .
C---------INITIALIZE ARRAYS
                .
                .
                .
C---------CALCULATE FLUXES AT GAUSS POINTS AND NODES IN EACH ELEMENT
        DO 100 IELNO=1,NUMEL
            ITYP = NTYPE(IELNO)
            NND = NNODES(ITYP)
            IF (ITYP.EQ.11) CALL ELEM11(FLUX)
                .
                .
                .
    100 CONTINUE
C---------CALCULATE NODAL FLUXES BY AVERAGING THE SUMMED ELEMENT
C             NODAL FLUXES FROM ELEMENT SUBROUTINES
        DO 150 I=1,NUMNP
            AIU = IU(I)
            IF (IU(I).GT.0) TAUXNP(I) = TAUXNP(I)/AIU
                .
                .
    150 CONTINUE
        RETURN
C----------------------------------------------------------------------
        END
```

IF (IU(I).GT.0) TAUXNP(I) = TAUXNP(I)/AIU ←——— Dividing $\Sigma\tilde{\tau}$ by n, in Eq. (7.43)

FIGURE 7.19 ● Subroutine CALFLX, a partial listing.

```
        SUBROUTINE ELEM11(OPT)
C
C          THIS SUBROUTINE IS THE MASTER ROUTINE FOR ELEMENT 11, A
C          2-NODE 1-D CO-LINEAR ELEMENT.
                .
                .
                .
C***************CALCULATE FLUXES*****************************************
C
    200 CONTINUE
        I = ICON(1,IELNO)
        J = ICON(2,IELNO)
        IPHY = NPHYS(IELNO)
C---------CALCULATE FLUXES AT THE GAUSS POINT (CENTER).
        XGP(1,IELNO) = (X(I)+X(J))/2.
        YGP(1,IELNO) = 0.
        XCENT = XGP(1,IELNO)
        YCENT = 0.
        CALL PROP(IPHY,XCENT,YCENT,ALPHA,ALPHAY,BETA,GAMMA,MU)
        TAUXGP(1,IELNO) = -ALPHA*(A(J)-A(I))/(X(J)-X(I))
        TAUYGP(1,IELNO) = 0.
C---------CALCULATE ELEMENT NODAL FLUXES AND SUM FOR NODAL FLUX
C             CALCULATION IN CALFLX (ELEMENT NODAL FLUXES SAME AS
C             GAUSS POINT FLUX)
        TAUXNP(I) = TAUXNP(I) + TAUXGP(1,IELNO)
        TAUXNP(J) = TAUXNP(J) + TAUXGP(1,IELNO)
        IU(I) = IU(I)+1
        IU(J) = IU(J)+1
        RETURN
C----------------------------------------------------------------------
        END
```

TAUXGP(1,IELNO) = -ALPHA*(A(J)-A(I))/(X(J)-X(I)) ←———Eq. (7.27)

TAUXNP(I) = TAUXNP(I) + TAUXGP(1,IELNO) ⎫ $\Sigma\tilde{\tau}$ in Eq. (7.43)
TAUXNP(J) = TAUXNP(J) + TAUXGP(1,IELNO) ⎭

FIGURE 7.20 ● Subroutine ELEM11, a partial listing showing only the flux calculations.

It is usually desirable to also know the flux at the nodes, especially those on the boundary of each element. There are several reasons for this. First, nodal values are generally easier to work with because if a flux value is desired at a specific location, it is easier to place a node at that point (during mesh generation) than to define an element such that its Gauss point coincides with the desired location.[17] Second, values at nodes on interelement boundaries exhibit the characteristic flux discontinuities that are useful for error estimation. Finally, it is easier to construct contour plots in 2-D and 3-D problems using nodal values.

Nodal fluxes can be calculated that have generally better accuracy than that obtainable from Eq. (7.27). The process involves two operations, called *flux smoothing* and *flux averaging*.[18] These operations are usually only applied to 2-D and 3-D elements, where it is considerably more effective (see Chapter 13). Nevertheless, the principle can easily be extended to 1-D elements where, as usual, it is easier to learn. Therefore we will introduce the basic ideas here, although it should be recognized at the outset that in 1-D problems the process can sometimes degenerate to a trivial or pointless exercise. It is expected that the reader will gain a fuller appreciation for the process after seeing the 2-D applications in Chapter 13.

Flux smoothing can be done in many ways (references are cited in Chapter 13), although there is one approach, called *local* smoothing, that is used in many commercial FE programs and will be used in UNAFEM. In this approach we construct a smoothing curve (also called a smoothing function) *within each element*. Each smoothing curve is a polynomial constructed by either interpolating or least-squares-fitting some or all of the Gauss-point flux values.[19] Flux values at the nodes are then determined by evaluating

---

[17] Probably the only exception to this argument is the present 1-D C^0-linear element. For this type of element it is just as easy to define an element in a mesh so that either its center or its nodes are at a particular point. For all other types of elements (see Chapters 8 and 13) the Gauss points are not located at such geometrically simple positions within the element. Therefore, when generating a mesh with those types of elements, it would be more difficult to locate the nodes for an element so that a Gauss point falls at a particular point.

[18] People working in solid mechanics usually refer to these operations as ''stress smoothing'' and ''stress averaging.''

[19] Interpolation may be viewed as a special case of least-squares-fitting. The former occurs when the number of Gauss-point values *equals* the number of coefficients in the smoothing polynomial, thereby forcing the polynomial to exactly equal the Gauss-point value at each Gauss point. The latter occurs when the number of Gauss-point values *exceeds* the number of coefficients, in which case the polynomial will generally not equal any of the Gauss-point values exactly, but will be close to all of them in a ''least-squares sense.'' Some program developers prefer interpolating; others prefer least-squares-fitting. There do not yet exist any theoretical arguments, based on error analyses, to favor either interpolation or least-squares-fitting for a general class of problems. The choice seems to be based on a combination of factors such as ease of coding, numerical experiments, and intuition. We will demonstrate both approaches by using interpolation with the 1-D elements and least-squares-fitting with the 2-D elements. When interpolating or least-squares-fitting, it is not necessary to use all the Gauss-point values within the element. In some of the more complicated types of 2-D and 3-D elements, which contain many Gauss points, satisfactory performance and a more efficient algorithm may be obtained by interpolating only some of the Gauss-point values. Again, theoretical guidelines are lacking.

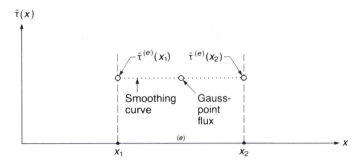

FIGURE 7.21 ● Determination of the element nodal fluxes, $\bar{\tau}^{(e)}(x_1)$ and $\bar{\tau}^{(e)}(x_2)$, in a 1-D C^0-linear element by interpolating the Gauss-point flux.

the smoothing curve at the nodes. These nodal flux values are commonly referred to as *element nodal fluxes*; the adjective "element" reminds us that these nodal values are derived from the solution *within a single element* and therefore represent only part of the total flux at a node that is shared by two or more elements (see the discussion of flux averaging below). In summary: *The element nodal fluxes are calculated by interpolating (or least-squares-fitting) the Gauss-point flux(es)* — rather than by using Eq. (7.27). The element nodal fluxes will generally be more accurate, as the element size becomes "small enough," than those calculated from Eq. (7.27), because the smoothing polynomial is defined by the more accurate Gauss-point fluxes.

Let's apply local smoothing to the present C^0-linear element. Since there is only one Gauss-point value, and since one point can only define a constant, the smoothing curve is a constant over the entire element, equal to the Gauss-point value, as shown in Fig. 7.21. The nodal values are therefore the same as the Gauss-point values for this type of element. (This is almost a trivial application of a smoothing curve; subsequent elements will be a bit more interesting. For example, the next-higher-order element in Chapter 8, the 1-D C^0-quadratic element, has two Gauss points, so its smoothing curve is a straight line interpolation through the two Gauss-point values.)

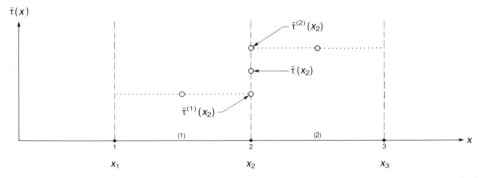

FIGURE 7.22 ● The nodal flux $\bar{\tau}(x_2)$ at an interelement node is the average of the two element nodal fluxes, $\bar{\tau}^{(1)}(x_2)$ and $\bar{\tau}^{(2)}(x_2)$.

We note that if $\alpha(x)$ is a constant, then $\tilde{\tau}^{(e)}(x)$ in Eq. (7.27) is a constant and therefore the smoothing curve is the same as the trial solution. In this special case, smoothing has accomplished nothing (the "degenerate" case referred to above). However, when the element nodal fluxes are subsequently averaged (see below), there will be a definite improvement in accuracy over Eq. (7.27). By way of contrast, in 2-D and 3-D elements the smoothing functions generally differ significantly from the trial solution (whether or not the material properties are constant), resulting in some cases in a dramatic improvement in the accuracy of the element nodal fluxes (see Chapter 13).

At interelement nodes there will be two element nodal fluxes, one from each element on either side of the node. The second operation, *flux averaging*, calculates the mean of these two fluxes, as shown in Fig. 7.22. Thus at node 2 between elements (1) and (2) there is an element nodal flux calculated from the element on the left, $\tilde{\tau}^{(1)}(x_2)$, and another element nodal flux from the element on the right, $\tilde{\tau}^{(2)}(x_2)$. The average value at node 2 is therefore

$$\tilde{\tau}(x_2) = \frac{1}{2}\left[\tilde{\tau}^{(1)}(x_2) + \tilde{\tau}^{(2)}(x_2)\right] \tag{7.42}$$

More generally, for problems in any dimension, if n elements share the node i at the location x_i, then the average flux will be

$$\tilde{\tau}(x_i) = \frac{1}{n}\sum_{(e)}\tilde{\tau}^{(e)}(x_i) \tag{7.43}$$

where the sum is over all elements sharing the node i.

The smoothed and averaged fluxes $\tilde{\tau}(x_2)$ and $\tilde{\tau}(x_1)$ are referred to simply as *nodal fluxes*. They are the total (complete) flux associated with a node, and are the nodal values we set out to calculate. They represent our best (most accurate) evaluation of the flux at a node.

Equations (7.42) and (7.43) are unweighted averages; they do not account for differences in sizes of the elements being averaged. Weighted averages have been used in some commercial FE programs; however, it requires more complicated programming and it represents more of an optimizing feature than a basic need, so UNAFEM will settle for Eq. (7.42). As a compromise, though, UNAFEM will *graphically* weight the element nodal fluxes in the flux plots, as explained below for the output subroutines.

The listing for ELEM11 (Fig. 7.20) shows the element nodal fluxes $\tilde{\tau}^{(e)}(x_i)$ in Eq. (7.43) being summed. Thus the element nodal flux from the current element is added to the Ith position in the TAUXNP array, corresponding to node I. In subsequent calls to ELEM11, if there are any other elements sharing node I, their element nodal flux at node I will also be added to position I in the TAUXNP array. In this way the TAUXNP array accumulates the sum of element nodal fluxes at each node. At the same time, the "counter" array IU accumulates for each node the number of element nodal fluxes added to TAUXNP; this corresponds to the number n in Eq. (7.43). After subroutine CALFLX (Fig. 7.19) completes the loop over all elements, each flux sum in the TAUXNP array is finally divided by its corresponding n value in the IU array.

The purpose of flux averaging is to provide a single value of flux at each interelement node, and thereby establish *continuity* in the flux at interelement boundaries. In other words, it removes the inherent discontinuities between elements. In addition, the averaged values are usually more accurate than any of the separate unaveraged values from each element.

Continuity is desirable because in most applications the exact flux solution is continuous. However, there are two important situations where the exact flux is discontinuous: (1) at concentrated loads in the interior of the domain (explained in Section 7.5.2) and (2) at interfaces between discontinuous physical or material properties in 2-D and 3-D problems, but not 1-D problems (explained in Section 7.4.3). Averaging would clearly not be desirable at points where such discontinuities occur.

The postprocessing phase • There are two types of postprocessing output: printed and graphical. The printed output employs standard ANSI Fortran conventions that should be compatible with a wide variety of computer installations. The graphical output uses DISSPLA [7.8], a commercially available graphical software package. DISSPLA is widely used in many industries although, unfortunately, it will probably not be readily available to every reader of this book. If it is not available, one can either remove the graphics from UNAFEM (a simple procedure) and run the program without it, or else replace the graphics with one's own in-house graphics software. The subroutine listings for the postprocessing may be found in the appendix.

Subroutine PRINTS is a small control routine that calls the appropriate printout routine for the type of problem being modeled. In particular, subroutine BIVPRI handles the printout for both boundary-value and initial-boundary-value problems. BIVPRI prints a listing of quantities associated with each element (the "element output") and a listing of quantities associated with each node (the "nodal output"). The element output lists for each element: the coordinate of each Gauss point, x, and the flux at each Gauss point, $\tilde{\tau}(x)$. The nodal output lists for each node: the coordinate x, the function $\tilde{U}(x)$, and the nodal flux $\tilde{\tau}(x)$. (The latter, of course, is the smoothed and averaged flux described previously).

Subroutine PLOTS is a small control routine, analogous to PRINTS, that calls the plotting routine BIVPLO for both boundary-value and initial-boundary-value problems. BIVPLO creates two plots: (1) the function solution, $\tilde{U}(x)$ versus x, and (2) the flux solution, $\tilde{\tau}(x)$ versus x. The function plot displays the solution $\tilde{U}^{(e)}(x)$ within each element; in fact, the curve is drawn element by element.

The flux plot is treated in a different manner from the function plot, a reflection of the special way that fluxes are calculated. Since the most accurate flux values are at the Gauss points, we plot only the Gauss-point values and then connect successive values with straight lines, proceeding from left to right across the plot, as shown in Fig. 7.23 for three adjacent elements.

It would have been reasonable to also plot the (smoothed and averaged) nodal fluxes. However, the straight line has the effect of providing a *weighted* average at the interelement nodes, as illustrated by the large and small elements in Fig. 7.23. Thus the straight line crosses the interelement node at a flux value that is closer to the value in the small element than to the value in the large element. This should be a better approximation since the node is closer to the Gauss point in the small element than it is to the Gauss point in the large element. When two adjacent elements are very different in size this graphically weighted nodal flux will differ appreciably from the analytically averaged (but nonweighted) value in the printed output. Not surprisingly, many numerical experiments have shown that the weighted average is almost always more accurate than the nonweighted average.

The plot also draws tiny circles at the element nodal fluxes on the boundary of each element — for example, $\tilde{\tau}^{(1)}(x_2)$ and $\tilde{\tau}^{(2)}(x_2)$ in Eq. (7.42). The spread in these values at

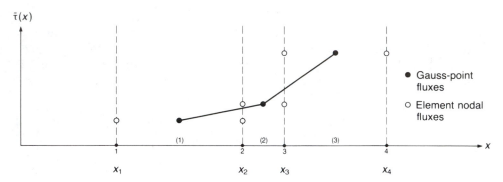

FIGURE 7.23 ● The method used to plot a flux solution in UNAFEM for 1-D C^0-linear elements.

the interelement nodes shows the interelement flux discontinuities and therefore affords a rough visual estimate of accuracy, as was illustrated in Chapter 5. However, in Chapter 5 we did not have the averaged values at the interelement nodes, and these are almost always of much better accuracy than the unaveraged values. Consequently, the spread in the unaveraged values, indicated by the tiny circles, will generally overestimate the error in the flux. A more reliable estimate of accuracy may be obtained by comparison with a second solution, one using a more refined (or a coarser) mesh.

There is no straight line drawn from the first and last Gauss-point values (the ones nearest the domain boundaries) to the boundary of the domain. There are a couple of reasons for this. First, the only value available on the boundary is an element nodal flux. This cannot be averaged with one from an adjacent element so it tends to be less accurate than a (averaged) nodal flux value or a Gauss-point value. Therefore, for the sake of consistency as regards accuracy, we prefer not to connect a Gauss-point value with an element nodal flux value. Second, this problem can be partly remedied by having a user tell the program how the solution is expected to approach the boundary. For example, a symmetry boundary (see the example in Chapter 13) would require a zero slope. This data could then be used along with one or more of the nearest Gauss-point values in order to extrapolate the latter values to the boundary. This approach, however, adds too much complication to a beginning instructional program. It would seem that a simple yet acceptable solution to this "boundary problem" is not to draw the Gauss-point curve to the boundary and settle instead for only plotting the tiny circle representing the (usually less accurate) element nodal flux on the boundary.

The smoothing curve itself is not displayed on the plots. Numerical experiments have shown that at most points within the element the smoothing curve is generally not as accurate as the Gauss-point fluxes or the (smoothed and averaged) nodal fluxes. This is especially evident with very coarse meshes. A secondary consideration is that plotting the smoothing curve complicates the programming. For these and other reasons dealing with program modularity, the smoothing curves are rarely plotted in FE programs.

In summary, it is generally sufficient to plot only point values of flux, i.e., Gauss-point fluxes and/or (smoothed and averaged) nodal fluxes. Numerical experiments show that these values are generally the most accurate in the mesh. For relatively coarse meshes this improvement in accuracy (over other locations) can frequently be quite significant.

In 1-D problems these point values can then be connected by straight lines to produce $\tilde{T}$ versus x plots. In 2-D and 3-D problems the point values can be used to construct contour plots.

It should be remembered that graphical plots are low-accuracy media that can be read to about two significant figures at best. Their value is in providing a bird's-eye view rather than a microscopic view. More specifically, they provide a compact presentation of a large amount of data, enabling the user to see overall patterns, trends, and possible trouble spots. It is usually the first output a user examines. Based on the plot information, the user may then wish to obtain more precise data in a particular location of the domain. This "microscopic" view may be obtained from the printed output, the purpose of which is to report the most precise representation possible of all the computed data.

7.4 Application of UNAFEM to a Heat Conduction Problem

7.4.1 Statement of the Problem

Figure 7.24 depicts a thin, cylindrical rod, 1 m long, composed of two different materials: A center section 20 cm long is made of copper, and two end sections, each 40 cm long, are made of steel. The cross section is circular, with a radius of 2 cm. Heat is flowing into the left end at a steady rate of 0.1 cal/sec-cm². The temperature of the right end is maintained at a constant 0°C. The rod is in contact with air at an ambient temperature of 20°C, so there is free convection from the lateral surface.

This type of problem was described in Section 6.1.2. The governing differential equation and domain [see Eq. (6.7b)] are[20]

$$-\frac{d}{dx}\left(k(x)\,\frac{dT(x)}{dx}\right) + \frac{hl}{A}\,T(x) = Q(x) + \frac{hlT_\infty}{A} \qquad 0 < x < 100 \text{ cm} \qquad \text{(7.44a)}$$

where

$$k(x) = \begin{cases} k_{\text{steel}} & 0 < x < 40 \\[2mm] k_{\text{copper}} & 40 < x < 60 \\[2mm] k_{\text{steel}} & 60 < x < 100 \end{cases} \qquad \begin{array}{l} \left(k_{\text{steel}} = 0.12\,\dfrac{\text{cal-cm}}{\text{sec-cm}^2\text{-}°\text{C}}\right) \\[4mm] \left(k_{\text{copper}} = 0.92\,\dfrac{\text{cal-cm}}{\text{sec-cm}^2\text{-}°\text{C}}\right) \end{array}$$

$$\frac{hl}{A} = \frac{h2\pi r}{\pi r^2} \qquad \left(h = 1.5 \times 10^{-4}\,\frac{\text{cal}}{\text{sec-cm}^2\text{-}°\text{C}}\right)$$

$$= 1.5 \times 10^{-4}\,\frac{\text{cal}}{\text{sec-cm}^3\text{-}°\text{C}}$$

$$\frac{hlT_\infty}{A} = 3.0 \times 10^{-3}\,\frac{\text{cal}}{\text{sec-cm}^3} \qquad\qquad\qquad \text{(7.44b)}$$

---

[20] This problem uses cgs units: centimeters, grams, seconds, calories (gram-calories), and degrees centigrade.

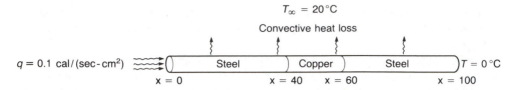

FIGURE 7.24 ● Heat conduction along a rod, with convection from the lateral surface.

$Q(x) = 0$

$T(x)$ = average temperature over cross section, at position x

$q(x)$ = average flux $-kdT/dx$ over cross section, at position x

The boundary conditions are

$$q(0) = 0.1 \frac{\text{cal}}{\text{sec-cm}^2} \qquad \text{natural BC}$$

$$T(100) = 0°C \qquad \text{essential BC} \qquad (7.45)$$

We want to use the UNAFEM computer program to calculate the distribution of temperature and flux, i.e., the functions $T(x)$ and $q(x)$, along the entire length of the rod.

7.4.2 One-Element Test of UNAFEM

Let's develop good habits right from the beginning. You the reader are about to use a program you have never seen or used before (presumably). The people who wrote it have their idiosyncrasies which may make parts of the input instructions or output format seem strange to you. And, being human, they made many mistakes while developing the program. Most of these mistakes were discovered and corrected before this book was published, but there is a reasonable likelihood that some bugs still remain.

In fact, the bigger the program, the more bugs that are likely to be present. This should not be surprising; the number of logic paths through a program grows perhaps exponentially with the number of lines of code. Many of the large FE programs contain over 100,000 lines of code. Although the developers verify the programs against a large number of test problems (typically several hundred) with known solutions, these can only confirm a small portion of the possible logic paths. Those programs under active development (i.e., new capabilities are being added) are, of course, continually acquiring new bugs.

One industry spokesman recently observed that working software of reasonable quality should not have more than about one bug per thousand lines of code [7.9]. Another spokesman stated, "In a large program the number of variables is so large, and the combinations of all possible variables so astronomical, that it is truly impossible to fully test all software. So it is essentially impossible to produce a large software system that is fault-free. No software system of any size goes out that doesn't have some bugs" [7.10].

The moral of this story is that the reader would be wise to assume that all FE programs contain bugs, and that there is a reasonable likelihood that one of your analyses will soon encounter one of them. In other words, *caveat emptor*. The author and his col-leagues have encountered over the years many bugs in commercial FE programs — some only a minor nuisance, others so obvious as to preclude any meaningful results, and some (the worst kind) giving wrong but reasonable results. These remarks are not made in a spirit of criticism. Quite the contrary, considering the sheer magnitude and complexity of the big commercial programs, it is almost a source of wonder that there are as few bugs as there are.

On the other hand, the author has encountered many inexperienced FE users who naively believe that commercial FE programs are (and should be) flawless. Bolstered by such unquestioning faith, many of these users plunge boldly into a big analysis without ever performing a single test of their own. This is certainly a risky proposition.

The author strongly encourages a user to perform as many tests as possible to verify a program he or she is about to use. These tests would involve doing simple problems with known answers. The more tests that one's time and budget permit, the greater the confidence one can have in the results of the actual analysis. It is not possible to have absolute certainty in a program, only varying levels of confidence. As increasing numbers of verification tests are performed, one's level of confidence approaches certainty asymp-totically! In the author's judgment, anywhere from one test to half a dozen tests would usually be appropriate, depending on the complexity of the problem and the importance of the project.

Perhaps the most important test of all is that of just a single element. After all, this is the basic building block of the analysis; any errors in a single element would be multiplied throughout the entire domain of the solution. It is therefore important for the user to be confident that the program can at least analyze a single element correctly. And it is also important that the user know how to interpret the program output for each element. Therefore, to gain both confidence and understanding, it is recommended that the user run one or more single-element tests before proceeding with an analysis.

Such tests consist of prescribing a simple load for which the exact element solution can be easily calculated, either by hand or by physical reasoning. We will illustrate with the one-element problem shown in Fig. 7.25. Assume $k=1$ cal-cm/sec-cm^2-°C, $h=0$, $Q=0$, and $Q_\infty=0$. So the governing equation, domain, and boundary conditions are

$$\frac{d^2T(x)}{dx^2} = 0 \qquad 0<x<10 \tag{7.46a}$$

$$T(0) = 1$$

$$q(10) = \left(-\frac{dT}{dx}\right)_{x=10} = -1 \qquad \text{heat flowing in } -x \text{ direction} \tag{7.46b}$$

The solution to Eq. (7.46a) is easy: $T=a+bx$. The constants a and b are determined from Eqs. (7.46b), yielding

$$T(x) = 1 + x$$

$$q(x) = -k\frac{dT(x)}{dx} = -1 \tag{7.47}$$

Since $T(x)$ is a linear polynomial and $q(x)$ is a constant, the C^0-linear element is capable of representing both exactly.

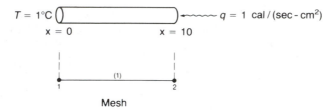

FIGURE 7.25 ● A one-element problem for testing the UNAFEM program.

The next step is to analyze this problem with UNAFEM and see if it will give the same result. At the same time we will begin to learn how to use the program. The data-input instructions may be found in the appendix. The instructions label the data sets by the general or mathematical descriptions used in the text, e.g., "interior loads," "essential boundary conditions," etc. To facilitate application to physical problems, Table 6, which accompanies the instructions, lists the physical meaning and corresponding physical dimensions for each of the data sets for several different types of problems — namely, thermal, mechanical, electrical, and acoustical. The input data for this problem is shown in Fig. 7.26. The reader is encouraged to verify this data.

One aspect of the input instructions deserves a careful explanation. In data set VIII the heat flux q is required to be a *positive* number if heat energy is flowing *into* the element. The reason for this sign convention is as follows.

```
1 2 3 4 5 6 7 8 9 10 11 12 13 14 15 16 17 18 19 20 21 22 23 24 25 26 27 28 29 30 31 32 33 34 35 36 37 38 39 40 41 42 43 44 45 46 47 48 49 50
1DIMBDVP          .ONE-ELEMENT TEST
NODE
             1             0.
             2            10.
ELEM
             1            11          1       1       2
PROP
             1        1.
                      0.
ILC
EBC
             1        1.
NBCC
             2        1.
PLOT

END
```

FIGURE 7.26 ● UNAFEM input data for the one-element test problem.

Recall the boundary flux vector on the RHS of the element equations [Eqs. (7.26)]:

$$
\left\{
\begin{array}{c}
\tilde{\tau}^{(e)}(x_1) \\
-\tilde{\tau}^{(e)}(x_2)
\end{array}
\right\}
\tag{7.48}
$$

noting in particular the difference in sign for the two fluxes. Substitute into Eq. (7.48) the expression for heat flux [Eqs. (7.44b) or Section 6.1]:

$$
\left\{
\begin{array}{c}
\tilde{\tau}^{(e)}(x_1) \\
-\tilde{\tau}^{(e)}(x_2)
\end{array}
\right\}
=
\left\{
\begin{array}{c}
\left(-k\dfrac{dT}{dx}\right)_{x_1} \\
-\left(-k\dfrac{dT}{dx}\right)_{x_2}
\end{array}
\right\}
=
\left\{
\begin{array}{c}
\left(-k\dfrac{dT}{dx}\right)_{x_1} \\
\left(k\dfrac{dT}{dx}\right)_{x_2}
\end{array}
\right\}
\tag{7.49}
$$

Figure 7.27 shows the physical meaning of these terms.

On the left end of the element, when $(-kdT/dx)_{x_1}$ is positive it represents heat flowing to the right (in positive x direction), which is therefore *into* the element. Similarly, on the right end of the element, when $(kdT/dx)_{x_2}$ is positive, it represents heat flowing to the left, which is again *into* the element. We could therefore write Eq. (7.49) as

$$
\left\{
\begin{array}{c}
\left(-k\dfrac{dT}{dx}\right)_{x_1} \\
\left(k\dfrac{dT}{dx}\right)_{x_2}
\end{array}
\right\}
=
\left\{
\begin{array}{c}
q_{in}(x_1) \\
q_{in}(x_2)
\end{array}
\right\}
\tag{7.50}
$$

indicating that if a heat flux is applied to either end of the element, directed into the element, then a positive number should be input in data set VIII.

This physical interpretation of the boundary flux terms is intuitively appealing because heat flowing *into* the element, at either end, is energy being *added* to the system, hence the positive sign. It will be seen in Part IV that this physical interpretation is especially appropriate for problems in 2-D (as well as 3-D) because the boundary of a 2-D element is a curve that may be oriented in any arbitrary direction in the x,y-plane. In such a case it is no longer meaningful to speak of a right or left end; it is instead more natural to describe the flux at any point on a boundary curve as flowing either into or out of the element (see Section 13.2). This physical interpretation is a widely used convention in FE programs.

It should be noted that in data set VIII one may specify heat fluxes at nodes on any of the element boundaries, not just the nodes on the boundary of the domain. This includes nodes at all interelement boundaries as well as on the domain boundary. The idea of specifying flux at interior nodes (rather than only at nodes on the domain boundary) is illustrated in the cable deflection problem in Section 7.5, and is explained more fully in Section 7.6.

Similarly, in data set VII one may specify temperatures at any nodes in the mesh, not just nodes on the boundary of the domain. This should not be surprising because specifying the temperature involves application of a constraint equation, and the row/column operations [see Eq. (5.29) or Section 4.4, step 9] are the same for any DOF in the system, not just those located on the boundary.

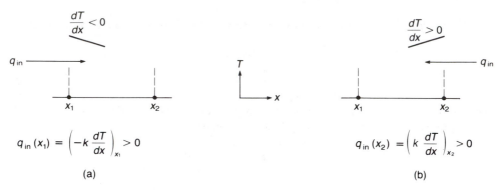

$$q_{in}(x_1) = \left(-k \frac{dT}{dx}\right)_{x_1} > 0 \qquad\qquad\qquad q_{in}(x_2) = \left(k \frac{dT}{dx}\right)_{x_2} > 0$$

(a)   (b)

FIGURE 7.27 ● The physical meaning of heat flux terms in the boundary load vector: (a) heat flowing into the left end of the element; (b) heat flowing into the right end of the element.

In short, although essential BCs (temperatures) and natural BCs (heat fluxes) are applied in data sets VII and VIII, respectively (and set IX for 2-D problems), one may also use these data sets to specify nodal temperatures and heat fluxes in the interior as well.

Part of the printed output from this problem is shown in Fig. 7.28, namely, the element output and the nodal output. We see that the function U was calculated to be 1 at node 1 ($x=0$) and 11 at node 2 ($x=10$), and the flux, $-$ALPHA*DU/DX, was calculated to be -1 at the center (the Gauss point) and therefore at the nodes too. Both results agree with the exact solution in Eqs. (7.47).

The program therefore passes this simple but important test, and we begin to gain some confidence in the program. We have also learned how to input data properly according to the input instructions and how to interpret the printed output.

Having performed only a single test, though, we can only have partial confidence in the program. Depending on the importance of the real problem we are about to analyze, we may want to increase our confidence level further. This would be done by modeling other simple problems with known solutions, involving one or several elements. The effort and dollars invested in this preliminary testing stage could pay significant dividends in avoiding expensive time spent later trying to explain strange-looking results from a complicated analysis.

```
ELEMENT OUTPUT

ELEM    GAUSS PT        X       -ALPHA*DU/DX

  1         1       5.0000      -0.1000E+01




NODAL OUTPUT

NODE          X            U        -ALPHA*DU/DX
  1        0.0000    0.1000E+01     -0.1000E+01
  2       10.0000    0.1100E+02     -0.1000E+01
```

FIGURE 7.28 ● UNAFEM printed output for the one-element test problem.

7.4.3 Dealing with Discontinuous Physical Properties

We would now like to begin analyzing the problem set forth in Section 7.4.1. However, the problem contains a special feature that needs some explanation since it will affect how the mesh should be constructed. The feature being referred to is the presence of two different materials (a commonly occurring situation). At the interface between the materials there is a step change, i.e., a discontinuity, in the material (physical) properties. Intuitively we would expect a sudden change in *some* aspect of the solution at such a discontinuity — if not in the temperature itself, perhaps in one of the derivatives. In other words, it is reasonable to expect that a discontinuity in the problem data is likely to produce some type of discontinuity in the solution. We will investigate the nature of this discontinuity from a mathematical viewpoint, from which we will then deduce a simple and general rule for dealing with such discontinuities.

The nature of the discontinuity is revealed by integrating the governing differential equation [Eq. (7.44a)] over an infinitesimally small interval surrounding the interface [as was done in Section 5.3.1, Eqs. (5.35) through (5.38)]. Consider the interface at $x=40$ cm, and integrate from $40-\epsilon$ to $40+\epsilon$:

$$\int_{40-\epsilon}^{40+\epsilon} \left[-\frac{d}{dx}\left(k\frac{dT}{dx}\right) + \frac{hl}{A}T \right] dx = \int_{40-\epsilon}^{40+\epsilon} (Q+Q_\infty)\, dx \tag{7.51}$$

$$\left[\left(-k\frac{dT}{dx}\right) \right]_{40-\epsilon}^{40+\epsilon} + \int_{40-\epsilon}^{40+\epsilon} \frac{hl}{A}T\, dx = \int_{40-\epsilon}^{40+\epsilon} (Q+Q_\infty)\, dx \tag{7.52}$$

In the limit as $\epsilon \to 0$, the integrals vanish since hl/A, T, Q, and Q_∞ are continuous at $x=40$. (They would vanish even if hl/A, T, Q, or Q_∞ had a step discontinuity.) Therefore Eq. (7.52) becomes

$$-\left(-k\frac{dT}{dx}\right)_{40^-} + \left(-k\frac{dT}{dx}\right)_{40^+} = 0 \tag{7.53}$$

where "40^-" means evaluate $-kdT/dx$ arbitrarily close to $x=40$ but on the left side (the $-$ sign). Similarly, "40^+" means evaluate on the right side of $x=40$ cm.

Figure 7.29 shows the materials on either side of the interface. The label "1" designates all quantities on one side (the left, say), and "2" all quantities on the other side (the right). Equation (7.53) may then be written

$$k_1 \frac{dT_1}{dx} = k_2 \frac{dT_2}{dx} \qquad \text{(at interface)} \tag{7.54}$$

The flux is therefore continuous at the interface. However, since $k_1 \neq k_2$, then $dT_1/dx \neq dT_2/dx$; that is, the temperature gradient must be discontinuous. For convenience, we rewrite Eq. (7.54) as

$$\frac{dT_1/dx}{dT_2/dx} = \frac{k_2}{k_1} \qquad \text{(at interface)} \tag{7.55}$$

which shows that the ratio of the temperature gradients at the interface must be equal to the inverse of the ratio of the thermal conductivities.

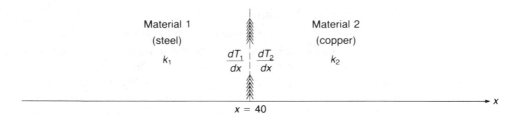

FIGURE 7.29 ● The regions on either side of the material interface.

Discontinuities in dT/dx (or any discontinuities in fact) cannot be modeled well inside elements because shape functions and their derivatives are very smooth functions. On the other hand, C^0-elements (e.g., element type 11) permit discontinuities in dT/dx on *inter-element boundaries*, because such elements constrain only T to be continuous on the boundaries. These observations lead to a general guideline:[21]

Never define an element so that it overlaps a material interface. Instead, always place the boundary of a C^0-element on the interface.

This is illustrated in Fig. 7.30 for the interface at $x=40$ cm, using, for convenience, element numbers (1) and (2), and node numbers 1, 2, and 3.

Equation (7.53) also tells us how to evaluate the natural interelement BC (Section 5.3.1) that arises during assembly of two elements on either side of an interface. Thus, when the element equations [Eqs. (7.26)] are assembled for the two elements in Fig. 7.30, they take the form,

$$
\begin{bmatrix}
K^{(1)}_{11} & K^{(1)}_{12} & \\
K^{(1)}_{21} & K^{(1)}_{22}+K^{(2)}_{22} & K^{(2)}_{23} \\
& K^{(2)}_{32} & K^{(2)}_{33}
\end{bmatrix}
\begin{Bmatrix}
a_1 \\
a_2 \\
a_3
\end{Bmatrix}
$$

$$
= \left\{
\begin{array}{c}
\dfrac{f^{(1)}L}{2} \\[2mm]
\dfrac{f^{(1)}L}{2} + \dfrac{f^{(2)}L}{2} \\[2mm]
\dfrac{f^{(2)}L}{2}
\end{array}
\right\}
+ \left\{
\begin{array}{c}
\left(-k_1 \dfrac{dT}{dx}\right)^{(1)}_{x_1} \\[2mm]
\left(k_1 \dfrac{dT}{dx}\right)^{(1)}_{x_2} + \left(-k_2 \dfrac{dT}{dx}\right)^{(2)}_{x_2} \\[2mm]
\left(k_2 \dfrac{dT}{dx}\right)^{(2)}_{x_3}
\end{array}
\right\}
\qquad (7.56)
$$

[21] This guideline holds for 2-D and 3-D problems as well. In these cases the temperature gradient (or, more generally, the gradient of the unknown function) has more than one component. Only the component *normal* to the boundary (see Part IV) obeys Eqs. (7.54) and (7.55). However, this is sufficient to require the guideline.

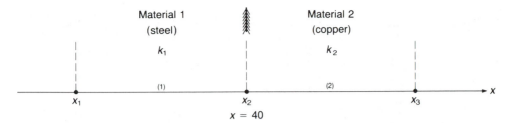

FIGURE 7.30 ● Element boundaries should be placed on a material interface.

We recognize the natural IBC in the second row of the boundary flux vector

$$\left(k_1 \frac{dT}{dx}\right)^{(1)}_{x_2} + \left(-k_2 \frac{dT}{dx}\right)^{(2)}_{x_2} \tag{7.57}$$

This is the same expression as seen in Eq. (7.53). Therefore the value zero must be substituted for the natural IBC. However, in practice no action need be taken at all since the load vector is zeroed at the beginning of the problem. The situation with regard to the natural IBC is therefore no different at material discontinuities than elsewhere in the middle of a material; i.e., since the flux is continuous, the natural IBC (which expresses the discontinuity in the flux) is zero.

7.4.4 UNAFEM Analysis

Let's continue now with the problem described in Section 7.4.1, and begin the analysis by constructing a mesh. We just learned that our principal constraint is to place an element boundary on each of the two material interfaces. Outside of this one restriction, we are free to construct nodes and elements any way we please.

It is generally a good idea to start with a coarse mesh, i.e., with a few large elements. This keeps the cost down if several runs are necessary to find and eliminate modeling errors. Let's try the five-element mesh shown in Fig. 7.31. The corresponding UNAFEM input data is shown in Fig. 7.32.

In order to minimize the data input, all the elements were made the same length, and the nodes and elements were numbered sequentially from one end of the domain to the other. The program will automatically generate node data (node numbers and coordinates) for a set of nodes that are uniformly separated and numbered in a repetitive pattern. The above sequential pattern therefore requires only two lines in data set II (see appendix). Sequential node numbering also has the advantage of reducing the bandwidth of the system

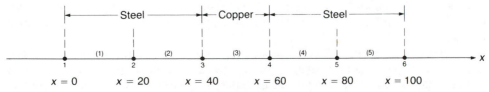

FIGURE 7.31 ● A five-element mesh using C^0-linear elements.

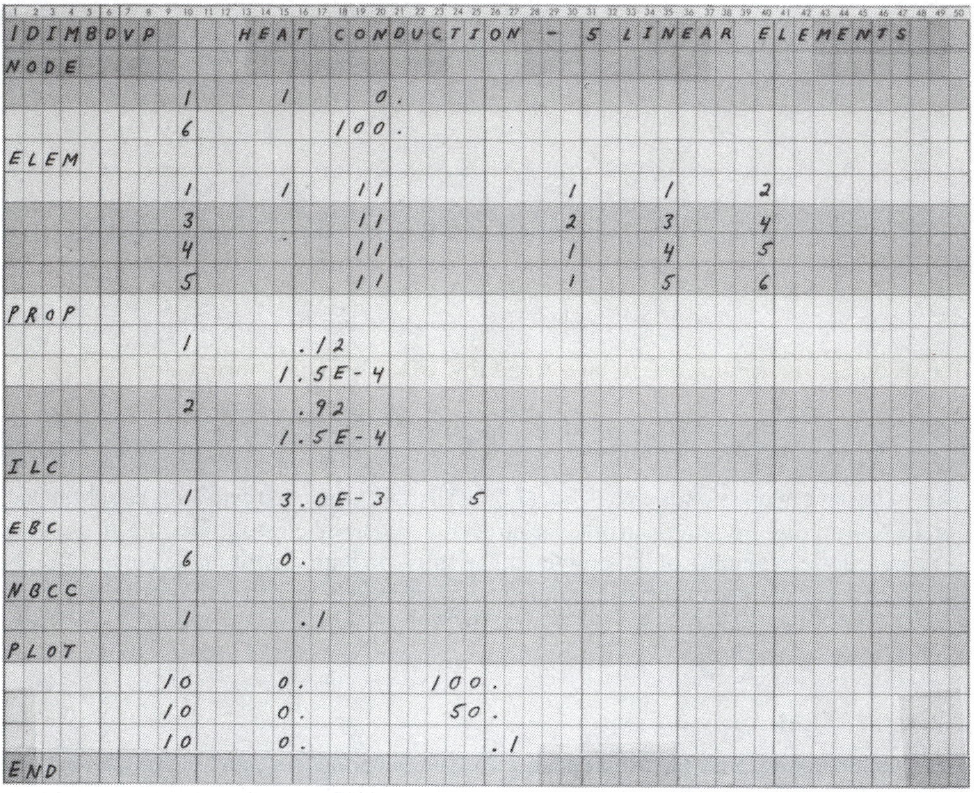

```
IDIMBDVP        HEAT  CONDUCTION  -  5  LINEAR  ELEMENTS
NODE
            1       1        0.
            6              100.
ELEM
            1       1       11        1       1       2
            3               11        2       3       4
            4               11        1       4       5
            5               11        1       5       6
PROP
            1             .12
                  1.5E-4
            2             .92
                  1.5E-4
ILC
            1           3.0E-3      5
EBC
            6           0.
NBCC
            1             .1
PLOT
           10        0.          100.
           10        0.           50.
           10        0.                .1
END
```

FIGURE 7.32 ● UNAFEM input data for the five-element mesh.

stiffness matrix (and therefore computation time and cost). In addition, it facilitates interpreting printed output.

The program will also automatically generate element data for a set of elements whose node numbers follow a repetitive pattern. Those elements that are generated are assigned the same physical property number as the first element in a set; therefore a set of generated elements must all have the same physical properties. This necessitates four lines in data set III — one for each of the different material sections, plus one for the last element.

Data set X defines the x- and y-axes for the plots. The values specified were obtained from a previous run (not shown) that input three blank lines so that the program would determine from the solution data the limits that would just enclose the plot. Normally these limits aren't known prior to the first run. Here we wish to show the plots on a set of axes that will be the same for all the different meshes (see Figs. 7.37 and 7.40) so that the reader can make a visual comparison between them — a procedure frequently done in practice.

Figures 7.33 and 7.34 show the printed and graphical output from UNAFEM. For heat conduction problems the headings U and $-\alpha dU/dx$ represent, respectively, temperature and heat flux. As usual we note that the essential BC, $T(100)=0\,°C$, is satisfied exactly, whereas the natural BC, $q(0)=0.1$ cal/sec-cm^2, is satisfied only approximately, namely,

```
ELEMENT OUTPUT

ELEM    GAUSS PT        X        -ALPHA*DU/DX

  1         1       10.0000       0.7483E-01

  2         1       30.0000       0.4712E-01

  3         1       50.0000       0.4180E-01

  4         1       70.0000       0.4607E-01

  5         1       90.0000       0.7216E-01


NODAL OUTPUT

NODE        X            U        -ALPHA*DU/DX
  1      0.0000     0.4094E+02     0.7483E-01
  2     20.0000     0.2847E+02     0.6097E-01
  3     40.0000     0.2061E+02     0.4446E-01
  4     60.0000     0.1971E+02     0.4394E-01
  5     80.0000     0.1203E+02     0.5912E-01
  6    100.0000     0.0000E+00     0.7216E-01
```

FIGURE 7.33 ● UNAFEM printed output for the five-element mesh.

$\tilde{q}(0) = 0.075$. A "ballpark" estimate of the flux error is afforded by the spread in the tiny circles, i.e., the discontinuities in the element nodal fluxes. They appear to differ by about 10 to 25% from the averaged fluxes. This is usually an overestimate of the error in the averaged fluxes. To better estimate, as well as improve, both the flux and temperature errors, we will generate a finer mesh.

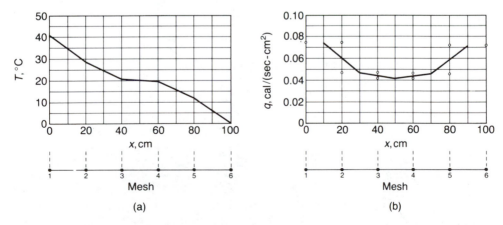

FIGURE 7.34 ● UNAFEM graphical output for the five-element mesh. (a) Function: temperature $\tilde{T}(x)$. Approximation-solution polynomial $\tilde{T}^{(e)}(x)$ plotted within each element. (b) Flux: heat flux $\tilde{q}(x)$ flowing to the right ($\tilde{q} = -kd\tilde{T}/dx$). Straight lines connect successive Gauss-point values; circles are element nodal values, interpolated from Gauss-point values.

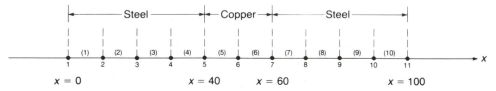

FIGURE 7.35 ● A 10-element mesh using C^0-linear elements.

Figures 7.35 through 7.37 show a 10-element mesh, the data input, and the resulting graphical output. A comparison of Figs. 7.37 and 7.34 provides an estimate of about 5% accuracy for the temperature and about 10 to 15% accuracy for the flux.

The small circles on the flux plot in Fig. 7.37(b) show that the greatest discontinuities occur near the ends of the rod where the slope of the flux is steepest. This is to be expected, since the flux in element type 11 is only a constant within each element, and a constant becomes a progressively poorer approximation as the slope increases. We therefore have better accuracy near the center than near the ends. To make the accuracy more uniform we will need more DOF in the trial solution near the ends. One way to do this is to employ more elements near the ends, a procedure called *local mesh refinement*. In our next mesh

IDIMBDVP			HEAT CONDUCTION		-	10	LINEAR	ELEMENTS	
NODE									
	1		1	0.					
	11			100.					
ELEM									
	1		1	11			1	1	2
	5		1	11			2	5	6
	7		1	11			1	7	8
	10			11			1	10	11
PROP									
	1			.12					
			1.	5E-4					
	2			.92					
			1.	5E-4					
ILC									
	1		3.	0E-3		10			
EBC									
	11		0.						
NBCC									
	1			.1					
PLOT									
	10		0.			100.			
	10		0.			50.			
	10		0.				.1		
END									

FIGURE 7.36 ● UNAFEM input data for the 10-element mesh.

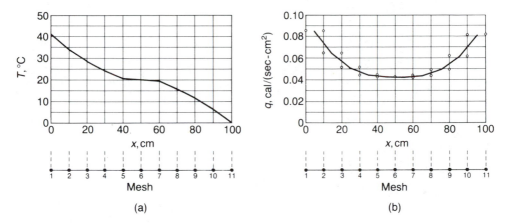

FIGURE 7.37 ● UNAFEM graphical output for the 10-element mesh. (a) Function: temperature $\tilde{T}(x)$. Approximate-solution polynomial $\tilde{T}^{(e)}(x)$ plotted within each element. (b) Flux: heat flux $\tilde{q}(x)$ flowing to the right ($\tilde{q} = -kd\tilde{T}/dx$). Straight lines connect successive Gauss-point values; circles are element nodal values, interpolated from Gauss-point values.

we will therefore refine the mesh near the ends (i.e., locally) to a greater extent than near the middle. (In Chapter 8 we will learn another way to add more DOF.)

Figure 7.38 shows a 44-element mesh that refines the 10-element mesh locally by subdividing elements (1) through (4) and (7) through (10) of that mesh into five elements each, and elements (5) and (6) into two elements each. The data input and resulting graphical output are shown in Figs. 7.39 and 7.40.

A comparison of Figs. 7.37 and 7.40 provides an estimate of much better than 1% accuracy for the temperature, and about 2 to 3% accuracy for the flux. If any further accuracy were desired, the reader should by now know what to do.

7.4.5 Are the Results Reasonable?

At the conclusion of an analysis, the results should always be closely examined for their reasonableness. Mistakes in the input data or a poor FE model (e.g., an inappropriate mesh or element type) can yield a solution that may appear correct under casual observation, but if examined critically will show unreasonable or at least suspicious behavior.

At this final and critical point in the analysis, good engineering judgment is needed. It is here that maturity and physical insight can make the difference between a reliable and useful solution or a useless computation. In the author's experience, most FE analyses

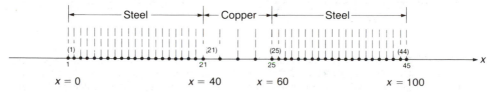

FIGURE 7.38 ● A 44-element mesh using C^0-linear elements. (Only those nodes and elements included in the input data are labeled).

IDIMBDVP   HEAT CONDUCTION - 44 LINEAR ELEMENTS

NODE
1	1	0.
21	1	40.
25	1	60.
45	1	100.

ELEM
1	1	11	1	1	2
21	1	11	2	21	22
25	1	11	1	25	26
44		11	1	44	45

PROP
1	.12
	1.5E-4
2	.92
	1.5E-4

ILC
1	3.0E-3	44

EBC
45	0.

NBCC
1	.1

PLOT
10	0.	100.
10	0.	50.
10	0.	.1

END

FIGURE 7.39 ● UNAFEM input data for the 44-element mesh.

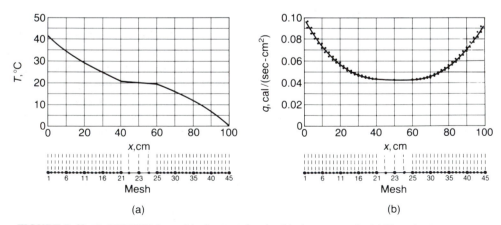

FIGURE 7.40 ● UNAFEM graphical output for the 44-element mesh. (a) Function: temperature $\tilde{T}(x)$. Approximate-solution polynomial $\tilde{T}^{(e)}(x)$ plotted within each element. (b) Flux: heat flux $\tilde{q}(x)$ flowing to the right ($\tilde{q} = -kd\tilde{T}/dx$). Straight lines connect successive Gauss-point values; circles are element nodal values, interpolated from Gauss-point values.

require several computer runs, or even several different models, before meaningful results are obtained. It would therefore seem quite irresponsible to take the first result of an FE analysis and, without further examination, assume that it is a correct solution to the engineering problem.

One of the easiest checks is to ensure that the essential BCs are satisfied exactly and that the natural BCs appear to be converging to the exact values. Other checks might involve simple balances between the applied loads and the computed reactions of the system. Different problems create different opportunities for the clever engineer to devise effective checks. The problems in this and subsequent chapters will illustrate some possible approaches.

In the present problem, Fig. 7.40 shows that the essential BC, $T(100)=0$, is correctly satisfied, and the natural BC, $q(0)=0.1$, is being approached very closely.

For another check, let's write the governing differential equation [Eq. (7.44a)] in the abbreviated form,

$$\frac{dq(x)}{dx} = -\frac{hl}{A}(T-T_\infty)$$

$$= -1.5\times10^{-4}(T-20) \tag{7.58}$$

Therefore, wherever dq/dx vanishes, T should equal 20 °C. Figure 7.40 indicates that Eq. (7.58) is approximately satisfied somewhere in the range $50<x<60$. More precision is available from the printed output (not shown), which indicates that dq/dx vanishes at $x=55$ cm, at which point the calculated value of T is 19.99 °C. This is encouraging.

We might use Eq. (7.58) to check one more point, say $x=0$ cm. From the printed output we have $T(0)=41.25$ °C, so Eq. (7.58) becomes

$$\left(\frac{dq(x)}{dx}\right)_{x=0} = -0.00319 \tag{7.59}$$

We can calculate dq/dx closely using the difference in values of q at the centers of elements (1) and (2):

$$\left(\frac{dq(x)}{dx}\right)_{x=0} \approx \frac{q(3)-q(1)}{3-1} = \frac{0.09100-0.09689}{2} = -0.00295 \tag{7.60}$$

which agrees to within about 8% with Eq. (7.59).

Yet another check would be the ratio of the temperature gradients across the material interface at $x=40$ cm, to verify Eq. (7.55). Thus

$$\frac{k_{copper}}{k_{steel}} = \frac{0.92}{0.12} = 7.67 \tag{7.61}$$

The temperature gradients can be calculated from adjacent nodal temperature values, obtained again from the printed output. Thus

$$\left(\frac{dT}{dx}\right)_{steel} \approx \frac{T(40)-T(38)}{40-38} = \frac{20.68-21.41}{2} = -0.365$$

$$\left(\frac{dT}{dx}\right)_{copper} \approx \frac{T(45)-T(40)}{45-40} = \frac{20.45-20.68}{5} = -0.046 \tag{7.62}$$

Hence,

$$\frac{(dT/dx)_{steel}}{(dT/dx)_{copper}} = \frac{-0.365}{-0.046} = 7.93 \tag{7.63}$$

which agrees to within about 3% with Eq. (7.61).

At this point, considering the one-element test of the program and these final numerical checks on the analysis, the reader can justifiably have some confidence (though not certainty) in the above approximate solution.

7.5 Application of UNAFEM to a Cable Deflection Problem

7.5.1 Statement of the Problem

Figure 7.41 depicts a 20-ft cable stretched horizontally with a tension of 100 lb and held at both ends. There are several applied loads on the cable. First is the self-weight due to gravity. The cable is of uniform construction with a lineal weight density of 0.5 lb/ft. This appears as a uniformly distributed load along the entire cable. Second, there is a 30-lb weight hanging at the position $x=16$ ft. This appears as a downward-acting concentrated force at $x=16$. Finally, there are a number of items suspended from the cable along an 8-ft span at the left end. The effect of these items has been modeled by the distributed load shown in the figure. As is usually the case, this load description is only an idealization of reality. Therefore when we later define the load to the program as an approximation to this idealization, the approximation might possibly be a more accurate representation of reality than this idealization.

This type of problem was described in Section 6.3. The governing differential equation and domain [see Eq. (6.17)] are as follows:

$$-\frac{d}{dx}\left(T(x)\frac{dW(x)}{dx}\right) = f(x) \qquad 0<x<20 \tag{7.64a}$$

where

$$T(x) = 100 \text{ lb} \qquad 0<x<20$$

$$f(x) = \begin{cases} -0.5 - 6\sin\frac{\pi x}{8} & 0<x<4 \\ -0.5 - 6 & 4<x<8 \\ -0.5 - 30\,\delta(x-16) & 8<x<20 \end{cases} \tag{7.64b}$$

Note that $f(x)$ is positive in the $+W$ direction.[22]

$W(x)$ = vertical displacement (deflection) of cable from unloaded horizontal position, positive upwards

$F(x)$ = vertical component of tension in cable, $-TdW/dx$, positive upwards when acting on the left end of a cable section. (For sign convention see Fig. 7.44.) $F(x)$ is the flux in this type of problem.

---

[22] See the explanation of $\delta(x-16)$ given with Eqs. (7.69) through (7.75).

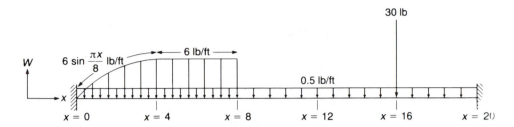

FIGURE 7.41 ● Vertical loads applied to a cable under tension.

The boundary conditions are

$$W(0) = 0 \qquad \text{essential BC}$$
$$W(20) = 0 \qquad \text{essential BC} \qquad (7.65)$$

Let's use UNAFEM to calculate the deflected shape of the cable. In addition, we want to know the displacement and forces at three specific locations:

1. $W(8)$, the vertical displacement at $x=8$
2. $|F(0)|$ and $|F(20)|$, the vertical force exerted on each end support

7.5.2 Dealing with Discontinuous Loads

This problem illustrates two types of load conditions that occur quite frequently in practice: a distributed load that has a step change (at $x=8$) and a concentrated load that acts at a point (at $x=16$). Both types of load involve a discontinuity in the load data and, as such, are convenient idealizations of the real world. As argued previously in the heat conduction problem, we can expect that discontinuities in the data will produce discontinuities in the solution (namely, in the function and/or some of its derivatives).

The nature of both types of discontinuity is revealed by integrating the governing differential equation [Eq. (7.64a)] over an infinitesimally small interval surrounding the interface, as was done for the heat conduction problem in Section 7.4.3. We treat each case separately.

Step change in load at $x=8$ ● The load surrounding the point $x=8$ is shown in Fig. 7.42. Integrate Eq. (7.64a) from $8-\epsilon$ to $8+\epsilon$:

$$\int_{8-\epsilon}^{8+\epsilon} \left[-\frac{d}{dx}\left(T\frac{dW}{dx} \right) \right] dx = \int_{8-\epsilon}^{8+\epsilon} f \, dx \qquad (7.66)$$

Substituting Eq. (7.64b) in the RHS, Eq. (7.66) reduces to

$$\left[\left(-T\frac{dW}{dx} \right) \right]_{8-\epsilon}^{8+\epsilon} = -6.5\epsilon - 0.5\epsilon \qquad (7.67)$$

In the limit as $\epsilon \to 0$, the RHS vanishes, leaving

$$-\left(-T\frac{dW}{dx} \right)_{8^-} + \left(-T\frac{dW}{dx} \right)_{8^+} = 0 \qquad (7.68)$$

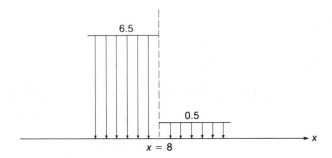

FIGURE 7.42 ● Loads in the vicinity of point $x=8$.

The flux (vertical force in the cable) is therefore continuous at $x=8$. On the other hand, it follows simply from an inspection of Eqs. (7.64a and b) that since the RHS is discontinuous at $x=8$, then so is the LHS; i.e., the derivative of the flux is discontinuous at $x=8$.

   We learned in Section 7.4.3 that discontinuities in the solution (of any kind) must be located on the boundaries between elements, not in the interior of elements. Therefore, during mesh construction we should always place an element boundary at a step change in the load.

   We also recognize Eq. (7.68) as the natural IBC that appears in the load vector after assembly of the two adjacent element equations. The situation here is exactly parallel to the heat conduction problem in Section 7.4.3; recall especially Eqs. (7.53), (7.56), and (7.57) and surrounding discussion. Since the value of the natural IBC in Eq. (7.68) is zero, and since the system load vector is initialized to zero, then nothing further needs to be done to the system load vector; i.e., no data needs to be input.

   In summary, our only concern when dealing with a step change (discontinuity) in the load is to place an element boundary at the discontinuity.

Concentratated load at $x=16$ ● A 30-lb force that is concentrated at a single point, $x=16$, and directed downwards (in the $-W$ direction), is represented mathematically as

$$-30\ \delta(x-16) \tag{7.69}$$

where $\delta(x-16)$ is called a *Dirac delta function*; it is shown in Fig. 7.43.

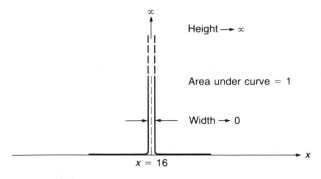

FIGURE 7.43 ● Representation of a concentrated load by a Dirac delta function.

This strange-looking function has an infinite height and an infinitesimal width; it is zero everywhere except at $x=16$. However, the area underneath it, by definition, is 1:

$$\int_{16-\epsilon}^{16+\epsilon} \delta(x-16) \, dx \; = \; 1$$

(7.70)

If the delta function is multiplied by any "ordinary," reasonably smooth, finite function, say $g(x)$, then we have the useful relation,

$$\int_{16-\epsilon}^{16+\epsilon} g(x) \, \delta(x-16) \, dx \; = \; g(16)$$

(7.71)

That is, the effect of the integral is simply to evaluate $g(x)$ at the concentration point. (Recall Fig. 3.11a.) In particular, if $g(x)$ is a constant, say G, then

$$\int_{16-\epsilon}^{16+\epsilon} G \, \delta(x-16) \, dx \; = \; G$$

(7.72)

Now let's integrate Eq. (7.64a) from $16-\epsilon$ to $16+\epsilon$, substituting Eq. (7.64b) in the RHS:

$$\int_{16-\epsilon}^{16+\epsilon} \left[-\frac{d}{dx} \left(T \frac{dW}{dx} \right) \right] dx \; = \; \int_{16-\epsilon}^{16+\epsilon} [-0.5 \; - \; 30 \, \delta(x-16)] dx$$

(7.73)

Applying Eq. (7.72) to the RHS, this reduces to

$$\left[\left(-T \frac{dW}{dx} \right) \right]_{16-\epsilon}^{16+\epsilon} \; = \; -0.5 \times 2\epsilon \; - \; 30$$

(7.74)

In the limit as $\epsilon \to 0$, only the first term on the RHS vanishes, leaving

$$-\left(-T \frac{dW}{dx} \right)_{16^-} \; + \; \left(-T \frac{dW}{dx} \right)_{16^+} \; = \; -30$$

(7.75)

The vertical force in the cable, i.e., the flux, is therefore *discontinuous* at $x=16$. Equation (7.75) states the physically obvious fact that the difference in vertical forces in the cable, on either side of the concentration, is the 30-lb applied force.

As always, a discontinuity in the solution must be located on the boundary between elements.

We want to note especially that the natural IBC represented by Eq. (7.75) is no longer zero, so we can't ignore this condition when inputting data as we did for the step discontinuity in the load or for the discontinuous material properties. As observed before, the difference expression on the LHS of Eq. (7.75) will appear in the boundary load vector as a result of assembling the adjacent element equations. Therefore we must input the value -30 lb for the DOF at the node at $x=16$. This treatment is therefore identical to a natural BC, which specifies an applied vertical force at a node on the boundary of the domain. It will be seen in Section 7.5.3 that this concentrated 30-lb load will therefore be input in data set VIII, just like a natural BC.

In summary, when we have a concentrated load, we must define our elements so that the load is acting on an element boundary node, and then we input the value of the load in the same way as a natural BC, i.e., just like a concentrated load on the boundary of the domain. See further comments in Section 7.6.

7.5.3 UNAFEM Analysis

The data input instructions for mechanical problems (see appendix) are the same as for the heat conduction problem except for the terminology and physical units, which may be determined with the help of Tables 6 and 7 accompanying the instructions.

In data set VIII the applied vertical force (flux) is required to be a *positive* number if the force is in the positive direction of transverse displacement. There is a physical motivation for this convention; similar to the heat conduction problem (see Section 7.4.2), that can be explained as follows.

Recall again the boundary flux vector on the RHS of the general element equations [Eqs. (7.26)], and substitute in the expression for flux below Eq. (7.64b):

$$
\left\{ \begin{array}{c} \tilde{\tau}^{(e)}(x_1) \\ -\tilde{\tau}^{(e)}(x_2) \end{array} \right\} = \left\{ \begin{array}{c} \left(-T\dfrac{dW}{dx}\right)_{x_1} \\ -\left(-T\dfrac{dW}{dx}\right)_{x_2} \end{array} \right\} = \left\{ \begin{array}{c} \left(-T\dfrac{dW}{dx}\right)_{x_1} \\ \left(T\dfrac{dW}{dx}\right)_{x_2} \end{array} \right\}
\tag{7.76}
$$

Figure 7.44 shows the physical meaning of these terms.

On the left end of the element, when $(-TdW/dx)_{x_1}$ is positive it represents an *upward*-directed force (in positive W direction). Similarly, on the right end of the element, when $(TdW/dx)_{x_2}$ is positive, it also represents an upward-directed force. We could therefore write Eq. (7.76) as

$$
\left\{ \begin{array}{c} \left(-T\dfrac{dW}{dx}\right)_{x_1} \\ \left(T\dfrac{dW}{dx}\right)_{x_2} \end{array} \right\} = \left\{ \begin{array}{c} F_{up}(x_1) \\ F_{up}(x_2) \end{array} \right\}
\tag{7.77}
$$

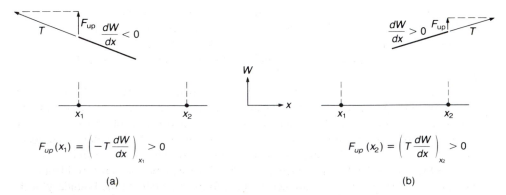

(a) (b)

FIGURE 7.44 ● The physical meaning of the boundary flux terms: (a) vertical force applied to the left end of the element; (b) vertical force applied to the right end of the element.

FIGURE 7.45 ● A five-element mesh using C^0-linear elements.

indicating that if an upward-directed force is applied to either end of the element, then a positive number should be input in data set VIII. This is analogous to the heat conduction relations in Eq. (7.50). Thus, an upward-directed force at either end is potential energy being *added* to the system, hence the positive sign.

We will begin the analysis with the five-element mesh shown in Fig. 7.45. Our only constraint is to place element boundaries at discontinuities in the interior load — namely, the step change at $x=8$ and the concentrated force at $x=16$. Otherwise we may define elements and nodes in any pattern we like. The five elements won't provide much accuracy, but we want to use a "cheap" model while we are familiarizing ourselves with the special features of the problem. As before, we number the nodes and elements sequentially so that we can take advantage of the automatic node and element generation in UNAFEM.

Figure 7.46 shows the UNAFEM input data. The first line after the keyword ILC needs an explanation. This is the interior load in element (1), i.e., a distributed vertical force. In element (1) the interior load $6 \sin (\pi x/8)$ is variable. Since UNAFEM only accepts a constant, we must input some representative constant value. It was argued in Section 7.2, step 5 [see comments following Eq. (7.24)] that the value at the *center* of the element would generally be the best value to use. This is because as the element size becomes small enough, the exact solution begins to "look like" (i.e., it can be approximated by) a straight line within a given element, in which case the value at the center is the average value. Therefore the load $6 \sin (\pi x/8)$ is evaluated at $x=2$, as indicated in the figure.

The three lines after the keyword PLOT define the axes for the plots. This data was obtained from a previous run (not shown) in which these lines were left blank, just as was done in the heat conduction problem.

Figure 7.47 shows the plots of displacement and vertical force for the five-element mesh.[23] The results look reasonable. The displacement BCs at $x=0$ and $x=20$ are satisfied exactly, and the vertical force plot indicates a change of about 30 lb near $x=16$, corresponding to the applied 30-lb load. Having satisfied ourselves that the input data appears to be correct and the results are reasonable (though not necessarily very accurate), we next construct a 10-element mesh to help estimate the accuracy, as well as improve it.

Figures 7.48 through 7.50 show the mesh, input data, and output plots for a 10-element model. Figure 7.51 lists the reaction forces at each end support and the displacement at

---

[23] Recall from Fig. 7.44 that the vertical force $-TdW/dx$ is positive upwards when acting on the left end of a cable section. Conversely, when $-TdW/dx$ is negative, the vertical force on the left end of a cable section is downward, and on the right end is upward. Therefore in Fig. 7.47 a negative value of $-TdW/dx$ means an upward force on the right end. This in turn means that the reaction force at $x=20$, which is a force acting on the right end of the cable, must be acting upward because Fig. 7.47 shows that $-TdW/dx=-36.9$ lb at $x=20$.

IDIMBDVP		1		CABLE DEFLECTION - 5 LINEAR ELEMENTS					
NODE									
		1	1	0.					
		6		20.					
ELEM									
		1	1	11		1	1	2	
		5		11		1	5	6	
PROP									
		1	100.						
			0.						
ILC									
		1	-4.7426 ◄─2── -6 sin($\frac{\pi}{4}$)-0.5						
		2	-6.5						
		3	-0.5		5				
EBC									
		1	0.						
		6	0.						
NBCC									
		5	-30.						
PLOT									
		10	0.		20.				
		10	-5.		0.				
		10	-50.		50.				
END									

FIGURE 7.46 ● UNAFEM input data for the five-element mesh.

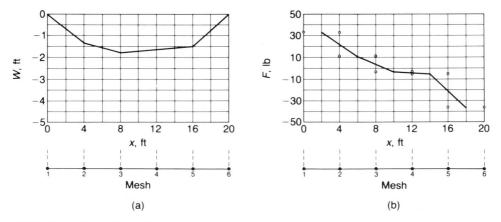

FIGURE 7.47 ● UNAFEM graphical output for the five-element mesh. (a) Function: vertical displacement $\tilde{W}(x)$. Approximate-solution polynomial $\tilde{W}^{(e)}(x)$ plotted within each element. (b) Flux: vertical force $\tilde{F}(x)$ acting on the section to the right of x ($\tilde{F} = -Td\tilde{W}/dx$). Straight lines connect successive Gauss-point values; circles are element nodal values, interpolated from Gauss-point values.

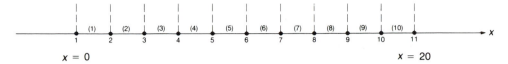

$x = 0$ $x = 20$

FIGURE 7.48 ● A 10-element mesh using C^0-linear elements.

```
IDIMBDVP    CABLE DEFLECTION  -  10 LINEAR ELEMENTS
NODE
        1           1           0.
        11                     20.
ELEM
        1           1       11          1           1       2
        10                  11          1           10      11
PROP
        1       100.
                  0.
ILC
        1           -2.7961    ← -6 sin(π/8) - 0.5
        2           -6.0433  ← -6 sin(3π/8) - 0.5
        3           -6.5            4
        5           -0.5            10
EBC
        1           0.
        11          0.
NBCC
        9           -30.
PLOT
        10          0.          20.
        10          -5.         0.
        10          -50.        50.
END
```

FIGURE 7.49 ● UNAFEM input data for the 10-element mesh.

$x=8$, for both the 5- and 10-element meshes. These values were read from the printed output, under the section entitled "Nodal Output."

A comparison of the 5- and 10-element results suggests the accuracy is higher in the middle region, say from $x=9$ to about $x=14$. We could therefore do one more mesh using local mesh refinement. However, a uniform mesh is easier and faster to construct and less likely to cause errors on data input. Weighing both computer costs and the engineer's time, the latter usually being much more expensive, the economical approach is simply to continue using uniform meshes, even though they may provide more accuracy than wanted in some regions of the model.

Note that the 10-element mesh has two elements in the variable load region $0<x<4$, whereas the five-element mesh had only one element in this region. The input data for

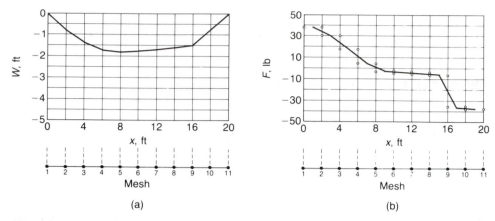

(a)   (b)

FIGURE 7.50 ● UNAFEM graphical output for the 10-element mesh. (a) Function: vertical displacement $\tilde{W}(x)$. Approximate-solution polynomial $\tilde{W}^{(e)}(x)$ plotted within each element. (b) Flux: vertical force $\tilde{F}(x)$ acting on the section to the right of x ($\tilde{F} = -Td\tilde{W}/dx$). Straight lines connect successive Gauss-point values; circles are element nodal values, interpolated from Gauss-point values.

Number of linear elements in mesh	$W(8)$	$\lvert F(0) \rvert$	$\lvert F(20) \rvert$
5	−1.788	33.6	36.9
10	−1.811	38.8	37.6

FIGURE 7.51 ● Vertical displacement at $x=8$ ft and vertical reaction forces at $x=0$ and 20 ft.

these two meshes shows that the variable load was approximated by one constant in the five-element mesh but by two constants in the 10-element mesh. Since the load data is part of the model definition (recall Section 7.2, step 5), we are therefore changing the model at the same time we are refining the mesh. The convergence theorems (see Section 5.6) are strictly valid only when the model definition does not change. However, it is common practice in FE analysis to improve the accuracy of the model definition (loads, properties, and/or geometry) as the mesh is refined. Neither the author nor the program developer[24] are aware of any instances when this practice has destroyed convergence;[25] on the contrary, it appears to always improve convergence (to the solution corresponding to the "exact" model, which the sequence of approximate models is converging to). This is surely not surprising.

As discussed earlier, if $f(x)$ is complicated enough that a piecewise-constant approximation seems to be a poor representation and, in addition, requires too much data handling, we can always rewrite the element routine to include more complicated functions (see Exercise 7.2). In today's sophisticated preprocessor programs, one can usually define

---

[24] See Special Acknowledgment at the front of this book.

[25] See the caveat in Section 13.6.3 regarding the "Babuska paradox."

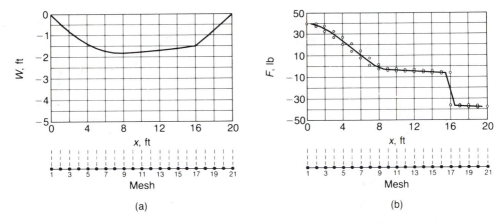

FIGURE 7.52 ● UNAFEM graphical output for the 20-element mesh. (a) Function: vertical displacement $\tilde{W}(x)$. Approximate-solution polynomial $\tilde{W}^{(e)}(x)$ plotted within each element. (b) Flux: vertical force $\tilde{F}(x)$ acting on the section to the right of x ($\tilde{F} = -Td\tilde{W}/dx$). Straight lines connect successive Gauss-point values; circles are element nodal values, interpolated from Gauss-point values.

loads independently from the mesh, and to any desired accuracy. Then as the mesh is refined the load definition can remain unchanged.

Figures 7.52 and 7.53 show the graphical plots produced from 20- and 40-element meshes, and Fig. 7.54 summarizes the calculated displacement and force values from all four meshes.

7.5.4 Are the Results Reasonable?

The values in Fig. 7.54 show a clearly converging trend. A comparison of the 20- and 40-element results yields error estimates of 0.06% for $W(8)$, 1.2% for $|F(0)|$ and 0.3% for $|F(20)|$. We would expect the accuracy of $|F(0)|$ to be less than that of $|F(20)|$ because the former is located next to the region $0<x<4$ in which the variable load was defined approximately by constant values.

A useful test for the present problem is as follows. Since the deflected cable is in static equilibrium, then the algebraic sum of all the applied vertical forces must equal the two vertical reaction forces at the end supports. From Fig. 7.41,

$$\Sigma f = \text{sum of applied forces}$$

$$= -\int_0^4 6 \sin \frac{\pi x}{8}\, dx - 6\times4 - 0.5\times20 - 30$$

$$= -79.3 \text{ lb} \tag{7.78}$$

Using the reaction forces from Fig. 7.54, Fig. 7.55 summarizes for each mesh the sum of the two reaction forces and the corresponding error relative to Σf.

These simple checks begin to provide some assurance that the results are meaningful. The reader can perhaps think of other ways to check these results.

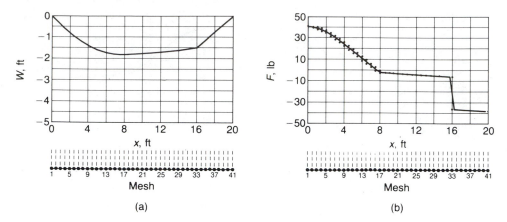

(a)   (b)

FIGURE 7.53 ● UNAFEM graphical output for the 40-element mesh. (a) Function: vertical displacement $\tilde{W}(x)$. Approximate-solution polynomial $\tilde{W}^{(e)}(x)$ plotted within each element. (b) Flux: vertical force $\tilde{F}(x)$ acting on the section to the right of x ($\tilde{F}= -Td\tilde{W}/dx$). Straight lines connect successive Gauss-point values; circles are element nodal values, interpolated from Gauss-point values.

Number of linear elements (and DOF) in mesh	$W(8)$	$\|F(0)\|$	$\|F(20)\|$
5 (6)	−1.788	33.6	36.9
10 (11)	−1.811	38.8	37.6
20 (21)	−1.816	40.4	37.9
40 (41)	−1.817	40.9	38.0

FIGURE 7.54 ● Vertical displacement at $x=8$ ft and vertical reaction forces at $x=0$ and 20 ft.

Number of linear elements (and DOF) in mesh	$\|F(0)\|+\|F(20)\|$	$\dfrac{\|\Sigma f\|-(\|F(0)\|+\|F(20)\|)}{\|\Sigma f\|}$
5 (6)	70.5	11.1%
10 (11)	76.4	3.7%
20 (21)	78.3	1.2%
40 (41)	78.9	0.5%

FIGURE 7.55 ● Sum of vertical reaction forces, and error relative to sum of applied forces.

7.6 The Physical Meaning of Interelement Boundary Conditions

The previous cable deflection problem demonstrated that a concentrated load applied to the interior of the domain is input to the program just like a natural BC, but it is applied to the boundary *between* two elements, i.e., a natural *interelement* BC (recall Section 5.3.1). The following discussion summarizes and generalizes several previous developments and provides a very meaningful and useful physical interpretation.

Recall the element equations [Eqs. (7.26)] for a C^0-linear element:

$$\begin{bmatrix} K_{11}^{(e)} & K_{12}^{(e)} \\ K_{21}^{(e)} & K_{22}^{(e)} \end{bmatrix} \begin{Bmatrix} a_1 \\ a_2 \end{Bmatrix} = \begin{Bmatrix} FI_1^{(e)} \\ FI_2^{(e)} \end{Bmatrix} + \begin{Bmatrix} \tilde{\tau}^{(e)}(x_1) \\ -\tilde{\tau}^{(e)}(x_2) \end{Bmatrix}$$

(7.79)

where the general symbols $K_{ij}^{(e)}$ and $FI_i^{(e)}$ represent stiffness and interior load, respectively, and $\tilde{\tau}^{(e)}(x)$ is the flux, given by

$$\tilde{\tau}^{(e)}(x) = -\alpha \frac{d\tilde{U}^{(e)}}{dx}$$

(7.80)

And consider the typical mesh shown in Fig. 7.56. The nodes and elements are numbered sequentially to simplify the discussion; any random numbering would also be acceptable.

After assembly of the element equations, the resulting system equations would have the form,

$$\begin{bmatrix} K_{11}^{(1)} & K_{12}^{(1)} & & & \\ K_{21}^{(1)} & K_{22}^{(1)}+K_{22}^{(2)} & K_{23}^{(2)} & & \\ & K_{32}^{(2)} & K_{33}^{(2)}+K_{33}^{(3)} & K_{34}^{(3)} & \\ & & K_{43}^{(3)} & K_{44}^{(3)}+K_{44}^{(4)} & \\ & & & & \ddots \\ & & & & & K_{N+1,N+1}^{(N)} \end{bmatrix} \begin{Bmatrix} a_1 \\ a_2 \\ a_3 \\ a_4 \\ \vdots \\ a_{N+1} \end{Bmatrix}$$

$$= \begin{bmatrix} FI_1^{(1)} \\ FI_2^{(1)}+FI_2^{(2)} \\ FI_3^{(2)}+FI_3^{(3)} \\ FI_4^{(3)}+ \\ \vdots \\ FI_{N+1}^{(N)} \end{bmatrix} + \begin{Bmatrix} \tilde{\tau}^{(1)}(x_1) \\ -\tilde{\tau}^{(1)}(x_2)+\tilde{\tau}^{(2)}(x_2) \\ -\tilde{\tau}^{(2)}(x_3)+\tilde{\tau}^{(3)}(x_3) \\ -\tilde{\tau}^{(3)}(x_4)+ \\ \vdots \\ -\tilde{\tau}^{(N)}(x_{N+1}) \end{Bmatrix}$$

(7.81)

FIGURE 7.56 ● A typical mesh.

Our concern here is the boundary flux vector,

$$
\left\{
\begin{array}{c}
\tilde{\tau}^{(1)}(x_1) \\
-\tilde{\tau}^{(1)}(x_2)+\tilde{\tau}^{(2)}(x_2) \\
-\tilde{\tau}^{(2)}(x_3)+\tilde{\tau}^{(3)}(x_3) \\
-\tilde{\tau}^{(3)}(x_4)+ \\
\vdots \\
-\tilde{\tau}^{(N)}(x_{N+1})
\end{array}
\right\}
\tag{7.82}
$$

The assembly operation has produced a sum of two flux terms at each interelement boundary, one from each element on either side of the boundary. For example, at node 2, which is the boundary between elements (1) and (2), the expression is

$$
-\tilde{\tau}^{(1)}(x_2) + \tilde{\tau}^{(2)}(x_2)
\tag{7.83a}
$$

or, substituting Eq. (7.80),

$$
-\left(-\alpha\,\frac{d\tilde{U}^{(1)}}{dx}\right)_{x_2} + \left(-\alpha\,\frac{d\tilde{U}^{(2)}}{dx}\right)_{x_2}
\tag{7.83b}
$$

To determine what numerical values should be specified for these pairs of flux terms, we can use either a mathematical approach or a physical approach. In previous sections (namely, Sections 5.3.1, 7.4.3, and 7.5.2), we used only the mathematical approach. In each of those sections we were working with an ordinary differential equation that was a particular application of the general equation [Eq. (7.1a)]. For the sake of generality, then, we will repeat the mathematical approach once more, this time for the general equation — Eq. (7.1a). Following this development we will then turn to the physical approach, which most readers will probably find more intuitively appealing and much easier to use.

The mathematical approach ● Consider again the general differential equation [Eq. (7.1a)]:

$$
-\frac{d}{dx}\left(\alpha(x)\,\frac{dU(x)}{dx}\right) + \beta(x)U(x) = f(x)
$$

To simplify the discussion, let's consider all the discontinuities to be located at x_2, the boundary between elements (1) and (2). Thus the material properties $\alpha(x)$ and $\beta(x)$ may both have a step change at x_2. The interior load $f(x)$ may have both a step change $f_s(x)$ and a concentrated load $f_c\delta(x-x_2)$; that is,

$$
f(x) = f_s(x) + f_c\delta(x-x_2)
\tag{7.84}
$$

In the same manner as was done in the above-mentioned sections, we now integrate Eq. (7.1a) over an infinitesimal interval from $x_2 - \epsilon$ to $x_2 + \epsilon$:

$$\int_{x_2-\epsilon}^{x_2+\epsilon} \left[-\frac{d}{dx}\left(\alpha(x) \frac{dU(x)}{dx} \right) + \beta(x)U(x) \right] dx = \int_{x_2-\epsilon}^{x_2+\epsilon} [f_s(x) + f_c\delta(x-x_2)]dx \tag{7.85}$$

which becomes

$$\left[\left(-\alpha(x) \frac{dU(x)}{dx} \right) \right]_{x_2-\epsilon}^{x_2+\epsilon} + \int_{x_2-\epsilon}^{x_2+\epsilon} \beta(x)U(x)dx = \int_{x_2-\epsilon}^{x_2+\epsilon} f_s(x)dx + f_c \tag{7.86}$$

In the limit as $\epsilon \to 0$, the integrals over $\beta(x)U(x)$ and $f_s(x)$ vanish (the integrands are finite, whereas the interval is infinitesimal), leaving

$$-\left(-\alpha \frac{dU}{dx} \right)_{x_2^-} + \left(-\alpha \frac{dU}{dx} \right)_{x_2^+} = f_c \tag{7.87}$$

Since x_2^- and x_2^+ indicate points just inside elements (1) and (2), respectively, Eq. (7.87) may be written instead as

$$-\left(-\alpha \frac{dU^{(1)}}{dx} \right)_{x_2} + \left(-\alpha \frac{dU^{(2)}}{dx} \right)_{x_2} = f_c \tag{7.88a}$$

or equivalently, using Eq. (7.80),

$$-\tau^{(1)}(x_2) + \tau^{(2)}(x_2) = f_c \tag{7.88b}$$

We conclude that the value to be substituted for each pair of interelement flux terms in Eq. (7.82) is the value of any concentrated interior load that is applied at each interelement boundary. If no concentrated load is applied (which is usually the case at most interelement boundaries), then the appropriate value is zero.

The physical approach • We consider first the heat conduction problem and then the cable deflection problem.

The heat conduction problem. Recall Fig. 7.27 and Eq. (7.50), which identify the element boundary flux terms as heat flowing *into* the ends of an element. Applying this interpretation to the first term in Eq. (7.83b) yields

$$-\left(-\alpha \frac{d\tilde{U}^{(1)}}{dx} \right)_{x_2} = \left(k \frac{d\tilde{T}^{(1)}}{dx} \right)_{x_2} = q_{in}^{(1)}(x_2) \tag{7.89}$$

which is the heat flowing into the right end of element (1) at x_2. Similarly for the second term,

$$\left(-\alpha \frac{d\tilde{U}^{(2)}}{dx} \right)_{x_2} = \left(-k \frac{d\tilde{T}^{(2)}}{dx} \right)_{x_2} = q_{in}^{(2)}(x_2) \tag{7.90}$$

which is the heat flowing into the left end of element (2) at x_2. Hence Eq. (7.83b) may be written

$$-\left(-\alpha \frac{d\tilde{U}^{(1)}}{dx} \right)_{x_2} + \left(-\alpha \frac{d\tilde{U}^{(2)}}{dx} \right)_{x_2} = q_{in}^{(1)}(x_2) + q_{in}^{(2)}(x_2) \tag{7.91}$$

Observe that the RHS of Eq. (7.91) is the heat flowing into element (1) plus the heat flowing into element (2), which is the *total heat flowing into the system at* x_2. We may therefore write Eq. (7.91) more simply as

$$-\left(-\alpha\,\frac{d\tilde{U}^{(1)}}{dx}\right)_{x_2} + \left(-\alpha\,\frac{d\tilde{U}^{(2)}}{dx}\right)_{x_2} = q_{in}(x_2) \tag{7.92}$$

where $q_{in}(x_2)$ represents the total heat flowing into the system at x_2. Figure 7.57 illustrates the physical interpretation of the two possibilities, $q_{in}(x_2)=0$ and $q_{in}(x_2)\neq 0$. In the first case the net heat flowing into the system is zero, which means only internal heat can be flowing across the interelement boundary. In the second case the net heat input is not zero, so it must be input from an external source. This externally injected flux, q_{ex}, is a concentrated load, corresponding to f_c in Eq. (7.88b).

We also note that the flux terms in Eq. (7.82) that occur on the boundary of the domain also have the same physical meaning:

$$\tilde{\tau}^{(1)}(x_1) = \left(-\alpha\,\frac{d\tilde{U}^{(1)}}{dx}\right)_{x_1} = q_{in}(x_1)$$

$$-\tilde{\tau}^{(N)}(x_{N+1}) = \left(\alpha\,\frac{d\tilde{U}^{(N)}}{dx}\right)_{x_{N+1}} = q_{in}(x_{N+1}) \tag{7.93}$$

We see, then, that the complete boundary flux vector in Eq. (7.82) may be written instead as

$$\begin{Bmatrix} q_{in}(x_1) \\ q_{in}(x_2) \\ q_{in}(x_3) \\ q_{in}(x_4) \\ \vdots \\ q_{in}(x_{N+1}) \end{Bmatrix} \tag{7.94}$$

and the assembled system equations in Eq. (7.81) may be written as Eqs. (7.95), see opposite page. Each of the $q_{in}(x_i)$ terms in Eq. (7.95) is a concentrated heat flux applied to a node on the boundary of an element — either on an interelement boundary or on the domain boundary. These concentrated flux values are specified in data set VIII in the data input instructions in the appendix. (The distributed interior loads, FI_i, on the RHS of Eq. (7.95) are specified in data set VI.) Since the system load vector is initially zeroed, one only needs to use data set VIII if a nonzero nodal heat flux is applied.

The cable-deflection problem. Recall Fig. 7.44 and Eq. (7.77), which identify the element boundary flux terms as an *upward*-directed force (i.e., in the positive W direction) on the ends of an element. Applying this interpretation to each term in Eq. (7.83b) yields

$$-\left(-\alpha\,\frac{d\tilde{U}^{(1)}}{dx}\right)_{x_2} = \left(T\,\frac{d\tilde{W}^{(1)}}{dx}\right)_{x_2} = F_{up}^{(1)}(x_2)$$

$$\left(-\alpha\,\frac{d\tilde{U}^{(2)}}{dx}\right)_{x_2} = \left(-T\,\frac{d\tilde{W}^{(2)}}{dx}\right)_{x_2} = F_{up}^{(2)}(x_2) \tag{7.96}$$

$$q_{in}(x_2) = -q + q = 0 \qquad\qquad q_{in}(x_2) = -q + q + q_{ex} = q_{ex}$$

FIGURE 7.57 ● Physical interpretation of a discontinuity in the heat flux as an externally applied, concentrated heat flux: (a) only internal heat flux—hence the flux is continuous at x_2; (b) internal heat flux plus externally injected flux—hence the flux is discontinuous at x_2.

$$
\begin{bmatrix}
K_{11}^{(1)} & K_{12}^{(1)} \\
K_{21}^{(1)} & K_{22}^{(1)}+K_{22}^{(2)} & K_{23}^{(2)} \\
 & K_{32}^{(2)} & K_{33}^{(2)}+K_{33}^{(3)} & K_{34}^{(3)} \\
 & & K_{43}^{(3)} & K_{44}^{(3)}+K_{44}^{(4)} \\
 & & & & \cdot \\
 & & & & & \cdot \\
 & & & & & & \cdot \\
 & & & & & & & K_{N+1,N+1}^{(N)}
\end{bmatrix}
\begin{Bmatrix}
a_1 \\ a_2 \\ a_3 \\ a_4 \\ \cdot \\ \cdot \\ \cdot \\ a_{N+1}
\end{Bmatrix}
$$

$$
=
\begin{bmatrix}
FI_1^{(1)} \\
FI_2^{(1)}+FI_2^{(2)} \\
FI_3^{(2)}+FI_3^{(3)} \\
FI_4^{(3)}+ \\
\cdot \\
\cdot \\
\cdot \\
FI_{N+1}^{(N)}
\end{bmatrix}
+
\begin{Bmatrix}
q_{in}(x_1) \\
q_{in}(x_2) \\
q_{in}(x_3) \\
q_{in}(x_4) \\
\cdot \\
\cdot \\
\cdot \\
q_{in}(x_{N+1})
\end{Bmatrix}
\qquad (7.95)
$$

where $F_{up}^{(1)}(x_2)$ is an upward-directed force applied to the end of element (1) at x_2 and $F_{up}^{(2)}(x_2)$ is an upward-directed force applied to the end of element (2) at x_2. Hence Eq. (7.83b) may be written

$$-\left(-\alpha \frac{d\tilde{U}^{(1)}}{dx}\right)_{x_2} + \left(-\alpha \frac{d\tilde{U}^{(2)}}{dx}\right)_{x_2} = F_{up}^{(1)}(x_2) + F_{up}^{(2)}(x_2) = F_{up}(x_2) \tag{7.97}$$

where $F_{up}(x_2)$ is the *total upward-directed force applied to the system at* x_2.

Figure 7.58 illustrates the physical interpretation of the two possibilities, $F_{up}(x_2) = 0$ and $F_{up}(x_2) \neq 0$. These two cases are directly analogous to the two heat conduction cases in Fig. 7.57.

Just as with the heat conduction problem, the flux terms in Eq. (7.82) that occur on the boundary of the domain have the same physical meaning as the pairs of terms at inter-element boundaries:

$$\tilde{\tau}^{(1)}(x_1) = \left(-T \frac{d\tilde{W}^{(1)}}{dx}\right)_{x_1} = F_{up}(x_1)$$

$$\tilde{\tau}^{(N)}(x_{N+1}) = \left(T \frac{d\tilde{W}^{(N)}}{dx}\right)_{x_{N+1}} = F_{up}(x_{N+1}) \tag{7.98}$$

Hence, in complete analogy with the heat conduction problem, the system boundary flux vector in Eq. (7.82) may be written

$$\begin{Bmatrix} F_{up}(x_1) \\ F_{up}(x_2) \\ F_{up}(x_3) \\ F_{up}(x_4) \\ \vdots \\ F_{up}(x_{N+1}) \end{Bmatrix} \tag{7.99}$$

Each of the $F_{up}(x_i)$ terms in Eq. (7.99) is an upward-directed, concentrated force applied to an element boundary node, either on the domain boundary or on an interelement

(a) $F_{up}(x_2) = -F + F = 0$

(b) $F_{up}(x_2) = -F + F + F_{ex} = F_{ex}$

FIGURE 7.58 ● Physical interpretation of a discontinuity in the vertical force as an externally applied, concentrated vertical force: (a) only internal forces—hence the vertical force is continuous at x_2; (b) internal forces plus external force—hence the vertical force is discontinuous at x_2.

boundary. These concentrated forces are specified in data set VIII in the data input instructions. This completes the discussion of the physical approach.

Looking back at both the mathematical and physical approaches, we see that the pair of flux terms at each interelement boundary represents a concentrated load, i.e., an interior load that is distributed over such a small region of the domain that it is effectively idealized as being concentrated at a single point — namely, the node between two elements. We first encountered the pair of interelement flux terms in Section 5.3.1 where it was given the name "natural IBC." For the particular illustration problem in that section, there were no concentrated loads so the natural IBC had the value zero, expressing the fact that the flux on either side of the interelement boundary was equal, i.e., continuous. Now we see that a concentrated load means a nonzero natural IBC and hence discontinuity of the flux.

We can summarize all the above observations in the following statements.

1. A concentrated interior load yields a nonzero natural IBC that corresponds to a discontinuity in the flux at the interelement boundary.[26] The discontinuity is precisely the value of the concentrated load. This value must be specified in the input data.

2. Conversely, the absence of a concentrated interior load (which is usually the case at most interelement boundaries) yields a zero natural IBC that corresponds to continuity in the flux at the interelement boundary. Since the system load vector is initially zeroed, no value needs to be specified in the input data.

7.7 Some Practical Guidelines

The above heat conduction and cable deflection problems illustrate several practical problem-solving techniques that can be used in all types of FE analysis. The following list is a convenient summary of these techniques. They are general guidelines that will be appropriate for most situations.

1. Before beginning an analysis with an unfamiliar FE program (or an unfamiliar aspect of a previously used program), perform one or more simple tests to gain confidence in the correctness of the program and to learn how to use it. Most important, test a single element.

2. Begin a new problem with a very simple mesh, perhaps using only a few elements. The resulting runs will be quick and cheap while you are becoming familiar with the problem and the input instructions. In other words, make your "learning curve" as short and inexpensive as possible.

   If there are bugs in the program or serious deficiencies in your model (e.g., a poorly posed problem), they will usually reveal themselves during these two preliminary steps. It is much easier to find and correct such problems when the analysis is small and uncomplicated; this will be greatly appreciated by anyone you might turn to for con-

---

[26] We recall from Section 7.5.2 that one must always construct elements so that any concentrated loads will act on the *boundaries* of elements.

sultation — such as the program developer! No one wants to be confronted with a thousand-element model and the complaint that "something's wrong."

3. Number the nodes in an orderly manner, keeping the numbers close together as you traverse the mesh.[27] For a 1-D problem this is trivial: Number sequentially from one end point to the other. For 2-D problems, see guidelines in Chapter 13. Orderly numbering will reduce the bandwidth, which means shorter computation time and lower costs. It also facilitates the use of automatic node generation, which means less preparation time for the engineer and hence lower costs.

4. When constructing a mesh, place elements so that discontinuities in the problem data are located on the element boundaries, not inside elements. Three types of discontinuities already encountered are

   a) Interfaces between different material or physical properties
   b) Step changes in loads
   c) Concentrated loads

5. After doing one or two analyses with coarse meshes, it will be clear (by examining flux discontinuities and/or by comparing results from two meshes) which regions of the domain are the least accurate. For the next mesh, place relatively more DOF in these critical regions, i.e., employ local mesh refinement. This may be accomplished by using smaller elements of the same type (h-refinement) or by using higher-order elements of the same size (p-refinement; see Chapters 8 and 9).

   This guideline is intended to optimize the mesh as regards computer efficiency. However, as demonstrated in the above cable problem, it may cost more for the engineer's time to construct an irregular mesh, even using a sophisticated preprocessor program, and the resulting output may be a bit more awkward to interpret, and hence may be another possible source of error. These are trade-offs that must be weighed in each problem.

6. After each mesh (or two), estimate the error by comparing the two solutions and/or examining the flux discontinuities. When an acceptable error estimate is reached, stop! This may require only one, two, or three meshes if you are working on a familiar type of problem with a familiar program. At the other extreme, it might involve dozens of meshes for a very complicated problem in an unfamiliar area.

7. At the end of the analysis, be able to explain the results. They must make sense physically. They must be reasonable.[28] Subject the results to simple but clever tests, using any available physical principles. Be resourceful. This is the stage that separates the mature from the immature engineer. Depending on the importance of the project, it may be worthwhile to seek an independent verification via an experiment or different type of analysis. (Recall the discussion at the end of Section 4.4.)

---

[27] If using a program with a wave-front equation solver, number the elements (not the nodes) in an orderly manner, for the same reasons given above. In this case, node numbering is essentially unimportant.

[28] On two separate and notable occasions, the author's FE analysis produced totally unexpected results. This forced a careful and critical examination, involving independent analysis and experiment. In both cases the original results turned out to be correct, resulting in new insight into the physical phenomenon.

Exercises

7.1   Section 7.2, step 6, argues that the optimal flux-sampling point for a 1-D C^0-linear element is generally at the center of the element. This is true only when the exact solution in the element "looks like" a linear polynomial — which generally occurs only as the element size $h \to 0$. The theory is based on the least-squares-fitting of a constant $(d\tilde{U}/dx)$ to a linear polynomial $(dU/dx$ as $h \to 0$, to $O(h^2))$.

Calculate the least-squares fit of a constant c to a linear polynomial, $f(x) = a + bx$, in an element spanning $[x_1, x_2]$. Show that the constant intersects $f(x)$ at the element center; that is,

$$c = f\left(\frac{x_1 + x_2}{2}\right)$$

Hint: Define

$$I(c) = \int_{x_1}^{x_2} [f(x) - c]^2 \, dx$$

and determine c from the minimizing condition $\partial I/\partial c = 0$.

7.2   Let $f(x)$ in Eq. (7.23) be represented by the function $A \sin \omega x$, in addition to a constant. Evaluate the interior load integral with closed-form expressions and recode subroutines ELEM11 and LOADDT. Then rerun the cable deflection problem in Section 7.5. Compare results using this exact definition (i.e., no modeling approximation) with those in Section 7.5 using the piecewise-constant approximation.

Exercises 7.3 through 7.9 were first analyzed in Chapters 3 and 4 using simple trial functions to facilitate hand calculations. They were repeated in Chapter 5 using elements, but again limited to only a few elements for hand calculations. Thus the number of DOF were chosen at the outset and the resulting accuracy was estimated. Here we have the benefit of a computer program so the order can be reversed: The desired accuracy will be decided at the outset and the process of mesh refinement and error estimation will determine the number of DOF needed. This is the usual procedure in practice.

In Exercises 7.3 through 7.9 use UNAFEM to obtain approximate solutions with an estimated pointwise error of about 1%. After observing the general nature of the solution from an initial mesh and making an error estimate with the help of a second mesh, you may want to locally refine for a third mesh to achieve the desired accuracy in certain regions. If the UNAFEM program is not available to you on a computer, you can still solve most or all of the problems using any other FE program capable of solving Eqs. (7.1a, b, and c), for example, elasticity or heat conduction programs (recall Table 6.1).

7.3   Use UNAFEM to find the deflection of a horizontal hanging cable (Exercise 3.4):

$$-\frac{d}{dx}\left(T\frac{dW}{dx}\right) = -\varrho g \qquad 0 < x < 10$$

with the BCs $W(0) = W(10) = 0$, $T = 98$ N, $\varrho = 1$ kg/m, and $g = 9.8$ m/sec^2.

7.4   Use UNAFEM to find the deflection of a horizontal hanging cable supported by an elastic foundation (Exercise 3.7):

$$-\frac{d}{dx}\left(T\frac{dW}{dx}\right) + kW = -\varrho g \qquad 0 < x < 10$$

with the BCs $W(0)=W(10)=0$, $k=24.5$ N/m-m, and the rest of the data the same as in Exercise 7.3.

7.5   Use UNAFEM to find the temperature and heat flux distributions in a heat-conducting rod with convection heat loss from the surface (Exercise 3.8):

$$-\frac{d}{dx}\left(k\frac{dT}{dx}\right) + \frac{2h}{r}T = \frac{2h}{r}T_\infty \qquad 0<x<10$$

with the BCs $T(0)=100\,°F$ and $q(10)=0.2$ BTU/sec-in^2, $k=22$ BTU/sec-in-°F, $h=0.193\times10^{-5}$ BTU/sec-in^2-°F, $r=0.5$ in., and $T_\infty=72\,°F$.

7.6   Use UNAFEM to obtain an approximate solution to the differential equation (Exercise 3.13):

$$\frac{d}{dx}\left[(x+1)\frac{dU}{dx}\right] = 0 \qquad 1<x<2$$

with the BCs $U(1)=1$ and $\tau(2)=1$.

7.7   Use UNAFEM to obtain an approximate solution to the differential equation (Exercise 3.14):

$$\frac{d}{dx}\left[(x+1)\frac{dU}{dx}\right] = 0 \qquad 1<x<2$$

with the BCs $U(1)=1$ and $U(2)=2$.

7.8   Use UNAFEM to obtain an approximate solution to the differential equation (Exercise 3.15):

$$\frac{d}{dx}\left(x^2\frac{dU}{dx}\right) = \frac{1}{12}(-30x^4+204x^3-351x^2+110x) \qquad 0<x<4$$

with the BCs $U(0)=1$ and $U(4)=0$.

7.9   Alucobond is a composite material consisting of two thin sheets of aluminum with a thermoplastic core. It is used as wall paneling for some of the small, outdoor enclosures protecting Bell System electronic equipment. Heat generated by the equipment will flow out through the wall panels. We can assume the heat flow over most of the panel area is essentially 1-D (see figure). It is estimated that the flux incident on the walls is 0.08 cal/sec-cm^2. The temperature of the outside surface of the walls is 20°C.

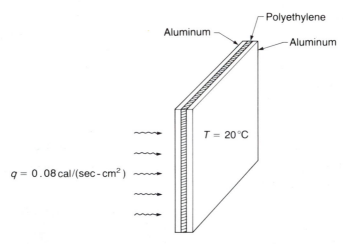

Using UNAFEM, calculate the temperature distribution through the panel. In particular, determine the temperature on the inside surface (at 0 cm). The aluminum layers are each 0.05 cm thick and the polyethylene is 0.3 cm. The thermal conductivities are $k_{alum}=0.49$ cal-cm/sec-cm^2- °C and $k_{poly}=0.001$ cal-cm/sec-cm^2- °C.

7.10 Using UNAFEM, obtain an approximate solution to the illustrative problem used in Chapters 3 through 5:

$$\frac{d}{dx}\left(x\,\frac{dU(x)}{dx}\right) = \frac{2}{x^2} \qquad 1<x<2$$

with the BCs $U(1)=2$ and $\tau(2)=1/2$. For the flux BCs, remember that on the left end of an element (lower x coordinate) we specify τ and on the right end of an element we specify $-\tau$. Do two or more meshes, of sufficient refinement to produce an estimated pointwise error of about 1% for both $\tilde{U}(x)$ and $\tilde{\tau}(x)$. [The exact solution is $U(x)=1/2 \ln x+2/x$.]

7.11 Using UNAFEM, obtain an approximate solution to the problem,

$$\frac{d^2U(x)}{dx^2} + 25U(x) = 0 \qquad 0<x<2\pi$$

with the BCs $U(0)=0$ and $\tau(2\pi)=-10$. See comment on flux BCs in Exercise 7.10. Do two or more meshes, of sufficient refinement to produce an estimated pointwise error of about 5% for both $\tilde{U}(x)$ and $\tilde{\tau}(x)$. Note: UNAFEM may fail at certain critical element sizes due to loss of positive-definiteness. See remarks in Exercise 5.5. [The exact solution is $U(x)=2 \sin 5x$.]

7.12 Using UNAFEM, obtain an approximate solution, with an estimated pointwise error of about 5%, to the problem,

$$\frac{d}{dx}\left(x\,\frac{dU(x)}{dx}\right) + xU(x) = 0 \qquad 0<x<10$$

with the BCs $U(0)=1$ and $U(10)=-0.2459358$. Note: UNAFEM may fail at certain critical element sizes due to loss of positive-definiteness. See remarks in Exercise 5.5. [The exact solution is the zero-order Bessel function, $J_0(x)$.]

7.13 Consider a 100-ft cable hanging *vertically* from a building. (See figure.) A 5-mph wind is blowing horizontally.

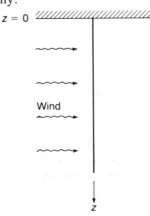

The tension in the cable, when hanging vertically, is

$$T(z) = \varrho_c g A (L-z)$$

where

ϱ_c = density of cable, 0.50 slugs/ft^3

$A = \dfrac{\pi D^2}{4}$ $(D$ = diam, 1 in.)

g = 32.2 ft/sec^2

L = 100 ft

The aerodynamic load exerted on the cable by the wind is a horizontal force f distributed uniformly along the length:

$$f = \frac{1}{2} \varrho_a C_n D V^2$$

where

f = distributed force, (lb/ft)

ϱ_a = density of air, 0.00251 slugs/ft^3 at 0 °C and 1 atm

C_n = coefficient of normal drag, 3.1 (with strumming)

V = wind velocity, 5 mph (7.3 ft/sec)

a) Use UNAFEM to calculate the horizontal displacement of the bottom of the cable (to within an estimated 5% accuracy).

b) Check the reasonableness of your solution by comparing the total horizontal force on the cable with the calculated horizontal reaction force at the top of the cable.

(Note that the solution can only be useful for small displacements since the theoretical model does not permit any rise in the cable as it moves horizontally.)

7.14 A boy is holding onto one end of a 10-ft rubber cord and whirling it in a horizontal circle over his head. Assume that his hand, acting as the pivot, does not move and that the rotational velocity ω is constant. The cord has a circular cross section of radius 0.2 in. There are three forces acting on the cord: centrifugal force acting radially, gravitational pull (weight) acting downward, and air resistance acting circumferentially. The first force will cause the cord to stretch radially, the second will cause it to sag vertically, and the third will cause it to lag circumferentially behind a straight radial line. We want to compute the displacement of the outer tip of the cord (relative to the straight radial configuration of an inextensible cord in a zero-gravity vacuum) due to these forces. For small displacements we can treat these effects as uncoupled.

a) Vertical sag: The centrifugal force on the cord produces a variation in tension:

$$T(r) = \frac{1}{2} \varrho \omega^2 (L^2 - r^2)$$

where

r = radial coordinate

p = lineal mass density, 0.01 slugs/ft

ω = rotational velocity, 10 rad/sec

L = 10 ft

Of course, the weight of the cord is a uniformly distributed load $-\varrho g$, where $g = 32.2$ ft/sec^2.

b) Circumferential lag: The air resistance is an aerodynamic load given by the expression in Exercise 7.13. Use those values for ϱ_a and C_n; note that V is the component normal to the cable of the relative velocity between the air (assumed still) and the cord.

c) Radial stretch: The cord behaves as an elastic rod, for which the governing equation is derived in Section 6.2. Young's modulus, E, is 500 lb/in^2. The interior load $f(r)$ is due to the centrifugal force:

$$f(r) = \varrho r \omega^2$$

Use UNAFEM to determine the three components of displacement of the tip of the cord. Compute your answers to an estimated accuracy of about 1%.

References

[7.1]   T. Moan, ''On the Local Distribution of Errors by Finite Element Approximations'' in Y. Yamada and R. H. Gallagher (eds.), *Theory and Practice in Finite Element Structural Analysis*, University of Tokyo Press, Tokyo, 1973.

[7.2]   E. Isaacson and H. B. Keller, *Analysis of Numerical Methods*, Wiley, New York, 1966.

[7.3]   U. W. Hochstrasser, ''Orthogonal Polynomials,'' Chapter 22 in M. Abramowitz and I. A. Stegun (eds.), *Handbook of Mathematical Functions*, National Bureau of Standards, Applied Mathematics Series, **55**, 1964.

[7.4]   G. Strang and G. J. Fix, *An Analysis of the Finite Element Method*, Prentice-Hall, Englewood Cliffs, N. J., 1973, p. 168.

[7.5]   K. J. Bathe and E. L. Wilson, *Numerical Methods in Finite Element Analysis*, Prentice-Hall, Englewood Cliffs, N. J., 1976.

[7.6]   O. Axelsson and V. A. Barker, *Finite Element Solution of Boundary Value Problems: Theory and Computation*, Academic Press, New York, 1984.

[7.7]   G. F. Carey and J. T. Oden, *Finite Elements: Computational Aspects*, vol. III, Prentice-Hall, Englewood Cliffs, N. J., 1984.

[7.8]   DISSPLA, a graphics software package developed by ISSCO, 4186 Sorrento Valley Blvd., San Diego, Calif., 92121.

[7.9]   J. S. Mayo, AT&T Bell Laboratories executive vice president, in remarks made at Software Quality Symposium, Bell Laboratories, Holmdel, N. J., June 22, 1982.

[7.10]   E. E. Sumner, AT&T Bell Laboratories vice president, ''Software and Society,'' *Bell Laboratories Record*, Feb. 1983 (interview by S. Aaronson).

8

**1-D Boundary-Value Problems:
Higher-Order Elements**

Introduction

This chapter has two principal objectives: (1) to develop three new elements, known as *higher-order* elements, that use polynomials of a higher-degree than needed to satisfy the completeness condition (Section 5.6), and (2) to introduce the concept of an *isoparametric* element. The two objectives overlap because the three higher-order elements will all be developed as isoparametric elements. Higher-order elements provide faster rates of convergence and therefore usually greater accuracy (see Chapter 9). In addition, the illustrative problems continue to provide practical guidelines for using FE programs.

This chapter continues to work with the general 1-D problem described in Chapter 6. Section 8.1 reviews steps 1 through 3 of the six-step procedure for deriving the element equations — these first three steps are identical for all the elements developed for this 1-D problem. Section 8.2 develops a C^0-quadratic element, using the same direct (non-isoparametric) approach as used in Chapter 7. Section 8.3 also develops a C^0-quadratic element, but uses the isoparametric approach. The isoparametric element is then added to UNAFEM, and the program is applied to the heat conduction and cable deflection problems in Chapter 7. Sections 8.4 and 8.5 develop C^0-cubic and C^0-quartic isoparametric elements, respectively, which are also added to UNAFEM. The program is then applied to the same two problems. Section 8.6 develops another kind of higher-order element, a C^1-cubic, in which interelement continuity is greater than the continuity condition requires.

8.1 The First Three Steps for Deriving the Element Equations

Before beginning the six-step procedure for each of our new elements, a review of Section 7.2 reminds us that the first three steps are independent of the specific form of the shape functions. These three steps deal only with the governing differential equation,

$$-\frac{d}{dx}\left(\alpha(x)\,\frac{dU(x)}{dx}\right) + \beta(x)U(x) = f(x) \qquad x_a < x < x_b \tag{8.1}$$

and the formal representation of the element trial solution as an expansion of shape functions,

$$\tilde{U}^{(e)}(x;a) = \sum_{j=1}^{n} a_j \phi_j^{(e)}(x) \tag{8.2}$$

The number of shape functions, n, and their specific form is not decided until step 4.

Therefore, at the end of step 3 we have a formal representation of the element equations in terms of the shape functions $\phi_j^{(e)}(x)$ and their first derivatives. These equations were derived in Section 7.2. For convenience, Eqs. (7.11) are reproduced here:

$$\begin{bmatrix} K_{11}^{(e)} & K_{12}^{(e)} & \cdots & K_{1n}^{(e)} \\ K_{21}^{(e)} & K_{22}^{(e)} & \cdots & K_{2n}^{(e)} \\ \vdots & \vdots & & \vdots \\ K_{n1}^{(e)} & K_{n2}^{(e)} & \cdots & K_{nn}^{(e)} \end{bmatrix} \begin{Bmatrix} a_1 \\ a_2 \\ \vdots \\ a_n \end{Bmatrix} = \begin{Bmatrix} F_1^{(e)} \\ F_2^{(e)} \\ \vdots \\ F_n^{(e)} \end{Bmatrix} \tag{8.3a}$$

where

$$K_{ij}^{(e)} = \int^{(e)} \frac{d\phi_i^{(e)}(x)}{dx}\,\alpha(x)\,\frac{d\phi_j^{(e)}(x)}{dx}\,dx + \int^{(e)} \phi_i^{(e)}(x)\beta(x)\phi_j^{(e)}(x)\,dx$$

$$F_i^{(e)} = \int^{(e)} f(x)\phi_i^{(e)}(x)\,dx - \left[\left(-\alpha(x)\frac{d\tilde{U}^{(e)}}{dx}\right)\phi_i^{(e)}(x)\right]^{(e)} \tag{8.3b}$$

This is the starting point for developing additional elements. Sections 8.2 through 8.5 each begin with the next step, step 4, and continue on through steps 5 and 6.

8.2 The 1-D C⁰-Quadratic Element: Direct Approach

Step 4: Develop specific expressions for the shape functions.

In Chapter 7 we employed a C^0-linear element, containing one node at each end, in which the element trial solution, $\tilde{U}^{(e)}(x;a)$, is expanded as a linear sum of two shape functions. Each shape function was a linear interpolation polynomial (Lagrange polynomial) equal to unity at one node and zero at the other (the δ_{ji} property). Recall the development in Section 5.2 and discussion in Section 5.8.

The C^0-quadratic element is developed in a completely analogous fashion. Since a quadratic polynomial has three terms $(a+bx+cx^2)$, it will require three nodes to uniquely define such a polynomial. One node must still be located at each end of the element, i.e., on the element boundary, in order to simplify assembly and to ensure that the resulting assembled trial functions are local. The third node may be located anywhere in the inte-

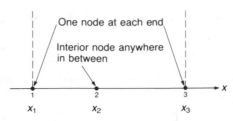

FIGURE 8.1 ● Placement of nodes for the 1-D C^0-quadratic element, using the direct (non-isoparametric) approach.

rior. Figure 8.1 shows such an element, using the local node numbers 1, 2, and 3. The middle node plays no role in establishing interelement continuity; its only purpose is to help define a quadratic polynomial.

The element trial solution is a sum of three shape functions,

$$\tilde{U}^{(e)}(x;a) = \sum_{j=1}^{3} a_j \phi_j^{(e)}(x)$$

(8.4)

where each shape function is a quadratic polynomial that satisfies the interpolation property,

$$\phi_j^{(e)}(x_i) = \delta_{ji}$$

(8.5)

Evaluating Eq. (8.4) at x_i, and using Eq. (8.5), yields

$$\tilde{U}^{(e)}(x_i;a) = a_i \qquad i = 1,2,3$$

(8.6)

Either Eq. (8.5) or (8.6) may be used to derive the expressions for the shape functions, as was demonstrated in Section 5.2 for the linear element. Let's use the first approach, Eq. (8.5).

Consider the first shape function, which we write as a quadratic polynomial:

$$\phi_1^{(e)}(x) = \alpha_1 + \alpha_2 x + \alpha_3 x^2$$

(8.7)

Apply Eq. (8.5) to Eq. (8.7) at each of the three node points:

$$\alpha_1 + \alpha_2 x_1 + \alpha_3 x_1^2 = 1$$

$$\alpha_1 + \alpha_2 x_2 + \alpha_3 x_2^2 = 0$$

$$\alpha_1 + \alpha_2 x_3 + \alpha_3 x_3^2 = 0$$

(8.8)

Solving Eqs. (8.8) for each of the α_i in terms of the x_i and then substituting the α_i back into Eq. (8.7) yields the following expression for $\phi_1^{(e)}(x)$:

$$\phi_1^{(e)}(x) = \frac{(x-x_2)(x-x_3)}{(x_1-x_2)(x_1-x_3)}$$

(8.9)

Repeating the same procedure for $\phi_2^{(e)}(x)$ and $\phi_3^{(e)}(x)$ yields (see Exercises 8.1 and 8.2)

$$\phi_2^{(e)}(x) = \frac{(x-x_1)(x-x_3)}{(x_2-x_1)(x_2-x_3)}$$

$$\phi_3^{(e)}(x) = \frac{(x-x_1)(x-x_2)}{(x_3-x_1)(x_3-x_2)}$$

(8.10)

The element trial solution for a typical element, $\tilde{U}^{(e)}(x; a)$, is a complete quadratic polynomial,

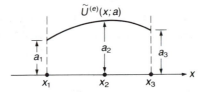

It is represented as the sum of 3 shape functions,

$$\tilde{U}^{(e)}(x; a) = \sum_{j=1}^{3} a_j \phi_j^{(e)}(x)$$

Each coefficient a_j is the value of $\tilde{U}^{(e)}(x; a)$ at the node at x_j. Each shape function, $\phi_j^{(e)}(x)$, is itself a quadratic polynomial satisfying the interpolation property, $\phi_j^{(e)}(x_i) = \delta_{ji}$:

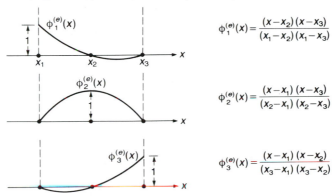

$$\phi_1^{(e)}(x) = \frac{(x - x_2)(x - x_3)}{(x_1 - x_2)(x_1 - x_3)}$$

$$\phi_2^{(e)}(x) = \frac{(x - x_1)(x - x_3)}{(x_2 - x_1)(x_2 - x_3)}$$

$$\phi_3^{(e)}(x) = \frac{(x - x_1)(x - x_2)}{(x_3 - x_1)(x_3 - x_2)}$$

FIGURE 8.2 ● The 1-D C⁰-quadratic element (nonisoparametric).

Figure 8.2 shows the three shape functions; in addition, it provides a convenient graphical summary of the interrelationship of the shape functions and the element trial solution for the 1-D C⁰-quadratic element.

We note in passing that these shape functions are second-degree (i.e., quadratic) *Lagrange interpolation polynomials*. This terminology was introduced in Section 5.2 where the shape functions for the C⁰-linear element were identified as first-degree (i.e., linear) Lagrange polynomials. Later in this chapter we will be using the third- and fourth-degree Lagrange polynomials. It is therefore instructive to see the expression for the 1-D Lagrange polynomials of general degree p, which may be found in almost any elementary numerical analysis book:

$$\phi_j(x) = \frac{(x - x_1)(x - x_2)\ldots(x - x_{i-1})(x - x_{i+1})\ldots(x - x_p)(x - x_{p+1})}{(x_j - x_1)(x_j - x_2)\ldots(x_j - x_{i-1})(x_j - x_{i+1})\ldots(x_j - x_p)(x_j - x_{p+1})}$$

$$j = 1, 2, \ldots, p+1 \qquad \textbf{(8.11)}$$

The polynomial $\phi_j(x)$ interpolates at the $p+1$ nodes $x_1, x_2, \ldots, x_{p+1}$; that is, it satisfies the interpolation property,

$$\phi_j(x_i) = \delta_{ji} \qquad i,j = 1,2,\ldots,p+1 \tag{8.12}$$

For example, setting $p=2$ in Eqs. (8.11) reproduces the quadratic shape functions in Eqs. (8.9) and (8.10). Setting $p=1$ reproduces the linear shape functions in Eqs. (5.9) in Section 5.2. And Sections 8.4 and 8.5 will set $p=3$ and $p=4$ to develop the cubic and quartic shape functions, respectively.

Step 5: Substitute the shape functions into the element equations, and transform the integrals into a form appropriate for numerical evaluation.

The element contains three DOF. Setting $n=3$ in Eqs. (8.3a) and using the local node numbers 1, 2, and 3 (Fig. 8.1), the element equations for the typical element may be written

$$\begin{bmatrix} K_{11}^{(e)} & K_{12}^{(e)} & K_{13}^{(e)} \\ K_{21}^{(e)} & K_{22}^{(e)} & K_{23}^{(e)} \\ K_{31}^{(e)} & K_{32}^{(e)} & K_{33}^{(e)} \end{bmatrix} \begin{Bmatrix} a_1 \\ a_2 \\ a_3 \end{Bmatrix} = \begin{Bmatrix} F_1^{(e)} \\ F_2^{(e)} \\ F_3^{(e)} \end{Bmatrix} \tag{8.13a}$$

where

$$K_{ij}^{(e)} = \int_{x_1}^{x_3} \frac{d\phi_i^{(e)}(x)}{dx}\, \alpha(x)\, \frac{d\phi_j^{(e)}(x)}{dx}\, dx + \int_{x_1}^{x_3} \phi_i^{(e)}(x)\beta(x)\phi_j^{(e)}(x)\, dx$$

$$F_i^{(e)} = \int_{x_1}^{x_3} f(x)\phi_i^{(e)}(x)\, dx - \left[\left(-\alpha(x)\frac{d\tilde{U}^{(e)}}{dx} \right) \phi_i^{(e)}(x) \right]_{x_1}^{x_3} \tag{8.13b}$$

To perform the integration using exact formulas, we must assume some specific expressions for $\alpha(x)$, $\beta(x)$, and $f(x)$. The choice of what kinds of expressions to use follows a line of reasoning similar to that used for the linear element in Section 7.2, step 5. In brief, the simplest choice is to represent $\alpha(x)$, $\beta(x)$, and $f(x)$ as constants within each element, that is, $\alpha^{(e)}$, $\beta^{(e)}$, and $f^{(e)}$, respectively. For problems in which $\alpha(x)$, $\beta(x)$, or $f(x)$ vary significantly, a coarse mesh would probably give poor results; however, as the mesh is refined the approximation would become progressively better.

Using constant values within each element, the stiffness and load terms in Eqs. (8.13b) are now straightforward to integrate, though a bit tedious. For example, using Eq. (8.9),

$$K_{11}^{(e)} = \alpha^{(e)} \int_{x_1}^{x_3} \left(\frac{2x-x_2-x_3}{(x_1-x_2)(x_1-x_3)} \right)^2 dx + \beta^{(e)} \int_{x_1}^{x_3} \left(\frac{(x-x_2)(x-x_3)}{(x_1-x_2)(x_1-x_3)} \right)^2 dx$$

$$= \frac{\alpha^{(e)}}{L}\left(1 + \frac{1}{3r^2} \right) + \frac{\beta^{(e)}L}{30r^2}(10r^2-5r+1)$$

$$F_1^{(e)} = f^{(e)} \int_{x_1}^{x_3} \frac{(x-x_2)(x-x_3)}{(x_1-x_2)(x_1-x_3)}\, dx - \left[\bar{\tau}^{(e)}(x)\phi_1^{(e)}(x) \right]_{x_1}^{x_3}$$

$$= \frac{f^{(e)}L}{6}\left(3 - \frac{1}{r} \right) + \bar{\tau}^{(e)}(x_1) \tag{8.14}$$

where $L=x_3-x_1$ and $r=(x_2-x_1)/L$. Integrating the remaining stiffness and load terms yields the following expressions for the element equations:

$$
\begin{bmatrix}
\dfrac{\alpha^{(e)}}{3L}\dfrac{3r^2+1}{r^2} & -\dfrac{\alpha^{(e)}}{3L}\dfrac{1}{r^2(1-r)} & \dfrac{\alpha^{(e)}}{3L}\dfrac{3r^2-3r+1}{r(1-r)} \\[2mm]
+\dfrac{\beta^{(e)}L}{30}\dfrac{10r^2-5r+1}{r^2} & +\dfrac{\beta^{(e)}L}{60}\dfrac{5r-2}{r^2(1-r)} & -\dfrac{\beta^{(e)}L}{60}\dfrac{10r^2-10r+3}{r(1-r)} \\[4mm]
 & \dfrac{\alpha^{(e)}}{3L}\dfrac{1}{r^2(1-r)^2} & -\dfrac{\alpha^{(e)}}{3L}\dfrac{1}{r(1-r)^2} \\[2mm]
 & +\dfrac{\beta^{(e)}L}{30}\dfrac{1}{r^2(1-r)^2} & -\dfrac{\beta^{(e)}L}{60}\dfrac{5r-3}{r(1-r)^2} \\[4mm]
\text{Symmetric} & & \dfrac{\alpha^{(e)}}{3L}\dfrac{3r^2-6r+4}{(1-r)^2} \\[2mm]
 & & +\dfrac{\beta^{(e)}L}{30}\dfrac{10r^2-15r+6}{(1-r)^2}
\end{bmatrix}
\begin{Bmatrix} a_1 \\[6mm] a_2 \\[6mm] a_3 \end{Bmatrix}
$$

$$
=\begin{Bmatrix}
\dfrac{f^{(e)}L}{6}\dfrac{3r-1}{r} \\[4mm]
\dfrac{f^{(e)}L}{6}\dfrac{1}{r(1-r)} \\[4mm]
\dfrac{f^{(e)}L}{6}\dfrac{2-3r}{1-r}
\end{Bmatrix}
+\begin{Bmatrix}
\tilde{\tau}^{(e)}(x_1) \\[4mm]
0 \\[4mm]
-\tilde{\tau}^{(e)}(x_3)
\end{Bmatrix}
\tag{8.15}
$$

Note that as r approaches 0 or 1, some of the stiffness and load terms become infinite, making Eqs. (8.15) useless. This corresponds to the interior node approaching either end of the element. The problem can be traced to the shape functions: Even though they are quadratic polynomials satisfying $\phi_j^{(e)}(x_i)=\delta_{ji}$ at the nodes, they will become arbitrarily large in between the nodes as the interior node approaches either end. Therefore in practical applications one should place the interior node of each element at or near the center (preferably at the center) in order to provide a balanced approximation within each element. We will encounter similar restrictions with the isoparametric elements in Sections 8.3 through 8.5.

Substituting $r=1/2$ in Eqs. (8.15) yields the following expressions for the element equations for an element with the interior node at the center:

$$
\begin{bmatrix}
\dfrac{7\alpha^{(e)}}{3L}+\dfrac{4\beta^{(e)}L}{30} & -\dfrac{8\alpha^{(e)}}{3L}+\dfrac{2\beta^{(e)}L}{30} & \dfrac{\alpha^{(e)}}{3L}-\dfrac{\beta^{(e)}L}{30} \\[2mm]
 & \dfrac{16\alpha^{(e)}}{3L}+\dfrac{16\beta^{(e)}L}{30}-\dfrac{8\alpha^{(e)}}{3L}+\dfrac{2\beta^{(e)}L}{30} & \\[2mm]
\text{Symmetric} & & \dfrac{7\alpha^{(e)}}{3L}+\dfrac{4\beta^{(e)}L}{30}
\end{bmatrix}
\begin{Bmatrix} a_1 \\[2mm] a_2 \\[2mm] a_3 \end{Bmatrix}
$$

$$
= \begin{Bmatrix} \dfrac{f^{(e)}L}{6} \\[3mm] \dfrac{4f^{(e)}L}{6} \\[3mm] \dfrac{f^{(e)}L}{6} \end{Bmatrix}
+ \begin{Bmatrix} \tilde{\tau}^{(e)}(x_1) \\[3mm] 0 \\[3mm] -\tau^{(e)}(x_3) \end{Bmatrix}
\tag{8.16}
$$

Note that the interior load vector assigns 1/6 of the total interior load $f^{(e)}L$ to each end node and the remaining 4/6 to the center node.

Assembling the element equations into the system equations follows the general assembly rule given in Section 5.7. This is illustrated for the present C^0-quadratic element by a three-element mesh with the nodes numbered sequentially from left to right (Fig. 8.3).

Using the assembly rule, the element stiffness and load terms from each of the three sets of element equations would be added to the rows and columns of the system matrices with the same numbers as the node numbers in the mesh; if two elements share a node (such as nodes 3 and 5), then their stiffness and load terms are added together. The resulting assembled system equations would appear as follows:

$$
\begin{bmatrix}
K_{11}^{(1)} & K_{12}^{(1)} & K_{13}^{(1)} & & & & \\
K_{21}^{(1)} & K_{22}^{(1)} & K_{23}^{(1)} & & & & \\
K_{31}^{(1)} & K_{32}^{(1)} & K_{33}^{(1)}+K_{33}^{(2)} & K_{34}^{(2)} & K_{35}^{(2)} & & \\
 & & K_{43}^{(2)} & K_{44}^{(2)} & K_{45}^{(2)} & & \\
 & & K_{53}^{(2)} & K_{54}^{(2)} & K_{55}^{(2)}+K_{55}^{(3)} & K_{56}^{(3)} & K_{57}^{(3)} \\
 & & & & K_{65}^{(3)} & K_{66}^{(3)} & K_{67}^{(3)} \\
 & & & & K_{75}^{(3)} & K_{76}^{(3)} & K_{77}^{(3)}
\end{bmatrix}
\begin{Bmatrix} a_1 \\ a_2 \\ a_3 \\ a_4 \\ a_5 \\ a_6 \\ a_7 \end{Bmatrix}
=
\begin{Bmatrix}
F_1^{(1)} \\ F_2^{(1)} \\ F_3^{(1)}+F_3^{(2)} \\ F_4^{(2)} \\ F_5^{(2)}+F_5^{(3)} \\ F_6^{(3)} \\ F_7^{(3)}
\end{Bmatrix}
\tag{8.17}
$$

FIGURE 8.3 ● A mesh of three C^0-quadratic elements.

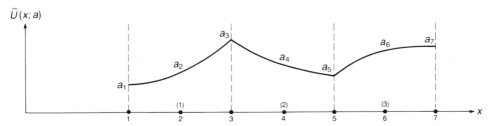

This 3-element trial solution is a sum of 7 trial functions,

$$\tilde{U}(x; a) = \sum_{j=1}^{7} a_j \phi_j(x)$$

Each trial function consists of either one shape function or a sum (assembly) of 2 shape functions:

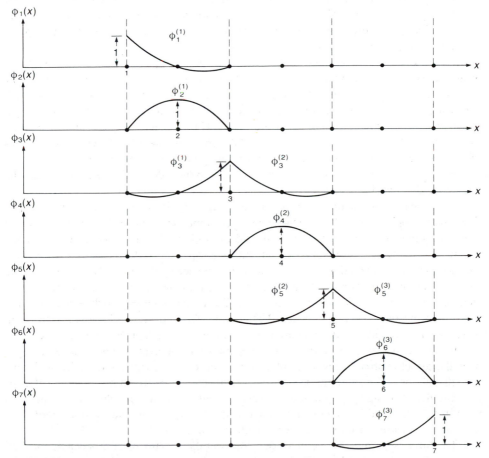

FIGURE 8.4 ● Assembled shape functions (trial functions) for a three-element mesh of 1-D C^0-quadratic elements.

As observed before (Fig. 5.15), assembly has the effect of joining shape functions from adjacent elements to form a single C^0-continuous trial function. This is illustrated in Fig. 8.4, which shows the shape functions after assembly for the same three-element mesh.

Step 6: Prepare expression for the flux.

$$\tilde{\tau}^{(e)}(x) = -\alpha(x) \frac{d\tilde{U}^{(e)}(x;a)}{dx}$$

$$= -\alpha(x) \sum_{j=1}^{3} a_j \frac{d\phi_j^{(e)}(x)}{dx}$$

$$= -\alpha(x) \left[a_1 \frac{2x-x_2-x_3}{(x_1-x_2)(x_1-x_3)} + a_2 \frac{2x-x_1-x_3}{(x_2-x_1)(x_2-x_3)} + a_3 \frac{2x-x_1-x_2}{(x_3-x_1)(x_3-x_2)} \right]$$

$$\tag{8.18}$$

This completes the six steps for deriving the element equations, their assembly rule, and the expression for flux. The development has directly paralleled that of the C^0-linear element. The procedure could easily be extended to elements with higher-degree polynomials such as cubic, quartic, etc.

The next logical step would be to add the element [Eqs. (8.15)] to the UNAFEM program. However, this will not be done (see Exercise 8.3). We will instead turn our attention to an alternate approach to the C^0-quadratic element. In the next section we will derive a different formulation for the element equations, using a very different technique which involves a transformation of coordinates. The element will be referred to as an *isoparametric* element. The resulting element equations will involve different expressions for the stiffness and load terms which, in turn, will generally produce different numerical values. (The two types of quadratic elements will produce identical results only when the interior node is at the center.)

The isoparametric approach is clearly not necessary for 1-D problems since we just finished developing two 1-D elements by the direct approach and could continue to develop more using the same approach. So why should we study it? There are a couple of reasons.

In the first place, isoparametric elements are especially effective (some would say indispensable) in 2-D (and 3-D) problems, as will be seen in Part IV. They represent the vast majority of elements in use today. It is therefore important to understand isoparametric elements and how to work with them. As usual, it is much easier to gain this understanding when working in a 1-D rather than a 2-D context. Consequently, there is a significant pedagogical advantage to introducing the isoparametric concept in a 1-D context.

In the second place, even in 1-D the isoparametric elements enjoy some advantages over the direct approach. For example, by the end of this chapter the reader will have seen that all the 1-D isoparametric elements are developed in essentially an identical manner, and the resulting programming is virtually identical for all. In addition, no integrals need to be integrated by hand [for example, Eq. (8.14)], a task that becomes more and more tedious as the order of the element increases.

The three isoparametric elements developed in the remainder of this chapter will all be incorporated into the UNAFEM program.

8.3 The 1-D C^0-Quadratic Isoparametric Element

8.3.1 Theoretical Development

Step 4: Develop specific expressions for the shape functions.

Isoparametric elements employ the standard form for the element trial solution, Eq. (8.2), and the shape functions satisfy the interpolation property, Eq. (8.5). However, the shape functions are generated indirectly, by first developing a master, or parent, set of shape functions using the direct approach employed in previous sections, and then mapping (transforming) the master set onto each of the real elements in a mesh. It will be seen below that the complexity of the mapping generally [but not always — see Eq. (8.27)] makes it impossible to write down explicit expressions for the resulting real shape functions, $\phi_j^{(e)}(x)$. Instead, we will have to be content with plotting pictures of the $\phi_j^{(e)}(x)$ from numerical data. Indeed, since the isoparametric approach results in shape functions being defined implicitly rather than explicitly, there is necessarily a greater reliance on numerical procedures.

We begin by defining a *master*, or *parent*, element along a separate ξ-axis, with the three nodes *uniformly* spaced at $\xi = -1$, 0, $+1$ (Fig. 8.5).

Next we define C^0-quadratic shape functions, $\phi_i(\xi)$, on the parent element. These can be copied from Fig. 8.2, substituting -1 for x_1, 0 for x_2, $+1$ for x_3, and ξ for x. The shape functions and their expressions are shown in Fig. 8.6. We will refer to them as the *parent shape functions*.

Finally, we define a *coordinate transformation*,

$$x = \chi^{(e)}(\xi) \tag{8.19}$$

which maps the parent element (i.e., the interval $\xi = -1$ to $\xi = +1$) onto each of the real elements in a mesh. A superscript (e) is used in Eq. (8.19) because a *separate mapping is used for each element*. Therefore ξ ranges from -1 to $+1$ in *each* of the real elements, as illustrated in Fig. 8.7.

An important consequence of this process is that the three parent shape functions, $\phi_i(\xi)$, are "carried along" by each mapping, resulting in *shape functions defined over each real element*. In other words, a set of three shape functions, $\phi_i^{(e)}(x)$, are created for each element. This, in fact, is the principal advantage of the entire parent-element/coordinate-transformation process: to facilitate the construction of shape functions on real elements. This advantage, though, is more significant in problems in two or more dimensions.[1] We will look at some typical mapped shape functions in a moment, but first we must construct an acceptable mapping.

For each element in the mesh it is desirable that the coordinate transformation [Eq. (8.19)] be *one to one*, i.e., each point in the parent element mapping onto one point in

---

[1] For example, in 2-D or 3-D problems, elements can have complicated shapes with curved boundaries. In these cases it would be very difficult, if not impossible, to define appropriate shape functions over such shapes. Therefore a parent element is constructed using the simplest possible shape, e.g., a bi-unit square in 2-D (corner nodes at ± 1, ± 1) or a bi-unit cube in 3-D (corner nodes at ± 1, ± 1, ± 1). It is much easier to construct shape functions on the parent square or cube and then to map them to the real curvilinear elements (see Chapter 13).

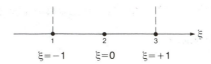

FIGURE 8.5 ● Parent element for the 1-D C^0-quadratic isoparametric element.

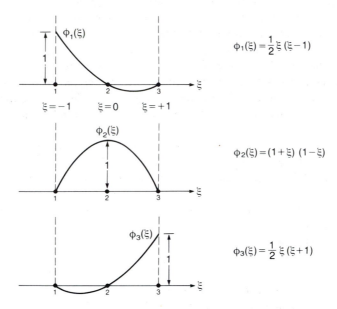

$$\phi_1(\xi) = \tfrac{1}{2}\xi\,(\xi-1)$$

$$\phi_2(\xi) = (1+\xi)\,(1-\xi)$$

$$\phi_3(\xi) = \tfrac{1}{2}\xi\,(\xi+1)$$

FIGURE 8.6 ● Parent shape functions for the 1-D C^0-quadratic isoparametric element.

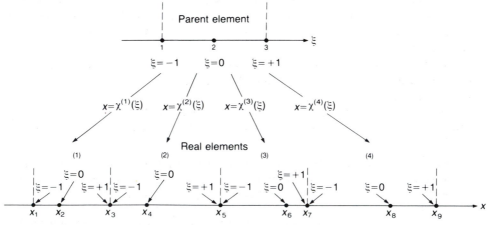

FIGURE 8.7 ● Mapping of ξ-coordinate in the parent element onto each real element, $x=\chi^{(e)}(\xi)$.

the real element, and vice versa. In particular, we want the boundary nodes of the parent element ($\xi = \pm 1$) to map onto the boundary nodes of the real element. This is necessary to ensure continuity of the mapped ξ-coordinate at interelement boundaries, i.e., no gaps or overlapping between adjacent elements. This in turn is necessary to ensure C^0-continuity in the mapped shape functions at the interelement boundaries.[2]

There are several ways to establish such a mapping. The most widely used at present is the isoparametric approach in which Eq. (8.19) takes the form,

$$x = \sum_{k=1}^{l} x_k^{(e)} \psi_k(\xi) \tag{8.20}$$

where $x_k^{(e)}$, $k = 1, 2, \ldots, l$, are the coordinates of the l nodes in the (e)th real element, and $\psi_k(\xi)$ are defined to be the parent shape functions,

$$\psi_k(\xi) = \phi_k(\xi) \qquad k = 1, 2, \ldots, l \tag{8.21}$$

Using the expressions in Fig. 8.6 for our C^0-quadratic element, we have

$$x = \sum_{k=1}^{3} x_k^{(e)} \phi_k(\xi)$$

$$= \frac{1}{2} \xi(\xi - 1) x_1^{(e)} + (1 + \xi)(1 - \xi) x_2^{(e)} + \frac{1}{2} \xi(\xi + 1) x_3^{(e)} \tag{8.22}$$

It is clear from inspection that Eq. (8.22) maps the parent nodes onto the real nodes; that is,

$$\xi = -1 \rightarrow x = x_1^{(e)}$$

$$\xi = 0 \rightarrow x = x_2^{(e)}$$

$$\xi = +1 \rightarrow x = x_3^{(e)} \tag{8.23}$$

The mapping in Eq. (8.22) may be used for each real element; we need only change the numerical values of $x_1^{(e)}$, $x_2^{(e)}$, and $x_3^{(e)}$ for each element.

The word "parametric" in the name isoparametric refers to the use of a mapping parameter, ξ; and the prefix "iso" (Greek, meaning "equal") refers to the fact that the mapping functions for the coordinates, $\psi_k(\xi)$, are chosen to be the same as the shape functions $\phi_k(\xi)$. In general the $\psi_k(\xi)$ need not be the same as the $\phi_k(\xi)$, and the $x_k^{(e)}$ may correspond to other points in the element, not necessarily the nodes. An element is called *subparametric* if the number of $\psi_k(\xi)$ is fewer than the $\phi_k(\xi)$,[3] and *superparametric* if there are more $\psi_k(\xi)$ than $\phi_k(\xi)$. Of the three types, isoparametric elements are in much greater use.

---

[2] Unacceptable mappings are easy to construct. For example, consider mapping the parent element in Fig. 8.7 onto element (1) by the function $x = x_2(3 + \xi)$. The boundary nodes at $\xi = -1$ and $\xi = +1$ would map onto the points $x = 2x_2$ and $4x_2$, respectively, which in general will not be at x_1 and x_3.

[3] An example of an acceptable subparametric mapping for the C^0-quadratic element is

$$x = \frac{1}{2}(1 - \xi) x_1^{(e)} + \frac{1}{2}(1 + \xi) x_3^{(e)}$$

Now let's look at how the parent shape functions $\phi_i(\xi)$ are mapped into shape functions on the real elements $\phi_i^{(e)}(x)$ as a result of the above coordinate mapping. We note at the outset that two of the three nodes in each real element are always at the ends, but the third node may be located anywhere in between. Therefore the nature of the coordinate transformation and of the mapped shape functions, is completely determined by the location of the interior node relative to the end nodes. This will become clear during the following discussion.

Let's consider first a real element in which the interior node $x_2^{(e)}$ is located at the *center*:

$$x_2^{(e)} = \frac{1}{2}\left(x_1^{(e)}+x_3^{(e)}\right) \tag{8.24}$$

This is the simplest situation to treat analytically. It is also the most interesting because locating the interior node at the center creates a "balanced" element, which is a desirable feature in almost all practical analyses (see later comments and application examples).

To derive the analytical expressions for the mapped shape functions $\phi_i^{(e)}(x)$, first substitute Eq. (8.24) into Eq. (8.22):

$$x = x^{(e)}(\xi) = \frac{1}{2}(1-\xi)x_1^{(e)} + \frac{1}{2}(1+\xi)x_3^{(e)} \tag{8.25}$$

Equation (8.25) is the coordinate transformation for this element. Note that Eq. (8.22) is quadratic with respect to ξ, but Eq. (8.25) is linear. We'll comment more on this later. For the moment, we note that because of this linearity we can invert Eq. (8.25), i.e., solve for ξ in terms of x:

$$\xi^{(e)}(x) = \frac{2x - x_3^{(e)} - x_1^{(e)}}{x_3^{(e)} - x_1^{(e)}} \tag{8.26}$$

Finally, we substitute Eq. (8.26) into the parent shape functions in Fig. 8.6. For example,

$$\phi_1^{(e)}(x) = \phi_1(\xi^{(e)}(x)) = \frac{1}{2}\left(\frac{2x-x_3^{(e)}-x_1^{(e)}}{x_3^{(e)}-x_1^{(e)}}\right)\left(\frac{2x-x_3^{(e)}-x_1^{(e)}}{x_3^{(e)}-x_1^{(e)}} - 1\right) \tag{8.27a}$$

Using Eq. (8.24), this can be simplified to

$$\phi_1^{(e)}(x) = \frac{\left(x-x_2^{(e)}\right)\left(x-x_3^{(e)}\right)}{\left(x_1^{(e)}-x_2^{(e)}\right)\left(x_1^{(e)}-x_3^{(e)}\right)} \tag{8.27b}$$

Similarly,

$$\phi_2^{(e)}(x) = \frac{\left(x-x_1^{(e)}\right)\left(x-x_3^{(e)}\right)}{\left(x_2^{(e)}-x_1^{(e)}\right)\left(x_2^{(e)}-x_3^{(e)}\right)}$$

$$\phi_3^{(e)}(x) = \frac{\left(x-x_1^{(e)}\right)\left(x-x_2^{(e)}\right)}{\left(x_3^{(e)}-x_1^{(e)}\right)\left(x_3^{(e)}-x_2^{(e)}\right)}$$

We recognize Eqs. (8.27b) as the Lagrange polynomial shape functions derived with the direct approach in Section 8.2 (recall Fig. 8.2). In other words, when the interior node is at the center, the isoparametric transformation maps the *quadratic* Lagrange polynomial *parent* shape functions into *quadratic* Lagrange polynomial *real* shape functions. These relations are summarized in Fig. 8.8.

Note that the mapped ξ-coordinate, shown under the real element, is *uniform* from -1 to $+1$. This is because the mapping in Eq. (8.25) is linear (more on this in a moment). Because of this uniformity, the real shape functions are simply a uniform expansion (or contraction) of the parent shape functions across the entire element, due to a change in length from 2 in the ξ-coordinate to $x_3^{(e)}-x_1^{(e)}$ in the x-coordinate. In short, the mapping in Eq. (8.25) has produced only a change in scale; it has not produced any distortion in the shape functions.

Now let's consider an element where the interior node is not located at the center; for example, if it's at the one-quarter point,

$$x_2^{(e)} = x_1^{(e)} + \frac{1}{4}\left(x_3^{(e)}-x_1^{(e)}\right)$$

(8.28)

Substituting Eq. (8.28) into Eq. (8.22) yields the following coordinate transformation:

$$x = x^{(e)}(\xi) = \frac{1}{4}(3+\xi)(1-\xi)x_1^{(e)} + \frac{1}{4}(1+\xi)^2 x_3^{(e)}$$

(8.29)

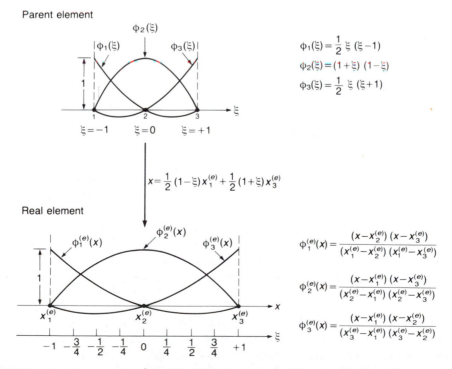

FIGURE 8.8 ● Isoparametric mapping of parent shape functions onto a real element in which the interior node is at the center.

Equation (8.29) is quadratic with respect to ξ, so it cannot be inverted to give an explicit expression for ξ in terms of x. Therefore we cannot obtain explicit expressions for the mapped shape functions like we did for the previous element. More generally, this problem occurs whenever the interior node of the real element is not at the center.

Fortunately the subsequent analysis doesn't need such expressions anyway. Nevertheless, it is quite instructive to see a plot of the mapped shape functions. This we can do by selecting several values for ξ in the interval -1 to $+1$ and then calculating the corresponding values of x and ϕ_i from Eq. (8.29) and Fig. 8.6, respectively. The results are shown in Fig. 8.9, which plots the mapped shape functions for several different real elements, starting with an element in which the interior node $x_2^{(e)}$ is at the center and gradually moving $x_2^{(e)}$ toward one end. Underneath each real element is the mapped ξ-axis.

Figure 8.9 illustrates an important principle: If the interior node in the real element is not at the center, then the mapping is nonuniform, i.e., distorted, resulting in the ξ-coordinate becoming compressed at one end of the element (toward $x_1^{(e)}$ in the figure) and stretched out at the other end. This in turn causes the mapped shape functions to be compressed toward one end — $x_1^{(e)}$ — and stretched out toward the other end — $x_3^{(e)}$ — thereby distorting the polynomial shape of the parent functions into nonpolynomial shapes. Note that the distortion is caused by a horizontal expansion and contraction, not a vertical; i.e., the range of values for each $\phi_i^{(e)}(x)$ remains the same as for its parent $\phi_i(\xi)$.

There is a limit to how far $x_2^{(e)}$ can approach either $x_1^{(e)}$ or $x_3^{(e)}$ before the ξ-coordinate becomes too compressed and hence the mapped shape functions become too distorted. This is shown by the sequence of four cases in Fig. 8.9. The only element with no distortion (Fig. 8.9a) was already illustrated in Fig. 8.8. In Fig. 8.9(b) the distortion is small, although the $\phi_i^{(e)}(x)$ are no longer polynomials with respect to x because the expansion or contraction is no longer uniform across the element. Nevertheless, the mapping is acceptable because it is everywhere one to one; i.e., each point in ξ maps onto one point in x and vice versa.

In Fig. 8.9(c) the distortion has become severe near $x_1^{(e)}$. The mapping is one to one everywhere except at $\xi = -1$, where it becomes infinitely compressed, resulting in the shape functions being *singular* at $x_1^{(e)}$, that is, having infinite slopes there (clearly not polynomial behavior). Such shape functions should generally be avoided.[4] In Fig. 8.9(d) the distortion is extreme; the ξ-coordinate maps outside the element, then wraps around. Consequently the shape functions also map outside the element, then wrap around, resulting in double-valued functions along the x-axis. Such severe distortion is no longer one to one over finite regions of x and would cause numerical problems, making this element unacceptable.[5]

Figure 8.9 also illustrates an important general property of isoparametric transformations. Since the parent shape functions satisfy the interpolation property at nodes in the parent element — that is, $\phi_j(\xi_i) = \delta_{ji}$ — and since the parent nodes map to the real nodes, then the mapped shape functions also satisfy the interpolation property in the

---

[4] These shape functions are only useful when modeling a region surrounding a point where it is known that the exact solution also has the same type of singular behavior.

[5] The reader will recall that the quadratic element developed by the direct approach in Section 8.2 also experiences difficulties if the interior node is located too close to the end nodes.

real element:

$$\phi_j^{(e)}(x_i) = \delta_{ji} \qquad (8.30)$$

This, of course, ensures C^0-continuity between the mapped shape functions.

Parent element and shape functions

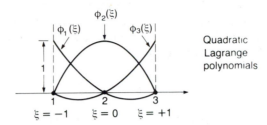

Real elements and shape functions

$$L = x_3^{(e)} - x_1^{(e)}$$

$x_2^{(e)}$ **at center:** $(x_2^{(e)} = x_1^{(e)} + \frac{1}{2} L)$:

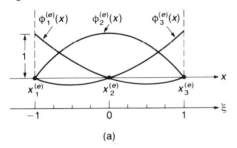

(a)

$x_2^{(e)}$ **at $\frac{3}{8}$ point:** $(x_2^{(e)} = x_1^{(e)} + \frac{3}{8} L)$:

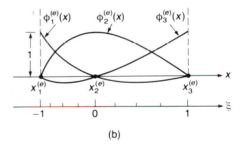

(b)

$x_2^{(e)}$ **at $\frac{1}{4}$ point:** $(x_2^{(e)} = x_1^{(e)} + \frac{1}{4} L)$:

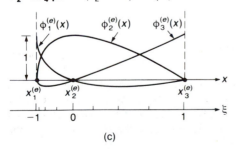

(c)

$x_2^{(e)}$ **at $\frac{1}{8}$ point:** $(x_2^{(e)} = x_1^{(e)} + \frac{1}{8} L)$:

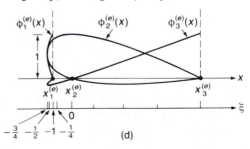

(d)

FIGURE 8.9 ● Mapping of parent shape functions to real shape functions for different locations of the interior node in the real element. (a) No distortion, only a uniform change of scale; $\phi_i^{(e)}(x)$ remain quadratic polynomials; an acceptable element. (b) Small distortion; $\phi_i^{(e)}(x)$ are not polynomials; an acceptable element. (c) Severe distortion; $d\phi_i^{(e)}(x)/dx \rightarrow \infty$ at x_1; not acceptable for most applications. (d) Extreme distortion; $\phi_i^{(e)}(x)$ are double-valued (mapping is not one to one); not acceptable .

The four cases shown in Fig. 8.9 are a convenient visual aid for instructional purposes, but constructing graphs is not a very practical way to determine whether a particular mapping is acceptable or not, especially with 2-D and 3-D elements. Fortunately there is an analytical test that can determine the acceptability of a mapping without actually calculating the mapping itself. It is based on a comparison of how an infinitesimal line segment $d\xi$ in the parent element is mapped onto a corresponding line segment dx in the real element. Their ratio, $dx/d\xi$, is, of course, the derivative of the mapping, and it tells us the amount of *local* expansion or contraction of the coordinates due to the mapping. When applied to a coordinate transformation, such as our present mapping, this derivative is called a *Jacobian*,[6] symbolized by $J^{(e)}(\xi)$:

$$J^{(e)}(\xi) = \frac{dx}{d\xi} \tag{8.31}$$

The Jacobian plays an important role in the FE analysis of isoparametric elements.

The criterion for an acceptable mapping is

$$J^{(e)}(\xi) > 0 \qquad \text{for } -1 \le \xi \le 1 \tag{8.32}$$

i.e., the Jacobian must be positive everywhere inside the element and on the boundary.[7] Let's apply this criterion to the mappings in Fig. 8.9. From Eq. (8.22),

$$J^{(e)}(\xi) = \sum_{k=1}^{3} x_k^{(e)} \frac{d\phi_k(\xi)}{d\xi} \tag{8.33}$$

and from Fig. 8.6,

$$\frac{d\phi_1(\xi)}{d\xi} = \xi - \frac{1}{2}$$

$$\frac{d\phi_2(\xi)}{d\xi} = -2\xi$$

$$\frac{d\phi_3(\xi)}{d\xi} = \xi + \frac{1}{2} \tag{8.34}$$

Therefore

$$J^{(e)}(\xi) = \left(\xi - \frac{1}{2}\right) x_1^{(e)} - 2\xi x_2^{(e)} + \left(\xi + \frac{1}{2}\right) x_3^{(e)} \tag{8.35}$$

---

[6] Actually, in 1-D transformations no special terminology is normally used. But 1-D is directly analogous to 2-D (or higher) in which the analogue to $dx/d\xi$ is a matrix called the *Jacobian matrix* (see Chapter 13), and the determinant of the Jacobian matrix is called the Jacobian. In 1-D, both the matrix and its determinant degenerate to the scalar $dx/d\xi$, which, by analogy, we will therefore refer to as the Jacobian.

[7] In principle the interval $-1 \le \xi \le 1$ could be changed to $-1 < \xi < 1$, thereby permitting $J^{(e)}(\xi) = 0$ on the element boundary. This corresponds to a special and unusual situation involving singular behavior on the boundary, as illustrated in Fig. 8.9(c). However, Eq. (8.32) is intended for general applications in which singular behavior is undesirable in an element.

Now consider each of the four cases in Fig. 8.9 by substituting the expression for $x_2^{(e)}$ above each figure into Eq. (8.35).

Figure 8.9(a) $J = \dfrac{L}{2}$ $J>0$ everywhere in element

Here J is a constant throughout the element because the mapping is *linear* with respect to ξ [recall Eq. (8.25)]. This illustrates an important general property of the Jacobian for elements in any dimension: The Jacobian is a constant if and only if the mapping is linear. In this case the mapping produces only a uniform change of scale; i.e., there is no distortion.

Figure 8.9(b) $J = \left(\dfrac{\xi+2}{2}\right)\dfrac{L}{2}$ $J>0$ everywhere in element

Here J varies over the element, resulting in a nonuniform mapping, i.e., distortion. However, the mapped shape functions are only mildly distorted.

Figure 8.9(c) $J = (\xi+1)\dfrac{L}{2}$ $J>0$ inside element, but $J=0$ on boundary at $\xi=-1$, i.e., at $x=x_1^{(e)}$

At the point $x_1^{(e)}$ a finite (nonzero) $d\xi$ maps onto a zero dx, which means that the ξ-coordinate becomes infinitely compressed when mapped onto the x-axis. Therefore the shape functions, which are carried along by the mapping, are similarly compressed, resulting in infinite slopes, i.e., a singularity. This type of element would be very effective if it were known that the exact solution has a similar singularity at that point; otherwise, it would give a very poor approximation near the point. The element is therefore unacceptable for general application.

Figure 8.9(d) $J = \left(\dfrac{3\xi+2}{2}\right)\dfrac{L}{2}$ $J>0$, $=0$, and <0 inside element

$$J > 0 \text{ for } -\frac{2}{3} < \xi \leq 1$$

$$J = 0 \text{ for } \xi = -\frac{2}{3}$$

$$J < 0 \text{ for } -1 \leq \xi < -\frac{2}{3}$$

$J<0$ means that a positive $d\xi$ maps onto a negative dx. This reversal in sign means that the mapped ξ-coordinate is folding back on itself, resulting in the shape functions also being folded over and therefore double-valued; i.e., the mapping is no longer one to one. Such extreme distortion does more damage than merely produce a poor approximation; it also produces such an ill-conditioned set of system equations that the equation-solving process breaks down. This type of element is obviously unacceptable.

Figure 8.9(a) through (d) has illustrated the movement of $x_2^{(e)}$ toward $x_1^{(e)}$. It should be clear from the symmetry of the parent shape functions about the center that if $x_2^{(e)}$ moves toward $x_3^{(e)}$, identical (mirror-image) results will be produced.

To conclude this discussion on mapping, we will use Eq. (8.32) to derive a simple rule for the acceptable range of placement of the interior node. Thus, applying Eq. (8.32) to Eq. (8.35) yields

$$J^{(e)}(\xi) = \left(\xi - \frac{1}{2}\right)x_1^{(e)} - 2\xi x_2^{(e)} + \left(\xi + \frac{1}{2}\right)x_3^{(e)} > 0 \qquad -1 \le \xi \le 1 \tag{8.36}$$

Note that $J^{(e)}(\xi)$ is a linear function of ξ. Therefore $J^{(e)}(\xi) > 0$ for $-1 \le \xi \le 1$ if and only if $J^{(e)}(\pm 1) > 0$, that is, $J^{(e)}(\xi) > 0$ at both boundary nodes, as shown in Fig. 8.10.

Applying the conditions $J^{(e)}(+1) > 0$ and $J^{(e)}(-1) > 0$ to Eq. (8.36) yields

$$J^{(e)}(-1) = -\frac{3}{2}x_1^{(e)} + 2x_2^{(e)} - \frac{1}{2}x_3^{(e)} > 0$$

$$x_2^{(e)} > x_1^{(e)} + \frac{1}{4}L$$

$$J^{(e)}(+1) = \frac{1}{2}x_1^{(e)} - 2x_2^{(e)} + \frac{3}{2}x_3^{(e)} > 0$$

$$x_2^{(e)} < x_1^{(e)} + \frac{3}{4}L \tag{8.37}$$

or, equivalently,

$$x_c^{(e)} - \frac{L}{4} < x_2^{(e)} < x_c^{(e)} + \frac{L}{4} \tag{8.38}$$

where $x_c^{(e)} = (x_1^{(e)} + x_3^{(e)})/2$ is the center of the element. This is illustrated in Fig. 8.11 and summarized by the following general guideline:

> *Guideline I.* For a 1-D C^0-quadratic isoparametric element, always locate the interior node within $L/4$ of the center. In practice, the best location is at the center unless special knowledge of the exact solution favors an alternate placement.

Step 5: Substitute the shape functions into the element equations, and transform the integrals into a form appropriate for numerical evaluation.

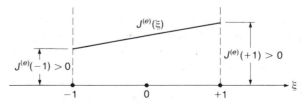

FIGURE 8.10 ● Conditions at the boundary nodes that ensure that $J^{(e)}(\xi) > 0$ over the entire element.

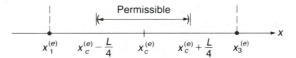

FIGURE 8.11 ● Permissible interval for locating the interior node of a 1-D C⁰-quadratic isoparametric element.

Consider the terms in the stiffness matrix in the element equations [Eqs. (8.13)]. Let $K\alpha_{ij}^{(e)}$ and $K\beta_{ij}^{(e)}$ represent the integrals over $\alpha(x)$ and $\beta(x)$, respectively. We treat the $K\alpha_{ij}^{(e)}$ integrals first:

$$K\alpha_{ij}^{(e)} = \int_{x_1}^{x_3} \frac{d\phi_i^{(e)}(x)}{dx} \, \alpha(x) \, \frac{d\phi_j^{(e)}(x)}{dx} \, dx \tag{8.39}$$

Recall from the previous discussion in step 4, especially Eq. (8.29), that the mapping may in general produce distorted shape functions $\phi_i^{(e)}(x)$, for which there are no explicit analytical expressions. Therefore we cannot use exact integration formulas in Eq. (8.39). We will instead use approximate numerical integration, which is also called numerical *quadrature*. It will be seen shortly that this has a decided advantage over exact integration because both the analysis and the programming are easier, and one general procedure can be used for all kinds of elements.

Numerical quadrature involves two steps:

1. Transform the integrals over the real element to integrals over the parent element, i.e., over the interval $-1 \le \xi \le +1$.

2. Use "classical" numerical quadrature formulas to evaluate the transformed integrals.

We can see right here that the isoparametric concept lends itself very nicely to numerical quadrature because the parent element has been deliberately defined over the same interval $(-1 \le \xi \le +1)$ that is used in the classical quadrature formulas.

Let's begin then with the first step by transforming each of the terms in Eq. (8.39).

a) The derivatives: $d\phi_i^{(e)}(x)/dx$

We cannot formally take the derivative of $\phi_i^{(e)}(x)$ with respect to x since the $\phi_i^{(e)}(x)$ are not explicit functions of x. Rather, the $\phi_i^{(e)}(x)$ are explicit functions of ξ (Fig. 8.6), and ξ in turn is implicitly a function of x via the coordinate transformation [Eq. (8.19) or (8.22)]. Therefore we must use the chain rule of differentiation:[8]

$$\frac{d\phi_i^{(e)}(x)}{dx} = \frac{d\phi_i^{(e)}(x^{(e)}(\xi))}{d\xi} \frac{d\xi}{dx} = \frac{1}{J^{(e)}(\xi)} \frac{d\phi_i(\xi)}{d\xi} \tag{8.40}$$

---

[8] In 1-D it is simple to invert $d\xi/dx$ to $dx/d\xi$, as in Eq. (8.40). This inversion is necessary in order to differentiate the coordinate transformation, Eq. (8.22). In 2-D, though, the analogue to Eq. (8.40) will involve a matrix of derivatives, and it will be seen in Chapter 13 that the chain rule of differentiation will be applied somewhat differently.

Here we have employed the fact that expressing $\phi_i^{(e)}(x)$ in terms of ξ, via the coordinate transformation, yields the parent shape function; that is,

$$\phi_i^{(e)}(x) = \phi_i^{(e)}(x^{(e)}(\xi)) = \phi_i(\xi)$$

b) The physical property: $\alpha(x)$

The mapping makes $\alpha(x)$ a function of ξ:

$$\alpha(x) = \alpha(x^{(e)}(\xi)) = \alpha^{(e)}(\xi) \tag{8.41}$$

c) The complete integral:

From elementary calculus, we know that under a general transformation of coordinates [for example, $x = x^{(e)}(\xi)$] an integral is transformed as follows:

$$\int_{x_1}^{x_3} I(x)\, dx = \int_{\xi(x_1)}^{\xi(x_3)} I(x^{(e)}(\xi))J^{(e)}(\xi)\, d\xi \tag{8.42}$$

where $I(x)$ represents the integrand and $dx = J^{(e)}(\xi)d\xi$. [Recall Eq. (8.31).]

Using Eqs. (8.40) through (8.42), Eq. (8.39) becomes

$$K\alpha_{ij}^{(e)} = \int_{-1}^{+1} \left(\frac{1}{J^{(e)}(\xi)} \frac{d\phi_i(\xi)}{d\xi}\right) \alpha^{(e)}(\xi) \left(\frac{1}{J^{(e)}(\xi)} \frac{d\phi_j(\xi)}{d\xi}\right) J^{(e)}(\xi)\, d\xi \tag{8.43}$$

Note that two of the $J^{(e)}(\xi)$ terms could cancel each other, leaving only one in the denominator; however, the integral will be left in this form to standardize the later numerical evaluation and program coding for all elements in 1-D and 2-D. Also note that the $J^{(e)}(\xi)$ terms in the denominator would cause problems if $J^{(e)}(\xi) = 0$ in the interval $-1 \leq \xi \leq 1$ [i.e., if it violates the acceptable mapping criterion of Eq. (8.32)].

The other integrals in the element equations do not contain derivatives, so their transformation is easier, involving only items b and c above. Thus

$$K\beta_{ij} = \int_{x_1}^{x_3} \phi_i^{(e)}(x)\beta(x)\phi_j^{(e)}(x)\, dx$$

$$= \int_{-1}^{+1} \phi_i(\xi)\beta^{(e)}(\xi)\phi_j(\xi)J^{(e)}(\xi)\, d\xi \tag{8.44}$$

$$Ff_i^{(e)} = \int_{x_1}^{x_3} f(x)\phi_i^{(e)}(x)\, dx$$

$$= \int_{-1}^{+1} f^{(e)}(\xi)\phi_i(\xi)J^{(e)}(\xi)\, d\xi \tag{8.45}$$

and

$$F\tau_i^{(e)} = -\left[\left(-\alpha(x)\frac{d\tilde{U}^{(e)}(x)}{dx}\right)\phi_i^{(e)}(x)\right]_{x_1}^{x_3}$$

$$= -\left[\tau^{(e)}(x)\phi_i^{(e)}(x)\right]_{x_1}^{x_3} \tag{8.46}$$

Using the interpolation property [Eq. (8.30)] for the shape functions, the boundary load terms in Eq. (8.46) simplify as follows:

$$\{F\tau\}^{(e)} = \left\{\begin{array}{c} \tau^{(e)}(x_1) \\ 0 \\ -\tau^{(e)}(x_3) \end{array}\right\} \tag{8.47}$$

The stiffness and load integrals [Eqs. (8.43) through (8.45)] have now all been transformed to the parent element, so we can turn to the second step, numerical quadrature. The integrals are each in the form

$$\int_{-1}^{+1} I(\xi)\,d\xi \tag{8.48}$$

where $I(\xi)$ represents all the terms in the integrand.

There are many well-established (classical) *interpolatory quadrature* formulas suitable for evaluating integrals in the form of Eq. (8.48) [8.1]. These formulas approximate the integral by a finite sum,

$$\int_{-1}^{+1} I(\xi)\,d\xi \simeq \sum_{l=1}^{n} w_{nl}I(\xi_{nl}) \tag{8.49}$$

where the w_{nl} are weight factors and the ξ_{nl} are points at which the integrand is evaluated. The ξ_{nl} are called *quadrature points, integration points, sampling points*, or *nodes* (or, in the special circumstances described below, *Gauss points*). Using the approximate formula in Eq. (8.49), the integrals may be written as follows:

$$K\alpha_{ij}^{(e)} \simeq \sum_{l=1}^{n} w_{nl}\left(\frac{1}{J^{(e)}(\xi)}\frac{d\phi_i(\xi)}{d\xi}\right)_{\xi_{nl}} \alpha^{(e)}(\xi_{nl})\left(\frac{1}{J^{(e)}(\xi)}\frac{d\phi_j(\xi)}{d\xi}\right)_{\xi_{nl}} J^{(e)}(\xi_{nl})$$

$$K\beta_{ij}^{(e)} \simeq \sum_{l=1}^{n} w_{nl}\phi_i(\xi_{nl})\beta^{(e)}(\xi_{nl})\phi_j(\xi_{nl})J^{(e)}(\xi_{nl})$$

$$Ff_i^{(e)} \simeq \sum_{l=1}^{n} w_{nl}f^{(e)}(\xi_{nl})\phi_i(\xi_{nl})J^{(e)}(\xi_{nl}) \tag{8.50}$$

The values for w_{nl} and ξ_{nl} depend on which quadrature formula we use. The simplest formulas are the equally spaced, closed Newton-Cotes rules, such as the Trapezoidal rule and Simpson's rule. If n is even, these rules will integrate exactly a polynomial of degree $n-1$; if n is odd, they will integrate exactly a polynomial of degree n.

The Gauss quadrature rules are more precise than the Newton-Cotes rules: For n points in Eq. (8.49), the Gauss rules will integrate exactly a polynomial of degree $2n-1$. There are an endless variety of Gauss rules; integrals in the form of Eq. (8.48) require the Gauss-*Lengendre* rules.[9] They are applicable to multiple integrals as well (see Chapter 13).

In Gauss-Legendre quadrature, the ξ_{nl} are the zeroes of the nth-degree Legendre polynomial and are usually referred to as *Gauss points*. The Legendre polynomials are an infinite set of polynomials that are orthogonal to each other over the interval $[-1,1]$. (Recall from Section 7.2, step 6, that the zeroes of the Legendre polynomials also occur in connection with the optimal flux-sampling points.) The weights w_{nl} and Gauss points ξ_{nl} are numbers calculated from the polynomials and have been tabulated in many numerical analysis textbooks and mathematical tables [8.2]. Figure 8.12 summarizes the Gauss points and weights for several orders of quadrature.[10] The second column indicates that the accuracy of the quadrature is $O(h^{2n})$, where the letter O is read ''order'' or ''of order'' and h is the length of the real element. This means that

$$\left| \int_{-1}^{+1} I(\xi)\, d\xi - \sum_{l=1}^{n} w_{nl} I(\xi_{nl}) \right| < Kh^{2n} \qquad \text{as } h \to 0$$

$$(8.51)$$

where $|\ |$ is the quadrature error and K is a constant. The condition ''as $h \to 0$'' reminds us that this is an asymptotic relationship, implying that it is only valid in the limit as the elements become small enough; conversely, Eq. (8.51) will generally be invalid for very coarse meshes.

Now that we have a table of values for w_{nl} and ξ_{nl}, the only remaining question is: What order quadrature rule should be used? Unfortunately, there are no simple answers, although 1-D elements are much less complicated than 2-D or 3-D elements. Some general guidelines for elements in any dimension are provided below, but it must be emphasized that there are frequent exceptions to these guidelines (mostly for 2-D and 3-D elements) because of numerical effects that are not accounted for by the underlying theory. In fact, many of the specific quadrature rules that are used for particular elements have evolved more from numerical experiments than from theoretical arguments (the latter frequently being created to explain the successes or failures of the former).

---

[9] The more general form of Eq. (8.48) is

$$\int_{a}^{b} I(\xi)\omega(\xi)\, d\xi$$

where $\omega(\xi)$ is a ''weighting function'' and a and b usually assume one of the values 0, ± 1, or $\pm \infty$. In FE work we usually encounter the special case of $a = -1$, $b = +1$, and $\omega(\xi) = 1$, which entails the Gauss-Legendre rules.

[10] It might be noted that for each integration order the sum of the weights equals 2, which is the length of the interval of integration; that is,

$$\sum_{l=1}^{n} w_{nl} = 2$$

$$\int_{-1}^{+1} I(\xi)\, d\xi \simeq \sum_{l=1}^{n} w_{nl}\, I(\xi_{nl})$$

Number of Gauss points, n	Accuracy of quadrature[†]	Gauss points,[‡] ξ_{nl}	Weights[‡] w_{nl}	Plot of ξ_{nl} (∗) and w_{nl} (○)
1	$O(h^2)$	$\xi_{11} = 0$	$w_{11} = 2$	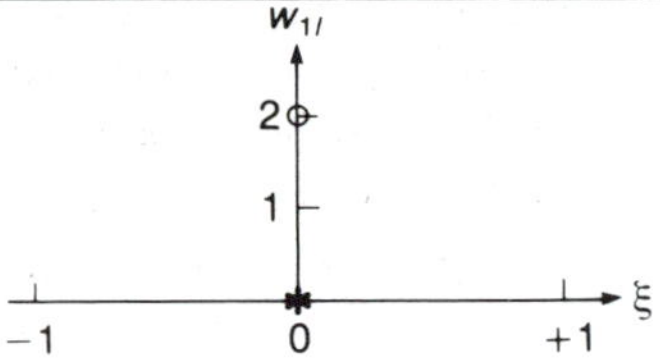
2	$O(h^4)$	$\xi_{21} = -\dfrac{1}{\sqrt{3}}$ $= -0.57735\ldots$ $\xi_{22} = -\xi_{21}$	$w_{21} = 1$ $w_{22} = w_{21}$	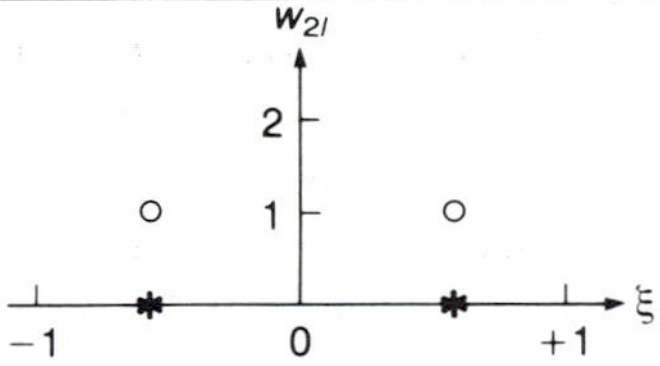
3	$O(h^6)$	$\xi_{31} = -\dfrac{\sqrt{3}}{\sqrt{5}}$ $= -0.77460\ldots$ $\xi_{32} = 0$ $\xi_{33} = -\xi_{31}$	$w_{31} = \dfrac{5}{9}$ $= 0.55555\ldots$ $w_{32} = \dfrac{8}{9}$ $= 0.88888\ldots$ $w_{33} = w_{31}$	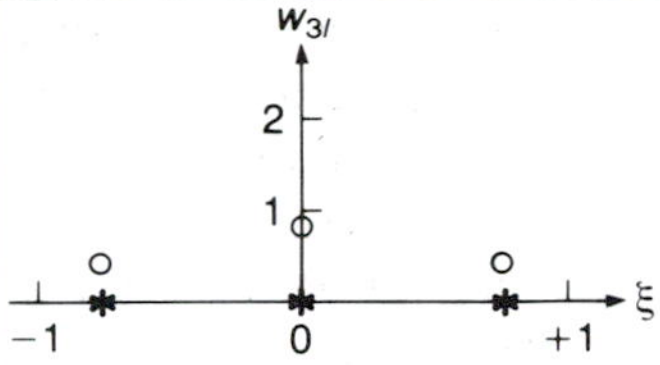
4	$O(h^8)$	$\xi_{41} = -\dfrac{\sqrt{15+2\sqrt{30}}}{\sqrt{35}}$ $= -0.86113\ldots$ $\xi_{42} = -\dfrac{\sqrt{15-2\sqrt{30}}}{\sqrt{35}}$ $= -0.33998\ldots$ $\xi_{43} = -\xi_{42}$ $\xi_{44} = -\xi_{41}$	$w_{41} = \dfrac{49}{6(18+\sqrt{30})}$ $= 0.34785\ldots$ $w_{42} = \dfrac{49}{6(18-\sqrt{30})}$ $= 0.65214\ldots$ $w_{43} = w_{42}$ $w_{44} = w_{41}$	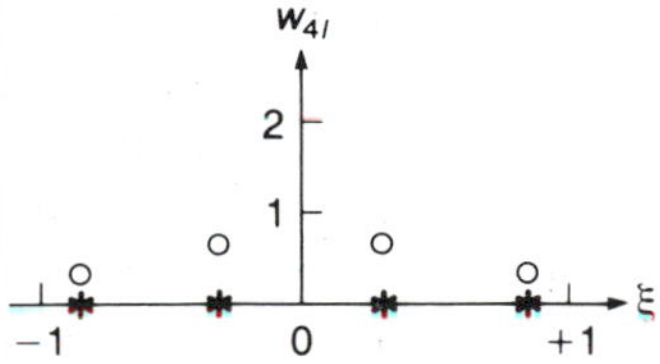
5	$O(h^{10})$	$\xi_{51} = -\dfrac{\sqrt{35+2\sqrt{70}}}{\sqrt{63}}$ $= -0.90617\ldots$ $\xi_{52} = -\dfrac{\sqrt{35-2\sqrt{70}}}{\sqrt{63}}$ $= -0.53846\ldots$ $\xi_{53} = 0$ $\xi_{54} = -\xi_{52}$ $\xi_{55} = -\xi_{51}$	$w_{51} = \dfrac{5103}{50(322+13\sqrt{70})}$ $= 0.23692\ldots$ $w_{52} = \dfrac{5103}{50(322-13\sqrt{70})}$ $= 0.47862\ldots$ $w_{53} = \dfrac{128}{225}$ $= 0.56888\ldots$ $w_{54} = w_{52}$ $w_{55} = w_{51}$	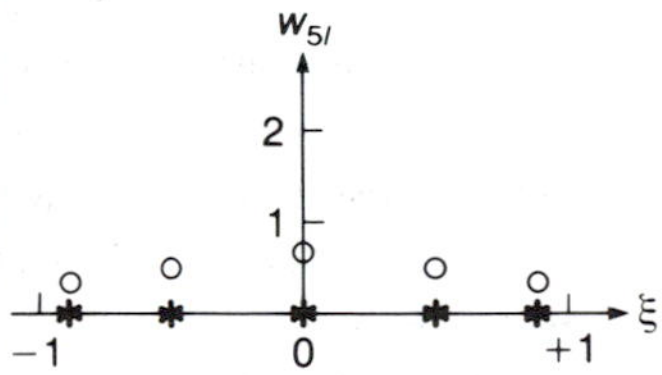

[†] Quadrature is exact if $I(\xi)$ is a polynomial of degree $2n-1$; e.g., a one-point rule integrates exactly a linear polynomial, a two-point rule integrates exactly a cubic polynomial, etc. An n-point rule is therefore said to have a degree of precision $2n-1$.

[‡] See Davis and Polonsky [8.2] for extensive tables of ξ_{nl}, w_{nl} to high order and high precision.

Note that $\displaystyle\sum_{l=1}^{n} w_{nl} = 2$, the interval of integration.

FIGURE 8.12 ● Gauss points and weights for Gauss-Legendre quadrature rules.

In selecting a quadrature rule, we are concerned with both cost and accuracy. On the one hand, we would like the order of the quadrature rule to be as low as possible in order to reduce the cost of computation. The cost is roughly proportional to the square of the number of DOF in an element since for n DOF there are $n(n+1)/2$ stiffness integrals to evaluate for each element (assuming a symmetric matrix). As the order decreases, though, the error introduced by the quadrature increases. If the order is too low, the quadrature error will exceed the discretization error (the error due to approximating the exact solution by a finite number of shape functions), resulting in a corresponding loss of accuracy to the FE solution, and possibly even preventing convergence of the solution. On the other hand, we can go to the other extreme and make the order so large that the quadrature error is several orders of magnitude less than the discretization error. The additional accuracy in the integrals would be wasted, however, because the overall accuracy would be limited by the discretization error.

Let's be more specific. Consider first the lowest-order rule possible that will merely preserve convergence (in a global energy sense, as predicted by the convergence theorems, which assume exact integration). Here it may be argued [8.3] that convergence should still occur if the volume of the element (i.e., the length, area, or volume of the 1-D, 2-D, or 3-D element, respectively) can be integrated exactly. This is because as $h \to 0$ the integrand approaches a constant value and hence in the limit could be moved outside the integral. In other words, the quadrature must be exact for a constant integrand.

Integrating a constant exactly requires an $O(h)$ rule which, in turn, requires only one quadrature point (see Fig. 8.12). This, however, sometimes creates a problem (one of the numerical effects alluded to above), referred to as a *zero-energy mode*. The concept draws upon the modal analysis techniques discussed in Chapter 10, so a technical explanation would be premature at this point. (See Cook [8.4], as well as the 2-D application in Section 16.3.) Nevertheless, a brief digression will be made here to indicate the nature of the problem.

An element possesses a zero-energy mode if there exists a set of values for the DOF (called a mode) that would yield zero flux (hence also zero energy) at all the quadrature points, while yielding a nontrivial (nonzero) solution elsewhere in the element. In other words, it might be possible for an element to have a nonzero response (with regard to the flux, or energy, expression in the stiffness integrals) over the entire element *except at one or more isolated points*. If we were to use a quadrature rule that sampled at precisely these points, we would find that the element stiffness matrix is singular. A mesh of such elements might lead to a singular system stiffness matrix that would preclude any solution. An easy way to resolve this problem is to use a higher-order quadrature rule in one or more of the elements; another way is to use "appropriate" BCs.[11]

---

[11] Element stiffness matrices are usually singular because of trivial solutions, referred to as "rigid-body modes" in structural mechanics. This is the case for $[K\alpha]^{(e)}$ (but not the sum $[K\alpha]^{(e)} + [K\beta]^{(e)}$) for both the 1-D and 2-D problem, as well as the elasticity stiffness matrices in Chapter 16. These singularities are eliminated when BCs necessary for a *well-posed* problem are imposed. The above zero-energy-mode singularities are in addition to the rigid-body-mode singularities. They too can be eliminated by BCs but only when such BCs happen to be part of the data for the problem.

In the present 1-D C^0-quadratic element the one-point quadrature rule yields a zero-energy mode. From Fig. 8.12, the single quadrature point is at the center of the element. The derivative of the second shape function, $d\phi_2/d\xi$, is zero at the center. Therefore all terms in the second row and column of $[K\alpha]^{(e)}$ are zero, producing a singularity in the element stiffness matrix (for problems in which $\beta=0$ so that $[K]^{(e)}=[K\alpha]^{(e)}$).

Zero-energy modes cannot usually be revealed as easily as in this example, since they generally involve several shape functions. The usual procedure would be to perform an eigenproblem analysis (see Chapter 10) on a single element. This ends the digression.

The "exact-volume" rule is usually unacceptable, for two reasons. In the first place, it frequently yields zero-energy modes (as with the present element). In addition, even though it might preserve convergence, it will result in a decrease in the *rate* of convergence for most types of elements, as explained below; the higher the order of the element, the greater the rate to be lost. In other words, the accuracy is too low for most applications.

At the other extreme of accuracy we might ask that the quadrature rule be capable of evaluating the entire integrand exactly. This is possible only when the integrand is a polynomial, which in turn means the Jacobian must be a constant, i.e., an undistorted element. In practical problems elements can generally be made undistorted (except those near to or touching a curved boundary in 2-D or 3-D problems). We will see that the rule recommended below (not based on exact integration) will, in fact, integrate the $K\alpha_{ij}$ integrals exactly for 1-D elements. However, for 2-D and 3-D elements, exact integration would require rules of much higher order (see Chapter 13). Such high-order rules are unnecessary (as shown below), very costly, and sometimes produce another type of numerical effect, called *locking*, which is relieved when a lower-order ("reduced") rule is used [8.5-8.7].

It would seem from these observations that the optimal-order rule should probably be somewhere in between the two extremes of exactly integrating only the volume or exactly integrating the entire integrand (if undistorted). For example, we might require that the quadrature error be comparable, in some sense, to the discretization error. This, in fact, is the rationale behind the conventional approach: One uses a rule with sufficient accuracy to at least preserve the *rate* of convergence due to the discretization error.

This raises questions as to which quantity we want to preserve the convergence rate for (function, flux, energy, etc.), and with respect to what measure of error (global, point-wise, etc.). The most natural quantity is perhaps the *global energy* since it is the quantity that is shown to converge by the fundamental convergence theorem [8.8]:[12]

$$\{a_0\}^T[K_0]\{a_0\} - \{a_h\}^T[K_h]\{a_h\} < Ch^{2(p-m+1)} \qquad \text{as } h_{\max} \to 0 \qquad \textbf{(8.52)}$$

Here C is a constant; p is the degree of the *complete* polynomial in the element trial solution $\tilde{U}^{(e)}$; and $2m$ is the order of the governing differential equation, which has been inte-

---

[12] Equation (8.52) is derived in Strang and Fix [8.8] using functional analysis. For those readers unfamiliar with this branch of mathematics, it is expressed here using matrix notation. The theorem is derived for linear, self-adjoint boundary-value problems (recall Sections 3.2 and 3.7). It may also apply to other problems in which a part of the problem is linear and self-adjoint, e.g., the spatial part of the mixed initial-value/boundary-value problem (Chapters 11, 12, and 15).

grated by parts (in the residual equations) m times, so that m is the order of the highest derivative of the shape functions in the stiffness integrals. $[K_h]$ and $\{a_h\}$ are the system stiffness matrix and the approximate-solution vector, respectively, for a mesh of element size h. $[K_0]$ and $\{a_0\}$ are the system stiffness matrix and solution vector, respectively, for a mesh of infinitesimally small elements, i.e., the exact solution.

Note that the product $\{a\}^T[K]\{a\}$ is a single number, not a vector or matrix; in physical applications it represents energy (e.g., in elasticity it represents strain energy). The LHS of Eq. (8.52) therefore represents the energy of the system corresponding to the exact solution, minus the energy of the system corresponding to a mesh of finite size h; in other words, the *error in the energy*.[13]

This error is due solely to the discretization error. It assumes that no other numerical errors are present; e.g., all integrals have been evaluated exactly and there are no computational round-off errors.

Equation (8.52) states that the approximate solution $\{a_h\}$ is converging in a global energy sense (not pointwise) to the exact solution $\{a_0\}$. It also tells us the *rate* of this convergence,[14] namely, $O(h^{2(p-m+1)})$. It is this rate that suggests a desirable order for a quadrature rule: If the stiffness integrals in Eq. (8.52) are integrated to an accuracy which is also $O(h^{2(p-m+1)})$, then the accuracy of the quadrature will improve at the same rate as the accuracy (with respect to energy) of the FE solution as $h \to 0$.

If we were to integrate less accurately than $O(h^{2(p-m+1)})$, then as $h \to 0$ the quadrature error would eventually exceed the discretization error (by an arbitrarily large amount for small enough h). Conversely, if we were to integrate more accurately, then as $h \to 0$ the discretization error would eventually exceed the quadrature error (by an arbitrarily large amount for small enough h); the additional quadrature accuracy would be wasted because it could not improve the inherent accuracy limitation of the FE approximation. If a quadrature rule with accuracy of $O(h^{2(p-m+1)})$ is not available, it is generally advisable to try a rule that is more accurate since a rule of lower accuracy may occasionally fail because of the zero-energy-mode problem.

These arguments have focused on the stiffness integrals, but they may also be applied to the load integrals. Thus, since $[K]\{a\} = \{F\}$, and hence $\{a\}^T[K]\{a\} = \{a\}^T\{F\}$, the same rate of convergence applies to $\{a\}^T\{F\}$. Therefore the load integrals should also be integrated to accuracy $O(h^{2(p-m+1)})$ so that both sides of the system equations converge at the same rate. In summary, we have the following guideline:

> *Guideline II.* To preserve the rate of convergence of the global energy due to the discretization error, integrate the stiffness and load integrals with a quadrature rule that has an accuracy of $O(h^{2(p-m+1)})$. If Gauss-Legendre quadrature is used (see Fig. 8.12), a $(p-m+1)$-point rule would be required.

The above argument is, of course, only heuristic; mathematically more rigorous analyses and numerical experiments are available [8.9-8.15]. Some of the evidence in the

---

[13] See Strang and Fix [8.8] to clarify important concepts such as ''error in the energy'' vis-à-vis ''energy in the error,'' the latter being the energy error discussed in Section 3.6.2.

[14] Note that if $p \le m-1$ the rate will be $O(h^0)$ or less, which implies no convergence at all. For example, if the mesh is refined by halving h, h^0 does not change. Therefore we must have $p \ge m$ for convergence (with respect to energy). This is the source of the completeness condition stated in Section 5.6.

cited references agrees with the quadrature rule given in the guideline, and some of the evidence suggests that a rule one order lower — $O(h^{2(p-m)+1})$ — may be adequate.[15] In either case, the evidence clearly supports the conclusion that a rule of $O(h^{2(p-m+1)})$ is sufficient to preserve the rate of convergence of the global energy.

We will refer to guideline II as *reduced integration*, a term which will be more meaningful after introducing guideline III, given below, for higher-order integration. The term "reduced" originated in the context of 2-D and 3-D elements where it was discovered that higher-order rules could in many instances be reduced to a lower order to yield better results. The term tends to be used somewhat loosely, and perhaps could be defined (somewhat circuitously) as any quadrature rule whose order is lower than that specified by guideline III for higher-order integration.

Now let's apply reduced integration to our C^0-quadratic isoparametric element. For this element $p=2$ and $m=1$; therefore $p-m+1=2$ and hence a two-point Gauss-Legendre rule is required. Each of the three quadrature expressions in Eqs. (8.50) should therefore be evaluated at the two Gauss points that use the values w_{21}, ξ_{21} and w_{22}, ξ_{22}.[16]

Observe that if the load is constant within an element, the F integral is integrated exactly by the two-point rule. This can be seen by examining the integrand in Eq. (8.45): $\phi_i(\xi)$ is a polynomial of degree 2, $f^{(e)}(\xi)$ is a constant, and $J^{(e)}(\xi)$ is a polynomial of maximum degree 1, so the integrand is a polynomial of maximum degree 3, which is integrated exactly by a two-point rule. Observe also that if the physical property $\alpha^{(e)}(\xi)$ is constant and the element is undistorted [i.e., the interior node $x_2^{(e)}$ is at the center of the real element, so $J^{(e)}(\xi)$ is a constant], then the two-point rule is also exact for the $K\alpha$ integral, since the integrand in Eq. (8.43) is then a polynomial of degree 2. If the element is distorted, then $K\alpha$ cannot be integrated exactly by any order rule because $J^{(e)}(\xi)$ is a polynomial

---

[15] The literature in this area, of which references [8.9-8.15] are only representative, is a bit confusing to this author. On the one hand, the rather abstract theoretical analysis in Zlamal [8.11] appears to agree with the above guideline. On the other hand, Fried [8.10], as well as Zienkiewicz and Hinton [8.7] citing Fix [8.9], claim that a rule with accuracy of $O(h^{2(p-m)+1})$ — i.e., one power of h lower than called for in the guideline — is sufficient to preserve the rate of convergence of the global energy. The author fails to see how reference [8.9] supports this one-power-lower rule. Indeed, such a one-power-lower rule would correspond to the rate of convergence of *pointwise* energy (which is the product of the two mth derivatives of shape functions, each containing complete polynomials of order p); and it is natural to expect a difference of one power of h between pointwise energy and global (integrated) energy.

In addition, Fried [8.13] offers some excellent numerical experiments, with the objective of demonstrating the sufficiency of the $O(h^{2(p-m)+1})$ rule, but they appear instead to demonstrate only the sufficiency of the $O(h^{2(p-m+1)})$ rule. Finally, Zienkiewicz [8.14] observes that if the quadrature points are chosen to be the same as the optimal flux-sampling points (which is an $O(h^{2(p-m+1)})$ rule — see guideline I in step 6), then the results show a "dramatic improvement" over using an $O(h^{2(p-m)+1})$ rule. In summary, the weight of the evidence, both theoretical and experimental, appears to lean towards the $O(h^{2(p-m+1)})$ rule in the guideline given above. In many practical applications, though, the difference between the two rules would probably not be readily apparent. Nevertheless, a more definitive and clearly written exposition is needed on this subject.

[16] For the C^0-linear element in Chapter 7 we approximated $\alpha(x)$, $\beta(x)$, and $f(x)$ by constants that were chosen at the *center* of the element. Note that the center location is the Gauss point for the one-point rule, the latter being appropriate for a linear element; that is, $p=1$ and $m=1$ so $p-m+1=1$.

in the denominator of $K\alpha$, making the integrand a nonpolynomial. The $K\beta$ integral cannot be integrated exactly by a two-point rule, although a three-point rule would integrate exactly even a distorted element.

It is important to remember that the above theory is asymptotic (like most FE theory), meaning that, in practice, numerical results will agree with the theory only when the elements become "small enough." What will happen for a coarse mesh of large elements in which the elements are considerably distorted (which may happen more frequently in 2-D and 3-D problems)? Experience shows that increasing the quadrature accuracy by one or two powers of h will sometimes improve the accuracy in these less-than-asymptotic cases. We therefore have the following additional guideline:

> *Guideline III*. For elements that are "considerably" distorted, accuracy may sometimes be improved by integrating the stiffness and load integrals with a quadrature rule that has an accuracy of at least $O(h^{2(p-m+1)+1})$. If Gauss-Legendre quadrature is used (see Fig. 8.12), a $(p-m+2)$-point rule, or higher, should be tried.

We will refer to this guideline as *higher-order integration*. Our present quadratic element would use at least a three-point rule $(p-m+2=3)$ for higher-order integration.

It is difficult to provide a general guideline, applicable to all types of problems, on whether to use reduced or higher-order integration. It depends on the type of physical problem being solved (i.e., the nature of the governing differential equation), the type of element being used, and the degree of mapping distortion in each element. There are some types of problems where reduced integration produces an extraordinarily greater accuracy than even the higher-order integration (the asymptotic physical theories mentioned below), and there are situations where reduced integration fails altogether (because of zero-energy modes).

As a rough rule-of-thumb, reduced integration is generally better suited for undistorted elements, whereas higher-order integration tends to do a better job as the element becomes more distorted.

An important class of exceptions to this rule-of-thumb consists of those problems described by a physical theory in which a parameter asymptotically approaches a limit. For example, in the field of solid mechanics [8.5-8.7] this includes some of the thin shell theories (thickness $\rightarrow$ 0) and the nearly-incompressible material formulations (ratio of shear modulus to bulk modulus $\rightarrow$ 0). In these cases reduced integration is generally superior for both undistorted and distorted elements, and in some instances it may be used selectively for only some terms in the stiffness matrix.

To complicate matters further, there are sometimes several different Gauss-type quadrature rules of the same accuracy, differing in number and placement of quadrature points as well as in weight factors. Numerical experiments usually resolve which one (if any) is the superior rule.

The reader has by now probably surmised that the issue of a proper quadrature rule is quite complicated, relying heavily on the collective experience of many previous applications. Commercial FE programs will therefore frequently offer more than one quadrature rule, sometimes accompanied by advice as to which one to use in a given application. In those circumstances where the analyst is uncertain which quadrature rule to use, both reduced and higher-order integration should be tested on a smaller but similar problem. If the solutions are comparable, then use the cheaper reduced integration. If the solutions differ significantly, testing against simple problems with known solutions can resolve which rule is more accurate. The documentation accompanying commercial FE programs will often include the results of many such test problems.

Step 6: Prepare expression for the flux.

$$\tilde{\tau}^{(e)}(x) = -\alpha(x) \frac{d\tilde{U}^{(e)}(x;a)}{dx}$$

$$= -\alpha(x) \sum_{j=1}^{3} a_j \frac{d\phi_j^{(e)}(x)}{dx}$$

$$= -\alpha(\chi^{(e)}(\xi)) \sum_{j=1}^{3} a_j \frac{d\phi_j^{(e)}(\chi^{(e)}(\xi))}{d\xi} \frac{d\xi}{dx}$$

or, simplifying the notation,

$$\tilde{\tau}^{(e)}(\xi) = -\alpha^{(e)}(\xi) \sum_{j=1}^{3} a_j \frac{d\phi_j(\xi)}{d\xi} \frac{1}{J^{(e)}(\xi)} \tag{8.53}$$

The values for a_j are available from the solution of the system equations, and the expressions for $d\phi_j(\xi)/d\xi$ and $J^{(e)}(\xi)$ are given in Eqs. (8.34) and (8.35) respectively.

Just as with the C^0-linear element in Chapter 7, there exist optimal flux-sampling points in the C^0-quadratic element (two points, in fact) at which the accuracy of Eq. (8.53) is one order better than anywhere else in the element. The theory underlying the existence of such points was presented in Section 7.2, step 6, with special emphasis on the 1-D C^0-linear element. We can apply that theory to the present element, although before doing so we'll generalize two important conclusions from that discussion so that the theory can be easily applied to other types of elements as well.

The first conclusion is that the discretization error of $\tilde{\tau}^{(e)}(\xi)$ for a *general* point in the element is $O(h^{p-m+1})$, where p is the degree of the complete polynomial in the element trial solution $\tilde{U}^{(e)}$ and m is the order of the highest derivative of $\tilde{U}^{(e)}$ in the expression for flux [which is also the highest derivative of the shape functions in the stiffness integrals]. The argument directly parallels that of the C^0-linear element [recall Eqs. (7.29) through (7.33)]: Since $d^m\tilde{U}^{(e)}/dx^m$ is a complete polynomial of degree $p-m$, it is capable of equaling a local Taylor series expansion of d^mU/dx^m to the same degree, leaving a remainder term proportional to h^{p-m+1}. Thus

$$|\tau(x) - \tilde{\tau}^{(e)}(x)| = |\alpha(x)| \left| \frac{d^mU}{dx^m} - \frac{d^m\tilde{U}^{(e)}}{dx^m} \right| = O(h^{p-m+1}) \qquad \text{as } h \to 0 \tag{8.54}$$

The second conclusion, based on the work by T. Moan [8.16] that was described in Section 7.2, pertains to the discretization error of $\tilde{\tau}^{(e)}(\xi)$ at *special* points in the element. Within each element, at the locations corresponding to the $p-m+1$ zeroes of the $(p-m+1)$th-degree Legendre polynomial, the error of $\tilde{\tau}^{(e)}(\xi)$ is $O(h^{p-m+2})$. This is one order better than given by Eq. (8.54) for general locations in the element. These special points are the optimal flux-sampling points, usually referred to as Gauss points because the $p-m+1$ zeroes of the $(p-m+1)$th-degree Legendre polynomial are also the integration points for the $(p-m+1)$-point Gauss-Legendre quadrature rule (see Fig. 8.12). We have, then, the following guideline:[17]

> *Guideline IV*. The optimal flux-sampling points are located at the Gauss points corresponding to the $(p-m+1)$-point Gauss-Legendre quadrature rule.

Note that if the integrals in the element equations are evaluated using reduced integration (see guideline II), then the optimal flux-sampling points are the same as the Gauss points used for the integration. This can sometimes yield efficient program coding. However, if higher-order integration is used (see guideline III), then the optimal flux-sampling points are not the same as the integration points (see numerical experiments in Section 9.2).

For the present C^0-quadratic element, $p-m+1=2$, so the optimal flux-sampling points are the Gauss points for the two-point rule. From Fig. 8.12, the two points are $\xi=\pm1/\sqrt{3}$. Figure 8.13 shows that these are the points where $d\tilde{U}^{(e)}/dx$ intersects dU/dx with accuracy $O(h^2)$ as $h\to0$. Recall from Section 7.2, step 6, that the intersections occur because the FE analysis is essentially providing a least-squares fit of $d\tilde{U}^{(e)}/dx$ to dU/dx (see Exercise 8.4) [8.17]. In summary, in order to calculate the flux in each element, Eq. (8.53) should be evaluated only at the two Gauss points $\xi=\pm1/\sqrt{3}$.[18]

Flux values at points other than the Gauss points may be obtained by interpolating from the Gauss-point values. In particular, we would also like to have flux values at the nodes, for the same reasons given in Section 7.3.2 for the C^0-linear element. Therefore, in a manner completely analogous to the C^0-linear element, we construct for each element a *smoothing curve* (an interpolation polynomial) through the most accurate flux values, namely, the two Gauss-point values. The interpolation polynomial is with respect to the ξ-coordinate, i.e., in the parent element.[19] The polynomial may then be evaluated at the nodes, i.e., at $\xi=-1, 0, +1$. These values are the desired nodal values in the real element, since nodes in the parent element map to nodes in the real element.

Since there are two Gauss-point values, the interpolation polynomial is a straight line with respect to ξ. This is most easily constructed by using the shape functions used for the C^0-linear element, but with the interpolating points being the Gauss points rather than the nodes [recall Eqs. (7.12)]. Thus, defining ξ_{21} as the first Gauss point $(-1/\sqrt{3})$ for the two-point rule and ξ_{22} as the second Gauss point $(+1/\sqrt{3})$, the smoothing curve for

---

[17] This guideline is applicable to 1-D elements as well as to most commonly used higher dimensional elements whose shapes are "rectangular products" of 1-D elements, e.g., quadrilaterals in 2-D (see Sections 13.4.1 and 13.4.2) and "brick" elements in 3-D. It is not applicable, for example, to 2-D triangles (see Sections 13.3.1 and 13.3.2) and 3-D tetrahedra.

[18] It is instructive to apply the above general conclusions to the 1-D C^0-linear element in Chapter 7 for which $p=1$ and $m=1$. Using these values, Eq. (8.53) reduces to Eq. (7.33). Since $p-m+1=1$, there is one Gauss point. From Fig. 8.12, the Gauss point for the one-point rule is at $\xi=0$ (the center). Finally, note that Fig. 7.6 is the linear analogue to Fig. 8.13.

[19] In 1-D elements we could just as easily interpolate with respect to the x variable in each real element. The results would be different from interpolating in the parent element if the element is distorted, although either approach appears to give satisfactory results. In 2-D and 3-D elements, however, there is a decided computational advantage to interpolating in the parent element (see Chapter 13). Therefore for consistency we choose to employ the same approach for 1-D elements.

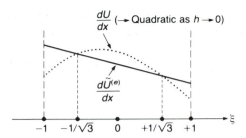

FIGURE 8.13 ● Least-squares fit of an approximate solution, $d\tilde{U}^{(e)}/dx$, to the exact solution, dU/dx, in a 1-D C^0-quadratic element. The intersections approach $\xi = \pm 1/\sqrt{3}$ as $h \to 0$.

our C^0-quadratic element $\tau_s(\xi)$ is

$$\tau_s(\xi) = \left(\frac{\xi_{22} - \xi}{\xi_{22} - \xi_{21}} \right) \tilde{\tau}^{(e)}(\xi_{21}) + \left(\frac{\xi - \xi_{21}}{\xi_{22} - \xi_{21}} \right) \tilde{\tau}^{(e)}(\xi_{22})$$

(8.55)

where $\tilde{\tau}^{(e)}(\xi_{21})$ and $\tilde{\tau}^{(e)}(\xi_{22})$ are the Gauss-point values. Evaluating $\tau_s(\xi)$ at the nodes yields the element nodal fluxes — that is, $\tau_s(-1)$, $\tau_s(0)$, and $\tau_s(+1)$.

We note in passing that Eq. (8.55) will be identical to Eq. (8.53) if and only if the element is undistorted and $\alpha(x)$ does not vary, i.e., if $J(\xi)$ and $\alpha^{(e)}(\xi)$ are both constants. In this case Eq. (8.53) is a linear polynomial which, by definition, equals the two Gauss-point values in Eq. (8.55). Therefore both linear polynomials must be identical.

Finally, at interelement nodes the element nodal fluxes $[\tau_s(-1)$ and $\tau_s(+1)]$ from adjacent elements are averaged to yield the nodal fluxes. We recall from Section 7.3.2 that averaging creates continuity in the flux at interelement boundaries and, in general, the resulting nodal fluxes are more accurate than the (unaveraged) element nodal fluxes. On the other hand, averaging is obviously inappropriate wherever the exact flux solution is expected to be discontinuous.

8.3.2 Implementation into the UNAFEM Program

It is not necessary to write a whole new FE program for the C^0-quadratic isoparametric element. We can use all of the UNAFEM code shown in the flowchart in Fig. 7.11 except for subroutine ELEM11, which deals with the C^0-linear element. We must add new subroutines that will do the calculations for the C^0-quadratic element. In other words, only operations at the element level are affected. This includes formation of the element equations and subsequent calculation of the element fluxes. All operations at the system level remain unchanged. This includes assembly, application of boundary conditions, and equation solving, as well as all pre- and postprocessing. This illustrates the inherent modularity of the FEM.

Figure 8.14 shows the flowchart for the improved UNAFEM program. Three new subroutines have been added: ELEM12, SHAP12, and JACOBI. These calculate the stiffness, load, and flux expressions for the C^0-quadratic element. In addition, only one extra line of code was added to subroutines FORM and CALFLX to call ELEM12. Let's examine each of these additions to the program.[20]

---

[20] Subroutine INITDT (see Appendix) has coded all the Gauss-point coordinates and weights from Fig. 8.12.

Figure 8.15 shows the new statement in subroutine FORM that calls ELEM12. Note that the statements calling ELEM11 and ELEM12 are IF statements, and they are located inside a loop (DO 500) over all the elements in a mesh. Thus each element is tested for its element type, which is determined by the value specified by the user (e.g., 11 or 12)

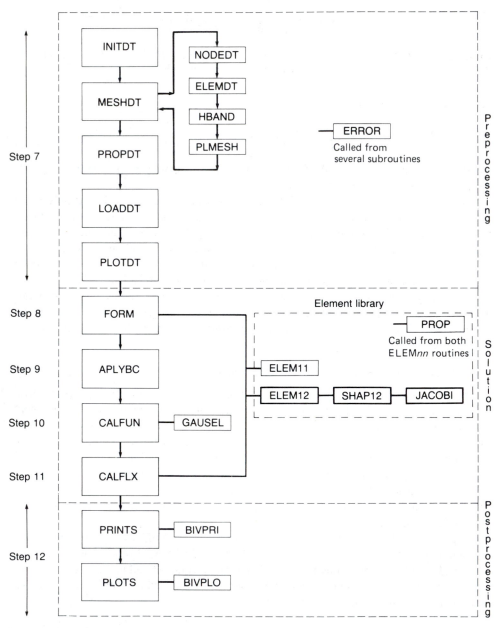

FIGURE 8.14 ● Subroutines (indicated by bold outline) added to UNAFEM to perform analysis for the 1-D C^0-quadratic isoparametric element.

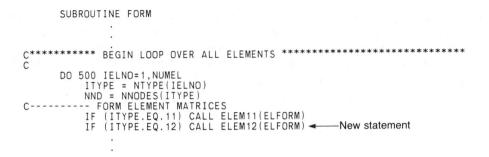

```
      SUBROUTINE FORM
              .
              .
              .
C*********** BEGIN LOOP OVER ALL ELEMENTS *****************************
C
      DO 500 IELNO=1,NUMEL
         ITYPE = NTYPE(IELNO)
         NND = NNODES(ITYPE)
C----------- FORM ELEMENT MATRICES
         IF (ITYPE.EQ.11) CALL ELEM11(ELFORM)
         IF (ITYPE.EQ.12) CALL ELEM12(ELFORM) ◄──New statement
              .
              .
              .
```

FIGURE 8.15 ● Subroutine FORM, a partial listing showing calls to both the C^0-linear element (ELEM11) and the C^0-quadratic element (ELEM12). (See Fig. 7.13.)

on each of the element cards. One may therefore construct a mesh using any mixture of the two element types. It should not be surprising that the two elements can be used in the same mesh since both are C^0-continuous, with a node (and DOF) at each end of the element (see Exercise 8.5).

The user may also specify the quadrature rule for each element — namely, any number from 1 to 5. The default is 3 (the lowest higher-order rule for the C^0-quadratic) rather than 2 (reduced) since this is the safer rule, although it is usually not as efficient as 2.

Assembly follows the usual rules. UNAFEM now has an *element library*, i.e., a collection of subroutines for two or more element types.

Figure 8.16 is a partial listing of ELEM12, showing only those statements relevant to forming the element equations. It calls the subroutine SHAP12 (Fig. 8.17), which evaluates the shape functions and their derivatives. SHAP12 in turn calls subroutine JACOBI (Fig. 8.18) to evaluate the Jacobian and its inverse. Annotations on the listings correlate lines of code with equations in the text.

Figure 8.19 is a partial listing of CALFLX, showing the new statement that calls ELEM12, requesting the flux calculation option in the element subroutine. Note again that the calls to ELEM11 and ELEM12 are inside a loop (DO 100) over all the elements in the mesh. Hence fluxes will be calculated for every element in the mesh.[21]

Figure 8.20 is another partial listing of ELEM12, this time showing the statements relevant to calculating the fluxes. The fluxes are first calculated at the optimal sampling points, which for this element are the two Gauss points corresponding to the two-point rule. They are then calculated at the three nodes by evaluating the smoothing curve (interpolating polynomial) that was constructed through the two Gauss points. Control then returns to subroutine CALFLX, which averages the (smoothed) element nodal values at all interelement nodes.

Postprocessing includes printed output from subroutine BIVPRI and graphical output from subroutine BIVPLO, in the manner described in Section 7.3.2. The isoparametric mapping presents a challenge in plotting the function $\tilde{U}^{(e)}(x)$. If an element is distorted,

---

[21] In principle it is not necessary to calculate fluxes in every element in the mesh. A more sophisticated program would let the user specify the elements in which to calculate fluxes. This could eliminate a considerable amount of unnecessary computation.

$\tilde{U}^{(e)}(x)$ is not a polynomial, so one cannot simply draw a quadratic polynomial through the three nodal values. As an alternative, we evaluate $\tilde{U}^{(e)}(x)$ at many closely spaced points in each element and then connect successive values by tiny straight-line segments. This requires reaching back into the solution phase to get the shape functions for $\tilde{U}^{(e)}(x)$. In large commercial FE programs this is not usually done, in order to keep solution and postprocessing phases separate. However, we do it in UNAFEM because the program

```
       SUBROUTINE ELEM12(OPT)
C
C          THIS SUBROUTINE IS THE MASTER ROUTINE FOR ELEMENT 12, A
C          3-NODE 1-D C0-QUADRATIC ELEMENT.
                 .
                 .
                 .
C*************FORM ELEMENT MATRICES AND VECTORS*************************
C
       YYY = 0.
       IPHY = NPHYS(IELNO)
       DO 20 I=1,3
          J = ICON(I,IELNO)
          XX(I) = X(J)      ◄── x-coordinates of nodes
          FE(I) = 0.        ⎫
          DO 10 J=1,3        ⎬ Initialize to zero
             KE(J,I) = 0.   ⎭
                 .
                 .
                 .
   10     CONTINUE
   20 CONTINUE
       NINT = NGPINT(IELNO)
C---------EVALUATE TERMS IN ELEMENT EQUATIONS
       DO 60 INTX=1,NINT
          XI = GP(NINT,INTX)◄─────── Select Gauss points for integration
          CALL SHAP12 ◄───────────── Evaluate dφᵢ(e)/dx [= DPHDX(I)]
          XXX = 0.
          DO 30 I=1,3
             XXX = XXX+PHI(I)*XX(I)
   30     CONTINUE
          CALL PROP(IPHY,XXX,YYY,ALPHA,ALPHAY,BETA,GAMMA,MU)
          DO 50 I=1,3
             FE(I) = FE(I)+FINT(IELNO)*PHI(I)*WT(NINT,INTX)*JAC ⎫
             DO 40 J=1,3                                         ⎬
                KE(I,J) = KE(I,J)+WT(NINT,INTX)*JAC*             ⎬ Eqs. (8.50)
     $                    (DPHDX(I)*ALPHA*DPHDX(J)+             ⎬
     $                    PHI(I)*BETA*PHI(J))                   ⎭
                 .
                 .
                 .
   40           CONTINUE
   50     CONTINUE
   60 CONTINUE
C---------FOR SYMMETRY
       DO 100 I=1,3
          DO 90 J=I,3
             KE(J,I) = KE(I,J)
                 .
                 .
                 .
   90     CONTINUE
  100 CONTINUE
                 .
                 .
                 .
```

FIGURE 8.16 ● Subroutine ELEM12, a partial listing showing evaluation of terms in the element equations.

```
      SUBROUTINE SHAP12
C
C          THIS ROUTINE EVALUATES SHAPE FUNCTIONS AT XI FOR THE 3-NODE
C                    1-D ELEMENT
                .
                .
                .
C*********SHAPE FUNCTIONS*****************************************
C
      PHI(1) = 1./2.*          XI*(XI-1.)  ⎫
      PHI(2) =        -(XI+1.)*    (XI-1.)  ⎬ Fig. 8.6
      PHI(3) = 1./2.*(XI+1.)*XI             ⎭
C
C*********DERIVATIVES OF THE SHAPE FUNCTIONS**********************
C
C----------DERIVATIVES WRT XI AT XI
      DPHDXI(1) = XI-1./2.      ⎫
      DPHDXI(2) = -2.*XI        ⎬Eqs. (8.34)
      DPHDXI(3) = XI+1./2.      ⎭
C----------DO JACOBIAN CALCULATIONS
      CALL JACOBI(1)
C----------DERIVATIVES WRT X AT XI
      DO 20 I=1,3
         DPHDX(I) = JACINV(1,1)*DPHDXI(I) ◄────Eq. (8.40)
   20 CONTINUE
      RETURN
C-------------------------------------------------------------------
      END
```

FIGURE 8.17 ● Subroutine SHAP12.

```
          SUBROUTINE JACOBI(NDIM)
C
C          THIS ROUTINE DOES THE JACOBIAN CALCULATIONS
                .
                .
                .
C----------FOR 1-D CASE
C---------COMPUTE JACOBIAN AT XI
      JAC = 0.
      DO 110 I=1,NND
         JAC = JAC+DPHDXI(I)*XX(I) ◄────Eq. (8.33)
  110 CONTINUE
C---------CHECK POSITIVENESS OF JACOBIAN
      IF (JAC.LT.1.E-8) CALL ERROR(19,IELNO)
C---------COMPUTE INVERSE OF THE JACOBIAN AT XI
      JACINV(1,1) = 1./JAC
                .
                .
                .
```

FIGURE 8.18 ● Subroutine JACOBI, a partial listing showing 1-D calculations.

```
      SUBROUTINE CALFLX
                .
                .
                .
C---------CALCULATE FLUXES AT GAUSS POINTS AND NODES IN EACH ELEMENT
      DO 100 IELNO=1,NUMEL
         ITYP = NTYPE(IELNO)
         NND = NNODES(ITYP)
         IF (ITYP.EQ.11) CALL ELEM11(FLUX)
         IF (ITYP.EQ.12) CALL ELEM12(FLUX) ◄────New statement
                .
                .
                .
```

FIGURE 8.19 ● Subroutine CALFLX, a partial listing showing calls to both the C^0-linear element (ELEM11) and the C^0-quadratic element (ELEM12). (See Fig. 7.19.)

```
      SUBROUTINE ELEM12(OPT)
C
C         THIS SUBROUTINE IS THE MASTER ROUTINE FOR ELEMENT 12, A
C         3-NODE 1-D C0-QUADRATIC ELEMENT.
            .
            .
            .
C**************CALCULATE FLUXES***************************************
C
  200 CONTINUE
      YYY = 0.
      IPHY = NPHYS(IELNO)
      DO 220 I=1,3
          J = ICON(I,IELNO)
          XX(I) = X(J)
  220 CONTINUE
C---------CALCULATE FLUXES AT THE 2 GAUSS POINTS
      DO 260 INT=1,2
          XI = GP(2,INT)
          CALL SHAP12
          XXI(INT) = 0.
          DUDX = 0.
          DO 240 I=1,3
              J = ICON(I,IELNO)
              XXI(INT) = XXI(INT)+PHI(I)*XX(I)       ◄──────────── Eq. (8.22)
              DUDX = DUDX+DPHDX(I)*A(J)
  240     CONTINUE
          XXX = XXI(INT)                                           Eq. (8.53)
          CALL PROP(IPHY,XXX,YYY,ALPHA,ALPHAY,BETA,GAMMA,MU)
          TAU(INT) = -ALPHA*DUDX
          XGP(INT,IELNO) = XXI(INT)
          YGP(INT,IELNO) = 0.
          TAUXGP(INT,IELNO) = TAU(INT)
          TAUYGP(INT,IELNO) = 0.
  260 CONTINUE
C---------CALCULATE ELEMENT NODAL FLUXES BY INTERPOLATING THE GAUSS
C             POINT FLUXES AND SUM FOR NODAL FLUX CALCULATION IN CALFLX
      XI12 = GP(2,1)-GP(2,2)
      XI21 = -XI12
      DO 300 I=1,3
          JNOD = ICON(I,IELNO)
          SII1 = SI(I)-GP(2,1)
          SII2 = SI(I)-GP(2,2)
          TAUXNP(JNOD) = TAUXNP(JNOD) + SII2/XI12*TAU(1)+    ◄──── Eq. (8.55)
      $                                 SII1/XI21*TAU(2)
          IU(JNOD) = IU(JNOD) + 1
  300 CONTINUE
      RETURN
C------------------------------------------------------------------
      END
```

FIGURE 8.20 ● Subroutine ELEM12, a partial listing showing flux calculations.

is relatively small and, most important, there is educational value in seeing the actual shape of $\tilde{U}^{(e)}(x)$, especially for distorted elements.

8.3.3 Application of UNAFEM

Let's now apply the improved UNAFEM program, using the new C^0-quadratic element, to the heat conduction and cable deflection problems treated previously in Sections 7.4 and 7.5.

The heat conduction problem. Since we are about to use a new element for the first time, it is good practice to begin with a one-element test problem, both to verify the correctness of the program and to familiarize ourselves with the details of data input and output for this type of element.

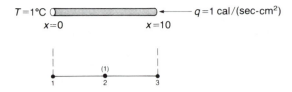

FIGURE 8.21 ● A one-element test problem.

We can use the same test problem analyzed in Section 7.4.2, which is shown in Fig. 8.21, along with a one-element mesh. The required input data, which is shown in Fig. 8.22, is generated using the data input instructions in the appendix. The printed output is shown in Fig. 8.23. Since the results agree with the exact solution ($T=1+x$, $q=-1$), we plunge into the real problem with just a bit more confidence in the program.

As usual we begin with a fairly coarse mesh so that any fundamental problems in the model can be easily and cheaply discovered. Let's try a mesh of five uniform elements, shown in Fig. 8.24. The interior nodes are placed at the center of each element. This is the recommended location in virtually all applications. One reason is that it creates no mapping distortion. Therefore reduced integration, rather than higher-order, will usually be appropriate. (Recall guideline III and accompanying discussion in Section 8.3.1.) Another

```
/DIMB DVP   ONE-ELEMENT TEST
NODE
            1       1       0.
            3              10.
ELEM
            1              12          1       1       2     3
PROP
            1       1.
                    0.
ILC
EBC
            1       1.
NBCC
            3       1.
PLOT


.
END
```

FIGURE 8.22 ● UNAFEM input data for the one-element test problem.

```
ELEMENT OUTPUT

ELEM    GAUSS PT        X        -ALPHA*DU/DX

  1         1        2.1132      -0.1000E+01
            2        7.8868      -0.1000E+01



NODAL OUTPUT

NODE         X           U        -ALPHA*DU/DX
  1       0.0000     0.1000E+01    -0.1000E+01
  2       5.0000     0.6000E+01    -0.1000E+01
  3      10.0000     0.1100E+02    -0.1000E+01
```

FIGURE 8.23 ● UNAFEM printed output for the one-element test problem.

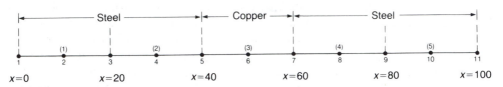

FIGURE 8.24 ● A five-element mesh using C^0-quadratic elements.

reason is that it simplifies node definition in the input data: Since the nodes are all uniformly spaced, we can take advantage of automatic node generation.

The input data is shown in Fig. 8.25. Note that in data set III the default integration order is used (columns 21-25 are blank). This is the higher-order integration, which is a three-point rule for the C^0-quadratic element. It is the default because, if in doubt, it is the safer (albeit more expensive) integration order to use. To help decide whether reduced integration would be adequate, we repeat the analysis using reduced integration (by putting a 2, for two-point rule, in column 25). A comparison of the two solutions[22] (results not shown) reveals that the temperatures agree to four significant figures and the heat fluxes agree to almost three significant figures. We will consider these differences to be negligible and, in addition, assume that the differences will remain negligible in finer meshes. Therefore we will use reduced integration in subsequent meshes.

Figure 8.26 shows the plots for the resulting FE solution for the temperature $T(x)$ and the heat flux $q(x)$. The element nodal fluxes show relatively small discontinuities, suggesting an error in the neighborhood of 5%, perhaps less for the Gauss-point values. The accuracy of $T(x)$ and $q(x)$ appears comparable to that of the 10-element mesh of C^0-linear elements in Fig. 7.37; both employ 11 DOF. Chapter 9 will demonstrate the relative performance of these two types of elements.

---

[22] This is an application of the wider principle that a solution should be independent of the mathematical technique used to produce the solution.

```
1DIM8DVP    HEAT  CONDUCTION  -  5  QUADRATIC  ELEMENTS

NODE
            1       1       0.
            11             100.

ELEM
            1   2   12          1   1   2   3
            3       12          2   5   6   7
            4       12          1   7   8   9
            5       12          1   9  10  11

PROP
            1            .12
                        1.5E-4
            2            .92
                        1.5E-4

ILC
            1           3.0E-3       5

EBC
            11          0.

NBCC
            1            .1

PLOT
            10      0.          100.
            10      0.           50.
            10      0.            .1

END
```

FIGURE 8.25 ● UNAFEM input data for the heat conduction problem, using five quadratic elements.

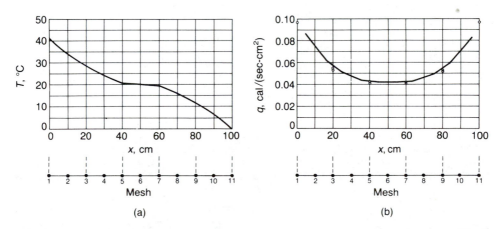

(a)   (b)

FIGURE 8.26 ● UNAFEM graphical output for the heat conduction problem, using five quadratic elements. (a) Function: temperature $\bar{T}(x)$. Approximate-solution polynomial $\bar{T}^{(e)}(x)$ plotted within each element. (b) Flux: heat flux $\tilde{q}(x)$ flowing to the right ($\tilde{q} = -k\,d\bar{T}/dx$). Straight lines connect successive Gauss-point values; circles are element nodal values, interpolated from Gauss-point values.

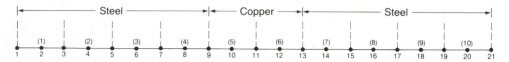

FIGURE 8.27 ● A 10-element mesh using C^0-quadratic elements.

Next consider a mesh of 10 elements, again of uniform length and undistorted (interior nodes at the centers), as shown in Fig. 8.27. The input data (using reduced integration) is shown in Fig. 8.28 and the output plots in Fig. 8.29.

A comparison of the temperature and flux values for the 5- and 10-element meshes (from the printed output) shows that the temperatures agree to at least four significant figures (about 0.01% difference), and the nodal fluxes agree to almost two significant figures (about 3% difference). Gauss-point flux values generally are difficult to compare because the Gauss points have different locations in different meshes. (This is one of the reasons stated previously for calculating nodal flux values.)

```
IDIMBDVP   HEAT  CONDUCTION  -  10  QUADRATIC  ELEMENTS
NODE
              1        1         0.
             21                 100.
ELEM
              1        2        12       2       1       1       2       3
              5        2        12       2       2       9      10      11
              7        2        12       2       1      13      14      15
             10                 12       2       1      19      20      21
PROP
              1                .12
                             1.5E-4
              2                .92
                             1.5E-4
ILC
              1               3.0E-3    10
EBC
             21              0.
NBCC
              1               .1
PLOT
             10      0.      100.
             10      0.       50.
             10      0.              .1
END
```

FIGURE 8.28 ● UNAFEM input data for the heat conduction problem, using 10 quadratic elements.

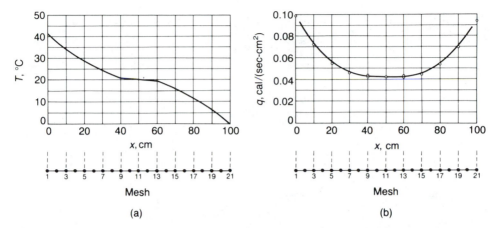

FIGURE 8.29 ● UNAFEM graphical output for the heat conduction problem, using 10 quadratic elements. (a) Function: temperature $\tilde{T}(x)$. Approximate-solution polynomial $\tilde{T}^{(e)}(x)$ plotted within each element. (b) Flux: heat flux $\tilde{q}(x)$ flowing to the right ($\tilde{q} = -kd\tilde{T}/dx$). Straight lines connect successive Gauss-point values; circles are element nodal values, interpolated from Gauss-point values.

Finally we consider a 22-element mesh consisting of 10 elements in each of the steel sections and two in the copper. This mesh is similar to the mesh of 44 C⁰-linear elements in Fig. 7.38 in that both meshes use the same 45 nodes. The results are shown in Fig. 8.30. A comparison of the nodal values of temperature and flux for the 10- and 22-element meshes shows that the temperatures agree to at least four significant figures (the maximum printed out) and the fluxes agree to at least two significant figures (about 1% difference). This 22-element mesh was reanalyzed using higher-order integration (a three-point rule). The temperatures and fluxes both agreed to four significant figures with the results using reduced integration. This lends further confidence in the overall analysis.

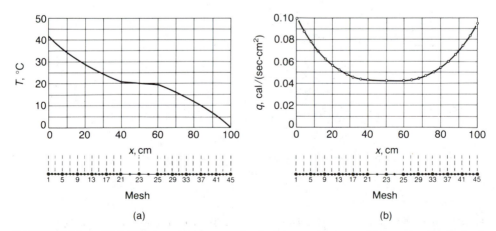

FIGURE 8.30 ● UNAFEM graphical output for the heat conduction problem, using 22 quadratic elements. (a) Function: temperature $\tilde{T}(x)$. Approximate-solution polynomial $\tilde{T}^{(e)}(x)$ plotted within each element. (b) Flux: heat flux $\tilde{q}(x)$ flowing to the right ($\tilde{q} = -kd\tilde{T}/dx$). Straight lines connect successive Gauss-point values; circles are element nodal values, interpolated from Gauss-point values.

	1 2 3 4 5 6 7 8 9 10 11 12 13 14 15 16 17 18 19 20 21 22 23 24 25 26 27 28 29 30 31 32 33 34 35 36 37 38 39 40 41 42 43 44 45 46 47 48 49 50
IDIMBDVP	*CABLE DEFLECTION - 5 QUADRATIC ELEMENTS*
NODE	
	1 *1* *0*.
	11 *20*.
ELEM	
	1 *2* *12* *2* *1* *1* *2* *3*
	5 *12* *2* *1* *9* *10* *11*
PROP	
	1 *100*.
	0.
ILC	
	1 *-4.7426*
	2 *-6.5*
	3 *-0.5* *5*
EBC	
	1 *0*.
	11 *0*.
NBCC	
	9 *-30*.
PLOT	
	10 *0*. *20*.
	10 *-5*. *0*.
	10 *-50*. *50*.
END	

FIGURE 8.31 ● UNAFEM input data for the cable deflection problem, using five quadratic elements.

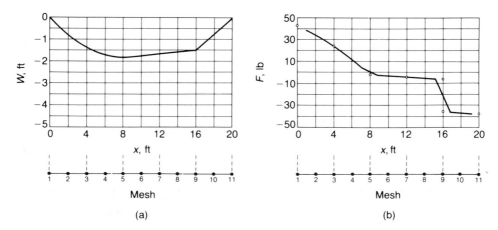

FIGURE 8.32 ● UNAFEM graphical output for the cable deflection problem, using five quadratic elements. (a) Function: vertical displacement $\tilde{W}(x)$. Approximate-solution polynomial $\tilde{W}^{(e)}(x)$ plotted within each element. (b) Flux: vertical force $\bar{F}(x)$ acting on the section to the right of x ($\bar{F} = -Td\tilde{W}/dx$). Straight lines connect successive Gauss-point values; circles are element nodal values, interpolated from Gauss-point values.

The cable deflection problem. The analysis here could follow a pattern very similar to that of the heat conduction problem. Thus we might begin with a coarse mesh of five uniform, undistorted elements and compare the relative performance of reduced and higher-order integration. If the differences were negligible, we could use reduced integration (otherwise, higher-order) in subsequent meshes. However, for this particular problem, if the elements are undistorted then reduced integration is exact. [Recall comments in Section

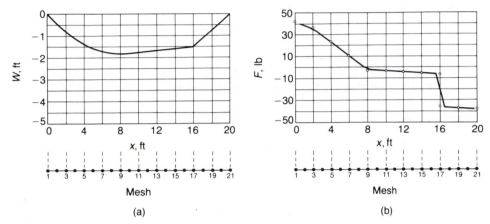

FIGURE 8.33 ● UNAFEM graphical output for the cable deflection problem, using 10 quadratic elements. (a) Function: vertical displacement $\tilde{W}(x)$. Approximate-solution polynomial $\tilde{W}^{(e)}(x)$ plotted within each element. (b) Flux: vertical force $\tilde{F}(x)$ acting on the section to the right of x ($\tilde{F} = -Td\tilde{W}/dx$). Straight lines connect successive Gauss-point values; circles are element nodal values, interpolated from Gauss-point values.

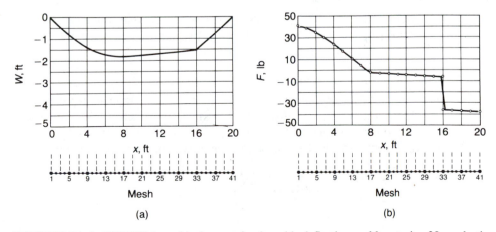

FIGURE 8.34 ● UNAFEM graphical output for the cable deflection problem, using 20 quadratic elements. (a) Function: vertical displacement $\tilde{W}(x)$. Approximate-solution polynomial $\tilde{W}^{(e)}(x)$ plotted within each element. (b) Flux: vertical force $\tilde{F}(x)$ acting on the section to the right of x ($\tilde{F} = -Td\tilde{W}/dx$). Straight lines connect successive Gauss-point values; circles are element nodal values, interpolated from Gauss-point values.

Number of quadratic elements (and DOF) in mesh	$W(8)$	$\|F(0)\|$	$\|F(20)\|$	$\|F(0)\|+\|F(20)\|$	$\dfrac{\|\Sigma f\|-(\|F(0)\|+\|F(20)\|)}{\|\Sigma f\|}$
5 (11)	-1.788	43.07	37.90	80.97	-2.13%
10 (21)	-1.811	41.59	38.09	79.68	-0.50%
20 (41)	-1.816	41.24	38.13	79.37	-0.11%

$$(\,|\Sigma f|=79.28)$$

FIGURE 8.35 ● Vertical displacement at $x=8$ ft, vertical reaction forces at $x=0$ and 20 ft, and error relative to applied forces.

8.3.1, step 5, noting that $\beta(x)=0$ for this problem.] Therefore we will use reduced integration, knowing that it will not introduce any integration error at all.

Figures 8.31 and 8.32 show the input data and resulting graphical plots for the five-element mesh. Figures 8.33 and 8.34 show the graphical plots for 10- and 20-element meshes.

Figure 8.35 summarizes the vertical displacements and reaction forces for the three meshes. They show a clearly converging trend. These values should be compared with those in Figs. 7.54 and 7.55 using the C^0-linear element. Note that the 5-, 10-, and 20-element meshes of quadratic elements employ 11, 21, and 41 DOF, respectively. These are the same numbers of DOF employed by the 10-, 20-, and 40-element meshes of linear elements.

8.4 The 1-D C^0-Cubic Isoparametric Element

8.4.1 Theoretical Development

The development of the C^0-cubic isoparametric element is analogous to the development of the C^0-quadratic isoparametric element. Therefore the following material will directly parallel Section 8.3.1, but in a more abbreviated form.

Step 4: Develop specific expressions for the shape functions.

The element trial solution *in the parent element* $\tilde{U}(\xi;a)$ is a cubic interpolation polynomial. A cubic has four terms $(1,\xi,\xi^2,\xi^3)$, hence four unknown parameters. We define four uniformly spaced nodes[23] in the interval $-1\le\xi\le+1$, as shown in Fig. 8.36.

Write $\tilde{U}(\xi;a)$ in the standard form, namely, as a sum of four shape functions:

$$\tilde{U}(\xi;a) = \sum_{j=1}^{4} a_j\phi_j(\xi)$$

$$(8.56)$$

---

[23] As a general rule, the nodes in any parent element are usually placed in a uniform or regular pattern in order to provide a balanced approximation throughout the element.

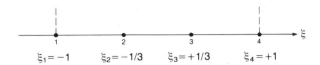

FIGURE 8.36 ● Parent element for the 1-D C^0-cubic isoparametric element.

Each $\phi_j(\xi)$ is a cubic polynomial:

$$\phi_j(\xi) = \alpha_{j0} + \alpha_{j1}\xi + \alpha_{j2}\xi^2 + \alpha_{j3}\xi^3 \qquad j = 1,2,3,4 \qquad (8.57)$$

which must satisfy the interpolation property

$$\phi_j(\xi_i) = \delta_{ji} \qquad (8.58)$$

The coefficients $\alpha_{j0},\ldots,\alpha_{j3}$ may be determined by applying Eq. (8.58) to Eq. (8.57), using the ξ_i values in Fig. 8.36, in exactly the same manner as for the C^0-quadratic element in Eqs. (8.7) through (8.10).

Figure 8.37 shows the four parent shape functions; in addition, it provides a convenient graphical summary of the interrelationship of the shape functions and the element trial solution for the 1-D C^0-cubic isoparametric element. The shape functions will be recognized as the cubic Lagrange interpolation polynomials; they can be obtained more directly by simply substituting $p=3$ into the expressions for the pth-degree Lagrange polynomials in Eq. (8.11).

Having defined the parent shape functions $\phi_j(\xi)$, we must next map them onto a typical real element, thereby creating the real (i.e., mapped) shape functions $\phi_j^{(e)}(x)$. To do this we define an isoparametric coordinate transformation which maps the ξ-coordinate in the parent element onto the x-coordinate in the (e)th real element:

$$x = \sum_{k=1}^{4} x_k^{(e)}\phi_k(\xi) \qquad (8.59)$$

where $x_k^{(e)}$, $k=1,2,3,4$, are the coordinates of the four nodes in the (e)th real element, and $\phi_k(\xi)$ are the mapping functions, which, for an isoparametric mapping, are identical to the parent shape functions in Fig. 8.37.

Just as with the C^0-quadratic [Eqs. (8.23)], nodes in the parent element are mapped by Eq. (8.59) onto nodes in each of the real elements:

$$\xi = \xi_1 \rightarrow x = x_1^{(e)}$$

$$\xi = \xi_2 \rightarrow x = x_2^{(e)}$$

$$\xi = \xi_3 \rightarrow x = x_3^{(e)}$$

$$\xi = \xi_4 \rightarrow x = x_4^{(e)} \qquad (8.60)$$

It follows from Eqs. (8.60) and (8.58) that the shape functions in the real element also satisfy the interpolation property at the nodes:

$$\phi_j^{(e)}(x_i) = \delta_{ji} \qquad (8.61)$$

The element trial soulution in the parent element, $\tilde{U}(\xi; a)$, is a complete cubic polynomial:

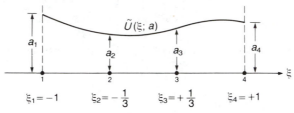

$$\xi_1 = -1 \qquad \xi_2 = -\frac{1}{3} \qquad \xi_3 = +\frac{1}{3} \qquad \xi_4 = +1$$

It is represented as the sum of 4 shape functions:

$$\tilde{U}(\xi; a) = \sum_{j=1}^{4} a_j\,\phi_j\,(\xi)$$

Each coefficient a_j is the value of $\tilde{U}(\xi; a)$ at the node at ξ_j.
Each parent shape function, $\phi_j(\xi)$, is itself a cubic polynomial satisfying the interpolation property, $\phi_j\,(\xi_i) = \delta_{ji}$:

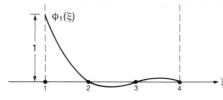

$$\phi_1(\xi) = \frac{(\xi - \xi_2)\,(\xi - \xi_3)\,(\xi - \xi_4)}{(\xi_1 - \xi_2)\,(\xi_1 - \xi_3)\,(\xi_1 - \xi_4)}$$

$$= -\frac{9}{16}\left(\xi + \frac{1}{3}\right)\left(\xi - \frac{1}{3}\right)(\xi - 1)$$

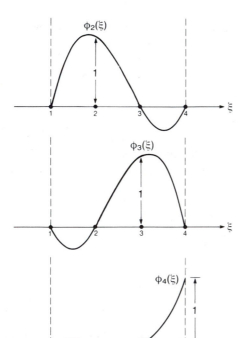

$$\phi_2(\xi) = \frac{(\xi - \xi_1)\,(\xi - \xi_3)\,(\xi - \xi_4)}{(\xi_2 - \xi_1)\,(\xi_2 - \xi_3)\,(\xi_2 - \xi_4)}$$

$$= \frac{27}{16}\,(\xi + 1)\left(\xi - \frac{1}{3}\right)(\xi - 1)$$

$$\phi_3(\xi) = \frac{(\xi - \xi_1)\,(\xi - \xi_2)\,(\xi - \xi_4)}{(\xi_3 - \xi_1)\,(\xi_3 - \xi_2)\,(\xi_3 - \xi_4)}$$

$$= -\frac{27}{16}\,(\xi + 1)\left(\xi + \frac{1}{3}\right)(\xi - 1)$$

$$\psi_4(\xi) = \frac{(\xi - \xi_1)\,(\xi - \xi_2)\,(\xi - \xi_3)}{(\xi_4 - \xi_1)\,(\xi_4 - \xi_2)\,(\xi_4 - \xi_3)}$$

$$= \frac{9}{16}\,(\xi + 1)\left(\xi + \frac{1}{3}\right)\left(\xi - \frac{1}{3}\right)$$

FIGURE 8.37 ● The 1-D C^0-cubic isoparametric element.

Let's examine the shape functions $\phi_k^{(e)}(x)$ to determine how their shape is affected by the placement of the two interior nodes. For this we need the Jacobian. From Eq. (8.59),

$$J^{(e)}(\xi) = \frac{dx}{d\xi} = \sum_{k=1}^{4} x_k^{(e)} \frac{d\phi_k(\xi)}{d\xi}$$

(8.62)

where the derivatives of the parent shape functions are obtained from Fig. (8.37):

$$\frac{d\phi_1(\xi)}{d\xi} = -\frac{9}{16}\left(3\xi^2 - 2\xi - \frac{1}{9}\right)$$

$$\frac{d\phi_2(\xi)}{d\xi} = \frac{27}{16}\left(3\xi^2 - \frac{2}{3}\xi - 1\right)$$

$$\frac{d\phi_3(\xi)}{d\xi} = -\frac{27}{16}\left(3\xi^2 + \frac{2}{3}\xi - 1\right)$$

$$\frac{d\phi_4(\xi)}{d\xi} = \frac{9}{16}\left(3\xi^2 + 2\xi - \frac{1}{9}\right)$$

(8.63)

First we determine the condition for no distortion; that is,

$$J^{(e)}(\xi) = \text{constant}$$

(8.64)

Substitute Eqs. (8.63) into Eq. (8.62) and combine the same powers of ξ:

$$J^{(e)}(\xi) = \frac{1}{16}\left(x_1^{(e)} - 27x_2^{(e)} + 27x_3^{(e)} - x_4^{(e)}\right)$$

$$+ \frac{9}{8}\left(x_1^{(e)} - x_2^{(e)} - x_3^{(e)} + x_4^{(e)}\right)\xi$$

$$- \frac{27}{16}\left(x_1^{(e)} - 3x_2^{(e)} + 3x_3^{(e)} - x_4^{(e)}\right)\xi^2$$

(8.65)

In order to satisfy Eq. (8.64), the coefficients of ξ and ξ^2 must vanish:

$$x_2^{(e)} + x_3^{(e)} = x_1^{(e)} + x_4^{(e)}$$

$$x_2^{(e)} - x_3^{(e)} = \frac{1}{3}x_1^{(e)} - \frac{1}{3}x_4^{(e)}$$

(8.66)

which yields

$$x_2^{(e)} = x_1^{(e)} + \frac{1}{3}L$$

$$x_3^{(e)} = x_1^{(e)} + \frac{2}{3}L$$

(8.67)

where $L = x_4^{(e)} - x_1^{(e)}$, the length of the (e)th real element. Hence the two interior nodes in the real element must be at the 1/3 and 2/3 points, just as in the parent element.

Substituting Eqs. (8.67) into Eq. (8.65) yields

$$J^{(e)}(\xi) = \frac{L}{2}$$

(8.68)

This is reasonable because when the mapping is uniform over the element,

$$\frac{dx}{d\xi} = \frac{\Delta x}{\Delta \xi} = \frac{x_4 - x_1}{(+1) - (-1)} = \frac{L}{2}$$

Substituting Eqs. (8.67) into Eq. (8.59) yields the following expression for the coordinate transformation:

$$x = x_c + \frac{L}{2} \xi$$

(8.69)

where $x_c = (x_1 + x_4)/2$ is the center of the element. The mapping is therefore linear, as it must be when there is no distortion. [Recall the discussion of Fig. 8.9(a) following Eq. (8.35).] When the mapping is linear, it can be inverted to give ξ explicitly as a function of x:

$$\xi = \frac{2}{L} (x - x_c)$$

(8.70)

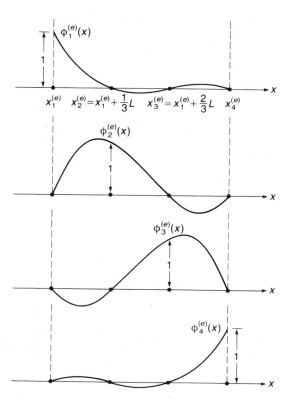

FIGURE 8.38 ● Shape functions in (e)th real element when interior nodes $x_2^{(e)}$ and $x_3^{(e)}$ are at the 1/3 and 2/3 points, respectively.

Substituting Eq. (8.70) into the parent shape functions $\phi_i(\xi)$ yields explicit algebraic expressions for the shape functions in the real element:

$$\phi_1^{(e)}(x) = \frac{(x-x_2)(x-x_3)(x-x_4)}{(x_1-x_2)(x_1-x_3)(x_1-x_4)}$$

$$\phi_2^{(e)}(x) = \frac{(x-x_1)(x-x_3)(x-x_4)}{(x_2-x_1)(x_2-x_3)(x_2-x_4)}$$

$$\phi_3^{(e)}(x) = \frac{(x-x_1)(x-x_2)(x-x_4)}{(x_3-x_1)(x_3-x_2)(x_3-x_4)}$$

$$\phi_4^{(e)}(x) = \frac{(x-x_1)(x-x_2)(x-x_3)}{(x_4-x_1)(x_4-x_2)(x_4-x_3)} \tag{8.71}$$

These expressions are, of course, the cubic Lagrange interpolation polynomials, which are illustrated in Fig. 8.38. Thus, when the interior nodes in the real element are at the 1/3 and 2/3 points (i.e., the same pattern as in the parent element), there is no distortion and the mapped shape functions remain cubic Lagrange polynomials. The only difference might be a uniform change of scale. This situation is exactly analogous to the C^0-quadratic element (Fig. 8.8).[24]

Having examined the condition on the placement of the interior nodes for no distortion, let's now look at the other extreme — infinite distortion. This happens when

$$J^{(e)}(\xi) = 0 \tag{8.72}$$

Recall from Eq. (8.32) that the condition for an acceptable mapping is $J^{(e)}(\xi) > 0$ for $-1 \le \xi \le +1$. Hence from Eqs. (8.62) and (8.63),

$$-\frac{9}{16}\left(3\xi^2 - 2\xi - \frac{1}{9}\right)x_1^{(e)} + \frac{27}{16}\left(3\xi^2 - \frac{2}{3}\xi - 1\right)x_2^{(e)}$$

$$-\frac{27}{16}\left(3\xi^2 + \frac{2}{3}\xi - 1\right)x_3^{(e)} + \frac{9}{16}\left(3\xi^2 + 2\xi - \frac{1}{9}\right)x_4^{(e)} > 0 \tag{8.73}$$

This is more complicated than the C^0-quadratic element because there are two interior node locations to adjust, namely, $x_2^{(e)}$ and $x_3^{(e)}$. By considering the special symmetrical case of both nodes located equidistant from the center, it can be shown (see Exercise 8.6) that the solution to Eq. (8.73) is

$$x_c^{(e)} - \frac{13}{27}\frac{L}{2} < x_2^{(e)} < x_c^{(e)} - \frac{1}{27}\frac{L}{2}$$

$$x_c^{(e)} + \frac{1}{27}\frac{L}{2} < x_3^{(e)} < x_c^{(e)} + \frac{13}{27}\frac{L}{2} \tag{8.74}$$

where $x_c^{(e)} = (x_1^{(e)} + x_4^{(e)})/2$ is the center of the element. These intervals are depicted graphically in Fig. 8.39.

---

[24] Also analogous is the fact that if the interior nodes are located anywhere else besides the 1/3 and 2/3 points, then the $\phi_i^{(e)}(x)$ are no longer polynomials. In such cases explicit expressions for the $\phi_i^{(e)}(x)$ cannot be obtained.

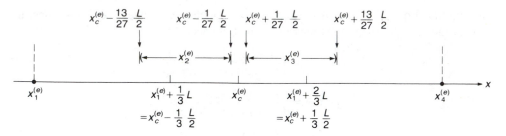

FIGURE 8.39 ● Permissible intervals for locating interior nodes of a 1-D C^0-cubic isoparametric element when both nodes are equidistant from the center.

We see that the real shape functions, $\phi_i^{(e)}(x)$, will not tolerate much displacement of the interior nodes from the $\pm 1/3$ points in the direction of the ends of the element, but they may approach the center quite closely. Figure 8.40 shows the infinite slopes of the $\phi_i^{(e)}(x)$ when $x_2^{(e)}$ and $x_3^{(e)}$ are at either end of their permissible intervals.

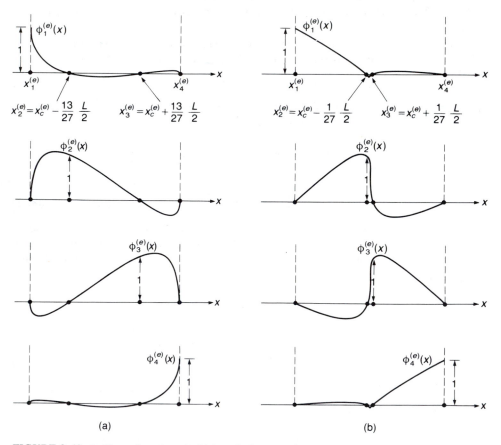

(a)   (b)

FIGURE 8.40 ● Shape functions in (e)th real element when symmetrically placed interior nodes are at limits of permissible intervals. (a) $J=0$ at the ends of the element $(x_2^{(e)}, x_4^{(e)})$; (b) $J=0$ at the center of the element $(x_c^{(e)})$.

FIGURE 8.41 ● A mesh of three C⁰-cubic elements.

The conclusion drawn from all this is as follows: When constructing a mesh of C⁰-cubic isoparametric elements, it is generally best to place the two middle nodes at the $\pm1/3$ points in each element. In practice, there is rarely any reason to do otherwise.[25]

Assembly of 1-D C⁰-cubic elements follows the general assembly rule in Section 5.7. This is illustrated for a three-element mesh with the nodes numbered sequentially from left to right (Fig. 8.41).

The element stiffness and load terms from each of the three sets of element equations are added to the rows and columns of the system matrices with the same numbers as the node numbers in the mesh; if two elements share a node (such as nodes 4 and 7), then their stiffness and load terms are added together. The resulting assembled system equations would appear as follows (see Exercise 8.7):

$$
\begin{bmatrix}
K_{11}^{(1)} & K_{12}^{(1)} & K_{13}^{(1)} & K_{14}^{(1)} \\
K_{21}^{(1)} & K_{22}^{(1)} & K_{23}^{(1)} & K_{24}^{(1)} \\
K_{31}^{(1)} & K_{32}^{(1)} & K_{33}^{(1)} & K_{34}^{(1)} \\
K_{41}^{(1)} & K_{42}^{(1)} & K_{43}^{(1)} & K_{44}^{(1)}+K_{44}^{(2)} & K_{45}^{(2)} & K_{46}^{(2)} & K_{47}^{(2)} \\
& & & K_{54}^{(2)} & K_{55}^{(2)} & K_{56}^{(2)} & K_{57}^{(2)} \\
& & & K_{64}^{(2)} & K_{65}^{(2)} & K_{66}^{(2)} & K_{67}^{(2)} \\
& & & K_{74}^{(2)} & K_{75}^{(2)} & K_{76}^{(2)} & K_{77}^{(2)}+K_{77}^{(3)} & K_{78}^{(3)} & K_{79}^{(3)} & K_{7,10}^{(3)} \\
& & & & & & K_{87}^{(3)} & K_{88}^{(3)} & K_{89}^{(3)} & K_{8,10}^{(3)} \\
& & & & & & K_{97}^{(3)} & K_{98}^{(3)} & K_{99}^{(3)} & K_{9,10}^{(3)} \\
& & & & & & K_{10,7}^{(3)} & K_{10,8}^{(3)} & K_{10,9}^{(3)} & K_{10,10}^{(3)}
\end{bmatrix}
\begin{Bmatrix}
a_1 \\ a_2 \\ a_3 \\ a_4 \\ a_5 \\ a_6 \\ a_7 \\ a_8 \\ a_9 \\ a_{10}
\end{Bmatrix}
=
\begin{Bmatrix}
F_1^{(1)} \\ F_2^{(1)} \\ F_3^{(1)} \\ F_4^{(1)}+F_4^{(2)} \\ F_5^{(2)} \\ F_6^{(2)} \\ F_7^{(2)}+F_7^{(3)} \\ F_8^{(3)} \\ F_9^{(3)} \\ F_{10}^{(3)}
\end{Bmatrix}
$$

$$(8.75)$$

As observed before (Fig. 8.4), assembly has the effect of joining shape functions from adjacent elements to form a single trial function that is C⁰-continuous. Figure 8.42 illustrates the assembled shape functions for the above three-element mesh (with interior nodes at the 1/3 and 2/3 points so the shape functions are undistorted).

Step 5: Substitute the shape functions into the element equations, and transform the integrals into a form appropriate for numerical evaluation.

---

[25] As with the quadratic element, the exception is when the user knows that the exact solution will be better approximated by a nonpolynomial trial function, e.g., one with very steep or singular behavior.

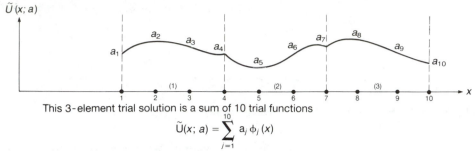

This 3-element trial solution is a sum of 10 trial functions

$$\tilde{U}(x; a) = \sum_{j=1}^{10} a_j \phi_j(x)$$

Each trial function consists of either one shape function or a sum (assembly) of 2 shape functions:

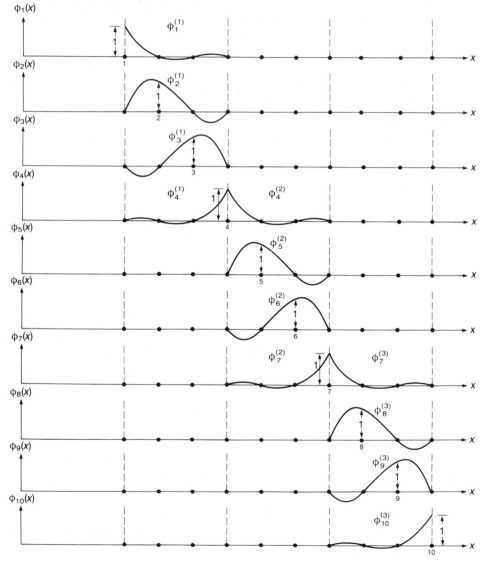

FIGURE 8.42 ● Assembled shape functions (trial functions) for a three-element mesh of 1-D C^0-cubic elements.

The development here is virtually identical to that in Section 8.3.1, step 5, for the C^0-quadratic element. The three integrals are

$$Ka_{ij} = \int_{x_1}^{x_4} \frac{d\phi_i^{(e)}(x)}{dx} \alpha(x) \frac{d\phi_j^{(e)}(x)}{dx} dx$$

$$= \int_{-1}^{+1} \left(\frac{1}{J^{(e)}(\xi)} \frac{d\phi_i(\xi)}{d\xi} \right) \alpha^{(e)}(\xi) \left(\frac{1}{J^{(e)}(\xi)} \frac{d\phi_j(\xi)}{d\xi} \right) J^{(e)}(\xi) d\xi$$

$$\simeq \sum_{l=1}^{n} w_{nl} \left(\frac{1}{J^{(e)}(\xi)} \frac{d\phi_i(\xi)}{d\xi} \right)_{\xi_{nl}} \alpha^{(e)}(\xi_{nl}) \left(\frac{1}{J^{(e)}(\xi)} \frac{d\phi_j(\xi)}{d\xi} \right)_{\xi_{nl}} J^{(e)}(\xi_{nl})$$

$$\tag{8.76}$$

$$K\beta_{ij}^{(e)} = \int_{x_1}^{x_4} \phi_i^{(e)}(x)\beta(x)\phi_j^{(e)}(x)dx$$

$$= \int_{-1}^{+1} \phi_i(\xi)\beta^{(e)}(\xi)\phi_j(\xi)J^{(e)}(\xi)d\xi$$

$$\simeq \sum_{l=1}^{n} w_{nl}\phi_i(\xi_{nl})\beta^{(e)}(\xi_{nl})\phi_j(\xi_{nl})J^{(e)}(\xi_{nl})$$

$$\tag{8.77}$$

$$Ff_i^{(e)} = \int_{x_1}^{x_4} f(x)\phi_i^{(e)}(x)dx$$

$$= \int_{-1}^{+1} f^{(e)}(\xi)\phi_i(\xi)J^{(e)}(\xi)d\xi$$

$$\simeq \sum_{l=1}^{n} w_{nl}f^{(e)}(\xi_{nl})\phi_i(\xi_{nl})J^{(e)}(\xi_{nl})$$

$$\tag{8.78}$$

Note that the quadrature expressions [Eqs. (8.76) through (8.78)] are identical to Eqs. (8.50).

The appropriate quadrature rule for reduced integration follows from guideline II: For our cubic element $p=3$ and $m=1$, so $p-m+1=3$; hence a three-point rule is required. From guideline III, higher-order integration requires at least a four-point rule.

We note that, just as with the quadratic element, if $f^{(e)}(\xi)$ is a constant (a uniform interior load over the element), then $Ff_i^{(e)}$ is integrated exactly by reduced integration. If $\alpha^{(e)}(\xi)$, $\beta^{(e)}(\xi)$, and $J^{(e)}(\xi)$ are constants (uniform physical properties over the element and no mapping distortion), then $Ka_{ij}^{(e)}$ is integrated exactly by reduced, but $K\beta_{ij}^{(e)}$ is integrated exactly only by higher-order.

As with the quadratic element, the flux vector can be simplified:

$$
F\tau_i^{(e)} = -\left[\left(-\alpha(x)\,\frac{d\tilde{U}^{(e)}(x)}{dx}\right)\phi_i^{(e)}(x)\right]_{x_1}^{x_4}
$$

$$
= -\left[\tau^{(e)}(x)\phi_i^{(e)}(x)\right]_{x_1}^{x_4} \tag{8.79}
$$

Using the interpolation property [Eq. (8.61)] yields

$$
\{F\tau\}^{(e)} = \left\{\begin{array}{c} \tau^{(e)}(x_1) \\ 0 \\ 0 \\ -\tau^{(e)}(x_4) \end{array}\right\} \tag{8.80}
$$

Step 6: Prepare expression for the flux.

$$
\tilde{\tau}^{(e)}(x) = -\alpha(x)\,\frac{d\tilde{U}^{(e)}(x;a)}{dx}
$$

$$
= -\alpha^{(e)}(x) \sum_{j=1}^{4} a_j\,\frac{d\phi_j^{(e)}(x)}{dx}
$$

$$
= -\alpha(\chi^{(e)}(\xi)) \sum_{j=1}^{4} a_j\,\frac{d\phi_j^{(e)}(\chi^{(e)}(\xi))}{d\xi}\,\frac{d\xi}{dx}
$$

or, simplifying the notation,

$$
\tilde{\tau}^{(e)}(\xi) = -\alpha^{(e)}(\xi) \sum_{j=1}^{4} a_j\,\frac{d\phi_j(\xi)}{d\xi}\,\frac{1}{J^{(e)}(\xi)} \tag{8.81}
$$

The values for a_j are available from the solution of the system equations, and the expressions for $d\phi_j(\xi)/d\xi$ and $J^{(e)}(\xi)$ are given in Eqs. (8.63) and (8.62), respectively.

From guideline IV in Section 8.3.1, step 6, the optimal flux-sampling points are the Gauss points corresponding to the $(p-m+1)$-point quadrature rule in Fig. 8.12. For the C^0-cubic element $p=3$ and $m=1$. This yields the three-point rule for which the Gauss points are $\xi_{31}=-\sqrt{3}/\sqrt{5}$, $\xi_{32}=0$, and $\xi_{33}=+\sqrt{3}/\sqrt{5}$. Equation (8.81) is evaluated at these three Gauss points within each element.

As with the linear and quadratic elements, in order to obtain nodal flux values in each element we construct a smoothing curve through the three Gauss-point values, consisting of an interpolation polynomial. Since three points define a quadratic, we can use the expressions from Fig. 8.2. The resulting smoothing curve, $\tau_s^{(e)}(\xi)$, is

$$
\tau_s^{(e)}(\xi) = \frac{(\xi-\xi_{32})(\xi-\xi_{33})}{(\xi_{31}-\xi_{32})(\xi_{31}-\xi_{33})}\,\tilde{\tau}^{(e)}(\xi_{31}) + \frac{(\xi-\xi_{31})(\xi-\xi_{33})}{(\xi_{32}-\xi_{31})(\xi_{32}-\xi_{33})}\,\tilde{\tau}^{(e)}(\xi_{32})
$$

$$
+ \frac{(\xi-\xi_{31})(\xi-\xi_{32})}{(\xi_{33}-\xi_{31})(\xi_{33}-\xi_{32})}\,\tilde{\tau}^{(e)}(\xi_{33}) \tag{8.82}
$$

where $\tilde{\tau}^{(e)}(\xi_{31})$, $\tilde{\tau}^{(e)}(\xi_{32})$, and $\tilde{\tau}^{(e)}(\xi_{33})$ are the Gauss-point values. Equation (8.82) is evaluated at the four nodes in each element. At interelement nodes the values are averaged.

8.4.2 Implementation into the UNAFEM Program

The element equations for the C^0-cubic isoparametric element, developed in the previous section, have been coded and added to UNAFEM as subroutines ELEM13 and SHAP13. Figure 8.43 shows the flowchart for the improved UNAFEM program. Note that SHAP13 can use the previously developed subroutine JACOBI to evaluate the Jacobian. The element library now contains three elements, enabling a mesh to be constructed from any combination of the three types of elements.

The FORM and CALFLX subroutines (see appendix) each have one extra statement that calls ELEM13. Figures 8.44 and 8.45 are the listings for ELEM13 and SHAP13. Annotations on the listings correlate lines of code with equations in the text.

Postprocessing includes printed output from subroutine BIVPRI and graphical output from subroutine BIVPLO, in the same manner as for the linear and quadratic elements.

8.4.3 Application of UNAFEM

Let's now apply the improved UNAFEM program, using the new C^0-cubic element, to the heat conduction and cable deflection problems treated previously in Sections 7.4, 7.5, and 8.3.3.

The heat-conduction problem. We can follow a procedure similar to the one in Section 8.3.3 with the C^0-quadratic element; i.e., first test the new element with a one-element test problem, then try both reduced and higher-order integration on a coarse mesh, and then refine the mesh. All elements should be undistorted; i.e., the interior nodes should be at the 1/3 and 2/3 points. Since the cubic element has more DOF than the quadratic or linear elements, and therefore more accuracy, fewer elements will be necessary to achieve a comparable accuracy. Figure 8.46 shows the input data for a three-element mesh, and Fig. 8.47 the resulting graphical output. Figure 8.48 shows the graphical output for a refined seven-element mesh. The results for the three cubic elements are comparable in accuracy to those for the five quadratic elements (Fig. 8.26) and the 10 linear elements (Fig. 7.37).

The cable-deflection problem. Figures 8.49 and 8.50 show the graphical output for a four- and seven-element mesh, respectively.

Figure 8.51 compares the values from both meshes. Note that these results are identical to the results for the quadratic elements in Fig. 8.35. This is because UNAFEM only accepts constant values for the interior loads within each element. The exact solution to the governing differential equation [Eq. (7.64a)], within an element with a constant interior load, is a quadratic polynomial. Any element that uses a quadratic or higher-degree polynomial can "approximate" a quadratic exactly. Therefore the quadratic and cubic elements (and quartic elements in Section 8.5) will yield the same solution if the meshes employ identical element patterns (although not necessarily the same nodes) and if the same (constant) interior load is used in each element. This, in fact, is the case for several of the meshes used in the cable problem.

This situation illustrates a wider principle: If the exact solution to the model can be represented exactly by the trial solution, then the FEM will produce that exact solution.

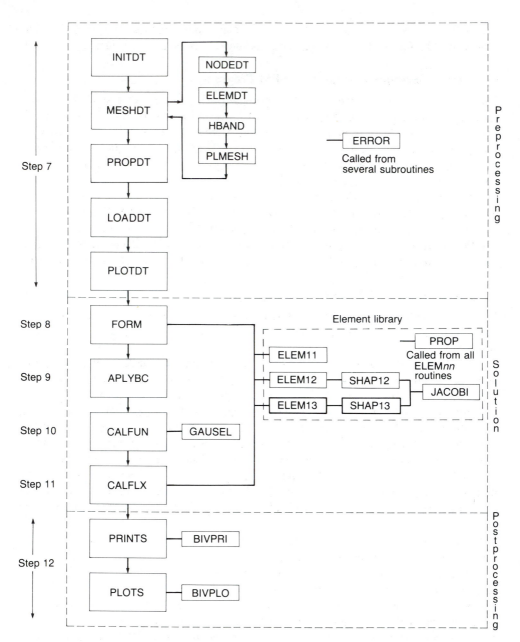

FIGURE 8.43 ● Subroutines (indicated by bold outline) added to UNAFEM to perform analysis for the 1-D C^0-cubic isoparametric element.

As we have frequently observed, the model defined to the program may frequently be different from the original problem statement. For example, in the present cable problem, the original statement — Eq. (7.64b) — defines an interior load of $-0.5-6 \sin (\pi x/8)$ in the range $0 < x < 4$, but the data input to UNAFEM represents it as piecewise constant.

Since the exact solution for a constant load is a quadratic polynomial, then a quadratic or higher-order element will produce that exact solution.

There is another lesson to be learned from this problem. If UNAFEM were to permit a more accurate definition of the interior load (by including, say, a sin ωx function rather

```
      SUBROUTINE ELEM13(OPT)
C
C         THIS SUBROUTINE IS THE MASTER ROUTINE FOR ELEMENT 13, A
C         4-NODE 1-D CO-CUBIC ELEMENT.
              .
              .
              .
C---------BRANCH TO FLUX CALCULATION IF OPT = 'FLUX'
C          FORM MATRICES IF OPT = 'FORM'
      IF (OPT .EQ. FLUX) GO TO 200
C
C*************FORM ELEMENT MATRICES AND VECTORS*************************
C
      YYY = 0.
      IPHY = NPHYS(IELNO)
      DO 20 I=1,4
          J = ICON(I,IELNO)
          XX(I) = X(J)     ◄──── x-coordinates of nodes
          FE(I) = 0.              ⎫
          DO 10 J=1,4            ⎬ Initialize to zero
              KE(J,I) = 0.       ⎭
                 .
                 .
                 .
  10      CONTINUE
  20  CONTINUE
      NINT = NGPINT(IELNO)
C---------EVALUATE TERMS IN ELEMENT EQUATIONS
      DO 60 INTX=1,NINT
          XI = GP(NINT,INTX)◄──── Select Gauss points for integration
          CALL SHAP13 ◄──────────Evaluate dφᵢ(e)/dx [= DPHDX(I)]
          XXX = 0.
          DO 30 I=1,4
              XXX = XXX+PHI(I)*XX(I)
  30      CONTINUE
          CALL PROP(IPHY,XXX,YYY,ALPHA,ALPHAY,BETA,GAMMA,MU)
          DO 50 I=1,4
              FE(I) = FE(I)+FINT(IELNO)*PHI(I)*WT(NINT,INTX)*JAC ◄── Eq. (8.78)
              DO 40 J=1,4
                  KE(I,J) = KE(I,J)+WT(NINT,INTX)*JAC*      ⎫
  $                           (DPHDX(I)*ALPHA*DPHDX(J)+     ⎬ Eqs. (8.76) and (8.77)
  $                           PHI(I)*BETA*PHI(J))            ⎭
                    .
                    .
                    .
  40          CONTINUE
  50      CONTINUE
  60  CONTINUE
C---------FOR SYMMETRY
      DO 100 I=1,4
          DO 90 J=I,4
              KE(J,I) = KE(I,J)
                 .
                 .
                 .
  90      CONTINUE
 100  CONTINUE
      RETURN
```

(a)

(Continued)

FIGURE 8.44(a) ● Subroutine ELEM13, a partial listing.

FIGURE 8.44(b) ● Continuation of subroutine ELEM13.

```
C
C*************CALCULATE FLUXES*****************************************
C
   200 CONTINUE
       YYY = 0.
       IPHY = NPHYS(IELNO)
       DO 220 I=1,4
           J = ICON(I,IELNO)
           XX(I) = X(J)
   220 CONTINUE
C--------CALCULATE FLUXES AT THE 3 GAUSS POINTS
       DO 260 INT=1,3
           XI = GP(3,INT)
           CALL SHAP13
           XXI(INT) = 0.
           DUDX = 0.
           DO 240 I=1,4
               J = ICON(I,IELNO)
               XXI(INT) = XXI(INT)+PHI(I)*XX(I)
               DUDX = DUDX+DPHDX(I)*A(J)
   240     CONTINUE
           XXX = XXI(INT)
           CALL PROP(IPHY,XXX,YYY,ALPHA,ALPHAY,BETA,GAMMA,MU)
           TAU(INT) = -ALPHA*DUDX
           XGP(INT,IELNO) = XXI(INT)
           YGP(INT,IELNO) = 0.
           TAUXGP(INT,IELNO) = TAU(INT)
           TAUYGP(INT,IELNO) = 0.
   260 CONTINUE
C--------CALCULATE ELEMENT NODAL FLUXES BY INTERPOLATING THE GAUSS
C            POINT FLUXES AND SUM FOR NODAL FLUX CALCULATION IN CALFLX
       XI12 = GP(3,1)-GP(3,2)
       XI13 = GP(3,1)-GP(3,3)
       XI21 = -XI12
       XI23 = GP(3,2)-GP(3,3)
       XI31 = -XI13
       XI32 = -XI23
       DO 300 I=1,4
           JNOD = ICON(I,IELNO)
           SII1 = SI(I)-GP(3,1)
           SII2 = SI(I)-GP(3,2)
           SII3 = SI(I)-GP(3,3)
           TAUXNP(JNOD) = TAUXNP(JNOD) + SII2/XI12*SII3/XI13*TAU(1)+
      $                                  SII1/XI21*SII3/XI23*TAU(2)+
      $                                  SII1/XI31*SII2/XI32*TAU(3)
           IU(JNOD) = IU(JNOD) + 1
   300 CONTINUE
       RETURN
C-------------------------------------------------------------------
       END
```

Eq. (8.81) (applies to the block from XXX = XXI(INT) through CALL PROP(...))

Eq. (8.82) (applies to the TAUXNP(JNOD) summation block)

(b)

than only constants), then the solutions using the quadratic and cubic elements (and quartic elements in Section 8.5) would not agree: The cubic elements would yield a more accurate solution, and the quartic elements an even more accurate one. Thus the low-accuracy definition of the interior load is dragging down the accuracy of the higher-order elements. The principal advantage of higher-order elements is that fewer of them need to be used to achieve a given accuracy. But fewer elements mean bigger elements, and in some problems the model data (physical properties, loads, and/or geometry) can vary appreciably over larger regions. Therefore in order to realize the full accuracy inherent in higher-order elements, for any kind of model data, a general-purpose program ought to also provide for more accurate definition of the model data.

```
              SUBROUTINE SHAP13
C
C              THIS ROUTINE EVALUATES SHAPE FUNCTIONS AT XI FOR THE 4-NODE
C                    1-D ELEMENT
                     .
                     .
                     .
C**********SHAPE FUNCTIONS*********************************************
C
          PHI(1) =  -9./16.*              (XI+1./3.)*(XI-1./3.)*(XI-1.)
          PHI(2) =  27./16.*(XI+1.)*              (XI-1./3.)*(XI-1.)
          PHI(3) = -27./16.*(XI+1.)*(XI+1./3.)*              (XI-1.)
          PHI(4) =   9./16.*(XI+1.)*(XI+1./3.)*(XI-1./3.)
C
C**********DERIVATIVES OF THE SHAPE FUNCTIONS************************
C
C----------DERIVATIVES WRT XI AT XI
          DPHDXI(1) =  -9./16.*(3.*XI*XI-2.*XI-1./9.)
          DPHDXI(2) =  27./16.*(3.*XI*XI-2./3.*XI-1.)
          DPHDXI(3) = -27./16.*(3.*XI*XI+2./3.*XI-1.)
          DPHDXI(4) =   9./16.*(3.*XI*XI+2.*XI-1./9.)
C----------DO JACOBIAN CALCULATIONS
          CALL JACOBI(1)
C----------DERIVATIVES WRT X AT XI
          DO 20 I=1,4
              DPHDX(I) = JACINV(1,1)*DPHDXI(I)
   20 CONTINUE
          RETURN
C----------------------------------------------------------------------
          END
```

Shape functions braces: } Fig (8.37)

Derivatives braces: } Eqs. (8.63)

DPHDX line arrow: ←———— Eq. (8.40)

FIGURE 8.45 ● Subroutine SHAP13.

	HEAT CONDUCTION - 3 CUBIC ELEMENTS							
IDIMBDVP								
NODE								
	1	1	0.					
	4	1	40.					
	7	1	60.					
	10		100.					
ELEM								
	1	13	3	1	1	2	3	4
	2	13	3	2	4	5	6	7
	3	13	3	1	7	8	9	10
PROP								
	1	.12						
		1.5E-4						
	2	.92						
		1.5E-4						
ILC								
	1	3.0E-3	3					
EBC								
	10	0.						
NBCC								
	1	.1						
PLOT								
	10	0.	100.					
	10	0.	50.					
	10	0.	.1					
END								

FIGURE 8.46 ● UNAFEM input data for the heat conduction problem, using three cubic elements.

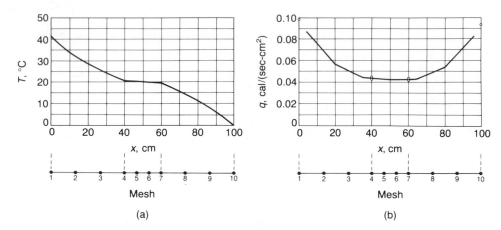

FIGURE 8.47 ● UNAFEM graphical output for the heat conduction problem, using three cubic elements. (a) Function: temperature $\tilde{T}(x)$. Approximate-solution polynomial $\tilde{T}^{(e)}(x)$ plotted within each element. (b) Flux: heat flux $\tilde{q}(x)$ flowing to the right ($\tilde{q} = -kd\tilde{T}/dx$). Straight lines connect successive Gauss-point values; circles are element nodal values, interpolated from Gauss-point values.

This suggests the following caution: When using higher-order elements in a problem where the model data is variable, do not expect a significant improvement in accuracy if the program does not permit a correspondingly greater accuracy in the definition of the data. This caution clearly applies to UNAFEM. Thus the variable load region in the cable problem, $0<x<4$, is more accurately modeled with several linear elements than with one high-order element (using the same number of DOF), since the former defines the load much more accurately (see Figs. 7.52 and 8.55, the latter in Section 8.5).

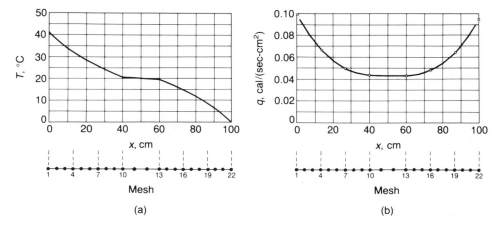

FIGURE 8.48 ● UNAFEM graphical output for the heat conduction problem, using seven cubic elements. (a) Function: temperatures $\tilde{T}(x)$. Approximate-solution polynomial $\tilde{T}^{(e)}(x)$ plotted within each element. (b) Flux: heat flux $\tilde{q}(x)$ flowing to the right ($\tilde{q} = -kd\tilde{T}/dx$). Straight lines connect successive Gauss-point values; circles are element nodal values, interpolated from Gauss-point values.

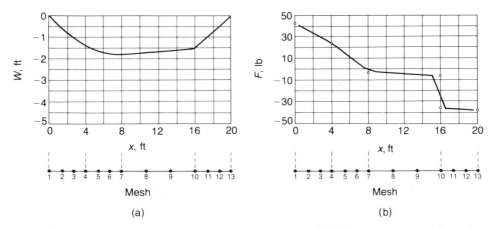

FIGURE 8.49 ● UNAFEM graphical output for the cable deflection problem, using four cubic elements. (a) Function: vertical displacement $\tilde{W}(x)$. Approximate-solution polynomial $\tilde{W}^{(e)}(x)$ plotted within each element. (b) Flux: vertical force $\tilde{F}(x)$ acting on the section to the right of x ($\tilde{F} = -Td\tilde{W}/dx$). Straight lines connect successive Gauss-point values; circles are element nodal values, interpolated from Gauss-point values.

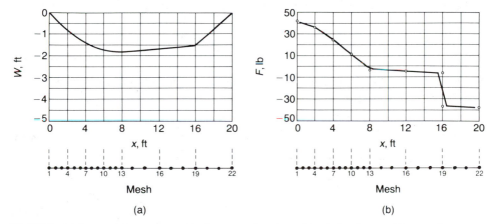

FIGURE 8.50 ● UNAFEM graphical output for the cable deflection problem, using seven cubic elements. (a) Function: vertical displacement $\tilde{W}(x)$. Approximate-solution polynomial $\tilde{W}^{(e)}(x)$ plotted within each element. (b) Flux: vertical force $\tilde{F}(x)$ acting on the section to the right of x ($\tilde{F} = -Td\tilde{W}/dx$). Straight lines connect successive Gauss-point values; circles are element nodal values, interpolated from Gauss-point values.

Number of cubic elements (and DOF) in mesh	$W(8)$	$\|F(0)\|$	$\|F(20)\|$	$\|F(0)\| + \|F(20)\|$	$\dfrac{\|\Sigma f\| - (\|F(0)\| + \|F(20)\|)}{\|\Sigma f\|}$
4 (13)	−1.788	43.07	37.90	80.97	−2.13%
7 (22)	−1.811	41.59	38.09	79.68	−0.50%

$(\|\Sigma f\| = 79.28)$

FIGURE 8.51 ● Vertical displacement at $x=8$ ft, vertical reaction forces at $x=0$ and 20 ft, and error relative to applied forces.

It should be noted that the source of these problems can be traced to the manner in which the load data is defined to the program; i.e., it is a preprocessing feature. In UNAFEM (as well as in some of the commercial FE programs) the loads are defined on an element-by-element basis. Therefore as the mesh is refined the load tends to be refined with it. The alternative is to define the load the same way we did the physical properties — i.e., over a given region, independent of the elements. Then the load may be defined once and for all, as accurately as desired, and subsequent meshes would all use the same load definition. This is a preferred approach and is used in many present-day preprocessor programs.

8.5 The 1-D C^0-Quartic Isoparametric Element

8.5.1 Theoretical Development

The development parallels the C^0-quadratic and C^0-cubic elements, so only the results will be summarized here.

Step 4: Develop specific expressions for the shape functions.

In the parent element the trial solution has the standard form,

$$\tilde{U}(\xi;a) = \sum_{j=1}^{5} a_j \phi_j(\xi)$$

(8.83)

where $\phi_j(\xi)$ are fourth-degree (quartic) Lagrange interpolation polynomials [see Eqs. (8.11)],

$$\phi_j(\xi) = \frac{(\xi-\xi_1)(\xi-\xi_2)\ldots(\xi-\xi_{j-1})(\xi-\xi_{j+1})\ldots(\xi-\xi_5)}{(\xi_j-\xi_1)(\xi_j-\xi_2)\ldots(\xi_j-\xi_{j-1})(\xi_j-\xi_{j+1})\ldots(\xi_j-\xi_5)} \quad j = 1,2,3,4,5$$

(8.84)

The nodes ξ_j are equally spaced,

$$\xi_1 = -1$$

$$\xi_2 = -\frac{1}{2}$$

$$\xi_3 = 0$$

$$\xi_4 = +\frac{1}{2}$$

$$\xi_5 = +1$$

(8.85)

Inserting the values from Eqs. (8.85) into Eqs. (8.84) yields the five parent shape functions.

Figure 8.52 shows the shape functions; in addition, it provides a convenient graphical summary of the interrelationship of the shape functions and the element trial solution for the 1-D C^0-quartic isoparametric element.

The isoparametric coordinate transformation (mapping) is

$$x = \sum_{k=1}^{5} x_k^{(e)} \phi_k(\xi)$$

(8.86)

The element trial solution in the parent element, $\tilde{U}(\xi; a)$, is a complete quartic polynomial:

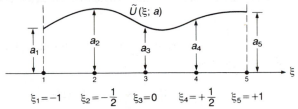

$$\xi_1 = -1 \qquad \xi_2 = -\frac{1}{2} \qquad \xi_3 = 0 \qquad \xi_4 = +\frac{1}{2} \qquad \xi_5 = +1$$

It is represented as the sum of 5 shape functions:

$$\tilde{U}(\xi; a) = \sum_{j=1}^{5} a_j \, \phi_j(\xi)$$

Each coefficient a_j is the value of $\tilde{U}(\xi; a)$ at the node at ξ_j.
Each parent shape function, $\phi_j(\xi)$, is itself a quartic polynomial satisfying the interpolation property, $\phi_j(\xi_i) = \delta_{ji}$:

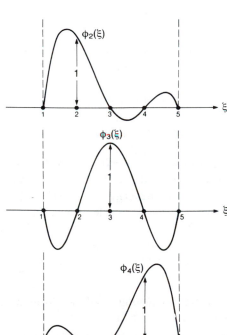

$$\phi_1(\xi) = \frac{(\xi-\xi_2)(\xi-\xi_3)(\xi-\xi_4)(\xi-\xi_5)}{(\xi_1-\xi_2)(\xi_1-\xi_3)(\xi_1-\xi_4)(\xi_1-\xi_5)}$$

$$= \frac{2}{3}\left(\xi+\frac{1}{2}\right)\xi\left(\xi-\frac{1}{2}\right)(\xi-1)$$

$$\phi_2(\xi) = \frac{(\xi-\xi_1)(\xi-\xi_3)(\xi-\xi_4)(\xi-\xi_5)}{(\xi_2-\xi_1)(\xi_2-\xi_3)(\xi_2-\xi_4)(\xi_2-\xi_5)}$$

$$= -\frac{8}{3}(\xi+1)\xi\left(\xi-\frac{1}{2}\right)(\xi-1)$$

$$\phi_3(\xi) = \frac{(\xi-\xi_1)(\xi-\xi_2)(\xi-\xi_4)(\xi-\xi_5)}{(\xi_3-\xi_1)(\xi_3-\xi_2)(\xi_3-\xi_4)(\xi_3-\xi_5)}$$

$$= 4(\xi+1)\left(\xi+\frac{1}{2}\right)\left(\xi-\frac{1}{2}\right)(\xi-1)$$

$$\phi_4(\xi) = \frac{(\xi-\xi_1)(\xi-\xi_2)(\xi-\xi_3)(\xi-\xi_5)}{(\xi_4-\xi_1)(\xi_4-\xi_2)(\xi_4-\xi_3)(\xi_4-\xi_5)}$$

$$= -\frac{8}{3}(\xi+1)\left(\xi+\frac{1}{2}\right)\xi(\xi-1)$$

$$\phi_5(\xi) = \frac{(\xi-\xi_1)(\xi-\xi_2)(\xi-\xi_3)(\xi-\xi_4)}{(\xi_5-\xi_1)(\xi_5-\xi_2)(\xi_5-\xi_3)(\xi_5-\xi_4)}$$

$$= \frac{2}{3}(\xi+1)\left(\xi+\frac{1}{2}\right)\xi\left(\xi-\frac{1}{2}\right)$$

FIGURE 8.52 ● The 1-D C^0-quartic isoparametric element.

where $x_k^{(e)}$, $k=1,2,\ldots,5$, are the coordinates of the five nodes in the (e)th real element. The Jacobian of this mapping is

$$J^{(e)}(\xi) = \frac{dx}{d\xi} = \sum_{k=1}^{5} x_k^{(e)} \frac{d\phi_k(\xi)}{d\xi}$$

(8.87)

where the derivatives of the shape functions are calculated from Fig. 8.52,

$$\frac{d\phi_1(\xi)}{d\xi} = \frac{2}{3}\left(4\xi^3 - 3\xi^2 - \frac{1}{2}\xi + \frac{1}{4}\right)$$

$$\frac{d\phi_2(\xi)}{d\xi} = -\frac{8}{3}\left(4\xi^3 - \frac{3}{2}\xi^2 - 2\xi + \frac{1}{2}\right)$$

$$\frac{d\phi_3(\xi)}{d\xi} = 4\left(4\xi^3 - \frac{5}{2}\xi\right)$$

$$\frac{d\phi_4(\xi)}{d\xi} = -\frac{8}{3}\left(4\xi^3 + \frac{3}{2}\xi^2 - 2\xi - \frac{1}{2}\right)$$

$$\frac{d\phi_5(\xi)}{d\xi} = \frac{2}{3}\left(4\xi^3 + 3\xi^2 - \frac{1}{2}\xi - \frac{1}{4}\right)$$

(8.88)

For no mapping distortion, the three middle nodes of the real element must be at the 1/4, 1/2, and 3/4 points, in which case Eq. (8.86) reduces to

$$x = x_c^{(e)} + \frac{L}{2}\xi \qquad \text{(no mapping distortion)}$$

(8.89)

where $x_c^{(e)} = (x_1^{(e)} + x_5^{(e)})/2$ is the center of the element. Equation (8.87) reduces to $J^{(e)}(\xi) = L/2$.

The conditions for unacceptable mapping distortion, $J^{(e)}(\xi) = 0$, are a bit complicated since there are three interior nodes that can be independently located. The curious reader may wish to derive the conditions for various symmetric patterns of these nodes. In addition to, or instead of, this approach one could do several one-element problems with UNAFEM to determine approximately how far the nodes can move from the 1/4, 1/2, and 3/4 positions before the program indicates $J^{(e)}(\xi) \leq 0$.

Assembly follows the same pattern as the other 1-D elements; recall the general assembly rule (Section 5.7).

Step 5: Substitute the shape functions into the element equations, and transform the integrals into a form appropriate for numerical evaluation.

The quadrature expressions are the same as for the quadratic elements, Eqs. (8.50), and the cubic elements, Eqs. (8.76) through (8.78). The appropriate quadrature rule for reduced integration is determined from guideline II: For the C^0-quartic element $p=4$ and $m=1$, so $p-m+1=4$ and hence a four-point rule is required. From guideline III higher-order integration requires at least a five-point rule.

As with the previous elements, we note that if $f^{(e)}(\xi)$ is a constant, then $Ff_i^{(e)}$ is integrated exactly by reduced integration. If $\alpha^{(e)}(\xi)$, $\beta^{(e)}(\xi)$, and $J^{(e)}(\xi)$ are constants, then $K\alpha_{ij}^{(e)}$ is integrated exactly by reduced, but $K\beta_{ij}^{(e)}$ is integrated exactly only by higher-order.

Step 6: Prepare expression for the flux.

The optimal flux-sampling points are the Gauss points for the four-point rule. The smoothing curve is a cubic Lagrange interpolation polynomial through the four Gauss-point values.

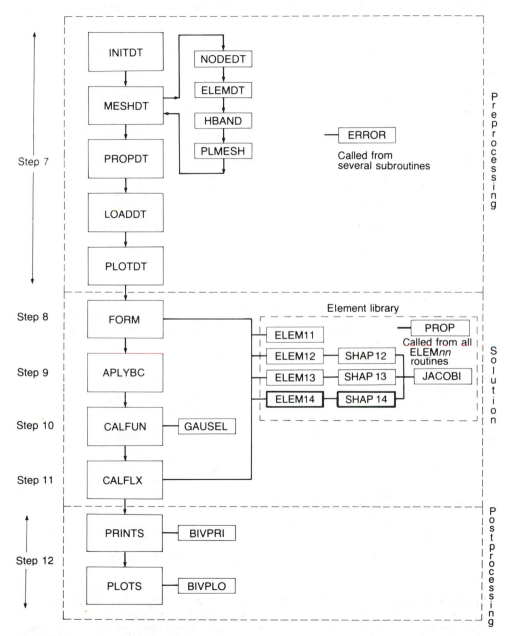

FIGURE 8.53 ● Subroutines (indicated by bold outline) added to UNAFEM to perform analysis for the 1-D C⁰-quartic isoparametric element.

8.5.2 Implementation into the UNAFEM Program

The element equations for the C^0-quartic isoparametric element have been coded and added to UNAFEM as subroutines ELEM14 and SHAP14 (see appendix). Figure 8.53 shows the flowchart for the improved UNAFEM program. The element library now contains four elements, enabling a mesh to be constructed from any combination of the four types of elements.

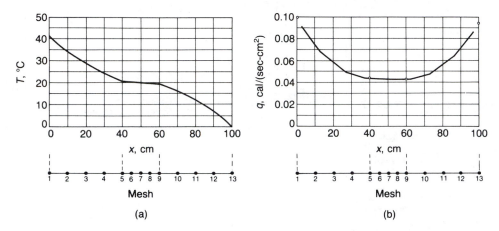

FIGURE 8.54 ● UNAFEM graphical output for the heat conduction problem, using three quartic elements. (a) Function: temperature $\tilde{T}(x)$. Approximate-solution polynomial $\tilde{T}^{(e)}(x)$ plotted within each element. (b) Flux: heat flux $\tilde{q}(x)$ flowing to the right ($\tilde{q} = -k\,d\tilde{T}/dx$). Straight lines connect successive Gauss-point values; circles are element nodal values, interpolated from Gauss-point values.

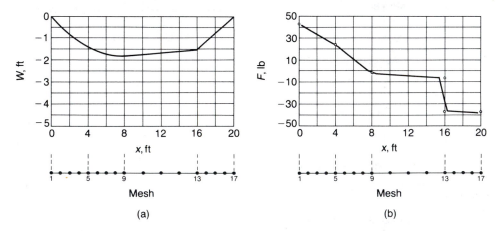

FIGURE 8.55 ● UNAFEM graphical output for the cable deflection problem, using four quartic elements. (a) Function: vertical displacement $\tilde{W}(x)$. Approximate-solution polynomial $\tilde{W}^{(e)}(x)$ plotted within each element. (b) Flux: vertical force $\tilde{F}(x)$ acting on the section to the right of x ($\tilde{F} = -T\,d\tilde{W}/dx$). Straight lines connect successive Gauss-point values; circles are element nodal values, interpolated from Gauss-point values.

Number of quartic elements (and DOF) in mesh	$W(8)$	$\|F(0)\|$	$\|F(20)\|$	$\|F(0)\|+\|F(20)\|$	$\dfrac{\|\Sigma f\|-(\|F(0)\|+\|F(20)\|)}{\|\Sigma f\|}$
4 (17)	-1.788	43.07	37.90	80.97	-2.13%

$$(\|\Sigma f\|=79.28)$$

FIGURE 8.56 ● Vertical displacement at $x=8$ ft, vertical reaction forces at $x=0$ and 20 ft, and error relative to applied forces.

A comparison of routines ELEM12, ELEM13, and ELEM14 reveals that they are all almost identical. The principal differences are merely the calls to different SHAPnn routines and the limits of DO loops, reflecting the different number of Gauss points in each element. Therefore, for more efficient coding, all three elements could have been easily incorporated into a single routine. However, this would have decreased the readability of the listing, especially for the reader who is studying the quadratic element for the first time and who is not yet aware of the treatment for the cubic and quartic elements. Since the purpose of this book is to teach finite element analysis, rather than to demonstrate tightly efficient computer code, the element routines were intentionally kept separate.

8.5.3 Application of UNAFEM

The new C^0-quartic element is applied to the previous heat conduction and cable deflection problems.

The heat-conduction problem. Figure 8.54 shows the graphical results for a three-element mesh using undistorted elements.

The cable-deflection problem. Figure 8.55 shows the graphical results for a four-element mesh using undistorted elements. Forces and deflections at particular points of the cable are shown in Fig. 8.56. The results are identical to the cubic and quadratic element analyses (see discussion accompanying Fig. 8.51).

In this and the previous chapter we have done several analyses of the same two problems, using four different types of elements. We have seen that the higher-order elements generally provide greater accuracy, so fewer of them are needed to achieve a given accuracy. Chapter 9 will examine in more detail the relative merits of these elements and will offer general guidelines for using higher-order elements.

8.6 The 1-D C¹-Cubic Element

The series of four C^0-continuous elements that we have developed so far could be continued indefinitely with the quintic, sextic, etc., up to any degree polynomial desired. As the degree increases much beyond the quartic, however, the practical value of the element begins to decrease because of the increased complexity and size of the element equations. Thus, since element stiffness matrices are fully populated (no bandedness), the number of element stiffness terms (to derive, and later compute) increases approximately as the square of the number of DOF in the element; i.e., for n DOF, a symmetric matrix has $n(n+1)/2$ independent terms. The *reductio ad absurdum* would be a single superhigh-order element for the entire problem; this would imply a global trial solution (as in Chapter 4), yielding a fully populated system stiffness matrix — clearly no longer the FEM.

The governing differential equation used in Chapters 6 through 8 [Eq. (7.1a)] is second order. According to the convergence conditions in Section 5.6, the minimum requirements placed on the element trial solution to ensure convergence for this equation (for which $m=1$) are (1) a complete polynomial of degree 1 and (2) C^0-continuity at interelement boundaries. This chapter has focused on the completeness condition by increasing the degree of the polynomial beyond the minimal first degree, while maintaining the minimal C^0-continuity. Now we want to consider an element that exceeds the minimal C^0-continuity condition, i.e., an element with C^1-continuity.

Let's derive the shape functions for a C^1-continuous element using the direct approach, as used in Section 8.2 for the linear element, i.e., no parametric mapping. C^1-continuity is enforced the same way as is C^0-continuity (using interpolation polynomials), only now we interpolate both the function and its derivative at each boundary node of the element. This requires two unknown parameters at each boundary node, hence a total of four for a 1-D element. Four parameters requires a cubic polynomial, which is therefore the lowest-degree polynomial capable of achieving C^1-continuity in a 1-D element. (More nodes could be placed inside the element if a higher-order C^1-continuous element were desired.)

The element trial solution takes the standard form,

$$\tilde{U}^{(e)}(x;a) = \sum_{j=1}^{4} a_j \phi_j^{(e)}(x)$$

$$(8.90)$$

but now only two of the parameters (a_1 and a_3) represent the value of $\tilde{U}^{(e)}$ at the nodes, and the other two parameters (a_2 and a_4) represent the value of $d\tilde{U}^{(e)}/dx$ at the nodes, as shown in Fig. 8.57.

The shape functions and their derivatives satisfy the following interpolation property:

$$\phi_1^{(e)}(x_1) = 1 \qquad \phi_2^{(e)}(x_1) = 0 \qquad \phi_3^{(e)}(x_1) = 0 \qquad \phi_4^{(e)}(x_1) = 0$$

$$\frac{d\phi_1^{(e)}(x_1)}{dx} = 0 \qquad \frac{d\phi_2^{(e)}(x_1)}{dx} = 1 \qquad \frac{d\phi_3^{(e)}(x_1)}{dx} = 0 \qquad \frac{d\phi_4^{(e)}(x_1)}{dx} = 0$$

$$\phi_1^{(e)}(x_2) = 0 \qquad \phi_2^{(e)}(x_2) = 0 \qquad \phi_3^{(e)}(x_2) = 1 \qquad \phi_4^{(e)}(x_2) = 0$$

$$\frac{d\phi_1^{(e)}(x_2)}{dx} = 0 \qquad \frac{d\phi_2^{(e)}(x_2)}{dx} = 0 \qquad \frac{d\phi_3^{(e)}(x_2)}{dx} = 0 \qquad \frac{d\phi_4^{(e)}(x_2)}{dx} = 1 \qquad (8.91)$$

From Eqs. (8.90) and (8.91),

$$\tilde{U}^{(e)}(x_1;a) = a_1$$

$$\frac{d\tilde{U}^{(e)}(x_1;a)}{dx} = a_2$$

$$\tilde{U}^{(e)}(x_2;a) = a_3$$

$$\frac{d\tilde{U}^{(e)}(x_2;a)}{dx} = a_4$$

$$(8.92)$$

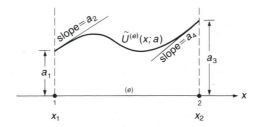

FIGURE 8.57 ● Element trial solution for the 1-D C^1-cubic element.

The shape functions are shown in Fig. 8.58 (see Exercise 8.8). They are called *Hermite interpolation polynomials*, and Eqs. (8.91) are the Hermite interpolation conditions for a cubic polynomial. The technique can clearly be extended to C^2-, C^3-, ..., continuity. Thus, a C^2-continuous element would interpolate $\tilde{U}$, $d\tilde{U}/dx$, and $d^2\tilde{U}/dx^2$ at both nodes, requiring six parameters and therefore a fifth-degree polynomial.

Assembly of two adjacent elements (see Fig. 8.59) follows the general assembly rule (Section 5.7). It is instructive, though, to once again rederive the rule by returning to basic principles and applying the essential IBCs (here, continuity of $\tilde{U}$ and $d\tilde{U}/dx$) to the adjacent element equations at the interelement boundary (recall Section 5.3.3, step 8). Thus, using the notation shown in Fig. 8.59, we first write down the trial solutions in each element.

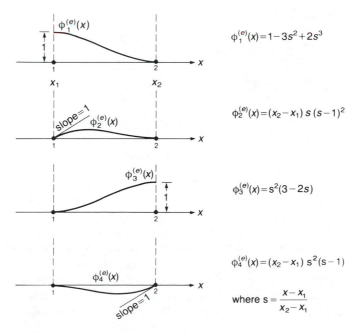

$$\phi_1^{(e)}(x) = 1 - 3s^2 + 2s^3$$

$$\phi_2^{(e)}(x) = (x_2 - x_1)\, s\, (s - 1)^2$$

$$\phi_3^{(e)}(x) = s^2(3 - 2s)$$

$$\phi_4^{(e)}(x) = (x_2 - x_1)\, s^2(s - 1)$$

where $s = \dfrac{x - x_1}{x_2 - x_1}$

FIGURE 8.58 ● Shape functions for the 1-D C^1-cubic element.

Element (1) •

$$\tilde{U}^{(1)}(x;a) = a_1\phi_1^{(1)}(x) + a_2\phi_2^{(1)}(x) + a_3^{(1)}\phi_3^{(1)}(x) + a_4^{(1)}\phi_4^{(1)}(x) \qquad (8.93)$$

Element (2) •

$$\tilde{U}^{(2)}(x;a) = a_3^{(2)}\phi_3^{(2)}(x) + a_4^{(2)}\phi_4^{(2)}(x) + a_5\phi_5^{(2)}(x) + a_6\phi_6^{(2)}(x) \qquad (8.94)$$

Enforcing continuity of $\tilde{U}$,

$$\tilde{U}^{(1)}(x_2;a) = \tilde{U}^{(2)}(x_2;a) \qquad (8.95)$$

which from Eqs. (8.90) and (8.91) yields the linear constraint equation

$$a_3^{(1)} = a_3^{(2)} \qquad (8.96)$$

Enforcing continuity of $d\tilde{U}/dx$,

$$\frac{d\tilde{U}^{(1)}(x_2;a)}{dx} = \frac{d\tilde{U}^{(2)}(x_2;a)}{dx} \qquad (8.97)$$

which yields the linear constraint equation[26]

$$a_4^{(1)} = a_4^{(2)} \qquad (8.98)$$

Applying the linear constraint equations — Eqs. (8.96) and (8.98) — to the four element equations that would be generated for each element (see the beam example below) would yield the following assembled system equations for the two-element mesh:

$$
\begin{bmatrix}
K_{11}^{(1)} & K_{12}^{(1)} & K_{13}^{(1)} & K_{14}^{(1)} & & & \\
K_{21}^{(1)} & K_{22}^{(1)} & K_{23}^{(1)} & K_{24}^{(1)} & & & \\
K_{31}^{(1)} & K_{32}^{(1)} & K_{33}^{(1)} + K_{33}^{(2)} & K_{34}^{(1)} + K_{34}^{(2)} & K_{35}^{(2)} & K_{36}^{(2)} \\
K_{41}^{(1)} & K_{42}^{(1)} & K_{43}^{(1)} + K_{43}^{(2)} & K_{44}^{(1)} + K_{44}^{(2)} & K_{45}^{(2)} & K_{46}^{(2)} \\
& & K_{53}^{(2)} & K_{54}^{(2)} & K_{55}^{(2)} & K_{56}^{(2)} \\
& & K_{63}^{(2)} & K_{64}^{(2)} & K_{65}^{(2)} & K_{66}^{(2)}
\end{bmatrix}
\begin{Bmatrix}
a_1 \\ a_2 \\ a_3 \\ a_4 \\ a_5 \\ a_6
\end{Bmatrix}
=
\begin{Bmatrix}
F_1^{(1)} \\ F_2^{(1)} \\ F_3^{(1)} + F_3^{(2)} \\ F_4^{(1)} + F_4^{(2)} \\ F_5^{(2)} \\ F_6^{(2)}
\end{Bmatrix}
\qquad (8.99)
$$

where $a_3^{(1)}$ and $a_3^{(2)}$ are replaced by the simpler notation, a_3; similarly, $a_4^{(1)}$ and $a_4^{(2)}$ are replaced by a_4. Assembly of 1-D C^1-continuous elements creates a 2-by-2 overlap of the element stiffness matrices, in contrast to only a 1-by-1 (one term) overlap for C^0-continuous elements. (C^2-continuous elements would overlap 3-by-3, etc.)

Figure 8.60 shows the assembled trial solution and shape functions for a three-element mesh. Note the C^1-continuity of all assembled shape functions.

The C^1-continuous trial solution looks attractive because of its smoothness, a property the C^0-continuous solutions lack. Because of this smoothness in the first derivative (which is proportional to flux), it might seem that the C^1-cubic element would yield more

---

[26] Recall Eq. (5.60) and accompanying discussion.

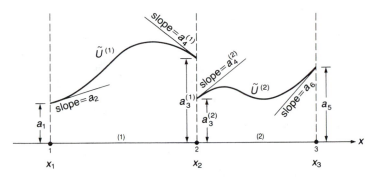

FIGURE 8.59 ● Element trial solutions in two adjacent C^1-cubic elements before assembly.

accurate flux values than the C^0-cubic element, and therefore would be preferred. However, this is not the case. The problem is this: Even though the C^1-cubic flux solution is continuous at the interelement boundary nodes, the optimal flux-sampling points (at which the flux is superconvergent) are still the same Gauss points as for the C^0-cubic for a second-order differential equation. In other words, after going to the trouble to establish continuity in the flux (first derivative) at the interelement boundaries, we end up ignoring the flux at these points.[27]

For this reason, it does not appear that C^1-continuous elements enjoy any advantage over C^0-continuous elements for second-order differential equations (that have been integrated by parts once). And, in general, it would appear that for all linear, self-adjoint governing equations (for which the optimal sampling-point theory applies), there is no advantage to enforcing continuity any higher than the minimum that is sufficient for convergence.

From these remarks it follows that C^1-continuous elements are better suited for problems governed by fourth-order differential equations (since $2m=4$, yielding $m=2$, and therefore C^{m-1}-continuity is C^1-continuity). Fourth-order equations occur in all the beam, plate, and shell theories in solid mechanics. As an example, let's derive the element equations for a 1-D beam, using the above C^1-cubic shape functions.

Figure 8.61 shows an elastic beam with a distributed lateral load.[28] The governing differential equation for the transverse deflection of the beam is [8.18]

$$\frac{d^2}{dx^2}\left(EI\,\frac{d^2W(x)}{dx^2}\right) = f(x) \tag{8.100}$$

---

[27] It might also be noted that a C^1-continuous element would be inappropriate wherever the first derivative *should* be discontinuous, such as at the interface between dissimilar materials. (Recall the guideline given in Section 7.4.3.)

[28] Observe that Fig. 8.61 for a beam is identical to Fig. 6.5(a) for a flexible cable. This illustrates that a particular *real* structure (typically long and slender) can be *modeled* as either a beam or a cable. The choice will depend on how we expect the structure to deform, which depends in turn on the loads and material properties. The beam theory includes bending stresses due to a loading-induced change in curvature of the beam's axis and assumes that any uniform axial tensile stresses (due to axial loads) are negligible relative to the bending stresses. The cable theory reverses these assumptions; i.e., it considers only uniform axial tensile stresses and ignores any bending stresses (recall Section 6.3).

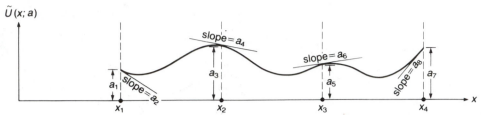

This 3 element trial solution is a sum of 8 trial functions,

$$\tilde{U}(x; a) = \sum_{j=1}^{8} a_j \phi_j(x)$$

Each trial function consists of one shape function or a sum (assembly) of two shape functions:

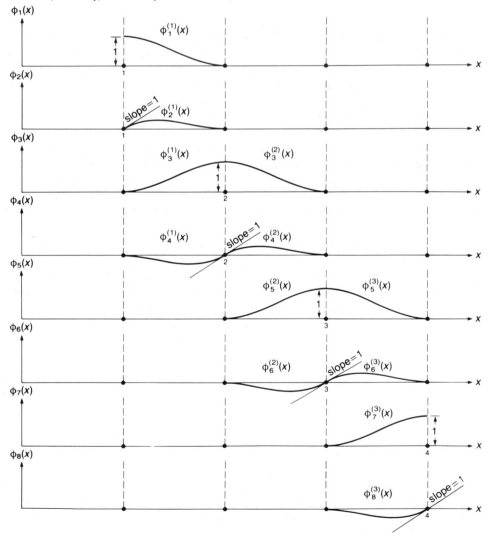

FIGURE 8.60 ● Assembled shape functions (trial functions) for a three-element mesh of 1-D C^1-cubic elements.

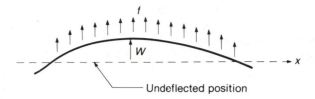

FIGURE 8.61 ● Transverse deflection of a beam due to a lateral load.

where $W(x)$ is the transverse displacement (deflection), E is Young's modulus, I is the moment of inertia of the beam's cross section, and $f(x)$ is the lateral load.[29] At each end two BCs must be specified. [Recall Table 4.2 (Section 4.5) for essential and natural BCs for fourth-order equations.] Most practical applications require any two of the following four:

Essential BCs ●

$$W = W_0 \qquad \text{displacement}$$

$$\frac{dW}{dx} = S_0 \qquad \text{slope} \tag{8.101}$$

Natural BCs ●

$$EI\frac{d^2W}{dx^2} = -M_0 \qquad \text{bending moment}$$

$$\frac{d}{dx}\left(EI\frac{d^2W}{dx^2}\right) = -V_0 \qquad \text{transverse shear force}$$

The element equations may be derived using steps 1 through 5 of the 12-step procedure (see Section 5.9). Thus writing the Galerkin residual equations for Eq. (8.100), integrating by parts *twice* (recall Section 4.5), and substituting the element trial solutions from Eq. (8.90) yields the following formal expression for the element equations:

$$[K]^{(e)}\{a\} = \{F\}^{(e)} \tag{8.102}$$

where

$$K_{ij}^{(e)} = \int_{x_1}^{x_2} \frac{d^2\phi_i^{(e)}(x)}{dx^2}\, EI\, \frac{d^2\phi_j^{(e)}(x)}{dx^2}\, dx$$

$$F_i^{(e)} = \int_{x_1}^{x_2} f(x)\phi_i^{(e)}(x)\, dx + \left[V_0\phi_i^{(e)}(x)\right]_{x_1}^{x_2} - \left[M_0\frac{d\phi_i^{(e)}(x)}{dx}\right]_{x_1}^{x_2}$$

and x_1, x_2 are the coordinates of the two end nodes.

---

[29] Equation (8.100) assumes no shearing deformation, no axial extension, small deflection, and a principal axis coinciding with the plane of bending.

The shape functions from Fig. 8.58 may next be substituted into the integrals in Eqs. (8.102). If we assume E, I, and f to be constant within an element (say the value at the center of the element), then the integration can be carried out exactly in closed form, yielding the following specific expressions for the terms in the element equations (see Exercise 8.9):

$$\frac{EI}{L}\begin{bmatrix} \dfrac{12}{L^2} & \dfrac{6}{L} & -\dfrac{12}{L^2} & \dfrac{6}{L} \\[2mm] & 4 & -\dfrac{6}{L} & 2 \\[2mm] & & \dfrac{12}{L^2} & \dfrac{6}{L} \\[2mm] \text{Symmetric} & & & 4 \end{bmatrix}\begin{Bmatrix} a_1 \\[2mm] a_2 \\[2mm] a_3 \\[2mm] a_4 \end{Bmatrix} = \begin{Bmatrix} \dfrac{fL}{2} - V_0(x_1) \\[2mm] \dfrac{fL^2}{12} + M_0(x_1) \\[2mm] \dfrac{fL}{2} + V_0(x_2) \\[2mm] -\dfrac{fL^2}{12} - M_0(x_2) \end{Bmatrix} \tag{8.103}$$

where $L = x_2 - x_1$. The reader trained in solid mechanics will appreciate the physical meaning in each of these terms. In fact, Eq. (8.103) could have been derived from engineering strength-of-materials relations, and, indeed, in the early days of the FEM, element stiffness and load matrices were frequently derived using such direct physical approaches [8.19].

For 1-D elements it is straightforward to apply Hermite interpolation for any level of continuity — that is, C^1, C^2, C^3, etc. However, in 2-D and 3-D (e.g., plate and shell models) it is quite difficult to obtain even C^1-continuity, so often the continuity is relaxed to only C^0-continuity along part of the element boundaries. As noted in Section 5.6, such elements are called *nonconforming*, or *incompatible*, because they do not satisfy the continuity condition. However, another convergence condition, called the *patch test*, is available [8.20-8.22]. There has been some controversy regarding the necessity and/or sufficiency of this test [8.23-8.25]. (Nonconforming elements have also been used for second-order differential equations, resulting in discontinuity in $\tilde{U}$ between elements [8.26].)

Nonconforming elements sometimes yield significantly better accuracy than conforming elements because relaxing the continuity constraint tends to ''soften'' the overall model; this may partially compensate for the ''stiffening'' inherent in the FE discretization (caused by using a finite, rather than infinite, number of DOF). Of course the softening may sometimes overcompensate too much. The behavior of nonconforming elements seems to depend on element type and problem type. Guidelines are presently more empirical than theoretical so one should be more cautious when using nonconforming elements.

Exercises

8.1 a) Solve Eqs. (8.8) for α_1, α_2, and α_3 and combine terms to yield Eq. (8.9) for $\phi_1^{(e)}(x)$.

   b) Set up the equations for $\phi_2^{(e)}(x)$ and $\phi_3^{(e)}(x)$, analogous to Eqs. (8.7) and (8.8) for $\phi_1^{(e)}(x)$, and solve to yield Eqs. (8.10).

8.2 As an alternative to the approach used in Exercise 8.1, the shape functions can also be derived by writing the element trial solution as a quadratic polynomial, $\tilde{U}^{(e)}(x;a) = \alpha + \beta x + \gamma x^2$, and imposing the three interpolatory conditions, Eqs. (8.6), to determine expressions for α, β, and γ. The terms may then be rearranged so that $\tilde{U}^{(e)}(x;a)$ is in the standard form of Eq. (8.4), thereby enabling the coefficients of the a_i to be identified as the shape functions. Use this approach to rederive Eqs. (8.9) and (8.10).

8.3 For the reader interested in the programming aspects of the FEM, a significant project would be to add to UNAFEM (listing in appendix) the C^0-quadratic element developed by the direct approach in Section 8.2. This will require creating a new element subroutine that evaluates Eqs. (8.15). As an additional benefit, it would be instructive to compare the performance of the direct and isoparametric elements, particularly when the middle node is moved away from the center.

8.4 Optimal flux-sampling points: Calculate the least-squares fit of a linear polynomial $(d\tilde{U}/dx)$ to a quadratic polynomial $(dU/dx$ as h $\rightarrow$ 0) in an undistorted 1-D C^0-quadratic element. Show that the intersection points are at $\xi = \pm 1/\sqrt{3}$. These are the zeroes of the second-degree Legendre polynomial, popularly referred to as Gauss points. *Hint:* Follow the same procedure as used in Exercise 7.1. *Note:* If the element is distorted (middle node not at the center), then $d\tilde{U}/dx$ is not a linear polynomial and the intersection will not, in general, be at the Gauss points.

8.5 Consider a two-element mesh. One element is a C^0-linear and the other is a C^0-quadratic:

a) Write the system equations in matrix form, using $K_{ij}^{(e)}$, $F_i^{(e)}$ notation. Use the standard assembly rule (Section 5.7).

b) Sketch each of the *assembled* shape functions, i.e., trial functions, in the manner of Fig. 8.4. (Assume node 3 is at the center of the quadratic element, so that there is no mapping distortion.)

8.6 Determine the permissible range of the two interior nodes for the 1-D C^0-cubic isoparametric element, assuming the nodes are equidistant from the center; i.e., reproduce Eqs. (8.74). *Hint:* consider two cases,

$$\frac{d^2J}{d\xi^2} > 0 \qquad \text{and} \qquad \frac{d^2J}{d\xi^2} < 0$$

8.7 Write the system equations in matrix form, using $K_{ij}^{(e)}$, $F_i^{(e)}$ notation as in Eqs. (8.75), for the following mesh of two 1-D C^0-cubic elements with nodes numbered nonsequentially. Use the standard assembly rule (Section 5.7).

8.8 Derive the expressions for each of the C^1-cubic shape functions shown in Fig. 8.58 by applying the interpolation conditions [Eqs. (8.91)] to a cubic polynomial (a different set of four conditions for each shape function).

8.9 Derive Eqs. (8.102) and (8.103) following the suggestions in the text.

The following problems were analyzed in Chapter 7 using only the C^0-linear element. Now these problems may be analyzed using higher-order elements. Use UNAFEM to obtain approximate solutions with an estimated pointwise error of about 1%. Try different element types and different quadrature rules. Mix different element types. Work especially with nonuniform meshes, comparing different patterns of DOF densities to see which give the best performance, i.e., the highest accuracy with the fewest DOF. Try distorting the isoparametric elements by moving the interior nodes, especially for those problems that have a steep gradient in their solutions. These kinds of numerical experiments will provide good insight into the behavior of elements and how to work with them effectively.

8.10 Use UNAFEM to determine the deflection of a horizontal hanging cable (Exercise 3.4):

$$-\frac{d}{dx}\left(T\frac{dW}{dx}\right) = -\varrho g \qquad 0 < x < 10$$

with the BCs $W(0) = W(10) = 0$, $T = 98$ N, $\varrho = 1$ kg/m, and $g = 9.8$ m/sec^2. Note that the exact solution (see Exercise 3.4) is a quadratic polynomial. Using quadratic elements, what would be the most efficient mesh to use?

8.11 Use UNAFEM to determine the deflection of a horizontal hanging cable supported by an elastic foundation (Exercise 3.7):

$$-\frac{d}{dx}\left(T\frac{dW}{dx}\right) + kW = -\varrho g \qquad 0 < x < 10$$

with the BCs $W(0) = W(10) = 0$, $k = 24.5$ N/m-m and the rest of the data the same as in Exercise 8.10.

8.12 Use UNAFEM to determine the heat conduction in a rod with convection heat loss from the surface (Exercise 3.8):

$$-\frac{d}{dx}\left(k\frac{dT}{dx}\right) + \frac{2h}{r}T = \frac{2h}{r}T_\infty \qquad 0 < x < 10$$

with the BCs $T(0) = 100\,°F$ and $q(10) = 0.2$ BTU/sec-in^2, $k = 22$ BTU/sec-in-°F, $h = 0.193 \times 10^{-5}$ BTU/sec-in^2-°F, $r = 0.5$ in., and $T_\infty = 72\,°F$.

8.13 Use UNAFEM to obtain an approximate solution to the differential equation (Exercise 3.13):

$$\frac{d}{dx}\left[(x+1)\frac{dU}{dx}\right] = 0 \qquad 1 < x < 2$$

with the BCs $U(1) = 1$ and $\tau(2) = 1$.

8.14 Use UNAFEM to obtain an approximate solution to the differential equation (Exercise 3.14):

$$\frac{d}{dx}\left[(x+1)\frac{dU}{dx}\right] = 0 \qquad 1 < x < 2$$

with the BCs $U(1) = 1$ and $U(2) = 2$.

8.15 Use UNAFEM to obtain an approximate solution to the differential equation (Exercise 3.15):

$$\frac{d}{dx}\left(x^2\,\frac{dU}{dx}\right) = \frac{1}{12}(-30x^4+204x^3-351x^2+110x) \qquad 0 < x < 4$$

with the BCs $U(0)=1$ and $U(4)=0$.

8.16 Use UNAFEM to determine the temperature distribution in an Alucobond wall panel (Exercise 7.9).

8.17 Use UNAFEM to obtain an approximate solution to the illustrative problem used in Chapters 3 through 5:

$$\frac{d}{dx}\left(x\,\frac{dU}{dx}\right) = \frac{2}{x^2} \qquad 1 < x < 2$$

with the BCs $U(1)=2$ and $\tau(2)=1/2$. [The exact solution is $U(x)=(\ln x)/2+2/x$.]

8.18 Use UNAFEM to obtain an approximate solution to the differential equation (Exercise 7.11):

$$\frac{d^2U}{dx^2} + 25U = 0 \qquad 0 < x < 2\pi$$

with the BCs $U(0)=0$ and $\tau(2\pi)=-10$. Recall the caveat in Exercise 7.11. [The exact solution is $U(x) = 2\sin 5x$.]

8.19 Use UNAFEM to obtain an approximate solution to the differential equation (Exercise 7.12):

$$\frac{d}{dx}\left(x\,\frac{dU}{dx}\right) + xU(x) = 0 \qquad 0 < x < 10$$

with the BCs $U(0)=1$ and $U(10)=-0.2459358$. Recall the caveat in Exercise 7.12. [The exact solution is the zero-order Bessel function, $J_0(x)$.]

8.20 Use UNAFEM to determine the horizontal deflection of a cable hanging vertically from a building and being blown by a 5-mph wind (Exercise 7.13).

8.21 Use UNAFEM to determine the tip displacement of a whirling rubber cord due to centrifugal force, air resistance, and weight (Exercise 7.14).

References

[8.1] E. Isaacson and H. B. Keller, *Analysis of Numerical Methods*, Wiley, New York, 1966, Chapter 7.

[8.2] P. J. Davis and I. Polonsky, ''Numerical Interpolation, Differentiation and Integration,'' Chapter 25 in M. Abramowitz and I. A. Stegun (eds.), *Handbook of Mathematical Functions*, National Bureau of Standards, Applied Mathematics Series, **55**, 1964.

[8.3] B. M. Irons, ''Engineering Applications of Numerical Integration in Stiffness Methods,'' *J Amer Inst Aero Astro*, **14**: 2035-2037 (1966).

[8.4]   R. D. Cook, *Concepts and Applications of Finite Element Analysis*, 2d ed., Wiley, New York, 1981, p. 135.

[8.5]   O. C. Zienkiewicz, R. L. Taylor, and J. M. Too, "Reduced Integration Technique in General Analysis of Plates and Shells," *Int J Num Meth Eng*, **3**: 275-290 (1971).

[8.6]   D. J. Naylor, "Stresses in Nearly Incompressible Materials for Finite Elements with Application to the Calculation of Excess Pore Pressures," *Int J Num Meth Eng*, **8**: 443-460 (1974).

[8.7]   O. C. Zienkiewicz and E. Hinton, "Reduced Integration, Function Smoothing and Non-Conformity in Finite Element Analysis (with Special Reference to Thick Plates)," *J Franklin Inst*, **302**: 443-461 (1976).

[8.8]   G. Strang and G. J. Fix, *An Analysis of the Finite Element Method*, Prentice-Hall, Englewood Cliffs, N. J., 1973, pp. 40, 106, and 165.

[8.9]   G. J. Fix, "On the Effects of Quadrature Errors in the Finite Element Method" in J. T. Oden, R. W. Clough, and Y. Yamamoto (eds.), *Advances in Computational Methods in Structural Mechanics and Design*, University of Alabama, 1972, pp. 55-68.

[8.10]   I. Fried, "Accuracy and Condition of Curved (Isoparametric) Finite Elements," *J Sound and Vib*, **31**(3): 345-355 (1973).

[8.11]   M. Zlamal, "Curved Elements in the Finite Element Method. II," *SIAM J Numer Anal*, **11**(2): 347-362 (1974).

[8.12]   G. Strang and G. J. Fix, op. cit., Sec. 4.3.

[8.13]   I. Fried, "Numerical Integration in the Finite Element Method," *Computers and Structures*, **4**: 921-932 (1974).

[8.14]   O. C. Zienkiewicz, *The Finite Element Method*, McGraw-Hill, New York, 1977, p. 284.

[8.15]   T. K. Hellen, "Numerical Integration Considerations in Two and Three Dimensional Isoparametric Finite Elements" in J. R. Whiteman (ed.), *The Mathematics of Finite Elements and Applications II*, Proc Brunel University Conf, April 1975, Academic Press, New York, 1976.

[8.16]   T. Moan, "On the Local Distribution of Errors by Finite Element Approximations" in Y. Yamada and R. H. Gallagher (eds.), *Theory and Practice in Finite Element Structural Analysis*, University of Tokyo Press, Tokyo, 1973.

[8.17]   J. Barlow, "Optimal Stress Locations in Finite Element Models," *Int J Num Meth Eng*, **10**: 243-251 (1976).

[8.18]   S. P. Timoshenko and J. M. Gere, *Theory of Elastic Stability*, McGraw-Hill, New York, 1961, pp. 1-2.

[8.19]   J. S. Przemieniecki, *Theory of Matrix Structural Analysis*, McGraw-Hill, New York 1968 (see especially Sec. 5.6, "Beam Elements").

[8.20]   G. P. Bazeley, Y. K. Cheung, B. M. Irons, and O. C. Zienkiewicz, "Triangular Elements in Bending — Conforming and Nonconforming Solutions," *Proc 1st Conf Matrix Meth Struct Mech*, Wright-Patterson AFB, Dayton, Ohio, 1965.

[8.21]   B. M. Irons and A. Razzaque, "Experience with the Patch Test for Convergence of Finite Elements" in A. K. Aziz (ed.), *The Mathematical Foundations of the Finite Element Method with Applications to Partial Differential Equations*, Academic Press, New York, 1972.

[8.22]   E. R. de Arantes e Oliveira, "The Patch Test and the General Convergence Criteria of the Finite Element Method," *Int J Solids Structures*, **13**: 159-178 (1977).

[8.23]   F. Stummel, "The Limitations of the Patch Test," *Int J Num Meth Eng*, **15**: 177-188 (1980).

[8.24]   J. Robinson, "The Patch Test — Is It or Isn't It?" *Finite Element News*, **1**: 30-34 (Feb. 1982).

[8.25]   B. Irons and M. Loikkanen, "An Engineers' Defence of the Patch Test," *Int J Num Meth Eng*, **19**: 1391-1401 (1983).

[8.26]   E. L. Wilson, R. L. Taylor, W. P. Doherty, and J. Ghaboussi, "Incompatible Displacement Models" in S. J. Fenves, N. Perrone, A. R. Robinson, and W. C. Schnobrich (eds.), *Numerical and Computer Methods in Structural Mechanics*, Academic Press, New York, 1973, pp. 43-57.

9

Relative Performance
of the Different Elements

Introduction

The two preceding chapters developed four elements of increasing complexity — namely, a linear, quadratic, cubic, and quartic. The illustrative problems revealed that as the order of the element increases, fewer elements are necessary to achieve a comparable accuracy — an intuitively reasonable result. This chapter will provide a theoretical explanation for this behavior, along with a body of numerical data illustrating and corroborating the theory. This will provide the analyst with a better understanding of the type of performance to be expected from such elements in any application problem.

Section 9.1 will develop an expression for the pointwise error at arbitrary points in an element. This expression is most naturally plotted as the logarithm of the error versus the logarithm of element size, so that the slope of the resulting curve corresponds to the *rate of convergence*. Such a log-log plot tells a great deal about the performance of an element. Section 9.2 will identify special points in each element that yield exceptionally high rates of convergence, and several sets of log-log plots will provide numerical data to illustrate and validate the theory for each element. Section 9.3 will redisplay some of the previous data in a way which will more clearly compare and contrast the performance of the four elements. Sections 9.4 and 9.5 will use the above approach to shed some light on two important situations.

9.1 Rate of Convergence

There are many different ways to measure the error of an FE solution. The convergence theorems generally employ global measures, such as the global energy [see Eq. (8.52)]. However, as argued in Section 3.6.2, it is the *pointwise* error that is usually of interest in practical applications, so that is the measure of error we will examine here. The pointwise error $E(x)$ is merely the difference between the exact solution $U(x)$ and the FE solu-

376

tion $\tilde{U}(x)$ at any point x:

$$E(x) = U(x) - \tilde{U}(x) \tag{9.1}$$

In addition, we will consider here only one source of error: the *discretization error*, which is due to the use of trial solutions that contain only a finite number of terms (see Section 3.6.2). This is the approximation that is at the heart of the FEM and hence is generally the most significant determinant of the accuracy of an FE solution.

Consider a typical problem for which a mesh has been constructed and an analysis performed. We want to examine the error at some point in the domain, x^*. Following the procedure employed for the flux in step 6 in Section 7.2, we begin by observing that the exact solution can be represented by a Taylor series about some point $\hat{x}$ in the element that contains the point x^*:

$$U(x) = c_0 + c_1(x-\hat{x}) + c_2(x-\hat{x})^2 + \dots + c_n(x-\hat{x})^n + \dots \tag{9.2}$$

where

$$c_n = \frac{1}{n!} \frac{d^2 U}{dx^2} \bigg|_{x=\hat{x}}$$

and $|x-\hat{x}| \leq h$, the size (length) of the element. This representation presumes that $U(x)$ is sufficiently smooth for the derivatives in the coefficients to exist (which may not be true near singularities; see Section 9.4). The approximate solution is represented by a polynomial that is complete to degree p:

$$\tilde{U}(x) = a_0 + a_1(x-\hat{x}) + \dots + a_p(x-\hat{x})^p \tag{9.3}$$

If the mesh is refined in a uniform manner,[1] then, as $h_{max} \to 0$, $\tilde{U}(x)$ in Eq. (9.3) will be able to approach arbitrarily close to the terms of degree p or lower in the Taylor series in Eq. (9.2). The remaining terms of degree $p+1$ and higher will therefore constitute the error at point x^*:

$$U(x^*) - \tilde{U}(x^*) \to c_{p+1}(x-\hat{x})^{p+1} + c_{p+2}(x-\hat{x})^{p+2} + \dots \qquad \text{as } h_{max} \to 0 \tag{9.4}$$

where, in the limit, $\hat{x} \to x^*$ and the derivatives in Eq. (9.2) are evaluated at x^*. For $|x-\hat{x}|$ small enough, the $p+1$ term will dominate, and Eq. (9.4) can be written:

$$\left| \frac{U-\tilde{U}}{U} \right| \leq Ch^{p+1} \qquad \text{as } h_{max} \to 0 \tag{9.5}$$

where we have normalized both sides by the constant $U(x^*)$ to yield a relative (or percent) error. Equation (9.5) may also be written equivalently as

$$\left| \frac{U-\tilde{U}}{U} \right| = O(h^{p+1}) \qquad \text{as } h_{max} \to 0 \tag{9.6}$$

---

[1] Recall comments in Section 5.6 regarding the requirement that the largest element, and hence all elements, must continually decrease in size. Section 9.5 discusses what happens when only part of the mesh is refined.

where the notation "$= O$" is read "is of order," a notation that was explained in Section 7.2, below Eqs. (7.32).

Equation (9.5) indicates that the *asymptotic rate* of convergence is of order[2] h^{p+1}. For example, if h is halved, then the error should decrease by the factor $(1/2)^{p+1}$. It must be emphasized that this is valid only in the limit of "small-enough" h; i.e., this is an asymptotic rate. What is meant by "small enough" will be amply illustrated in the data below.

For values of h that are not small, the coefficient C is a function of h since the RHS of Eq. (9.5) is replacing all the terms on the RHS of Eq. (9.4). As the element becomes smaller, though, C approaches a constant and the $\leq$ sign approaches an equality. Taking the logarithm of both sides would then yield[3]

$$\log \left| \frac{U - \tilde{U}}{U} \right| = \log C + (p+1) \log h \qquad \text{as } h_{\max} \to 0$$

$$(9.7)$$

Equation (9.7) is a straight line with slope $p+1$ when plotted as $\log \left| \dfrac{U - \tilde{U}}{U} \right|$ versus $\log h$. If numerical experiments are performed for a sequence of meshes and the data is plotted on such log-log axes, *the slope of the curve is the rate of convergence*. Because of the asymptotic nature of Eq. (9.7), we might expect such an experimental curve to be a straight line only for very fine meshes, and to exhibit nonlinear behavior for coarse meshes. This is indeed what usually happens, as will be seen below.

All the above arguments can be applied in the same manner to the flux, as, in fact, was done already in Section 7.2. In essence, since the flux is the first derivative of the function, its asymptotic rate should be one order lower than that of the function, that is, $O(h^p)$.

The above theory describes the rate of convergence that is observed at arbitrary points in an element; such points will be referred to here as "ordinary" points. However, there usually exist a few "special" points in each type of element at which the convergence rates are greater, so-called *superconvergent points*. We will defer discussion of these points until Section 9.2.

To demonstrate Eq. (9.7) we will employ the following sample problem, which is representative of the general problem studied in Chapters 6 through 8 [see Eq. (6.25)]:

$$-\frac{d^2 U(x)}{dx^2} + U(x) = f(x) \qquad 0 < x < 5$$

$$(9.8)$$

where

$$f(x) = -x^3 + 6x^2 + x - 12$$

---

[2] As a point of diction, it would be better to say that the asymptotic rate of convergence is $p+1$ (rather than "of order h^{p+1}") since the rate refers only to the exponent of h. It is the error that is of order h^{p+1}. Nevertheless, the language used above is not uncommon, probably because "of order" (or O) is a reminder of the asymptotic nature of the convergence and the h is a reminder of which parameter is being varied to produce the convergence.

[3] The logarithms may employ any base. The log-log plots in this chapter will use base 10 (rather than, say, the natural base e) because of the convenience of interpreting the axis values in terms of integers.

and the BCs are

$$U(0) = 10$$

$$\tau(5) = 5$$

Thus $\alpha(x) = \beta(x) = 1$ in Eq. (6.25), and there is an essential BC on the left end and a natural BC on the right end. The interior load $f(x)$ was chosen to be a polynomial of a degree high enough to provide the exact solution with a couple of "wiggles" in order to simulate, albeit modestly, a bit of the complexity found typically in 2-D and 3-D problems.[4] The exact solution for the function $U(x)$ and flux $\tau(x)$ is

$$U(x) = Ae^x + Be^{-x} - x(x-1)(x-5)$$

$$\tau(x) = -Ae^x + Be^{-x} + 3x^2 - 12x + 5 \qquad (9.9)$$

where

$$A = \frac{15 + 10e^{-5}}{e^5 + e^{-5}}$$

$$B = 10 - A$$

In constructing a sequence of meshes for such a convergence study, it is important to impose some degree of *uniformity* on the refinement; i.e., *all* of the elements must decrease in size from one mesh to the next. This is implied by the requirement $h_{max} \to 0$, not just $h \to 0$. This would preclude a local refinement in which only part of the mesh is refined (see Section 9.5). The simplest approach would be perfectly uniform refinement in which every element decreases by exactly the same ratio. However, this is generally impossible to do because every mesh in the sequence should evaluate the error at the same point in the domain and the same intraelement location[5] (e.g., a Gauss point, a nodal point, an ordinary point, etc.), and these requirements cannot be satisfied simultaneously with a uniform refinement for most intraelement locations. Fortunately, a perfectly uniform refinement is not necessary. It is sufficient to employ a *quasiuniform* refinement [9.1] in which the ratio of the largest to smallest element is bounded as $h_{max} \to 0$. This is the approach used here. Figure 9.1 illustrates the strategy employed.

The point x^* is the *error point*, i.e., the point in the domain at which the error is evaluated for every mesh in the sequence. The element containing x^* will be called the *error element*. The location of the error element (x_l and x_r) and the construction of the

---

[4] In contrast, the solution to the illustrative problem in Chapters 3 through 5 is only a shallow curve with no inflection points; consequently, even a "coarse" mesh of only two linear elements is close to the asymptotic range. Indeed, it is something of a challenge to create a *well-posed* 1-D problem in the form of Eq. (6.25) that has an uncomplicated formulation but a solution with a complexity similar to 2-D and 3-D problems, e.g., several irregularly spaced inflection points. (Formulations which are the same as their associated eigenproblems, such as those which yield the circular or Bessel functions, must be excluded for a boundary-value problem because $\alpha(x)$ and $\beta(x)$ are of opposite sign, making the stiffness matrix singular at critical element sizes.)

[5] Several figures below demonstrate how the error curves can differ from point to point in the domain (Fig. 9.3) and for different intraelement locations (Figs. 9.4 through 9.7).

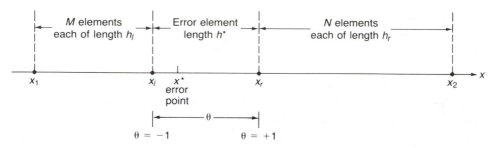

FIGURE 9.1 ● Definition of terms used in quasiuniform mesh refinement.

rest of the mesh are determined by the following four conditions, which refer to Fig. 9.1:

1. The intraelement location of x^* in the error element is the same in all meshes. The location is specified by the dimensionless parameter θ, which varies linearly from -1 at the left end of the element, x_l, to $+1$ at the right end, x_r. Specifying $\theta=0$ would position the error element so that its center is at x^*; $\theta=\pm1$ would position it so that one of the end nodes is at x^*. In other words, θ determines the type of intraelement point (ordinary, nodal, Gauss, etc.) at which the error will be evaluated.

2. Define a uniform mesh in the regions on either side of the error element:

$$h_l = \frac{x_l-x_1}{M}$$

$$h_r = \frac{x_2-x_r}{N} \tag{9.10}$$

3. Maintain a constant ratio in the number of elements on either side of the error element:

$$R = \frac{M}{N} \text{ (held constant during refinement)} \tag{9.11}$$

4. Define the length of the error element to be the mean of the element lengths on either side:

$$h^* = \frac{1}{2}\,(h_l+h_r) \tag{9.12}$$

From Eq. (9.10), as $M,N\to\infty$, $h_l,h_r\to0$. Hence from Eq. (9.12), $h^*\to0$, which implies $x_l,x_r\to x^*$. Therefore

$$\frac{h_l}{h_r} \to \frac{x^*-x_1}{x_2-x^*}\frac{1}{R} \qquad \text{as } M,N \to \infty \tag{9.13}$$

Since there are only three element sizes (h_l,h_r,h^*) and h^* is the mean, then h_l/h_r equals h_{max}/h_{min} or its inverse, and Eq. (9.13) implies h_{max}/h_{min} is bounded. Therefore the refinement is quasiuniform. (The exception is when $x^*=x_1$ or x_2, and this can be handled with a minor modification of the above conditions.)

To construct a sequence of meshes, one need only specify x^*, θ, and a sequence of pairs of M,N values. The above conditions will then uniquely determine each of the meshes.

The parameters M and N are arbitrary, excepting only the condition in Eq. (9.11) that their ratio R remain constant. They were generally chosen so that R would be as close as possible to the fraction $(x^*-x_1)/(x_2-x^*)$ since, from Eq. (9.13), this would make the meshes as uniform as possible; that is, $h_l/h_r \approx 1$.[6]

All the subsequent log-log error plots were created by a special version of the UNAFEM program, which requires as input the usual problem data, along with x^*, θ, and a sequence of M,N values. Before proceeding with the bulk of the data in the next two sections, let's first examine a typical plot.

Figure 9.2 shows a function and flux curve corresponding to a sequence of meshes of linear elements (ELEM11), using the sample problem in Eq. (9.8). The error point is at the center of the domain ($x^*=2.5$) and at an ordinary point in the error element ($\theta=-0.5$). Each dot on the curves corresponds to a mesh that was analyzed. The dots are connected by straight lines, and the slopes of some of the lines are indicated underneath. Below the error plots are a few UNAFEM plots of both the exact and approximate solutions for some of the coarser meshes. By seeing the actual FE solutions associated with points on the error curves, one can perhaps get a better feel for such curves.

The vertical scale to the right of each error plot provides a convenient interpretation of the logarithmic scale in terms of a percentage error. Only the range 0.01 to 100% is labeled because this is the range of error most frequently encountered in practical applications. Indeed, the 1 to 10% range is sufficient for many engineering applications.

The horizontal axis uses $\log_{10}(h_{ave}/L)$, rather than $\log_{10} h$ as indicated by Eq. (9.7). The average element length, h_{ave}, is defined by the relation

$$\frac{h_{ave}}{L} = \frac{1}{N_{elem}} \tag{9.14}$$

where $L=x_2-x_1$ and N_{elem} is the number of elements in the mesh. The factor L is introduced into Eq. (9.7) by multiplying the $(p+1)$st term in Eq. (9.4) by $(L/L)^{p+1}$, thereby making the coefficient c_{p+1} nondimensional too. The variable h in Eq. (9.7) is the size of the error element h^*. Because of the above refinement strategy, it can be replaced by h_{ave}, which decreases at essentially the same rate as h^* (equaling h^* in the limit) since h^* is bounded between h_l and h_r. Indeed, when x^* is at the center of the domain and $M=N$ (as is the case for most of the data that follows), $h_{ave}=h^*$. Equation (9.14) is the basis for the secondary horizontal axis on the plots.

Since the linear element uses a polynomial of degree $p=1$, the asymptotic rate of convergence at an ordinary point should be $O(h^2)$ for the function and $O(h)$ for the flux. The curves in Fig. 9.2 do indeed exhibit slopes of 2 and 1, respectively, for *very fine meshes*. It should be noted, though, that the slopes for the coarser meshes differ significantly from the asymptotic values. And the asymptotic slopes are effectively not reached until below the practical range of accuracy.

---

[6] As noted previously, perfect uniformity, $h_l=h_r=h^*$, is impossible for most values of θ. However, it can be approached in the limit if x^* is located so that $(x^*-x_1)/(x_2-x^*)$ is a ratio of small integers, in which case M and N can be chosen so that R equals the same ratio, making $h_l/h_r \to 1$. For example, if x^* were located at the 1/4 point of the domain, that is, $x^*=x_1+(x_2-x_1)/4$, then $(x^*-x_1)/(x_2-x^*)=1/3$, so all meshes would have three times as many elements to the right of the error element as to the left.

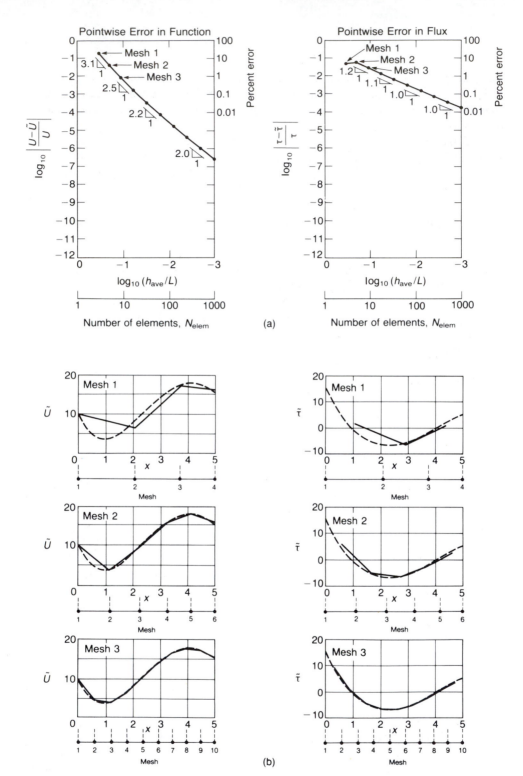

FIGURE 9.2 ● (a) Error in function and flux at an ordinary point in a 1-D linear element; rates of convergence are indicated by slopes. (b) Exact and FE solutions for the first three meshes.

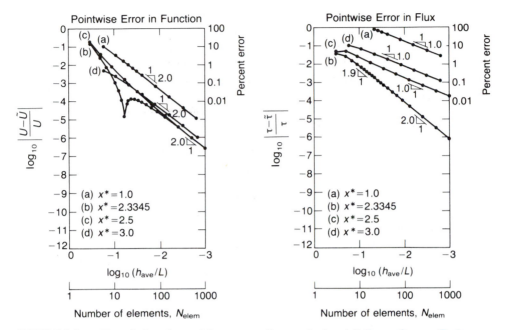

FIGURE 9.3 ● Error in function and flux at an ordinary point in a 1-D linear element. Each curve corresponds to a different point in the domain. Rates of convergence are indicated by slopes.

Figure 9.3 shows several curves corresponding to different error points x^*. The curves demonstrate that *pointwise* error can exhibit different behavior from point to point in a given problem, though the asymptotic slopes will all be the same. This is because the coefficients in the *local* Taylor series representation will vary from point to point. Thus, the log C term in Eq. (9.7), which determines the vertical position of the curve, contains the $(p+1)$st derivative evaluated at or near to x^*. As x^* varies, the curves will shift up or down.

In addition, the shape of the curves in the preasymptotic region (before the slope is close to the asymptotic slope, say within about 10% of the latter) can vary dramatically. For example, if the $(p+1)$st derivative happens to be very small at x^*, then the next-higher-degree term in Eq. (9.4), i.e., the $p+2$ term, will dominate until h becomes small enough, resulting in a dip in the convergence curve. This is demonstrated by curve b, for which x^* was deliberately chosen at a point where $d^2U/dx^2 = -0.0001$ (and $d^3U/dx^3 = -5.91$). Note that the function eventually reaches its asymptotic slope of 2, whereas the flux (in data not shown here) runs into machine-precision limits before it can approach its asymptotic slope of 1. The author has encountered many such dips (as well as spikes) in a variety of different problems. In many of those cases the convergence curves are very irregular in the realm of practical errors and don't approach the asymptotic slope until extremely low errors are reached (e.g., $10^{-4}\%$, or sometimes much lower, as in the flux for curve b).

Such anomalous behavior in the preasymptotic region is more characteristic of point-wise error than of any global measure of error since the latter averages the local fluctuations over the entire domain. This means that in practical applications one can expect to occasionally meet strange convergence behavior when observing a sequence of values at a given point. In the next two sections the data presented in the log-log error plots should be examined primarily for the asymptotic slopes, as well as part of the smooth portion in the preasymptotic region, because it is this behavior that is problem-independent and that should be used as general guidelines for predicting the typical behavior of each element type.

9.2 Ordinary and Superconvergent Points

The theoretical rate of convergence studied in the previous section was based solely on polynomial approximation theory, i.e., fitting the FE trial-solution polynomial to a local Taylor series representation of the exact solution. This is what is observed at most points in an element. However, as we've seen already in Chapters 7 and 8 in regard to the flux, there usually exist special points in each type of element at which the rate exceeds that predicted by approximation theory alone. Such points are said to be *superconvergent*, and their locations can be predicted theoretically (i.e., they are known a priori).

The existence of such special points should not be surprising when we recall Fig. 3.1. There it was observed that an FE solution is generated from two ingredients: a trial solution (which is usually a polynomial) and an optimizing criterion. The above theoretical rate of convergence based on approximation theory looked at only the first ingredient and ignored the optimizing criterion. Hence it represents a *minimum* rate; i.e., we would expect an FE solution to converge *at least* at this rate at any point. It seems reasonable to expect that the optimizing criterion might possibly improve upon this rate. And indeed it does.

We list here several superconvergence theorems applicable to the four 1-D elements when used in linear, self-adjoint, 2mth-order boundary-value problems. The theorems specify the locations of the superconvergent points within an arbitrary element.

1. The flux converges at a rate $O(h^{p+1})$ at the zeroes of the Legendre polynomial of degree $p-m+1$ [9.2, 9.3]. This is one order higher than approximation theory. (Chapters 7 and 8, — see especially guideline IV in Chapter 8 — covered this type of superconvergence in some detail, and it was noted that the zeroes of the Legendre polynomials are also the Gauss points used in the Gauss-Legendre quadrature formulas tabulated in Fig. 8.12.)

2. The function converges at a rate $O(h^{2p})$ at the interelement nodes (at each end of an element), for second-order ($m=1$) problems [9.4, 9.5]. This is *twice* the order of approximation theory. Such an extraordinary rate could perhaps be called "super-superconvergence."

3. The function converges at a rate $O(h^{p+2})$ at the zeroes of the Jacobi polynomial $P_{p+1-2m}^{m,m}$ [9.6, 9.7]. This is one order higher than approximation theory. This may be more of academic interest since the above superconvergence at the nodes is faster, and nodal values, being the primary FE variables, require no additional processing.

These theorems deal with the function and flux as they are evaluated directly from the FE solution. In addition, it has been shown [9.8] that if a small amount of postprocessing is performed on the flux, it too will exhibit "super-superconvergence" at a rate $O(h^{2p})$,

TABLE 9.1 ● Rates of convergence and locations of ordinary and superconvergent points for the four 1-D elements in second-order problems: $-1 \leq \theta \leq +1$ over element.

		Function			Flux	
		Superconvergent				Superconvergent
	Ordinary	Interelement nodes	Zeroes of Jacobi polynomials	Ordinary		Zeroes of Legendre polynomials† (Gauss points)
Linear (ELEM11)	$O(h^2)$	$O(h^2)$ at $\theta = \pm 1$	None	$O(h)$		$O(h^2)$ at $\theta = 0$
Quadratic (ELEM12)	$O(h^3)$	$O(h^4)$ at $\theta = \pm 1$	$O(h^4)$ at $\theta = 0$	$O(h^2)$		$O(h^3)$ at $\theta = \pm \dfrac{1}{\sqrt{3}}$
Cubic (ELEM13)	$O(h^4)$	$O(h^6)$ at $\theta = \pm 1$	$O(h^5)$ at $\theta = \pm \dfrac{1}{\sqrt{5}}$	$O(h^3)$		$O(h^4)$ at $\theta = 0, \pm \dfrac{\sqrt{3}}{\sqrt{5}}$
Quartic (ELEM14)	$O(h^5)$	$O(h^8)$ at $\theta = \pm 1$	$O(h^6)$ at $\theta = 0, \pm \dfrac{\sqrt{3}}{\sqrt{7}}$	$O(h^4)$		$O(h^5)$ at $\theta = \pm \dfrac{\sqrt{15+2\sqrt{30}}}{\sqrt{35}},$ $\pm \dfrac{\sqrt{15-2\sqrt{30}}}{\sqrt{35}}$

†Bakker [9.6] demonstrates that the superconvergent points for the flux occur at the zeroes of the Jacobi polynomial $P_{p+2-2m}^{m-1,m-1}$. For second-order problems $(m=1)$, this becomes $P_2^{0,0}$ which is the Legendre polynomial of degree p discussed in Chapters 7 and 8.

the same as exhibited by the function at the interelement nodes. Since the postprocessing entails the use of already computed data, the technique appears quite promising. To the author's knowledge, it has not yet been implemented in commercial programs.

Table 9.1 lists the locations and asymptotic rates of convergence for the superconvergent points predicted by the above three theorems, for the four 1-D elements in Chapters 7 and 8, when used in second-order $(m=1)$ problems. The locations are specified by the variable θ defined in the previous section; that is, $-1 \leq \theta \leq +1$ over an element. The table also includes the rates of convergence at ordinary (nonsuperconvergent) points; of course, their locations are all points other than superconvergent points.

Figures 9.4 through 9.7 display error curves for the points listed in Table 9.1, using the sample problem and mesh-refinement strategy described in Section 9.1. It is clear from Table 9.1 that when there are two or more superconvergent points of the same type for a given element, they are always located symmetrically with respect to the center of the element. Therefore, to avoid a cluttered appearance in the plots, curves are displayed for only one point of each symmetric pair. All the curves evaluate the error at the center of the domain, $x^* = 2.5$. The three isoparametric elements use reduced integration. Other approaches were also studied and their results are discussed following the plots.

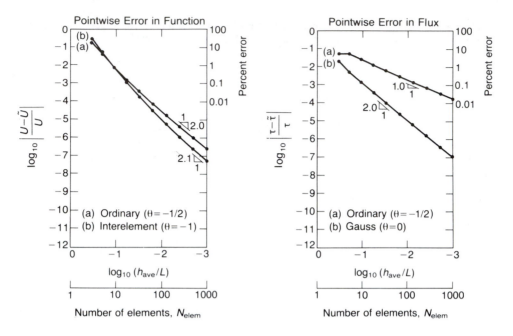

FIGURE 9.4 ● Error in function and flux at ordinary and superconvergent points in 1-D linear elements; rates of convergence are indicated by slopes.

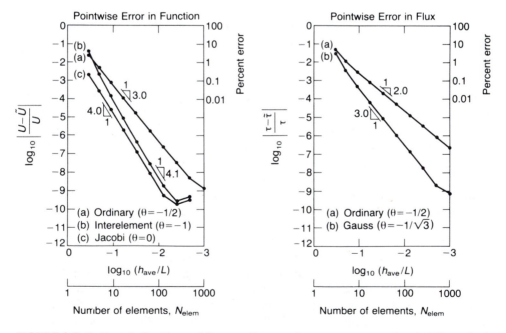

FIGURE 9.5 ● Error in function and flux at ordinary and superconvergent points in 1-D quadratic elements; rates of convergence are indicated by slopes.

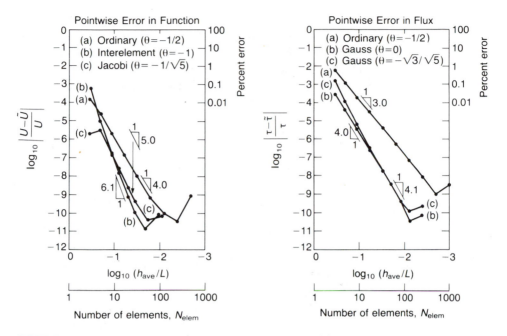

FIGURE 9.6 ● Error in function and flux at ordinary and superconvergent points in 1-D cubic elements; rates of convergence are indicated by slopes.

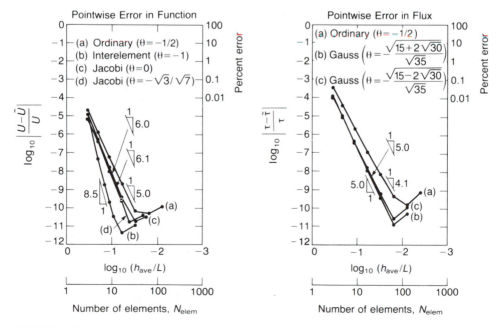

FIGURE 9.7 ● Error in function and flux at ordinary and superconvergent points in 1-D quartic elements; rates of convergence are indicated by slopes.

When the errors decrease to about 10^{-9} or 10^{-10}, the curves turn upward; i.e., the error increases. This is where the computation is limited by the single-precision accuracy of about 14 decimal digits on the Cray-1, the computer used to generate these curves. For the size meshes involved here, one can expect to lose about three or four significant figures because of propagation of round-off error, which leaves about 10 significant figures.

A large number of other curves, not shown here, were also generated during this study. They included testing all the other θ values from the symmetric pairs mentioned above; sampling ''random'' θ values for a variety of ordinary points; placing x^* at different locations in the domain; and repeating all the above curves using higher-order integration instead of reduced. In all cases, the asymptotic rates of convergence conformed to the theory. As would be expected (see Section 9.1), the curves exhibited some differences in vertical positioning (see especially Figs. 9.3 and 9.8) and occasional irregularities in the preasymptotic region.

An example of a predictable irregularity is when θ is chosen very close to, though not exactly at, a superconvergent point. For coarse meshes the separation distance is insignificant so the slope of the curve initially approaches the superconvergence rate. As the error decreases, the points begin to appear significantly separated so the slope decreases to the ordinary rate, which it then approaches asymptotically.

Figure 9.8 addresses an issue about which there has been some confusion (see the last footnote in Section 7.2). The author has frequently encountered the belief that the optimal flux-sampling points occur at whatever Gauss points are used to perform the numerical integration of the stiffness terms. It was argued in Chapters 7 and 8, however, that the location of these points is determined by a theory that is independent of any

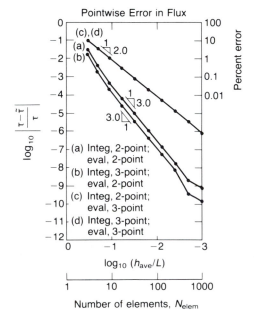

FIGURE 9.8 ● Error in flux in 1-D quadratic elements, using different combinations of Gauss points from two- and three-point quadrature rules for integration and evaluation points; rates of convergence are indicated by slopes.

numerical integration concepts. (Indeed, the linear element, which is integrated exactly, has one optimal flux point.) It is hoped that the figure will help to clarify this issue.

Figure 9.8 exhibits two pairs of curves for the quadratic element. (Two of the four curves happen to be coincident.) One pair evaluates the flux at a Gauss point for the two-point rule; the other pair evaluates the flux at a Gauss point for the three-point rule. Within each pair, one curve uses the two-point rule for the integration of the stiffness terms and the other curve uses the three-point rule. Thus the four curves cover all combinations for using the two- and three-point rules for integration points and for sampling points.

The curves show that superconvergence occurs only at the locations corresponding to the Gauss points for the two-point rule, irrespective of which set of points were used for the integration. The Gauss points for the three-point rule are merely ordinary points for evaluating the flux, even when the integration uses the same points.

In both pairs, the asymptotic rates of convergence are independent of the order of integration, which is consistent with all the previous data. However, at the superconvergent points the accuracy (vertical position of curve) noticeably improves with higher-order integration. This is another manifestation of the fact that the theory for such points presumes exact integration.

Similar results were obtained for the cubic and quartic elements.

9.3 Comparison of the Elements

Having studied the performance of each element separately, let's now take a look at how they perform relative to each other. Figure 9.9 shows the error curves for each of the four elements at an ordinary point. These curves were selected from Figs. 9.4 through 9.7 and replotted on a single set of axes.

Of course, the accuracy steadily improves as the order of the element increases, as would be expected. However, this is not really a fair comparison because the variable along the horizontal axis is the number of elements N_{elem}, so points on the separate curves that are aligned vertically represent meshes with the same number (and size) of elements. But the higher-order elements have more DOF per element, so their meshes are employing an increasingly greater number of DOF.

A more objective approach would be to compare elements on the basis of meshes that use the same number of DOF, since computational costs are more closely related to DOF. For example, consider a 1-D mesh containing two linear elements. The interesting refinement question becomes: If an extra node (DOF) is added in the middle of each element, will greater accuracy be achieved by defining four linear elements or two quadratic elements (both meshes increasing from three to five DOF)? The former represents h-refinement, i.e., increasing the number of elements while maintaining the same degree of the polynomial trial solution within each element. The latter represents p-refinement, i.e., increasing the degree of the polynomial while maintaining the same number and size of elements (recall Section 5.6).

We can transform the horizontal axis from N_{elem} to N_{DOF}, the number of DOF in the mesh, simply by multiplying and dividing in the log-log equation by the ratio N_{DOF}/N_{elem}. This ratio is essentially (neglecting an additive constant of 1) the number of DOF per element. The ratio equals 2, 3, 4, and 5 for the four 1-D elements, respectively. This transformation will shift each of the curves in Fig. 9.9 by different amounts, without altering

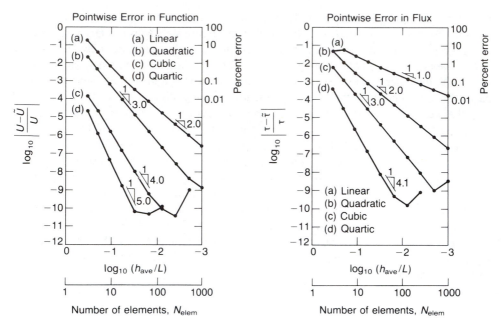

FIGURE 9.9 ● Error in function and flux at an ordinary point in each of the four 1-D elements; rates of convergence are indicated by slopes.

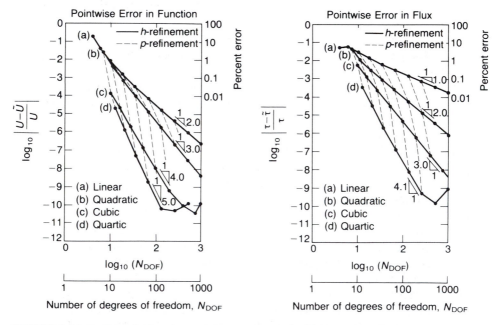

FIGURE 9.10 ● Error in function and flux at ordinary points in each of the four 1-D elements. The same data is used here as in Fig. 9.9, though it is plotted as a function of N_{DOF} instead of N_{elem}; rates of h-convergence are indicated by slopes.

the asymptotic rate of convergence. Of course, it is precisely these shifts that will provide the "fairer" comparison.

Figure 9.10 replots the same curves from Fig. 9.9, using N_{DOF} instead of N_{elem} on the horizontal axis. The dashed lines connect points corresponding to meshes with the same element geometry (same number, size, and distribution of elements) and hence represent paths of p-refinement. Since the solid curves are paths of h-refinement, we see that p-refinement yields a faster rate of convergence than h-refinement. Thus, in the above example of two linear elements, a greater improvement in accuracy is achieved by refining to two quadratic elements rather than to four linear elements.

As a general principle, the rate of convergence with p-refinement is at least as fast as (and, in practice, usually faster than) h-refinement. In certain situations, e.g., near singularities caused by corners, the p-refinement rate can be much faster than h-refinement, merely because the h-refinement rate may deteriorate [9.9, 9.10]; however, the latter can be recovered by using an optimally graded mesh (see Section 9.4), which is a form of local mesh refinement (see Section 9.5). As a practical matter, though, finding the optimal grading strategy, especially in 2-D and 3-D problems, may not be a simple task. The fastest rate of convergence can theoretically be achieved by a combination of both p- and h-refinement.

Figure 9.11 compares the four elements at the interelement nodes for the function and at the Gauss points for the flux, these being superconvergent points and therefore of greater practical value.

These curves demonstrate the general computational superiority of higher-order elements, particularly when used in conjunction with p-refinement. This superiority comes

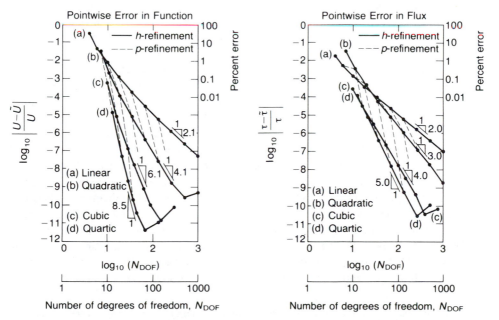

FIGURE 9.11 ● Error in function and flux at superconvergent points in each of the four 1-D elements: function values at interelement nodes; flux values at Gauss points. Rates of h-convergence are indicated by slopes.

at a cost of greater programming complexity (which implies a higher risk of program bugs) and usually a greater complexity in pre- and postprocessing such elements (which frequently means greater difficulty for the analyst). These costs become more pronounced in 2-D and 3-D problems.

Commercial FE programs have traditionally offered only the capability to do *h*-refinement, namely, a choice of a linear or a quadratic element (rarely a cubic). In these cases, considering the above trade-offs, the author's personal preference has been to work with the quadratic element and locally grade the meshes where the solution appears to be less smooth (see Sections 9.4 and 9.5).

Significant advances have been made recently in implementing the *p*-method, resulting in at least one such commercial program [9.11]. The method has been made more practicable by the use of *hierarchical elements* [9.12]. These are elements that increase the polynomial degree by adding higher-order shape functions that do not disturb the lower-order ones. This very promising technique greatly improves the computational efficiency of the *p*-method and claims other practical advantages as well.

9.4 Behavior Near Singularities

Singularities are points where the function or its derivatives above some order become infinite. They generally occur where there are sudden or sharp irregularities in the geometry, physical properties, or loads. For example, in 2-D and 3-D problems the geometry may contain sharp or reentrant corners, cracks, or slits. Properties and loads may have discontinuities.

Of course, in the real world such infinities are impossible. They only occur in our analyses because of mathematical simplifications that are made in the problem data or the governing differential equations. Thus it is much easier to describe the corner of an object (in 2-D, say) by two straight lines meeting at an angle (resulting in a perfectly sharp vertex) rather than to give a more precise and very complicated description of all the tiny, but perhaps functionally smoother, bumps and crevices actually present in a real material. And it is much easier to use a linear material-property law (e.g., linear elasticity) than a more accurate nonlinear law (plasticity) that would prevent the derivatives (stresses) from becoming infinite. The price for these simplifying modeling idealizations is having to make an extra effort to handle the resulting singular behavior.

Since polynomials cannot become infinite (for finite values of the variable), it should not be surprising that polynomials provide an exceptionally poor approximation in the neighborhood of a singularity (as will be demonstrated below). The theorems in Section 9.2 presume that the exact solution has no singularities; i.e., it is "smooth." If the exact solution contains singularities, then the theorems, as stated above, must be modified [9.13-9.15]. The modified theorems show that the rates of convergence for quasiuniform mesh refinement may be governed by the "strength" of the singularity[7] rather than by the degree of the trial-solution polynomial, resulting in rates that may be much lower than

---

[7] In the neighborhood of the singularity the exact solution is generally of the form $r^{\alpha} + f(x)$, where r is a radial coordinate with the singularity as origin, $0 < \alpha < 1$, and $f(x)$ is a "well-behaved" function (representable by polynomials). Thus the singularity is in the first derivative, i.e., the flux. The exponent α is the strength of the singularity.

the rates demonstrated in Sections 9.2 and 9.3. Therefore low-degree polynomials might be preferable, since the increased accuracy of higher-degree polynomials would merely be wasted.

The higher convergence rates associated with smooth solutions can often be recovered, though, by one of the following strategies:

1. *Adding a singular trial function, of the same strength as the singularity, to the FE trial solution*   This approach is a direct and obvious one. The singular trial function can provide the appropriate approximation of the singularity, permitting the remaining polynomial trial functions to approximate the nonsingular part of the solution. The strength of the singularity can usually be determined by classical theoretical methods.

   The singular trial function can be global (see Section 5.8) or local. The latter approach is more common [9.16-9.19], partly because it is easier to implement with already existing programs. Indeed, UNAFEM contains such elements: Recall from Chapter 8 that if the interior nodes of isoparametric elements are placed at certain locations (e.g., the middle node of the quadratic element placed at the 1/4 point), the shape functions will be singular (their derivatives being infinite). The cited references also describe other techniques for developing special singular elements.

2. *Grading the mesh in the neighborhood of the singular point so that the DOF density increases as the point is approached*   Although it is intuitively obvious that increasing the number of DOF near the singularity should help to improve the accuracy, it is not so obvious — it is perhaps even remarkable — that the *rate* of convergence can be fully restored using only polynomial trial functions. In this approach the elements are made progressively smaller as they get closer to the singularity. Appropriate grading strategies have been devised, based on the strength of the singularity, the dimension of the problem, and the degree of the trial solution polynomial [9.20-9.23].

The discussion so far has focused on singularities which lie *in* the domain or *on* the boundary. A closely related and very common situation is when the function or flux rises very steeply near a boundary, reaching a sharp maximum (but not an infinity) on the boundary. Typical examples are concentrations of stress, heat flux, electric field, etc., in the vicinity of curved boundaries with small radii of curvature. In these cases the theoretical expressions for the exact solutions involve a singularity located just *outside* the boundary of the domain. (If the radius decreased to zero, i.e., a sharp corner, the singularity would move to the boundary.)

In these "quasisingular" situations the solution is smooth (nonsingular) everywhere *in* the domain, so the asymptotic *rates* of convergence are unaltered. Nevertheless, the *accuracy* in the vicinity of the singularity can seriously deteriorate, much as it could if the singularity were on the boundary. Because this quasisingular behavior is quite similar to the singular behavior described above, it can be handled in a similar manner, namely, using special trial functions or graded meshes.

Figures 9.12 and 9.13 demonstrate the extent to which the accuracy can deteriorate. These figures use the illustrative problem in Chapters 3 through 5 for which the exact solution is

$$U(x) = \frac{1}{2} \ln x + \frac{2}{x} \qquad\qquad (9.15)$$

and

$$\tau(x) = -\frac{1}{2} + \frac{2}{x}$$

both of which have a singularity at $x=0$. The domain used in Chapters 3 through 5 is $1<x<2$, which is sufficiently far removed from the origin that the effect of the singularity on the smoothness of the solution is negligible. The convergence curves for this nonsingular problem are shown in Fig. 9.12 in order to provide a reference for comparison with Fig. 9.13.

For Fig. 9.13 the domain was enlarged to $0.1<x<10$. The BCs were altered so that the exact solution would remain the same as Eqs. (9.14). The evaluation point was also

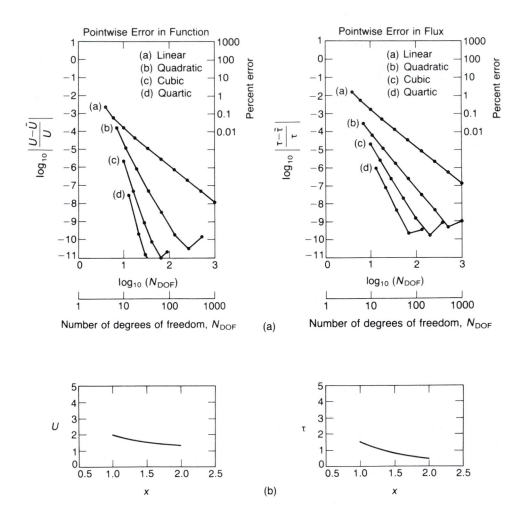

FIGURE 9.12 ● (a) Error curves for the illustrative problem in Chapters 3 through 5. The domain is $1<x<2$. There is no singularity in or near the domain. The solution is evaluated at $x^*=1.5$. Function values are at interelement nodes, and flux values are at Gauss points. (b) Exact solution.

kept the same, namely, $x^*=1.5$. Thus the only change in the problem is merely an extension of the previous domain and solution in order to include the quasisingular behavior near $x=0$.

These figures reveal a decrease in accuracy of quite a few orders of magnitude. In addition, the curves are bunched more closely together, almost overlapping at the coarser meshes. This is to be expected since the log C term in Eq. (9.7), which determines the vertical shift, contains the $(p+1)$st derivative. Near a singularity this constant can increase dramatically as p is increased; hence, the higher-order curves will shift upwards more than the lower-order curves. This behavior is perhaps closely related to the previous observation that when the singularity is in the domain, the rate of convergence may be governed by the strength of the singularity, so higher-degree polynomials would be wasteful,

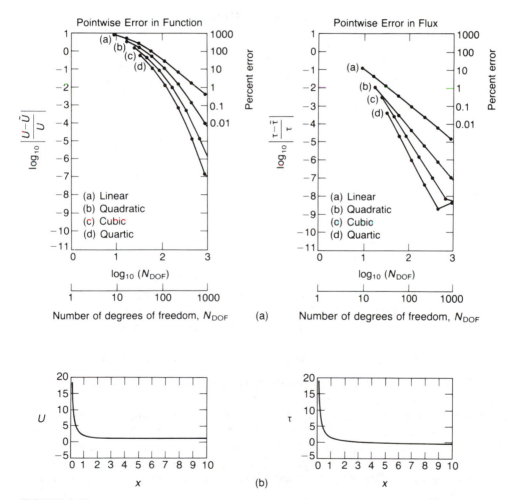

FIGURE 9.13 ● (a) Error curves for the illustrative problem in Chapters 3 through 5, with the domain enlarged to $0.1 < x < 10$. There is a singularity at $x=0$. The solution is evaluated at $x^*=1.5$. Function values are at interelement nodes, and flux values are at Gauss points. (b) Exact solution.

unless one of the aforementioned strategies is employed. Based on these observations, this author prefers the use of lower-degree polynomials (linears or quadratics) in regions of singular or quasisingular behavior, unless an appropriate grading strategy can be determined to recover the higher convergence rates associated with smooth solutions.

The use of graded meshes can be an effective strategy in regions of singular or quasisingular behavior.[8] However, one frequently doesn't know the steepness of the singular or quasisingular solution (i.e., the strength of the singularity, either inside or just outside the domain). Therefore a naive but usually effective approach is to employ *some* degree of grading, such as a geometric spacing of nodes, which most commercial preprocessors can easily generate. Two or three coarse meshes with different spacing ratios would help to determine a good grading.

9.5 Local Mesh Refinement

One is usually interested in a solution only in local regions of the domain, so it is common practice to provide additional DOF in those regions to increase the local accuracy, a process called *local mesh refinement*. This is an effective and recommended technique. However, there are some precautions that must be observed, which the example below will illustrate.

Recall that all the previous error curves used a quasiuniform mesh refinement. Such a refinement satisfies the important condition $h_{max} \to 0$, not just $h \to 0$. This, of course, implies that *all* the elements in the mesh must become arbitrarily small, not just a few within a local region. Local mesh refinement violates this condition. The consequences of the violation may or may not be important, depending on the extent to which the local refinement is pursued. The following example demonstrates what happens errorwise during a local refinement, especially when the refinement is pursued ''too far.''

We will consider once again the sample problem in Eq. (9.8) and will use the linear element. (The other elements exhibit very similar behavior.) The mesh refinement will consist of the following two stages:

- *An initial global, quasiuniform refinement for several meshes.* This is the same strategy used in Section 9.2, so the rate of convergence should approach a slope of 2 (evaluating at an interelement node for the function and at a Gauss point for the flux).
- *Local mesh refinement for all subsequent meshes.* The mesh in the interval $1 < x < 5$ will remain fixed at the last mesh used in that interval in the first stage. The mesh in the remainder of the domain, $0 < x < 1$, will continue with quasiuniform refinement.

Figure 9.14 shows the error curve corresponding to the evaluation point $x^* = 0.5$, which is in the middle of the locally refined region. This might simulate a typical situation in which the analyst is interested in the solution only at or near a particular point, so he or she locally refines the mesh only in the vicinity of that point. Also shown, for reference, is the curve corresponding to a global refinement (i.e., no local refinement).

The local refinement curve departs from the global refinement curve in two ways: an initial sharp dip, followed by a horizontal plateau. These features can be better understood

---

[8] An alternate approach for regions of quasisingular behavior would be to use trial functions that are almost singular, e.g., isoparametric elements with the interior nodes placed close to, but not at, the locations that yield singular shape functions. Knowing just where to place the nodes is the challenge here. There seems to be little use made of this strategy.

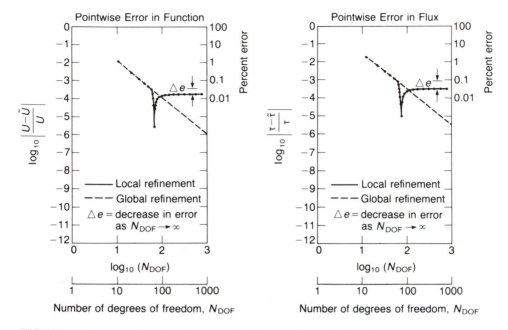

FIGURE 9.14 ● Local mesh refinement in the interval $0<x<1$ in the sample problem in Eq. (9.8). The evaluation point is at $x^*=0.5$.

if we simply plot the function $\tilde{U}$ — rather than $\log_{10}|(U-\tilde{U})/U|$ — versus $\log_{10} N_{DOF}$, as shown in Fig. 9.15. The latter displays the same values of $\tilde{U}(0.5)$ used in Fig. 9.14, one for each mesh, excepting only the values for the first two meshes in the first stage, which do not fall within the range shown.

The essential feature in Fig. 9.15 is that, after local refinement begins, the sequence of approximate solutions *approaches a different limit* from the exact solution. This is a general characteristic of any local refinement process. By fixing a portion of the mesh, a certain amount of error is locked into that region, and that contaminates the rest of the domain to one degree or another. Stated more precisely, the approximate solution and error within the fixed region will continue to change slightly as the remainder of the mesh is locally refined. However, the error in the fixed region necessarily approaches a nonzero limit since the finite number of trial functions is incapable of representing the exact solution. The limiting error in the fixed region in turn produces a limiting error in the rest of the domain because of the continuity in the FE solution across elements. Thus the local refinement process *does* have a limit that it converges to, but it is an *approximate* solution, not the exact solution. (Recall Fig. 3.13 and Section 5.6.)

In the present example the convergence is monotonic,[9] and the new limit is on the other side of the exact solution. Therefore the local refinement curve passes through the

---

[9] Several of the meshes (not all of them) form an *imbedded sequence*; i.e., all the nodes in one mesh are contained in the next mesh, which implies (assuming the same types of shape functions are used) that all trial solutions of one mesh are contained in the next mesh. This ensures that the global energy of the solution, though not necessarily the pointwise values, will converge monotonely. Frequently, though, pointwise values will also converge monotonely for such sequences.

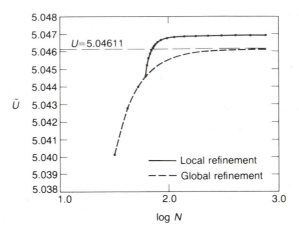

FIGURE 9.15 ● Local mesh refinement, using the same values of $\tilde{U}(0.5)$ as used in Fig. 9.14 (excepting the first two points).

exact solution, resulting in some meshes yielding values very close to the exact solution. This accounts for the initial sharp dip in Fig. 9.14. In other examples, where the new limit is on the same side of the exact solution, there is no dip.

Figure 9.14 reveals only a small improvement in accuracy as a result of the local refinement. This might seem rather discouraging. However, the amount of improvement in accuracy that is possible depends on the proximity of the evaluation point to the fixed

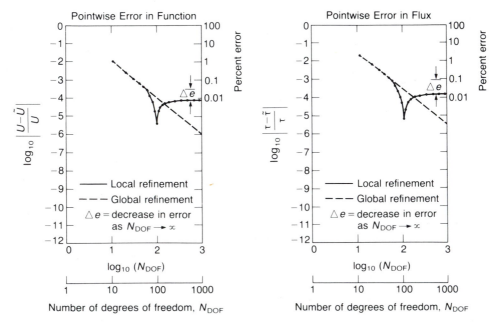

FIGURE 9.16 ● Local mesh refinement in the interval $0<x<2.5$ in the sample problem in Eq. (9.8). The evaluation point is at $x^*=0.5$.

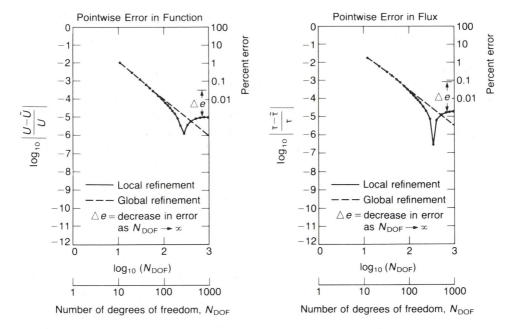

FIGURE 9.17 ● Local mesh refinement in the interval $0<x<4$ in the sample problem in Eq. (9.8). The evaluation point is at $x^*=0.5$.

portion of the mesh (assuming the evaluation point to be within the locally refined portion). To demonstrate this, let's repeat the above two-stage process, only this time move the boundary between the fixed and locally refined regions further away from the evaluation point. The latter will be kept at $x^*=0.5$. Figures 9.16 and 9.17 show the results for progressively larger separations.

A comparison of Figs. 9.14, 9.16, and 9.17 reveals that the closer the evaluation point is to the boundary of the fixed region, the greater is the contamination by that region.[10] Stated another way, one cannot expect high accuracy immediately adjacent to a low-accuracy region. It would therefore not be computationally efficient to juxtapose a very fine and very coarse mesh, since the extra DOF close to the coarse mesh would essentially be wasted. A more efficient strategy would be to employ a graded mesh between the fine and coarse regions (see Section 13.6.4 for a 2-D illustration).[11]

The above results show that local refinement can generally improve accuracy to some extent, usually to an appreciable extent, and that is the *raison d'etre* for the technique.[12]

---

[10] This behavior cannot be attributed merely to the fact that the fixed region is changing in size. In numerical experiments not shown here, the fixed region was held constant while the evaluation point was moved progressively further away. The results were very similar to those reported above.

[11] In 2-D and 3-D problems analyzed with commercial FE programs, it may sometimes be significantly easier, depending on the capabilities of the preprocessor, to construct a uniform rather than a graded mesh. Here the cost savings of the analyst's time might outweigh the extra cost of wasted computation.

[12] Occasionally local refinement can yield an increase in the pointwise error. This may be illustrated with the sample problem in Eq. (9.8), using $2.5 \leq x \leq 5$ for the fixed region and an evaluation point $x^*=2$.

However, the curves show that there is a limit to the amount of improvement. In addition, the error can no longer be estimated by comparing the differences between successive solutions, because these differences will be approaching zero, while the true error will not be approaching zero.

There is another consideration that can limit the extent to which local refinement can be pursued. This is the problem of *ill-conditioning* of the stiffness matrix [9.24, 9.25]. The terms in an element stiffness matrix contain the dimensions of the element; hence the presence of very large and very small elements in a mesh results in very large and very small *eigenvalues* (see Chapter 10) for the system stiffness matrix. Because of finite machine precision, the resulting accumulation of round-off errors causes loss of significant figures. Ill-conditioning can be circumvented by performing a separate analysis on just the refined portion of the mesh, as illustrated for a 2-D mesh in Section 13.6.4.

Exercises

It is recommended that the reader continue to work with the exercises described in Chapters 7 and 8, but here the emphasis should be on studying mesh-refinement strategies and observing performance with regard to accuracy and rate of convergence. A suggested approach is to work with just one exercise and perform the following investigation:

9.1 First try *h*-refinement, using any one of the element types. Do a sequence of uniform meshes, calculating the pointwise error for each mesh and plotting the error versus *h* on log-log axes. Calculate the rate of convergence and compare with the asymptotic rates shown in the text. In a classroom setting, several students might work on the same exercise, each using a different element type. This would afford a comparative study of the different elements.

9.2 Next try *p*-refinement. Begin with a mesh of uniform linear elements; then change the element type to quadratic, then cubic, and finally quartic. Again calculate the pointwise error for each mesh and plot the error versus N_{DOF} on log-log axes. On the same axes replot the data from the *h*-refinement study in Exercise 9.1 and observe the difference in performance between *p*- and *h*-refinement.

References

[9.1]   I. Babuska and B. Szabo, "On the Rates of Convergence of the Finite Element Method," *Int J Num Meth Eng*, **18**: 323-341 (1982).

[9.2]   T. Moan, "On the Local Distribution of Errors by Finite Element Approximations" in Y. Yamada and R. H. Gallagher (eds.), *Theory and Practice in Finite Element Structural Analysis*, University of Tokyo Press, Tokyo, 1973.

[9.3]   V. G. Korneev, "Superconvergence of Solutions of the Finite-Element Method in Mesh Norms" (in English), *U.S.S.R. Comput Maths Math Phys*, **22**(5): 115-131 (1982).

[9.4]   T. J. Oden and J. N. Reddy, *An Introduction to the Mathematical Theory of Finite Elements*, Wiley, New York, 1976.

[9.5]   G. F. Carey, D. Humphrey, and M. F. Wheeler, "Galerkin and Collocation-
Galerkin Methods with Superconvergence and Optimal Fluxes," *Int J Num Meth
Eng*, **17**:, 939-950 (1981).

[9.6]   M. Bakker, "One-Dimensional Galerkin Methods and Superconvergence at
Interior Nodal Points," *SIAM J Numer Anal*, **21**(1): 101-110 (1984).

[9.7]   U. W. Hochstrasser, "Orthogonal Polynomials," Chapter 22 in M. Abramowitz
and I. Stegun (eds.), *Handbook of Mathematical Functions*, National Bureau of
Standards, Applied Mathematics Series, **55**, 1964.

[9.8]   G. F. Carey, "Derivative Calculation from Finite Element Solutions," *Comp
Meth App Mech Eng*, **35**: 1-14 (1982).

[9.9]   I. Babuska and B. Szabo, op. cit.

[9.10] I. Babuska, B. A. Szabo, and I. N. Katz, "The *p*-Version of the Finite Element
Method," *SIAM J Numer Anal*, **18**(3): 515-545 (1981).

[9.11] A. G. Peano and J. W. Walker, "Modeling of Solid Continua by the *P*-Version
of the Finite Element Method," Paper 39 in J. Robinson (ed.), *New and Future
Developments in Commercial Finite Element Methods* (Proc Third World Congress
and Exhibition on Finite Element Methods, Beverly Hills, Calif.), Robinson and
Associates, Devon, England, 1981.

[9.12] O. C. Zienkiewicz, J. P. de S. R. Gago, and D. W. Kelly, "The Hierarchical
Concept in Finite Element Analysis," *Computers and Structures*, **16**(1-4): 53-65
(1983).

[9.13] P. Tong and T. H. H. Pian, "On the Convergence of the Finite Element Method
for Problems with Singularity," *Int J Solids and Structures*, **9**: 313-321 (1973).

[9.14] G. Strang and G. J. Fix, *An Analysis of the Finite Element Method*, Prentice-
Hall, Englewood Cliffs, N. J., 1973, Chapter 8.

[9.15] G. F. Carey and J. T. Oden, *Finite Elements: A Second Course*, vol. II, Prentice-
Hall, Englewood Cliffs, N. J., 1983, Section 5.3.

[9.16] R. D. Henshell and K. G. Shaw, "Crack Tip Finite Elements are Unnecessary,"
Int J Num Meth Eng, **9**: 495-507 (1975).

[9.17] R. Barsoum, "On the Use of Isoparametric Finite Elements in Linear Fracture
Mechanics," *Int J Num Meth Eng*, **10**: 25-37 (1976).

[9.18] J. E. Akin, "The Generation of Elements with Singularities," *Int J Num Meth
Eng*, **10**: 1249-1259 (1976).

[9.19] M. Stern, "Families of Consistent Conforming Elements with Singular Derivative
Fields," *Int J Num Meth Eng*, **14**: 409-421 (1979).

[9.20] I. Fried and S. K. Yang, "Best Finite Elements Distribution Around a Singularity," *AIAA J*, **10**(9): 1244-1246 (1972).

[9.21] G. Strang and G. J. Fix, op. cit.

[9.22] D. M. Parks, "On the Convergence of Finite Element Meshes in the Presence
of Singularities," *Int J Num Meth Eng*, **14**(5): 780-784 (1979).

[9.23] J. A. Gregory, D. Fishelov, B. Schiff, and J. R. Whiteman, "Local Mesh Refinement with Finite Elements for Elliptic Problems," *J Comp Phys*, **29**: 133-140
(1978).

[9.24] G. Strang and G. J. Fix, op. cit., Chapter 5.

[9.25] G. E. Forsythe and C. B. Moler, *Computer Solution of Linear Algebraic Systems*,
Prentice-Hall, Englewood Cliffs, N. J., 1967, Chapter 8.

10

1-D Eigenproblems

Introduction

Most boundary-value problems, including the one analyzed in Chapters 6 through 8, have an associated eigenproblem. The two are closely related, both in physical meaning and in mathematical expression. This chapter will analyze the 1-D eigenproblem associated with Eq. (6.25). This will require only a minor change to the formation of the element equations. The resulting assembled system equations will be in the form of an algebraic eigenproblem, which is a standard problem that can be solved by a wide assortment of conventional numerical analysis algorithms.

10.1 Mathematical and Physical Description of the Problem

We will analyze a general class of 1-D eigenproblems, governed by the following second-order ordinary differential equation:

$$-\frac{d}{dx}\left[\alpha(x)\frac{dU(x)}{dx}\right] + \beta(x)U(x) - \lambda\gamma(x)U(x) = 0 \tag{10.1}$$

The domain is any interval $x_a < x < x_b$ along the x-axis. A boundary condition must be applied at each of the two boundary points x_a and x_b. The condition may be one of the following:

At x_a

$$U(x_a) = 0 \qquad \text{or} \qquad \tau(x_a) = 0 \tag{10.2a}$$

At x_b

$$U(x_b) = 0 \qquad \text{or} \qquad \tau(x_b) = 0 \tag{10.2b}$$

where $\tau(x) = -\alpha(x)dU/dx$ is the flux.[1]

---

[1] Equation (10.1), with the boundary conditions given in Eq. (10.2), is the classic *Sturm-Liouville* eigenproblem, which occurs in a broad spectrum of physical applications [10.1]. Its solutions include many well-known and well-tabulated special functions, e.g., trigonometric, Bessel, Legendre, etc.

Equation (10.1) is the same as the boundary-value problem in Eq. (6.25) and Eq. (7.1a) except for (1) the presence of an additional term $-\lambda\gamma(x)U(x)$ on the LHS and (2) the vanishing of $f(x)$ on the RHS; that is, $f(x)=0$. $\gamma(x)$ is a known function describing a physical property of the system, similar to $\alpha(x)$ and $\beta(x)$. For example, in mechanical applications it is a mass or mass density. The coefficient λ is an unknown scalar-valued parameter, called an *eigenvalue*. In most physical applications it represents a frequency or energy level.

Clearly there is an uninteresting trivial solution, $U(x)=0$, which satisfies Eq. (10.1) and Eqs. (10.2) for any value of λ. Of much greater interest, though, is the fact that there exist nontrivial solutions (many different ones, in fact) called *eigenfunctions* (also *eigenvectors, eigenmodes, natural modes, normal modes, mode shapes*, or simply, *modes*). Corresponding to each eigenfunction is a critical value of the eigenvalue λ (usually a different eigenvalue for each different eigenfunction, but not always). We will refer to each eigenfunction/eigenvalue pair as an *eigensolution*. A solution to Eqs. (10.1) and (10.2) will consist of many eigensolutions. The exact solution always comprises an infinity of eigensolutions. An FE approximate solution will comprise a large but finite number, equal, in fact, to the number of active DOF in the FE model.[2] Fortunately, most practical problems require calculating only a few of the eigensolutions.

There are two essential features that distinguish the eigenproblem from the boundary-value problem. First is the fact that λ *is an unknown quantity*; its critical values $\lambda_1, \lambda_2,...$ must be solved for, along with the corresponding solutions $U_1(x), U_2(x),...$. Note that if λ were a known quantity, we would be able to simply write the last two terms in Eq. (10.1) as $\beta'(x)U(x)$, with $\beta'(x)=\beta(x)-\lambda\gamma(x)$, and then solve it as a boundary-value problem. It is important, therefore, that these two terms be kept separate.

The second feature is that the interior loads [RHS of Eq. (10.1)] and boundary loads [RHS of Eqs. (10.2)] are all zero; i.e., *no loads are driving the system*. Yet there exist many nonzero solutions. In physical terms, the state of the system changes despite the absence of any (externally applied) loads. This is characteristic of a *natural* (or *resonant*) condition in which energy internal to the system is oscillating back and forth between different forms (e.g., kinetic and potential), but no energy is being exchanged with the surroundings. Familiar examples from the physical world abound everywhere, e.g., resonant vibrations of buildings and machinery, resonant acoustic modes in rooms and auditoriums, etc. (Of course, this is an idealization because in real applications some energy is always lost from the system by a variety of mechanisms, such as friction or radiation, which are referred to generically as *damping*.)

To illustrate the application of Eq. (10.1) to a specific problem, consider again the cable deflection problem (recall Fig. 6.5). Let's derive the governing equation from physical principles. Here we want to describe the actual time-dependent motion of the cable, rather than a single static equilibrium position. Therefore the transverse displacement[3] w becomes a function of both position x and time t. In accordance with Newton's second law (force

---

[2] An inactive DOF is one that has been constrained by a constraint equation such as an essential BC.

[3] Previous chapters used an uppercase W for transverse displacement of the cable. Here we use a lowercase w for the same transverse displacement (although here it is time-dependent), and we reserve the uppercase W for the displacement amplitude, introduced later in Eq. (10.7).

= mass × acceleration), we must add an inertial acceleration term, $\varrho dx(\partial^2 w/\partial t^2)$, to the RHS of the balance of forces in Eq. (6.13), resulting in

$$\frac{\partial F(x,t)}{\partial x} + k(x)w(x,t) = f(x) - \varrho(x)\frac{\partial^2 w(x,t)}{\partial t^2} \tag{10.3}$$

where $\varrho(x)$ is the lineal mass density (mass per unit length) and $\partial^2 w/\partial t^2$ is the transverse acceleration.

For *free* motion (i.e., no applied loads) the interior load $f(x)$ must be zero, so Eq. (10.3) becomes

$$\frac{\partial F(x,t)}{\partial x} + k(x)w(x,t) = -\varrho(x)\frac{\partial^2 w(x,t)}{\partial t^2} \tag{10.4}$$

The constitutive Eq. (6.16) remains the same, although the total derivative becomes partial:

$$F(x,t) = -T(x)\frac{\partial w(x,t)}{\partial x} \tag{10.5}$$

Inserting Eq. (10.5) into Eq. (10.4) yields

$$-\frac{\partial}{\partial x}\left[T(x)\frac{\partial w(x,t)}{\partial x}\right] + k(x)w(x,t) = -\varrho(x)\frac{\partial^2 w(x,t)}{\partial t^2} \tag{10.6}$$

which is the *wave equation* for transverse waves in the cable.

A *free-vibration* solution to Eq. (10.6) is characterized by a motion that varies sinusoidally in time [10.2, 10.3] (called *time harmonic, simple harmonic,* or just *harmonic*):

$$w(x,t) = W(x)\sin \omega t \tag{10.7}$$

where $W(x)$ is the amplitude of vibration at each location x and $\omega = 2\pi f$ is the circular frequency. Since the x and t variables are in separate functions in Eq. (10.7), the partial derivatives of $w(x,t)$ can be expressed as total derivatives of either $W(x)$ or $\sin \omega t$; that is,

$$\frac{\partial}{\partial x}\left[T(x)\frac{\partial w(x,t)}{\partial x}\right] = \frac{d}{dx}\left[T(x)\frac{dW(x)}{dx}\right]\sin \omega t \tag{10.8a}$$

and

$$\frac{\partial^2 w(x,t)}{\partial t^2} = W(x)\frac{d^2(\sin \omega t)}{dt^2}$$

$$= -W(x)\omega^2 \sin \omega t \tag{10.8b}$$

Inserting Eqs. (10.7) and (10.8) into Eq. (10.6) and canceling $\sin \omega t$ from both sides yields

$$-\frac{d}{dx}\left[T(x)\frac{dW(x)}{dx}\right] + k(x)W(x) - \omega^2\varrho(x)W(x) = 0 \tag{10.9}$$

Equation (10.9) is the governing equation for free transverse vibrations of a flexible cable; it has the same form as the general eigenproblem in Eq. (10.1), where $U(x) = W(x)$, $\alpha(x) = T(x)$, $\beta(x) = k(x)$, $\gamma(x) = \varrho(x)$, and $\lambda = \omega^2$. Note that the eigenvalue is proportional to the square of the frequency; that is, $\lambda = (2\pi f)^2$. Thus, if λ_n is the nth eigenvalue (corresponding to the nth eigensolution), then the natural (or resonant) frequencies are

$$f_n = \frac{1}{2\pi}\sqrt{\lambda_n} \qquad n = 1,2,\dots \tag{10.10}$$

If the cable is of length L and its ends are held fixed for all time, then the boundary conditions are

$$w(0,t) = 0$$

$$w(L,t) = 0 \tag{10.11a}$$

which, from Eq. (10.7), become

$$W(0) = 0$$

$$W(L) = 0 \tag{10.11b}$$

corresponding to the first of the two conditions in Eqs. (10.2).

There are eigenproblems from many other fields of application which are also described by Eq. (10.1) and the boundary conditions in Eq. (10.2). Table 10.1 summarizes four different applications, including the above cable vibration problem, using notation that

TABLE 10.1 ● One-dimensional eigenproblems

Application	Unknown	Physical/material properties			Eigenvalue	Governing equation
General mathematical formulation	U	α	β	γ	λ	$-\dfrac{d}{dx}\left(\alpha\dfrac{dU}{dx}\right)$ $+\beta U-\lambda\gamma U=0$
Cable deflection: Transverse vibrations	W (transverse displacement amplitude)	T (tension in cable)	k (elastic modulus of foundation)	ϱ (mass density of cable)	ω^2 (ω=circular frequency)	$-\dfrac{d}{dx}\left(T\dfrac{dW}{dx}\right)$ $+kW-\omega^2\varrho W=0$
Elasticity: Longitudinal vibrations in rod	U (longitudinal displacement amplitude)	E (Young's modulus)		ϱ (mass density of rod)	ω^2 (ω=circular frequency)	$-\dfrac{d}{dx}\left(E\dfrac{dU}{dx}\right)$ $-\omega^2\varrho U=0$
Acoustics: Plane waves	Φ (acoustic potential)	1		1	k^2 ($k=\omega/c$; ω=circ. freq.; c=sound veloc.)	$-\dfrac{d^2\Phi}{dx^2}-k^2\Phi=0$
Cylindrical waves	Φ	r (radial coordinate)	$\dfrac{m^2}{r}$ $m=0,1,2,...$	r	k^2	$-\dfrac{d}{dr}\left(r\dfrac{d\Phi}{dr}\right)+\dfrac{m^2}{r}\Phi$ $-k^2r\Phi=0$
Spherical waves	Φ	r^2	$m(m+1)$ $m=0,1,2,...$	r^2	k^2	$-\dfrac{d}{dr}\left(r^2\dfrac{d\Phi}{dr}\right)$ $+m(m+1)\Phi-k^2r^2\Phi=0$
Quantum mechanics: Time-independent Schroedinger equation for single particle	ψ (probability amplitude function)	$\dfrac{\hbar^2}{2m}$ $\hbar=h/2\pi$; h=Planck's constant; m=mass of particle	V (potential energy of particle)	1	E (total energy of particle)	$-\dfrac{\hbar^2}{2m}\dfrac{d^2\psi}{dx^2}$ $+V\psi-E\psi=0$

is conventional and/or meaningful to each application. Sections 10.5 and 10.6 will apply the UNAFEM program to both a cable vibration problem and a classical problem in quantum mechanics.

10.2 Derivation of Element Equations

Steps 1 through 6 are virtually the same as in Section 7.2 since the governing Eqs. (10.1) and (7.1a) are the same except for the new $-\lambda\gamma(x)U(x)$ term, which is mathematically manipulated in the same way as is the $\beta(x)U(x)$ term. That is, we could simply copy those equations and add a new term identical to the $\beta(x)$ term in which $-\lambda\gamma(x)$ is substituted for $\beta(x)$. However, it is worthwhile going through the steps once again in order to reinforce our understanding of the procedure and to see explicitly the additional details created by the $-\lambda\gamma(x)U(x)$ term.

Step 1: Write the Galerkin residual equations for a typical element.

$$\int^{(e)} \left[-\frac{d}{dx}\left(\alpha(x)\, \frac{d\tilde{U}^{(e)}(x;a)}{dx} \right) + \beta(x)\tilde{U}^{(e)}(x;a) - \lambda\gamma(x)\tilde{U}^{(e)}(x;a) \right] \phi_i^{(e)}(x)\, dx = 0$$

$$i = 1,2,\ldots,n \qquad \textbf{(10.12)}$$

where n is the number of DOF (and therefore shape functions) in the element, which depends on the type of element developed in step 4. For notational consistency we should put a tilde over the λ as a reminder that this quantity, like $\tilde{U}$ and $\tilde{\tau}$, is an unknown quantity that the FE solution will provide approximate values for. However, in order to avoid an unconventional appearance to many of the standard eigenvalue expressions later in the chapter, we will omit the tilde.

Step 2: Integrate by parts.

The second derivative term is integrated by parts once:

$$\int^{(e)} \frac{d\phi_i^{(e)}(x)}{dx}\, \alpha(x)\, \frac{d\tilde{U}^{(e)}(x;a)}{dx}\, dx + \int^{(e)} \phi_i^{(e)}(x)\beta(x)\tilde{U}^{(e)}(x;a)\, dx$$

$$-\lambda \int^{(e)} \phi_i^{(e)}(x)\gamma(x)\tilde{U}^{(e)}(x;a)\, dx = -\left[\left(-\alpha(x)\, \frac{d\tilde{U}^{(e)}(x;a)}{dx} \right) \phi_i^{(e)}(x) \right]_{x_1}^{x_n}$$

$$i = 1,2,\ldots n \qquad \textbf{(10.13)}$$

where x_1, x_n are the two boundary nodes of the element.

The boundary term contains, as usual, the flux,

$$\left[\left(-\alpha(x)\, \frac{d\tilde{U}^{(e)}(x;a)}{dx} \right) \phi_i^{(e)}(x) \right]_{x_1}^{x_n} = \left[\tilde{\tau}(x;a)\phi_i^{(e)}(x) \right]_{x_1}^{x_n} \qquad \textbf{(10.14)}$$

For the eigenproblem, this term must vanish from the system equations. To see this, recall that the boundary term occurs in two different ways in the system equations: as the expression shown above — that is, $\tilde{\tau}\phi_i^{(e)}$ — at each node on the boundary of the domain, and as the difference of two such expressions at each node on interelement boundaries. In the first case, Eqs. (10.2) require that the term vanish at the domain boundary nodes. Thus, imposing $\tau(x)=0$ obviously eliminates the term; and imposing $U(x)=0$ eliminates the entire equation in which the term appears (recall the application of essential BCs by constraint equations, Section 4.4). In the second case, a nonzero difference in flux at an interelement boundary represents an applied concentrated load (Section 7.6); however, the eigenproblem does not permit applied loads.

Since the boundary term must vanish from the system equations, we will ignore it right at the outset by eliminating it from the element equations. Therefore Eq. (10.13) may be written as follows:

$$\int^{(e)} \frac{d\phi_i^{(e)}(x)}{dx}\, \alpha(x)\, \frac{d\tilde{U}^{(e)}(x;a)}{dx}\, dx + \int^{(e)} \phi_i^{(e)}(x)\beta(x)\tilde{U}^{(e)}(x;a)\, dx$$

$$- \lambda \int^{(e)} \phi_i^{(e)}(x)\gamma(x)\tilde{U}^{(e)}(x;a)\, dx = 0 \qquad i = 1,2,\ldots,n \qquad \textbf{(10.15)}$$

Step 3: Substitute the general form of the element trial solution into interior integrals in residual equations.

The element trial solution expansion has the general form,

$$\tilde{U}^{(e)}(x;a) = \sum_{j=1}^{n} a_j \phi_j^{(e)}(x) \qquad\qquad \textbf{(10.16)}$$

Inserting Eq. (10.16) into Eq. (10.15) yields

$$\sum_{j=1}^{n} \left[\int^{(e)} \frac{d\phi_i^{(e)}(x)}{dx}\, \alpha(x)\, \frac{d\phi_j^{(e)}(x)}{dx}\, dx \right] a_j + \sum_{j=1}^{n} \left[\int^{(e)} \phi_i^{(e)}(x)\beta(x)\phi_j^{(e)}(x)\, dx \right] a_j$$

$$- \lambda \sum_{j=1}^{n} \left[\int^{(e)} \phi_i^{(e)}(x)\gamma(x)\phi_j^{(e)}(x)\, dx \right] a_j = 0 \qquad i = 1,2,\ldots,n \qquad \textbf{(10.17)}$$

These are the element equations for a typical element.

Equation (10.17) may be written in conventional matrix form:

$$[K]^{(e)}\{a\} - \lambda[M]^{(e)}\{a\} = \{0\} \qquad\qquad \textbf{(10.18)}$$

where

$$[K]^{(e)} = [K\alpha]^{(e)} + [K\beta]^{(e)}$$

and

$$
Ka_{ij}^{(e)} = \int^{(e)} \frac{d\phi_i^{(e)}(x)}{dx} \, \alpha(x) \, \frac{d\phi_j^{(e)}(x)}{dx} \, dx
$$

$$
K\beta_{ij}^{(e)} = \int^{(e)} \phi_i^{(e)}(x)\beta(x)\phi_j^{(e)}(x) \, dx
$$

$$
M_{ij}^{(e)} = \int^{(e)} \phi_i^{(e)}(x)\gamma(x)\phi_j^{(e)}(x) \, dx \tag{10.19}
$$

The symbol M is employed for the $\gamma(x)$ integral because $\gamma(x)$ represents mass, or mass density, in so many physical applications. Because of this, the matrix is usually referred to as the *mass matrix*, irrespective of the actual physical meaning (analogous to calling $[K]$ the stiffness matrix).

Step 4: Develop specific expressions for the shape functions, $\phi_j^{(e)}(x)$.

Here our work is already done. We can use any of the four 1-D elements already developed in Chapters 7 and 8.

Step 5: Substitute the shape functions into the element equations, and transform the integrals into a form appropriate for numerical evaluation.

For the four 1-D elements in UNAFEM, the $Ka_{ij}^{(e)}$ and $K\beta_{ij}^{(e)}$ integrals have already been treated in Chapters 7 and 8. Here we only need to treat the $M_{ij}^{(e)}$ integral.

Consider first the C^0-linear element (Chapter 7) which was developed by the direct (nonisoparametric) approach. The integrals were evaluated exactly with closed-form expressions. Using the expressions for $\phi_1^{(e)}(x)$ and $\phi_2^{(e)}(x)$ in Eqs. (7.12b), we have

$$
M_{11}^{(e)} = \int_{x_1}^{x_2} \phi_1^{(e)}(x)\gamma(x)\phi_1^{(e)}(x) \, dx
$$

$$
= \int_{x_1}^{x_2} \frac{x_2 - x}{x_2 - x_1} \, \gamma^{(e)} \, \frac{x_2 - x}{x_2 - x_1} \, dx
$$

$$
= \frac{1}{3} \, \gamma^{(e)}L \tag{10.20a}
$$

where $L = x_2 - x_1$ is the length of the element. Similarly,

$$
M_{12}^{(e)} = \frac{1}{6} \, \gamma^{(e)}L
$$

$$
M_{22}^{(e)} = \frac{1}{3} \, \gamma^{(e)}L \tag{10.20b}
$$

Since $M_{ij}^{(e)}$ is symmetric, $M_{21}^{(e)} = M_{12}^{(e)}$.

The physical property $\gamma(x)$ is treated in the same manner as are $\alpha(x)$ and $\beta(x)$. [Recall Eqs. (7.17) through (7.19) and Figs. 7.3 through 7.5.] Thus $\gamma(x)$ is represented as a constant, $\gamma^{(e)}$, within each element, where $\gamma^{(e)}$ is the value of $\gamma(x)$ at the center of each element:

$$\gamma^{(e)} = \gamma(x_c) \tag{10.21a}$$

where

$$x_c = \frac{1}{2}(x_1 + x_2) \tag{10.21b}$$

UNAFEM will calculate values for $\gamma^{(e)}$ from Eq. (10.21), using expressions for $\gamma(x)$ supplied by the user. Just as with $\alpha(x)$ and $\beta(x)$, we will let $\gamma(x)$ be defined by a piecewise-quadratic polynomial over the domain. The user may specify from 1 to 20 different quadratics:

$$\gamma(x) = \gamma_0 + \gamma_1 x + \gamma_2 x^2 \tag{10.22}$$

each one being applicable to a different region of the domain. If $\gamma(x)$ in the original problem statement, Eq. (10.1), cannot be represented exactly by a piecewise-quadratic, then this method will introduce a modeling error (see Section 7.2, step 5).

The expressions in Eqs. (10.20) are physically meaningful. For example, in the cable problem $\gamma(x) = \varrho(x)$, the mass per unit length. Therefore $\gamma^{(e)}L$ is the mass of element (e). Noting that $M_{11}^{(e)} + M_{12}^{(e)} = \varrho L/2$ and $M_{21}^{(e)} + M_{22}^{(e)} = \varrho L/2$, we see that half the element mass is associated with node 1 and the other half with node 2, an intuitively appealing result.

The C^0-quadratic, -cubic, and -quartic elements are all isoparametric elements requiring numerical integration. The integrals over the real element are transformed to integrals over the parent element, and then approximated by a finite sum. [Recall Eqs. (8.43), (8.44), and (8.50) for the quadratic element and Eqs. (8.76) and (8.77) for the cubic element.] The $M_{ij}^{(e)}$ integral is transformed in an identical fashion:

$$M_{ij}^{(e)} = \int^{(e)} \phi_i^{(e)}(x)\gamma(x)\phi_j^{(e)}(x)\,dx$$

$$= \int_{-1}^{+1} \phi_i(\xi)\gamma^{(e)}(\xi)\phi_j(\xi)J^{(e)}(\xi)\,d\xi$$

$$\simeq \sum_{l=1}^{n} W_{nl}\phi_i(\xi_{nl})\gamma^{(e)}(\xi_{nl})\phi_j(\xi_{nl})J^{(e)}(\xi_{nl}) \tag{10.23}$$

where n is the number of quadrature points used in the numerical integration. The value of n is determined by the desired accuracy, and we observe that the order of the quadrature rule for the mass integrals may be the same as for the stiffness and load integrals. (Recall guidelines II and III in Chapter 8.)

Step 6: Derive expression for the flux.

The expression for the flux (as well as the location of the optimal flux-sampling points) is unaffected by the mass matrix. [Recall Eq. (7.27) for the linear element, Eq. (8.53) for the quadratic, and Eq. (8.81) for the cubic.]

This completes the six steps for deriving the element equations and the expression for flux.

10.3 The Algebraic Eigenproblem

Consider briefly the first few steps in the numerical computation phase. Step 7 is the preprocessing step in which numerical data for a particular problem is input to the program; there are only minor differences between the eigenproblem and boundary-value problem, and they will be illustrated in Sections 10.5 and 10.6.

Step 8 involves the assembly of the element Eqs. (10.18) to form the system equations:

$$[K]\{a\} - \lambda[M]\{a\} = \{0\} \tag{10.24}$$

It should be almost obvious that the element mass matrices assemble into the system mass matrix by exactly the same procedure that the element stiffness matrices assemble into the system stiffness matrix. (Recall the general assembly rule in Section 5.7.)

Step 9 involves the application of boundary conditions and concentrated loads. However, as argued in step 2 above, these are all zero-valued; i.e., there can be no applied fluxes or concentrated loads, and any essential BCs must specify a zero value. Hence this step will preserve the zero vector on the RHS of Eq. (10.24).

This brings us to step 10, which involves the solution of a system of equations in the form of Eq. (10.24). Since λ and $\{a\}$ are both unknowns that must be solved for, the existing Gaussian elimination algorithm, GAUSEL, is inappropriate. We will have to use a different type of equation solver. The purpose of the present section is to discuss the available techniques for solving Eq. (10.24).

Equation (10.24) is a system of algebraic equations, referred to as an *algebraic eigenproblem* (or, algebraic eigenvalue problem). Our FE analysis has essentially transformed the original eigenproblem in differential form, Eq. (10.1), into an equivalent eigenproblem in algebraic form. By "equivalent," we mean that the latter is a discretized approximation to the former. Thus Eq. (10.1) has an infinite number of DOF (the continuum of x-values in the interval $x_a < x < x_b$), whereas Eq. (10.24) has only a finite number (the parameters a_i, $i=1,2,\ldots,N$). As the number of DOF in our FE model increases indefinitely ($N \rightarrow \infty$), the solution to Eq. (10.24) should converge to the solution of Eq. (10.1), assuming the convergence conditions are satisfied in the FE model.

The algebraic eigenproblem is a classical mathematical problem, predating the FEM by over a century. A great many solution techniques have evolved, and there is an extensive body of literature [10.4-10.6]. The reader wishing to delve more deeply into this field is advised to begin with the references cited here, which in turn contain more extensive bibliographies.

In the author's opinion, an engineer using any of the commercially available FE programs does not need to understand the mathematical intricacies of an eigensolution algorithm in order to use the program effectively. (Similarly, detailed knowledge of algorithms for solving systems of linear algebraic equations, e.g., Gaussian elimination, is not necessary either.) Therefore Sections 10.3.1 through 10.3.3 are intended to provide only a brief summary of a few salient features of the problem.

10.3.1 Some Mathematical Properties

Equation (10.24) is called the *generalized eigenproblem*. If $[M]=[I]$, the identity matrix, Eq. (10.24) becomes

$$[K]\{a\} = \lambda\{a\} \tag{10.25}$$

which is the *standard eigenproblem*.[4] The standard eigenproblem occurs more frequently than the generalized eigenproblem in many scientific applications, so most of the literature and available solution algorithms pertain to the standard eigenproblem. However, FE applications usually produce the generalized form, so the following material will deal explicitly with Eq. (10.24). It might be noted that Eq. (10.24) includes Eq. (10.25) as a special case, so all the mathematical results for Eq. (10.24) will also apply to Eq. (10.25) but in simpler form.

We will assume $[K]$ and $[M]$ are real, symmetric, positive-definite matrices of order N; this is the case for most FE applications. *Positive-definite* means

$$\{v\}^T[K]\{v\} > 0 \tag{10.26a}$$

$$\{v\}^T[M]\{v\} > 0 \tag{10.26b}$$

for all nonzero vectors $\{v\}$; *that is,* $\{v\}^T\{v\} > 0$. The superscript T denotes the transpose. Sometimes $[M]$ may be positive-semidefinite [the ">" sign in Eq. (10.26b) is replaced by "$\geq$"], which requires adjustments to some algorithms.

There exist N nontrivial ($\{a\} \neq \{0\}$) solutions to Eq. (10.24). Each solution consists of an eigenvalue λ_i and a corresponding eigenvector $\{v\}_i$, satisfying

$$[K]\{v\}_i = \lambda_i[M]\{v\}_i \tag{10.27}$$

The complete set of N eigenpairs $(\lambda_1, \{v\}_1), (\lambda_2, \{v\}_2), \ldots, (\lambda_N, \{v\}_N)$ constitutes the *eigensystem* for eq. (10.24). In addition,

$$0 < \lambda_1 \leq \lambda_2 \leq \cdots \leq \lambda_N \tag{10.28}$$

The positive-definiteness of $[K]$ and $[M]$ ensures that $\lambda_i > 0$, $i = 1, 2, \ldots, N$. The $\{v\}_i$ are only defined to within an arbitrary constant c, since the vector $c\{v\}_i$ is also a solution of Eq. (10.27):

$$[K](c\{v\}_i) = \lambda_i[M](c\{v\}_i) \tag{10.29}$$

It is easily shown that the λ_i are real (not complex-valued) and the $\{v\}_i$ are *orthogonal* with respect to both $[K]$ and $[M]$:

$$\{v\}_i^T[K]\{v\}_j = \lambda_i\delta_{ij}$$

$$\{v\}_i^T[M]\{v\}_j = \delta_{ij} \tag{10.30}$$

where δ_{ij} is the Kronecker delta. It is helpful to write out Eq. (10.30):

$$\left.\begin{array}{l}\{v\}_i^T[K]\{v\}_j = 0 \\[2mm] \{v\}_i^T[M]\{v\}_j = 0\end{array}\right\} \quad i \neq j \tag{10.31a}$$

$$\left.\begin{array}{l}\{v\}_i^T[K]\{v\}_i = \lambda_i \\[2mm] \{v\}_i^T[M]\{v\}_i = 1\end{array}\right\} \quad i = j \tag{10.31b}$$

---

[4] Equation (10.25) is often written in the form $([K] - \lambda[I])\{a\} = \{0\}$.

We see that the arbitrary constant c in Eq. (10.29) is determined by Eqs. (10.31b). It is said that $\{v\}_i$ is *orthonormal* with respect to $[M]$, where "normal" refers to the value unity on the RHS of the second of Eqs. (10.31b). (There are other normalizing procedures for determining the arbitrary constant c. Thus, one could use the condition $\{v\}_i^T\{v\}_i=1$, rather than $\{v\}_i^T[M]\{v\}_i=1$, which would yield $\{v\}_i^T[M]\{v\}_i=\mu_i$ and $\{v\}_i^T[K]\{v\}_i=\mu_i\lambda_i$, where μ_i is the "generalized mass" of the ith mode. For hand calculations, an easier procedure is to merely divide all terms of $\{v\}_i$ by the maximum-valued term, thereby making the maximum value unity.)

Equation (10.24) has the form $[C]a=0$, where the coefficient $[C]$ equals $[K]-\lambda[M]$. This is a system of N linear algebraic equations with a zero RHS. A nontrivial solution exists if and only if the determinant of the coefficient matrix vanishes [10.7]:

$$\det |[K] - \lambda[M]| = 0 \tag{10.32}$$

that is,

$$\begin{vmatrix} K_{11}-\lambda M_{11} & K_{12}-\lambda M_{12} & \cdots & K_{1N}-\lambda M_{1N} \\ K_{21}-\lambda M_{21} & K_{22}-\lambda M_{22} & \cdots & K_{2N}-\lambda M_{2N} \\ \cdot & \cdot & & \cdot \\ \cdot & \cdot & & \cdot \\ \cdot & \cdot & & \cdot \\ K_{N1}-\lambda M_{N1} & K_{N2}-\lambda M_{N2} & \cdots & K_{NN}-\lambda M_{NN} \end{vmatrix} = 0 \tag{10.33}$$

Expanding Eq. (10.33) in the conventional way in terms of minors will clearly produce a polynomial in λ of degree N:

$$p_N(\lambda) = \det |[K] - \lambda[M]| = 0 \tag{10.34}$$

which is the *characteristic polynomial* associated with Eq. (10.24). The solutions to Eq. (10.34) are the eigenvalues λ_i, which are therefore the roots of the characteristic polynomial,

$$p_N(\lambda_i) = 0 \qquad i = 1,2,\ldots,N \tag{10.35}$$

If Eq. (10.35) could be solved for λ_i, then each λ_i could be substituted into Eq. (10.27) and solved for its corresponding eigenvector $\{v\}_i$. This is illustrated by the following simple example with $N=2$:

$$[K] = \begin{bmatrix} 3 & 2 \\ 2 & 5 \end{bmatrix} \qquad [M] = \begin{bmatrix} 4 & 1 \\ 1 & 2 \end{bmatrix} \tag{10.36}$$

$[K]$ and $[M]$ are both real, symmetric, and positive-definite (see Exercise 10.1).

$$\begin{aligned} p_2(\lambda) &= \begin{vmatrix} 3-4\lambda & 2-1\lambda \\ 2-1\lambda & 5-2\lambda \end{vmatrix} \\ &= (3-4\lambda)(5-2\lambda) - (2-\lambda)^2 \\ &= 7\lambda^2 - 22\lambda + 11 \end{aligned} \tag{10.37}$$

Therefore

$$p_2(\lambda_i) = 7\lambda_i^2 - 22\lambda_i + 11 = 0 \tag{10.38}$$

The roots of Eq. (10.38) are the two eigenvalues,

$$\lambda_1 = \frac{\sqrt{11}}{7} (\sqrt{11} - 2) = 0.624$$

$$\lambda_2 = \frac{\sqrt{11}}{7} (\sqrt{11} + 2) = 2.519 \tag{10.39}$$

Note that $0 < \lambda_1 < \lambda_2$.

To determine the eigenvector corresponding to λ_1, substitute λ_1 back into Eq. (10.27) and solve the resulting system of linear algebraic equations. Thus

$$([K] - \lambda_1[M])\{v\}_1 = \{0\}$$

$$\begin{bmatrix} 3-0.624\times4 & 2-0.624\times1 \\ 2-0.624\times1 & 5-0.624\times2 \end{bmatrix} \begin{Bmatrix} v_1 \\ v_2 \end{Bmatrix}_1 = \begin{Bmatrix} 0 \\ 0 \end{Bmatrix}$$

$$\begin{bmatrix} 0.504 & 1.376 \\ 1.376 & 3.752 \end{bmatrix} \begin{Bmatrix} v_1 \\ v_2 \end{Bmatrix}_1 = \begin{Bmatrix} 0 \\ 0 \end{Bmatrix} \tag{10.40}$$

These are two equations in two unknowns, but both equations are identical. (The vanishing of the determinant ensures that the equations are not linearly independent.) Therefore we can solve either equation by itself, yielding values for v_1 and v_2 multiplied by an arbitrary constant:

$$\begin{Bmatrix} v_1 \\ v_2 \end{Bmatrix}_1 = c \begin{Bmatrix} -2.730 \\ 1 \end{Bmatrix} \tag{10.41}$$

The arbitrary constant c is determined by using Eqs. (10.31b) to normalize $\{v\}_1$ with respect to $[M]$:

$$c\{-2.730 \quad , \quad 1\} \begin{bmatrix} 4 & 1 \\ 1 & 2 \end{bmatrix} c \begin{Bmatrix} -2.730 \\ 1 \end{Bmatrix} = 1 \tag{10.42a}$$

which yields

$$c = 0.195 \tag{10.42b}$$

Therefore

$$\{v\}_1 = \begin{Bmatrix} -0.532 \\ 0.195 \end{Bmatrix} \tag{10.43a}$$

In a similar fashion,

$$\{v\}_2 = \begin{Bmatrix} -0.054 \\ 0.730 \end{Bmatrix} \tag{10.43b}$$

The complete eigensystem consists of the two eigenpairs $(\lambda_1, \{v\}_1)$ and $(\lambda_2, \{v\}_2)$ in Eqs. (10.39) and (10.43). The orthogonality relations in Eqs. (10.31) can be verified by direct multiplication (see Exercise 10.2).

10.3.2 Different Techniques for Solution

In the previous example the two eigenvalues were calculated from an explicit formula for the two roots of a second-degree polynomial. Explicit formulas exist only for $N \leq 4$; for $N \geq 5$ there are no known explicit formulas to solve for the λ_i and $\{v\}_i$ solutions to either the standard or generalized eigenproblem. Hence all available solution techniques for general N are *iterative*, requiring approximate numerical methods. This means the calculations proceed in a cyclic fashion, gradually improving the accuracy in each successive cycle. The accuracy must be repeatedly tested to determine when to stop the calculations. In short, the calculations gradually converge until some acceptable level of accuracy is reached. Some problems may converge quickly, others very slowly or not at all.[5]

There are three basic approaches to solving the algebraic eigenproblem, depending on which of the properties in Section 10.3.1 are principally utilized. Each approach encompasses a variety of particular methods, some of which are only minor variants of each other.

Vector-iteration methods. These use Eq. (10.27). The general technique can be illustrated by the following example. Let's assume a convenient starting value for λ_i, say 1, and a convenient starting vector for $\{v\}_i$, call it $\{u\}_1$; for example, $u_{i1} = 1$, $i = 1, 2, \ldots, N$. Substitute these values for $\lambda_i \{v\}_i$ on the RHS of Eq. (10.27):

$$[K]\{u\}_2 = [M]\{u\}_1 \qquad (10.44)$$

where $\{u\}_2$ on the LHS is an unknown vector. Equation (10.44) is a system of linear algebraic equations that can be solved for $\{u\}_2$ (using Gaussian elimination, say). Next put $\{u\}_2$ on the RHS and solve again to obtain $\{u\}_3$ on the LHS. In general, the lth step is

$$[K]\{u\}_{l+1} = [M]\{u\}_l \qquad (10.45)$$

As $l \to \infty$ we would observe that $\{u\}_l \to c\{v\}_i$, a multiple of one of the eigenvectors (if the vectors are appropriately normalized after each iteration). Therefore, from Eqs. (10.29) and (10.45),

$$\{u\}_{l+1} \to \frac{1}{\lambda_i} c\{v\}_i = \frac{1}{\lambda_i} \{u\}_l \qquad (10.46)$$

that is

$$\lambda_i = \lim_{l \to \infty} \frac{(u_j)_l}{(u_j)_{l+1}} \qquad j = 1, 2, \ldots, N$$

where $(u_j)_l$ is the lth iteration of the jth element of $\{v\}_i$,

The eigenvector to which the method first converges depends on which variant of the method is used and the nature of the starting vector $\{u\}_1$. One variant is to start the

---

[5] It should be clear that for iterative methods there does not exist a predetermined number of arithmetic steps to be performed. This is in contrast to the Gaussian elimination method of solution for linear algebraic equations which calculates the exact solution (limited only by the finite precision of the numbers) within a predetermined number of steps.

iteration putting $\{u\}_1$ on the LHS of Eq. (10.27) and then solve for $[M]\{u\}_2$ on the RHS. Convergence to particular eigenvectors, as well as the rate of convergence, can be controlled by a strategy called *shifting*.

Popular methods in this category include the power (or Stodolla) method, simultaneous iteration, and inverse iteration [10.5]. The latter exhibits particularly rapid convergence to an eigenvector if one already has an approximation to the corresponding eigenvalue, say by one of the methods described below.

Characteristic-polynomial methods. These methods compute the roots, i.e., the eigenvalues, of the characteristic polynomial, Eq. (10.34). This involves working with a sequence of polynomials related to Eq. (10.34), known as a *Sturm sequence*, and applying the method of bisection [10.5]. The latter is a traditional technique of successively halving an interval in order to define an increasingly smaller subinterval enclosing a root of an equation. After computing in this manner several of the eigenvalues (usually not all of them), the corresponding eigenvectors must then be computed by a separate method. Here inverse iteration with shifting is a generally recommended procedure.

Transformation methods. These methods work with Eq. (10.24). They transform $[K]$ and $[M]$ (or a matrix related to both) to a simpler form by pre- and postmultiplying by a sequence of (usually orthogonal) matrices. The Givens [10.5], Householder [10.5], and Lanczos [10.8, 10.9] methods require that Eq. (10.24) first be converted to an equivalent standard eigenproblem [Eq. (10.25)]. These methods then transform the converted stiffness matrix into a *tridiagonal* matrix. The eigenvalues of the tridiagonal matrix may then be very efficiently computed by a separate method, such as the Sturm sequence/bisection method mentioned above or the popular and very powerful QR method with shifting [10.5]. Corresponding eigenvectors may then be computed by inverse iteration with shifting and, finally, transformed back to eigenvectors of the original generalized eigenproblem.

The generalized Jacobi method [10.10-10.14] works directly with the generalized eigenproblem, Eq. (10.24). It transforms both $[K]$ and $[M]$ to *diagonal* matrices, yielding both the eigenvalues and eigenvectors directly and therefore not requiring any further methods. Since the generalized Jacobi method was chosen for use in UNAFEM (see Section 10.3.3), let's take a closer look at how it works.

We first observe that Eqs. (10.30) may be written in the more general form

$$[V]^T[K][V] = [\Lambda]$$

$$[V]^T[M][V] = [I] \tag{10.47}$$

where $[V]$ is a square matrix of order N containing eigenvector $\{v\}_i$ in column i:

$$[V] = [\{v\}_1, \{v\}_2, \ldots, \{v\}_N] \tag{10.48}$$

$[\Lambda]$ is a square diagonal *matrix* of order N containing the eigenvalues λ_i along the main diagonal:

$$[\Lambda] = \begin{bmatrix} \lambda_1 & & & & \\ & \lambda_2 & & \mathbf{0} & \\ & & \cdot & & \\ & & & \cdot & \\ \mathbf{0} & & & & \cdot \\ & & & & \lambda_N \end{bmatrix} \tag{10.49}$$

and $[I]$ is the diagonal unit matrix. A moment's reflection should convince the reader that Eqs. (10.47) contain the relations for all $i,j=1,2,...,N$ in Eqs. (10.30).

The matrix $[V]$ that satisfies Eqs. (10.47) is unique. We therefore try to construct it by pre- and postmultiplying $[K]$ and $[M]$ by a sequence of orthogonal ("rotation") matrices, gradually transforming $[K]$ and $[M]$ into diagonal matrices. Thus, defining $[K]_1=[K]$ and $[M]_1=[M]$, we have

$$[K]_2 = [P]_1^T[K]_1[P]_1$$
$$[M]_2 = [P]_1^T[M]_1[P]_1$$

$$[K]_3 = [P]_2^T[K]_2[P]_2$$
$$[M]_3 = [P]_2^T[M]_2[P]_2$$

$$\cdot$$
$$\cdot$$
$$\cdot$$

$$[K]_{l+1} = [P]_l^T[K]_l[P]_l$$
$$[M]_{l+1} = [P]_l^T[K]_l[P]_l \tag{10.50}$$

After l iterations,

$$[K]_{l+1} = [V]_l^T[K][V]_l$$
$$[M]_{l+1} = [V]_l^T[M][V]_l \tag{10.51a}$$

where

$$[V]_l = [P]_1[P]_2...[P]_l \tag{10.51b}$$

Each $[P]_i$ is constructed so that it will annihilate (reduce to zero) an off-diagonal term of $[K]_i$ and simultaneously the same off-diagonal term of $[M]_i$. A given $[P]_i$ will also affect other off-diagonal terms, causing some of the previously zeroed terms to become nonzero again. It is therefore necessary to make more than one sweep through all the off-diagonal terms. The process could be very time-consuming and inefficient if the annihilation operation were automatically applied to every off-diagonal term, irrespective of how close to zero a term might already be. This problem is resolved by the *threshold* technique.

At the beginning of each sweep a threshold is established (which is lowered each sweep) and only those terms whose magnitude exceeds the threshold are annihilated. Once the off-diagonal terms become "small," convergence is quite rapid. Convergence theorems demonstrate that as $l \to \infty$ in Eqs. (10.51), then

$$[V]_l \to c[V]$$
$$[K]_l \to [\Lambda]$$
$$[M]_l \to [I] \tag{10.52}$$

where c is a scalar. In practice, the procedure is terminated when these limits are approached to within an accuracy specified by the analyst. Typically only a few sweeps are required for high accuracy.

10.3.3 Available Computer Algorithms for Big and Small Problems

A wide variety of algorithms exist, each utilizing one or more of the above methods or their variants. Some of these may be further specialized to match the requirements of particular problems:

1. Is the problem in standard form or generalized form?
2. Is the $[K]$ (and perhaps $[M]$) matrix real or complex-valued; symmetric or nonsymmetric; banded or full; positive-definite, semidefinite, or indefinite?
3. Will the entire eigensystem be calculated or only a few eigenpairs?
4. Are only eigenvalues desired, or both eigenvalues and eigenvectors?

To choose from what must seem a rather bewildering array of algorithms, perhaps the first and most important and practical consideration should be the size of the problem. The solution to an algebraic eigenproblem is generally much more complicated (and hence more expensive) than the solution to the corresponding boundary-value problem of the same size. Very large problems therefore demand efficient solvers; conversely, small problems will tolerate relatively inefficient solvers.

What is large and what is small? Based on the cost of computing power available to most industrial users in the 1980s, I would estimate a "small" eigenproblem to be less than 100 to 200 DOF. In writing an FE program for a problem of this size, it would be adequate to select from any of the widely available eigensolver routines in the public domain such as EISPACK [10.6], or from numerical analysis computer program libraries at one's company or university. Time (and therefore money) saved by this simple, direct approach usually more than compensates for any additional computational costs incurred as a result of using a less-than-optimal solver.

Most of the publicly available routines are for solving the standard eigenproblem, Eq. (10.25). One could use these routines for the generalized eigenproblem, Eq. (10.24), by converting the latter to standard form. Even those routines in Garbow et al. [10.6] for the generalized eigenproblem include a preliminary conversion to standard form. If $[M]$ is not diagonal, then the conversion involves a *Cholesky decomposition*, which is a bit involved though straightforward. The transformed $[K]$ matrix would be full, even if originally banded. If, however, $[M]$ is diagonal, then the conversion to standard form is trivial and the bandedness of $[K]$ is preserved.

There are techniques for diagonalizing a banded $[M]$ matrix, called *mass lumping* (see Section 11.3.2), which can sometimes be quite effective.[6] Mass lumping is still an imprecisely understood topic. There are a variety of techniques, yielding variable performances depending on problem or element type. A mass matrix that is not lumped, i.e., one left in the form derived from (consistent with) the weighted residual or variational integrals, is called a *consistent* mass matrix. Neither lumped nor consistent matrices yield superior accuracy in all situations. Chapters 11 and 15 discuss lumping procedures as applied to the capacity matrix in initial-boundary-value problems; the capacity and mass matrices are identical so the procedures are the same for both.

---

[6] Example 1.5 (see Chapter 1) used this approach. The engineer wrote an entire FE program in 67 lines of FORTRAN, employing mass lumping to enable a trivial conversion to standard form and then using two routines from the EISPACK library.

Practical FE problems usually fall in the medium-to-large category: hundreds or thousands of DOF, sometimes tens of thousands. Calculating the entire eigensystem of a very large problem could be prohibitively expensive. Furthermore, it would generally be wasteful because of two observations: (1) In many practical applications, only a few eigenvectors determine the significant response of the system, and they are usually the ones corresponding to the lowest eigenvalues; and (2) the higher eigenvalues and corresponding eigenvectors of an FE-generated algebraic eigenproblem have much less accuracy [relative to the exact solution of the original eigenproblem in Eq. (10.1)] than do the lower eigenvalues and corresponding eigenvectors. This will be illustrated in Section 10.5. Hence an efficient solver for large problems should be capable of calculating only a few eigenpairs, usually the ones with the lowest eigenvalues. In many problems, 5 to 10 eigenpairs will suffice; on occasion, 50 to 100 may be necessary.

There are presently several methods that are appropriate for the medium-to-large generalized eigenproblems that occur in FE applications. They take full advantage of the symmetric, positive-definite, banded properties of the [K] and [M] matrices, and they allow the user some degree of control in selecting the number of eigenpairs to be calculated. One or more of the following methods will be found in most commercial FE programs that solve eigenproblems [10.15]:

1. Generalized Jacobi, usually preceded by an eigenvalue economizer (see below)

2. Householder, preceded by conversion to standard form and sometimes an eigenvalue economizer, and followed by additional routines (see below)

3. Givens, following the same procedure noted in method 2

4. Lanczos, following the same procedure noted in method 2 [10.16]

5. Subspace iteration [10.13, 10.14, 10.16, 10.17]

6. Determinant search [10.13, 10.14]

The first method is based on the classical (1846) Jacobi method [10.10] for the standard eigenproblem. It was modified during the 1960s [10.11, 10.12] to handle the generalized eigenproblem. This modification, referred to as the generalized Jacobi method, was described at the end of Section 10.3.2. The algorithm is very stable (converging for a wide range of problems), and is one of the easiest to understand and code.

The generalized Jacobi method calculates the entire eigensystem (i.e., all eigenvalues and eigenvectors) and therefore, by itself, would be appropriate only for small problems. In order to extend its usefulness to large problems, it is necessary to first reduce the large eigenproblem to a much smaller eigenproblem (typically a few tens of DOF) by eliminating the DOF associated with the higher eigenvalues. The resulting smaller problem models only the lowest eigenvalues and corresponding eigenvectors.

An algorithm that performs such a reduction is sometimes called an *eigenvalue economizer* [10.18, 10.19]. The most widely used one at present appears to be *Guyan reduction* [10.20]. In this technique one must select those DOF that contribute most to the eigenvectors with the lowest eigenvalues. Those DOF are called the *master* DOF, and the others are called the *slave* DOF. Deciding which DOF to choose as masters can be difficult for inexperienced users; fortunately, many FE programs now automatically perform this selection [10.21]. Following the reduction, the generalized Jacobi method is then used to calculate the complete eigensystem of the reduced problem. It should be noted that Guyan reduction introduces an approximation error of its own [10.22], although the error can be made progressively smaller by choosing more and more DOF as masters.

The generalized Jacobi algorithm, without Guyan reduction, was chosen for the UNAFEM program because of its stability, simplicity of programming, and relative ease of being understood. In addition, the code is publicly available [10.13, 10.14]. When used by itself, without Guyan reduction, it is particularly well suited for small problems.

Methods (2), (3), and (4) are not in themselves eigensolvers. Their only function is to transform the stiffness matrix (after the problem has been converted to standard form) to tridiagonal form. They must be followed by additional techniques to extract eigenvalues (e.g., QR for all the eigenvalues, or Sturm sequences with bisection for only a few) and eigenvectors (e.g., inverse iteration with shifting). Methods 2 through 6 are mathematically more involved than the generalized Jacobi method and their programming is more complicated. The reader is referred to the cited references for further details.

10.4 Implementation into the UNAFEM Program

Two principal additions must be made to UNAFEM to incorporate the eigenproblem capability: (1) additional statements in each of the four ELEM*ij* subroutines to evaluate the mass matrix integrals and (2) a new subroutine, called GENJAC, containing the generalized Jacobi algorithm for solving the assembled system equations [Eq. (10.24)]. Figure 10.1 shows the modified flowchart for UNAFEM. Note the addition of GENJAC as well as two new output routines, EIGPRI and EIGPLO.

Figure 10.2 shows the additional lines of coding added to each of the four 1-D element routines. Note that only a single statement is needed for the three isoparametric elements, and it is identical for all of them — an attractive feature of parametric elements. (Recall the observation in Section 8.5 that we could have written the three isoparametric element routines as a single routine.)

All mass matrices were left in consistent form rather than lumped. The interested reader might wish to modify the code to include lumping (in the same manner as the [C] matrix in Chapter 11). The eigensolver GENJAC would not need to be altered since it is capable of treating the general case of a nondiagonal (and hence also diagonal) mass matrix.

The listing for the GENJAC subroutine is in the appendix; it is reproduced, with slight modification, from Bathe [10.13]. The algorithm in Bathe is limited to a positive-definite [K] matrix, thereby not permitting zero eigenvalues. In order to permit zero eigenvalues, several statements were modified so that a positive-*semi*definite [K] would be acceptable.

The GENJAC subroutine does not shift the rows of the banded [K] and [M] matrices to the left.[7] The reason for this is that the zero terms outside the band will generally become nonzero during the course of the generalized Jacobi calculations, and then in the final itera-

---

[7] Recall that the Gaussian elimination solver GAUSEL performs shifting in order to avoid wastefully storing all the zero terms outside the band.

tions will approach zero again. In other words, bandedness is temporarily destroyed, resulting in a full matrix at intermediate stages. Therefore the full $[K]$ and $[M]$ matrices must be stored. This means that the FORM subroutine needs new assembly statements that assemble the element matrices $[K]^{(e)}$ and $[M]^{(e)}$ without shifting. These are shown

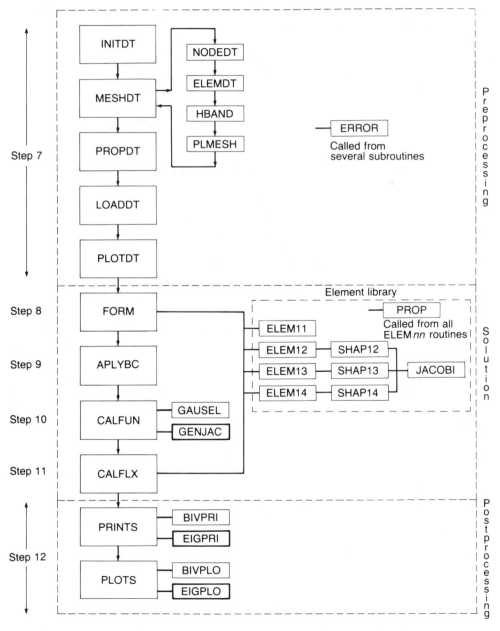

FIGURE 10.1 ● Subroutines (indicated by bold outline) added to UNAFEM to perform analysis for 1-D eigenproblems.

in Fig. 10.3, along with the previous shifted statements for the boundary-value problem to provide a comparison of shifted and nonshifted assembly. Note that there is no load vector to assemble for the eigenproblem.

```
      SUBROUTINE ELEM11(OPT)
C
C         THIS SUBROUTINE IS THE MASTER ROUTINE FOR ELEMENT 11, A
C         2-NODE 1-D CO-LINEAR ELEMENT.
             .
             .
      ME(1,1) = GAMMA*L/3.
      ME(1,2) = GAMMA*L/6.
      ME(2,1) = ME(1,2)
      ME(2,2) = ME(1,1)

             .
             .
```

```
      SUBROUTINE ELEM12(OPT)
C
C         THIS SUBROUTINE IS THE MASTER ROUTINE FOR ELEMENT 12, A
C         3-NODE 1-D CO-QUADRATIC ELEMENT.
             .
             .
             IF (PROB.EQ.EIGP) ME(I,J)=ME(I,J)+WT(NINT,INTX)*JAC*
     $                         PHI(I)*GAMMA*PHI(J)
             .
             .
```

```
      SUBROUTINE ELEM13(OPT)
C
C         THIS SUBROUTINE IS THE MASTER ROUTINE FOR ELEMENT 13, A
C         4-NODE 1-D CO-CUBIC ELEMENT.
             .
             .
             IF (PROB.EQ.EIGP) ME(I,J)=ME(I,J)+WT(NINT,INTX)*JAC*
     $                         PHI(I)*GAMMA*PHI(J)
             .
             .
```

```
      SUBROUTINE ELEM14(OPT)
C
C         THIS SUBROUTINE IS THE MASTER ROUTINE FOR ELEMENT 14, A
C         5-NODE 1-D CO-QUARTIC ELEMENT.
             .
             .
             IF (PROB.EQ.EIGP) ME(I,J)=ME(I,J)+WT(NINT,INTX)*JAC*
     $                         PHI(I)*GAMMA*PHI(J)
             .
             .
```

FIGURE 10.2 ● Statements added to the element subroutines to evaluate the element mass matrix.

```
   SUBROUTINE FORM
                 .
                 .
                 .
C**************** ASSEMBLE MATRICES FOR BOUNDARY-VALUE PROBLEM ***********
                 .
                 .
                 .
         DO 140 I=1,NND
             II = ICON(I,IELNO)
             F(II) = F(II)+FE(I)
             DO 120 J=1,NND
                 JJ = ICON(J,IELNO)
                 JSHIFT = JJ-(II-1)
                 IF (JJ.GE.II) K(II,JSHIFT) = K(II,JSHIFT)+KE(I,J)
  120        CONTINUE
  140    CONTINUE
                 .
                 .
                 .
C**************** ASSEMBLE MATRICES FOR EIGENPROBLEM ********************
                 .
                 .
                 .
         DO 240 I=1,NND
             II = ICON(I,IELNO)
             DO 220 J=1,NND
                 JJ = ICON(J,IELNO)
                 KEIG(II,JJ) = KEIG(II,JJ)+KE(I,J)
                 MEIG(II,JJ) = MEIG(II,JJ)+ME(I,J)
  220        CONTINUE
  240    CONTINUE
                 .
                 .
                 .
```

FIGURE 10.3 ● Subroutine FORM, a partial listing showing shifted assembly for boundary-value problems and nonshifted assembly for eigenproblems.

In the postprocessing phase, subroutine EIGPRI prints the solution for the first five eigenpairs, i.e., the lowest five eigenvalues and their corresponding eigenmodes (eigenvectors).[8] The printed output for each eigenmode consists of a list of the nodal values of the function at every node. Subroutine EIGPLO plots the mode shapes for these first five eigenmodes. We recall that a mode is arbitrary up to a multiplicative constant; i.e., it can be scaled by any constant factor. Subroutine GENJAC chooses the constant so that each mode is normalized with respect to the mass matrix, using the relation $\{v_i\}^T[M]\{v_i\}=1$ from Eqs. (10.31b). This ensures that each mode $\{v_i\}$ is unique, excepting only an ambiguity in the algebraic sign. Thus, as a mesh is refined, a given mode may possibly reverse its sign from one mesh to the next.

10.5 Application to Free Vibration of a Flexible Cable

We want to examine the natural frequencies and modes (eigenvalues and eigenmodes) of transverse vibration of a flexible cable that is suspended horizontally with tension T and held fixed at both ends. The governing equation, Eq. (10.9), and boundary conditions,

---

[8] Regarding terminology, there seems to be a general tendency to use the term eigen*vector* during the theoretical analysis and discussion, but then to switch to terms such as eigen*mode*, *mode*, and/or *mode shape* to describe the output from the numerical computation. At the risk of oversimplifying or overgeneralizing, ''mode'' seems to be more engineering-(application) oriented and ''vector'' more mathematics-(theory) oriented.

Eqs. (10.11b), were derived in Section 10.1. Let's consider the same cable used in the boundary-value (equilibrium) problem in Section 7.5, namely, a length of 20 ft, a tension of 100 lb, and a lineal weight density of 0.5 lb/ft. The latter is the weight of the cable in a gravitational field. From this we calculate the lineal mass density ϱ to be 0.01553 slug/ft (using $g=32.2$ ft/sec^2). Note that this does not imply the presence of a gravitational field for the vibration problem. On the contrary, the problem assumes no gravitational field nor any applied loads of any kind, since it is a free-vibration problem.[9] In summary, the physical properties in Eq. (10.9) are the following:

$$T(x) = 100 \text{ lb}$$

$$k(x) = 0$$

$$\varrho(x) = 0.01553 \text{ slug/ft} \tag{10.53}$$

As usual, we begin with a very simple mesh to familiarize ourselves with the details of program input and output. Let's use six C^0-linear elements, all of uniform length, as shown in Fig. 10.4. This mesh has seven nodes and hence seven DOF, but the boundary conditions constrain the first and seventh DOF, leaving only five "active" DOF. In other words, after applying the essential BCs to the assembled system equations (using constraint equations), the resulting algebraic eigenproblem will have only five DOF. This will yield five modes. (We could use a mesh with fewer DOF, and therefore produce fewer modes, but it will be interesting to see at least five modes.)

Figure 10.5 shows the input data for the six-element mesh, using the input instructions in the appendix. Note that the keywords ILC and NBCC and associated data are omitted for the eigenproblem, since nonzero loads are not permitted and zero values do not affect the system equations. On the other hand, the EBC keyword must be included because, even though only zero values are permitted, an EBC does affect the system equations (via application of a constraint equation).

The printed output (Fig. 10.6) shows the lowest five eigenvalues — λ_1, λ_2, λ_3, λ_4, and λ_5 — and the nodal values for the corresponding modes. The eigenvalues may be converted to natural frequencies using Eq. (10.10).[10]

Figure 10.7 shows the mode shapes and natural frequencies for both the FE solution and the exact solution.[11] These results reveal an important feature common to all eigenproblem solutions: The accuracy of both the modes and frequencies decreases as the frequency (and therefore mode number) increases. This should not be surprising. As the frequency increases, the mode shapes become more complicated. If one thinks of a mode

---

[9] The distinction between weight and mass is important. *Weight* is an externally applied load. *Mass* is a physical property, namely, an inertial resistance to acceleration.

[10] UNAFEM calculates only the eigenvalues, not the natural frequencies, the difference between the two being a multiplicative constant that has different physical dimensions for different problems. Thus, Eq. (10.10) does not apply to all physical applications of Eq. (10.1) [see, for example, Eq. (10.62)].

[11] Using the properties in Eqs. (10.53), the governing Eq. (10.9) becomes $d^2W/dx^2+\varkappa^2W=0$, where $\varkappa=\omega/c$, and $c=\sqrt{T/\varrho} = 80.2443$ ft/sec is the propagation velocity of transverse waves in the cable. The BCs in Eqs. (10.11b) become $W(0)=W(20)=0$. The exact solution for the modes and eigenvalues, respectively, is $W_n(x)=\sin(n\pi x/20)$ and $\lambda_n=(n\pi c/20)^2$, $n=1,2,3,\dots$. From Eq. (10.10) the natural frequencies are $f_n=2.006n$ Hz (cycles/sec).

FIGURE 10.4 ● A simple starting mesh for the cable vibration problem.


```
| D I M E I G P    C A B L E   V I B R A T I O N  -  6  L I N E A R   E L E M E N T S
N O D E
           1         1        0 .
           7                 2 0 .
E L E M
           1         1        1 1              1       1        2
           6                  1 1              1       6        7
P R O P
           1       1 0 0 .
                     0 .
                   1 5 5 . 3 E - 4
E B C
           1         0 .
           7         0 .
P L O T
         1 0         0 .             2 0 .
         1 0       - 5 .              5 .
E N D
```

FIGURE 10.5 ● UNAFEM input data for the six-element mesh.

```
EIGENPROBLEM-LOWEST 5 EIGENVALUES & EIGENVECTORS

                          EIGENVALUES

              1           2           3           4           5
         0.1625E+03  0.6954E+03  0.1739E+04  0.3477E+04  0.5722E+04


                          EIGENVECTORS

NODE          1            2            3            4            5
  1     0.0000E+00   0.0000E+00   0.0000E+00   0.0000E+00   0.0000E+00
  2     0.1298E+01  -0.2407E+01   0.3108E+01  -0.3108E+01  -0.2064E+01
  3     0.2248E+01  -0.2407E+01   0.4037E-12   0.3108E+01   0.3574E+01
  4     0.2596E+01  -0.4401E-12  -0.3108E+01  -0.5624E-12  -0.4127E+01
  5     0.2248E+01   0.2407E+01  -0.4479E-12  -0.3108E+01   0.3574E+01
  6     0.1298E+01   0.2407E+01   0.3108E+01   0.3108E+01  -0.2064E+01
  7     0.0000E+00   0.0000E+00   0.0000E+00   0.0000E+00   0.0000E+00
```

FIGURE 10.6 ● UNAFEM printed output for the six-element mesh.

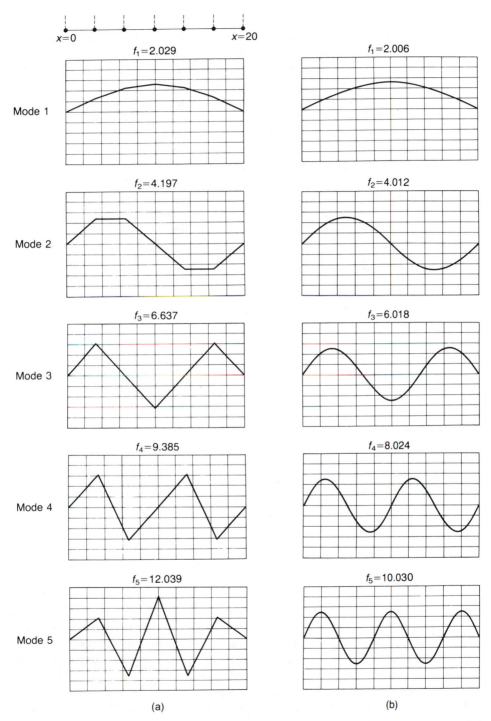

FIGURE 10.7 ● (a) The first five mode shapes and natural frequencies for the mesh with six C^0-linear elements; (b) The exact solution.

as a standing wave pattern containing several half-wavelengths (e.g., a sequence of ⌒ and ⌣ segments), then higher modes have a greater number of half-wavelengths. Since a given mesh has a fixed number of DOF, then as the frequency increases, the number of half-wavelengths increases, so the number of DOF/half-wavelength decreases. Thus each half-wavelength segment is modeled with fewer and fewer DOF.

For example, mode 1 in Fig. 10.7 consists of one half-wavelength segment. It is being modeled by seven DOF: five are active and two are constrained on the boundary. This yields a pretty decent approximation. Mode 5, however, consists of five half-wavelength segments, so each half-wavelength is being modeled by only about one DOF, which is clearly a much poorer approximation. In order to model mode 5 more accurately, we will need to refine the mesh.

Figure 10.8 shows the solutions generated by a sequence of refined meshes in which the element order remains the same (linear) but the number of elements increases (i.e., h-refinement). Figure 10.9 shows the solutions generated by a sequence of refined meshes in which the number (and size) of elements remains the same, but the element order increases (p-refinement). A careful perusal of these figures will provide some insight into the effect of mesh refinement on accuracy.

The rate of convergence of the mode shapes is the same as for the function in the corresponding boundary-value problem. Thus, recalling Section 9.1, the rate at ordinary points x is controlled simply by approximation theory,

$$\left| \frac{U_n(x) - \tilde{U}_n(x)}{U_n(x)} \right| = O(h^{p+1}) \qquad \text{as } h_{\max} \to 0$$

$$(10.54a)$$

where n is the mode number and p is the degree of the approximating polynomial. The rate at superconvergent points x^* is as shown in Table 9.1; thus, at interelement nodes,

$$\left| \frac{U_n(x^*) - \tilde{U}_n(x^*)}{U_n(x^*)} \right| = O(h^{2p}) \qquad \text{as } h_{\max} \to 0$$

$$(10.54b)$$

The rate of convergence of the eigenvalues for linear, self-adjoint problems is [10.23]:

$$\frac{\left| \dfrac{\lambda_n - \tilde{\lambda}_n}{\lambda_n} \right|}{\dfrac{\lambda_n^{p-m+1}}{m}} = O(h^{2(p-m+1)}) \qquad \text{as } h_{\max} \to 0$$

$$(10.55)$$

where $2m$ is the order of the governing differential equation. The factor $\lambda_n^{(p-m+1)/m}$ means the higher eigenvalues will require smaller h values (more-refined meshes) before showing these rates. A comparison of Eqs. (10.55) and (8.52) shows that the eigenvalues converge at the same rate as the global energy. This is not surprising, since in physically motivated problems the eigenvalue either *is* the global energy (e.g., the quantum-mechanical oscillator problem below) or else is the ratio of global potential energy to global kinetic energy (e.g., the square of the frequency in the present vibration problem — also referred to as *Rayleigh's quotient*).

The author has verified the rates in Eqs. (10.54) and (10.55) for the 1-D elements in Chapters 7 and 8 by performing numerical experiments similar to those in Chapter 9.

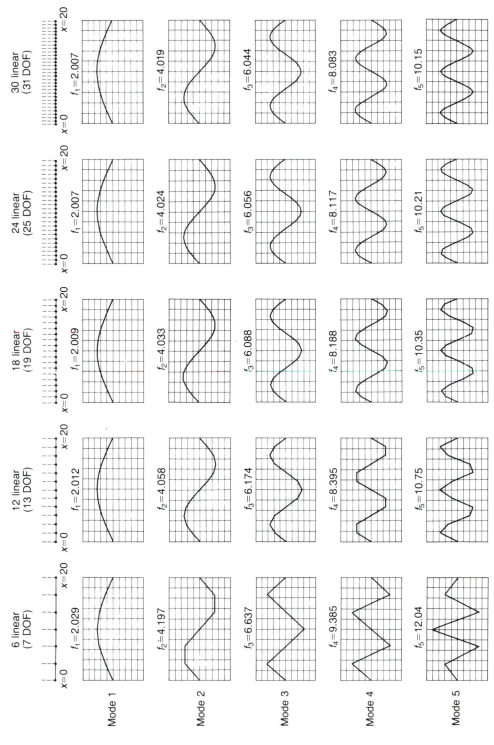

FIGURE 10.8 ● The first five mode shapes and natural frequencies for the cable vibration problem, using five meshes with 6, 12, 18, 24, and 30 C^0-linear elements, respectively (*h*-refinement).

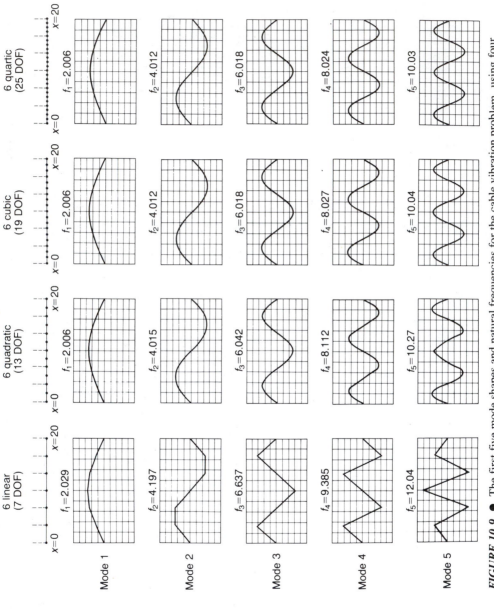

FIGURE 10.9 ● The first five mode shapes and natural frequencies for the cable vibration problem, using four meshes with six C^0-linear, -quadratic, -cubic, and -quartic elements, respectively (p-refinement).

It is interesting that for second-order governing equations ($m=1$), the convergence rate for the modes at interelement nodes is the same as for the eigenvalues. This contrasts with traditional (pre-FEM) analytical methods that use global trial functions, which only exhibit ordinary convergence rates for the modes, as in Eq. (10.54a). It is well known that these traditional methods compute the mode shapes less accurately than the eigenvalues. The superconvergence properties of the FEM, however, are able to increase the accuracy of the modes to that of the eigenvalues.

Before leaving this problem, we can use Figs. 10.8 and 10.9 to generate a rough guideline for determining a reasonable number of DOF to use in constructing a mesh. An examination of the printed output (not shown here) for all the mode shapes and frequencies reveals an error in the neighborhood of 1% if there are q active DOF/half-wavelength, where q is about 2 to 5, the smaller integers being appropriate for higher-order elements. This is the error in the frequencies as well as the error in the mode shapes at the interelement nodes, both of which, as indicated above, converge at the same rate. Therefore, if one wants to calculate the nth mode and frequency to about 1% accuracy (and hence the lower modes to better than 1% accuracy), then a 1-D uniform mesh would require about $q \times n$ DOF.

More generally, for problems in D dimensions, if one wants to calculate the lth mode and frequency to about 1% accuracy, where $l=n^D$, then a uniform mesh needs about $q \times n$ DOF in each direction, or a total of $(qn)^D$ DOF, where q is about 2 to 5 if the higher-dimensional elements exhibit similar superconvergence properties; otherwise q would be a larger number. This guideline presumes that the geometry is not too complicated; it would be less reliable for domains with very complex geometries. Nevertheless, we can see that an FE model needs a much greater number of DOF than the number of desired modes, and this disparity increases with the dimensionality.

10.6 Application to a Harmonic Oscillator in Quantum Mechanics

Figure 10.10 illustrates the classic 1-D harmonic oscillator, one of the earliest quantum mechanical systems ever studied and a useful model for a variety of quantum mechanical applications. A point mass m is constrained to move along a straight line (the x-axis) and is attracted toward a point on this line ($x=0$) with a force proportional to the distance kx. The mass is shown at two different positions on either side of the attracting point.

The governing equation is the time-independent *Schroedinger equation* [10.24, 10.25] (see Table 10.1),

$$-\frac{\hbar^2}{2m}\frac{d^2\psi(x)}{dx^2} + V(x)\psi(x) - E\psi(x) = 0$$

(10.56)

where $\psi(x)$ is the probability amplitude function. Here $\hbar=h/2\pi$ and h is Planck's constant, representing a "quantum of action" and having the value 6.6256×10^{-34} Joule-

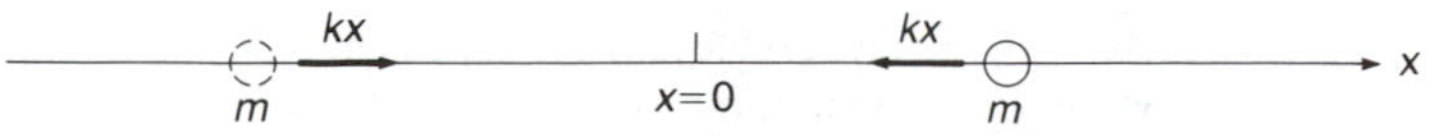

FIGURE 10.10 ● A 1-D harmonic oscillator.

sec. The term $(\hbar^2/2m)(d^2\psi/dx^2)$ is the kinetic energy of the particle, and $V(x)$ is the potential energy. Since the force on the particle is a restoring force kx, then

$$V(x) = \frac{1}{2} kx^2$$

(10.57)

E is the total energy of the particle (kinetic plus potential) which is a constant. Combining Eqs. (10.56) and (10.57) yields

$$-\frac{\hbar^2}{2m}\frac{d^2\psi(x)}{dx^2} + \frac{1}{2} kx^2\psi(x) - E\psi(x) = 0$$

(10.58)

The particle is not constrained to lie within any finite limits, so the *domain of the problem is infinite:*

$$-\infty < x < \infty$$

(10.59)

There are many engineering and scientific problems that are theoretically modeled with infinite domains. How to analyze a problem with an infinite domain will be the principal focus of this problem. Quantum mechanical principles require that[12]

$$\lim_{x \to \pm\infty} \psi(x) = 0$$

(10.60)

Hence the boundaries of the problem are at $x = \pm\infty$ and the boundary conditions are

$$\psi(-\infty) = 0$$
$$\psi(+\infty) = 0$$

(10.61)

Equations (10.58) and (10.61) are the statement of our problem, which is in the form of the general 1-D eigenproblem in Eqs. (10.1) and (10.2). Here the eigenvalue is the total energy E. Eigenvalues E_1, E_2,... represent discrete (quantized) energy levels corresponding to natural quantum states (eigenmodes) of the particle, $\psi_1(x)$, $\psi_2(x)$,... .

Because of the typically very small values for $\hbar$, m, and k, it is easier to first convert Eq. (10.58) to nondimensional form by a change of scale in the x variable. Letting

$$\xi = \mu x \qquad \text{(where } \mu^4 = km/\hbar^2)$$

(10.62a)

and

$$\lambda = 2E/\hbar \sqrt{k/m}$$

(10.62b)

Eq. (10.58) becomes[13]

$$-\frac{d^2\psi(\xi)}{d\xi^2} + \xi^2\psi(\xi) - \lambda\psi(\xi) = 0$$

(10.63)

---

[12] The probability amplitude function is normalized by the integral

$$\int_{-\infty}^{\infty} \psi^2 d\xi = 1$$

The quantity $\psi^2 d\xi$ is the probability that the particle may be found in the interval $d\xi$. Equation (10.60) is necessary to ensure the existence of the integral.

[13] This ξ has no relationship to the ξ used in Chapter 8 for isoparametric elements.

The boundary conditions remain the same as in Eqs. (10.61). Comparing Eq. (10.63) to our standard eigenproblem in Eq. (10.1), we have $\alpha(\xi)=1$, $\beta(\xi)=\xi^2$, and $\gamma(\xi)=1$; the eigenvalue γ is nondimensional.

There are three principal approaches to analyzing a problem with an infinite domain.

1. Model a finite portion of the domain with standard elements and model the remaining infinite portion with *infinite elements* [10.26-10.32]. An infinite element covers a sector of the domain from a finite radius out to infinity.[14]

2. Model a finite portion of the domain with standard elements and use the *boundary integral equation method* (BIEM), also called the boundary integral method (BIM) or *boundary element method* (BEM), to model the remaining infinite portion [10.34-10.39], coupling the two solutions along their common boundary (as illustrated by Example 1.9 in Section 1.6).[15]

3. Model a finite portion of the domain with standard elements and ignore the remaining infinite portion. One must choose the finite portion large enough so that the contribution from the infinite portion will be insignificant.

The first approach is a very natural one, requiring only the development of a new element. However, the evolution of these elements is fairly recent, so they are not yet available in the popular commercial FE programs (to this author's knowledge). The second approach is also not generally available in commercial FE programs, although there are some programs that formulate the entire problem with the BIEM [10.40]. The third approach is actually the oldest of the three, having been used almost universally since the inception of the FEM. It still remains an effective, easy-to-implement approach for many applications since it doesn't require any new elements or new theoretical formulations. We will use this approach with the present problem (as well as the problem in Chapter 15).

We know from Eq. (10.60) that the solution must decay towards zero as $\xi \to \pm\infty$. Therefore, as suggested by Fig. 10.11, there must be a finite portion of the domain, let's

---

[14] "Finite" in "finite element method" does not mean "not infinite"; it means "not infinitesimal" [10.33]. Therefore, in principle, elements may also be infinitely large, in which case their shape functions must decay to zero as $|x| \to \infty$.

[15] The theoretical formulation of the BIEM is quite different, in some respects, from that of the FEM. An essential feature of the BIEM is that it discretizes only the boundary of the domain, not the interior, thereby reducing the dimensionality of the analysis by one. This greatly reduces the mesh generation effort, as well as the number of DOF. These might seem to be compelling advantages. However, from a practical standpoint, they may not be that significant. Thus, in judging the relative merits of a BIEM or FEM approach, the following observations should be considered.

First, virtually all practical problems require several analyses, either for error estimation or, quite frequently, for design iterations and parametric studies. The cost of preparing a complicated FE mesh, compared to a less complicated BIEM mesh, is therefore only a small part of the total analysis cost, and it is becoming even less significant as the sophistication and automation of preprocessors continue to improve. Second, the use of superelements (substructures) in the interior of the domain can reduce the active DOF in an FE model to only a few interior DOF plus those on the boundary, comparable to the BIEM. The cost, too, becomes comparable if one performs more than one analysis or uses repetitive superelements within one analysis. Third, on the accuracy side of the accuracy-versus-cost trade-off, one should compare the rates of convergence, noting especially any superconvergence properties, such as those demonstrated in this text for FE trial functions.

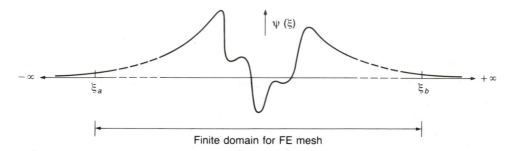

FIGURE 10.11 ● Typical behavior of a solution over an infinite domain.

1 DIME IGP			QUANTUM MECHANICAL OSCILLATOR — 12 QUADRATIC ELEMENTS								
NODE											
	1	1	-3.								
	25		3.								
ELEM											
	1	2	12	2	1	1	2	3			
	12		12	2	1	23	24	25			
PROP											
	1	1.									
		1.				1.					
EBC											
	1	0.									
	25	0.									
PLOT											
	6	-3.		3.							
	6	-.75		.75							
END											

FIGURE 10.12 ● UNAFEM input data for the quantum mechanical oscillator, using 12 C^0-quadratic elements and a finite domain $-3 < \xi < +3$.

call it $\xi_a < \xi < \xi_b$, inside of which all the significant activity takes place, and outside of which the solution is insignificant, i.e., close enough to zero.

What values should we use for ξ_a and ξ_b? This will have to be discovered, in addition to the solution itself. In practice, a reasonable first guess can usually be made, based on whatever insight we already have regarding the general nature of the solution.[16] Having selected an initial ξ_a and ξ_b, we can then construct an FE mesh over the finite domain $\xi_a < \xi < \xi_b$, and the boundary conditions in Eqs. (10.61) would be applied instead to these finite boundaries; that is,

$$\psi(\xi_a) = 0$$

$$\psi(\xi_b) = 0 \qquad\qquad (10.64)$$

---

[16] It would be generally inadvisable for an engineer to begin an FE analysis without having some minimal understanding of the general nature of the solution. However, even assuming complete naiveté, two or three coarse-mesh analyses with different ξ_a and ξ_β should identify the significant region.

This approach introduces a further approximation to the FE model. But the additional errors introduced can be made arbitrarily small by moving ξ_a toward $-\infty$ and ξ_b toward $+\infty$ in a sequence of meshes until the effect (of moving the boundary) on the solution is as small as desired.

In summary, we must construct a sequence of refined meshes with two goals in mind: (1) to achieve a desired accuracy within an assumed finite domain and (2) to make the accuracy in the finite domain relatively unaffected by the arbitrary location of its boundaries. Both goals can be condensed into a single guideline:

> We seek a mesh that is sufficiently refined and a finite domain sufficiently large that the accuracy is not significantly affected by any further refinement of the mesh or enlargement of the domain.

Let's begin with two meshes of 6 and 12 quadratic elements, respectively, and a domain from $\xi_a = -3$ to $\xi_b = +3$. We choose the quadratic element because, as demonstrated by the previous cable vibration problem (as well as the data in Chapter 9), the higher-order elements generally perform better than the linear elements. Figure 10.12 shows the input data for the 12-element mesh, using the input instructions in the appendix. Note that reduced integration is used.

Figure 10.13 shows the first five wave functions (eigenmodes) and the corresponding nondimensional energy levels (eigenvalues).[17] All five wave functions appear to be converging fairly well, the lower ones better than the higher ones, of course. However, they could be converging to a solution considerably different from the true solution if the finite domain isn't large enough to encompass all the significant behavior. Indeed, wave functions 3, 4, and 5, and to some extent 2, indicate that the domain is probably too small because the *slopes* of the wave functions are not approaching zero near the boundaries of the domain. For any problem in an infinite domain, Eq. (10.60) implies that not only $\psi(\xi)$ but also all its derivatives must approach zero as $\xi \to \pm\infty$, that is, $\psi(\xi)$ becomes identically zero "at" infinity.

Let's test the size of the domain by doubling it to $-6 < \xi < +6$. We won't disturb the mesh in the previous domain, $-3 < \xi < +3$, so that we can discern the effect on the solution in this region caused solely by movement of the boundary. Therefore, to the previous six-element mesh we'll add six more elements of the same size. Similarly, we'll add 12 more elements to the previous 12-element mesh. The results are shown in Fig. 10.14.

All five wave functions in Fig. 10.14 exhibit two important features:

1. They are showing very good convergence. A comparison of the tabular printout of the solutions from the two meshes indicates accuracies ranging from about 0.2% for the first wave function to about 4% for the fifth, for both the wave functions and the eigenvalues. Note that the higher wave functions were significantly affected by enlarging the domain.

2. The wave functions and their slopes are both decaying rapidly to zero near the boundaries of the domain.

---

[17] The generalized Jacobi routine GENJAC normalizes the eigenvectors relative to the mass matrix according to Eqs. (10.30). Since $\gamma(\xi) = 1$ in Eq. (10.63), the normalizing relation $\{\psi\}_i[M]\{\psi\}_i = 1$ is the discretization of the normalizing integral $\int_{-\infty}^{\infty} \psi^2 d\xi = 1$. [Recall the footnote to Eq. (10.60)].

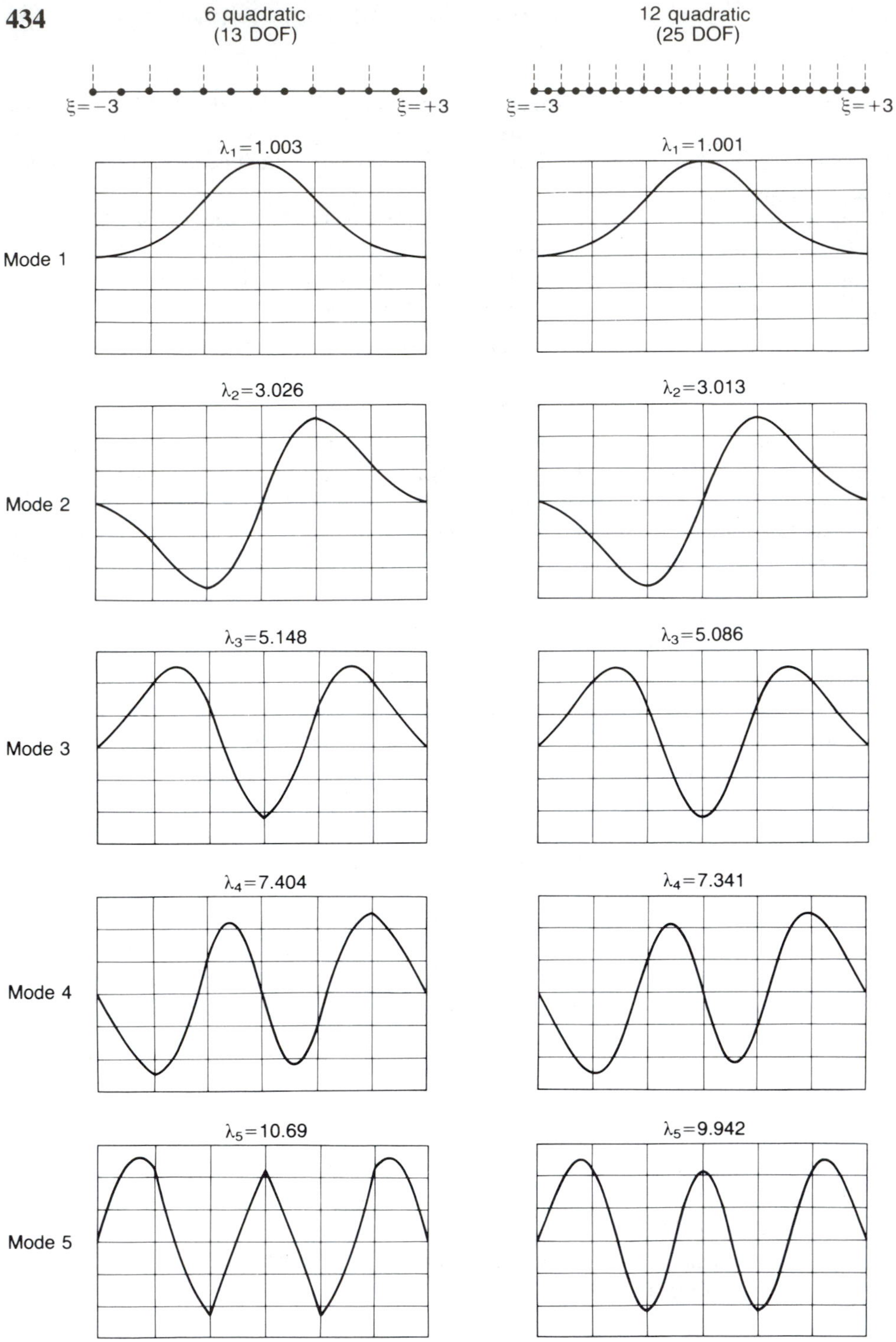

FIGURE 10.13 ● The first five wave functions and energy levels (nondimensional) for the quantum mechanical oscillator, using two meshes of 6 and 12 C^0-quadratic elements, respectively, and a finite domain $-3 < \xi < +3$.

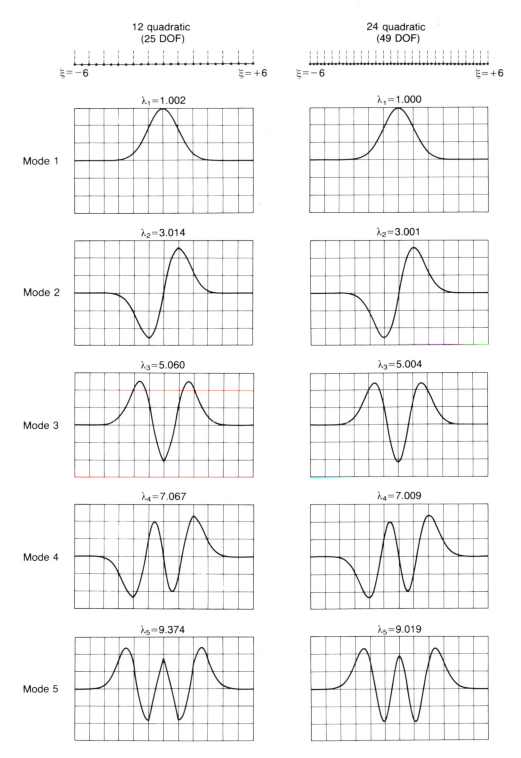

FIGURE 10.14 ● The first five wave functions and energy levels (nondimensional) for the quantum mechanical oscillator, using two meshes of 12 and 24 C^0-quadratic elements, respectively, and a finite domain $-6 < \xi < +6$.

436

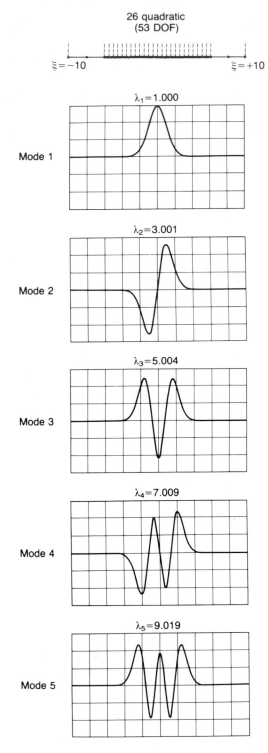

FIGURE 10.15 ● The first five wave functions and energy levels (nondimensional) for the quantum mechanical oscillator, using a nonuniform mesh of 26 C^0-quadratic elements and a finite domain $-10 < \xi < +10$.

The second feature suggests that the interval $-6<\xi<6$ is probably adequate to model all the significant behavior of these first five wave functions. If it isn't, then any error is probably being controlled by the size of the domain, and therefore the solution would change significantly if the domain were enlarged (even without altering the mesh in the present domain). To test for this possibility, let's add one large element to either end of the domain $-6<\xi<+6$, enlarging it to $-10<\xi<+10$. The results for this final 26-element mesh are shown in Fig. 10.15. A comparison of the tabular printout of the solutions from the 24- and 26-element meshes shows agreement to four significant figures in both the energy levels and the wave functions. Therefore, by the guideline following Eq. (10.64), we have an acceptable solution.[18]

Exercises

10.1   Use Eqs. (10.26) to show that $[K]$ and $[M]$ in Eqs. (10.36) are both positive-definite.

10.2   Verify that $\{v\}_1$ and $\{v\}_2$ in Eqs. (10.43) are both orthogonal to $[K]$ and $[M]$ in Eqs. (10.36), i.e., that they both satisfy Eqs. (10.31).

Exercises 10.3 through 10.8 examine the buckling of long, slender rods under axial compressive loads. They illustrate a classic static buckling analysis and the use of elastic beam theory. Exercise 10.9 considers the transverse vibration of a beam, and Exercise 10.10 combines the buckling and vibration concepts into a dynamic buckling analysis.

10.3   Consider the buckling of a long, slender rod compressed axially by a force P:

$$P \longrightarrow \fbox{} \longleftarrow P$$

One or both ends of the rod must be constrained. There are several types of constraints, the usual ones being clamps and/or hinges.

Below a certain critical load P_{cr} (which depends on the physical properties of the rod and the type of end constraints), the rod will remain straight, merely compressing axially; it is the only equilibrium configuration and it is stable. However, for $P>P_{cr}$, there are two equilibrium configurations: the straight one which is now unstable, and a deflected (buckled) one, which is stable. The problem is to determine P_{cr} and the shape of the buckled configuration. This is the classic "Euler buckling" problem. An extensive treatment may be found in the work by Timoshenko and Gere [10.43].

The governing differential equation is

$$\frac{d^2}{dx^2}\left(EI\,\frac{d^2W}{dx^2}\right) + \frac{d}{dx}\left(P\,\frac{dW}{dx}\right) = 0 \qquad \text{(E10.1)}$$

---

[18] The exact solution for the wave functions and nondimensional energy levels may be found in Leighton [10.41] and Pauling and Wilson [10.42]. The expression for the energy levels is particularly simple:

$$\lambda_n = 2n-1 \qquad n = 1,2,3,\ldots$$

where W is the deflection from the straight configuration. We will treat E, I, and P as constants. Equation (E10.1) is the elastic beam equation, with the addition of the compressive axial load term and the absence of any lateral loads. It was treated in Section 8.6 [see Eq. (8.100)] in connection with the C^1-cubic element. Since the end constraints (see exercises below) all involve zero-valued essential or natural BCs, Eq. (E10.1) is an eigenproblem with the eigenvalue P.

Using the C^1-cubic element in Section 8.6, and following steps 1 through 5 of the 12-step procedure, show that the element equations have the form,

$$([K]-P[\varkappa])\{a\} = \{F\} \tag{E10.2a}$$

where

$$[\varkappa] = \begin{bmatrix} \dfrac{6}{5L} & \dfrac{1}{10} & -\dfrac{6}{5L} & \dfrac{1}{10} \\[2mm] & \dfrac{2L}{15} & -\dfrac{1}{10} & -\dfrac{L}{30} \\[2mm] & & \dfrac{6}{5L} & -\dfrac{1}{10} \\[2mm] \text{Symmetric} & & & \dfrac{2L}{15} \end{bmatrix}$$

$$\{F\} = \begin{Bmatrix} -V(x_1) \\[2mm] M(x_1) \\[2mm] V(x_2) \\[2mm] -M(x_2) \end{Bmatrix} \tag{E10.2b}$$

Here $L\ (= x_2-x_1)$ is the length of the element and M and V are the bending moment and transverse shear force, respectively. The expression for M is the same as in Eq. (8.101), but V becomes instead

$$-\frac{d}{dx}\left(EI\frac{d^2W}{dx^2}\right) - P\frac{dW}{dx}$$

The expression for $[K]$ is given in Eq. (8.103).

10.4   Calculate the critical buckling load P_{cr} and the corresponding buckled mode shape (deflected configuration) for a rod hinged at both ends:

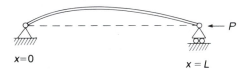

$x=0$

$x = L$

The figure is only a suggestion of a possible buckled mode shape. A hinge, by definition, prevents lateral displacement but allows free rotation, the latter implying that no bending moment can be imparted by the hinge.

a) What are the BCs that must be specified at each end? Which are essential and which are natural? (See Section 8.6.)

b) Using the element equations [Eqs. (E10.2)] in Exercise 10.3, analyze a one-element model.

The exact solution is $P_{cr} = \pi^2 EI/L^2$, with the mode shape $W(x) = A \sin (\pi x/L)$, where A is an arbitrary amplitude. (The analysis will yield two eigensolutions, which are an approximation to the first two of an infinite family for the exact solution. The higher modes are generally of no practical interest since a structure would buckle as soon as the lowest buckling load is reached unless that mode shape were prevented by appropriately positioned constraints.)

10.5 Calculate the critical buckling load P_{cr} and the corresponding buckled mode shape for a rod clamped at one end and free at the other:

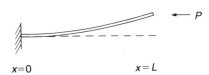

$$x = 0 \qquad\qquad\qquad x = L$$

A clamp, by definition, prevents both lateral deflection and rotation. Of course, there are no bending moments or transverse shear forces at a free end.

a) What are the BCs that must be specified at each end? Which are essential and which are natural?

b) Using the element equations [Eqs. (E10.2)] in Exercise 10.3, analyze a one-element model. The exact solution is $P_{cr} = \pi^2 EI/4L^2$, with the mode shape $W(x) = A[1 - \cos (\pi x/2L)]$.

c) Comparing the end conditions in this and the previous exercise, it can be seen that this buckled mode is the same as *half* the previous one. (The clamped end is analogous to the center of the hinged/hinged rod.) Thus, if we change L to $L/2$ for the clamped/free rod, P_{cr} would be the same as for the hinged/hinged rod. We have therefore effectively accomplished a *mesh refinement*; i.e., the present exercise is equivalent to having done a two-element analysis of the hinged/hinged rod.

Using the exact solutions to determine error, calculate the *rate of convergence* of the critical buckling loads. *Hint:* Use $2^\alpha = R$, where α is the rate, the base 2 corresponds to halving the element size, and R is the ratio of the percent errors in P_{cr}. Compare with the *asymptotic* rate predicted by Eq. (10.55). (Recall from Chapter 9 that the rates for coarse meshes may vary appreciably from the asymptotic rate.) The substantial decrease in percent error (more than an order of magnitude) as a result of only halving the element size is therefore to be expected for higher-order elements, e.g., cubic C^1 or quadratic C^0 elements.

10.6   Calculate the critical buckling load P_{cr} for a rod clamped at one end and hinged at the other:

$$x=0 \qquad\qquad\qquad\qquad x=L$$

a)   What are the BCs that must be specified at each end? Note that this exercise has one more essential BC than the previous two exercises. How will this affect the number of active DOF and eigenvalues in the FE model?

b)   Using the element equations [Eqs. (E10.2)] in Exercise 10.3, analyze a one-element model. The "exact" solution is $P_{cr}=\pi^2 EI/(0.69916L)^2$, the number 0.69916 being an approximate solution to a transcendental equation. (See Timoshenko and Gere [10.43].)

10.7   Analyze the clamped/hinged rod in Exercise 10.6 using a *refined* mesh of two elements, each of length $L/2$. (The procedure for assembling C^1 elements is explained in Section 8.6.)

a)   Calculate the critical buckling load P_{cr}. As discussed above, P_{cr} is the lowest eigenvalue.

b)   Using the values of P_{cr} calculated for the one-element mesh (Exercise 10.6) and the two-element mesh, calculate the rate of convergence α from the relation $2^\alpha=R$, where R is the ratio of percent errors in P_{cr}. Compare with the asymptotic rate predicted by Eq. (10.55).

10.8   Analyze the clamped/hinged rod in Exercise 10.6 using a mesh of two elements of *unequal* length:

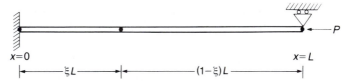

$$x=0 \qquad\qquad\qquad\qquad\qquad\qquad x=L$$
$$\vdash\!\!-\!\!-\!\!\xi L\!\!-\!\!-\!\!\dashv\!\!\vdash\!\!-\!\!-\!\!-\!\!(1-\xi)L\!\!-\!\!-\!\!-\!\!\dashv$$

where $0<\xi<1$. Determine the *optimum* value of ξ, namely, that which minimizes the error in P_{cr}. (Use the "exact" solution in Exercise 10.6 to determine error.) One approach would be to calculate P_{cr} for several values of ξ and interpolate graphically or numerically. *Suggestion*: Before starting the analysis, make a "guesstimate" of ξ_{opt} by asking yourself which end of the rod has the more complex part of the expected mode shape and hence needs a greater DOF density. It will be interesting to see how close your intuitive expectation comes to the actual solution. In more complex problems an experienced analyst can often construct a good nonuniform mesh (high accuracy for a given number of DOF) by using an intuitive "feel" for the expected qualitative nature of the solution.

10.9   Transverse vibrations of a beam with no axial loads are governed by the differential equation,

$$\frac{\partial^2}{\partial x^2}\left(EI\,\frac{\partial^2 w}{\partial x^2}\right)=-\varrho\,\frac{\partial^2 w}{\partial t^2} \qquad\qquad\text{(E10.3)}$$

This follows from Eq. (8.100) where the lateral load $f(x)$ becomes the inertial acceleration $-\varrho \partial^2 w/\partial t^2$. [Recall Eq. (10.3).] For free vibrations, $w(x,t)=W(x) \sin \omega t$, and Eq. (E10.3) becomes

$$\frac{d^2}{dx^2}\left(EI \frac{d^2W}{dx^2}\right) - \varrho \omega^2 W = 0 \tag{E10.4}$$

a) Using the C^1-cubic element in Section 8.6, and following steps 1 through 5 of the 12-step procedure, derive the element equations. [The stiffness matrix is given in Eq. (8.103); it remains only to derive the mass matrix.]

b) Using a one-element model, calculate the fundamental (lowest) natural frequency for the hinged/hinged beam described in Exercise 10.4. The exact solution is $\omega_1 = \pi^2 c/L^2$, where $c=\sqrt{EI/\varrho}$. *Hint:* Considering the essential BCs, is it necessary in the first part of this exercise to derive all the terms in the mass matrix for this one-element problem?

10.10 We can combine the preceding vibration and buckling analyses to perform a *dynamic buckling* analysis. The governing differential equation for transverse vibrations of a beam with a compressive axial load uses the terms from both Eqs. (E10.1) and (E10.4):

$$\frac{d^2}{dx^2}\left(EI \frac{d^2W}{dx^2}\right) + \frac{d}{dx}\left(P \frac{dW}{dx}\right) - \varrho \omega^2 W = 0 \tag{E10.5}$$

a) Using a one-element model, calculate the fundamental frequency for the hinged/hinged beam. Make use of previous derivations for the mass matrix and $[k]$. If you have not done the previous exercises, note the hint in Exercise 10.9. The exact solution is $\omega_1=(\pi^2 c/L^2)\sqrt{1-(PL^2/\pi^2 EI)}$, where $c=\sqrt{EI/\varrho}$.

b) Buckling occurs when $\omega_1=0$. Thus, as P increases, ω_1 decreases, until eventually, at $\omega_1=0$, the beam does not return to the straight configuration; i.e., it buckles. Apply this condition to your result from the first part of this exercise to determine the buckling load P_{cr}. (If you have done Exercise 10.4, compare that solution with the present one.)

c) The vibrating beam under compression can be converted to a vibrating flexible cable under tension, analyzed in Section 10.5, by changing P to $-T$ and assuming $PL^2/EI \gg 1$. Apply these conditions to your result in the first part of this exercise to determine the fundamental frequency of the flexible cable. The exact solution is $\omega_1=\pi c/L$, where $c=\sqrt{T/\varrho}$.

Use UNAFEM, or any other FE program capable of solving Eq. (10.1), for the following exercises.

10.11 Consider the 100-ft cable hanging vertically from a building, as described in Exercise 7.13. Using the same physical properties but neglecting aerodynamic drag (i.e., there is no wind, and no air resistance during vibration), calculate the first five natural frequencies and mode shapes for transverse vibration of the cable. [Note the conversion from eigenvalues to frequencies in Eq. (10.10).]

10.12 Consider the 10-ft rubber cord whirling in a circle, as described in Exercise 7.14. Using the same physical properties and considering only the effect of the centrifugal force on the tension (i.e., neglecting aerodynamic drag, gravity, and stretch), calculate the first five natural frequencies and mode shapes for vibration of the cord transverse to the plane of rotation. (The exact solution is the family of Legendre functions.)

10.13 The free, transverse vibration of a flexible membrane is a 2-D eigenproblem. [Recall Section 2.4, especially Eq. (2.19).] For a *circular* membrane it is natural to use polar coordinates (r,θ) rather than Cartesian. Then one can use the classic *separation of variables* technique and write the vibration amplitude as $W(r,\theta)=U(r)V(\theta)$, which enables the 2-D governing differential equation to be converted into separate 1-D equations for $U(r)$ and $V(\theta)$. The circumferential equation is $d^2V/d\theta^2+m^2V=0$, which has as its solutions $\sin m\theta$ and $\cos m\theta$, where $m=0,1,2,\ldots$. The radial equation is

$$-\frac{d}{dr}\left(r\frac{dU}{dr}\right)+\frac{m^2}{r}U-k^2rU=0 \tag{E10.6}$$

The eigenvalue is $k^2=\omega^2/c^2$ and c is the propagation velocity of transverse waves. For each of the above integral values of m there is an infinite family of eigenfunctions, $U_{mn}(r)$, $n=1,2,\ldots$. Thus there is a doubly infinite family of eigenfunctions (mode shapes) for $W(r,\theta)$.

Consider a circular membrane of radius a, fastened at its edge, $r=a$. The boundary conditions for Eq. (E10.6) are

$$r\frac{dU(0)}{dr}=0 \qquad \text{(no transverse force applied at center)}$$

$$U(a)=0 \qquad \text{(fastened at edge)} \tag{E10.7}$$

Calculate the first five natural frequencies and mode shapes for the membrane.

Note that the first five frequencies may occur for different values of m, necessitating solutions to Eq. (E10.6) for several values of m and then ranking the frequencies in increasing order. The analysis for $m=0$ requires very little input data for UNAFEM. However, for $m\geq1$, the $1/r$ term must be approximated by one or more polynomials in the domain, with the approximation becoming "infinitely bad" as r approaches zero. It is instructive to compare the computed results with the exact solution given below to observe the effect of the local modeling error introduced by approximating a singularity by a polynomial. [A simple alternative is to change one line of code in UNAFEM to define $\beta(r)=m^2/r$.] The exact solution to the radial equation is the family of *Bessel* functions: $U_{mn}(r)=J_m(k_{mn}r)$, $n=1,2,\ldots$, where $k_{mn}=\nu_{mn}/a$ and ν_{mn} are the zeroes of J_m; that is, $J_m(\nu_{mn})=0$.

References

[10.1]   P. M. Morse and H. Feshbach, *Methods of Theoretical Physics*, McGraw-Hill, New York, 1953, p. 719.

[10.2]   Y. Chen, *Vibrations: Theoretical Methods*, Addison-Wesley, Reading, Mass., 1966. See Section 7.6, "Free Vibration and Normal Modes of a Finite String."

[10.3]   P. M. Morse, *Vibration and Sound*, McGraw-Hill, New York, 1948. See Chapter III, "The Flexible String."

[10.4]   J. H. Wilkinson, *The Algebraic Eigenvalue Problem*, Oxford University Press, New York, 1965.

[10.5]   A. R. Gourlay and G. A. Watson, *Computational Methods for Matrix Eigenproblems*, Wiley, New York, 1973.

[10.6]   B. S. Garbow, J. M. Boyle, J. J. Dongarra, and C. B. Moler, *Matrix Eigensystem Routines — EISPACK Guide Extension*, Springer-Verlag, New York, 1976. (Volume 51 in the Lecture Notes in Computer Science series. See also vol. 6 by B. T. Smith, et al.)

[10.7]   F. B. Hildebrand, *Methods of Applied Mathematics*, Prentice-Hall, Englewood Cliffs, N. J., 1952. See Chapter 1, "Matrices, Determinants, and Linear Equations."

[10.8]   C. Lanczos, "An Iteration Method for the Solution of the Eigenvalue Problem of Linear Differential in an Integral Operation," *J Res Natl Bureau of Standards*, **45**: 255-282 (1950).

[10.9]   V. I. Weingarten, R. K. Ramanathan, and C. N. Chen, "Lanczos Eigenvalue Algorithm for Large Structures on a Minicomputer," *Computers and Structures*, **16**(1-4): 253-257 (1983).

[10.10] C. G. J. Jacobi, "Über ein leichtes Verfahren die in der Theorie der Säculärstörungen vorkommenden Gleichungen numerisch aufzulösen," *Crelle's Journal*, **30**: 51-94 (1846).

[10.11] S. Falk and P. Langemeyer, "Das Jacobische Rotationsverfahren für Reellsymmetrische Matrizenpaare," *Elektronische Datenverarbeitung*, 1960, pp. 30-34.

[10.12] K. J. Bathe, "Solution Methods for Large Generalized Eigenvalue Problems in Structural Engineering," Report UC SESM 71-20, Civil Engineering Department, University of California, Berkeley, 1971.

[10.13] K. J. Bathe and E. L. Wilson, *Numerical Methods in Finite Element Analysis*, Prentice-Hall, Englewood Cliffs, N. J., 1976, Chapters 10-12.

[10.14] K. J. Bathe, *Finite Element Procedures in Engineering Analysis*, Prentice-Hall, Engelwood Cliffs, N. J., 1982, Chapters 10-12. (Chapters 10 through 12 appear to be identical in this reference and the one above. They present a thorough treatment of several popular methods of solving the algebraic eigenproblem in FE applications.)

[10.15] H. H. Fong, "An Evaluation of Eight U.S. General Purpose Finite Element Computer Programs," Part I, *Proc* 23rd *AIAA/ASME/ASCE/AHS Structures, Structural Dynamics and Materials Conference*, New Orleans, May 10-12, 1982, pp. 145-160.

[10.16] B. Nour-Omid, B. N. Parlett, and R. L. Taylor, "Lanczos Versus Subspace Iteration for Solution of Eigenvalue Problems," *Int J Num Meth Eng*, **19**: 859-871 (1983).

[10.17] E. L. Wilson and T. Itoh, "An Eigensolution Strategy for Large Systems," *Computers and Structures*, **16**(1-4): 259-265 (1983).

[10.18] B. Irons, "Eigenvalue Economisers in Vibration Problems," *J Royal Aero Soc*, **67**: 526-528 (1963).

[10.19] B. Irons, "Structural Eigenvalue Problems: Elimination of Unwanted Variables," *AIAAJ*, **3**(5): 961-962 (1965).

[10.20] R. J. Guyan, "Reduction of Stiffness and Mass Matrices," *AIAAJ*, **3**(2): 380 (1965).

[10.21] R. D. Henshell and J. H. Ong, "Automatic Masters for Eigenvalue Economization," *Earthquake Eng Struc Dyn*, **3**: 375-383 (1975).

[10.22] D. L. Thomas, "Errors in Natural Frequency Calculations Using Eigenvalue Economization," *Int J Num Meth Eng*, **18**: 1521-1527 (1982).

[10.23] G. Strang and G. J. Fix, *An Analysis of the Finite Element Method*, Prentice-Hall, Englewood Cliffs, N. J., 1973, p. 232.

[10.24] R. B. Leighton, *Principles of Modern Physics*, McGraw-Hill, New York, 1959, p. 112.

[10.25] L. Pauling and E. B. Wilson, *Introduction to Quantum Mechanics*, McGraw-Hill, New York, 1935, p. 57. (A venerable text and still one of the most readable.)

[10.26] P. Bettess, "Infinite Elements," *Int J Num Meth Eng*, **11**: 53-64 (1977).

[10.27] P. Bettess, "More on Infinite Elements," *Int J Num Meth Eng*, **15**: 1613-1626 (1980).

[10.28] G. Beer and J. L. Meek, "'Infinite Domain' Elements," *Int J Num Meth Eng*, **17**: 43-52 (1981).

[10.29] F. Medina and R. L. Taylor, "Finite Element Techniques for Problems of Unbounded Domains," *Int J Num Meth Eng*, **19**: 1209-1226 (1983).

[10.30] S. Pissanetzky, "An Infinite Element and a Formula for Numerical Quadrature over an Infinite Interval," *Int J Num Meth Eng*, **19**: 913-927 (1983).

[10.31] J. M. M. C. Marques and D. R. J. Owen, "Infinite Elements in Quasi-Static Materially Nonlinear Problems," *Computers and Structures*, **18**(4): 739-751 (1984).

[10.32] O. C. Zienkiewicz, K. Bando, P. Bettess, C. Emson, and T. C. Chiam, "Mapped Infinite Elements for Exterior Wave Problems," *Int J Num Meth Eng*, **21**: 1229-1251 (1985).

[10.33] R. W. Clough, "The Finite Element Method in Plane Stress Analysis," *Proc Second ASCE Conference Electronic Computation*, Pittsburgh, Sept. 1960.

[10.34] O. C. Zienkiewicz, *The Finite Element Method*, McGraw-Hill, New York, 1977. See Chapter 23, "'Boundary Solution' Processes and the Finite Element Method. Infinite Domains; Singularity in Fracture Mechanics."

[10.35] O. C. Zienkiewicz, D. W. Kelly, and P. Bettess, "The Coupling of the Finite Element Method and Boundary Solution Procedure," *Int J Num Meth Eng*, **11**(2): 355-375 (1977). (This is a survey of the field, citing 96 references.)

[10.36] C. A. Brebbia, *The Boundary Element Method for Engineers*, Halsted (Wiley), New York, 1978.

[10.37] P. K. Banerjee and R. Butterfield, *Boundary Element Methods in Engineering Science*, McGraw-Hill, New York, 1981.

[10.38] G. Beer, "Finite Element, Boundary Element and Coupled Analysis of Unbounded Problems in Elastostatics," *Int J Num Meth Eng,* **19**: 567-580 (1983).

[10.39] C. A. Brebbia, J. C. F. Telles, and L. C. Wrobel, *Boundary Element Techniques: Theory and Applications in Engineering*, Springer-Verlag, Berlin, 1984.

[10.40] C. A. Brebbia, ed., *Finite Element Systems*, Springer-Verlag, Berlin, 1985.

[10.41] R. B. Leighton, op. cit., pp. 133-135.

[10.42] L. Pauling and E. B. Wilson, op. cit., pp. 73-76.

[10.43] S. P. Timoshenko and J. M. Gere, *Theory of Elastic Stability*, McGraw-Hill, New York, 1961. (See "Elastic Buckling of Bars and Frames," Chapter 2.)

11

1-D Initial-Boundary-Value Problems: Diffusion

Introduction

Chapter 11 treats the mixed initial-value/boundary-value problem, also referred to as the initial-boundary-value problem. An additional term, involving a derivative with respect to time, is added to the boundary-value problem in Chapters 6 through 8, making the unknown U a function of both space and time. We therefore look for a solution which is changing in time, called a *transient* solution, which begins at some instant t_0 and proceeds indefinitely into the future $(t \to \infty)$.

Sections 11.1 through 11.5 present a complete analysis of a particular class of initial-boundary-value problems, namely, *diffusion* phenomena. As in the previous chapters we begin with a mathematical and physical description of the problem (Section 11.1), followed by a derivation of the element equations (Section 11.2), then implementation of the equations into the UNAFEM program (Section 11.4), and finally application of the program to a transient heat conduction problem (Section 11.5). Inserted into the middle of this standard sequence is an additional section (Section 11.3) that explains how to treat the time-varying part of the problem.

11.1 Mathematical and Physical Description of the Problem

11.1.1 The Initial-Value Problem

Before examining the mixed problem, let's first consider the "pure" initial-value problem in which the unknown U is a function only of time.[1] One of the simplest such problems is the differential equation,

$$c \, \frac{dU(t)}{dt} + kU(t) = f(t) \qquad t > t_0 \tag{11.1a}$$

---

[1] In principle, the variable t may represent any physical quantity, not necessarily time. However, in practice, it almost always represents time, so we will employ that meaning here.

FIGURE 11.1 ● Semi-infinite domain characteristic of an initial-value problem, with initial condition at time t_0.

with the initial condition

$$U(t_0) = U_0 \qquad (11.1b)$$

Here U is the unknown function, c and k are known physical properties, and f is a known applied load.[2] If this were a heat conduction problem (see below) these variables would represent temperature, heat capacity, convection loss coefficient, and interior heat source, respectively.

Equation (11.1a) is a first-order ordinary differential equation — first order because the highest derivative of U is of order 1. The domain is *infinite* (more precisely, semi-infinite), as shown in Fig. 11.1. The problem begins at time t_0 and "marches forward" indefinitely into the future. To get the problem started, we must specify a value for U at t_0, called an *initial condition*. It is not necessary to specify data at any other time $t > t_0$. (Contrast this with the boundary-value problem which requires data at both ends of the domain, whether it is finite or infinite.)

Considerable insight may be gained by examining the *free response* of the system, namely, when the applied load $f(x)$ vanishes:

$$c\frac{dU(t)}{dt} + kU(t) = 0 \qquad t > t_0 \qquad (11.2a)$$

$$U(t_0) = U_0 \qquad (11.2b)$$

As shown in any introductory textbook on ordinary differential equations [11.1-11.3], the solution must have the form,

$$U(t) = Ae^{-\lambda t} \qquad (11.3)$$

Substituting Eq. (11.3) into Eq. (11.2a) yields

$$A(-c\lambda + k)e^{-\lambda t} = 0 \qquad (11.4)$$

or

$$k - \lambda c = 0 \qquad (11.5)$$

---

[2] c and k are considered here to be constants. In general, they could be functions of t as well as U, the latter making the problem nonlinear. Both cases are straightforward to handle by the FEM, but they would only add unnecessary generality to the present introductory remarks.

which is called the *characteristic equation*. The solution to Eq. (11.5) is the *characteristic value*, or *eigenvalue*,

$$\lambda = \frac{k}{c} \tag{11.6}$$

where $\lambda > 0$ since the physical properties k and c are generally positive-valued. Substituting Eq. (11.6) into Eq. (11.3) yields

$$U(t) = Ae^{-(k/c)t} \tag{11.7}$$

The constant A is determined by applying the initial condition [Eq. (11.2b)] to Eq. (11.7), resulting in the following expression for the free response:

$$U(t) = U_0 e^{-(k/c)(t-t_0)} \qquad t \geq t_0 \tag{11.8}$$

which is shown in Fig. 11.2.

If the applied load is any constant (steady) value f, the solution is just as easily shown to be

$$U(t) = \frac{f}{k} + \left(U_0 - \frac{f}{k} \right) e^{-(k/c)(t-t_0)} \tag{11.9}$$

which again is exponential decay. As $t \to \infty$ in Eq. (11.9), $U(t) \to f/k$ which is the *steady-state* (equilibrium) solution, corresponding to $dU/dt = 0$ in Eq. (11.1a). In other words, if the loads are held steady for any interval of time, the solution decays exponentially during that interval, and if they are held steady indefinitely, the solution ultimately "settles down" (exponentially) to a final equilibrium state, i.e., the steady-state solution. The important point here is that the free response, as well as the response to any steady load, is an exponential decay. This is the hallmark of all diffusion phenomena.[3]

The above discussion has focused on a system with only one unknown function, $U(t)$. More generally, if there are n unknown functions $U_1(t), U_2(t), \ldots, U_n(t)$, there will be a system of n coupled ordinary differential equations, written in matrix form as

$$[c] \left\{ \frac{dU(t)}{dt} \right\} + [k]\{U(t)\} = \{f(t)\} \tag{11.10}$$

Approximate numerical solutions to Eqs. (11.10) are generally obtained by *time-stepping* methods, in which the differential equations are approximated by *recurrence relations*. Such relations can be derived by a variety of techniques, for example, finite differences or finite elements, as explained in Section 11.3. There exists an extensive body of literature for solving initial-value problems, most of it written prior to or independent of the FEM. A sampling of that literature [11.5-11.10] is listed at the end of this chapter.

If our only interest were in pure initial-value problems we could simply refer to the above literature. Our interest, however, is in the *mixed* initial-boundary-value problem

---

[3] Mathematically speaking, diffusion phenomena are described by differential equations in which the time derivatives are *first* order. Such equations are said to be *parabolic* [11.4]. (The equations for boundary-value problems are *elliptic*.) Another very important class of initial-value problems (*hyperbolic*) is characterized by *second*-order time derivatives: $md^2U/dt^2 + cdU/dt + kU = f(t)$. For those kinds of problems (see Chapter 12) the free response is *oscillatory*, describing wave propagation and vibration phenomena.

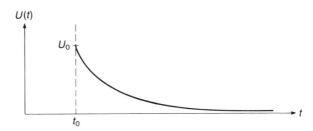

FIGURE 11.2 ● An exponentially decaying free response, which is characteristic of diffusion phenomena.

in which the unknown U is a function of both time (an initial-value problem) and space (a boundary-value problem). As we will see shortly, the spatial part is dealt with most effectively by the FEM (in precisely the manner presented in Chapters 6 through 8), and the temporal part is dealt with very effectively by time-stepping methods.

11.1.2 A Mixed Initial-Boundary-Value Problem

Sections 11.2 through 11.5 will analyze a class of problems described by the following governing differential equation and associated boundary conditions and initial condition:

Differential equation ●

$$\mu(x)\,\frac{\partial U(x,t)}{\partial t} - \frac{\partial}{\partial x}\left(\alpha(x)\,\frac{\partial U(x,t)}{\partial t}\right) + \beta(x)U(x,t) = f(x,t) \tag{11.11a}$$

Domain (see Fig. 11.3) ●

$$x_a < x < x_b$$

$$t > t_0 \tag{11.11b}$$

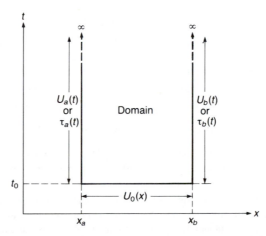

FIGURE 11.3 ● Domain for initial-boundary-value problem, showing initial condition and boundary conditions.

Boundary conditions •

At x_a $(t > t_0)$

$$U(x_a,t) = U_a(t) \qquad \text{essential BC}$$

or

$$\left(-\alpha(x) \frac{\partial U(x,t)}{\partial x}\right)_{x_a} = \tau_a(t) \qquad \text{natural BC}$$

At x_b $(t > t_0)$

$$U(x_b,t) = U_b(t) \qquad \text{essential BC}$$

or

$$\left(-\alpha(x) \frac{\partial U(x,t)}{\partial x}\right)_{x_b} = \tau_b(t) \qquad \text{natural BC} \qquad \text{(11.11c)}$$

Initial condition •

At t_0 $(x_a < x < x_b)$

$$U(x,t_0) = U_0(x) \qquad \text{(11.11d)}$$

We see in Eq. (11.11a) all the same terms that were present in the boundary-value problem in Chapters 6 through 8 [Eqs. (6.25) and (7.1a)], but now there is an additional time derivative term, $\mu(x)\partial U/\partial t$. The unknown U becomes a function of both x and t, and the previous ordinary derivatives become partial derivatives. In addition, the interior load f may now be a function of both x and t.[4] The boundary conditions are the same as for the boundary-value problem, i.e., essential and natural, although now they may be functions of t.

In a strict mathematical sense, this is a 2-D problem since there are two independent variables, x and t. However, within the context of a finite element analysis, it is appropriate to refer to it as a 1-D problem since we will treat only the spatial variable x by the FEM, and therefore we will be using all of the previously developed 1-D elements.

11.1.3 Transient Heat Conduction and Other Physical Applications

To demonstrate the application of Eqs. (11.11) to a 1-D transient heat conduction problem, let's repeat the derivation of the equations in Section 6.1.1, using the same notation in that section, but adding the extra time derivative term. And for simplicity we'll consider the usual case of a cylindrical rod with constant cross-sectional area (since the case of a variable area is just as easily derived, following the example in Section 6.1.1).

Balance of energy • (for the differential element of volume $A dx$ shown in Fig. 6.1)

$$\begin{array}{ccc} \text{Rate of heat} & \text{rate of heat} & \text{rate of increase of} \\ \text{energy added} & \text{energy lost} & \text{internal heat energy} \end{array}$$

(the equation shows: Rate of heat energy added − rate of heat energy lost = rate of increase of internal heat energy)

---

[4] More generalities could have been added to the problem, such as making α and β also functions of t. It was felt that such embellishments do not entail important new concepts, and the additional analysis and coding due to their implementation would only obscure the above features, which the author believes to be the most important ones.

TABLE 11.1 • One-dimensional initial-boundary-value problems (diffusion phenomena).

Application	Unknown	Physical/material properties			Interior load	Governing equation
		μ	α	β		
General mathematical formulation	U	μ	α	β	f	$\mu \dfrac{\partial U}{\partial t} - \dfrac{\partial}{\partial x}\left(\alpha \dfrac{\partial U}{\partial x}\right) + \beta U = f$
Heat conduction: diffusion of heat in a solid	T (temperature)	ϱc (ϱ = mass density, c = specific heat)	k (thermal conductivity)	$\dfrac{hl}{A}$ (convection loss coefficient)	$Q + \dfrac{hlT_\infty}{A}$ Q = heat source $\dfrac{hlT_\infty}{A}$ = part of ambient convection	$\varrho c \dfrac{\partial T}{\partial t} - \dfrac{\partial}{\partial x}\left(k \dfrac{\partial T}{\partial x}\right) + \dfrac{hl}{A}T = Q + \dfrac{hlT_\infty}{A}$
Transient seepage: flow of fluid through a porous medium	H (hydraulic head)	S (storage coefficient)	T (hydraulic transmissivity)		Q (inflow from external source)	$S \dfrac{\partial H}{\partial t} - \dfrac{\partial}{\partial x}\left(T \dfrac{\partial H}{\partial x}\right) = Q$
Soil consolidation: diffusion of fluid through pores of an elastic medium	P (pore pressure of fluid)	C (compressibility)	$\dfrac{\varkappa}{\varrho}$ $\varkappa$ = permeability ϱ = density			$C \dfrac{\partial P}{\partial t} - \dfrac{\partial}{\partial x}\left(\dfrac{\varkappa}{\varrho} \dfrac{\partial P}{\partial x}\right) = 0$
Atomic and nuclear physics: diffusion of electrons in a gas, or neutrons in matter	ϱ (spatial density of particles)	m (mass of particles)	a^2 (diffusion constant - proportional to P_a, the average momentum, and λ_a, the mean free path)	μ (mean rate of absorption of particles)	r (number of particles created per second per volume)	$m \dfrac{\partial \varrho}{\partial t} - \dfrac{\partial}{\partial x}\left(a^2 \dfrac{\partial \varrho}{\partial x}\right) + \mu\varrho = r$

Thus Eq. (6.1) becomes instead

$$qA + QAdx - (q+dq)A = \varrho c \frac{\partial T}{\partial t} Adx \tag{11.12}$$

or

$$\varrho(x)c(x) \frac{\partial T(x,t)}{\partial t} + \frac{\partial q(x,t)}{\partial x} = Q(x,t) \tag{11.13}$$

where

$\varrho(x)$ = mass density,[5] M/L^3

$c(x)$ = specific heat,[6] E/MT

Constitutive equation • This is Fourier's law of heat conduction, which remains unchanged. Thus, repeating Eq. (6.3) but using partial derivatives,

$$q(x,t) = -k(x) \frac{\partial T(x,t)}{\partial x} \tag{11.14}$$

Substituting Eq. (11.14) into Eq. (11.13) yields

$$\varrho(x)c(x) \frac{\partial T(x,t)}{\partial t} - \frac{\partial}{\partial x}\left(k(x) \frac{\partial T(x,t)}{\partial x} \right) = Q(x,t) \tag{11.15}$$

Equation (11.15) has the same form as Eq. (11.11a), where $U(x,t)=T(x,t)$, $\mu(x)=\varrho(x)c(x)$, $\alpha(x)=k(x)$, $\beta(x)=0$, and $f(x,t)=Q(x,t)$.

Similarly, Eq. (6.7b) for the 1-D model of 2-D heat flow in Section 6.1.2 has the corresponding transient form,

$$\varrho(x)c(x) \frac{\partial T(x,t)}{\partial t} - \frac{\partial}{\partial x}\left(k(x) \frac{\partial T(x,t)}{\partial x} \right) + \frac{hl}{A} T(x,t) = Q(x,t) + \frac{hlT_\infty}{A} \tag{11.16}$$

There are diffusion phenomena that occur in other fields of application which are also described by eqs. (11.11). Several of these are summarized in Table 11.1.

11.2 Derivation of Element Equations

11.2.1 Separation of Space and Time

Since the unknown U is a function of two variables, x and t, it would seem quite natural to write the element trial solution in the standard form used in all the previous chapters but with the shape functions now a function of both x and t; that is,

$$\tilde{U}^{(e)}(x,t;a) = \sum_{j=1}^{n} a_j \phi_j^{(e)}(x,t) \tag{11.17}$$

---

[5] See Table 7 in the appendix for the physical dimensions corresponding to these symbols.

[6] If the material is permitted to expand freely, use the specific heat at constant pressure, c_p. If expansion is not permitted, use the specific heat at constant volume, c_v. (See Carslaw and Jaeger [11.11].)

This approach would require the use of 2-D elements (such as, but not limited to, those discussed in Part IV) and the construction of a mesh of such elements over the 2-D domain shown in Fig. 11.3. However, the infinite size of the domain, and the concomitant lack of "boundary" (or "final") conditions at $t = \infty$, causes a problem.

One way to resolve this problem would be to first calculate the steady-state solution at $t = \infty$.[7] This would generate the missing "boundary" conditions at $t = \infty$. Then the infinite domain could be approximated by a finite domain; i.e., we would apply the steady-state condition to a finite time t_f, where t_f is chosen so that the solution is insignificantly affected if t_f is made any larger (as was demonstrated in Section 10.6).

Another resolution is to use 2-D infinite elements, which are rectangular strips of width Δx and infinitely long in the t-direction [11.12]. This method also requires a prior calculation of the steady-state solution.

Certainly these two methods suffer from the added burden of having to first calculate the steady-state solution (which one may not be interested in). In any case, methods based on Eq. (11.17) have never been popular, probably because there exists an alternative approach that has been used almost universally from the inception of the FEM to the present day, and which will probably continue to dominate commercial FE programs for the foreseeable future. We now turn our attention to this very popular and well-established approach, which will be used in the remainder of this chapter.

Instead of putting all the independent variables in the shape functions, as in Eq. (11.17), we will include only the spatial variables, namely, those that correspond to the boundary-value part of the problem. The parameters a_j will then be made functions of time:

$$\tilde{U}^{(e)}(x,t;a) = \sum_{j=1}^{n} a_j(t)\phi_j^{(e)}(x)$$

$$(11.18)$$

This is the classical *separation of variables* technique [11.13], sometimes also referred to as the *method of Kantorovich* [11.14]. For any given value of t, Eq. (11.18) has the same form as the standard 1-D FE approximation used in previous chapters. The only difference here is that the numerical values for the a_j may vary from one instant to the next.

It will be seen in the next section that the time variation of the a_j in Eq. (11.18) does not disturb in any way the procedures involved in steps 1 through 6 of the theoretical analysis. Since the ϕ_j are now functions only of x, we will be able to use all of our previously developed 1-D elements. The principal effect of the a_j being functions of t is that the element equations, and therefore the assembled system equations, will be *ordinary differential equations* in time, rather than algebraic equations. [Recall Eqs. (11.10).] In other words, our usual FE procedure will transform the initial-boundary-value problem into a pure initial-value problem. The latter will then be solved by time-stepping techniques, as discussed previously. With this "preview of coming attractions," let's now derive the element equations for the general problem described in Eqs. (11.11), using the element trial solution in Eq. (11.18).

---

[7] The steady-state solution is a boundary-value problem that was treated in Chapters 7 and 8. It is Eq. (11.11a) without the $\mu \partial U/\partial t$ term, and using the boundary conditions [Eqs. (11.11c)] and interior load $f(x,t)$ evaluated at $t = \infty$.

11.2.2 Steps 1 through 6: Theoretical Development

It is important to realize at the outset that because the shape functions are functions only of x, we are essentially performing a *one-dimensional* FE analysis with respect to the variable x, just as in Chapters 7 through 9. Thus the Galerkin integrals and integration by parts, for example, are done in the usual manner with respect to x. The variable t, for the most part, is merely "carried along" during the six steps; its turn will come later in step 10.

Step 1: Write the Galerkin residual equations for a typical element.

$$\int\limits^{(e)} \left[\mu(x) \frac{\partial \tilde{U}^{(e)}(x,t;a)}{\partial t} - \frac{\partial}{\partial x}\left(\alpha(x) \frac{\partial \tilde{U}^{(e)}(x,t;a)}{\partial x} \right) \right.$$

$$\left. + \beta(x)\tilde{U}^{(e)}(x,t;a) - f(x,t) \right] \phi_i^{(e)}(x)\, dx = 0 \qquad i = 1,2,\ldots,n \tag{11.19}$$

where n is the number of DOF (and therefore shape functions) in the element.

Step 2: Integrate by parts.

The second derivative term (with respect to x) is integrated by parts once. The fact that the derivatives are partial rather than ordinary does not alter the standard integration-by-parts formula [Eq. (4.9)]:

$$\int\limits^{(e)} \phi_i^{(e)}(x)\mu(x) \frac{\partial \tilde{U}^{(e)}(x,t;a)}{\partial t}\, dx + \int\limits^{(e)} \frac{d\phi_i^{(e)}(x)}{dx} \alpha(x) \frac{\partial \tilde{U}^{(e)}(x,t;a)}{\partial x}\, dx$$

$$+ \int\limits^{(e)} \phi_i^{(e)}(x)\beta(x)\tilde{U}^{(e)}(x,t;a)\, dx$$

$$= \int\limits^{(e)} f(x,t)\phi_i^{(e)}(x)\, dx - \left[\left(-\alpha(x) \frac{\partial \tilde{U}^{(e)}(x,t;a)}{\partial x} \right) \phi_i^{(e)}(x) \right]_{x_1}^{x_n}$$

$$i = 1,2,\ldots n \tag{11.20}$$

Note that the boundary term contains, as usual, the flux,

$$\left[\left(-\alpha(x) \frac{\partial \tilde{U}^{(e)}(x,t;a)}{\partial x} \right) \phi_i^{(e)}(x) \right]_{x_1}^{x_n} = \left[\tilde{\tau}^{(e)}(x,t;a)\phi_i^{(e)}(x) \right]_{x_1}^{x_n} \tag{11.21}$$

Step 3: Substitute the general form of the element trial solution into interior integrals in residual equations.

Because of the separation of variables in the element trial solution, the partial derivatives of $\tilde{U}^{(e)}$ with respect to both x and t in Eqs. (11.20) revert to ordinary derivatives. Thus, from Eq. (11.18),

$$\frac{\partial \tilde{U}^{(e)}(x,t;a)}{\partial x} = \sum_{j=1}^{n} a_j(t) \frac{d\phi_j^{(e)}(x)}{dx}$$

$$\frac{\partial \tilde{U}^{(e)}(x,t;a)}{\partial t} = \sum_{j=1}^{n} \frac{da_j(t)}{dt} \phi_j^{(e)}(x) \tag{11.22}$$

Substituting Eqs. (11.18), (11.21), and (11.22) into Eqs. (11.20) yields

$$\sum_{j=1}^{n} \left[\int^{(e)} \phi_i^{(e)}(x)\mu(x)\phi_j^{(e)}(x) \ dx \right] \frac{da_j(t)}{dt} + \sum_{j=1}^{n} \left[\int^{(e)} \frac{d\phi_i^{(e)}(x)}{dx} \ \alpha(x) \ \frac{d\phi_j^{(e)}(x)}{dx} \ dx \right] a_j(t)$$

$$+ \sum_{j=1}^{n} \left[\int^{(e)} \phi_i^{(e)}(x)\beta(x)\phi_j^{(e)}(x) \ dx \right] a_j(t)$$

$$= \int^{(e)} f(x,t)\phi_i^{(e)}(x) \ dx - \left[\tilde{\tau}^{(e)}(x,t;a)\phi_i^{(e)}(x) \right]_{x_1}^{x_n}$$

$$i = 1,2,\dots,n \qquad \textbf{(11.23)}$$

Equations (11.23) are the element equations for a typical element. They may also be written in the usual matrix form:

$$[C]^{(e)} \left\{ \frac{da(t)}{dt} \right\} + [K]^{(e)}\{a(t)\} = \{F(t)\}^{(e)} \qquad \textbf{(11.24a)}$$

where

$$C_{ij}^{(e)} = \int^{(e)} \phi_i^{(e)}(x)\mu(x)\phi_j^{(e)}(x) \ dx$$

$$K_{ij}^{(e)} = K\alpha_{ij}^{(e)} + K\beta_{ij}^{(e)}$$

$$= \int^{(e)} \frac{d\phi_i^{(e)}(x)}{dx} \ \alpha(x) \ \frac{d\phi_j^{(e)}(x)}{dx} \ dx + \int^{(e)} \phi_i^{(e)}(x)\beta(x)\phi_j^{(e)}(x) \ dx$$

$$F_i^{(e)}(t) = Ff_i^{(e)}(t) + F\tau_i^{(e)}(t)$$

$$= \int^{(e)} f(x,t)\phi_i^{(e)}(x) \ dx - \left[\tilde{\tau}^{(e)}(x,t;a)\phi_i^{(e)}(x) \right]_{x_1}^{x_n} \qquad \textbf{(11.24b)}$$

The stiffness and load terms are identical to those in Chapters 7 through 9 (although the load terms are now a function of time). The symbol C is conventionally used for the $\mu(x)$ integral, and is frequently referred to as the *capacity* or *heat capacity* integral, in obvious reference to heat conduction applications. Although this terminology is not very appropriate for some of the other applications in Table 11.1, we will nevertheless accept the convention and hereafter refer to $[C]$ as the *capacity matrix*. As "advertised" previously, we see that the element equations are no longer algebraic equations but rather ordinary differential equations. [Recall Eqs. (11.10).]

Step 4: Develop specific expressions for the shape functions, $\phi_j^{(e)}(x)$.

Just as in Chapter 10, our work here is already done. We can use any of the four 1-D elements developed in Chapters 7 and 8.

Step 5: Substitute the shape functions into the element equations, and transform the integrals into a form appropriate for numerical evaluation.

The $K\alpha_{ij}^{(e)}$, $K\beta_{ij}^{(e)}$, and $Ff_i^{(e)}(t)$ integrals have already been treated in Chapters 7 and 8 for all four 1-D elements. The fact that the load integral may be a function of time does not alter those expressions; it only means that the expressions will be continually reevaluated at many different instants of time during the time-stepping solution (see Section 11.4).

Here we only need to treat the $C_{ij}^{(e)}$ integral. However, our job is once again simplified because we note from Eqs. (11.24b) that the $C_{ij}^{(e)}$ integral is identical in form to the $K\beta_{ij}^{(e)}$ integral as well as the $M_{ij}^{(e)}$ integral in Chapter 10 (Section 10.2, step 5). Therefore, for the C^0-linear element we can write down the following expressions from Eqs. (10.20):

$$C_{11}^{(e)} = C_{22}^{(e)} = \frac{1}{3}\mu^{(e)}L$$

$$C_{12}^{(e)} = C_{21}^{(e)} = \frac{1}{6}\mu^{(e)}L \tag{11.25}$$

where

$$\mu^{(e)} = \mu(x_c) \tag{11.26}$$

and

$$x_c = \frac{1}{2}(x_1+x_2) \text{ (the center of the element)} \tag{11.27}$$

The UNAFEM program evaluates $\mu(x_c)$ from a piecewise-quadratic polynomial. Just as with $\alpha(x)$, $\beta(x)$, and $\gamma(x)$, the user may specify from 1 to 20 different quadratics:

$$\mu(x) = \mu_0 + \mu_1 x + \mu_2 x^2 \tag{11.28}$$

each one being applicable to a different region of the domain. If $\mu(x)$ in the original problem statement, Eq. (11.11a), cannot be represented exactly by a piecewise-quadratic, then this method will introduce a modeling error (recall Section 7.2, step 5).

The integrals for the C^0-quadratic, -cubic, and -quartic isoparametric elements all involve the same numerical quadrature expression, which we can write down directly from Eq. (10.23):

$$C_{ij}^{(e)} \simeq \sum_{l=1}^{n} W_{nl}\phi_i(\xi_{nl})\mu^{(e)}(\xi_{nl})\phi_j(\xi_{nl})J^{(e)}(\xi_{nl}) \tag{11.29}$$

where n is the number of quadrature points used in the numerical integration. The value of n is determined by the desired accuracy. Here the situation is the same as for the mass integrals in Chapter 10. Thus the order of the quadrature rule for the capacity integrals may be the same as for the stiffness and load integrals. (Recall guidelines II and III in Section 8.3.1.)

Step 6: Derive expression for the flux.

The expression for the flux (as well as the location of the optimal flux-sampling points) is unaffected by the capacity matrix. Thus, for the C^0-linear element [recall Eq. (7.27)],

$$\tilde{\tau}^{(e)}(x,t) = -\alpha(x)\frac{a_2(t)-a_1(t)}{x_2-x_1} \tag{11.30}$$

and for the isoparametric elements [recall Eqs. (8.53) and (8.81)],

$$\bar{\tau}^{(e)}(\xi,t) = -\alpha^{(e)}(\xi) \sum_{j=1}^{n} a_j(t) \frac{d\phi_j(\xi)}{d\xi} \frac{1}{J^{(e)}(\xi)}$$

(11.31)

where $n=3$, 4, and 5 for the quadratic, cubic, and quartic elements, respectively. The flux is now a function of time since the a_j are functions of time.

This completes the six steps for deriving the element equations and the expression for flux. During the subsequent numerical computation phase (steps 7 through 12), assembly of the element equations in step 8 is performed according to the usual rule [recall the general rule for assembly given in Section 5.7]. The resulting system equations,

$$[C]\left\{\frac{da(t)}{dt}\right\} + [K]\{a(t)\} = \{F(t)\}$$

(11.32)

are a system of coupled ordinary differential equations, identical in form to the pure initial-value problem in Eqs. (11.10). The next section addresses the problem of how to solve Eqs. (11.32).[8]

11.3 Time-Stepping Methods for Solving Initial-Value Problems

11.3.1 Some Basic Concepts

In all time-stepping methods the time axis is divided into a succession of time steps Δt_i, $i=1,2,\ldots$, beginning at time t_0, as shown in Fig. 11.4. Some methods permit the steps to be of different lengths, as suggested in the figure; others require uniform steps. Then, instead of seeking a solution for $\{a(t)\}$ over the continuous domain of t, we look for an approximate solution consisting of *discrete* values for $\{a(t)\}$ at the end of each step — that is, $\{a\}_1$ at time t_1, $\{a\}_2$ at time t_2, etc., starting from the known initial value $\{a\}_0$ at time t_0, as indicated in the figure for the ith component of $\{a(t)\}$.

The discrete values $\{a\}_n$, $n=1,2,\ldots$, are computed from a *recurrence* relation, which is an algebraic equation that relates the values $\{a\}_n$ at two or more successive times. The recurrence relation is an approximation to the differential equation. In principle, there is an unlimited variety of possible recurrence relations for any given differential equation, and any particular recurrence relation can usually be derived by several different methods. In practice, though, only a dozen or so different recurrence relations, i.e., time-stepping methods, have actually been used in commercial FE programs, and almost all of these can be classified as *linear multistep (LMS)* methods [11.15].[9] The adjective "linear" refers to the nature of the recurrence relation, not to the differential equation; thus, LMS methods are used for both linear and nonlinear differential equations. LMS methods encompass

---

[8] Methods that solve initial-value problems, by time-stepping or any other technique, are frequently called "time-integration" methods. (However, methods that solve boundary-value problems are not analogously called "space-integration" methods.)

[9] The author is not aware of any commercial or private FE programs that use non-LMS methods to solve first-order initial-value problems.

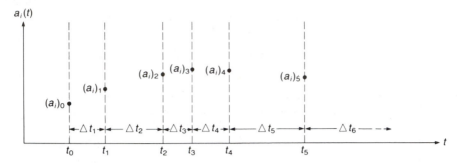

FIGURE 11.4 ● Division of the time axis into steps Δt_1, Δt_2, ..., for time-stepping methods, and computed discrete values $(a_i)_n$ at times t_n, $n=1,2,\ldots$.

a broad class of classic time-stepping methods. In the remainder of this chapter we will examine several of the most popular LMS methods used in FE applications.

A *one-step* method relates the discrete values at both ends of a single step. For example, for the nth step Δt_n, an LMS recurrence relation for Eq. (11.32) would have the form,

$$[P]\{a\}_n + [Q]\{a\}_{n-1} = p\{F\}_n + q\{F\}_{n-1} \tag{11.33}$$

where the coefficient matrices $[P]$ and $[Q]$ are known constants, related to $[C]$ and $[K]$, and are determined from the data of the problem. The coefficients p and q are also known constants. The vectors $\{F\}_n$ and $\{F\}_{n-1}$ are the load vector $\{F(t)\}$ evaluated at times t_n and t_{n-1}, respectively, and hence are also known. Sections 11.3.2 and 11.3.3 will derive specific expressions for these coefficients for several different recurrence relations.

To use Eq. (11.33) we would begin with the first step, Δt_1, and substitute $n=1$ into the equation:

$$[P]\{a\}_1 = p\{F\}_1 + q\{F\}_0 - [Q]\{a\}_0 \tag{11.34}$$

The initial conditions can be substituted for $\{a\}_0$, the solution at time t_0. The entire RHS is therefore known. Equation (11.34) is a system of linear algebraic equations [linear because Eq. (11.32) is linear] which can be solved for $\{a\}_1$, say by Gaussian elimination.

For the second step, substitute $n=2$ into Eq. (11.33):

$$[P]\{a\}_2 = p\{F\}_2 + q\{F\}_1 - [Q]\{a\}_1 \tag{11.35}$$

The vector $\{a\}_1$ on the RHS is now known from Eq. (11.34) so the entire RHS is known and we can again solve the system of linear algebraic equations by Gaussian elimination to obtain $\{a\}_2$.

The process can obviously be repeated indefinitely to obtain $\{a\}_3$, $\{a\}_4$, $\{a\}_5$, etc. Phrases such as "time-stepping," "time-marching," "marching forward," and "stepping forward" are used to describe this process.

A *multistep* method, also called a *k-step* method, relates the discrete values corresponding to k successive time steps. Thus, for the nth step an LMS recurrence relation for Eq. (11.32) would have the form,

$$[P]\{a\}_n + [Q]\{a\}_{n-1} + \ldots + [R]\{a\}_{n-k} = p\{F\}_n + q\{F\}_{n-1} + \ldots + r\{F\}_{n-k}$$

$$\tag{11.36}$$

Here one would begin with the kth step, substituting $n=k$ into Eq. (11.36). Initial values would have to be specified for $\{a\}_0, \{a\}_1,\ldots,\{a\}_{k-1}$. The values for $\{a\}_0$ are the given initial conditions; the other values at times $t_1, t_2,\ldots,t_{k-1}$ could be obtained by one-step methods or other strategies.

For FE applications, one- and two-step methods are used almost universally for diffusion problems, i.e., first-order initial-value problems (for example, see Kohnke [11.16] and Peeters [11.17]; also see Wood [11.18]). The one-step methods include three classic finite difference formulas:

1. The backward difference method, also known as the backward Euler rule

2. The mid-difference method, also known as the Crank-Nicolson method or the trapezoidal rule

3. The forward difference method, also known as Euler's rule

These three are special cases of a more general formula, referred to as

4. The θ-method

Section 11.3.2 will derive the recurrence relations for each of these four methods using the finite difference technique. Section 11.3.3 will rederive the θ-method using the weighted residual technique. These one-step methods are then implemented into the UNAFEM program in Section 11.4.

From a program-application standpoint, all time-stepping methods are applied using similar considerations (see Section 11.5). Proficiency with any one method should make it relatively easy to learn and use any other method. For this reason, only one-step methods are explicitly developed here. Derivation of a two-step formula will be left as an exercise (see Section 11.3.2).

11.3.2 Finite Difference Methods

As regards notation, each of the following recurrence relations will be derived for the nth time step, Δt_n, which carries the solution from time t_{n-1} to time t_n. As illustrated symbolically in Fig. 11.5 for a typical DOF $a_i(t)$, we have already stepped the solution forward through the first $n-1$ time steps. We therefore know the solution at time t_{n-1} but not at t_n.

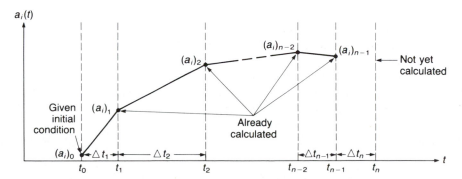

FIGURE 11.5 ● Solution has been computed at the end of the first $n-1$ time steps, $(a_i)_1$, $(a_i)_2,\ldots(a_i)_{n-1}$; next we want to compute the solution $(a_i)_n$ at time t_n.

Backward difference method • In this method we "evaluate" each of the terms in the differential Eq. (11.32) at the forward end of the time step, t_n; i.e., we simply write Eq. (11.32) with a subscript n on each of the terms,

$$[C]\left\{\frac{da}{dt}\right\}_n + [K]\{a\}_n = \{F\}_n$$

(11.37)

The $[C]$ and $[K]$ matrices don't need a subscript since they are constant matrices for the linear problems being considered in this chapter.

The time derivative is then approximated by a *backward* difference over the time step:

$$\left\{\frac{da}{dt}\right\}_n \simeq \frac{\{a\}_n - \{a\}_{n-1}}{\Delta t_n}$$

(11.38)

where $\Delta t_n = t_n - t_{n-1}$. Note that Eq. (11.38) is a vector relation, indicating that the difference expression applies separately to each DOF in the vector. Thus, for the *i*th DOF,

$$\left(\frac{da_i}{dt}\right)_n \simeq \frac{(a_i)_n - (a_i)_{n-1}}{\Delta t_n} \qquad i = 1,2,\ldots,N$$

(11.39)

Substituting Eq. (11.38) into Eq. (11.37) and placing all the known terms on the RHS yields

$$\left[\frac{1}{\Delta t_n}[C] + [K]\right]\{a\}_n = \{F\}_n + \frac{1}{\Delta t_n}[C]\{a\}_{n-1}$$

(11.40)

Equation (11.40) is the desired recurrence relation. Note that it has the form of the general one-step relation, Eq. (11.33), where $[P] = (1/\Delta t_n)[C] + [K]$, $[Q] = -(1/\Delta t_n)[C]$, $p = 1$, and $q = 0$.

Equation (11.40) is a system of algebraic equations in the standard form

$$[K_{\text{eff}}]\{a\}_n = \{F_{\text{eff}}\}$$

(11.41a)

where $[K_{\text{eff}}]$ is an effective stiffness matrix and $\{F_{\text{eff}}\}$ is an effective load vector, that is,

$$[K_{\text{eff}}] = \frac{1}{\Delta t_n}[C] + [K]$$

$$\{F_{\text{eff}}\} = \{F\}_n + \frac{1}{\Delta t_n}[C]\{a\}_{n-1}$$

(11.41b)

Equation (11.41a) can be solved by the same Gaussian elimination method used for the boundary-value problems, i.e., subroutine GAUSEL.

Since $[K_{\text{eff}}]$ is a nondiagonal matrix,[10] Eqs. (11.41a) are *coupled*, necessitating the use of an equation solver such as GAUSEL to "invert" $[K_{\text{eff}}]$. Because of this coupling (and consequent need for an equation solver), the backward difference method is said to be *implicit*, meaning that the unknown, $\{a\}_n$, is defined implicitly by Eq. (11.40) or (11.41a). (In contrast, the forward difference method explained below can be made explicit.)

It can be shown [11.19] that the accuracy of the backward difference method (or, more precisely, the asymptotic accuracy) is $O(\Delta t)$, meaning that the error at a given time,

---

[10] Although the matrix $[C]$ can be diagonalized (as explained below for the forward difference method), the matrix $[K]$ cannot.

in the limit as $\Delta t \to 0$, is proportional to Δt. This is equivalent to saying that the asymptotic *rate* of convergence is Δt. For example, if the analysis is repeated, halving the time step and using twice as many steps, the error at a given time will in general decrease by approximately one-half if Δt is small enough.

It is important to note that in practice this rate may not always be realized, simply because Δt is not "small enough" for this asymptotic formula to be valid. That is, higher-order terms such as Δt^2, Δt^3, etc., may predominate for "not-so-small" values of Δt because of the relative magnitudes of the coefficients of these terms at the given time, similar to the effects observed in Chapter 9 for "not-so-small" values of h.

As will be shown in Section 11.3.4, the backward difference method is "well behaved." The author and his colleagues have used it for almost 15 years to solve a variety of industrial problems in heat conduction, both linear and nonlinear, and it is used in many of the large commercial FE programs (for example, see Kohnke [11.16] and Peeters [11.17]). Although not always optimal from a computational efficiency standpoint, it is nevertheless a reliable and robust algorithm and relatively easy to program.

Mid-difference method • In this method, Eq. (11.32) is evaluated at the center of the time step (call it $t_{n-1/2}$):

$$[C]\left\{\frac{da}{dt}\right\}_{n-1/2} + [K]\{a\}_{n-1/2} = \{F\}_{n-1/2} \tag{11.42}$$

The time derivative is then approximated by a *mid*-difference over the time step:[11]

$$\left\{\frac{da}{dt}\right\}_{n-1/2} \simeq \frac{\{a\}_n - \{a\}_{n-1}}{\Delta t_n} \tag{11.43}$$

The function value $\{a\}_{n-1/2}$ is approximated by an average over the step:

$$\{a\}_{n-1/2} \simeq \frac{\{a\}_{n-1} + \{a\}_n}{2} \tag{11.44}$$

Equation (11.44) is consistent with Eq. (11.43). Both make the approximation that $\{a(t)\}$ varies linearly between t_{n-1} and t_n; that is,

$$\{a(t)\} \simeq (1-\theta)\{a\}_{n-1} + \theta\{a\}_n \qquad \theta = \frac{t - t_{n-1}}{\Delta t_n} \tag{11.45}$$

Evaluating Eq. (11.45) at $t_{n-1/2}$ ($\theta = 1/2$) reproduces Eq. (11.44), and taking the derivative reproduces Eq. (11.43). We will meet Eq. (11.45) again in the θ-method explained below.

Substituting Eqs. (11.43) and (11.44) into Eq. (11.42) and placing all the known terms on the RHS yields

$$\left(\frac{1}{\Delta t_n}[C] + \frac{1}{2}[K]\right)\{a\}_n = \{F\}_{n-1/2} + \left(\frac{1}{\Delta t_n}[C] - \frac{1}{2}[K]\right)\{a\}_{n-1} \tag{11.46}$$

---

[11] Note that the difference expressions in Eqs. (11.38) and (11.43) are identical. In fact, we will meet this expression once again in Eq. (11.50) for the forward difference method. In other words, the first derivative has only one difference expression over a single time step, but the name we call it (backward, mid, or forward) depends on the point to which we assign it in the step.

Equation (11.46) is the desired recurrence relation. It is again a system of algebraic equations in the standard form $[K_{eff}]\{a\}_n = \{F_{eff}\}$, but here the effective stiffness matrix and load vector are, respectively,

$$[K_{eff}] = \frac{1}{\Delta t_n} [C] + \frac{1}{2} [K]$$

$$\{F_{eff}\} = \{F\}_{n-1/2} + \left(\frac{1}{\Delta t_n} [C] - \frac{1}{2} [K] \right) \{a\}_{n-1} \tag{11.47}$$

Since $[K_{eff}]$ is nondiagonal, this is an implicit method, programmed very similarly to the backward difference method.

The applied load vector, $\{F\}_{n-1/2}$, involves the interior and natural boundary loads, which are known (specified) at all times and therefore could be evaluated at time $t_{n-1/2}$. However, it is usually easier to input data at the ends of each time step, in which case $\{F\}_{n-1/2}$ may be averaged in the same manner as $\{a\}_{n-1/2}$:

$$\{F\}_{n-1/2} \simeq \frac{\{F\}_{n-1} + \{F\}_n}{2} \tag{11.48}$$

The mid-difference method has also been a popular one. Its accuracy is $O(\Delta t^2)$; i.e., its asymptotic rate of convergence is Δt^2. This is one order better than the backward difference method. However, for typical time-step sizes the solutions are frequently characterized by annoying oscillations; although sometimes quite severe, they will always die out (for linear problems) as the solution steps forward. Because of the oscillations, the asymptotic accuracy is frequently not realized. The accuracy can be improved, however, by averaging adjacent time-step values. (See additional comments in Section 11.3.4.)

Forward difference method • In this method Eq. (11.32) is evaluated at the backward end of the time step, t_{n-1}:

$$[C] \left\{ \frac{da}{dt} \right\}_{n-1} + [K]\{a\}_{n-1} = \{F\}_{n-1} \tag{11.49}$$

The time derivative is then approximated by a *forward* difference over the time step [recall the footnote to Eq. (11.43)]:

$$\left\{ \frac{da}{dt} \right\}_{n-1} \simeq \frac{\{a\}_n - \{a\}_{n-1}}{\Delta t_n} \tag{11.50}$$

Substituting Eq. (11.50) into Eq. (11.49) and placing all the known terms on the RHS yields

$$\frac{1}{\Delta t_n} [C]\{a\}_n = \{F\}_{n-1} + \left(\frac{1}{\Delta t_n} [C] - [K] \right) \{a\}_{n-1} \tag{11.51}$$

Equation (11.51) is the desired recurrence relation. It is once again a system of algebraic equations in the standard form $[K_{eff}]\{a\}_n = \{F_{eff}\}$, although now the effective stiffness matrix and load vector are, respectively,

$$[K_{eff}] = \frac{1}{\Delta t_n} [C]$$

$$\{F_{eff}\} = \{F\}_{n-1} + \left(\frac{1}{\Delta t_n} [C] - [K] \right) \{a\}_{n-1} \tag{11.52}$$

Unlike the previous two methods, the stiffness matrix $[K]$ no longer appears on the LHS, so $[K_{eff}]$ now consists of only the capacity matrix $[C]$. The latter is normally not a diagonal matrix [recall Eqs. (11.25) and (11.29)], making Eq. (11.51) implicit. However, if $[C]$ could be diagonalized, then Eq. (11.51) would be *uncoupled* and $\{a\}_n$ could be evaluated explicitly and therefore *very rapidly*. That is, each component of $\{a\}_n$ could be evaluated separately by merely dividing the corresponding RHS term by the LHS coefficient on the diagonal of $[K_{eff}]$.

Happily, there are techniques for diagonalizing $[C]$, known as *lumping*. (The same techniques are used for lumping mass matrices; recall the brief comments in Section 10.3.3.) There are several different techniques used for lumping [11.20-11.23], each yielding a different lumped matrix. The computational efficiency of the resulting system equations (e.g., accuracy and rate of convergence) varies from one technique to another and seems to depend on the type of governing differential equations and the type of polynomial approximation used in the element shape functions. Here we describe one of the earliest and still widely used techniques, one that is easy to understand and program.

Lumping is applied to the element capacity matrix $[C]^{(e)}$ prior to assembly.[12] In each row of $[C]$ all the terms are added together and placed on the diagonal; then the off-diagonal terms are zeroed. Thus, denoting the lumped capacity matrix by $[CL]$, its components are

$$CL_{ii}^{(e)} = \sum_{j=1}^{n} C_{ij}^{(e)} \qquad i = 1,2,\dots,n$$

$$CL_{ij}^{(e)} = 0 \qquad\qquad i \neq j \tag{11.53}$$

As a point of terminology, the nonlumped $[C]^{(e)}$ matrix is called the *consistent* capacity matrix since it was generated with the original shape functions in a manner consistent with the standard Galerkin formulation. The technique described in Eq. (11.53) does not work well for some types of 2-D and 3-D elements. Chapter 15 will describe an alternative technique to handle those situations.

Lumping can be interpreted as using a different ("effective") set of shape functions for just the capacity integrals. The shape functions would be equal to 1 over part of the element touching a given node, and zero everywhere else, as illustrated in Fig. 11.6. Note that the shape functions (1) satisfy the interpolation property $\phi_j(x_i) = \delta_{ji}$, (2) are discontinuous within the element (which is acceptable for convergence since the ϕ_j do not appear as derivatives in the capacity integrals), and (3) do not overlap (which is the essential feature that diagonalizes the matrix).

The lumping technique in Eq. (11.53) conserves the total capacity of the consistent capacity matrix:

$$\sum_{i=1}^{n} \sum_{j=1}^{n} CL_{ij}^{(e)} = \sum_{i=1}^{n} \sum_{j=1}^{n} C_{ij}^{(e)} \tag{11.54}$$

This is necessary to preserve convergence of the FE discretization of the spatial variable [11.24]. The *rate* of convergence, though, may be affected. Thus $[C]^{(e)}$ preserves the

---

[12] The method described here is independent of the order of lumping and assembly (i.e., the two processes are commutative); this is because the lumping involves only the *addition* of terms, just like assembly. Thus we could lump the element capacity matrices and then assemble them, or assemble first and lump the resulting system capacity matrix. The first approach (lumping at the element level) is generally easier to program.

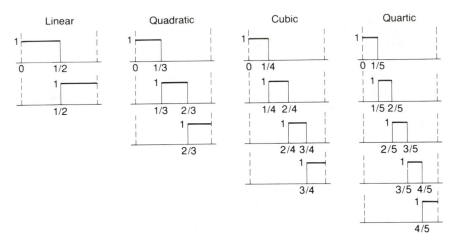

FIGURE 11.6 ● "Effective" shape functions $\phi_j(x)$ for lumping the element capacity matrix for each of the 1-D elements. [$\mu(x)$ is assumed to be constant.]

same rate as the $[K]$ matrix, whereas $[CL]^{(e)}$ may sometimes lower the rate, depending on the type of problem, element, etc. Interestingly, lumping sometimes yields a significant improvement in accuracy [11.22, 11.23].

In the UNAFEM program we will use $[CL]^{(e)}$ for the explicit forward difference rule because of the considerable increase in efficiency of the equation solving process. The implicit rules will use $[C]^{(e)}$ because the program allows for a variable $\mu(x)$ in the element capacity integrals; since the above cited research assumes a constant $\mu(x)$, it would be risky to use $[CL]^{(e)}$ with a variable $\mu(x)$.

Replacing the (assembled) consistent capacity matrix by the (assembled) lumped capacity matrix on both sides of Eq. (11.51) yields

$$\{a\}_n = \{a\}_{n-1} + \Delta t_n [CL]^{-1} \left(\{F\}_{n-1} - [K]\{a\}_{n-1} \right) \tag{11.55a}$$

where $[CL]^{-1}$ is a diagonal matrix,

$$[CL]^{-1} = \begin{bmatrix} \dfrac{1}{CL_{11}} & & & & \\ & \dfrac{1}{CL_{22}} & & & \\ & & \cdot & & \\ & & & \cdot & \\ & & & & \dfrac{1}{CL_{NN}} \end{bmatrix} \tag{11.55b}$$

The use of an equation solver ("matrix inversion") is therefore unnecessary and we can evaluate $\{a\}_n$ explicitly, the only matrix operation being the multiplication $[K]\{a\}_{n-1}$ on the RHS. This method is therefore much faster than the previous two methods. Its accuracy is $O(\Delta t)$; that is, its asymptotic rate of convergence is Δt.

The outstanding advantage of the forward difference method over the other methods is that it is explicit and therefore computationally much faster. Unfortunately it also has

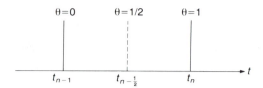

FIGURE 11.7 ● Variation of parameter θ over the nth time step.

a serious disadvantage: It is potentially *unstable*. Section 11.3.4 will explain the concept of stability and what needs to be done to ensure stability when using the forward difference method.

θ-method ● This method is the natural generalization of the three previous methods. Here we evaluate Eq. (11.32) at a general location in the time step denoted by the dimensionless parameter θ:

$$[C]\left\{\frac{da}{dt}\right\}_\theta + [K]\{a\}_\theta = \{F\}_\theta$$

(11.56a)

where

$$\theta = \frac{t - t_{n-1}}{\Delta t_n} \qquad (\Delta t_n = t_n - t_{n-1})$$

(11.56b)

The parameter θ varies from 0 to 1 over the time step,[13] as shown in Fig. 11.7.

Approximate expressions for $\{da/dt\}_\theta$, $\{a\}_\theta$, and $\{F\}_\theta$ may be obtained by approximating $\{a(t)\}$ and $\{F(t)\}$ by linear polynomials over the step.[14] We can make use of θ to write the polynomials as interpolation polynomials in exactly the same manner as for the C^0-linear element [recall Eqs. (7.12)]. Thus

$$\{a\}_\theta \simeq (1-\theta)\{a\}_{n-1} + \theta\{a\}_n$$

(11.57)

and

$$\{F\}_\theta \simeq (1-\theta)\{F\}_{n-1} + \theta\{F\}_n$$

(11.58)

which are shown in Fig. 11.8 for the ith components of the vectors.

Differentiating Eq. (11.57) yields

$$\left\{\frac{da}{dt}\right\}_\theta = \frac{1}{\Delta t_n}\frac{d\{a\}_\theta}{d\theta} = \frac{\{a\}_n - \{a\}_{n-1}}{\Delta t_n}$$

(11.59)

which is the same difference expression used in the previous three methods, only now it is assigned to the point θ. Substituting Eqs. (11.57) through (11.59) into Eq. (11.56a)

---

[13] Some authors define a theta, call it Θ, that is the "reflection" of the above theta; that is, $\Theta = 1 - \theta$. There is no compelling mathematical reason to favor one over the other. The author prefers the above definition merely for the aesthetic appeal of having θ increase from 0 to 1 as time increases from t_{n-1} to t_n.

[14] This approach can obviously be extended to the higher-degree interpolation polynomials to generate multistep methods: a quadratic polynomial for a two-step method (see Exercise 11.1), a cubic polynomial for a three-step method, etc.

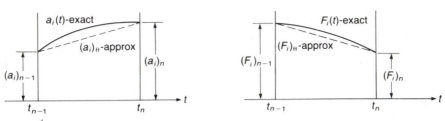

FIGURE 11.8 ● $\{a(t)\}$ and $\{F(t)\}$ approximated by linear variation over nth step.

and placing all the known terms on the RHS yields

$$\left[\frac{1}{\Delta t_n}[C] + \theta[K]\right]\{a\}_n$$

$$= (1-\theta)\{F\}_{n-1} + \theta\{F\}_n + \left[\frac{1}{\Delta t_n}[C] - (1-\theta)[K]\right]\{a\}_{n-1} \qquad (11.60)$$

Equation (11.60) is the desired recurrence relation. It is a system of algebraic equations in the standard form,

$$[K_{\text{eff}}]\{a\}_n = \{F_{\text{eff}}\} \qquad (11.61a)$$

where

$$[K_{\text{eff}}] = \frac{1}{\Delta t_n}[C] + \theta[K] \qquad (11.61b)$$

$$\{F_{\text{eff}}\} = (1-\theta)\{F\}_{n-1} + \theta\{F\}_n + \left[\frac{1}{\Delta t_n}[C] - (1-\theta)[K]\right]\{a\}_{n-1} \qquad (11.61c)$$

Equation (11.60) includes the previous three methods as special cases:

$$\theta = 0 : \text{forward difference [Eq. (11.51)]}$$
$$\theta = 1/2 : \text{mid-difference [Eq. (11.46)]}$$
$$\theta = 1 : \text{backward difference [Eq. (11.40)]}$$

More important, however, is the fact that we can now set θ equal to values other than 0, 1/2, or 1 in order to create additional recurrence relations that might perform better

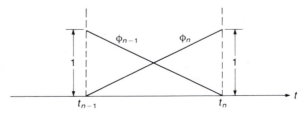

FIGURE 11.9 ● C^0-linear shape functions in time, used for each spatial DOF, $a_i(t)$, in the nth time step Δt_n.

in some respects than these three methods. We must therefore turn our attention to two important questions regarding θ:

1. What effect does the value of θ have on the performance of the recurrence relation (11.60)?

2. Is there an optimum value of θ?

Section 11.3.4 will address these questions. (Section 11.3.3 presents a rederivation of the θ-method which the reader may wish to skip over temporarily in order to proceed directly to Section 11.3.4.)

11.3.3 Weighted Residual Finite Element Methods[15]

In this section we will rederive the recurrence relations for the θ-method, using instead the weighted residual FEM in the time domain. The purpose is to demonstrate an alternative approach to deriving recurrence relations that may be applied to virtually any kind of initial value problem. The following development will focus on only the one-step θ-method, but the principles discussed may easily be extended to multistep methods [11.25].

Recalling the basic weighted residual concepts in Chapter 3, let's first write down the residual, $\{R(t)\}$, for Eq. (11.32):

$$\{R(t)\} = [C]\left\{\frac{d\tilde{a}(t)}{dt}\right\} + [K]\{\tilde{a}(t)\} - \{F(t)\} \tag{11.62}$$

where $\{\tilde{a}(t)\}$ is some approximation to $\{a(t)\}$. Since Eq. (11.32) is a system of N equations, the residual is a vector quantity; i.e., there are N residuals.

Since we seek recurrence relations for a single time step, we only need to apply the weighted residual method to a single time step, say the nth step. This means that we want to minimize the residual over only the nth step. In addition, it means that trial solutions for $\{\tilde{a}(t)\}$ need to be constructed for only one step. Here we can use the C^0-linear shape functions, as shown in Fig. 11.9. Thus

$$\{\tilde{a}(t)\} = \{a\}_{n-1}\phi_{n-1}(t) + \{a\}_n\phi_n(t) \tag{11.63a}$$

where

$$\phi_{n-1}(t) = \frac{t_n - t}{\Delta t_n} \qquad (\Delta t_n = t_n - t_{n-1})$$

$$\phi_n(t) = \frac{t - t_{n-1}}{\Delta t_n} \tag{11.63b}$$

Note that Eq. (11.63a) is a vector equation, although ϕ_n and ϕ_{n-1} are scalar-valued functions, indicating that the same shape functions are used for every DOF. In analogy with the weighted residual process for the spatial boundary-value problem, the "temporal element" here is the nth time step Δt_n, the "temporal nodes" are at t_{n-1} and t_n, and the "temporal DOF" are the values of the ith spatial DOF $a_i(t)$ at the temporal nodes,

---

[15] This section may be omitted without any loss of continuity to the chapter.

that is, $a_i(t_{n-1})$ and $a_i(t_n)$, which are abbreviated $(a_i)_{n-1}$ and $(a_i)_n$, respectively (recall Fig. 11.5).

Substituting Eq. (11.63a) into Eq. (11.60) yields the weighted residual equation,

$$\int_{t_{n-1}}^{t_n} \left[[C] \left(\{a\}_{n-1} \frac{d\phi_{n-1}(t)}{dt} + \{a\}_n \frac{d\phi_n(t)}{dt} \right) \right.$$
$$\left. + [K] \left(\{a\}_{n-1}\phi_{n-1}(t) + \{a\}_n\phi_n(t) \right) - \{F(t)\} \right] W(t)dt = 0 \qquad \textbf{(11.64)}$$

where $W(t)$ is a weighting function. Observe that there is only one residual equation (for each spatial DOF). This is in marked contrast to the boundary-value problem, which would require two residual equations, one for each DOF in the element trial solution. The reason stems from the fact that the time dimension has only one "boundary," namely, at t_0; it is open at the other end. By using only one residual equation, we will generate a *recurrence* relation that determines $\{a\}_n$ in terms of $\{a\}_{n-1}$.

To simplify the remainder of the analysis we make the usual change of variable, which changes the interval of integration to $[0,1]$:

$$\xi = \frac{t-t_{n-1}}{\Delta t_n} \qquad \textbf{(11.65)}$$

Therefore

$$\phi_{n-1} = 1 - \xi$$
$$\phi_n = \xi$$
$$\{\tilde{a}(t)\} = (1-\xi)\{a\}_{n-1} + \xi\{a\}_n$$
$$\left(\frac{d\tilde{a}(t)}{dt} \right) = - \frac{1}{\Delta t_n} \{a\}_{n-1} + \frac{1}{\Delta t_n} \{a\}_n \qquad \textbf{(11.66)}$$

It would seem "consistent" to approximate $\{F(t)\}$ in the same manner as $\{a(t)\}$ [recall Eqs. (11.57) and (11.58)], and indeed, the form of the resulting equations would support this. Hence

$$\{F(t)\} \simeq (1-\xi)\{F\}_{n-1} + \xi\{F\}_n \qquad \textbf{(11.67)}$$

Equation (11.64) becomes

$$\int_0^1 \left[[C] \left(- \frac{1}{\Delta t_n} \{a\}_{n-1} + \frac{1}{\Delta t_n} \{a\}_n \right) \right.$$
$$\left. + [K] \left[(1-\xi)\{a\}_{n-1} + \xi\{a\}_n \right] - \left[(1-\xi)\{F\}_{n-1} + \xi\{F\}_n \right] \right] W(\xi)d\xi = 0$$

$$\textbf{(11.68)}$$

which can be written more concisely as

$$\left(\frac{1}{\Delta t_n} [C] + \theta[K] \right) \{a\}_n$$

$$= (1-\theta)\{F\}_{n-1} + \theta\{F\}_n + \left(\frac{1}{\Delta t_n} [C] - (1-\theta)[K] \right) \{a\}_{n-1} \qquad \textbf{(11.69a)}$$

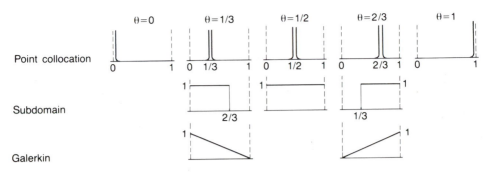

FIGURE 11.10 ● A few examples of weighting functions $W(\xi)$ and their corresponding θ values.

where

$$\theta = \frac{\int_0^1 W(\xi)\xi \, d\xi}{\int_0^1 W(\xi) \, d\xi}$$

(11.69b)

Equation (11.69a) is the desired recurrence relation. It is identical to Eq. (11.60). The parameter θ, as defined by Eq. (11.69b), may be interpreted as the centroid of the weighting function $W(\xi)$. If $W(\xi) > 0$ for $0 \le \xi \le 1$, as it usually is, then $0 \le \theta \le 1$. The specific form that $W(\xi)$ might assume is academic since only θ itself appears in Eq. (11.69a), and for any given value of θ there is an unlimited number of different $W(\xi)$ with centroids at θ. Figure 11.10 shows just a few examples of some familiar weighting functions and their corresponding θ values. Clearly any value of θ can be produced from an endless variety of different $W(\xi)$. We can therefore forget about $W(\xi)$ and turn our attention to the parameter θ and its effect on the performance of the recurrence relation.

11.3.4 Comparison of the Performance of Different Methods

From a practical standpoint, perhaps the most important and dramatic performance characteristic is *stability*. In words, stability is concerned with the behavior of the solution *as $t \to \infty$*, while keeping the step size Δt constant.[16] Stable or unstable behavior can be investigated by examining the *free response* of a system. Recall from Section 11.1.1 that the free response of a 1-DOF system is an exponential decay. (It will be shown below that the free response for a multi-DOF system is a sum of exponential decays.) As illustrated in Fig. 11.11, an unstable method[17] will yield a solution that oscillates about the exponen-

---

[16] This should be contrasted with the notion of convergence (accuracy) referred to previously. For convergence we require that t be held fixed, at t^* say. Then two or more solutions are computed, each with progressively smaller Δt (analogous to mesh refinement). One then observes whether the two or more computed values at t^* are converging (and at what rate) or diverging.

[17] It should be realized that stability, like convergence, is a characteristic of the *method* (e.g., forward difference, backward difference, etc.), not the particular problem the method is applied to.

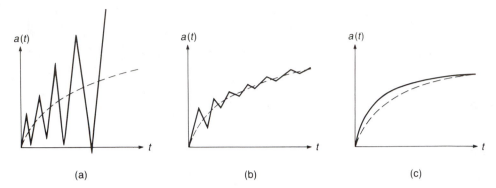

FIGURE 11.11 ● Typical behavior of solutions produced by stable and unstable time-stepping methods: (a) unstable — oscillatory divergence; (b) stable — oscillatory decay; (c) stable — monotonic decay.

tial decay with increasingly larger amplitude from one step to the next until eventually the "solution" becomes worthless (Fig. 11.11a). A stable solution, on the other hand, either experiences oscillations that decay as $t \to \infty$ (Fig. 11.11b) or else no oscillations at all (Fig. 11.11c). Clearly we cannot accept any methods that are unstable.

Having made these general observations, lets now examine the specific form of the free response. For this task we will find it much simpler and more informative to deal with only a 1-DOF system. Such a system may occur either as an actual 1-DOF system or, as is more often the case, in a multi-DOF system in which each mode behaves as a 1-DOF system. It is the latter that we are more interested in because it is applicable to every real problem, no matter how complicated. The former, however, is conceptually simpler so we consider it first.

An actual 1-DOF system ● The system of Eq. (11.32) reduces to a single equation,

$$C \frac{da(t)}{dt} + Ka(t) = F(t) \tag{11.70}$$

For free response,

$$C \frac{da(t)}{dt} + Ka(t) = 0 \tag{11.71}$$

which is identical to Eq. (11.2a). Hence, from Eqs. (11.2) through (11.7), the exact solution is

$$a(t) = Ae^{-\lambda t} \tag{11.72}$$

where $\lambda = K/C$ is the eigenvalue.

The recurrence relation [Eq. (11.60)], which was derived from Eq. (11.32), similarly reduces to a single equation:

$$\left(\frac{1}{\Delta t} C + \theta K \right) a_n = (1-\theta)F_{n-1} + \theta F_n + \left(\frac{1}{\Delta t} C - (1-\theta)K \right) a_{n-1} \tag{11.73}$$

which for free response becomes

$$\left(\frac{1}{\Delta t}\,C + \theta K\right)a_n = \left(\frac{1}{\Delta t}\,C - (1-\theta)K\right)a_{n-1} \tag{11.74}$$

Multiplying both sides by $\Delta t/C$ and substituting the eigenvalue λ for K/C yields

$$\frac{a_n}{a_{n-1}} = \frac{1-(1-\theta)\lambda\Delta t}{1+\theta\lambda\Delta t} \tag{11.75}$$

Equation (11.75) is the desired relation that will tell us the stability behavior of the approximate (time-stepping) solution; it will answer the two questions posed at the end of Section 11.3.2 regarding the influence of θ on stability. Before analyzing Eq. (11.75), though, we want to rederive the equation in a much more general and therefore more meaningful context.

Each mode in a multi-DOF system • The response of any linear multi-DOF system can be represented by a linear *superposition of its modes*. We demonstrate below that the free response of each mode has the same mathematical form as for the 1-DOF system described above. The following development incorporates several of the relationships from Section 10.3 ("The Algebraic Eigenproblem").

For the free response of a multi-DOF system, Eq. (11.32) becomes

$$[C]\left\{\frac{da(t)}{dt}\right\} + [K]\{a(t)\} = \{0\} \tag{11.76}$$

A solution must have the form

$$\{a(t)\} = \{v\}e^{-\lambda t} \tag{11.77}$$

Substituting Eq. (11.77) into Eq. (11.76) yields

$$(-\lambda[C] + [K])\{v\}e^{-\lambda t} = \{0\} \tag{11.78}$$

Nontrivial solutions for $\{v\}$ exist if and only if the determinant of the coefficient matrix vanishes:[18]

$$\det|[K] - \lambda[C]| = 0 \tag{11.79}$$

Equation (11.79) is the characteristic equation and its roots $\lambda_1, \lambda_2, \ldots, \lambda_N$ are the characteristic values, or eigenvalues. The matrices $[K]$ and $[C]$ are both symmetric; they are also positive-definite for any physically meaningful (well-posed) problem. Therefore all the λ_i are real and positive. To each λ_i corresponds an eigenvector (or mode) $\{v\}_i$. The $\{v\}_i$ satisfy the usual orthogonality relations [recall Eqs. (10.30)]:

$$\{v\}_i^T[K]\{v\}_j = \lambda_i\delta_{ij}$$

$$\{v\}_i^T[C]\{v\}_j = \delta_{ij} \tag{11.80}$$

where the $\{v\}_i$ have been normalized with respect to $[C]$. (If desired, the $\{v\}_i$ could be calculated using any of the procedures discussed in Chapter 10.)

---

[18] Note that Eq. (11.79) is identical to Eq. (10.32) if $[C]$ is replaced by $[M]$.

The general solution to Eq. (11.76) is a linear superposition (sum) of solutions of the form of Eq. (11.77), one for each mode:

$$\{a(t)\} = \sum_{j=1}^{N} A_j \{v\}_j e^{-\lambda_j t}$$

(11.81)

where the coefficients A_j are determined from the initial conditions. Since all the λ_j are real and positive, the general solution for free response is a sum of exponentially decaying modes.

The general solution to the forced problem, i.e., Eq. (11.32) with $\{F(t)\} \neq \{0\}$, is also a linear superposition of the modes:

$$\{a(t)\} = \sum_{j=1}^{N} A_j(t) \{v\}_j$$

(11.82)

but the time dependence is no longer exponential.[19] Instead, each mode has a general time-varying amplitude, $A_j(t)$. It is, in fact, these amplitude functions that behave like separate, i.e., uncoupled, 1-DOF systems. To see this, substitute Eq. (11.82) into Eq. (11.32):

$$[C]\left(\sum_{j=1}^{N} \frac{dA_j(t)}{dt} \{v\}_j\right) + [K]\left(\sum_{j=1}^{N} A_j(t)\{v\}_j\right) = \{F(t)\}$$

(11.83)

Premultiplying both sides of Eq. (11.83) by $\{v\}_i^T$ and using relations (11.80) yields

$$\frac{dA_i(t)}{dt} + \lambda_i A_i(t) = f_i(t) \qquad i = 1,2,\dots,N$$

(11.84)

where $f_i(t) = \{v\}_i^T \{F(t)\}$.

Equation (11.84) is our effective 1-DOF system (one for each mode). It is identical to Eq. (11.70) in which C, K, a, and F are replaced by 1, λ_i, A_i, and f_i, respectively. Therefore the recurrence relation corresponding to Eq. (11.84) is obtained simply by making these same replacements in Eq. (11.73):

$$\left(\frac{1}{\Delta t} + \theta\lambda_i\right)(A_i)_n = (1-\theta)(f_i)_{n-1} + \theta(f_i)_n + \left(\frac{1}{\Delta t} - (1-\theta)\lambda_i\right)(A_i)_{n-1}$$

$$i = 1,2,\dots,N \qquad (11.85)$$

For free response this becomes

$$\left(\frac{1}{\Delta t} + \theta\lambda_i\right)(A_i)_n = \left(\frac{1}{\Delta t} - (1-\theta)\lambda_i\right)(A_i)_{n-1} \qquad i = 1,2,\dots,N$$

(11.86)

or

$$\frac{(A_i)_n}{(A_i)_{n-1}} = \frac{1-(1-\theta)\lambda_i\Delta t}{1+\theta\lambda_i\Delta t} \qquad i = 1,2,\dots,N$$

(11.87)

---

[19] Gallagher and Mallett [11.26] solve a transient heat conduction problem using modal superposition.

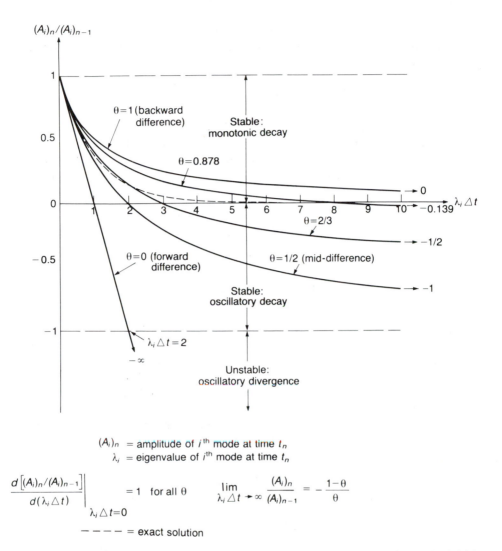

$(A_i)_n = $ amplitude of ith mode at time t_n

$\lambda_i = $ eigenvalue of ith mode at time t_n

$$\left. \frac{d\left[(A_i)_n/(A_i)_{n-1}\right]}{d(\lambda_i \Delta t)} \right|_{\lambda_i \Delta t=0} = 1 \quad \text{for all } \theta \qquad \lim_{\lambda_i \Delta t \to \infty} \frac{(A_i)_n}{(A_i)_{n-1}} = -\frac{1-\theta}{\theta}$$

$---- = $ exact solution

FIGURE 11.12 ● Stability behavior of one-step time-integration methods for first-order initial-value problems.

Equation (11.87) is identical in form to Eq. (11.75), but with the more general interpretation that λ_i and A_i are the eigenvalue and amplitude, respectively, of the ith mode. Since the total system response is the sum of the responses of all the modes, the system is stable if and only if every mode is stable. We can therefore determine the stability of the system by investigating the stability of each mode, using Eq. (11.87).

Let's now proceed with the stability analysis by examining the behavior of Eq. (11.87). Essentially the entire story is summarized in Fig. 11.12, which is a plot of Eq. (11.87) with θ as a parameter. This figure deserves a careful perusal because it contains a considerable amount of information. It offers a practical guide to the type of behavior one can expect of a solution when using any one of the methods. The dashed curve is derived from the exact solution: Since the exact solution is $A_i(t)=ce^{-\lambda_i t}$ [setting $f_i(t)=0$ in Eq. (11.84)], then

$$\frac{(A_i)_n}{(A_i)_{n-1}} = \frac{A_i(t_n)}{A_i(t_n-\Delta t)} = e^{-\lambda_i \Delta t} \qquad i = 1,2,...,N \tag{11.88}$$

The dashed curve is therefore a plot of Eq. (11.88).

The most important feature is the condition for stability. Since $(A_i)_n/(A_i)_{n-1}$ is the ratio of two successive time-step values, we must require for stability that the magnitude of this ratio be less than 1, otherwise successive values will continually grow larger. Thus

$$\left| \frac{(A_i)_n}{(A_i)_{n-1}} \right| < 1 \qquad i = 1,2,...,N \tag{11.89}$$

Applying this inequality[20] to Eq. (11.87) yields the following conditions.

$$\text{For } 0 \le \theta < 1/2: \quad \lambda_i \Delta t < \frac{2}{1-2\theta} \qquad i = 1,2,...,N \tag{11.90a}$$

$$\text{For } \theta \ge 1/2: \quad \lambda_i \Delta t > \frac{-2}{2\theta-1} \qquad i = 1,2,...,N \tag{11.90b}$$

Equation (11.90a) places restrictions (conditions) on the step size. For example, for $\theta=0$ we require $\lambda_i \Delta t<2$ (as indicated in Fig. 11.12); for $\theta=1/3$ we require $\lambda_i \Delta t<6$, etc. Methods for which $0\le\theta<1/2$ are therefore said to be *conditionally stable*. Equation (11.90b), however, is satisfied for all $\lambda_i \Delta t>0$. Since $\lambda_i>0$ for $i=1,2,...,N$, the equation is satisfied for all $\Delta t>0$. Thus there are no restrictions (conditions) on the step size. Therefore methods for which $\theta\ge1/2$ are said to be *unconditionally stable*. Let's look at both types of stability in more detail.

Conditional stability • Figure 11.13 shows three approximate solutions calculated from the recurrence relation [Eq. (11.87)] for the forward difference method ($\theta=0$); that is,

$$\frac{(A_i)_n}{(A_i)_{n-1}} = 1 - \lambda_i \Delta t \qquad i = 1,2,3 \tag{11.91}$$

The three solutions correspond to three modes for which the eigenvalues were arbitrarily chosen to be $\lambda_1=1$, $\lambda_2=3$, and $\lambda_3=5$. Using a time step $\Delta t=0.5$ yields $\lambda_1 \Delta t=0.5$, $\lambda_2 \Delta t=1.5$, and $\lambda_3 \Delta t=2.5$. The exact solutions $e^{-\lambda_i t}$ are shown by dashed lines. [For convenience we assume the initial condition for each mode is $(A_i)_0=1$ at $t_0=0$.] As predicted by the $\theta=0$ curve in Fig. 11.12, the first mode is stable, showing monotone decay; the second mode is also stable but showing oscillatory decay; and the third mode is unstable, diverging in an oscillatory fashion.

This example illustrates the principal problem inherent in all conditionally stable methods: The inequality in Eq. (11.90a) must be satisfied by *every* mode in the system.

---

[20] Equation (11.89) may be expressed as two inequalities: $(A_i)_n/(A_i)_{n-1}< +1$ and $(A_i)_n/(A_i)_{n-1}>-1$. The former yields $\lambda_i \Delta t >0$ (which is always true), while the latter yields Eqs. (11.90).

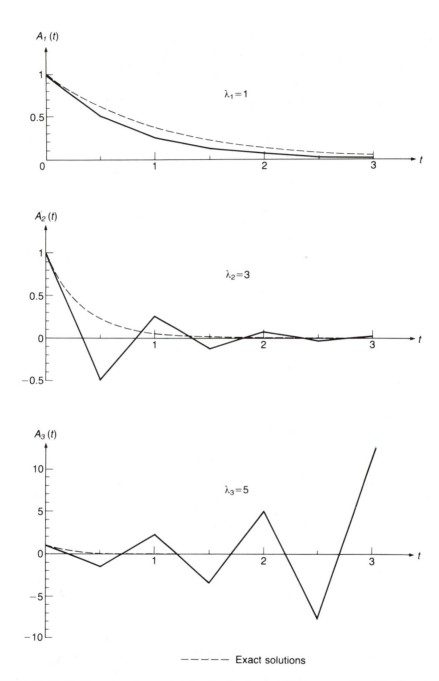

FIGURE 11.13 ● Three modes computed using the forward difference method ($\theta=0$) and a time step $\Delta t=0.5$.

If only one mode is unstable, the method will fail because eventually, as $t \to \infty$, the unbounded oscillations of that mode would dominate all the other modes.[21] Since those modes with larger λ_i's require proportionately shorter Δt's, the largest λ_i of the system, λ_{max}, determines the smallest Δt. It is this smallest Δt that is the *critical time step*, Δt_{crit}, for the entire system. Equation (11.90a) can therefore be replaced by a single condition,

$$\Delta t < \Delta t_{crit}$$

where

$$\Delta t_{crit} = \frac{2}{1-2\theta} \frac{1}{\lambda_{max}} \tag{11.92}$$

This is the stability condition for the conditionally stable methods ($0 \le \theta < 1/2$). It should be emphasized that λ_{max} is the maximum eigenvalue of the *FE model* (not the exact problem, which has an infinity of eigenvalues that are unbounded).

As shown below, for many problems λ_{max} can be very large, thereby requiring very small time steps. This in turn usually requires a very large number of such steps, and therefore large amounts of computation, to reach a given time. For this reason conditionally stable methods are generally not attractive unless they are explicit (e.g., the forward difference method), and therefore computationally fast enough to possibly compensate for the large number of steps. They can be particularly effective when analyzing a very short duration transient solution due to a sudden change in loading, in which case the desired time-step size for the physical model may be comparable to Δt_{crit} anyway.

To determine Δt_{crit} we must determine λ_{max}. One way would be to calculate λ_{max} directly by solving the eigenproblem for the system, Eq. (11.77), using the eigensolver routine in UNAFEM or any other eigensolver program. However, this is not a practical suggestion, since few analysts would care to spend the time to perform an additional computer analysis simply to determine λ_{max}. What is needed is a simple formula that can easily be calculated by hand and that provides a reasonable approximation to λ_{max}. The author has derived such a formula in a rather nonrigorous fashion, yet it has given consistently good results in a variety of applications. The derivation will be presented here to provide insight into the diffusion problem and also to provide a plausible explanation for the formula. We begin by deriving the exact eigensystem for the following problem:

$$\alpha \frac{\partial^2 U(x,t)}{\partial x^2} - \mu \frac{\partial U(x,t)}{\partial t} = 0 \qquad 0 < x < L \tag{11.93}$$

The BCs are $U(0,t) = U(L,t) = 0$ and α and μ are assumed constant. The free response has the standard form

$$U(x,t) = v(x)e^{-\lambda t} \tag{11.94}$$

Substituting Eq. (11.94) into Eq. (11.93) yields the eigenproblem

$$\frac{d^2 v(x)}{dx^2} + \lambda \frac{\mu}{\alpha} v(x) = 0 \tag{11.95}$$

---

[21] Even if an unstable mode were absent from the *exact* solution, it would probably be present in the approximate solution because of truncation and/or round-off errors.

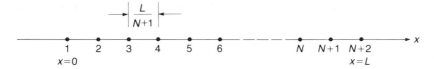

FIGURE 11.14 ● A uniform mesh.

with the BCs $v(0)=v(L)=0$. The solution to Eq. (11.95) has the form

$$v(x) = a \sin \sqrt{\lambda \frac{\mu}{\alpha}} x + b \cos \sqrt{\lambda \frac{\mu}{\alpha}} x \qquad (11.96)$$

Applying the BCs to Eq. (11.96) yields the following eigensystem:

$$\text{Eigenvalues:} \quad \lambda_n = \left(\frac{n\pi}{L}\right)^2 \frac{\alpha}{\mu} \qquad (11.97a)$$
$$n = 1,2,\ldots$$

$$\text{Eigenvectors:} \quad v_n(x) = \sin \frac{n\pi}{L} x \qquad (11.97b)$$

Now consider an FE analysis for this problem using a uniform mesh with $N+2$ nodes (Fig. 11.14). Nodes 1 and $N+2$ are both constrained, leaving only N free DOF in the FE model. An eigensolution will therefore calculate the first N modes. The corresponding N FE eigenvalues will be an approximation to the lowest N exact eigenvalues. In particular, the maximum FE eigenvalue, λ_{max} $(=\lambda_N)$ will be an approximation to the Nth exact eigenvalue:[22]

$$\lambda_{max} \simeq \left(\frac{N\pi}{L}\right)^2 \frac{\alpha}{\mu} \simeq \left(\frac{(N+1)\pi}{L}\right)^2 \frac{\alpha}{\mu} = \left(\frac{\pi}{\delta}\right)^2 \frac{\alpha}{\mu} \qquad (11.98)$$

where $\delta = L/(N+1)$ is the distance between two adjacent nodes.

Equation (11.98) provides an approximation for λ_{max}, but it is not a very useful formula since it can only be used for a uniform 1-D mesh. We want to generalize it to a nonuniform mesh in any number of dimensions. For this purpose we observe that [11.27]

$$\lambda_{max} \leq \lambda_{max}^{(e)} \qquad (11.99)$$

where $\lambda_{max}^{(e)}$ is the largest *element* eigenvalue of any element in the mesh. [Equation (11.99) is valid in any number of dimensions.] Assuming that we can treat the upper bound in Eq. (11.99) as an approximation (the accuracy of our final formula will show the reasonableness of this assumption), our search for λ_{max} can therefore be reduced to finding the largest eigenvalue of any single element.

Here again we could calculate $\lambda_{max}^{(e)}$ exactly by solving the eigenproblem of Eq. (11.79) for a single element, generally the smallest element if α and μ are constants. However, we can estimate $\lambda_{max}^{(e)}$ by performing the same analysis that led to Eq. (11.98),

---

[22] Recall Fig. 10.7, which depicts the first five modes of the eigensystem in Eqs. (11.97) as well as an FE solution for $N=5$. Note that λ_{max} $(=\lambda_5)$ is in error by 20%.

only this time for a single element. For a nonuniform 1-D mesh this would yield

$$\lambda_{max} \simeq \frac{\pi^2}{\left[(\mu/\alpha)\delta^2 \right]^{(e)}_{min}} \tag{11.100}$$

where $[(\mu/\alpha)\delta^2]^{(e)}_{min}$ is the smallest value that can be found in any element in the mesh, and δ is the distance between two adjacent nodes in the element (the smallest distance if the nodes are nonuniformly spaced). If α and μ are constant over the entire mesh, as is frequently the case, then

$$\lambda_{max} \simeq \frac{\pi^2}{(\mu/\alpha)\delta^2_{min}} \qquad (\alpha,\mu \text{ are constants}) \tag{11.101}$$

where δ_{min} is the smallest distance between any two nodes in the mesh.

For problems in more than one dimension, the arguments are analogous to the 1-D arguments. For example, for a cubic domain in a 3-D problem the eigenvalues are [11.28].

$$\lambda_{lmn} = \left(\frac{\pi}{L} \right)^2 [l^2 + m^2 + n^2] \frac{\alpha}{\mu} \tag{11.102}$$

Using a uniform mesh with $N+2$ nodes in each direction, the maximum FE eigenvalue, $\lambda_{max}(=\lambda_{NNN})$, would be approximately

$$\lambda_{max} \simeq 3 \left(\frac{N\pi}{L} \right)^2 \frac{\alpha}{\mu} \simeq 3 \left(\frac{\pi}{\delta} \right)^2 \frac{\alpha}{\mu} \tag{11.103}$$

which is three times the value in Eq. (11.98). Expression (11.100) for nonuniform meshes can therefore be generalized as follows:

$$\lambda_{max} \simeq \frac{d\pi^2}{\left[(\mu/\alpha)\delta^2 \right]^{(e)}_{min}} \tag{11.104}$$

where d is the dimension of the problem.

Substituting (11.104) into Eq. (11.92) yields the following very simple guideline for estimating Δt_{crit}:

$$\Delta t_{crit} \simeq \frac{2}{d(1-2\theta)\pi^2} \left[(\mu/\alpha)\delta^2 \right]^{(e)}_{min} \qquad 0 \leq \theta < 1/2 \tag{11.105}$$

As mentioned previously, the only conditionally stable one-step method of practical value seems to be the explicit $\theta=0$ method. Also, in most applications α and μ are constants. Therefore Eq. (11.105) will usually be used in the following form:

$$\Delta t_{crit} \simeq \frac{2}{d\pi^2} (\mu/\alpha)\delta^2_{min} \qquad \left(\begin{array}{l} \theta = 0 \\ \alpha,\mu \text{ are constants} \end{array} \right) \tag{11.106}$$

The above derivation made several approximations that could either increase or decrease the exact values, so it is difficult to place upper or lower bounds on Eq. (11.105). Nevertheless, in a variety of 1-D and 2-D numerical experiments, the author has found Eq. (11.105) to be within a factor of about 5 of the exact value, and often a factor of only 2 or 3. This affords a very useful first estimate considering that in many diffusion

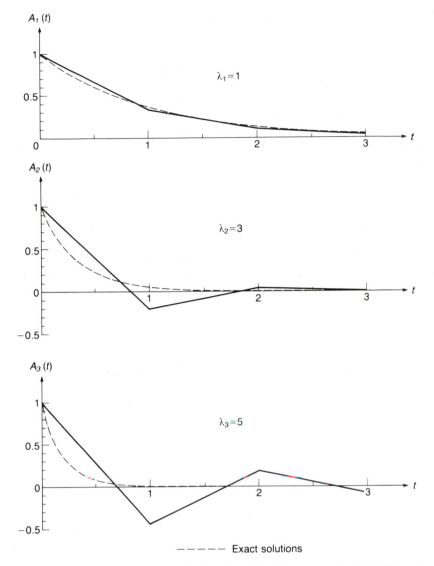

FIGURE 11.15 ● Three modes computed using the mid-difference method ($\theta=1/2$) and a time step $\Delta t = 1$.

One way to handle this type of situation would be to use very small time steps for the short duration of the change and for a short interval immediately following the change, i.e., long enough for the higher modes to decay [11.29]. The step size should therefore satisfy $\lambda_{max}\Delta t \approx 2$. From Fig. 11.12 this yields the same value for Δt as Δt_{crit} for the $\theta=0$ method (the latter becoming unstable at $\lambda_i\Delta t=2$). Hence Eq. (11.105) with $\theta=0$ would yield a good estimate for a Δt that would prevent oscillations in the mid-difference method. This might be computationally expensive and it would make more sense to use the explicit forward difference method ($\theta=0$) as long as the time steps are small enough to ensure stability anyway. Another approach, as indicated in Section 11.3.2, is to numerically smooth the oscillations by using an averaging technique [11.30].

problems time-step sizes may vary over several orders of magnitude in progressing from an initial shock load to nearly steady-state conditions. Interestingly, Eq. (11.105) has always estimated on the low side, suggesting that the inequality in Eq. (11.99) is the dominant source of error. Therefore, multiplying Eq. (11.105) by an experimental "correction factor" of 2 or 3 will usually provide a close estimate for Δt_{crit}, which can then be tried using only a few time steps (see Section 11.5). If unstable modes are present, they will generally be revealed very quickly, and Δt can be halved once or twice until they become stable.

Because Δt_{crit} is proportional to the *square* of δ_{min}, we would like to keep δ_{min} as large as possible, consistent with a desired spatial accuracy. This argument would favor using a few large higher-order elements. However, in quasisingular regions (e.g., a shock response), we might prefer a locally uniform mesh of lower-order elements (see Section 9.4); higher-order elements generally require a graded mesh to realize a significantly better performance, and the resulting smallest element might yield an even smaller Δt_{crit} than for the uniform mesh of lower-order elements. One other implication of the *square* of δ_{min}: During mesh refinement, each time δ_{min} is halved, Δt_{crit} is reduced by a factor of 1/4.

Unconditional stability • Figure 11.15 shows three approximate solutions calculated from the recurrence relation (11.87) for the mid-difference method ($\theta = 1/2$); that is,

$$\frac{(A_i)_n}{(A_i)_{n-1}} = \frac{1 - \lambda_i \Delta t/2}{1 + \lambda_i \Delta t/2} \qquad i = 1,2,3 \tag{11.107}$$

Figure 11.16 shows three approximate solutions for the backward difference method ($\theta = 1$), that is,

$$\frac{(A_i)_n}{(A_i)_{n-1}} = \frac{1}{1 + \lambda_i \Delta t} \qquad i = 1,2,3 \tag{11.108}$$

The three eigenvalues are the same as before: $\lambda_1 = 1$, $\lambda_2 = 3$, $\lambda_3 = 5$. Using a time step $\Delta t = 1$ yields $\lambda_1 \Delta t = 1$, $\lambda_2 \Delta t = 3$, and $\lambda_3 \Delta t = 5$. Initial conditions are again assumed to be $(A_i)_0 = 1$ at $t_0 = 0$. As expected, all the curves are stable, and they show the type of decay (monotone or oscillatory) predicted by Fig. 11.12.

These two methods illustrate the two extremes of the unconditionally stable methods. The mid-difference method is the most oscillatory. Figure 11.12 shows that $(A_i)_n/(A_i)_{n-1} \rightarrow -1$ for $\lambda_i \Delta t \gg 2$, indicating that if λ_i and/or Δt is very large, the solution may oscillate for many time steps with very slow decay in the oscillations. This would be a problem if one used a very large Δt in order to go directly to a steady-state solution. In this case it would be advisable to use the backward difference method. (See comments in Section 11.5.)

Large oscillations also usually occur when the highest modes are a significant component of the overall solution, even though Δt is not large (so that $\lambda_i \Delta t \gg 2$ for only the highest modes). This happens when there are sudden changes in the loading.[23]

---

[23] A sudden load change produces a correspondingly sudden change in the system response which can only be described accurately by those modes with short "time constants" (the reciprocals of the eigenvalues).

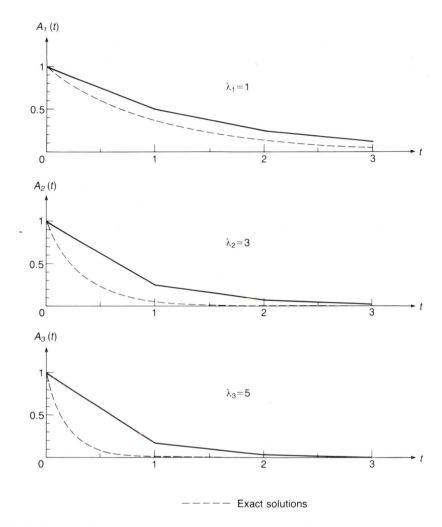

FIGURE 11.16 ● Three modes computed using the backward difference method ($\theta=1$) and a time step $\Delta t=1$.

The backward difference method shows no oscillation at all. In fact Fig. 11.12 shows that $(A_i)_n/(A_i)_{n-1}>0$ for all $\lambda_i\Delta t$. Since all modes decay monotonely, it can be observed in practice that solutions always approach steady-state monotonely from one side of the exact solution.

It is tempting to draw an analogy with vibrating systems and suggest that the backward difference method acts "overdamped" and the mid-difference method acts "under-damped." Perhaps it is the "overdamping" of the backward difference method that accounts for its generally reliable behavior even for nonlinear problems.

The mid- and backward difference methods have dominated most FE applications since the early years of the FEM. The θ-method, however, demonstrates that they represent only two special cases at the extremes of a continuum, with the strong implication that there might exist an optimum somewhere between the overdamped backward method ($\theta=1$) and the under-damped mid-difference method ($\theta=1/2$). Indeed, the value $\theta=2/3$ [11.31] has been demonstrated to provide a superior performance [11.32], and the author's own numerical experiments also suggest that $\theta=2/3$ is near-optimal. Thus, Fig. 11.12 shows that for $\theta=2/3$, $(A_i)_n/(A_i)_{n-1} \to -1/2$ as $\lambda_i \Delta t \to \infty$, indicating that oscillations of the higher modes will be smaller (than mid-difference) and will die out more quickly. And for the lower modes, which are usually the most significant ones, the $\theta=2/3$ curve in Fig. 11.12 is a close approximation to the exact curve.

Figure 11.12 also shows a curve for $\theta=0.878$ [11.33]. This value of θ minimizes the maximum absolute value of the difference between the exact solution and the time-stepping solution over the interval $0\leq\lambda_i \Delta t\leq\infty$, that is,

$$\min\left(\max_{0\leq\lambda_i\Delta t\leq\infty}\left|e^{-\lambda_i\Delta t}-\frac{1-(1-\theta)\lambda_i\Delta t}{1+\theta\lambda_i\Delta t}\right|\right) \tag{11.109}$$

This is certainly a plausible model, perhaps deserving more numerical experimentation.[24]

The recurrence relation for the θ-method, Eq. (11.60), has been implemented into the UNAFEM program so that the user may experiment with the full range of possibilities from $\theta=0$ to $\theta=1$. The next section will discuss the programming.

11.4 Implementation into the UNAFEM Program

A mixed initial-boundary-value problem can potentially require a substantial amount of input data. In addition to the usual FE data for the boundary-value part of the problem, initial conditions must be specified for every DOF in the mesh. Time steps must also be defined. Typical problems will involve dozens or even hundreds of steps, frequently of different sizes. If any of the loads (interior or boundary) vary with time, they may have to be defined every step, or every few steps.

Postprocessing can likewise generate a voluminous amount of data. The usual function and flux values over the entire mesh could be printed and plotted not just at the end of the problem but also at the end of every time step. Function and flux values at several DOF could also be plotted as a function of time.

In the UNAFEM program an attempt has been made to keep the input and output data to a minimum while retaining a reasonable degree of program versatility. To facilitate defining the time steps (and associated time-varying loads), the user defines *intervals* of time, as shown in Fig. 11.17. Within each interval there may be an arbitrary number of time steps, all the same size. Each interval is therefore defined by two numbers: the number of steps, n_s, and the size of the steps, Δt. Most problems require about 1 to 10 intervals. (The user does not need to define an initial time t_0 since the variable t appears in the analysis only as the time difference Δt. The output data therefore arbitrarily sets $t_0=0$.)

At the beginning of each interval the user defines all loads (boundary and interior) that are to be applied at the *end* of the interval, i.e., at the end of the last step in the interval

---

[24] The value $\theta=3/4$ would seem to be an intriguing possibility, if for no other reason than its being almost the mean of the four meaningful values: 1/2, 2/3, 0.878, and 1!

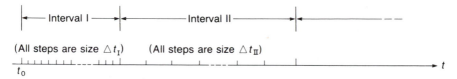

FIGURE 11.17 ● Definition of intervals and time steps for UNAFEM.

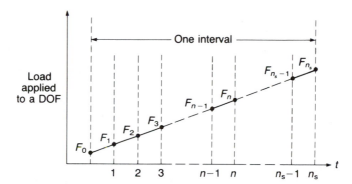

FIGURE 11.18 ● Ramping the loads between the user-defined values F_0 and F_{n_s}.

(step n_s). The loads at the ends of all the intermediate steps are calculated by the program by linearly interpolating from the loads at the end of the previous interval, as illustrated in Fig. 11.18. The linear interpolation is often referred to as *ramping*. For example, letting F_0 and F_{n_s} designate loads applied to a given DOF at the beginning and end, respectively, of an interval containing n_s steps, the ramping formula is as follows:

$$F_n = F_0 + \frac{n}{n_s}\left(F_{n_s} - F_0\right) \tag{11.110}$$

where F_n is the load applied at the end of the nth step. Here F represents a natural or essential boundary load or an interior load. Note that F_0 is the user-defined load for the end of the preceding interval (or the initial load if this is the first interval).

Ramping is merely a convenient form of automatic load generation. If in the above example one preferred a "sudden" (i.e., step) load[25] rather than a ramped load, this could be accomplished by applying the entire load to a very short one-step interval, then applying the same load to a much longer subsequent interval, as illustrated in Fig. 11.19.

A particularly attractive feature of the program is that the user may define a different θ in every interval, thereby making it possible to use the time-stepping method that is best

---

[25] The recurrence relations are all based on finite (nonzero) time steps so it is impossible to define a true step load. We can only approximate it arbitrarily closely by using very small time steps. It should be recognized, though, that a true step load is a discontinuity and therefore can only be an idealization of the real world where changes occur continuously, however rapidly. Consequently, the types of loading changes described above are capable of approximating with arbitrary accuracy any real time-varying loads.

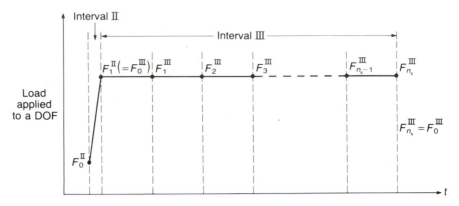

FIGURE 11.19 ● Applying a sudden load (ideally, a step load) by using two intervals.

suited for the current loads. This technique will be demonstrated in the illustrative problem in Section 11.5.

Now let's see how these ideas have been implemented into the coding of the program. All the significant coding for the initial-boundary-value problem was added to already existing routines; consequently, the flowchart remains the same as in Chapter 10, Fig. 10.1. Several of the modified routines are discussed below.

Figure 11.20 shows the MAIN routine, emphasizing the loop over each interval. Note that the mesh data and physical property data are outside the loop since they are defined only once for a given problem.[26] The load data, however, is inside the loop so that it may be redefined every interval. (Although PLOTDT is also inside the loop, it is used only once for a given problem, namely, in the first interval.)

Figure 11.21 shows the extra statements needed in the element routines to calculate the capacity integrals. As usual, the statements for the isoparametric elements are all identical.

Figure 11.22 shows the assembly of the system matrices in subroutine FORM. For the explicit problem there is no stiffness matrix on the LHS of the recurrence relation (11.55a), but we still need to assemble $[K]$ for the RHS. The element capacity matrices are lumped and assembled, and the resulting lumped system matrix is stored in the one-dimensional array CLUMP(I). The interior loads for the *end* of the interval are stored in the usual F(I) array; they will later be ramped in subroutine CALFUN.

---

[26] If one wanted to include the capability of time-dependent properties, the loop could be moved up to include subroutine PROPDT. It might be noted that if the properties were also temperature-dependent, thereby making the analysis nonlinear, subroutine PROPDT would again be inside the loop, and, interestingly, only minor changes would need to be made to the rest of the program, since the time-stepping techniques naturally accommodate the incremental techniques necessary for a nonlinear solution. [Brahmanandam and Chatterji [11.34] analyze such a nonlinear version of Eq. (11.11), in which $\alpha = \alpha(U,x,t)$.]

As regards the mesh data, it is conceivable that in the near future programs will allow the mesh to adaptively change during the time-stepping so as to best suit the changing complexity of a solution as it progresses from an initial shock response, say, to a more gentle steady-state response.

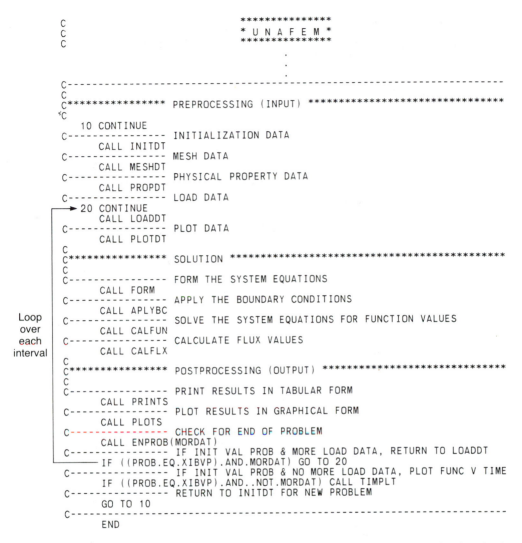

```
        C                          ***************
        C                          * U N A F E M *
        C                          ***************
                                          .
                                          .
                                          .
        C----------------------------------------------------------------
        C
        C*************** PREPROCESSING (INPUT) ***************************
       *C
           10 CONTINUE
        C---------------- INITIALIZATION DATA
              CALL INITDT
        C---------------- MESH DATA
              CALL MESHDT
        C---------------- PHYSICAL PROPERTY DATA
              CALL PROPDT
        C---------------- LOAD DATA
           20 CONTINUE
              CALL LOADDT
        C---------------- PLOT DATA
              CALL PLOTDT
        C
        C*************** SOLUTION **************************************
        C
        C---------------- FORM THE SYSTEM EQUATIONS
              CALL FORM
        C---------------- APPLY THE BOUNDARY CONDITIONS
              CALL APLYBC
        C---------------- SOLVE THE SYSTEM EQUATIONS FOR FUNCTION VALUES
              CALL CALFUN
        C---------------- CALCULATE FLUX VALUES
              CALL CALFLX
        C
        C*************** POSTPROCESSING (OUTPUT) ***********************
        C
        C---------------- PRINT RESULTS IN TABULAR FORM
              CALL PRINTS
        C---------------- PLOT RESULTS IN GRAPHICAL FORM
              CALL PLOTS
        C---------------- CHECK FOR END OF PROBLEM
              CALL ENPROB(MORDAT)
        C---------------- IF INIT VAL PROB & MORE LOAD DATA, RETURN TO LOADDT
              IF ((PROB.EQ.XIBVP).AND.MORDAT) GO TO 20
        C---------------- IF INIT VAL PROB & NO MORE LOAD DATA, PLOT FUNC V TIME
              IF ((PROB.EQ.XIBVP).AND..NOT.MORDAT) CALL TIMPLT
        C---------------- RETURN TO INITDT FOR NEW PROBLEM
              GO TO 10
        C----------------------------------------------------------------
              END
```

Loop
over
each
interval

FIGURE 11.20 ● MAIN routine of UNAFEM, showing a loop over each interval for the mixed initial-boundary-value problem.

For the implicit problem $[C]$ and $[K]$ are assembled directly into the effective stiffness matrix [Eq. (11.61b)] and stored in array KEFF(I,J). Note that KEFF(I,J) is equivalenced to K(I,J). This is done for two reasons. The system matrices (stiffness, mass, etc.) are the largest arrays in the program so the saving in storage is significant. Second, KEFF(I,J) for the initial-boundary-value problem and K(I,J) for the boundary-value problem, once assembled, are treated identically throughout the remainder of the program; hence using a single array is logically meaningful and the coding is simplified. FORM also assembles the terms $(1/\Delta t_n)[C] - (1-\theta)[K]$ for the effective load vector [Eq. (11.61c)] and stores them in array CK(I,J). The interior loads are stored in array F(I) just as for the explicit problem.

```
         SUBROUTINE ELEM11(OPT)
C
C            THIS SUBROUTINE IS THE MASTER ROUTINE FOR ELEMENT 11, A
C            2-NODE 1-D C0-LINEAR ELEMENT.
                          .
                          .
                          .
         CE(1,1) = MU*L/3.
         CE(1,2) = MU*L/6.
         CE(2,1) = CE(1,2)
         CE(2,2) = CE(1,1)
                          .
                          .
                          .
```

```
         SUBROUTINE ELEM12(OPT)
C
C            THIS SUBROUTINE IS THE MASTER ROUTINE FOR ELEMENT 12, A
C            3-NODE 1-D C0-QUADRATIC ELEMENT.
                          .
                          .
                 IF (PROB.EQ.XIBVP)CE(I,J)=CE(I,J)+WT(NINT,INTX)*JAC*
      $                      PHI(I)*MU*PHI(J)
                          .
                          .
```

```
         SUBROUTINE ELEM13(OPT)
C
C            THIS SUBROUTINE IS THE MASTER ROUTINE FOR ELEMENT 13, A
C            4-NODE 1-D C0-CUBIC ELEMENT.
                          .
                          .
                 IF (PROB.EQ.XIBVP)CE(I,J)=CE(I,J)+WT(NINT,INTX)*JAC*
      $                      PHI(I)*MU*PHI(J)
                          .
                          .
                          .
```

```
         SUBROUTINE ELEM14(OPT)
C
C            THIS SUBROUTINE IS THE MASTER ROUTINE FOR ELEMENT 14, A
C            5-NODE 1-D C0-QUARTIC ELEMENT.
                          .
                          .
                 IF (PROB.EQ.XIBVP)CE(I,J)=CE(I,J)+WT(NINT,INTX)*JAC*
      $                      PHI(I)*MU*PHI(J)
                          .
                          .
                          .
```

FIGURE 11.21 ● Statements added to the element subroutines to evaluate the element capacity matrix.

FIGURE 11.22 ● Subroutine FORM, a ▶ partial listing showing assembly of system matrices for initial-boundary-value problems.

```
      SUBROUTINE FORM
C
C           THIS ROUTINE FORMS THE SYSTEM MATRICES BY PERFORMING
C               TWO TASKS FOR EACH ELEMENT IN THE MESH:
C                   1.  FORMING THE ELEMENT MATRICES BY CALLING THE
C                       APPROPRIATE ELEMENT ROUTINES, AND
C                   2.  ASSEMBLING THE SYSTEM MATRICES FROM THE
C                       ELEMENT MATRICES.
                       .
                       .
                       .
      EQUIVALENCE (K,KEFF)
                       .
                       .
                       .
C----------- INITIALIZE ARRAYS FOR THE SYSTEM MATRICES
                       .
                       .
                       .
C*********** BEGIN LOOP OVER ALL ELEMENTS ******************************
C
      DO 500 IELNO=1,NUMEL
              ITYPE = NTYPE(IELNO)
              NND = NNODES(ITYPE)
C----------- FORM ELEMENT MATRICES
                       .
                       .
                       .
C----------- ASSEMBLE THE SYSTEM MATRICES FROM THE ELEMENT MATRICES
                       .
                       .
                       .
              IF (PROB.EQ.XIBVP.AND.THETA.EQ.0.) GO TO 300
              IF (PROB.EQ.XIBVP.AND.THETA.NE.0.) GO TO 400
                       .
                       .
                       .
C************** ASSEMBLE MATRICES FOR
C                   EXPLICIT INITIAL-BOUNDARY-VALUE PROBLEM *************
C
  300     CONTINUE
                       .
                       .
                       .
          DO 350 I=1,NND
              II = ICON(I,IELNO)
              F(II) = F(II)+FE(I)
              DO 340 J=1,NND
                  JJ = ICON(J,IELNO)
                  JSHIFT = JJ-(II-1)
                  IF (JJ.GE.II) K(II,JSHIFT)=K(II,JSHIFT)+KE(I,J)
C------------------(LUMP THE CAPACITY MATRIX)
                  IF (DIMEN.EQ.ONED) CLUMP(II) = CLUMP(II)+CE(I,J) ◄─── Eq. (11.53)
  340             CONTINUE
                       .
                       .
                       .
  350     CONTINUE
          GO TO 500
C
C************** ASSEMBLE MATRICES FOR
C                   IMPLICIT INITIAL-BOUNDARY-VALUE PROBLEM *************
C
  400     CONTINUE
          DO 440 I=1,NND
              II = ICON(I,IELNO)
              F(II) = F(II)+FE(I)
              DO 420 J=1,NND
                  JJ = ICON(J,IELNO)
                  JSHIFT = JJ-(II-1)
                  IF (JJ.GE.II) KEFF(II,JSHIFT)=KEFF(II,JSHIFT)+    ◄─── Eq. (11.61b)
     $                CE(I,J)/DT+THETA*KE(I,J)
                  IF (JJ.GE.II) CK(II,JSHIFT)=CK(II,JSHIFT)+       ◄─── in Eq. (11.61c)
     $                CE(I,J)/DT-ONEMTH*KE(I,J)
  420             CONTINUE
  440     CONTINUE
  500 CONTINUE
      RETURN
C--------------------------------------------------------------------
      END
```

```
          SUBROUTINE APLYBC
C
C         THIS ROUTINE APPLIES THE NATURAL AND ESSENTIAL BOUNDARY
C              CONDITIONS TO THE SYSTEM MATRICES.
                         .
                         .
                         .
C-----------------------------------------------------------------------
C
C******** APPLY THE NONZERO NATURAL BOUNDARY CONDITIONS, IF ANY, *******
C                 BY ADDING TERMS TO LOAD VECTOR
C
C
          IF (NUMNBC.EQ.0) GO TO 90
C
          DO 80 N=1,NUMNBC
             II = INP(N)
             JJ = JNP(N)
             KK = KNP(N)
C--------- IF JJ EQ 0 WE HAVE A CONCENTRATED FLUX AT NODE II
             IF (JJ.NE.0) GO TO 20
             F(II) = F(II)+TAUNBC(N)
             GO TO 80
      20     CONTINUE
                .
                .
                .
      80 CONTINUE
      90 CONTINUE
C********* APPLY THE ESSENTIAL BOUNDARY CONDITIONS, IF ANY, ************
C                BY IMPOSING CONSTRAINT EQUATIONS
C
                    .
                    .
                    .
          IF (PROB.EQ.XIBVP.AND.THETA.EQ.0.) GO TO 300
          IF (PROB.EQ.XIBVP.AND.THETA.NE.0.) GO TO 400
                    .
                    .
                    .
C********** MODIFY EXPLICIT INITIAL-BOUNDARY-VALUE PROBLEM
C
  300 CONTINUE
C--------- CALCULATE STEP CHANGE IN ESSENTIAL BOUNDARY CONDITIONS
C------------- (FOR RAMPING IN SUBROUTINE CALFUN)
          DO 310 I=1,NUMNP
             DELU(I) = 0.
  310 CONTINUE
          RNS = NSTEPS
          DO 320 IEB=1,NUMEBC
             I = IUEBC(IEB)
             DELU(I) = (UEBC(IEB)-A(I))/RNS  ◄——— Eq. (11.114b)
  320 CONTINUE
          RETURN
C
C********** MODIFY IMPLICIT INITIAL-BOUNDARY-VALUE PROBLEM
C
  400 CONTINUE
C--------- CALCULATE STEP CHANGE IN ESSENTIAL BOUNDARY CONDITIONS
C------------- (FOR RAMPING IN SUBROUTINE CALFUN)
          DO 410 I=1,NUMNP
             U0(I) = 0.
             DELU(I) = 0.
```

FIGURE 11.23 ● Subroutine APLYBC, a partial listing showing application of natural boundary conditions and part of application of essential boundary conditions.

Figure 11.23 ● Continued

```
            KEFFDU(I) = 0.
            KEFFUO(I) = 0.
  410 CONTINUE
        RNS = NSTEPS
        DO 420 IEB=1,NUMEBC
            I = IUEBC(IEB)
            UO(I) = A(I)
            DELU(I) = (UEBC(IEB)-A(I))/RNS  ◄────── Eq. (11.114b)
  420 CONTINUE
C--------------- MODIFY THE EFFECTIVE STIFFNESS MATRIX (KEFF)
        DO 450 IEB=1,NUMEBC
            I = IUEBC(IEB)
            DO 440 J=2,MBAND
                IR1 = I+(J-1)
                IR2 = I-(J-1)
                IF (IR1.LE.NUMNP) KEFFDU(IR1)=KEFFDU(IR1)+
     $                     KEFF(I,J)*DELU(I)
                IF (IR1.LE.NUMNP) KEFFUO(IR1)=KEFFUO(IR1)+
     $                     KEFF(I,J)*UO(I)
                IF (IR1.LE.NUMNP) KEFF(I,J)=0.
                IF (IR2.GT.0) KEFFDU(IR2)=KEFFDU(IR2)+
     $                     KEFF(IR2,J)*DELU(IR2)
                IF (IR2.GT.0) KEFFUO(IR2)=KEFFUO(IR2)+
     $                     KEFF(IR2,J)*UO(IR2)
                IF (IR2.GT.0) KEFF(IR2,J)=0.
  440       CONTINUE
            KEFF(I,1) = 1.
  450 CONTINUE
        RETURN
C-------------------------------------------------------------------
        END
```

Application of the boundary conditions in subroutine APLYBC is shown in Fig. 11.23. The natural boundary loads for the *end* of the interval, stored in array TAUNBC(I) are added to the interior loads already assembled in array F(I). For the remainder of the problem, F(I) will contain only the natural boundary loads and interior loads in order to facilitate ramping these loads later in subroutine CALFUN.

The essential boundary conditions must be handled differently for the implicit and explicit problems. The explicit problem is easier because there is no (effective) stiffness matrix on the LHS to modify. In fact, the procedure is almost trivial; we merely need to assign the value of the essential BC to its appropriate parameter a_i. However, the value to be assigned has to first be calculated from the ramping formula [Eq. (11.110)], and ramping is done more efficiently in subroutine CALFUN (inside a loop over all the time steps in the interval). All that can be done here in APLYBC is to calculate the *change* in the value during each time step. Then later CALFUN will calculate the value to apply in each step by adding the change to the value applied in the previous step.

To be more specific, consider an essential BC for the ith DOF, a_i, at the end of the current interval, i.e., at the end of step n_s:

$$(a_i)_{n_s} = (U_i)_{n_s} \tag{11.111}$$

where $(U_i)_{n_s}$ is the user-defined value. Applying the ramping formula [Eq. (11.110)] to $(U_i)_{n_s}$ yields

$$(U_i)_n = (U_i)_0 + \frac{n}{n_s}\left[(U_i)_{n_s}-(U_i)_0\right] \tag{11.112}$$

where $(U_i)_0$ is the user-defined value at the beginning of the interval (end of previous interval), and $(U_i)_n$ is the ramped value to be applied at the end of step n; that is,

$$(a_i)_n = (U_i)_n \tag{11.113}$$

Note that Fig. 11.18, substituting U_i for F, illustrates Eq. (11.112).

It will be convenient to write Eq. (11.112) in the form

$$(U_i)_n = (U_i)_0 + n(\Delta U_i) \tag{11.114a}$$

where

$$(\Delta U_i) = \frac{1}{n_s}\left[(U_i)_{n_s} - (U_i)_0\right] \tag{11.114b}$$

Here (ΔU_i) is the change in the value of the essential BC during each time step. This quantity is the same for all time steps in the interval, so it is calculated just once at the beginning of the interval. It is the only part of the essential BC application performed in subroutine APLYBC. As shown in Fig. 11.23, the values $(\Delta U)_i$ are stored in array DELU(I). The user-defined values $(U_i)_{n_s}$ were read into array UEBC(I) in subroutine LOADDT. The values $(U_i)_0$ are available in the system solution vector $\{a\}$ from the end of the previous interval.

The implicit problem has an effective stiffness matrix on the LHS, $\{K_{\text{eff}}\}$, which is modified for the essential BCs in the conventional manner. Here again, though, we save part of the job for subroutine CALFUN. As shown in Fig. 11.23, the appropriate rows and columns of KEFF(I,J) are deleted, and 1s are substituted for the diagonal terms. Prior to the deletion the column terms must be multiplied by the value of the appropriate essential BC and then subtracted from the RHS load vector. Just as with the explicit problem, the value of the essential BC is determined by the ramping Eq. (11.112) in subroutine CALFUN, so here in subroutine APLYBC we can only form the appropriate load vectors and then save them.

To be more specific, consider an essential BC, $(U_i)_n$, applied to the ith DOF at the end of step n. Using Eq. (11.114a), the generated load vector subtracted from the RHS is

$$\{K_{\text{eff}_i}\}(U_i)_n = \{K_{\text{eff}_i}\}(U_i)_0 + n\{K_{\text{eff}_i}\}(\Delta U_i) \tag{11.115}$$

where $\{K_{\text{eff}_i}\}$ denotes the ith column of $[K_{\text{eff}}]$. If there are essential BCs at several different DOF, it is convenient to write Eq. (11.115) in matrix form:

$$[K_{\text{eff}}]\{U\}_n = [K_{\text{eff}}]\{U\}_0 + n[K_{\text{eff}}]\{\Delta U\} \tag{11.116}$$

where the vectors $\{U\}_n$, $\{U\}_0$, and $\{\Delta U\}$ have zeroes at all DOF at which there is no essential BC. As shown in Fig. 11.23, the vectors $\{U\}_0$, $\{\Delta U\}$, $[K_{\text{eff}}]\{U\}_0$, and $[K_{\text{eff}}]\{\Delta U\}$ are stored in arrays U0(I), DELU(I), KEFFU0(I), and KEFFDU(I), respectively.

Figure 11.24(a) and (b) shows the statements added to subroutine CALFUN to solve the recurrence relations for the explicit and implicit problems for each time step in the interval. The explicit problem solves Eqs. (11.55). Note the ramping of the natural and interior load vector according to Eq. (11.110). The ramping of the essential BCs was begun in subroutine APLYBC by calculating the step change (ΔU_i) from Eq. (11.114b). Here

```
      SUBROUTINE CALFUN
C
C        THIS ROUTINE SOLVES THE SYSTEM EQUATIONS FOR THE FUNCTION VALUES
            .
            .
            .
C--------------- BRANCH TO PROPER EQUATION-SOLVING METHOD
            .
            .
            .
      IF (PROB.EQ.XIBVP.AND.THETA.EQ.0.) GO TO 300
      IF (PROB.EQ.XIBVP.AND.THETA.NE.0.) GO TO 400
            .
            .
            .
C*************** EXPLICIT INITIAL-BOUNDARY-VALUE PROBLEM
C
  300 CONTINUE
C--------------- DEFINE SOLUTION VECTOR AT BEGINNING OF INTERVAL
      DO 310 I=1,NUMNP
          ANM1(I) = AN(I)
  310 CONTINUE
C-------------- SOLVE RECURRENCE RELATION OVER ALL TIME STEPS
      RNS = NSTEPS
      DO 380 N = 1, NSTEPS
          TIME = TIME+DT
          RNM1 = N-1
          DO 350 I=1,NUMNP
              KA = K(I,1)*ANM1(I)
              DO 340 J=2,MBAND
                  IR1 = I+(J-1)
                  IR2 = I-(J-1)
                  IF (IR1.LE.NUMNP) KA=KA+K(I,J)*ANM1(IR1)
                  IF (IR2.GT.0) KA=KA+K(IR2,J)*ANM1(IR2)
  340             CONTINUE
C----------------- CALCULATE NAT +INT LOADS AT STEP N-1 BY RAMPING
              FNM1(I) = FO(I)+(RNM1/RNS)*(FNS(I)-FO(I))
C----------------- SOLVE FOR FUNCTION AT STEP N (IGNORE ESSENTIAL BC)
              IF (CLUMP(I).EQ.0.) CALL ERROR(33,I)
              AN(I) = ANM1(I)+(DT/CLUMP(I))*(FNM1(I)-KA)
  350         CONTINUE
C------------ MODIFY FUNCTION FOR RAMPED ESSENTIAL BC
          DO 360 IEB=1,NUMEBC
              I = IUEBC(IEB)
              AN(I) = ANM1(I)+DELU(I)
  360         CONTINUE
C------------ SAVE FUNCTION VALUES FOR NEXT TIME STEP
          DO 370 I=1,NUMNP
              ANM1(I) = AN(I)
  370         CONTINUE
          IF (NTIME.GE.501) GO TO 380
          NTIME = NTIME+1
C----------------- SAVE TIME AND FUNCTION VALUES FOR TIME PLOTS
          TPLOT(NTIME) = TIME
          DO 375 L=1,3
              NNDE = NNODEI(L)
              IF (NNDE.NE.0) FTPLOT(L,NTIME) = AN(NNDE)
  375         CONTINUE
  380 CONTINUE
C---------- SAVE NAT + INT LOAD AT END OF INTERVAL
      DO 390 I=1,NUMNP
          FO(I) = FNS(I)
  390 CONTINUE
      RETURN                                              (Continued)
```

Annotations:
- Multiplies $[K]\{a\}_{n-1}$ in Eq. (11.55a)
- Eq. (11.110), for step $n-1$
- Eq. (11.55a)
- Eq. (11.117)

FIGURE 11.24a ● Subroutine CALFUN, a partial listing showing evaluation of recurrence relation for *explicit* initial-boundary-value problems.

FIGURE 11.24b ● Continuation of subroutine CALFUN, showing evaluation of recurrence relation for *implicit* initial-boundary-value problems.

```
C
C*************** IMPLICIT INITIAL-BOUNDARY-VALUE PROBLEM
C
  400 CONTINUE
C---------- SAVE LOAD VECTOR AND FUNCTION AT START OF INTERVAL
      DO 410 I=1,NUMNP
          FNM1(I) = FO(I)
          ANM1(1) = A(1)
  410 CONTINUE
      ONEMTH = 1.-THETA
      RNS = NSTEPS
C---------- FORWARD REDUCTION OF EFFECTIVE STIFFNESS MATRIX (KEFF)
      CALL GAUSEL(NUMNP,MBAND,KEFF,FEFF,AN,1)
C---------- SOLVE RECURRENCE RELATIONS OVER ALL TIME STEPS
      DO 480 N=1,NSTEPS
      TIME = TIME+DT
      RN = N
C------------- CALCULATE NAT + INT LOADS AT STEP N BY RAMPING
          DO 420 I=1,NUMNP
          FN(I) = FO(I)+(RN/RNS)*(FNS(I)-FO(I))  ◀──── Eq. (11.110), for step n
  420     CONTINUE
C------------- CALCULATE EFFECTIVE LOAD VECTOR (FEFF) AT STEP N
          DO 440 I=1,NUMNP
              CKA = CK(I,1)*ANM1(I)                    ⎫
              DO 430 J=2,MBAND                         ⎪
                  IR1 = I+(J-1)                        ⎬ Multiplies [CK]{a}ₙ₋₁
                  IR2 = I-(J-1)                        ⎪ ([CK] was formed in
                  IF (IR1.LE.NUMNP)CKA=CKA+CK(I,J)*ANM1(IR1)  ⎪ subroutine FORM)
                  IF (IR2.GT.0) CKA=CKA+CK(IR2,J)*ANM1(IR2)   ⎭
  430         CONTINUE
              FEFF(I) = ONEMTH*FNM1(I)+THETA*FN(I)+CKA  ◀── Eq. (11.61c)
  440     CONTINUE
C------------- MODIFY EFFECTIVE LOAD VECTOR FOR RAMPED ESSENTIAL BC
          DO 445 I=1,NUMNP
          FEFF(I) = FEFF(I)-KEFFUO(I)-RN*KEFFDU(I)  ◀── Subtracts [Kₑff]{U}ₙ,
  445     CONTINUE                                       Eq. (11,116), from {Fₑff}
          DO 450 IEB=1,NUMEBC
          I = IUEBC(IEB)
          FEFF(I) = ANM1(I)+DELU(I)                 ◀──── Eq. (11.117)
  450     CONTINUE
C------------- REDUCTION OF LOAD VECTOR: BACKSUBSTITUTION
          CALL GAUSL2(NUMNP,MBAND,KEFF,FEFF,AN,0)
C------------- SAVE VALUES FOR NEXT STEP
          DO 460 I=1,NUMNP
              ANM1(I) = AN(I)
              FNM1(I) = FN(I)
  460     CONTINUE
          IF (NTIME.GE.501) GO TO 480
          NTIME = NTIME+1
C------------- SAVE TIME AND FUNCTION VALUES FOR TIME PLOTS
          TPLOT(NTIME) = TIME
          DO 470 L=1,3
              NNDE = NNODEI(L)
              IF (NNDE.NE.0) FTPLOT(L,NTIME) = AN(NNDE)
  470     CONTINUE
  480     CONTINUE
C--------------- SAVE NAT + INT LOADS AT END OF INTERVAL
      DO 490 I=1,NUMNP
          FO(I) = FNS(I)
  490 CONTINUE
      RETURN
C---------------------------------------------------------------------
      END
```

we complete the ramping with the expression

$$(U_i)_n = (U_i)_{n-1} + (\Delta U_i) \tag{11.117}$$

which follows from Eq. (11.114a) applied to steps n and $n-1$.

The implicit problem solves Eqs. (11.61). Ramping of the natural and interior loads is the same as for the explicit problem. Ramping of the essential BCs is also the same; however, instead of applying the ramped value directly to the DOF (as in the explicit problem), it is substituted for the load vector term on the RHS, in the usual manner of applying essential BCs.

Note that the programming for the implicit case takes advantage of the fact that the effective stiffness matrix, $[K_{eff}] = (1/\Delta t_n)[C] + \theta[K]$, is the same for every time step in an interval since Δt_n is the same for every step. Therefore the forward reduction phase of Gaussian elimination (which is the most time consuming and expensive phase) needs to be done only once, during the first step, and the resulting triangularized effective stiffness matrix saved. Then, in each of the subsequent time steps in that interval, the remaining phases (forward reduction of the load vector and back substitution) are performed by entering the GAUSEL routine at the secondary entrance point GAUSL2 (see the appendix).

The postprocessing calculations are so similar to the boundary-value problem that they are incorporated in subroutines BIVPRI and BIVPLO (*b*oundary- and *i*nitial-*v*alue *pri*nt and *plo*t). At the end of each interval the function and flux are each tabularly printed and graphically plotted as a function of x (the same plots as for the boundary-value problem). At the end of the problem the function is plotted versus time for three DOF chosen by the user.[27]

11.5 Application to a Transient Heat Conduction Problem

Consider the problem shown in Fig. 11.25. This is the same problem described in Section 7.4.1 and analyzed in Section 7.4.4 and again in Chapter 8, although here we apply the loads "suddenly" and then hold them constant thereafter. This will produce a time-varying response that will gradually approach a steady-state solution, the latter being the solution in Chapters 7 and 8.

The rod is initially in thermal equilibrium with its surroundings at a temperature of 20 °C (and therefore zero heat flux). At $t=0$ a flux of 0.1 cal/sec-cm^2 is suddenly applied to the left end, and the temperature is suddenly dropped 20 °C at the right end. It is estimated (from an experimental procedure, say) that these sudden changes both occur in about 0.1 sec; i.e., the "rise" time (or time of application of the full load) is assumed to be 0.1 sec. Such a rapid application represents a thermal "shock" to the system and we can expect correspondingly rapid changes in the temperature distribution immediately following the shock and in the vicinity of the shock.

---

[27] Flux-versus-time data is not output because in real problems it tends to be used less frequently than function-versus-time data. In addition, the coding is a bit messy, thereby further diminishing its value to the program.

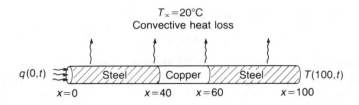

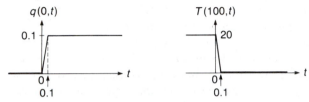

Initial conditions: $T(x,0) = 20°C$ $0 \le x \le 100$

FIGURE 11.25 ● Sudden loads applied to a rod that is initially in thermal equilibrium.

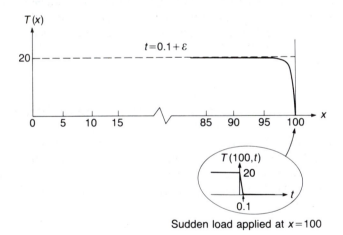

Sudden load applied at $x=100$

FIGURE 11.26 ● A typical response immediately following application of a sudden load, such as the change in temperature at the right end of the rod in Fig. 11.25.

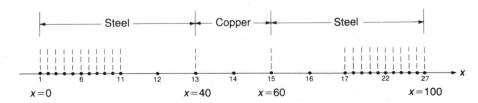

FIGURE 11.27 ● A mesh appropriate for the initial shock response, and therefore more than adequate for all subsequent responses out to steady state.

The density and specific heat for steel and copper are as follows:

Steel	Copper
$\varrho = 7.8$ gm/cm^3 $c = 0.11$ cal/gm-°C	$\varrho = 8.5$ gm/cm^3 $c = 0.092$ cal/gm-°C

We would like to know the temperature-time histories at the points $x=10$ cm and $x=90$ cm, as well as the temperature distribution along the rod at several instants of time, including both the initial transient (shock) response and the final steady-state response.

Our next task is to select a mesh and a time-stepping strategy, i.e., a sequence of time steps. Let's consider the mesh first. Since the temperature distribution will vary with time, the "appropriate" mesh will also vary. Thus, during the initial shock response we would expect steep temperature gradients near the ends, as well as a rapid decay in these gradients, as suggested in Fig. 11.26. This would require a finer mesh in these regions. At the other extreme, as steady state is approached the temperature distribution will become smoother, thereby allowing a coarser mesh.

For small-to-medium sized problems (such as the present one) it is usually expedient to choose one mesh that can provide the desired accuracy at all times of interest.[28] Since the shock response is the most complicated, it usually dictates the necessary fineness of the mesh. Such a mesh would provide too much accuracy near steady state, but the cost of the excessive computation may be compensated by the savings in time taken to construct only one mesh.

In the present problem we are interested in observing the shock response, so let's try the mesh shown in Fig. 11.27. The choice of type and size of elements is determined in the same manner as for a boundary-value problem (which this is, with respect to the variable x). Thus we usually have some rough idea of what the exact solution looks like, e.g., where the solution tends to be "wiggly" and where it tends to be more-or-less smooth. We can then mentally picture how many straight lines (linear elements) or parabolic lines (quadratic elements), etc., would be required to provide a meaningful approximation. A subsequent finer or coarser mesh will then reveal the approximate accuracy achieved, and it will suggest how many more (or less) DOF will provide the desired accuracy.[29] Based on our expectations from Figs. 11.25 and 11.26, the number and distribution of nodes shown in Fig. 11.27 would seem reasonable.

As regards the type of element, recall the discussion in Section 9.4. In regions where the solution has a very steep, quasisingular behavior, lower-degree polynomials are generally more computationally efficient if the mesh is not graded. Since UNAFEM has an unsophisticated mesh generator, it is much easier to define a uniform mesh. Therefore we will employ a locally uniform mesh of linear elements at each end of the rod. Subsequent meshes would then employ h-refinement in these regions.

---

[28] For large problems with shock loads it may be more economical to separate the analysis into a shock phase and a postshock phase, and construct a different mesh for each phase. See comments following this problem, as well as the illustration problem in Chapter 15.

[29] If we have no feeling at all about the exact solution we could try virtually any type of mesh, but the results might be so poor that several meshes would be required before acceptable accuracy is obtained.

Interval	θ	Δt, sec	Number of steps, n_s	Time span, sec
I	0	0.1	1	0 – 0.1
II	0	0.1	19	0.1 – 2
III	2/3	2	9	2 – 20
IV	2/3	20	9	20 – 200
V	2/3	200	9	200 – 2000
VI	2/3	1000	8	2000 – 10,000
VII	1	10^6	1	$10{,}000 - 1.01 \times 10^6$
VIII	1	10^6	1	$1.01 \times 10^6 - 2.01 \times 10^6$

FIGURE 11.28 ● A suggested time-stepping strategy for the problem in Figure 11.25, using the mesh in Figure 11.27.

Throughout the remainder of the rod we can expect smoother behavior, so a few higher-order elements should provide the best accuracy. We'll use quadratics for our first mesh, which is usually low accuracy, while we get the bugs out of our input data. Subsequent meshes would converge most rapidly in this region using *p*-refinement, i.e., changing the quadratics to cubics or quartics. Of course, *h*-refinement could also be used; it might require a few more DOF to achieve the same accuracy as *p*-refinement, but for such a small problem the difference is unimportant.

Now let's consider the time-stepping strategy. Just as with mesh generation, there is essentially an unlimited number of different possible strategies, many of which will do the job satisfactorily, but with varying degrees of efficiency, i.e., with different costs to achieve a given accuracy. Figure 11.28 shows just one of many reasonable strategies that could be employed.

Several observations can be made regarding the selection of intervals and values for θ, Δt, and n_s in Fig. 11.28. These observations illustrate several principles and techniques which are applicable, either directly or with suitable modification, to all time-stepping methods (such as those described in Section 12.2 for second-order initial-boundary-value problems in dynamics). In other words, the basic techniques of time-stepping extend to all types of initial value problems.

1. The interesting temperature changes cover time spans from as short as a few tenths of a second (during and immediately following the shock) to as long as a few hours (to reach steady state), a range of several orders of magnitude. This, in turn, requires time steps that similarly vary from tenths of a second to hours. We therefore use several intervals, beginning with small values of Δt for the shock response and increasing in size (by multiples of about 5 to 10)[30] as steady state is approached.

2. The value of θ changes as the time steps increase. Some values are better suited for short-duration shock response, other values for near steady state, and still other values for the broad transition region between these two extremes. The remaining observa-

---

[30] One-step methods can tolerate large differences in step size in adjacent steps since their recurrence relations involve only a single step. Multi-step methods, however, generally require more gradual changes in step size to avoid ill-conditioning of their recurrence relations.

tions will comment further on the particular θ choices used in each of the intervals in Fig. 11.28.[31]

3. The first interval applies the full load in a single step. It is the total rise time (here, 0.1 sec) that determines the shock to the system, i.e., the relative amplitudes of the higher modes. If the higher modes are significantly excited (due to a short rise time), then there may be considerable numerical oscillation, depending on the $\lambda_i \Delta t$ values in Fig. 11.12 for the θ-method being used. Oscillations are usually most severe for $0 \leq \theta \leq 1/2$ since $(A_i)_n/(A_i)_{n-1}$ can be close to -1 for the higher modes.

There are several ways to reduce or eliminate the oscillations: (1) Use the averaging techniques referred to previously; (2) lengthen the rise time (to reduce the shock); (3) use $\theta > 1/2$ (preferably $\theta \geq 2/3$), or (4) use very short steps so that $\lambda_i \Delta t$ for even the highest mode is small enough that $(A_i)_n/(A_i)_{n-1} \geq -1/2$ (preferably > 0). The first method is an option only if the program being used employs such techniques. The second can always be easily implemented, but changing the total rise time is tantamount to changing the physical model, which might not be desirable. The third method is very effective, with $\theta \approx 2/3$ being close to optimum. The fourth is also very effective, but usually only with the explicit $\theta = 0$ method, as explained next.

Tracking the shock response usually requires very short time steps. If the steps are comparable to Δt_{crit}, as determined by Eq. (11.105), then it is most efficient to use the explicit $\theta = 0$ method. For the present problem, Eq. (11.105) yields $\Delta t_{\text{crit}} \approx 6$ sec.[32] We also note that the $\theta = 0$ line in Fig. 11.12 crosses the $(A_i)_n/(A_i)_{n-1} = 0$ axis at $(1/2)\Delta t_{\text{crit}}$, which is ≈ 3 sec. The zero axis is the boundary between oscillatory and nonoscillatory decay. Therefore, recalling that Eq. (11.105) tends to underestimate by a factor of 2 or 3,[33] we conclude that a time step of a few seconds will probably ensure stability, as well as no oscillations. Since the rise time is only 0.1 sec, we can safely use the $\theta = 0$ method with $\Delta t = 0.1$ sec and apply the full load in just one step, as shown for interval I in Fig. 11.28. (Using several smaller steps, such as 10 steps at $\Delta t = 0.01$, will improve accuracy during the shock application, but will not essentially change the computed relative magnitudes of the higher modes following the shock.)

In interval II the load is maintained constant (as it is in all subsequent intervals), while continuing with the $\theta = 0$ method and using many steps of $\Delta t = 0.1$ sec to follow the rapidly changing solution immediately following application of the load. Figure 11.28 shows only 19 steps in interval II because this initial mesh is intended to have low

---

[31] In principle, one can use any θ value at any time during the analysis. For example, one could use a single θ for the entire analysis. Most commercial FE programs, in fact, still offer only a single one-step method, usually $\theta = 1/2$ or $\theta = 1$. Many of these programs, though, also offer a more accurate two-step method (see Exercise 11.1).

[32] The minimum value of $[(\mu/\alpha)\delta^2]^{(e)}$ occurs in the linear elements in Fig. 11.27, for which $\delta = 2$ cm and $\mu/\alpha = 7.17$ sec/cm^2.

[33] For the mesh in Fig. 11.27, the actual value of Δt_{crit} is close to 14.5 sec, as evidenced by a stable solution using $\Delta t = 14$ sec and an unstable solution using $\Delta t = 15$ sec.

accuracy. However, for a high-accuracy final mesh one might employ as many as several hundred time steps because of the efficiency of the explicit method.

4. Following the shock response, and up until steady state is almost reached, there is a broad transition region in which the time steps can gradually increase as the rate of change of the response gradually slows down.[34] During this transition the appropriate time-step sizes will usually increase by one or more orders of magnitude. Intervals III through VI in Fig. 11.28 provide a convenient way to increase Δt from 2 to 1,000 sec using only a few time steps. The choice of step size and number of steps is somewhat arbitrary (just like element sizes) since subsequent step refinement (like mesh refinement) will improve the accuracy and reveal where further refinement, if any, is necessary.

In the transition region, step sizes are generally much larger than Δt_{crit}, so for stability we must use θ in the range $1/2 \le \theta \le 1$. The comments at the end of Section 11.3.4 indicate that $\theta = 2/3$ seems to be near to an optimum value. For this value oscillations decay rapidly because, as shown by the $\theta = 2/3$ curve in Fig. 11.12, $(A_i)_n/(A_i)_{n-1} \to -1/2$ (rather than -1) as $\lambda_i \Delta t \to \infty$.

The mid-difference ($\theta = 1/2$) method seems attractive because of its higher rate of convergence, namely, Δt^2 rather than Δt. However, if Δt is not "small enough," its accuracy may actually be worse (and frequently is) because of the numerical oscillations. We have already noted that averaging techniques can help to remedy this problem. But if averaging is not done, then one would have to use small enough Δt values such that $\lambda_i \Delta t$ for the highest *significant* modes yields $(A_i)_n/(A_i)_{n-1}$ greater than about $-1/2$. As steady state is approached, the higher modes become less significant. Therefore Δt can gradually increase (up to some finite limit determined by the lowest modes) without producing oscillations. Such generalities and caveats make the mid-difference method (without averaging) a bit risky for the inexperienced user. As a general rule, the more reliable approach is to use θ in the range $2/3 \le \theta \le 1$. As noted in Sections 11.3.2 and 11.3.4, the backward difference method ($\theta = 1$) is a very reliable method for any time step size.

5. Interval VII calculates the steady-state solution. This can be obtained most efficiently by using the backward difference method ($\theta = 1$) and a single very large time step. This can be demonstrated theoretically by letting $\Delta t_n \to \infty$ in Eq. (11.60):

$$\theta[K]\{a\}_n = (1-\theta)\{F\}_{n-1} + \theta\{F\}_n - (1-\theta)[K]\{a\}_{n-1} \qquad (\Delta t_n \to \infty)$$

$$(11.118)$$

Setting $\theta = 1$ yields

$$[K]\{a\}_n = \{F\}_n \qquad (11.119)$$

Since $\Delta t_n \to \infty$, $\{F\}_n$ must be the steady-state value $\{F\}_{ss}$. Consequently $\{a\}_n$ in Eq. (11.119) is the steady-state solution, $\{a\}_{ss}$. In summary, for $\theta = 1$ and Δt_n extremely large, the recurrence relation [Eq. (11.60)] reduces to

$$[K]\{a\}_{ss} = \{F\}_{ss} \qquad (11.120)$$

---

[34] Some of the large commercial FE programs automatically increase (or decrease) Δt if some measure of change during the previous steps, such as the second time derivative for a one-step method or the third time derivative for a two-step method, is calculated to be below (or above) a given threshold.

This is a pure boundary-value problem, corresponding to Eq. (11.11a) with the time-varying term $\mu \partial U / \partial t$ omitted since U is not a function of time at steady state. It is therefore the same problem solved in Chapters 7 and 8. Note that the single step can be taken from any point in the time history, including the initial time.

6. Interval VIII is simply an extra very large time step to check for convergence of the steady-state solution, i.e., to verify that the previous step was indeed large enough.

7. The steady-state solution may also be calculated using $1/2 \leq \theta < 1$, but oscillations may necessitate several (large) steps. Thus Fig. 11.12 shows that the θ curves approach the limit $-(1-\theta)/\theta$ as $\Delta t \rightarrow \infty$. For θ in the range $1/2 \leq \theta < 1$, the limit is negative (between -1 and 0); i.e., all modes exhibit oscillatory decay. (For $\theta < 1/2$ the methods are unstable since the limit is less than -1.) The nature of the oscillation is easily shown if we consider the usual situation in which we have stepped the solution forward to a point where the loads are no longer changing, namely, $\{F\}_{n-1} = \{F\}_n = \{F\}_{ss}$, and we are taking very large time steps after step $n-1$, so that Eq. (11.118) effectively applies for steps n, $n+1$, etc. It then follows from Eq. (11.118),

$$\{a\}_n = \{a\}_{ss} + \frac{1-\theta}{\theta} (\{a\}_{ss} - \{a\}_{n-1}) \qquad (\Delta t_n \rightarrow \infty) \qquad \textbf{(11.121)}$$

(See Exercise 11.2.) The solution therefore oscillates about $\{a\}_{ss}$, the amplitude of the oscillation being proportional to the difference between $\{a\}_{ss}$ and $\{a\}_{n-1}$, where $\{a\}_{n-1}$ is the solution just prior to the very large step. The amplitude decays at the rate $(1-\theta)/\theta$ during subsequent large steps. Clearly then, if the difference between $\{a\}_{ss}$ and $\{a\}_{n-1}$ is large or θ is close to $1/2$, it may require many (large) steps to converge to $\{a\}_{ss}$. It is therefore preferable to make θ close to 1 or equal to 1 when calculating the steady-state solution.

This completes the observations regarding the choice of time-stepping strategy. Continuing now with the analysis, Fig. 11.29 shows the UNAFEM input data corresponding to the mesh in Fig. 11.27 and the time-stepping strategy in Fig. 11.28, using the data input instructions in the appendix. Note the use of reduced integration for the C^0-quadratic elements (a "2" in column 25) since they are undistorted.

Figure 11.30 shows the temperature distributions at the end of intervals II through VII. Judging from the general shape of the curves it would seem that the three large quadratic elements in the middle and the twenty small linear elements at the ends are probably providing an approximately balanced error throughout the rod. We will be able to judge this better, though, after analyzing a second mesh. The printed output shows that the temperatures and fluxes at the end of intervals VII and VIII agree to almost four significant figures, indicating good steady-state convergence *for the given mesh*; a finer mesh could change these values considerably. Figure 11.31 shows the temperature-time histories at $x = 10$ cm and $x = 90$ cm.

To estimate accuracy we follow the usual procedure of performing a second analysis using a different discretization. Unlike the previous boundary-value and eigenvalue problems, the present initial-boundary-value problem involves a discretization of both the x and t variables, the former by finite elements and the latter by finite time steps. To improve (as well as estimate) accuracy we must therefore refine the mesh or refine the steps or both. The decision as to whether and how to refine the time steps is made in much the

```
IDIMI BVP    TRANSIENT HEAT CONDUCTION IN ROD
NODE
             1      1       0.
            11      1      20.
            17      1      80.
            27            100.
ELEM
             1      1      11             1      1       2
            11             12      2      1     11      12     13
            12             12      2      2     13      14     15
            13             12      2      1     15      16     17
            14      1      11             1     17      18
            23             11             1     26      27
PROP
             1           .12
                       1.5E-4
                        .86
             2          .92
                       1.5E-4
                        .78
STEP
             1           .1       0.           20.
ILC
             1          3.0E-3    23
```

```
EBC
            27      0.
NBCC
             1            .1
PLOT
            10      0.          100.
             5      0.           50.
             5      0.            1.
             5      0.        10000.            1      6     22
STEP
            19            .1       0.
ILC
             1          3.0E-3    23
EBC
            27      0.
NBCC
             1            .1
STEP
             9      2.            .67
ILC
             1          3.0E-3    23
EBC
            27      0.
NBCC
```

(Continued)

FIGURE 11.29 ● UNAFEM input data for the transient heat conduction problem.

```
         /              . /
STEP
         9      2 0 .              . 6 7
ILC
         /           3 . 0 E - 3      2 3
EBC
      2 7         0 .
NBCC
         /              . /
STEP
         9      2 0 0 .            . 6 7
ILC
         /           3 . 0 E - 3      2 3
EBC
      2 7         0 .
NBCC
         /              . /
STEP
         8    1 0 0 0 .            . 6 7
ILC
         /           3 . 0 E - 3      2 3
EBC
      2 7         0 .
NBCC


         /              . /
STEP
         /           1 . 0 E + 6       / .
ILC
         /           3 . 0 E - 3      2 3
EBC
      2 7         0 .
NBCC
         /              . /
STEP
         /           1 . 0 E + 6       / .
ILC
         /           3 . 0 E - 3      2 3
EBC
      2 7         0 .
NBCC
         /              . /
END
```

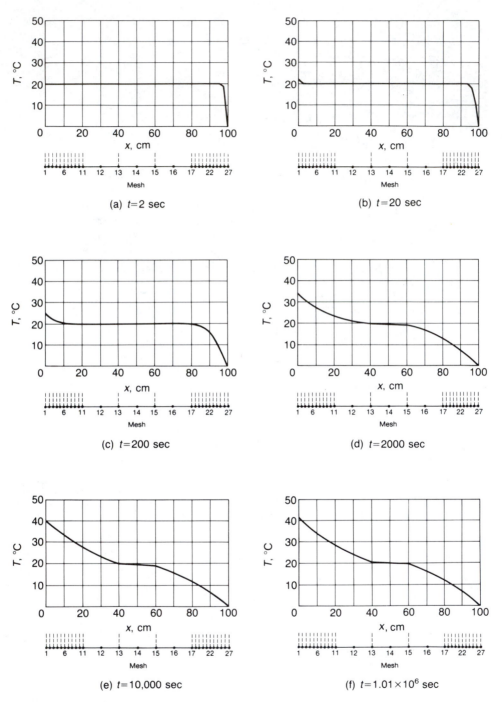

FIGURE 11.30 ● Temperature distributions at several times, using the coarse mesh and time-stepping data in Figs. 11.27 and 11.28.

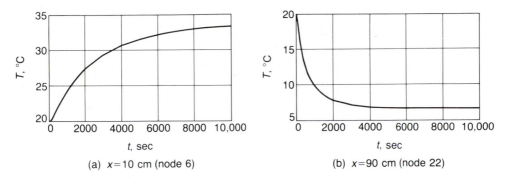

(a) $x=10$ cm (node 6)

(b) $x=90$ cm (node 22)

FIGURE 11.31 ● Temperature-time histories at two locations, using the coarse mesh and time-stepping data in Figs. 11.27 and 11.28.

same way as for mesh refinement — namely, an examination of the temperature-time plot to see where the FE solution is "least smooth," or, if two analyses are available, a comparison of the relative differences from the graphical or printed output.

The plots in Fig. 11.30(a) and (b) indicate more spatial refinement is probably needed near the ends of the rod for the shock response. Figure 11.31 indicates more temporal refinement may be needed after 2000 sec. We might therefore choose to locally refine in these areas. However, this is only the first mesh, which is presumably a coarse one, so we're really not too sure what the error is at any value of x or t. Therefore we can obtain a better overall estimate of error by refining globally, i.e., over the entire mesh and the entire time history. A subsequent comparison of two such analyses, one "coarse" and one "medium" discretization, will indicate where further local refinement, if any, might be needed for a final analysis.

Figures 11.32 and 11.33 show, respectively, the globally refined mesh and time-stepping strategy. Note the use of h-refinement near the rod ends to handle the very steep shock response, and the use of p-refinement over the broad middle section, where the solution at all times seems to have gentler slopes, curvatures, etc., within each element.

It should also be observed that Δt_{crit} for the refined mesh will be decreased by a factor of 1/4 because, according to Eq. (11.105), Δt_{crit} is proportional to δ_{min}^2, the square of the minimum separation between nodes. A comparison of Figs. 11.27 and 11.32 shows that δ_{min} has decreased by a factor of 1/2 for the linear elements (and a factor of only 2/3 for the three higher-order elements). Our estimate of Δt_{crit} from Eq. (11.105) has therefore decreased from approximately 6 sec to approximately 1 1/2 sec. However, Fig.

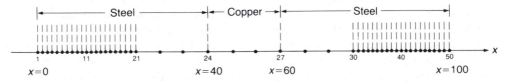

FIGURE 11.32 ● A refined mesh. (A coarser mesh is shown in Fig. 11.27.)

Interval	θ	Δt, sec	Number of steps, n_s	Time span, sec
I	0	0.05	2	0 – 0.1
II	0	0.05	38	0.1 – 2
III	2/3	1	18	2 – 20
IV	2/3	10	18	20 – 200
V	2/3	100	18	200 – 2000
VI	2/3	500	16	2000 – 10,000
VII	1	10^6	1	$10,000 – 1.01 \times 10^6$
VIII	1	10^6	1	$1.01 \times 10^6 – 2.01 \times 10^6$

FIGURE 11.33 ● A refined time-stepping strategy. (A coarser time-stepping strategy is shown in Figure 11.28).

11.33 shows that we have also halved Δt from 0.1 to 0.05 sec, which is still well below the new Δt_{crit}. Hence we can expect a stable solution for the explicit method in intervals I and II.

Figures 11.34 and 11.35 show the temperature distributions and time histories, respectively, analogous to Figs. 11.30 and 11.31. A comparison of the graphical plots for the coarse and refined analyses shows very good convergence at almost all locations and times, i.e., the plotted curves appear to be almost identical, suggesting a maximum pointwise error of about 1%. (Apparently the first mesh and time steps weren't so coarse after all.)

An exception to this error estimate is the shock response in the region $90 \leq x \leq 100$, as can be seen by comparing Figs. 11.30(a) and (b) and 11.34(a) and (b). If better accuracy is desired in this region, we could do a third analysis to decrease h locally, and probably Δt as well. Such a local refinement will generally improve the local accuracy. However, we must always be careful not to locally refine "too much," i.e., not to let the mesh refinement in one region vastly exceed that in nearby regions. We recall from Chapter 9 that local refinement can yield only limited improvement in accuracy, and the additional accuracy cannot always be reliably estimated by comparing successive solutions.

We conclude this section by describing an alternative procedure for analyzing the above problem which, for larger problems (e.g., in 2-D and 3-D), can significantly reduce the computational cost. It was observed in a footnote at the beginning of this section that one could analyze a shock-loading problem by separating the analysis into a shock phase and a postshock phase and constructing a different mesh for each phase. The results of the above problem illustrate why this is possible. Figure 11.34(b) shows that the effect of the sudden temperature drop at $x=100$ is not "felt" (to any significant degree) at $x=90$, say, until more than 20 sec later. Consequently, the solution during the first 20 sec in the region $90 \leq x \leq 100$cm — that is — the shock phase, could be obtained much more efficiently by modeling only the region $90 \leq x \leq 100$cm and ignoring the remainder of the rod. In other words, during the first 20 sec the region $0 \leq x \leq 90$ has no appreciable effect on the solution in the region $90 \leq x \leq 100$, so carrying along that portion of the model (the region $0 \leq x \leq 90$) only entails wasteful computation. The boundary condition at $x=90$ cm would be the initial temperature, held indefinitely: $T(90,t)=20\,°C$. One could now afford to use a much smaller element and time step.

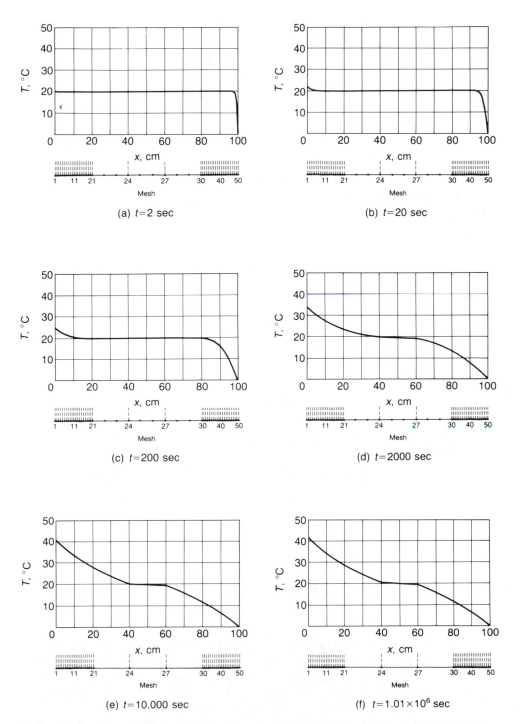

FIGURE 11.34 ● Temperature distributions at several times, using the refined mesh and time-stepping data in Figs. 11.32 and 11.33.

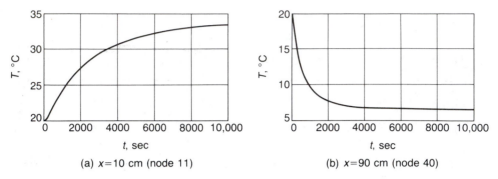

(a) $x=10$ cm (node 11)   (b) $x=90$ cm (node 40)

FIGURE 11.35 ● Temperature-time histories at two locations, using the refined mesh and time-stepping data in Figs. 11.32 and 11.33.

For the postshock phase, say $t>20$ sec, one would model the entire rod, using a much coarser mesh in the region $90 \leq x \leq 100$. The analysis would begin at $t=0$ with application of the load in 0.1 sec, but followed by larger time steps chosen to provide the desired accuracy only for $t>20$ sec. This two-phase approach could significantly reduce computational costs in complicated 2-D and 3-D problems involving localized application of a shock load, as will be demonstrated in Chapter 15.

Exercises

11.1   Derive the following *two-step* backward difference recurrence relation [11.16] using the quadratic interpolation polynomials in Fig. 8.2 (changing x to t):

$$\left(\frac{2\Delta t_n + \Delta t_{n-1}}{(\Delta t_n + \Delta t_{n-1})\Delta t_n} [C] + [K] \right) \{a\}_n = \{F\}_n + \frac{\Delta t_n + \Delta t_{n-1}}{\Delta t_n \Delta t_{n-1}} [C]\{a\}_{n-1}$$

$$- \frac{\Delta t_n}{(\Delta t_n + \Delta t_{n-1})\Delta t_{n-1}} [C]\{a\}_{n-2}$$

(E11.1)

where the variables are defined in Fig. 11.5.

11.2   Derive Eq. (11.121) from Eq. (11.118). *Hint:* Use the relation $[K]\{a\}_{ss} = \{F\}_{ss}$.

Exercises 11.3 through 11.12 form a set of 10 exercises that treat several different aspects of the 1-D transient heat conduction problem described in Exercise 11.3. Solutions to several of the exercises are shown, enabling subsequent exercises to reference these solutions and thus be done independently. The complete set covers most of the material in the chapter and provides a comparative study of various approaches. The problem has only two active DOF so that time-stepping calculations by hand will require only a modest effort. Nevertheless, if a programmable pocket calculator is available, even this effort is minimized, permitting one to easily extend the time-stepping to longer time intervals and/or shorter time steps.

11.3   A thin cylindrical rod with the lateral surface insulated is initially at $0\,°C$. (See (a) below.) An interior heat source Q, uniform along the length, is turned on at $t=0$ and remains on indefinitely. The end faces are maintained at $0\,°C$.

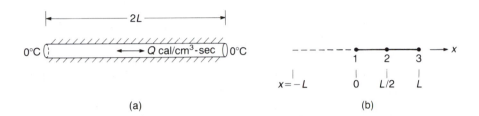

(a)   (b)

The problem has bilateral symmetry about the center so only half the rod needs to be analyzed. We will use a uniform mesh of two C^0-linear elements. (See (b) above.) Assuming constant density ϱ, specific heat c, and thermal conductivity k, show that the element stiffness matrix, consistent capacity matrix, and load vector are as follows:

$$[K]^{(e)} = \frac{k}{l}\begin{bmatrix} 1 & -1 \\ -1 & 1 \end{bmatrix} \qquad [C]^{(e)} = \frac{\varrho c l}{6}\begin{bmatrix} 2 & 1 \\ 1 & 2 \end{bmatrix}$$

$$\{F\}^{(e)} = \frac{Ql}{2}\begin{Bmatrix} 1 \\ 1 \end{Bmatrix} + \begin{Bmatrix} q(x_1) \\ -q(x_2) \end{Bmatrix} \qquad \text{(E11.2)}$$

where $l=x_2-x_1$ is the length of an element.

Show that the assembled system equations after application of the boundary conditions are as follows, depending on whether one uses a consistent or lumped capacity matrix:

Consistent •

$$\frac{\varrho c L}{12}\begin{bmatrix} 2 & 1 \\ 1 & 4 \end{bmatrix}\begin{Bmatrix} \dot{a}_1 \\ \dot{a}_2 \end{Bmatrix} + \frac{2k}{L}\begin{bmatrix} 1 & -1 \\ -1 & 2 \end{bmatrix}\begin{Bmatrix} a_1 \\ a_2 \end{Bmatrix} = \frac{QL}{4}\begin{Bmatrix} 1 \\ 2 \end{Bmatrix} \qquad \text{(E11.3)}$$

Lumped •

$$\frac{\varrho c L}{12}\begin{bmatrix} 3 & 0 \\ 0 & 6 \end{bmatrix}\begin{Bmatrix} \dot{a}_1 \\ \dot{a}_2 \end{Bmatrix} + \frac{2k}{L}\begin{bmatrix} 1 & -1 \\ -1 & 2 \end{bmatrix}\begin{Bmatrix} a_1 \\ a_2 \end{Bmatrix} = \frac{QL}{4}\begin{Bmatrix} 1 \\ 2 \end{Bmatrix} \qquad \text{(E11.4)}$$

where $a_1(t)=T(0,t)$, $a_2(t)=T(L/2,t)$, $\dot{a}_i=da_i/dt$, and $T(x,t)$ is the temperature.

11.4   Use the forward difference method to calculate several steps of the transient solution to the heat conduction problem in Exercise 11.3. Assume the values $L=10$ cm,

$\varrho = 10$ gm/cm^3, $c = 0.1$ cal/gm-°C, $k = 1$ cal/sec-cm-°C, and $Q = 2$ cal/cm^3-sec. Using Eq. (E11.4) based on the *lumped* capacity matrix, show that the recurrence relation is as follows:

$$\left\{ \begin{array}{c} a_1 \\ a_2 \end{array} \right\}_n = \left\{ \begin{array}{c} 2\Delta t \\ 2\Delta t \end{array} \right\} + \left[\begin{array}{cc} 1 - \dfrac{2\Delta t}{25} & \dfrac{2\Delta t}{25} \\ \dfrac{\Delta t}{25} & 1 - \dfrac{2\Delta t}{25} \end{array} \right] \left\{ \begin{array}{c} a_1 \\ a_2 \end{array} \right\}_{n-1}$$

(E11.5)

Estimate the critical time step for a stable solution [see Eq. (11.106)]. Using Eq. (E11.5), calculate four or five steps with $\Delta t = 10$. Does the solution appear to be stable? Repeat the analysis with $\Delta t = 15$, asking yourself the same question. You may wish to also try $\Delta t = 20$. Based on these results, determine a range that the critical time step lies within.

11.5 Repeat Exercise 11.4 but use Eq. (E11.3) based on the *consistent* capacity matrix. Show that the recurrence relation can be put in the following form:

$$\left\{ \begin{array}{c} a_1 \\ a_2 \end{array} \right\}_n = \left\{ \begin{array}{c} \dfrac{12\Delta t}{7} \\ \dfrac{18\Delta t}{7} \end{array} \right\} + \left[\begin{array}{cc} 1 - \dfrac{30\Delta t}{175} & \dfrac{36\Delta t}{175} \\ \dfrac{18\Delta t}{175} & 1 - \dfrac{30\Delta t}{175} \end{array} \right] \left\{ \begin{array}{c} a_1 \\ a_2 \end{array} \right\}_{n-1}$$

(E11.6)

The effective stiffness matrix, $(1/\Delta t)[C]$, has been inverted in Eq. (E11.6). Although this is easy for a 2-DOF problem, note that in general $[C]$ only needs to be inverted once if it does not change in time, thereby leaving an explicit recurrence relation for the remainder of the problem.

If you did not do Exercise 11.4, estimate the critical time step for a stable solution [see Eq. (11.106)]. Using Eq. (E11.6), calculate four or five steps with $\Delta t = 10$. Does the solution appear to be stable? Repeat the analysis with one or two other time steps to determine a range within which the critical time step lies.

If you did Exercise 11.4, observe the differences in the (approximate) critical time steps using the consistent and lumped capacity matrices, and note which shows better agreement with the estimated value from Eq. (11.106). The modal analysis below will shed more light on this.

11.6 Calculate the temperature at the center of the rod 10 sec after the source is turned on. Use the forward difference recurrence relation, Eq. (E11.5) in Exercise 11.4, with a time step $\Delta t = 5$ sec. Check the *convergence* of your solution at $t = 10$ sec by repeating the analysis with half the time step, $\Delta t = 2$ 1/2 sec. (Of course, this will require twice as many steps to reach $t = 10$ sec.) Does your solution appear to be converging? If you're not sure, repeat again with $\Delta t = 1$ 1/4 sec. Bear in mind that the limit as $\Delta t \to 0$ (assuming it exists) is not, in general, the exact solution but rather the approximate solution corresponding to the given *spatial mesh*. Both the mesh and the time step would have to be refined to converge to the exact solution.

The exact solution to this transient heat conduction problem [11.11, p. 130] is represented by the following Fourier series:

$$T(x,t) = \frac{Q}{2k}(L^2 - x^2) - \frac{QL^2}{k}\sum_{n=0}^{\infty}\frac{2(-1)^n}{v_n^3}e^{-(\sigma v_n^2/L^2)t}\cos(v_n x/L) \qquad \text{(E11.7)}$$

where $\sigma = k/\varrho c$ is the diffusivity and $v_n = (n+1/2)\pi$. This series converges very rapidly for $t \geq L^2/10\sigma v_0^2$, requiring only one or two terms for "engineering accuracy." Evaluate $T(0,10)$ from Eq. (E11.7) and compare with your FE solutions.

11.7 Let's use the *modal* approach [see Eqs. (11.76) through (11.85)] to analyze the 1-D heat conduction problem in Exercise 11.3. This approach usually provides considerable physical insight to a problem. Beginning with the system Eqs. (E11.3) based on the consistent capacity matrix, show that the two normalized modes and corresponding eigenvalues are as follows:

Mode 1   Mode 2

$$\{v\}_1 = \sqrt{\frac{1}{\varrho cL}\frac{6}{4+\sqrt{2}}}\begin{Bmatrix}\sqrt{2}\\1\end{Bmatrix} \qquad \{v\}_2 = \sqrt{\frac{1}{\varrho cL}\frac{6}{4-\sqrt{2}}}\begin{Bmatrix}-\sqrt{2}\\1\end{Bmatrix}$$

$$\lambda_1 = \frac{24(5-3\sqrt{2})\sigma}{7L^2} \qquad\qquad \lambda_2 = \frac{24(5+3\sqrt{2})\sigma}{7L^2} \qquad \text{(E11.8)}$$

The Fourier series representation of the exact solution in Eq. (E11.7) is an expansion in terms of the thermal modes. Hence the two modes in Eq. (E11.8) are the FE approximation to the first two terms in that infinite series. Using the Fourier terms, compare the accuracies of the eigenvalues and mode shapes of the above two FE modes.

11.8 Using the physical properties in Exercise 11.4, calculate the maximum eigenvalue, λ_{max}, from Eq. (E11.8). What is the critical time step for the forward difference method? Note that this is the exact value for Δt_{crit}, and it is based on a *consistent* capacity matrix. Compare this with the estimated value from Eq. (11.106).

Recalculate λ_{max} in the same manner as was done above, but this time use the system Eqs. (E11.4) based on the *lumped* capacity matrix. Now what is Δt_{crit} for the forward difference method? Note that Δt_{crit} using the consistent capacity matrix is in much closer agreement with the approximate formula in Eq. (11.106), since the latter is based on the original (consistent) FE formulation.

These results explain the regimes of stable and unstable behavior observed in Exercises 11.4 and 11.5. In general, lumping tends to decrease the eigenvalues of the FE model, especially the higher ones. This frequently yields greater accuracy since an FE model is inherently too "stiff" because of the discretization, especially the highest modes. Thus, lumping is often said to "soften" the model. Of course, a lower λ_{max} means a larger Δt_{crit}.

11.9 Complete the modal analysis begun in Exercises 11.7 and 11.8, employing the forward difference method to compute the temperature history at the center of the rod ($x=0$). Use the eigenvalues and eigenvectors from Eq. (E11.8) (which are based on the consistent capacity matrix) in Eq. (11.85) to calculate the modal amplitudes and in Eq. (11.82) to calculate the temperature.

If you did not do Exercise 11.8, calculate the critical time step. Then try $\Delta t = 10$ sec. This is greater than Δt_{crit} so the solution will be unstable. Which mode is the source of the instability? Recall that the nonmodal analysis in Exercise 11.5 with $\Delta t = 10$ was unstable. Here we can clearly see the source of the instability.

Now try $\Delta t = 5$ sec, which is less than Δt_{crit}. Calculate the temperature at 20 sec. Repeat the analysis, halving Δt, to check convergence and estimate accuracy.

11.10 Calculate the temperature history at the center of the rod ($x=0$) from 0 to 100 sec. Use the *mid*-difference method with a 20-sec time step. Repeat the analysis using the *backward* difference method with the same time step. Also evaluate the exact solution at the end of each time step from Eq. (E11.7). Plot both FE solutions as well as the exact solution. Compare the accuracies of the two FE solutions; recall the observation on their relative accuracies following Eq. (11.48). Note that the mid-difference solution is not oscillatory, although, as observed in the text, it often is. This is due to the "gentleness" in the application of the thermal loads, which results in a relatively insignificant excitation of the higher thermal modes, the latter being the usual source of oscillations. (Exercise 11.9 illustrates the relative role played by the two modes in this problem. Exercise 11.12 illustrates how this role changes when the thermal loads are applied suddenly rather than gently.)

11.11 Use the *two-step* backward difference recurrence relation in Exercise 11.1 to calculate the temperature history at the center of the rod ($x=0$) from 0 to 100 sec. Employ 20-sec time steps. Recall from the discussion following Eq. (11.36) that the two-step method requires values for $\{a\}_0$ and $\{a\}_1$ in order to begin, although the initial conditions can only supply values for $\{a\}_0$. Therefore use the one-step backward difference relation to determine $\{a\}_1$. If you did Exercise 11.10, compare the relative accuracies of the one-step and two-step backward difference methods.

Optional: The two-step method is of higher order than the one-step, so it will converge at a faster rate. Repeat the analysis using 10-sec time steps and observe the improvement in accuracy at $t=40$, 60, 80, and 100 sec, relative to the solution using 20-sec steps.

11.12 Consider the same cylindrical rod as in the previous exercises, i.e., having the same geometry and physical properties, but change the thermal loads to the following:

$$\text{Initial conditions: } T(x,0) = 100\,°C$$
$$\text{Boundary conditions: } T(\pm L,t) = 0\,°C \qquad \text{(E11.9)}$$

Thus, at $t=0$ the temperature of the end faces is suddenly dropped from 100 to $0\,°C$. There is no interior heat source. Calculate the temperature history at the center of the rod ($x=0$) from 0 to 100 sec, using the mid-difference method with 20-sec time steps. Compare your answers with the exact solution [11.11, p. 97, Exercise 11.6]:

$$T(x,t) = 100 \sum_{n=0}^{\infty} \frac{2(-1)^n}{v_n} e^{-(\sigma v_n^2/L^2)t} \cos (v_n x/L) \qquad \text{(E11.10)}$$

where $\sigma = k/\varrho c$ and $v_n = (n+1/2)\pi$.

N.B.: Since the geometry and physical properties are the same as before, the natural modes for this problem are the same as before, as evidenced by the identical eigenvalues and eigenmodes in Eqs. (E11.7) and (E11.10). By using the same FE mesh, you can therefore use the FE eigenvalues and eigenmodes derived previously in Exercise 11.7, thereby simplifying your computational work. This is one of the principal motivations for doing a modal analysis: Once the modes are derived, they can be used repeatedly for different loading conditions.

The loads in this problem are applied suddenly and hence they excite the higher modes to a greater degree than a gently applied load. [Compare the amplitudes of corresponding modes in Eqs. (E11.7) and (E11.10).] Note that, as a result, the mid-difference solution shows more oscillation than in Exercise 11.10.

This completes the set of 10 exercises associated with the problem described in Exercise 11.3.

11.13 Derive the following governing differential equation for 1-D transient heat conduction in the *radial* direction in cylindrical coordinates:

$$\varrho c \frac{\partial T}{\partial t} - \frac{1}{r} \frac{\partial}{\partial r} \left(kr \frac{\partial T}{\partial r} \right) + \frac{h}{\Delta z} T = Q + \frac{h}{\Delta z} T_\infty \qquad \text{(E11.11)}$$

For the derivation (recall Section 11.1.3) apply balance of energy to a differential element of volume $\Delta z \Delta \theta \, r \, dr$:

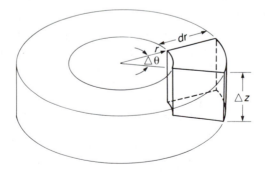

Include convective heat loss from the z-surfaces, using h to represent the sum of the upper and lower convection coefficients. (This will permit application to a thin slab of material, as in the illustrative example in Section 15.4. See especially the shock response in Section 15.4.3.) The constitutive equation (Fourier's law) is $q_r = -k\partial T/\partial r$, where q_r is the flux in the radial direction.

11.14 Develop a C^0-linear element for radial flow of heat in cylindrical coordinates:

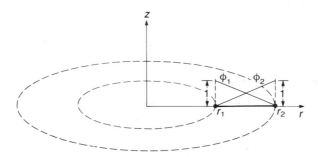

Using the governing differential equation in Exercise 11.13, show that the element equations are as follows:

$$
\varrho c \frac{\Delta r}{6}
\begin{bmatrix}
r_1 + r_m & r_m \\
r_m & r_2 + r_m
\end{bmatrix}
\begin{Bmatrix}
\dot{a}_1 \\
\dot{a}_2
\end{Bmatrix}
+ k \frac{r_m}{\Delta r}
\begin{bmatrix}
1 & -1 \\
-1 & 1
\end{bmatrix}
\begin{Bmatrix}
a_1 \\
a_2
\end{Bmatrix}
$$

$$
+ \frac{h}{\Delta z} \frac{\Delta r}{6}
\begin{bmatrix}
r_1 + r_m & r_m \\
r_m & r_2 + r_m
\end{bmatrix}
\begin{Bmatrix}
a_1 \\
a_2
\end{Bmatrix}
= Q' \frac{\Delta r}{6}
\begin{Bmatrix}
r_1 + 2r_m \\
r_2 + 2r_m
\end{Bmatrix}
+
\begin{Bmatrix}
(qr)_{r_1} \\
-(qr)_{r_2}
\end{Bmatrix}
$$

$$\textbf{(E11.12)}$$

where $\Delta r = r_2 - r_1$, $r_m = (r_1 + r_2)/2$, and $Q' = Q + \dfrac{h}{\Delta z} T_\infty$.

Note that even though the finite element is one-dimensional, it may be thought of as representing a solid of revolution, as suggested by the dashed lines in the figure, with the nodes being nodal circles. Thus it is an axisymmetric element (sometimes referred to as a *ring element*). This provides a clue for the above element derivation: Integrate the differential equation residual over the three-dimensional solid of revolution.

11.15 Consider a thin circular disc of radius 10 cm, insulated on the upper and lower surfaces. (See (a) below.) The disc is initially at $0\,°C$. A uniform interior heat source Q is turned on at $t=0$ and remains on indefinitely. The edge of the disc is maintained at $0\,°C$. The physical properties are $\varrho=10$ gm/cm^3, $c=0.1$ cal/gm-$°C$, $k=1$ cal/sec-cm-$°C$, and $Q=2$ cal/cm^3-sec. (These properties are deliberately made the same as for the rod in Exercises 11.3 through 11.12 so that the effects of linear and radial heat flow can be easily compared.)

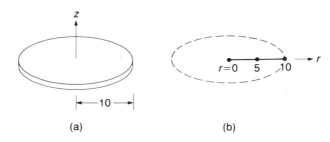

(a)   (b)

Using the element equations in Eq. (E11.12) for axisymmetric elements, employ a mesh of two uniform elements. (See (b) above.) Use the mid-difference method with time steps $\Delta t = 20$ sec to calculate the temperature at the center of the disc at $t = 20$ sec and $t = 40$ sec. Then jump to the steady-state solution with a step $\Delta t = 1000$ sec. Do another 1000-sec step to check convergence. Recall the discussion associated with Eq. (11.121). Repeat the analysis using the backward difference method and the same time-stepping strategy. Which method is better suited for jumping directly to the steady-state solution in one large step?

The exact solution [11.11, p. 204, Exercise 11.6] is

$$T(r,t) = \frac{Q(a^2 - r^2)}{4k} - \frac{Q}{ak} \sum_{n=1}^{\infty} \frac{2}{\zeta_n^3} e^{-\sigma \zeta_n^2 t} \frac{J_0(r\zeta_n)}{J_1(a\zeta_n)} \tag{E11.13}$$

where $\sigma = k/\varrho c$, a is the radius of the disc, J_0 and J_1 are the Bessel functions of order 0 and 1, respectively, and ζ_n are the roots of $J_0(a\zeta_n) = 0$.

References

[11.1] G. J. Etgen and W. L. Morris, *Introduction to Ordinary Differential Equations: with Difference Equations, Numerical Methods, and Applications*, Harper & Row, New York, 1977.

[11.2] N. Finizio and G. Ladas, *Ordinary Differential Equations with Modern Applications*, Wadsworth, Belmont, Calif., 1982.

[11.3] W. Kaplan, *Ordinary Differential Equations*, Addison-Wesley, Reading, Mass., 1958.

[11.4] A. N. Tikhonov and A. A. Samarskii, *Equations of Mathematical Physics*, Pergamon, New York, 1963, Chapter 1.

[11.5] J. D. Lambert, *Computational Methods in Ordinary Differential Equations*, Wiley, New York, 1973.

[11.6] G. Dahlquist and A. Bjorck, *Numerical Methods*, Prentice-Hall, Englewood Cliffs, N. J., 1974.

[11.7] P. Henrici, *Discrete Variable Methods in Ordinary Differential Equations*, Wiley, New York, 1962.

[11.8] R. D. Richtmyer and K. W. Morton, *Difference Methods for Initial-Value Problems*, Wiley-Interscience, New York, 1967.

[11.9] I. Fried, *Numerical Solution of Differential Equations*, Academic Press, New York, 1979.

[11.10] G. J. Etgen and W. L. Morris, op. cit.

[11.11] H.S. Carslaw and J.C. Jaeger, *Conduction of Heat in Solids*, 2d ed., Oxford University Press, New York, 1959.

[11.12] J. C. Cavendish, C. A. Hall, and O. C. Zienkiewicz, "Blended Infinite Elements for Parabolic Boundary-Value Problems," *Int J Num Meth Eng*, **12**: 1841-1851 (1978).

[11.13]   J. Mathews and R. L. Walker, *Mathematical Methods of Physics*, Benjamin, New York, 1965, p. 218.

[11.14]   L. V. Kantorovitch and V. I. Krylov, *Approximate Methods of Higher Analysis*, Interscience, New York, 1964.

[11.15]   J. D. Lambert, op. cit., Chapters 2 and 3.

[11.16]   P. C. Kohnke, *ANSYS Engineering Analysis System Theoretical Manual*, Swanson Analysis Systems, Inc., Houston, Pa., 1977 (Revised 1983).

[11.17]   F. J. H. Peeters, "Thermal Analysis by Means of Finite Element Methods," *Seminar on Advanced Topics in Nonlinear Finite Element Analysis*, MARC Analysis Research Corp., Palo Alto, Calif., 1981.

[11.18]   W. L. Wood, "On the Zienkiewicz Three- and Four-Time Level Schemes Applied to the Numerical Integration of Parabolic Equations," *Int J Num Meth Eng*, **12**: 1717-1726 (1978).

[11.19]   J. D. Lambert, op. cit., pp. 13-15, 28, 60.

[11.20]   O. C. Zienkiewicz, *The Finite Element Method*, McGraw-Hill, New York, 1977, p. 535.

[11.21]   P. Tong, T. H. H. Pian, and L. L. Bucciarelli, "Mode Shapes and Frequencies by Finite Element Method Using Consistent and Lumped Masses," *Computers and Structures*, **1**: 623-638 (1971).

[11.22]   I. Fried and D. S. Malkus, "Finite Element Mass Matrix Lumping by Numerical Integration with No Convergence Rate Loss," *Int J Solids Structures*, **11**: 461-466 (1975).

[11.23]   E. Hinton, T. Rock, and O. C. Zienkiewicz, "A Note on Mass Lumping and Related Processes in the Finite Element Method," *Earthquake Eng Struct Dyn*, **4**: 245-249 (1976).

[11.24]   R. W. Clough, "Analysis of Structural Vibrations and Dynamic Response" in R. H. Gallagher, Y. Yamada and J. T. Oden (eds.), *Recent Advances in Marix Methods of Structural Analysis and Design*, University of Alabama Press, University, Alabama, 1971, pp. 441-482.

[11.25]   O. C. Zienkiewicz, op. cit., Chapter 21.

[11.26]   R. H. Gallagher and R. H. Mallett, *Efficient Solution Processes for Finite Element Analysis of Transient Heat Conduction*, Bell Aerosystems, Buffalo, N. Y, 1969.

[11.27]   B. Irons and S. Ahmad, *Techniques of Finite Elements*, Halsted (Wiley), New York, 1980, p. 349.

[11.28]   J. Mathews and R. L. Walker, op. cit., p. 225.

[11.29]   M. E. Weber, "Improving the Accuracy of Crank-Nicolson Numerical Solutions to the Heat-Conduction Equation," *J Heat Transfer, Trans ASME*, February 1969, pp. 189-191.

[11.30]   W. L. Wood and R. W. Lewis, "A Comparison of Time Marching Schemes for the Transient Heat Conduction Equation", *Int J Num Meth Eng*, **9**: 679-689 (1975).

[11.31]   O. C. Zienkiewicz, *The Finite Element Method in Engineering Science*, McGraw-Hill, New York, 1971, p. 336. [Equation (16.49) has several errors in earlier printings.]

[11.32] M. Zlamal, ''Finite Element Methods in Heat Conduction Problems'' in J. Whiteman (ed.), *The Mathematics of Finite Elements and Applications II*, Academic Press, New York, 1977, pp. 85-104.

[11.33] W. Liniger, ''Optimization of a Numerical Integration Method for Stiff Systems of Ordinary Differential Equations,'' *IBM Research Report RC2198*, 1968.

[11.34] M. B. Brahmanandam and B. N. Chatterji, ''Lumped Model of a Non-Linear Distributed System by Finite Element Method,'' *Int J Systems Sci*, **13**(10): 1117-1123 (1982).

12

1-D Initial-Boundary-Value Problems: Dynamics

Introduction

The previous chapter dealt with diffusion problems, which are characterized by governing differential equations that contain a first-order time derivative, $\partial U/\partial t$. This chapter deals with dynamics problems, which are characterized by governing differential equations that contain a *second*-order time derivative, $\partial^2 U/\partial t^2$. The latter term is due to *inertial* forces, which are an essential feature of a dynamics problem. Inertial forces are the source of kinetic energy. If the structure or medium can also store potential energy (e.g., by means of elastic deformation and/or changing position in a gravitational field), then energy can oscillate between kinetic and potential, yielding motion referred to as *wave propagation* or *vibration*.[1] The motion of solids is usually referred to simply as *dynamics*, which may include solids idealized as nondeformable ("rigid-body dynamics") or deformable ("elastodynamics"). Terms such as "hydrodynamics," "fluid dynamics," "aerodynamics," and "acoustics" are used for various theories of motion of liquids and gases.

The purpose of this chapter is to present an overview that simply acquaints the reader with some of the concepts and methodologies currently used in the FE dynamic analysis of solid structures. A more detailed development would exceed the scope of this book, since dynamics is a many-faceted subject that could easily fill a separate tome. Indeed, many books and a vast number of journal articles have been written on the topic. The references cited herein are only a small sample of all the published work; these references contain their own lists of references, which collectively provide a more substantial bibliography.

As shown in Fig. 12.1, there are a variety of different, though mathematically related methods that are used to solve dynamics problems, depending primarily on the nature of

---

[1] The words "wave propagation" and "vibration" refer to the same physical phenomenon but with different emphasis. The former focuses on how energy travels in *space* (through the structure or medium), whereas the latter focuses on how the position of the structure or medium changes in *time*.

the loading. And within the different methods there may be a variety of specific techniques (as shown, for example, in Section 12.2).

Perhaps the first and broadest level of classification distinguishes *deterministic* and *nondeterministic* (or *random*) loads. The former deal with any load for which the variation in time (the *time history*) can be specified, whereas the latter deal with loads that can only be characterized by statistical properties — e.g., mean, variance, root-mean-square (RMS), and power spectral density. When using these adjectives, it is important to distinguish between the actual physical process and how that process is being modeled. Thus, even though a physical process may be random in nature, one could experimentally measure a particular event and use the resulting data in a deterministic analysis.

For example, an earthquake or the surface of the sea are random processes, meaning that the variables which determine the processes are not under control of the observer, and therefore the time histories are neither predictable nor repeatable from one earthquake or sea state to the next. However, if one has experimentally recorded the time history of a particular earthquake or sea state, then the loads (e.g., accelerations) produced by the earthquake or sea state on a building or ship are a known function of time and one can perform a deterministic analysis (see Example 1.2 in Chapter 1). Such an analysis can provide insight into the physics of a problem. For design purposes, though, one may wish to also perform a random analysis, using the statistical properties of several recorded time histories in order to predict statistical properties of future events (as was done subsequent to the work shown in Example 1.2). Since random-vibration analysis is a special-

Loads		
Deterministic: known variation in time		Nondeterministic: unknown variation in time, but statistical properties are known; usually called *random* loads

Nonperiodic: arbitrary variation in time
 	
• Direct integration; time-stepping using recurrence relations
 	
• Mode superposition; transformation to natural modes, then direct integration
 	
• Fast Fourier transform; uses modes and frequency-response functions
	Periodic: variation in time described by one or more sinusoids
	
	• Harmonic analysis (1 sinusoid at 1 frequency);
	  • frequency-response function
	
	• Nonharmonic analysis; sum of two or more sinusoids at different frequencies
	  • Fourier series: sum of two or more frequency-response functions
		Typical calculations include:
		
		• Power spectral densities
		• Autocorrelation functions
		• Root-mean-square values
		
		Use modes and frequency-response functions

FIGURE 12.1 ● Classification of loads, and corresponding methods of analysis.

ized topic that is beyond the scope of this text [12.1-12.4], we turn our attention now to deterministic methods.

Deterministic loads may be further classified as either *nonperiodic* or *periodic*. The three nonperiodic analysis methods shown in Fig. 12.1 are used to calculate the *transient* response, i.e., a time history of the solution, $\tilde{U}(x,t)$. (Sections 11.3 and 11.5 dealt with the transient response for diffusion problems.) Of these, direct integration and mode superposition are the most widely used in current commercial FE programs, and hence will be discussed in Sections 12.2 and 12.3, respectively. The Fast Fourier Transform (FFT) approach [12.4, 12.5] has not generally caught on in FE structural analysis, although there have been some impressive applications [12.6, 12.7]. Note the overlap in the three methods; i.e., the FFT uses modes, and mode superposition uses direct integration.

Periodic loads are typically generated by devices of human origin, such as machinery containing rotating or reciprocating parts. The simplest of all periodic loads is one that varies sinusoidally in time at a single frequency, called a *harmonic* (or *simple-harmonic*) load. The response of a (linear) structure to a harmonic load, as a function of frequency, is described by a *frequency-response function*. A harmonic load rarely occurs in actual applications, although it is often used in experiments. Its importance, both theoretically and experimentally, lies in the fact that any arbitrary load can be represented as a sum (Fourier series) or integral (Fourier transform) of harmonic loads at different frequencies. And if a structure behaves linearly, then the structural response to an arbitrary load is the sum (or integral) of frequency-response functions. The frequency-response function therefore plays an important role in other types of analyses, including even random analysis, as shown in Fig. 12.1. Section 12.4 will treat harmonic analysis.

Based on this overall perspective, it would appear that the fundamentally important topics are direct integration, mode superposition, and harmonic analysis. These topics will therefore be explained in Sections 12.2 through 12.4, following a preliminary development in Section 12.1, which applies the usual FE discretization to convert the governing differential equation to a system of coupled ordinary differential equations in time.

12.1 Derivation of Element and System Equations

The development here is virtually identical to that in Sections 11.1.2 and 11.2, the only difference being the addition of an extra term containing the second derivative with respect to time. Thus the governing differential equation and associated boundary conditions and initial conditions are as follows:

Differential equation •

$$\varrho(x)\,\frac{\partial^2 U(x,t)}{\partial t^2} + \mu(x)\,\frac{\partial U(x,t)}{\partial t} - \frac{\partial}{\partial x}\left(\alpha(x)\,\frac{\partial U(x,t)}{\partial x}\right) + \beta(x)U(x,t) = f(x,t)$$

$$(12.1a)$$

Domain •

$$x_a < x < x_b$$

$$t > t_0 \qquad\qquad (12.1b)$$

Boundary conditions •

At x_a $(t > t_0)$

$$U(x_a,t) = U_a(t) \qquad \text{essential BC}$$

or

$$\left(-\alpha(x)\ \frac{\partial U(x,t)}{\partial x}\right)_{x_a} = \tau_a(t) \qquad \text{natural BC} \tag{12.1c}$$

At x_b $(t > t_0)$

$$U(x_b,t) = U_b(t) \qquad \text{essential BC}$$

or

$$\left(-\alpha(x)\ \frac{\partial U(x,t)}{\partial x}\right)_{x_b} = \tau_b(t) \qquad \text{natural BC}$$

Initial conditions •

At t_0 $(x_a < x < x_b)$

$$U(x,t_0) = U_0(x)$$

$$\left(\frac{\partial U(x,t)}{\partial t}\right)_{t_0} = V_0(x) \tag{12.1d}$$

These equations are the same as Eqs. (11.11) for the diffusion problem except for the extra term, $\varrho \partial^2 U/\partial t^2$, and the extra initial condition,[2] $\partial U/\partial t = V_0$.

Regarding terminology, in solid mechanics applications U, $\partial U/\partial t$, and $\partial^2 U/\partial t^2$ represent displacement, velocity, and acceleration, respectively. ϱ is a density (mass/volume) and $\varrho \partial^2 U/\partial t^2$ an inertial force (per volume); μ is a viscosity and $\mu \partial U/\partial t$ a damping force. U may have other meanings in other applications (e.g., pressure or velocity potential in acoustics), but $\partial U/\partial t$ and $\partial^2 U/\partial t^2$, multiplied perhaps by physical constants, usually retain the meaning of velocity and acceleration, and $\mu \partial U/\partial t$ and $\varrho \partial^2 U/\partial t^2$ usually refer to damping and inertial effects, respectively. Consequently, in the remainder of this chapter we will associate $\partial/\partial t$, $\partial^2/\partial t^2$, μ, and ϱ with velocity, acceleration, damping, and mass, respectively.

To derive the element equations we begin with the usual element trial-solution formulation and separate the space and time variables as in Eq. (11.18); that is,

$$\tilde{U}^{(e)}(x,t;a) = \sum_{j=1}^{n} a_j(t)\phi_j^{(e)}(x) \tag{12.2}$$

We then apply steps 1 through 6 to the differential Eq. (12.1a). The results are the same as in Eqs. (11.19) through (11.32) except for the additional mass term, as the reader can

---

[2] More generally, if the highest-order time derivative in Eq. (12.1a) were of order n, then n initial conditions would be needed, specifying U, $\partial U/\partial t, \dots, \partial^{n-1} U/\partial t^{n-1}$ at time t_0. (Derivatives of order n and greater at time t_0 are calculable from the differential equation itself by repeatedly differentiating the latter and evaluating at t_0.)

easily verify. For example, at the end of step 3 the element equations are as follows:

$$[M]^{(e)}\left\{\frac{d^2a(t)}{dt^2}\right\} + [C]^{(e)}\left\{\frac{da(t)}{dt}\right\} + [K]^{(e)}\{a(t)\} = \{F(t)\}^{(e)} \qquad (12.3a)$$

where the new mass matrix has the form,

$$M_{ij}^{(e)} = \int^{(e)} \phi_i^{(e)}(x)\varrho(x)\phi_j^{(e)}(x)\,dx \qquad (12.3b)$$

and $C_{ij}^{(e)}$, $K_{ij}^{(e)}$, $F_i^{(e)}$ remain the same as in Eq. (11.24b) (although $[C]^{(e)}$ is now called a *damping matrix* rather than a capacity matrix). Note that the mass and damping matrices both have the same form with respect to the shape functions.

In step 4 one could use any of the four 1-D elements developed in Chapters 7 and 8. In step 5, the expressions for the mass matrix terms would be the same as in Eqs. (11.25) to (11.29), exchanging μ for ϱ, since $C_{ij}^{(e)}$ and $M_{ij}^{(e)}$ have identical integral expressions. And, of course, the flux expressions in step 6 are unaffected by the time derivatives.

Assembly of the element equations proceeds as usual, resulting in the following system equations:

$$[M]\{\ddot{a}\} + [C]\{\dot{a}\} + [K]\{a\} = \{F\} \qquad (12.4)$$

Here we have adopted the standard shorthand notation of representing time differentiation by a dot; that is, $\{\dot{a}\} = \{da/dt\}$ and $\{\ddot{a}\} = \{d^2a/dt^2\}$. Equation (12.4) is a system of coupled second-order ordinary differential equations. The initial conditions on $\{a\}$ follow directly from Eqs. (12.1d) and (12.2); they may be expressed symbolically as $\{a\}_0$ and $\{\dot{a}\}_0$, which represent $\{a(t_0)\}$ and $\{\dot{a}(t_0)\}$, respectively. Sections 12.2 through 12.4 present methods for obtaining solutions to Eq. (12.4).

12.2 Direct Integration

This and the next section will deal with two common methods for obtaining a transient solution: direct integration and mode superposition. Both methods must, of course, yield the same solution in the limit of arbitrarily fine discretization, any differences in practice being attributable to discretization and round-off errors. The choice between the two is primarily one of computational efficiency and depends on the nature of the particular prob-

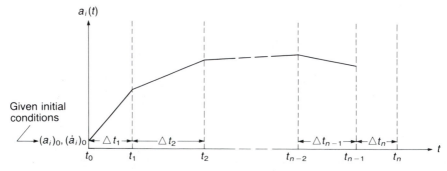

FIGURE 12.2 ● Notation used for time-stepping recurrence relations.

lem, namely, the system properties, loads, and desired information. Guidelines will be given at the end of Section 12.3.

Direct integration subsumes a broad spectrum of time-stepping methods. Section 11.3 discussed in some detail a particular class of time-stepping methods suitable for the first order initial-value (diffusion) problem, and the reader will recognize in the material below many similarities to the diffusion problem. It may be worthwhile at this point for the reader to review the basic ideas of time-stepping in Section 11.3.1 since they apply to the present section as well. Figure 12.2 defines the notation; it is the same as shown in Fig. 11.5 except for the extra initial condition.

We seek a recurrence relation that relates the unknown values at t_n to the known values at one or more previous times — t_{n-1}, t_{n-2}, etc. Four of the most popular recurrence relations currently used in commercial FE programs [12.8] are explained below. These four have been the workhorses for FE dynamics problems for many years [12.9-12.23].

12.2.1 The Central Difference Method

This is conceptually the simplest recurrence relation. It may be found in almost any elementary numerical analysis text dealing with finite difference techniques. As shown in Fig. 12.3, this is a two-step method involving three times: t_{n-2}, t_{n-1}, and t_n.

The system equations [Eqs. (12.4)] are evaluated at the central time, t_{n-1}:

$$[M]\{\ddot{a}\}_{n-1} + [C]\{\dot{a}\}_{n-1} + [K]\{a\}_{n-1} = \{F\}_{n-1} \qquad (12.5)$$

The two derivatives are then approximated by central difference expressions,[3] and it is assumed that both steps are of the same length; that is, $\Delta t = t_n - t_{n-1} = t_{n-1} - t_{n-2}$. Thus

$$\{\dot{a}\}_{n-1} \simeq \frac{\{a\}_n - \{a\}_{n-2}}{2\Delta t}$$

$$\{\ddot{a}\}_{n-1} \simeq \frac{\{\dot{a}\}_{n-1/2} - \{\dot{a}\}_{n-3/2}}{\Delta t}$$

$$= \frac{\dfrac{\{a\}_n - \{a\}_{n-1}}{\Delta t} - \dfrac{\{a\}_{n-1} - \{a\}_{n-2}}{\Delta t}}{\Delta t}$$

$$= \frac{\{a\}_n - 2\{a\}_{n-1} + \{a\}_{n-2}}{\Delta t^2} \qquad (12.6)$$

Substituting Eqs. (12.6) into Eq. (12.5) and placing all known terms on the RHS yields the central difference recurrence relation:

$$\left(\frac{1}{\Delta t^2}[M] + \frac{1}{2\Delta t}[C]\right)\{a\}_n =$$

$$\{F\}_{n-1} - \left([K] - \frac{2}{\Delta t^2}[M]\right)\{a\}_{n-1} - \left(\frac{1}{\Delta t^2}[M] - \frac{1}{2\Delta t}[C]\right)\{a\}_{n-2} \qquad (12.7)$$

---

[3] This is the traditional and simplest derivation. Another approach is to use an interpolation polynomial, as shown in Eq. (12.13).

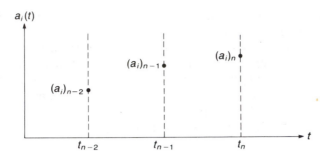

FIGURE 12.3 ● Data used in the central difference recurrence relation; all three values are interpolated.

Equation (12.7) is a system of linear algebraic equations with an effective stiffness matrix

$$[K_{eff}] = \frac{1}{\Delta t^2}[M] + \frac{1}{2\Delta t}[C] \tag{12.8}$$

Since $[K]$ is not present in $[K_{eff}]$, we can diagonalize $[K_{eff}]$ by lumping both $[M]$ and $[C]$ (recall the forward difference method in Section 11.3.2), thereby uncoupling the equations and making the method *explicit*.

The explicit central difference method is computationally very fast. This efficiency advantage is offset by its being only conditionally stable; i.e., the method becomes unstable unless the time-step size is less than a critical value:

$$\Delta t \leq \Delta t_{crit} = \frac{2}{\omega_{max}} \qquad \text{for stability} \tag{12.9}$$

where $f_{max} = \omega_{max}/2\pi$ is the maximum natural frequency of the FE model. Recalling Eq. (10.10), $\omega_{max} = \sqrt{\lambda_{max}}$, where λ_{max} is the maximum eigenvalue of the FE model. Hence

$$\Delta t_{crit} = \frac{2}{\sqrt{\lambda_{max}}} \tag{12.10}$$

We can estimate Δt_{crit} by using the approximate formula in either Eq. (11.100), (11.101), or (11.104), replacing μ by ϱ. For example, if α and ϱ are constant over the problem domain, then for a 1-D problem Eqs. (11.101) and (12.10) would yield

$$\Delta t_{crit} \simeq \frac{2}{\pi}\sqrt{\frac{\varrho}{\alpha}}\,\delta_{min} \tag{12.11}$$

Comparing Eqs. (12.11) and (11.106), it should be noted that Δt_{crit} is proportional to δ_{min} for the dynamics problem but to δ_{min}^2 for the diffusion problem, making mesh refinement less costly for the dynamics problem.

The central difference method is two-step, but initial conditions are specified at only one time t_0, namely, $\{a\}_0$ and $\{\dot{a}\}_0$. Therefore a starting procedure is required. For this, one could use the relation

$$\{a\}_{-1} = \{a\}_0 - \Delta t\{\dot{a}\}_0 \tag{12.12}$$

and then substitute the values $\{a\}_{-1}$ and $\{a\}_0$ into Eq. (12.7) evaluated at $n=1$.

The reader has probably noted the strong parallels between the explicit forward difference method for the diffusion problem and the explicit central difference method for the dynamics problem. Indeed, the latter has been most effectively used for calculating the immediate transient response to shock loads. Underwood and Park [12.24] describe a program for implementing the method.

It is instructive to note that the central difference recurrence relation [Eq. (12.7)] can also be derived by using a quadratic interpolation polynomial through the values of $\{a\}$ at times t_{n-2}, t_{n-1}, and t_n; that is,

$$\{a(t)\} = \{a\}_{n-2}\phi_{n-2}(t) + \{a\}_{n-1}\phi_{n-1}(t) + \{a\}_n\phi_n(t) \tag{12.13}$$

where $\phi_{n-2}(t)$, $\phi_{n-1}(t)$, and $\phi_n(t)$ are the quadratic shape functions (replacing x with t) in Fig. 8.2 (see Exercise 12.1). This permits different time-step sizes; that is, $\Delta t_n = t_n - t_{n-1}$ and $\Delta t_{n-1} = t_{n-1} - t_{n-2}$. It also introduces a general approach that can be extended to many other applications. [Recall Eqs. (11.63).] For example, evaluating the system equations [Eqs. (12.4)] at time t_n, rather than t_{n-1}, and calculating $\{a\}_n$, $\{\dot{a}\}_n$, and $\{\ddot{a}\}_n$ from Eq. (12.13) will yield an unconditionally stable two-step backward difference formula (see Exercise 12.2). The interpolation polynomial approach will also be used below for the Houbolt and Wilson methods and for part of the Newmark method.

12.2.2 The Houbolt Method

As shown in Fig. 12.4, this is a three-step method [12.9] in which $\{a(t)\}$ is approximated by a cubic polynomial that interpolates $\{a(t)\}$ at times t_{n-3}, t_{n-2}, t_{n-1}, and t_n; that is,

$$\{a(t)\} = \{a\}_{n-3}\phi_{n-3}(t) + \{a\}_{n-2}\phi_{n-2}(t) + \{a\}_{n-1}\phi_{n-1}(t) + \{a\}_n\phi_n(t) \tag{12.14}$$

where $\phi_{n-3}(t),\dots,\phi_n(t)$ are the cubic shape functions in Fig. 8.37 (replacing ξ with t). The system equations [Eqs. (12.4)] are evaluated at the forward time t_n:

$$[M]\{\ddot{a}\}_n + [C]\{\dot{a}\}_n + [K]\{a\}_n = \{F\}_n \tag{12.15}$$

Differentiating Eq. (12.14) once and twice with respect to t, then evaluating at t_n, and finally substituting into Eq. (12.15) and placing all known terms on the RHS yields the

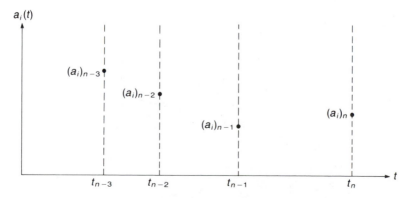

FIGURE 12.4 ● Data used in the Houbolt recurrence relation; all four values are interpolated.

Houbolt recurrence relation:

$$\left(\frac{2}{\Delta t^2}[M] + \frac{11}{6\Delta t}[C] + [K]\right)\{a\}_n = \{F\}_n$$

$$+ \left(\frac{5}{\Delta t^2}[M] + \frac{3}{\Delta t}[C]\right)\{a\}_{n-1} - \left(\frac{4}{\Delta t^2}[M] + \frac{3}{2\Delta t}[C]\right)\{a\}_{n-2} + \left(\frac{1}{\Delta t^2}[M] + \frac{1}{3\Delta t}[C]\right)\{a\}_{n-3}$$

$$\text{(12.16)}$$

Since Eq. (12.14) is a cubic polynomial, the velocity and acceleration are quadratic and linear polynomials, respectively. For example, differentiating Eq. (12.14) twice [using the cubic shape functions for $\phi_i(t)$] yields

$$\{\ddot{a}(t)\} = \{\ddot{a}\}_{n-2}\left(\frac{t_{n-1}-t}{\Delta t}\right) + \{\ddot{a}\}_{n-1}\left(\frac{t-t_{n-2}}{\Delta t}\right) \qquad \text{(12.17)}$$

where

$$\{\ddot{a}\}_{n-1} = \frac{\{a\}_n - 2\{a\}_{n-1} + \{a\}_{n-2}}{\Delta t^2}$$

$$\{\ddot{a}\}_{n-2} = \frac{\{a\}_{n-1} - 2\{a\}_{n-2} + \{a\}_{n-3}}{\Delta t^2}$$

Equation (12.17) linearly interpolates the accelerations at times t_{n-1} and t_{n-2}, as shown in Fig. 12.5. Observe that $\{\ddot{a}\}_{n-1}$ and $\{\ddot{a}\}_{n-2}$ use the central difference expression in Eq. (12.6).

The recurrence relation in Eq. (12.16) assumes uniform time steps — i.e., $\Delta t = t_n - t_{n-1} = t_{n-1} - t_{n-2} = t_{n-2} - t_{n-3}$ — and in this form it is identical to the original Houbolt recurrence relation. However, uniform steps are clearly not necessary, since the above derivation easily accommodates variable time steps. Such a "generalized Houbolt" relation is used in a large commercial FE program (see Exercise 12.3).

Since the system equations are evaluated at the forward time t_n, all the difference expressions for the derivatives at time t_n involve previous time steps. Hence this is sometimes described as a backward difference method. [Recall the footnote below Eq. (11.42).] In fact, it behaves rather similarly to the one-step backward difference method

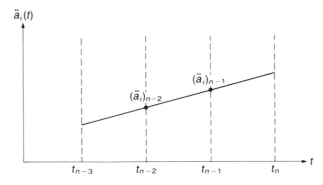

FIGURE 12.5 ● Linear acceleration assumed by the Houbolt method.

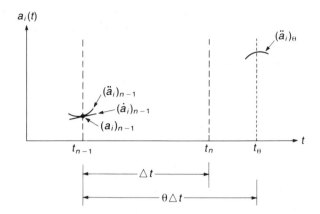

FIGURE 12.6 ● Data used in the Wilson recurrence relation; all four values are interpolated.

for the diffusion problem: It is implicit, unconditionally stable for linear problems, quite "robust" even for nonlinear problems, and tends to act somewhat "overdamped."

This method also needs a starting procedure. Here one could use expressions similar to Eq. (12.12) or else one of the one-step methods described next.

12.2.3 The Wilson Method[4]

This is a one-step method [12.12, 12.16] in which $\{a(t)\}$ is approximated by a cubic polynomial:

$$\{a(t)\} = \{c_0\} + \{c_1\}t + \{c_2\}t^2 + \{c_3\}t^3 \tag{12.18}$$

The polynomial is defined over not only the current step but part of the next one as well, as shown in Fig. 12.6. The parameter θ measures how far beyond t_n the interval of definition extends:

$$t_\theta = t_{n-1} + \theta\Delta t \qquad \theta \geq 1 \tag{12.19}$$

The coefficients $\{c_i\}$, $i=1,2,3,4$, are determined by interpolating $\{a(t)\}$, $\{\dot{a}(t)\}$, and $\{\ddot{a}(t)\}$ at t_{n-1}, and $\{\ddot{a}(t)\}$ at t_θ. Therefore Eq. (12.18) becomes

$$\{a(t)\} = \{a\}_{n-1} + \{\dot{a}\}_{n-1}(t-t_{n-1}) + \frac{1}{2}\{\ddot{a}\}_{n-1}(t-t_{n-1})^2 + \frac{1}{6}\left(\frac{\{\ddot{a}\}_\theta - \{\ddot{a}\}_{n-1}}{\theta\Delta t}\right)(t-t_{n-1})^3 \tag{12.20}$$

This is the basic approximating polynomial from which all the remaining relations are derived.

Differentiating Eq. (12.20) once and twice yields, respectively,

$$\{\dot{a}(t)\} = \{\dot{a}\}_{n-1} + \{\ddot{a}\}_{n-1}(t-t_{n-1}) + \frac{1}{2}\left(\frac{\{\ddot{a}\}_\theta - \{\ddot{a}\}_{n-1}}{\theta\Delta t}\right)(t-t_{n-1})^2 \tag{12.21}$$

$$\{\ddot{a}(t)\} = \{\ddot{a}\}_{n-1} + \left(\frac{\{\ddot{a}\}_\theta - \{\ddot{a}\}_{n-1}}{\theta\Delta t}\right)(t-t_{n-1}) \tag{12.22}$$

---

[4] Also called the Wilson-θ method.

Equations (12.20) through (12.22) clearly satisfy the interpolation conditions. The acceleration is linear (like the Houbolt method) since $\{a(t)\}$ is cubic; it interpolates the values at t_{n-1} and t_θ, as shown in Fig. 12.7.

The system equations [Eqs. (12.4)] are now evaluated at time t_θ:

$$[M]\{\ddot{a}\}_\theta + [C]\{\dot{a}\}_\theta + [K]\{a\}_\theta = \{F\}_\theta \tag{12.23a}$$

where $\{F\}_\theta$ is linearly interpolated:

$$\{F\}_\theta = \{F\}_{n-1} + \theta(\{F\}_n - \{F\}_{n-1}) \tag{12.23b}$$

Expressions for $\{\ddot{a}\}_\theta$ and $\{\dot{a}\}_\theta$ are obtained by evaluating Eqs. (12.20) and (12.21) at t_θ, solving the former for $\{\ddot{a}\}_\theta$, and substituting into the latter for $\{\dot{a}\}_\theta$. These expressions may then be substituted into Eq. (12.23a) to yield the following result:

$$\left(\frac{6}{\theta^2 \Delta t^2}[M] + \frac{3}{\theta \Delta t}[C] + [K]\right)\{a\}_\theta = \{F\}_{n-1} + \theta(\{F\}_n - \{F\}_{n-1})$$

$$+ \left(\frac{6}{\theta^2 \Delta t^2}[M] + \frac{3}{\theta \Delta t}[C]\right)\{a\}_{n-1} + \left(\frac{6}{\theta \Delta t}[M] + 2[C]\right)\{\dot{a}\}_{n-1} + \left(2[M] + \frac{\theta \Delta t}{2}[C]\right)\{\ddot{a}\}_{n-1}$$

$$\tag{12.24}$$

Equation (12.24) is the Wilson recurrence relation. After solving it for $\{a\}_\theta$, the latter is substituted into the equation for $\{\ddot{a}\}_\theta$ described above (but not shown). Then $\{\ddot{a}\}_\theta$ is substituted into Eqs. (12.20) through (12.22) evaluated at time t_n to obtain $\{a\}_n$, $\{\dot{a}\}_n$, and $\{\ddot{a}\}_n$.

The Wilson method is obviously implicit. It is unconditionally stable for linear problems only if $\theta \geq 1.37$; a value of $\theta = 1.40$ is normally used in practice. Because it is only one step, no special starting procedure is needed. Thus the first step is obtained using $n=1$ in Eq. (12.24). On the RHS $\{a\}_0$ and $\{\dot{a}\}_0$ are given, and $\{\ddot{a}\}_0$ may be calculated from the system equations [Eqs. (12.4)] at time t_0.

12.2.4 The Newmark Method

This is also a one-step method [12.11] in which $\{a(t)\}$ is approximated by a cubic polynomial:

$$\{a(t)\} = \{c_0\} + \{c_1\}t + \{c_2\}t^2 + \{c_3\}t^3 \tag{12.25}$$

which is defined only over the current time step. The coefficients $\{c_0\}$ and $\{c_1\}$ are determined by interpolating $\{a(t)\}$ and $\{\dot{a}(t)\}$ at time t_{n-1}. Therefore Eq. (12.25) becomes

$$\{a(t)\} = \{a\}_{n-1} + \{\dot{a}\}_{n-1}(t - t_{n-1}) + \{c_2\}(t - t_{n-1})^2 + \{c_3\}(t - t_{n-1})^3 \tag{12.26}$$

The other two coefficients are determined by the following two conditions on $\{a(t)\}$ and $\{\dot{a}(t)\}$ at time t_n (Fig. 12.8):

$$\{a(t_n)\} = \{a\}_n = \{a\}_{n-1} + \{\dot{a}\}_{n-1}\Delta t + \frac{1}{2}\left[(1-2\beta)\{\ddot{a}\}_{n-1} + 2\beta\{\ddot{a}\}_n\right]\Delta t^2$$

$$\{\dot{a}(t_n)\} = \{\dot{a}\}_n = \{\dot{a}\}_{n-1} + \left[(1-\gamma)\{\ddot{a}\}_{n-1} + \gamma\{\ddot{a}\}_n\right]\Delta t \tag{12.27}$$

The unknowns $\{a(t_n)\}$ and $\{\dot{a}(t_n)\}$ are thus approximated by what looks like the first few terms of a Taylor series about t_{n-1} (they would be if $\beta = \gamma = 0$), although the coefficients

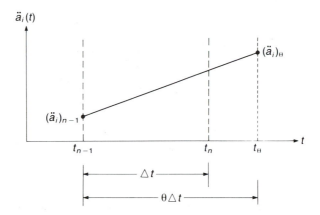

FIGURE 12.7 ● Linear acceleration assumed by the Wilson method.

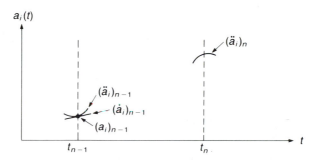

FIGURE 12.8 ● Data used in the Newmark recurrence relation; only $(a_i)_{n-1}$ and $(\dot{a}_i)_{n-1}$ are interpolated.

of the acceleration terms are expressed as a linear interpolation between $\{\ddot{a}\}_{n-1}$ and $\{\ddot{a}\}_n$. Different interpolations are used in the expressions for $\{a(t)\}$ and $\{\dot{a}(t)\}$; the parameter β controls the former, and γ the latter.

Applying conditions (12.27) to Eq. (12.26) yields the following cubic polynomial:

$$\{a(t)\} = \{a\}_{n-1} + \{\dot{a}\}_{n-1}(t-t_{n-1})$$

$$+ \frac{1}{2}\left[\left(\frac{\{\ddot{a}\}_{n-1}+\{\ddot{a}\}_n}{2}\right) + \left(6\beta-2\gamma-\frac{1}{2}\right)(\{\ddot{a}\}_n-\{\ddot{a}\}_{n-1})\right](t-t_{n-1})^2$$

$$+ (\gamma-2\beta)\left(\frac{\{\ddot{a}\}_n-\{\ddot{a}\}_{n-1}}{\Delta t}\right)(t-t_{n-1})^3 \qquad \textbf{(12.28)}$$

This is the basic approximating polynomial from which all the remaining relations are derived.[5] Note that it degenerates to a quadratic polynomial if $\gamma-2\beta=0$, as will be illustrated below.

---

[5] To the author's knowledge, no previous publications have expressed the basic Newmark approximating polynomials in terms of a single cubic polynomial as shown in Eq. (12.28).

Differentiating Eq. (12.28) once and twice yields, respectively,

$$\{\dot{a}(t)\} = \{\dot{a}\}_{n-1} + \left[\left(\frac{\{\ddot{a}\}_{n-1}+\{\ddot{a}\}_n}{2}\right) + \left(6\beta-2\gamma-\frac{1}{2}\right)(\{\ddot{a}\}_n-\{\ddot{a}\}_{n-1})\right](t-t_{n-1})$$

$$+3(\gamma-2\beta)\left(\frac{\{\ddot{a}\}_n-\{\ddot{a}\}_{n-1}}{\Delta t}\right)(t-t_{n-1})^2 \qquad \text{(12.29)}$$

$$\{\ddot{a}(t)\} = \left(\frac{\{\ddot{a}\}_{n-1}+\{\ddot{a}\}_n}{2}\right) + \left(6\beta-2\gamma-\frac{1}{2}\right)(\{\ddot{a}\}_n-\{\ddot{a}\}_{n-1})$$

$$+6(\gamma-2\beta)\left(\frac{\{\ddot{a}\}_n-\{\ddot{a}\}_{n-1}}{\Delta t}\right)(t-t_{n-1}) \qquad \text{(12.30)}$$

If $\gamma-2\beta\neq 0$, the acceleration is linear (like the Houbolt and Wilson methods) since $\{a(t)\}$ is cubic. If $\gamma-2\beta=0$, the acceleration is constant since $\{a(t)\}$ is quadratic.

The Newmark method is a two-parameter family of algorithms. For particular values of the parameters γ and β we can sometimes provide meaningful interpretations of these expressions. There are two cases that are especially popular.

1. $\gamma = \dfrac{1}{2}, \ \beta = \dfrac{1}{4}$

$$\gamma - 2\beta = 0$$

$$6\beta - 2\gamma - \frac{1}{2} = 0 \qquad \text{(12.31)}$$

Equations (12.28) through (12.30) become

$$\{a(t)\} = \{a\}_{n-1} + \{\dot{a}\}_{n-1}(t-t_{n-1}) + \frac{1}{2}\left(\frac{\{\ddot{a}\}_{n-1}+\{\ddot{a}\}_n}{2}\right)(t-t_{n-1})^2$$

$$\{\dot{a}(t)\} = \{\dot{a}\}_{n-1} + \left(\frac{\{\ddot{a}\}_{n-1}+\{\ddot{a}\}_n}{2}\right)(t-t_{n-1})$$

$$\{\ddot{a}(t)\} = \frac{\{\ddot{a}\}_{n-1}+\{\ddot{a}\}_n}{2} \qquad \text{(12.32)}$$

The general cubic polynomial for $\{a(t)\}$ has degenerated to a quadratic, so the acceleration is a constant, namely, the average value, as shown in Figure 12.9.

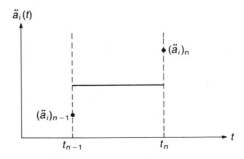

FIGURE 12.9 ● Constant acceleration assumed by the Newmark method when $\gamma=1/2$, $\beta=1/4$.

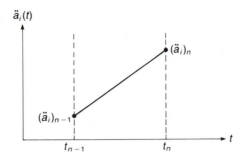

FIGURE 12.10 ● Linear acceleration assumed by the Newmark method when $\gamma = 1/2$, $\beta = 1/6$.

2. $\gamma = \dfrac{1}{2}$, $\beta = \dfrac{1}{6}$

$$\gamma - 2\beta = \frac{1}{6}$$

$$6\beta - 2\gamma - \frac{1}{2} = -\frac{1}{2} \tag{12.33}$$

Equations (12.28) through (12.30) become

$$\{a(t)\} = \{a\}_{n-1} + \{\dot{a}\}_{n-1}(t-t_{n-1}) + \frac{1}{2}\{\ddot{a}\}_{n-1}(t-t_{n-1})^2 + \frac{1}{6}\left(\frac{\{\ddot{a}\}_n - \{\ddot{a}\}_{n-1}}{\Delta t}\right)(t-t_{n-1})^3$$

$$\{\dot{a}(t)\} = \{\dot{a}\}_{n-1} + \{\ddot{a}\}_{n-1}(t-t_{n-1}) + \frac{1}{2}\left(\frac{\{\ddot{a}\}_n - \{\ddot{a}\}_{n-1}}{\Delta t}\right)(t-t_{n-1})^2$$

$$\{\ddot{a}(t)\} = \{\ddot{a}\}_{n-1} + \left(\frac{\{\ddot{a}\}_n - \{\ddot{a}\}_{n-1}}{\Delta t}\right)(t-t_{n-1}) \tag{12.34}$$

Here $\{a(t)\}$ is a Taylor series expansion about t_{n-1}, since $(\{\ddot{a}\}_n - \{\ddot{a}\}_{n-1})/\Delta t = \{\dddot{a}\}$ is a constant over the time step and hence is the value at t_{n-1}. We also see that $\{\ddot{a}(t)\}$ linearly interpolates $\{\ddot{a}\}_{n-1}$ and $\{\ddot{a}\}_n$, as shown in Fig. 12.10. Note that Fig. 12.10 is identical to Fig. 12.7 when $\theta = 1$ in the latter. Thus the Newmark method with $\gamma = 1/2$, $\beta = 1/6$ is identical to the Wilson method with $\theta = 1$; this can also be seen by substituting $\theta = 1$ into Eq. (12.24) and $\gamma = 1/2$, $\beta = 1/6$ into Eq. (12.36) below.[6]

To derive the recurrence relation, the system equations [Eqs. (12.4)] are evaluated at time t_n:

$$[M]\{\ddot{a}\}_n + [C]\{\dot{a}\}_n + [K]\{a\}_n = \{F\}_n \tag{12.35}$$

---

[6] It is perhaps worth noting that the Houbolt, Wilson, and Newmark methods all employ cubic approximating polynomials for $\{a(t)\}$; hence all assume linear acceleration. (Newmark, as observed above, degenerates to a quadratic polynomial, and therefore constant acceleration, if $\gamma - 2\beta = 0$.) The difference between the methods is that the linear acceleration for Houbolt is defined over the current and two previous time steps; for Wilson it is over the current step plus part of the next step; and for Newmark it is over only the current step.

Expressions for $\{\ddot{a}\}_n$ and $\{\dot{a}\}_n$ are obtained by solving the first of Eqs. (12.27) for $\{\ddot{a}\}_n$, then substituting $\{\ddot{a}\}_n$ into the second equation for $\{\dot{a}\}_n$. These in turn are substituted into Eq. (12.35) to yield the following result:

$$
\left(\frac{1}{\beta\Delta t^2}[M] + \frac{\gamma}{\beta\Delta t}[C] + [K]\right)\{a\}_n = \{F\}_n + \left(\frac{1}{\beta\Delta t^2}[M] + \frac{\gamma}{\beta\Delta t}[C]\right)\{a\}_{n-1}
$$

$$
+ \left[\frac{1}{\beta\Delta t}[M] + \left(\frac{\gamma}{\beta} - 1\right)[C]\right]\{\dot{a}\}_{n-1} + \left[\left(\frac{1}{2\beta} - 1\right)[M] + \left(\frac{\gamma}{2\beta} - 1\right)\Delta t[C]\right]\{\ddot{a}\}_{n-1}
$$

$$
\tag{12.36}
$$

Equation (12.36) is the Newmark recurrence relation. After solving it for $\{a\}_n$, the latter is substituted into the equations for $\{\dot{a}\}_n$ and $\{\ddot{a}\}_n$ described above (but not shown).

The Newmark method is obviously implicit. It is unconditionally stable for linear problems only if $\gamma \geq 1/2$ and $\beta \geq (\gamma + 1/2)^2/4$. Thus the previously described special case using $\gamma = 1/2$, $\beta = 1/4$ is unconditionally stable, although the other case using $\gamma = 1/2$, $\beta = 1/6$ (equivalent to $\theta = 1$ in the Wilson method) is only conditionally stable. Being only one step, the method is self-starting, using, for example, the technique explained for the Wilson method.

More detailed analyses of the stability and accuracy of the above four methods may be found in most of the previously cited references [12.9-12.23]. Of particular importance are the algorithmic errors introduced by *amplitude decay* and *period elongation*. Although these four methods have performed quite well for many years over a broad spectrum of applications, there has been a continuing search to find even better methods [12.25-12.37].

A couple of final observations may be made regarding all four methods. First, for large problems it is generally desirable to reduce the size of the system equations [Eqs. (12.4)] before beginning the time-stepping. This can be accomplished by employing *Guyan reduction*. This technique was described briefly in Section 10.3.3 as a means to reduce the number of DOF in an eigenproblem. The technique can also be used to reduce the number of DOF in Eqs. (12.4) [12.38]. The two applications are intimately related since, in essence, Guyan reduction eliminates from Eqs. (12.4) those DOF that do not contribute (significantly) to the important modes of the response, a comment that will be better appreciated after studying Section 12.3.

Second, the system damping matrix $[C]$ is usually difficult to construct by the standard procedure of assembling element damping matrices because damping in most structures is caused by several mechanisms such as internal friction in materials, surface friction at interfaces and joints, viscous damping from lubricated surfaces, aerodynamic effects, etc. In addition, damping effects are usually sensitive functions of frequency (or velocity). For these reasons, local damping properties, such as the damping coefficient $\mu(x)$ in the governing differential equation, are not as easy to measure experimentally as are stiffness and mass.

To circumvent these difficulties, it is common practice to simply express $[C]$ as a linear combination of the system mass and stiffness matrices:

$$
[C] = \alpha[M] + \beta[K] \tag{12.37}
$$

an approximation referred to as *Rayleigh damping*. The coefficients α and β can be related to the *modal damping ratios* (see Section 12.3), which measure the percent of critical damping in each mode. These ratios are experimentally easier to obtain, and they essentially sum up all of the damping effects over the entire structure for each particular mode. We see that once again the concept of modes has entered the discussion, so further explanation of this procedure must be deferred to the next section. [See the discussion involving Eqs. (12.55) through (12.59).]

12.3 Mode Superposition

The system equations [Eqs. (12.4)] are generally *coupled*, with $[K]$ banded and $[M]$ and $[C]$ either banded or diagonal. They would be uncoupled if and only if all three matrices were diagonal. A system of uncoupled ordinary differential equations is considerably less expensive to integrate than a coupled system. The purpose of mode superposition is to transform Eqs. (12.4) to an uncoupled system in order to realize these cost savings. Of course, the transformation itself is an added cost, but there are many situations (to be discussed later) when this cost is more than offset by the subsequent savings in dealing with a computationally more efficient system.

There are two approaches to uncoupling Eqs. (12.4). Both involve the calculation of the system of eigenvectors (modes) corresponding to Eqs. (12.4), as explained in Chapter 10. One approach calculates the *undamped* modes and the other approach the *damped* modes. The latter can always be done exactly (with no simplifying assumptions regarding the damping) but is more complicated and hence more expensive. The former approach must make simplifying assumptions regarding the damping, but the assumptions are appropriate for many common structural applications.[7] Consequently, the undamped-modes approach is currently the standard approach in virtually all commercial FE structural programs and will be described below. (References to the other approach will be cited later.) The following development makes use of the material in Section 10.3 on the generalized algebraic eigenproblem.

The undamped modes for Eqs. (12.4) are the time-harmonic (sinusoidal) solutions when there are no loads ($\{F\} = \{0\}$) and damping is neglected ($[C] = [0]$); that is,

$$[M]\{\ddot{a}\} + [K]\{a\} = \{0\} \tag{12.38}$$

Let

$$\{a\} = \{v\}e^{i\omega t} \tag{12.39}$$

where $e^{i\omega t} = \cos \omega t + i \sin \omega t$ is complex notation for a sinusoidal time variation. Substituting Eq. (12.39) into Eq. (12.38) yields

$$[K]\{v\} - \omega^2[M]\{v\} = \{0\} \tag{12.40}$$

---

[7] The undamped-modes approach does not mean damping is neglected in the analysis. It means that the modes are first calculated for the system assuming no damping. These modes are then used in a time-integration procedure which includes the damping, as explained in the text. This approach is effective whenever the damping is small enough that the differences between the damped and undamped modes and frequencies are negligible.

which is the generalized algebraic eigenproblem in Eq. (10.24) where the eigenvalue λ is the square of the circular frequency, that is,

$$\lambda = \omega^2 \tag{12.41}$$

Since $[K]$ and $[M]$ are in general real, symmetric, and positive-definite, all the results in Section 10.3.1 carry over to the present problem. Thus, if the system has N active DOF (those not constrained by zero-valued essential BCs), then there are N eigensolutions:

$$\left(\omega_1^2, \{v\}_1\right), \left(\omega_2^2, \{v\}_2\right), \ldots, \left(\omega_N^2, \{v\}_N\right) \tag{12.42a}$$

where

$$0 < \omega_1^2 \leq \omega_2^2 \leq \cdots \leq \omega_N^2 \tag{12.42b}$$

If we recall Eq. (10.30), we see the eigenvectors are *orthogonal* with respect to $[K]$ and $[M]$, and they are *orthonormal* with respect to $[M]$; that is,

$$\{v\}_i^T[K]\{v\}_j = \omega_i^2 \delta_{ij}$$

$$\{v\}_i^T[M]\{v\}_j = \delta_{ij} \tag{12.43}$$

It is convenient to define an N-by-N square matrix $[V]$ that contains the N eigenvectors as columns:

$$[V] = \left[\ \{v\}_1, \{v\}_2, \ldots, \{v\}_N\ \right] \tag{12.44}$$

and an N-by-N diagonal matrix $[\Omega^2]$ that contains the N eigenvalues on the diagonal:

$$[\Omega^2] = \begin{bmatrix} \omega_1^2 & & & & \\ & \omega_2^2 & & & \\ & & \cdot & & \\ & & & \cdot & \\ & & & & \cdot \\ & & & & & \omega_N^2 \end{bmatrix} \tag{12.45}$$

With this notation we may compactly write all N solutions to Eqs. (12.40) as follows:[8]

$$[K][V] = [M][V][\Omega^2] \tag{12.46}$$

and the orthogonality relations [Eqs. (12.43)] may be written

$$[V]^T[K][V] = [\Omega^2]$$

$$[V]^T[M][V] = [I] \tag{12.47}$$

where $[I]$ is the identity matrix, i.e., a diagonal matrix with 1s on the diagonal.

---

[8] Some of the matrix operations in the subsequent material are a bit more advanced than those used in the rest of the book. Explanations of these operations may be found in almost any introductory text on matrix algebra [12.39].

This system of N modes can now be used to transform the system equations [Eqs. (12.4)]. The general solution to Eqs. (12.4) may be written as a linear superposition of the N modes, each multiplied by a general time-varying amplitude:[9]

$$\{a(t)\} = \sum_{j=1}^{N} A_j(t)\{v\}_j$$

$$= [V]\{A(t)\} \tag{12.48}$$

Substituting Eq. (12.48) into eq. (12.4), premultiplying the latter by $[V]^T$, and using Eqs. (12.47) yields

$$\{\ddot{A}\} + [V]^T[C][V]\{\dot{A}\} + [\Omega^2]\{A\} = [V]^T\{F\} \tag{12.49}$$

The initial conditions $\{a\}_0$ and $\{\dot{a}\}_0$ are transferred to $\{A\}$ by premultiplying Eq. (12.48) by $[V]^T[M]$; that is,

$$\{A\}_0 = [V]^T[M]\{a\}_0$$

$$\{\dot{A}\}_0 = [V]^T[M]\{\dot{a}\}_0 \tag{12.50}$$

Equations (12.49) are the transformed system equations. *The DOF are now the amplitudes of each mode.* The stiffness and mass terms have been uncoupled, but the damping term remains coupled for arbitrary $[C]$.

For problems in which damping can be neglected, Eqs. (12.49) become

$$\{\ddot{A}\} + [\Omega^2]\{A\} = [V]^T\{F\} \tag{12.51}$$

Equations (12.51) are completely uncoupled so they can be written as N separate equations:

$$\ddot{A}_i(t) + \omega_i^2 A_i(t) = f_i(t) \qquad i = 1,2,\ldots,N \tag{12.52}$$

where[10] $f_i(t) = \{v\}_i^T\{F(t)\}$. Equations (12.52) can be numerically solved using any of the time-stepping methods in Section 12.2.[11] Finally, the separate solutions for each $A_i(t)$ are recombined according to Eq. (12.48) to produce $\{a(t)\}$.

---

[9] Note the direct parallel with the modal solution to the diffusion problem in Section 11.3.4, Eqs. (11.82) through (11.84).

[10] Note that $\{v\}_i^T\{F(t)\}$ is the *inner product* of the ith mode shape with the applied load. This measures how $\{F(t)\}$ is spatially distributed over the mode. $\{F(t)\}$ is said to be *orthogonal* to the mode when the inner product vanishes. This occurs when $\{F(t)\}$ is some places in phase with the mode (inputting energy) and other places out of phase (outputting energy) so that the net effect over the entire mode is to input zero energy, in which case the mode cannot be excited and therefore $A_i(t)=0$. An example would be trying to excite the second (or fourth or sixth, etc.) mode of a transversely vibrating string (see Fig. 10.8) with a uniform force along the length of the string. Such a force is orthogonal to the even-order modes, so only the odd-order modes would be excited.

[11] Equations (12.52) could also be solved using a classical integral formula called the *Duhamel integral* [12.4, 12.40, 12.41]. However, for general $f_i(t)$ the integral would have to be numerically evaluated, so there is little incentive to adopt this approach if an FE program already contains one or more of the direct-integration methods.

When damping cannot be neglected, it is obviously desirable to try to diagonalize the transformed damping matrix, $[V]^T[C][V]$, in order to completely uncouple the system equations and thereby realize the computational simplifications exhibited above for the zero-damping case. For many structures this may be accomplished merely by making the simplifying assumption that $[V]^T[C][V]$ *is* diagonal; that is,

$$[V]^T[C][V] = [C_D] \tag{12.53a}$$

where

$$[C_D] = \begin{bmatrix} 2\omega_1\xi_1 & & & & \\ & 2\omega_2\xi_2 & & & \\ & & \cdot & & \\ & & & \cdot & \\ & & & & \cdot \\ & & & & & 2\omega_N\xi_N \end{bmatrix} \tag{12.53b}$$

Here ξ_i are the *modal damping ratios*, which are the percent of critical damping in each mode. This assumption is referred to as *proportional damping*, for reasons explained below.

Substituting Eqs. (12.53) into Eqs. (12.49) uncouples the latter, enabling them to be written as N separate equations:

$$\ddot{A}_i(t) + 2\omega_i\xi_i\dot{A}_i(t) + \omega_i^2 A_i(t) = f_i(t) \qquad i = 1,2,\ldots,N \tag{12.54}$$

As before, Eqs. (12.54) can be numerically solved using one of the time-stepping methods, and then each $A_i(t)$ summed according to Eq. (12.48) to produce $\{a(t)\}$.

The principal advantage of mode superposition should now be apparent. In many vibration problems (e.g., earthquake excitation of buildings, wave action on ships, narrow-band excitation of machinery) only a few modes of the structure (say p, where $p \ll N$) are excited. These few are capable of accurately describing the time history of the response. Therefore we need to calculate only p modes and then solve only p of the equations in Eqs. (12.52) or (12.54).

At this point we return to the discussion on damping in the context of the direct-integration methods at the end of Section 12.2 . It was observed that it is much easier to obtain experimental data for the modal damping ratios ξ_i, which appear in $[C_D]$, than it is to obtain data for the damping coefficient $\mu(x)$, which is used to construct $[C]$. Since the direct-integration methods deal with $[C]$ rather than $[C_D]$, we would like to devise a method for constructing $[C]$ from the modal damping ratios rather than from $\mu(x)$. This may be done by assuming that $[C]$ is linearly *proportional* to $[M]$ and/or $[K]$ and/or various products thereof, as expressed by the following *Caughey series* [12.42]:

$$[C] = [M] \sum_l b_l([M]^{-1}[K])^l \qquad l = \ldots,-2,-1,0,1,2,\ldots \tag{12.55}$$

By writing out a few terms in this series, that is,

$$[C] = \ldots + b_{-1}[M][K]^{-1}[M] + b_0[M] + b_1[K] + b_2[K][M]^{-1}[K] + \ldots$$

$$(12.56)$$

we see that it includes *Rayleigh damping* as a special case if we use only the terms for $l=0$ and $l=1$; that is,

$$[C] = \alpha[M] + \beta[K] \tag{12.57}$$

where the conventional notation α, β is used in place of b_0, b_1. [Recall Eq. (12.37).] Substituting Eq. (12.57) into Eq. (12.53a) and using Eqs. (12.47) yields

$$[C_D] = \alpha[I] + \beta[\Omega^2] \tag{12.58}$$

(see Exercise 12.10) which is a system of N uncoupled equations,

$$\xi_i = \frac{\alpha}{2\omega_i} + \frac{\beta\omega_i}{2} \qquad i = 1, 2, \ldots, N \tag{12.59}$$

Values for α and β can then be determined from eqs. (12.59) [12.22, 12.43]. For example, if values for ξ_i are available for a few modes, say $p(p \geq 2)$, then eqs. (12.59) provide p equations in the two unknowns α and β, which can be solved using a least-squares fit. Since Eqs. (12.59) are applicable to all the modes in the structure, and clearly $\xi_i \to \infty$ as $\omega_i \to \infty$, then the higher modes will be excessively damped. This is generally desirable, though, since, as discussed in Chapter 10, the accuracy of the modes in an FE model decreases with increasing frequency. The excessive damping at the higher frequencies therefore effectively eliminates from the analysis the higher, less-accurate modes.

Having determined numerical values for α and β, Eq. (12.57) provides the desired expression for $[C]$ for use in the direct-integration methods. If one has data for several modal damping ratios, it might be tempting to include more terms from the Caughey series (e.g., p terms for p different ξ_i). However, the Rayleigh damping terms are the only ones in the series that yield a banded $[C]$ if $[M]$ and $[K]$ are also banded; all the other terms will in general yield a full $[C]$. It might therefore be preferable to stay with Rayleigh damping and determine α and β by a least-squares fit to the data as just described.[12]

In the event that proportional damping is a poor model (e.g., for structures composed of very different materials), then the damping matrix cannot be uncoupled using undamped modes. Here there are a number of options. First, as mentioned at the beginning of this section, an alternate approach is to calculate the *damped* modes since all three matrices $[M]$, $[K]$, and $[C]$ can be simultaneously diagonalized with respect to these modes [12.6, 12.44, 12.45]. The computations are more expensive since there are $2N$ such modes rather than N and they are complex-valued. A second option is to perform mode superposition with undamped modes as described above, leading to Eqs. (12.49), and assume the total response may be represented by only a few modes. Then directly integrate this smaller,

---

[12] In Example 1.2 in Chapter 1 the author used similar curve-fitting techniques to determine the Rayleigh parameter α so that *nonlinear* damping due to hydrodynamic drag could be approximately modeled by the linear Rayleigh damping $[C] = \alpha[M]$, thereby enabling harmonic analyses (Section 12.4) to be performed more efficiently.

though coupled, system. A third option is to employ one of several approximate methods for diagonalizing the transformed damping matrix [12.46]. A final option is, of course, to ignore mode superposition altogether and instead use direct integration.[13]

How does one choose whether to use direct integration or mode superposition? This depends on how many modes will be needed to accurately represent the solution (e.g., a couple dozen vis-à-vis a couple hundred), as well as how long a transient solution is desired. Mode superposition incurs an initial start-up cost to calculate the modes the analyst thinks will be necessary. To recoup this initial investment, one should anticipate performing a long transient solution or, equivalently, performing many separate transient solutions using the same modes (with presumably different loads).

Two extreme cases illustrate these points. A sudden shock load will excite many modes. If only a brief transient response needs to be calculated, then direct integration would be much cheaper than the calculation of a large eigensystem. At the other extreme, an earthquake tends to excite only a few of the lowest modes of typical structures, and the loading history is usually much longer than a shock load. Here mode superposition would be particularly effective since the initial eigensystem calculation would be relatively cheap (only a few modes needing to be calculated), and the resulting small system of decoupled equations would probably be used for thousands or tens of thousands of time steps.

A final point: Mode superposition is a *linear* superposition, so it is only effective for linear systems. Direct integration, however, is valid for both linear and nonlinear systems. Therefore nonlinear problems should use direct integration (unless the nonlinearities are sufficiently weak that the problem can be effectively "linearized").

12.4 Harmonic Analysis

A *harmonic* (or *time-harmonic* or *simple-harmonic*) load is one whose time variation is sinusoidal and therefore is characterized by a single frequency, ω. The load vector $\{F(t)\}$ in the system equations [Eqs. (12.4)] may be written[14]

$$\{F(t)\} = \{\bar{F}\}e^{i\omega t} \tag{12.60}$$

where the amplitude $\{\bar{F}\}$ may be real or complex-valued, $e^{i\omega t} = \cos \omega t + i \sin \omega t$, and $i = \sqrt{-1}$. If the system equations are linear, then the solution vector $\{a(t)\}$ must also vary sinusoidally with the same frequency:

$$\{a(t)\} = \{\bar{a}\}e^{i\omega t} \tag{12.61}$$

---

[13] Wilson et al. [12.47] describe an alternative and intriguing mode-superposition approach that calculates a set of orthogonal vectors, different from the modes discussed above. These vectors are the only ones that are significantly excited by the loading, and hence they form an efficient basis with which to represent the solution. Any of the above direct-integration or mode-superposition methods may then be used to analyze this greatly reduced system. This approach takes advantage of a fundamental property of any forced linear system: the response of a given mode is proportional to the inner product (i.e., degree of orthogonality) of the load distribution and the mode shape [12.48, 12.49].

[14] Some authors use the notation $e^{-i\omega t}$ instead of $e^{i\omega t}$. The difference is only a change in phase of π radians that can be absorbed into $\{F\}$.

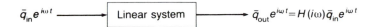

FIGURE 12.11 ● Frequency-response function $H(i\omega)$.

where the amplitude $\{\bar{a}\}$ is generally complex-valued. Note that *phase* information is contained in the complex values. Thus different components of $\{\bar{F}\}$ (or $\{\bar{a}\}$) may be out of phase with each other, and $\{\bar{F}\}$ will generally be out of phase with $\{\bar{a}\}$. All components, though, will be oscillating at frequency ω.

Substituting Eqs. (12.60) and (12.61) into Eqs. (12.4) and canceling the $e^{i\omega t}$ factors on both sides yields

$$[K_{\text{eff}}]\{\bar{a}\} = \{\bar{F}\} \tag{12.62}$$

where

$$[K_{\text{eff}}] = -\omega^2[M] + i\omega[C] + [K]$$

Equations (12.62) are a system of linear algebraic equations with a complex-valued effective stiffness matrix and load vector. They can be solved by Gaussian elimination, using complex arithmetic. The resulting solution, $\{\bar{a}\}$, may then be multiplied by $e^{i\omega t}$ to obtain $\{a(t)\}$, according to eq. (12.61). Since $\{a(t)\}$ varies sinusoidally 'indefinitely' (for as long as $\{F(t)\}$ varies sinusoidally), this is called a *steady-state* solution.

Since $[K_{\text{eff}}]$ is a function of the excitation frequency ω, then $\{\bar{a}\}$ will also be a function of ω, that is, $\{\bar{a}(\omega)\}$. The variation with frequency may be determined by solving Eqs. (12.62) at many different frequencies. This data can then be used to construct a *frequency-response function $H(i\omega)$*, which is the ratio of the amplitudes of any two specified variables, loosely called ''input'' and ''output,'' when the load is harmonic and the system is linear.[15] Thus, as shown in Fig. 12.11,

$$H(i\omega) = \frac{\bar{q}_{\text{out}}(\omega)}{\bar{q}_{\text{in}}(\omega)} \tag{12.63}$$

where $\bar{q}_{\text{in}}(\omega)$ and $\bar{q}_{\text{out}}(\omega)$ may be frequency-dependent complex-valued quantities.

For a given structure one can define a great variety of different frequency-response functions, depending on what variables are chosen for input and output, e.g., force, displacement, velocity, acceleration, or linear combinations thereof. In addition, one must define the *locations* on the structure where the input and output are calculated, since the frequency-response function depends (usually strongly) on the choice of input and output locations.

Special names have been given to the frequency-response function, or its absolute value, for a few commonly used definitions of $\bar{q}_{\text{in}}$ and $\bar{q}_{\text{out}}$. Letting $\bar{a}_i$, $\dot{\bar{a}}_i$, $\ddot{\bar{a}}_i$, and $\bar{F}_i$ denote the (possibly complex-valued) amplitudes of displacement, velocity, acceleration, and force, respectively, at the ith DOF, we have the following definitions:

---

[15] $H(i\omega)$ is often called the *complex frequency-response function*, or the *transfer function*, and it is conventional to write the argument as $i\omega$ rather than ω as a reminder that it is complex-valued.

Transmissibility, T(ω) •

$$T(\omega) = \left| \frac{\bar{a}_i(\omega)}{\bar{a}_j(\omega)} \right| = \left| \frac{\dot{\bar{a}}_i(\omega)}{\dot{\bar{a}}_j(\omega)} \right| = \left| \frac{\ddot{\bar{a}}_i(\omega)}{\ddot{\bar{a}}_j(\omega)} \right|$$ (12.64)

which is the *dimensionless* ratio of dimensionally similar quantities at two different points in the structure, e.g., the ith and jth DOF. Since the motion is harmonic, the ratio is identical for displacement, velocity, and acceleration. One can also define a force transmissibility in the same manner, namely, $T_F(\omega) = |\bar{F}_i(\omega)/\bar{F}_j(\omega)|$, which may or may not be the same as $T(\omega)$, so one should always state which type of transmissibility one is referring to.

Impedance, Z($i\omega$) •

$$Z(i\omega) = \frac{\bar{F}_i(\omega)}{\dot{\bar{a}}_j(\omega)}$$ (12.65)

This is the complex-valued ratio of force to velocity. If the input and output locations are the same ($i=j$), $Z(i\omega)$ is called the *driving-point impedance*. If the locations are different ($i \neq j$), $Z(i\omega)$ is called the *transfer impedance*.

Admittance (or *mobility*), *Y($i\omega$)* •

$$Y(i\omega) = \frac{1}{Z(i\omega)}$$ (12.66)

Dynamic magnification factor, D(ω) •

$$D(\omega) = \left| \frac{\bar{a}_i(\omega)/\bar{F}_j}{\bar{a}_i(0)/\bar{F}_j} \right| = \left| \frac{\bar{a}_i(\omega)}{\bar{a}_i(0)} \right|$$ (12.67)

The numerator, $|\bar{a}_i(\omega)/\bar{F}_j|$, is the frequency-response function, giving the ratio of the amplitude of the displacement response to the amplitude of the force excitation. It is normalized (made dimensionless) by dividing by the same ratio evaluated at $\omega=0$, which yields $D(0)=1$.

To illustrate, the above frequency-response functions are calculated for the single-DOF mechanical oscillator shown in Fig. 12.12. The system equation is

$$M\ddot{a} + C\dot{a} + Ka = Fe^{i\omega t}$$ (12.68)

This may also be written as

$$\ddot{a} + 2\omega_0\xi\dot{a} + \omega_0^2 a = \frac{F}{M}e^{i\omega t}$$ (12.69)

where $\omega_0 = \sqrt{K/M}$ is the resonant frequency and $\xi = C/C_c = C/(2\sqrt{MK})$ is the damping ratio.

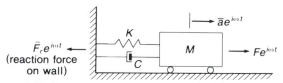

FIGURE 12.12 • A single-DOF mechanical oscillator excited by a harmonic force $Fe^{i\omega t}$.

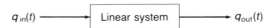

FIGURE 12.13 ● General time-varying input and output for a linear system.

Equation (12.69) is in the same form as each of the modal-response equations [Eqs. (12.54)], and therefore the following results could be applied to each mode in a multi-DOF system.

Writing the steady-state solution as $a = \bar{a}e^{i\omega t}$, we obtain from Eq. (12.69)

$$\bar{a}(\omega) = \frac{F/K}{1 - v^2 + i2v\xi} \tag{12.70}$$

where $v = \omega/\omega_0$. This yields the following frequency-response functions (see Exercise 12.11):

$$T_F(\omega) = \left| \frac{\bar{F}_r(\omega)}{F} \right| = \frac{1 + (2v\xi)^2}{(1 - v^2)^2 + (2v\xi)^2}$$

$$Z(i\omega)_{\substack{\text{driving} \\ \text{point}}} = \frac{F}{\bar{a}(\omega)} = \frac{F}{i\omega\bar{a}} = C_c \left(\xi + i\frac{v^2 - 1}{2v} \right)$$

$$D(\omega) = \left| \frac{\bar{a}(\omega)}{\bar{a}(0)} \right| = \frac{1}{\sqrt{(1 - v^2)^2 + (2v\xi)^2}} \tag{12.71}$$

A frequency-response function $H(i\omega)$ may be computed using an input of any magnitude. It should be obvious, though, that the resulting *ratio* may be interpreted as the output produced *per unit* input. Therefore, once $H(i\omega)$ is computed, the response $\bar{q}_{\text{out}}(\omega)$ to any given harmonic input $\bar{q}_{\text{in}}(\omega)$, consistent with the particular definition of $H(i\omega)$ being used, is simply the product of $H(i\omega)$ times the input, as indicated by Eq. (12.63); that is,

$$\bar{q}_{\text{out}}(\omega) = H(i\omega)\bar{q}_{\text{in}}(\omega) \tag{12.72}$$

For this reason, a quick visual scan of a plot of $|H(i\omega)|$ versus ω can tell a designer a great deal about the vibration characteristics of a given design [12.50].

As explained in connection with Fig. 12.1, the frequency-response function plays a ubiquitous role in vibration analysis. By virtue of the principle of *linear superposition*, the response of a linear system to any general time-varying load can be synthesized by summing or integrating harmonic responses. This includes periodic, nonperiodic, and random loads. Defining $q_{\text{in}}(t)$ and $q_{\text{out}}(t)$ as the time-varying input and output variables (Fig. 12.13), the pertinent relations are as follows [12.1-12.7, 12.41]:

Periodic loads ●
Fourier series:

$$q_{\text{in}}(t) = \sum_{n=-\infty}^{\infty} \bar{C}_n e^{in\omega t}$$

$$q_{\text{out}}(t) = \sum_{n=-\infty}^{\infty} H(in\omega)\bar{C}_n e^{in\omega t} \tag{12.73}$$

Nonperiodic loads •
Fourier transform:

$$q_{in}(t) = \frac{1}{2\pi} \int_{-\infty}^{\infty} \bar{C}(i\omega)e^{i\omega t} \, d\omega$$

$$q_{out}(t) = \frac{1}{2\pi} \int_{-\infty}^{\infty} H(i\omega)\bar{C}(i\omega)e^{i\omega t} \, d\omega \tag{12.74}$$

Random loads •
Power spectral density (PSD):

$$S_{out}(\omega) = |H(i\omega)|^2 S_{in}(\omega) \tag{12.75}$$

where $S_{in}(\omega)$ and $S_{out}(\omega)$ are the PSDs, respectively, of $q_{in}(t)$ and $q_{out}(t)$.

This completes the overview of the dynamics problem. Those readers with a special interest in the FE analysis of dynamics problems could add significantly to their understanding by implementing into the UNAFEM program one or more of the methods described in this chapter. Such an undertaking would be ambitious, but the rewards would be great.

Exercises

12.1   Rederive the central difference recurrence relation, Eq. (12.7), using the quadratic interpolation polynomial in eq. (12.13) and the quadratic shape functions in Fig. 8.2.

12.2   Derive a two-step backward difference recurrence relation by writing the system equations [Eqs. (12.4)] at the forward time t_n and using Eq. (12.12) to evaluate $\{a\}_n$, $\{\dot{a}\}_n$, and $\{\ddot{a}\}_n$. (See Kohnke [12.51] for use of this relation in a large general-purpose FE program.)

12.3   Derive the "generalized Houbolt" recurrence relation, following the steps described after Eq. (12.15) but allowing for a variable time step; that is, $\Delta t_n = t_n - t_{n-1}$, $\Delta t_{n-1} = t_{n-1} - t_{n-2}$, $\Delta t_{n-2} = t_{n-2} - t_{n-3}$. (This relation is also used in Kohnke [12.51].) Show that the relation reduces to Eq. (12.14) when $\Delta t_n = \Delta t_{n-1} = \Delta t_{n-2} = \Delta t$.

12.4   Verify Eq. (12.20) by applying conditions (12.19) to Eq. (12.18).

12.5   Derive the Wilson recurrence relation, Eq. (12.24), following the steps described after Eqs. (12.23).

12.6   Complete the derivation of the Newmark cubic approximating polynomial, Eq. (12.28), by applying conditions (12.27) to Eq. (12.26).

12.7   Derive the Newmark recurrence relation, Eq. (12.36), following the steps described after Eq. (12.35).

12.8   Class or group project: Add a dynamics capability to the UNAFEM program by coding one (or more) of the four recurrence relations in Section 12.2. Mass matrix calculations will have to be added to one (or more) of the element routines, and a starting procedure (imposition of initial conditions) and input-output procedures will also have to be included.

12.9 Consider a single-DOF oscillator with no damping:

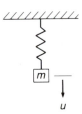

The governing differential equation is $m\ddot{u}+ku=f$. Let $m=k=1, f=0$, and assume the initial conditions at $t=0$ to be $u_0=0$, $\dot{u}_0=1$. (The exact solution is $u = \sin t$.)

a) Calculate the transient response using the central difference method. What is the value of Δt_{crit}? First use $\Delta t=\Delta t_{crit}/2$. Repeat the calculations using $\Delta t=2\Delta t_{crit}$.

b) Calculate the transient response using one or more of the other methods. Observe the varying amounts of *amplitude decay* ("numerical damping") and *period elongation*, depending on method, parameter values, and size of Δt. These are errors introduced by the algorithms [12.9-12.23]. Amplitude decay is desirable for the highest modes in an FE model since they are unwanted because their accuracy is much lower than that of the lower modes.

12.10 Consider a Caughey series [Eq. (12.55)] containing the terms from $l=0$ to $l=p$:

$$[C] = [M] \sum_{l=0}^{p} b_l([M]^{-1}[K])^l \tag{E12.10a}$$

Using Eqs. (12.47) and (12.53), derive the following generalization of Eq. (12.59):

$$\xi_i = \frac{1}{2}\left(\frac{b_0}{\omega_i} + b_1\omega_i + b_2\omega_i^3 + \dots + b_p\omega_i^{2p-1}\right) \qquad i = 1,2,\dots,N \tag{E12.10b}$$

12.11 For the single-DOF oscillator in Fig. 12.12, verify Eqs. (12.71), using Eqs. (12.64) through (12.67) and Eq. (12.70).

References

[12.1] J. D. Robson, *An Introduction to Random Vibration*, Elsevier Publishing Company, Amsterdam, 1964.

[12.2] S. H. Crandall and W. D. Mark, *Random Vibration in Mechanical Systems*, Academic Press, New York, 1963.

[12.3] R. H. Lyon, *Statistical Energy Analysis of Dynamical Systems: Theory and Applications*, MIT Press, Cambridge, Mass., 1975.

[12.4] R. W. Clough and J. Penzien, *Dynamics of Structures*, McGraw-Hill, New York, 1975.

[12.5] A. D. Fagan, "An Introduction to the Fast Fourier Transform," *Marconi Review*, **42**: 38-47 (1979).

[12.6]   T. Itoh, "Damped Vibration Mode Superposition Method for Dynamic Response Analysis," *Earthquake Eng Struct Dyn*, **2**: 47-57 (1973).

[12.7]   J. Penzien and S. Tseng, "Three-Dimensional Dynamic Analysis of Fixed Offshore Platforms" in O. C. Zienkiewicz, R. W. Lewis, and K. G. Stagg (eds.), *Numerical Methods in Offshore Engineering*, Wiley, New York, 1978.

[12.8]   H. H. Fong, "An Evaluation of Eight U.S. General Purpose Finite Element Computer Programs," Part I, *Proc 23d AIAA/ASME/ASCE/AHS Structures, Structural Dynamics and Materials Conference*, New Orleans, May 10-12, 1982, pp. 145-160.

[12.9]   John C. Houbolt, "A Recurrence Matrix Solution for the Dynamic Response of Elastic Aircraft," *J Aero Sci*, **17**: 540-550 (1950).

[12.10] S. Levy and W. D. Kroll, "Errors Introduced by Finite Space and Time Increments in Dynamic Response Computation," *Proc 1st U.S. Nat Cong App Mech*, Illinois Inst. of Tech., Chicago, 1951, pp. 1-8.

[12.11] N. M. Newmark, "A Method of Computation for Structural Dynamics," *J Eng Mech Div, Proc Amer Soc Civil Eng*, **85**(EM3): 67-94 (1959).

[12.12] E. L. Wilson, "A Computer Program for the Dynamic Stress Analysis of Underground Structures," *Report UCSESM 68-1*, University of California, Berkeley, 1968.

[12.13] R. E. Nickell, "On the Stability of Approximation Operators in Problems of Structural Dynamics," *Int J Solids Structures*, **7**: 301-319 (1971).

[12.14] G. L. Goudreau and R. L. Taylor, "Evaluation of Numerical Integration Methods in Elastodynamics," *Comp Meth App Mech Eng*, **2**: 69-97 (1972).

[12.15] R. S. Dunham, R. E. Nickell, and D. C. Stickler, "Integration Operators for Transient Structural Response," *Computers and Structures*, **2**: 1-15 (1972).

[12.16] E. L. Wilson, I. Farhoomand, and K. J. Bathe, "Nonlinear Dynamic Analysis of Complex Structures," *Earthquake Eng Struct Dyn*, **1**: 241-252 (1973).

[12.17] K. J. Bathe and E. L. Wilson, "Stability and Accuracy Analysis of Direct Integration Methods," *Earthquake Eng Struct Dyn*, **1**: 283-291 (1973).

[12.18] R. D. Krieg and S. W. Key, "Transient Shell Response by Numerical Time Integration," *Int J Num Meth Eng*, **7**: 273-286 (1973).

[12.19] Ted Belytschko, "Transient Analysis" in W. Pilkey, K. Saczalski, and H. Schaeffer (eds.), *Structural Mechanics Computer Programs*, University Press of Virginia, Charlottsville, 1974.

[12.20] T. Belytschko, R. L. Chiapetta, and H. D. Bartel, "Efficient Large Scale Non-Linear Transient Analysis by Finite Elements," *Int J Num Meth Eng*, **10**: 579-596 (1976).

[12.21] D. Shantram, D. R. J. Owen, and O. C. Zienkiewicz, "Dynamic Transient Behavior of Two- and Three-Dimensional Structures Including Plasticity, Large Deformation Effects and Fluid Interaction," *Earthquake Eng Struct Dyn*, **4**: 561-578 (1976).

[12.22] K. J. Bathe and E. L. Wilson, *Numerical Methods in Finite Element Analysis*, Prentice-Hall, Englewood Cliffs, N. J., 1976, Chapter 9.

[12.23] S. Levy and J. P. D. Wilkinson, *The Component Element Method in Dynamics*, McGraw-Hill, New York, 1976, Chapter 1.

[12.24] P. Underwood and K. C. Park, "STINT/CD: A Stand-Alone Explicit Time Integration Package for Structural Dynamics Analysis," *Int J Num Meth Eng*, **17**: 1285-1312 (1981).

[12.25] H. M. Hilber, T. J. R. Hughes, and R. L. Taylor, "Improved Numerical Dissipation for Time Integration Algorithms in Structural Dynamics," *Earthquake Eng Struct Dyn*, **5**: 283-292 (1977).

[12.26] O. C. Zienkiewicz, "A New Look at the Newmark, Houbolt, and Other Time Stepping Formulas: A Weighted Residual Approach," *Earthquake Eng Struct Dyn*, **5**: 413-418 (1977).

[12.27] O. C. Zienkiewicz, *The Finite Element Method*, McGraw-Hill, New York, 1977, Chapter 21.

[12.28] G. F. Howard and J. E. T. Penny, "The Accuracy and Stability of Time Domain Finite Element Solutions," *J Sound Vibration*, **61**(4): 585-595 (1978).

[12.29] H. M. Hilber and T. J. R. Hughes, "Collocaton, Dissipation and 'Overshoot' for Time Integration Schemes in Structural Dynamics," *Earthquake Eng Struct Dyn*, **6**: 99-117 (1978).

[12.30] T. J. R. Hughes and W. K. Liu, "Implicit-Explicit Finite Elements in Transient Analysis: Stability Theory," *J App Mech*, **45**: 371-374 (1978).

[12.31] T. J. R. Hughes and W. K. Liu, "Implicit-Explicit Finite Elements in Transient Analysis: Implementation and Numerical Examples," *J App Mech*, **45**: 375-378 (1978).

[12.32] J. H. Cushman, "Difference Schemes or Element Schemes?" *Int J Num Meth Eng*, **14**: 1643-1651 (1979).

[12.33] A. G. Collings and G. J. Tee, "The Solution of Structural Dynamics Problems by the Generalized Euler Method," *Computers and Structures*, **10**: 505-515 (1979).

[12.34] W. L. Wood, M. Bossak, and O. C. Zienkiewicz, "An Alpha Modification of Newmark's Method," *Int J Num Meth Eng*, **15**: 1562-1566 (1980).

[12.35] J. Braekhus and J. O. Aasen, "Experiments with Direct Integration Algorithms for Ordinary Differential Equations in Structural Dynamics," *Computers and Structures*, **13**: 91-96 (1981).

[12.36] K. C. Park and J. M. Housner, "Semi-Implicit Transient Analysis Procedures for Structural Dynamics Analysis," *Int J Num Meth Eng*, **18**: 609-622 (1982).

[12.37] M. Chi and J. Tucker, "Integration Methods for Stiff Systems," in N. Perrone and W. Pilkey (eds.), *Structural Mechanics Software Series*, University Press of Virginia, Charlottesville, 1982.

[12.38] C. R. Rogers, "Substructuring as an Everyday Design Tool," ASME Paper No. 77-PVP-27, September, 1977.

[12.39] F. B. Hildebrand, *Methods of Applied Mathematics*, Prentice-Hall, Englewood Cliffs, N. J., 1952, Chapter 1.

[12.40] Y. Chen, *Vibrations: Theoretical Methods*, Addison-Wesley, Reading, Mass., 1966.

[12.41] R. R. Craig, Jr., *Structural Dynamics: An Introduction to Computer Methods*, Wiley, New York, 1981.

[12.42] T. K. Caughey, "Classical Normal Modes in Damped Linear Dynamic Systems," *J Appl Mech*, **27**: 269-271 (1960).

[12.43]   E. L. Wilson and J. Penzien, ''Evaluation of Orthogonal Damping Matrices,'' *Int J Num Meth Eng*, **4**: 5-10 (1972).

[12.44]   K. A. Foss, ''Coordinates Which Uncouple the Equations of Motion of Damped Linear Dynamic Systems,'' *J Appl Mech*, **25**: 361-364 (1958).

[12.45]   W. C. Hurty and M. F. Rubinstein, *Dynamics of Structures*, Prentice-Hall, Englewood Cliffs, N. J., 1964.

[12.46]   W. T. Thomson, T. Calkins, and P. Caravani, ''A Numerical Study of Damping,'' *Earthquake Eng Struct Dyn*, **3**: 97-103 (1974).

[12.47]   E. L. Wilson, M. W. Yuan, and J. M. Dickens, ''Dynamic Analysis by Direct Superposition of Ritz Vectors,'' *Earthquake Eng Struct Dyn*, **10**: 813-821 (1982).

[12.48]   P. M. Morse and H. Feshbach, *Methods of Theoretical Physics, Part I*, McGraw-Hill, New York, 1953, Chapter 7.

[12.49]   G. E. Shilov, *An Introduction to Linear Spaces*, Prentice-Hall, Englewood Cliffs, N. J. 1961, Chapter 12.

[12.50]   J. C. Snowdon, *Vibration and Shock in Damped Mechanical Systems*, Wiley, New York, 1968.

[12.51]   P. C. Kohnke, *ANSYS Engineering Analysis System Theoretical Manual*, Swanson Analysis Systems, Inc., Houston, Pa., 1977 (revised 1983).

P A R T

IV

Two-Dimensional Problems: Theory, Programming, and Applications

13

![2-D Boundary-Value Problems banner]

2-D Boundary-Value Problems

Introduction

This chapter begins the analysis of 2-D problems. It will be assumed that the reader is familiar with basic concepts from multiple-variable and vector calculus — e.g., gradient, normal derivative, divergence, divergence theorem, and line integrals.

This chapter is similar in structure to Chapters 6, 7, and 8 combined. The mathematical and physical description in Section 13.1 is the 2-D analogue of the material in Chapter 6. In addition, there is a brief explanation of the concept of anisotropy, which arises more in 2-D problems than in 1-D. Section 13.2 performs steps 1, 2, and 3 of the theoretical analysis (see Section 5.9), but employs the 2-D analogue of integration by parts, namely, the 2-D divergence theorem. Sections 13.3 and 13.4 develop four 2-D elements. The first one is the simplest possible: a C^0-linear that is integrated exactly, analogous to the 1-D C^0-linear. The other three are isoparametric elements using numerical integration, and here the proper quadrature rules and optimal flux-sampling points are determined with the help of the guidelines and theory presented in Chapters 7 and 8. This completes the theoretical derivation of the element equations.

The remaining sections follow the usual procedure of implementing the theory into a program and then applying the program to an illustrative problem. Thus Section 13.5 adds the element equations to the UNAFEM program, Section 13.6 presents guidelines for 2-D mesh construction, and finally Section 13.7 works through a complete illustrative problem in elasticity, namely, the torsion of an I-beam.

13.1 Description of the Problem

13.1.1 Mathematical Description

This chapter will treat boundary-value problems that are the 2-D analogue of the 1-D problems described in Chapter 6. The general governing differential equation has the form

$$-\frac{\partial}{\partial x}\left(\alpha_x(x,y)\frac{\partial U(x,y)}{\partial x}\right) - \frac{\partial}{\partial y}\left(\alpha_y(x,y)\frac{\partial U(x,y)}{\partial y}\right) + \beta(x,y)U(x,y) = f(x,y) \tag{13.1}$$

where $U(x,y)$ is the unknown function (state variable); $\alpha_x(x,y)$, $\alpha_y(x,y)$, and $\beta(x,y)$ are known physical properties; and $f(x,y)$ is a known interior load. The various physical meanings of these quantities are similar to those in the 1-D problem; specific examples are cited below in Section 13.1.3.

The domain is typically a finite,[1] closed region in the x,y-plane, possibly containing interior holes, as suggested by Fig. 13.1. The vector $\{n\}$ is the *outward* unit normal to the boundary:

$$\{n\} = \begin{Bmatrix} n_x \\ n_y \end{Bmatrix}$$

(13.2)

where the components n_x and n_y are the direction cosines of the vector (see inset in Fig. 13.1). For a curved boundary, the values of n_x and n_y will vary along the boundary. Hence, in general, $n_x = n_x(s)$ and $n_y = n_y(s)$, where s is a coordinate measuring position along the boundary.

We will consider the two usual types[2] of BCs.

Essential BC • Specify the function $U(s)$ along part or all of the boundary.

Natural BC • Specify the component of flux normal to the boundary, $\tau_n(s)$, along part or all of the boundary.

(13.3)

The flux vector $\{\tau\}$ has components τ_x and τ_y:

$$\{\tau\} = \begin{Bmatrix} \tau_x \\ \tau_y \end{Bmatrix}$$

(13.4)

The normal component, τ_n, is the projection of $\{\tau\}$ along the outward unit normal, $\{n\}$, as shown in Fig. 13.2. This projection is expressed by the inner product,[3] $\{\tau\}^T\{n\}$:

$$\tau_n = \{\tau\}^T\{n\} = \{\tau_x, \tau_y\}\begin{Bmatrix} n_x \\ n_y \end{Bmatrix}$$

$$= \tau_x n_x + \tau_y n_y$$

(13.5)

How many BCs must be specified along the boundary? Recall from Section 4.5 that if a differential equation is of order $2m$, then m BCs must be specified at each boundary

---

[1] Infinite regions may also be treated, using any of the techniques in Section 10.6.

[2] As noted in footnotes in Sections 6.1.1 and 6.5, there is also a third type, called a mixed BC, which is a linear combination of the other two. Although of some practical importance, such as in heat conduction problems where it represents a convective BC, it was felt that its inclusion in the theoretical development and the UNAFEM program would be a distracting detail, an unnecessary embellishment, rather than a fundamentally important addition. In fact, from a programming standpoint, it is treated similarly to a natural BC. For these reasons, mixed BCs are not included in the above treatment.

[3] Some readers, like the author, may be more familiar with the vector notation that employs unit vectors $\boldsymbol{i}$ and $\boldsymbol{j}$ in the x and y directions, respectively. The flux and normal vectors are then expressed as $\tau = \tau_x \boldsymbol{i} + \tau_y \boldsymbol{j}$ and $\boldsymbol{n} = n_x \boldsymbol{i} + n_y \boldsymbol{j}$, and the inner product is written as the "dot product," $\tau \cdot \boldsymbol{n}$. The matrix notation used above is a more natural notation for computer programming.

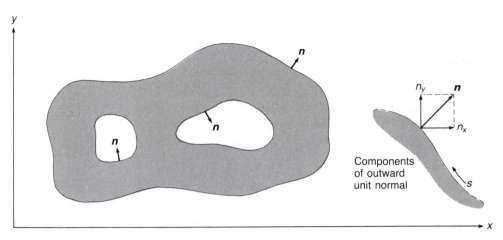

FIGURE 13.1 ● General domain for a 2-D boundary-value problem. The inset defines components of the outward unit normal vector.

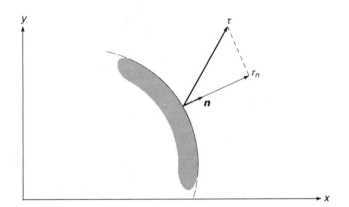

FIGURE 13.2 ● Component of flux normal to the boundary, τ_n.

point.[4] Since Eq. (13.1) is of second order (hence $m=1$), then *one* BC (an essential or natural, say, or a mixed if we were to include that type) must be specified at every point along the boundary. The type of BC may vary from point to point, subject only to the requirement that the problem be "well posed."

The requirements for a well-posed problem, as well as the underlying mathematical and physical rationale, are the same as for the 1-D problem discussed in Section 6.6. Thus, if $\beta(x)=0$, an essential BC must be specified at at least one point along the boundary. This is because if only natural BCs were specified along the entire boundary, then the total problem formulation (differential equation and BCs) would contain only derivatives of U, so the solution could only be determined up to an arbitrary additive constant, i.e., a nonunique solution. If $\beta(x)\neq0$, then this nonuniqueness problem cannot arise, so any mix of BCs is permissible.

---

[4] More generally, if there are n governing differential equations, each of order $2m$, then $n\times m$ BCs must be specified at each boundary point (see Chapter 16).

Equation (13.1) is the result of combining two separate equations (analogous to the 1-D problem in Section 6.5).

A balance equation (or conservation principle) •

$$\{\nabla\}^T\{\tau(x,y)\} + \beta(x,y)U(x,y) = f(x,y) \tag{13.6}$$

where $\{\nabla\}$, pronounced "del," is the *gradient* operator:

$$\{\nabla\} = \left\{ \begin{array}{c} \dfrac{\partial}{\partial x} \\[2mm] \dfrac{\partial}{\partial y} \end{array} \right\} \tag{13.7}$$

and the inner product $\{\nabla\}^T\{\tau\}$ is the *divergence* of the flux:[5]

$$\{\nabla\}^T\{\tau\} = \left\{ \dfrac{\partial}{\partial x}, \dfrac{\partial}{\partial y} \right\} \left\{ \begin{array}{c} \tau_x \\[1mm] \tau_y \end{array} \right\}$$

$$= \dfrac{\partial \tau_x}{\partial x} + \dfrac{\partial \tau_y}{\partial y} \tag{13.8}$$

Substituting Eq. (13.8) into Eq. (13.6) yields

$$\dfrac{\partial \tau_x(x,y)}{\partial x} + \dfrac{\partial \tau_y(x,y)}{\partial y} + \beta(x,y)U(x,y) = f(x,y) \tag{13.9}$$

A constitutive equation (or physical law) •

$$\{\tau(x,y)\} = -[\alpha(x,y)]\{\nabla U(x,y)\} \tag{13.10}$$

where $[\alpha]$ is a matrix of material (or physical) properties:[6]

$$[\alpha] = \begin{bmatrix} \alpha_x & \alpha_{xy} \\ \alpha_{xy} & \alpha_y \end{bmatrix} \tag{13.11}$$

and $\{\nabla U\}$ is the gradient of U:

$$\{\nabla U\} = \left\{ \begin{array}{c} \dfrac{\partial U}{\partial x} \\[2mm] \dfrac{\partial U}{\partial y} \end{array} \right\} \tag{13.12}$$

Using Eqs. (13.4), (13.11), and (13.12), Eq. (13.10) may be written out in full as follows:

$$\tau_x = -\alpha_x \dfrac{\partial U}{\partial x} - \alpha_{xy} \dfrac{\partial U}{\partial y}$$

$$\tau_y = -\alpha_{xy} \dfrac{\partial U}{\partial x} - \alpha_y \dfrac{\partial U}{\partial y} \tag{13.13}$$

---

[5] Using the vector notation mentioned in a previous footnote, the gradient operator would be written $\nabla = (\partial/\partial x)i + (\partial/\partial y)j$, and the divergence would be expressed as the dot product $\nabla \cdot \tau$.

[6] The $[\alpha]$ matrix in Eq. (13.11) is symmetric. This is a manifestation of an empirically observed *reciprocity* between gradient and flux (or, loosely speaking, a symmetry between cause and effect). This, in turn, will make Eq. (13.15) self-adjoint (recall Section 3.2).

Substituting Eq. (13.10) into Eq. (13.6) yields[7]

$$-\{\nabla\}^T([\alpha]\{\nabla U\}) + \beta U = f \tag{13.14}$$

Carrying out the matrix multiplications in Eq. (13.14) [or, equivalently, substituting Eqs. (13.13) into Eq. (13.9)] yields

$$-\frac{\partial}{\partial x}\left(\alpha_x \frac{\partial U}{\partial x}\right) - \frac{\partial}{\partial x}\left(\alpha_{xy} \frac{\partial U}{\partial y}\right) - \frac{\partial}{\partial y}\left(\alpha_{xy} \frac{\partial U}{\partial x}\right) - \frac{\partial}{\partial y}\left(\alpha_y \frac{\partial U}{\partial y}\right) + \beta U = f \tag{13.15}$$

Equation (13.14), or its expanded form, Eq. (13.15), is called the *quasiharmonic* equation.[8]

There are several special forms of the quasiharmonic equation that occur quite frequently.

1. $[\alpha]$ is a diagonal matrix; that is, $\alpha_{xy}(x,y)=0$:

$$[\alpha] = \begin{bmatrix} \alpha_x & 0 \\ 0 & \alpha_y \end{bmatrix} \tag{13.16}$$

so that

$$\tau_x = -\alpha_x \frac{\partial U}{\partial x}$$

$$\tau_y = -\alpha_y \frac{\partial U}{\partial y} \tag{13.17}$$

Note that the two directions are *uncoupled*. Thus a gradient in the x-direction produces a flux only in the x-direction, i.e., parallel to the gradient. Similarly, a gradient in the y-direction produces a flux only in the y-direction.

Equation (13.15) becomes

$$-\frac{\partial}{\partial x}\left(\alpha_x \frac{\partial U}{\partial x}\right) - \frac{\partial}{\partial y}\left(\alpha_y \frac{\partial U}{\partial y}\right) + \beta U = f \tag{13.18}$$

This is our original Eq. (13.1). Section 13.1.2 will discuss the physical meaning of a diagonal $[\alpha]$ matrix and explain why this chapter will analyze Eq. (13.18) rather than the more general Eq. (13.15).

2. $[\alpha]$ is a (nonzero) constant times the identity matrix; that is, $\alpha_{xy}(x,y)=0$ and $\alpha_x(x,y)=\alpha_y(x,y)=\alpha$:

$$[\alpha] = \alpha \begin{bmatrix} 1 & 0 \\ 0 & 1 \end{bmatrix} \tag{13.19}$$

This too has a special physical meaning that will be explained in Section 13.1.2.

---

[7] Again, using the vector notation described in previous footnotes, Eq. (13.14) would be written $-\nabla\cdot(\alpha\nabla U)+\beta U=f$.

[8] Equations (13.14) and (13.15) are the 2-D form of the quasiharmonic equation. The 3-D form follows rather obviously by letting $[\alpha]$ be a 3-by-3 matrix and

$$\{\nabla\}^T = \left\{\frac{\partial}{\partial x}, \frac{\partial}{\partial y}, \frac{\partial}{\partial z}\right\}$$

Equation (13.15) becomes

$$-\alpha \nabla^2 U + \beta U = f \tag{13.20}$$

where

$$\nabla^2 U = \frac{\partial^2 U}{\partial x^2} + \frac{\partial^2 U}{\partial y^2}$$

Here $\nabla^2 = \{\nabla\}^T \{\nabla\} = \partial^2/\partial x^2 + \partial^2/\partial y^2$ is the *Laplacian* operator.

3. This is the same case as in 2, but in addition $\beta(x,y)=0$, so Eq. (13.20) becomes

$$\nabla^2 U = -\frac{f}{\alpha} \tag{13.21}$$

which is *Poisson's* equation, one of the most famous of the classic field equations.

4. This is the same case as in 3, but in addition $f(x,y)=0$, so Eq. (13.21) becomes

$$\nabla^2 U = 0 \tag{13.22}$$

which is the *Laplace*, or *harmonic*, equation, and perhaps *the* most famous of the classic field equations.

13.1.2 Anisotropic Materials

Anisotropy is a concept that refers to the "directionality" of a material property. Being a material (physical) property, it affects the structure of the constitutive equation, not the balance equation. In particular, it determines the *relative* magnitudes of the terms in the $[\alpha]$ matrix. For example, in Eq. (13.13), if $\alpha_{xy}=0$ and $\alpha_x > \alpha_y$, and if there is an equal gradient in both directions — that is, $\partial U/\partial x = \partial U/\partial y$, then the resulting flux would be greater in the x-direction. In other words, a gradient of constant magnitude produces different responses in different directions.

Such a material is said to be *anisotropic*, or *directional*, or to have a *preferred orientation*. Conversely, a material whose properties are independent of direction ($\alpha_{xy}=0$ and $\alpha_x = \alpha_y$; see case 2 above) is said to be *isotropic*.[9]

The following discussion is only a brief introduction to this topic, sufficient to explain the meaning and limitations of Eq. (13.1). The discussion is restricted to 2-D, although the comments extend naturally to 3-D. Indeed, there is some advantage to developing the general theory in a 3-D context and then specializing it to 2-D. Such an approach provides a more complete picture and also provides guidelines (see below) on how, or whether, a problem with anisotropic materials can be modeled as 2-D (or 1-D). For a more complete mathematical treatment of this kind, the reader is referred elsewhere [13.1, 13.2].

For an anisotropic material, the set of α coefficients will assume different values when the constitutive equations [Eqs. (13.13)] are written with respect to a set of axes that have a different orientation from the x,y-axes. To see this, consider a set of x',y'-axes rotated

---

[9] *Isotropic* comes from the Greek "iso-" and "-tropos," meaning "the same" and "turning." Thus the same response is observed as one turns (in different directions). The prefix "an-" means "not" or "without."

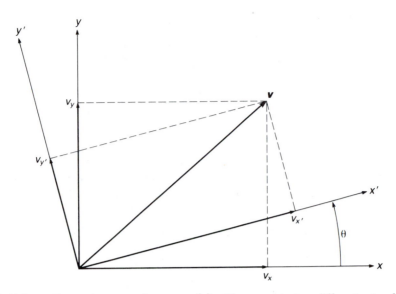

FIGURE 13.3 ● The components of a vector $\{v\}$ with respect to two different sets of axes.

an angle θ from the x,y-axes, as shown in Fig. 13.3. An arbitrary vector $\{v\}$ has different components with respect to each set of axes. By simple trigonometry, the components are related as follows:

$$\left\{\begin{matrix} v_{x'} \\ v_{y'} \end{matrix}\right\} = \begin{bmatrix} \cos\theta & \sin\theta \\ -\sin\theta & \cos\theta \end{bmatrix} \left\{\begin{matrix} v_x \\ v_y \end{matrix}\right\}$$

(13.23)

or,

$$\{v'\} = [R]\{v\}$$

where $[R]$ is a *rotation matrix*.

Next let's write the constitutive equations with respect to the x',y'-axes. Formally, we merely copy Eqs. (13.13), putting a prime over every x and y:

$$\tau_{x'} = -\alpha_{x'}\frac{\partial U}{\partial x'} - \alpha_{x'y'}\frac{\partial U}{\partial y'}$$

$$\tau_{y'} = -\alpha_{x'y'}\frac{\partial U}{\partial x'} - \alpha_{y'}\frac{\partial U}{\partial y'}$$

(13.24)

or,

$$\{\tau'\} = -[\alpha']\{\nabla'U\}$$

Then we use the general vector transformation equation [Eq. (13.23)] to transfer $\{\tau'\}$ and $\{\nabla'U\}$ to their components with respect to the x,y-axes:

$$\{\tau'\} = [R]\{\tau\}$$

$$\{\nabla'U\} = [R]\{\nabla U\}$$

(13.25)

Finally, substitute Eqs. (13.25) into Eqs. (13.24):

$$[R]\{\tau\} = -[\alpha'][R]\{\nabla U\} \tag{13.26}$$

and premultiply both sides by $[R]^T$ to yield[10]

$$\{\tau\} = -[R]^T[\alpha'][R]\{\nabla U\} \tag{13.27}$$

Comparing Eqs. (13.27) and (13.10) yields the desired relation:

$$[\alpha] = [R]^T[\alpha'][R] \tag{13.28}$$

Equation (13.28) shows how the α coefficients with respect to the x,y-axes are related to the coefficients with respect to any other axes. Carrying out the multiplications on the RHS of Eq. (13.28), and using trigonometric identities, yields the following transformation equations:

$$\alpha_x = \alpha_{x'} \cos^2 \theta - \alpha_{x'y'} \sin 2\theta + \alpha_{y'} \sin^2 \theta$$

$$\alpha_y = \alpha_{y'} \cos^2 \theta + \alpha_{x'y'} \sin 2\theta + \alpha_{x'} \sin^2 \theta$$

$$\alpha_{xy} = \alpha_{x'y'} \cos 2\theta + \frac{1}{2}\left(\alpha_{x'} - \alpha_{y'}\right) \sin 2\theta \tag{13.29}$$

We next make a very important observation: Since $[\alpha]$ is a real symmetric matrix, there exists at least one set of mutually perpendicular axes with respect to which the matrix is diagonal [13.4]. Such axes are called *principal axes*, and they will be denoted by a circumflex ($\hat{\ }$). The corresponding coefficients are the principal coefficients. Thus

$$[\hat{\alpha}] = \begin{bmatrix} \alpha_{\hat{x}} & 0 \\ 0 & \alpha_{\hat{y}} \end{bmatrix} \tag{13.30}$$

where $\alpha_{\hat{x}\hat{y}} = 0$ by definition.

In most materials the orientation of the principal axes bears an obvious relation to directions of geometric symmetry in the molecular structure or in the components of a composite material. (Less obvious orientations occur in some crystalline materials [13.5].) For example, in stratified or laminated materials the principal axes are usually parallel and perpendicular to the layers[11] as suggested by Fig. 13.4.

The concept of principal axes plays a central role in the treatment of anisotropic materials, both analytically and experimentally. This is because the number of α coefficients is reduced to a minimum when using principal axes, and this in turn yields considerable economization. For example, if the global x,y-axes for the quasiharmonic equation [Eq. (13.15)] are chosen parallel to the principal axes, then $\alpha_{xy}=0$, and the equation simplifies to Eq. (13.18), which is case 1 in Section 13.1.1. This, of course, means that fewer stiffness integrals will need to be evaluated.

We therefore have a useful guideline: If a problem uses an anisotropic material, and if the principal axes have the same orientation everywhere in the domain, then define the

---

[10] Here we make use of the fact that the rotation matrix is orthogonal [13.3], meaning that $[R]^T=[R]^{-1}$. Hence $[R]^T[R]=[I]$, the identity matrix, as may easily be verified.

[11] This assumes that the material *within* each layer, if anisotropic, has its own principal axes also aligned parallel and perpendicular to the layers.

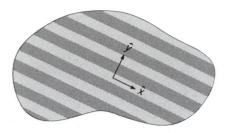

FIGURE 13.4 ● The usual direction of principal axes $\hat{x}$ and $\hat{y}$ in a stratified material.

global x,y-axes to be parallel with the principal axes (Fig. 13.5). Any other orientation would involve wasteful calculations [involving the conversion of the principal coefficients to some other set — see Eqs. (13.31) below — and then wasteful calculation of additional integrals because of the extra α_{xy} term].

If the problem employs materials with principal axes oriented in two or more different directions (Fig. 13.6), then it is impossible to align the global axes with all of them. In this case we are forced to solve Eq. (13.15), and the orientation of the global axes is unimportant. The coefficients α_x, α_y, α_{xy} for each material could be calculated from Eqs. (13.29) if they are known with respect to some other (x',y') orientation. They are usually known with respect to the principal axes, in which case Eqs. (13.29) simplify to

$$\alpha_x = \alpha_{\hat{x}} \cos^2 \theta + \alpha_{\hat{y}} \sin^2 \theta$$

$$\alpha_y = \alpha_{\hat{y}} \cos^2 \theta + \alpha_{\hat{x}} \sin^2 \theta$$

$$\alpha_{xy} = \frac{1}{2} \left(\alpha_{\hat{x}} - \alpha_{\hat{y}} \right) \sin 2\theta \tag{13.31}$$

Of course, the simplest situation to deal with is the use of isotropic materials. Fortunately, this occurs frequently in practice. For an isotropic material the principal coeffi-

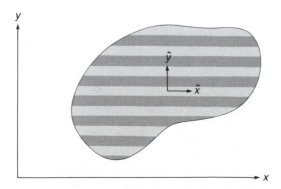

FIGURE 13.5 ● Align the global axes (x,y) with the principal axes $(\hat{x},\hat{y})$ for efficient analysis.

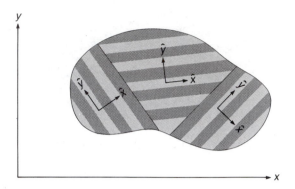

FIGURE 13.6 ● The orientation of the global axes is unimportant when the direction of the principal axes varies over the domain.

cients are identical: $\alpha_{\hat{x}}=\alpha_{\hat{y}}$, so $[\hat{\alpha}]$ is the identity matrix times a constant (α, say):

$$[\hat{\alpha}] = \alpha \begin{bmatrix} 1 & 0 \\ 0 & 1 \end{bmatrix} \tag{13.32}$$

Substituting Eq. (13.32) for $[\hat{\alpha}]$ in Eqs. (13.28) or (13.29) shows that $[\alpha]=\alpha[I]$. Thus $[\alpha]$ has the same diagonal form in all orientations. Consequently, all axes are principal, and the orientation of the global axes is unimportant. The isotropic material corresponds to case 2 in Section 13.1.1.

As noted previously, the above discussion of principal axes extends readily to 3-D problems. The $[\alpha]$ matrix is 3-by-3 so there are three principal coefficients: $\alpha_{\hat{x}}$, $\alpha_{\hat{y}}$, and $\alpha_{\hat{z}}$. With respect to nonprincipal axes there are six coefficients: α_x, α_y, α_z, α_{xy}, α_{xz}, and α_{yz}. Since real materials exist in three dimensions and therefore always have three principal axes, all of the above discussion of 2-D anisotropy implicitly made the following assumption: *One of the three principal axes is perpendicular to the plane of the 2-D model.* This is necessary to uncouple the third dimension from the other two. If the principal axes

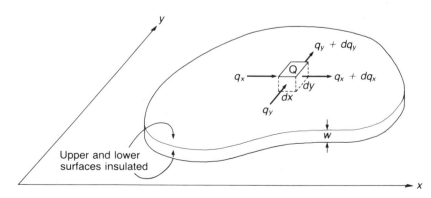

FIGURE 13.7 ● A model for 2-D heat conduction.

were skewed, then a gradient in the plane would produce a flux perpendicular to the plane, thereby invalidating the 2-D assumption.[12]

Let's now summarize the assumptions (i.e., restrictions) that are implied by Eq. (13.1), since this is the governing differential equation that will be analyzed in the remainder of this chapter. The material may be isotropic or anisotropic. If isotropic, there are no special restrictions. If anisotropic, the restrictions are as follows:

1. **2-D requirement:** One of the three principal axes must be perpendicular to the plane of the 2-D model.

2. The other two principal axes (which necessarily lie in the 2-D plane) must have the same orientation throughout the domain of the problem — i.e., as shown in Fig. 13.5, not Fig. 13.6.[13]

3. The global x,y-axes must be aligned with the principal axes, as shown in Fig. 13.5.

13.1.3 Physical Applications

Table 2.4 in Section 2.4 listed just a few of the many areas of applications of Eq. (13.1). The following examples provide a physical interpretation of the terms in Eq. (13.1) for four different physical applications. The reader should note the similarities (and differences) between these 2-D models and the 1-D models in Chapter 6.

Heat conduction • Figure 13.7 shows a thin layer of material, meaning that the x and y dimensions are much greater than the thickness t. The upper and lower surfaces are insulated so that heat is constrained to flow essentially parallel to the plane of the material. In analogy with the 1-D problems in Chapter 6, we could allow the thickness to vary gradually over the material — that is, $t(x,y)$ — and the resulting equations would directly parallel the 1-D equations. However, for simplicity, we will consider below the usual case of a constant thickness. Figure 13.7 is the 2-D analogue of Fig. 6.1(a). [A similar 2-D analogy could be made for Fig. 6.1(b).]

Using the infinitesimal element in Fig. 13.7, the balance-of-energy equation becomes

$$\frac{\partial q_x}{\partial x} + \frac{\partial q_y}{\partial y} = Q \tag{13.33}$$

---

[12] Before leaving this section, a final point of clarification may be helpful. Mathematically speaking, anisotropy is a *point* property. That is, within an infinitesimal neighborhood of any given point in the material, anisotropy refers to how the constitutive quantities (flux and gradient) vary in different *directions* emanating from that point. In most applications-oriented discussions, however, it is more natural to draw pictures of finite regions, as in the above figures. In these cases it is tacitly assumed that the anisotropic properties are uniform over the illustrated regions.

[13] It is this restriction that enables us to simplify the completely general quasiharmonic equation [Eq. (13.15)] to the form of Eq. (13.1). This was done so that the theoretical development in Sections 13.2 through 13.4 would be less cluttered. Thus only two stiffness integrals will need to be evaluated instead of four. Since all four integrals have the same structure, it was felt that the extra two added no additional insight, and hence constituted an unnecessary embellishment. Yet the resulting UNAFEM program will still permit modeling anisotropic materials. The ambitious student might want to modify UNAFEM to be able to handle Eq. (13.15).

where

q_x, q_y = components of heat flux in the x- and y-directions, respectively,[14] E/tL^2

Q = interior (volume) heat source, E/tL^3

The constitutive equation (Fourier's law) written with respect to principal axes is

$$
\begin{Bmatrix} q_x \\ q_y \end{Bmatrix} = - \begin{bmatrix} k_x & 0 \\ 0 & k_y \end{bmatrix} \begin{Bmatrix} \dfrac{\partial T}{\partial x} \\ \dfrac{\partial T}{\partial y} \end{Bmatrix}
\tag{13.34}
$$

where

T = temperature, T

k_x, k_y = thermal conductivities in the x- and y-directions, respectively, E/tLT

Substituting Eq. (13.34) into Eq. (13.33) yields the governing differential equation:

$$
- \frac{\partial}{\partial x} \left(k_x \frac{\partial T}{\partial x} \right) - \frac{\partial}{\partial y} \left(k_y \frac{\partial T}{\partial y} \right) = Q
\tag{13.35}
$$

which has the same form as Eq. (13.1).[15]

The 1-D model of the 2-D convection problem in Section 6.1.2 can be similarly extended to a 2-D model of a 3-D convection problem. By removing the insulation from the upper and lower surfaces in Fig. 13.7 and permitting convection from these surfaces, the balance-of-energy equation [Eq. (13.33)] becomes

$$
\frac{\partial q_x}{\partial x} + \frac{\partial q_y}{\partial y} + \frac{h}{t} (T - T_\infty) = Q
\tag{13.36}
$$

where

h = effective convective heat transfer coefficient, i.e., the sum of the two coefficients for the upper and lower surfaces, E/tL^2T

Substituting Eq. (13.34) into Eq. (13.36) yields

$$
- \frac{\partial}{\partial x} \left(k_x \frac{\partial T}{\partial x} \right) - \frac{\partial}{\partial y} \left(k_y \frac{\partial T}{\partial y} \right) + \frac{h}{t} T = Q + \frac{h T_\infty}{t}
\tag{13.37}
$$

which again has the same form as Eq. (13.1).

Transverse deflection of a flexible membrane • This problem is the 2-D analogue of the 1-D flexible cable in Section 6.3, and thus it makes the same small-deflection assumption. (Recall

---

[14] See Table 7 in the appendix for the physical dimensions corresponding to these symbols.

[15] It may sometimes be more convenient, and more meaningful in a 2-D context, to multiply the quantities q_x, q_y, k_x, k_y, and Q by the thickness. Then, for example, the dimensions of Q would be E/tL^2, i.e., power per unit area rather than per unit volume.

also Fig. 2.6 in Section 2.4.) The derivation proceeds in the same manner as for the above heat conduction problem (see Exercise 13.1). The resulting equation for balance of transverse forces is

$$\frac{\partial F_x}{\partial x} + \frac{\partial F_y}{\partial y} + kW = f \tag{13.38}$$

where

W = transverse displacement, L

F_x = transverse component of the x-component of tension, per unit length in y-direction, F/L

F_y = transverse component of the y-component of tension, per unit length in x-direction, F/L

k = elastic modulus of foundation, i.e., the transverse stiffness, F/L, per area of foundation, L^2

f = applied transverse force, distributed over area of membrane, F/L^2

The constitutive equation with respect to principal axes is

$$\begin{Bmatrix} F_x \\ \\ F_y \end{Bmatrix} = - \begin{bmatrix} T_x & 0 \\ \\ 0 & T_y \end{bmatrix} \begin{Bmatrix} \dfrac{\partial W}{\partial x} \\ \\ \dfrac{\partial W}{\partial y} \end{Bmatrix} \tag{13.39}$$

where

T_x = tension in x-direction, per unit length in y-direction, F/L

T_y = tension in y-direction, per unit length in x-direction, F/L

Note that the property of anisotropy in the constitutive equation is determined by the applied membrane tensions, T_x and T_y, and not by a material property (e.g., elastic stiffness) of the membrane.[16] Thus the membrane behaves isotropicly if and only if $T_x = T_y$, otherwise anisotropicly.

Substituting Eq. (13.39) into Eq. (13.38) yields the governing differential equation,

$$- \frac{\partial}{\partial x}\left(T_x \frac{\partial W}{\partial x} \right) - \frac{\partial}{\partial y}\left(T_y \frac{\partial W}{\partial y} \right) + kW = f \tag{13.40}$$

---

[16] For *small* transverse deflections, the primary "stiffness" that is resisting deflection is due to the transverse component of the applied membrane tension. This component of force is linearly proportional to the slope of the membrane [as expressed by Eq. (13.39) and illustrated in Fig. 6.5 for the cable]. Such a resistance is therefore like a linear spring, but it is caused by a geometric effect (projected component) from an applied tension. It is therefore sometimes referred to as *geometric stiffening* or *stress* (tension) *stiffening*. The intrinsic elastic stiffness of the membrane material is a secondary (second order) effect because it requires an in-plane stretching of the membrane material. Such a stretching is second order relative to the transverse deflection, and therefore is neglected in a small-displacement model.

Electrostatic field • The equation for conservation of electric charge (Gauss' law) is

$$\frac{\partial D_x}{\partial x} + \frac{\partial D_y}{\partial y} = \varrho \tag{13.41}$$

where

D_x, D_y = displacement of electric charge in the x- and y-directions, respectively, Q/L^2

ϱ = charge density, Q/L^3

The constitutive equation with respect to principal axes is

$$\left\{ \begin{matrix} D_x \\ \\ D_y \end{matrix} \right\} = - \begin{bmatrix} \epsilon_x & 0 \\ \\ 0 & \epsilon_y \end{bmatrix} \left\{ \begin{matrix} \dfrac{\partial \Phi}{\partial x} \\ \\ \dfrac{\partial \Phi}{\partial y} \end{matrix} \right\} \tag{13.42}$$

where

Φ = electrostatic potential, V

ϵ_x, ϵ_y = permittivities in the x- and y-directions, respectively, C/L

Substituting Eq. (13.42) into Eq. (13.41) yields the governing differential equation,

$$-\frac{\partial}{\partial x}\left(\epsilon_x \frac{\partial \Phi}{\partial x} \right) - \frac{\partial}{\partial y}\left(\dot{\epsilon}_y \frac{\partial \Phi}{\partial y} \right) = \varrho \tag{13.43}$$

In the very common situation that the dielectric medium is isotropic, Eq. (13.43) reduces to

$$\nabla^2 \Phi = -\frac{\varrho}{\epsilon} \tag{13.44}$$

which is Poisson's equation [Eq. (13.21)].

Flow of incompressible, inviscid fluid (hydrodynamics) • Conservation of mass is expressed by the equation of continuity. For an incompressible fluid the equation simplifies to the vanishing of the divergence of the velocity field:

$$\frac{\partial v_x}{\partial x} + \frac{\partial v_y}{\partial y} = 0 \tag{13.45}$$

where

v_x, v_y = components of fluid velocity in the x- and y-directions, respectively, L/t

The constitutive equation expresses the fact that the fluid is inviscid. Mathematically speaking, an inviscid fluid must be "irrotational," and this property in turn implies that the velocity vector is the gradient of a scalar [13.6]; that is,

$$\left\{ \begin{matrix} v_x \\ \\ v_y \end{matrix} \right\} = - \begin{bmatrix} 1 & 0 \\ \\ 0 & 1 \end{bmatrix} \left\{ \begin{matrix} \dfrac{\partial \Phi}{\partial x} \\ \\ \dfrac{\partial \Phi}{\partial y} \end{matrix} \right\} \tag{13.46}$$

where

$$\Phi = \text{velocity potential function, } L^2/t$$

The coefficient matrix in Eq. (13.46) is the identity matrix, reflecting the fact that the fluid is isotropic.

Substituting Eq. (13.46) into Eq. (13.45) yields the governing differential equation,

$$\nabla^2 \Phi = 0 \tag{13.47}$$

which is Laplace's equation [Eq. (13.22)].

13.2 Derivation of Element Equations: Steps 1 through 3

In this and the next section we will apply the first six steps of the 12-step procedure to Eq. (13.1). These steps were applied to 1-D problems in several of the previous chapters, and are summarized in Section 5.9. As before, we can always write down at the outset the general form for the typical element trial solution,

$$\tilde{U}^{(e)}(x,y;a) = \sum_{j=1}^{n} a_j \phi_j^{(e)}(x,y) \tag{13.48}$$

where n is the number of DOF in the element and $\phi_j^{(e)}(x,y)$ are the shape functions, which are now functions of both x and y for 2-D problems. We recall that the first three steps are short, formal operations using only the general form in Eq. (13.48), and it is not until step 4 (see Section 13.3) that we decide on the value of n and the specific form of each of the shape functions.

Step 1: Write the Galerkin residual equations for a typical element.

The residual for Eq. (13.1) is

$$R(x,y,;a) = -\frac{\partial}{\partial x}\left(\alpha_x(x,y)\frac{\partial \tilde{U}^{(e)}(x,y;a)}{\partial x}\right) - \frac{\partial}{\partial y}\left(\alpha_y(x,y)\frac{\partial \tilde{U}^{(e)}(x,y;a)}{\partial y}\right)$$
$$+ \beta(x,y)\tilde{U}^{(e)}(x,y;a) - f(x,y) \tag{13.49}$$

We need one weighted residual equation for each DOF in Eq. (13.48):

$$\int\int^{(e)} R(x,y;a)\phi_i^{(e)}(x,y) \, dx \, dy = 0 \qquad i = 1,2,\dots,n \tag{13.50}$$

where the double integral $\int\int^{(e)} dx \, dy$ means integrate over the area of element (e). We will see in step 4 below that 2-D elements are triangles or quadrilaterals (as illustrated in several figures in Chapter 1), so the integration will be over areas of these shapes.

Substituting Eq. (13.49) into Eq. (13.50) yields the desired residual equations,

$$\int\int^{(e)} \left[-\frac{\partial}{\partial x}\left(\alpha_x(x,y)\frac{\partial \tilde{U}^{(e)}(x,y;a)}{\partial x}\right) - \frac{\partial}{\partial y}\left(\alpha_y(x,y)\frac{\partial \tilde{U}^{(e)}(x,y;a)}{\partial y}\right) \right.$$
$$\left. + \beta(x,y)\tilde{U}^{(e)}(x,y;a) - f(x,y) \right] \phi_i^{(e)}(x,y) \, dx \, dy = 0 \qquad i = 1,2,\dots,n \tag{13.51}$$

Step 2: Integrate by parts.

The reader should note how the following development parallels the 1-D development in step 2 in Section 4.3. In particular, we must first apply the "chain rule" (for differentiating the product of two functions) to each of the derivative terms in Eq. (13.51).

$$\frac{\partial}{\partial x}\left(\alpha_x \frac{\partial \tilde{U}^{(e)}}{\partial x}\right)\phi_i^{(e)} = \frac{\partial}{\partial x}\left(\alpha_x \frac{\partial \tilde{U}^{(e)}}{\partial x}\phi_i^{(e)}\right) - \left(\alpha_x \frac{\partial \tilde{U}^{(e)}}{\partial x}\right)\frac{\partial \phi_i^{(e)}}{\partial x}$$

$$\frac{\partial}{\partial y}\left(\alpha_y \frac{\partial \tilde{U}^{(e)}}{\partial y}\right)\phi_i^{(e)} = \frac{\partial}{\partial y}\left(\alpha_y \frac{\partial \tilde{U}^{(e)}}{\partial y}\phi_i^{(e)}\right) - \left(\alpha_y \frac{\partial \tilde{U}^{(e)}}{\partial y}\right)\frac{\partial \phi_i^{(e)}}{\partial y} \qquad \text{(13.52)}$$

Substituting Eq. (13.52) into Eq. (13.51) yields

$$-\iint\limits^{(e)}\left[\frac{\partial}{\partial x}\left(\alpha_x \frac{\partial \tilde{U}^{(e)}}{\partial x}\phi_i^{(e)}\right) + \frac{\partial}{\partial y}\left(\alpha_y \frac{\partial \tilde{U}^{(e)}}{\partial y}\phi_i^{(e)}\right)\right] dx\, dy$$

$$+\iint\limits^{(e)}\left[\left(\alpha_x \frac{\partial \tilde{U}^{(e)}}{\partial x}\right)\frac{\partial \phi_i^{(e)}}{\partial x} + \left(\alpha_y \frac{\partial \tilde{U}^{(e)}}{\partial y}\right)\frac{\partial \phi_i^{(e)}}{\partial y} + \beta\tilde{U}^{(e)}\phi_i^{(e)} - f\phi_i^{(e)}\right] dx\, dy = 0$$

$$i = 1,2,\ldots,n \qquad \text{(13.53)}$$

The first integral in Eq. (13.53) is the 2-D version of a perfect differential in 1-D [recall Eq. (4.8)]; i.e., it can be reduced to an integral over the *boundary* of the area. This is accomplished with the aid of the 2-D *divergence theorem* [13.6].[17] This theorem states that if $F(x,y)$ and $G(x,y)$ are two functions defined in a region in the x,y-plane, say the (e)th element, then

$$\iint\limits^{(e)}\left(\frac{\partial F}{\partial x} + \frac{\partial G}{\partial y}\right) dx\, dy = \oint\limits^{(e)}\left(Fn_x^{(e)} + Gn_y^{(e)}\right) ds \qquad \text{(13.54)}$$

where the single integral $\oint\limits^{(e)} ds$ is a line integral that integrates along the boundary of the element.[18] The closed-circle-with-arrow symbol, $\circlearrowleft$, indicates that the integration should be around the complete boundary and in the counterclockwise direction (from positive x towards positive y), as illustrated, for example, by the dashed line for the triangular element in Fig. 13.8. Here $n_x^{(e)}$ and $n_y^{(e)}$ are the direction cosines, i.e., x and y components, of the outward unit normal to the element boundary, and s is a coordinate along the boundary.

Using Eq. (13.54), the first integral in Eq. (13.53) may be written

$$-\oint\limits^{(e)}\left(\alpha_x \frac{\partial \tilde{U}^{(e)}}{\partial x}\phi_i^{(e)}n_x^{(e)} + \alpha_y \frac{\partial \tilde{U}^{(e)}}{\partial y}\phi_i^{(e)}n_y^{(e)}\right) ds \qquad \text{(13.55)}$$

---

[17] This is a corollary of *Green's theorem* for line integrals. Its extension to 3-D is called simply the *divergence theorem* or *Gauss' theorem*.

[18] Letting F and G be the x and y components of a vector v — that is, $v = Fi + Gj$ — Eq. (13.54) in vector notation would be $\iint\limits^{(e)} \nabla\cdot v\, dx\, dy = \oint\limits^{(e)} v\cdot n\, ds$.

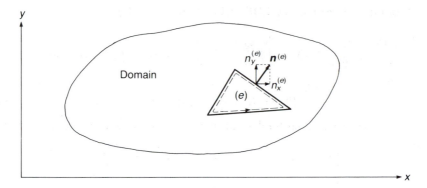

FIGURE 13.8 ● The dashed line is the path of integration for the boundary integral in Eq. (13.54), illustrated here for a triangular element.

This can be further simplified by using expressions (13.17) for the x and y flux components:

$$\tilde{\tau}_x^{(e)} = -\alpha_x \frac{\partial \tilde{U}^{(e)}}{\partial x}$$

$$\tilde{\tau}_y^{(e)} = -\alpha_y \frac{\partial \tilde{U}^{(e)}}{\partial y} \tag{13.56}$$

as well as relation (13.5) for the outward-normal component of the flux,

$$\tilde{\tau}_n^{(e)} = \tilde{\tau}_x^{(e)} n_x^{(e)} + \tilde{\tau}_y^{(e)} n_y^{(e)} \tag{13.57}$$

Substituting Eqs. (13.56) and (13.57) into Eq. (13.55) reduces the boundary integral to the relatively simple form,

$$\oint^{(e)} \tilde{\tau}_n^{(e)} \phi_i^{(e)} \, ds \tag{13.58}$$

This, of course, is the 2-D analogy of the boundary flux that we encountered in all the 1-D analyses.

Replacing the first integral in Eq. (13.53) with Eq. (13.58), the residual equations become

$$\iint^{(e)} \left[\left(\alpha_x \frac{\partial \tilde{U}^{(e)}}{\partial x} \right) \frac{\partial \phi_i^{(e)}}{\partial x} + \left(\alpha_y \frac{\partial \tilde{U}^{(e)}}{\partial y} \right) \frac{\partial \phi_i^{(e)}}{\partial y} + \beta \tilde{U}^{(e)} \phi_i^{(e)} \right] dx \, dy$$

$$= \iint^{(e)} f \phi_i^{(e)} \, dx \, dy - \oint^{(e)} \tilde{\tau}_n^{(e)} \phi_i^{(e)} \, ds \qquad i = 1, 2, \dots, n \tag{13.59}$$

As usual we have placed all load terms on the RHS; this includes the interior loads and the boundary fluxes. The reader should note the direct parallel between the 1-D and 2-D residual equations in Eqs. (7.8) and (13.59), respectively.

Step 3: Substitute the general form of the element trial solution into interior integrals in residual equations.

Substituting Eq. (13.48) into the LHS of Eq. (13.59) yields

$$\sum_{j=1}^{n} \left(\int^{(e)}\!\!\!\int \frac{\partial \phi_i^{(e)}}{\partial x} \alpha_x \frac{\partial \phi_j^{(e)}}{\partial x} \, dx \, dy + \int^{(e)}\!\!\!\int \frac{\partial \phi_i^{(e)}}{\partial y} \alpha_y \frac{\partial \phi_j^{(e)}}{\partial y} \, dx \, dy + \int^{(e)}\!\!\!\int \phi_i^{(e)} \beta \phi_j^{(e)} \, dx \, dy \right) a_j$$

$$= \int^{(e)}\!\!\!\int f \phi_i^{(e)} \, dx \, dy + \oint \tilde{\tau}_{-n}^{(e)} \phi_i^{(e)} \, ds \qquad i = 1,2,...,n \qquad (13.60)$$

These are the element equations for the typical element. Writing them also in matrix form, we have

$$
\begin{bmatrix}
K_{11}^{(e)} & K_{12}^{(e)} & \cdots & K_{1n}^{(e)} \\
K_{21}^{(e)} & K_{22}^{(e)} & \cdots & K_{2n}^{(e)} \\
\cdot & \cdot & & \cdot \\
\cdot & \cdot & & \cdot \\
\cdot & \cdot & & \cdot \\
K_{n1}^{(e)} & K_{n2}^{(e)} & \cdots & K_{nn}^{(e)}
\end{bmatrix}
\begin{Bmatrix}
a_1 \\ a_2 \\ \cdot \\ \cdot \\ \cdot \\ a_n
\end{Bmatrix}
=
\begin{Bmatrix}
F_1^{(e)} \\ F_2^{(e)} \\ \cdot \\ \cdot \\ \cdot \\ F_n^{(e)}
\end{Bmatrix}
\qquad (13.61a)
$$

where the expressions for the stiffness and load terms are, respectively,

$$K_{ij}^{(e)} = \int^{(e)}\!\!\!\int \frac{\partial \phi_i^{(e)}}{\partial x} \alpha_x \frac{\partial \phi_j^{(e)}}{\partial x} \, dx \, dy + \int^{(e)}\!\!\!\int \frac{\partial \phi_i^{(e)}}{\partial y} \alpha_y \frac{\partial \phi_j^{(e)}}{\partial y} \, dx \, dy + \int^{(e)}\!\!\!\int \phi_i^{(e)} \beta \phi_j^{(e)} \, dx \, dy$$

$$F_i^{(e)} = \int^{(e)}\!\!\!\int f \phi_i^{(e)} \, dx \, dy + \oint \tilde{\tau}_{-n}^{(e)} \phi_i^{(e)} ds \qquad (13.61b)$$

Note the change in sign from minus to plus in front of the boundary flux integral and the corresponding change in notation from $\tilde{\tau}_n^{(e)}$ to $\tilde{\tau}_{-n}^{(e)}$. The subscript $-n$ indicates the

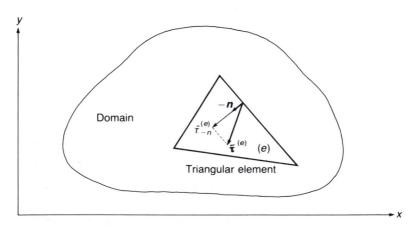

FIGURE 13.9 ● Component of flux along the inward normal vector of an element.

inward unit normal, as shown in Fig. 13.9, and consequently

$$\tilde{\tau}_{-n}^{(e)} = -\tilde{\tau}_n^{(e)} \qquad (13.62)$$

The quantity $\tilde{\tau}_{-n}^{(e)}$ is the *inward*-normal component of flux and thus represents energy being *added* to the system (which is consonant with the plus sign in front of the integral). This physical interpretation was discussed in the context of 1-D problems in Section 7.4.2 [recall Fig. 7.27 and Eqs. (7.48) through (7.50)].

13.3 Triangular Elements
Theoretical Development (Steps 4 through 6)

In Sections 13.3 and 13.4 we will complete the derivation of the element equations for four different types of elements. This will include developing the shape functions (step 4 of the 12-step procedure), preparing the integrals for numerical evaluation (step 5), and preparing expressions for evaluating the flux (step 6).

   The conditions for convergence are the same for all four elements. As explained in Section 5.6, those conditions are determined by the form of the element equations at the end of step 2. Thus, since the governing differential equation [Eq. (13.1)] is of second order $(2m=2)$, the order of the highest derivative of $\tilde{U}^{(e)}$ in Eq. (13.59) is $m=1$. Hence, for a conforming element, the element trial solution $\tilde{U}^{(e)}(x,y;a)$ will be represented by a polynomial which must be (1) complete at least to degree 1, that is, linear, (recall the "Completeness condition using polynomials" in Section 5.6), and (2) C^0-continuous across interelement boundaries. Of course, these are the same requirements used in Part III since that also dealt with a second-order equation.

   It might seem that 2-D elements could have an infinite variety of shapes, e.g., polygons (straight- or curved-sided), circles, ellipses, etc. However, of all such possible geometric figures, only three- and four-sided polygons (i.e., triangles and quadrilaterals, with straight or curved sides, as shown in Fig. 13.10) seem to be well suited. To the author's knowledge, these are the only shapes used in FE programs. Other shapes are more difficult to construct shape functions on, and most would make it difficult or impossible to subdivide a domain into a mesh. The four elements developed in this section correspond to each of the four shapes in Fig. 13.10.

13.3.1 The C^0-Linear Triangle

Step 4: Develop specific expressions for the shape functions.

The simplest of all elements that will just satisfy the above convergence conditions is a C^0-linear 2-D polynomial in a straight-sided triangular-shaped element. The reason for the triangular shape will become apparent in a moment.

   First let's examine the completeness condition. A 2-D polynomial has the general (power series) form,

$$\underbrace{a}_{\text{Constant}} \quad + \underbrace{bx + cy}_{\substack{\text{Linear} \\ \text{terms}}} + \underbrace{dx^2 + exy + fy^2}_{\substack{\text{Quadratic} \\ \text{terms}}} + \underset{\text{etc.}}{\cdots}$$

$$(13.63)$$

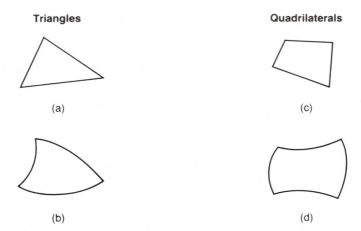

FIGURE 13.10 ● The four standard shapes used for 2-D elements: (a) straight-sided triangle; (b) curved-sided triangle; (c) straight-sided quadrilateral; (d) curved-sided quadrilateral.

A more meaningful way to display these terms is in a triangular array called *Pascal's triangle*, shown in Fig. 13.11. We see that every term has the form $x^r y^s$, where $r,s = 0,1,2,3,\ldots$. A term is said to be of degree p if $r + s = p$. Each row in Pascal's triangle contains all the terms of a given degree. Thus, the second row contains all the first-degree (linear) terms, the third row all the second-degree (quadratic) terms, etc. A 2-D polynomial is complete to degree p if it contains all terms up to and including those of degree p, i.e., all terms in the first $p+1$ rows of Pascal's triangle, as indicated in Fig. 13.11.

A complete linear polynomial therefore contains at least the three terms $1, x, y$, so the simplest form that our element trial solution can take is the following:

$$\tilde{U}^{(e)} = a + bx + cy \tag{13.64}$$

Now let's examine the continuity condition. It is here that we encounter a special problem that cannot arise in 1-D analyses. Recall that for 1-D elements interelement continuity is most efficiently achieved by using interpolation polynomials and placing an

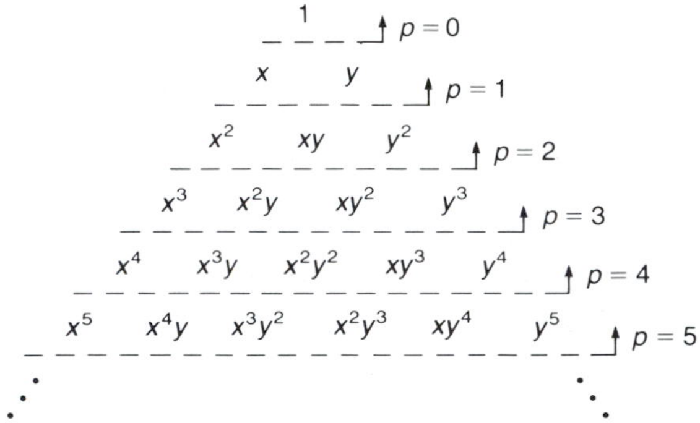

FIGURE 13.11 ● Pascal's triangle: A 2-D polynomial is complete to degree p if it contains all terms in the first $p+1$ rows of the triangle.

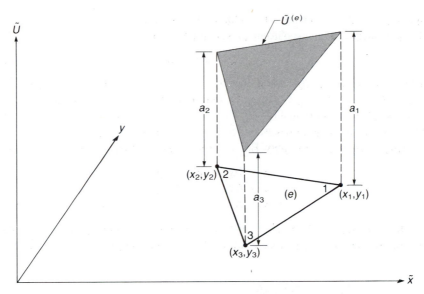

FIGURE 13.12 ● The C^0-linear triangular element and element trial solution.

interpolation node at each of the two boundary points. This ensures continuity at the nodes, which therefore ensures continuity across the *entire boundary*, since the boundary consists of only the two points. In other words, the nodes completely cover the boundary.

For 2-D (and higher-dimensional) elements, interelement continuity is also most efficiently achieved by using interpolation polynomials and placing interpolation nodes on the boundary. However, now the boundary is a line (or a surface in 3-D, etc.) and it obviously cannot be covered by a finite number of nodes, so we have the additional problem of ensuring interelement continuity along the entire boundary *between* the nodes. Rather than trying to explain in abstract terms how to accomplish this, it is probably easier to simply present the standard derivation of the shape functions with a minimum of explanation. Then afterwards we will look back and identify a general principle, applicable to all elements, that enabled these shape functions to achieve the desired continuity.

We begin by observing that the element trial solution in Eq. (13.64) has three DOF. This, of course, is because it is a complete 2-D linear polynomial which has three terms and therefore three coefficients (DOF). Just as with the 1-D elements, we will rewrite this polynomial in the form of an interpolation polynomial, and this will require three interpolation points, i.e., nodes. We will use a straight-sided triangle and place one node at each of the three vertices,[19] as shown in Fig. 13.12. The nodes are numbered counterclockwise to facilitate subsequent programming for the counterclockwise boundary integrals, and local node numbers 1, 2, 3 are employed.[20]

The element trial solution $\tilde{U}^{(e)}$ is shown in Fig. 13.12 by the shaded triangular surface above the x,y-plane. Since $\tilde{U}^{(e)}$ is linear, this surface is flat, though generally tilted

---

[19] It is this choice of a triangular shape and the placement of the nodes at the vertices that achieves the interelement continuity, as will be explained shortly.

[20] Only the local nodes in the typical (master) element ought to be numbered counterclockwise. Global nodes in a mesh of elements may be numbered in a random pattern (see Section 13.6.5).

at some angle to the x,y-plane. The value of $\tilde{U}^{(e)}$ at node i is the interpolatory DOF a_i (recall the principle of interpolation in Section 5.2), that is,

$$\tilde{U}^{(e)}(x_i, y_i; a) = a_i \tag{13.65}$$

The remainder of the derivation directly parallels the derivation for the 1-D C^0-linear element in Section 5.2, step 4. (In particular, note that Fig. 5.2 is identical to any one of the sides of the triangle in Fig. 13.12.) As before, we wish to express $\tilde{U}^{(e)}$ in Eq. (13.64) in terms of the parameters a_1, a_2, and a_3 rather than the parameters a, b, and c. One way to do this is to apply Eq. (13.65) to Eq. (13.64) [recall Eqs. (5.4)]:

$$\begin{aligned}
a + bx_1 + cy_1 &= a_1 \\
a + bx_2 + cy_2 &= a_2 \\
a + bx_3 + cy_3 &= a_3
\end{aligned} \tag{13.66}$$

Solving Eqs. (13.66) for a, b, and c, then substituting the resulting expressions into Eq. (13.64) and rearranging terms (see Exercise 13.2), yields the desired interpolatory form for $\tilde{U}^{(e)}$:

$$\tilde{U}^{(e)}(x,y;a) = \sum_{j=1}^{3} a_j \phi_j^{(e)}(x,y) \tag{13.67}$$

where

$$\phi_j^{(e)}(x,y) = \frac{a_j + b_j x + c_j y}{2\Delta} \qquad j = 1,2,3$$

and

$$a_j = x_k y_l - x_l y_k$$

$$b_j = y_k - y_l$$

$$c_j = x_l - x_k$$

$$\Delta = \frac{1}{2} \begin{vmatrix} 1 & x_1 & y_1 \\ 1 & x_2 & y_2 \\ 1 & x_3 & y_3 \end{vmatrix}$$

$$= \frac{1}{2} \left[(x_2 y_3 - x_3 y_2) - (x_1 y_3 - x_3 y_1) + (x_1 y_2 - x_2 y_1) \right]$$

$$= \text{area of element}$$

The subscripts j, k, l have the values 1, 2, 3 for $\phi_1^{(e)}(x,y)$ and are permuted cyclically for $\phi_2^{(e)}(x,y)$ and $\phi_3^{(e)}(x,y)$.

The functions $\phi_j^{(e)}(x,y)$ are the shape functions, and it is easily verified from Eqs. (13.67) that they possess the requisite interpolation property,

$$\phi_j^{(e)}(x_i, y_i) = \delta_{ji} \tag{13.68}$$

Figure 13.13 illustrates $\phi_1^{(e)}(x,y)$, the shape function associated with node 1, which equals 1 at node 1 and 0 at the other nodes. The other two shape functions have the same triangular shape but with the 1 and 0 nodal values cyclically permuted.

As explained in Section 5.2, we could alternatively have derived the shape functions by writing each one as a 2-D linear polynomial, as in Eq. (13.64), and applying the interpolation property [Eq. (13.68)] at each of the nodes.

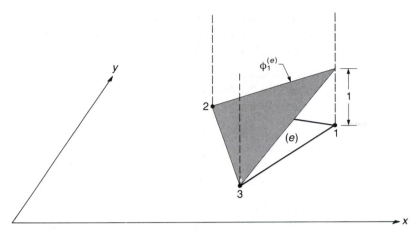

FIGURE 13.13 ● Shape function $\phi_1^{(e)}(x,y)$ associated with node 1 for the C^0-linear triangular element.

Now let's return to the interelement-continuity question and ask whether this element will provide C^0-continuity along its entire boundary. Consider one of the sides of the triangle in Fig. 13.12, say side $\overline{23}$. Note that $\tilde{U}^{(e)}$ is a straight line along side $\overline{23}$, i.e., between the parameters a_2 and a_3. This, of course, is because $\tilde{U}^{(e)}$ is a linear polynomial. Since a straight line is uniquely determined by two points, then $\tilde{U}^{(e)}$ *is uniquely determined along the entire side $\overline{23}$ by the values of a_2 and a_3.* Consequently, if the element trial solution $\tilde{U}^{(f)}$ in an adjacent element (f) (see Fig. 13.14) should assume the same values for a_2 and a_3, then $\tilde{U}^{(f)}$ would be the same straight line along side $\overline{23}$, thereby establishing interelement continuity. Of course, the purpose of assembly is to enforce con-

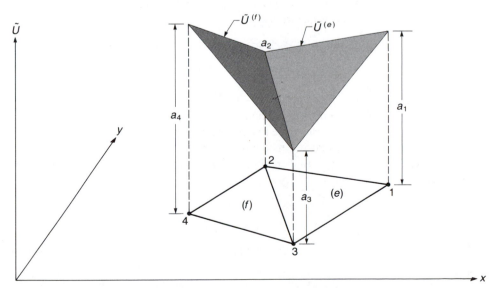

FIGURE 13.14 ● For the C^0-linear triangular element, continuity in $\tilde{U}^{(e)}$ and $\tilde{U}^{(f)}$ at the nodes (produced by assembly) ensures continuity along the sides.

tinuity at the nodes. Therefore, when elements (e) and (f) are assembled, the resulting continuity at the nodes will also produce continuity along the entire side.

From this example we can identify the following general guideline:

To establish interelement continuity along any side of an element, place q nodes along the side, where q is the number of nodes that will *uniquely* determine the element trial solution *along the side*.

The value of q is determined by the degree of the polynomial that represents the element trial solution on the element *boundary*. In the present triangular element the polynomial is linear (degree 1) everywhere in the element. In Sections 13.4.1 and 13.4.2, however, we will see examples where the degree of the polynomial is lower on the boundary than in the interior.

We can now see why the triangular-shaped element was chosen for the linear trial solution in Eq. (13.64). Since there are three DOF and therefore only three nodes, placing one node at each of the three vertices satisfies the continuity requirement for two nodes on each side.

Figure 13.15 illustrates the typical form of a complete trial solution over a mesh of several C^0-linear triangular elements. The solution looks like a surface composed of "hinged" triangular plates, suggestive of a geodesic dome.

Step 5: Substitute the shape functions into the element equations, and transform the integrals into a form appropriate for numerical evaluation.

We must substitute the $\phi_j^{(e)}(x,y)$ from Eq. (13.67) into the stiffness and load integrals in Eq. (13.61b).

Stiffness integrals • The physical properties $\alpha_x(x,y)$, $\alpha_y(x,y)$, and $\beta(x,y)$ will be treated in the same manner as for the 1-D elements (see Section 7.2, step 5). Thus we will let these properties be defined by a piecewise-quadratic (2-D) polynomial. The user may specify from 1 to 20 different sets of quadratics:

$$\alpha_x(x,y) = \alpha_{x0} + \alpha_{x1}x + \alpha_{x2}y + \alpha_{x3}x^2 + \alpha_{x4}xy + \alpha_{x5}y^2$$
$$\alpha_y(x,y) = \alpha_{y0} + \alpha_{y1}x + \alpha_{y2}y + \alpha_{y3}x^2 + \alpha_{y4}xy + \alpha_{y5}y^2$$
$$\beta(x,y) = \beta_0 + \beta_1 x + \beta_2 y + \beta_3 x^2 + \beta_4 xy + \beta_5 y^2 \qquad (13.69)$$

each set being applicable to a different region of the domain. If α_x, α_y, or β in the original problem statement, Eq. (13.1), cannot be represented exactly by such a piecewise quadratic, then this definition will introduce a modeling error.

Within each element we may then approximate the properties by a constant, equal to the value at the centroid:

$$\alpha_x^{(e)} = \alpha_x(x_c,y_c)$$
$$\alpha_y^{(e)} = \alpha_y(x_c,y_c)$$
$$\beta^{(e)} = \beta(x_c,y_c) \qquad (13.70)$$

where

$$x_c = \frac{1}{3}(x_1+x_2+x_3) \qquad y_c = \frac{1}{3}(y_1+y_2+y_3) \qquad (13.71)$$

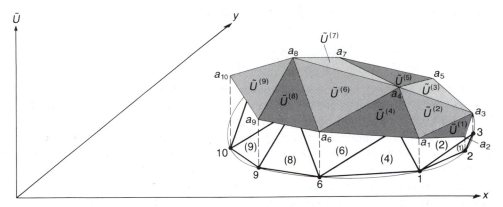

FIGURE 13.15 ● A mesh of several C^0-linear triangular elements and the corresponding trial solution.

and x_i, y_i, $i = 1, 2, 3$, are the coordinates of the nodes.[21] This assumption provides suffi-cient accuracy to the integrals to preserve the rate of convergence of the global energy due to the discretization error. Thus, for the present C^0-linear element, $m = 1$ and $p = 1$, so by guideline II in Chapter 8 (Section 8.3.1.) we must integrate to an accuracy of $O(h^2)$. The triangle quadrature rules in Section 13.3.2 indicate that this accuracy is obtained by evaluating the integrand at the centroid.

Consider first the α_x and α_y integrals. From Eq. (13.67),

$$\frac{\partial \phi_j^{(e)}(x,y)}{\partial x} = \frac{b_j}{2\Delta}$$

$$\frac{\partial \phi_j^{(e)}(x,y)}{\partial y} = \frac{c_j}{2\Delta} \tag{13.72}$$

which are both constants. Therefore, from Eq. (13.61b),

$$K\alpha_{ij}^{(e)} = \int\int^{(e)} \frac{\partial \phi_i^{(e)}(x,y)}{\partial x} \alpha_x(x,y) \frac{\partial \phi_j^{(e)}(x,y)}{\partial x} \, dx \, dy + \int\int^{(e)} \frac{\partial \phi_i^{(e)}(x,y)}{\partial y} \alpha_y(x,y) \frac{\partial \phi_j^{(e)}(x,y)}{\partial y} \, dx \, dy$$

$$\approx \frac{b_i}{2\Delta} \alpha_x^{(e)} \frac{b_j}{2\Delta} \int\int^{(e)} dx \, dy + \frac{c_i}{2\Delta} \alpha_y^{(e)} \frac{c_j}{2\Delta} \int\int^{(e)} dx \, dy$$

$$= \frac{\alpha_x^{(e)}}{4\Delta} b_i b_j + \frac{\alpha_y^{(e)}}{4\Delta} c_i c_j \tag{13.73}$$

Next consider the β integral:

$$K\beta_{ij}^{(e)} = \int\int^{(e)} \phi_i^{(e)}(x,y)\beta(x,y)\phi_j^{(e)}(x,y) \, dx \, dy$$

$$\approx \beta^{(e)} \int\int^{(e)} \phi_i^{(e)}(x,y)\phi_j^{(e)}(x,y) \, dx \, dy \tag{13.74}$$

---

[21] The reader should note the strong parallel between the 1-D and 2-D elements, by comparing Eqs. (7.17) through (7.19) with Eqs. (13.69) through (13.71).

This may be evaluated with the aid of the following triangle integration formula:

$$\int\int^{(e)} \zeta_1^l \zeta_2^m \zeta_3^n \, dx \, dy = \frac{l!m!n!}{(l+m+n+2)!} 2\Delta \tag{13.75}$$

where the integrand is the product of the three *area coordinates*, ζ_1, ζ_2, ζ_3, raised to the powers l, m, n, respectively, and the integration is over the area of the triangular element. The RHS involves factorials of the powers ($0!=1$), and the area of the triangle, Δ.

Area coordinates (also known as triangular, simplex, or barycentric coordinates) are a set of three coordinates naturally related to triangular geometry (Fig. 13.16). Each coordinate varies linearly from the value 0 on one of the sides to the value 1 at the opposite vertex. Geometrically, ζ_i is a fractional area, Δ_i/Δ, (hence the name) and also the fractional distance from the side opposite node i, as indicated by the coordinate lines for ζ_1 shown in the figure. It is immediately apparent from the geometry that the area coordinates are not independent, but in fact satisfy the relation

$$\zeta_1 + \zeta_2 + \zeta_3 = 1 \tag{13.76}$$

We can apply Eq. (13.75) to the β integral in Eq. (13.74) by observing that the shape functions for the C^0-linear triangle are identical with the area coordinates:

$$\phi_i^{(e)}(x,y) = \zeta_i \qquad i = 1,2,3 \qquad \text{for } C^0\text{-linear triangle} \tag{13.77}$$

Therefore, substituting Eq. (13.77) into Eq. (13.74), and using Eq. (13.75), yields

$$K\beta_{ij}^{(e)} = \beta^{(e)} \int\int^{(e)} \zeta_i \zeta_j \, dx \, dy = \begin{cases} \dfrac{\beta^{(e)}\Delta}{6} & i = j \\[2mm] \dfrac{\beta^{(e)}\Delta}{12} & i \neq j \end{cases} \tag{13.78}$$

Load integrals • Consider first the interior load integral. This will be treated in the same manner as for the 1-D elements (see Section 7.2, step 5). Thus we can approximate the

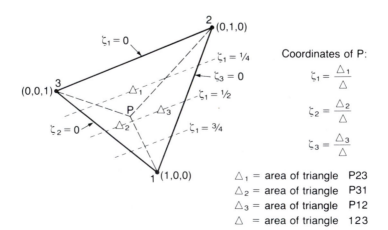

FIGURE 13.16 ● Area coordinates for a triangle.

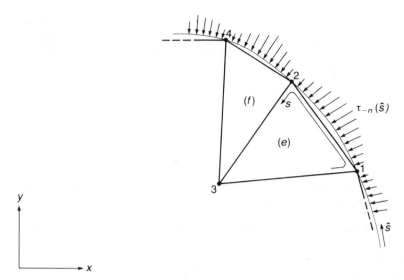

FIGURE 13.17 ● Elements lying along a boundary with an applied flux (natural BC).

distributed load $f(x,y)$ as a constant $f^{(e)}$ over each element, equal to the value of the load at the centroid,

$$f^{(e)} = f(x_c, y_c) \tag{13.79}$$

This will provide the $O(h^2)$ accuracy appropriate for this linear element (recall a similar argument for the stiffness integrals) because this is equivalent to "sampling" the integrand at the centroid for an $O(h^2)$ triangular quadrature rule (see Section 13.3.2). We can therefore move $f^{(e)}$ outside the integral:

$$Ff_i^{(e)} \simeq f^{(e)} \int\!\!\int^{(e)} \phi_i^{(e)}(x,y) \, dx \, dy \tag{13.80}$$

and the remaining integral can be evaluated exactly. Here we can use formula (13.75), where $l,m,n=1,0,0$ or $0,1,0$ or $0,0,1$ which yields $\Delta/3$.[22] Hence

$$Ff_i^{(e)} = \frac{f^{(e)} \Delta}{3} \tag{13.81}$$

In 2-D physical applications (see Section 13.1.3) $f^{(e)}$ is a load per unit area, so $f^{(e)}\Delta$ is the total load acting on the element. Therefore Eq. (13.81) assigns one-third of the total load to each of the three nodes, which is an intuitively appealing result.

Now consider the boundary flux integral,

$$F_{T_i^{(e)}} = \oint^{(e)} \bar{\tau}_{-n}^{(e)} \phi_i^{(e)} \, ds \tag{13.82}$$

Figure 13.17 shows two adjacent elements, (e) and (f), which have sides along the boundary of the domain. An inward-normal component of applied flux, $\tau_{-n}(\hat{s})$, is shown acting

---

[22] Referring to Fig. 13.13, the integral in Eq. (13.80) is the volume of a pyramid of unit height. Hence from solid geometry we know that volume = 1/3 × area of base × height = $\Delta/3$.

on the domain boundary; its magnitude may in general vary with position, as suggested by the arrows. Here $\hat{s}$ is a coordinate along the boundary, which may be a slightly different path from the boundary of the element.[23] We recognize $\tau_{-n}(\hat{s})$ as the prescribed value for the natural BCs:

$$\tilde{\tau}^{(e)}_{-n} = \tau_{-n}(\hat{s})$$

$$\tilde{\tau}^{(f)}_{-n} = \tau_{-n}(\hat{s})$$

$$\vdots$$

etc. for any other elements   (13.83)

In the following development it will be convenient to distinguish between external and internal sides of an element:

- An external side lies on (or approximately on) the boundary of the domain and hence is common to only one element. Examples in Fig. 13.17 are sides $\overline{12}$ and $\overline{24}$.
- An internal side lies inside the domain and is shared by two adjacent elements. Examples in Fig. 13.17 are sides $\overline{31}$, $\overline{32}$, and $\overline{34}$.

Every element side must be either external or internal.

The integral in Eq. (13.82) is around the entire periphery of an element and hence is the sum of three integrals, one along each side. Thus, for element (e) in Fig. 13.17,

$$\oint^{(e)} = \int_1^2 + \int_2^3 + \int_3^1 \tag{13.84}$$

where $\int_i^j$ means integrate from node i to node j. The boundary flux vector may therefore be written in an expanded form as follows:

$$\{F_T\}^{(e)} = \left\{ \begin{array}{c} \displaystyle\int_1^2 \tilde{\tau}^{(e)}_{-n}\phi^{(e)}_1\, ds + \int_2^3 \tilde{\tau}^{(e)}_{-n}\phi^{(e)}_1\, ds + \int_3^1 \tilde{\tau}^{(e)}_{-n}\phi^{(e)}_1\, ds \\[2ex] \displaystyle\int_1^2 \tilde{\tau}^{(e)}_{-n}\phi^{(e)}_2\, ds + \int_2^3 \tilde{\tau}^{(e)}_{-n}\phi^{(e)}_2\, ds + \int_3^1 \tilde{\tau}^{(e)}_{-n}\phi^{(e)}_2\, ds \\[2ex] \displaystyle\int_1^2 \tilde{\tau}^{(e)}_{-n}\phi^{(e)}_3\, ds + \int_2^3 \tilde{\tau}^{(e)}_{-n}\phi^{(e)}_3\, ds + \int_3^1 \tilde{\tau}^{(e)}_{-n}\phi^{(e)}_3\, ds \end{array} \right\} \tag{13.85}$$

Since each shape function is zero along the side opposite to its node, this simplifies to the following:

---

[23] If the domain boundary is curved, then of course a straight-sided element can only approximately conform to the boundary (see Section 13.6).

$$\{F_T\}^{(e)} = \left\{ \begin{array}{l} \displaystyle\int_1^2 \tilde{\tau}_{-n}^{(e)}\phi_1^{(e)}\,ds \qquad\qquad + \int_3^1 \tilde{\tau}_{-n}^{(e)}\phi_1^{(e)}\,ds \\[2em] \displaystyle\int_1^2 \tilde{\tau}_{-n}^{(e)}\phi_2^{(e)}\,ds + \int_2^3 \tilde{\tau}_{-n}^{(e)}\phi_2^{(e)}\,ds \\[2em] \displaystyle\int_2^3 \tilde{\tau}_{-n}^{(e)}\phi_3^{(e)}\,ds + \int_3^1 \tilde{\tau}_{-n}^{(e)}\phi_3^{(e)}\,ds \end{array} \right\} \begin{array}{l} \leftarrow \text{Node 1} \\[2em] \leftarrow \text{Node 2} \\[2em] \leftarrow \text{Node 3} \end{array}$$

$$\begin{array}{ccc} \uparrow & \uparrow & \uparrow \\ \text{External} & \text{Internal integrals} & \\ \text{integrals} & & \end{array} \qquad\qquad \textbf{(13.86)}$$

We will examine the external integrals first.

Boundary flux integrals along an external side • From Eq. (13.86) we have two integrals to evaluate:

$$\int_1^2 \tilde{\tau}_{-n}^{(e)}\phi_1^{(e)}\,ds$$

$$\int_1^2 \tilde{\tau}_{-n}^{(e)}\phi_2^{(e)}\,ds \qquad\qquad \textbf{(13.87)}$$

Substituting the value of the natural BC from Eq. (13.83) yields

$$\int_1^2 \tilde{\tau}_{-n}^{(e)}\phi_1^{(e)}\,ds \simeq \int_1^2 \tau_{-n}(s)\phi_1^{(e)}\,ds$$

$$\int_1^2 \tilde{\tau}_{-n}^{(e)}\phi_2^{(e)}\,ds \simeq \int_1^2 \tau_{-n}(s)\phi_2^{(e)}\,ds \qquad\qquad \textbf{(13.88)}$$

where the approximation sign allows for the possibility of the side of the element not coinciding exactly with the domain boundary, leaving a "sliver" of domain between them, i.e., a modeling error (see Section 13.6.3). In this case the analyst might approximate $\tau_{-n}(\hat{s})$ by some other function, $\tau_{-n}(s)$, applied to the element. In practice, however, the sliver may be made narrow enough that one can simply apply the given flux directly to the side of the element.

As with all the previous integrals we only need $O(h^2)$ accuracy. For this we can approximate $\tau_{-n}(s)$ as a constant, $\tilde{\tau}^{(e)}$, evaluated at the midpoint of the side:

$$\tau^{(e)} = \tau_{-n}(x_m, y_m) \qquad\qquad \textbf{(13.89)}$$

where

$$x_m = \frac{1}{2}(x_1 + x_2)$$

$$y_m = \frac{1}{2}(y_1 + y_2)$$

Again, the choice of midpoint is because, for a 1-D integral, this is equivalent to "sampling" the integrand at the midpoint for an $O(h^2)$ 1-D quadrature rule (see Fig. 8.12). We can therefore move $\tau^{(e)}$ outside the integrals:

$$\tau^{(e)} \int_1^2 \phi_1^{(e)} \, ds$$

$$\tau^{(e)} \int_1^2 \phi_2^{(e)} \, ds \tag{13.90}$$

The remaining integrals can be evaluated exactly with the aid of another triangle integration formula, this one for a side of the triangle:

$$\int_i^j \varsigma_i^l \varsigma_j^m \, ds = \frac{l! \, m!}{(l+m+1)!} L_{ij} \tag{13.91}$$

where $L_{ij} = \int_i^j ds$ is the length of the side between nodes i and j. Applying Eq. (13.91) to the integrals in Eq. (13.90) yields $L_{12}/2$ for each one.[24] The external boundary flux integrals are therefore as follows:

$$\int_1^2 \tilde{\tau}_{-n}^{(e)} \phi_1^{(e)} \, ds = \frac{\tau^{(e)} L_{12}}{2}$$

$$\int_1^2 \tilde{\tau}_{-n}^{(e)} \phi_2^{(e)} \, ds = \frac{\tau^{(e)} L_{12}}{2} \tag{13.92}$$

In 2-D physical applications (see Section 13.1.3) $\tau^{(e)}$ is a load per unit length, so $\tau^{(e)} L_{12}$ is the total load acting on the side of the element. Therefore Eq. (13.92) assigns one-half of the total load to each of the two nodes, which is an intuitively appealing result.

As usual we observe that the manner in which the values of $\tau^{(e)}$ are calculated is basically a programming consideration. They could either be calculated by hand and then input directly (as in UNAFEM), or else a general function $\tau_{-n}(s)$ could be input and the program would then evaluate it at the midpoints of the element sides.

Figure 13.17 showed the applied flux $\tau_{-n}(\hat{s})$ distributed over a "significant" length of the boundary. Sometimes $\tau_{-n}(\hat{s})$ may be concentrated in a "very small" region, as

---

[24] Referring to Fig. 13.13, each integral in Eq. (13.90) is the area of a triangle of unit height and base L_{12}, that is, $L_{12}/2$.

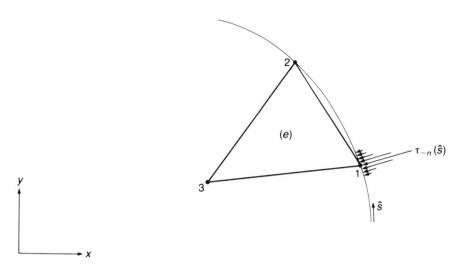

FIGURE 13.18 ● A concentrated applied flux.

shown in Fig. 13.18. We then have a choice. If we want to obtain a detailed solution in the immediate vicinity of the concentrated flux, we would use expressions (13.92) and employ many tiny elements near the concentration. On the other hand, we frequently don't want such a detailed, local solution but only want the solution further away throughout the rest of the domain. In this latter case a more efficient approach would be to model the concentrated flux as a *point* flux. Then we could use larger elements and apply the point flux to one of the nodes, as suggested in the figure.

This approach is analogous to how we dealt with concentrated loads in 1-D (see Section 7.5.2). Mathematically we represent $\tau_{-n}(\hat{s})$ as a Dirac delta function (see Fig. 7.43), concentrated, say, at node 1 on element (e) (Fig. 13.19):

$$\tau_{-n}(\hat{s}) = \tau_1 \delta(\hat{s} - \hat{s}_1) \tag{13.93}$$

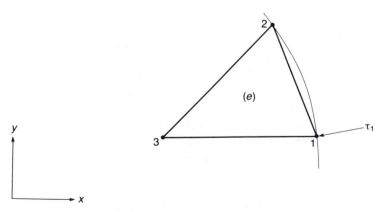

FIGURE 13.19 ● Modeling the concentrated flux in Fig. 13.18 as a point flux (a Dirac delta function).

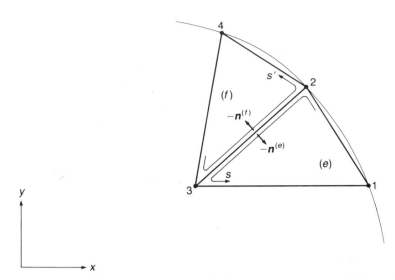

FIGURE 13.20 ● Notation for interelement boundary flux integrals.

where $\hat{s}_1$ represents the coordinate value at node 1, and τ_1 is the value of the point flux.[25]
Substituting Eq. (13.93) into Eqs. (13.88) yields

$$\int_1^2 \bar{\tau}_{-n}^{(e)} \phi_1^{(e)} \, ds = \tau_1$$

$$\int_1^2 \bar{\tau}_{-n}^{(e)} \phi_2^{(e)} \, ds = 0 \tag{13.94}$$

which therefore assigns the entire flux to node 1, as expected. We note in passing that
such a flux would be input to the UNAFEM program as a nodal flux, just as in the 1-D
problems.

Boundary flux integrals along an internal side ● Let's now look at the internal integrals in
the boundary flux vector in Eq. (13.86). The two integrals along side $\overline{13}$ are handled in
the same manner as the two integrals along side $\overline{23}$ so we need only examine one side.
Considering side $\overline{23}$, Fig. 13.17 shows that this side is shared by element (f), so the boundary flux vector for that element [Eq. (13.95)] will also contribute an integral along that
side. The boundary coordinate in element (f) is s', as shown in Fig. 13.20.

When elements (e) and (f) are assembled, the terms corresponding to nodes 2 and
3 are added together. Hence, from Eqs. (13.86) and (13.95), the assembled boundary flux
vector is the expression shown in Eq. (13.96).

---

[25] The value of τ_1 is essentially the area under the $\tau_{-n}(\hat{s})$ distribution in Fig. 13.18, that is, $\tau_1 \simeq$
$\int \tau_{-n}(\hat{s}) \, d\hat{s}$. Since $\tau_{-n}(\hat{s})$ is a load per unit length, τ_1 is simply a load. In practice, however, such an
integration is rarely performed since concentrated fluxes are usually available from experimental data
directly in the form of (point) loads rather than as loads per unit length.

$$\{F_T\}^{(f)} = \begin{cases} \displaystyle\int_2^4 \tilde{\tau}_{-n}^{(f)}\phi_2^{(f)}\,ds' & +\ \displaystyle\int_3^2 \tilde{\tau}_{-n}^{(f)}\phi_2^{(f)}\,ds' & \leftarrow \text{Node 2} \\[3mm] \displaystyle\int_4^3 \tilde{\tau}_{-n}^{(f)}\phi_3^{(f)}\,ds' + \int_3^2 \tilde{\tau}_{-n}^{(f)}\phi_3^{(f)}\,ds' & & \leftarrow \text{Node 3} \\[3mm] \displaystyle\int_2^4 \tilde{\tau}_{-n}^{(f)}\phi_4^{(f)}\,ds' + \int_4^3 \tilde{\tau}_{-n}^{(f)}\phi_4^{(f)}\,ds' & & \leftarrow \text{Node 4} \end{cases}$$

(13.95)

$$\{F_T\}^{(e)+(f)} = \begin{cases} \displaystyle\int_1^2 \tilde{\tau}_{-n(e)}^{(e)}\phi_1^{(e)}\,ds + \int_3^1 \tilde{\tau}_{-n(e)}^{(e)}\phi_1^{(e)}\,ds & \leftarrow \text{Node 1} \\[3mm] \displaystyle\int_1^2 \tilde{\tau}_{-n(e)}^{(e)}\phi_2^{(e)}\,ds + \boxed{\int_2^3 \tilde{\tau}_{-n(e)}^{(e)}\phi_2^{(e)}\,ds + \int_3^2 \tilde{\tau}_{-n(f)}^{(f)}\phi_2^{(f)}\,ds'} + \int_2^4 \tilde{\tau}_{-n(f)}^{(f)}\phi_2^{(f)}\,ds' & \leftarrow \text{Node 2} \\[3mm] \displaystyle\int_3^1 \tilde{\tau}_{-n(e)}^{(e)}\phi_3^{(e)}\,ds + \boxed{\int_2^3 \tilde{\tau}_{-n(e)}^{(e)}\phi_3^{(e)}\,ds + \int_3^2 \tilde{\tau}_{-n(f)}^{(f)}\phi_3^{(f)}\,ds'} + \int_4^3 \tilde{\tau}_{-n(f)}^{(f)}\phi_3^{(f)}\,ds' & \leftarrow \text{Node 3} \\[3mm] \displaystyle\int_2^4 \tilde{\tau}_{-n(f)}^{(f)}\phi_4^{(f)}\,ds' + \int_4^3 \tilde{\tau}_{-n(f)}^{(f)}\phi_4^{(f)}\,ds' & \leftarrow \text{Node 4} \end{cases}$$

(13.96)

Here it is important to attach the appropriate (e) and (f) superscripts to the normal vectors to distinguish $n^{(e)}$ from $n^{(f)}$ (Fig. 13.20).

The two boxed-in pairs of terms in Eq. (13.96) are the complete boundary flux terms along side $\overline{23}$. Both pairs are handled in the same manner so we only need to examine one. (The other terms are either external integrals, i.e., $\int_1^2$ and $\int_2^4$, which we have already examined, or only one internal integral of a similar pair associated with another side, i.e., $\int_3^1$ and $\int_4^3$.)

Consider the pair of terms associated with node 2:

$$\int_2^3 \tilde{\tau}_{-n(e)}^{(e)}\phi_2^{(e)}\,ds + \int_3^2 \tilde{\tau}_{-n(f)}^{(f)}\phi_2^{(f)}\,ds'$$

(13.97)

This can be rewritten more meaningfully if we use the following relationships from Fig. 13.20:

$$ds = -ds'$$

$$n^{(e)} = -n^{(f)}$$

$$\tilde{\tau}^{(f)}_{-n(f)} = -\tilde{\tau}^{(f)}_{-n(e)} \tag{13.98}$$

The third relation, although obvious from the figure, follows from the second relation and Eq. (13.62). In addition,

$$\phi_2^{(e)}(\text{side } \overline{23}) = \phi_2^{(f)}(\text{side } \overline{23}) \tag{13.99}$$

This, of course, is the C^0-continuity condition we built into the shape functions; it is also evident from a plot of these functions such as the one in Fig. 13.13.

Applying relations (13.98) and (13.99) to the second integral in Eq. (13.97) enables the pair of integrals to be written as follows:

$$\int_2^3 \left(\tilde{\tau}^{(e)}_{-n(e)} - \tilde{\tau}^{(f)}_{-n(e)} \right) \phi_2^{(e)} \, ds \tag{13.100}$$

We recognize this as the 2-D analogue of the flux-difference expressions across inter-element boundaries (i.e., nodes) in 1-D problems, as discussed in Section 7.6. All of that development can be appropriately extended to 2-D.

Thus, in order to evaluate Eq. (13.100), we must know what value to assign to the flux difference $\tilde{\tau}^{(e)}_{-n(e)} - \tilde{\tau}^{(f)}_{-n(e)}$. Analogous to the 1-D approach, we integrate the original governing differential equation [Eq. (13.1)] around a rectangle (Fig. 13.21) whose thickness

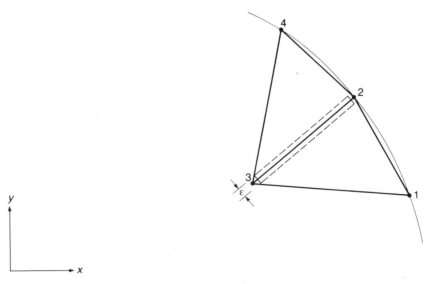

FIGURE 13.21 ● An integration path (dashed rectangle) to derive interelement natural BCs.

$\epsilon \to 0$. If the interior load $f(x,y)$ is finite along side $\overline{23}$ (e.g., is continuous or has a step discontinuity) then

$$\tilde{\tau}^{(e)}_{-n^{(e)}} - \tilde{\tau}^{(f)}_{-n^{(e)}} = 0 \qquad \left(\begin{array}{c} f(x,y) \text{ not concentrated} \\ \text{along interelement boundary} \end{array} \right) \qquad \textbf{(13.101)}$$

i.e., the flux for the exact solution should be continuous across side $\overline{23}$. This is the usual condition along interelement boundaries. If the interior load has a concentrated distribution along side $\overline{23}$ (a Dirac delta function along a line), then

$$\tilde{\tau}^{(e)}_{-n^{(e)}} - \tilde{\tau}^{(f)}_{-n^{(e)}} = \tau_{23} \qquad \left(\begin{array}{c} f(x,y) \text{ concentrated} \\ \text{along interelement boundary} \end{array} \right) \qquad \textbf{(13.102)}$$

where τ_{23} is the line integral of the concentrated interior load and thus has the same physical units as flux. In practical applications τ_{23} would normally be part of the given load data for the problem (i.e., the aforementioned integration around an ϵ-rectangle is generally not necessary).

Equations (13.101) and (13.102) are the 2-D *natural interelement BCs*. We recall from all of the 1-D analysis (recall especially Sections 5.3.1 and 7.6) that natural interelement BCs in the interior behave in all respects like natural BCs on the boundary. For example, we can specify distributed fluxes along interior element sides[26] and concentrated point fluxes at interior nodes[27].

As noted above, the natural interelement BC in Eq. (13.101), *continuity of flux*, is the usual condition in practical problems and therefore, as in 1-D analyses, the internal boundary flux integrals can be ignored when inputting data to a program, since adding a zero to an array accomplishes nothing. Only nonzero values, corresponding to Eq. (13.102), need to be input. We also recall that the resulting approximate solution will only satisfy the natural interelement BCs approximately; i.e., there will generally be some discontinuity in the normal component of flux across interelement boundaries.

In summary, the boundary flux integral in Eq. (13.61b) reduces to two possible types of fluxes that can be specified: a distributed flux along any element side, Eqs. (13.92) or Eq. (13.102) [with Eq. (13.100)], or a nodal flux (point load) at any node, Eqs. (13.94).

Step 6: Prepare expressions for the flux.

From Eqs. (13.17),

$$\tilde{\tau}^{(e)}_x(x,y) = -\alpha_x(x,y) \frac{\partial \tilde{U}^{(e)}(x,y;a)}{\partial x}$$

$$\tilde{\tau}^{(e)}_y(x,y) = -\alpha_y(x,y) \frac{\partial \tilde{U}^{(e)}(x,y;a)}{\partial y} \qquad \textbf{(13.103)}$$

---

[26] To complete the story, τ_{23} in Eq. (13.102) could in general vary along side $\overline{23}$, that is, $\tau_{23}(s)$. Substituting $\tau_{23}(s)$ into Eq. (13.100) yields an integral in the same form as the external boundary integrals in Eq. (13.87). It would therefore be evaluated in the same manner, yielding either Eqs. (13.92) or (13.94).

[27] Note that nodal fluxes (point loads) can be mathematically described as either concentrated boundary fluxes or "doubly concentrated" interior loads.

and from Eq. (13.67),

$$\frac{\partial \tilde{U}^{(e)}(x,y;a)}{\partial x} = \sum_{j=1}^{3} a_j \frac{\partial \phi_j^{(e)}(x,y)}{\partial x} = \sum_{j=1}^{3} a_j \frac{b_j}{2\Delta}$$

$$\frac{\partial \tilde{U}^{(e)}(x,y;a)}{\partial y} = \sum_{j=1}^{3} a_j \frac{\partial \phi_j^{(e)}(x,y)}{\partial y} = \sum_{j=1}^{3} a_j \frac{c_j}{2\Delta} \qquad \textbf{(13.104)}$$

Hence,

$$\tilde{\tau}_x^{(e)}(x,y) = -\alpha_x(x,y) \sum_{j=1}^{3} a_j \frac{b_j}{2\Delta}$$

$$\tilde{\tau}_y^{(e)}(x,y) = -\alpha_y(x,y) \sum_{j=1}^{3} a_j \frac{c_j}{2\Delta} \qquad \textbf{(13.105)}$$

For this C^0-linear triangle element there is one optimal flux-sampling point: at the centroid. There the accuracy is $O(h^2)$. [At other points the accuracy is only $O(h)$.] The location of this point (recall the footnote to guideline IV in Section 8.3.1) may be derived in the same manner as outlined in Exercises 7.1 and 8.4, i.e., by determining the least-squares fit of a constant ($\partial \tilde{U}/\partial x$ and $\partial \tilde{U}/\partial y$) to a 2-D linear polynomial [$\partial U/\partial x$ and $\partial U/\partial y$ as $h \to 0$, to $O(h^2)$]. See Exercise 13.3. Thus the optimal flux values are

$$\tilde{\tau}_x^{(e)} = -\alpha_x(x_c,y_c) \sum_{j=1}^{3} a_j \frac{b_j}{2\Delta}$$

$$\tilde{\tau}_y^{(e)} = -\alpha_y(x_c,y_c) \sum_{j=1}^{3} a_j \frac{c_j}{2\Delta} \qquad \textbf{(13.106)}$$

where x_c, y_c are the coordinates of the centroid [Eqs. (13.71)].

We also want to calculate nodal fluxes. These are generally easier to work with than the optimal fluxes because the locations of nodes are easier to identify than the locations of the optimal points (especially for higher-order, curved elements), and it is easier to construct flux contour plots using nodal values. Nodal fluxes are calculated by using a smoothing and averaging procedure similar to that used for 1-D elements (see Sections 7.3.2 and 8.3.1). Thus we first construct in each element a smoothing function for each component of flux, i.e., one for $\tilde{\tau}_x^{(e)}(x,y)$ and another one for $\tilde{\tau}_y^{(e)}(x,y)$. In 1-D these functions are curves; in 2-D they are surfaces. These functions are generally polynomials, and they are determined by either interpolating or least-squares-fitting (depending on whether one chooses the number of DOF in the polynomial to be equal to or less than, respectively, the number of selected fluxes) some or all of the optimal fluxes. The choice is largely one of computational convenience. The element nodal fluxes are then determined by evaluating the smoothing functions at the nodes. Finally, the nodal flux at a given node is the mean of all the element nodal fluxes at that node.

For our present C^0-linear element each flux component has only one optimal flux (at the centroid) so the smoothing functions in each element can only be a constant, equal

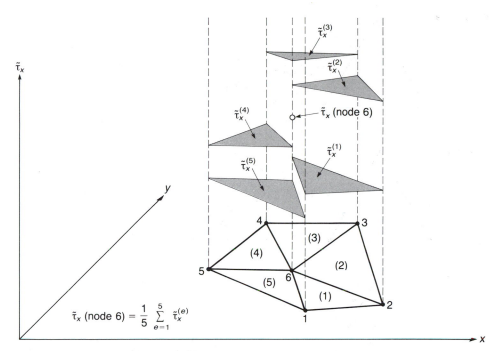

FIGURE 13.22 ● An example of nodal flux computation: flux at node 6, $\tilde{\tau}_x$ (node 6), is the mean of the element nodal fluxes from the five elements sharing that node.

to the optimal values $\tilde{\tau}_x^{(e)}$ and $\tilde{\tau}_y^{(e)}$. Therefore the element nodal fluxes at each of the nodes must be $\tilde{\tau}_x^{(e)}$ for the x-component and $\tilde{\tau}_y^{(e)}$ for the y-component. The nodal flux is then the mean of such values, as illustrated in Fig. 13.22 for $\tilde{\tau}_x$.

13.3.2 The C⁰-Quadratic Isoparametric Triangle

This particular element is just one of an infinite hierarchy of higher-order C⁰-triangular elements. Therefore, before delving into the details of this element, let's first examine the general hierarchy and thereby gain a better appreciation of why this element is a useful one to study. We recall that the three-node C⁰-linear element in the previous section is the simplest possible 2-D element, employing only the first three terms in Pascal's triangle (recall Fig. 13.11). If desired, we could now develop a sequence of higher-order C⁰-triangular elements, each containing one more term from Pascal's triangle — namely, four-node triangles, five-node triangles, etc. However, most such elements would involve incomplete polynomials. There is nothing wrong with this, but we will see in a moment that those triangular elements employing complete polynomials are generally the most useful, for reasons other than the completeness. Those with incomplete polynomials have important but limited application.

From Pascal's triangle, the complete 2-D polynomials contain 3, 6, 10, 15, 21, etc., terms. Of course, the previous three-node linear element was the first in this series. Figure 13.23 shows the next four elements in this series, namely, the C⁰-quadratic, -cubic, -quartic, and -quintic. Also shown are the terms in the corresponding element trial solutions. Each trial solution is a polynomial complete to degree p (e.g., $p=2$ for the quadratic,

C⁰-quadratic

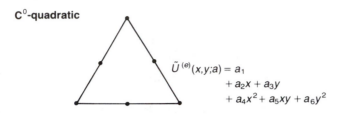

$$\tilde{U}^{(e)}(x,y;a) = a_1$$
$$+ a_2 x + a_3 y$$
$$+ a_4 x^2 + a_5 xy + a_6 y^2$$

C⁰-cubic

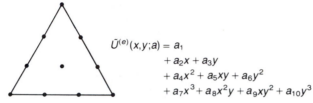

$$\tilde{U}^{(e)}(x,y;a) = a_1$$
$$+ a_2 x + a_3 y$$
$$+ a_4 x^2 + a_5 xy + a_6 y^2$$
$$+ a_7 x^3 + a_8 x^2 y + a_9 xy^2 + a_{10} y^3$$

C⁰-quartic

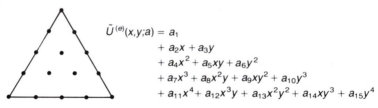

$$\tilde{U}^{(e)}(x,y;a) = a_1$$
$$+ a_2 x + a_3 y$$
$$+ a_4 x^2 + a_5 xy + a_6 y^2$$
$$+ a_7 x^3 + a_8 x^2 y + a_9 xy^2 + a_{10} y^3$$
$$+ a_{11} x^4 + a_{12} x^3 y + a_{13} x^2 y^2 + a_{14} xy^3 + a_{15} y^4$$

C⁰-quintic

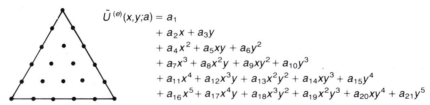

$$\tilde{U}^{(e)}(x,y;a) = a_1$$
$$+ a_2 x + a_3 y$$
$$+ a_4 x^2 + a_5 xy + a_6 y^2$$
$$+ a_7 x^3 + a_8 x^2 y + a_9 xy^2 + a_{10} y^3$$
$$+ a_{11} x^4 + a_{12} x^3 y + a_{13} x^2 y^2 + a_{14} xy^3 + a_{15} y^4$$
$$+ a_{16} x^5 + a_{17} x^4 y + a_{18} x^3 y^2 + a_{19} x^2 y^3 + a_{20} xy^4 + a_{21} y^5$$

FIGURE 13.23 ● The first four higher-order C⁰-triangular elements employing complete 2-D polynomials.

$p=3$ for the cubic, etc.) and therefore it uses all the terms in the first $p+1$ rows of Pascal's triangle.

Note that the overall geometric pattern of the node locations in each element is triangular, just like the array of terms in Pascal's triangle. For example, the cubic element has 10 DOF and therefore 10 nodes; these nodes are placed so that there is one at each vertex, two along each side, and one in the interior, which is the same pattern as the first 10 terms in Pascal's triangle.[28] This is principally why the triangular-shaped

---

[28] This is purely a similarity in the overall patterns and does not imply any association between a particular node and a particular term in the polynomial.

elements lend themselves naturally to complete 2-D polynomials (and, conversely, why quadrilateral-shaped elements work better, in some respects, with incomplete polynomials, as will be seen in Sections 13.4.1 and 13.4.2).

The triangles in Fig. 13.23 are shown straight-sided with approximately equal-length sides and the nodes in a uniformly spaced pattern. This was only to simplify these diagrams. In practice, the sides may be curved and of different lengths, and the nodes may be nonuniformly distributed. Just as with the 1-D elements, though, extreme deviations from these "regular-shaped" elements will generally cause numerical problems, as will be discussed later.

We observe two important practical properties of the complete triangular elements. First, the trial solutions have *geometric isotropy*, meaning that they are balanced with respect to the x and y independent variables. Thus, for each term of the form $x^r y^s$ there is also a term $x^s y^r$, so that x and y could be interchanged and the form of the trial solution would not change. Although this property can easily be imposed on any trial solution, it occurs automatically for complete polynomials. Geometric isotropy is not necessary for convergence, but it is nevertheless a desirable property. Lack of isotropy means that one variable is more accurately represented than the other, and hence the performance of the element could vary significantly from one point in the mesh to another.

The other very useful property is that each side of the element has the same number of nodes. This yields the same order of approximation along each side, and hence the performance of the element is independent of its orientation. This in turn makes it easy to combine large numbers of such elements into any kind of mesh pattern, since any side of a given element is compatible with any side of another (similar) one.

The incomplete polynomial triangles have a different number of nodes along different sides and hence cannot be combined in large clusters in a generally meaningful way. For example, the four-node triangle in Fig. 13.24(a) is linear along two sides and quadratic along the third. (Note that the fourth term in the element trial solution, which needs to be a quadratic term, is chosen as xy rather than x^2 or y^2, in order to provide for geometric isotropy — see Exercise 13.4.) A mesh of only four-noded triangles would yield alter-

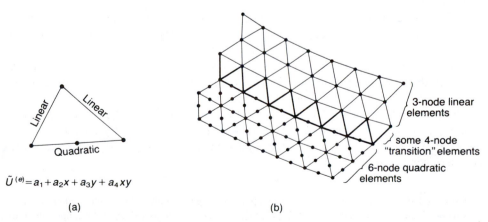

$$\bar{U}^{(e)} = a_1 + a_2 x + a_3 y + a_4 xy$$

(a)   (b)

FIGURE 13.24 ● (a) A four-node triangular element employing an incomplete quadratic polynomial; (b) a mesh using four-node triangles as transition elements.

nating bands of linear and quadratic accuracy, which would not appear to be of any general practical value. However, as illustrated in Fig. 13.24(b), a single row of such elements could provide a transition between a region of complete linear elements and a region of complete quadratic elements. In general, elements with different numbers of nodes along different sides are limited to functioning mainly as *transition elements*.

It can thus be seen that, following the three-node linear triangle, the six-node quadratic triangle is the next most generally useful triangular element to develop. So let's proceed with steps 4, 5, and 6 for this element.

Step 4: Develop specific expressions for the shape functions.

We can develop the element either by the direct approach or by parametric (sub-, super-, or iso-) mapping. Recall that the direct approach constructs the shape functions directly on the real elements in x,y-space (see Section 8.2), whereas parametric mapping constructs the shape functions on a parent element and then maps them to the real elements (see Section 8.3).

For 2-D (and 3-D) elements the direct approach is limited to straight-sided elements because it is incapable of producing interelement continuity on curved sides. The reason is as follows. Although there is no problem constructing the standard interpolatory polynomial shape functions for curved-sided elements, the functions are *all* generally nonzero along a curved side (see Exercise 13.5). Therefore the element trial solution along a curved side is determined by all the DOF rather than just the DOF associated with the nodes on the curved side. In other words, the trial solution along the curved side is not uniquely determined by only the DOF on that side. Therefore assembly of two adjacent elements at the nodes does not force their trial solutions to also be identical along the entire side (recall the guideline in Section 13.3.1 for establishing interelement continuity).

Curved-sided elements are quite useful in practice, principally for modeling curved boundaries. Therefore the above observation provides a strong motivation to use parametric mapping rather than the direct approach. In particular, we will use isoparametric mapping. The following development of a 2-D C^0-quadratic isoparametric element has many similarities to the development of the 1-D C^0-quadratic isoparametric element in Section 8.3.

Figure 13.25(a) shows the parent element in ξ, η-space; Fig. 13.25(b) shows the parent element mapped onto a curved-sided real element in x,y-space. Let's first examine the parent element, then we'll develop the shape functions on the parent element, and finally we'll develop the mapping of the parent element onto the real element. The latter, of course, results in the parent shape functions being mapped into shape functions on the real element, which is the goal of this entire procedure.

The sides of the parent element must be straight, even though the real elements may be curved, for the reason just argued above — namely, to provide for interelement continuity.[29] The fact that straight sides (rather than curved sides) greatly simplifies the algebra is a peripheral benefit.

---

[29] Recall from Chapter 8 that the subsequent mapping of the parent element to a real element involves only a coordinate transformation (from ξ,η to x,y), not a change in the values of the shape functions. Therefore, if the parent shape functions yield interelement continuity, then the mapped shape functions will retain the continuity, provided the mapping produces no overlapping or gaps between the real elements.

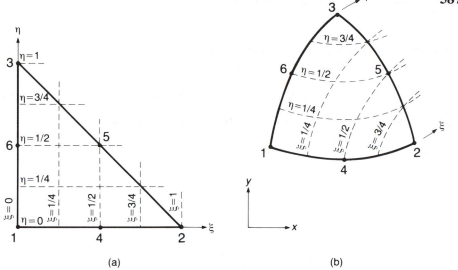

FIGURE 13.25 ● (a) The parent element for the six-node C^0-quadratic isoparametric triangle; (b) a curved-sided real element showing the parent element mapped onto it.

The parent element may have any triangular shape. However, a fairly regular shape (similar-length sides and similar angles) is desirable because real elements that are severely distorted from the parent shape cause numerical problems, and regularly shaped real elements are generally most useful. A right triangle or an equilateral triangle are therefore good candidates, but, of course, not the only ones. We choose a right triangle because two of the sides can then be aligned with the Cartesian ξ,η-axes, thereby simplifying the algebra. In addition, having angles of both 45 and 90° makes it a bit more versatile for constructing meshes than an equilateral with only 60° angles. (Remember that even though the real elements need not use these angles, we nevertheless don't want them to deviate "too much" from the angles in the parent element.)

Note that the two sides of the parent element along the ξ,η-axes are both of unit length. The coordinate ξ has the value zero along side $\overline{361}$, and it varies linearly to the value unity at node 2. Similarly, the coordinate η is zero along side $\overline{142}$ and varies linearly to the value unity at node 3. Therefore ξ and η are area coordinates (see Fig. 13.16).[30] This will be convenient later on when applying quadrature formulas. We see that there are definite computational advantages to defining the parent element as a unit right triangle aligned with the ξ,η-axes, no matter what the order of the element may be — i.e., quadratic, cubic, quartic, etc.

Let's develop the parent shape functions using the interpolation property, $\phi_i(\xi_j, \eta_j) = \delta_{ij}$. Since the trial solution in the parent element is a complete 2-D polynomial, then each shape function may contain any of the terms in a complete 2-D polynomial. Thus, for node 1,

$$\phi_1(\xi,\eta) = \alpha_1 + \alpha_2\xi + \alpha_3\eta + \alpha_4\xi^2 + \alpha_5\xi\eta + \alpha_6\eta^2 \qquad (13.107)$$

---

[30] A third area coordinate, call it ζ, would be zero on side $\overline{253}$ and unity at the origin. There is no need to introduce this coordinate, however, since it is dependent on the other two ($\xi + \eta + \zeta = 1$), so we can always work in terms of only ξ and η.

where the interpolation property requires that

$$\phi_1(\xi_1,\eta_1) = 1$$

$$\phi_1(\xi_j,\eta_j) = 0 \quad j = 2,3,4,5,6 \tag{13.108}$$

Applying conditions (13.108) to Eq. (13.107) yields six equations:

$$\alpha_1 \qquad\qquad\qquad\qquad = 1$$

$$\alpha_1 \quad + \alpha_2 \qquad\qquad + \alpha_4 \qquad\qquad = 0$$

$$\alpha_1 \qquad + \alpha_3 \qquad\qquad\quad + \alpha_6 \quad = 0$$

$$\alpha_1 \quad + \frac{1}{2}\alpha_2 \qquad\quad + \frac{1}{2}\alpha_4 \qquad\qquad = 0$$

$$\alpha_1 \quad + \frac{1}{2}\alpha_2 + \frac{1}{2}\alpha_3 + \frac{1}{4}\alpha_4 + \frac{1}{4}\alpha_5 + \frac{1}{4}\alpha_6 = 0$$

$$\alpha_1 \qquad\quad + \frac{1}{2}\alpha_3 \qquad\qquad + \frac{1}{4}\alpha_6 = 0 \tag{13.109}$$

Solving Eqs. (13.109) yields

$$\phi_1(\xi,\eta) = [1-(\xi+\eta)][1-2(\xi+\eta)] \tag{13.110a}$$

and in a similar fashion we can show

$$\phi_2(\xi,\eta) = \xi(2\xi-1)$$

$$\phi_3(\xi,\eta) = \eta(2\eta-1)$$

$$\phi_4(\xi,\eta) = 4\xi[1-(\xi+\eta)]$$

$$\phi_5(\xi,\eta) = 4\xi\eta$$

$$\phi_6(\xi,\eta) = 4\eta[1-(\xi+\eta)] \tag{13.110b}$$

Figure 13.26 shows $\phi_1(\xi,\eta)$ and $\phi_4(\xi,\eta)$. The other two corner shape functions, $\phi_2(\xi,\eta)$ and $\phi_3(\xi,\eta)$, have the same form as $\phi_1(\xi,\eta)$, and the other two midside shape functions, $\phi_5(\xi,\eta)$ and $\phi_6(\xi,\eta)$, have the same form as $\phi_4(\xi,\eta)$.

Mapping the parent element onto each of the real elements follows the same procedure as in Section 8.3.1 [see Eqs. (8.19) through (8.22)], only here we define a coordinate transformation for two coordinates instead of one. For an isoparametric transformation the mapping functions are chosen to be the same as the shape functions. Thus, in analogy with Eq. (8.22), we have

$$x = \sum_{k=1}^{6} x_k^{(e)}\phi_k(\xi,\eta)$$

$$y = \sum_{k=1}^{6} y_k^{(e)}\phi_k(\xi,\eta) \tag{13.111}$$

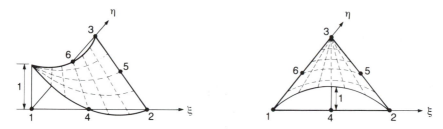

FIGURE 13.26 ● Corner and midside parent shape functions for the C^0-quadratic isoparametric triangle.

where $x_k^{(e)}$, $y_k^{(e)}$, $k=1,2,\ldots,6$ are the coordinates of the six nodes in the (e)th real element. Equations (13.111) clearly map the parent nodes onto the real nodes, for example,

$$\text{Node 1: } \xi = 0, \ \eta = 0 \ \to \ x_1^{(e)}, y_1^{(e)}$$

$$\text{Node 2: } \xi = 1, \ \eta = 0 \ \to \ x_2^{(e)}, y_2^{(e)}$$

$$\vdots$$

$$\text{Node 6: } \xi = 0, \ \eta = \frac{1}{2} \ \to \ x_6^{(e)}, y_6^{(e)} \qquad \textbf{(13.112)}$$

Equations (13.111) map the ξ,η-coordinates of the parent element onto each real element, as illustrated in Fig. 13.27. Only the numerical values for $x_k^{(e)}$, $y_k^{(e)}$ differ from one element to the next. For clarity, the figure shows the ξ,η coordinate lines in only two of the elements. In element (5), for example, parent nodes 1 to 6 are shown mapped onto global nodes 18, 9, 7, 15, 8, 14, respectively. We could just as well have mapped parent node 1 onto global nodes 9 or 7 (instead of 18 as shown), and the ξ,η lines would have been rotated correspondingly.

The parent shape functions are carried by these coordinate transformations to each of the real elements, resulting in shape functions being defined on each real element. Figure 13.28 shows how the element trial solutions might appear in two adjacent elements. Note that the C^0-continuity along the interelement boundary implies continuity of all derivatives *parallel* to the boundary. However, the derivative *normal* to the boundary (and therefore all higher-order derivatives normal to the boundary) will in general be discontinuous for the approximate FE solution.[31]

Just as with the 1-D isoparametric elements, we must be careful not to make our 2-D isoparametric elements too distorted. We recall from Section 8.3.1 that an isoparametric element is acceptable if and only if its mapping from the parent element is 1 to 1; i.e., each point in the parent element maps onto one point in the real element, and vice versa.

---

[31] Recall from Eq. (13.101) that the normal component of flux for the *exact* solution is continuous across the interelement boundary (if there are no concentrated loads there). Hence the normal derivative of the *exact* solution is also continuous across the boundary (if the properties α_x and α_y are continuous). This, of course, is the natural interelement BC, and, as we have learned, it is satisfied only approximately by the FE solution.

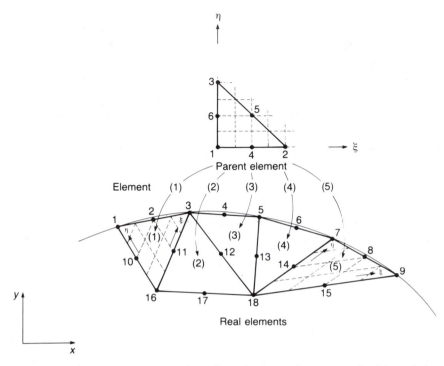

FIGURE 13.27 ● Mapping of ξ,η-coordinates from the parent element to each of the real elements in a mesh.

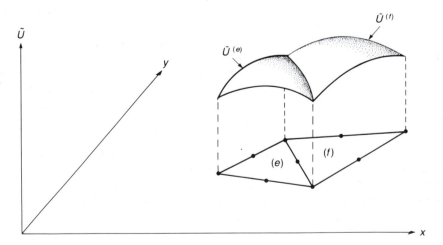

FIGURE 13.28 ● Typical form of element trial solutions in two adjacent elements.

The analytical test for acceptability is the 2-D analogue of Eq. (8.32), namely,

$$|J^{(e)}(\xi,\eta)| > 0 \qquad \text{everywhere in element} \tag{13.113}$$

where $|J^{(e)}(\xi,\eta)|$ is the Jacobian, which is the determinant of the Jacobian matrix. The latter is given by

$$J^{(e)}(\xi,\eta) = \begin{bmatrix} \dfrac{\partial x}{\partial \xi} & \dfrac{\partial y}{\partial \xi} \\ \dfrac{\partial x}{\partial \eta} & \dfrac{\partial y}{\partial \eta} \end{bmatrix} = \begin{bmatrix} J_{11}^{(e)}(\xi,\eta) & J_{12}^{(e)}(\xi,\eta) \\ J_{21}^{(e)}(\xi,\eta) & J_{22}^{(e)}(\xi,\eta) \end{bmatrix} \tag{13.114}$$

Hence

$$|J^{(e)}(\xi,\eta)| = \frac{\partial x}{\partial \xi}\frac{\partial y}{\partial \eta} - \frac{\partial x}{\partial \eta}\frac{\partial y}{\partial \xi} \tag{13.115}$$

The derivatives of the coordinate transformation are obtained from Eqs. (13.111):

$$J_{11}^{(e)}(\xi,\eta) = \frac{\partial x}{\partial \xi} = \sum_{k=1}^{6} x_k^{(e)} \frac{\partial \phi_k(\xi,\eta)}{\partial \xi} \qquad J_{12}^{(e)}(\xi,\eta) = \frac{\partial y}{\partial \xi} = \sum_{k=1}^{6} y_k^{(e)} \frac{\partial \phi_k(\xi,\eta)}{\partial \xi}$$

$$J_{21}^{(e)}(\xi,\eta) = \frac{\partial x}{\partial \eta} = \sum_{k=1}^{6} x_k^{(e)} \frac{\partial \phi_k(\xi,\eta)}{\partial \eta} \qquad J_{22}^{(e)}(\xi,\eta) = \frac{\partial y}{\partial \eta} = \sum_{k=1}^{6} y_k^{(e)} \frac{\partial \phi_k(\xi,\eta)}{\partial \eta}$$

$$\tag{13.116}$$

and the derivatives of the shape functions are obtained from Eqs. (13.110),

$$\frac{\partial \phi_1(\xi,\eta)}{\partial \xi} \quad -3+4\xi+4\eta \qquad\qquad \frac{\partial \phi_1(\xi,\eta)}{\partial \eta} \quad -3+4\xi+4\eta$$

$$\frac{\partial \phi_2(\xi,\eta)}{\partial \xi} = -1+4\xi \qquad\qquad \frac{\partial \phi_2(\xi,\eta)}{\partial \eta} = 0$$

$$\frac{\partial \phi_3(\xi,\eta)}{\partial \xi} = 0 \qquad\qquad \frac{\partial \phi_3(\xi,\eta)}{\partial \eta} = -1+4\eta$$

$$\frac{\partial \phi_4(\xi,\eta)}{\partial \xi} = 4(1-2\xi-\eta) \qquad\qquad \frac{\partial \phi_4(\xi,\eta)}{\partial \eta} = -4\xi$$

$$\frac{\partial \phi_5(\xi,\eta)}{\partial \xi} = 4\eta \qquad\qquad \frac{\partial \phi_5(\xi,\eta)}{\partial \eta} = 4\xi$$

$$\frac{\partial \phi_6(\xi,\eta)}{\partial \xi} = -4\eta \qquad\qquad \frac{\partial \phi_6(\xi,\eta)}{\partial \eta} = 4(1-\xi-2\eta) \tag{13.117}$$

Analogous to Eq. (8.31) in 1-D, the Jacobian in 2-D is the ratio of an infinitesimal area in the parent element to the corresponding infinitesimal area in the real element that it is mapped into:

$$dx\,dy = |J^{(e)}(\xi,\eta)|\,d\xi\,d\eta \tag{13.118}$$

The value of $|J^{(e)}(\xi,\eta)|$ tells us the amount of local expansion or contraction of the coordinates due to the mapping. For example, if $|J^{(e)}(\xi^*,\eta^*)| = 0$, then a nonzero area $d\xi\,d\eta$ in the neighborhood of the point (ξ^*,η^*) in the parent element maps to a zero area in the real element (which is not a 1-to-1 mapping), resulting in real shape functions with infinite slopes, and hence an unacceptable element.

There is a much greater diversity of shapes and node patterns in 2-D than in 1-D elements, so there are no simple rules to cover all possible types of distortion. However, there are some guidelines that are sufficient for most practical applications. The most frequently used shape (see Section 13.6) employs straight sides, with the midside nodes at the centers, that is,

$$x_4^{(e)} = \frac{1}{2}\left(x_1^{(e)}+x_2^{(e)}\right)$$

$$x_5^{(e)} = \frac{1}{2}\left(x_2^{(e)}+x_3^{(e)}\right)$$

$$x_6^{(e)} = \frac{1}{2}\left(x_3^{(e)}+x_1^{(e)}\right) \tag{13.119}$$

Substituting Eqs. (13.110) and (13.119) into Eqs. (13.111) yields

$$x = x_1^{(e)} + \left(x_2^{(e)}-x_1^{(e)}\right)\xi + \left(x_3^{(e)}-x_1^{(e)}\right)\eta$$

$$y = y_1^{(e)} + \left(y_2^{(e)}-y_1^{(e)}\right)\xi + \left(y_3^{(e)}-y_1^{(e)}\right)\eta \tag{13.120}$$

which is a *linear* mapping. Substituting Eqs. (13.120) into Eq. (13.115) yields

$$|J^{(e)}(\xi,\eta)| = \left(x_2^{(e)}-x_1^{(e)}\right)\left(y_3^{(e)}-y_1^{(e)}\right) - \left(x_3^{(e)}-x_1^{(e)}\right)\left(y_2^{(e)}-y_1^{(e)}\right)$$

$$= 2\Delta \text{ (twice the area of the triangle)} \tag{13.121}$$

The Jacobian is a constant over the entire element, implying no distortion. That is, the real shape functions are quadratic polynomials, having the same form as the parent shape functions (see Fig. 13.26), though possibly skewed or perhaps only enlarged or diminished by a change of scale. (Note the strong parallel with the 1-D C^0-quadratic element.)

The next most frequently used shape employs two straight sides, with centered midside nodes, and one curved side, as shown in Fig. 13.29. This would be useful along a curved boundary of a domain, since only the side along the boundary needs to be curved. It can be shown [13.7] that the Jacobian is nonzero in the triangle if the midside node on the curved side lies anywhere in the indicated sector.

For any general shape, the interior angle at each vertex should equal neither 0 nor 180°. For straight-sided triangles this is obvious, since the area of the triangle would then be zero. For curved-sided triangles, as shown in Fig. 13.30, an angle of 0 or 180° implies a locally zero area in the neighborhood of the vertex. It should also be evident that near such vertices the ξ- and η-coordinate lines would become parallel, which is an obvious collapse of the 2-D coordinate system.

Figures 13.29 and 13.30 describe the extreme limits of acceptable distortion, i.e., the conditions in which the element would *always* destroy a calculation. Using an element that is close to these limits is therefore clearly inviting trouble, since the element is likely

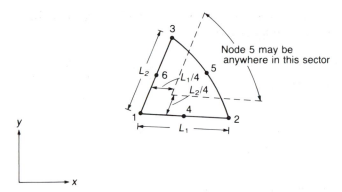

FIGURE 13.29 ● Acceptable location of the midside node along the one curved side of a C^0-quadratic isoparametric triangle.

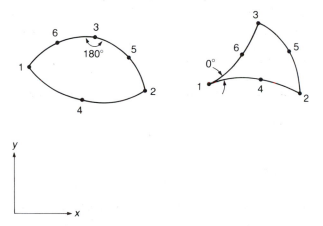

FIGURE 13.30 ● Unacceptable vertex angles for a C^0-quadratic isoparametric element.

to be numerically *ill-conditioned* (involving loss of significant figures), which would cause sporadic and unpredictable problems. There is never any practical reason to use such severely distorted elements anyway (see Section 13.6).

In summary, then, we should always use elements whose sides are straight or only "moderately" curved, the vertex angles somewhat similar to the parent angles (certainly not near 0 or 180°, and the midside nodes at or near the centers of the sides (not near the 1/4 points of the sides). In any case, a good FE program should evaluate the Jacobian at several points in an element to warn the user of a potentially badly distorted element.

Step 5: Substitute the shape functions into the element equations, and transform the integrals into a form appropriate for numerical evaluation.

Since the real shape functions may possibly be distorted, we cannot use closed-form integral formulas; we must use numerical integration (quadrature) instead. As with the 1-D elements, we first transform the integrals over the real element to integrals over the parent element, and then use appropriate quadrature formulas on the transformed integrals. Thus, using

Eq. (13.118), the stiffness and load integrals in Eqs. (13.61b) are transformed to the following integrals over the parent element:

$$K_{ij}^{(e)} = \int_0^1 \int_0^{1-\eta} \frac{\partial \phi_i^{(e)}}{\partial x} \alpha_x \frac{\partial \phi_j^{(e)}}{\partial x} |J^{(e)}| d\xi \, d\eta + \int_0^1 \int_0^{1-\eta} \frac{\partial \phi_i^{(e)}}{\partial y} \alpha_y \frac{\partial \phi_j^{(e)}}{\partial y} |J^{(e)}| d\xi \, d\eta$$

$$+ \int_0^1 \int_0^{1-\eta} \phi_i^{(e)} \beta \phi_j^{(e)} |J^{(e)}| d\xi \, d\eta$$

$$F_i^{(e)} = \int_0^1 \int_0^{1-\eta} f\phi_i^{(e)} |J^{(e)}| d\xi \, d\eta + \oint \tilde{\tau}_{-n}^{(e)} \phi_i^{(e)} \, ds$$

$$\tag{13.122}$$

The quadrature formulas will require evaluating the integrands at various quadrature points (ξ_l, η_l) and this will require a bit more manipulation of some of the terms in Eqs. (13.122).

Consider first the $K\alpha$ integrals. The physical properties $\alpha_x(x,y)$ and $\alpha_y(x,y)$ may be evaluated at a point (ξ_l, η_l) by substituting ξ_l and η_l into the coordinate transformation, Eqs. (13.111), to determine the corresponding x and y, which in turn are used to evaluate $\alpha_x(x,y)$ and $\alpha_y(x,y)$ from the functions defined by the program input data. The Jacobian $|J^{(e)}|$ may be evaluated using Eqs. (13.115) through (13.117).

The derivatives $\partial \phi_i^{(e)}/\partial x$ and $\partial \phi_i^{(e)}/\partial y$ require using the chain rule of differentiation:[32]

$$\frac{\partial \phi_i}{\partial \xi} = \frac{\partial \phi_i}{\partial x} \frac{\partial x}{\partial \xi} + \frac{\partial \phi_i}{\partial y} \frac{\partial y}{\partial \xi}$$

$$i = 1,2,\ldots,6$$

$$\frac{\partial \phi_i}{\partial \eta} = \frac{\partial \phi_i}{\partial x} \frac{\partial x}{\partial \eta} + \frac{\partial \phi_i}{\partial y} \frac{\partial y}{\partial \eta}$$

$$\tag{13.123}$$

which, when written in matrix form, contains the Jacobian matrix on the RHS:

$$\left\{ \begin{array}{c} \dfrac{\partial \phi_i}{\partial \xi} \\ \dfrac{\partial \phi_i}{\partial \eta} \end{array} \right\} = [J^{(e)}] \left\{ \begin{array}{c} \dfrac{\partial \phi_i}{\partial x} \\ \dfrac{\partial \phi_i}{\partial y} \end{array} \right\} \qquad i = 1,2,\ldots,6$$

$$\tag{13.124}$$

Inverting Eq. (13.124) yields

$$\left\{ \begin{array}{c} \dfrac{\partial \phi_i}{\partial x} \\ \dfrac{\partial \phi_i}{\partial y} \end{array} \right\} = [J^{(e)}]^{-1} \left\{ \begin{array}{c} \dfrac{\partial \phi_i}{\partial \xi} \\ \dfrac{\partial \phi_i}{\partial \eta} \end{array} \right\} \qquad i = 1,2,\ldots,6$$

$$\tag{13.125}$$

---

[32] It might seem more natural to write the chain rule in the form,

$$\frac{\partial \phi_i}{\partial x} = \frac{\partial \phi_i}{\partial \xi} \frac{\partial \xi}{\partial x} + \frac{\partial \phi_i}{\partial \eta} \frac{\partial \eta}{\partial x}$$

(with a similar expression for $\partial \phi_i/\partial y$), since the unknown quantity $\partial \phi_i/\partial x$ is given *explicitly* in terms of other quantities, rather than *implicitly* as in Eqs. (13.123) which requires inverting the equations. This approach, unfortunately, produces the derivatives $\partial \xi/\partial x$, $\partial \eta/\partial x$, etc., which we can't evaluate using Eqs. (13.111); we need their inverted forms, $\partial x/\partial \xi$, $\partial x/\partial \eta$, etc., which appear in Eqs. (13.123).

where

$$[J^{(e)}]^{-1} = \frac{1}{|J^{(e)}|} \begin{bmatrix} \dfrac{\partial y}{\partial \eta} & -\dfrac{\partial y}{\partial \xi} \\[2ex] -\dfrac{\partial x}{\partial \eta} & \dfrac{\partial x}{\partial \xi} \end{bmatrix} \tag{13.126}$$

Therefore

$$\frac{\partial \phi_i}{\partial x} = (J_{11})^{-1} \frac{\partial \phi_i}{\partial \xi} + (J_{12})^{-1} \frac{\partial \phi_i}{\partial \eta}$$

$$i = 1,2,\ldots,6$$

$$\frac{\partial \phi_i}{\partial y} = (J_{21})^{-1} \frac{\partial \phi_i}{\partial \xi} + (J_{22})^{-1} \frac{\partial \phi_i}{\partial \eta} \tag{13.127a}$$

where

$$(J_{11})^{-1} = \frac{1}{|J^{(e)}|} \frac{\partial y}{\partial \eta} \qquad\qquad (J_{12})^{-1} = -\frac{1}{|J^{(e)}|} \frac{\partial y}{\partial \xi}$$

$$(J_{21})^{-1} = -\frac{1}{|J^{(e)}|} \frac{\partial x}{\partial \eta} \qquad\qquad (J_{22})^{-1} = \frac{1}{|J^{(e)}|} \frac{\partial x}{\partial \xi} \tag{13.127b}$$

Equations (13.127) are the desired expressions since all the terms are known functions of ξ and η, using Eqs. (13.115) through (13.117).

The $K\beta$ integral requires no further manipulation. Thus $\beta(x,y)$ becomes a function of ξ and η by using Eqs. (13.111). The shape functions and Jacobian use Eqs. (13.110) and Eqs. (13.115) through (13.117), respectively.

The interior load integral Ff is treated in the same manner as the $K\beta$ integral. In addition, it was stated in Section 13.3.1 that $f(x,y)$ will be defined in the UNAFEM program as a constant $f^{(e)}$ in each element [a procedure that will introduce a modeling error if $f(x,y)$ is a more complicated function]. Therefore $f^{(e)}$ can be factored outside the integral.

Before examining the boundary flux integral, which is a 1-D line integral, let's first complete the numerical integration of the 2-D area integrals by developing the appropriate quadrature formulas. The 1-D Gaussian quadrature formulas in Chapter 8 are not appropriate for triangles (although they can be used for the 2-D quadrilateral elements in Sections 13.4.1 and 13.4.2).

Figure 13.31 tabulates Gauss points and weights for several formulas of increasing levels of accuracy [13.8]. These formulas have two important properties. They are Gaussian type rather than Newton-Cotes and thus produce greater accuracy for a given number of quadrature points. And they are triangularly symmetric, which avoids biasing any one vertex. Thus points occur in groups of one (at the centroid), three (along bisectors of the vertices), or six (in pairs symmetrically offset from each bisector).

Figure 13.31 shows only the most frequently used formulas from the more comprehensive list in Cowper [13.8]. A more recent attempt to improve on these formulas is given

$$\int_0^1 \int_0^{1-\eta} I(\xi,\eta)\, d\xi\, d\eta \simeq \frac{1}{2} \sum_{l=1}^n w_{nl}\, I(\xi_{nl},\eta_{nl})$$

Area of triangle

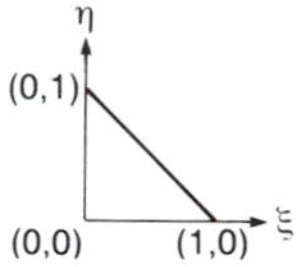

Number of Gauss points, n	Accuracy of quadrature[†]	Gauss points[‡] ξ_{nl}	η_{nl}	Weights[‡] w_{nl}	Illustration of Gauss-point locations (∗)
1	$O(h^2)$	$\xi_{11}=1/3$	$\eta_{11}=1/3$	$w_{11}=1$	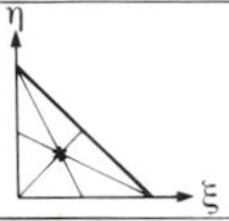
3	$O(h^3)$	$\xi_{31}=1/6$ $\xi_{32}=2/3$ $\xi_{33}=1/6$	$\eta_{31}=1/6$ $\eta_{32}=1/6$ $\eta_{33}=2/3$	$w_{31}=1/3$ $w_{32}=1/3$ $w_{33}=1/3$	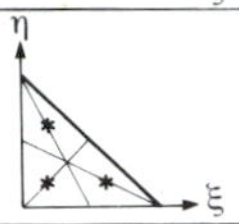
4	$O(h^4)$	$\xi_{41}=1/3$ $\xi_{42}=1/5$ $\xi_{43}=3/5$ $\xi_{44}=1/5$	$\eta_{41}=1/3$ $\eta_{42}=1/5$ $\eta_{43}=1/5$ $\eta_{44}=3/5$	$w_{41}=-27/48$ $w_{42}=25/48$ $w_{43}=25/48$ $w_{44}=25/48$	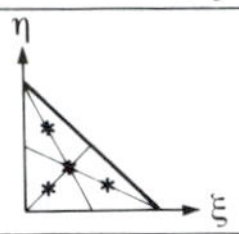
6	$O(h^5)$	$\xi_{61}=0.09157\,62135\,09771$ $\xi_{62}=0.81684\,75729\,80459$ $\xi_{63}=\xi_{61}$ $\xi_{64}=0.44594\,84909\,15965$ $\xi_{65}=0.10810\,30181\,68070$ $\xi_{66}=\xi_{64}$	$\eta_{61}=\xi_{61}$ $\eta_{62}=\xi_{61}$ $\eta_{63}=\xi_{62}$ $\eta_{64}=\xi_{64}$ $\eta_{65}=\xi_{64}$ $\eta_{66}=\xi_{65}$	$w_{61}=0.10995\,17436\,55322$ $w_{62}=w_{61}$ $w_{63}=w_{61}$ $w_{64}=0.22338\,15896\,78011$ $w_{65}=w_{64}$ $w_{66}=w_{64}$	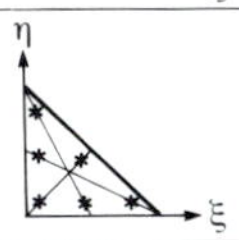
7	$O(h^6)$	$\xi_{71}=1/3$ $\xi_{72}=0.10128\,65073\,23456$ $\xi_{73}=0.79742\,69853\,53087$ $\xi_{74}=\xi_{72}$ $\xi_{75}=0.47014\,20641\,05115$ $\xi_{76}=0.05971\,58717\,89770$ $\xi_{77}=\xi_{75}$	$\eta_{71}=1/3$ $\eta_{72}=\xi_{72}$ $\eta_{73}=\xi_{72}$ $\eta_{74}=\xi_{73}$ $\eta_{75}=\xi_{75}$ $\eta_{76}=\xi_{75}$ $\eta_{77}=\xi_{76}$	$w_{71}=0.225$ $w_{72}=0.12593\,91805\,44827$ $w_{73}=w_{72}$ $w_{74}=w_{72}$ $w_{75}=0.13239\,41527\,88506$ $w_{76}=w_{75}$ $w_{77}=w_{75}$	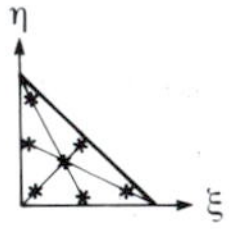

[†] An accuracy of $O(h^{p+1})$ means that $I(\xi,\eta)$ will be integrated exactly if it is a 2-D polynomial of degree $\le p$. Such a rule is said to have a degree of precision p.

[‡] ξ and η are area coordinates. Therefore the above **Gauss** points and weights are for triangles of any shape, not just the unit right triangle shown. See comments in text.

FIGURE 13.31 ● Gauss points and weights for Gauss quadrature rules on triangles [13.8].

in Reddy and Shippy [13.9]. Other contributions to this topic are cited in both these references.

The stiffness and interior load integrals in Eqs. (13.122) may now be written in terms of the quadrature formula in Fig. 13.31 as follows:

$$K_{ij}^{(e)} = \frac{1}{2} \sum_{l=1}^n w_{nl} \left[\left(\frac{\partial \phi_i^{(e)}}{\partial x} \alpha_x \frac{\partial \phi_j^{(e)}}{\partial x} + \frac{\partial \phi_i^{(e)}}{\partial y} \alpha_y \frac{\partial \phi_j^{(e)}}{\partial y} + \phi_i^{(e)} \beta \phi_j^{(e)} \right) |J^{(e)}| \right]_{(\xi_{nl},\eta_{nl})}$$

$$Ff_i^{(e)} = \frac{1}{2} \sum_{l=1}^n w_{nl} \left[f\phi_i^{(e)} |J^{(e)}| \right]_{(\xi_{nl},\eta_{nl})}$$

$$(13.128)$$

where all the terms inside the brackets may be evaluated at the Gauss points (ξ_{nl}, η_{nl}) using the expressions developed previously.

To determine which order quadrature rule to use, we may use guidelines II and III of Section 8.3.1. Thus for the C^0-quadratic triangle $m=1$ and $p=2$; reduced integration calls for a quadrature accuracy of $O(h^4)$ which is the four-point rule. Higher-order integration calls for an accuracy of $O(h^5)$, which is the six-point rule.

Now let's examine the boundary flux integral,

$$F_{T_i}^{(e)} = \oint^{(e)} \tilde{\tau}_{-n}^{(e)} \phi_i^{(e)} \, ds \tag{13.129}$$

As explained in Section 13.3.1, step 5 [see Eqs. (13.84) through (13.86) and Eq. (13.96)], Eq. (13.129) may be written as the sum of three integrals, one on each side of the triangle. If a side is interior to the domain, it will pair with a similar integral from an adjacent element, resulting in interelement boundary flux integrals that are evaluated in the same manner as those on the domain boundary. Therefore we need only evaluate Eq. (13.129) for a side lying on the domain boundary.

Figure 13.32 shows such an element, with the side $\overline{142}$ on the domain boundary, Γ. In general, this side may possibly be curved and the midside node need not be located at the center. The following derivation will be based on side $\overline{142}$. However, because of the interpolation property of the shape functions and their (collective) triangular symmetry, the final expressions derived below for side $\overline{142}$ will also be applicable to sides $\overline{253}$ and $\overline{361}$, simply by changing the indices 1, 4, 2 to 2, 5, 3 and 3, 6, 1, respectively. (We will comment on this again at the end of the derivation.)

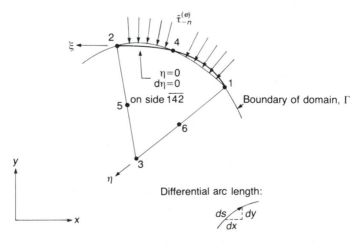

FIGURE 13.32 ● Evaluating the boundary flux integral on an element side lying on (or approximately on) the domain boundary.

The portion of the boundary integral in Eq. (13.129) on side $\overline{142}$ may be written as follows:

$$F\tau_i^{(e)}(\Gamma) = \int_{\overline{142}} \bar{\tau}_{-n}^{(e)}\phi_i(\xi,0) \, ds \qquad i = 1,4,2$$

(13.130)

where we have used the fact that $\eta=0$ on side $\overline{142}$ and hence the shape functions in Eqs. (13.110) become

$$\phi_1(\xi,0) = (1-\xi)(1-2\xi)$$

$$\phi_2(\xi,0) = \xi(2\xi-1)$$

$$\phi_3(\xi,0) = 0$$

$$\phi_4(\xi,0) = 4\xi(1-\xi)$$

$$\phi_5(\xi,0) = 0$$

$$\phi_6(\xi,0) = 0$$

(13.131)

Therefore the index i in Eq. (13.130) may be limited to the values 1, 4, 2 since $F\tau_3^{(e)}(\Gamma)$ $= F\tau_5^{(e)}(\Gamma) = F\tau_6^{(e)}(\Gamma) = 0$, and we know that zero-valued natural BCs can be ignored (assuming the natural BC array in the computer program is initially zeroed).

The next step is to express the differential arc length ds in the ξ,η-coordinate system. From the basic Pythagorean relation with dx and dy (see Fig. 13.32), the coordinate transformation in Eqs. (13.111), and the definition of the Jacobian components in Eq. (13.116), we have the following general expression for ds in terms of $\xi,\eta,d\xi$, and $d\eta$:

$$ds = \sqrt{dx^2 + dy^2}$$

$$= \sqrt{\left(\frac{\partial x}{\partial \xi} d\xi + \frac{\partial x}{\partial \eta} d\eta\right)^2 + \left(\frac{\partial y}{\partial \xi} d\xi + \frac{\partial y}{\partial \eta} d\eta\right)^2}$$

$$= \sqrt{\left(J_{11}^{(e)}(\xi,\eta) \, d\xi + J_{21}^{(e)}(\xi,\eta) \, d\eta\right)^2 + \left(J_{12}^{(e)}(\xi,\eta) \, d\xi + J_{22}^{(e)}(\xi,\eta) \, d\eta\right)^2} \qquad (13.132)$$

On side $\overline{142}$ $\eta = 0$ and $d\eta = 0$ (see Fig. 13.32), so Eq. (13.132) simplifies to

$$ds = J_\Gamma^{(e)}(\xi,0) \, d\xi \qquad (13.133a)$$

where

$$J_\Gamma^{(e)}(\xi,0) = \sqrt{\left(J_{11}^{(e)}(\xi,0)\right)^2 + \left(J_{12}^{(e)}(\xi,0)\right)^2} \qquad (13.133b)$$

The quantity $J_\Gamma^{(e)}$ could be called a *boundary Jacobian* since it is the ratio of differential arc lengths in the two coordinate systems. The boundary integral in Eq. (13.130) becomes

$$F\tau_i^{(e)}(\Gamma) = \int_0^1 \tilde{\tau}_{-n}^{(e)}\phi_i(\xi,0)J_\Gamma^{(e)}(\xi,0) \ d\xi \qquad i = 1,4,2$$

$$\text{(13.134)}$$

The boundary Jacobian can be evaluated using Eqs. (13.116) and (13.131). Thus

$$J_{11}^{(e)}(\xi,0) = \frac{\partial x}{\partial \xi}(\xi,0) = \sum_{k=1,4,2} x_k^{(e)} \frac{\partial \phi_k(\xi,0)}{\partial \xi}$$

$$= (4\xi-3)x_1^{(e)} - (8\xi-4)x_4^{(e)} + (4\xi-1)x_2^{(e)}$$

$$J_{12}^{(e)}(\xi,0) = \frac{\partial y}{\partial \xi}(\xi,0)$$

$$= \sum_{k=1,4,2} y_k^{(e)} \frac{\partial \phi_k(\xi,0)}{\partial \xi}$$

$$= (4\xi-3)y_1^{(e)} - (8\xi-4)y_4^{(e)} + (4\xi-1)y_2^{(e)} \qquad \text{(13.135)}$$

All the terms in the integrand of Eq. (13.134) are now known functions of ξ. Thus $\tilde{\tau}_{-n}^{(e)}$ is a prescribed function of x and y but is implicitly a function of ξ from Eqs. (13.111) and (13.131), $\phi_i(\xi,0)$ is defined by Eqs. (13.131), and $J_\Gamma^{(e)}$ is defined by Eqs. (13.133b) and (13.135).

If $\tilde{\tau}_{-n}^{(e)}$ is constant (or assumed constant, as in UNAFEM), Eq. (13.134) can be integrated in closed form, although the expressions are quite complicated because of the square root in $J_\Gamma^{(e)}$. For a more complicated $\tilde{\tau}_{-n}^{(e)}$, a closed-form expression is generally not possible at all. The simpler and straightforward solution is to use numerical integration.

In order to use the 1-D Gauss-Legendre rules in Fig. 8.12, the integration interval $0 \le \xi \le 1$ in Eq. (13.134) must be changed to the bi-unit interval $-1 \le \xi' \le 1$. This can be accomplished by the transformation

$$\xi = \frac{1}{2}(\xi'+1)$$

$$\text{(13.136)}$$

which changes the shape functions and Jacobian components [Eqs. (13.131) and (13.135)] to the following expressions:

$$\phi_1(\xi',0) = -\frac{1}{2}\xi'(1-\xi')$$

$$\phi_2(\xi',0) = \frac{1}{2}\xi'(1+\xi')$$

$$\phi_4(\xi',0) = (1+\xi')(1-\xi')$$

$$J_{11}^{(e)}(\xi',0) = (2\xi'-1)x_1^{(e)} - 4\xi'x_4^{(e)} + (2\xi'+1)x_2^{(e)}$$

$$J_{12}^{(e)}(\xi',0) = (2\xi'-1)y_1^{(e)} - 4\xi'y_4^{(e)} + (2\xi'+1)y_2^{(e)} \qquad \text{(13.137)}$$

where

$$J_\Gamma^{(e)}(\xi',0) = \sqrt{ \left[J_{11}^{(e)}(\xi',0) \right]^2 + \left[J_{12}^{(e)}(\xi',0) \right]^2 }$$

Eq. (13.134) becomes

$$F\tau_i^{(e)}(\Gamma) = \frac{1}{2} \int_{-1}^{1} \tilde{\tau}_{-n}^{(e)} \phi_i(\xi',0) J_\Gamma^{(e)}(\xi',0) \, d\xi' \qquad i = 1,4,2$$

(13.138)

Using the 1-D Gauss-Legendre quadrature formula in Fig. 8.12, Eq. (13.138) becomes

$$F\tau_i^{(e)}(\Gamma) \simeq \frac{1}{2} \sum_{l=1}^{n} w_{nl} \left[\tilde{\tau}_{-n}^{(e)} \phi_i(\xi',0) J_\Gamma^{(e)}(\xi',0) \right]_{\xi'_{nl}} \qquad i = 1,4,2$$

(13.139)

To simplify data input to UNAFEM, we will require $\tilde{\tau}_{-n}^{(e)}$ to be a constant along each element. This will provide a piecewise-constant approximation of $\tilde{\tau}_{-n}^{(e)}$ along the entire boundary, thereby possibly introducing a modeling error that would decrease as the mesh is refined.

The same order of accuracy should be used for $F\tau_i^{(e)}(\Gamma)$ as for all the other integrals in the element equations. Thus reduced integration calls for accuracy of $O(h^4)$, which is the 1-D two-point rule, and higher-order integration calls for an accuracy of at least $O(h^5)$, which requires the 1-D three-point rule.

It was observed at the beginning of this derivation that the final integration expressions, Eqs. (13.137) and (13.139), would also be applicable to the other two sides of the element, simply by changing the indices 1, 4, 2 to 2, 5, 3 and 3, 6, 1, respectively. Thus, if side $\overline{361}$ were on the boundary, then $\xi = 0$, and Eqs. (13.110) become

$$\phi_1(0,\eta) = (1-\eta)(1-2\eta)$$

$$\phi_2(0,\eta) = 0$$

$$\phi_3(0,\eta) = \eta(2\eta-1)$$

$$\phi_4(0,\eta) = 0$$

$$\phi_5(0,\eta) = 0$$

$$\phi_6(0,\eta) = 4\eta(1-\eta)$$

In addition, $d\xi=0$, so the boundary Jacobian [recall Eqs. (13.132) and (13.133)] becomes

$$J_\Gamma^{(e)}(0,\eta) = \sqrt{ \left[J_{21}^{(e)}(0,\eta) \right]^2 + \left[J_{22}^{(e)}(0,\eta) \right]^2 }$$

The counterclockwise integration path means η is integrated from 1 to 0 (not 0 to 1), so a transformation different from Eq. (13.136) is needed to convert to the bi-unit interval $-1 \le \xi' \le 1$. The reader can easily verify (see Exercise 13.6) that this will lead to Eqs. (13.137) and (13.139), with the indices 1, 4, 2 replaced by 3, 6, 1. Similar results pertain to side $\overline{253}$.

One particular situation deserves special attention because it occurs so frequently —
namely, the element side on the boundary is straight and its midside node is at its center:

$$x_4^{(e)} = \frac{1}{2} \left(x_1^{(e)} + x_2^{(e)} \right)$$

$$y_4^{(e)} = \frac{1}{2} \left(y_1^{(e)} + y_2^{(e)} \right) \tag{13.140}$$

Substituting Eqs. (13.140) into Eqs. (13.137) yields $J_{11}^{(e)}(\xi,0) = x_2^{(e)} - x_1^{(e)}$, $J_{12}^{(e)}(\xi,0) = y_2^{(e)} - y_1^{(e)}$ and therefore $J_\Gamma^{(e)} = L$, the length of the side.
The integral in Eq. (13.138) simplifies to

$$F_{T_i}^{(e)}(\Gamma) = \frac{L}{2} \int_{-1}^{1} \tilde{\tau}_{-n}^{(e)} \phi_i(\xi',0) \, d\xi' \qquad i = 1,4,2 \tag{13.141}$$

If $\tilde{\tau}_{-n}^{(e)}$ is constant (or assumed constant, as in UNAFEM), then the integral in Eq. (13.141)
can be evaluated exactly, yielding

$$F_{T_1}^{(e)}(\Gamma) = \frac{1}{6} L \tilde{\tau}_{-n}^{(e)}$$

$$F_{T_4}^{(e)}(\Gamma) = \frac{4}{6} L \tilde{\tau}_{-n}^{(e)}$$

$$F_{T_2}^{(e)}(\Gamma) = \frac{1}{6} L \tilde{\tau}_{-n}^{(e)} \tag{13.142}$$

Equations (13.142) have appeared widely in the FE literature. The above derivation
has shown that they are strictly valid only under three assumptions: (1) the boundary
side of the element is straight; (2) the midside node is at the center; and (3) the specified
boundary flux is constant along the side of the element. If any of these assumptions is
not true, then the quadrature formula, Eq. (13.139), should be used instead.

Step 6: Prepare expressions for the flux.

As usual, we have two sets of points at which to calculate the flux: first, at the optimal
flux-sampling points (Gauss points), and second, at the nodes, using smoothing and
averaging.

The general expressions for the components of flux in terms of the shape functions
follow directly from the basic flux formula, Eq. (13.17), and the element trial solution,
Eq. (13.48) with $n=6$:

$$\tilde{\tau}_x^{(e)}(x,y) = -\alpha_x(x,y) \sum_{j=1}^{6} a_j \frac{\partial \phi_j^{(e)}(x,y)}{\partial x}$$

$$\tilde{\tau}_y^{(e)}(x,y) = -\alpha_y(x,y) \sum_{j=1}^{6} a_j \frac{\partial \phi_j^{(e)}(x,y)}{\partial y} \tag{13.143}$$

Expressions (13.143) may be evaluated at any point (ξ,η) using Eqs. (13.111) and (13.127) which in turn use Eqs. (13.110) and Eqs. (13.115) through (13.117).

The optimal flux-sampling points can be determined using the least-squares technique illustrated in the exercises in Chapters 7 and 8 and also in this chapter. This yields a system of nonlinear equations which possesses only complex-valued roots for ξ and η. These roots are very close to the real values $\sqrt{6}/[4(1+\sqrt{6})]$ (or 0.178) and $(4+\sqrt{6})/10$ (or 0.645).

Using an alternative technique, Moan [13.10] demonstrates that the bound for the pointwise error in the flux is minimal at the points $\xi=(8-\sqrt{10})/27$ (or 0.179) and $\eta=(11+2\sqrt{10})/27$ (or 0.642), as well as the other two points that are symmetrically positioned with respect to area coordinates. However, Moan observes that "these points are located near to the sampling points [Gauss points] of the four point integration formula of order four" (i.e., the values 0.2 and 0.6, shown in Fig. 13.31), and gives a numerical example that exhibits an error at each of the four Gauss points that is one to two orders of magnitude less than the maximum error in the element (the latter occurring at element vertices).

Indeed, the Gauss points for the four-point rule are generally used for the flux-sampling points for this element. Whether these points are optimal or close to optimal is unclear to this author.[33] Nevertheless, the general practice will be followed here (Figure 13.33). This will yield four optimal values for both $\tilde{\tau}_x^{(e)}$ and $\tilde{\tau}_y^{(e)}$, which we write as vectors for use in the subsequent smoothing analysis:

$$\{\tilde{\tau}_x\}_{opt}^{(e)} = \left\{ \begin{array}{l} \tilde{\tau}_x^{(e)}(\xi_I,\eta_I) \\ \tilde{\tau}_x^{(e)}(\xi_{II},\eta_{II}) \\ \tilde{\tau}_x^{(e)}(\xi_{III},\eta_{III}) \\ \tilde{\tau}_x^{(e)}(\xi_{IV},\eta_{IV}) \end{array} \right\}$$

$$\{\tilde{\tau}_y\}_{opt}^{(e)} = \left\{ \begin{array}{l} \tilde{\tau}_y^{(e)}(\xi_I,\eta_I) \\ \tilde{\tau}_y^{(e)}(\xi_{II},\eta_{II}) \\ \tilde{\tau}_y^{(e)}(\xi_{III},\eta_{III}) \\ \tilde{\tau}_y^{(e)}(\xi_{IV},\eta_{IV}) \end{array} \right\}$$

(13.144)

where, for example, $\tilde{\tau}_x^{(e)}(\xi_I,\eta_I)$ is shorthand for $\tilde{\tau}_x^{(e)}[x(\xi_I,\eta_I),y(\xi_I,\eta_I)]$.

We next construct a smoothing function for each component of flux, i.e., one for $\tilde{\tau}_x^{(e)}(x,y)$ and another one for $\tilde{\tau}_y^{(e)}(x,y)$. As noted in Section 13.3.1, step 6, the smoothing functions are determined by either interpolating or least-squares-fitting some or all of the optimal fluxes. The choice is primarily one of convenience since there appears to be no theoretical or numerical evidence showing superior accuracy of any one choice for a given class of problems. Both interpolation and least-squares-fitting, using some or all of the optimal fluxes, have been used in practice [13.11-13.13].

In the following expressions τ will represent either $\tilde{\tau}_x^{(e)}$ or $\tilde{\tau}_y^{(e)}$ since the development is identical for both components. Our goal is to derive a *transformation matrix* $[TR]$ that

---

[33] If the flux-sampling points are close to optimal, but not optimal, a $\log|error|$ versus $\log h$ plot, similar to those in Chapter 9, would have an initial slope corresponding to an optimal point. As h decreases, the slope would eventually decrease to the value corresponding to a nonoptimal point.

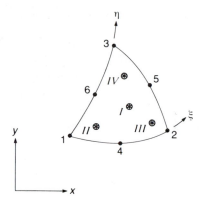

FIGURE 13.33 ● Optimal flux-sampling points for the C^0-quadratic isoparametric triangle. Optimal points (⊛) are the Gauss points for the four-point rule in Fig. 13.31.

relates the four optimal fluxes in an element to the six element nodal fluxes,

$$\{\tau\}_{\text{nodes}} = [TR]\{\tau\}_{\text{opt}} \qquad\qquad (13.145)$$

where

$$\{\tau\}_{\text{nodes}} = \begin{Bmatrix} \tau_1 \\ \tau_2 \\ \tau_3 \\ \tau_4 \\ \tau_5 \\ \tau_6 \end{Bmatrix} \qquad \{\tau\}_{\text{opt}} = \begin{Bmatrix} \tau_I \\ \tau_{II} \\ \tau_{III} \\ \tau_{IV} \end{Bmatrix}$$

and $[TR]$ is a 6-by-4 matrix. The vector $\{\tau\}_{\text{opt}}$ is either one of the vectors in Eqs. (13.144). Roman numerals are used to designate Gauss points and quantities related to those points, e.g., the optimal fluxes.

   There is a definite computational advantage to constructing the smoothing function τ_s in the parent element rather than in the real elements, since then there is only one $[TR]$ applicable to all real elements. It can be evaluated in advance, as shown below, and stored in the program. Conversely, if τ_s were constructed in the real elements, then the shape functions would be needed in order to calculate a different $[TR]^{(e)}$ for each element, resulting not only in increased computation but also decreased modularity of the post-processing phase. The two approaches would differ only for distorted elements $[J^{(e)} \neq$ constant]. There is no evidence yet to show which approach is more accurate; once again, convenience becomes the guiding factor.

   For a smoothing function we use a 2-D polynomial such that the number of coefficients is equal to or less than the number of optimal flux values. The former would require an interpolatory fit and the latter a least-squares fit. We'll do the latter in order to demonstrate the procedure. (Higher-order elements probably perform better with least-squares-fitting since they have larger numbers of optimal values, yet low-degree polynomials are generally better smoothing functions since they are less oscillatory.) Given four optimal fluxes, we will therefore use a complete first-degree polynomial:

$$\tau_s(\xi,\eta) = c_1 + c_2\xi + c_3\eta \tag{13.146}$$

which has three unknown coefficients to be determined by the least-squares procedure.

Equation (13.146) describes a flat plane (possibly tilted) which cannot in general coincide with all four optimal fluxes. We therefore seek to orient the plane so that it is as close as possible, in a least-squares sense, to the four optimal fluxes. Let $e(c_1,c_2,c_3)$ be the least-squares error, i.e., the sum of the squares of the distances between the plane and each of the optimal fluxes:

$$e(c_1,c_2,c_3) = \sum_{i=I}^{IV} [\tau_s(\xi_i,\eta_i) - \tau_i]^2$$

$$= \sum_{i=I}^{IV} [c_1 + c_2\xi_i + c_3\eta_i - \tau_i]^2 \tag{13.147}$$

where each of the terms is illustrated graphically in Fig. 13.34.

The error is minimized by imposing the conditions

$$\frac{\partial e}{\partial c_i} = 0 \qquad i = 1,2,3 \tag{13.148}$$

This yields the following three equations:

$$\frac{\partial e}{\partial c_1} = 0 : \sum_{i=I}^{IV} [c_1 + c_2\xi_i + c_3\eta_i - \tau_i] = 0$$

$$\frac{\partial e}{\partial c_2} = 0 : \sum_{i=I}^{IV} [c_1 + c_2\xi_i + c_3\eta_i - \tau_i]\xi_i = 0$$

$$\frac{\partial e}{\partial c_3} = 0 : \sum_{i=I}^{IV} [c_1 + c_2\xi_i + c_3\eta_i - \tau_i]\eta_i = 0 \tag{13.149}$$

which may be written in matrix form as

$$
\underbrace{\begin{bmatrix} \sum\limits_{i=I}^{IV} 1 & \sum\limits_{i=I}^{IV} \xi_i & \sum\limits_{i=I}^{IV} \eta_i \\[2ex] \sum\limits_{i=I}^{IV} \xi_i & \sum\limits_{i=I}^{IV} \xi_i\xi_i & \sum\limits_{i=I}^{IV} \xi_i\eta_i \\[2ex] \sum\limits_{i=I}^{IV} \eta_i & \sum\limits_{i=I}^{IV} \xi_i\eta_i & \sum\limits_{i=I}^{IV} \eta_i\eta_i \end{bmatrix}}_{[P]} \begin{Bmatrix} c_1 \\[2ex] c_2 \\[2ex] c_3 \end{Bmatrix} = \underbrace{\begin{bmatrix} 1 & 1 & 1 & 1 \\[2ex] \xi_I & \xi_{II} & \xi_{III} & \xi_{IV} \\[2ex] \eta_I & \eta_{II} & \eta_{III} & \eta_{IV} \end{bmatrix}}_{[Q]} \begin{Bmatrix} \tau_I \\ \tau_{II} \\ \tau_{III} \\ \tau_{IV} \end{Bmatrix} \tag{13.150}
$$

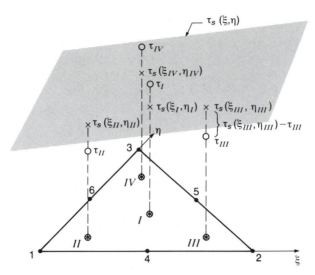

FIGURE 13.34 ● Least-squares-fitting of a flat plane to the four optimal flux values; the fitting is done in the parent element.

or in more abbreviated form,[34]

$$[P]\{c\} = [Q]\{\tau\}_{\text{opt}} \tag{13.151}$$

Hence

$$\{c\} = [S]\{\tau\}_{\text{opt}} \tag{13.152}$$

where

$$[S] = [P]^{-1}[Q] \tag{13.153}$$

The *smoothing matrix* $[S]$ determines the coefficients $\{c\}$ of the smoothing function that gives the least-squares fit.

From Fig. 13.31,

$$\xi_I = \frac{1}{3} \qquad \xi_{II} = \frac{1}{5} \qquad \xi_{III} = \frac{3}{5} \qquad \xi_{IV} = \frac{1}{5}$$

$$\eta_I = \frac{1}{3} \qquad \eta_{II} = \frac{1}{5} \qquad \eta_{III} = \frac{1}{5} \qquad \eta_{IV} = \frac{3}{5} \tag{13.154}$$

---

[34] In approximation theory (a branch of numerical analysis) Eqs. (13.150) are called the *normal equations*. It should be apparent that the above least-squares procedure can be readily generalized to fitting a surface with m unknowns — for example, $\tau_s(\xi, \eta) = c_1\psi_1(\xi, \eta) + \dots + c_m\psi_m(\xi, \eta)$ — to a set of $n(n > m)$ optimal fluxes. The normal equations would then have the form of Eq. (13.151) where

$P_{jk} = \sum_{i=1}^{n} \psi_j(\xi_i, \eta_i)\psi_k(\xi_i, \eta_i)$, $Q_{jk} = \psi_j(\xi_k, \eta_k)$, and $[P]$ and $[Q]$ are m-by-m and m-by-n matrices,

respectively.

Hence

$$[P] = \frac{4}{3}\begin{bmatrix} 3 & 1 & 1 \\ 1 & \dfrac{31}{75} & \dfrac{22}{75} \\ 1 & \dfrac{22}{75} & \dfrac{31}{75} \end{bmatrix}$$

$$[Q] = \begin{bmatrix} 1 & 1 & 1 & 1 \\ \dfrac{1}{3} & \dfrac{1}{5} & \dfrac{3}{5} & \dfrac{1}{5} \\ \dfrac{1}{3} & \dfrac{1}{5} & \dfrac{1}{5} & \dfrac{3}{5} \end{bmatrix}$$

(13.155)

and

$$[S] = \frac{1}{12}\begin{bmatrix} 3 & 23 & -7 & -7 \\ 0 & -30 & 30 & 0 \\ 0 & -30 & 0 & 30 \end{bmatrix}$$

(13.156)

Having determined the smoothing function, it remains only to evaluate it at the six nodes. Using Eq. (13.146) and the known ξ, η values at each of the nodes (see Fig. 13.25), we can construct (by inspection) an evaluation matrix $[E]$ relating the element nodal fluxes to the coefficients $\{c\}$:

$$\begin{Bmatrix} \tau_1 \\ \tau_2 \\ \tau_3 \\ \tau_4 \\ \tau_5 \\ \tau_6 \end{Bmatrix} = \begin{bmatrix} 1 & 0 & 0 \\ 1 & 1 & 0 \\ 1 & 0 & 1 \\ 1 & \dfrac{1}{2} & 0 \\ 1 & \dfrac{1}{2} & \dfrac{1}{2} \\ 1 & 0 & \dfrac{1}{2} \end{bmatrix} \begin{Bmatrix} c_1 \\ c_2 \\ c_3 \end{Bmatrix}$$

$$\uparrow$$
$$[E]$$

(13.157)

or in matrix form,

$$\{\tau\}_{\text{nodes}} = [E]\{c\}$$

(13.158)

Substituting Eq. (13.152) into Eq. (13.158) yields

$$\{\tau\}_{\text{nodes}} = [E][S]\{\tau\}_{\text{opt}}$$

(13.159)

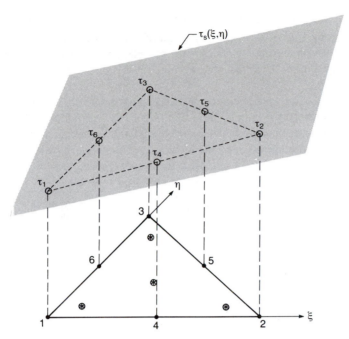

FIGURE 13.35 ● The element nodal fluxes are obtained by evaluating the smoothing function $\tau_s(\xi,\eta)$ at each of the nodes.

which is in the same form as Eq. (13.145) and hence our desired transformation matrix is

$$[TR] = [E][S] \tag{13.160}$$

From Eqs. (13.156) and (13.157),

$$[TR] = \frac{1}{12}
\begin{bmatrix}
3 & 23 & -7 & -7 \\
3 & -7 & 23 & -7 \\
3 & -7 & -7 & 23 \\
\\
3 & 8 & 8 & -7 \\
3 & -7 & 8 & 8 \\
3 & 8 & -7 & 8
\end{bmatrix} \tag{13.161}$$

The evaluation of the element nodal fluxes is illustrated graphically in Fig. 13.35. Thus the element nodal flux at node i is simply the value of $\tau_s(\xi,\eta)$ at node i.

In summary, Eq. (13.145) is used for both of the flux components in each element in the mesh; that is,

$$\{\tilde{\tau}_x\}^{(e)}_{\text{nodes}} = [TR]\{\tilde{\tau}_x\}^{(e)}_{\text{opt}}$$
$$\{\tilde{\tau}_y\}^{(e)}_{\text{nodes}} = [TR]\{\tilde{\tau}_y\}^{(e)}_{\text{opt}} \tag{13.162}$$

where $[TR]$ is given by Eq. (13.161), and $\{\tilde{\tau}_x\}^{(e)}_{opt}$ and $\{\tilde{\tau}_y\}^{(e)}_{opt}$ are given by Eqs. (13.144) and evaluated using Eqs. (13.143). The calculated values for $\{\tilde{\tau}_x\}^{(e)}_{nodes}$ and $\{\tilde{\tau}_y\}^{(e)}_{nodes}$ are then assigned to the corresponding nodes in the real elements. Finally, these element nodal fluxes are averaged in the usual manner (see Fig. 13.22) at each node in the mesh to yield the nodal fluxes.

It is worth remarking once again that the smoothing and averaging procedure for computing nodal fluxes from the optimal fluxes introduces several additional approximations into the FE model, — namely the least-squares procedure, the fitting in the parent element rather than in the real elements, and the averaging at the nodes. Although the effect of

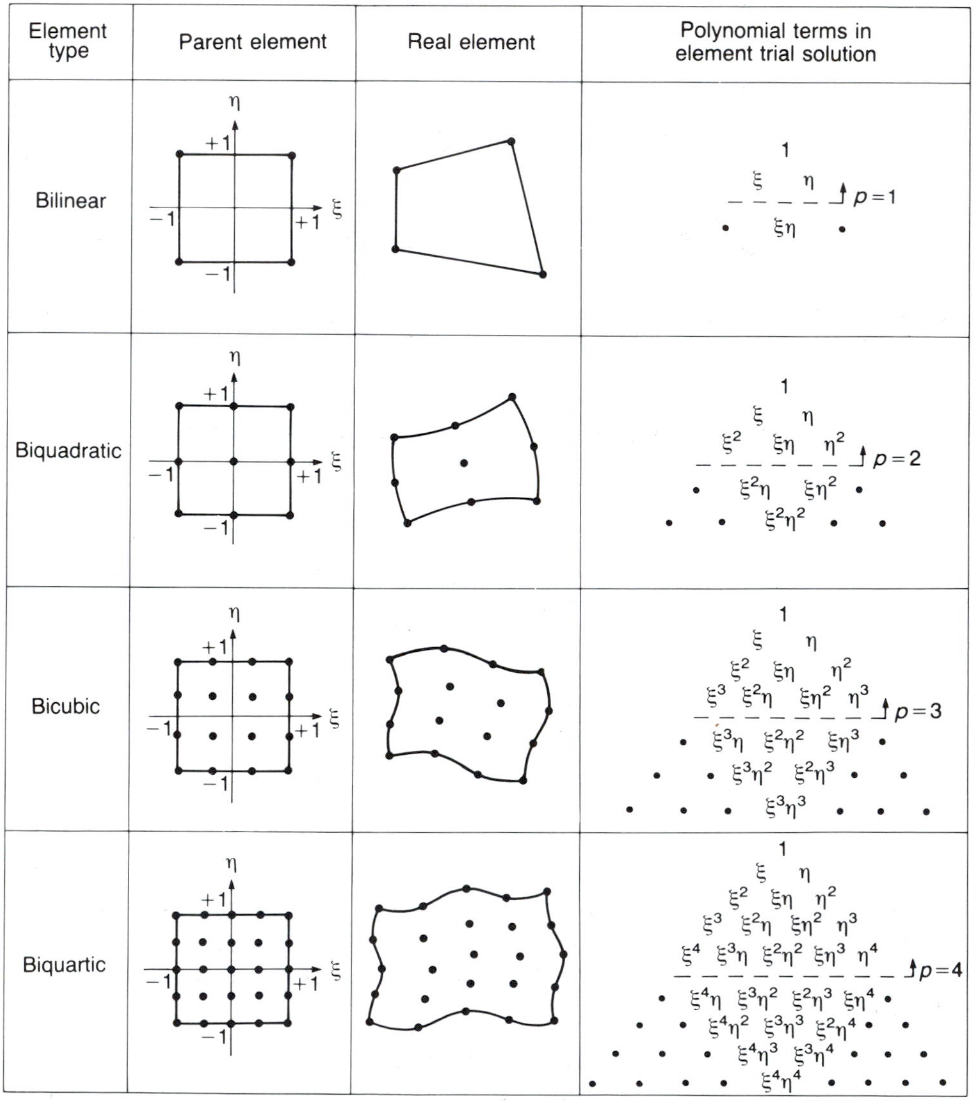

Element type	Parent element	Real element	Polynomial terms in element trial solution
Bilinear			1 $\xi \quad \eta$ $\xi\eta$ $p=1$
Biquadratic			1 $\xi \quad \eta$ $\xi^2 \quad \xi\eta \quad \eta^2$ $\xi^2\eta \quad \xi\eta^2$ $\xi^2\eta^2$ $p=2$
Bicubic			1 $\xi \quad \eta$ $\xi^2 \quad \xi\eta \quad \eta^2$ $\xi^3 \quad \xi^2\eta \quad \xi\eta^2 \quad \eta^3$ $\xi^3\eta \quad \xi^2\eta^2 \quad \xi\eta^3$ $\xi^3\eta^2 \quad \xi^2\eta^3$ $\xi^3\eta^3$ $p=3$
Biquartic			1 $\xi \quad \eta$ $\xi^2 \quad \xi\eta \quad \eta^2$ $\xi^3 \quad \xi^2\eta \quad \xi\eta^2 \quad \eta^3$ $\xi^4 \quad \xi^3\eta \quad \xi^2\eta^2 \quad \xi\eta^3 \quad \eta^4$ $\xi^4\eta \quad \xi^3\eta^2 \quad \xi^2\eta^3 \quad \xi\eta^4$ $\xi^4\eta^2 \quad \xi^3\eta^3 \quad \xi^2\eta^4$ $\xi^4\eta^3 \quad \xi^3\eta^4$ $\xi^4\eta^4$ $p=4$

FIGURE 13.36 ● The first four members of the 2-D Lagrange family of quadrilateral elements.

these approximations on the accuracy of the final solution is not yet well understood theoretically, there is nevertheless a vast amount of numerical evidence (e.g., comparison with exact solutions [13.14, 13.15]) showing that nodal fluxes calculated this way are virtually always more accurate than simply evaluating the element trial solutions at the nodes. Naylor [13.15] shows that in some situations (nearly incompressible elastic materials) the improvement in accuracy can be several orders of magnitude.

13.4 Quadrilateral Elements:
Theoretical Development (Steps 4 through 6)

Conceptually the simplest and most direct way to develop the shape functions for a C^0-quadrilateral element is merely to form the product of two 1-D Lagrange interpolation polynomials, one with respect to x (or ξ if in a parent element) and the other with respect to y (or η), resulting in so-called bilinear, biquadratic, etc., elements. Figure 13.36 illustrates the first four elements in this family, and indicates the terms in Pascal's triangle that are present in the element trial solutions.

For example, the nine parent shape functions for the biquadratic element are the products $\phi_i(\xi)\phi_j(\eta)$ for $i,j=1,2,3$ where ϕ_1, ϕ_2, and ϕ_3 are the three 1-D C^0-quadratic shape functions in Fig. 8.6. The shape function corresponding to a corner node is shown in Fig. 13.37. It clearly satisfies the interpolation property, namely, unity at the corner node and zero at the other eight nodes. This, of course, is a direct consequence of forming the product of two 1-D interpolatory functions. Along each side common to the unity node, one of the ϕ_i is identically unity, so the 2-D shape function reduces to one of the 1-D shape functions.

The 2-D Lagrangian elements have a couple of disadvantages. First, they become computationally less efficient as the order of the element increases. We recall that the rate of convergence of an element depends on the degree of the highest *complete* polynomial in the element trial solution. As indicated by the Pascal triangles in Fig. 13.36, these elements contain many extra higher-degree terms that do not contribute to completeness. For example, the bicubic element is complete to degree 3 (by definition), but it also con-

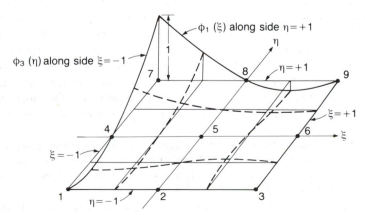

FIGURE 13.37 ● Illustration of a corner shape function (in the parent element) for a biquadratic Lagrange element.

tains six additional terms of degrees 4, 5, and 6. In other words, the element trial solution is a complete 2-D cubic but an incomplete 2-D quartic (or quintic or sextic). In general, a 2-D Lagrange element complete to degree p contains $(p+1)^2$ terms of which only $(p+1)(p+2)/2$ of the terms are needed for completeness, and the remaining $p(p+1)/2$ terms do not improve the rate of convergence.[35]

A second disadvantage is that the interior nodes can create a problem when constructing meshes. It requires additional computation to locate the interior nodes of a curved-sided element so that mapping distortion is minimized.

These problems are effectively resolved by the *serendipity* family of elements[36] [13.16-13.19]. Figure 13.38 illustrates the first four members of this family. The most conspicuous feature of this family is the absence of interior nodes and, correspondingly, the relatively few extra incomplete polynomial terms in the element trial solutions. Each of these elements has an equal number of nodes on all sides. As noted previously, such elements are the "basic" elements since they provide the same level of accuracy in all directions. However, we can also construct serendipity transition elements, i.e., elements with different numbers of nodes on different sides which therefore provide transitions between regions of higher-order and lower-order elements, analogous to the triangular elements in Fig. 13.24 (see Section 13.4.2).

The linear serendipity element in Fig. 13.38 is, in fact, a bilinear element, identical to the bilinear Lagrange element in Fig. 13.36. These are the only two identical elements in the two families. To progress from the linear serendipity to the quadratic serendipity we want to add four more nodes (one per side). This correspondingly requires adding four more terms to Pascal's triangle, namely, the two missing quadratic terms (ξ^2 and η^2) plus two cubic terms ($\xi^2\eta$ and $\xi\eta^2$) that are symmetric in ξ and η in order to preserve geometric isotropy. Similarly, to progress from the quadratic to the cubic we want to add four more nodes (one per side); hence we add the two missing cubic terms (ξ^3 and η^3) plus two quartic terms ($\xi^3\eta$ and $\xi\eta^3$) that are symmetric in ξ and η.

To progress from the cubic to the quartic we run into a problem. Now we need to add five more terms: three to complete the quartic (ξ^4, $\xi^2\eta^2$, and η^4), plus two (rather than one) quintic terms to preserve geometric isotropy (say $\xi^4\eta$ and $\xi\eta^4$). A fifth node would preferably be located at the center to avoid biasing any side of the element. As noted above, though, interior nodes are difficult to work with in applications. The solution is to use a *nodeless* shape function. This would be $(1-\xi^2)(1-\eta^2)$, which introduces the above $\xi^2\eta^2$ term. This shape function is intentionally constructed to be zero at all the other nodes (by vanishing on all four sides), thereby preserving their interpolation property. However, the other shape functions are not constrained to be zero at the center or any other interior point, so the parameter (DOF) multiplying this nodeless shape function cannot be equated to the value of the trial solution at any point (hence the name "nodeless"). We note that when the element equations are formed, the equation corresponding to this

---

[35] These $p(p+1)/2$ extra incomplete terms may improve the accuracy at random points in a problem (if the local Taylor series expansion of the exact solution at those points does not contain the missing terms). However, there is generally no way to predict a priori such behavior, so we can't rely on these extra terms to supply the accuracy wherever it's needed.

[36] This name refers to the original development of these elements, which was mainly by inspection and intuition.

Element type	Parent element	Real element	Polynomial terms in element trial solution
Linear	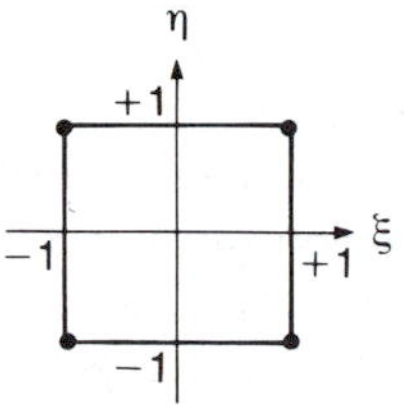		1 ξ η $\;p=1$ $\bullet$ $\xi\eta$ $\bullet$
Quadratic	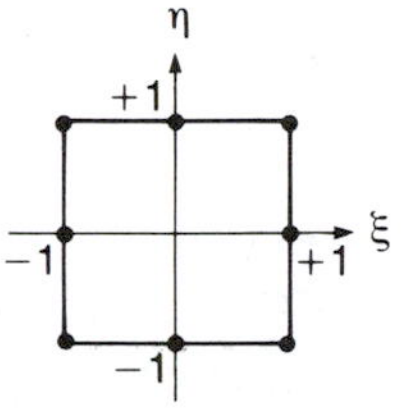		1 ξ η ξ^2 $\xi\eta$ η^2 $\;p=2$ $\bullet$ $\xi^2\eta$ $\xi\eta^2$ $\bullet$
Cubic	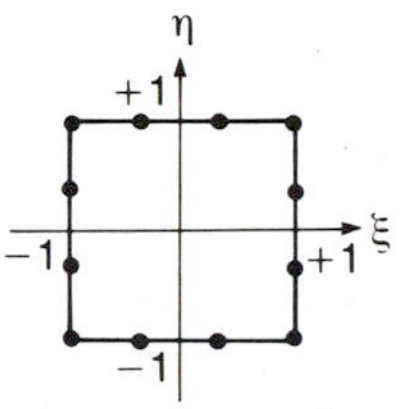		1 ξ η ξ^2 $\xi\eta$ η^2 ξ^3 $\xi^2\eta$ $\xi\eta^2$ η^3 $\;p=3$ $\bullet$ $\xi^3\eta$ $\bullet$ $\xi\eta^3$ $\bullet$
Quartic	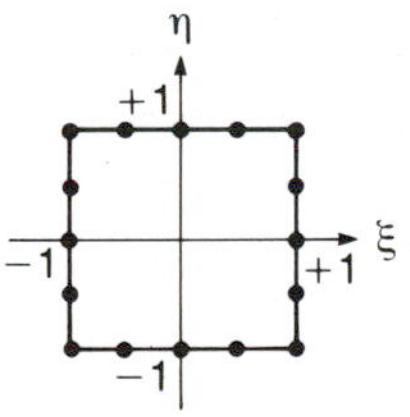		1 ξ η ξ^2 $\xi\eta$ η^2 ξ^3 $\xi^2\eta$ $\xi\eta^2$ η^3 ξ^4 $\xi^3\eta$ $\xi^2\eta^2$ $\xi\eta^3$ η^4 $\;p=4$ $\bullet$ $\xi^4\eta$ $\bullet$ $\bullet$ $\xi\eta^4$ $\bullet$

FIGURE 13.38 ● The first four members of the 2-D serendipity family of quadrilateral elements.

interior DOF, i.e., nodeless shape function, can be eliminated by *static condensation*,[37] leaving a 16-by-16 element stiffness matrix that involves only the sixteen DOF on the element boundary.

---

[37] Static condensation is a technique for reducing the size of the element stiffness matrix by eliminating those DOF whose shape functions are zero at all the boundary nodes. The elimination procedure is nothing more than the forward elimination that is normally performed on the system equations during Gaussian elimination. However, it can be performed instead at the element level for those DOF whose shape functions do not couple with the DOF in any other elements. The technique can also be applied to groups of elements, in which case it is more commonly referred to as *substructuring* (see remarks at the end of Section 13.6.4).

To form the quintic element we would add the four missing quintic terms to the quartic element — namely, ξ^5, $\xi^3\eta^2$, $\xi^2\eta^3$, and η^5. This would create a complete quintic (21 terms) with no extra higher-degree terms. Twenty of the terms would be associated with boundary nodes and the twenty-first would be the same interior nodeless function as in the quartic.

Above the quintic element the pattern of adding nodes and/or nodeless shape functions can become less regular and the analysis more complicated, depending on which terms in Pascal's triangle one chooses to add. For example, the sextic would need to add at least seven additional terms to the quintic.

The very high order elements (say quintic and above) are at present mostly of academic interest. They have potential for use primarily as hierarchic elements with the *p*-refinement method (see Section 9.3).

The linear, quadratic, and cubic elements (and their related transition elements) have been the principal ones used in conventional (*h*-refinement) FE programs. Of these, the linear and quadratic elements have definitely been the most popular and hence will be developed in the next two sections.

13.4.1 The C^0-Linear Isoparametric Quadrilateral

Step 4: Develop specific expressions for the shape functions.

This is the first member of the 2-D serendipity family (see Fig. 13.38). Figure 13.39(a) shows the parent element in ξ,η-space, and Fig. 13.39(b) shows the parent element mapped onto a real element in x,y-space. The parent element for quadrilaterals is chosen to be a bi-unit (rather than unit) square for convenience in later applying the Gauss-Legendre quadrature formulas (which integrate from -1 to $+1$). As suggested in the figure, the real element may have arbitrary corner angles and side lengths (within allowable distortion limits), but the sides must be straight since the isoparametric mapping (described below) is linear along each side.

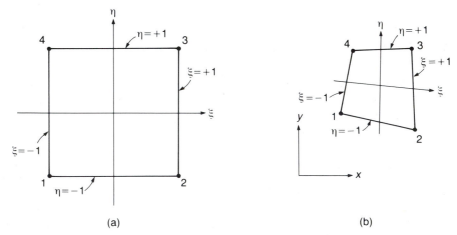

(a)   (b)

FIGURE 13.39 ● (a) The parent element for the four-node C^0-linear isoparametric quadrilateral; (b) a real element with the parent element mapped onto it.

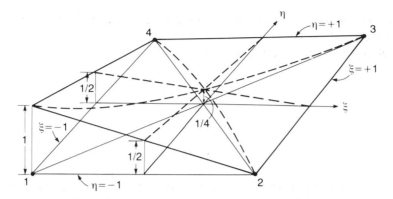

FIGURE 13.40 ● The parent shape function $\phi_1(\xi,\eta)$ for the C^0-linear isoparametric quadrilateral.

As indicated in Fig. 13.38, the element trial solution is a 2-D bilinear polynomial, using the terms 1, ξ, η, and $\xi\eta$ — that is, a complete 2-D linear polynomial plus an extra quadratic term (hence an incomplete quadratic polynomial). We could derive the parent shape functions in the usual manner, using the interpolation property, but for this simple element they can almost be written down by inspection:

$$\phi_1(\xi,\eta) \ = \ \frac{1}{4} \ (1-\xi)(1-\eta)$$

$$\phi_2(\xi,\eta) \ = \ \frac{1}{4} \ (1+\xi)(1-\eta)$$

$$\phi_3(\xi,\eta) \ = \ \frac{1}{4} \ (1+\xi)(1+\eta)$$

$$\phi_4(\xi,\eta) \ = \ \frac{1}{4} \ (1-\xi)(1+\eta) \tag{13.163}$$

They all have the same form, as illustrated by $\phi_1(\xi,\eta)$ in Fig. 13.40.

The shape functions are linear *along each side*. This ensures C^0-continuity between elements since there are two nodes on a side which, of course, uniquely determine a straight line (see the guideline for establishing interelement continuity given in Section 13.3.1). Note, however, that inside the element the ϕ_i may vary quadratically in some directions, because of the extra $\xi\eta$ quadratic term. Thus, parallel to the ξ,η-coordinate lines the ϕ_i vary linearly, but in any other direction they vary quadratically. For example, along the $\overline{13}$ diagonal $\xi=\eta=s$ and $\phi_1(s)=(1-s)^2/4$, as shown in Fig. 13.40. This extra quadratic term may occasionally yield a quadratic-like rate of convergence at some point in a problem if the local Taylor series expansion of the exact solution happens to contain only this one quadratic term. But such behavior can't be predicted in general. As noted previously, we can only guarantee rates of convergence according to the highest degree *complete* polynomial. Thus we should always think of this as essentially a linear element (while not forgetting that there is an incomplete quadratic term present which might yield occa-sionally exceptional behavior).[38]

For an isoparametric mapping we have

$$x = \sum_{k=1}^{4} x_k^{(e)} \phi_k(\xi, \eta)$$

$$y = \sum_{k=1}^{4} y_k^{(e)} \phi_k(\xi, \eta) \tag{13.164}$$

which provides continuity between adjacent elements (i.e., no gaps or overlapping) because of the C^0-continuity of the ϕ_k. To calculate the Jacobian for this mapping [see Eqs. (13.114) through (13.116)], the program will need the shape function derivatives,

$$\frac{\partial \phi_1(\xi, \eta)}{\partial \xi} = -\frac{1}{4}(1-\eta) \qquad \frac{\partial \phi_1(\xi, \eta)}{\partial \eta} = -\frac{1}{4}(1-\xi)$$

$$\frac{\partial \phi_2(\xi, \eta)}{\partial \xi} = \frac{1}{4}(1-\eta) \qquad \frac{\partial \phi_2(\xi, \eta)}{\partial \eta} = -\frac{1}{4}(1+\xi)$$

$$\frac{\partial \phi_3(\xi, \eta)}{\partial \xi} = \frac{1}{4}(1+\eta) \qquad \frac{\partial \phi_3(\xi, \eta)}{\partial \eta} = \frac{1}{4}(1+\xi)$$

$$\frac{\partial \phi_4(\xi, \eta)}{\partial \xi} = -\frac{1}{4}(1+\eta) \qquad \frac{\partial \phi_4(\xi, \eta)}{\partial \eta} = \frac{1}{4}(1-\xi) \tag{13.165}$$

Hence

$$J_{11}^{(e)}(\xi, \eta) = \frac{\partial x}{\partial \xi} = \sum_{k=1}^{4} x_k^{(e)} \frac{\partial \phi_k(\xi, \eta)}{\partial \xi}$$

$$= \frac{1}{4} \left[\left(x_2^{(e)} - x_1^{(e)} \right)(1-\eta) + \left(x_3^{(e)} - x_4^{(e)} \right)(1+\eta) \right] \tag{13.166a}$$

Similarly,

$$J_{12}^{(e)}(\xi, \eta) = \frac{1}{4} \left[\left(y_2^{(e)} - y_1^{(e)} \right)(1-\eta) + \left(y_3^{(e)} - y_4^{(e)} \right)(1+\eta) \right]$$

$$J_{21}^{(e)}(\xi, \eta) = \frac{1}{4} \left[\left(x_4^{(e)} - x_1^{(e)} \right)(1-\xi) + \left(x_3^{(e)} - x_2^{(e)} \right)(1+\xi) \right]$$

$$J_{22}^{(e)}(\xi, \eta) = \frac{1}{4} \left[\left(y_4^{(e)} - y_1^{(e)} \right)(1-\xi) + \left(y_3^{(e)} - y_2^{(e)} \right)(1+\xi) \right] \tag{13.166b}$$

We recall that the criterion for an acceptable element is that the mapping be 1 to 1 — that is, $|J^{(e)}(\xi, \eta)| > 0$ — everywhere in the element. Using this it can be shown (see

---

[38] The element can be made into a complete quadratic by adding the missing ξ^2 and η^2 terms as nodeless interior shape functions, though C^0-continuity would be lost, i.e., the modified element would be incompatible [13.20].

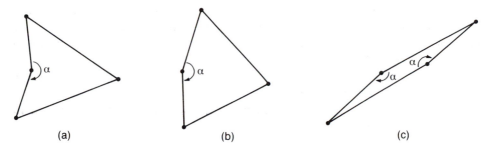

(a)   (b)   (c)

FIGURE 13.41 ● Poorly shaped C^0-linear isoparametric quadrilateral elements: (a) Reentrant corner ($\alpha > 180°$) unacceptable; mapping not 1 to 1; (b) mapping 1 to 1, but ill-conditioned since α is close to $180°$; (c) no distortion (parallelogram), but ill-conditioned because α is close to $180°$ and $J^{(e)}$ is close to zero everywhere.

Exercise 13.7) that the only limitation on the shape of the real element is that each interior corner angle must be $< 180°$; i.e., the element must be convex (Fig. 13.41a). Of course, angles that are less than but close to $180°$ (Fig. 13.41b) will be ill-conditioned and should be avoided.

It can also be shown (see Exercise 13.8) that $|J^{(e)}(\xi,\eta)| = $ (area of element)$/4$ if the element is a parallelogram; i.e., there is no mapping distortion. However, an extremely narrow parallelogram (Fig. 13.41c) will be ill-conditioned because two of the angles are approaching $180°$. In addition, the area of such an element, and hence its Jacobian, is approaching zero, so it would have a much greater stiffness than other, less-distorted elements, and this would be a source of ill-conditioning in an assemblage of elements.

As remarked previously, there is no practical reason to ever use shapes approaching these extremes. One can and should use elements that are not too dissimilar from the parent shape, i.e., having sides and angles of comparable values. An aspect ratio of 5 to 1 or 10 to 1 is a prudent limit in most applications. Interior corner angles ought to stay within about 20 or $30°$ of a right angle; otherwise a triangular element would be better suited.

Step 5: Substitute the shape functions into the element equations, and transform the integrals into a form appropriate for numerical evaluation.

In the usual manner we write the stiffness and load integrals in Eqs. (13.61b) over the parent element, using Eq. (13.118) to transform the differential area:

$$K_{ij}^{(e)} = \int_{-1}^{1} \int_{-1}^{1} \frac{\partial \phi_i^{(e)}}{\partial x} \alpha_x \frac{\partial \phi_j^{(e)}}{\partial x} |J^{(e)}| \, d\xi \, d\eta + \int_{-1}^{1} \int_{-1}^{1} \frac{\partial \phi_i^{(e)}}{\partial y} \alpha_y \frac{\partial \phi_j^{(e)}}{\partial y} |J^{(e)}| \, d\xi \, d\eta$$

$$+ \int_{-1}^{1} \int_{-1}^{1} \phi_i^{(e)} \beta \phi_j^{(e)} |J^{(e)}| \, d\xi \, d\eta$$

$$F_i^{(e)} = \int_{-1}^{1} \int_{-1}^{1} f\phi_i^{(e)} |J^{(e)}| \, d\xi \, d\eta + \oint^{(e)} \tilde{\tau}_{-n}^{(e)} \phi_i^{(e)} \, ds$$

(13.167)

The integrands are all known functions of ξ and η: $\phi_i^{(e)}$ from Eq. (13.163) and $|J^{(e)}|$, $\partial\phi_i^{(e)}/\partial x$, and $\partial\phi_i^{(e)}/\partial y$ from Eqs. (13.164) through (13.166), (13.114), and (13.127). The quantities α_x, α_y, β, f, $\bar{\tau}_{-n}^{(e)}$ are prescribed functions of x and y that are transformed to ξ and η with Eqs. (13.164).

Since the area integrals are over the bi-unit square, they may be treated as two separate 1-D integrals, each over an interval from -1 to $+1$, and hence the 1-D Gauss-Legendre quadrature formulas in Fig. 8.12 can be used for each integration. Thus, letting $I(\xi,\eta)$ represent the integrand, any of the above area integrals may be evaluated as follows:

$$\int_{-1}^{1}\int_{-1}^{1} I(\xi,\eta) \, d\xi \, d\eta \approx \int_{-1}^{1} \left[\sum_{l=1}^{n} w_{nl}I(\xi_{nl},\eta) \right] d\eta$$

$$\approx \sum_{k=1}^{n}\sum_{l=1}^{n} w_{nk}w_{nl}I(\xi_{nl},\eta_{nk}) \tag{13.168}$$

The sampling points now form a 2-D array, as illustrated in Fig. 13.42, and the weights are the products of the 1-D weights. (Extension to 3-D and higher dimensions follows the same pattern.)

As indicated by the accuracy comment in the figure, these product-type 2-D Gauss-Legendre rules are not optimally efficient for the serendipity elements (excepting only the linear element that is bilinear). As shown in Fig. 13.43, these rules integrate exactly all the terms associated with the Lagrange family of elements (which involve the products of 1-D polynomials), whereas some of these terms have been intentionally excluded from the serendipity elements. More efficient rules are available for some types of elements [13.21-13.25], especially the more complex ones where the improvement would be greatest (e.g., 3-D elements). For the lower-order elements, especially in 2-D, these product rules are still used almost universally.

The stiffness and interior load integrals in Eqs. (13.167) may now be written as the following quadrature expressions:

$$K_{ij}^{(e)} = \sum_{k=1}^{n}\sum_{l=1}^{n} w_{nk}w_{nl} \left[\left(\frac{\partial\phi_i^{(e)}}{\partial x} \alpha_x \frac{\partial\phi_j^{(e)}}{\partial x} + \frac{\partial\phi_i^{(e)}}{\partial y} \alpha_y \frac{\partial\phi_j^{(e)}}{\partial y} + \phi_i^{(e)}\beta\phi_j^{(e)} \right) |J^{(e)}| \right]_{(\xi_{nl},\eta_{nk})}$$

$$Ff_i^{(e)} = \sum_{k=1}^{n}\sum_{l=1}^{n} w_{nk}w_{nl} \left[f\phi_i^{(e)}|J^{(e)}| \right]_{(\xi_{nl},\eta_{nk})} \tag{13.169}$$

To determine which order quadrature rule to use, we may use guidelines II and III in Section 8.3.1. Thus, for the C^0-linear quadrilateral $m=1$ and $p=1$, so reduced integration calls for a quadrature accuracy of $O(h^2)$ which is the one-point rule. Higher-order integration calls for an accuracy of at least $O(h^3)$, which requires the 2-by-2-point rule. Remember that these are asymptotic accuracies, i.e., valid in the limit of "small-enough" h. For coarse meshes a 2-by-2 rule sometimes provides better accuracy than a one-point rule, probably because it can more accurately integrate the extra quadratic term.

The treatment of the boundary flux integral in Eq. (13.167) directly parallels the treatment in Section 13.3.2, step 5. Thus we need to consider only that portion of the integral

$$\int\limits_{-1}^{+1}\int\limits_{-1}^{+1} I(\xi,\eta)\ d\xi\ d\eta \approx \sum_{k=1}^{n}\ \sum_{l=1}^{n}\ w_{nk}\ w_{nl}\ I(\xi_{nk},\eta_{nl})$$

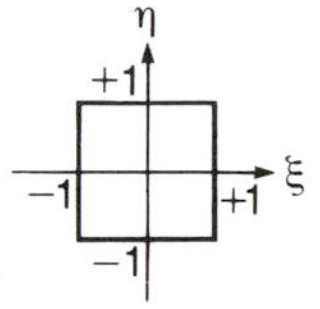

n	Number of Gauss points, $n \times n$	Accuracy of quadrature[†]	Gauss points[‡] ξ_{nl},η_{nk}	Weights[‡] $w_{nk} \times w_{nl}$
1	1 (1×1)	$O(h^2)$	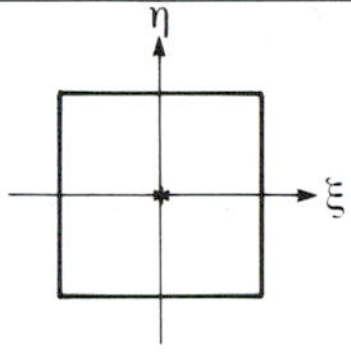	4 $(=2\times2)$ at center
2	4 (2×2)	$O(h^4)$	$\xi = -\dfrac{1}{\sqrt3}$ $\xi = +\dfrac{1}{\sqrt3}$ $\eta = +\dfrac{1}{\sqrt3}$ $\eta = -\dfrac{1}{\sqrt3}$ 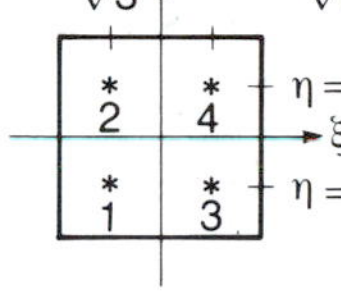	1 $(=1\times1)$ at points 1,2,3,4
3	9 (3×3)	$O(h^6)$	$\xi = -\dfrac{\sqrt3}{\sqrt5}$ $\xi = +\dfrac{\sqrt3}{\sqrt5}$ $\eta = +\dfrac{\sqrt3}{\sqrt5}$ $\eta = -\dfrac{\sqrt3}{\sqrt5}$ 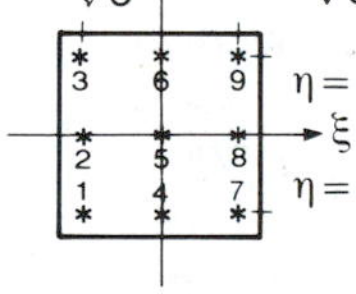	$\dfrac{25}{81}\left(=\dfrac59 \times \dfrac59\right)$ at points 1,3,7,9 $\dfrac{40}{81}\left(=\dfrac59 \times \dfrac89\right)$ at points 2,4,6,8 $\dfrac{64}{81}\left(=\dfrac89 \times \dfrac89\right)$ at point 5

[†] An accuracy of $O(h^{2n})$ means that integration is exact for any polynomial term $\xi^r\eta^s$ where $r\le 2n-1$ and $s \le 2n-1$. Therefore the product of two 1-D polynomials, each of degree $2n-1$ or less, is integrated exactly. This implies (see Fig. 13.43) that a 2-D polynomial complete up to degree $2n-1$ (terms with $r+s\le 2n-1$) is also integrated exactly.

[‡] Gauss points and weights for ξ and η separately are identical to values in 1-D formulas, so the table can be extended to any rule for which the corresponding 1-D formula is available.

FIGURE 13.42 ● Gauss points and weights for 2-D product-type Gauss-Legendre quadrature rules.

along an element side that is lying on the domain boundary (since the integrals along interior sides pair off with adjacent elements to produce similar integrals involving the *difference* in flux across the interelement boundaries). And, as before, we need only develop the expressions for any one side since they will be the same on all four sides.

Consider an element adjacent to the boundary, with side $\overline{12}$ lying on (or approximately on) the domain boundary, as shown in Fig. 13.44. On side $\overline{12}$ $\eta = -1$ and $d\eta = 0$. The

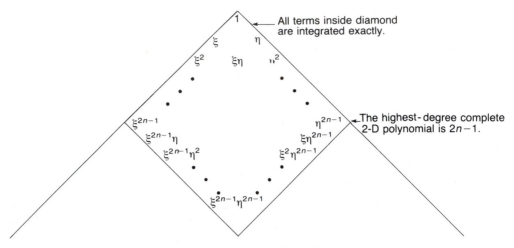

FIGURE 13.43 ● Polynomial terms (shown here in Pascal's triangle) that are integrated exactly by an n-by-n-point 2-D Gauss-Legendre rule.

shape functions in Eqs. (13.163) reduce to

$$\phi_1(\xi, -1) = \frac{1}{2}(1 - \xi)$$

$$\phi_2(\xi, -1) = \frac{1}{2}(1 + \xi)$$

$$\phi_3(\xi, -1) = 0$$

$$\phi_4(\xi, -1) = 0 \tag{13.170}$$

and from Eq. (13.132) the boundary Jacobian becomes

$$J_\Gamma^{(e)}(\xi, -1) = \sqrt{\left[J_{11}^{(e)}(\xi, -1)\right]^2 + \left[J_{12}^{(e)}(\xi, -1)\right]^2} \tag{13.171}$$

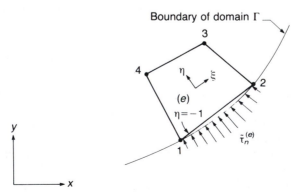

FIGURE 13.44 ● Evaluating the boundary flux integral on an element side lying on (or approximately on) the domain boundary.

Therefore the boundary flux integral along side $\overline{12}$ may be written

$$F\tau_i^{(e)}(\Gamma) = \int_{-1}^{1} \tilde{\tau}_{-n}^{(e)} \phi_i(\xi, -1) J_{\Gamma}^{(e)}(\xi, -1) \, d\xi \qquad i = 1,2$$

(13.172)

From Eqs. (13.116) and (13.170),

$$J_{11}^{(e)}(\xi, -1) = \sum_{k=1,2} x_k^{(e)} \frac{\partial \phi_k(\xi, -1)}{\partial \xi} = \frac{1}{2} \left(x_2^{(e)} - x_1^{(e)} \right)$$

$$J_{12}^{(e)}(\xi, -1) = \sum_{k=1,2} y_k^{(e)} \frac{\partial \phi_k(\xi, -1)}{\partial \xi} = \frac{1}{2} \left(y_2^{(e)} - y_1^{(e)} \right)$$

(13.173)

Therefore

$$J_{\Gamma}^{(e)}(\xi, -1) = \frac{1}{2} \sqrt{ \left(x_2^{(e)} - x_1^{(e)} \right)^2 + \left(y_2^{(e)} - y_1^{(e)} \right)^2 } = \frac{L}{2}$$

(13.174)

where L is the length of side $\overline{12}$ and Eq. (13.172) becomes

$$F\tau_i^{(e)}(\Gamma) = \frac{L}{2} \int_{-1}^{1} \tilde{\tau}_{-n}^{(e)} \phi_i(\xi, -1) \, d\xi \qquad i = 1,2$$

(13.175)

If $\tilde{\tau}_{-n}^{(e)}$ is permitted to be an arbitrary function along an element, then the 1-D Gauss-Legendre quadrature rules would be used to evaluate Eq. (13.175). For this linear element ($p=1$), reduced integration requires a one-point rule, and higher-order integration a two-point rule.

However, as with the triangular elements, we will simplify data input to UNAFEM by requiring $\tilde{\tau}_{-n}^{(e)}$ to be constant along each element. This will provide a piecewise-constant approximation of $\tilde{\tau}_{-n}$ along the entire boundary, thereby possibly introducing a modeling error that would decrease as the mesh is refined. In this case,

$$F\tau_i^{(e)}(\Gamma) = \frac{L}{2} \tilde{\tau}_{-n}^{(e)} \int_{-1}^{1} \phi_i(\xi, -1) \, d\xi \qquad i = 1,2$$

(13.176)

which can be evaluated exactly using Eqs. (13.170), yielding finally

$$F\tau_1^{(e)}(\Gamma) = \frac{1}{2} L \tilde{\tau}_{-n}^{(e)}$$

$$F\tau_2^{(e)}(\Gamma) = \frac{1}{2} L \tilde{\tau}_{-n}^{(e)}$$

(13.177)

This is an intuitively appealing result, assigning half the total flux ($L\tilde{\tau}_{-n}^{(e)}$) to each node. As noted previously, since the shape functions have the same form on each side, Eqs. (13.177) are valid on all four sides.

Step 6: Prepare expressions for the flux.

The general expressions for the components of flux in terms of the shape functions follow directly from the basic flux formula, Eqs. (13.17), and the element trial solution, Eq.

(13.48) with $n=4$:

$$\tilde{\tau}_x^{(e)}(x,y) = -\alpha_x(x,y) \sum_{j=1}^{4} a_j \frac{\partial \phi_j^{(e)}(x,y)}{\partial x}$$

$$\tilde{\tau}_y^{(e)}(x,y) = -\alpha_y(x,y) \sum_{j=1}^{4} a_j \frac{\partial \phi_j^{(e)}(x,y)}{\partial y} \qquad \textbf{(13.178)}$$

Expressions (13.178) may be evaluated at any point (ξ,η) using Eqs. (13.127) and Eqs. (13.164) through (13.166), the general procedure being the same for all the isoparametric elements.

There is one optimal point for this element, which is at the center of the element (see Exercise 13.9), coinciding with the quadrature point for the one-point Gauss rule. Therefore expressions (13.178) are evaluated at only this point, i.e., at $\xi=\eta=0$.

Since there is only one optimal flux value (for each flux component), the smoothing function for each component can only be a constant, equal to the optimal value at the center. Hence the element nodal fluxes (which are determined by evaluating the smoothing function at each of the nodes) are all equal to the optimal value. (Note the strong parallel with the 1-D C^0-linear element in Chapter 7.) Finally, the element nodal fluxes are averaged in the usual manner (see Fig. 13.22) at each node in the mesh to yield the nodal fluxes.

13.4.2 The C^0-Quadratic Isoparametric Quadrilateral

Step 4: Develop specific expressions for the shape functions.

This is the second member of the 2-D serendipity family (see Fig. 13.38). Figure 13.45(a) shows the parent element in ξ,η-space, and Fig. 13.45(b) shows the parent element mapped onto a real element in x,y-space. As suggested in the figure, the real element may have arbitrary corner angles and side lengths (within allowable distortion limits), but the sides must be a quadratic since the *iso*parametric mapping (described below) is quadratic along each side. The numbering of the corner nodes first, followed by that of the midside nodes, allows for a consistent numbering pattern when dealing with transition elements that lack some of the midside nodes.

As indicated in Fig. 13.38 the element trial solution uses the eight terms 1, ξ, η, ξ^2, $\xi\eta$, η^2, $\xi^2\eta$, and $\xi\eta^2$, which is a complete 2-D quadratic polynomial plus two extra cubic terms (hence an incomplete cubic polynomial). We could derive the parent shape functions in the usual manner, using the interpolation property. However, we will instead demonstrate a simpler procedure that can be used to generate other serendipity elements as well [13.26]. The procedure consists of first generating the midside-node shape functions from a simple prescription, and then generating the corner-node shape functions by subtracting suitable multiples of the midside functions from a starting function.

Figure 13.46(a) shows that the shape function for midside node 5, $\phi_5(\xi,\eta)$, can be written down "by inspection" as the product of a quadratic in ξ along side $\overline{152}$ times a linear in η. The quadratic provides the necessary 1 and 0 nodal values along side $\overline{152}$, as well as all zero nodal values along sides $\overline{263}$ and $\overline{481}$. The linear in η does not disturb these values, and in addition provides zero nodal values along the fourth side, $\overline{374}$. The figure also shows $\phi_8(\xi,\eta)$, which has an identical shape (rotated $90°$) and a similar

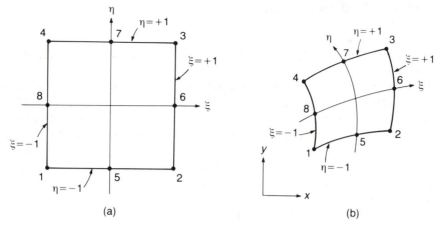

FIGURE 13.45 ● (a) The parent element for the eight-node C^0-quadratic isoparametric quadrilateral; (b) a real element with the parent element mapped onto it.

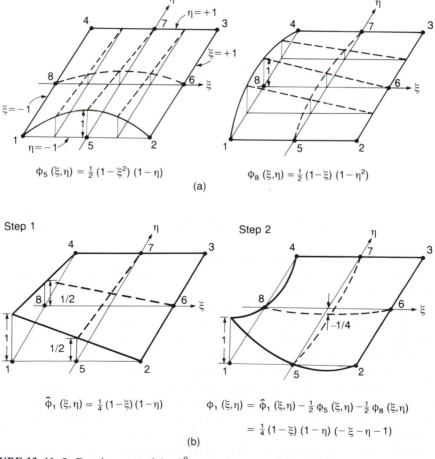

$\phi_5 (\xi,\eta) = \tfrac{1}{2} (1 - \xi^2)(1 - \eta)$

$\phi_8 (\xi,\eta) = \tfrac{1}{2} (1 - \xi)(1 - \eta^2)$

(a)

Step 1

Step 2

$\hat{\phi}_1 (\xi,\eta) = \tfrac{1}{4} (1 - \xi)(1 - \eta)$

$\phi_1 (\xi,\eta) = \hat{\phi}_1 (\xi,\eta) - \tfrac{1}{2} \phi_5 (\xi,\eta) - \tfrac{1}{2} \phi_8 (\xi,\eta)$

$$= \tfrac{1}{4} (1 - \xi)(1 - \eta)(-\xi - \eta - 1)$$

(b)

FIGURE 13.46 ● Development of the C^0-quadratic parent shape functions for (a) midside nodes and (b) corner nodes.

expression. Of course, $\phi_6(\xi,\eta)$ and $\phi_7(\xi,\eta)$ also have identical shapes and similar expressions.

The corner node functions are constructed in two steps, as illustrated in Fig. 13.46(b) for $\phi_1(\xi,\eta)$. In the first step we write down the *bilinear* function [see Eqs. (13.163)], which has the value 1 at node 1 and 0 at the other corner nodes. This provides the appropriate 1 and 0 values at all nodes except the two midside nodes 5 and 8; at these two nodes $\hat{\phi}_1(\xi,\eta)$ equals 1/2. Therefore, in the second step we correct this deficiency by subtracting $\phi_5(\xi,\eta)/2$ and $\phi_8(\xi,\eta)/2$. These subtractions do not disturb the correct values at the other six nodes. The shape functions for the other three corner nodes are developed in the same manner.

The resulting expressions for the shape functions are as follows:

$$\phi_1(\xi,\eta) = \frac{1}{4}(1-\xi)(1-\eta)(-\xi-\eta-1)$$

$$\phi_2(\xi,\eta) = \frac{1}{4}(1+\xi)(1-\eta)(\xi-\eta-1)$$

$$\phi_3(\xi,\eta) = \frac{1}{4}(1+\xi)(1+\eta)(\xi+\eta-1)$$

$$\phi_4(\xi,\eta) = \frac{1}{4}(1-\xi)(1+\eta)(-\xi+\eta-1)$$

$$\phi_5(\xi,\eta) = \frac{1}{2}(1-\xi^2)(1-\eta)$$

$$\phi_6(\xi,\eta) = \frac{1}{2}(1+\xi)(1-\eta^2)$$

$$\phi_7(\xi,\eta) = \frac{1}{2}(1-\xi^2)(1+\eta)$$

$$\phi_8(\xi,\eta) = \frac{1}{2}(1-\xi)(1-\eta^2) \tag{13.179}$$

The shape functions are quadratic *along each side*. This ensures C^0-continuity between elements since there are three nodes on a side, and three nodes uniquely determine a quadratic (see the guideline for interelement continuity given in Section 13.3.1). Inside the element the ϕ_i vary cubically in some directions because of the incomplete cubic terms. As noted previously, we can only guarantee rates of convergence according to the highest degree *complete* polynomial. Thus we should always think of this as essentially a quadratic element (while not forgetting that there are incomplete cubic terms present which might yield occasionally exceptional behavior).

The above method can be generalized to higher-order elements. Thus shape functions for midside nodes are products of an mth-degree polynomial in one variable and a linear in the other variable; shape functions for corner nodes are bilinear polynomials minus multiples of the midside node functions (see Exercise 13.10). This procedure makes it particularly easy to generate transition elements (see Exercise 13.11). However, the procedure only produces boundary-node shape functions that contain terms of the form ξ^p, $\xi^p\eta$, $\xi\eta^p$, and η^p, i.e., those terms along the two outer diagonals of Pascal's triangle (Fig.

13.47), resulting in incomplete elements for the quartic and above. As explained at the beginning of Section 13.4, the missing term for the quartic is provided by the nodeless shape function $(1-\xi^2)(1-\eta^2)$, but the quintic would perhaps not be optimally constructed by this procedure.

For an isoparametric mapping we have

$$x = \sum_{j=1}^{8} x_k^{(e)} \phi_k(\xi,\eta)$$

$$y = \sum_{j=1}^{8} y_k^{(e)} \phi_k(\xi,\eta) \tag{13.180}$$

which provides continuity between adjacent elements (i.e., no gaps or overlapping) because of the C^0-continuity of the ϕ_k. To calculate the Jacobian for this mapping [see Eqs. (13.114) through (13.116)], we need the shape function derivatives,

$$\frac{\partial \phi_1(\xi,\eta)}{\partial \xi} = \frac{1}{4}(1-\eta)(2\xi+\eta) \qquad \frac{\partial \phi_1(\xi,\eta)}{\partial \eta} = \frac{1}{4}(1-\xi)(\xi+2\eta)$$

$$\frac{\partial \phi_2(\xi,\eta)}{\partial \xi} = \frac{1}{4}(1-\eta)(2\xi-\eta) \qquad \frac{\partial \phi_2(\xi,\eta)}{\partial \eta} = \frac{1}{4}(1+\xi)(-\xi+2\eta)$$

$$\frac{\partial \phi_3(\xi,\eta)}{\partial \xi} = \frac{1}{4}(1+\eta)(2\xi+\eta) \qquad \frac{\partial \phi_3(\xi,\eta)}{\partial \eta} = \frac{1}{4}(1+\xi)(\xi+2\eta)$$

$$\frac{\partial \phi_4(\xi,\eta)}{\partial \xi} = \frac{1}{4}(1+\eta)(2\xi-\eta) \qquad \frac{\partial \phi_4(\xi,\eta)}{\partial \eta} = \frac{1}{4}(1-\xi)(-\xi+2\eta)$$

$$\frac{\partial \phi_5(\xi,\eta)}{\partial \xi} = -\xi(1-\eta) \qquad \frac{\partial \phi_5(\xi,\eta)}{\partial \eta} = -\frac{1}{2}(1-\xi^2)$$

$$\frac{\partial \phi_6(\xi,\eta)}{\partial \xi} = \frac{1}{2}(1-\eta^2) \qquad \frac{\partial \phi_6(\xi,\eta)}{\partial \eta} = -\eta(1+\xi)$$

$$\frac{\partial \phi_7(\xi,\eta)}{\partial \xi} = -\xi(1+\eta) \qquad \frac{\partial \phi_7(\xi,\eta)}{\partial \eta} = \frac{1}{2}(1-\xi^2)$$

$$\frac{\partial \phi_8(\xi,\eta)}{\partial \xi} = -\frac{1}{2}(1-\eta^2) \qquad \frac{\partial \phi_8(\xi,\eta)}{\partial \eta} = -\eta(1-\xi) \tag{13.181}$$

The first component of the Jacobian is thus

$$J_{11}^{(e)}(\xi,\eta) = \frac{\partial x}{\partial \xi} = \sum_{k=1}^{8} x_k^{(e)} \frac{\partial \phi_k(\xi,\eta)}{\partial \xi}$$

$$= \frac{1}{4}(1-\eta)(2\xi+\eta)x_1^{(e)} + \frac{1}{4}(1-\eta)(2\xi-\eta)x_2^{(e)}$$

$$+ \frac{1}{4}(1+\eta)(2\xi+\eta)x_3^{(e)} + \frac{1}{4}(1+\eta)(2\xi-\eta)x_4^{(e)} - \xi(1-\eta)x_5^{(e)}$$

$$+ \frac{1}{2}(1-\eta^2)x_6^{(e)} - \xi(1+\eta)x_7^{(e)} - \frac{1}{2}(1-\eta^2)x_8^{(e)} \tag{13.182}$$

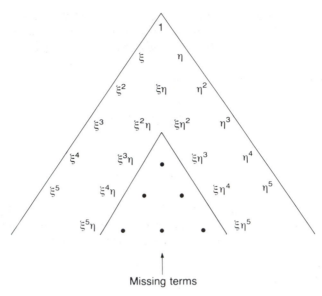

Missing terms

FIGURE 13.47 ● Terms generated by the procedure described in the text for serendipity elements, resulting in incomplete elements for quartic and above. Missing terms must be generated by other means.

The other three components

$$J_{12}^{(e)} = \frac{\partial y}{\partial \xi} \qquad J_{21}^{(e)} = \frac{\partial x}{\partial \eta} \qquad J_{22}^{(e)} = \frac{\partial y}{\partial \eta}$$

may be written down in a similar manner using Eqs. (13.180) and (13.181).

As usual, the criterion for an acceptable element is that the mapping be 1 to 1; that is, $|J^{(e)}(\xi,\eta)| > 0$ everywhere in the element. This imposes two restrictions on an element (Fig. 13.48): Interior corner angles must be $< 180°$, and midside nodes must be within $\pm L/4$ of their centers. (Note the similarity to the 1-D quadratic element.) In addition, numerical experiments [13.27] reveal an extreme loss of accuracy when the curvature of a side is so great that the circular arc defined by the three nodes subtends an angle of $180°$.

The same remarks and caveats made for the linear element in Section 13.4.1, step 4, also apply to this element. Thus these restrictions represent *extreme* limits[39], so being even close to these limits is likely to cause problems. (For example, Cook and Zhao-hua [13.27] report very large errors when corner angles reach $169°$ or, similarly, only $11°$.) In practice there is no need to use such extremely distorted elements. In the interior of a domain one can and should use straight-sided elements with centered midside nodes. (Parallelograms with centered midside nodes yield no distortion.) Aspect ratios should usually be less than about 5 to 1 or 10 to 1 for isotropic physical properties ($\alpha_x = \alpha_y$). If there is considerable anisotropy, say $\alpha_x \gg \alpha_y$, then it would be advisable to make the x-dimension of the element comparable to or greater than the y-dimension in order to avoid

---

[39] If a midside node is exactly at the $L/4$ point, a square root singularity is produced (analogous to the 1-D quadratic element) which might be desired if it is known that the exact solution possesses a similar singularity, e.g., a crack tip in elasticity theory.

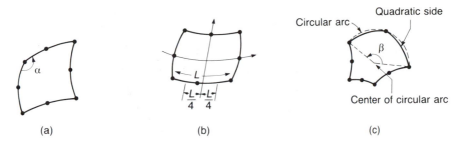

FIGURE 13.48 ● Limitations on C^0-quadratic isoparametric quadrilateral elements to avoid mapping problems: (a) $\alpha < 180°$; (b) midside node should be within $\pm L/4$ of the center of the side; (c) $\beta \le 45°$ is a reasonable guideline.

ill-conditioning (due to the components of K_{ij} in the x-direction being much greater than in the y-direction).

Curved sides are needed only along a curved boundary. However, the greater the curvature of the boundary, the greater in general will be the complexity of the exact solution in its vicinity, necessitating many DOF and hence many elements. Consequently, there is generally no need for the curved side of any element to subtend more than a few tens of degrees, say $\approx 45°$.

Step 5: Substitute the shape functions into the element equations, and transform the integrals into a form appropriate for numerical evaluation.

The quadrature formulas for the stiffness and interior load integrals (the area integrals) are the same as for the linear quadrilateral — namely, Eqs. (13.169). The expressions for $\phi_i^{(e)}$, $\partial\phi_i^{(e)}/\partial x$, $\partial\phi_i^{(e)}/\partial y$, and $|J^{(e)}|$ are available in Eqs. (13.179) through (13.182), along with the basic relations in Eqs. (13.114) and (13.127). For the C^0-quadratic quadrilateral, $m=1$ and $p=2$, so guidelines II and III in Section 8.3.1 call for 2-by-2 and 3-by-3 Gauss rules, respectively, for reduced and higher-order integration. As with the linear quadrilateral, the 3-by-3 rule may sometimes provide better accuracy with coarse meshes because of the extra incomplete cubic terms which the 3-by-3 rule will integrate more accurately.

The treatment of the boundary flux integral [see Eqs. (13.167) or (13.129)] follows the same pattern as for the previous 2-D isoparametric elements in Sections 13.3.2 and 13.4.1. Thus, for the reasons given earlier, we need only develop the integration expressions for a side of an element lying on the boundary, and any side may be used for the development since the resulting expressions will be identical for all four sides.

Consider an element adjacent to the boundary, with side $\overline{152}$ lying on (or approximately on) the domain boundary, as shown in Fig. 13.49. On side $\overline{152}$, $\eta=-1$ and $d\eta=0$. The shape functions in Eqs. (13.179) reduce to

$$\phi_1(\xi, -1) = -\frac{1}{2}\,\xi(1-\xi)$$

$$\phi_2(\xi, -1) = \frac{1}{2}\,\xi(1+\xi)$$

$$\phi_5(\xi, -1) = (1+\xi)(1-\xi) \qquad\qquad (13.183)$$

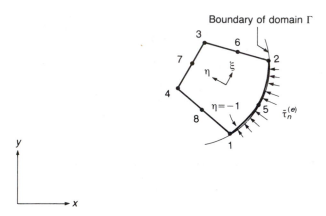

FIGURE 13.49 ● Evaluating the boundary flux integral on an element side lying on (or approximately on) the domain boundary.

and $\phi_i(\xi, -1) = 0$ for $i = 3,4,6,7,8$. From Eq. (13.132) the boundary Jacobian becomes

$$J_\Gamma^{(e)}(\xi, -1) = \sqrt{\left[J_{11}^{(e)}(\xi, -1)\right]^2 + \left[J_{12}^{(e)}(\xi, -1)\right]^2} \qquad (13.184)$$

and hence the boundary flux integral along side $\overline{152}$ may be written

$$F\tau_i^{(e)}(\Gamma) = \int_{-1}^{1} \bar{\tau}_{-n}^{(e)}\phi_i(\xi, -1)J_\Gamma^{(e)}(\xi, -1) \, d\xi \qquad i = 1,5,2 \qquad (13.185)$$

From Eqs. (13.116) and (13.183),

$$J_{11}^{(e)}(\xi, -1) = \sum_{k=1,5,2} x_k^{(e)} \frac{\partial \phi_k(\xi, -1)}{\partial \xi} = \left(\xi - \frac{1}{2}\right) x_1^{(e)} - 2\xi x_5^{(e)} + \left(\xi + \frac{1}{2}\right) x_2^{(e)}$$

$$J_{12}^{(e)}(\xi, -1) = \sum_{k=1,5,2} y_k^{(e)} \frac{\partial \phi_k(\xi, -1)}{\partial \xi} = \left(\xi - \frac{1}{2}\right) y_1^{(e)} - 2\xi y_5^{(e)} + \left(\xi + \frac{1}{2}\right) y_2^{(e)}$$

$$(13.186)$$

Using the 1-D Gauss-Legendre quadrature formulas in Fig. 8.12, Eq. (13.185) becomes

$$F\tau_i^{(e)}(\Gamma) \simeq \sum_{l=1}^{n} w_{nl}\left[\bar{\tau}_{-n}^{(e)}\phi_i(\xi, -1)J_\Gamma^{(e)}(\xi, -1)\right]_{\xi_{nl}} \qquad (13.187)$$

For this quadratic element ($p=2$), reduced integration requires a two-point rule, and higher-order integration a three-point rule. As with the previous elements we will simplify data input to UNAFEM by requiring $\bar{\tau}_{-n}^{(e)}$ to be constant along each element (thereby possibly introducing a modeling error that would decrease as the mesh is refined).

A comparison of Eqs. (13.183) through (13.187) with Eqs. (13.137) through (13.139) for the C^0-quadratic triangle reveals that the *final integration expressions for the boundary flux integrals are identical for the C^0-quadratic triangle and C^0-quadratic quadrilateral*. (The shape functions along the boundary side are identical; the boundary

Jacobian for the triangle is twice that of the quadrilateral, but there is a compensating factor of 1/2 outside the integral.) This is not a coincidence, but a consequence of the fact that the element trial solutions are 1-D quadratic polynomials along the sides of the parent elements. Therefore these integration expressions will apply to the side of any element along which the element trial solution is quadratic in the parent element, i.e., three-node sides. Examples would be transition elements containing one or more quadratic sides.

It should be obvious now that if (1) the side is straight, (2) the midside node is at the center, and (3) the flux is constant, then the boundary flux integral reduces to the same 1/6, 4/6, and 1/6 expressions as for the quadratic triangle in Eqs. (13.142) — namely,

$$F_{T_1}^{(e)}(\Gamma) = \frac{1}{6} L\tilde{\tau}_{-n}^{(e)} \qquad F_{T_5}^{(e)}(\Gamma) = \frac{4}{6} L\tilde{\tau}_{-n}^{(e)} \qquad F_{T_2}^{(e)}(\Gamma) = \frac{1}{6} L\tilde{\tau}_{-n}^{(e)}$$

This 1/6, 4/6, 1/6 weighting of a total load is characteristic of undistorted quadratic elements, having also occurred previously with the interior load integral in the 1-D C^0-quadratic element [Eq. (8.45) with constant $f^{(e)}$ and $J^{(e)}(\xi) = L/2$].

Step 6: Prepare expressions for the flux.

As before, the general flux expressions follow directly from the basic flux formula, Eqs. (13.17), and the element trial solution, Eq. (13.48) with $n = 8$:

$$\tilde{\tau}_x^{(e)}(x,y) = -\alpha_x(x,y) \sum_{j=1}^{8} a_j \frac{\partial \phi_j^{(e)}(x,y)}{\partial x}$$

$$\tilde{\tau}_y^{(e)}(x,y) = -\alpha_y(x,y) \sum_{j=1}^{8} a_j \frac{\partial \phi_j^{(e)}(x,y)}{\partial y} \tag{13.188}$$

Expressions (13.188) may be evaluated at any point (ξ,η) using Eqs. (13.127) and Eqs. (13.179) through (13.182), in the same manner as the previous isoparametric elements.

It can be shown (see Exercise 13.12) that there are four optimal flux-sampling points for this element, coinciding with the quadrature points for the 2-by-2 Gauss rule (Fig. 13.50). Therefore expressions (13.188) are evaluated at only these four points, yielding

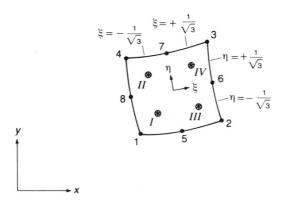

FIGURE 13.50 ● Optimal flux-sampling points for the C^0-quadratic isoparametric quadrilateral. Optimal points (⊛) are the four Gauss points for the 2-by-2 rule in Fig. 13.42.

four optimal values for both $\tilde{\tau}_x^{(e)}$ and $\tilde{\tau}_y^{(e)}$:

$$\{\tilde{\tau}_x\}_{\text{opt}}^{(e)} = \left\{ \begin{array}{c} \tilde{\tau}_x^{(e)}(\xi_I, \eta_I) \\[2mm] \tilde{\tau}_x^{(e)}(\xi_{II}, \eta_{II}) \\[2mm] \tilde{\tau}_x^{(e)}(\xi_{III}, \eta_{III}) \\[2mm] \tilde{\tau}_x^{(e)}(\xi_{IV}, \eta_{IV}) \end{array} \right\}$$

$$\{\tilde{\tau}_y\}_{\text{opt}}^{(e)} = \left\{ \begin{array}{c} \tilde{\tau}_y^{(e)}(\xi_I, \eta_I) \\[2mm] \tilde{\tau}_y^{(e)}(\xi_{II}, \eta_{II}) \\[2mm] \tilde{\tau}_y^{(e)}(\xi_{III}, \eta_{III}) \\[2mm] \tilde{\tau}_y^{(e)}(\xi_{IV}, \eta_{IV}) \end{array} \right\} \tag{13.189}$$

As with the previous elements, we next fit a smoothing function through the optimal flux values. Here we will least-squares-fit a 2-D linear polynomial,

$$\tau_s(\xi, \eta) = c_1 + c_2\xi + c_3\eta \tag{13.190}$$

to the above four values, in exactly the same manner as was done for the C^0-quadratic triangle in Section 13.3.2. (Figures 13.34 and 13.35 are applicable here if the six-node triangle is changed to an eight-node quadrilateral.) Thus we seek a transformation matrix $[TR]$ that relates the four optimal fluxes [for either $\tilde{\tau}_x^{(e)}$ or $\tilde{\tau}_y^{(e)}$] to eight element nodal fluxes:

$$\{\tau\}_{\text{nodes}} = [TR]\{\tau\}_{\text{opt}} \tag{13.191}$$

where

$$\{\tau\}_{\text{nodes}} = \left\{ \begin{array}{c} \tau_1 \\ \tau_2 \\ \tau_3 \\ \tau_4 \\ \tau_5 \\ \tau_6 \\ \tau_7 \\ \tau_8 \end{array} \right\} \qquad \{\tau\}_{\text{opt}} = \left\{ \begin{array}{c} \tau_I \\ \tau_{II} \\ \tau_{III} \\ \tau_{IV} \end{array} \right\}$$

and $[TR]$ is an 8-by-4 matrix. Here τ represents either $\tilde{\tau}_x^{(e)}$ or $\tilde{\tau}_y^{(e)}$.

The remainder of the development exactly parallels the development in Section 13.3.2. For example, Eq. (13.150) remains the same; only the numerical values for ξ and η differ, resulting in

$$
\underbrace{\begin{bmatrix} 4 & 0 & 0 \\ 0 & \dfrac{4}{3} & 0 \\ 0 & 0 & \dfrac{4}{3} \end{bmatrix}}_{[P]} \begin{Bmatrix} c_1 \\ c_2 \\ c_3 \end{Bmatrix} = \underbrace{\begin{bmatrix} 1 & 1 & 1 & 1 \\ -\dfrac{1}{\sqrt{3}} & -\dfrac{1}{\sqrt{3}} & \dfrac{1}{\sqrt{3}} & \dfrac{1}{\sqrt{3}} \\ -\dfrac{1}{\sqrt{3}} & \dfrac{1}{\sqrt{3}} & -\dfrac{1}{\sqrt{3}} & \dfrac{1}{\sqrt{3}} \end{bmatrix}}_{[Q]} \underbrace{\begin{Bmatrix} \tau_I \\ \tau_{II} \\ \tau_{III} \\ \tau_{IV} \end{Bmatrix}}_{\{\tau\}_{\text{opt}}} \quad \textbf{(13.192)}
$$

Hence the smoothing matrix $[S]$ becomes

$$
[S] = [P]^{-1}[Q] = \frac{1}{4}\begin{bmatrix} 1 & 1 & 1 & 1 \\ -\sqrt{3} & -\sqrt{3} & \sqrt{3} & \sqrt{3} \\ -\sqrt{3} & \sqrt{3} & -\sqrt{3} & \sqrt{3} \end{bmatrix} \quad \textbf{(13.193)}
$$

The evaluation matrix $[E]$ follows directly from Eq. (13.190) and the coordinates of the eight nodal points:

$$
\underbrace{\begin{Bmatrix} \tau_1 \\ \tau_2 \\ \tau_3 \\ \tau_4 \\ \tau_5 \\ \tau_6 \\ \tau_7 \\ \tau_8 \end{Bmatrix}}_{\{\tau\}_{\text{nodes}}} = \underbrace{\begin{bmatrix} 1 & -1 & -1 \\ 1 & 1 & -1 \\ 1 & 1 & 1 \\ 1 & -1 & 1 \\ 1 & 0 & -1 \\ 1 & 1 & 0 \\ 1 & 0 & 1 \\ 1 & -1 & 0 \end{bmatrix}}_{[E]} \begin{Bmatrix} c_1 \\ c_2 \\ c_3 \end{Bmatrix} \quad \textbf{(13.194)}
$$

The resulting transformation matrix [see Eqs. (13.158) through (13.160)] is therefore

$$[TR] = [E][S] = \frac{1}{4} \begin{bmatrix} 1+2\sqrt{3} & 1 & 1 & 1-2\sqrt{3} \\ 1 & 1-2\sqrt{3} & 1+2\sqrt{3} & 1 \\ 1-2\sqrt{3} & 1 & 1 & 1+2\sqrt{3} \\ 1 & 1+2\sqrt{3} & 1-2\sqrt{3} & 1 \\ 1+\sqrt{3} & 1-\sqrt{3} & 1+\sqrt{3} & 1-\sqrt{3} \\ 1-\sqrt{3} & 1-\sqrt{3} & 1+\sqrt{3} & 1+\sqrt{3} \\ 1-\sqrt{3} & 1+\sqrt{3} & 1-\sqrt{3} & 1+\sqrt{3} \\ 1+\sqrt{3} & 1+\sqrt{3} & 1-\sqrt{3} & 1-\sqrt{3} \end{bmatrix} \quad \textbf{(13.195)}$$

In summary, the element nodal fluxes are evaluated from Eq. (13.191), applied to each of the flux components,

$$\{\tilde{\tau}_x\}_{\text{nodes}}^{(e)} = [TR]\{\tilde{\tau}_x\}_{\text{opt}}^{(e)}$$

$$\{\tilde{\tau}_y\}_{\text{nodes}}^{(e)} = [TR]\{\tilde{\tau}_y\}_{\text{opt}}^{(e)} \quad \textbf{(13.196)}$$

where $[TR]$ is given by Eq. (13.195), and $\{\tilde{\tau}_x\}_{\text{opt}}^{(e)}$ and $\{\tilde{\tau}_y\}_{\text{opt}}^{(e)}$ are given by Eqs. (13.189) and evaluated using Eqs. (13.188). The calculated values for $\{\tilde{\tau}_x\}_{\text{nodes}}^{(e)}$ and $\{\tilde{\tau}_y\}_{\text{nodes}}^{(e)}$ are then assigned to the corresponding nodes in the real elements. Finally, these element nodal fluxes are then averaged in the usual manner (see Fig. 13.22) at each node in the mesh to yield the nodal fluxes. (Recall the comments at the end of Section 13.3.2, step 6, regarding the above procedure of calculating nodal fluxes by smoothing and averaging the optimal fluxes.)

13.5 Implementation into the UNAFEM Program

Figure 13.51 shows the flowchart for UNAFEM after adding the 2-D boundary-value capability developed in this chapter to the previous 1-D program developed in Part III (see Fig. 10.1). The principal enhancements are the four new 2-D elements added to the element library: ELEM21 for the linear triangle, ELEM22 and SHAP22 for the quadratic triangle, ELEM23 and SHAP23 for the linear quadrilateral, and ELEM24 and SHAP24 for the quadratic quadrilateral. There is also a new subroutine, BDYFLX, to evaluate the boundary flux integrals. In addition, several of the other subroutines have 2-D calculations added to the previous 1-D calculations. Following is a brief description of the changes made to each of the three phases of the program — namely, preprocessing, solution, and postprocessing.

Most of the changes in the preprocessing phase are relatively minor so only a few comments are necessary to guide the reader who may be interested in examining the program listing in the appendix. The MAIN routine, which calls the subroutines in the left-most vertical column of Fig. 13.51, remains the same (except for the banner comments). Subroutine INITDT now includes the initialization of the Gauss points and weights for the triangle quadrature rules in Fig. 13.31. The quadrature rules for the quadrilaterals

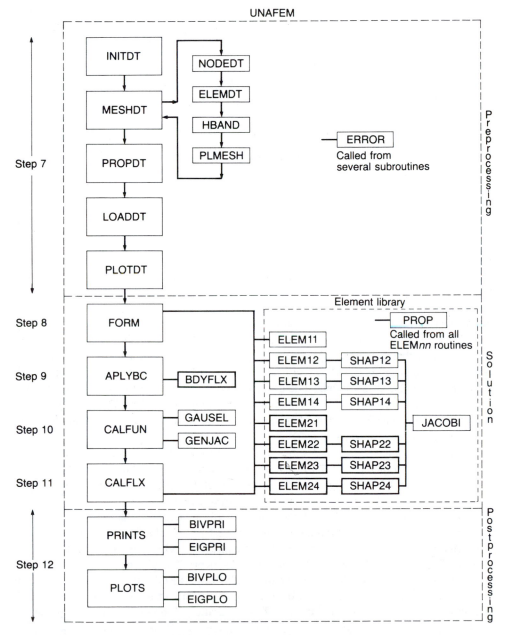

FIGURE 13.51 ● Subroutines (indicated by bold outline) added to UNAFEM to perform analysis for 2-D boundary-value problems; 2-D calculations are also added to some other subroutines.

```
        SUBROUTINE FORM
                      .
                      .
                      .
C*********** BEGIN LOOP OVER ALL ELEMENTS ******************************
C
        DO 500 IELNO=1,NUMEL
              ITYPE = NTYPE(IELNO)
              NND = NNODES(ITYPE)
C---------- FORM ELEMENT MATRICES
              IF (ITYPE.EQ.11) CALL ELEM11(ELFORM) ⎫
              IF (ITYPE.EQ.12) CALL ELEM12(ELFORM) ⎬ 1-D elements
              IF (ITYPE.EQ.13) CALL ELEM13(ELFORM) ⎪
              IF (ITYPE.EQ.14) CALL ELEM14(ELFORM) ⎭
              IF (ITYPE.EQ.21) CALL ELEM21(ELFORM) ⎫
              IF (ITYPE.EQ.22) CALL ELEM22(ELFORM) ⎬ 2-D elements
              IF (ITYPE.EQ.23) CALL ELEM23(ELFORM) ⎪
              IF (ITYPE.EQ.24) CALL ELEM24(ELFORM) ⎭
                      .
                      .
                      .
```

FIGURE 13.52 ● Subroutine FORM, a partial listing showing calls to all eight element subroutines.

in Fig. 13.42 don't need to be added because they use the values from the 1-D rules. INITDT also initializes the constants in the transformation matrices $[TR]$, which transform the Gauss-point fluxes to element nodal fluxes [see Eqs. (13.161) and (13.195)].

Subroutine NODEDT needs only an extra line here and there for the new y-coordinate. Subroutine ELEMDT can now read in up to eight (instead of five) nodes per element (eight are needed for ELEM24); it also defines the dimensionality of the problem (variable NUMDIM) as either 1-D or 2-D. Subroutine HBAND remains the same. Subroutine PLMESH adds 2-D mesh plotting. Subroutine PROPDT reads in up to six (instead of three) coefficients so that each physical property may be defined as either a 1-D or 2-D quadratic polynomial. Subroutine LOADDT remains the same for interior loads (a constant in each element), essential BCs (defined nodally), and concentrated natural BCs (also defined nodally). A few lines are added for reading in distributed natural BCs — namely, a constant value of flux and the node numbers on an element side (two nodes for linear elements, three nodes for quadratic elements).

The solution phase implements most of the theory developed in this chapter, so parts of the listing from the appendix are reproduced here to illustrate the coding of the equations. Subroutine FORM (Fig. 13.52) now has a total of eight element routines that it can call. The assembly procedures remain the same since they are matrix addition operations and therefore independent of the source of the matrices.

Figure 13.53 is a partial listing of subroutine ELEM21, the linear triangle, showing the formation of the element stiffness matrix and interior load vector. Note the similarities between ELEM21 and ELEM11 (the 1-D linear element). The physical properties $\alpha_x(x,y)$, $\alpha_y(x,y)$, and $\beta(x,y)$ are evaluated by a call to subroutine PROP, shown in Fig. 13.54. It defines the properties as either 1-D quadratic polynomials in the variable x (using only the first three terms in each equation) or 2-D quadratic polynomials in x and y (using all

```
      SUBROUTINE ELEM21(OPT)
C
C         THIS SUBROUTINE IS THE MASTER ROUTINE FOR ELEMENT 21, A
C         3-NODE 2-D C0-LINEAR TRIANGLE.
             .
             .
      I  = ICON(1,IELNO)
      J  = ICON(2,IELNO)
      L  = ICON(3,IELNO)
      IPHY = NPHYS(IELNO)
      B(1) = Y(J)-Y(L)
      B(2) = Y(L)-Y(I)
      B(3) = Y(I)-Y(J)
      C(1) = X(L)-X(J)                        Eqs. (13.67)
      C(2) = X(I)-X(L)
      C(3) = X(J)-X(I)
      AREA = (B(2)*C(3)-B(3)*C(2))/2.
      XCENT = (X(I)+X(J)+X(L))/3.             Eqs. (13.71)
      YCENT = (Y(I)+Y(J)+Y(L))/3.
      CALL PROP(IPHY,XCENT,YCENT,ALPHAX,ALPHAY,BETA,GAMMA,MU)
             .
             .
C************FORM ELEMENT MATRICES AND VECTORS************************
C
      IF (AREA.LE..0) CALL ERROR(20,IELNO)
      DO 20 I=1,3
          FE(I) = FINT(IELNO)*AREA/3.
          DO 10 J=I,3
              IF (I.EQ.J) FACTOR=6.
              IF (I.NE.J) FACTOR=12.
              KE(I,J) = (B(I)*ALPHAX*B(J)+            Eqs. (13.73)
     $                   C(I)*ALPHAY*C(J))/(4.*AREA)+  and (13.78)
     $                   BETA*AREA/FACTOR
             .
             .
   10     CONTINUE
   20 CONTINUE
C------------FOR SYMMETRY
      DO 40 I=1,3
          DO 30 J=I,3
              KE(J,I) = KE(I,J)
             .
             .
   30     CONTINUE
   40 CONTINUE
             .
             .
```

FIGURE 13.53 ● Subroutine ELEM21, a partial listing showing evaluation of terms in the element stiffness matrix and the interior load vector.

six terms). The property $\gamma(x,y)$ is used for eigenproblems (Chapters 10 and 14) and $\mu(x,y)$ for initial-boundary-value problems (Chapters 11 and 15).

Figure 13.55 is a partial listing of subroutine ELEM24, the quadratic quadrilateral, showing the formation of the element stiffness matrix and interior load vector. Again, note the similarities between ELEM24 and ELEM12 (the 1-D quadratic element). ELEM24

```
      SUBROUTINE PROP(IPHY,X,Y,AX,AY,B,G,M)
C
C             THIS ROUTINE CALCULATES THE PHYSICAL PROPERTIES
C
C------------ INPUT
C             IPHY - PHYSICAL PROPERTY NUMBER
C             X      - X COORDINATE
C             Y      - Y COORDINATE
C------------ OUTPUT
C             AX - ALPHA-X AT X,Y
C             AY - ALPHA-Y AT X,Y
C             B  - BETA    AT X,Y
C             G  - GAMMA   AT X,Y.
C             M  - MU      AT X,Y`
C
      COMMON /PHYS/   ALPHAX(6,20),ALPHAY(6,20),BETA(6,20),
     $                GAMMA(6,20),MU(6,20)
      REAL MU
      REAL M
C-----------------------------------------------------------------
      AX =  ALPHAX(1,IPHY)    + ALPHAX(2,IPHY)*X   + ALPHAX(3,IPHY)*X*X +
     $      ALPHAX(4,IPHY)*Y  + ALPHAX(5,IPHY)*Y*Y + ALPHAX(6,IPHY)*X*Y
      AY =  ALPHAY(1,IPHY)    + ALPHAY(2,IPHY)*X   + ALPHAY(3,IPHY)*X*X +
     $      ALPHAY(4,IPHY)*Y  + ALPHAY(5,IPHY)*Y*Y + ALPHAY(6,IPHY)*X*Y
      B  =    BETA(1,IPHY)    +   BETA(2,IPHY)*X   +   BETA(3,IPHY)*X*X +
     $        BETA(4,IPHY)*Y  +   BETA(5,IPHY)*Y*Y +   BETA(6,IPHY)*X*Y
      G  =   GAMMA(1,IPHY)    +  GAMMA(2,IPHY)*X   +  GAMMA(3,IPHY)*X*X +
     $       GAMMA(4,IPHY)*Y  +  GAMMA(5,IPHY)*Y*Y +  GAMMA(6,IPHY)*X*Y
      M  =      MU(1,IPHY)    +     MU(2,IPHY)*X   +     MU(3,IPHY)*X*X +
     $          MU(4,IPHY)*Y  +     MU(5,IPHY)*Y*Y +     MU(6,IPHY)*X*Y
      RETURN
C-----------------------------------------------------------------
      END
```

$$\left. \begin{array}{c} \\ \\ \\ \\ \\ \end{array} \right\} \text{Eqs. (13.69)}$$

FIGURE 13.54 ● Subroutine PROP.

calls subroutine SHAP24 (Fig. 13.56) which in turn calls subroutine JACOBI (Fig. 13.57). The listings for elements 22 and 23 are not reproduced here (see the appendix) since they are so similar to element 24.

The only addition to subroutine APLYBC (Fig. 13.58) is the application of distributed natural BCs (fluxes) along the sides of 2-D elements. (Application of concentrated natural BCs and essential BCs are all nodal operations and hence the same in 1-D or 2-D.) The evaluation of the boundary flux integral is performed in subroutine BDYFLX (Fig. 13.59). Recall that UNAFEM assumes the boundary flux to be constant along each element.

Subroutines CALFUN, GAUSEL, and GENJAC are unchanged because they involve only matrix operations. All that matters for these routines are the properties of the matrices (e.g., symmetric, positive-definite, banded), not where the matrices came from (e.g., a 1-D, 2-D, or 3-D problem).

Subroutine CALFLX (Fig. 13.60) has minor additions, namely, four more element calls (as in subroutine FORM) and a couple of extra arrays for the y-component of flux and y-coordinate of Gauss points. This subroutine calls the element subroutines to calculate the optimal fluxes and element nodal fluxes in each element (Figs. 13.61 and 13.62), then returns to finish calculating the nodal fluxes by averaging the element nodal fluxes.

In the postprocessing phase, subroutine BIVPRI prints out the data in tabular form. Only two extra format statements are needed to handle the extra columns of 2-D data. Subroutine BIVPLO has a major addition for generating 2-D contour plots for both the function and flux. The reader interested in contour-plotting techniques is referred to the listing in the appendix.

```
        SUBROUTINE ELEM24(OPT)
C
C          THIS SUBROUTINE IS THE MASTER ROUTINE FOR ELEMENT 24, AN
C          8-NODE 2-D CO-QUADRATIC QUADRILATERAL.
                                      .
                                      .
                                      .
C**********FORM ELEMENT MATRICES AND VECTORS**************************
C
        IPHY = NPHYS(IELNO)
        DO 20 I=1,8
             J = ICON(I,IELNO)
             XX(I) = X(J)
             YY(I) = Y(J)
             FE(I) = 0.
             DO 10 J=1,8
                  KE(J,I) = 0.
                             .
                             .
                             .
     10     CONTINUE
     20 CONTINUE
        NINT = NGPINT(IELNO)
C----------INTEGRATE TERMS
        DO 100 INTX=1,NINT
             XI = GP(NINT,INTX)
             DO 90 INTY=1,NINT
                  ETA = GP(NINT,INTY)
                  WGT = WT(NINT,INTX)*WT(NINT,INTY)
                  CALL SHAP24
                  XXX = 0.
                  YYY = 0.
                  DO 30 I=1,8
                       XXX = XXX+PHI(I)*XX(I) }
                       YYY = YYY+PHI(I)*YY(I) }  Eqs. (13.180)
     30           CONTINUE
                  CALL PROP(IPHY,XXX,YYY,ALPHAX,ALPHAY,BETA,GAMMA,MU)
                  DO 80 I=1,8
                       FE(I) = FE(I)+FINT(IELNO)*PHI(I)*WGT*JAC  ⎤
                       DO 70 J=I,8                               ⎥
                            KE(I,J) = KE(I,J)+WGT*JAC*           ⎥
     $                          (DPHDX(I)*ALPHAX*DPHDX(J)+      ⎬ Eqs. (13.169)
     $                           DPHDY(I)*ALPHAY*DPHDY(J)+      ⎥
     $                           PHI(I)*BETA*PHI(J))            ⎦
                                 .
                                 .
                                 .
     70                     CONTINUE
     80                 CONTINUE
     90        CONTINUE
    100 CONTINUE
C----------FOR SYMMETRY
        DO 150 I=1,8
             DO 140 J=I,8
                  KE(J,I) = KE(I,J)
                       .
                       .
                       .
    140        CONTINUE
    150 CONTINUE
               .
               .
               .
```

FIGURE 13.55 ● Subroutine ELEM24, a partial listing showing evaluation of terms in the element stiffness matrix and interior load vector.

```
      SUBROUTINE SHAP24
C
C         EVALUATE SHAPE FUNCTIONS AND DERIVATIVES AT POINT (XI,ETA)
C                 FOR 8-NODE QUADRILATERAL
                  .
                  .
                  .
C----------CONSTANTS
      XIP = 1.+XI
      XIM = 1.-XI
      ETP = 1.+ETA
      ETM = 1.-ETA
      XI2 = XIP*XIM
      ET2 = ETP*ETM
C
C**********SHAPE FUNCTIONS********************************************
C
      PHI(1) = XIM*ETM*(-XI-ETA-1.)/4.
      PHI(2) = XIP*ETM*( XI-ETA-1.)/4.
      PHI(3) = XIP*ETP*( XI+ETA-1.)/4.
      PHI(4) = XIM*ETP*(-XI+ETA-1.)/4.     Eqs. (13.179)
      PHI(5) = XI2*ETM/2.
      PHI(6) = XIP*ET2/2.
      PHI(7) = XI2*ETP/2.
      PHI(8) = XIM*ET2/2.
C
C**********DERIVATIVES OF THE SHAPE FUNCTIONS*************************
C
C--------DERIVATIVES WRT XI AT (XI,ETA)
      DPHDXI(1) = ETM*( 2.*XI+ETA)/4.
      DPHDXI(2) = ETM*( 2.*XI-ETA)/4.
      DPHDXI(3) = ETP*( 2.*XI+ETA)/4.
      DPHDXI(4) = ETP*( 2.*XI-ETA)/4.
      DPHDXI(5) = -XI*ETM
      DPHDXI(6) =   ET2/2.
      DPHDXI(7) = -XI*ETP
      DPHDXI(8) = -ET2/2.
C--------DERIVATIVES WRT ETA AT (XI,ETA)     Eqs. (13.181)
      DPHDET(1) = XIM*( XI+2.*ETA)/4.
      DPHDET(2) = XIP*(-XI+2.*ETA)/4.
      DPHDET(3) = XIP*( XI+2.*ETA)/4.
      DPHDET(4) = XIM*(-XI+2.*ETA)/4.
      DPHDET(5) = -XI2/2.
      DPHDET(6) = -XIP*ETA
      DPHDET(7) =  XI2/2.
      DPHDET(8) = -XIM*ETA
C----------DO JACOBIAN CALCULATIONS
      CALL JACOBI(2)
C----------DERIVATIVES WRT X AND Y AT (XI,ETA)
      DO 50 I=1,8
         DPHDX(I) = JACINV(1,1)*DPHDXI(I)+JACINV(1,2)*DPHDET(I)
         DPHDY(I) = JACINV(2,1)*DPHDXI(I)+JACINV(2,2)*DPHDET(I)     Eqs. (13.127a)
   50 CONTINUE
      RETURN
C-------------------------------------------------------------------
      END
```

FIGURE 13.56 ● Subroutine SHAP24.

```
      SUBROUTINE JACOBI(NDIM)
C
C        THIS ROUTINE DOES THE JACOBIAN CALCULATIONS
                .
                .
                .
C----------COMPUTE JACOBIAN AT (XI,ETA)
      DXDXI = 0.
      DXDET = 0.
      DYDXI = 0.
      DYDET = 0.
      DO 210 I=1,NND
          DXDXI = DXDXI+DPHDXI(I)*XX(I)  }
          DXDET = DXDET+DPHDET(I)*XX(I)  }
          DYDXI = DYDXI+DPHDXI(I)*YY(I)  }  Eqs. (13.116)
          DYDET = DYDET+DPHDET(I)*YY(I)  }
  210 CONTINUE
      JAC = DXDXI*DYDET-DXDET*DYDXI <------ Eq.(13.115)
C----------CHECK POSITIVENESS OF JACOBIAN
      AVG = (ABS(DXDXI)+ABS(DXDET)+ABS(DYDXI)+ABS(DYDET))/4.
      IF (JAC.LT.AVG*1.E-8) CALL ERROR(19,IELNO)
C----------COMPUTE INVERSE OF THE JACOBIAN AT (XI,ETA)
      JACINV(1,1) = DYDET/JAC   }
      JACINV(1,2) = -DYDXI/JAC  }
      JACINV(2,1) = -DXDET/JAC  }  Eqs. (13.127b)
      JACINV(2,2) = DXDXI/JAC   }
      RETURN
          .
          .
          .
```

FIGURE 13.57 ● Subroutine JACOBI, a partial listing showing 2-D calculations.

```
      SUBROUTINE APLYBC
C
C        THIS ROUTINE APPLIES THE NATURAL AND ESSENTIAL BOUNDARY
C             CONDITIONS TO THE SYSTEM MATRICES.
                .
                .
                .
C******** APPLY THE NONZERO NATURAL BOUNDARY CONDITIONS, IF ANY, *******
C                BY ADDING TERMS TO LOAD VECTOR
C
      IF (NUMNBC.EQ.0) GO TO 90
C
      DO 80 N=1,NUMNBC
          II = INP(N)
          JJ = JNP(N)
          KK = KNP(N)
C--------- IF JJ EQ 0 WE HAVE A CONCENTRATED FLUX AT NODE II
          IF (JJ.NE.0) GO TO 20
          F(II) = F(II)+TAUNBC(N)
          GO TO 80
   20     CONTINUE
C--------- IF KK EQ 0 WE HAVE A DISTRIBUTED FLUX ON A SIDE
C             OF A 3-NODE TRIANGLE OR A 4-NODE QUADRILATERAL
          IF (KK.NE.0) GO TO 40
          L = SQRT((X(JJ)-X(II))**2+(Y(JJ)-Y(II))**2)  }  Eqs. (13.92)
          F(II) = F(II)+TAUNBC(N)*L/2.                  }  or (13.177)
          F(JJ) = F(JJ)+TAUNBC(N)*L/2.                  }
          GO TO 80
   40     CONTINUE
C--------- WE HAVE A DISTRIBUTED FLUX ON A SIDE OF A 6-NODE TRIANGLE
C                  OR AN 8-NODE QUADRILATERAL
          XX(1) = X(II)
          XX(2) = X(JJ)
          XX(3) = X(KK)
          YY(1) = Y(II)
          YY(2) = Y(JJ)
          YY(3) = Y(KK)
          CALL BDYFLX(N,XX,YY,BDYTAU)
          F(II) = F(II) + BDYTAU(1)
          F(JJ) = F(JJ) + BDYTAU(2)
          F(KK) = F(KK) + BDYTAU(3)
   80 CONTINUE
   90 CONTINUE
          .
          .
          .
```

FIGURE 13.58 ● Subroutine APLYBC, a partial listing showing application of natural BCs, both concentrated and distributed.

```
      SUBROUTINE BDYFLX(N,XX,YY,BDYTAU)
C
C         THIS ROUTINE CALCULATES THE BOUNDARY FLUX INTEGRAL FOR
C             A DISTRIBUTED FLUX ON THE BOUNDARY OF A 6-NODE
C             TRIANGLE OR AN 8-NODE QUADRILATERAL, USING A
C             3-POINT GAUSSIAN QUADRATURE FORMULA.  THE
C             FLUX IS ASSUMED CONSTANT.
              .
              .
              .
C-----------------------------------------------------------------
      DO 20 I=1,3
         BDYTAU(I) = 0.
   20 CONTINUE
      DO 50 IPT=1,3
         XI = GP(3,IPT)
         PHI(1) = -XI*(1.-XI)/2.  ⎫
         PHI(2) = 1.-XI*XI        ⎬ Eqs. (13.183)
         PHI(3) = XI*(1.+ XI)/2.  ⎭
         DPH1 = XI - 1./2.        ⎫
         DPH2 = -2.*XI            ⎬ in Eqs. (13.186)
         DPH3 = XI + 1./2.        ⎭
         BDYJAC = SQRT((DPH1*XX(1)+DPH2*XX(2)+DPH3*XX(3))**2 +  ◄─── Eq. (13.184)
      $              (DPH1*YY(1)+DPH2*YY(2)+DPH3*YY(3))**2)
         DO 40 I=1,3
            BDYTAU(I) = BDYTAU(I)+WT(3,IPT)*(TAUNBC(N)*PHI(I)*BDYJAC) ◄──Eq. (13.187)
   40    CONTINUE
   50 CONTINUE
      RETURN
C-----------------------------------------------------------------
      END
```

FIGURE 13.59 ● Subroutine BDYFLX.

```
      SUBROUTINE CALFLX
C
C         THIS ROUTINE CONTROLS THE CALCULATION OF FLUXES IN EACH
C             ELEMENT, AND CALCULATES THE NODAL FLUXES.
              .
              .
              .
C---------INITIALIZE ARRAYS
              .
              .
              .
C---------CALCULATE FLUXES AT GAUSS POINTS AND NODES IN EACH ELEMENT
      DO 100 IELNO=1,NUMEL
         ITYP = NTYPE(IELNO)
         NND = NNODES(ITYP)
         IF (ITYP.EQ.11) CALL ELEM11(FLUX)
         IF (ITYP.EQ.12) CALL ELEM12(FLUX)
         IF (ITYP.EQ.13) CALL ELEM13(FLUX)
         IF (ITYP.EQ.14) CALL ELEM14(FLUX)
         IF (ITYP.EQ.21) CALL ELEM21(FLUX)
         IF (ITYP.EQ.22) CALL ELEM22(FLUX)
         IF (ITYP.EQ.23) CALL ELEM23(FLUX)
         IF (ITYP.EQ.24) CALL ELEM24(FLUX)
  100 CONTINUE
C---------CALCULATE NODAL FLUXES BY AVERAGING THE SUMMED ELEMENT
C             NODAL FLUXES FROM ELEMENT SUBROUTINES
      DO 150 I=1,NUMNP
         AIU = IU(I)
         IF (IU(I).GT.0) TAUXNP(I) = TAUXNP(I)/AIU ⎫ Fig. 13.22
         IF (IU(I).GT.0) TAUYNP(I) = TAUYNP(I)/AIU ⎭
  150 CONTINUE
      RETURN
C-----------------------------------------------------------------
      END
```

FIGURE 13.60 ● Subroutine CALFLX, a partial listing.

```
      SUBROUTINE ELEM21(OPT)
C
C           THIS SUBROUTINE IS THE MASTER ROUTINE FOR ELEMENT 21, A
C           3-NODE 2-D CO-LINEAR TRIANGLE.
                   .
                   .
C**************CALCULATE FLUXES*****************************************
C
  200 CONTINUE
C--------- CALCULATE FLUXES AT THE GAUSS POINT (CENTER)
      XGP(1,IELNO) = XCENT
      YGP(1,IELNO) = YCENT
      TAUXGP(1,IELNO) = -ALPHAX*(A(I)*B(1)+A(J)*B(2)+A(L)*B(3))   ⎫ Eqs. (13.106)
      TAUYGP(1,IELNO) = -ALPHAY*(A(I)*C(1)+A(J)*C(2)+A(L)*C(3))   ⎭
C--------- CALCULATE ELEMENT NODAL FLUXES AND SUM FOR NODAL FLUX
C                 CALCULATION IN CALFLX (ELEMENT NODAL FLUXES SAME AS
C                 GAUSS POINT FLUX)
      DO 220 II=1,3
          JNOD = ICON(II,IELNO)
          IU(JNOD) = IU(JNOD)+1
          TAUXNP(JNOD) = TAUXNP(JNOD)+TAUXGP(1,IELNO)
          TAUYNP(JNOD) = TAUYNP(JNOD)+TAUYGP(1,IELNO)
  220 CONTINUE
C--------- CALCULATE FUNCTION, FLUXES, AND COORDINATES AT
C                     CENTROID FOR CONTOUR PLOTTING
      UC(IELNO) = (A(I)+A(J)+A(L))/3.
      TAUXC(IELNO) = TAUXGP(1,IELNO)
      TAUYC(IELNO) = TAUYGP(1,IELNO)
      XC(IELNO) = XGP(1,IELNO)
      YC(IELNO) = YGP(1,IELNO)
      RETURN
C--------------------------------------------------------------------
      END
```

FIGURE 13.61 ● Subroutine ELEM21, a partial listing showing flux calculations.

```
      SUBROUTINE ELEM24(OPT)
C
C           THIS SUBROUTINE IS THE MASTER ROUTINE FOR ELEMENT 24, AN
C           8-NODE 2-D CO-QUADRATIC QUADRILATERAL.
                   .
                   .
C***********CALCULATE FLUXES*******************************************
C
  200 CONTINUE
      IPHY = NPHYS(IELNO)
      DO 220 I=1,8
          J = ICON(I,IELNO)
          XX(I) = X(J)
          YY(I) = Y(J)
  220 CONTINUE
C--------- CALCULATE FLUXES AT THE 4 GAUSS POINTS
      IGP = 0
      DO 280 IR=1,2
          XI = GP(2,IR)
          DO 260 IS=1,2
              IGP = IGP+1
              ETA = GP(2,IS)
              CALL SHAP24
              XXX = 0.
              YYY = 0.
```

(Continued)

FIGURE 13.62 ● Subroutine ELEM24, a partial listing showing flux calculations.

FIGURE 13.62 ● *(Continued)*

```
                DUDX = 0.
                DUDY = 0.
                DO 240 I=1,8
                    J = ICON(I,IELNO)
                    XXX = XXX+PHI(I)*XX(I)        } Eqs. (13.180)
                    YYY = YYY+PHI(I)*YY(I)
                    DUDX = DUDX+DPHDX(I)*A(J)
                    DUDY = DUDY+DPHDY(I)*A(J)
    240             CONTINUE
                CALL PROP(IPHY,XXX,YYY,ALPHAX,ALPHAY,BETA,GAMMA,MU)  } Eqs. (13.188)
                XGP(IGP,IELNO) = XXX
                YGP(IGP,IELNO) = YYY
                TAUXGP(IGP,IELNO) = -ALPHAX*DUDX
                TAUYGP(IGP,IELNO) = -ALPHAY*DUDY
    260     CONTINUE
    280 CONTINUE
C-------- CALCULATE ELEMENT NODAL FLUXES USING A LEAST-SQUARES FIT
C              THROUGH GAUSS POINTS AND SUM FOR NODAL FLUX CALCULATION
C              IN CALFLX
      DO 320 I=1,8
          JNOD = ICON(I,IELNO)
          IU(JNOD) = IU(JNOD)+1                    Eq. (13.195)
          DO 300 L=1,4
              TAUXNP(JNOD) = TAUXNP(JNOD)+TR2(I,L)*TAUXGP(L,IELNO)  } Eqs. (13.196)
              TAUYNP(JNOD) = TAUYNP(JNOD)+TR2(I,L)*TAUYGP(L,IELNO)
    300     CONTINUE
    320 CONTINUE
C-------- CALCULATE FUNCTION, FLUXES, AND COORDINATES AT
C                  CENTROID FOR CONTOUR PLOTTING
      XI = 0.
      ETA = 0.
      CALL SHAP24
      UC(IELNO) = 0.
      DO 340 I=1,8
         J = ICON(I,IELNO)
         UC(IELNO) = UC(IELNO)+PHI(I)*A(J)
    340 CONTINUE
C------------------CALCULATE FLUXES AND COORDINATES AT CENTROID
C                  BY AVERAGING GAUSS PT VALUES (SAME AS FITTING
C                  A PLANE THROUGH GAUSS PT VALUES)
      TAUXC(IELNO) = 0.
      TAUYC(IELNO) = 0.
      XC(IELNO) = 0.
      YC(IELNO) = 0.
      DO 360 I=1,4
          TAUXC(IELNO) = TAUXC(IELNO)+TAUXGP(I,IELNO)/4.
          TAUYC(IELNO) = TAUYC(IELNO)+TAUYGP(I,IELNO)/4.
          XC(IELNO) = XC(IELNO)+XGP(I,IELNO)/4.
          YC(IELNO) = YC(IELNO)+YGP(I,IELNO)/4.
    360 CONTINUE
      RETURN
C-----------------------------------------------------------------------
      END
```

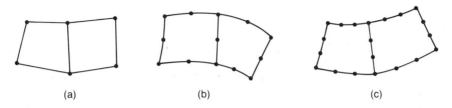

(a)   (b)   (c)

FIGURE 13.63 ● A correct (and usual) manner of combining elements: matching corner and midside nodes along sides with the same number of nodes.

13.6 Mesh Construction Guidelines

13.6.1 Individual Elements

Step 4, as discussed in Sections 13.3.2, 13.4.1, and 13.4.2, described both the extreme limits and prudent limits for acceptable shapes of individual elements. It was observed that for all parametric elements numerical problems gradually increase as the shapes and midside-node locations deviate further and further from that of the parent element.

In practice, there is no reason to ever use severely distorted elements; they can always be replaced by two or three mildly distorted elements. Throughout the interior of a domain, elements can be constructed with no distortion at all, e.g., eight-node quadrilaterals as straight-sided parallelograms (usually rectangles). Curved sides should be used only on curved boundaries. Here again, only a mild curvature in each element is necessary. As the curvature of the domain boundary becomes more severe, so will the complexity of the solution in its vicinity, thereby requiring more DOF, i.e., many elements, each one subtending only a small arc (say 20 to 30°) of the boundary.

13.6.2 Combining Elements

The different types of 1-D elements in Part III were easily combined since they connected only at nodes. We must be more careful in combining different types of 2-D (or 3-D) elements in order to preserve interelement continuity across the sides (or faces).[40]

Elements should have corresponding nodes connected, i.e., corner nodes to corner nodes and midside nodes to midside nodes (Fig. 13.63). Thus, a side with n nodes should normally be connected only to another element side that also has n nodes (unless a linear constraint equation is imposed, as described below).

If this guideline is not observed, then interelement continuity is generally destroyed, since the DOF defining the trial solution on the side of one element are not the same DOF defining the trial solution on the side of the adjacent element. Figure 13.64 illustrates some typical combinations that should generally be avoided.

There are two ways to combine elements of different orders. One approach is to use *transition elements*, discussed previously in Sections 13.3 and 13.4. Figure 13.65 illustrates

---

[40] Interelement continuity is not necessary for convergence (see Section 5.6); however, if discontinuity is permitted, then a patch test (see Section 8.6) should be performed to assure convergence.

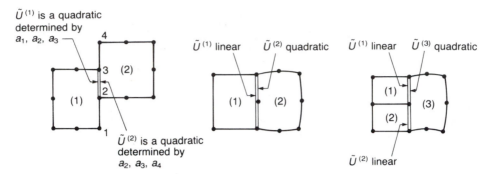

FIGURE 13.64 ● Some element combinations that generally destroy interelement continuity (and hence are incorrect if continuity is desired).

the use of a linear/quadratic element to make the transition from eight-node quadratics to four-node linears (see Exercise 13.13).

Another approach is to impose a linear constraint equation that constrains the midside DOF of the higher-order element to equal the element trial solution on the side of the lower-order element. Figure 13.66 illustrates the procedure for combining a linear element and a quadratic element (see Exercise 13.14).

In principle, elements of different dimensions can also be combined (Fig. 13.67). This is done routinely in commercial FE programs. For example, if the mesh in Fig. 13.67 were for a heat conduction problem, the 1-D element might represent a thin rod that is connecting two plates, permitting heat to flow back and forth between the two plates. This would be a numerically better conditioned and more efficient model than trying to model the thin rod with long, thin 2-D elements, which in turn would require several tiny 2-D elements in each plate at the connection points. (This purely 2-D model *would* be appropriate if we wanted a detailed analysis right in the local vicinity of the connection points.)

For an m-dimensional element to combine with an n-dimensional element $(m<n)$, the nodal coordinates of the former should have n components so that the element can be positioned arbitrarily in an n-dimensional space. In addition, postprocessing routines,

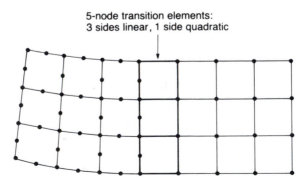

FIGURE 13.65 ● Use of linear/quadratic transition elements.

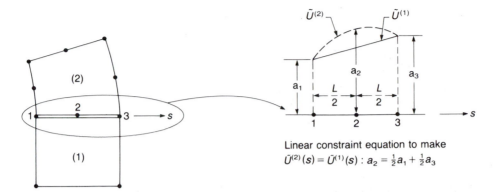

FIGURE 13.66 ● Combining a linear element and a quadratic element by using a linear constraint equation.

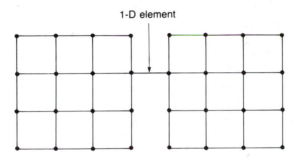

FIGURE 13.67 ● Combining 1-D and 2-D elements (both C^0-linear).

especially graphical routines such as contour plotters, should be able to accommodate both types of elements in the same mesh. The UNAFEM program is not intended to mix 1-D and 2-D elements since the coding would become a bit more complicated for the 1-D elements, and the primary purpose of the program is beginning-level instruction, not general-purpose industrial production. Nevertheless, the program would require only minor modifications.[41]

13.6.3 Modeling Errors

Sections 7.2 and 7.5 commented on modeling errors associated primarily with physical properties and loads. The present section looks at modeling errors caused by curved boundaries (*geometric* errors), and comments further on modeling loads.

---

[41] The UNAFEM 1-D elements are oriented parallel to the x-axis, so 1-D elements oriented only in this direction could be mixed with 2-D elements (after removing the NUMDIM variable, initially defined in subroutine ELEMDT). Subroutine BIVPLO would also have to be modified. For arbitrary orientation, the 1-D elements should have both x and y coordinates defined for the nodes in order to calculate the length correctly.

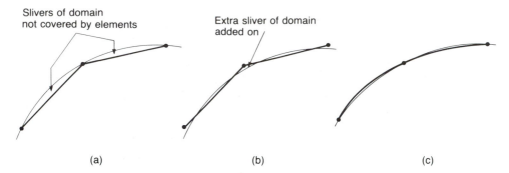

(a)   (b)   (c)

FIGURE 13.68 ● Boundary mismatches when using polynomial elements to model a nonpolynomial curved boundary (light line): (a) linear elements, nodes on the boundary; (b) linear elements, nodes off the boundary; (c) quadratic element, nodes on the boundary.

If the original problem statement defines the boundary of the domain to be a polygon — i.e., all sides are straight — then elements of any type can match the boundary exactly since all elements can have straight sides (linear polynomials). Similarly, if the boundary is defined by higher-degree polynomials (quadratics, cubic, etc.), then corresponding higher-order elements can also match the boundary exactly. However, nonpolynomial curves will be impossible to match exactly by polynomial elements, so there will be small "slivers" of the domain not covered by an element or extra slivers added on, depending on the types of elements used and the placement of their nodes. This changes the domain of the problem since, in effect, the domain is defined by the outer boundary of the mesh of elements.

Figure 13.68 illustrates some typical boundary mismatches. Figure 13.68(b) provides better accuracy than Fig. 13.68(a) by approximately preserving the correct area of the domain, but its drawback is the difficulty in placing the nodes to accomplish this. Figure 13.68(c) generally provides the best accuracy (for the same number of nodes), even with the nodes conveniently placed on the boundary.

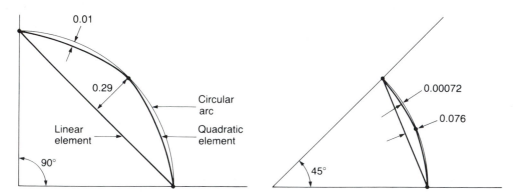

FIGURE 13.69 ● Amount of boundary mismatch introduced by fitting linear and quadratic interpolatory curves to a unit radius circular arc (light line).

The most common nonpolynomial curve is a circular arc. It can be shown [13.28] that the maximum radial error between a circular arc and a quadratic interpolating curve (i.e., the side of a quadratic element with nodes on the arc) is about 1% for a 90° arc and less than 0.1% for a 45° arc (Fig. 13.69). Corresponding values for a linear element are 29 and 7.6%, respectively. There is clearly a significant improvement (namely, radial error decreases from 7.6 to 1%) fitting one quadratic, rather than two linears, through three nodes.

Boundary mismatches produce corresponding errors in the FE solution, called *geometric* errors. The convergence theorems strictly require that the domain not change during the course of mesh refinement; i.e., every mesh in a sequence of refined meshes should have the same boundary. This would imply that a coarse approximation of a curved boundary cannot be improved during refinement, or else that we should begin the sequence with a very fine mesh along the curved boundaries and not change that part of the mesh thereafter. In practice, neither is done, and analysts routinely refine the modeling of the curved boundary during a sequence of meshes. This would seem reasonable, and, in fact, the expected convergence usually occurs.

However, it has been reported [13.29-13.35] that the use of straight-sided elements on curved boundaries may sometimes (not always) cause convergence to the wrong answer in the vicinity of the boundary for second-order problems, or possibly no convergence at all for fourth-order problems, a phenomenon referred to as the *Babuska paradox* [13.29, 13.30]. According to the paradox, within a couple or so elements of the boundary (a "boundary-layer" effect) the normal components of flux may have an exceptionally high error. This has been demonstrated both theoretically [13.31, 13.32] and numerically [13.33-13.35], yet other numerical experiments [13.36] have failed to produce such an error. In any case, curved-sided isoparametric elements (quadratic or higher order) do not exhibit these convergence problems.[42]

In view of these potential problems, as well as the above modeling considerations, this author prefers to model curved boundaries with curved-sided elements — quadratics or higher order — whenever a solution is sought in the vicinity of the curved boundary.

A geometric error may also be introduced when a distributed boundary flux is applied to a curved boundary (see Section 13.3.1, step 5). As illustrated in Fig. 13.70 for a straight-sided element, the flux (which is normal to the boundary) will be applied to a different location, with generally a different orientation. Here again curved-sided elements should produce considerably less error than straight-sided elements.

Another type of modeling error arises when a boundary flux (or any applied load) is distributed over a very small region (see Fig. 13.18). Such a localized load will produce a complicated solution in the immediate vicinity of the load, requiring many DOF (e.g., many tiny elements), as illustrated in Fig. 13.71(a). However, frequently the analyst has no interest in this local behavior and is concerned with only the more distant effects. In these cases the distributed load can be approximated by a point (concentrated) load of the same total strength, applied to a point that is located at or near (as best as can be

---

[42] A possible explanation for the paradox is that the curvature of the boundary does not converge when approximated by piecewise linears, but does converge when approximated by piecewise quadratics, cubics, etc.

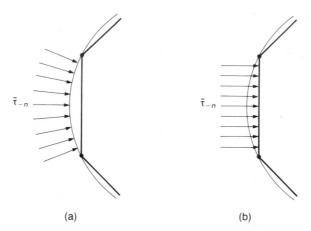

(a)   (b)

FIGURE 13.70 ● A boundary mismatch (e.g., a straight-sided element on a curved boundary) shifts the location and orientation of the boundary flux: (a) the definition of $\tilde{\tau}_{-n}$ in the problem statement; (b) actual application in the FE model.

estimated) the centroid of the distribution. Then much larger elements can be used, with a node (corner nodes are better than midside nodes) placed at the point of application of the load (Fig. 13.71b). The error produced by this approximation will decrease with distance from the load. A rough guideline (from St. Venant's principle in elasticity theory) is that the error is generally "small" when the distance is at least five times the width of the original distributed load.

The above discussion on modeling errors has implicitly assumed that the original problem statement is the "correct" model, and that any deviations from this model represent approximations that produce errors. However, the original problem statement is itself a mathematical idealization of some real-world situation, containing approximations of its own, some of which may be poorly understood and not easily quantifiable. After all, in most engineering applications it is the real-world situation we are trying to approximate as closely as possible, not the mathematical idealization. It can therefore be difficult at

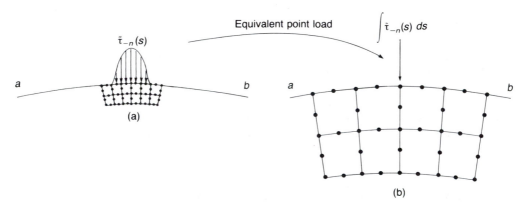

FIGURE 13.71 ● Modeling a narrowly distributed load (a) with a fine mesh for a locally detailed solution or (b) as an equivalent point load with a coarser mesh for distant effects only.

times to assess whether the modeling errors discussed above have decreased or improved the accuracy *relative to the real-world situation*, since they may partially offset other errors in the mathematical idealization. It is because of uncertainties like these that the credibility and reliability of an FE analysis is strengthened by independent corroboration, such as experimental data. (Recall the comments in Sections 3.6.2 and 4.4 under step 12.)

13.6.4 Local Refinement

Sections 13.6.1 through 13.6.3 have provided guidelines for dealing with local problems involving only a few elements. We are now ready to look at the bigger picture — namely, at large quantities of elements covering substantial areas of the domain.

A uniform mesh, consisting of elements all about the same size and shape and repeated in a fairly regular pattern, is the easiest type to construct and needs no special words of advice. Such meshes provide a more-or-less uniform *DOF density* (the number of DOF per area of domain), and hence they are generally most appropriate in the large central areas of a domain where the solution tends to be least complicated, i.e., more or less uniform. In those portions of the domain where the solution is more complicated (e.g., near sharply contoured boundaries and concentrated loads), a greater DOF density is necessary in order to maintain a given level of accuracy; that is, the mesh must be locally refined.

The relationship of accuracy to DOF density was amply explained and demonstrated in Part III, especially in Chapter 9. Those principles apply just as well to 2-D (and 3-D) problems. Here we need only examine a few mesh construction techniques, appropriate to 2-D (and 3-D) problems, that will achieve local refinement, i.e., a change in DOF density.

There are basically four different methods for changing the DOF density. These are illustrated in Fig. 13.72 and, for clarity, the change is shown in one direction only. (Changes

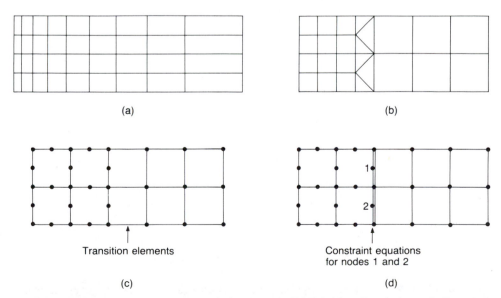

FIGURE 13.72 ● Four different methods for changing DOF density. For clarity, only linear and quadratic elements are shown, but the methods are applicable to elements of any order.

in more than one direction are discussed below.) The method shown in Fig. 13.72(a) is the most straightforward; it uses only one type of element and gradually increases (or decreases) its size. Present-day preprocessors will automatically generate such a mesh with only a couple of commands by generating a sequence of geometrically spaced nodes (i.e., the spacings increase by a constant ratio). Figure 13.72(b) employs triangles as transitions from smaller to larger quadrilaterals. This type of mesh is a bit more irregular than that shown in Fig. 13.72(a) and hence can be a bit more time-consuming to generate (especially if the change is in both directions, as shown below). Figure 13.72(c) uses transition elements and Fig. 13.72(d) uses linear constraint equations. Note that the methods shown in Fig. 13.72(a) and (b) change the size of the elements but not their order, whereas the methods shown in Fig. 13.72(c) and (d) change the order of the elements without changing their size. By mixing the methods together, some very efficient and effective meshes can be created.

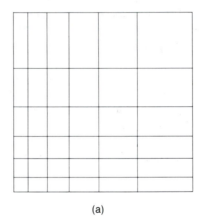

(a)

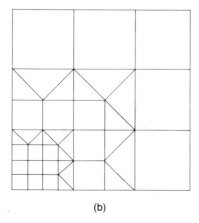

(b)

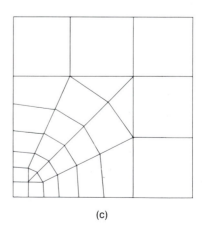

(c)

FIGURE 13.73 ● Some possible element patterns for changing DOF density. (Any ''compatible'' nodal patterns may be employed.)

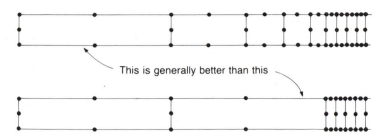

FIGURE 13.74 ● Generally avoid placing very fine and very coarse meshes adjacent to each other; a gradual transition is usually more efficient.

Figure 13.73 demonstrates how these methods can be used to change the DOF density in all directions. The three patterns of elements illustrate changing element sizes by geometrically spacing nodes (Fig. 13.73a), using triangles as transition from smaller to larger quadrilaterals (Fig. 13.73b), and a combination of both (Fig. 13.73c). Of course these are only three of the endless variety of element patterns that are possible. In addition, the reader can visualize a great variety of nodal patterns on these elements. For example, the gradation in the DOF density can be made especially "steep" by making the small elements higher order than the large elements, using the methods shown in Fig. 13.72(c) and (d).

These methods can be applied to domains of any shape, as will be demonstrated by the illustrative problems in this and the next two chapters (see also examples in Chapter 1). Domains of complicated shape are first divided into several subdomains with simpler shapes and the latter meshed separately. The meshes along the subdomain boundaries are made compatible so that they can be easily joined together.

Elements in a mesh should not be locally refined so much that the size of the smallest element is "tiny" in comparison to the largest element, as this tends to produce an ill-conditioned stiffness matrix. The degree of ill-conditioning is very problem-dependent, and even rough guidelines are difficult. Ratios of 100 to 1 (in linear dimensions, hence 10,000 to 1 in area) between the largest and smallest element can sometimes cause trouble in medium-size problems (a few hundred DOF) on low-precision (32-bit) computers.

We also recall from Chapter 9 that abrupt changes in mesh refinement from very fine to very coarse should generally be avoided (Fig. 13.74) because the error in the immediate vicinity of the coarse mesh is locked in to being comparable to the error in the coarse mesh. (Exceptions would be shock-front phenomena.) Recall from Section 9.5 that the solution in the adjacent, locally refined mesh, if refined indefinitely, would converge to the wrong value.

To help minimize the total DOF in an analysis, as well as minimize the ill-conditioning problem due to range of element sizes, the analysis can be separated into two or more stages (Fig. 13.75). The first stage uses a coarse mesh over the entire domain. The second stage uses a fine mesh and models only a portion of the domain surrounding an area needing local refinement; the boundary conditions are determined from the output of the first stage. In line with the previous discussion, the second stage mesh ought to use the same coarse mesh on the new boundaries, and then gradually become more refined further away from these boundaries. In other words, the second-stage portion of the domain needs a buffer

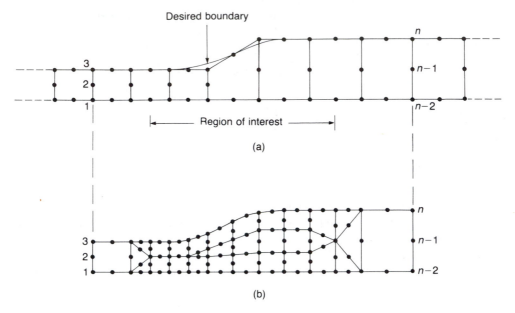

FIGURE 13.75 ● Separating an analysis into two stages, using output from the first stage as boundary conditions for the second stage: (a) first stage — coarse mesh over the entire domain; (b) second stage — fine mesh over part of the domain. Values calculated at nodes 1, 2, 3, $n-2$, $n-1$, n using coarse mesh are imposed as BCs on fine mesh.

zone to make the transition from the lower accuracy in the surrounding coarse mesh.[43] In this staged approach the number of DOF in each stage is, of course, less than if the combined coarse and fine meshes were analyzed together. This approach can be especially economical if the second stage needs to be reanalyzed many times (for a parametric study, say). In this regard, the technique of *substructuring*[44] affords similar economies.

---

[43] Some of the more sophisticated commercial programs enable the analyst to generate a very refined mesh immediately adjacent to the new boundary, using newly created nodes along the boundary. Boundary values for the new nodes are obtained by interpolating from the old nodal values. This type of automation greatly facilitates the local refinement procedure since it is usually faster and cheaper to generate a uniformly fine mesh throughout the entire local region. However, the same inherent error limitations still exist near the boundary so one should use the results from the refined mesh only in areas somewhat removed from the boundary.

[44] It was observed in a footnote in the opening remarks of Section 13.4 that substructuring is the extension of static condensation to a group of elements. Thus, a typical application would be the elimination (at the element level) of all the DOF in a group except those on the boundary of the group. The resulting condensed group, commonly referred to as a *superelement*, could then be treated like any other element as regards assembly and solution. Substructuring can also be applied to dynamic problems. It is a practical, efficient technique for analyzing large problems that have some degree of repetitiveness, either in the structural geometry or the analytical procedure (e.g., a parametric study) [13.37].

Some final remarks are in order regarding a general meshing strategy. Consider an analyst about to construct a mesh for a given problem. If the analyst has absolutely no feeling for what the general qualitative nature of the solution is likely to be (a situation that is not recommended for performing FE analyses), he or she could proceed naively by constructing a more-or-less uniform mesh, observing the solution, and then refining locally wherever it seems necessary. However, a good analyst will usually have a sufficient understanding of the physics of the problem to anticipate the approximate behavior of the overall solution, including the approximate locations of extreme values. Indeed, in most engineering applications the purpose of the FE analysis is primarily to provide quantitative precision to an a priori qualitative understanding. Any such a priori knowledge should be used to help design a more computationally efficient mesh. This will generally result in a nonuniform mesh, one in which the DOF density varies from one area to another.

Probably the two most important factors in determining a global meshing strategy are (1) the variable complexity of the solution from area to area and (2) the analyst's interest in the solution in only certain areas. These factors are usually interrelated. The first factor implies that complicated areas will need more DOF to achieve an accuracy comparable to less complicated areas. However, this may be tempered by the second factor, which indicates that a globally uniform accuracy may not be desired. If the analyst has no interest in a particular area, then a locally coarse mesh with a correspondingly large local error would be acceptable because the error would not significantly affect the solution far away. This is a general property of boundary-value problems that was illustrated in Section 9.5. We recall that there can be a wide range in accuracy in a given mesh; the farther two areas are separated, the greater can be the difference in their accuracies. Hence, the mesh can be coarser farther away from the areas of interest.

A computationally efficient mesh would be one that achieves a desired accuracy in the areas of interest, while using a minimum number of DOF throughout the rest of the domain. This generally implies a greater DOF density in the areas of interest and lesser DOF densities elsewhere. Of course, it is usually the complicated areas of the solution that are of greatest interest. Such areas will therefore require considerable local refinement: the complexity factor demanding extra DOF to bring the accuracy up to that of the surroundings, and the interest factor demanding still more DOF to make the accuracy somewhat greater than that of the surroundings. As noted previously, excessive local refinement leads to ill-conditioning, although staging the refinements as in Fig. 13.74 can help eliminate that problem.

In 2-D (and 3-D) problems there is an endless variety of ways to construct meshes for any given problem, all providing desired accuracies but with varying degrees of efficiency (and hence cost). It would seem a safe bet that if 100 skilled FE analysts were to separately model the same (moderately complicated) problem, no two meshes would be identical.

There has been considerable effort devoted to the development of automatic mesh generators, i.e., computer algorithms that will generate a computationally efficient mesh, requiring the user to supply only the geometric data for defining the shape, i.e., the boundary of the domain. Such generators would conceivably make it possible to perform an FE analysis without ever seeing a mesh. There is some evidence that automatic mesh generators may become a reality in the not-too-distant future [13.38].

13.6.5 Bandwidth Minimization

Computer processing and storage costs decrease significantly as the bandwidth of the system stiffness matrix decreases [see Fig. 5.25]. For a given mesh topology[45] the bandwidth is determined by the *node-numbering pattern* (see Section 5.7). There always exists at least one pattern that minimizes the bandwidth. It is not always practical to seek this special (and perhaps elusive) pattern, but by following the few special guidelines given below, it is usually easy to number the nodes in a near-optimal pattern (or, at the very least, to avoid very badly numbered meshes).

There exist many algorithms [13.39-13.47] that will automatically renumber a mesh to minimize (or nearly minimize) the bandwidth if the program uses a banded equation solver, or minimize the wavefront if a frontal solver is used. The latter requires an optimal numbering of the elements rather than the nodes, and the principles used are similar for either elements or nodes. These algorithms have been incorporated in most commercial FE programs. They sometimes require the user to supply a few starting nodes; these can be chosen judiciously using the guidelines below. On one occasion the author observed a mesh numbered by a colleague that gave a lower bandwidth than the computer algorithm. Although this process will someday soon probably be totally automated and 99.9% reliable, these observations suggest there is still a need for analysts to understand some simple rules for optimal node- (or element-) numbering strategies.

First let's review how the bandwidth is calculated. For symmetric matrices (which characterize most FE analyses) we need to deal with only the half-bandwidth; this was defined in Section 5.5 (see Fig. 5.25). Letting hB represent the half-bandwidth of the system stiffness matrix and $hB^{(e)}$ the half-bandwidth of the element stiffness matrix for element (e), then

$$hB = \underset{\substack{\text{(over all} \\ \text{elements)}}}{\text{maximum}: hB^{(e)}} \tag{13.197}$$

where

$$hB^{(e)} = N_{max}^{(e)} - N_{min}^{(e)} + 1$$

and $N_{max}^{(e)}$ and $N_{min}^{(e)}$ are the largest and smallest node numbers,[46] respectively, in element (e). Thus $hB^{(e)}$ must be calculated for every element in the mesh, and the maximum value is hB. In UNAFEM these calculations are done in subroutine HBAND.

To illustrate the calculation of $hB^{(e)}$, consider the quadrilateral element and corresponding assembled stiffness matrix shown in Fig. 13.76. In this example $N_{max}^{(e)}=33$ and $N_{min}^{(e)}=28$, so $hB^{(e)}=33-28+1=6$. The value $hB^{(e)}=6$ indicates that along any one row of the system stiffness matrix the largest distance of a nonzero term from the main diagonal (for this element) is six terms, counting the diagonal; this occurs along rows $N_{min}^{(e)}$ and $N_{max}^{(e)}$.

---

[45] The topology refers to the pattern of nodes and the lines (element boundaries) connecting the nodes. The coordinates of the nodes do not affect the topology.

[46] For elements that have more than one DOF per node (see Chapter 16), $N_{max}^{(e)}$ and $N_{min}^{(e)}$ are the largest and smallest DOF numbers.

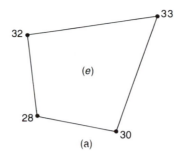

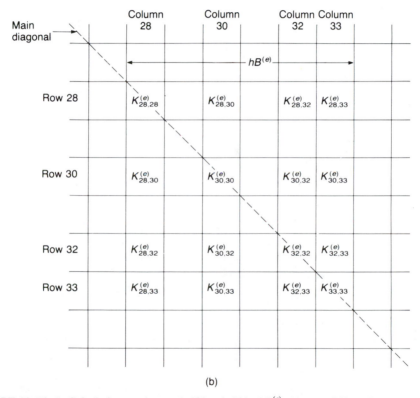

FIGURE 13.76 ● Calculating an element half-bandwidth, $hB^{(e)}$: (a) a quadrilateral element; (b) its assembled stiffness matrix.

To illustrate the calculation of hB, consider the mesh shown in Fig. 13.77. The corresponding assembled stiffness matrix shows all the nonzero terms by x's. Element (4) has the maximum $hB^{(e)}$ so its terms are labeled by (4)s. (We will see in a moment a better way to number this mesh.)

To keep hB small we therefore strive to keep $hB^{(e)}$ as small as possible in every element, which means we want the node numbers in each element to be as close together

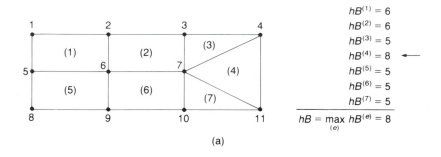

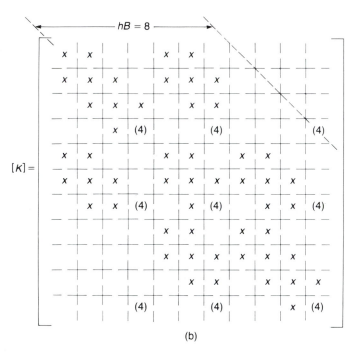

(b)

FIGURE 13.77 ● Calculating a system half-bandwidth, hB: (a) a mesh; (b) its assembled system stiffness matrix.

as possible. This may be accomplished by the following basic guideline for numbering a mesh:

> Number the nodes along paths that contain the least number of nodes *from one boundary to another*, proceeding from one element to an adjacent element, i.e., without skipping over elements.

For a mesh in which the spacing between nodes is comparable in all directions, we should number across the shortest dimension of the domain (Fig. 13.78). If the internodal spacings are not comparable, then numbering across the shortest dimension might yield the larger half-bandwidth (Fig. 13.79). Thus, as stated in the guideline above, it is the number

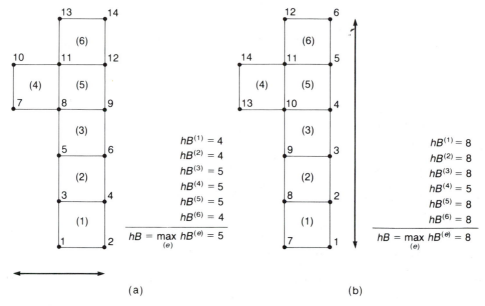

FIGURE 13.78 ● Numbering across the shortest dimension of the domain will decrease the half-bandwidth if the internodal spacings are comparable in all directions: (a) numbering across the short dimension minimizes hB; (b) numbering across the long dimension yields a larger hB.

of nodes (or DOF) from one side of the domain to the other that is the determining factor, not the dimensions of the domain.

It is important to number the nodes from boundary to boundary. Otherwise, as with the middle row in Fig. 13.77(a), an element will span more than two rows and the band-

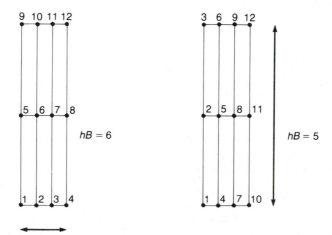

FIGURE 13.79 ● Numbering across the shortest dimension will yield a larger half-bandwidth if there are more nodes in this direction.

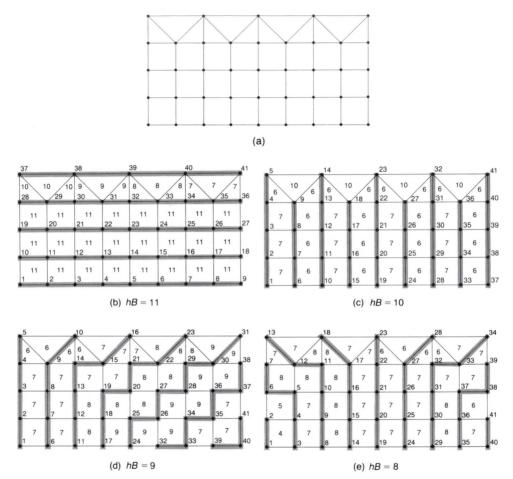

FIGURE 13.80 ● (a) Mesh used to illustrate strategies for minimizing the half-bandwidth; (b) a poor pattern in which the nodes are numbered along the long dimension; (c) a better pattern in which the nodes are numbered along the short dimension; however, some numbering lines do not extend from boundary to boundary; (d) a still better pattern in that all numbering lines extend from boundary to boundary, but the lines on the right are much longer than the lines on the left; (e) the best (?) pattern; all numbering lines extend from boundary to boundary and they are balanced in length from left to right. The number inside each element is the half-bandwidth for that element. The number below each mesh is the system half-bandwidth for that mesh.

width may increase significantly. Figure 13.80 illustrates these ideas by showing several numbering patterns, each an improvement over the former for the specific reason stated. If a program uses a frontal solver instead of a bandwidth solver, then strategies similar to those shown in Fig. 13.80 would be applied to the element numbers (rather than node numbers) to minimize the wave front.

It should be remembered that the basic reason for trying to minimize bandwidth is to decrease computational costs. However, the analyst's time is often much more expensive than the computer fees (even with a very poorly numbered mesh). Hence the "best" numbering pattern is perhaps one that is reasonably (not optimally) efficient for the computer and also easy to construct and use by the analyst. Exotic numbering patterns can be more time-consuming to generate and more confusing to interpret during postprocessing. In this wider perspective, the "best" mesh might not be that of Fig. 13.80(e) but rather that of Fig. 13.80(c). Many of the commercial FE programs that employ the aforementioned bandwidth (or wave-front) optimizers resolve this human/machine conflict by using the optimal renumbering only for internal processing, and then converting back to the user's numbering scheme for postprocessing.

13.7 Application of UNAFEM to a Torsion Problem

13.7.1 Description of the Problem

A classic problem in the linear theory of elasticity is the torsion of a cylindrical bar with an arbitrary cross-sectional shape and isotropic material properties [13.48-13.50]. Figure 13.81 shows a cylinder being twisted about its axis by torsional moments M applied in opposite directions at each end. Cross sections a distance Δz apart rotate relative to each other through an angle $\theta\Delta z$, where θ is the angle of twist per unit length. The twisting produces shear stresses on the cross-sectional planes which, in turn, cause the planes to warp. The shear stresses and warping are identical in every cross section; only the angular rotation of the cross sections varies (linearly) along the length.

For small rotations a useful engineering formula is $M=D\theta$, where D is the torsional rigidity of the cylinder, a quantity that may be calculated theoretically (using, say, an FE approach as described here) or experimentally (by measuring M and θ). In a practical problem concerning a cylinder (bar, beam, etc.) with a specific cross-sectional shape, the quantities one is usually interested in calculating are the shear stresses, torsional rigidity, and/or warping.

The governing equations in the general linear theory of elasticity (see Chapter 16) are a set of coupled partial differential equations. However, for the torsion problem those equations simplify to a single 2-D quasiharmonic equation, the domain being the cross section of the cylinder. Therefore the problem may be analyzed with UNAFEM.

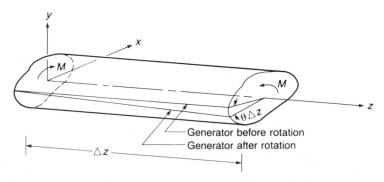

FIGURE 13.81 ● Torsion of a cylinder with an arbitrary cross-sectional shape.

Here we merely present the relevant governing equation and boundary conditions and briefly explain the quantities involved. A more complete explanation of basic elasticity concepts may be found in Chapter 16, and a fuller treatment of the torsion problem may be found in the cited references, as well as in many other textbooks on elasticity.

Governing differential equation[47] •

$$\nabla^2 \Psi = -2 \qquad \text{[Poisson's equation, Eq. (13.21)]} \qquad \textbf{(13.198)}$$

Domain • The 2-D cross section (in the x,y-plane)

Boundary conditions •

$$\Psi = 0 \qquad \text{on entire perimeter of cross section}$$

The function Ψ is called the *Prandtl stress function*. The two components of stress acting on the plane of the cross section may be calculated from Ψ using the relations

$$\tau_{zx} = \mu\theta \frac{\partial \Psi}{\partial y}$$

$$\tau_{zy} = -\mu\theta \frac{\partial \Psi}{\partial x} \qquad \textbf{(13.199)}$$

where μ is the shear modulus (force/area) and τ_{zx}, τ_{zy} are the two components of shear stress (force/area); that is, τ_{zx} and τ_{zy} are the x- and y-components, respectively, of the shear force acting on a unit area of the z-plane (cross section).

Defining i and j as unit vectors in the x and y directions, respectively, the shear stress vector is

$$\tau = \tau_{zx}i + \tau_{zy}j$$

$$= \mu\theta \frac{\partial \Psi}{\partial y}i - \mu\theta \frac{\partial \Psi}{\partial x}j \qquad \textbf{(13.200)}$$

Since the gradient of Ψ (see Section 13.1.1) is

$$\nabla\Psi = \frac{\partial \Psi}{\partial x}i + \frac{\partial \Psi}{\partial y}j \qquad \textbf{(13.201)}$$

it follows that τ is everywhere normal to $\nabla\Psi$, or, equivalently, τ is parallel to lines of constant Ψ. Thus, Ψ-contours are lines of shear stress. We also have

$$|\tau| = \mu\theta|\nabla\Psi|$$

$$= \mu\theta\sqrt{\left(\frac{\partial \Psi}{\partial x}\right)^2 + \left(\frac{\partial \Psi}{\partial y}\right)^2} \qquad \textbf{(13.202)}$$

---

[47] There are three 2-D formulations of the torsion problem, referred to as the *Poisson, Neumann*, and *Dirichlet problems*. They are closely interrelated mathematically, each having advantages and disadvantages regarding ease of analysis and usefulness of the form of the solution. Here we use the Poisson formulation because the UNAFEM function and flux plots can both be used to interpret the stresses, as explained above. In the Neumann formulation, on the other hand, the function represents the warped shape of the cross section. All three formulations can be analyzed by UNAFEM, using the same meshes but different BCs and interior loads.

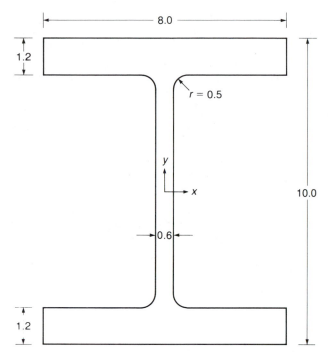

FIGURE 13.82 ● Cross section of the I-beam to be analyzed (dimensions in inches).

Hence the maximum shear stress, $|\tau|_{max}$, occurs where $|\nabla\Psi|$ is a maximum. This property can be used to determine (reasonably closely) the location of $|\tau|_{max}$ since the UNAFEM function plots will show Ψ-contours, and they will be closest together where $|\nabla\Psi|$ is a maximum.

Comparing Eq. (13.198) with Eq. (13.18), we see that the x- and y-components of flux are $-\partial\Psi/\partial x$ and $-\partial\Psi/\partial y$, respectively. From Eqs. (13.199),

$$x\text{-component of flux} = \frac{1}{\mu\theta}\, \tau_{zy}$$

$$y\text{-component of flux} = -\frac{1}{\mu\theta}\, \tau_{zx} \tag{13.203}$$

that is, the UNAFEM flux components are the shear stress components normalized by $\mu\theta$. Hence, after determining the location of the maximum shear stress from the function plot, its value (normalized by $\mu\theta$) can be determined from Eq. (13.202) by reading the component flux values from either the flux plots or the printout.

Having laid the necessary groundwork, let's now take a look at a specific problem: the torsion of an I-beam (Fig. 13.82). Our objective will be to calculate the maximum shear stress in the beam, to an estimated accuracy of about 5%.

13.7.2 Taking Advantage of Symmetry

At the outset of any problem one should look for every opportunity to simplify the analysis as much as possible. Many practical problems contain points, lines, or planes of *sym-*

metry which divide the domain into two or more identically shaped subdomains, and *the solution is identical in each subdomain*. Therefore only one of the subdomains needs to be analyzed.

To identify a symmetry, we note that the solution is determined by three types of data: geometry (the shape of the domain), physical properties (the coefficients in the governing equations), and loads (interior loads and all boundary conditions). For the solution to be symmetrical, all three groups of data must possess the same symmetry. It is usually geometric symmetry that first catches the analyst's eye; one must then check the physical properties and loads to see if they also possess the same symmetry.

In our I-beam problem we notice immediately (Fig. 13.83a) that the geometric shape has two lines of symmetry — the x- and y-axes — which divide the domain into four identically shaped quadrants. From Eq. (13.198) we see that the physical properties are constants $[\alpha_x(x,y)=\alpha_y(x,y)=1, \; \beta(x,y)=0]$, so they have the same distribution in each quadrant. The interior load is also a constant $[f(x,y)=2]$ and so are the boundary conditions ($\Psi=0$). Since all the data is distributed in an identical manner in each quadrant, we conclude that the x- and y-axes are indeed lines of symmetry. Therefore we only need to analyze one of the quadrants, say the upper-right one (Fig. 13.83b).

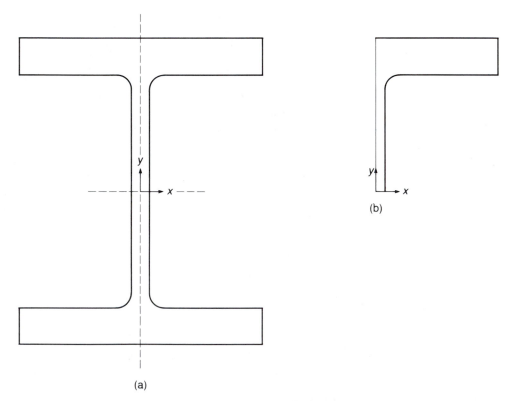

(a)

(b)

FIGURE 13.83 ● (a) The x- and y-axes are lines of symmetry for the I-beam shape; (b) symmetry considerations reduce the analysis to only one quadrant of the I-beam.

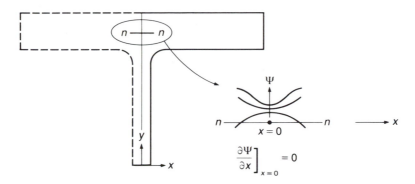

FIGURE 13.84 ● Appropriate boundary conditions on a symmetry boundary (y-axis).

There remains one problem: The removal of a quadrant from the original domain has created new boundaries along the x- and y-axes. The original problem data, though, did not specify boundary conditions along these axes (since they were interior to the original domain). We must therefore determine appropriate boundary conditions for these new boundaries.

Consider first the y-axis. It is the line of symmetry between the right and left quadrants (see Fig. 13.84). Since the left quadrant is geometrically a mirror reflection of the right quadrant, the solutions in the two quadrants must also be mirror reflections. This implies that the slope of Ψ normal to the y-axis and just to the left of the axis must be equal to the *negative* of the slope just to the right of the axis:

$$\left.\frac{\partial \Psi}{\partial x}\right]_{x=0_-} = -\left.\frac{\partial \Psi}{\partial x}\right]_{x=0_+} \tag{13.204}$$

We also know that the exact solution (function and all derivatives) is smooth throughout the interior of the original domain since the problem data is smooth. In particular, this implies continuity of the normal derivative across the y-axis; i.e., the normal derivatives on the left and right sides of the axis, as well as on the axis, must be equal:

$$\left.\frac{\partial \Psi}{\partial x}\right]_{x=0_-} = \left.\frac{\partial \Psi}{\partial x}\right]_{x=0_+} = \left.\frac{\partial \Psi}{\partial x}\right]_{x=0} \tag{13.205}$$

Clearly Eqs. (13.204) and (13.205) can only be satisfied if

$$\left.\frac{\partial \Psi}{\partial x}\right]_{x=0} = 0 \tag{13.206}$$

These relations are illustrated graphically in Fig. 13.84, which shows a sketch of some typical variations of Ψ along a path normal to the y-axis, demonstrating that symmetry and smoothness jointly necessitate a horizontal slope at the axis.

A similar argument applied to the symmetry boundary on the x-axis yields

$$\left.\frac{\partial \Psi}{\partial y}\right]_{y=0} = 0 \tag{13.207}$$

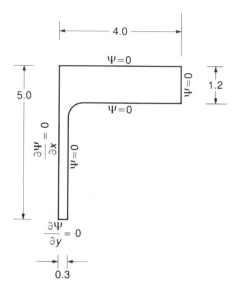

FIGURE 13.85 ● Simplification of the I-beam problem after taking advantage of symmetries (dimensions in inches).

Figure 13.85 shows the reduced domain and appropriate boundary conditions. This is now a well-posed problem, ready for an FE analysis. It is certainly much simpler than the total I-beam in Fig. 13.82. By exploiting all the symmetries, we have reduced the size of the problem fourfold. In fact, some of the computational costs, namely, solution of the system equations, will be reduced by an even greater factor.

13.7.3 Verification of UNAFEM

Presumably the reader has never performed a 2-D analysis with the UNAFEM program. Therefore, before jumping directly into the I-beam problem, it is important to first establish a reasonable level of *confidence* (not certainty) that the program can indeed do what it claims it can do. Recalling the discussion in Section 7.4.2, let's begin with a simple one-element test. This will provide a measure of confidence in the element itself and, at the same time, develop a familiarity with the input-output procedures peculiar to this program.

For the I-beam analysis we will use the quadratic elements ELEM22 and ELEM24 rather than the linear elements. There is a curved boundary on the fillet, and it was recommended in Section 13.6.3 that quadratic or higher-order elements be used along curved boundaries. In the rest of the domain, it is the author's preference to work with quadratic elements because of their faster rate of convergence (see Section 9.3).

Figure 13.86(a) describes a simple test of the quadrilateral. All types of loads are applied: essential BCs, natural BCs, and an interior load. The exact solution (Fig. 13.86b) is a quadratic polynomial, which the element should be able to reproduce exactly. Figure 13.86(c) and (d) show the mesh and printed output. Happily, the results are correct. It would be wise to repeat the same test in the *y*-direction, as well as perform similar tests on the triangular element.

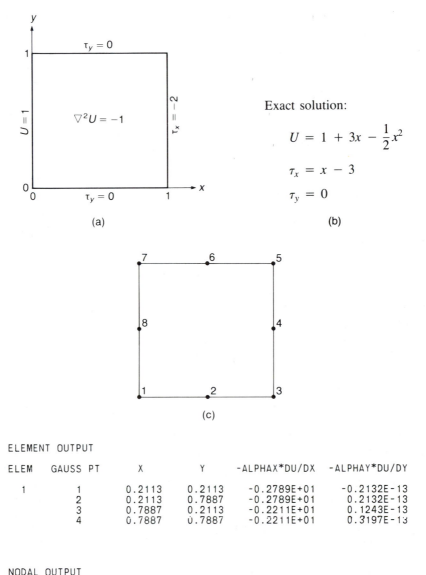

FIGURE 13.86 ● A one-element test of the quadrilateral element ELEM24 in UNAFEM: (a) test problem; (b) exact solution; (c) mesh; (d) computed values.

The test described here is not sufficient to verify the element under all situations. That would take several carefully designed tests to accomplish. There has been an ongoing effort in the FE community to establish objective standards for verifying elements [13.51-13.54]. Such standards will likely prove quite valuable for commercial program developers. Users can learn to apply these tests themselves or, as here, simply be resourceful and devise one or two tests of their own which should provide *some* level of confidence.

Passing one-element tests provides a minimum level of confidence. However, much greater confidence can be gained by also analyzing a simple problem with a known solution. The test problem should employ a mesh of several elements and have as many features as possible similar to the real problem, with the principal difference usually being a simpler geometry. It is usually not difficult to find such problems in reference books, textbooks, or the open literature. The difference between the real problem and a well-chosen test problem is primarily just additional computation using the same algorithmic sections of the program. Therefore, if the test problem succeeds, there is small likelihood of encountering program bugs when doing the real problem.

Commercial program developers generally perform a large number of such test problems to qualify their programs, and the documentation of these tests is often made available to users. Such documentation can be a rich source of test problems. Even when using carefully qualified commercial programs, it is still advisable to perform test problems frequently (either theirs or yours) to assure yourself that everything is working correctly throughout the computational network, which includes not only the FE program but also all the supporting software and hardware.

When working with previously qualified programs (e.g., commercial codes), it would be reasonable to skip the one-element tests and perform only simple test problems (with several elements), since a successful test problem usually (not always) implies a valid element. Of course, two or three test problems will provide even greater confidence. If the results are suspicious, then one-element tests would be advisable.

For our present I-beam problem, a simple torsion test problem would be a rectangular cross section (see Fig. 13.87a). A mesh of two (or four, six, eight, etc.) elements is easy to construct and an exact solution is available [13.55]. A rectangle has two lines of symmetry, which can be used to reduce the problem to a quadrant of the rectangle (see Fig. 13.87b). Although the shape remains a rectangle, the principal purpose of this reduction is to include symmetry BCs in the test since they are present in the I-beam problem.

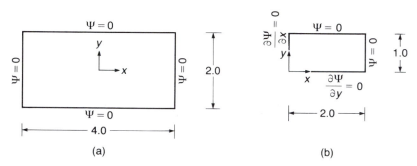

FIGURE 13.87 ● Torsion test problem: (a) rectangular cross section; (b) reduction of the problem, because of symmetries, to one quadrant (dimensions in inches).

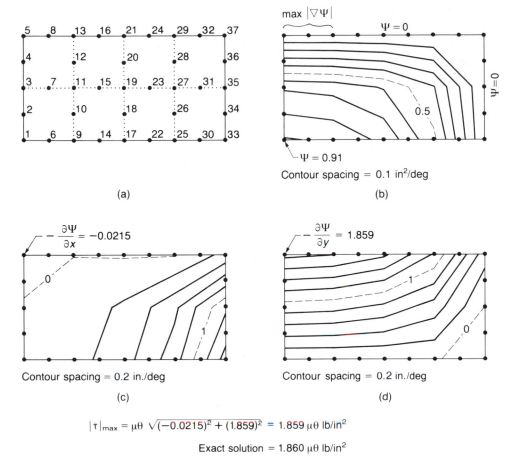

$$|\tau|_{max} = \mu\theta \sqrt{(-0.0215)^2 + (1.859)^2} = 1.859 \; \mu\theta \; lb/in^2$$

$$\text{Exact solution} = 1.860 \; \mu\theta \; lb/in^2$$

FIGURE 13.88 ● (a) Mesh for the torsion test problem; (b) contour plot of the Prandtl stress function, Ψ; (c) contour plot of $-\partial\Psi/\partial x$; (d) contour plot of $-\partial\Psi/\partial y$. Values of dashed contours are indicated.

Figure 13.88(a) shows a mesh of 2-by-4 elements for the rectangular quadrant. UNAFEM draws all *exterior* element sides (those common to only one element) as solid lines, and all *interior* element sides (those common to two elements) as dotted lines. A correctly generated mesh will therefore have solid lines along the boundary of the domain, and all interior element sides will be dotted. This feature is an aid for verifying a correctly generated mesh, since any (inadvertently) undefined interior elements can be easily detected by the presence of solid lines in the interior of the domain[48]. In the remainder of this book, however, all 2-D mesh plots will be displayed with only solid lines on all element sides, since dotted lines are sometimes difficult to perceive in photo-reduced pictures, especially in refined meshes with short element sides.

---

[48] Another technique for identifying missing interior elements, used by many commercial programs, is to shrink each element a small amount.

The function and flux contour plots in Fig. 13.88 show that $|\tau|_{max}$ occurs at the upper-left corner, i.e., at the center of the long side of the whole rectangle.[49] Its error is less than 0.1%. This certainly increases our confidence in the program. At the same time, we begin to get a feeling for the approximate level of accuracy obtainable with a given DOF density in problems of this type. This will be helpful in constructing a first mesh for the I-beam problem.

13.7.4 Analysis of the I-Beam

We can now turn our attention to the construction of a mesh for the L-shaped domain in Fig. 13.85, and we will apply the general meshing strategy discussed in Section 13.6.4.

As stated previously, our objective is to calculate the maximum shear stress to an estimated accuracy of about 5%. This means we will be interested in the solution primarily within a small region surrounding the maximum, which in turn implies that we need to find the location of the maximum.

An examination of the geometry of the domain draws our attention immediately to the small-radius fillet. We know that concave sections of a boundary with small radii of curvature will locally produce sharp increases in the solution. Indeed, a radius of zero, i.e., a reentrant corner, produces a singularity. Therefore we might expect the maximum to occur along the boundary of the fillet. (Of course, we will watch for extreme values anywhere else in case our expectations turn out to be wrong.)

Since the fillet is a quasisingular region (see Section 9.4), a graded mesh would be appropriate. It is not necessary to employ an optimal grading strategy (which is not easy to determine). Virtually any form of grading that involves modest changes in size from element to element will improve the accuracy.[50]

Farther away from the fillet the solution is expected to be less complicated (another expectation that should be verified when examining the actual solution), so a coarser mesh may be used in those areas.

Figure 13.89(a) shows a mesh that incorporates these features. The grading is not very pronounced since this is only a first mesh and there are only a few elements to work with. In addition, too much local refinement in a first mesh might be wasteful; we need to examine the first mesh solution to better determine where the refinement is really needed. For the sake of clarity, only a few node numbers are shown. Note that they number across the shorter dimensions to lower the bandwidth.

The resulting Prandtl stress function plot (Fig. 13.89b) indicates two regions with steep gradients: one on the fillet and one on the "corner" opposite the fillet (which is the midpoint of the surface of the flange). Using Eq. (13.202) and the printed tabular listing of computed nodal values for $-\partial\psi/\partial x$ and $-\partial\psi/\partial y$, it is determined that $|\tau|_{max}$ occurs

---

[49] The UNAFEM contour-plotting algorithm uses only the corner nodes on all 2-D elements; midside nodes in element types 22 and 24 are ignored.

[50] In many commercial preprocessors, grading can be easily accomplished by specifying a nonuniform (usually geometric) spacing of nodes. When using higher-order isoparametric elements, it is advisable to grade only the corner nodes, then fill in the midside nodes at their undistorted positions within each element, in order to avoid unnecessary, and generally undesirable, distortion of the mapped elements.

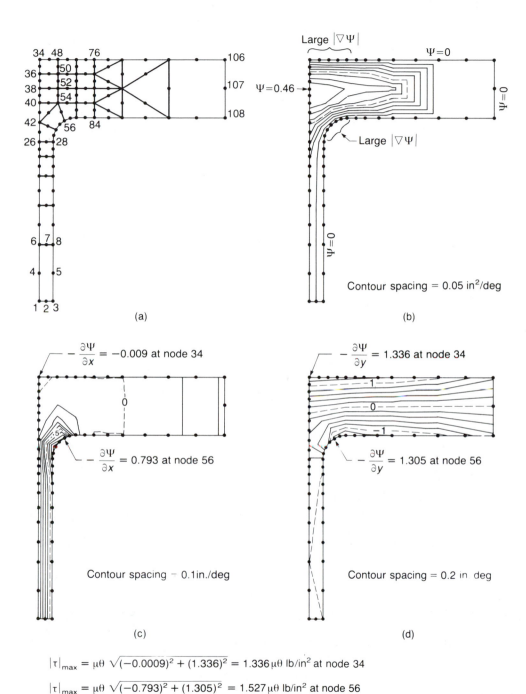

$$|\tau|_{max} = \mu\theta \sqrt{(-0.0009)^2 + (1.336)^2} = 1.336\,\mu\theta \text{ lb/in}^2 \text{ at node 34}$$

$$|\tau|_{max} = \mu\theta \sqrt{(-0.793)^2 + (1.305)^2} = 1.527\,\mu\theta \text{ lb/in}^2 \text{ at node 56}$$

FIGURE 13.89 ● First mesh and solution for the I-beam analysis: (a) mesh; (b) contour plot of the Prandtl stress function, Ψ; (c) contour plot of $-\partial\Psi/\partial x$; (d) contour plot of $-\partial\Psi/\partial y$. Values of dashed contours are indicated.

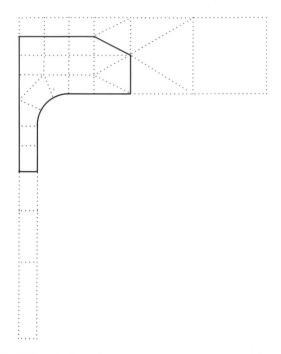

FIGURE 13.90 ● Bold lines indicate the local region near the fillet which will be isolated from the rest of the domain and modeled with a finer mesh. Dotted lines indicate the previous global mesh.

at node 56 on the fillet, as indicated in Fig. 13.89(c) and (d).[51] All other aspects of the solution appear to be reasonable.

Since the maximum shear stress is located on the fillet, we are now interested in only the local region surrounding the fillet. This calls for a local mesh refinement. We will employ the technique described in Section 13.6.4 (see Fig. 13.75 and accompanying discussion) wherein we isolate a small portion of the domain from the rest of the domain, and for boundary conditions on the new "boundary" we specify the values computed from the previous coarser analysis. Figure 13.90 shows the portion of the domain that will be isolated.

Figure 13.91(a) shows the locally refined mesh. Along the new boundary between the global and local meshes the same nodes are retained to facilitate using the previously computed values at these nodes (indicated in the figure). In addition, one row of elements inside the new boundary is left the same as before, since the error immediately adjacent to the unrefined global mesh can't improve much because of its proximity to a larger error

---

[51] Neither of the separate maxima for $|-\partial\psi/\partial x|$ and $|-\partial\psi/\partial y|$ occurs at node 56, but rather at nearby nodes on the edge of the fillet. Hence Eq. (13.202) must be evaluated at several nodes throughout the region to determine the local maximum. For a small program like UNAFEM with an unsophisticated postprocessor, this operation must be done manually. However, large commercial programs will generally automate such a procedure by providing the capability of performing arithmetic (and sometimes calculus) operations on the nodal values, as well as searches for extreme values.

(see Section 9.5). In other words, there is a buffer zone just inside the new boundary that ought to be ignored when interpreting output from the new mesh, and therefore it would be computationally wasteful to refine the mesh in this zone. Again, only a few node numbers are indicated. The numbering pattern attempts to keep the bandwidth low while, at the same time, enabling the program to automatically generate as many nodes and elements as possible.

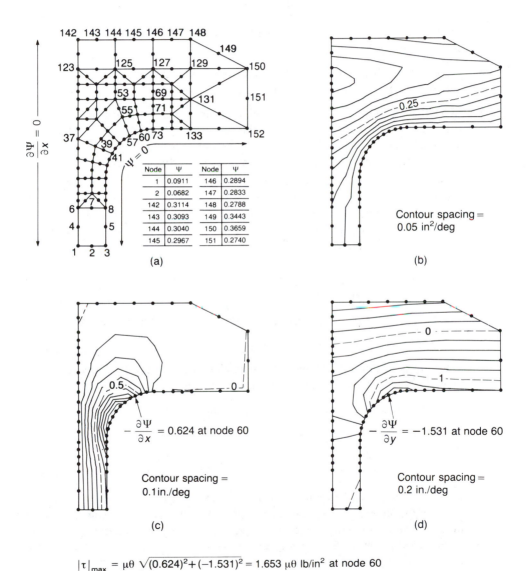

Node	Ψ	Node	Ψ
1	0.0911	146	0.2894
2	0.0682	147	0.2833
142	0.3114	148	0.2788
143	0.3093	149	0.3443
144	0.3040	150	0.3659
145	0.2967	151	0.2740

$$|\tau|_{max} = \mu\theta \sqrt{(0.624)^2 + (-1.531)^2} = 1.653 \; \mu\theta \; \text{lb/in}^2 \text{ at node 60}$$

FIGURE 13.91 ● Isolated, locally refined mesh and solution for the I-beam analysis: (a) mesh, showing specified values of Ψ from previous global analysis; (b) contour plot of the Prandtl stress function, Ψ; (c) contour plot of $-\partial\Psi/\partial x$; (d) contour plot of $-\partial\Psi/\partial y$. Values of dashed contours are indicated.

As before, Eq. (13.202) is evaluated at several nodes along the fillet. The maximum value occurs at node 60, which is very close (7.5° clockwise) to the previous maximum location at node 56 in the global mesh. A comparison of the results from the global mesh and the locally refined mesh indicates an accuracy of about 7 to 8% in the maximum shear stress.

Exercises

13.1   Derive the balance equation [Eq. (13.38)] for small transverse deflections of a flexible membrane.

13.2   Derive the expressions in Eq. (13.67) for the C^0-linear triangle shape functions, following the procedure described in the text.

13.3   Show that the optimal flux-sampling point for the 2-D C^0-linear triangular element is at the centroid, i.e., at $\xi = \eta = 1/3$. The coordinates ξ and η may be interpreted as cartesian coordinates for a unit right triangle, as shown in Fig. 13.31 (or, equivalently, as area coordinates for a triangle of arbitrary shape).

Comment: The proof follows the same least-squares methodology used for the 1-D problems in Chapters 7 and 8. However, for 2-D (and 3-D) problems the physical property α has two or more components so we must use the complete variational integral given in the paper by Moan [8.16, Chapter 8]. For our problem this integral takes the following form:

$$I = \int_0^1 \int_0^{1-\eta} \left[\alpha_x \left(\frac{\partial U}{\partial \xi} - \frac{\partial \tilde{U}^{(e)}}{\partial \xi} \right)^2 + \alpha_y \left(\frac{\partial U}{\partial \eta} - \frac{\partial \tilde{U}^{(e)}}{\partial \eta} \right)^2 \right] d\xi \, d\eta$$

$$(E13.1)$$

The remainder of the procedure is analogous to the 1-D approach. Thus, let

$$\tilde{U}^{(e)}(\xi, \eta) = c_1 + c_2 \xi + c_3 \eta \qquad (E13.2)$$

Since this is a complete 2-D linear polynomial, expand the exact solution as a complete 2-D quadratic polynomial (one degree higher):

$$U(\xi, \eta) = b_1 + b_2 \xi + b_3 \eta + b_4 \xi^2 + b_5 \xi \eta + b_6 \eta^2 \qquad (E13.3)$$

Substitute these expressions into the above integral and apply the minimizing conditions:

$$\frac{\partial I}{\partial c_2} = 0 \qquad \frac{\partial I}{\partial c_3} = 0 \qquad (E13.4)$$

Use the triangle integration formula [Eq. (13.75)] and assume α_x and α_y are constants in the element (which they approach in the limit as $h \to 0$, consonant with the asymptotic nature of the theory). Solve the resulting two equations for c_2 and c_3 in terms of the b_i. Finally, determine the values of ξ and η that satisfy the conditions

$$\frac{\partial U}{\partial \xi} - \frac{\partial \tilde{U}^{(e)}}{\partial \xi} = 0$$

$$\frac{\partial U}{\partial \eta} - \frac{\partial \tilde{U}^{(e)}}{\partial \eta} = 0 \qquad (E13.5)$$

which is the point where the approximate flux will equal the exact flux to one degree higher, i.e., the optimal flux-sampling point.

13.4 Develop the four shape functions for the four-node straight-sided C^0-triangular element shown in Fig. 13.24, using the same technique as used for the six-node parent (straight-sided) shape functions in Section 13.3.2.

13.5 Prove the claim in Section 13.3.2, step 4, that the shape functions developed by the direct approach are generally *all* nonzero along a curved side of an element, and hence interelement continuity cannot be established along the curved side. *Hint*: Consider the following right triangle with two straight sides and a curved hypotenuse that is a quadratic defined by nodes 2, 5, and 3,

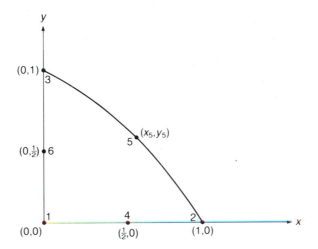

Note that the hypotenuse is a straight line if and only if $x_5+y_5=1$.

a) Develop the six shape functions using the interpolation property as in Section 13.3.2.
b) Write the expression for the quadratic curve through nodes 2, 5, and 3.
c) Substitute the expression from part (b) into the shape functions in part (a) and show that $\phi_1(x,y)$, $\phi_4(x,y)$, and $\phi_6(x,y)$ are zero along the hypotenuse only if $x_5+y_5=1$.

13.6 The boundary flux integral in Section 13.3.2, step 5, was evaluated for side $\overline{142}$ of the triangular element, resulting in Eqs. (13.137) and (13.139). Show that these same equations are produced along sides $\overline{253}$ and $\overline{361}$, excepting only a change in indices. [Note the discussion following Eq. (13.139).]

13.7 Show that the condition for an acceptable mapping of a C^0-linear isoparametric quadrilateral element is that the element must be convex, i.e., each interior angle must be less than $180°$ (see Strang and Fix [13.7], pp. 157-158). *Hint*: Using Eqs. (13.164) through (13.166), show that $|J^{(e)}(\xi,\eta)|$ is linear (rather than bilinear), and therefore $|J^{(e)}(\xi,\eta)|>0$ everywhere in the element if it is greater than 0 at each of the four corners.

13.8 Show that if a C^0-linear isoparametric quadrilateral element is a parallelogram, then the Jacobian is a constant, that is, $|J^{(e)}(\xi,\eta)|=A/4$, where A is the area of the parallelogram. Hence there is no mapping distortion. Use Eqs. (13.164) through (13.166).

13.9 Show that the C^0-linear isoparametric quadrilateral element in Section 13.4.1 has only one optimal flux sampling point, located at the center of the element, i.e., at $\xi=\eta=0$. *Hint*: Follow the procedure described in Exercise 13.3, altering those equations where appropriate. Show that there are two lines ($\xi=0$ and $\eta=0$) along which the ξ and η flux components are separately optimal. Hence their intersection is where both components are optimal, thereby forming the optimal flux-sampling point for any orientation of the real element.

13.10 Develop the parent shape functions for the C^0-cubic isoparametric element (third member of serendipity family in Fig. 13.38), using the procedure described in Section 13.4.2, step 4.

13.11 Develop the parent shape functions for a 5-node isoparametric C^0-linear/quadratic transition element,

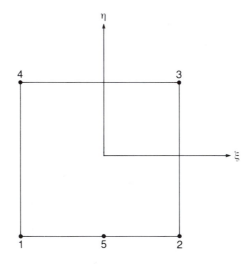

using the procedure described in Section 13.4.2, step 4.

13.12 Show that the C^0-quadratic isoparametric quadrilateral element in Section 13.4.2 has four optimal flux-sampling points, located at the four sampling points of the 2-by-2 Gauss quadrature rule (Fig. 13.50). *Hint*: Follow the same procedure as in Exercise 13.3, but use the eight serendipity quadratic terms for $\tilde{U}(\xi,\eta)$ and a complete 2-D cubic polynomial for $U(\xi,\eta)$. Show that there are two pairs of lines ($\xi=\pm 1/\sqrt{3}$ and $\eta=\pm 1/\sqrt{3}$) along which the ξ and η flux components are separately optimal, and hence their four intersection points are where both components are optimal.

13.13 Using the parent shape functions developed in Exercise 13.11 for the five-node linear/quadratic transition element, complete steps 4 through 6 of the theoretical development and implement the equations into the UNAFEM program. Such a transition element will enable UNAFEM to mix linear and quadratic elements in the same mesh.

13.14 Derive the linear constraint equations to enable a quadratic element to combine with a cubic element (recall Fig. 13.66).

13.15 Using the relations in Section 13.3.1 for the C^0-linear triangular element, calculate the stiffness matrix and load vector for each of the following three elements. Assume that $\alpha_x = \alpha_y = \alpha$ and that α, β, and f are constants.

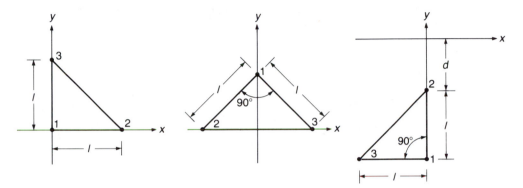

All three elements have identical geometry (i.e., identical size and shape) and identical physical properties. They differ only in their location and orientation in the x,y-plane. In other words, they differ by only a rigid-body translation and rotation. What is the most conspicuous feature of the three sets of stiffness matrices and load vectors?

Repeat the derivations for the following two anisotropic elements whose principal axes are indicated by the parallel lines. These two elements also have identical geometry and physical properties; this is because the physical properties have maintained the same orientation with respect to the element geometry as the element is rotated.

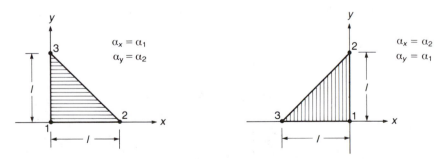

What general conclusion can be drawn about the element equations for elements whose geometry and physical properties differ only by a rigid-body translation and/or rotation? This is a very useful feature that can frequently save considerable computation when generating the system equations.

Caveat: Although this feature applies to all elements with one DOF per node, it generally requires additional considerations for elements with multiple DOF per node. For example, when two or more DOF at a node represent the components of a vector function (e.g., displacements in elasticity, as in Chapter 16), then the directions of the DOF must also rotate as a rigid body with the element to preserve the same element equations.

13.16 Consider the torsion of a beam with a square cross section and constant, isotropic material properties. Recalling Section 13.7, especially Section 13.7.2, we see that the problem has four lines of symmetry — the vertical and horizontal bisectors of the square and both diagonals — permitting us to analyze only one eighth of the domain, i.e., a triangular region:

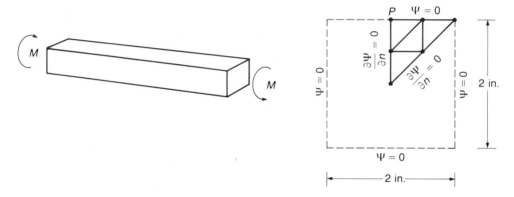

Using a mesh of four C^0-linear triangular elements as shown, and the relations in Section 13.3.1, calculate the maximum shear stress, which occurs at the point *P* at the midpoint of the sides of the square. (The exact solution [13.55] is $|\tau|_{max} = 1.350\mu\theta$.)

Hints: (1) Can the stiffness matrix and load vector for one of the elements be used for any of the other elements? See Exercise 13.15. (2) What effect will the zero essential BCs have on the assembled system equations?

13.17 For the C^0-linear rectangular element shown in the figure,

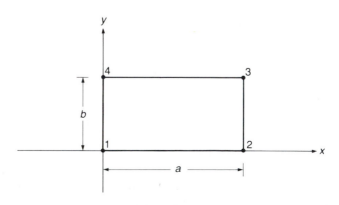

derive the following expressions for the stiffness matrix and load vector:

$$[K]^{(e)} = \frac{\alpha}{3}\frac{b}{a}\begin{bmatrix} 1 & -1 & -\frac{1}{2} & \frac{1}{2} \\ & 1 & \frac{1}{2} & -\frac{1}{2} \\ & & 1 & -1 \\ & & & 1 \end{bmatrix} + \frac{\alpha}{3}\frac{a}{b}\begin{bmatrix} 1 & \frac{1}{2} & -\frac{1}{2} & -1 \\ & 1 & -1 & -\frac{1}{2} \\ & & 1 & \frac{1}{2} \\ & & & 1 \end{bmatrix}$$

$$\{F\}^{(e)} = \frac{fab}{4}\begin{Bmatrix} 1 \\ 1 \\ 1 \\ 1 \end{Bmatrix}$$

(E13.6)

where $\alpha_x = \alpha_y = \alpha$ is a constant, $\beta = 0$, and f is a constant. Either the direct or isoparametric approach may be used; if the latter, note from Exercise 13.8 that the Jacobian may be written down merely by inspection.

13.18 Analyze the torsion of a square-section beam, as described in Exercise 13.16, but this time employ a mesh of four C^0-linear quadrilateral elements covering one quadrant of the square section:

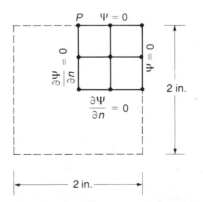

Use the element stiffness matrix and load vector from Exercise 13.17, and note the hints in Exercise 13.16. This mesh takes advantage of the symmetry with respect to the vertical and horizontal bisectors of the square, but ignores the symmetry with respect to the diagonals. However, the latter can be taken into account by an appropriate *linear constraint equation*. What is the equation? Apply the constraint equation to the system equations (recall Section 5.3.3); this will reduce the number of system equations and hence simplify their solution. Compare your solution for $|\tau|_{max}$ with the approximate and exact solutions in Exercise 13.16.

13.19   Analyze the torsion of a square-section beam, as described in Exercise 13.16, but this time employ a mesh of one C^0-quadratic quadrilateral element covering one quadrant of the square section:

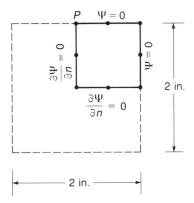

Use the relations in Section 13.4.2 to calculate stiffness and load terms. Although the shape functions are a bit complicated, the amount of work will be relatively small *if* you heed the suggestions in Exercises 13.16 and 13.18 regarding the zero essential BCs and the diagonal symmetry. Be alert! Calculate the nodal flux components at point P using the optimal flux values and the transformation (smoothing) matrix. Compare your solution for $|\tau|_{max}$ with the approximate and exact solutions in Exercises 13.16 and 13.18.

13.20   The parent element for a C^0-quadratic isoparametric triangle is mapped onto the two different elements shown below. Nodes 2, 5, and 3 lie on a $90°$ circular arc of radius r in element (a) and on a $180°$ circular arc in element (b). Nodes 4, 5, and 6 are at the midpoints of their sides.

a)   Show that the isoparametric mappings yield the following equations for the curved boundaries:

$$90° \text{ element: } \frac{\sqrt{2}-1}{\sqrt{2}}\left(\frac{x}{r}-\frac{y}{r}\right)^2 + \frac{1}{\sqrt{2}}\left(\frac{x}{r}+\frac{y}{r}\right) - 1 = 0$$

$$180° \text{ element: } y = r - \frac{x^2}{r} \tag{E13.7}$$

These are parabolas, not circular arcs, as indicated in the computer-generated figure.

b)   Show that the Jacobians for the mappings are as follows:

$$90° \text{ element: } |J(\xi,\eta)| = r^2[2(\sqrt{2}-1)(\xi+\eta)+1]$$

$$180° \text{ element: } |J(\xi,\eta)| = 4r^2(\xi-\eta) \tag{E13.8}$$

c)   For each mapping show graphically on the parent element how the value of $|J(\xi,\eta)|$ varies over the element. Note particularly any regions where

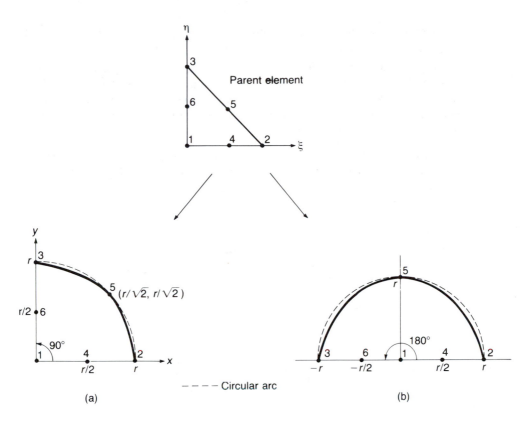

--- Circular arc

(a)   (b)

$|J(\xi,\eta)| \leq 0$. Also note the ratio of the maximum and minimum magnitudes of $|J(\xi,\eta)|$.

d) Is either mapping acceptable? If so, does the distortion seem severe or modest? If not acceptable, why not?

13.21 The stiffness matrix and load vector for the 90° "circular" arc triangular element in Exercise 13.20 are as follows (to five-decimal-place accuracy):

$$
[K\alpha]^{(e)} = \alpha
\begin{bmatrix}
0.75430 & 0.14423 & 0.14423 & -0.46646 & -0.10984 & -0.46646 \\
 & 0.74903 & 0.08374 & -0.61385 & -0.19427 & -0.16888 \\
 & & 0.74903 & -0.16888 & -0.19427 & -0.61385 \\
 & & & 2.01436 & -0.55625 & -0.20891 \\
 & & & & 1.61088 & -0.55625 \\
 & \text{Symmetric} & & & & 2.01436
\end{bmatrix}
$$

$$
\{Ff\}^{(e)} = fr^2
\begin{Bmatrix}
-0.01381 \\
0.00690 \\
0.00690 \\
0.24951 \\
0.27712 \\
0.24951
\end{Bmatrix}
$$

(E13.9)

where $\alpha_x=\alpha_y=\alpha$ and α and f are assumed constant. The integrals were evaluated with the four-point Gauss quadrature rule in Fig. 13.31.

Using the isoparametric relations in Section 13.3.2, calculate any one of the stiffness and load terms, using the four-point quadrature rule. Use the above values to verify your answer. (This exercise reviews all the steps in generating stiffness and load terms for distorted isoparametric elements. One term is sufficient for this purpose; any more would be tedious.)

13.22   Consider the torsion of a circular rod with constant, isotropic material properties. (See (a) below.) Radial symmetry makes it a 1-D problem. However, let's use the 90° "circular" arc element treated in the previous two exercises and model one quadrant with a single element. (See (b) below.)

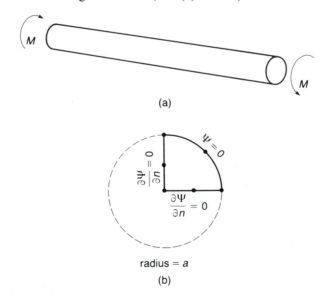

(a)

(b)

radius $= a$

Calculate the maximum shear stress, which occurs on the circumference. What is the linear constraint equation that expresses the radial symmetry? Apply this constraint equation to the system equations to reduce their size. Use Eq. (13.203) and the relations in Section 13.3.2, step 6, to evaluate the flux at one of the nodes on the circumference.

The exact solution is:

$$\Psi = -\frac{1}{2}(x^2+y^2-a^2)$$

$$\tau_{zx} = -\mu\theta y$$

$$\tau_{zy} = \mu\theta x$$

$$|\tau|_{max} = \mu\theta a \qquad\qquad \text{(E13.10)}$$

Note that the exact solution is a 2-D quadratic polynomial, and we have used a C^0-quadratic element. Yet the FE solution is not equal to the exact solution. Identify two sources of error.

13.23 A thin rectangular plate is insulated on both surfaces so that heat can only flow parallel to the surfaces. (See (a) below.) On three of the edges the temperature is 0 °C. On the fourth edge the temperature varies according to a piecewise-linear profile. (See (b) below.) Calculate the temperature at the center of the plate.

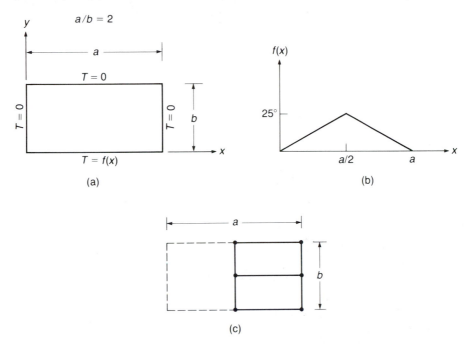

(a)

(b)

(c)

Exploit symmetry by modeling only half the plate, and employ a mesh of two C^0-linear quadrilateral elements. (See (c) above.) Use the stiffness matrix from Exercise 13.17. Can the same matrix be used for both elements? (Recall Exercise 13.15.)

Note that the loading (temperature profile) is modeled exactly by the linear elements. Since the geometry (rectangular domain) and physical properties are also modeled exactly, there are no modeling errors. There are also no quadrature errors since the stiffness matrix (Exercise 13.17) is integrated exactly. Hence the only source of error is the discretization error. Do you think this error will make the temperatures computed by this FE model too high or too low? *Hint*: The FE solution is constrained to yield the exact temperature at $(a/2,0)$, but throughout the rest of the domain the bilinear shape functions cannot conform to the curvature of the exact solution; i.e., they are "too stiff." In your mind picture these shape functions trying to "bend" to fit the exact solution.

The exact solution (using separation of variables — see also Carslaw and Jaeger [13.2], p. 166) is

$$T(x,y) = 50 \sum_{n=1}^{\infty} \left[\frac{2}{(2n-1)\pi} \right]^2 \frac{\sin[(2n-1)\pi x/a] \, \sinh\,[(2n-1)\pi(b-y)/a]}{\sinh[(2n-1)\pi b/a]}$$

(E13.11)

At $x=a/2$, $y=b/2$ the series converges very rapidly; using only the first two terms yields $T(a/2,b/2)= 7.45\,°C$. Was your intuition correct in predicting a high or low temperature?

13.24 Consider the problem of heat conduction in the rectangular plate described in Exercise 13.23. Replace the piecewise-linear temperature profile with the following parabolic profile:

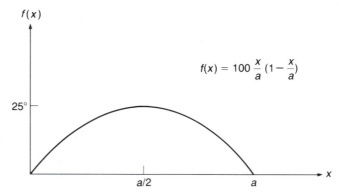

$$f(x) = 100\,\frac{x}{a}\,(1 - \frac{x}{a})$$

As before, calculate the temperature at the center of the plate, $T(a/2,b/2)$, using the same two-element mesh. *Hint:* If you already did Exercise 13.23, how much of that analysis can be used, without alteration, in this problem?

Note that in this problem there is a modeling error. Where? Do you think this error will make the calculated temperature too high or too low? *Hint:* Consider whether the temperature profile being input to the calculation represents a positive or negative error in the total heat input.

The exact solution is the same as Eq. (E13.11) except that the factor $50[2/(2n-1)\pi]^2$ is replaced by $100[2/(2n-1)\pi]^3$. This yields $T(a/2,b/2)=9.65\,°C$.

13.25 Consider again the heat conduction problem treated in the previous two exercises. Use the parabolic temperature profile described in Exercise 13.24 but employ a mesh of one C^0-quadratic quadrilateral element:

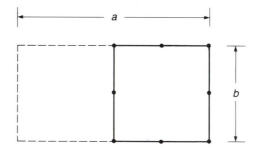

As before, calculate the temperature at the center of the plate, $T(a/2,b/2)$. *Hint:* Note the similarities with the torsion problem in Exercise 13.19. How much of that analysis can be used, without alteration, in this problem?

Is there a modeling error in this problem? The exact solution is given in Exercise 13.24. It is instructive to compare the errors in the solutions to Exercises 13.23 through 13.25.

13.26 Consider once again the problem of 2-D heat conduction in a rectangular plate. (See (a) below.) On two of the edges the temperature is 100 °C. On the other two there is heat convection to the surroundings which are at an ambient temperature of 0 °C.

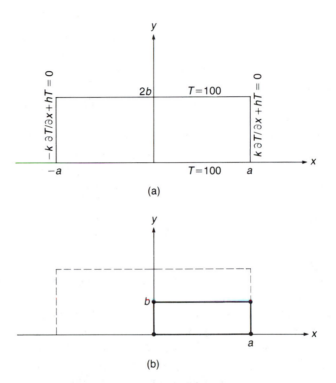

(a)

(b)

There are clearly two lines of symmetry which reduce the problem to one quadrant. What are the appropriate BCs on the quadrant? Model the quadrant with a single C^0-linear quadrilateral element (see (b) above) and use the stiffness matrix from Exercise 13.17.

Calculate the temperature at the center of the plate, $T(0,b)$. The dimensions are $a=30$ cm and $b=10$ cm; the material properties are $k=0.2$ cal/sec-cm- °C and $h=0.0001$ cal/sec-cm²- °C. (See Exercise 4.6 for treatment of convective BCs.) The exact solution (see Carslaw and Jaeger [13.2], p. 167) is

$$T(x,y) = 200 \frac{h}{k} \sum_{n=1}^{\infty} \frac{\cos \alpha_n x \cosh \alpha_n(b-y)}{\left[\left(\alpha_n^2 + (h^2/k^2) \right) + (h/k) \right] \cos \alpha_n a \cosh \alpha_n b} \tag{E13.12}$$

where α_n are the roots of $\alpha_n \tan \alpha_n a = h/k$.

References

[13.1]   M. N. Özisik, *Heat Conduction*, Wiley, New York, 1980, Chapter 15.

[13.2]   H. S. Carslaw and J. C. Jaeger, *Conduction of Heat in Solids*, Oxford University Press, New York, 1959, Sections 1.17-1.20.

[13.3]   F. B. Hildebrand, *Methods of Applied Mathematics*, Prentice-Hall, Englewood Cliffs, N. J., 1952, Sections 1.13, 1.19-1.21.

[13.4]   Ibid., Section 1.11.

[13.5]   Carslaw and Jaeger, op. cit.

[13.6]   W. Kaplan, *Advanced Calculus*, Addison-Wesley, Reading, Mass., 1984, Chapter 5.

[13.7]   G. Strang and G. J. Fix, *An Analysis of the Finite Element Method*, Prentice-Hall, Englewood Cliffs, N. J., 1973, pp. 158-159.

[13.8]   G. R. Cowper, "Gaussian Quadrature Formulas for Triangles," *Int J Num Meth Eng*, **7**: 405-408 (1973).

[13.9]   C. T. Reddy and D. J. Shippy, "Alternative Integration Formulae for Triangular Finite Elements," *Int J Num Meth Eng*, **17**: 133-139 (1981).

[13.10] T. Moan, "Experiences with Orthogonal Polynomials and 'Best' Numerical Integration Formulas on a Triangle; with Particular Reference to Finite Element Approximations," *ZAMM*, **54**: 501-508 (1974).

[13.11] E. Hinton and J. S. Campbell, "Local and Global Smoothing of Discontinuous Finite Element Functions Using a Least Squares Method," *Int J Num Meth Eng*, **8**: 461-480 (1974).

[13.12] E. Hinton, F. C. Scott, and R. E. Ricketts, "Local Least Squares Stress Smoothing for Parabolic Isoparametric Elements," *Int J Num Meth Eng*, **9**: 235-238 (1975).

[13.13] P. C. Kohnke, *ANSYS, Engineering Analysis System Theoretical Manual*, Swanson Analysis Systems, Inc., Houston, Pa., 1983, Section 2.0.1.

[13.14] L. L. Durocher, A. Gasper, and G. Rhoades, "A Numerical Comparison of Axisymmetric Finite Elements," *Int J Num Meth Eng*, **12**: 1415-1427 (1978).

[13.15] D. J. Naylor, "Stresses in Nearly Incompressible Materials by Finite Elements with Application to the Calculation of Excess Pore Pressures," *Int J Num Meth Eng*, **8**: 443-460 (1974).

[13.16] J. Ergatoudis, B. M. Irons, and O. C. Zienkiewicz, "Curved, Isoparametric, 'Quadrilateral' Elements for Finite Element Analysis," *Int J Solids Struct*, **4**: 31-42 (1968).

[13.17] O. C. Zienkiewicz, B. M. Irons, J. G. Ergatoudis, S. Ahmad, and F. C. Scott, "Iso-Parametric and Associated Element Families for Two- and Three-Dimensional Analysis" in I. Holland and K. Bell (eds.), *Finite Element Methods in Stress Analysis*, Tapir forlag, Trondheim, Norway, 1972, Chapter 13.

[13.18] R. L. Taylor, "On Completeness of Shape Functions for Finite Element Analysis," *Int J Num Meth Eng*, **4**: 17-22 (1972).

[13.19] A. A. Ball, "The Interpolation Function of a General Serendipity Rectangular Element," *Int J Num Meth Eng*, **15**(5): 773-778 (1980).

[13.20] E. L. Wilson, R. L. Taylor, W. P. Doherty, and J. Ghaboussi, "Incompatible Displacement Models" in S. J. Fenves, N. Perrone, A. R. Robinson, and

W. C. Schnobrich (eds.), *Numerical and Computer Methods in Structural Mechanics*, Academic Press, New York, 1973, pp. 43-57.

[13.21] B. M. Irons, "Quadrature Rules for Brick Based Finite Elements," *Int J Num Meth Eng*, **3**: 293-294 (1971).

[13.22] T. K. Hellen, "Effective Quadrature Rules for Quadratic Solid Isoparametric Finite Elements," *Int J Num Meth Eng*, **4**: 597-599 (1972).

[13.23] W. G. Gray and M. T. van Genuchten, "Economical Alternatives to Gaussian Quadrature over Isoparametric Quadrilaterals," *Int J Num Meth Eng*, **12**(9): 1478-1484 (1978).

[13.24] A. H. Stroud and D. Secrest, *Gaussian Quadrature Formulas*, Prentice-Hall, Englewood Cliffs, N. J., 1966.

[13.25] A. H. Stroud, *Approximate Calculation of Multiple Integrals*, Prentice-Hall, Englewood Cliffs, N. J., 1971.

[13.26] O. C. Zienkiewicz, *The Finite Element Method*, McGraw-Hill, New York, 1977, p. 158.

[13.27] R. D. Cook and F. Zhao-hua, "Control of Spurious Modes in the Nine-Node Quadrilateral Element," *Int J Num Meth Eng*, **18**(10): 1576-1580 (1982).

[13.28] R. D. Henshell, "Differences Between Isoparametric Assumptions and True Circles," *Int J Num Meth Eng*, **10**(5): 1193-1196 (1976).

[13.29] I. Babuska, "Stability of Domains of Definition with Respect to the Fundamental Problems in the Theory of Partial Differential Equations, Primarily in Relation to the Theory of Elasticity" (in Russian), *Czech Math J*, **11**(86): 1961. Part I on pp. 76-105 and Part II on pp. 162-203.

[13.30] I. Babuska, "The Rate of Convergence for the Finite Element Method," *SIAM J Num Analysis*, **8**(2): 304-315 (1971).

[13.31] G. Strang and G. J. Fix, op. cit., pp. 192-204.

[13.32] M. Zlamal, "Curved Elements in the Finite Element Method," *SIAM J Num Anal*, Part I in **10**(1): 229-240 (1973); Part II in **11**(2): 347-362 (1974).

[13.33] A. K. Rao and K. Rajaiah, "Polygon-Circle Paradox of Simply Supported Thin Plates under Uniform Pressure," *AIAA J*, **6**(1): 155-156 (1968).

[13.34] M. Zlamal, "The Finite Element Method in Domains with Curved Boundaries," *Int J Num Meth Eng*, **5**: 367-373 (1973).

[13.35] T. Krauthammer, "Accuracy of the Finite Element Method Near a Curved Boundary," *Int J Comp Struct*, **10**: 921-929 (1979).

[13.36] F. S. Kelley, Swanson Analysis Systems, Inc., Houston, Pa., private communication.

[13.37] C. R. Rogers, "Substructuring as an Everyday Design Tool," ASME, Paper No. 77-PVP-27, September 1977.

[13.38] M. S. Shephard and M. A. Yerry, "Approaching the Automatic Generation of Finite Element Meshes," *Computers in Mech Eng*, April 1983, pp. 49-55.

[13.39] E. Cuthill and J. M. McKee, "Reducing the Bandwidth of Sparse Symmetric Matrices," *Proc 24th Nat Conf Assn Comp Machinery*, ACM Pub. P69, 1969, pp. 157-172.

[13.40] H. R. Grooms, "Algorithm for Matrix Bandwidth Reduction," *J Struct Div*, ASCE, **98**: 203-214 (1972).

[13.41] N. E. Gibbs, W. G. Poole, Jr., and P. K. Stockmeyer, "A Comparison of Several Bandwidth and Profile Reduction Algorithms," *ACM Trans Math Software*, **2**: 322-330 (1976).

[13.42] G. Akhras and G. Dhatt, "An Automatic Node Relabelling Scheme for Minimizing a Matrix or Network Bandwidth," *Int J Num Meth Eng*, **10**: 787-797 (1976).

[13.43] G. Everstine, "A Comparison of Three Resequencing Algorithms for the Reduction of Matrix Profile and Wavefront," *Int J Num Meth Eng*, **14**: 837-853 (1979). (Thirty test meshes are shown; an extensive bibliography is included.)

[13.44] A. Razzaque, "Automatic Reduction of Frontwidth for Finite Element Analysis," *Int J Num Meth Eng*, **14**: 1315-1324 (1980).

[13.45] G. Beer and W. Haas, "A Partitioned Frontal Solver for Finite Element Analysis," *Int J Num Meth Eng*, **18**: 1623-1654 (1982).

[13.46] J. Puttonen, "Simple and Effective Bandwidth Reduction Algorithm," *Int J Num Meth Eng*, **19**: 1139-1152 (1983).

[13.47] S. W. Sloan and M. F. Randolph, "Automatic Element Reordering for Finite Element Analysis with Frontal Solution Schemes," *Int J Num Meth Eng*, **19**(8): 1153-1181 (1983).

[13.48] I. S. Sokolnikoff, *Mathematical Theory of Elasticity*, McGraw-Hill, New York, 1956, Chapter 4.

[13.49] S. Timoshenko and J. N. Goodier, *Theory of Elasticity*, 2d ed., McGraw-Hill, New York, 1951, Chapter 11.

[13.50] C. Wang, *Applied Elasticity*, McGraw-Hill, New York, 1953, Chapter 5.

[13.51] J. Robinson, "Element Evaluation — A Set of Assessment Points and Standard Tests" in J. Robinson (ed.), *Finite Element Methods in the Commercial Environment* (Proc Second World Congress and Exhibition on Finite Element Methods, Bournemouth, Dorset, England), Robinson and Associates, Devon, England, 1978.

[13.52] H. H. Fong, "Standards for Finite Element Codes: The Long Road Ahead," *Computers in Mech Eng*, November 1984, pp. 10-14.

[13.53] R. H. MacNeal and R. L. Harder, "A Proposed Standard Set of Problems to Test Finite Element Accuracy," presented at Finite Element Validation Forum, AIAA/ASME/ASCE/AHS 25th Structures, Structural Dynamics, and Materials Conference, May 14, 1984, Palm Springs, Calif.

[13.54] K. J. Forsberg and H. H. Fong, cochairmen, Finite Element Standards Forum, AIAA/ASME/ASCE/AHS 26th Structures, Structural Dynamics, and Materials Conference, April 15, 1985, Orlando, Fla. Briefing Books 1 and 2 produced by PDA Engineering, Santa Ana, Calif.

[13.55] C. Wang, op. cit., p. 90.

2-D Eigenproblems

Introduction

This chapter will analyze the eigenproblem associated with the boundary-value problem treated in Chapter 13. The development closely parallels that of the 1-D eigenproblem in Chapter 10. Thus the FE discretization will transform the governing partial differential equation into an algebraic eigenproblem, and to solve the latter we can use the GENJAC eigensolver and all the techniques developed in Chapter 10. In fact, the only task left for this chapter is to develop the mass matrix, since all other aspects of the problem have been dealt with in Chapters 10 and 13. Practical modeling considerations will be demonstrated by using UNAFEM to solve an acoustics problem.

14.1 Mathematical and Physical Description of the Problem

The eigenproblem associated with the boundary-value problem in Section 13.1 is governed by the following second-order partial differential equation:

$$-\frac{\partial}{\partial x}\left(\alpha_x(x,y)\frac{\partial U(x,y)}{\partial x}\right) - \frac{\partial}{\partial y}\left(\alpha_y(x,y)\frac{\partial U(x,y)}{\partial y}\right) + \beta(x,y)U(x,y)$$

$$- \lambda\gamma(x,y)U(x,y) = 0 \qquad (14.1a)$$

The domain is typically a finite, closed region in the x,y-plane (see Fig. 13.1), although infinite regions may also be treated, using the techniques in Section 10.6. As is characteristic of eigenproblems, there cannot be any loads. Thus the interior load, $f(x,y)$, is zero, and the BCs must all be zero;[1] i.e.,

$$U = 0 \qquad \text{or} \qquad \tau_n = 0 \qquad \text{at every boundary point} \qquad (14.1b)$$

---

[1] We are including in UNAFEM both essential and natural BCs, but not mixed BCs, for the reasons given in Section 13.1.1.

The first three terms in Eq. (14.1a) are the same as in Eq. (13.1) and are fully described in Section 13.1. The fourth term is the addition that changes the boundary-value problem to an eigenproblem; it is the 2-D analogue of the 1-D term, $\lambda\gamma(x)U(x)$, described in Section 10.1. Thus λ is an unknown eigenvalue, usually representing a resonant frequency or energy level, and $\gamma(x,y)$ is a known material or physical property, usually representing a mass density or other inertia-like quantity.

Some typical physical applications of Eq. (14.1a) are listed in Table 2.5. Of course these are the natural 2-D extensions of related 1-D problems (see Tables 2.2 and 10.1). For example, the vibrating membrane is the 2-D extension of the 1-D vibrating cable [see Eqs. (2.18) through (2.20), Eqs. (10.3) through (10.11) and Section 10.5).

When $\beta=0$, Eq. (14.1a) is the classic *Helmholtz* equation. In this case U is the amplitude of the time-harmonic solution to the wave equation, the latter arising in several fields of application, e.g., acoustics, electromagnetism, and hydrodynamics. The acoustics problem in Section 14.4 will analyze the Helmholtz equation.

14.2 Derivation of Element Equations

The first three terms in Eq. (14.1a) were treated in Sections 13.2 through 13.4. Here we need only consider the additional term, $-\lambda\gamma(x,y)U(x,y)$. However, this has the same form as the term $\beta(x,y)U(x,y)$, so we need only copy the β terms from Chapter 13, exchanging $\beta(x,y)$ for $-\lambda\gamma(x,y)$. The situation is directly parallel to the 1-D situation in Chapter 10. By now, though, the reader should be quite familiar with steps 1 through 6 of the theoretical development, so we will only summarize the results here.

At the end of step 3, the formal expression for the element equations is as follows:

$$\left([K]^{(e)} - \lambda[M]^{(e)} \right)\{a\} = \{0\} \tag{14.2}$$

where

$$K_{ij}^{(e)} = \int\int^{(e)} \frac{\partial\phi_i^{(e)}}{\partial x}\alpha_x\frac{\partial\phi_j^{(e)}}{\partial x}\,dx\,dy + \int\int^{(e)} \frac{\partial\phi_i^{(e)}}{\partial y}\alpha_y\frac{\partial\phi_j^{(e)}}{\partial y}\,dx\,dy + \int\int^{(e)} \phi_i^{(e)}\beta\phi_j^{(e)}\,dx\,dy$$

$$M_{ij}^{(e)} = \int\int^{(e)} \phi_i^{(e)}\gamma\phi_j^{(e)}\,dx\,dy$$

Of course, the expression for $K_{ij}^{(e)}$ remains the same as in Eq. (13.61b) for the boundary-value problem, and the expression for $M_{ij}^{(e)}$ has the same form as its 1-D counterpart in Eqs. (10.19).

Step 4, which develops the element shape functions, needs no further work. Here we can use any of the four 2-D elements from Chapter 13.

In step 5, the evaluation of the γ-integrals (mass matrix) and β-integrals is identical:

C^0-*linear triangle* [see Eq. (13.78)] •

$$M_{ij}^{(e)} = \begin{cases} \dfrac{\gamma^{(e)}\Delta}{6} & i = j \\[3mm] \dfrac{\gamma^{(e)}\Delta}{12} & i \neq j \end{cases} \tag{14.3}$$

```
      SUBROUTINE ELEM21(OPT)
C
C        THIS SUBROUTINE IS THE MASTER ROUTINE FOR ELEMENT 21, A
C        3-NODE 2-D CO-LINEAR TRIANGLE.
                     .
                     .
             IF (I.EQ.J) FACTOR=6.
             IF (I.NE.J) FACTOR=12.
                     .
                     .
             IF (PROB.EQ.EIGP) ME(I,J) = GAMMA*AREA/FACTOR ◄────Eq. (14.3)
                     .
                     .
                     .
```

```
      SUBROUTINE ELEM22(OPT)
C
C        THIS SUBROUTINE IS THE MASTER ROUTINE FOR ELEMENT 22, A
C        6-NODE 2-D CO-QUADRATIC TRIANGLE.
                     .
                     .
         WGT = WGHT(N1,INT)/2.
                     .
                     .
                IF(PROB.EQ.EIGP) ME(I,J) = ME(I,J)+WGT*JAC* ◄────Eq. (14.4)
      $                          PHI(I)*GAMMA*PHI(J)
                     .
                     .
```

```
      SUBROUTINE ELEM23(OPT)
C
C        THIS SUBROUTINE IS THE MASTER ROUTINE FOR ELEMENT 23, A
C        4-NODE 2-D CO-LINEAR QUADRILATERAL.
                     .
                     .
         WGT = WT(NINT,INTX)*WT(NINT,INTY)
                     .
                     .
                IF(PROB.EQ.EIGP) ME(I,J) = ME(I,J)+WGT*JAC* ◄────Eq. (14.5)
      $                          PHI(I)*GAMMA*PHI(J)
                     .
                     .
                     .
```

```
      SUBROUTINE ELEM24(OPT)
C
C        THIS SUBROUTINE IS THE MASTER ROUTINE FOR ELEMENT 24, AN
C        8-NODE 2-D CO-QUADRATIC QUADRILATERAL.
                     .
                     .
         WGT = WT(NINT,INTX)*WT(NINT,INTY)
                     .
                     .
                IF(PROB.EQ.EIGP) ME(I,J) = ME(I,J)+WGT*JAC* ◄────Eq. (14.5)
      $                          PHI(I)*GAMMA*PHI(J)
                     .
                     .
                     .
```

FIGURE 14.1 ● Addition of mass matrix calculations to the 2-D element subroutines.

where Δ is the area of the triangle, and $\gamma^{(e)} = \gamma(x_c, y_c)$ is the value at the centroid [see Eqs. (13.70) and (13.71)].[2]

C^0-quadratic isoparametric triangle [see Eq. (13.128)] •

$$M_{ij}^{(e)} = \frac{1}{2} \sum_{l=1}^{n} w_{nl} \left[\phi_i^{(e)} \gamma \phi_j^{(e)} |J^{(e)}| \right]_{(\xi_{nl}, \eta_{nl})}$$

(14.4)

C^0-linear and C^0-quadratic isoparametric quadrilaterals [see Eqs. (13.169)] •

$$M_{ij}^{(e)} = \sum_{k=1}^{n} \sum_{l=1}^{n} w_{nk} w_{nl} \left[\phi_i^{(e)} \gamma \phi_j^{(e)} |J^{(e)}| \right]_{(\xi_{nl}, \eta_{nl})}$$

(14.5)

Step 6 derives expressions for the optimal and nodal fluxes. These are unaffected by the mass matrix, so the results given in Chapter 13 are applicable here.

14.3 Implementation into the UNAFEM Program

The flowchart remains the same as in Fig. 13.51; i.e., there are no new subroutines. The principal changes are the addition of the mass matrix calculations, Eqs. (14.3) through (14.5), to the four 2-D element subroutines, as shown in Fig. 14.1. The function $\gamma(x,y)$ may be specified in UNAFEM as a piecewise-quadratic (2-D) polynomial, in the same manner as are $\alpha_x(x,y)$, $\alpha_y(x,y)$, and $\beta(x,y)$ in Eqs. (13.69).

The eigensolver, GENJAC, needs no change since it solves an algebraic eigenproblem. The algorithm does not depend on the origin of the $[K]$ and $[M]$ matrices; they might originate from a 1-D, 2-D, or n-D problem, or perhaps not even from an FE problem at all. The postprocessors EIGPRI and EIGPLO need changes similar to BIVPRI and BIVPLO for the boundary-value problem.

14.4 Application of UNAFEM to an Acoustics Problem

14.4.1 Description of the Problem

We will examine the propagation of sound through a vehicular/pedestrian tunnel, e.g., a tunnel cut through a hill or mountain for the passage of automobiles, buses, trucks, and pedestrians. Figure 14.2 shows a typical cross section of such a tunnel, with the characteristic arched roof and a raised pedestrian walkway along one side. In order to better understand the noise levels created by the traffic, we will determine the resonant frequencies and mode shapes associated with this cross-sectional geometry.

---

[2] Note that $\sum_{i=1}^{3} \sum_{j=1}^{3} M_{ij}^{(e)} = \gamma^{(e)} \Delta$, the total mass of the triangle. This conservation-of-mass property holds true for the isoparametric elements as well, as can be verified from the above quadrature expressions; it follows from the interpolation property, $\sum_{i} \phi_i^{(e)}(x,y) = 1$.

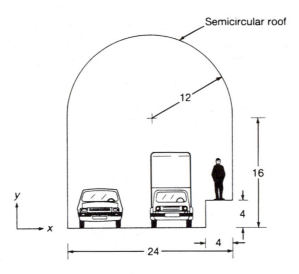

FIGURE 14.2 ● Cross section of a vehicular/pedestrian tunnel (dimensions in feet).

Propagation of sound in a 3-D region is governed by the *wave equation*,

$$\nabla^2 \Phi - \frac{1}{c^2} \frac{\partial^2 \Phi}{\partial t^2} = 0 \tag{14.6}$$

where $\nabla^2 (= \partial^2/\partial x^2 + \partial^2/\partial y^2 + \partial^2/\partial z^2)$ is the Laplacian operator (see Section 13.1.1), $\Phi(x,y,z,t)$ is the velocity potential function, and c is the speed of sound. Additional relations are

$$v = -\nabla \Phi$$

$$p = \varrho \frac{\partial \Phi}{\partial t} \tag{14.7}$$

where v is the particle velocity of the air, $\nabla (= \partial/\partial x i + \partial/\partial y j + \partial/\partial z k)$ is the gradient operator, p is the acoustic pressure (i.e., the change in the ambient pressure due to the sound waves), and ϱ is the ambient air density.

For free vibrations, the time variation is sinusoidal, $e^{\pm i\omega t}$, where $e^{\pm i\omega t} = \cos \omega t \pm i \sin \omega t$, $\omega = 2\pi f$ is the circular frequency, and $i = \sqrt{-1}$. Thus

$$\Phi(x,y,z,t) = \Psi(x,y,z)e^{\pm i\omega t} \tag{14.8}$$

Substituting Eq. (14.8) into Eq. (14.6) reduces the wave equation to the Helmholtz equation:

$$\nabla^2 \Psi + k^2 \Psi = 0 \tag{14.9}$$

where

$$k = \frac{\omega}{c} = \frac{2\pi}{\lambda} \qquad (f\lambda = c)$$

The quantity k is called the *wavenumber*; it is inversely proportional to the wavelength, λ.

The tunnel is essentially a long cylinder, parallel to the z-axis, so we would expect the x,y-variation of $\Psi(x,y,z)$ to be the same at every cross section (except near the ends of the tunnel). Employing the classical separation of variables technique [14.1] we define

$$\Psi(x,y,z) = U(x,y)W(z) \tag{14.10}$$

Substituting Eq. (14.10) into Eq. (14.9) and carrying out the standard separation of variables operations yields the following relations for $U(x,y)$ and $W(z)$:

$$\nabla^2_{x,y}U(x,y) + \varkappa^2 U(x,y) = 0 \tag{14.11}$$

$$W(z) = e^{ik_z z} \tag{14.12}$$

where

$$\nabla^2_{x,y} = \frac{\partial^2}{\partial x^2} + \frac{\partial^2}{\partial y^2} \tag{14.13}$$

and

$$\varkappa^2 = \frac{\omega^2}{c^2} - k_z^2 \tag{14.14}$$

The quantity $\nabla^2_{x,y}$ is the 2-D Laplacian operator, $k_z = 2\pi/\lambda_z$ is the z-component of the wave number, and λ_z is the wavelength along the tunnel axis. Equation (14.11) is Eq. (14.1a) with $\alpha_x(x,y) = \alpha_y(x,y) = 1$; $\beta(x,y) = 0$; $\gamma(x,y) = 1$; and $\lambda = \varkappa^2$.

Substituting Eqs. (14.12) and (14.10) into Eq. (14.8) yields for the velocity potential function,

$$\Phi(x,y,z,t) = U(x,y)e^{i(k_z z \pm \omega t)} \tag{14.15}$$

Equation (14.15) describes the propagation of a sound wave of amplitude $U(x,y)$ along the z-axis of the tunnel (in the negative or positive z-direction), with a wavelength λ_z along the tunnel and a phase velocity ω/k_z. It remains only to determine the amplitude, $U(x,y)$, which is the solution to Eq. (14.11).

What are the appropriate boundary conditions? In reality, some sound energy will penetrate the tunnel walls. However, we shall treat all surfaces as rigid.[3] Therefore the normal component of particle velocity v_n must vanish on the boundary:

$$v_n = \boldsymbol{n}\cdot\boldsymbol{v} = 0 \qquad \text{on the boundary} \tag{14.16}$$

where $\boldsymbol{n} = n_x\boldsymbol{i} + n_y\boldsymbol{j}$. From Eqs. (14.7) and (14.15),

$$\boldsymbol{n}\cdot\boldsymbol{v} = -\boldsymbol{n}\cdot\nabla\Phi = -\boldsymbol{n}\cdot\nabla_{x,y}\Phi = -(\boldsymbol{n}\cdot\nabla_{x,y}U)e^{i(k_z z \pm \omega t)} \tag{14.17}$$

where $\nabla_{x,y}(=\partial/\partial x\boldsymbol{i} + \partial/\partial y\boldsymbol{j})$ is the 2-D gradient operator. Therefore Eq. (14.16) becomes

$$-\boldsymbol{n}\cdot\nabla_{x,y}U = 0 \qquad \text{on the boundary} \tag{14.18}$$

In Eq. (14.11) the flux is $\tau = -\nabla_{x,y}U$, so Eq. (14.16) becomes finally,

$$\tau_n = \boldsymbol{n}\cdot\boldsymbol{\tau} = 0 \qquad \text{on the boundary} \tag{14.19}$$

---

[3] An alternative treatment might consider the acoustic impedance of the walls, which would be applied with mixed boundary conditions (a linear relation between pressure and normal velocity). More rigorously, one could model the coupled fluid/solid interaction between the air and the solid materials of the wall and earthen exterior.

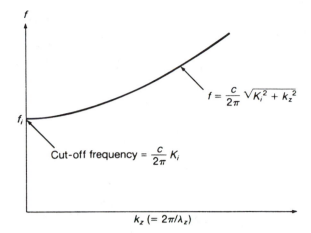

FIGURE 14.3 ● A typical dispersion curve for the ith mode.

Thus the boundary conditions are natural (not essential), vanishing at all points on the boundary.

In summary, we seek solutions to the following 2-D eigenproblem:

$$\nabla^2_{x,y} U + \varkappa^2 U = 0$$

Domain: Fig. 14.2

BCs: $\tau_n = -\boldsymbol{n}\cdot\nabla_{x,y} U = 0$ $\qquad\qquad$ **(14.20)**

There are an infinity of solutions, consisting of the modes $U_i(x,y)$ and corresponding eigenvalues $\varkappa_i^2$, $i=1,2,\ldots$.

Note that Eq. (14.14) yields, for each eigenvalue, a relation between ω and k_z or, equivalently, between f and k_z:

$$f = \frac{c}{2\pi}\sqrt{\varkappa_i^2 + k_z^2} \qquad i = 1,2,\ldots$$
$$\qquad\qquad\qquad\qquad\qquad\text{(14.21)}$$

This is known as a *dispersion* relation.[4] Figure 14.3 shows a typical dispersion curve for this type of acoustics problem. It is a hyperbola that intersects the f-axis at a frequency called the *cut-off frequency*, f_i:

$$f_i = \frac{c}{2\pi}\varkappa_i \qquad i = 1,2,\ldots$$
$$\qquad\qquad\qquad\qquad\qquad\text{(14.22)}$$

---

[4] A dispersion relation is so named because a "packet" of waves composed of the same mode shapes at different frequencies will spread apart, or disperse, as it travels along, because each component wave will travel at a different phase velocity, ω/k_z. This follows from Eq. (14.21), for which ω/k_z is, in general, a function of k_z (or, equivalently, ω). An important exception is a mode with a zero cut-off frequency, for which $\omega/k_z=c$, a constant; hence, a packet of these waves is "dispersionless."

These concepts have played an important role in quantum mechanics and, on a larger engineering scale, in any applications involving a waveguide that propagates energy, e.g., electromagnetic, acoustic, or stress waves. (See Fig. 1.13.) Of course, the field of fiber optics is an especially important and current practical application.

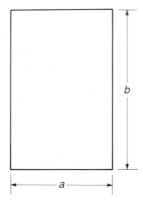

FIGURE 14.4 ● Tunnel with a rectangular cross section, used for the test problem.

This is the lowest frequency at which the ith mode can propagate freely along the tunnel. At this frequency the axial wavelength λ_z and phase velocity ω/k_z become infinite (except for dispersionless modes, as explained in the footnote).
     We will use UNAFEM to determine the first few cut-off frequencies and corresponding mode shapes for the tunnel.

14.4.2 Verification of UNAFEM

Because of the curved roof, as well as previously discussed accuracy/performance considerations, we will use the quadratic elements, ELEM22 and ELEM24. We have already performed some tests on these elements in Section 13.7.3 — namely, single-element tests as well as a rectangular-beam torsion problem (and an I-beam torsion problem for which the results appeared to be reasonable). However, these were all boundary-value problems, so they tested parts of the code common to both boundary-value and eigenproblems (e.g., the 2-D stiffness matrices), but they didn't test those parts used only by eigenproblems (e.g., the 2-D mass matrices). Therefore it would be wise to test these elements on a simple eigenproblem with a known solution.
     In the previous illustrative example (Section 13.7.3) it was suggested that an excellent type of test problem is one which is as similar as possible to the real problem, the principal difference being usually a much simpler geometry. A good test for our tunnel problem would therefore be a tunnel with a *rectangular* cross section (Fig. 14.4). For such a shape the exact solution for the cut-off frequencies and mode shapes is[5]

$$f^{(mn)} = \frac{c}{2} \sqrt{\left(\frac{m}{a}\right)^2 + \left(\frac{n}{b}\right)^2}$$

$$m,n = 0,1,2,\ldots$$

$$U^{(mn)}(x,y) = A \cos\left(\frac{\pi m x}{a}\right) \cos\left(\frac{\pi n y}{b}\right)$$

$$(14.23)$$

---

[5] Equations (14.23) are easily derived from Eq. (14.11) using the aforementioned separation of variables technique [14.1].

where A is an arbitrary amplitude. [A is determined in subroutine GENJAC by normalizing the eigenvectors with respect to the mass matrix. Recall Eqs. (10.31b).]

Let's use data for the test problem that is similar to that in the real problem, namely,
$a=24$ ft, $b=28$ ft, and $c=1127$ ft/sec (the speed of sound in air at $20\,^\circ$C and 1 atm pressure).
Substituting this data into Eqs. (14.23) yields the following values for the first 11 cut-off
frequencies:[6]

$$f_1 = f^{(00)} = 0.000 \text{ Hz}$$
$$f_2 = f^{(01)} = 20.125 \text{ Hz}$$
$$f_3 = f^{(10)} = 23.479 \text{ Hz}$$
$$f_4 = f^{(11)} = 30.924 \text{ Hz}$$
$$f_5 = f^{(02)} = 40.250 \text{ Hz}$$
$$f_6 = f^{(12)} = 46.598 \text{ Hz}$$
$$f_7 = f^{(20)} = 46.958 \text{ Hz}$$
$$f_8 = f^{(21)} = 51.089 \text{ Hz}$$
$$f_9 = f^{(03)} = 60.375 \text{ Hz}$$
$$f_{10} = f^{(22)} = 61.848 \text{ Hz}$$
$$f_{11} = f^{(13)} = 64.780 \text{ Hz} \tag{14.24}$$

Figure 14.5(a) shows a mesh that incorporates some of the guidelines in Chapter 10
for choosing an appropriate number of DOF. Thus, recalling the discussion at the end
of Section 10.5 and using the values $D=2$ (for two dimensions, $n=3$ [for the (n^D)th, or
9th, mode], and $q=3$ or 4 (for a higher-order element, although not very high order),
then about 9 to 12, or $q \times n$, active DOF in each direction should yield an accuracy in
the neighborhood of 1% for the 9th mode (and, of course, even higher accuracy for the
lower modes). The DOF density in Fig. 14.5(a) is 13-by-15 active DOF,[7] or about 4/3
as dense, so we might expect better than 1% accuracy for the 9th mode. Indeed, from
Eqs. (10.54b) or (10.55), the error is $O(h^4)$ for quadratic elements, so if the DOF density
is 4/3 greater, then $h \rightarrow 3h/4$ and the error decreases by $(3/4)^4$, or 0.3. Hence we might
expect an accuracy in the neighborhood of 0.3% for the 9th mode.

Figure 14.5 also shows the computed mode shapes and corresponding cut-off frequencies. The latter were calculated using Eq. (14.22) and the square root of the computed eigenvalues, $\varkappa_i$. Compared with Eq. (14.24), the error in f_9 is 0.2%, which agrees
well with the error predicted by the above guideline. The errors in f_{10} and f_{11} are 0.1%
and 0.2%, respectively. The lower frequencies, of course, have smaller errors.[8] The con-

---

[6] The fundamental frequency is zero, corresponding to a dispersionless plane wave. We are mainly
interested in the first 10 nonzero frequencies and their mode shapes.

[7] For the purpose of this guideline we can count an imaginary extra node in the center of each element
since the rate of convergence is the same as for a 3-by-3 node biquadratic Lagrange element.

[8] Note that the computed frequencies are all greater than the exact frequencies. This is because the variational principle corresponding to Eq. (14.1a) minimizes the frequency, so the approximate frequencies will approach the exact frequencies (which are nonnegative) from above. Recall the discussion
in Chapter 3 on the equivalence of the variational and Galerkin formulations.

FIGURE 14.5 ● UNAFEM analysis of acoustic modes for the rectangular-tunnel test problem: (a) mesh; (b) through (l) first 11 modes and corresponding cut-off frequencies. Solid contours, $U > 0$; dotted contours, $U < 0$; contour spacing = 0.01 ft^2/sec.

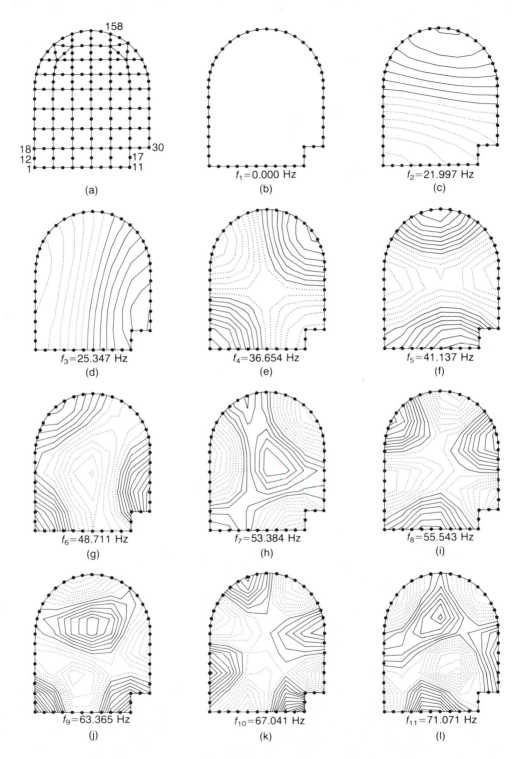

FIGURE 14.6 ● UNAFEM analysis of acoustic modes for the vehicular/pedestrian tunnel: (a) mesh; (b) through (l) first 11 modes and corresponding cut-off frequencies. Solid contours, $U > 0$; dotted contours, $U < 0$; contour spacing = 0.01 ft²/sec.

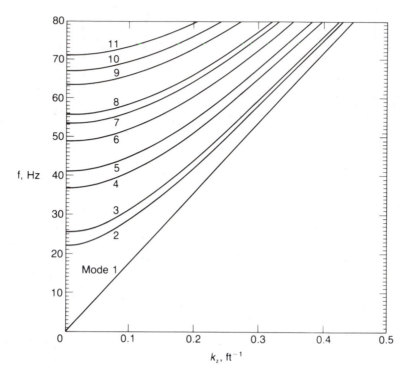

FIGURE 14.7 ● Dispersion curves corresponding to the 11 modes in Fig. 14.6.

tour plots of the modes appear to agree very well with the theoretical expression in Eqs. (14.23). Very similar results were obtained using ELEM22.

14.4.3 Analysis of the Tunnel

We can now proceed with the analysis of the real tunnel problem having quite a bit more confidence in the program. And we also have a good idea of what type of accuracy to expect if we use a similar mesh. Therefore we will construct a mesh that uses essentially the same DOF density as in the test problem.[9] This is shown in Fig. 14.6, along with the computed mode shapes and corresponding cut-off frequencies.

   Figure 14.7 plots the dispersion curves, Eq. (14.21), for each mode. The first mode is a plane wave with a zero cut-off frequency, i.e., a dispersionless mode. This is to be

---

[9] This mesh and the test problem mesh use 158 and 153 DOF, respectively. Recalling the discussion in Section 10.3.3, these sizes are approaching the practical limit for the Jacobi algorithm if it is not preceded by an eigenvalue economizer (such as Guyan reduction). This is because the algorithm computes the *complete* eigensystem, in this case all 158 (or 153) modes. The array sizes in UNAFEM had to be increased about 70,000 more words. If one were to use an eigenvalue economizer in front of GENJAC, then one would select a few master DOF (typically 1 1/2 to 2 times the number of desired frequencies), and GENJAC would only need to solve a much smaller problem involving just the master DOF. For example, 15 to 20 master DOF would suffice for calculating the first 10 modes in the above problems.

expected because the particle velocity of a plane wave is everywhere parallel to the direction of propagation (the tunnel axis), so it can satisfy the BCs on the rigid walls for any geometric shape. Therefore its phase velocity should be independent of the axial wavelength. The next several mode shapes look like distorted versions of the corresponding rectangular mode shapes. The distortions are what one might expect if one were to round off the end of a rectangle and "perturb" one corner. The results therefore seem reasonable.

There is no need to do a refined mesh since the rectangular test problem is sufficiently similar to provide a good accuracy estimate. The only important difference between the test problem and the real problem that would affect accuracy is the local complexity in the mode shapes surrounding the walkway in the corner. One could do a local refinement to improve the accuracy in that region. The accuracy of the frequencies, however, will not be significantly affected by local errors in the mode shapes since frequencies are a global property; i.e., they represent an average response over the entire domain. Hence we can expect the computed frequencies to have about the same accuracies as in the test problem.

Exercises

14.1 Consider the following C^0-linear rectangular element:

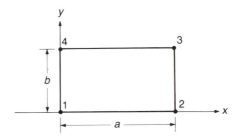

Assuming γ is a constant, show that the mass matrix is as follows:

$$[M] = \frac{\gamma ab}{9} \begin{bmatrix} 1 & \frac{1}{2} & \frac{1}{4} & \frac{1}{2} \\ & 1 & \frac{1}{2} & \frac{1}{4} \\ & & 1 & \frac{1}{2} \\ \text{Symmetric} & & & 1 \end{bmatrix}$$

(E14.1)

Hint: Only three different integrals need to be evaluated. (Also, recall Exercise 13.8.)

14.2 Consider the transverse vibration of a square membrane that is fastened along all four sides. (See (a) below.) Its area density is ϱ, and it is stretched with a uniform tension T (force per unit length). Calculate the fundamental natural frequency.

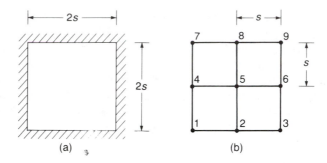

(a)

(b)

Employ a mesh of four C^0-linear square elements. (See (b) above.) Use the stiffness and mass matrices from Exercises 13.17 and 14.1, respectively. If you follow the hints in Exercise 13.16, this becomes a simple "back-of-the-envelope" exercise. The exact solution is

$$f_0 = \frac{1}{2\sqrt{2}} \frac{1}{s} \sqrt{\frac{T}{\varrho}}$$

(E14.2)

14.3   The C^0-linear quadrilateral element used in the previous two exercises is the first member in both the Lagrange and serendipity families of elements. (Recall Section 13.4.) The second member of the Lagrange family is the biquadratic Lagrange element, shown here both as a parent element in the ξ,η-plane and a rectangular element in the x,y-plane.

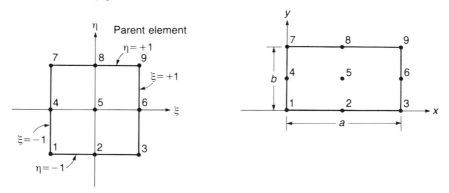

The shape functions in the parent element are products of the 1-D C^0-quadratic shape functions. In particular, the parent shape function for the center node is $\phi_5(\xi,\eta)=(1-\xi^2)(1-\eta^2)$. In evaluating the mass matrix for the rectangular element, the integral expression for the diagonal term associated with the center node takes the usual form,

$$M_{55} = \int_{-1}^{+1} \int_{-1}^{+1} \phi_5 \varrho \phi_5 |J|\, d\xi\, d\eta$$

(E14.3)

To gain familiarity with the quadrature rules and an understanding of their accuracy, evaluate the M_{55} integral using Eq. (14.5) and the quadrature rules in Fig. 13.42.

What is the expression for the Jacobian? [Consider Eq. (13.118) and Exercise 13.8.] Evaluate M_{55} using the 1, 2-by-2, and 3-by-3 rules. Which rule should integrate M_{55} exactly? The exact expression, easily verified, is $M_{55} = \varrho ab(8/15)^2$.

14.4 Calculate the fundamental frequency of vibration of the square membrane in Exercise 14.2 using a single biquadratic Lagrange element:

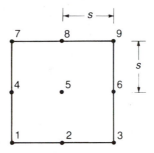

Simplify the analysis by considering at the outset the effect that the zero essential BCs will have on the system equations, and make use of the results of Exercise 14.3. Compare your solution with the approximate and exact solutions in Exercise 14.2.

14.5 Consider the transverse vibration of a triangular membrane that is fastened along all three sides. (See (a) below.) Its area density is ϱ, and it is stretched with a uniform tension T (force per unit length). Calculate the fundamental natural frequency.

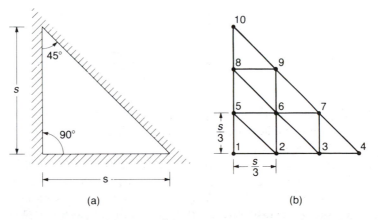

(a)   (b)

Employ a mesh of nine identical C^0-linear triangular elements. (See (b) above.) As in previous exercises, the analysis will be greatly simplified if you evaluate and assemble only those terms that will remain in the final system equations. (Recall Exercises 13.15 and 13.16.) The exact solution is

$$f = \frac{\sqrt{5}}{2}\frac{1}{s}\sqrt{\frac{T}{\varrho}} \tag{E14.4}$$

14.6   Repeat Exercise 14.5 but employ the following mesh, which uses both triangular and square elements:

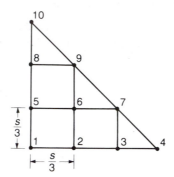

The stiffness and mass terms for the square elements are available from Exercises 13.17 and 14.1, respectively. Compare your solution with the approximate and exact solutions in Exercise 14.5.

14.7   Repeat Exercise 14.5 but employ the following mesh of one C^0-cubic triangular element:

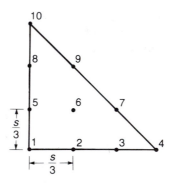

It should be evident from the previous exercises that only the stiffness and mass terms for the middle node are needed, i.e., K_{66} and M_{66}, which in turn require the shape function ϕ_6.

a)   Derive the shape function $\phi_6(x,y)$ or, if you prefer, $\phi_6(\xi,\eta)$ for a unit isosceles right triangle. Recall from Fig. 13.23 that the element uses a complete 2-D cubic polynomial. *Hint:* Make use of the symmetry with respect to the line $y=x$ (or $\xi=\eta$) to eliminate 4 of the 10 coefficients. Then apply the zero essential BCs at five nodes to eliminate five more coefficients. (Observe that this exercise is equivalent to applying the theoretical method in Chapter 4 for developing a trial solution that satisfies the BCs.) *Another hint:* Since ϕ_6 will be zero along the entire boundary (why?), you could actually write the expression for ϕ_6 merely by inspection.

b) Evaluate K_{66} and M_{66} either by exact or numerical integration, the latter using one of the Gauss quadrature rules in Fig. 13.31. Which rule would yield exact integration for K_{66}? for M_{66}?

c) Calculate the fundamental frequency of vibration. Compare your solution with the approximate and exact solutions in Exercises 14.5 and 14.6.

14.8 Consider the $90°$ ''circular'' arc element that was treated in Exercises 13.20 through 13.22. This is a C^0-quadratic isoparametric triangle in which nodes 2, 5, and 3 on the hypotenuse of the parent element are mapped onto a $90°$ circular arc of radius r.

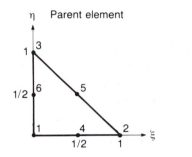
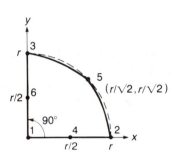

The mass matrix for this ''circular'' arc element is as follows (to five-decimal-place accuracy):

$$[M]^{(e)} = \gamma r^2 \begin{bmatrix} 0.01207 & -0.01038 & -0.01038 & 0.00828 & -0.02167 & 0.00828 \\ & 0.01207 & -0.01287 & 0.01822 & 0.01490 & -0.01504 \\ & & 0.01207 & -0.01504 & 0.01490 & 0.01822 \\ & & & 0.10449 & 0.07341 & 0.06015 \\ & & & & 0.12217 & 0.07341 \\ \text{Symmetric} & & & & & 0.10449 \end{bmatrix}$$

(E14.5)

where γ is assumed constant, and the integrals were evaluated with the four-point Gauss quadrature rule in Fig. 13.31. Calculate any one of the mass terms, using the four-point rule. Verify your answer by comparison with the above values.

14.9 Consider the transverse vibration of a circular membrane that is fastened along the circumference. (See (a) below.) Its area density is ϱ, and it is stretched with a uniform tension T (force per unit length). Calculate the lowest natural frequency for the family of radially symmetric modes. You will recall from Exercise 10.13 that these are the modes that have no variation with respect to the circumferential angle θ ($m=0$ in Exercise 10.13).

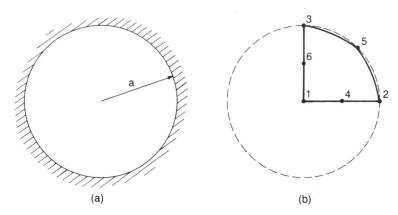

(a)   (b)

Model only one quadrant of the circle with a single 90° "circular" arc element.
(See (b) above.) Use the stiffness and mass matrices in Exercises 13.21 and 14.8.
Express the radial symmetry property as a linear constraint equation and apply
it to the system equations. The exact solution is

$$f_{01} = \frac{\nu_{01}}{2\pi a} \sqrt{\frac{T}{\varrho}}$$

(E14.6)

where $\nu_{01}=2.405$, which is the first zero of the Bessel function of order 0; that
is, $J_0 (\nu_{01})=0$.

14.10 Repeat Exercise 14.9 using the same mesh but calculate the lowest natural frequency for both of the first two families of radially nonsymmetric modes. Recall
from Exercise 10.13 that the modes in the first of these two families have a cos
θ (or sin θ) variation $(m=1$ in Exercise 10.13). Figure (a) below shows that these
modes have a "nodal diameter" along which the mode has zero transverse displacement; the cos θ function is positive on one side and negative on the other. The
modes in the second family have a cos 2θ (or sin 2θ) variation $(m=2)$, and hence
two nodal diameters, as indicated in Figure (b).

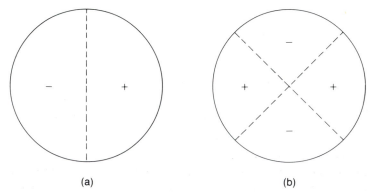

(a)   (b)

Determine the appropriate BCs and/or linear constraint equations for each family
of modes. Observe that it is possible to solve for these higher modes modeling
only one quadrant of the domain. This illustrates how it is sometimes possible to

simplify an analysis by being alert to ways that one can apply any a priori knowledge of the solution that is available. The exact solution for the $\cos\theta$ family is

$$f_{11} = \frac{v_{11}}{2\pi a} \sqrt{\frac{T}{\varrho}}$$ (E14.7)

where $v_{11}=3.832$, corresponding to $J_1(v_{11})=0$. The exact solution for the $\cos 2\theta$ family is

$$f_{21} = \frac{v_{21}}{2\pi a} \sqrt{\frac{T}{\varrho}}$$ (E14.8)

where $v_{21}=5.135$, corresponding to $J_2(v_{21})=0$.

14.11 Calculate the first nonzero cut-off frequency for the propagation of radially symmetric sound waves through a circular cylindrical tube of radius a with rigid walls. Use the 90° "circular" arc element; the stiffness and mass matrices are in Exercises 13.21 and 14.8. Model one quadrant of the cross section with a single element, as in Exercise 14.9. What are the linear constraint equations that express the radial symmetry property? The exact solution is $f=v_{11}c/2\pi a$, where c is the sound speed and $v_{11}=3.832$, corresponding to $J_1(v_{11})=0$.

Reference

[14.1]   J. Mathews and R. L. Walker, *Mathematical Methods of Physics*, Benjamin, New York, 1965, p. 218.

15

2-D Initial-Boundary-Value Problems

Introduction

This chapter will analyze the initial-boundary-value problem associated with the boundary-value problem treated in Chapter 13. The development closely parallels that of the 1-D initial-boundary-value problem in Chapter 11. Thus the FE discretization will transform the governing partial differential equation into a system of ordinary differential equations, and to solve the latter we can use all the time-stepping methods developed in Chapter 11. In fact, the only task left for this chapter is to develop the capacity matrix, since all other aspects of the problem have been dealt with in Chapters 11 and 13. However, the capacity matrix is identical to the mass matrix, which was developed in Chapter 14, so no new development work need be done at all. Nevertheless, a different way to lump the capacity matrix will be shown. Practical modeling considerations will be demonstrated by using UNAFEM to solve a transient heat conduction problem.

15.1 Mathematical and Physical Description of the Problem

The initial-boundary-value problem associated with the boundary-value problem in Section 13.1 is governed by the following partial differential equation, which is second-order in space and first-order in time:

$$
\mu(x,y)\frac{\partial U(x,y,t)}{\partial t} - \frac{\partial}{\partial x}\left(\alpha_x(x,y)\frac{\partial U(x,y,t)}{\partial x}\right) - \frac{\partial}{\partial y}\left(\alpha_y(x,y)\frac{\partial U(x,y,t)}{\partial y}\right)
$$
$$
+ \beta(x,y)U(x,y,t) = f(x,y,t) \qquad (15.1)
$$

704

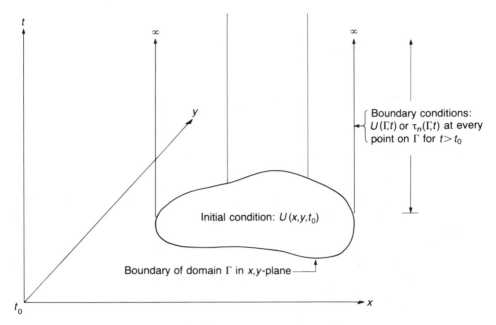

FIGURE 15.1 ● Domain for 2-D initial-boundary-value problem, showing initial condition and boundary conditions.

The domain is a region in x,y,t-space (Fig. 15.1) that is usually (but not always) bounded in the x,y-plane and unbounded (semi-infinite) in the t-direction. Boundary conditions are of the same type as for the boundary-value problem, namely, specifying either $U(\Gamma,t)$ or $\tau_n(\Gamma,t)$ at every point on the boundary. Here, though, the boundary values are a function of time. The initial condition, $U(x,y,t_0)$, specifies the value of U over the 2-D spatial domain at the initial time t_0. The situation is analogous to the 1-D problem in Chapter 11 (see Fig. 11.3), the only difference being that the spatial dimensions have increased from 1 to 2.

Physical applications of Eq. (15.1) include any of the diffusion phenomena listed in Table 11.1.

15.2 Derivation of Element Equations

As noted in the introduction to this chapter, all of the development work has been done in Chapters 11 and 13, so only the results will be summarized here. Thus, employing the separation of variables technique as in Section 11.2 yields the following expression for the element equations at the end of step 3:

$$[C]^{(e)}\left\{\frac{da(t)}{dt}\right\} + [K]^{(e)}\{a(t)\} = \{F(t)\}^{(e)}$$

$$(15.2)$$

where

$$C_{ij}^{(e)} = \int\int^{(e)} \phi_i^{(e)} \mu \phi_j^{(e)} \, dx \, dy$$

$$K_{ij}^{(e)} = \int\int^{(e)} \frac{\partial \phi_i^{(e)}}{\partial x} \alpha_x \frac{\partial \phi_j^{(e)}}{\partial x} \, dx \, dy + \int\int^{(e)} \frac{\partial \phi_i^{(e)}}{\partial y} \alpha_y \frac{\partial \phi_j^{(e)}}{\partial y} \, dx \, dy + \int\int^{(e)} \phi_i^{(e)} \beta \phi_j^{(e)} \, dx \, dy$$

$$F_i^{(e)} = \int\int^{(e)} f \phi_i^{(e)} \, dx \, dy + \oint^{(e)} \bar{\tau}_{-n}^{(e)} \phi_i^{(e)} \, ds$$

Of course, the expressions for $K_{ij}^{(e)}$ and $F_i^{(e)}$ remain the same as in Eqs. (13.61b) for the boundary-value problem. The expression for $C_{ij}^{(e)}$ has the same form as its 1-D counterpart in Eqs. (11.24b); it is also identical to the β-integral above or the mass integral in Eq. (14.2).

Step 4 needs no work since we can use any of the four 2-D elements developed in Chapter 13. In step 5, the evaluation of the capacity integrals is identical to the evaluation of the mass integrals in Chapter 14. Therefore the expressions for $C_{ij}^{(e)}$ for each of the four types of elements are given by Eqs. (14.3) through (14.5), substituting C_{ij} for M_{ij} and μ for γ. These expressions yield the *consistent* capacity matrix.

For the explicit forward difference time-stepping method (see Section 11.3.2), we need a *lumped* (i.e., diagonalized), rather than a consistent, capacity matrix. A particular lumping technique was described in Section 11.3.2 and implemented in UNAFEM for all the 1-D elements. However, for 2-D and 3-D elements with midside nodes, that technique makes some of the diagonal terms negative. This has caused numerical problems in several different situations [15.1]. These problems disappear, though, when a different lumping technique is used [15.2]. We will therefore describe this alternative technique, and implement it into UNAFEM for all the 2-D elements (including the ones without midside nodes, for which both techniques would be satisfactory).

The terms of the lumped element capacity matrix, $CL_{ij}^{(e)}$, are calculated as follows [15.2]:

$$CL_{ii}^{(e)} = S \times C_{ii}^{(e)} \qquad i = 1,2,\ldots n$$

$$CL_{ij}^{(e)} = 0 \qquad i \neq j \tag{15.3}$$

Here S is a scaling factor,

$$S = \frac{C^{(e)}}{\sum_{i=1}^{n} C_{ii}^{(e)}} \tag{15.4}$$

where $C^{(e)}$ is the total capacity of the element. $C^{(e)}$ may be calculated in either of two ways:[1]

$$C^{(e)} = \int\int^{(e)} \mu(x) \, dx \, dy \tag{15.5a}$$

---

[1] Both yield the same value because of the interpolation property of the shape functions:

$$\sum_{i=1}^{n} \phi_i(x,y) = 1 \qquad \text{and therefore} \qquad \sum_{i=1}^{n}\sum_{j=1}^{n} \phi_i(x,y)\phi_j(x,y) = 1$$

or

$$C^{(e)} = \sum_{i=1}^{n} \sum_{j=1}^{n} C_{ij}^{(e)}$$

(15.5b)

whichever is more convenient. (UNAFEM will use the second way.)

The essential difference between this lumping technique and the one in Section 11.3.2 [see Eq. (11.53)] is that here only the diagonal terms of the consistent matrix, which are always positive, are used to calculate the lumped matrix. (The off-diagonal terms are needed if Eq. (15.5b) is used, but the resulting value of $C^{(e)}$ is always positive.) Note that the total capacity is conserved [recall Eq. (11.54) and accompanying discussion]; that is,

$$\sum_{i=1}^{n} \sum_{j=1}^{n} CL_{ij}^{(e)} = S \times \sum_{i=1}^{n} C_{ii}^{(e)} = \frac{C^{(e)}}{\sum_{i=1}^{n} C_{ii}^{(e)}} \times \sum_{i=1}^{n} C_{ii}^{(e)} = C^{(e)}$$

(15.6)

After the element capacity matrices are lumped, they are then assembled into a (lumped) system capacity matrix, $[CL]$.[2] Being diagonal, it is trivially inverted [see Eq. (11.55b)]. The resulting system recurrence relation has the same form as before [see Eq. (11.55a)] and therefore is solved in the same manner.

Step 6 derives expressions for the optimal and nodal fluxes. These are unaffected by the capacity matrix so the results in Chapter 13 are applicable here.

15.3 Implementation into the UNAFEM Program

The principal changes to the program are the addition of the capacity matrix calculations and the new lumping technique. The expressions for the (consistent) capacity matrix terms are identical to the mass matrix terms (see Fig. 14.1), as the reader can verify from the listing in the appendix. The function $\mu(x,y)$ may be specified in UNAFEM as a piecewise-quadratic (2-D) polynomial, similar to $\alpha_x(x,y)$, $\alpha_y(x,y)$, and $\beta(x,y)$ in Eqs. (13.69). Figure 15.2 shows the implementation of the lumping procedure (for the 2-D elements) for the explicit initial-boundary-value problem.

The time-stepping algorithms for the system recurrence relations need no changes since they are general matrix operations and hence do not depend on the origin of the matrices; i.e., they might originate from a 1-D, 2-D, or n-D problem, or perhaps not even from an FE problem at all.

15.4 Application of UNAFEM to a Transient Heat Conduction Problem

15.4.1 Description of the Problem

Soldering and welding are techniques commonly used to join metallic parts. Examples are ubiquitous, ranging from delicate electronic components to very large structures. In

---

[2] This lumping technique involves a multiplicative scaling factor, so lumping and assembly do not commute. It is recommended that lumping be performed before assembly, i.e., at the element level, because (1) this yields the improved results cited above, and (2) it is easier to program.

```
      SUBROUTINE FORM
                    .
                    .
                    .
C************** ASSEMBLE MATRICES FOR
C                    EXPLICIT INITIAL-BOUNDARY-VALUE PROBLEM **************
C
  300     CONTINUE
C------ COMPUTE SCALE FACTOR
C          FOR LUMPING 2-D CAPACITY MATRIX
      IF (DIMEN.NE.TWOD) GO TO 330
      CEDIAG = 0.
      CESUM = 0.
      DO 320 I=1,NND
          CEDIAG = CEDIAG+CE(I,I)
          DO 310 J=1,NND
              CESUM = CESUM+CE(I,J) ◄────Eq. (15.5b)
  310     CONTINUE
  320 CONTINUE
      SCALE = CESUM/CEDIAG ◄────Eq. (15.4)
C
  330 CONTINUE
      DO 350 I=1,NND
          II = ICON(I,IELNO)
          F(II) = F(II)+FE(I)
          DO 340 J=1,NND
              JJ = ICON(J,IELNO)
              JSHIFT = JJ-(II-1)
              IF (JJ.GE.II) K(II,JSHIFT)=K(II,JSHIFT)+KE(I,J)
C----------------(LUMP THE CAPACITY MATRIX)
              IF (DIMEN.EQ.ONED) CLUMP(II) = CLUMP(II)+CE(I,J)
  340         CONTINUE
              IF (DIMEN.EQ.TWOD) CLUMP(II) = CLUMP(II)+SCALE*CE(I,I) ◄──── Eq. (15.3)
  350     CONTINUE
                    .
                    .
                    .
```

FIGURE 15.2 ● Subroutine FORM, a partial listing showing lumping and assembly of element capacity matrices.

a typical soldering or welding operation an iron, gun, or torch is "suddenly" applied to a small spot on a structure, injecting a fairly steady, concentrated amount of thermal energy for the duration of the operation. Of course, the temperature in the immediate vicinity of the heat source will rise very rapidly. At points farther away it will rise less rapidly. It may be of interest to know *how long* it will take for the temperature to rise to a certain critical level at a nearby point on the structure, e.g., where a sensitive component is located. This is a transient heat conduction problem.

In the following illustrative example we want to focus on the techniques for performing a 2-D (or 3-D) transient analysis. We will examine a typical soldering/welding operation, looking in particular at the temperature-versus-time histories at several points on the structure, as well as the spatial temperature distribution at several instants of time.

Figure 15.3 shows a rectangular sheet of copper. All physical properties (thickness, density, thermal conductivity, specific heat, and convective heat transfer coefficient) will be assumed uniform over the sheet, and the thermal conductivity will be assumed isotropic.[3]

---

[3] Of course, variable properties and anisotropy could be handled just as easily by UNAFEM. The above assumptions are mainly to avoid cluttering the problem with a lot of unimportant input data. Nevertheless, uniform, isotropic properties are frequently realistic assumptions. In a similar spirit, the geometry will be kept simple so that we can focus principally on the transient analysis techniques.

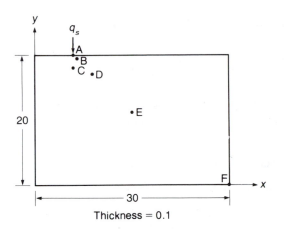

Locations of points:

	x	y
A	6	20
B	6.4	19.6
C	6	18
D	9	17
E	15	11
F	30	0

Thickness = 0.1

FIGURE 15.3 ● Sheet of copper with a heat source applied at point A. Temperature histories are monitored at points B, C, D, E, and F (dimensions and coordinates in centimeters).

The soldering iron or welding torch is applied at point A on the edge of the sheet. We will be interested in the temperature histories at points B, C, D, E, and F.

The tip of the soldering iron or flame of the torch is best modeled as a concentrated (i.e., point) source since we are not interested in extremely close-up details of the temperature distribution directly underneath the point of application. We will model the iron or torch as a steady source of heat energy, $q_s = 100$ watts, applied to the sheet at time $t=0$. Not having any idea what the actual ''rise time'' is (which could vary considerably, depending on who or what is doing the soldering or welding), we will let the source be applied suddenly at time $t=0$, as shown in Fig. 15.4. Different rise times would only affect the solution immediately following application and in the immediate vicinity of the point of application.

There will be a convective heat loss from the top and bottom surfaces as well as the edges. We can model the surface loss by the approximation technique described in Section 13.1.3 [see Eq. (13.37)]. The convective loss along the edges would normally be modeled directly with convective (i.e., mixed) BCs. Since UNAFEM doesn't accept this type of BC, we will instead specify the heat flux to be zero along the edges. This should be a reasonable approximation because the edge loss will be small relative to the surface loss, and it will affect the resulting temperature distribution only near the edges.

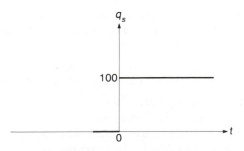

FIGURE 15.4 ● Idealized time history of heat source q_s at point A.

The governing differential equation is Eq. (13.37) with the addition of a time-derivative term,

$$\varrho c \frac{\partial T}{\partial t} - k\nabla^2 T + \frac{h}{w}T = \frac{h}{w}T_\infty \tag{15.7}$$

This is Eq. (15.1) with $U(x,y,t)=T(x,y,t)$, the temperature; $\mu(x,y)=\varrho c$; $\alpha_x(x,y)=\alpha_y(x,y)=k$; $\beta(x,y)=h/w$; and $f(x,y,t)=(h/w)T_\infty$. The physical properties have the following values:

$$\text{Mass density } \varrho = 8.5 \text{ gm/cm}^3$$

$$\text{Specific heat } c = 0.092 \text{ cal/gm-}°\text{C}$$

$$\text{Thermal conductivity } k = 0.92 \text{ cal-cm/sec-cm}^2\text{-}°\text{C}$$

$$\text{Convective heat transfer coefficient } h = 0.0004 \text{ cal/sec-cm}^2\text{-}°\text{C}$$
$$\text{(sum of upper \& lower surface values)}$$

$$\text{Thickness of sheet } w = 0.1 \text{ cm} \tag{15.8}$$

The boundary conditions are zero flux along all edges:

$$q_x(0,y,t) = 0 \qquad 0 \le y \le 20 \; , \; t \ge 0$$
$$q_x(30,y,t) = 0 \qquad 0 \le y \le 20 \; , \; t \ge 0$$
$$q_y(x,0,t) = 0 \qquad 0 \le x \le 30 \; , \; t \ge 0$$
$$q_y(x,20,t) = 0 \qquad 0 \le x \le 30 \; , \; t \ge 0 \tag{15.9a}$$

with the exception of the concentrated heat source at point A:

$$[q_y(6,20,t)]_{conc} = q_s \qquad t > 0$$
$$= 23.9 \text{ cal/sec (100 watts)} \tag{15.9b}$$

The quantity $[q_y]_{conc}$ corresponds to τ_1 in Eq. (13.94); that is, it is an integral of the actual distributed 100-watt source over its small area of application. [See Figs. 13.17 and 13.18 as well as the footnote at the end of Section 15.4.3.] The sheet is initially in thermal equilibrium at the ambient temperature:

$$T(x,y,0) = T_\infty = 20°\text{C} \tag{15.10}$$

We want to compute the temperature histories at points B, C, D, E, and F, and temperature contour plots at several different times during the transient solution, including the steady-state solution.

15.4.2 Splitting the Analysis into Shock and Postshock Phases

Since the load is applied suddenly, there will be an immediate response consisting of a very steep temperature profile in the immediate vicinity of the load (see Fig. 11.26). During this brief interval, distant boundaries have no significant effect on the solution. Therefore we will only need to analyze a small portion of the domain surrounding the

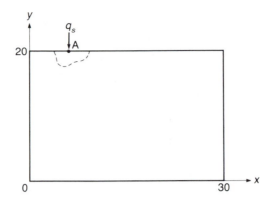

FIGURE 15.5 ● An arbitrarily shaped small region surrounding the heat source (dashed line) is adequate for analyzing the shock response (dimensions in centimeters).

heat source. The solution during this primary phase — within a local region for an initially brief interval — will be referred to as the *shock response*.

At later times the temperature profile will become smoother as the heat diffuses throughout the structure and distant boundaries start to become significant. The solution during this secondary phase — throughout the rest of the domain at times beyond the initial shock — will be referred to as the *postshock response*.

Considerable computational efficiency will be gained by analyzing the two phases separately.[4] This is because the phases should use different element types, radically different element and time-step sizes, and different time-stepping (time-integration) methods.

15.4.3 Shock Response

As observed above, we only need to model a small portion of the domain surrounding the heat source. The size or shape of the region is not crucial; it depends mainly on how far into the domain we want to analyze the shock response (Fig. 15.5).

It is important to note at the outset that the response from the concentrated (point) source will initially propagate in essentially a *radial* direction away from the source. More generally, when a point source is on a boundary and the local boundary is insulated and straight (or a straight-sided angle, with the source at the vertex of the angle), and the material properties are isotropic, then the initial propagation is "exactly" radial and we really have a 1-D problem. This is the case in our present problem, so we could really solve this using the 1-D radial heat conduction equation in cylindrical coordinates (see Exercises 11.13 and 11.14).

In real applications, though, there may be local variations in geometry or properties that would destroy the locally radial symmetry and necessitate a 2-D analysis. Therefore we will solve this as a 2-D problem in order to demonstrate several useful techniques for

---

[4] In the 1-D thermal shock problem studied in Section 11.5 we did not split the analysis apart because the efficiency gains would not have been significant for such a small problem. The gains are more significant in 2-D, and especially in 3-D.

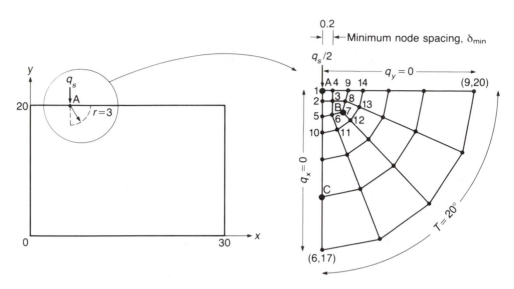

FIGURE 15.6 ● A local mesh for the shock response (dimensions and coordinates in centimeters).

general situations. Even in these cases, though, an auxiliary 1-D analysis would provide a quick and easy check on the reasonableness of the 2-D solution.

These comments suggest that in most situations an appropriate mesh would be similar to a *polar* coordinate pattern (resembling the mesh in Fig. 13.73c, minus, perhaps, the outside row of rectangles). Figure 15.6 shows such a mesh, indicating the portion of the overall domain that it occupies. Here we have modeled a 90° arc surrounding point A. Because of the locally radial symmetry in this geometrically simple problem, we could have modeled an arc of any number of degrees. However, the 90° arc is likely to be appropriate in more complicated applications that sometimes possess at least a locally bilateral symmetry.

The appropriate BCs are indicated in the figure. The horizontal straight edge along the top is part of the original boundary, so it uses the specified conditions, $q_y=0$. The vertical side is a line of symmetry (locally) so the normal component of flux must vanish, $q_x=0$ (see Section 13.7.2). Along the curved side we don't know the correct "boundary conditions," i.e., the exact solution as a function of time. However, we know the solution at $t=0$. We can therefore specify the initial condition ($T=20$°C) and hold this condition constant.[5] Of course, this will introduce an error into the solution that will be zero initially and will gradually increase with time. We will see below how to determine this error and hence the maximum time of usefulness for this shock model. The heat source applied at point A on the mesh in Fig. 15.6 is half the total source since, by symmetry, only half the energy flows into the 90° area.

Recalling the recommendations in Chapters 9 and 11 for very steep solutions in the vicinity of a singularity, we will employ low-order elements, namely, the linear elements. (The initial temperature profile will be close to a step function. A quadratic interpolating polynomial trying to approximate such a step will badly overshoot, whereas a linear will

---

[5] We could, alternatively, specify the initial condition $q_n=0$.

not.) In addition, numerical experiments indicate the linear quadrilateral, ELEM23, yields consistently better accuracy than the linear triangle, ELEM21. (Perhaps the incomplete quadratic term in ELEM23 improves the accuracy, yet avoids the overshooting associated with additional midside interpolation points.) Therefore the mesh in Fig. 15.6 employs all linear quadrilaterals. The circumferential node numbering minimizes bandwidth. Note that node 7 at point B is common to only three quadrilaterals. To avoid excessive mapping distortion in any one of them (due to departure from the 90° angles in the parent element), the surrounding mesh is adjusted to make all three vertex angles approximately 120°. The mesh is graded only in the radial direction since the steep slopes will be only in the radial direction.

ELEM23 has not been used in previous chapters so, of course, several tests should be performed, including one-element tests as well as a simple problem with a known solution (e.g., 1-D transient heat flow along a row of elements). These tests will not be demonstrated here since it is hoped that by now the reader has acquired this important habit of program ''verification.''

Having the spatial part of the model established, i.e., a mesh, we must now turn to the temporal part, i.e., a time-integration method and a time-step size. Here we will make use of many of the guidelines established in Chapter 11.

An explicit method is very efficient for a shock response since the very small time steps required for stability are usually comparable to the step size necessary to characterize the short duration response. And we recall that explicit methods involve only matrix multiplication, whereas implicit methods require matrix inversion so the former is much faster and cheaper than the latter.

We will use the forward difference ($\theta=0$) method. Since this is conditionally stable we must choose a time step Δt that is less than Δt_{crit}. The latter can be estimated by the approximate formula in Eq. (11.104), which for this problem has the form

$$\Delta t_{crit} \simeq \frac{\varrho c}{k \pi^2} \delta^2_{min}$$

$$(15.11)$$

Using the values for ϱ, c, and k from Eqs. (15.8), and $\delta_{min}=0.2$ cm from Fig. 15.6, Eq. (15.11) yields $\Delta t_{crit} \approx 0.0034$ sec. Chapter 11 indicated this estimate is always low, usually by a factor of about 3 to 5, so we expect Δt_{crit} to be in the neighborhood of 0.01 or 0.02 sec.

To minimize costs it is desirable to use as large a Δt as possible, especially since we will decrease Δt to one-fourth in a subsequent analysis in order to ascertain convergence. Therefore we must perform a few short experiments with our mesh, testing several Δt values. Starting with a value near 0.01 or 0.02, we observe the time history at any node of interest, watching for stable or unstable behavior. The latter will quickly reveal itself, so the analyses are simple and cheap, and generally only two, three, or four are necessary. Some results are shown in Fig. 15.7. Clearly, Δt_{crit} is between 0.019 and 0.020 sec since we see the characteristic diverging and converging oscillations associated with a Δt just to either side of Δt_{crit} (recall Fig. 11.12). A value of $\Delta t=0.01$ sec seems to be reasonable and quite safe.

Next we must determine the effect of the location of the arbitrary boundary. Here we can use the technique applied to the arbitrary finite boundary in the quantum mechanical oscillator problem in Section 10.6. Thus we repeat the analysis with a different boundary location and observe the resulting effect on the response. To enlarge the mesh, new elements

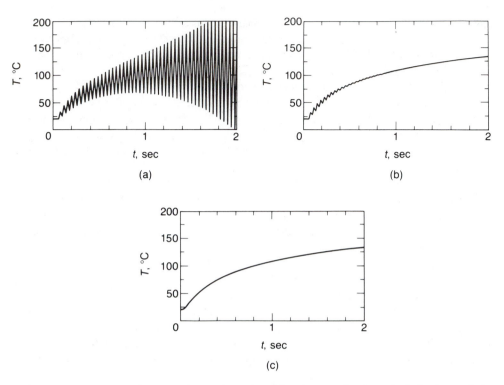

FIGURE 15.7 ● Determination of Δt_{crit} from time histories at point B (node 7) in the mesh in Fig. 15.6, using different time steps: (a) $\Delta t = 0.020$ sec; (b) $\Delta t = 0.019$ sec; (c) $\Delta t = 0.010$ sec.

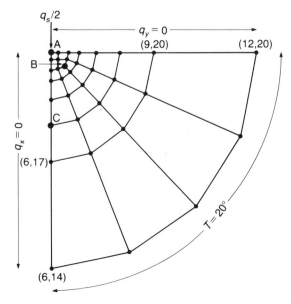

FIGURE 15.8 ● Enlargement of the mesh in Fig. 15.6 to determine the effect of the location of the arbitrary boundary on the response (coordinates in centimeters).

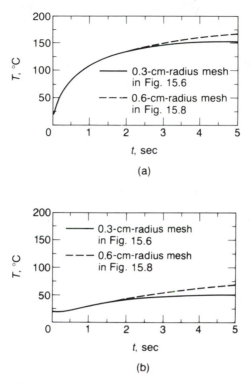

FIGURE 15.9 ● Effect of the location of the arbitrary boundary: (a) response at point B; (b) response at point C.

are added to the previous mesh, without disturbing the latter, so that the only change is to move the boundary, and its boundary conditions, farther out.

Figure 15.8 shows the enlarged mesh. Note the addition of only one extra arc of four elements. Their large size continues the radial grading of the mesh, and the bandwidth is unaffected. This change to the mesh is a trivial effort.

Figure 15.9 shows the time histories at points B and C for both meshes. The difference is less than 2 °C up until about 2 sec at point B and up until about 1 1/2 sec at point C. The latter, of course, is shorter since point C is closer to the boundary. Therefore, for points within 0.2 cm of the heat source, locating the boundary at a radius of 0.3 cm will introduce an error of less than about 2 °C during the first 1 1/2 sec of the shock response.

Having determined an appropriate time step and a maximum useful time limit for the shock analysis, we can now run the complete analysis. Figure 15.10 shows the resulting temperature and flux contour plots at several different times, as well as the temperature histories at points B and C.

To verify convergence and estimate error we refine both the mesh and the time step. Note from Eq. (15.11) that when the node spacing δ_{min} is halved, Δt must be decreased to one-fourth. Figure 15.11 shows the refined mesh. The element sizes are halved both radially and circumferentially. This made mesh generation easy. However, as we've discussed above, most shock responses from concentrated loads vary predominantly in the radial direction so usually mesh refinement only needs to be done radially. In the present

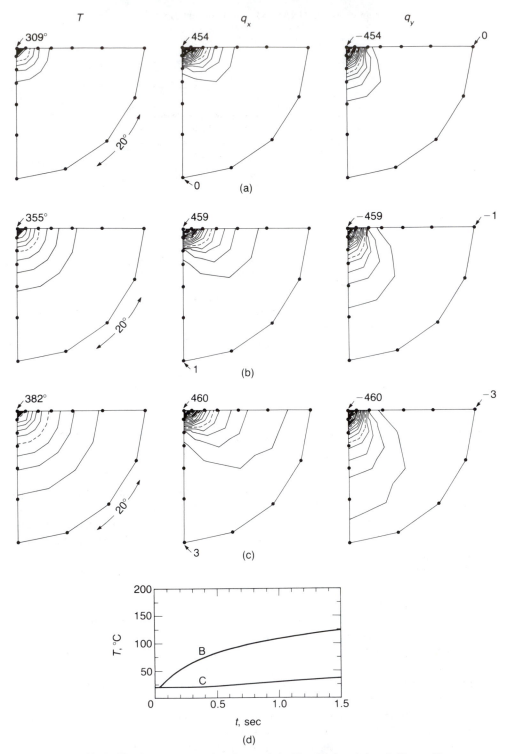

FIGURE 15.10 ● Shock response, using the mesh in Fig. 15.6 and $\Delta t = 0.01$ sec. Temperature and heat flux contours at (a) $t = 1/4$ sec; (b) $t = 3/4$ sec; (c) $t = 1\ 1/2$ sec. Temperature contour spacings, $20\,°C$; dashed contours at multiples of $100\,°C$. Flux contour spacings $20\ cal/cm^2$-sec; dashed contours at multiples of $100\ cal/cm^2$-sec. (d) Temperature histories at points B and C.

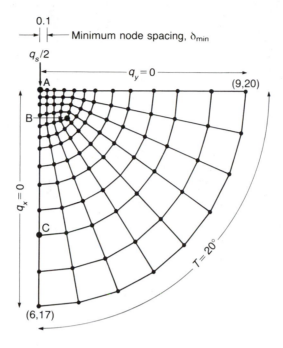

FIGURE 15.11 ● Refined mesh for shock response (coordinates in centimeters).

problem the explicit solver is so cheap that the wasted refinement in the circumferential direction was offset by the savings in mesh-generation effort.

The resulting contour and time plots are shown in Fig. 15.12. A comparison of the coarse and refined results indicates the error at points B and C is about 1 or 2°.

The error at point A (the point of application of the heat source) is much greater: The temperatures in the refined mesh are 60° higher than in the coarse mesh, and the fluxes are essentially double. The problem is that the *point* source is a *singularity*. There is a constant power input (100 watts) applied to a point (zero area). Therefore the flux, which is power *per unit area* (the area perpendicular to the 2-D domain), is infinite at the point, as is the temperature. We would consequently expect a significant deterioration in accuracy in the immediate vicinity of the point unless we were to incorporate in the model one of the remedies suggested in Section 9.4.

Despite these local inaccuracies, the approximate *doubling* of the flux at point A as the mesh is refined is to be expected (providing a convenient check on the correctness of the analysis). To see this, we first observe that in the immediate vicinity of point A the flux flows essentially radially and is inversely proportional to r, the radius from A.[6]

---

[6] The integral of the flux over any surface enclosing the source should equal the power input. Consider a semicircular path of radius r around the source. For small r the flux is essentially radial and constant along the path, say q_r, and the surface heat loss perpendicular to the sheet is negligible compared to the radial flow. (Within $r=0.5$ cm, surface loss is 0.13 cal/sec for a temperature rise of 400 °C, which is 0.5% of the input power.) Therefore $q_r \pi r w = 23.9$ cal/sec (100 watts), where w is the thickness of the sheet. Hence q_r is inversely proportional to r.

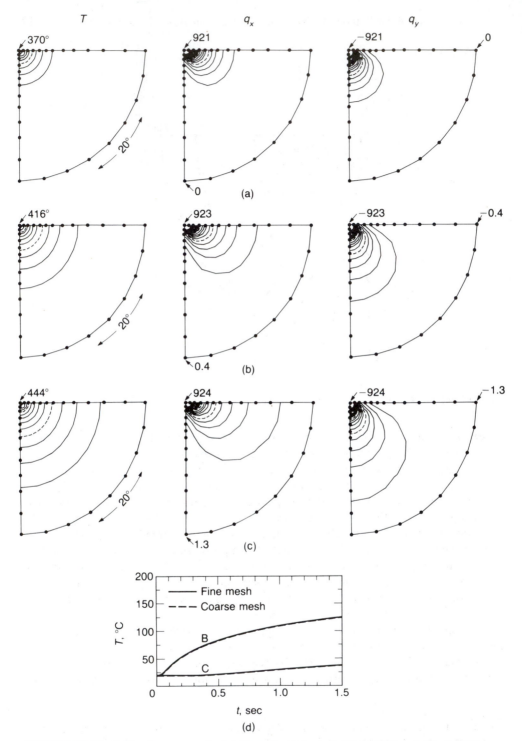

FIGURE 15.12 ● Shock response, using the refined mesh in Fig. 15.11 and $\Delta t = 0.0025$ sec. Temperature and heat flux contours at (a) $t = 1/4$ sec; (b) $t = 3/4$ sec; (c) $t = 1\ 1/2$ sec. Temperature contour spacings, $20\,°C$; dashed contours at multiples of $100\,°C$. Flux contour spacings, 20 cal/cm²-sec; dashed contours at multiples of 100 cal/cm²-sec. (d) Temperature histories at points B and C.

Therefore if r is halved, the flux should double. We then recall that the flux computed at the node at point A is equal to the (finite) Gauss-point value at the center of the element containing that node. Consequently, when the element size is halved, the radial position of the Gauss point is halved. We thus conclude that the computed flux at point A should be about double for the refined mesh.

15.4.4 Postshock Response

This analysis is more straightforward than the shock response since there are no artificial boundary effects or critical time steps to determine. We model the entire domain, for all times $t > 0$, using implicit time-integration methods and much larger elements and time steps. The solution near the heat source for $t \le 1\ 1/2$ sec we ignore since we have already computed it. In any case, our large element and time steps would yield very inaccurate results in that region. Here we are interested instead in the temperatures at points D, E, and F, which are further removed from the heat source (see Fig. 15.3).

Since the postshock phase involves a smoother response, higher-order elements are appropriate. Figure 15.13 shows a mesh using the quadratic quadrilaterals (ELEM24). The mesh is nonuniform since we expect the solution to be more complicated near the heat source.

In this phase larger time steps, and therefore implicit, unconditionally stable methods, are best suited. For a first analysis one may not yet have any idea how long it will take for most of the significant effects to take place. Therefore one could employ a time-stepping strategy similar to the one in Fig. 11.33; i.e., cover several decades of time-step sizes, using only a few steps in each interval and $\theta = 2/3$, and end with two steady-state solutions using $\theta = 1$ (the second to verify convergence). This was done, and it was determined that most of the temperature changes occurred in a few hundred seconds, reaching to within

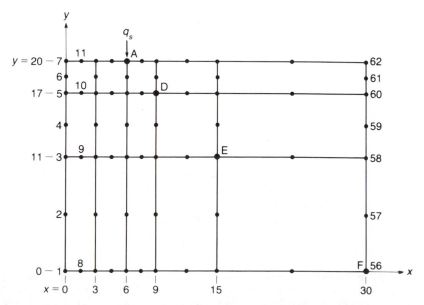

FIGURE 15.13 ● Global mesh for postshock response (coordinates in centimeters).

Interval	θ	Δt, sec	Number of steps, n_s	Time span, sec
I	2/3	0.1	1	0–0.1
II	2/3	0.1	9	0.1–1
III	2/3	1	9	1–10
IV	2/3	10	4	10–50
V	2/3	50	3	50–200
VI	2/3	100	8	200–1000
VII	1	10^6	1	$1000–10^6$

FIGURE 15.14 ● Time-stepping strategy used with the mesh in Fig. 15.13.

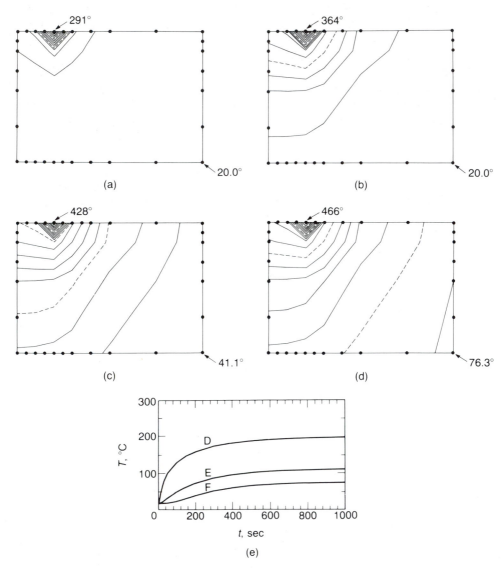

FIGURE 15.15 ● Postshock response, using the mesh in Fig. 15.13 and time-stepping strategy in Fig. 15.14. Temperature contours at (a) $t=10$ sec; (b) $t=50$ sec; (c) $t=200$ sec; (d) $t=10^6$ sec (steady state). Contour spacings, 20 °C; dashed contours at multiples of 100 °C. (e) Temperature histories at points D, E, and F.

1% of steady-state values by 1000 sec. Therefore the analysis was rerun using the time-stepping strategy shown in Fig. 15.14. The resulting temperature contour and time plots are shown in Fig. 15.15.

Note that the rise time of the "suddenly" applied heat source is 0.1 sec, which is much longer than the 0.01 and 0.0025 sec used in the shock response. However, this is unimportant on the present scale of tens and hundreds of seconds. Indeed, if the rise time is decreased, these results show no perceptible change.

A refined mesh is shown in Fig. 15.16. The elements have simply been halved in both directions, making mesh generation easy. A coarser mesh would be more appropriate in the lower-left corner, but again, as with the shock analysis, the extra computational cost is offset by the savings in mesh-generation effort. As a general rule, one should try to estimate how many times a given model is likely to be analyzed (e.g., perhaps dozens of times for a series of parametric studies). If the expectation is for many repeated analyses, then the higher computational costs would justify a greater initial effort to generate a more computationally efficient mesh.

A refined time-stepping strategy is shown in Fig. 15.17. Here the time step can be simply halved (rather than decreased to one-fourth) since there is no critical time step.

The resulting temperature contour and time plots are shown in Fig. 15.18. Note that the contours are approaching normality with the boundary (within the expected accuracy of the contour-plotting algorithm which ignores midside node values). This is necessary for the zero-flux boundary conditions, and hence affords a simple check on the reasonableness of the solution. A comparison of the coarse and refined results indicates the error at point D is about 5 °C and at points E and F is about 1 °C. (The error at point A is much greater, as discussed previously in regard to the shock response.)

Further discussion of the transient heat conduction problem, including practical modeling considerations, may be found in Damjanic and Owen [15.3, 15.4].

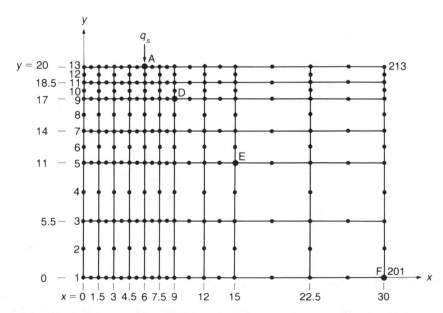

FIGURE 15.16 ● Refined (global) mesh for postshock response (coordinates in centimeters).

Interval	θ	Δt, sec	Number of steps, n_s	Time span, sec
I	2/3	0.05	2	0–0.1
II	2/3	0.05	18	0.1–1
III	2/3	0.5	18	1–10
IV	2/3	5	8	10–50
V	2/3	25	6	50–200
VI	2/3	50	16	200–1000
VII	1	10^6	1	$1000–10^6$

FIGURE 15.17 ● Refined time-stepping strategy used with the refined mesh in Fig. 15.16.

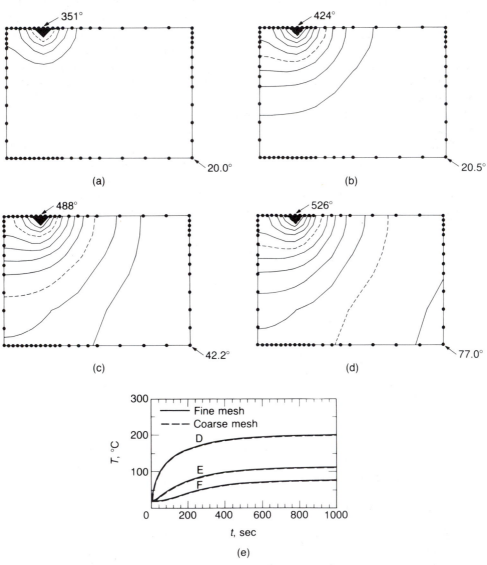

FIGURE 15.18 ● Postshock response, using the refined mesh in Fig. 15.16 and refined time-stepping strategy in Fig. 15.17. Temperature contours at (a) $t=10$ sec; (b) $t=50$ sec; (c) $t=200$ sec; (d) $t=10^6$ sec (steady state). Contour spacings, 20 °C; dashed contours at multiples of 100 °C. (e) Temperature histories at points D, E, and F.

Exercises

15.1   In the illustrative example in Section 15.4, the flux and temperature were infinite at the point of application of the concentrated heat source. This is because the dimension of the point, i.e., zero, is at least two less than that of the domain of the problem. Recall that the concentrated force in the 1-D hanging cable example in Chapters 7 and 8 yielded only a discontinuity in the force, not an infinity — because there the force was only one dimension less than that of the problem.

Conventional, nonsingular elements will give very poor results at or near such singularities (see Section 9.4), but farther away the accuracy is generally acceptable. How far away is "far enough"? This and the next exercise will provide a feel for how conventional elements behave near singularities.

Consider a thin circular disc, insulated on the upper and lower surfaces, initially at 0°C. (See (a) below.) At $t=0$ a point heat source, P cal/sec, is applied to the center and held indefinitely. (If we were to assume a line source, perpendicular to the plane of the disc, of uniform strength P cal/sec-cm, the problem would be exactly 2-D for any thickness of the disc. By considering only a "thin" disc, similar to the copper sheet in Section 15.4, we can assume a point source and ignore 3-D details of the temperature field in the immediate vicinity of the point. Thus, the field will be essentially 2-D throughout the entire disc, excepting only a small region, of diameter comparable to the thickness, surrounding the point.) The edge of the disc is maintained at 0°C. The exact solution for the steady-state temperature distribution is $T(r) = -(P/2\pi k\Delta z)\ln(r/a)$, where k is the thermal conductivity.

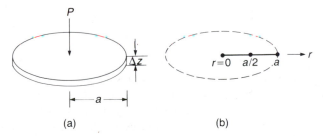

(a)   (b)

Employ a mesh of two uniform C^0-linear axisymmetric elements (shown in (b) above), using the element equations in Exercise 11.14. (*Hint:* Recall the footnote at the end of Section 15.4.3 relating radial flux to power input.) Calculate the steady-state nodal temperatures and plot them along with the curve for the exact solution. Refine the mesh to four uniform elements and again plot the nodal temperatures. Observe the accuracies as a function of radius.

15.2   Consider again the circular disc with the point heat source at the center, as described in Exercise 15.1. Again calculate the steady-state solution, but this time employ a mesh of one C^0-quadratic axisymmetric ring element, using the same three nodal points shown in (b) in Exercise 15.1.

Plot the nodal temperatures as well as the curve for the exact solution. Compare the accuracy, as a function of radius, with that of the mesh of two linear elements in the previous exercise. (*Hint:* Follow the method described in Exercise 11.14

for deriving the element equations, noting especially that the steady-state condition and the boundary conditions simplify the derivation so that only three integrals
need to be evaluated.)

15.3   Use the C^0-linear axisymmetric elements in Exercise 11.14 to model the shock
response in the illustrative example in Section 15.4.3. Employ a mesh of two
nonuniform elements, as shown below. The radius of node 2 (from point A) is
the distance between points A and B (see Fig. 15.3). The radius of node 3 is
3 cm (see Fig. 15.6).

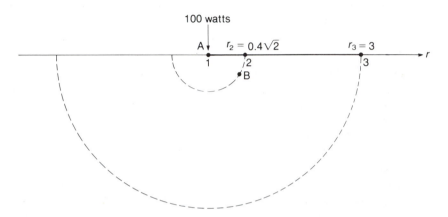

[Recall the footnote at the end of Section 15.4.3 relating radial flux to power input. However, be careful in translating the power input at the edge of the sheet
(180°) to the power input to an axisymmetric element (360°).] Compare your
results with those in Fig. 15.10.

The following set of eight exercises analyzes a 2-D *nonlinear* transient heat conduction
problem. Nonlinear analysis is a many-faceted topic that in general is outside the scope
of this book. However, as mentioned in the footnote related to Fig. 11.20, the incremental
time-stepping technique in transient analyses easily accommodates nonlinear physical properties, so it is possible to provide a brief introduction to nonlinear analysis in these exercises.

Exercise 15.4 defines the physical problem and derives the system equation (there
is only one!). Exercise 15.5 analyzes a closely related linear system in order to provide
a comparison with the nonlinear analyses in Exercises 15.6 through 15.11. There is usually
a variety of methods for solving nonlinear problems, no one method being best for all
problems. However, the backward difference method explained below has been used suc-
cessfully by the author and his colleagues for a variety of nonlinear heat conduction problems over a span of 15 years. It therefore seems to be fairly robust, i.e., able to handle
a wide class of mild-to-severe nonlinearities.

15.4   Consider a thin square sheet of material (shown in (a) below), insulated on both
surfaces and initially at 0°C. An interior heat source Q, distributed uniformly
throughout the sheet, is turned on suddenly at $t=0$ and remains on indefinitely.
The four edges of the sheet are maintained at 0°C. The thermal conductivity varies
linearly with temperature. (See (b) below.) The other properties have the following constant values: $L=10$ cm, $\varrho c=0.9$ cal/cm^3-°C, and $Q=4$ cal/sec-cm^3.

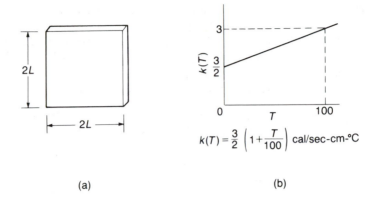

$$k(T) = \frac{3}{2}\left(1+\frac{T}{100}\right) \text{ cal/sec-cm-°C}$$

(a)   (b)

Model the sheet with four C^0-linear square elements (shown in (c) below), using the element matrices in Exercises 13.17 and 14.1. Those matrices assumed constant material properties. Hence, using the same rationale as used in Chapters 7 and 13, we can approximate the thermal conductivity in each element as a constant. It would be calculated from the expression for $k(T)$, using for T the average temperature in the element T_{ave}. The latter may be defined as the mean of the element's four nodal temperatures. (See (d) below.)

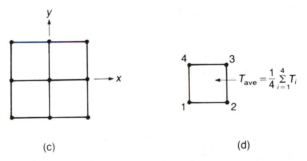

(c)   (d)

Considering the symmetry in the problem, as well as the zero essential BCs, show that the system equations consist of the following single differential equation:

$$\frac{\varrho c L^2}{9}\frac{dT(t)}{dt} + \frac{2}{3}k(T_{ave})T(t) = \frac{QL^2}{4} \tag{E15.1}$$

where $T(t)$ is the nodal temperature at the center node in the mesh. Using the above property and load data, show that Eq. (E15.1) reduces to the following expression:

$$10\frac{dT(t)}{dt} + \left(1+\frac{T}{400}\right)T(t) = 100 \tag{E15.2}$$

15.5 Before beginning the nonlinear analyses in the subsequent exercises, it will be instructive to look at a closely related linear problem. This will provide some physical insight into the effects of the nonlinearities. Thus, for this exercise only, let $k=3/2$ cal/sec-cm-°C, i.e., a constant, independent of temperature.

a) Using $k=3/2$, along with the other property and load data given in Exercise 15.4, solve Eq. (E15.1) with the backward difference method. Use 10-sec time steps and calculate the temperature history until it is within 5% of steady state.

b) What is the expression for the exact solution? [Recall Eqs. (11.1a) and (11.9).] Compare the FE solution with the exact solution.

c) The rate of convergence of the one-step backward difference method is $O(\Delta t)$. Show this by calculating the temperature at, say, $t=4$ sec, using first $\Delta t=2$ sec, then $\Delta t=1$ sec, and then $\Delta t=1/2$ sec. Observe the change in computed temperatures as Δt is progressively halved.

15.6 To solve the nonlinear problem [Eq. (E15.2)] with the backward difference method, we need to modify slightly the recurrence relation in order to accommodate a temperature-dependent conductivity matrix, $[K(T)]$, which will generally change in value from step to step. Thus, in Eq. (11.37) we need only include a subscript n on $[K]$, so that the recurrence relation in Eq. (11.40) becomes

$$\left(\frac{1}{\Delta t_n}[C]+[K]_n\right)\{a\}_n = \{F\}_n + \frac{1}{\Delta t_n}[C]\{a\}_{n-1} \tag{E15.3}$$

(If ϱc were temperature-dependent, we would do the same with the capacity matrix.)

The difficulty in solving this recurrence relation for each time step is that the coefficient matrix depends on the unknown temperature at time t_n, making Eq. (E15.3) nonlinear. This is resolved in the following manner. At the beginning of the step we *predict* the unknown temperature (in each element, and therefore at each node) as closely as possible, based on one or more of the preceding temperatures. (Several different prediction strategies will be illustrated below.) The predicted temperature is used to evaluate $[K]_n$, and then Eq. (E15.3) can be solved for $\{a\}_n$, the nodal temperatures at time t_n. The latter are referred to as *corrected* values; they will generally not agree with the predicted values. One could then reevaluate $[K]_n$ using the corrected values and solve Eq. (E15.3) again to obtain an improved set of corrected values. This *iterative* process can be repeated until some error criterion is satisfied or, alternatively, terminated after a predetermined number of steps, accepting whatever error remains.

Applying the recurrence relation [Eq. (E15.3)] and the above ideas to Eq. (E15.2) yields the following iterative recurrence relation for our thermal problem:

$$\left(\frac{10}{\Delta t_n}+1+\frac{T_n^{(i)}}{400}\right)T_n^{(i+1)} = 100 + \frac{10}{\Delta t_n}T_{n-1} \qquad i = 0,1,2,\cdots \tag{E15.4}$$

where i is the iteration number and $T_n^{(0)}$ is the predicted value which begins the iterative sequence. The reader is encouraged to verify Eq. (E15.4).

For "mild" nonlinearities (smoothly varying properties, without abrupt changes), experience has shown the following simple strategy to be quite effective:

> *Predictor:* $T_n^{(0)} = T_{n-1}$
>
> > that is, use the temperature at the end of the previous step to evaluate $[K(T)]$
>
> *Corrector:* Eq. (E15.4)
>
> *Iterations:* none

This simplifies Eq. (E15.4) to the following:

$$\left(\frac{10}{\Delta t_n}+1+\frac{T_{n-1}}{400}\right)T_n = 100 + \frac{10}{\Delta t_n}T_{n-1} \tag{E15.5}$$

Use Eq. (E15.5) to perform the following analyses.

a) Calculate the temperature history from 0 to 50 sec using 10-sec steps. Plot your results.

b) Check for convergence by repeating the analysis from 0 to 50 sec using 5-sec steps. Calculate a third value at $t=10$ sec using 2 1/2-sec steps. Show these results on the plot used in part (a). Do your approximate solutions appear to be converging?

c) The exact solution to the nonlinear differential equation (E15.2) is

$$T(t) = \frac{200}{1+\sqrt{2}} \left[\frac{1-e^{-(\sqrt{2}/10)t}}{1+(3-2\sqrt{2})e^{-(\sqrt{2}/10)t}} \right] \tag{E15.6}$$

Evaluate Eq. (E15.6) at $t=0$, 10, 20, 30, 40, and 50 sec, and add these values to the above plot. This will give you a feel for the accuracy of the one-step backward difference method in a nonlinear application.

Note: It is much more difficult to obtain closed-form exact solutions to nonlinear differential equations than to linear differential equations. Indeed, for almost any practical problem it is impossible, hence the need for numerical techniques like the FEM. The present thermal problem is tractable only because of the simplicity of Eq. (E15.2).

d) Verify that the *rate* of convergence is still $O(\Delta t)$ for this nonlinear problem, just as it was for the linear problem in Exercise 15.5. As with the spatial discretization, we speak of *asymptotic* rates, so a fairly constant rate may not emerge until Δt is "small enough." Thus, if you examine the three temperatures calculated at $t=10$ sec in parts (a) and (b) above, you may wish to calculate a fourth or fifth value to clearly reveal a trend.

15.7 Repeat part (a) of Exercise 15.6 but this time use Eq. (E15.4) and perform three iterations in each step. Compare your calculated values of T at 10, 20, 30, 40, and 50 sec with the noniterative and exact values calculated in parts (a) and (c), respectively, of Exercise 15.6. Surprised? Remember that the purpose of the iteration is to converge to the exact solution of the *recurrence* relation, not to the exact solution of the differential equation. These two solutions will generally agree only in the limit as $\Delta t \to 0$. In the present problem we see that the error introduced by the noniterative (approximate) solution to the recurrence relation partially offsets the error due to the recurrence relation itself (i.e., the backward difference error).

15.8 Calculate the *steady-state* solution by setting $\Delta t = \infty$ in Eq. (E15.4) and iterating until you estimate the error is less than 1%. Compare your solution with the exact solution from Eq. (E15.6).

Note that you are using only one (very large) time step so the error after predicting and correcting only once is likely to be significant. Hence iterating is generally necessary. If, however, the nonlinearities are severe (e.g., sudden, almost discontinuous, changes in the material properties), this iterative approach may not converge; in such cases one should instead step through the transient history using time steps small enough to follow the rapidly changing response.

15.9 The previous three exercises used the temperature at the end of the previous time step for the predicted temperature. Here we will consider *alternative strategies* *for the predictor*. For example, we can use the *two* previous temperatures to determine a linear variation of temperature with time which can then be extrapolated forward into the current time step, as shown below.

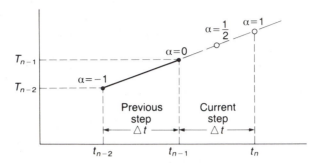

The dimensionless parameter α determines the time at which the extrapolated temperature is to be evaluated:

$$T_n^{(0)} = T_{n-1} + \alpha(T_{n-1} - T_{n-2}) \qquad \text{(E15.7)}$$

where $T_n^{(0)}$ is the predicted temperature for the current step. This expression assumes a constant time step, Δt; it would have to be modified slightly for a variable step. For the first time step at the beginning of the analysis ($n = 1$), it is convenient to define T_{-1} to be the same as T_0.

Note that $\alpha = 0$ is the predictor used in the previous exercises. Also note that $\alpha = 1/2$, which corresponds to evaluating $k(T)$ at the middle of the current step, is equivalent to taking the average $k(T)$ over the current step if $k(T)$ varies linearly with T (as it does in our problem).

a) Calculate the temperature history from 0 to 50 sec using 10-sec steps and $\alpha = 1$. Do not iterate, since that would eliminate most of the error produced by the predictor, and we wish to make a comparison of the relative accuracies of the different predictors.

b) Repeat the analysis using $\alpha = -1$.

c) If you did Exercise 15.6, add the above values to the plot constructed in that exercise. If you didn't, then (1) repeat the analysis once more for $\alpha = 0$, (2) evaluate Eq. (E15.6) for the exact solution, and (3) plot the results for $\alpha = -1$, 0, 1, and the exact solution.

Which predictor gives more accurate results? Before drawing any general conclusions, consider the problem in Exercise 15.10.

15.10 As a second comparative study of different predictors, let's alter the original problem in the following manner: change the thermal conductivity to $k(T) = (3/2)[1 - (T/100)]$, i.e., from a positive to a negative slope. This is not physically meaningful for $T \geq 100\,^{\circ}C$. However, the recurrence relation for our simple problem uses $1 - (T_{ave}/100)$, where $T_{ave} = T/4$, so it remains stable until $T \geq 400\,^{\circ}C$. (Both positive and negative slopes occur in real materials; e.g., aluminum has a positive slope and nickel has a negative slope.)

As was done in Exercise 15.9, calculate the temperature history from 0 to 50 sec using 10-sec steps. Use the predictor in Eq. (E15.7) with $\alpha = -1$, then $\alpha = 0$, and finally $\alpha = 1$. Do not iterate. Also calculate values from the following exact solution:

$$T(t) = \frac{200t}{20+t} \tag{E15.8}$$

Plot the three approximate solutions and the exact solution. Which predictor gives more accurate results? Considering the results from both Exercises 15.9 and 15.10, which predictor would you use for other problems that have more general types of nonlinear $k(T)$ functions?

15.11 The ambitious reader may wish to explore the *two-step* backward difference method defined in Exercise 11.1, performing analyses similar to those performed in the above exercises with the one-step method. The two-step method is, of course, of higher order than the one-step, so the rate of convergence will be faster. The predictor in Eq. (E15.7) could again be tested for a ''best'' α value, and these results compared with the effects of iterating. In addition, one might experiment with a higher-order predictor, e.g., fitting a *quadratic* polynomial to the *three* previous temperatures and then evaluating it at various times in the current or previous time steps.

References

[15.1]   P. C. Kohnke, Swanson Analysis Systems, Inc., Houston, Pa., private communication.

[15.2]   E. Hinton, T. Rock, and O. C. Zienkiewicz, ''A Note on Mass Lumping and Related Processes in the Finite Element Method,'' *Earthquake Eng Struct Dyn*, **4**: 245-249 (1976).

[15.3]   F. Damjanic and D. R. J. Owen, ''Practical Considerations for Thermal Transient Finite Element Analysis Using Isoparametric Elements,'' *Nuclear Eng Design*, **69**: 109-126 (1982).

[15.4]   D. R. J. Owen and F. Damjanic, ''Reduced Numerical Integration in Thermal Transient Finite Element Analysis,'' *Computers and Structures*, **17**(2): 261-276 (1983).

16

Elasticity

Introduction

As noted previously (in the preface and again in Section 1.6), the field of structural mechanics, of which elasticity is the central discipline, provided the initial impetus to the practical development of the FEM. And today, more than 25 years later, elasticity is still the field where the great bulk of FE analysis is performed. In fact most of the FE literature is devoted to this field. It is therefore fitting that this final chapter be devoted to elasticity.

Since this field of FE application is quite mature, a single chapter can't begin to cover the detailed developments that have evolved over a quarter century; indeed, most FE text-books are devoted predominantly or exclusively to the topic. The primary purpose here is to demonstrate how the methods of analysis developed in the previous chapters are applied to this field. Secondarily, for those who wish to pursue this topic in greater depth, this treatment (along with the rest of the book) will, it is hoped, provide an easy springboard to the more advanced, specialized literature. Thirdly, the method shown here for handling more than one unknown function is applicable to other fields, e.g., electromagnetism, fluid dynamics, etc.

For those readers unfamiliar with elasticity theory, Section 16.1 will review the basic relations in 3-D elasticity [16.1-16.3]. With these at hand, it will be easier to understand the assumptions that simplify these relations to 2-D applications. We will have frequent recourse to these relations in the ensuing sections.

The remainder of the chapter will examine three types of 2-D elasticity problems: plane stress, plane strain, and axisymmetric bodies of revolution. The first two employ rectangular Cartesian coordinates. Their elasticity expressions are very similar and their FE formulations virtually identical, so it is natural to treat both types together. Sections 16.2 and 16.3 will therefore analyze the two plane problems.

Axisymmetric bodies of revolution employ cylindrical coordinates, and this alters the elasticity expressions a bit. The FE formulation, though, is remarkably similar to the plane problems. This third type of 2-D problem will be treated separately in Section 16.4.

Unlike the previous chapters, we will not implement the theory into a program. It is presumed that any reader interested in elasticity will very likely already have available

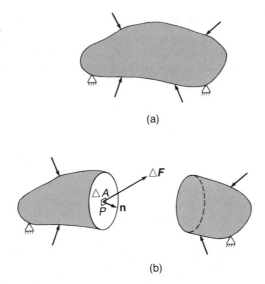

FIGURE 16.1 ● (a) A body with applied loads and (b) an imaginary planar cut through the body.

an elasticity FE program (with probably much more capability than developed here). Hence there is little incentive to develop yet another elasticity program that is not likely to be used very much, if at all. Therefore Section 16.5 will only briefly discuss some programming considerations.

16.1 Basic Relations in Linear Static Elasticity

The following relations are presented in 3-D rectangular Cartesian coordinates, which will be denoted by x_1, x_2, x_3 in some expressions and x, y, z in others. The former is more useful for theoretical expressions and the latter for practical applications. Vectors will be indicated by bold type.

Stress ● Consider a body with applied loads (Fig. 16.1a), and an imaginary plane cut through the body that passes through an interior point P (Fig. 16.1b). Here n is the unit normal to the plane, ΔA is an element of area on the plane surrounding P, and ΔF is the resultant force on ΔA due to the part of the body on the RHS of the cut exerting forces on the LHS.

The stress vector $\overset{(n)}{\tau}$ is defined as[1]

$$\overset{(n)}{\tau} = \lim_{\Delta A \to 0} \frac{\Delta F}{\Delta A} \tag{16.1}$$

---

[1] The vector symbol n is placed above the τ, rather than as a sub- or superscript, since it is not an integer-valued index and hence is incompatible with the integer-valued sub- and superscripts that commonly occur in the tensor notation used in elasticity theory.

The stress vector depends on the location of the point as well as the orientation of the surface through the point.

If the cutting plane is perpendicular to one of the axes, say x_1, (Fig. 16.2), then the components of the stress vector in each of the three coordinate directions are denoted by doubly subscripted letters:

$$\overset{(i)}{T_1} = \tau_{11}$$

$$\overset{(i)}{T_2} = \tau_{12}$$

$$\overset{(i)}{T_3} = \tau_{13} \tag{16.2}$$

The first subscript refers to the positive direction of the normal to the surface; the second subscript refers to the positive direction of a component of the stress vector acting on the surface.

Similar expressions apply to cutting planes perpendicular to directions x_2 and x_3, resulting in a total of nine stresses acting on three mutually perpendicular coordinate planes (Fig. 16.3). Thus, τ_{ij} is the jth component of the stress vector that acts on the positive side of the ith coordinate plane. The nine stresses are usually written in the matrix form,

$$\begin{bmatrix} \tau_{11} & \tau_{12} & \tau_{13} \\ \tau_{21} & \tau_{22} & \tau_{23} \\ \tau_{31} & \tau_{32} & \tau_{33} \end{bmatrix} \tag{16.3}$$

The terms with identical subscripts (which lie on the main diagonal), τ_{11}, τ_{22}, τ_{33}, are called the *normal stresses*, and the terms with mixed subscripts, τ_{12}, τ_{13}, τ_{32}, etc., are called the *shear stresses*.

We next construct a small tetrahedral element at the point P in which three of the sides are coordinate planes and the fourth side is in the general direction n (Fig. 16.4). Applying balance of linear momentum to the element and letting the volume shrink to zero, one can derive the important relation,

$$\overset{(n)}{T_i} = \tau_{ji} n_j \qquad i = 1,2,3 \tag{16.4}$$

where the summation convention is used — namely, a pair of repeated indices in a monomial expression means sum from 1 to 3. For example, $\overset{(n)}{T_1} = \tau_{11}n_1 + \tau_{21}n_2 + \tau_{31}n_3$. (This convention will be used henceforth.) Equation (16.4) states that the components of the stress vector acting on *any* plane through a point are a linear combination of the components of the stress vectors acting on the three coordinate planes at that point. In other words, the array τ_{ji} completely describes the state of stress at a point. Equation (16.4) is useful, for example, in relating stresses acting on a boundary surface to the normal and shear stresses.

Consider a second set of rectangular Cartesian axes x_1', x_2', x_3' (Fig. 16.5). Let l_{kj} be the cosine of the angle between x_k' and x_j. Using Eq. (16.4) it can be shown [16.4, 16.5] that the nine stresses in Eq. (16.3) transform as a *tensor*:

$$\tau_{km}' = l_{kj} l_{mi} \tau_{ji} \qquad k,m = 1,2,3 \tag{16.5}$$

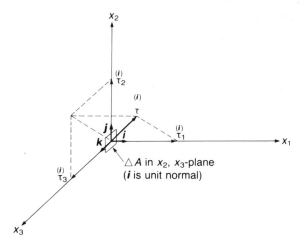

FIGURE 16.2 ● Cutting plane perpendicular to the x_1-axis.

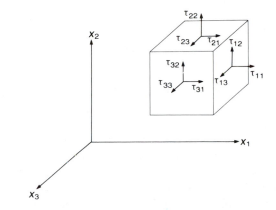

FIGURE 16.3 ● Normal and shear stresses at a point.

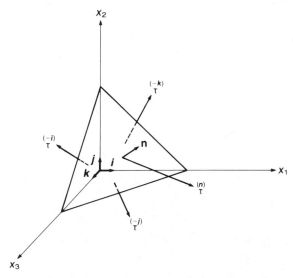

FIGURE 16.4 ● The stress vector on any plane at a point is a linear combination of the normal and shear stresses on the coordinate planes.

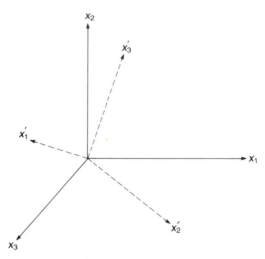

FIGURE 16.5 ● Two coordinate systems for showing transformation of stresses.

where, for example, τ'_{23} is the x'_3-component of the stress vector that acts on the positive side of the x'_2-coordinate plane. (Because of the summation convention, the two pairs of repeated indices in the term on the RHS imply a double sum over i and j, each sum from 1 to 3, i.e., a total of nine terms.) The 3-by-3 array in Eq. (16.3) is therefore referred to as the *stress tensor*, and the individual terms are the components of the stress tensor.

There is only one stress tensor at a point, but it may, according to Eq. (16.5), have different components with respect to different coordinate systems. In particular, it is always possible to find one set of orthogonal coordinate axes such that the stress tensor with respect to these axes is diagonal; i.e., all shear stresses vanish. These are the *principal axes of stress*. The normal stresses along these axes are called the *principal stresses*.

The equations of static equilibrium (balance equations) may be derived by considering a *balance of forces* on an infinitesimal volume element, as shown in Fig. 16.6. To prevent crowding of the figure, only forces in the x_1-direction are shown; similar forces

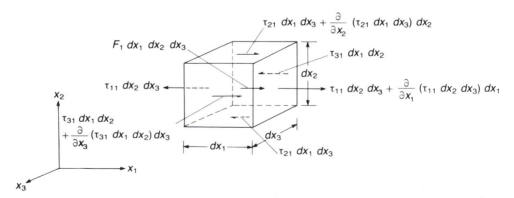

FIGURE 16.6 ● Forces in the x_1-direction on an infinitesimal volume element.

act in the x_2- and x_3-directions. Here f_1 is the x_1-component of the body force, i.e., a force per unit volume. A balance of forces in the x_1-direction yields

$$\frac{\partial \tau_{11}}{\partial x_1} + \frac{\partial \tau_{21}}{\partial x_2} + \frac{\partial \tau_{31}}{\partial x_3} = -f_1 \tag{16.6a}$$

Similarly for the other directions,

$$\frac{\partial \tau_{12}}{\partial x_1} + \frac{\partial \tau_{22}}{\partial x_2} + \frac{\partial \tau_{32}}{\partial x_3} = -f_2$$

$$\frac{\partial \tau_{13}}{\partial x_1} + \frac{\partial \tau_{23}}{\partial x_2} + \frac{\partial \tau_{33}}{\partial x_3} = -f_3 \tag{16.6b}$$

These three equations may be written in the abbreviated form,

$$\frac{\partial \tau_{ji}}{\partial x_j} = -f_i \qquad i = 1,2,3 \tag{16.7}$$

(Note the summation convention on the index j.) Equations (16.6) are the *equations of (stress) equilibrium* in rectangular Cartesian coordinates.

In a similar manner, a *balance of moments* about each axis yields

$$\tau_{ij} = \tau_{ji} \tag{16.8}$$

Thus, the stress tensor is symmetric.

Deformation •

1. *General (nonlinear) theory.* Figure 16.7 shows a material body in two configurations: an initial undeformed configuration (say at time t_0), before any loads are applied, and a subsequent deformed configuration (at time t), after loads have been applied. The displacement vector u is defined as the vector that extends from a material point P in the undeformed configuration to the same material point in the deformed configuration. u has three components, each a function of the position of $P(t_0)$:[2]

$$u(x_1,x_2,x_3) = u_1(x_1,x_2,x_3)i + u_2(x_1,x_2,x_3)j + u_3(x_1,x_2,x_3)k \tag{16.9}$$

The displacement field (i.e., the displacement vector, viewed as a function over the entire domain) may contain rigid-body motions as well as pure deformations. Rigid-body motions include translations and rotations; they are characterized by the property that the distance between any pair of material points remains unchanged. Conversely, any quantity that measures the *change* in length (and only the change in length) between any pair of neighboring material points is a measure of pure deformation only. One such measure is the *Lagrangian strain* tensor E_{ij}, which in rectangular Cartesian coordinates has the form

$$E_{ij} = \frac{1}{2}\left(\frac{\partial u_i}{\partial x_j} + \frac{\partial u_j}{\partial x_i} + \frac{\partial u_k}{\partial x_i}\frac{\partial u_k}{\partial x_j} \right) \qquad i,j = 1,2,3 \tag{16.10}$$

---

[2] Equation (16.9) expresses u as a function of the coordinates of the *un*deformed configuration, sometimes called the *material coordinates*. This will lead to a measure of deformation called the *Lagrangian strain*. Alternatively we could express u as a function of the coordinates of the deformed configuration, called the *spatial coordinates*. This would lead to an *Eulerian strain*. In the limit of an infinitesimal deformation, the two strains become identical.

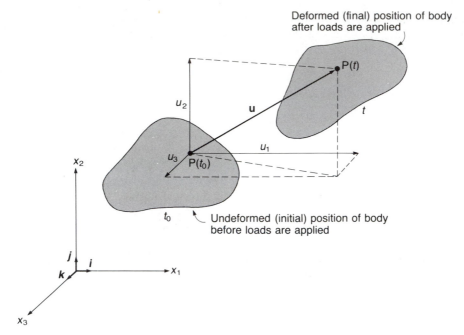

FIGURE 16.7 ● Displacement **u** from undeformed to deformed configuration.

For practical applications we will use the conventional notation u,v,w in place of u_1,u_2,u_3, respectively. If x_1,x_2,x_3 are also replaced by x,y,z, respectively, then each of the nine strain components in Eq. (16.10) may be written in the following conventional form:

$$E_{xx} = \frac{\partial u}{\partial x} + \frac{1}{2}\left[\left(\frac{\partial u}{\partial x}\right)^2 + \left(\frac{\partial v}{\partial x}\right)^2 + \left(\frac{\partial w}{\partial x}\right)^2\right]$$

$$E_{yy} = \frac{\partial v}{\partial y} + \frac{1}{2}\left[\left(\frac{\partial u}{\partial y}\right)^2 + \left(\frac{\partial v}{\partial y}\right)^2 + \left(\frac{\partial w}{\partial y}\right)^2\right]$$

$$E_{zz} = \frac{\partial w}{\partial z} + \frac{1}{2}\left[\left(\frac{\partial u}{\partial z}\right)^2 + \left(\frac{\partial v}{\partial z}\right)^2 + \left(\frac{\partial w}{\partial z}\right)^2\right]$$

$$E_{xy} = \frac{1}{2}\left(\frac{\partial u}{\partial y} + \frac{\partial v}{\partial x}\right) + \frac{1}{2}\left[\frac{\partial u}{\partial x}\frac{\partial u}{\partial y} + \frac{\partial v}{\partial x}\frac{\partial v}{\partial y} + \frac{\partial w}{\partial x}\frac{\partial w}{\partial y}\right]$$

$$E_{yz} = \frac{1}{2}\left(\frac{\partial v}{\partial z} + \frac{\partial w}{\partial y}\right) + \frac{1}{2}\left[\frac{\partial u}{\partial y}\frac{\partial u}{\partial z} + \frac{\partial v}{\partial y}\frac{\partial v}{\partial z} + \frac{\partial w}{\partial y}\frac{\partial w}{\partial z}\right]$$

$$E_{zx} = \frac{1}{2}\left(\frac{\partial w}{\partial x} + \frac{\partial u}{\partial z}\right) + \frac{1}{2}\left[\frac{\partial u}{\partial z}\frac{\partial u}{\partial x} + \frac{\partial v}{\partial z}\frac{\partial v}{\partial x} + \frac{\partial w}{\partial z}\frac{\partial w}{\partial x}\right]$$

$$E_{yx} = E_{xy}$$

$$E_{zy} = E_{yz} \qquad \text{(the strain tensor is symmetric)}$$

$$E_{xz} = E_{zx} \qquad\qquad\qquad\qquad\qquad\qquad\qquad \text{(16.11)}$$

The presence of the squared terms in Eqs. (16.11) makes the strains a nonlinear function of the displacements. Any theoretical formulation utilizing Eqs. (16.11) would constitute a nonlinear theory of elasticity.

It is clear from Eqs. (16.11) that the strain is defined in terms of nine independent *displacement gradients*. They are the components of a 3-by-3 displacement gradient tensor,

$$
\left[\frac{\partial u_i}{\partial x_j} \right] =
\begin{bmatrix}
\dfrac{\partial u}{\partial x} & \dfrac{\partial u}{\partial y} & \dfrac{\partial u}{\partial z} \\[2mm]
\dfrac{\partial v}{\partial x} & \dfrac{\partial v}{\partial y} & \dfrac{\partial v}{\partial z} \\[2mm]
\dfrac{\partial w}{\partial x} & \dfrac{\partial w}{\partial y} & \dfrac{\partial w}{\partial z}
\end{bmatrix}
\tag{16.12}
$$

Since a general tensor may always be expressed as the sum of symmetric and skew-symmetric parts, we may write

$$
\frac{\partial u_i}{\partial x_j} = e_{ij} + \omega_{ij}
\tag{16.13}
$$

where

$$
e_{ij} = \frac{1}{2} \left(\frac{\partial u_i}{\partial x_j} + \frac{\partial u_j}{\partial x_i} \right)
$$

$$
\omega_{ij} = \frac{1}{2} \left(\frac{\partial u_i}{\partial x_j} - \frac{\partial u_j}{\partial x_i} \right)
$$

The displacement gradients are dimensionless numbers in Cartesian coordinates. Their values may be large relative to unity (nonlinear theory) or small relative to unity (linear theory).

2. **Linear theory.** If all nine displacement gradients are small relative to unity, then the products of the gradients are small relative to the gradients themselves, and Eq. (16.10) can be written approximately as

$$
E_{ij} \simeq \frac{1}{2} \left(\frac{\partial u_i}{\partial x_j} + \frac{\partial u_j}{\partial x_i} \right)
$$

$$
= e_{ij}
\tag{16.14}
$$

These are the kinematic assumptions of the classical *linear* (or, infinitesimal) *elasticity* theory, which will be used in the remainder of this chapter.

The symmetric tensor e_{ij} is the classical linear strain tensor,

$$
e_{11} = \frac{\partial u}{\partial x} \qquad e_{12} = \frac{1}{2} \left(\frac{\partial u}{\partial y} + \frac{\partial v}{\partial x} \right)
$$

$$
e_{22} = \frac{\partial v}{\partial y} \qquad e_{23} = \frac{1}{2} \left(\frac{\partial v}{\partial z} + \frac{\partial w}{\partial y} \right)
$$

$$
e_{33} = \frac{\partial w}{\partial z} \qquad e_{31} = \frac{1}{2} \left(\frac{\partial w}{\partial x} + \frac{\partial u}{\partial z} \right)
\tag{16.15}
$$

When the displacement gradients are small, the components e_{11}, e_{22}, e_{33} may be given the physical interpretation of *extension*, i.e., the change in length per unit length of an infinitesimal ''fiber'' that is initially parallel to one of the coordinate axes. The components e_{12}, e_{23}, e_{31} may be interpreted as half the *shear*, where shear is the decrease in angle of two infinitesimal fibers that are initially parallel to two perpendicular coordinate axes.[3]

The skew-symmetric tensor ω_{ij} is the classical linear rotation tensor,

$$\omega_{12} = \frac{1}{2}\left(\frac{\partial v}{\partial x} - \frac{\partial u}{\partial y}\right)$$

$$\omega_{23} = \frac{1}{2}\left(\frac{\partial w}{\partial y} - \frac{\partial v}{\partial z}\right)$$

$$\omega_{31} = \frac{1}{2}\left(\frac{\partial u}{\partial z} - \frac{\partial w}{\partial x}\right) \qquad (16.16)$$

When the displacement gradients are small, the components ω_{ij} may be given the physical interpretation of *rigid-body rotation*, i.e., the change in angle of a fiber relative to the coordinate axes.

The assumption that all nine displacement gradients are small is equivalent to the assumption that all six independent e_{ij} and all three independent ω_{ij} are small. When performing a linear elastic analysis, it is wise to check the results to ensure that e_{ij} and ω_{ij} are indeed small.

We note in passing that the so-called linear ''strain'' tensor e_{ij} is not really a measure of strain (i.e., pure deformation) because it also measures, to second order, effects of rotation. Similarly, ω_{ij} measures second-order effects of strain. Of course, these second-order effects are precisely the nonlinear terms in Eq. (16.10) that were neglected (see Exercises 16.1 and 16.2).

Constitutive (stress-strain) relations • The above relations for static equilibrium (involving stresses) and deformation (strains) are valid for all materials, whether solid, liquid, or gas. In addition, a constitutive equation is needed that relates stress and strain for a particular type of material. A *linear elastic* material is described by the generalized Hooke's law,

$$\tau_{ij} = C_{ijkl} e_{kl} \qquad (16.17)$$

where C_{ijkl} is a fourth-rank tensor of 81 (=3×3×3×3) elastic moduli. However, by utilizing the symmetry of τ_{ij} and e_{kl}, and by assuming the existence of a strain-energy density function, the maximum number of independent elastic moduli reduces to 21, which represents the maximum possible degree of anisotropy (see discussion on anisotropy in Section 13.1.2). It is then more convenient for engineering applications to write τ_{ij} and

---

[3] For extension, the linear theory deals only with the projection of the final fiber along its initial direction. In this context, extension becomes synonymous with *elongation* (a different measure of strain in nonlinear theory). For shear, the linear theory deals only with the projection of the final angle onto the plane defined by the initial perpendicular fibers. Truesdell and Toupin [16.3] rigorously define the various nonlinear and linearized measures of strain.

e_{kl} as vectors (rather than second-rank tensors) and C_{ijkl} as a matrix, so that Eq. (16.17) becomes

$$
\begin{Bmatrix}
\sigma_x \\
\sigma_y \\
\sigma_z \\
\tau_{xy} \\
\tau_{yz} \\
\tau_{zx}
\end{Bmatrix}
=
\begin{bmatrix}
C_{11} & C_{12} & C_{13} & C_{14} & C_{15} & C_{16} \\
       & C_{22} & C_{23} & C_{24} & C_{25} & C_{26} \\
       &   & C_{33} & C_{34} & C_{35} & C_{36} \\
       &   &   & C_{44} & C_{45} & C_{46} \\
       & \text{Symmetric} & &   & C_{55} & C_{56} \\
       &   &   &   &   & C_{66}
\end{bmatrix}
\begin{Bmatrix}
\epsilon_x \\
\epsilon_y \\
\epsilon_z \\
\gamma_{xy} \\
\gamma_{yz} \\
\gamma_{zx}
\end{Bmatrix}
\tag{16.18}
$$

A more engineering-oriented notation has been introduced in Eq. (16.18). It is related to the tensor notation shown in Table 16.1.

TABLE 16.1 ●

	Tensor	Engineering		Tensor	Engineering
Normal stresses	$\tau_{11} =$ $\tau_{22} =$ $\tau_{33} =$	σ_x σ_y σ_z	Normal strains	$e_{11} =$ $e_{22} =$ $e_{33} =$	ϵ_x ϵ_y ϵ_z
Shear stresses	$\tau_{12} =$ $\tau_{23} =$ $\tau_{31} =$	τ_{xy} τ_{yz} τ_{zx}	Shear strains	$2e_{12} =$ $2e_{23} =$ $2e_{31} =$	γ_{xy} γ_{yz} γ_{zx}

Note the factor of 2 in the shear strains; γ_{xy} physically represents shear, so e_{12} is half the shear. We also note that the engineering stress and strain vectors are not (first-rank) tensors; i.e., they do not transform according to the rules of tensor transformation when there is a change of coordinates. Such a transformation is performed by using the tensor rules with the doubly subscripted stress and strain tensors, as in Eq. (16.5), and then subsequently changing back to the engineering notation (see Sections 16.2.2 and 16.4.1).

For the important and frequently occurring case of *isotropic* materials, Eq. (16.18) simplifies to the following:

$$
\begin{Bmatrix}
\sigma_x \\
\sigma_y \\
\sigma_z \\
\tau_{xy} \\
\tau_{yz} \\
\tau_{zx}
\end{Bmatrix}
=
\frac{E}{(1+v)(1-2v)}
\begin{bmatrix}
1-v & v & v & 0 & 0 & 0 \\
    & 1-v & v & 0 & 0 & 0 \\
    &   & 1-v & 0 & 0 & 0 \\
    &   &   & \frac{1}{2}-v & 0 & 0 \\
    & \text{Symmetric} & & & \frac{1}{2}-v & 0 \\
    &   &   &   &   & \frac{1}{2}-v
\end{bmatrix}
\begin{Bmatrix}
\epsilon_x \\
\epsilon_y \\
\epsilon_z \\
\gamma_{xy} \\
\gamma_{yz} \\
\gamma_{zx}
\end{Bmatrix}
\tag{16.19}
$$

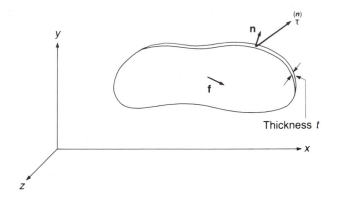

FIGURE 16.8 ● Typical geometry for a plane stress problem.

where E is Young's modulus and ν is Poisson's ratio. Other engineering moduli may be defined in terms of these two, e.g., bulk modulus, shear modulus, or the Lamé constants [16.6]. A full discussion of linear elastic moduli for anisotropic materials may be found in Lekhnitskii [16.7].

16.2 Plane Stress and Plane Strain

16.2.1 Mathematical and Physical Description

For the plane stress state, consider a thin plate of thickness t, where t is small relative to the other two dimensions (Fig. 16.8). The middle plane lies halfway between both faces. For convenience, we define the x,y-axes to lie in the middle plane. Let the faces be free of applied loads. Any surface stresses $\overset{(n)}{\tau}$ (sometimes called *tractions*), applied to the edges, and any body forces **f**, applied to the interior, must be parallel to the middle plane and distributed symmetrically with respect to it (in the z-direction). Anisotropic materials must be limited to orthotropic symmetry [16.7] (nine independent elastic moduli) and must be oriented such that two of the three principal axes are "in plane" (in the x,y-plane), and hence the third axis is perpendicular to the plane (in the z-direction). This is necessary to decouple in-plane normal components of stress and strain from out-of-plane shear components.

The resulting problem is not exactly two-dimensional but, for practical purposes, can be made so by averaging all quantities over the thickness.[4] The (generalized) plane stress state can then be defined by the properties

$$\sigma_z = 0$$

$$\tau_{xz} = 0$$

$$\tau_{yz} = 0 \tag{16.20}$$

---

[4] This is called *generalized plane stress*. See Sokolnikoff [16.8], who attributes the averaging idea to Filon in 1903.

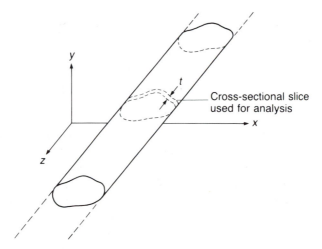

Cross-sectional slice
used for analysis

FIGURE 16.9 ● Typical geometry for a plane strain problem.

It also follows for the displacement field that

$$u = u(x,y)$$

$$v = v(x,y) \tag{16.21}$$

The average value of the z-component of displacement w is zero, but the average normal strain in the z-direction, ϵ_z, is not zero. Given the above requirement on the orientation of the principal axes of material anisotropy, it follows from Eqs. (16.20) that $\gamma_{xz}=\gamma_{yz}=0$.[5]
In summary, the stresses and strains in a plane stress problem are as follows:

$$\sigma_x(x,y) \qquad \epsilon_x(x,y)$$

$$\sigma_y(x,y) \qquad \epsilon_y(x,y)$$

$$\tau_{xy}(x,y) \qquad \gamma_{xy}(x,y)$$

$$\sigma_z = 0 \qquad \epsilon_z(x,y) \quad (\epsilon_z \text{ is related to } \epsilon_x \text{ and } \epsilon_y \text{ by}$$
$$\text{stress-strain relations, using } \sigma_z{=}0)$$
$$\tau_{xz} = 0 \qquad \gamma_{xz} = 0$$

$$\tau_{yz} = 0 \qquad \gamma_{yz} = 0 \tag{16.22}$$

For the plane strain state, consider a long, slender, cylindrical body with a cross section of arbitrary shape (Fig. 16.9). For convenience, we define the z-axis to be parallel to the generators of the cylinder, so that the cross section is in the x,y-plane. Applied surface tractions and body forces must be parallel to the x,y-plane (''in-plane'') and independent of the z-direction, i.e., applied along its entire length. Displacement in the z-direction must be constant (no loss of generality if assumed zero). Anisotropic materials

---

[5] Vanishing shear stresses imply vanishing shear strains for most material anisotropies encountered in practice.

must be oriented so that two principal axes are in-plane. The resulting plane strain state can then be defined by the properties

$$u = u(x,y)$$

$$v = v(x,y)$$

$$w = \text{constant} \tag{16.23}$$

It follows from this that $\gamma_{xz} = \gamma_{yz} = \epsilon_z = 0$. And from the assumption on orientation of the principal axes of material anisotropy, we have $\tau_{xz} = \tau_{yz} = 0$. However, in general, σ_z does not vanish. All quantities are independent of z. Hence, for analysis one needs to examine only a cross-sectional slice of uniform, but arbitrary, thickness t.

In summary, the stresses and strains in a plane strain problem are as follows:

$$\sigma_x(x,y) \qquad \epsilon_x(x,y)$$

$$\sigma_y(x,y) \qquad \epsilon_y(x,y)$$

$$\tau_{xy}(x,y) \qquad \gamma_{xy}(x,y)$$

$$\sigma_z(x,y) \qquad \epsilon_z = 0 \quad (\sigma_z \text{ is related to } \epsilon_x \text{ and } \epsilon_y \text{ by}$$

$$\tau_{xz} = 0 \qquad \gamma_{xz} = 0 \qquad \text{stress-strain relations, using } \epsilon_z = 0)$$

$$\tau_{yz} = 0 \qquad \gamma_{yz} = 0 \tag{16.24}$$

Note that these differ from the plane stress problem, Eqs. (16.22), only with respect to σ_z and ϵ_z.

As with the plane stress state, these conditions are usually only approximated in reality. If a soft material is constrained between two relatively rigid, flat end-blocks, then plane strain will essentially be achieved for any length of cylinder, even a thin plate (as in Fig. 16.8). If such external constraints aren't available, then requiring the cylinder to be long and slender will approximately create a plane strain state at some distance from the ends, since the end regions effectively provide a constraint for the center region.

The equations of equilibrium are the same for both plane stress and plane strain. Thus, using either Eqs. (16.22) or (16.24), the first two equations in Eqs. (16.6) reduce to the following:

$$\frac{\partial \sigma_x}{\partial x} + \frac{\partial \tau_{xy}}{\partial y} = -f_x$$

$$\frac{\partial \tau_{xy}}{\partial x} + \frac{\partial \sigma_y}{\partial y} = -f_y \tag{16.25}$$

The third equation is identically satisfied.

16.2.2 Stress-Strain Relations

We begin with the general 3-D relations, Eq. (16.18), and reduce them appropriately. We don't need to consider the most general form of anisotropy (21 independent moduli) because such materials are incompatible with a 2-D analysis (since effects in all three directions would be interrelated). In any case, one rarely encounters materials so highly anisotropic. *Orthotropic* material symmetry (nine independent moduli) is compatible with

2-D and sufficiently anisotropic to represent most practical materials. For orthotropic symmetry, Eq. (16.18) becomes

$$
\begin{Bmatrix} \sigma_{\hat{x}} \\ \sigma_{\hat{y}} \\ \sigma_{\hat{z}} \\ \tau_{\hat{x}\hat{y}} \\ \tau_{\hat{y}\hat{z}} \\ \tau_{\hat{z}\hat{x}} \end{Bmatrix} =
\begin{bmatrix}
C_{\hat{x}\hat{x}} & C_{\hat{x}\hat{y}} & C_{\hat{x}\hat{z}} & 0 & 0 & 0 \\
 & C_{\hat{y}\hat{y}} & C_{\hat{y}\hat{z}} & 0 & 0 & 0 \\
 & & C_{\hat{z}\hat{z}} & 0 & 0 & 0 \\
 & & & G_{\hat{x}\hat{y}} & 0 & 0 \\
 & \text{Symmetric} & & & G_{\hat{y}\hat{z}} & 0 \\
 & & & & & G_{\hat{z}\hat{x}}
\end{bmatrix}
\begin{Bmatrix} \epsilon_{\hat{x}} \\ \epsilon_{\hat{y}} \\ \epsilon_{\hat{z}} \\ \gamma_{\hat{x}\hat{y}} \\ \gamma_{\hat{y}\hat{z}} \\ \gamma_{\hat{z}\hat{x}} \end{Bmatrix}
\tag{16.26a}
$$

or, in abbreviated form,

$$
\{\hat{\sigma}\} = [\hat{C}] \{\hat{\epsilon}\}
\tag{16.26b}
$$
$$
{}_{6\times1} \qquad {}_{6\times6}\ {}_{6\times1}
$$

where the circumflex indicates local principal coordinates, i.e., aligned with the principal axes of the material (Fig. 16.10). As suggested in the figure, the orientation of the principal axes might vary over the domain. At each point in the domain, Eq. (16.26) is with respect to the local principal axes at that point. As noted before, the $\hat{z}$-axis must everywhere be perpendicular to the x,y-plane (parallel to the z-axis) in order to decouple in-plane effects from the z-direction.

It is frequently useful to include in the stress-strain relations the effects of thermal expansion (and contraction) by subtracting thermal strains from total strains [16.9], the latter including both stress-induced and thermally induced strains. Equation (16.26b) becomes

$$
\{\hat{\sigma}\} = [\hat{C}](\underbrace{\{\hat{\epsilon}\}}_{\text{total}} - \underbrace{\{\hat{\epsilon}_T\}}_{\text{thermal}})
\tag{16.27}
$$

where

$$
\{\hat{\epsilon}_T\} = \begin{Bmatrix} \alpha_{\hat{x}}\Delta T \\ \alpha_{\hat{y}}\Delta T \\ \alpha_{\hat{z}}\Delta T \\ 0 \\ 0 \\ 0 \end{Bmatrix}
$$

Here $\alpha_{\hat{x}}$, $\alpha_{\hat{y}}$, and $\alpha_{\hat{z}}$ are the thermal expansion coefficients with respect to the local principal axes, and ΔT is the change in temperature from a stress-free reference temperature. The last three components of $\{\hat{\epsilon}_T\}$ are zero because free thermal expansion only produces a change in volume (normal strains), not a change in shape (shear strains).[6]

---

[6] The thermal strain vector $\{\hat{\epsilon}_T\}$ is a special case of so-called *initial strains* that may be produced by a variety of different physical phenomena, e.g., swelling or crystal growth.

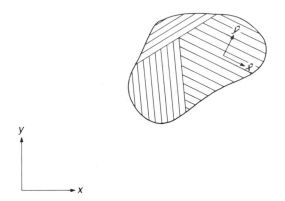

y
↑
|
|
|
└─────→ x

FIGURE 16.10 ● Local principal axes for an orthotropic material.

To perform a 2-D FE analysis using the global x,y-coordinate system, we will need to transform the set of six 3-D equations with respect to local principal axes, Eq. (16.27), to a set of (three) 2-D equations with respect to the global x,y-axes. The desired stress-strain equations will have the following form:

$$
\left\{
\begin{array}{c}
\sigma_x \\
\sigma_y \\
\tau_{xy}
\end{array}
\right\} =
\left[
\begin{array}{ccc}
C_{xx} & C_{xy} & C_{x,xy} \\
 & C_{yy} & C_{y,xy} \\
\text{Symmetric} & & C_{xy,xy}
\end{array}
\right]
\left(
\left\{
\begin{array}{c}
\epsilon_x \\
\epsilon_y \\
\gamma_{xy}
\end{array}
\right\} -
\left\{
\begin{array}{c}
\alpha_x \Delta T \\
\alpha_y \Delta T \\
\alpha_{xy} \Delta T
\end{array}
\right\}
\right)
\tag{16.28}
$$

The transformation from Eq. (16.27) to (16.28) will be done in two stages: (1) reduction of the 3-D formulation to 2-D and (2) rotation from local $\hat{x},\hat{y}$-axes to global x,y-axes.

1. *Reduction of six 3-D equations to three 2-D equations.* Recall from Eqs. (16.22) and (16.24) that the out-of-plane shear stresses and strains are zero for both plane stress and plane strain. Hence the last two equations in Eqs. (16.26) or (16.27) are identically satisfied and we can eliminate them immediately, leaving the following four equations:

$$
\left\{
\begin{array}{c}
\sigma_{\hat{x}} \\
\sigma_{\hat{y}} \\
\sigma_{\hat{z}} \\
\tau_{\hat{x}\hat{y}}
\end{array}
\right\} =
\left[
\begin{array}{cccc}
C_{\hat{x}\hat{x}} & C_{\hat{x}\hat{y}} & C_{\hat{x}\hat{z}} & 0 \\
 & C_{\hat{y}\hat{y}} & C_{\hat{y}\hat{z}} & 0 \\
 & & C_{\hat{z}\hat{z}} & 0 \\
\text{Symmetric} & & & G_{\hat{x}\hat{y}}
\end{array}
\right]
\left(
\left\{
\begin{array}{c}
\epsilon_{\hat{x}} \\
\epsilon_{\hat{y}} \\
\epsilon_{\hat{z}} \\
\gamma_{\hat{x}\hat{y}}
\end{array}
\right\} -
\left\{
\begin{array}{c}
\alpha_{\hat{x}} \Delta T \\
\alpha_{\hat{y}} \Delta T \\
\alpha_{\hat{z}} \Delta T \\
0
\end{array}
\right\}
\right)
\tag{16.29}
$$

We can next eliminate the third equation from Eq. (16.29) but we must proceed differently for plane stress than for plane strain, since $\sigma_{\hat{z}}=0$ for the former, whereas $\epsilon_{\hat{z}}=0$ for the latter. Considering first plane stress, the third equation becomes

$$0 = C_{\hat{x}\hat{z}}(\epsilon_{\hat{x}}-\alpha_{\hat{x}}\Delta T) + C_{\hat{y}\hat{z}}(\epsilon_{\hat{y}}-\alpha_{\hat{y}}\Delta T) + C_{\hat{z}\hat{z}}(\epsilon_{\hat{z}}-\alpha_{\hat{z}}\Delta T) \tag{16.30}$$

Solving Eq. (16.30) for $\epsilon_{\hat{z}}-\alpha_{\hat{z}}\Delta T$ and substituting into the first, second, and fourth equations, yields Eqs. (16.33) below.

For plane strain, simply substitute $\epsilon_{\hat{z}}=0$ into Eqs. (16.29). The third equation becomes

$$\sigma_{\hat{z}} = C_{\hat{x}\hat{z}}(\epsilon_{\hat{x}}-\alpha_{\hat{x}}\Delta T) + C_{\hat{y}\hat{z}}(\epsilon_{\hat{y}}-\alpha_{\hat{y}}\Delta T) - C_{\hat{z}\hat{z}}\alpha_{\hat{z}}\Delta T \tag{16.31}$$

and the other three equations become

$$
\left\{
\begin{array}{c}
\sigma_{\hat{x}} \\
\sigma_{\hat{y}} \\
\tau_{\hat{x}\hat{y}}
\end{array}
\right\}
=
\left[
\begin{array}{ccc}
C_{\hat{x}\hat{x}} & C_{\hat{x}\hat{y}} & 0 \\
 & C_{\hat{y}\hat{y}} & 0 \\
\text{Symmetric} & & G_{\hat{x}\hat{y}}
\end{array}
\right]
\left(
\left\{
\begin{array}{c}
\epsilon_{\hat{x}} \\
\epsilon_{\hat{y}} \\
\gamma_{\hat{x}\hat{y}}
\end{array}
\right\}
-
\left\{
\begin{array}{c}
\alpha_{\hat{x}}\Delta T \\
\alpha_{\hat{y}}\Delta T \\
0
\end{array}
\right\}
\right)
-
\left\{
\begin{array}{c}
C_{\hat{x}\hat{z}}\alpha_{\hat{z}}\Delta T \\
C_{\hat{y}\hat{z}}\alpha_{\hat{z}}\Delta T \\
0
\end{array}
\right\}
$$

$$\tag{16.32}$$

A bit of algebraic manipulation will permit both thermal vectors in Eqs. (16.32) to be combined and premultiplied by the stiffness matrix, resulting in Eqs. (16.33) below.

Thus the 2-D orthotropic stress-strain relations for both plane stress and plane strain, relative to local principal axes, have the following form:

$$
\left\{
\begin{array}{c}
\sigma_{\hat{x}} \\
\sigma_{\hat{y}} \\
\tau_{\hat{x}\hat{y}}
\end{array}
\right\}
=
\left[
\begin{array}{ccc}
\bar{C}_{\hat{x}\hat{x}} & \bar{C}_{\hat{x}\hat{y}} & 0 \\
 & \bar{C}_{\hat{y}\hat{y}} & 0 \\
\text{Symmetric} & & \bar{G}_{\hat{x}\hat{y}}
\end{array}
\right]
\left(
\left\{
\begin{array}{c}
\epsilon_{\hat{x}} \\
\epsilon_{\hat{y}} \\
\gamma_{\hat{x}\hat{y}}
\end{array}
\right\}
-
\left\{
\begin{array}{c}
\bar{\alpha}_{\hat{x}}\Delta T \\
\bar{\alpha}_{\hat{y}}\Delta T \\
0
\end{array}
\right\}
\right) \tag{16.33}
$$

with the stiffness and thermal expansion coefficients given by the following expressions:

	Plane stress	Plane strain
$\bar{C}_{\hat{x}\hat{x}} =$	$C_{\hat{x}\hat{x}} - \dfrac{C_{\hat{x}\hat{z}}^2}{C_{\hat{z}\hat{z}}}$	$C_{\hat{x}\hat{x}}$
$\bar{C}_{\hat{x}\hat{y}} =$	$C_{\hat{x}\hat{y}} - \dfrac{C_{\hat{x}\hat{z}}C_{\hat{y}\hat{z}}}{C_{\hat{z}\hat{z}}}$	$C_{\hat{x}\hat{y}}$
$\bar{C}_{\hat{y}\hat{y}} =$	$C_{\hat{y}\hat{y}} - \dfrac{C_{\hat{y}\hat{z}}^2}{C_{\hat{z}\hat{z}}}$	$C_{\hat{y}\hat{y}}$
$\bar{G}_{\hat{x}\hat{y}} =$	$G_{\hat{x}\hat{y}}$	$G_{\hat{x}\hat{y}}$
$\bar{\alpha}_{\hat{x}} =$	$\alpha_{\hat{x}}$	$\alpha_{\hat{x}} + \left(\dfrac{C_{\hat{y}\hat{y}}C_{\hat{x}\hat{z}}-C_{\hat{x}\hat{y}}C_{\hat{y}\hat{z}}}{C_{\hat{x}\hat{x}}C_{\hat{y}\hat{y}}-C_{\hat{x}\hat{y}}^2}\right)\alpha_{\hat{z}}$
$\bar{\alpha}_{\hat{y}} =$	$\alpha_{\hat{y}}$	$\alpha_{\hat{y}} + \left(\dfrac{C_{\hat{x}\hat{x}}C_{\hat{y}\hat{z}}-C_{\hat{x}\hat{y}}C_{\hat{x}\hat{z}}}{C_{\hat{x}\hat{x}}C_{\hat{y}\hat{y}}-C_{\hat{x}\hat{y}}^2}\right)\alpha_{\hat{z}}$

$$\tag{16.34}$$

We note that Eqs. (16.30) and (16.31) have been eliminated from the analysis. However, after completing the FE analysis, the calculated values for $\epsilon_{\hat{x}}$ and $\epsilon_{\hat{y}}$ can be substituted back into either Eq. (16.30) or (16.31), the former to obtain $\epsilon_{\hat{z}}$ for a plane stress problem, the latter to obtain $\sigma_{\hat{z}}$ for a plane strain problem.[7] This is not generally done for plane stress since $\epsilon_{\hat{z}}$ is usually insignificant. In plane strain, though, $\sigma_{\hat{z}}$ is usually significant, comparable in magnitude to $\sigma_{\hat{x}}$ and $\sigma_{\hat{y}}$.

For isotropic materials these expressions simplify considerably. From Eq. (16.19),

$$C_{\hat{x}\hat{x}} = C_{\hat{y}\hat{y}} = C_{\hat{z}\hat{z}} = \frac{E(1-v)}{(1+v)(1-2v)}$$

$$C_{\hat{x}\hat{y}} = C_{\hat{x}\hat{z}} = C_{\hat{y}\hat{z}} = \frac{Ev}{(1+v)(1-2v)}$$

$$G_{\hat{x}\hat{y}} = \frac{E}{2(1+v)} \tag{16.35}$$

Also,

$$\alpha_{\hat{x}} = \alpha_{\hat{y}} = \alpha_{\hat{z}} = \alpha \tag{16.36}$$

Substituting Eqs. (16.36) and (16.35) into Eqs. (16.34), the 2-D stress-strain relations in Eqs. (16.33) simplify to the following expressions:

Plane stress (isotropic) •

$$\begin{Bmatrix} \sigma_x \\ \sigma_y \\ \tau_{xy} \end{Bmatrix} = \frac{E}{1-v^2} \begin{bmatrix} 1 & v & 0 \\ v & 1 & 0 \\ 0 & 0 & \frac{1}{2}(1-v) \end{bmatrix} \left(\begin{Bmatrix} \epsilon_x \\ \epsilon_y \\ \gamma_{xy} \end{Bmatrix} - \begin{Bmatrix} \alpha\Delta T \\ \alpha\Delta T \\ 0 \end{Bmatrix} \right) \tag{16.37}$$

Plane strain (isotropic) •

$$\begin{Bmatrix} \sigma_x \\ \sigma_y \\ \tau_{xy} \end{Bmatrix} = \frac{E}{(1+v)(1-2v)} \begin{bmatrix} 1-v & v & 0 \\ v & 1-v & 0 \\ 0 & 0 & \frac{1}{2}-v \end{bmatrix} \left(\begin{Bmatrix} \epsilon_x \\ \epsilon_y \\ \gamma_{xy} \end{Bmatrix} - \begin{Bmatrix} (1+v)\alpha\Delta T \\ (1+v)\alpha\Delta T \\ 0 \end{Bmatrix} \right)$$

$$\tag{16.38}$$

It should be noted that Eqs. (16.37) and (16.38) are written with respect to the global x,y-axes. This, of course, is because an isotropic material has no directionality; i.e., all directions are "principal." Therefore it makes no difference whether the circumflex is added to or removed from the x,y-subscripts; Eqs. (16.37) and (16.38) are valid for any orientation of the axes. Consequently these isotropic relations do not need to be subjected to the axes-rotation operations in the next stage; they are already in appropriate form [see Eqs. (16.28)] for the FE analysis in Section 16.3.

2. *Rotation from local principal axes to global axes.* Equation (16.33) may be written in the abbreviated form,

$$\{\hat{\sigma}\} = [\hat{C}](\{\hat{\epsilon}\} - \{\hat{\epsilon}_T\}) \tag{16.39a}$$

---

[7] One would first calculate the strains relative to the global axes, ϵ_x, ϵ_y, γ_{xy}, and then transform them to $\epsilon_{\hat{x}}$ and $\epsilon_{\hat{y}}$ using Eq. (16.42).

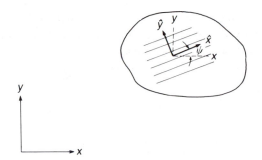

FIGURE 16.11 ● Orientation of local principal axes relative to the global axes.

where

$$\{\hat{\epsilon}_T\} = \left\{ \begin{array}{c} \bar{\alpha}_{\hat{x}}\Delta T \\ \bar{\alpha}_{\hat{y}}\Delta T \\ 0 \end{array} \right\} = \{\hat{\bar{\alpha}}\}\Delta T$$

$$(16.39b)$$

This is valid only with respect to a set of local principal $\hat{x},\hat{y}$-axes that are oriented at some angle ψ relative to the global x,y-axes (Fig. 16.11). We therefore must transform (i.e., rotate) the stress and strain components from one set of axes to the other.

It was observed in Section 16.1 that stresses and strains are tensors in their doubly subscripted form (but not in their engineering form; see Table 16.1), and, as tensors, they transform according to the rules of tensor transformation — Eq. (16.5). Writing Eq. (16.5) in matrix form, we have for both stresses and strains,

$$[\hat{\tau}] = [l] \ [\tau] \ [l]^T \qquad (16.40a)$$
$$\scriptstyle 3\times 3 \qquad 3\times 3 \ 3\times 3$$

$$[\hat{e}] = [l] \ [e] \ [l]^T \qquad (16.40b)$$
$$\scriptstyle 3\times 3 \qquad 3\times 3$$

where $[\tau]$ and $[e]$ are the 3-by-3 stress and strain matrices (tensors) discussed in Section 16.1, and $[l]$ is the rotation matrix for the axes shown in Fig. 16.11 (z and $\hat{z}$ are identical)[8]:

$$[l] = \begin{bmatrix} \cos\psi & \sin\psi & 0 \\ -\sin\psi & \cos\psi & 0 \\ 0 & 0 & 1 \end{bmatrix}$$

$$(16.41)$$

Consider first the strains. Because $[\hat{e}]$ and $[e]$ are symmetric, Eq. (16.40b) yields six (not nine) independent strain equations. For our 2-D plane problems we only need to retain the three equations for $\hat{e}_{11}$, $\hat{e}_{22}$, and $\hat{e}_{12}$. Then, replacing the tensor strains

---

[8] A vector V, with components V_x, V_y, V_z and $V_{\hat{x}}, V_{\hat{y}}, V_{\hat{z}}$ in the two systems, would be transformed by the rule $\{\hat{V}\} = [l]\{V\}$.

$\hat{e}_{11}$, $\hat{e}_{22}$, and $\hat{e}_{12}$ with the engineering strains $\epsilon_{\hat{x}}$, $\epsilon_{\hat{y}}$, and $\gamma_{\hat{x}\hat{y}}/2$, respectively (similarly for the strains relative to the global axes), the 3-D transformation in Eq. (16.40b) reduces to the following 2-D transformation:

$$\{\hat{\epsilon}\} = [L]\{\epsilon\} \tag{16.42}$$

where

$$\{\hat{\epsilon}\} = \left\{ \begin{array}{c} \epsilon_{\hat{x}} \\ \epsilon_{\hat{y}} \\ \gamma_{\hat{x}\hat{y}} \end{array} \right\} \qquad \{\epsilon\} = \left\{ \begin{array}{c} \epsilon_{x} \\ \epsilon_{y} \\ \gamma_{xy} \end{array} \right\}$$

and

$$[L] = \begin{bmatrix} \cos^2 \psi & \sin^2 \psi & \cos \psi \sin \psi \\ \sin^2 \psi & \cos^2 \psi & -\cos \psi \sin \psi \\ -2 \cos \psi \sin \psi & 2 \cos \psi \sin \psi & \cos^2 \psi - \sin^2 \psi \end{bmatrix} \tag{16.43}$$

The $[L]$ matrix is a rotation matrix, but it is not orthogonal — that is, $[L]^{-1} \neq [L]^{T}$ (the inverse does not equal the transpose) — because the engineering strain vectors are not tensors.[9] Equations (16.42) and (16.43) also apply to the thermal strain vector,

$$\{\hat{\epsilon}_T\} = [L]\{\epsilon_T\} \tag{16.44}$$

Substituting Eqs. (16.42) and (16.44) into Eq. (16.39a) converts the stress-strain relation to the following form:

$$\{\hat{\sigma}\} = [\hat{C}][L](\{\epsilon\} - \{\epsilon_T\}) \tag{16.45}$$

It remains to transform the stresses. We proceed in the same manner as for the strains, although it is convenient to first invert Eq. (16.40a), using the orthogonality of $[l]$:

$$[\tau] = [l]^{T}[\hat{\tau}][l] \tag{16.46}$$

Writing out Eq. (16.46) as six equations, retaining only the three for τ_{11}, τ_{22}, and τ_{12}, and replacing the tensor notation with engineering notation (see Table 16.1), yields

$$\{\sigma\} = [L]^{T}\{\hat{\sigma}\} \tag{16.47}$$

where

$$\{\sigma\} = \left\{ \begin{array}{c} \sigma_{x} \\ \sigma_{y} \\ \tau_{xy} \end{array} \right\} \qquad \{\hat{\sigma}\} = \left\{ \begin{array}{c} \sigma_{\hat{x}} \\ \sigma_{\hat{y}} \\ \tau_{\hat{x}\hat{y}} \end{array} \right\}$$

Substituting Eq. (16.45) into Eq. (16.47) yields

$$\{\sigma\} = [C](\{\epsilon\} - \{\epsilon_T\}) \tag{16.48}$$

---

[9] Rotation operations on tensors always involve orthogonal matrices. The classic situation, usually described in textbooks on matrix algebra, is rotation of a *position* vector. (The latter is a tensor quantity.)

where

$$[C] = [L]^T[\hat{C}][L]$$

$$= \begin{bmatrix} C_{xx} & C_{xy} & C_{x,xy} \\ & C_{yy} & C_{y,xy} \\ \text{Symmetric} & & C_{xy,xy} \end{bmatrix}$$

(16.49)

Expanding Eq. (16.49) yields

$$C_{xx} = \bar{C}_{\hat{x}\hat{x}} \cos^4 \psi + \bar{C}_{\hat{y}\hat{y}} \sin^4 \psi + 2(\bar{C}_{\hat{x}\hat{y}} + 2\bar{G}_{\hat{x}\hat{y}}) \cos^2 \psi \sin^2 \psi$$

$$C_{xy} = \bar{C}_{\hat{x}\hat{y}} (\cos^4 \psi + \sin^4 \psi) + (\bar{C}_{\hat{x}\hat{x}} + \bar{C}_{\hat{y}\hat{y}} - 4\bar{G}_{\hat{x}\hat{y}}) \cos^2 \psi \sin^2 \psi$$

$$C_{x,xy} = [(\bar{C}_{\hat{x}\hat{x}} - \bar{C}_{\hat{x}\hat{y}} - 2\bar{G}_{\hat{x}\hat{y}}) \cos^2 \psi - (\bar{C}_{\hat{y}\hat{y}} - \bar{C}_{\hat{x}\hat{y}} - 2\bar{G}_{\hat{x}\hat{y}}) \sin^2 \psi] \cos \psi \sin \psi$$

$$C_{yy} = \bar{C}_{\hat{x}\hat{x}} \sin^4 \psi + \bar{C}_{\hat{y}\hat{y}} \cos^4 \psi + 2(\bar{C}_{\hat{x}\hat{y}} + 2\bar{G}_{\hat{x}\hat{y}}) \cos^2 \psi \sin^2 \psi$$

$$C_{y,xy} = [(\bar{C}_{\hat{x}\hat{x}} - \bar{C}_{\hat{x}\hat{y}} - 2\bar{G}_{\hat{x}\hat{y}}) \sin^2 \psi - (\bar{C}_{\hat{y}\hat{y}} - \bar{C}_{\hat{x}\hat{y}} - \bar{G}_{\hat{x}\hat{y}}) \cos^2 \psi] \cos \psi \sin \psi$$

$$C_{xy,xy} = \bar{G}_{\hat{x}\hat{y}} (\cos^2 \psi - \sin^2 \psi)^2 + (\bar{C}_{\hat{x}\hat{x}} - 2\bar{C}_{\hat{x}\hat{y}} + \bar{C}_{\hat{y}\hat{y}}) \cos^2 \psi \sin^2 \psi$$

(16.50)

where the expressions for $\bar{C}_{\hat{x}\hat{x}}$, $\bar{C}_{\hat{x}\hat{y}}$, $\bar{C}_{\hat{y}\hat{y}}$, and $\bar{G}_{\hat{x}\hat{y}}$ are given by Eqs. (16.34).

It is more convenient from the standpoint of inputting data to a program to define $\{\epsilon_T\}$ in terms of the thermal expansion coefficients along the principal axes (the form in which the data is usually available). Therefore let's transform $\{\epsilon_T\}$ back to $\{\hat{\epsilon}_T\}$. From Eqs. (16.44) and (16.39b), the result is

$$\{\epsilon_T\} = [L]^{-1}\{\hat{\epsilon}_T\} = [L]^{-1}\{\hat{\alpha}\}\Delta T$$
$$= \{\alpha\}\Delta T$$

(16.51)

Thus

$$\{\alpha\} = [L]^{-1}\{\hat{\alpha}\}$$

(16.52)

where

$$\{\alpha\} = \begin{Bmatrix} \alpha_x \\ \alpha_y \\ \alpha_{xy} \end{Bmatrix} \qquad \{\hat{\alpha}\} = \begin{Bmatrix} \bar{\alpha}_{\hat{x}} \\ \bar{\alpha}_{\hat{y}} \\ 0 \end{Bmatrix}$$

and

$$[L]^{-1} = \begin{bmatrix} \cos^2 \psi & \sin^2 \psi & -\cos \psi \sin \psi \\ \sin^2 \psi & \cos^2 \psi & \cos \psi \sin \psi \\ 2 \cos \psi \sin \psi & -2 \cos \psi \sin \psi & \cos^2 \psi - \sin^2 \psi \end{bmatrix}$$

(16.53)

Writing out Eqs. (16.52) and (16.53) yields

$$\alpha_x = \bar{\alpha}_{\hat{x}} \cos^2 \psi + \bar{\alpha}_{\hat{y}} \sin^2 \psi$$

$$\alpha_y = \bar{\alpha}_{\hat{x}} \sin^2 \psi + \bar{\alpha}_{\hat{y}} \cos^2 \psi$$

$$\alpha_{xy} = 2(\bar{\alpha}_{\hat{x}} - \bar{\alpha}_{\hat{y}})\cos \psi \sin \psi \tag{16.54}$$

where the expressions for $\bar{\alpha}_{\hat{x}}$ and $\bar{\alpha}_{\hat{y}}$ are given by Eqs. (16.34).

Substituting Eq. (16.51) into Eq. (16.48) yields the final form for the stress-strain relations [see Eqs. (16.28)]:

$$\{\sigma\} = [C](\{\epsilon\} - \{\alpha\}\Delta T) \tag{16.55}$$

This completes the transformation of the six 3-D stress-strain relations to a set of three 2-D relations with respect to the global x,y-axes.

16.2.3 Summary of Important Relations: Conditions for a Well-Posed Problem

Let's bring together all those relations from Sections 16.2.1 and 16.2.2 that will be needed for a meaningful problem.

Strain-displacement (kinematic) relations [Eqs. (16.15) with Table 16.1] •

$$\epsilon_x = \frac{\partial u}{\partial x}$$

$$\epsilon_y = \frac{\partial v}{\partial y}$$

$$\gamma_{xy} = \frac{\partial u}{\partial y} + \frac{\partial v}{\partial x} \tag{16.56}$$

Stress-strain (constitutive) relations [Eq. (16.55)] •

$$\left\{ \begin{array}{c} \sigma_x \\ \sigma_y \\ \tau_{xy} \end{array} \right\} = \left[\begin{array}{ccc} C_{xx} & C_{xy} & C_{x,xy} \\ & C_{yy} & C_{y,xy} \\ \text{Symmetric} & & C_{xy,xy} \end{array} \right] \left(\left\{ \begin{array}{c} \epsilon_x \\ \epsilon_y \\ \gamma_{xy} \end{array} \right\} - \left\{ \begin{array}{c} \alpha_x \Delta T \\ \alpha_y \Delta T \\ \alpha_{xy} \Delta T \end{array} \right\} \right) \tag{16.57}$$

The stiffness and thermal expansion coefficients are given by Eqs. (16.50) and (16.54), respectively, which in turn require Eqs. (16.34).

Equations of stress equilibrium [Eqs. (16.25)] •

$$\frac{\partial \sigma_x}{\partial x} + \frac{\partial \tau_{xy}}{\partial y} = -f_x$$

$$\frac{\partial \tau_{xy}}{\partial x} + \frac{\partial \sigma_y}{\partial y} = -f_y \tag{16.58}$$

The above three sets of equations constitute the governing equations for the problem. There are eight equations involving eight unknown functions: three stresses, three strains, and two displacements. These can be reduced to two equations involving only the two displacements by first substituting Eqs. (16.56) into Eqs. (16.57), yielding stresses in terms of displacements:

$$
\left\{\begin{array}{c} \sigma_x \\ \\ \sigma_y \\ \\ \tau_{xy} \end{array}\right\} = \left[\begin{array}{ccc} C_{xx} & C_{xy} & C_{x,xy} \\ \\ & C_{yy} & C_{y,xy} \\ \\ \text{Symmetric} & & C_{xy,xy} \end{array}\right] \left(\left\{\begin{array}{c} \dfrac{\partial u}{\partial x} \\[6pt] \dfrac{\partial v}{\partial y} \\[6pt] \dfrac{\partial u}{\partial y} + \dfrac{\partial v}{\partial x} \end{array}\right\} - \left\{\begin{array}{c} \alpha_x \Delta T \\ \\ \alpha_y \Delta T \\ \\ \alpha_{xy} \Delta T \end{array}\right\}\right)
$$
(16.59)

and then substituting the stresses from Eqs. (16.59) into Eqs. (16.58):

$$
\frac{\partial}{\partial x}\left[C_{xx}\left(\frac{\partial u}{\partial x} - \alpha_x \Delta T\right) + C_{xy}\left(\frac{\partial v}{\partial y} - \alpha_y \Delta T\right) + C_{x,xy}\left(\frac{\partial u}{\partial y} + \frac{\partial v}{\partial x} - \alpha_{xy} \Delta T\right)\right]
$$

$$
+ \frac{\partial}{\partial y}\left[C_{x,xy}\left(\frac{\partial u}{\partial x} - \alpha_x \Delta T\right) + C_{y,xy}\left(\frac{\partial v}{\partial y} - \alpha_y \Delta T\right) + C_{xy,xy}\left(\frac{\partial u}{\partial y} + \frac{\partial v}{\partial x} - \alpha_{xy} \Delta T\right)\right]
$$

$$
= -f_x
$$

$$
\frac{\partial}{\partial x}\left[C_{x,xy}\left(\frac{\partial u}{\partial x} - \alpha_x \Delta T\right) + C_{y,xy}\left(\frac{\partial v}{\partial y} - \alpha_y \Delta T\right) + C_{xy,xy}\left(\frac{\partial u}{\partial y} + \frac{\partial v}{\partial x} - \alpha_{xy} \Delta T\right)\right]
$$

$$
+ \frac{\partial}{\partial y}\left[C_{xy}\left(\frac{\partial u}{\partial x} - \alpha_x \Delta T\right) + C_{yy}\left(\frac{\partial v}{\partial y} - \alpha_y \Delta T\right) + C_{y,xy}\left(\frac{\partial u}{\partial y} + \frac{\partial v}{\partial x} - \alpha_{xy} \Delta T\right)\right]
$$

$$
= -f_y
$$

(16.60)

For a material that is isotropic and homogeneous (i.e., the elastic moduli are constant) and for isothermal conditions ($\Delta T = 0$), Eqs. (16.60) simplify to the two equations already given in Section 1.3, Eqs. (1.7) [and repeated in Section 2.4, Eqs. (2.22)].

The important point here is that the problem is described by two coupled, second-order partial differential equations in which there are *two* unknown functions, namely, the two displacement components,

$$u(x,y)$$

$$v(x,y)$$
(16.61)

In the FE analysis in Section 16.3, $u(x,y)$ and $v(x,y)$ will each be approximated by trial solutions in the same manner as in Chapter 13.

There are important parallels between the present 2-D elasticity problem and the 2-D quasiharmonic problem in Chapter 13. For example, the constitutive equations [Eqs. (13.13) and (16.59)] have a similar form. Here stresses play the role of fluxes, and they are proportional to gradients of the displacement components. For the quasiharmonic equation there were two flux components related to one unknown function, $U(x,y)$. Here there are three flux components related to two unknown functions, $u(x,y)$ and $v(x,y)$. In addition, both equations in Eqs. (16.60) are similar to the quasiharmonic Eq. (13.15); i.e., they are all second-order differential equations, and their terms have a similar form.

Perhaps one of the most important differences between the quasiharmonic and elasticity equations (from an analysis standpoint) is simply that there are now two unknown functions instead of one, and this will mean that things will tend to occur in pairs, e.g., two unknown functions, two boundary conditions at each boundary point, two DOF at each node, etc.

The domain for the 2-D plane elasticity problem has the same limitations/characteristics as for the 2-D quasiharmonic problem: It is typically a finite,[10] closed region in the x,y-plane, possibly containing interior holes (see Fig. 13.1). For the plane stress problem this domain would typically be the middle surface of a thin plate; for a plane strain problem it would typically be the cross section of a long cylinder (or a short cylinder with end faces constrained against axial displacement).

The static elasticity problem is a boundary-value problem, requiring *two* boundary conditions at each boundary point. They may be specified in one of the following ways:

1. Essential BCs: Specify both components of displacement (Fig. 16.12a).

2. Natural BCs: Specify both components of stress (Fig. 16.12b).

3. Mixed BCs: Specify displacement in one direction and stress perpendicular to it (Fig. 16.12c).

Figure 16.12 shows the specified displacement and stress components aligned with the global x,y-axes. This is not necessary; the data may be specified for any two mutually perpendicular directions, not necessarily aligned with the global axes. For essential or natural BCs one would simply resolve a given displacement or stress vector into its components along the global axes. However, mixed BCs require a local transformation of axes if the displacement and stress components are not aligned with the global axes; this will be explained in Section 16.5.

For a well-posed problem we must specify enough displacements (essential BCs) to *prevent rigid-body motion (translation or rotation) of the entire structure*. This is because the governing equations only describe deformation, i.e., relative motion between points in the body. If the structure is not sufficiently constrained to prevent rigid-body motions, then there is an infinity of solutions to the governing equations, each having the same deformation but a different location in space. (Recall similar arguments in Sections 6.6 and 13.1.1.)[11]

In many problems the boundary loads contain sufficient displacements to prevent rigid-body motion. In other problems, though, there may not be enough, and in these cases the analyst must add extra displacements. This requires care to ensure that the extra displacements prevent rigid-body motion without *over*constraining the structure, thereby producing unwanted deformation.

As a guideline, we observe that in order to prevent rigid-body motion in a plane, at least three pointwise (e.g., nodal) displacement components must be specified as zero,

---

[10] Infinite regions may also be treated, using any of the techniques discussed in Section 10.6.

[11] We note in passing that the prevention of rigid-body motion is relevant only to the *static* elasticity problem. The dynamic problem (Chapter 12) involves both relative deformation and absolute motion with respect to an external (inertial) reference frame, e.g., the x,y-axes. Rigid-body motions are therefore an appropriate part of the solution in dynamics problems.

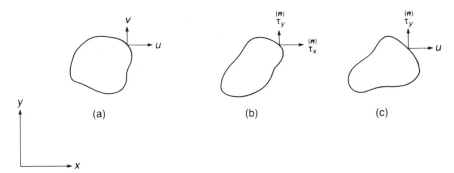

FIGURE 16.12 ● Different ways to specify BCs for plane elasticity problems: (a) essential BCs specify u and v; (b) natural BCs specify $\overset{(n)}{\tau_x}$ and $\overset{(n)}{\tau_y}$; (c) mixed BCs specify one displacement component — for example, u and $\overset{(n)}{\tau_y}$.

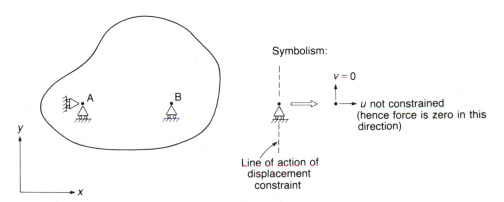

FIGURE 16.13 ● A recommended procedure for preventing rigid-body motion of a structure that has only applied stresses.

i.e., constrained.[12] The displacement constraints may be applied at any points in the body, and their lines of action must not be all parallel (permitting translation) nor all passing through the same point (permitting rotation).

To illustrate, consider the common situation of a structure with only stress BCs (e.g., the bar in simple tension, discussed in Section 1.3). Here we must arbitrarily create at least three extra displacement BCs. The usual procedure, which avoids overconstraining or underconstraining, is shown in Fig. 16.13. One point (A) is "pinned," i.e., constrained from displacement in any direction. This specifies two displacement constraints ($u_A=0$ and $v_A=0$) that prevent rigid-body translation but not rigid-body rotation about the point. A second point (B) is constrained against displacement in only one direction. This specifies the third displacement constraint (say $v_B=0$), which is normally chosen so that its line

---

[12] In principle, the three displacement components could be given nonzero values if they are small enough to be compatible with the linear theory and if they do not produce any deformation by themselves. Such generality, however, is rarely needed.

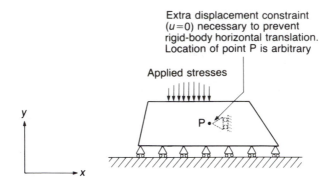

FIGURE 16.14 ● The given boundary data prevent a rigid-body rotation but not a rigid-body horizontal translation, necessitating the specification of a single horizontal displacement at any point in the structure.

of action is approximately at right angles to the line between A and B. The reader should convince himself or herself that this procedure will prevent rigid-body motion without overconstraining, irrespective of the locations of points A and B.

In each particular application, the analyst must examine the specified loads and decide whether they are capable of preventing rigid-body translations as well as rigid-body rotations. In some cases, one type of rigid-body motion may be prevented but not the other, thereby requiring an extra displacement or two. For example, the bottom edge of the structure in Fig. 16.14 is constrained by the given load data to slide along a flat surface. Here there are specified boundary displacements $(v=0)$ all along the bottom, which are obviously more than enough to prevent a rigid-body rotation, but they are insufficient to prevent a rigid-body horizontal translation. Therefore, a single horizontal displacement must be specified (usually zero) at any point in the structure.

16.3 Derivation of Element Equations

As in the preceding chapters we begin the six-step theoretical development by first writing the standard form for the element trial solution. This time, however, we have *two* unknown functions — the two components of displacement, $u(x,y)$ and $v(x,y)$. Therefore we will need a trial solution for each one:

$$\tilde{u}^{(e)}(x,y;a) = \sum_{j=1}^{n} u_j \, \phi_j^{(e)}(x,y)$$

$$\tilde{v}^{(e)}(x,y;a) = \sum_{j=1}^{n} v_j \, \phi_j^{(e)}(x,y)$$

$$(16.62a)$$

Here u_j and v_j are the unknown parameters (DOF). Since DOF are always associated with nodal values of the unknown function (except for nodeless trial functions), u_i and v_i are the x- and y-components of displacement at node i, usually referred to as *nodal displacements*. These are illustrated in Fig. 16.15 for a triangular element, and in step 4 below for several other elements.

Equation (16.62a) may also be written in matrix form:

$$\{\tilde{U}\}^{(e)} = [\Phi]^{(e)}\{a\} \qquad\qquad\qquad (16.62b)$$

where

$$\{\tilde{U}\}^{(e)}_{2\times1} = \left\{ \begin{array}{c} \tilde{u}^{(e)} \\ \tilde{v}^{(e)} \end{array} \right\}$$

$$[\Phi]^{(e)}_{2\times2n} = \left[\begin{array}{cccccc} \phi_1^{(e)} & 0 & \phi_2^{(e)} & 0 & \cdots & \phi_n^{(e)} & 0 \\ 0 & \phi_1^{(e)} & 0 & \phi_2^{(e)} & \cdots & 0 & \phi_n^{(e)} \end{array} \right]$$

$$\{a\}_{2n\times1} = \left\{ \begin{array}{c} u_1 \\ v_1 \\ u_2 \\ v_2 \\ \cdot \\ \cdot \\ \cdot \\ u_n \\ v_n \end{array} \right\} = \left\{ \begin{array}{c} a_1 \\ a_2 \\ a_3 \\ a_4 \\ \cdot \\ \cdot \\ \cdot \\ a_{2n-1} \\ a_{2n} \end{array} \right\}$$

Here we have introduced an alternative notation for the DOF: a_j. Because of the consecutive numbering and single character, the a_j notation is better suited to matrix formulations and the arrays used in computer implementation (see Section 16.5). The u_j and v_j notation, on the other hand, is more physically meaningful and hence is better suited to the theoretical development, since it is easier to distinguish which displacement compo-

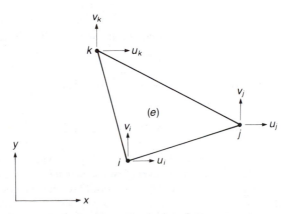

FIGURE 16.15 ● Nodal displacements (DOF) for a C^0-linear triangular elasticity element.

nent a DOF is associated with. Both notations will be used interchangeably in the ensuing material.[13]

Observe that each trial solution uses the same set of trial functions but a different set of DOF. No generality is lost by using the same trial functions since they are merely a "basis" in terms of which the trial solutions are expanded (similar to the terms 1, x, x^2, ..., in a Taylor series expansion or sin $n\theta$, cos $n\theta$, $n=1,2,...$, in a Fourier series expansion). It is the numerical values of the coefficients of these functions, i.e., the values of the DOF, that determine the resulting trial solution. Therefore different DOF must be used in each trial solution.

Sometimes it is desirable to represent one (or more) of the trial solutions by polynomials of a higher degree than in the other trial solution(s). This would be the case if the governing differential equations are of different orders, necessitating different degrees of inter-element continuity for the different trial solutions. For example, an elastic bar/beam element in 3-D [16.10] typically approximates the axial displacement and rotation (governed by second-order extension and torsion equations) by linear polynomials; and the transverse displacements (governed by fourth-order bending equations, see Section 8.6) are approximated by cubic polynomials. For the 2-D elasticity problems in this chapter, both governing differential equations are of second order, with u and v appearing in similar terms in each equation [see Eqs. (16.60)], so it is natural to represent both $\tilde{u}^{(e)}$ and $\tilde{v}^{(e)}$ by the same set of trial functions.

Step 1: Write the Galerkin residual equations for a typical element.

Equations (16.60) are the two governing differential equations. However, for the sake of clarity, and to avoid a lot of messy algebra, it will be more convenient in steps 1 and 2 to work with only the two stress equilibrium equations [Eqs. (16.58)]. Of course, these *are* the governing differential equations, but without the constitutive and kinematic relations [Eqs. (16.56) and (16.57)] substituted into them. The latter will be introduced in step 3, along with a conventional matrix notation that will obviate the messy algebra.

From Eqs. (16.58), we have the following *two* residuals:

$$R_x = \frac{\partial \tilde{\sigma}_x^{(e)}}{\partial x} + \frac{\partial \tilde{\tau}_{xy}^{(e)}}{\partial y} + f_x$$

$$R_y = \frac{\partial \tilde{\tau}_{xy}^{(e)}}{\partial x} + \frac{\partial \tilde{\sigma}_y^{(e)}}{\partial y} + f_y \tag{16.63}$$

As usual, we need one weighted residual equation for each DOF in Eqs. (16.62a), that is, $2n$ equations,

$$\iint\limits^{(e)} R_x \phi_i^{(e)} t \, dx \, dy = 0 \qquad i = 1,2,\dots,n$$

$$\iint\limits^{(e)} R_y \phi_i^{(e)} t \, dx \, dy = 0 \qquad i = 1,2,\dots,n \tag{16.64}$$

---

[13] It is a universal convention to order the DOF in the vector $\{a\}$ in the manner shown, i.e., grouped together by nodes, rather than, say, listing all the u_i first and then the v_i. Grouping by nodes yields a more consistent formulation between related elements with different numbers of nodes (e.g., a family of transition elements).

where the weighting functions in the R_x equation are the trial functions for $\tilde{u}^{(e)}$, and those in the R_y equation are the trial functions for $\tilde{v}^{(e)}$. From Eqs. (16.62a), both sets of trial functions, and hence weighting functions, are the same.

Here t represents the thickness of the plate for a plane stress problem (Fig. 16.8), or the thickness of a cross-sectional slice of a cylinder for a plane strain problem (Fig. 16.9). The integration is over the 3-D volume of the material but the integral over the z-direction reduces to $\int dz = t$ since the integrand is independent of z. In practice, t is usually constant, in which case it could be factored out of Eqs. (16.64) and eliminated from the remainder of the analysis. [If we wanted to restrict t to always being a constant, we could have ignored it at the outset, and written Eqs. (16.64) directly as 2-D integrals.] In some problems, though, t may vary slightly,[14] say $t = t(x,y)$, and we can still effectively approximate them as plane problems. In the following material, then, the thickness t may vary "mildly" over the domain.

Substituting Eqs. (16.63) into Eqs. (16.64) yields the desired residual equations:

$$\iint^{(e)} \left[\frac{\partial \tilde{\sigma}_x^{(e)}}{\partial x} + \frac{\partial \tilde{\tau}_{xy}^{(e)}}{\partial y} + f_x \right] \phi_i^{(e)} t \, dx \, dy = 0 \qquad i = 1, 2, \dots, n$$

$$\iint^{(e)} \left[\frac{\partial \tilde{\tau}_{xy}^{(e)}}{\partial x} + \frac{\partial \tilde{\sigma}_y^{(e)}}{\partial y} + f_y \right] \phi_i^{(e)} t \, dx \, dy = 0 \qquad i = 1, 2, \dots, n \qquad \textbf{(16.65)}$$

Step 2: Integrate by parts.

All the stress derivative terms in Eqs. (16.65) involve second-order derivatives of the displacements [see Eqs. (16.60)] so they will be integrated by parts once. Here the procedure is the same as for the quasiharmonic equation in Section 13.2 — namely, apply the chain rule of differentiation and then the 2-D divergence theorem.

Applying the chain rule, Eqs. (16.65) become

$$\iint^{(e)} \left[\tilde{\sigma}_x^{(e)} \frac{\partial \phi_i^{(e)}}{\partial x} + \tilde{\tau}_{xy}^{(e)} \frac{\partial \phi_i^{(e)}}{\partial y} \right] t \, dx \, dy = \iint^{(e)} \left[\frac{\partial}{\partial x} \left(\tilde{\sigma}_x^{(e)} \phi_i^{(e)} \right) + \frac{\partial}{\partial y} \left(\tilde{\tau}_{xy}^{(e)} \phi_i^{(e)} \right) \right] t \, dx \, dy$$

$$+ \iint^{(e)} f_x \phi_i^{(e)} t \, dx \, dy \qquad i = 1, 2, \dots, n$$

$$\iint^{(e)} \left[\tilde{\tau}_{xy}^{(e)} \frac{\partial \phi_i^{(e)}}{\partial x} + \tilde{\sigma}_y^{(e)} \frac{\partial \phi_i^{(e)}}{\partial y} \right] t \, dx \, dy = \iint^{(e)} \left[\frac{\partial}{\partial x} \left(\tilde{\tau}_{xy}^{(e)} \phi_i^{(e)} \right) + \frac{\partial}{\partial y} \left(\tilde{\sigma}_y^{(e)} \phi_i^{(e)} \right) \right] t \, dx \, dy$$

$$+ \iint^{(e)} f_y \phi_i^{(e)} t \, dx \, dy \qquad i = 1, 2, \dots, n \qquad \textbf{(16.66)}$$

The integrand in the first integral on the RHS of each equation is the 2-D analogue of a perfect differential. Hence these integrals may be reduced to boundary integrals with

---

[14] If t varies substantially, the problem is better modeled as 3-D.

the aid of the 2-D divergence theorem [see Eq. (13.54)]. Therefore Eqs. (16.66) become

$$
\overset{(e)}{\iint} \left[\tilde{\sigma}_x^{(e)} \frac{\partial \phi_i^{(e)}}{\partial x} + \tilde{\tau}_{xy}^{(e)} \frac{\partial \phi_i^{(e)}}{\partial y} \right] t \; dx \; dy = \overset{(e)}{\oint} \left(\tilde{\sigma}_x^{(e)} n_x^{(e)} + \tilde{\tau}_{xy}^{(e)} n_y^{(e)} \right) \phi_i^{(e)} t \; ds
$$

$$
+ \overset{(e)}{\iint} f_x \phi_i^{(e)} t \; dx \; dy \qquad i = 1,2,\ldots,n
$$

$$
\overset{(e)}{\iint} \left[\tilde{\tau}_{xy}^{(e)} \frac{\partial \phi_i^{(e)}}{\partial x} + \tilde{\sigma}_y^{(e)} \frac{\partial \phi_i^{(e)}}{\partial y} \right] t \; dx \; dy = \overset{(e)}{\oint} \left(\tilde{\tau}_{xy}^{(e)} n_x^{(e)} + \tilde{\sigma}_y^{(e)} n_y^{(e)} \right) \phi_i^{(e)} t \; ds
$$

$$
+ \overset{(e)}{\iint} f_y \phi_i^{(e)} t \; dx \; dy \qquad i = 1,2,\ldots,n \qquad \textbf{(16.67)}
$$

The integrands in the boundary integrals are the x- and y-components of a stress vector acting on a surface with normal $\mathbf{n}$. Thus, recalling Eq. (16.4), and noting the change in notation in Table 16.1 and the symmetry of the stress components,

$$
\overset{(n)}{\tilde{\tau}_x^{(e)}} = \tilde{\sigma}_x^{(e)} n_x^{(e)} + \tilde{\tau}_{xy}^{(e)} n_y^{(e)}
$$

$$
\overset{(n)}{\tilde{\tau}_y^{(e)}} = \tilde{\tau}_{xy}^{(e)} n_x^{(e)} + \tilde{\sigma}_y^{(e)} n_y^{(e)} \qquad \textbf{(16.68)}
$$

where $n_z^{(e)}=0$ for our 2-D plane problems. As in previous chapters, we recognize $\overset{(n)}{\tilde{\tau}_x^{(e)}}$ and $\overset{(n)}{\tilde{\tau}_y^{(e)}}$ as boundary fluxes and hence they may be *prescribed*, either by natural BCs on external element sides [see Eqs. (13.83)] or by natural interelement BCs on internal element sides [see Eqs. (13.101) and (13.102)]. It is only along element sides with essential BCs (displacements prescribed) that the boundary fluxes are unknown and hence can't be prescribed; however, these unknown boundary terms drop out of the system equations when the essential BCs are applied (see Section 4.4, step 9). We can therefore replace $\overset{(n)}{\tilde{\tau}_x^{(e)}}$ and $\overset{(n)}{\tilde{\tau}_y^{(e)}}$ with $\overset{(n)}{\tau_x(s)}$ and $\overset{(n)}{\tau_y(s)}$, respectively, representing known boundary data (see Fig. 16.12b). It should be understood by this notation that data is available for $\overset{(n)}{\tau_x(s)}$ and $\overset{(n)}{\tau_y(s)}$ for natural BCs, or for differences (pairs) of these terms for natural interelement BCs; for essential BCs (on external or internal sides) no data is available, but these unknown terms eventually disappear anyway.

Making the indicated changes to the boundary integrals, Eqs. (16.67) become

$$
\overset{(e)}{\iint} \left[\tilde{\sigma}_x^{(e)} \frac{\partial \phi_i^{(e)}}{\partial x} + \tilde{\tau}_{xy}^{(e)} \frac{\partial \phi_i^{(e)}}{\partial y} \right] t \; dx \; dy = \overset{(e)}{\int} \overset{(n)}{\tau_x} \phi_i^{(e)} t \; ds + \overset{(e)}{\iint} f_x \phi_i^{(e)} t \; dx \; dy
$$

$$
i = 1,2,\ldots,n
$$

$$
\overset{(e)}{\iint} \left[\tilde{\tau}_{xy}^{(e)} \frac{\partial \phi_i^{(e)}}{\partial x} + \tilde{\sigma}_y^{(e)} \frac{\partial \phi_i^{(e)}}{\partial y} \right] t \; dx \; dy = \overset{(e)}{\int} \overset{(n)}{\tau_y} \phi_i^{(e)} t \; ds + \overset{(e)}{\iint} f_y \phi_i^{(e)} t \; dx \; dy
$$

$$
i = 1,2,\ldots,n \qquad \textbf{(16.69)}
$$

Step 3: Substitute the general form of the element trial solution into interior integrals in residual equations.

Equations (16.69) include only the equations of stress equilibrium, Eqs. (16.58). In order to substitute the expressions for $\tilde{u}^{(e)}$ and $\tilde{v}^{(e)}$ from Eqs. (16.62) into Eqs. (16.69), we will also need to include the constitutive and kinematic relations [Eqs. (16.55) and (16.56)].
    Let's first substitute $\tilde{u}^{(e)}$ and $\tilde{v}^{(e)}$ into the kinematic relations:

$$\tilde{\epsilon}_x^{(e)} = \frac{\partial \tilde{u}^{(e)}}{\partial x} = \sum_{j=1}^{n} u_j \frac{\partial \phi_j^{(e)}}{\partial x}$$

$$\tilde{\epsilon}_y^{(e)} = \frac{\partial \tilde{v}^{(e)}}{\partial y} = \sum_{j=1}^{n} v_j \frac{\partial \phi_j^{(e)}}{\partial y}$$

$$\tilde{\gamma}_{xy}^{(e)} = \frac{\partial \tilde{u}^{(e)}}{\partial y} + \frac{\partial \tilde{v}^{(e)}}{\partial x} = \sum_{j=1}^{n} \left(u_j \frac{\partial \phi_j^{(e)}}{\partial y} + v_j \frac{\partial \phi_j^{(e)}}{\partial x} \right)$$

(16.70)

or, in matrix form,

$$\{\tilde{\epsilon}\}^{(e)} = [B]^{(e)}\{a\}$$

(16.71)

where

$$\{\tilde{\epsilon}\}^{(e)} = \left\{ \begin{array}{c} \tilde{\epsilon}_x^{(e)} \\ \tilde{\epsilon}_y^{(e)} \\ \tilde{\gamma}_{xy}^{(e)} \end{array} \right\}$$

$$[B]^{(e)}_{3 \times 2n} = \begin{bmatrix} \dfrac{\partial \phi_1^{(e)}}{\partial x} & 0 & \dfrac{\partial \phi_2^{(e)}}{\partial x} & 0 & \cdots & \dfrac{\partial \phi_n^{(e)}}{\partial x} & 0 \\[2mm] 0 & \dfrac{\partial \phi_1^{(e)}}{\partial y} & 0 & \dfrac{\partial \phi_2^{(e)}}{\partial y} & \cdots & 0 & \dfrac{\partial \phi_n^{(e)}}{\partial y} \\[2mm] \dfrac{\partial \phi_1^{(e)}}{\partial y} & \dfrac{\partial \phi_1^{(e)}}{\partial x} & \dfrac{\partial \phi_2^{(e)}}{\partial y} & \dfrac{\partial \phi_2^{(e)}}{\partial x} & \cdots & \dfrac{\partial \phi_n^{(e)}}{\partial y} & \dfrac{\partial \phi_n^{(e)}}{\partial x} \end{bmatrix}$$

and $\{a\}$ is given in Eq. (16.62b).
    Next we substitute the kinematic relations, Eq. (16.71) into the constitutive relations, Eq. (16.55):

$$\{\tilde{\sigma}\}^{(e)} = [C]^{(e)} \left[[B]^{(e)}\{a\} - \{\alpha\}^{(e)}\Delta T \right]$$

(16.72)

    Finally, we want to substitute Eq. (16.72) into Eqs. (16.69), and this will require writing Eqs. (16.69) in matrix form. The appropriate matrices can be readily identified if we arrange the equations in the same order as the DOF in the $\{a\}$ vector — namely, alternating between the x- and y-components:

$$
\text{Node 1}
\begin{cases}
\overset{(e)}{\iint} \left[\tilde{\sigma}_x^{(e)} \frac{\partial \phi_1^{(e)}}{\partial x} + \tilde{\tau}_{xy}^{(e)} \frac{\partial \phi_1^{(e)}}{\partial y} \right] t \, dx \, dy = \overset{(e)}{\oint} \overset{(n)}{T_x} \phi_1^{(e)} t \, ds + \overset{(e)}{\iint} f_x \phi_1^{(e)} t \, dx \, dy \\[2em]
\overset{(e)}{\iint} \left[\tilde{\tau}_{xy}^{(e)} \frac{\partial \phi_1^{(e)}}{\partial x} + \tilde{\sigma}_y^{(e)} \frac{\partial \phi_1^{(e)}}{\partial y} \right] t \, dx \, dy = \overset{(e)}{\oint} \overset{(n)}{T_y} \phi_1^{(e)} t \, ds + \overset{(e)}{\iint} f_y \phi_1^{(e)} t \, dx \, dy
\end{cases}
$$

$$
\text{Node 2}
\begin{cases}
\overset{(e)}{\iint} \left[\tilde{\sigma}_x^{(e)} \frac{\partial \phi_2^{(e)}}{\partial x} + \tilde{\tau}_{xy}^{(e)} \frac{\partial \phi_2^{(e)}}{\partial y} \right] t \, dx \, dy = \overset{(e)}{\oint} \overset{(n)}{T_x} \phi_2^{(e)} t \, ds + \overset{(e)}{\iint} f_x \phi_2^{(e)} t \, dx \, dy \\[2em]
\overset{(e)}{\iint} \left[\tilde{\tau}_{xy}^{(e)} \frac{\partial \phi_2^{(e)}}{\partial x} + \tilde{\sigma}_y^{(e)} \frac{\partial \phi_2^{(e)}}{\partial y} \right] t \, dx \, dy = \overset{(e)}{\oint} \overset{(n)}{T_y} \phi_2^{(e)} t \, ds + \overset{(e)}{\iint} f_y \phi_2^{(e)} t \, dx \, dy
\end{cases}
$$

$$\vdots \qquad\qquad\qquad\qquad\qquad\qquad\qquad\qquad\qquad \vdots$$

$$
\text{Node } n
\begin{cases}
\overset{(e)}{\iint} \left[\tilde{\sigma}_x^{(e)} \frac{\partial \phi_n^{(e)}}{\partial x} + \tilde{\tau}_{xy}^{(e)} \frac{\partial \phi_n^{(e)}}{\partial y} \right] t \, dx \, dy = \overset{(e)}{\oint} \overset{(n)}{T_x} \phi_n^{(e)} t \, ds + \overset{(e)}{\iint} f_x \phi_n^{(e)} t \, dx \, dy \\[2em]
\overset{(e)}{\iint} \left[\tilde{\tau}_{xy}^{(e)} \frac{\partial \phi_n^{(e)}}{\partial x} + \tilde{\sigma}_y^{(e)} \frac{\partial \phi_n^{(e)}}{\partial y} \right] t \, dx \, dy = \overset{(e)}{\oint} \overset{(n)}{T_y} \phi_n^{(e)} t \, ds + \overset{(e)}{\iint} f_y \phi_n^{(e)} t \, dx \, dy
\end{cases}
$$

$$\textbf{(16.73)}$$

Using the relation for $[B]^{(e)}$ in Eq. (16.71), the integrands on the LHS of Eqs. (16.73) may be written in the following matrix form:

$$
\left\{
\begin{array}{c}
\tilde{\sigma}_x^{(e)} \frac{\partial \phi_1^{(e)}}{\partial x} + \tilde{\tau}_{xy}^{(e)} \frac{\partial \phi_1^{(e)}}{\partial y} \\[1em]
\tilde{\tau}_{xy}^{(e)} \frac{\partial \phi_1^{(e)}}{\partial x} + \tilde{\sigma}_y^{(e)} \frac{\partial \phi_1^{(e)}}{\partial y} \\[1em]
\tilde{\sigma}_x^{(e)} \frac{\partial \phi_2^{(e)}}{\partial x} + \tilde{\tau}_{xy}^{(e)} \frac{\partial \phi_2^{(e)}}{\partial y} \\[1em]
\tilde{\tau}_{xy}^{(e)} \frac{\partial \phi_2^{(e)}}{\partial x} + \tilde{\sigma}_y^{(e)} \frac{\partial \phi_2^{(e)}}{\partial y} \\[1em]
\vdots \\[1em]
\tilde{\sigma}_x^{(e)} \frac{\partial \phi_n^{(e)}}{\partial x} + \tilde{\tau}_{xy}^{(e)} \frac{\partial \phi_n^{(e)}}{\partial y} \\[1em]
\tilde{\tau}_{xy}^{(e)} \frac{\partial \phi_n^{(e)}}{\partial x} + \tilde{\sigma}_y^{(e)} \frac{\partial \phi_n^{(e)}}{\partial y}
\end{array}
\right\}
=
\begin{bmatrix}
\frac{\partial \phi_1^{(e)}}{\partial x} & 0 & \frac{\partial \phi_1^{(e)}}{\partial y} \\[1em]
0 & \frac{\partial \phi_1^{(e)}}{\partial y} & \frac{\partial \phi_1^{(e)}}{\partial x} \\[1em]
\frac{\partial \phi_2^{(e)}}{\partial x} & 0 & \frac{\partial \phi_2^{(e)}}{\partial y} \\[1em]
0 & \frac{\partial \phi_2^{(e)}}{\partial y} & \frac{\partial \phi_2^{(e)}}{\partial x} \\[1em]
\vdots & \vdots & \vdots \\[1em]
\frac{\partial \phi_n^{(e)}}{\partial x} & 0 & \frac{\partial \phi_n^{(e)}}{\partial y} \\[1em]
0 & \frac{\partial \phi_n^{(e)}}{\partial y} & \frac{\partial \phi_n^{(e)}}{\partial x}
\end{bmatrix}
\left\{
\begin{array}{c}
\tilde{\sigma}_x^{(e)} \\[1em]
\tilde{\sigma}_y^{(e)} \\[1em]
\tilde{\tau}_{xy}^{(e)}
\end{array}
\right\}
= [B]^{(e)^T} \{\tilde{\sigma}\}^{(e)}
$$

$$\textbf{(16.74)}$$

Using the relation for $[\Phi]^{(e)}$ in Eq. (16.62b), the integrands on the RHS of Eqs. (16.73) may also be written in matrix form:

$$
\begin{Bmatrix}
\overset{(n)}{\tau_x} \phi_1^{(e)} \\[4pt]
\overset{(n)}{\tau_y} \phi_1^{(e)} \\[4pt]
\overset{(n)}{\tau_x} \phi_2^{(e)} \\[4pt]
\overset{(n)}{\tau_y} \phi_2^{(e)} \\[4pt]
\vdots \\[4pt]
\overset{(n)}{\tau_x} \phi_n^{(e)} \\[4pt]
\overset{(n)}{\tau_y} \phi_n^{(e)}
\end{Bmatrix}
=
\begin{bmatrix}
\phi_1^{(e)} & 0 \\[4pt]
0 & \phi_1^{(e)} \\[4pt]
\phi_2^{(e)} & 0 \\[4pt]
0 & \phi_2^{(e)} \\[4pt]
\vdots & \vdots \\[4pt]
\phi_n^{(e)} & 0 \\[4pt]
0 & \phi_n^{(e)}
\end{bmatrix}
\begin{Bmatrix}
\overset{(n)}{\tau_x} \\[4pt]
\overset{(n)}{\tau_y}
\end{Bmatrix}
= [\Phi]^{(e)^T} \{\overset{(n)}{\tau}\}
\tag{16.75}
$$

and

$$
\begin{Bmatrix}
f_x \phi_1^{(e)} \\[4pt]
f_y \phi_1^{(e)} \\[4pt]
f_x \phi_2^{(e)} \\[4pt]
f_y \phi_2^{(e)} \\[4pt]
\vdots \\[4pt]
f_x \phi_n^{(e)} \\[4pt]
f_y \phi_n^{(e)}
\end{Bmatrix}
=
\begin{bmatrix}
\phi_1^{(e)} & 0 \\[4pt]
0 & \phi_1^{(e)} \\[4pt]
\phi_2^{(e)} & 0 \\[4pt]
0 & \phi_2^{(e)} \\[4pt]
\vdots & \vdots \\[4pt]
\phi_n^{(e)} & 0 \\[4pt]
0 & \phi_n^{(e)}
\end{bmatrix}
\begin{Bmatrix}
f_x \\[4pt]
f_y
\end{Bmatrix}
= [\Phi]^{(e)^T} \{f\}
\tag{16.76}
$$

Equations (16.73) may therefore be written in the following matrix form:

$$
\overset{(e)}{\iint} [B]^{(e)^T} \{\bar{\sigma}\}^{(e)} t \, dx \, dy = \overset{(e)}{\oint} [\Phi]^{(e)^T} \{\overset{(n)}{\tau}\} t \, ds + \overset{(e)}{\iint} [\Phi]^{(e)^T} \{f\} t \, dx \, dy
\tag{16.77}
$$

We can now substitute Eqs. (16.72) into Eqs. (16.77):

$$
\overset{(e)}{\iint} [B]^{(e)^T} [C]^{(e)} [B]^{(e)} \{a\} t \, dx \, dy = \overset{(e)}{\oint} [\Phi]^{(e)^T} \{\overset{(n)}{\tau}\} t \, ds + \overset{(e)}{\iint} [\Phi]^{(e)^T} \{f\} t \, dx \, dy
$$

$$
+ \overset{(e)}{\iint} [B]^{(e)^T} [C]^{(e)} \{\alpha\}^e \Delta T t \, dx \, dy
\tag{16.78}
$$

These are the element equations ($2n$ of them), which may be written in the usual condensed matrix form,

$$
[K]^{(e)} \{a\} = \{F\}^{(e)}
\tag{16.79a}
$$

where the element stiffness matrix is given by

$$[K]^{(e)} = \int\int^{(e)} [B]^{(e)^T} [C]^{(e)} [B]^{(e)} t \; dx \; dy \tag{16.79b}$$

and the element load vector is a sum of three load vectors:

$$\{F\}^{(e)} = \{F_\tau\}^{(e)} + \{F_f\}^{(e)} + \{F_T\}^{(e)} \tag{16.79c}$$

where

$$\{F_\tau\}^{(e)} = \oint^{(e)} [\Phi]^{(e)^T} \{\tau\}^{(n)} t \; ds$$

$$\{F_f\}^{(e)} = \int\int^{(e)} [\Phi]^{(e)^T} \{f\} t \; dx \; dy$$

$$\{F_T\}^{(e)} = \int\int^{(e)} [B]^{(e)^T} [C]^{(e)} \{\alpha\}^{(e)} \Delta T t \; dx \; dy$$

Here $\{\overset{(n)}{\tau}(s)\}$, $\{f(x,y)\}$, and $\Delta T(x,y)$ are the prescribed boundary stresses, interior loads, and temperature changes, respectively.

Step 4: Develop specific expressions for the shape functions.

The governing differential equations [Eqs. (16.60)] are second order ($2m=2$) which are integrated by parts once, yielding element equations in which the highest-order derivatives are first order ($m=1$).[15] Therefore the completeness and continuity conditions for a conforming element (Section 5.6) require that each trial solution $\tilde{u}^{(e)}$ and $\tilde{v}^{(e)}$ be a complete polynomial to at least first degree (i.e., linear) and C^0-continuous across interelement boundaries.

Of course, these are exactly the same requirements as for the 2-D quasiharmonic equation in Chapter 13 (since both are second-order equations). Therefore, *2-D plane elasticity problems can use essentially the same elements as 2-D quasiharmonic problems.* That is, an elasticity element will employ two identical sets of shape functions, one for $\tilde{u}^{(e)}$ and one for $\tilde{v}^{(e)}$, and each set is identical to the set used for a quasiharmonic element. Graphically speaking, both types of problems could employ identical meshes, but each element in the elasticity mesh would have two trial solutions plotted over it (each one constructed from the same shape functions used for the quasiharmonic element).

Therefore all the triangular and quadrilateral elements discussed in Chapter 13 are applicable here, it being understood that the set of shape functions described for each of those elements would be used for both $\tilde{u}^{(e)}$ and $\tilde{v}^{(e)}$ in the present elements. For example, the element trial solutions for the C^0-linear triangular elasticity element are shown in Fig. 16.16. Each is identical in form to the quasiharmonic trial solution shown in Fig. 13.12, i.e., a complete 2-D linear polynomial, with the shape functions given by Eq. (13.67).

---

[15] Recall the discussion in Section 4.5 in which we agreed to adopt in this book the standard convention of integrating by parts m times governing differential equations of order $2m$, thereby producing derivatives of maximum order m in the element equations at the end of step 2.

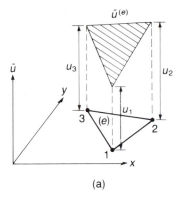

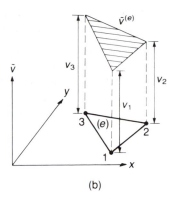

(a)   (b)

FIGURE 16.16 ● The element trial solutions $\tilde{u}^{(e)}$ and $\tilde{v}^{(e)}$ for the C^0-linear triangular elasticity element.

The six DOF illustrated in Fig. 16.16 were shown in a simpler fashion in Fig. 16.15. In a similar manner, Fig. 16.17 shows the DOF for each of the four types of elements developed in Chapter 13, but used as elasticity elements.

The coordinate transformations for the isoparametric elements are the same as for the quasiharmonic elements; i.e., they use the shape functions for either $\tilde{u}^{(e)}$ or $\tilde{v}^{(e)}$, both being the same. Therefore the expressions for the Jacobian remain the same, and hence all the distortion guidelines also remain the same.

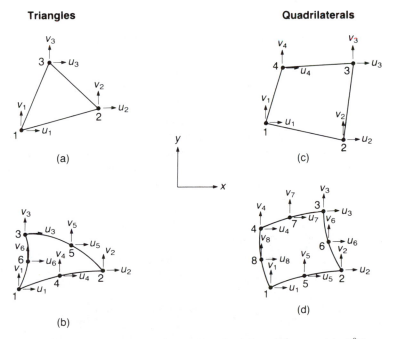

FIGURE 16.17 ● DOF for four popular 2-D elasticity elements: (a) C^0-linear triangle; (b) C^0-quadratic triangle; (c) C^0-linear quadrilateral; (d) C^0-quadratic quadrilateral. Shape functions for $\tilde{u}^{(e)}$ and $\tilde{v}^{(e)}$ are the same as for corresponding 2-D quasiharmonic elements in Chapter 13.

Step 5: Substitute the shape functions into the element equations, and transform the integrals into a form appropriate for numerical evaluation.

As might be expected, this step is also essentially the same as for the quasiharmonic problem. For example, consider the C^0-linear triangular element. This element is simple enough that we can evaulate all the integrals in the element equations [Eqs. (16.79)] exactly, using closed-form integration formulas, just as we did for the quasiharmonic element. Thus, for the stiffness matrix, substituting the shape functions from Eqs. (13.67) into Eq. (16.71) yields

$$\underset{3\times 6}{[B]^{(e)}} = \begin{bmatrix} \dfrac{b_1}{2\Delta} & 0 & \dfrac{b_2}{2\Delta} & 0 & \dfrac{b_3}{2\Delta} & 0 \\[2mm] 0 & \dfrac{c_1}{2\Delta} & 0 & \dfrac{c_2}{2\Delta} & 0 & \dfrac{c_3}{2\Delta} \\[2mm] \dfrac{c_1}{2\Delta} & \dfrac{b_1}{2\Delta} & \dfrac{c_2}{2\Delta} & \dfrac{b_2}{2\Delta} & \dfrac{c_3}{2\Delta} & \dfrac{b_3}{2\Delta} \end{bmatrix} \tag{16.80}$$

where b_j and c_j are given in Eqs. (13.67) and Δ is the area of the element. The terms in the $[B]^{(e)}$ matrix are all constants. In addition, the elastic moduli in the $[C]^{(e)}$ matrix are also constants, since for this linear element the physical properties may be assumed constant throughout the element, equal to the centroidal values (recall the argument in Section 13.3.1, step 5, regarding the preservation of the rate of convergence). The same argument applies to the thickness t if it varies slightly over the element. Hence

$$\underset{6\times 6}{[K]^{(e)}} = \underset{6\times 3}{[B]^{(e)}}^{T} \underset{3\times 3}{[C]^{(e)}} \underset{3\times 6}{[B]^{(e)}} t\Delta \qquad \text{(for } C^0\text{-linear triangle)} \tag{16.81}$$

where $[B]^{(e)}$ is evaluated from Eq. (16.80), $[C]^{(e)}$ from the expressions in Section 16.2.2, and the matrices then multiplied to yield $[K]^{(e)}$. The integrals in the thermal load vector $\{F_T\}^{(e)}$ are evaluated in essentially the same manner as the $[K]^{(e)}$ integrals. The integrals in the other two load vectors, $\{F_\tau\}^{(e)}$ and $\{F_f\}^{(e)}$, are handled in the same manner as in Chapter 13.

The isoparametric elements are also handled in essentially the same manner as in Chapter 13, i.e., by transforming to the ξ,η-coordinates, evaluating the Jacobian terms, etc. For example, consider the C^0-linear isoparametric quadrilateral element (see Fig. 16.17(c) and Section 13.4.1). In terms of the ξ,η-coordinates, the stiffness matrix in Eq. (16.79b) may be written in the usual form [see Eqs. (13.167)],

$$\underset{8\times 8}{[K]^{(e)}} = \int_{-1}^{1}\int_{-1}^{1} \underset{8\times 3}{[B]^{(e)}}^{T} \underset{3\times 3}{[C]^{(e)}} \underset{3\times 8}{[B]^{(e)}} t\,|J^{(e)}|\,d\xi\,d\eta \tag{16.82}$$

The integral is then evaluated by a Gauss-Legendre n-by-n-point rule [see Eqs. (13.169) and Fig. 13.42],

$$K^{(e)} \simeq \sum_{k=1}^{n}\sum_{l=1}^{n} w_{nk}w_{nl}\left([B]^{(e)}{}^{T}[C]^{(e)}[B]^{(e)}t\,|J^{(e)}|\right)_{(\xi_{nl},\eta_{nk})} \tag{16.83}$$

At each Gauss point in succession, the terms in $[B]^{(e)}$, $[C]^{(e)}$, and $|J^{(e)}|$ are numerically evaluated [e.g., using Eqs. (13.127) for $\partial\phi_i^{(e)}/\partial x$ and $\partial\phi_i^{(e)}/\partial y$ in $[B]^{(e)}$] and the matrices then multiplied. (This is more efficient than performing the matrix products first in analytical form and then numerically evaluating.) The value of each $K_{ij}^{(e)}$ term is then the sum of contributions from each Gauss point.

The area integrals in the $\{F_f\}^{(e)}$ and $\{F_T\}^{(e)}$ vectors are treated in the same way as $[K]^{(e)}$. The line integrals in $\{F_\tau\}^{(e)}$ are also treated as in Chapter 13 [see Eq. (13.172)]. Thus along side 12 of the element,

$$\{F_\tau\}_{8\times1}^{(e)} = \int_{-1}^{1} \left[\underset{8\times2}{[\Phi]^{(e)^T}} \underset{2\times1}{\{\tau\}} tJ_\Gamma^{(e)} \right]_{(\xi,-1)} d\xi \tag{16.84}$$

where only the first four terms in $\{F_\tau\}^{(e)}$ are nonzero since $\phi_3(\xi,-1)=\phi_4(\xi,-1)=0$. The 1-D Gauss-Legendre rules are then used to evaluate Eq. (16.84) at the Gauss points for ξ:

$$\{F_\tau\}^{(e)} \simeq \sum_{l=1}^{n} w_{nl} \left[[\Phi]^{(e)^T} \{\tau\} tJ_\Gamma^{(e)} \right]_{(\xi_{nl},-1)} \tag{16.85}$$

The choice of the appropriate-order quadrature rule follows the usual procedure. We first determine the minimal order that preserves the rate of convergence with respect to energy, using guidelines II and III in Section 8.3.1. Since the quasiharmonic and elasticity equations are both second order ($2m=2$, hence $m=1$), the minimal order is the same for both. Thus, for reduced integration, the C^0-linear and C^0-quadratic quadrilaterals require a one-point rule and 2-by-2-point rule, respectively (see Fig. 13.42), and the C^0-quadratic triangle requires a four-point rule (see Fig. 13.31).

The one-point rule with the C^0-linear quadrilateral and the 2-by-2-point rule with the C^0-quadratic quadrilateral both exhibit zero-energy modes [16.11]. Figure 16.18 shows the shapes into which these elements can deform, producing zero strain at the indicated quadrature points, and therefore making the element stiffness matrix singular. This would destroy a one-element problem if the element did not have sufficient displacement constraints imposed to prevent such a deformation.

What about a mesh of several elements? Figure 16.18(a) and (b) indicates that the zero-energy mode shapes for the linear element can form a compatible (C^0-continuous) indefinite repeating pattern. Therefore the assembled system stiffness matrix would also be singular unless at least one of the elements were constrained to prevent such a shape. A simple resolution of the problem is to use the next-higher-order quadrature rule (2-by-2) in at least one element (which would constrain that one element, and hence the entire mesh).

The C^0-quadratic zero-energy mode in Fig. 16.18(c) cannot form a compatible pattern with two or more elements; i.e., they mutually constrain each other. Therefore the 2-by-2 rule is appropriate for this element (for reduced integration) in a mesh of two or more elements.

Step 6: Prepare expressions for the flux (i.e., stresses).

As usual, we first evaluate the fluxes (stresses) at the optimal flux (stress)-sampling points, using the constitutive and kinematic relations and element trial solutions. These expressions were derived in step 3, resulting in Eq. (16.72), namely,

$$\{\tilde{\sigma}\}_{3\times1}^{(e)} = \underset{3\times3}{[C]^{(e)}} \left(\underset{3\times2n}{[B]^{(e)}} \underset{2n\times1}{\{a\}} - \underset{3\times1}{\{\alpha\}^{(e)}\Delta T} \right) \tag{16.86}$$

Zero strain at
Gauss point

Zero strain at each
of the four Gauss points

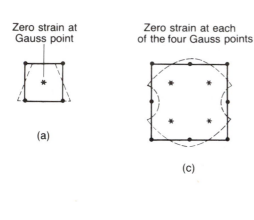

(a)

(c)

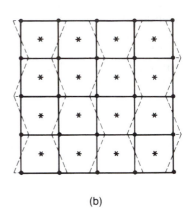

(b)

FIGURE 16.18 ● Zero-energy modes in 2-D plane elasticity elements. (a) Zero-energy mode for a C^0-linear quadrilateral using a one-point Gauss integration rule. This deformation pattern is compatible with adjacent elements deformed in a similar manner, such as shown in (b). Therefore a mesh of such elements (integrated with a one-point rule) may yield a zero-energy mode for the entire system, making the system stiffness matrix singular. (c) Zero-energy mode for a C^0-quadratic quadrilateral using a 2-by-2-point Gauss integration rule. This deformation pattern is not compatible with an adjacent element deformed in a similar manner. Therefore a mesh of two or more such elements (integrated with a 2-by-2-point rule) cannot yield a zero-energy mode for the entire system.

The terms in the matrices are evaluated using the same expressions as in step 5. The strains are also evaluated at the optimal points, using the same expressions. The locations of the optimal flux (stress)-sampling points are the same for both the elasticity and quasiharmonic elements, since both involve the least-squares-fitting of derivatives of the same trial solutions (see Sections 13.3 and 13.4). Evaluating Eq. (16.86) at the optimal points yields the optimal stresses for each of the three stress components.

The next step is to calculate nodal stresses by smoothing and averaging the optimal values for each of the three stress components. Here the procedure is identical to the quasiharmonic element in Chapter 13.

16.4 The Axisymmetric Problem

16.4.1 Mathematical and Physical Description

This class of 2-D problems deals with solid bodies of revolution, i.e., structures whose geometry possesses an axis of rotational symmetry (Fig. 16.19). Cylindrical coordinates r, θ, z are naturally suited to such problems, with the z-axis being the axis of rotational symmetry. The shape of the body depends on r and z but is independent of θ. Therefore we need only examine a cross section in the r,z-plane ($r \geq 0$), as suggested by the bold outline in the figure.

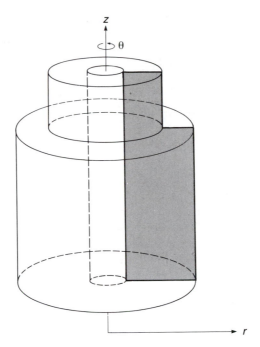

FIGURE 16.19 ● A solid body of revolution; cross section in the r,z-plane is indicated by the bold outline.

An axisymmetric problem must be independent of the θ-coordinate in *all* respects. Thus, not only must the geometry be independent of θ but so must the physical properties and the loads. If materials are anisotropic, two of the principal axes must lie in the r,z-plane; hence the third principal axis is in the θ-direction. All variables (displacement, stress, strain, etc.) are a function of only the two coordinates r and z, so the resulting analysis is 2-D.[16]

The three components of displacement along the directions r,θ,z, respectively, are

$$u = u(r,z)$$
$$v = 0$$
$$w = w(r,z) \tag{16.87}$$

---

[16] If the geometry and physical properties are axisymmetric but the loads are not, the problem can still be analyzed in 2-D by using Fourier analysis with respect to θ. The nonaxisymmetric loads are resolved into Fourier components — for example, $f_0 + \sum_n (f'_n \cos n\theta + f''_n \sin n\theta)$. The displacements are likewise expanded, resulting in a modification to the element equations derived in this section. Such problems are referred to as *axisymmetric bodies with nonaxisymmetric loads*, and many commercial structural analysis FE programs provide this type of analysis capability. In a given application, a program would perform a separate 2-D analysis for each Fourier component present in the load, and the results added together (linear superposition). If there are only a few Fourier components, this approach can be much cheaper than performing a 3-D analysis.

The six components of stress and strain are as follows:

$$\sigma_r(r,z) \qquad \epsilon_r(r,z)$$

$$\sigma_\theta(r,z) \qquad \epsilon_\theta(r,z)$$

$$\sigma_z(r,z) \qquad \epsilon_z(r,z)$$

$$\tau_{rz}(r,z) \qquad \gamma_{rz}(r,z)$$

$$\tau_{\theta r} = 0 \qquad \gamma_{\theta r} = 0$$

$$\tau_{\theta z} = 0 \qquad \gamma_{\theta z} = 0 \tag{16.88}$$

The properties shown in Eq. (16.88) should be compared with those shown in Eqs. (16.22) and (16.24) for the plane stress and plane strain problems. For both the plane and axisymmetric problems the two pairs of out-of-plane shear stresses and shear strains vanish, and the three pairs of in-plane stresses and strains (two normal, one shear) are nonvanishing. The difference is in the out-of-plane normal stress and strain: In the plane problems either the stress or the strain component was zero, enabling the other component to be eliminated from the analysis. Here both components (σ_θ and ϵ_θ) are nonzero, so they cannot be eliminated. Hence the axisymmetric problem will have four pairs of stress-strain components to work with (as compared with only three for the plane problems).

The following relations may be found in many elementary elasticity textbooks (e.g., Timoshenko and Goodier [16.1]).

Strain-displacement (kinematic) relations • The six components of strain with respect to cylindrical coordinates are

$$\epsilon_r = \frac{\partial u}{\partial r} \qquad\qquad \gamma_{rz} = \frac{\partial w}{\partial r} + \frac{\partial u}{\partial z}$$

$$\epsilon_\theta = \frac{1}{r}\frac{\partial v}{\partial \theta} + \frac{u}{r} \qquad\qquad \gamma_{\theta r} = \frac{1}{r}\frac{\partial u}{\partial \theta} + \frac{\partial v}{\partial r} - \frac{v}{r}$$

$$\epsilon_z = \frac{\partial w}{\partial z} \qquad\qquad \gamma_{\theta z} = \frac{\partial v}{\partial z} + \frac{1}{r}\frac{\partial w}{\partial \theta} \tag{16.89}$$

Using Eqs. (16.87) these reduce to the following four nonzero components for the axisymmetric problem:

$$\{\epsilon\} = \left\{\begin{array}{c} \epsilon_r \\[4pt] \epsilon_\theta \\[4pt] \epsilon_z \\[4pt] \gamma_{rz} \end{array}\right\} = \left\{\begin{array}{c} \dfrac{\partial u}{\partial r} \\[8pt] \dfrac{u}{r} \\[8pt] \dfrac{\partial w}{\partial z} \\[8pt] \dfrac{\partial w}{\partial r} + \dfrac{\partial u}{\partial z} \end{array}\right\} \tag{16.90}$$

Stress-strain (constitutive) relations • The treatment here is similar to the plane problems presented in Section 16.2.2 but is less complicated since the fourth equation does not need to be eliminated. As before, we consider orthotropic material symmetry since it is sufficiently anisotropic to represent most practical materials. The complete 3-D stress-strain relations with respect to principal axes $\hat{r},\hat{\theta},\hat{z}$ are therefore Eqs. (16.26), or Eq. (16.27) for thermoelastic behavior, after changing the notation from $\hat{x},\hat{y},\hat{z}$ to $\hat{r},\hat{z},\hat{\theta}$, respectively. (Note that z in the plane problem is analogous to θ in the axisymmetric problem, both being the out-of-plane coordinate.) The two equations relating $\tau_{\hat{\theta}\hat{r}}$ to $\gamma_{\hat{\theta}\hat{r}}$ and $\tau_{\hat{\theta}\hat{z}}$ to $\gamma_{\hat{\theta}\hat{z}}$ are identically zero so we are left with the following four stress-strain equations, relative to principal axes:

$$\left\{ \begin{array}{c} \sigma_{\hat{r}} \\ \sigma_{\hat{\theta}} \\ \sigma_{\hat{z}} \\ \tau_{\hat{r}\hat{z}} \end{array} \right\} = \left[\begin{array}{cccc} C_{\hat{r}\hat{r}} & C_{\hat{r}\hat{\theta}} & C_{\hat{r}\hat{z}} & 0 \\ & C_{\hat{\theta}\hat{\theta}} & C_{\hat{\theta}\hat{z}} & 0 \\ & & C_{\hat{z}\hat{z}} & 0 \\ \text{Symmetric} & & & G_{\hat{r}\hat{z}} \end{array} \right] \left(\left\{ \begin{array}{c} \epsilon_{\hat{r}} \\ \epsilon_{\hat{\theta}} \\ \epsilon_{\hat{z}} \\ \gamma_{\hat{r}\hat{z}} \end{array} \right\} - \left\{ \begin{array}{c} \alpha_{\hat{r}} \\ \alpha_{\hat{\theta}} \\ \alpha_{\hat{z}} \\ 0 \end{array} \right\} \Delta T \right) \qquad \textbf{(16.91)}$$

or, in abbreviated form,

$$\{\hat{\sigma}\} = [\hat{C}](\{\hat{\epsilon}\} - \{\hat{\alpha}\}\Delta T)$$

There are a variety of matrix notation conventions in use, which can sometimes be confusing, so we digress briefly for a word of explanation. In principle, the order in which the stress and strain components are listed in their vectors is arbitrary. In Eqs. (16.90) and (16.91) we have chosen to list them in the order σ_r, σ_θ, σ_z, τ_{rz} (and the same for strains).[17]

However, from a programming standpoint, other orders are generally preferred. The problem arises when one writes a program to analyze both the plane problem and the axisymmetric problem. In order to write compact code that uses the same equations for both types of problems, it is necessary to correlate the positions of the in-plane and out-of-plane components in the vectors in both problems. Thus, if the stress vector for the plane problem uses the order σ_x, σ_y, σ_z, then the order for the axisymmetric problem should be σ_r, σ_z, σ_θ or σ_z, σ_r, σ_θ; that is, the first two components are in-plane and the third component out-of-plane. In addition, the stress-strain relations for the plane problem should be expanded to include the fourth equation. As a result, both types of problem can then be described by the same four-component stress and strain vectors and a 4-by-4 elastic modulus matrix. Since this book is not developing an elasticity program, we have chosen the order shown in Eqs. (16.90) and (16.91) because it conforms to the usual r,θ,z listing of cylindrical coordinates and it is a widely used convention in elasticity theory textbooks. This order may therefore seem more natural to most readers. This ends the digression.

Equations (16.91) must be rotated from local principal axes to the global r,θ,z-axes. The rotation is only in the r,z-plane (same as $\hat{r},\hat{z}$-plane) since θ and $\hat{\theta}$ are identical (Fig. 16.20). The procedure is the same as for the plane problem, although the algebra is different because the axisymmetric stress and strain vectors have four components instead of three.

---

[17] Any other order would also be acceptable, the only restriction being that the stresses and strains should both be listed in the same order so that the symmetry of the elastic modulus matrix $[C]$ is preserved.

As before, one begins with the basic tensor transformation rules given by Eqs. (16.40a) and (16.40b), which use the doubly subscripted stress and strain tensors. The rotation matrix $[l]$ in Eq. (16.41) is the same for both the plane and axisymmetric problem. Therefore Eqs. (16.40a) and (16.40b) yield the same six independent stress equations and six strain equations, respectively, as for the plane problem. However, for the axisymmetric problem we must retain the four equations corresponding to $\sigma_{\hat{r}}$, $\sigma_{\hat{\theta}}$, $\sigma_{\hat{z}}$, and $\tau_{\hat{r}\hat{z}}$ and the four equations corresponding to $\epsilon_{\hat{r}}$, $\epsilon_{\hat{\theta}}$, $\epsilon_{\hat{z}}$, and $\gamma_{\hat{r}\hat{z}}$. The rest of the matrix manipulations remain the same so we merely record the results here.

The orthotropic stress-strain relations for the axisymmetric problem, relative to global axes, are as follows:

$$
\left\{ \begin{array}{c} \sigma_r \\ \sigma_\theta \\ \sigma_z \\ \tau_{rz} \end{array} \right\} = \left[\begin{array}{cccc} C_{rr} & C_{r\theta} & C_{rz} & C_{r,rz} \\ & C_{\theta\theta} & C_{\theta z} & C_{\theta,rz} \\ & & C_{zz} & C_{z,rz} \\ \text{Symmetric} & & & C_{rz,rz} \end{array} \right] \left(\left\{ \begin{array}{c} \epsilon_r \\ \epsilon_\theta \\ \epsilon_z \\ \gamma_{rz} \end{array} \right\} - \left\{ \begin{array}{c} \alpha_r \\ \alpha_\theta \\ \alpha_z \\ \alpha_{rz} \end{array} \right\} \Delta T \right)
\tag{16.92}
$$

or, in abbreviated form,

$$\{\sigma\} = [C](\{\epsilon\} - \{\alpha\}\Delta T)$$

Also,

$$[C] = [L]^T[\hat{C}][L]$$
$$\{\alpha\} = [L]^{-1}\{\hat{\alpha}\}$$

and

$$
[L] = \left[\begin{array}{cccc} \cos^2\psi & 0 & \sin^2\psi & \cos\psi\sin\psi \\ 0 & 1 & 0 & 0 \\ \sin^2\psi & 0 & \cos^2\psi & -\cos\psi\sin\psi \\ -2\cos\psi\sin\psi & 0 & 2\cos\psi\sin\psi & \cos^2\psi - \sin^2\psi \end{array} \right]
$$

$$
[L]^{-1} = \left[\begin{array}{cccc} \cos^2\psi & 0 & \sin^2\psi & -\cos\psi\sin\psi \\ 0 & 1 & 0 & 0 \\ \sin^2\psi & 0 & \cos^2\psi & \cos\psi\sin\psi \\ 2\cos\psi\sin\psi & 0 & -2\cos\psi\sin\psi & \cos^2\psi - \sin^2\psi \end{array} \right]
\tag{16.93}
$$

where $[\hat{C}]$ and $[\hat{\alpha}]$ are, respectively, the elastic moduli and thermal expansion coefficients relative to principal axes, and $[L]$ is the rotation matrix (nonorthogonal) between principal and global axes.

For isotropic materials all directions are "principal" so rotation isn't necessary. We merely substitute into Eqs. (16.91) the expressions (16.35) and (16.36) (replacing $\hat{x},\hat{y},\hat{z}$

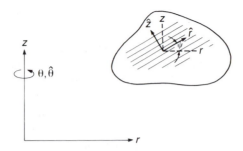

FIGURE 16.20 ● Orientation of local principal axes relative to the global axes.

with $\hat{r},\hat{z},\hat{\theta}$. Hence the isotropic stress-strain relations for the axisymmetric problem are as follows:

$$
\begin{Bmatrix} \sigma_r \\ \sigma_\theta \\ \sigma_z \\ \tau_{rz} \end{Bmatrix} = \frac{E}{(1+v)(1-2v)} \begin{bmatrix} 1-v & v & v & 0 \\ & 1-v & v & 0 \\ & & 1-v & 0 \\ \text{Symmetric} & & & \frac{1}{2}-v \end{bmatrix} \left(\begin{Bmatrix} \epsilon_r \\ \epsilon_\theta \\ \epsilon_z \\ \gamma_{rz} \end{Bmatrix} - \begin{Bmatrix} \alpha \\ \alpha \\ \alpha \\ 0 \end{Bmatrix} \Delta T \right)
$$

(16.94)

Equations of stress equilibrium ● The three equations of static equilibrium [compare with Eqs. (16.6)] take a different form with respect to cylindrical coordinates:

$$
\frac{\partial \sigma_r}{\partial r} + \frac{1}{r}\frac{\partial \tau_{r\theta}}{\partial \theta} + \frac{\partial \tau_{rz}}{\partial z} + \frac{1}{r}(\sigma_r - \sigma_\theta) = -f_r
$$

$$
\frac{\partial \tau_{r\theta}}{\partial r} + \frac{1}{r}\frac{\partial \sigma_\theta}{\partial \theta} + \frac{\partial \tau_{\theta z}}{\partial z} + \frac{2}{r}\tau_{r\theta} = -f_\theta
$$

$$
\frac{\partial \tau_{rz}}{\partial r} + \frac{1}{r}\frac{\partial \tau_{\theta z}}{\partial \theta} + \frac{\partial \sigma_z}{\partial z} + \frac{1}{r}\tau_{rz} = -f_z
$$

(16.95)

Using the properties shown in Eq. (16.88), the r and z equilibrium equations reduce to the following:

$$
\frac{\partial \sigma_r}{\partial r} + \frac{\partial \tau_{rz}}{\partial z} + \frac{1}{r}(\sigma_r - \sigma_\theta) = -f_r
$$

$$
\frac{\partial \tau_{rz}}{\partial r} + \frac{\partial \sigma_z}{\partial z} + \frac{1}{r}\tau_{rz} = -f_z
$$

(16.96)

The θ equilibrium equation is satisfied identically; i.e., all the stress terms vanish, and out-of-plane loads are not permitted ($f_\theta = 0$).

The stress vector acting on a boundary obeys Eq. (16.4), written with respect to the r,θ,z-axes:

$$
\overset{(n)}{T_r} = \sigma_r n_r + \tau_{rz} n_z \qquad \overset{(n)}{T_z} = \tau_{rz} n_r + \sigma_z n_z
$$

(16.97)

where $n_\theta = 0$ on the surface of a body of revolution, and $\overset{(n)}{\tau}_\theta = 0$ since out-of-plane loads are not permitted.

Combining Eqs. (16.90), (16.92), and (16.96) would yield two second-order ($2m = 2$) coupled partial differential equations, quite similar in form to the equations for the plane problems, although with some important differences that will be noted in the next section.

For a well-posed problem we note that rigid-body rotations in the r,z-plane and translations in the r direction are necessarily accompanied by circumferential strain ($\epsilon_\theta \neq 0$); that is, the axisymmetry provides a constraint against these motions. Therefore it is only necessary to ensure that rigid-body translations in the z-direction are prevented. This requires that a z-displacement be specified at at least one point on the structure.

16.4.2 Derivation of Element Equations

The two trial solutions take the standard form,

$$\tilde{u}^{(e)}(r,z;a) = \sum_{j=1}^{n} u_j \phi_j^{(e)}(r,z)$$

$$\tilde{w}^{(e)}(r,z;a) = \sum_{j=1}^{n} w_j \phi_j^{(e)}(r,z)$$

$$(16.98a)$$

where u_j and w_j are nodal displacements. Alternatively, in matrix form,

$$\{\tilde{U}\}^{(e)} = [\Phi]^{(e)}\{a\} \qquad (16.98b)$$

where

$$\{\tilde{U}\}^{(e)}_{2\times 1} = \left\{ \begin{array}{c} \tilde{u}^{(e)} \\ \tilde{w}^{(e)} \end{array} \right\}$$

$$[\Phi]^{(e)}_{2\times 2n} = \left[\begin{array}{ccccccc} \phi_1^{(e)} & 0 & \phi_2^{(e)} & 0 & \cdots & \phi_n^{(e)} & 0 \\ 0 & \phi_1^{(e)} & 0 & \phi_2^{(e)} & \cdots & 0 & \phi_n^{(e)} \end{array} \right]$$

$$\{a\}_{2n\times 1} = \left\{ \begin{array}{c} u_1 \\ w_1 \\ u_2 \\ w_2 \\ \cdot \\ \cdot \\ \cdot \\ u_n \\ w_n \end{array} \right\} = \left\{ \begin{array}{c} a_1 \\ a_2 \\ a_3 \\ a_4 \\ \cdot \\ \cdot \\ \cdot \\ a_{2n-1} \\ a_{2n} \end{array} \right\}$$

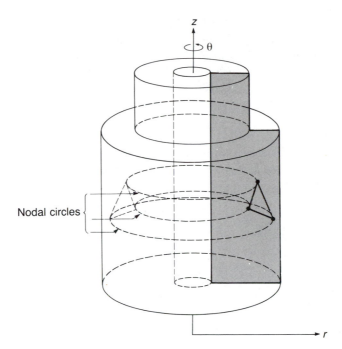

FIGURE 16.21 ● Axisymmetric elements (ring elements) are solids of revolution, just like the structure itself, as illustrated by this C^0-linear triangular element.

Here we see an important difference from the plane problem: Since the axisymmetric structure is a solid of revolution, the *elements are also solids of revolution*, as illustrated by the triangular element in Fig. 16.21. Similarly, the nodes are *nodal circles* rather than points. Axisymmetric elements are sometimes referred to as *ring elements*.

Nodal displacements in the r- or z-direction occur around the entire nodal circle (obviously, since the problem is independent of θ). Although it is important to keep this 3-D picture in mind, the ensuing analysis is all 2-D in the r,z-plane. Therefore we will find it convenient to speak of nodes as though they were points in the r,z-plane, and to speak of elements as triangles or quadrilaterals, by which we mean the shape of the cross sections of the ring elements. Indeed, when a mesh is constructed for an axisymmetric problem, it is a 2-D pattern (in the r,z-plane) of nodal points and 2-D element shapes, just like for a plane problem (recall Fig. 1.11).

The following development closely parallels that for the plane problem in Section 16.3.

Step 1: Write the Galerkin residual equations for a typical element.

From Eqs. (16.96),

$$\iint\limits^{(e)} \left[\frac{\partial \tilde{\sigma}_r^{(e)}}{\partial r} + \frac{\partial \tilde{\tau}_{rz}^{(e)}}{\partial z} + \frac{1}{r}\left(\tilde{\sigma}_r^{(e)} - \tilde{\sigma}_\theta^{(e)} \right) + f_r \right] \phi_i^{(e)} 2\pi r \, dr \, dz = 0 \qquad i = 1,2,\ldots,n$$

$$\iint\limits^{(e)} \left[\frac{\partial \tilde{\tau}_{rz}^{(e)}}{\partial r} + \frac{\partial \tilde{\sigma}_z^{(e)}}{\partial z} + \frac{1}{r}\tilde{\tau}_{rz}^{(e)} + f_z \right] \phi_i^{(e)} 2\pi r \, dr \, dz = 0 \qquad i = 1,2,\ldots,n \qquad \textbf{(16.99)}$$

As usual, the integration is over the domain of the element. Since the element is a 3-D solid of revolution, we must use a volume integration. The differential volume element in cylindrical coordinates is $dV = r\, d\theta\, dr\, dz$. The integrand is independent of θ so the integration from 0 to 2π merely leaves the constant 2π, which can be eliminated from Eqs. (16.99) and the remainder of the analysis.

Step 2: Integrate by parts.

Before applying the chain rule of differentiation and the 2-D divergence theorem in the usual manner, we combine some of the terms in Eq. (16.99) so that the derivatives are multiplied only by $\phi_i^{(e)}$ (not by factors of r):

$$\overset{(e)}{\iint} \left[\frac{\partial}{\partial r}\left(r\tilde{\sigma}_r^{(e)} \right) + \frac{\partial}{\partial z}\left(r\tilde{\tau}_{rz}^{(e)} \right) - \tilde{\sigma}_\theta^{(e)} + rf_r \right] \phi_i^{(e)} dr\, dz = 0 \qquad i = 1, 2, \ldots, n$$

$$\overset{(e)}{\iint} \left[\frac{\partial}{\partial r}\left(r\tilde{\tau}_{rz}^{(e)} \right) + \frac{\partial}{\partial z}\left(r\tilde{\sigma}_z^{(e)} \right) + rf_z \right] \phi_i^{(e)} dr\, dz = 0 \qquad i = 1, 2, \ldots, n \tag{16.100}$$

The chain rule can now be applied to the first two terms in each integral, followed by the 2-D divergence theorem [see Eq. (13.54)]. The latter produces boundary stress integrals containing the expressions (16.97). Performing these operations, Eqs. (16.100) become

$$\overset{(e)}{\iint} \left[\tilde{\sigma}_r^{(e)} \frac{\partial \phi_i^{(e)}}{\partial r} + \tilde{\tau}_{rz}^{(e)} \frac{\partial \phi_i^{(e)}}{\partial z} + \tilde{\sigma}_\theta^{(e)} \frac{\phi_i^{(e)}}{r} \right] r\, dr\, dz = \oint \overset{(n)}{\tilde{\tau}}_r \phi_i^{(e)} r\, ds + \overset{(e)}{\iint} f_r \phi_i^{(e)} r\, dr\, dz$$
$$i = 1, 2, \ldots, n$$

$$\overset{(e)}{\iint} \left[\tilde{\tau}_{rz}^{(e)} \frac{\partial \phi_i^{(e)}}{\partial r} + \tilde{\sigma}_z^{(e)} \frac{\partial \phi_i^{(e)}}{\partial z} \right] r\, dr\, dz = \oint \overset{(n)}{\tilde{\tau}}_z \phi_i^{(e)} r\, ds + \overset{(e)}{\iint} f_z \phi_i^{(e)} r\, dr\, dz$$
$$i = 1, 2, \ldots, n \tag{16.101}$$

Step 3: Substitute the general form of the element trial solution into interior integrals in residual equations.

In the same manner as for the plane problem [see Eqs. (16.70) and (16.71)], we first substitute the displacement trial solutions, Eqs. (16.98), into the strain-displacement equations, Eqs. (16.90). This yields

$$\{\tilde{\epsilon}\}^{(e)} = [B]^{(e)}\{a\} \tag{16.102}$$

where

$$\{\tilde{\epsilon}\}^{(e)} = \begin{Bmatrix} \tilde{\epsilon}_r^{(e)} \\ \tilde{\epsilon}_\theta^{(e)} \\ \tilde{\epsilon}_z^{(e)} \\ \tilde{\gamma}_{rz}^{(e)} \end{Bmatrix}$$

and

$$[B]^{(e)}_{4 \times 2n} = \begin{bmatrix} \dfrac{\partial \phi_1^{(e)}}{\partial r} & 0 & \dfrac{\partial \phi_2^{(e)}}{\partial r} & 0 & \cdots & \dfrac{\partial \phi_n^{(e)}}{\partial r} & 0 \\[2mm] \dfrac{1}{r}\phi_1^{(e)} & 0 & \dfrac{1}{r}\phi_2^{(e)} & 0 & \cdots & \dfrac{1}{r}\phi_n^{(e)} & 0 \\[2mm] 0 & \dfrac{\partial \phi_1^{(e)}}{\partial z} & 0 & \dfrac{\partial \phi_2^{(e)}}{\partial z} & \cdots & 0 & \dfrac{\partial \phi_n^{(e)}}{\partial z} \\[2mm] \dfrac{\partial \phi_1^{(e)}}{\partial z} & \dfrac{\partial \phi_1^{(e)}}{\partial r} & \dfrac{\partial \phi_2^{(e)}}{\partial z} & \dfrac{\partial \phi_2^{(e)}}{\partial r} & \cdots & \dfrac{\partial \phi_n^{(e)}}{\partial z} & \dfrac{\partial \phi_n^{(e)}}{\partial r} \end{bmatrix}$$

Next we substitute Eqs. (16.102) into the stress-strain equations, Eqs. (16.92), yielding

$$\{\tilde{\sigma}\}^{(e)} = [C]^{(e)} \left[[B]^{(e)}\{a\} - \{\alpha\}^{(e)}\Delta T \right] \tag{16.103}$$

Finally we substitute Eqs. (16.103) into Eqs. (16.101), after the latter are reordered and written in matrix form in the same manner as before [see Eqs. (16.73) through (16.76)]. This yields the following $2n$ element equations, written in matrix form:

$$[K]^{(e)}\{a\} = \{F_\tau\}^{(e)} + \{F_f\}^{(e)} + \{F_T\}^{(e)}$$
$$= \{F\}^{(e)} \tag{16.104}$$

where

$$[K]^{(e)} = \int\!\!\int^{(e)} [B]^{(e)^T}[C]^{(e)}[B]^{(e)} r \, dr \, dz$$

$$\{F_\tau\}^{(e)} = \oint^{(e)} [\Phi]^{(e)^T}\{\overset{(n)}{\tau}\} r \, ds$$

$$\{F_f\}^{(e)} = \int\!\!\int^{(e)} [\Phi]^{(e)^T}\{f\} r \, dr \, dz$$

$$\{F_T\}^{(e)} = \int\!\!\int^{(e)} [B]^{(e)^T}[C]^{(e)}\{\alpha\}^{(e)}\Delta T r \, dr \, dz$$

and

$$\overset{(n)}{\{\tau\}} = \begin{Bmatrix} \overset{(n)}{\tau_r} \\ \overset{(n)}{\tau_z} \end{Bmatrix} \qquad \{f\} = \begin{Bmatrix} f_r \\ f_z \end{Bmatrix}$$

Here $\{\overset{(n)}{\tau}(s)\}$, $\{f(r,z)\}$, and $\Delta T(r,z)$ are the prescribed boundary stresses, interior loads, and temperature changes, respectively.

Step 4: Develop specific expressions for the shape functions.

The axisymmetric problem employs all the same elements as the plane problem, i.e., the same 2-D triangular and quadrilateral shapes, nodal patterns, and shape functions described in Chapter 13 and Section 16.3, step 4. It bears repeating that even though the elements are solids of revolution (recall Fig. 16.21), the analysis is only 2-D in the r,z-plane. Therefore the mathematical development of the elements is strictly 2-D, treating the nodal circles as points, and using all of the analysis developed previously for 2-D elements.

Step 5: Substitute the shape functions into the element equations, and transform the integrals into a form appropriate for numerical evaluation.

In most respects, this step is the same as for the plane problem. However, there are a few differences, associated primarily with the axis of symmetry itself, i.e., at $r=0$.[18] Consider the C^0-linear triangular element. Substituting the shape functions from Eqs. (13.67) (replacing x and y with r and z) into the strain-displacement equations, Eq. (16.102), yields

$$
\left\{
\begin{array}{c}
\tilde{\epsilon}_r^{(e)} \\[2mm]
\tilde{\epsilon}_\theta^{(e)} \\[2mm]
\tilde{\epsilon}_z^{(e)} \\[2mm]
\tilde{\gamma}_{rz}^{(e)}
\end{array}
\right\}
=
\left[
\begin{array}{cccccc}
\dfrac{b_1}{2\Delta} & 0 & \dfrac{b_2}{2\Delta} & 0 & \dfrac{b_3}{2\Delta} & 0 \\[3mm]
\dfrac{a_1+b_1 r+c_1 z}{2\Delta r} & 0 & \dfrac{a_2+b_2 r+c_2 z}{2\Delta r} & 0 & \dfrac{a_3+b_3 r+c_3 z}{2\Delta r} & 0 \\[3mm]
0 & \dfrac{c_1}{2\Delta} & 0 & \dfrac{c_2}{2\Delta} & 0 & \dfrac{c_3}{2\Delta} \\[3mm]
\dfrac{c_1}{2\Delta} & \dfrac{b_1}{2\Delta} & \dfrac{c_2}{2\Delta} & \dfrac{b_2}{2\Delta} & \dfrac{c_3}{2\Delta} & \dfrac{b_3}{2\Delta}
\end{array}
\right]
\left\{
\begin{array}{c}
a_1 \\[2mm]
a_2 \\[2mm]
a_3 \\[2mm]
a_4 \\[2mm]
a_5 \\[2mm]
a_6
\end{array}
\right\}
$$

$$(16.105)$$

The first observation is that the circumferential strain $\tilde{\epsilon}_\theta^{(e)}$ is not constant within the element like the other components. Since each stress component is a linear combination of the strains, then the stresses will also vary over the element. This situation should be contrasted with the plane problems in which the stresses and strains are constant in each element [recall Eq. (16.80)]. Thus, even though the same linear shape functions are used in both cases, the element stresses and strains are constant for the plane problems, but ϵ_θ and the normal stresses are variable for the axisymmetric problem.[19] (The other elements show a similar difference between the axisymmetric and plane problems.)

A second observation is that the $1/r$ factor makes $\tilde{\epsilon}_\theta^{(e)}$ (and the related stresses) infinite at $r=0$. It might seem that this would cause a problem for elements with nodes on the axis of symmetry. The problem is automatically resolved, however, when one applies

---

[18] This is to be expected in any curvilinear coordinate system, where the origin of one of the coordinates (r, in cylindrical coordinates) corresponds to an "infinite concentration" of one of the other coordinates (θ), giving rise to singular behavior at the origin ($r=0$).

[19] The three normal stresses always depend on ϵ_θ. The shear stress τ_{rz} depends on ϵ_θ only if the material is anisotropic and the principal axes are not aligned with the global axes.

(in step 9) the necessary BC

$$u(0,z) = 0 \qquad \qquad (16.106)$$

This BC preserves material continuity by preventing any radial displacement on the axis of symmetry. (For example, $u(0,z) > 0$ would create a cylindrical hole surrounding the axis.)

The reason that Eq. (16.106) eliminates the singular behavior can be seen by recalling the equivalence of the theoretical and numerical methods for applying BCs (see Section 3.3 and Exercise 4.1). Thus, one could use the theoretical method to develop (in step 4) a special "core" element that touches the axis of symmetry. Applying Eq. (16.106) to the element trial solution would yield

$$\tilde{U}^{(e)}(r,z) = rP(r,z) \qquad \qquad (16.107)$$

where $P(r,z)$ is a polynomial in r and z.[20] For the C^0-linear triangular element, this would eliminate the a_i and $c_i z$ terms from the shape functions for $\tilde{u}^{(e)}$ [Eqs. (13.67), replacing x and y with r and z] and hence would eliminate these terms from $\tilde{\epsilon}_{\theta}^{(e)}$ and the c_i terms from $\tilde{\gamma}_{rz}^{(e)}$ in Eqs. (16.105). In other words, the terms a_i and $c_i z$ are inappropriate for $\tilde{u}^{(e)}$ in the neighborhood of the axis and are the source of the singularity. However, because of the equivalence of the theoretical and numerical methods, these undesirable terms can be left in the element trial solution, and then in step 9, when Eq. (16.106) is applied, they will be eliminated anyway — the final result being the same as if a core element had been used.

It is recommended that all axisymmetric elements be integrated numerically, using the formulas in Figs. 13.31 or 13.42. The Gauss points for all these formulas are *inside* the elements, so no terms are evaluated at $r=0$. This avoids creating any infinities during evaluation of the stiffness integrals in step 8, before they are eliminated in step 9. Alternatively, the code could be modified in step 8 to circumvent any singular terms. (If exact integration formulas were used, the $1/r$ factor would create stiffness terms containing ln r and, as just observed, the code would have to be modified in step 8 to avoid evaluating these terms at $r=0$.)

For the C^0-linear triangle the three-point rule has proved to be adequate. The four-point rule is appropriate for the C^0-quadratic triangle (for reduced integration). For the C^0-linear quadrilateral, the one-point rule is sufficient for reduced integration but it would yield the zero-energy modes shown in Fig. 16.18(a) and (b) (although only when oriented so that the radial displacement, and hence ϵ_θ, is zero). Therefore this element should use the 2-by-2-point rule. The C^0-quadratic quadrilateral would also use the 2-by-2-point rule (for reduced integration).

Step 6: Prepare expressions for the flux (i.e., stresses).

---

[20] The author used this approach in developing a special purpose FE program for acoustically induced, long-wavelength stress wave propagation in cables (related to Example 1.3 in Section 1.4). It was shown from elasticity theory that a Taylor series expansion of the exact solution for $u(r,z)$ about $r=0$ contains only odd powers of r [i.e., $P(r,z)$ in Eq. (16.107) contains only even powers of r]. Therefore a core element was developed that contained only odd powers of r, yielding extraordinarily high accuracy with only a few DOF. Because of the long-wavelength condition, a single such element was adequate to model the entire cable cross section.

Here the procedure is identical to the plane problem, i.e., evaluate the stresses and strains at the optimal points (using the trial solution expressions), then smooth the optimal values to obtain element nodal values, and finally average the latter to obtain nodal values. For both triangular and quadrilateral elements the optimal stress-sampling points are interior to the element. Hence the nodal stresses and strains calculated on the z-axis ($r=0$) do not encounter a singularity problem, since they are extrapolated from off-axis (optimal point) values.

16.5 Program Implementation

Since we are not developing an elasticity program in this textbook (see the introduction to this chapter), we will only comment briefly here on some programming considerations not covered in previous chapters.

Step 7: Specify numerical data for a particular problem.

When constructing a mesh for elasticity problems (or any type of problem involving more than one DOF per node), it should only be necessary for the user of a program to specify one number per node. The program would then internally assign the appropriate number of DOF to each node. Figure 16.22 illustrates the internal DOF numbering that would be used for our present 2-D elasticity problems. Thus, at node n the nodal displacements in the x- and y-directions for plane problems (or r- and z-directions for axisymmetric problems) would be a_{2n-1} and a_{2n}, respectively, as previously indicated in Eqs. (16.62b) and (16.98b).

Labeling of nodes and elements therefore remains the same as for the one-DOF-per-node quasiharmonic problem, and all the guidelines for numbering strategies given in Section 13.6.5 are applicable to elasticity and other multi-DOF-per-node problems.

It should be remembered that *two* BCs are required at every boundary node (recall Fig. 16.12). A stress-free boundary corresponds to zero-valued natural BCs. As in previous chapters, zero-valued natural BCs do not require any data input (assuming the load vector is initially zeroed), since it would only involve adding zeroes. The only data input required (as usual) are essential BCs (displacements) and nonzero natural BCs (nonzero stresses). In the author's experience, most BCs in most stress-analysis problems are stress-free conditions. Thus the amount of data input required is usually not as voluminous as it might at first seem.

Mixed BCs specify a displacement in one direction and a stress perpendicular to it. If the stress is zero, then only the one displacement need be specified. If the directions are rotated with respect to the global axes, then the DOF must first be redefined, as explained in step 9 below.

For a well-posed problem we recall that enough (but not too many) nodal displacements must be specified to prevent rigid-body translations and rotations. Guidelines for the plane stress problem were given in Section 16.2.3. and for the axisymmetric problem in Section 16.4.1. In addition, for an axisymmetric problem all nodes on the z-axis should be given zero radial displacement [see Eq. (16.106)].

Step 8: Evaluate the interior terms in the element equations for each element, and assemble the terms into system equations.

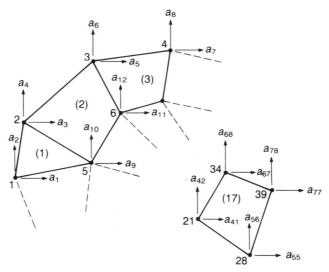

FIGURE 16.22 ● The user specifies one number at each node; the program internally assigns two DOF to each node.

The evaluation of the integrals requires no further comments beyond those already made in step 5 in previous sections. The programming of the element subroutines would be very similar to UNAFEM. Assembly follows the usual rules, allowing for the internal reassignment of DOF numbers, as illustrated in Fig. 16.23.

Step 9: Apply the BCs, including the natural interelement BCs, to the system equations.

Displacements (essential BCs) and stresses (natural BCs and natural interelement BCs) are applied to the system equations in the standard way described in previous chapters. Of course, displacements are specified only at nodes since they correspond to the DOF at each node. Stresses, on the other hand, occur as functions in the integrands of boundary stress integrals — see Eqs. (16.79c) and (16.104). When these integrals are evaluated over an area along an element side ($t \, ds$ for plane stress and plane strain or $r \, ds$ for axisymmetry), the result is concentrated forces applied to the nodes, i.e., the F_i terms on the RHS of the element (and system) equations.

We saw in Chapter 13 with UNAFEM that it is desirable for the user to be able to input stresses (fluxes) in two ways: distributed along the element side or concentrated at the nodes. If the former, then the program evaluates the integral and assigns the resulting concentrated forces to the nodes. In either case, all applied stresses end up as concentrated nodal forces.

In practice it is quite common to have data available as "very localized" forces, in which case it is usually appropriate to approximate the local distribution as concentrated and apply the total force to a node, rather than to try to model the minute details of the distribution (see Fig. 13.70). Indeed, such approximations are frequently obvious from merely examining an engineering drawing or an actual structure.

It is also quite common that the applied concentrated forces, as well as displacements, are in directions rotated from the global axes. This is no problem if only a force or displace-

$$
\begin{bmatrix}
K_{11}^{(1)} & K_{12}^{(1)} & K_{13}^{(1)} & K_{14}^{(1)} & & & & & K_{19}^{(1)} & K_{1,10}^{(1)} & & \\
 & K_{22}^{(1)} & K_{23}^{(1)} & K_{24}^{(1)} & & & & & K_{29}^{(1)} & K_{2,10}^{(1)} & & \\
 & & K_{33}^{(1)}+K_{33}^{(2)} & K_{34}^{(1)}+K_{34}^{(2)} & K_{35}^{(2)} & K_{36}^{(2)} & & & K_{39}^{(1)}+K_{39}^{(2)} & K_{3,10}^{(1)}+K_{3,10}^{(2)} & K_{3,11}^{(2)} & K_{3,12}^{(2)} \\
 & & & K_{44}^{(1)}+K_{44}^{(2)} & K_{45}^{(2)} & K_{46}^{(2)} & & & K_{49}^{(1)}+K_{49}^{(2)} & K_{4,10}^{(1)}+K_{4,10}^{(2)} & K_{4,11}^{(2)} & K_{4,12}^{(2)} \\
 & & & & K_{55}^{(2)} & K_{56}^{(2)} & & & K_{59}^{(2)} & K_{5,10}^{(2)} & K_{5,11}^{(2)} & K_{5,12}^{(2)} \\
 & & & & & K_{66}^{(2)} & & & K_{69}^{(2)} & K_{6,10}^{(2)} & K_{6,11}^{(2)} & K_{6,12}^{(2)} \\
 & & & & & & & & & & & \\
 & & & & \text{Symmetric} & & & & & & & \\
 & & & & & & & & K_{99}^{(1)}+K_{99}^{(2)} & K_{9,10}^{(1)}+K_{9,10}^{(2)} & K_{9,11}^{(2)} & K_{9,12}^{(2)} \\
 & & & & & & & & & K_{10,10}^{(1)}+K_{10,10}^{(2)} & K_{10,11}^{(2)} & K_{10,12}^{(2)} \\
 & & & & & & & & & & K_{11,11}^{(2)} & K_{11,12}^{(2)} \\
 & & & & & & & & & & & K_{12,12}^{(2)}
\end{bmatrix}
\begin{Bmatrix}
a_1 \\ a_2 \\ a_3 \\ a_4 \\ a_5 \\ a_6 \\ a_7 \\ a_8 \\ a_9 \\ a_{10} \\ a_{11} \\ a_{12} \\ \vdots
\end{Bmatrix}
=
\begin{Bmatrix}
F_1^{(1)} \\ F_2^{(1)} \\ F_3^{(1)}+F_3^{(2)} \\ F_4^{(1)}+F_4^{(2)} \\ F_5^{(2)} \\ F_6^{(2)} \\ \\ \\ F_9^{(1)}+F_9^{(2)} \\ F_{10}^{(1)}+F_{10}^{(2)} \\ F_{11}^{(2)} \\ F_{12}^{(2)} \\ \vdots
\end{Bmatrix}
$$

FIGURE 16.23 ● Assembly of element equations for elements (1) and (2) in Fig. 16.22.

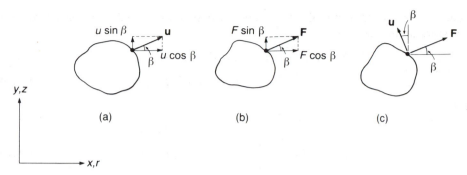

FIGURE 16.24 ● Specified displacement *or* force can be resolved into global components: (a) displacement only; (b) force only. (c) Combination of displacement *and* force cannot be resolved.

ment is specified (Fig. 16.24a and b) since either vector can easily be expressed in terms of its components along the global axes, which are then input to the program. However, if mixed, rotated BCs are specified, i.e., a mutually perpendicular force and displacement rotated from the global axes (Fig. 16.24c), then the situation is more complicated. A simple resolution of both the force and the displacement into their components along the global axes is not permissible because this would yield a known force and displacement along each axis. One cannot in general specify both a force and a displacement in the same direction since their ratio is the stiffness of the structure at that point (due to elastic properties and geometry) and that is determined by the governing differential equations.

For mixed, rotated BCs at a given node, we must define a new set of nodal displacements (DOF) and nodal forces along a local set of axes that are aligned with the directions of the applied force and displacement (Fig. 16.25). The displacement and force

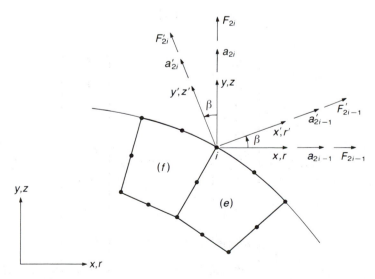

FIGURE 16.25 ● Local x',y'-axes (or r',z'-axes, if axisymmetric) for applying mixed BCs at node i.

components with respect to the global and local axes are related by the standard vector transformation rules:

$$a_{2i-1} = a'_{2i-1} \cos \beta - a'_{2i} \sin \beta$$

$$a_{2i} = a'_{2i-1} \sin \beta + a'_{2i} \cos \beta \qquad (16.108)$$

and

$$F_{2i-1} = F'_{2i-1} \cos \beta - F'_{2i} \sin \beta$$

$$F_{2i} = F'_{2i-1} \sin \beta + F'_{2i} \cos \beta \qquad (16.109)$$

Using Eqs. (16.108) and (16.109) we can now transform the element equations (and system equations) so that the unprimed variables are replaced by the primed variables. Thus, consider the element equations for element (e):

$$\underset{2n \times 2n}{[K]^{(e)}} \underset{2n \times 1}{\{a\}} = \underset{2n \times 1}{\{F\}^{(e)}} \qquad (16.110)$$

Substitute a_{2i-1} and a_{2i} from Eqs. (16.108) into Eqs. (16.110) and collect common coefficients of a_{2i-1} and a_{2i}. The result is equivalent to performing the following column operations on $[K]^{(e)}$: Column $2i-1$ is replaced by $\cos \beta$ times column $2i-1$ plus $\sin \beta$ times column $2i$; column $2i$ is replaced by $-\sin \beta$ times column $2i-1$ plus $\cos \beta$ times column $2i$. The DOF a'_{2i-1} and a'_{2i} replace a_{2i-1} and a_{2i}, respectively. Equations (16.109) produce the same set of operations on the rows of $[K]^{(e)}$ and $\{F\}^{(e)}$.

These row-and-column operations can also be described using matrix notation. Some readers may find this approach more meaningful. We rewrite Eqs. (16.108) and (16.109) in matrix form, expanding the vectors to element size, i.e., with $2n$ rows:

$$\{a\} = [R]\{a'\}$$

$$\{F\}^{(e)} = [R]\{F'\}^{(e)} \qquad (16.111)$$

where

$$\{a'\} = \begin{Bmatrix} a_1 \\ a_2 \\ \vdots \\ a'_{2i-1} \\ a'_{2i} \\ \vdots \\ a_{2n-1} \\ a_{2n} \end{Bmatrix} \qquad \{F'\}^{(e)} = \begin{Bmatrix} F_1 \\ F_2 \\ \vdots \\ F'_{2i-1} \\ F'_{2i} \\ \vdots \\ F_{2n-1} \\ F_{2n} \end{Bmatrix}$$

and

$$[R] = \begin{bmatrix} 1 \\ & 1 \\ & & \cdot \\ & & & \cdot & & & & 0 \\ & & & & \cdot \\ & & & & & 1 \\ & & & & & & c & -s \\ & & & & & & s & c \\ & & & & & & & & 1 \\ & & & & & & & & & \cdot \\ & & 0 & & & & & & & & \cdot \\ & & & & & & & & & & & \cdot \\ & & & & & & & & & & & & 1 \\ & & & & & & & & & & & & & 1 \end{bmatrix} \qquad \begin{array}{l} c = \cos \beta \\ s = \sin \beta \end{array}$$

Here $[R]$ is an orthogonal rotation matrix; that is, $[R]^{-1}=[R]^{T}$ (the inverse equals the transpose); hence

$$[R]^{T}[R] = [I] \tag{16.112}$$

where $[I]$ is the identity matrix. Since Eqs. (16.111) are only a rewriting of Eqs. (16.108) and (16.109), it should be clear that they only alter the two components associated with node i; all other components are left unchanged. The formality is only to facilitate the following matrix operations.

Substitute Eqs. (16.111) into Eq. (16.110):

$$[K]^{(e)}[R]\{a'\} = [R]\{F'\}^{(e)} \tag{16.113}$$

Premultiply both sides of Eq. (16.113) by $[R]^{T}$ and use relation (16.112). This yields the following transformed element equations:

$$[K']^{(e)}\{a'\} = \{F'\}^{(e)} \tag{16.114}$$

where

$$[K']^{(e)} = [R]^{T}[K]^{(e)}[R]$$

$$\{F'\}^{(e)} = [R]^{T}\{F\}^{(e)}$$

The only terms in $[K']^{(e)}$ that differ from $[K]^{(e)}$ are those in rows and columns $2i-1$ and $2i$. Similarly, only rows $2i-1$ and $2i$ in $\{F'\}^{(e)}$ differ from $\{F\}^{(e)}$.

An adjacent element (f) that shares node i (see Fig. 16.25) would, of course, have to have the same transformation applied to its stiffness matrix and load vector before assembling. Since the identical rotation is applied to both elements, it does not matter whether we first transform the separate element equations and then assemble into the system equations, or assemble first and then transform the system equations. The end result is the same.[21] (In mathematical language, the operations of assembly and local axes rotation commute.) It is usually more convenient to transform at the element level, before assembling.

If other nodes in an element must be transformed, then a series of similar transformations is applied to $[K']^{(e)}$; that is, $[K'']^{(e)} = [R']^T[K']^{(e)}[R']$, $[K''']^{(e)} = [R'']^T[K'']^{(e)}[R'']$, etc.

Following assembly, the new (rotated) DOF are the active variables in the resulting system equations. The mixed BCs can then be applied to these nodes in the standard manner for applying nodal displacements and forces.

Solution of the system equations yields *both* components of displacement at node i — namely, a'_{2i-1} and a'_{2i}. Hence they can then be transformed back to the global components a_{2i-1} and a_{2i}, using Eqs. (16.111), for use in stress calculations with the original trial solutions.

Step 10: Solve the system equations.

Here we could use the GAUSEL subroutine used in previous chapters. However, when we introduce multiple DOF per node and/or increase the dimensionality from 1-D to 2-D or 3-D, we run into a size problem; i.e., the amount of data quickly overwhelms the available main storage space in even the largest of computers.

For example, consider a square domain (Fig. 16.26) with a uniform mesh of 20 eight-node 2-D quadrilateral elements in each direction. The system stiffness matrix would require just a little less than $2562 \times 130 = 333{,}060$ words. Using a 32-bit (4-byte) word length (which is minimal for most FE computations), storage for just the stiffness matrix would require more than 1.3 million bytes. And this is only a modest-size problem. Analyses up to 10 times this size are run routinely. (Much larger analyses than this are also run occasionally, but they generally employ substructuring — see the end of Section 13.6.4 — to break the problem into smaller more manageable pieces.)

There exists a variety of data-handling procedures to resolve this problem. The interested reader is referred to the general literature [16.12-16.18]. They all exploit the sparsity characteristic of FE matrices, which requires only a small subset of the system equations to be in main storage at any given moment during their processing. The remainder can then reside on a less expensive secondary storage medium, such as a disk. The number of equations (DOF) "active" at any moment in main storage is approximately equal to the half-bandwidth (for banded solvers) or the wave front (for frontal solvers). A buffer of a few extra equations may be desirable to help decrease the amount of data handling due to swapping back and forth between main and secondary storage. This is an area that requires good file structure and database management techniques.

Step 11: Evaluate the flux (stresses and strains).

---

[21] To apply the above matrix operations to two or more elements, we would need to expand the force and displacement vectors to include all elements (or the entire system, for that matter). The operations would still apply to each element separately, and the $[R]$ matrix would be identical for each element. In such expanded notation we could write

$$[R]^T[K]^{(e)}[R] + [R]^T[K]^{(f)}[R] + \ldots + [R]^T[K]^{(g)}[R] = [R]^T \left[[K]^{(e)} + [K]^{(f)} + \ldots + [K]^{(g)} \right] [R]$$

Thus transforming and then assembling is the same as assembling and then transforming.

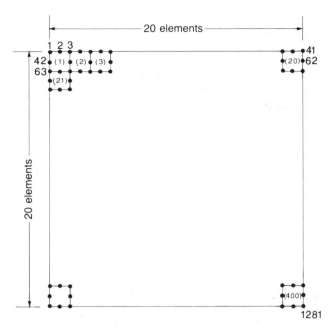

FIGURE 16.26 ● DOF-count for a modest-size problem: Total DOF=1281×2=2562; half-bandwidth=65×2=130.

No further comments are necessary beyond those made in step 6 in previous sections.

Step 12: Display the solution and estimate its accuracy.

The output of the solution phase (steps 8 through 11) consists of the displacements at the nodes and the stresses and strains at the nodes and optimal (Gauss) points.[22] Any further manipulation of these quantities (e.g., arithmetic or calculus operations, as described in item 4 in Fig. 7.10) is done in postprocessing (step 12). Following are some typical calculations that might be performed. The equations are written using the Cartesian x,y,z-coordinates for the 2-D plane problems. The equations are also applicable to the axisymmetric problem if x,y,z are replaced by r,z,θ, respectively.

Principal stresses and strains ● For any general state of stress (at a point) there are three mutually perpendicular planes on which the shear stresses vanish [16.1-16.3]. The normals to these planes define the principal directions, and the normal stresses on these planes are the principal stresses. Similarly, there also exist three mutually perpendicular principal strains at the point. The directions of the principal strains are the same as the principal stresses only if the material is isotropic.

By definition of the plane problem ($\tau_{zx}=\tau_{zy}=0$ and $\gamma_{zx}=\gamma_{zy}=0$), σ_z and ϵ_z must be a principal stress and principal strain. Therefore the other two principal stresses, σ_1 and σ_2, and principal strains, ϵ_1 and ϵ_2, must lie in the x,y-plane. They are calculated from

---

[22] These calculations use the shape functions which, for the sake of modularity, are usually kept separate from postprocessing.

the following formulas [16.19]:

$$\sigma_1 = \frac{1}{2}(\sigma_x + \sigma_y) \pm \sqrt{\left(\frac{\sigma_x - \sigma_y}{2}\right)^2 + \tau_{xy}^2} \tag{16.115}$$

where σ_1 is defined to be the algebraically larger of the two stresses. The angle between the direction of σ_1 and the x-axis, θ_{σ_1}, is given by

$$\theta_{\sigma_1} = \frac{1}{2}\tan^{-1}\left(\frac{\tau_{xy}}{(\sigma_x - \sigma_y)/2}\right) \tag{16.116}$$

The direction of σ_2 is rotated 90° from the direction of σ_1. Similarly,

$$\epsilon_1 = \frac{1}{2}(\epsilon_x + \epsilon_y) \pm \sqrt{\left(\frac{\epsilon_x - \epsilon_y}{2}\right)^2 + \left(\frac{\gamma_{xy}}{2}\right)^2} \tag{16.117}$$

where ϵ_1 is defined to be the algebraically larger of the two strains, and

$$\theta_{\epsilon_1} = \frac{1}{2}\tan^{-1}\left(\frac{\gamma_{xy}}{\epsilon_x - \epsilon_y}\right) \tag{16.118}$$

The direction of ϵ_2 is rotated 90° from the direction of ϵ_1.

Stresses acting on a rotated surface • It is frequently desirable to know the stresses acting on a surface other than those aligned with the global x- and y-axes (Fig. 16.27). The expressions for the rotated stress components σ_Θ and τ_Θ are obtained directly from the tensor transformation relations, Eq. (16.5), which were subsequently applied to the plane problem [see Eqs. (16.47) and (16.53)]. Thus

$$\sigma_\Theta = \sigma_x \cos^2\Theta + \sigma_y \sin^2\Theta + 2\tau_{xy} \sin\Theta \cos\Theta$$
$$\tau_\Theta = \tau_{xy}(\cos^2\Theta - \sin^2\Theta) + (\sigma_y - \sigma_x)\sin\Theta \cos\Theta \tag{16.119}$$

Strain energy density W •

$$W = \frac{1}{2}\{\sigma\}^T(\{\epsilon\} - \{\alpha\}\Delta T) \tag{16.120}$$

Equivalent stress σ_E •

$$\sigma_E = \sqrt{\frac{1}{2}\left[(\sigma_x - \sigma_y)^2 + (\sigma_y - \sigma_z)^2 + (\sigma_z - \sigma_x)^2\right] + 3\tau_{xy}^2} \tag{16.121}$$

or, in terms of principal stresses,

$$\sigma_E = \sqrt{\frac{1}{2}\left[(\sigma_1 - \sigma_2)^2 + (\sigma_2 - \sigma_3)^2 + (\sigma_3 - \sigma_1)^2\right]} \tag{16.122}$$

where $\sigma_3 = \sigma_z$ for the plane problems and $\sigma_3 = \sigma_\theta$ for the axisymmetric problems. For isotropic materials, the onset of plastic yielding is predicted by the Mises yield criterion,[23]

$$\sigma_E = Y \tag{16.123}$$

---

[23] Equation (16.123) is also valid for anisotropic materials, but a more general expression than Eq. (16.121) or (16.122) is needed for σ_E [16.20].

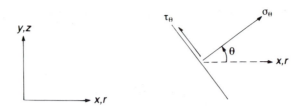

FIGURE 16.27 ● Normal stress σ_θ and shear stress τ_θ acting on a surface with the normal rotated an angle Θ from the x-axis.

where Y is the experimentally determined yield strength in simple tension. Though Eq. (16.123) is generally limited to ductile metals, the latter constitute a structurally important class of materials.

If the theory in this chapter were to be implemented into a computer program, we would be able to analyze the plane stress problem of the bar with a hole, under tension, that was illustrated in Chapter 1.

This completes the presentation. It is hoped that the reader now has a solid grasp of fundamental concepts and has also acquired some practical modeling skills.

Exercises

16.1   Consider a thin bar ("infinitesimal fiber," if you prefer) that rotates rigidly through an angle θ in the x,y-plane:

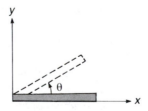

Clearly there is no deformation, only a rigid-body rotation. Calculate the linear "strain" e_{xx} [e_{11} in Eqs. (16.15)], and show that it is not zero but includes second-order terms in θ. (It is therefore not a measure of pure deformation because it includes effects of rigid-body rotation.) Calculate the nonlinear Lagrangian strain E_{xx} [the first equation in Eqs. (16.11)], and show that it is zero for any value of θ. (Hence it is a measure of pure deformation.)

16.2 Consider now a thin plate that rotates rigidly through an angle θ in the x,y-plane:

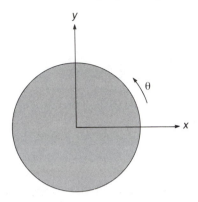

Calculate the linear "rotation" ω_{xy} [ω_{12} in Eqs. (16.16)], and show that it does not equal the angle θ because of second-order terms in θ.

References

[16.1] S. Timoshenko and J. N. Goodier, *Theory of Elasticity*, McGraw-Hill, New York, 1951. (An elementary, readable treatment.)

[16.2] I. S. Sokolnikoff, *Mathematical Theory of Elasticity*, McGraw-Hill, New York, 1956. (An intermediate-level text.)

[16.3] C. Truesdell and R. Toupin, "The Classical Field Theories" in S. Flugge (ed.), *Principles of Classical Mechanics and Field Theory*, Vol. III/1 of *Encyclopedia of Physics*, Springer-Verlag, New York, 1960. (An advanced, abstract treatise.)

[16.4] H. Jeffreys, *Cartesian Tensors*, Cambridge University Press, New York, 1961.

[16.5] I. S. Sokolnikoff, *Tensor Analysis: Theory and Applications to Geometry and Mechanics of Continua*, 2d ed., Wiley, New York, 1964.

[16.6] I. S. Sokolnikoff, *Mathematical Theory of Elasticity*, op. cit., p. 71.

[16.7] S. G. Lekhnitskii, *Theory of Elasticity of an Anisotropic Elastic Body*, Holden-Day, San Francisco, 1963.

[16.8] I. S. Sokolnikoff, *Mathematical Theory of Elasticity*, op. cit. p. 254.

[16.9] Ibid., p. 358.

[16.10] J. S. Przemieniecki, *Theory of Matrix Structural Analysis*, McGraw-Hill, New York, 1968, Section 5.6.

[16.11] R. D. Cook, *Concepts and Applications of Finite Element Analysis*, 2d ed., Wiley, New York, 1981, p. 135.

[16.12] B. M. Irons, "A Frontal Solution Program," *Int J Num Meth Eng*, **2**: 5-32 (1970).

[16.13] D. P. Mondkar and G. H. Powell, "Towards Optimal In-Core Equation Solving," *Int J Comp Struct*, **4**: 531-548 (1974).

[16.14] D. P. Mondkar and G. H. Powell, "Large Capacity Equation Solver for Structural Analyses," *Int J Comp Struct*, **4**: 699-728 (1974).

[16.15] C. Meyer, "Special Problems Related to Linear Equation Solvers," *J Struct Div A.S.C.E.*, **101**(4): 869-890 (1975).

[16.16] J. A. Bettess, "A Data Structure for FE Analysis," *Int J Num Meth Eng,* **11**(12): 1779-1799 (1977).

[16.17] E. Hinton and D. R. J. Owen, *Finite Element Programming*, Academic Press, New York, 1977.

[16.18] C. A. Felippa, "Database Management in Scientific Computing — I. General Description," *Int J Comp Struct,* **10**: 53-62 (1979).

[16.19] E. P. Popov, *Mechanics of Materials*, Prentice-Hall, Englewood Cliffs, N. J., 1952, Chapter 8.

[16.20] F. A. McClintock and A. S. Argon, *Mechanical Behavior of Materials*, Addison-Wesley, Reading, Mass., 1966, pp. 276-278.

Glossary of Abbreviated Terms

BC(s)	Boundary condition(s)	NBC(s)	Natural boundary condition(s)
DOF	Degree(s) of freedom	ODE(s)	Ordinary differential equation(s)
EBC(s)	Essential boundary condition(s)	O(h)	(of) Order h
FDM	Finite difference method	1-D	One dimensional
FE	Finite element(s)	PDE(s)	Partial differential equation(s)
FEA	Finite element analysis	RHS	Right-hand side
FEM	Finite element method	RVM	Ritz variational method
IBC(s)	Interelement boundary condition(s)	2-D	Two dimensional
LHS	Left-hand side	UNAFEM	UNderstanding and Applying the FEM (name of the computer program developed in this book)
LMS	Linear multistep methods		
MWR	Method(s) of weighted residuals		

Appendix:
Data Input Instructions and
Listing for the UNAFEM
Program

Data Input Instructions for UNAFEM

The input data is composed of 11 data sets: problem description; nodes; elements; physical properties; time-stepping; interior loads; essential boundary conditions; natural boundary conditions, concentrated; natural boundary conditions, distributed; postprocessing plots; and problem termination. The sets must be input in the order shown. Each set consists of two parts: a keyword on a line by itself, followed by lines of data, i.e.,

All keywords begin in column 1. Keywords serve a dual purpose: They terminate the previous set and they indicate the nature of the data to follow. Therefore the first set has no keyword and the last set (problem terminator) has no data.

In the instructions for each data set, the keyword is presented first, followed by one or more tables listing the type and format of the data to be input. Each table lists the items for one line of data (the equivalent of an 80-column punch card). Use as many lines as necessary. The format symbols A, F, and I refer to alphanumeric, floating point, and integer, respectively. Integers must be right-justified in their fields. Floating point numbers must have a decimal point and may optionally have an exponent. Blanks (omitted data) are interpreted as zeroes. UNAFEM supplies default values for some of the data; in these cases, if an item is left blank, the default value will be used.

If a data set is meaningless for a certain type of problem (e.g., interior loads for an eigenproblem), the instructions will indicate to omit the set. If a set involves data that is *necessary* for a type of problem (e.g., nodes and elements for all problems) or that *might* be used (e.g., natural boundary conditions for boundary-value and initial-boundary-value problems), then the keyword will be required in either case, whereas the data will be required for the former but optional for the latter.

Superscripts in parentheses refer to explanatory notes following the data sets. A table of physical units for several types of problems follows the explanatory notes.

I. Problem Description
Keyword: (none)
Data: Required for all problems

Columns	Format	Data
1-4	A	'1DIM' if problem is 1-dimensional
		'2DIM' if problem is 2-dimensional
5-8	A	'BDVP' if problem is boundary value
		'EIGP' if problem is eigenproblem
		'IBVP' if problem is initial-boundary value
9-80	A	Title of problem (any meaningful description)

II. Nodes[1]
Keyword: NODE (required for all problems)
Data: Required for all problems

Columns	Format	Data
6-10	I	Node number
11-15	I	Increment in node numbers between this node and node read on next line (Use only for program generation of nodes in between.)
16-25	F	x-coordinate
26-35	F	y-coordinate (use only for 2-D problems)

III. Elements[2]
Keyword: ELEM (required for all problems)
Data: Required for all problems

Columns	Format	Data
6-10	I	Element number
11-15	I	Increment in node numbers between this element and element read on next line; default = 1 (Use only for program generation of elements in between.)
16-20	I	Element-type number[3]
21-25	I	Number of Gauss points, n, for n-point (or n-by-n-point) quadrature rule[4]
26-30	I	Identification number for group of physical properties associated with this element (defined in data set IV); default = 1
31-35	I	Node number of first node in element[5]
36-40	I	Node number of second node in element
⋮	⋮	⋮
65-70	I	Node number of eighth node in element

IV. Physical Properties[6]

Keyword: PROP (required for all problems)

Data: Some required, as explained below

Line 1: Required for all problems

Columns	Format	Data
6-10	I	Identification number for group of properties defined on lines 1-5
11-20	F	Property $\alpha(x)$ (if 1-D problem) or $\alpha_x(x,y)$ (if 2-D problem)
21-30	F	Coefficients of quadratic polynomial, defined in the order 1, x, x^2, y, y^2, xy
⋮	⋮	
		Use only first three terms for 1-D problems, all six terms
61-70	F	for 2-D problems

Line 2: Required only for 2-D problems

Columns	Format	Data
11-20	F	
21-30	F	Property $\alpha_y(x,y)$
⋮	⋮	Coefficients of quadratic polynomial, defined in the order 1, x, x^2, y, y^2, xy
61-70	F	

Line 3: Required for all problems

Columns	Format	Data
11-20	F	
21-30	F	Property $\beta(x)$ or $\beta(x,y)$
⋮	⋮	Coefficients of quadratic polynomial, defined in the order 1, x, x^2, y, y^2, xy
61-70	F	

Line 4: Required only for eigenproblems

Columns	Format	Data
11-20	F	
21-30	F	Property $\gamma(x)$ or $\gamma(x,y)$
⋮	⋮	Coefficients of quadratic polynomial, defined in the order 1, x, x^2, y, y^2, xy
61-70	F	

Line 5: Required only for initial-boundary-value problems

Columns	Format	Data
11-20	F	
21-30	F	Property $\mu(x)$ or $\mu(x,y)$
⋮	⋮	Coefficients of quadratic polynomial, defined in the order 1, x, x^2, y, y^2, xy
61-70	F	

V. Time-Stepping[7]
  Omit this data set for boundary-value problems and eigenproblems.
Keyword: STEP (required for initial-boundary-value problems)
Data: Required for initial-boundary-value problems

Columns	Format	Data
6-10	I	Number of time steps in this interval, n_s
11-20	F	Duration of each time step in this interval, Δt
21-30	F	Value of θ in this interval, $0 \le \theta \le 1$
31-40	F	Initial value at all nodes (Define only for first interval.)

VI. Interior Loads
  Omit this data set for eigenproblems.
Keyword: ILC (required for boundary-value and initial-boundary-value problems)
Data: Optional

Columns	Format	Data
6-10	I	First element number acted on by load
11-20	F	Value of interior load
21-25	I	Last element number acted on by load
26-30	I	Increment in element numbers between these first and last elements; default = 1 (Use columns 21-30 only for program generation of loads on elements in between.[8])

VII. Essential Boundary Conditions[9]
Keyword: EBC (required for all problems)
Data: Optional

Columns	Format	Data
6-10	I	First node number acted on by essential boundary load
11-20	F	Value of load
21-25	I	Last node number acted on by essential boundary load
26-30	I	Increment in node numbers between these first and last nodes; default = 1 (Use columns 21-30 only for program generation of loads on nodes in between.[10])

VIII. Natural Boundary Conditions, Concentrated[11]

Omit this data set for eigenproblems.

Keyword: NBCC (required for boundary-value and initial-boundary-value problems)

Data: Optional

Columns	Format	Data
6-10	I	Node number acted on by concentrated natural boundary load
11-20	F	Value of load

IX. Natural Boundary Conditons, Distributed[12]

Omit this data set for 1-D problems and/or eigenproblems.

Keyword: NBCD (required for 2-D boundary-value and 2-D initial-boundary-value problems)

Data: Optional

Columns	Format	Data
6-10	I	First node of side of element acted on by distributed natural boundary load
11-15	I	Second node of side of element acted on by distributed natural boundary load
16-20	I	Third node of side of element acted on by distributed natural boundary load (Use only for element types 22 and 24.)
21-30	F	Value of load

X. Postprocessing Plots[13]

Keywords: PLOT, if plots wanted (One or the other keyword required for all problems)

NPLOT, if plots not wanted

Data: Optional for PLOT; omitted for NPLOT

Line 1: Use for 1-D problems.

Columns	Format	Data
6-10	I	Number of divisions along x-axis of plot
11-20	F	Minimum x-value on plot
21-30	F	Maximum x-value on plot

Line 2: Use for 1-D problems and 2-D initial-boundary-value problems.

Columns	Format	Data
6-10	I	Number of divisions along U-axis for U-versus-x and U-versus-t plots
11-20	F	Minimum U-value on plot
21-30	F	Maximum U-value on plot

Line 3: Use for 1-D boundary-value and 1-D initial-boundary-value problems.

Columns	Format	Data
6-10	I	Number of divisions along flux axis for flux-versus-x plot
11-20	F	Minimum flux value on plot
21-30	F	Maximum flux value on plot

Line 4: Use for 1-D and 2-D initial-boundary-value problems.

Columns	Format	Data
6-10	I	Number of divisions along t-axis for U-versus-t plot
11-20	F	Minimum time value on plot
21-30	F	Maximum time value on plot
31-35	I	First node number at which U-versus-t is plotted
36-40	I	Second node number at which U-versus-t is plotted (optional)
41-45	I	Third node number at which U-versus-t is plotted (optional)

XI. Problem Termination[14]

Keyword: END (required for all problems)

Data: (none)

Explanatory Notes

1. The nodes in a mesh may be numbered in any pattern, the only restriction being that the numbers must be in the range $1 \leq n \leq n_{max}$, where n_{max} is the number of nodes in the mesh. It is recommended that a pattern be chosen which reduces the bandwidth.

   In data set II the nodes may be defined in any order. UNAFEM has a limited node-generation capability. If three or more nodes are uniformly spaced along a straight line, UNAFEM can generate the data for all the nodes lying in between the two end nodes; in such a case the user need only input data for the two end nodes. This is illustrated by the following example of eight uniformly spaced nodes, showing the two lines of input data that define all eight.

   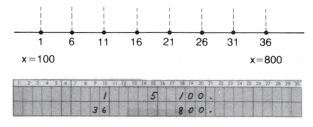

   Nodes that are program-generated must use a positive node number increment. The first and last nodes (numbers 1 and n_{max}) must be input; i.e., they cannot be generated.

2. The elements in a mesh may be numbered in any pattern, the only restriction being that the numbers must be in the range $1 \le e \le e_{max}$, where e_{max} is the number of elements in the mesh. It is recommended that a pattern be chosen which enables the user to make maximum use of program generation of elements, as described below.

   In data set III the elements must be defined in numerical order, beginning with number 1, i.e., $1,2,3,\ldots,e_{max}$. UNAFEM has a limited element-generation capability. If two or more elements have node number patterns that differ by a constant (which may be specified in columns 11-15) and the other element properties are the same in each element, then UNAFEM can generate the data for all the elements following the first one. This is illustrated by the following example of four elements, numbered (6) through (9), showing the single line of input data that defines all four.

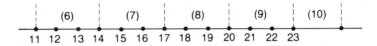

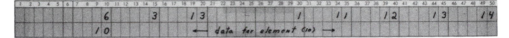

   Generated elements are given the same element-type number, number of Gauss points, and physical property number as the element from which they are generated. The first and last elements (numbers 1 and e_{max}) must be input; i.e., they cannot be generated.

3. The element-type number is a two-digit number. The first digit is the dimension of the element (1 or 2), and the second digit indicates the order in which the element was developed in the text, as shown in Table 1.

4. The number of Gauss points determines the accuracy of the quadrature rule. UNAFEM offers several different quadrature rules for each type of element. Table 2 lists the permissible and default values for number of Gauss points.

5. For the eight elements in UNAFEM, there may be anywhere from two to eight nodes per element. Use only the number of fields necessary for each type of element. The global node numbers should be entered in the same *order* as the local node numbers used for the typical (parent) element. Thus 1-D elements must list the node numbers

TABLE 1

Element-type number	Element type	UNAFEM name
11	1-D 2-node C^0-linear	ELEM11
12	1-D 3-node C^0-quadratic	ELEM12
13	1-D 4-node C^0-cubic	ELEM13
14	1-D 5-node C^0-quartic	ELEM14
21	2-D 3-node C^0-linear triangle	ELEM21
22	2-D 6-node C^0-quadratic triangle	ELEM22
23	2-D 4-node C^0-linear quadrilateral	ELEM23
24	2-D 8-node C^0-quadratic quadrilateral	ELEM24

TABLE 2

Element-type number	Number of Gauss points	
	Permissible	Default
11	NA†	NA†
12	1,2,3,4,5	3
13	1,2,3,4,5	4
14	1,2,3,4,5	5
21	NA†	NA†
22	1,3,4,6,7	6
23	1-by-1,...,5-by-5	2-by-2
24	1-by-1,...,5-by-5	3-by-3

†Leave field blank.

from one end of the element to the other, in the positive coordinate direction. 2-D elements must list the numbers counterclockwise (for a right-handed coordinate system), corner nodes first and then midside nodes, if any (see Figs. 13.12, 13.25, 13.39, and 13.45).

6. The physical property set defines a group of properties: $\alpha(x)$, $\beta(x)$, $\gamma(x)$, and $\mu(x)$ for 1-D problems, or $\alpha_x(x,y)$, $\alpha_y(x,y)$, $\beta(x,y)$, $\gamma(x,y)$, and $\mu(x,y)$ for 2-D problems. An arbitrary identification number (from 1 to 20) is assigned to the group.

Each property is defined on a different line, by specifying the coefficients of a quadratic polynomial [for example, see Eqs. (7.17) and (13.69)]. For 1-D problems, the coefficients of 1, x, x^2 are defined, in that order; for 2-D problems, 1, x, x^2, y, y^2, xy.

Up to 20 groups may be defined, each requiring a separate set of lines 1 to 5 and a unique identification number. Note that the identification number is used in the element data set (columns 26-30 in data set III) to associate a group of physical properties with each element.

7. For initial-boundary-value problems, UNAFEM requires the analyst to divide the time dimension into one or more *intervals*, as explained in Section 11.4. Data sets V through VIII for 1-D problems, or V through IX for 2-D problems, must be input *for each interval*, as illustrated in Section 11.5. This allows for changing any of the loads, as well as the time-stepping parameters, from interval to interval.

Data set X (plotting) is input only once, at the end of the first interval.

The type of time-stepping algorithm is controlled by the parameter θ. It may assume any value in the range $0 \le \theta \le 1$. Table 3 lists several interesting values.

8. Program generation of interior loads may be illustrated by the following example in which a load of 500 is applied to elements 8, 12, 16, and 20.

If an interior load acts on only one element, leave columns 21-30 blank, e.g.,

TABLE 3

θ	Common names for algorithm	Comments
0	Forward difference or Euler's rule	Conditonally stable Requires short time steps Well suited for brief, shocklike response
$\frac{1}{2}$	Mid-difference or Crank-Nicolson method	Unconditionally stable Tends to be very oscillatory
$\frac{2}{3}$	(none)	Unconditionally stable Near-optimal for most responses
1	Backward difference or backward Euler's rule	Unconditionally stable No oscillations; acts overdamped Well suited for approaching steady state

9. Essential boundary conditions can be applied to any node in the mesh, not just those on the boundary of the domain.

10. Program generation of essential boundary loads follows the same structure as for interior loads. (See explanatory note 8.)

11. Concentrated natural boundary conditions can be applied to any node in the mesh, not just those on the boundary of the domain.

12. Distributed natural boundary conditions can be applied to any side of any element in the mesh, not just those on the boundary of the domain.

    Data set IX pertains only to 2-D elements. Each data line defines a constant load (inward-normal component of flux) acting on *one side* of an element. Additional lines would be needed to apply a load to other sides of the same element, or to other elements. For element types 21 and 23, i.e., linear elements, which have only two nodes per side, node numbers should be entered in the first two fields but not the third. For element types 22 and 24, i.e., quadratic elements, which have three nodes per side, node numbers should be entered in all three fields; the nodes should be listed in counterclockwise order along the side. (Since the load is assumed to be uniform along the entire side, it is not permitted to define a load acting between only two nodes on the side of a quadratic element.)

13. Data set X controls the postprocessing plots. (It does not control mesh plots, which are part of preprocessing and are always produced.) If no plots are wanted, use the keyword NPLOT, followed by no data. If plots are wanted, use the keyword PLOT; it may optionally be followed by data.

    Table 4 lists the types of plots produced by UNAFEM (if requested).

    The user has the choice of defining the plotting parameters, or else accepting the default values in UNAFEM.

    - If the choice is to define the parameters, then one should input the lines of data specified in Table 5.
    - If the choice is to accept the default values, then one may omit all data lines (the

TABLE 4

	Boundary-value problems	Eigenproblems	Initial-boundary-value problems
1-D	Function plots: $\tilde{U}(x)$ versus x $\tilde{\tau}(x)$ versus x	Function plots of lowest 5 modes: $\tilde{U}(x)$ versus x	Function plots at end of each interval: $\tilde{U}(x,t)$ versus x $\tilde{\tau}(x,t)$ versus x Function plots at 1-3 selected nodes: $\tilde{U}(x,t)$ versus t
2-D	Contour plots: $\tilde{U}(x,y)$ $\tilde{\tau}_x(x,y)$ $\tilde{\tau}_y(x,y)$	Contour plots of lowest 5 modes: $\tilde{U}(x,y)$	Contour plots at end of each interval: $\tilde{U}(x,y,t)$ $\tilde{\tau}_x(x,y,t)$ $\tilde{\tau}_y(x,y,t)$ Function plots at 1-3 selected nodes: $\tilde{U}(x,y,t)$ versus t

easiest method), or else input blank lines, using the number of lines specified in Table 5 for each problem type. All plots shown in Table 4 will be produced except the time plots for initial-boundary-value problems. (Default parameters cannot be set for time plots since there are no general guidelines to determine default nodes at which to generate the plots.)

For initial-boundary-value problems, data set X is input at the end of the first interval; it is not input after any subsequent intervals.

14. Two or more problems may be analyzed in the same computer run merely by concatenating in one file (i.e., listing sequentially) data sets I through XI for each problem.

To ensure that your input data is dimensionally correct, Table 6 lists the physical dimensions for several different types of problems. An empty box indicates that no data is input, either because the item is not applicable to that type of problem or, in the case of natural boundary conditions for eigenproblems, only zero values are permitted and don't need to be input. The α column describes either a single α for 1-D problems or both α_x and α_y for 2-D problems. Table 7 is a list of abbreviations used in Table 6, along with the corresponding SI units (the *Système International d'Unités*, which is based on mks units). The SI units are shown merely for illustration; any consistent system would be acceptable.

TABLE 5

	Boundary-value problems	Eigenproblems	Initial-boundary-value problems
1-D	Lines 1, 2, 3	Lines 1, 2	Lines 1, 2, 3, 4
2-D	None	None	Lines 2, 4

TABLE 6 ● Physical dimensions for several types of problems

		Data Sets							
		IV				VI	VII	VIII	IX
		Physical properties				Interior loads	Essential boundary conditions	Natural boundary conditions, concentrated	Natural boundary conditions, distributed
		α	β	γ	μ	f	U	τ	τ_{-n}
1-D	Heat conduction†	Thermal conductivity E/tLT	Convection loss coefficient E/tL^3T		Heat capacity E/L^3T	Heat source + part of convection E/tL^3	Temperature T	Heat flux E/tL^2	
	Cable deflection	Tension F	Foundation modulus F/L^2			Distributed transverse force F/L	Transverse displacement L	Transverse force F	
	Elongation of elastic rod†	Young's modulus F/L^2	0			Interior force F/L^3	Longitudinal displacement L	Stress F/L^2	
	Electrostatic field†	Permittivity C/L	0			Charge density Q/L^3	Electrostatic potential V	Electric displacement Q/L^2	
	Vibration of flexible cable	Tension F	Foundation modulus F/L^2	Lineal mass density M/L			Transverse displacement (can only be 0) L		
2-D	Heat conduction‡	Thermal conductivity E/tLT	Convection loss coefficient E/tL^3T		Heat capacity E/L^3T	Heat source + part of convection E/tL^3	Temperature T	Heat flux E/tL	Inward normal component of heat flux E/tL^2
	Membrane deflection	Tension F/L	Foundation modulus F/L^3			Distributed pressure F/L^2	Transverse displacement L	Transverse force F	Transverse force per unit length F/L
	Torsion of elastic cylinder	1	0			2	Prandtl stress function (0 on entire boundary) L^2/A		
	Electrostatic field‡	Permittivity C/L	0			Charge density Q/L^3	Electrostatic potential V	Electric displacement Q/L	Inward normal component of electric disp. Q/L^2
	Acoustic eigenproblem	1	0	1			Amplitude of velocity potential function (can only be 0) L^2/t		

†All terms except essential boundary conditions can be multiplied by cross-sectional area, L^2.

‡All terms except essential boundary conditions can be multiplied by thickness, L.

TABLE 7

Quantity	Abbreviation	SI unit
Length	L	meter
Mass	M	kilogram
Time	t	second
Force	F	newton
Energy	E	joule
Temperature	T	degree Kelvin
Voltage	V	volt
Electric charge	Q	coulomb
Electric capacitance	C	farad
Angle	A	radian

The UNAFEM Program

The complete listing for the UNAFEM program is shown photoreduced on pages 806–828. Below is a table of contents giving the pages on which each subroutine appears, followed by a flowchart of the program on page 805.

UNAFEM

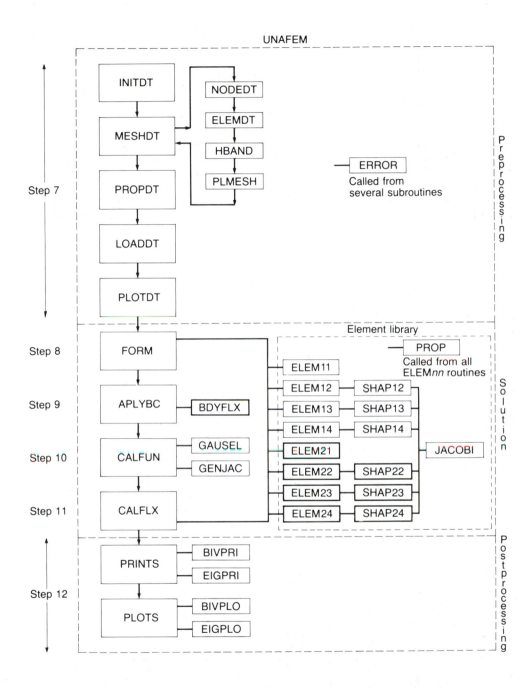

Step 7

Step 8

Step 9

Step 10

Step 11

Step 12

```fortran
C     **************
C     * U N A F E M *
C     **************
C
C     (UN)DERSTANDING AND (A)PPLYING THE (F)INITE (E)LEMENT (M)ETHOD
C
C     THIS PROGRAM WAS WRITTEN BY W. JOHN DENKMANN, OF AT&T
C     CONSUMER PRODUCT LABORATORIES, IN COLLABORATION WITH
C     DAVID S. BURNETT, OF AT&T BELL LABORATORIES, FOR THE
C     TEXTBOOK "FINITE ELEMENT ANALYSIS: FROM CONCEPTS TO
C     APPLICATIONS."
C
C     THE PROGRAM SOLVES THREE TYPES OF ONE- AND TWO-DIMENSIONAL
C     PROBLEMS USING THE FINITE ELEMENT METHOD:
C
C     1.  BOUNDARY-VALUE PROBLEMS
C         (INCLUDES LAPLACE AND POISSON EQUATIONS)
C
C     1-D
C         -D/DX(ALPHA(X)*DU/DX) + BETA(X)*U = F(X)
C     2-D
C         -D/DX(ALPHAX(X,Y)*DU/DX) - D/DY(ALPHAY(X,Y)*DU/DY)
C            + BETA(X,Y)*U = F(X,Y)
C
C     2.  EIGENPROBLEMS
C         (INCLUDES STURM-LIOUVILLE, HELMHOLTZ, AND
C          TIME-INDEPENDENT SCHROEDINGER EQUATIONS)
C
C     1-D
C         -D/DX(ALPHA(X)*DU/DX) + BETA(X)*U - LAMBDA*GAMMA(X)*U = 0
C     2-D
C         -D/DX(ALPHAX(X,Y)*DU/DX) - D/DY(ALPHAY(X,Y)*DU/DY)
C            + BETA(X,Y)*U - LAMBDA*GAMMA(X,Y)*U = 0
C
C     3.  INITIAL-BOUNDARY-VALUE PROBLEMS
C         (INCLUDES DIFFUSION EQUATION)
C
C     1-D
C         MU(X)*DU/DT - D/DX(ALPHA(X)*DU/DX) + BETA(X)*U = F(X,T)
C     2-D
C         MU(X,Y)*DU/DT - D/DX(ALPHAX(X,Y)*DU/DX)
C            - D/DY(ALPHAY(X,Y)*DU/DY) + BETA(X,Y)*U = F(X,Y,T)
C
C     ALL LIMITS ARE SET IN SUBROUTINE INITDT
C
      COMMON /PROBLM/ DIMEN,PROB,TITLE(18)
      DATA XIBVP/'IBVP'/
      LOGICAL MORDAT
C
C********* PREPROCESSING (INPUT) **********
C
C------     INITIALIZATION DATA
10    CONTINUE
      CALL INITDT
C------     MESH DATA
      CALL MESHDT
C------     PHYSICAL PROPERTY DATA
      CALL PRPDT
C------     LOAD DATA
20    CONTINUE
      CALL LOADDT
C------     PLOT DATA
      CALL PLOTDT
C

C********* SOLUTION **********
C
C------     FORM THE SYSTEM EQUATIONS
      CALL FORM
C------     APPLY THE BOUNDARY CONDITIONS
      CALL APLYBC
C------     SOLVE THE SYSTEM EQUATIONS FOR FUNCTION VALUES
      CALL CALFUN
C------     CALCULATE FLUX VALUES
      CALL CALFLX
C
C********* POSTPROCESSING (OUTPUT) **********
C
C------     PRINT RESULTS IN TABULAR FORM
      CALL PRINTS
C------     PLOT RESULTS IN GRAPHICAL FORM
      CALL PLOTS
C
C------     CHECK FOR END OF PROBLEM
      CALL ENPROB(MORDAT)
C
C------     IF INIT VAL PROB & MORE LOAD DATA, RETURN TO LOADDT
      IF ((PROB.EQ.XIBVP) AND MORDAT) GO TO 20
C------     IF INIT VAL PROB & NO MORE LOAD DATA, PLOT FUNC V TIME
      IF ((PROB.EQ.XIBVP) AND..NOT.MORDAT) CALL TIMPLT
C------     RETURN TO INITDT FOR NEW PROBLEM
      GO TO 10
C
      END
      SUBROUTINE ENPROB(MORDAT)
C
C     THIS ROUTINE CHECKS FOR THE END OF DATA FOR A PROBLEM
C
      COMMON /PROBLM/ DIMEN,PROB,TITLE(18)
      DATA XIBVP/'IBVP'/
      LOGICAL MORDAT
      COMMON /INVAL/ INTRVL,NSTEPS,NTIME,TIME,DT,THETA,AO
      COMMON /KEYWRD/ AKEY,PLTKEY
      DATA ENDA/' END'/,ENDB/'END '/,STEP/'STEP'/
C
      IF (AKEY.EQ.ENDA.OR.AKEY.EQ.ENDB) MORDAT=.FALSE.
      IF (AKEY.EQ.STEP) MORDAT=.TRUE.
      IF (.NOT.MORDAT.AND.(PROB.NE.XIBVP)) WRITE (6,3000)
      RETURN
1000  FORMAT(A4)
3000  FORMAT(///,'*****POSTPROCESSING COMPLETE.  RETURN FOR NEXT PROB.')
      END
      SUBROUTINE INITDT
C
C     THIS ROUTINE READS THE INITIAL DATA, SETS LIMITS, AND SETS
C         INTEGRATION POINTS
C
      COMMON /LIMITS/ MAXNP,MAXEL,MAXPP,MAXIL,MAXEBC,MAXNBC,MAXBND
      COMMON /ELSPEC/ NNODES(30),INTDEF(30),IOPTAU(30)
      COMMON /GLINT/ GP(5,5),WT(5,5)
      COMMON /TRINT/ NTIP(7),GPXI(7,7),GPET(7,7),WGHT(7,7)
      COMMON /INVAL/ INTRVL,NSTEPS,NTIME,TIME,DT,THETA,AO
      COMMON /TRANS/ TR1(6,4),TR2(8,4)
      COMMON /PROBLM/ DIMEN,PROB,TITLE(18)
      DATA EIGP/'EIGP'/
C
C********** TITLE CARD **********
C
5     CONTINUE
      READ (5,1100,END=500) DIMEN,PROB,TITLE
C------     SET MAXIMUMS

      MAXNP = 300
      MAXEL = 200
      MAXPP = 20
      MAXIL = 200
      MAXEBC = 100
      MAXNBC = 100
      MAXBND = 50
      IF (PROB.EQ.EIGP) MAXBND = 61
      IF (PROB.EQ.EIGP) MAXNP = 61
      IF (PROB.EQ.EIGP) MAXEL = 61
      IF (PROB.EQ.EIGP) MAXIL = 61
      WRITE (6,2000)
      WRITE (6,2100) DIMEN,PROB,TITLE
      WRITE (6,2200) MAXNP,MAXEL,MAXPP,MAXIL,MAXEBC,MAXNBC,MAXBND
C------     SET NUMBER OF NODES PER ELEMENT, DEFAULT INTEGRATION ORDER AND
C               NUMBER OF OPTIMAL FLUX POINTS
      DO 10 I=1,30
      NNODES(I) = 0
      INTDEF(I) = 0
      IOPTAU(I) = 0
10    CONTINUE
      NNODES(11) = 2
      NNODES(12) = 3
      NNODES(13) = 3
      NNODES(14) = 5
      NNODES(21) = 3
      NNODES(22) = 6
      NNODES(23) = 4
      NNODES(24) = 8
      INTDEF(11) = 3
      INTDEF(12) = 3
      INTDEF(13) = 5
      INTDEF(14) = 5
      INTDEF(21) = 6
      INTDEF(22) = 6
      INTDEF(23) = 4
      INTDEF(24) = 4
      IOPTAU(11) = 1
      IOPTAU(12) = 2
      IOPTAU(13) = 3
      IOPTAU(14) = 4
      IOPTAU(21) = 1
      IOPTAU(22) = 3
      IOPTAU(23) = 2
      IOPTAU(24) = 4
C------     SET INTERVAL COUNTER FOR INITIAL-BOUNDARY-VALUE PROBLEMS
      INTRVL = 0
C------     GAUSS (INTEGRATION) POINTS AND WEIGHTS FOR
C               GAUSS-LEGENDRE QUADRATURE
      DO 50 N=1,5
      DO 40 L=1,5
      GP(N,L) = 0.
      WT(N,L) = 0.
40    CONTINUE
50    CONTINUE
C------     FOR ONE GAUSS POINT
      GP(1,1) = 0.
      WT(1,1) = 2.
C------     FOR TWO GAUSS POINTS
      GP(2,1) = -1./SQRT(3.)
      GP(2,2) = -GP(2,1)
      WT(2,1) = 1.
      WT(2,2) = 1.
C------     FOR THREE GAUSS POINTS
      GP(3,1) = -SQRT(3./5.)
      GP(3,2) = 0.
```

(Continued)

```
      GP(3,3) = -GP(3,1)
      WT(3,1) = 5./9.
      WT(3,2) = 8./9.
      WT(3,3) = WT(3,1)
C----------FOR FOUR GAUSS POINTS
      GP(4,1) = -SQRT((15.+2.*SQRT(30.))/35.)
      GPET(4,1) = -SQRT((15.-2.*SQRT(30.))/35.)
      GP(4,2) = -GP(4,2)
      GP(4,3) = -GP(4,1)
      GP(4,4) = -GP(4,1)
      WT(4,1) = 49./16.*(18.+SQRT(30.)))
      WT(4,2) = 49./16.*(18.-SQRT(30.)))
      WT(4,3) = WT(4,2)
      WT(4,4) = WT(4,1)
C----------FOR FIVE GAUSS POINTS
      GP(5,1) = -SQRT((35.+2.*SQRT(70.))/63.)
      GP(5,2) = -SQRT((35.-2.*SQRT(70.))/63.)
      GP(5,3) = 0.
      GP(5,4) = -GP(5,2)
      GP(5,5) = -GP(5,1)
      WT(5,1) = 5103./50.*(322.+13.*SQRT(70.)))
      WT(5,2) = 5103./50.*(322.-13.*SQRT(70.)))
      WT(5,3) = 128./225.
      WT(5,4) = WT(5,2)
      WT(5,5) = WT(5,1)
C----------GAUSS (INTEGRATION) POINTS AND WEIGHTS FOR GAUSSIAN
C          QUADRATURE OF TRIANGLES
C          (FROM G. R. COWPER, "GAUSSIAN QUADRATURE FORMULAS
C          FOR TRIANGLES," INT. J. NUM. METH. ENGNG .
C          VOL 7, 1973, PP405-408)
C
      NTIP(1) = 1
      NTIP(3) = 3
      NTIP(4) = 4
      NTIP(6) = 6
      NTIP(7) = 7
C
      DO 70 N=1,7
      DO 60 L=1,7
         GPXI(N,L) = 0.
         GPET(N,L) = 0.
         WGHT(N,L) = 0.
60    CONTINUE
70    CONTINUE
C----------FOR ONE GAUSS POINT
      GPXI(1,1) = 1./3.
      GPET(1,1) = GPXI(1,1)
      WGHT(1,1) = 1.
C----------FOR THREE GAUSS POINTS
      GPXI(3,1) = 1./6.
      GPXI(3,2) = 2./3.
      GPXI(3,3) = GPXI(3,1)
      GPET(3,1) = GPXI(3,1)
      GPET(3,2) = GPXI(3,1)
      GPET(3,3) = GPXI(3,1)
      WGHT(3,1) = 1./3.
      WGHT(3,2) = WGHT(3,1)
      WGHT(3,3) = WGHT(3,1)
C----------FOR FOUR GAUSS POINTS
      GPXI(4,1) = 1./3.
      GPXI(4,2) = 1./5.
      GPXI(4,3) = 3./5.
      GPXI(4,4) = GPXI(4,2)
      GPET(4,1) = GPXI(4,1)
      GPET(4,2) = GPXI(4,1)
      GPET(4,3) = GPXI(4,2)
      GPET(4,4) = GPXI(4,3)
      WGHT(4,1) = -27./48.


      WGHT(4,2) = 25./48.
      WGHT(4,3) = WGHT(4,2)
      WGHT(4,4) = WGHT(4,1)
C----------FOR SIX GAUSS POINTS
      GPXI(6,1) = 0.09157621350771
      GPXI(6,2) = 0.81684757298C459
      GPXI(6,3) = GPXI(6,1)
      GPXI(6,4) = 0.445948490915965
      GPXI(6,5) = 0.108103018168070
      GPXI(6,6) = GPXI(6,4)
      GPET(6,1) = GPXI(6,1)
      GPET(6,2) = GPXI(6,1)
      GPET(6,3) = GPXI(6,4)
      GPET(6,4) = GPXI(6,4)
      GPET(6,5) = GPXI(6,4)
      GPET(6,6) = GPXI(6,5)
      WGHT(6,1) = 0.109951743655322
      WGHT(6,2) = WGHT(6,1)
      WGHT(6,3) = WGHT(6,1)
      WGHT(6,4) = 0.223381589678011
      WGHT(6,5) = WGHT(6,4)
      WGHT(6,6) = WGHT(6,4)
C----------FOR SEVEN GAUSS POINTS
      GPXI(7,1) = 1./3.
      GPXI(7,2) = 0.101286507323456
      GPXI(7,3) = 0.797426985353087
      GPXI(7,4) = GPXI(7,2)
      GPXI(7,5) = 0.470142064105115
      GPXI(7,6) = 0.059715871789770
      GPXI(7,7) = GPXI(7,5)
      GPET(7,1) = GPXI(7,1)
      GPET(7,2) = GPXI(7,2)
      GPET(7,3) = GPXI(7,2)
      GPET(7,4) = GPXI(7,3)
      GPET(7,5) = GPXI(7,5)
      GPET(7,6) = GPXI(7,5)
      GPET(7,7) = GPXI(7,6)
      WGHT(7,1) = 0.225
      WGHT(7,2) = 0.125939180544822
      WGHT(7,3) = WGHT(7,2)
      WGHT(7,4) = WGHT(7,2)
      WGHT(7,5) = 0.132394152788506
      WGHT(7,6) = WGHT(7,5)
      WGHT(7,7) = WGHT(7,5)
C----------TRANSFORMATION MATRIX - TO TRANSFORM GAUSS POINT FLUXES TO
C          ELEMENT NODAL FLUXES FOR 6-NODE TRIANGULAR ELEMENT
C
      TR1(1,1) = 3./12.
      TR1(2,1) = TR1(1,1)
      TR1(3,1) = TR1(1,1)
      TR1(4,1) = TR1(1,1)
      TR1(5,1) = TR1(1,1)
      TR1(6,1) = TR1(1,1)
      TR1(1,2) = 23./12.
      TR1(2,2) = -7./12.
      TR1(3,2) = 8./12.
      TR1(4,2) = TR1(2,2)
      TR1(5,2) = TR1(1,2)
      TR1(6,2) = TR1(4,2)
      TR1(1,3) = TR1(1,2)
      TR1(2,3) = TR1(1,2)
      TR1(3,3) = TR1(2,2)
      TR1(4,3) = TR1(1,2)
      TR1(5,3) = TR1(6,3)
      TR1(6,3) = TR1(2,2)
      TR1(1,4) = TR1(2,2)
      TR1(2,4) = TR1(2,2)


      TR1(3,4) = TR1(1,2)
      TR1(4,4) = TR1(2,2)
      TR1(5,4) = TR1(4,2)
      TR1(6,4) = TR1(4,2)
C
      SR3 = SQRT(3.)
      TR2(1,1) = (1.+2.-SR3)/4.
      TR2(2,1) = 1./4.
      TR2(3,1) = (1.-2.-SR3)/4.
      TR2(4,1) = TR2(2,1)
      TR2(5,1) = (1.-SR3)/4.
      TR2(6,1) = (1.-SR3)/4.
      TR2(7,1) = TR2(6,1)
      TR2(8,1) = TR2(5,1)
      TR2(1,2) = TR2(2,1)
      TR2(2,2) = TR2(3,1)
      TR2(3,2) = TR2(2,1)
      TR2(4,2) = TR2(1,1)
      TR2(5,2) = TR2(6,1)
      TR2(6,2) = TR2(5,1)
      TR2(7,2) = TR2(5,1)
      TR2(8,2) = TR2(6,1)
      TR2(1,3) = TR2(2,1)
      TR2(2,3) = TR2(2,1)
      TR2(3,3) = TR2(2,1)
      TR2(4,3) = TR2(2,1)
      TR2(5,3) = TR2(5,1)
      TR2(6,3) = TR2(5,1)
      TR2(7,3) = TR2(6,1)
      TR2(8,3) = TR2(6,1)
      TR2(1,4) = TR2(2,1)
      TR2(2,4) = TR2(2,1)
      TR2(3,4) = TR2(2,1)
      TR2(4,4) = TR2(2,1)
      TR2(5,4) = TR2(6,1)
      TR2(6,4) = TR2(5,1)
      TR2(7,4) = TR2(5,1)
      TR2(8,4) = TR2(6,1)
      RETURN
C
500   CONTINUE
      CALL ERROR(27,0)
C
1100  FORMAT(20A4)
2000  FORMAT('1',20A4)
   $
2100  FORMAT(//' MAXIMUM SIZE LIMITS ON VARIABLES IN UNAFEM'/)
2200  FORMAT(//' MAXIMUM SIZE LIMITS ON VARIABLES IN UNAFEM'/
   $   10X,'NODAL POINTS (MAXNP) ---------------------',I5/
   $   10X,'ELEMENTS (MAXEL) -------------------------',I5/
   $   10X,'PHYSICAL PROPERTIES (MAXPP) --------------',I5/
   $   10X,'INTERNAL LOADS (MAXIL) -------------------',I5/
   $   10X,'ESSENTIAL BOUNDARY LOADS (MAXEBC) --------',I5/
   $   10X,'NATURAL BOUNDARY LOADS (MAXNBC) ----------',I5/
   $   10X,'SYSTEM HALF-BANDWIDTH (MAXBND) -----------',I5)
      END
      SUBROUTINE ERROR(N,M)
C
C     THIS ROUTINE PRINTS ERROR MESSAGES FOR FATAL ERRORS
C     ENCOUNTERED DURING EXECUTION AND TERMINATES THE PROGRAM.
C
      COMMON /LIMITS/ MAXNP,MAXEL,MAXPP,MAXIL,MAXEBC,MAXNBC,MAXBND
C
      WRITE (6,2000)
      IF (N.EQ.1) WRITE (6,2001) MAXNP
      IF (N.EQ.2) WRITE (6,2002) MAXEL
```

(Continued)

807

```
      IF (N.EQ.3)  WRITE (6,2003) MAXPP
      IF (N.EQ.4)  WRITE (6,2004) MAXIL
      IF (N.EQ.5)  WRITE (6,2005) MAXEBC
      IF (N.EQ.6)  WRITE (6,2006) MAXNBC
      IF (N.EQ.7)  WRITE (6,2007) M
      IF (N.EQ.8)  WRITE (6,2008) M
      IF (N.EQ.9)  WRITE (6,2009) M
      IF (N.EQ.10) WRITE (6,2010) M
      IF (N.EQ.11) WRITE (6,2011) M
      IF (N.EQ.12) WRITE (6,2012) M
      IF (N.EQ.13) WRITE (6,2013) M
      IF (N.EQ.14) WRITE (6,2014) M
      IF (N.EQ.15) WRITE (6,2015) M
      IF (N.EQ.16) WRITE (6,2016) M
      IF (N.EQ.17) WRITE (6,2017) M
      IF (N.EQ.18) WRITE (6,2018) M,MAXBND
      IF (N.EQ.19) WRITE (6,2019) M
      IF (N.EQ.20) WRITE (6,2020) M
      IF (N.EQ.21) WRITE (6,2021) MAXBND
      IF (N.EQ.22) WRITE (6,2022)
      IF (N.EQ.23) WRITE (6,2023)
      IF (N.EQ.24) WRITE (6,2024)
      IF (N.EQ.25) WRITE (6,2025)
      IF (N.EQ.26) WRITE (6,2026) M
      IF (N.EQ.27) WRITE (6,2027)
      IF (N.EQ.28) WRITE (6,2028) M
      IF (N.EQ.29) WRITE (6,2029) M
      IF (N.EQ.30) WRITE (6,2030) M
      IF (N.EQ.31) WRITE (6,2031) M
      IF (N.EQ.32) WRITE (6,2032) M
      IF (N.EQ.33) WRITE (6,2033) M
      IF (N.EQ.34) WRITE (6,2034)
      IF (N.EQ.35) WRITE (6,2035)
      WRITE (6,2050)
      STOP
 2000 FORMAT('1',  UNAFEM-UNDERSTANDING AND APPLYING THE FEM'/
     $  ' ************************************************* ')
 2001 FORMAT(' NUMBER OF NODAL POINTS EXCEEDS ',I5,
     $  '  THE MAXIMUM ALLOWABLE.')
 2002 FORMAT(' NUMBER OF ELEMENTS EXCEEDS ',I5,
     $  '  THE MAXIMUM ALLOWABLE.')
 2003 FORMAT(' NUMBER OF PHYSICAL PROPERTIES EXCEEDS ',I5,
     $  '  THE MAXIMUM ALLOWABLE.')
 2004 FORMAT(' NUMBER OF INTERNAL LOADS EXCEEDS ',I5,
     $  '  THE MAXIMUM ALLOWABLE.')
 2005 FORMAT(' NUMBER OF ESSENTIAL BOUNDARY LOADS EXCEEDS ',I5,
     $  '  THE MAXIMUM ALLOWABLE.')
 2006 FORMAT(' NUMBER OF NATURAL BOUNDARY LOADS EXCEEDS ',I5,
     $  '  THE MAXIMUM ALLOWABLE.')
 2007 FORMAT(' NODE NUMBER ',I5,
     $  '  IS NOT IN PERMISSABLE RANGE OF NODE NUMBERS.')
 2008 FORMAT(' NODE NUMBER ',I5,' IS OUT OF ORDER.')
 2009 FORMAT(' ELEMENT NUMBER ',I5,
     $  '  IS NOT IN PERMISSABLE RANGE OF ELEMENT NUMBERS.')
 2010 FORMAT(' ELEMENT TYPE SPECIFIED IN ELEMENT ',I5,
     $  '  IS NOT IN PERMISSABLE RANGE OF TYPE NUMBERS.')
 2011 FORMAT(' PHYSICAL PROPERTY SPECIFIED IN ELEMENT ',I5,
     $  '  IS NOT IN PERMISSABLE RANGE OF PROPERTY NUMBERS.')
 2012 FORMAT(' ONE OF NODES ON ELEMENT ',I5,
     $  '  IS NOT IN PERMISSABLE RANGE OF NODE NUMBERS.')
 2013 FORMAT(' ELEMENT NUMBER ',I5,' IS OUT OF ORDER.')
 2014 FORMAT(' PHYSICAL PROPERTY NUMBER ',I5,
     $  '  IS NOT IN PERMISSABLE RANGE OF PROPERTY NUMBERS.')
 2015 FORMAT(' INTERNAL LOAD NUMBER ',I5,
     $  '  IS NOT IN PERMISSABLE RANGE OF LOAD NUMBERS.')
```

```
 2016 FORMAT(' NODE NUMBER ',I5,
     $  '  FOR ESSENTIAL BOUNDARY LOADS IS NOT IN PERMISSABLE RANGE'
     $  '  OF NODE NUMBERS.')
 2017 FORMAT(' NATURAL BOUNDARY LOAD ',I5,
     $  '  IS NOT IN PERMISSABLE RANGE OF NBC NUMBERS.')
 2018 FORMAT(' MAXIMUM BANDWIDTH AS CALCULATED ',I5,
     $  '  EXCEEDS MAXIMUM PERMISSABLE BANDWIDTH.',I5)
 2019 FORMAT(' ZERO OR NEGATIVE DETERMINANT OF THE JACOBIAN MATRIX ',
     $  '  IS FOUND FOR ELEMENT ',I5)
 2020 FORMAT(' ZERO OR NEGATIVE AREA FOUND IN ELEMENT ',I5)
 2021 FORMAT(' FOR AN EIGEN PROBLEM, THE NUMBER OF NODES EXCEEDS',
     $  I5/10X,' THE MAXIMUM ALLOWABLE FOR EIGEN PROBLEMS.')
 2022 FORMAT(' THE PRECEDING NODES WERE NOT READ.')
 2023 FORMAT(' THE PRECEDING NODES WERE NOT USED IN ANY ELEMENT.')
 2024 FORMAT(' DUPLICATE ESSENTIAL BOUNDARY LOADS INPUT.')
 2025 FORMAT(' NO NODE CARDS READ.')
 2026 FORMAT(' NUMBER OF NODE CARDS READ EQUALS ',I5/
     $  10X,' THERE MUST BE TWO OR MORE NODES FOR EACH PROBLEM.')
 2027 FORMAT(' END OF INPUT DATA REACHED.')
 2028 FORMAT(' PHYSICAL PROPERTIES FOR PROPERTY NO.',I3,' INCOMPLETE.')
 2029 FORMAT(' INTERNAL LOAD DATA OUT OF ORDER FOR TIME INTERVAL',I3)
 2030 FORMAT(' NUMBER OF TIME STEPS INCORRECT IN TIME INTERVAL',I3)
 2031 FORMAT(' TIME STEP INCORRECT IN TIME INTERVAL',I3)
 2032 FORMAT(' THETA INCORRECT IN TIME INTERVAL',I3)
 2033 FORMAT(' DIAGONAL TERM OF LUMPED CAPACITY MATRIX = 0'//
     $  ' FOR NODE ',I3)
 2034 FORMAT(' KEYWORD SEQUENCE INCORRECT.')
 2035 FORMAT(' ERROR IN CONTOUR PLOT.')
 2050 FORMAT(' EXECUTION HALTED.')
      END
      SUBROUTINE MESHDT
C
C-------------------------------------------------------------------
C
C   THIS ROUTINE CALLS THE ROUTINES WHICH GENERATE AND PLOT THE MESH
C
C-------------------------------------------------------------------
C--------- NODAL POINT DATA
      CALL NODEDT
C--------- ELEMENT DATA
      CALL ELEMDT
C--------- CALCULATE HALF BANDWIDTH
      CALL HBAND
C--------- PLOT THE MESH
      CALL PLMESH
C
      RETURN
      END
      SUBROUTINE NODEDT
C
C-------------------------------------------------------------------
C
C   THIS ROUTINE READS AND GENERATES NODE POINT INFORMATION
C
C-------------------------------------------------------------------
      COMMON /LIMITS/ MAXNP,MAXEL,MAXPP,MAXIL,MAXEBC,MAXNBC,MAXBND
      COMMON /GENRL/  NUMNP,NUMEL,NUMPP,NUMILC,NUMEBC,NUMNBC,MBAND
      COMMON /PROBLM/ DIMEN,PROB,TITLE(18)
      COMMON /NODES/  X(300),Y(300)
      REAL MISS
      DATA ELLM/'ELEM'/,MISS/'MISS'/,NODD/'NODE'/
      DATA IBLANK/'    '/,BLANK/'    '/
      DATA ONED/'1D1M'/,TWOD/'2D1M'/
      DATA BDVP/'BDVP'/,EIGP/'EIGP'/,IBVP/'IBVP'/
C
      NUMNP = 0
      WRITE (6,2100) DIMEN,PROB,TITLE
      READ (5,1300) KEY
      IF (KEY.NE.NODD) CALL ERROR(25,NUMNP)
C--------- PREPARE CHECK FOR MISSING NODES
```

```
      DO 50 I=1,300
         X(I) = MISS
   50 CONTINUE
C--------- READ FIRST NODE
      READ (5,1300) AKEY,N1,N1INC,XN1,YN1
      IF (AKEY.EQ.BLANK) GO TO 80
      IF (AKEY.EQ.ELLM) CALL ERROR(25,NUMNP)
      CALL ERROR(34,1)
   80 CONTINUE
      NUMNP = N1
      IF (N1.GT.MAXNP .OR. N1.LE.0) CALL ERROR(7,N1)
      X(N1) = XN1
      Y(N1) = YN1
C--------- BEGIN LOOP TO READ REST OF NODAL DATA
  100 CONTINUE
C--------- READ NEXT CARD (BRANCH OUT OF LOOP IF DELIMITER IS READ)
      READ (5,1300) AKEY,N2,N2INC,XN2,YN2
      IF (AKEY.EQ.BLANK) GO TO 120
      IF (AKEY.EQ.ELLM) GO TO 200
      CALL ERROR(34,1)
  120 CONTINUE
      IF (N2.GT.NUMNP) NUMNP = N2
      IF (N2.GT.MAXNP .OR. N2.LE.0) CALL ERROR(7,N2)
      X(N2) = XN2
      Y(N2) = YN2
C--------- CHECK IF GENERATION NECESSARY-BRANCH AROUND IF NOT NEEDED
      IF (N1INC.LE.0 .OR. N2.EQ.(N1+N1INC)) GO TO 160
C--------- GENERATION
      DIFF = N2-N1
      DINC = N1INC
      DX = (X(N2)-X(N1))/DIFF*DINC
      DY = (Y(N2)-Y(N1))/DIFF*DINC
      N11 = N1+N1INC
      N22 = N2-N1INC
      IF (((N2-N1)/N1INC)*N1INC.NE.(N2-N1)) WRITE (6,3000) N1,N2
      IF (N11.GT.N22) N22 = N11
      IF (N11.GT.N2) WRITE (6,3100) N1,N2
      IF (N11.GT.N2) GO TO 160
      DO 140 N=N11,N22,N1INC
         X(N) = X(N-N1INC)+DX
         Y(N) = Y(N-N1INC)+DY
  140 CONTINUE
  160 CONTINUE
C--------- SHIFT INDEX
      N1 = N2
      N1INC = N2INC
      GO TO 100
C--------- WRITE NODES IN ORDER
  200 CONTINUE
      IF (DIMEN.EQ.ONED) WRITE (6,2500) (I,X(I),I=1,NUMNP)
      IF (DIMEN.EQ.TWOD) WRITE (6,2550) (I,X(I),Y(I),I=1,NUMNP)
      IF (NUMNP.LE.1) CALL ERROR(26,NUMNP)
      IF ((PROB.EQ.EIGP).AND.(NUMNP.GT.MAXBND)) CALL ERROR(21,MAXBND)
C--------- CHECK FOR MISSING NODES
      IMISS = 0
      DO 250 I=1,NUMNP
         IF (X(I).EQ.MISS) WRITE (6,3200) I
         IF (X(I).EQ.MISS) IMISS=IMISS+1
  250 CONTINUE
      IF (IMISS.EQ.1) CALL ERROR(22,NUMNP)
      RETURN
 1300 FORMAT(A4,1X,2I5,2F10.5)
 2000 FORMAT('1',  UNAFEM-UNDERSTANDING AND APPLYING THE FEM'/
     $  ' ... ')
 2100 FORMAT(// ...,2044)
```

(Continued)

```fortran
2500 FORMAT(//' NODAL POINT DATA'//)
   $  8X,'NODE NO',7X,'X-COORD',7X,'Y-COORD'//(10X,I5,E14.3))
2550 FORMAT(//' NODAL POINT DATA'//)
   $  8X,'NODE NO',7X,'X-COORD',7X,'Y-COORD'//(10X,I5,2E14.3))
3000 FORMAT(10X,'BETWEEN THESE NODES GENERATION WILL NOT'/)
   $  15X,'EQUALLY SPACED.'/15X,215/10X,'CHECK INCREMENT'/)
3100 FORMAT(10X,'BETWEEN THESE NODES GENERATION CANNOT BE DONE.'/
   $  10X,215)
3200 FORMAT(' THE FOLLOWING NODE WAS NOT READ OR GENERATED-',
   $  15)
     END
     SUBROUTINE ELEMDT
C
C    THIS ROUTINE READS AND GENERATES ELEMENT INFORMATION
C
     COMMON /PROBLM/ DIMEN,PROB,TITLE(18)
     COMMON /LIMITS/ MAXNP,MAXEL,MAXIL,MAXEBC,MAXNBC,MAXBND
     COMMON /GENRL/ NUMNP,NUMEL,NUMPP,NUMILC,NUMEBC,NUMNBC,MBAND
     COMMON /ECMNTS/ NTYPE(200),NGPINT(200),NPHYS(200),ICON(8,200)
     COMMON /ELSPEC/ NNODES(30),INTDEF(30),IDPTAU(30)
     DIMENSION IC(8),IMISS(300)
     DATA PPROP/'PROP'/,MISS/'MISS'/,IPRES/'PRES'/
     DATA IBLANK /'    '/,BLANK/' '/
C
     WRITE (6,2000)
     WRITE (6,2100) DIMEN,PROB,TITLE
     WRITE (6,2400)
     NUMEL = 0
     NO = 0
100  CONTINUE
     READ (5,1400) AKEY,N1,INC,N2,INTG,NPH,(IC(I),I=1,8)
C---- CHECK FOR DELIMITER, RETURN IF END
     IF (AKEY.EQ.BLANK) GO TO 110
     IF (AKEY.EQ.PPROP) GO TO 200
     CALL ERROR(34,1)
110  CONTINUE
     IF (N1.GT.NUMEL) NUMEL=N1
     IF (N1.LE.O.OR.N1.GT.MAXEL) CALL ERROR(9,N1)
     IF (NO.EQ.O.AND.N1.NE.1) CALL ERROR(9,N1)
     IF (INC.LE.O) INC=1
     IF (NTY.LE.1O.OR.NTY.GT.30) CALL ERROR(10,N1)
     NND = NNODES(NTY)
     IF (NND.LE.O) CALL ERROR(10,N1)
     IF (INTG.LE.O) INTG=INTDEF(NTY)
     IF (INTG.LE.21.OR.NTY.GE.23).AND.INTG.GT.5) INTG=INTDEF(NTY)
     IF (NTY.EQ.22.AND.(INTG.EQ.2.OR.INTG.EQ.5.OR.INTG.GT.7))
   $  INTG=INTDEF(NTY)
     NTYPE(N1) = NTY
     NGPINT(N1) = INTG
     NPHYS(N1) = NPH
     DO 120 I=1,NND
     IF (IC(I).GT.NUMNP.OR.IC(I).LE.O) CALL ERROR(12,N1)
     ICON(I,N1) = IC(I)
120  CONTINUE
C---- CHECK IF GENERATION REQUIRED, BRANCH AROUND IF NOT NEEDED
     IF (N1.EQ.NO+1) GO TO 180
C---- GENERATION
     N11 = NO+1
     N22 = N1-1
     DO 160 N=N11,N22
     NTYPE(N) = NTYPE(N-1)
     NGPINT(N) = NGPINT(N-1)
     NPHYS(N) = NPHYS(N+1)
     DO 140 I=1,8
     ICON(I,N) = O
```

```fortran
     IF (ICON(I,N-1).NE.O) ICON(I,N)=ICON(I,N-1)+INCO
     IF (ICON(I,N).GT.NUMNP) CALL ERROR(12,N)
140  CONTINUE
160  CONTINUE
180  CONTINUE
C---- SHIFT AND RETURN FOR NEXT ELEMENT
     NO = N1
     INCO = INC
     GO TO 100
200  CONTINUE
C---- OUTPUT ELEMENT DATA
     DO 220 I=1,NUMEL
     JTYP = NTYPE(I)
     NND = NNODES(JTYP)
     WRITE (6,2500) I,NTYPE(I),NGPINT(I),NPHYS(I),
   $  (ICON(J,I),J=1,NND)
220  CONTINUE
C---- CHECK ELEMENTS TO SEE IF ALL NODES ARE USED
     DO 280 I=1,NUMNP
     IMISS(I) = MISS
280  CONTINUE
     DO 320 IELNO=1,NUMEL
     ITYPE = NTYPE(IELNO)
     NND = NNODES(ITYPE)
     DO 300 I=1,NND
     INODE = ICON(I,IELNO)
     IF (INODE.GT.O.AND.INODE.LE.NUMNP)
   $  IMISS(INODE) = IPRES
300  CONTINUE
320  CONTINUE
     JMISS = O
     DO 340 I=1,NUMNP
     IF (IMISS(I).EQ.MISS) WRITE (6,2600) I
     IF (IMISS(I).EQ.MISS) JMISS=+1
340  CONTINUE
     IF (JMISS.EQ.1) CALL ERROR(23,NUMEL)
     RETURN
C
1400 FORMAT(A4,7X,1315)
2000 FORMAT(' UNAFEM-UNDERSTANDING AND APPLYING THE FEM'//
   $  '------------------------------')
2100 FORMAT(//  20A4)
2400 FORMAT(//' ELEMENT DATA'//
   $  8X,'ELEM   ELEM   GAUSS     PROP        NODES'/
   $  9X,'NO   TYPE   POINTS'/)
2500 FORMAT(8X,I3,4X,I3,5X,I3,5X,I3,1X,8I4)
2600 FORMAT(' THE FOLLOWING NODE WAS NOT USED IN ANY ELEMENT-',
   $  15)
     END
     SUBROUTINE HBAND
C
C    THIS ROUTINE COMPUTES THE SYSTEM HALF-BANDWIDTH OF THE PROBLEM
C
     COMMON /GENRL/ NUMNP,NUMEL,NUMPP,NUMILC,NUMEBC,NUMNBC,MBAND
     COMMON /LIMITS/ MAXNP,MAXEL,MAXIL,MAXEBC,MAXNBC,MAXBND
     COMMON /ELMTS/ NTYPE(200),NGPINT(200),NPHYS(200),ICON(8,200)
     COMMON /ELSPEC/ NNODES(30),INTDEF(30),IDPTAU(30)
C
     MBAND = O
     DO 260 IELNO=1,NUMEL
     ITYPE = NTYPE(IELNO)
     NND = NNODES(ITYPE)
     NNDMIN = MAXNP+1
     NNDMAX = MAXNP-1
     DO 240 I=1,NND
     IF (ICON(I,IELNO).GT.NNDMAX) NNDMAX=ICON(I,IELNO)
```

```fortran
     IF (ICON(I,IELNO).LT.NNDMIN) NNDMIN=ICON(I,IELNO)
240  CONTINUE
     MBE = NNDMAX-NNDMIN+1
     IF (MBE.GT.MBAND) MBAND=MBE
260  CONTINUE
     WRITE (6,2000)
     WRITE (6,2100) MBAND
     IF (MBAND.GT.MAXBND) CALL ERROR(18,MBAND)
     RETURN
C
2000 FORMAT(' UNAFEM-UNDERSTANDING AND APPLYING THE FEM'/)
2100 FORMAT(//' SYSTEM HALF-BANDWIDTH =',I4)
     END
     SUBROUTINE PLMESH
C
C    THIS ROUTINE PLOTS THE INITIAL MESH
C
     COMMON /NODES/  XI(300),YI(300)
     COMMON /ELMTS/ NTYPE(200),NGPINT(200),NPHYS(200),ICON(8,200)
     COMMON /GENRL/ NUMNP,NUMEL,NUMPP,NUMILC,NUMEBC,NUMNBC,MBAND
     COMMON /ELSPEC/ NNODES(30),INTDEF(30),IDPTAU(30)
     COMMON /PROBLM/ DIMEN,PROB,TITLE(18)
     DATA TWOD/'2DIM'/
C
     IF (DIMEN.EQ.TWOD) GO TO 200
C---- PLOT 1-D MESH
     CALL MAXMIN(NUMNP,X,XMAX,XMIN,JMAX,JMIN)
     CALL BGNPL(O)
     CALL PAGE(8.,6.)
     CALL TITLE('MESH PLOT#',',','  ',7.,4.)
     XSTEP=(XMAX-XMIN)/7.
     CALL MESH1D(XMAX,XMIN)
     CALL ENDPL(O)
     RETURN
C---- PLOT 2-D MESH
200  CONTINUE
     CALL SETUP2(XMIN,XMAX,YMIN,YMAX,HAX,DAX,XOR,XSTEP,YSTEP)
     CALL TITLE('MESH PLOT#',',','  ',HAX,DAX)
     CALL MESH2D(XOR,XMAX,YMAX,XSTEP,YSTEP)
     CALL ENDPL(O)
     RETURN
C
     END
     SUBROUTINE SETUP2(XMIN,XMAX,YMIN,YMAX,DAX,XOR,XSTEP,YSTEP)
C
C    THIS SUBROUTINE SETS UP THE PARAMETERS FOR 2-D PLOTS
C
     COMMON /NODES/  XI(300),YI(300)
     COMMON /GENRL/ NUMNP,NUMEL,NUMPP,NUMILC,NUMEBC,NUMNBC,MBAND
C
     CALL MAXMIN(NUMNP,X,XMAX,XMIN,JMAX,JMIN)
     CALL MAXMIN(NUMNP,Y,YMAX,YMIN,JMAX,JMIN)
     DX = XMAX-XMIN
     DY = YMAX-YMIN
     SPACE = O.02*AMAX1(DX,DY)
     XAX = DX + 2.*SPACE
     YAX = DY + 2.*SPACE
     XSTEP = XAX
     YSTEP = YAX
     XOR = XMIN - SPACE
     YOR = YMIN - SPACE
     XMAX = XMAX + SPACE
     YMAX = YMAX + SPACE
     CALL BGNPL(O)
```

(Continued)

```fortran
      CALL MAXPAG(PAGEX,PAGEY)
      ALX = PAGEX-1.
      ALY = PAGEY-2
      P1 = ALX/ALY
      Q1 = XAX/YAX
      IF (P1.GE.Q1) GO TO 200
      HAX = ALX
      DAX = ALX/Q1
      GO TO 220
  200 CONTINUE
      DAX = ALY
      HAX = Q1*ALY
  220 CONTINUE
      CALL PAGE(PAGEX,PAGEY)
      RETURN
C-----
      END
      SUBROUTINE PROPDT
C
C     THIS ROUTINE READS THE PHYSICAL PROPERTIES
C
      COMMON /LIMITS/ MAXNP,NUMEL,MAXEL,MAXIL,MAXPP,NUMILC,NUMEBC,MAXNBC,MAXBND
      COMMON /GENRL/ NUMNP,NUMEL,NUMPP,NUMILC,NUMEBC,NUMNBC,MBAND
      COMMON /PHYS/   ALPHAY(6,20),ALPHAX(6,20),BETA(6,20),
     $                GAMMA(6,20),MU(6,20)
      REAL MU
      COMMON /PROBLM/ DIMEN,PROB,TITLE(18)
      DIMENSION PPAX(6),PPAY(6),PPB(6),PPG(6),PPM(6),ICOEF(6)
      DATA ILC1/'ILC '/, ILC2/' ILC2'/, IBLANK/'     '/
      DATA NBC1/'EBC '/, IEBC2/' EBC '/, ISTEP/'STEP '/, TWOD/'2D1M '/
      DATA ICOEF/' 1','  X','  X*X','  Y','  Y*Y','  X*Y'/
      DATA BDVP/'BDVP'/, EIGP/'EIGP'/, XIBVP/'IBVP'/
C
      WRITE (6,2000)
      WRITE (6,2100) DIMEN,PROB,TITLE
      NUMPP = 0
      DO 20 I=1,6
        PPG(I) = 0.
        PPM(I) = 0.
        PPAY(I) = 0.
   20 CONTINUE
      WRITE (6,2500)
  100 CONTINUE
      READ (5,1500) KEY,IP1,(PPAX(I),I=1,6)
      IF (KEY.NE.IBLANK) RETURN
      IF (IP1.GT.NUMPP) NUMPP=IP1
      IF (DIMEN.EQ.TWOD) READ (5,1500) KEY,IP2,(PPAY(I),I=1,6)
      IF (KEY.NE.IBLANK) CALL ERROR(28,IP1)
      READ (5,1500) KEY,IP2,(PPB(I),I=1,6)
      IF (KEY.NE.IBLANK) CALL ERROR(28,IP1)
      IF (PROB.EQ.EIGP) READ (5,1500) KEY,IP2,(PPG(I),I=1,6)
      IF (KEY.NE.IBLANK) CALL ERROR(28,IP1)
      IF (PROB.EQ.XIBVP) READ (5,1500) KEY,IP2,(PPM(I),I=1,6)
      IF (KEY.NE.IBLANK) CALL ERROR(28,IP1)
      IF (NUMPP.GT.MAXPP .OR.IP.LE.0) CALL ERROR(14,IP1)
      DO 120 I=1,6
        ALPHAX(I,IP1) = PPAX(I)
        ALPHAY(I,IP1) = PPAY(I)
        BETA(I,IP1) = PPB(I)
        GAMMA(I,IP1) = PPG(I)
        MU(I,IP1) = PPM(I)
  120 CONTINUE
      WRITE (6,2600) IP1,(ICOEF(I),ALPHAX(I,IP1),ALPHAY(I,IP1),
     $       BETA(I,IP1),GAMMA(I,IP1),MU(I,IP1),I=1,6)
      GO TO 100
C-----
```

```fortran
 1500 FORMAT(A4,1X,I5,6F10.5)
 2000 FORMAT('1    UNAFEM--UNDERSTANDING AND APPLYING THE FEM'/
     $ '    ------------------------------------------------')
 2100 FORMAT(//'  ',20A4)
 2500 FORMAT(//'  PHYSICAL PROPERTIES:'/
     $1X,'NO.',2X,'COEF',4X,'ALPHA-X',4X,'ALPHA-Y',8X,'BETA',7X,
     $ 'GAMMA',10X,'MU')
 2600 FORMAT(/I4,2X,A4,4E12.4/(6X,A4,5E12.4))
      END
      SUBROUTINE LOADDT
C
C
C     THIS ROUTINE READS THE INTERIOR AND BOUNDARY LOADS
C
      COMMON /GENRL/ NUMNP,NUMEL,NUMPP,NUMILC,NUMEBC,NUMNBC,MBAND
      COMMON /INTLD/ FINT(200)
      COMMON /LIMITS/ MAXNP,MAXEL,MAXPP,MAXIL,MAXEBC,NUMNBC,MAXBND
      COMMON /ESSBC/ IUEBC(100),UEBC(100)
      COMMON /NATBC/ INP(100),JNP(100),KNP(100),TAUNBC(100)
      COMMON /SYSEQS/ K(300,50),F(300),A(300)
      REAL K
      COMMON /INVAL/ INTRVL,NSTEPS,NTIME,TIME,DT,THETA,AO
      COMMON /IVEQS/ CKA(300,50),ANM1(300),FNM1(300),
     $               FN(300),FO(300),CLUMP(300),DELU(300),
     $               UO(300),KEFFDU(300),KEFFUO(300)
      REAL KEFFDU, KEFFUO
      COMMON /TTMPLT/ TPLOT(501),FTPLOT(3,501),NNODEI(3)
      COMMON /PROBLM/ DIMEN,PROB,TITLE(18)
      COMMON /KEYWRD/ AKEY,PLTKEY
      DATA EIGP/'EIGP'/, BDVP/'BDVP'/, ONED/'1D1M'/
      DATA NBC1/'NBCC'/, NBC2/'NBCD'/, PLOT/'PLOT'/,XNPLT/'NPLD'/
      DATA IPLOT/'PLOT'/, ILC1/'ILC1 '/, ILC2/' ILC2'/
      DATA NEND1/'END '/, NEND2/' END '/, ENBC2/'NBCD'/
      DATA EEND1/'END '/, EEND2/' END '/, IBLANK/'     '/, BLANK/'    '/
C
      WRITE (6,2000)
      WRITE (6,2100) DIMEN,PROB,TITLE
      INTRVL = INTRVL+1
      IF (PROB.EQ.BDVP) GO TO 40
      IF (PROB.EQ.EIGP) GO TO 200
      READ (5,1500) NSTEPS,DT,THETA,AO
C----- SET INITIAL CONDITIONS AT BEGINNING OF FIRST INTERVAL ONLY
      WRITE (6,2200) INTRVL,NSTEPS,DT,THETA
      IF (INTRVL.NE.1) GO TO 30
      WRITE (6,2250) AO
      NTIME = 1
      TPLOT(1) = 0.
      TIME = 0.
      DO 10 L=1,3
        FTPLOT(L,1) = AO
   10 CONTINUE
      DO 20 I=1,NUMNP
        A(I) = AO
        FO(I) = 0.
   20 CONTINUE
   30 CONTINUE
      READ (5,1600) KEY
      IF (KEY.NE.ILC1.AND.KEY.NE.ILC2) CALL ERROR(29,INTRVL)
      IF (NSTEPS.LE.0) CALL ERROR(30,INTRVL)
      IF (DT .LE.0) CALL ERROR(31,INTRVL)
      IF (THETA.LT.0.)CALL ERROR(32,INTRVL)
   40 CONTINUE
C----- INTERIOR LOADS
C
      NUMILC = 0
C-----
```

```fortran
      DO 50 IELND=1,NUMEL
        FINT(IELNO) = 0.
   50 CONTINUE
      WRITE (6,2600)
  100 CONTINUE
      READ (5,1600) AKEY,IEL1,FV,IEL2,INC
      IF (AKEY.EQ.BLANK) GO TO 120
      IF (AKEY.EQ.EBC1.OR AKEY.EQ EBC2) GO TO 200
      CALL ERROR(34,1)
  120 CONTINUE
      IF (IEL1.LE.O) CALL ERROR(15,IEL1)
      IF (IEL2.LT.IEL1) IEL2=IEL1
      IF (INC.LE.O) INC=1
      DO 150 IEL=IEL1,IEL2,INC
        NUMILC = NUMILC+1
        IF (NUMILC GT MAXIL) CALL ERROR(4,MAXIL)
        IF (IEL.GT.NUMEL) CALL ERROR(15,IEL)
        FINT(IEL) = FV
  150 CONTINUE
      GO TO 100
  200 CONTINUE
C----- READ ESSENTIAL BOUNDARY LOADS
      NUMEBC = 0
      WRITE (6,2750)
  300 CONTINUE
      READ (5,1700) KEY,JEBC,TEBC,JEBC2,INC
      IF (KEY.EQ.NOPLT) AKEY=XNOPLT
      IF (KEY.EQ.IBLANK) GO TO 320
      IF (KEY.EQ NBC1).OR.
     $   (KEY.EQ NEND1).OR.
     $   (KEY.EQ NEND2).OR.
     $   (KEY.EQ NOPLT).OR.
     $   (KEY.EQ.IPLOT)) GO TO 340
      CALL ERROR(34,1)
  320 CONTINUE
      IF (JEBC1.LE.O) CALL ERROR(16,JEBC1)
      IF (JEBC2.LT.JEBC1) JEBC2=JEBC1
      IF (INC.LE.O) INC=1
      DO 330 JEBC=JEBC1,JEBC2,INC
        NUMEBC = NUMEBC+1
        WRITE (6,2700) JEBC,TEBC
        IF (NUMEBC GE MAXEBC) CALL ERROR(5,NUMEBC)
        IUEBC(NUMEBC) = JEBC
        UEBC(NUMEBC) = TEBC
        IF (JEBC.GT.NUMNP) CALL ERROR(16,JEBC)
  330 CONTINUE
      GO TO 300
  340 CONTINUE
C----- CHECK FOR DUPLICATE SPECIFICATIONS
      IF (NUMEBC .LE.1) GO TO 380
      NUMEB1 = NUMEBC-1
      NBLTOT = NUMEBC
      JERROR = 1
      DO 370 I=1,NUMEB1
        I1 = I+1
        DO 360 L=I1,NUMEBC
          IF (IUEBC(I).NE.IUEBC(J)) GO TO 360
          NBLTOT = NBLTOT-1
          WRITE (6,2850) J,I
          DO 350 L=I1,NUMEB1
            IUEBC(L) = IUEBC(L+1)
            UEBC(L) = UEBC(L+1)
  350     CONTINUE
          IUEBC(NUMEBC) = 0
```

```
        UEBC(NUMEBC) = 0.
360     CONTINUE
370     CONTINUE
        NUMEBC = NBLTOT
        IF (JERROR.EQ.1) CALL ERROR(24,NUMEBC)
380     CONTINUE
C------- READ NATURAL BOUNDARY LOADS
        NUMNBC = 0
        IF (PROB.EQ.EIGP) RETURN
C------- CONCENTRATED NATURAL BOUNDARY LOADS
        WRITE (6,2900)
400     CONTINUE
        READ (5,1800) AKEY,INNP,TAUC
        IF (AKEY.EQ.BLANK) GO TO 440
        IF (AKEY.EQ.ENBC2.OR.AKEY.EQ.PPLOT.OR.AKEY.EQ.STEP.OR
     $   AKEY.EQ.EEND1.OR.AKEY.EQ.EEND2.OR.AKEY.EQ.XNOPLT) GO TO 480
        CALL ERROR(34,1)
440     CONTINUE
        NUMNBC = NUMNBC+1
        WRITE (6,2950) TAUC
        IF (NUMNBC.GT.MAXNBC) CALL ERROR(6,NUMNBC)
        IF (INNP.GT.NUMNP.OR.INNP.LE.0) CALL ERROR(17,NUMNBC)
        INP(NUMNBC) = INNP
        KNP(NUMNBC) = 0
        TAUNBC(NUMNBC) = TAUC
        GO TO 400
480     CONTINUE
C------- DISTRIBUTED NATURAL BOUNDARY LOADS
        IF (DIMEN.EQ.ONED) RETURN
        WRITE (6,3000)
500     CONTINUE
        READ (5,1900) AKEY,INNP,JNNP,KNNP,TAUN
        IF (AKEY.EQ.BLANK) GO TO 520
        IF (AKEY.EQ.PPLOT.OR.AKEY.EQ.STEP.OR.AKEY.EQ.STEP.OR.
     $   AKEY.EQ.EEND1.OR.AKEY.EQ.EEND2.OR.AKEY.EQ.XNOPLT) RETURN
        CALL ERROR(34,1)
520     CONTINUE
        NUMNBC = NUMNBC+1
        WRITE (6,3050) INNP,JNNP,KNNP,TAUN
        IF (NUMNBC.GT.MAXNBC) CALL ERROR(6,NUMNBC)
        IF (INNP.GT.NUMNP.OR.INNP.LE.0) CALL ERROR(17,NUMNBC)
        INP(NUMNBC) = INNP
        JNP(NUMNBC) = JNNP
        KNP(NUMNBC) = KNNP
        TAUNBC(NUMNBC) = TAUN
        GO TO 500
C------------------------------------
1500    FORMAT(5X,I5,3F10.5)
1600    FORMAT(A4,1X,I5,F10.5,2I5)
1700    FORMAT(A4,1X,I5,F10.5,2I5)
1800    FORMAT(A4,1X,I5,F10.5)
1900    FORMAT(A4,1X,3I5,F10.5)
2000    FORMAT(' UNAFEM-UNDERSTANDING AND APPLYING THE FEM'/
     $   ' ...........................................')
2100    FORMAT(//' ',20X4)
2200    FORMAT(///' INITIAL-BOUNDARY-VALUE PROBLEM'//
     $   20X,' NUMBER INTEGRATION PARAMETERS FOR INTERVAL ',I3/
     $   20X,' NUMBER OF TIME STEPS = ',I5/
     $   20X,' TIME STEP = ',E12.5/
     $   20X,' THETA = ',E12.5/)
2250    FORMAT(20X,' INITIAL VALUE FOR ALL DOF = ',E12.5/)
2600    FORMAT(///' INTERIOR LOADS'//)
2700    FORMAT(10X,'ELEM NO',7X,'VALUE'//)
2750    FORMAT(///' ESSENTIAL BOUNDARY LOADS'//)
```

```
$  10X,'NODE NO',7X,'VALUE'/)
2850    FORMAT(//' ESSENTIAL BOUNDARY LOAD',I3,' - DUPLICATE OF LOAD',I3)
2900    FORMAT(///' NATURAL BOUNDARY LOADS - CONCENTRATED'//
$  10X,'NODE NO',7X,'VALUE'/)
2950    FORMAT(10X,I7,3X,E12.3)
3000    FORMAT(///' NATURAL BOUNDARY LOADS - DISTRIBUTED'//
$  10X,'NODE NO',3X,'NODE -J',3X,'NODE -K',7X,'VALUE'/)
3050    FORMAT( 10X,I7,3X,I7,3X,I7 E12.3)
        END
        SUBROUTINE PLOTDT
C
C       THIS ROUTINE READS THE PLOTTING DATA
C
        COMMON /PROBLM/ DIMEN,PROB,TITLE(18)
        COMMON /INVAL/ INTRVL,NSTEPS,NTIME,TIME,DT,THETA,AO
        COMMON /GRAFIC/ XXMIN,XXMAX,NDIVX,YYMIN,YYMAX1,NDIVY1,
$  YYMIN2,YYMAX2,NDIVY2,TTMIN,TTMAX,NDIVT
        COMMON /TTMPLT/ TPLOT(501),FTPLOT(3,501),NNODEII(3)
        COMMON /GENRL/ NUMNP,NUMEL,NUMPP,NUMIC,NUMEBC,NUMNBC,MBAND
        DATA ONED/'1DIM'/,TWOD/'2DIM'/
        DATA BLANK/'    '/,ONED/'1DIM'/,TWOD/'2DIM'/
        DATA BDVP/'BDVP'/,EIGP/'EIGP'/,XIBVP/'1BVP'/
        DATA EEND1/'END '/,EEND2/'END2'/
        DATA XNOPLT/'NPLO'/,YESPLT/'PLOT'/
C-----------------------------------------------
C------- READ ONLY FOR THE FIRST INTERVAL
        IF (INTRVL.NE.1) GO TO 100
        PLTKEY = XNOPLT
        IF (AKEY.EQ.XNOPLT) RETURN
        PLTKEY = YESPLT
        XXMIN = 0.
        XXMAX = 0.
        NDIVX = 0.
        YYMIN1 = 0.
        YYMAX1 = 0.
        NDIVY1 = 0.
        YYMIN2 = 0.
        YYMAX2 = 0.
        NDIVY2 = 0.
        NDIVT = 10
        TTMIN = 0.
        TTMAX = 0.
        DO 20 I=1,3
        NNODEII(I) = 0
20      CONTINUE
        READ (5,1000) AKEY,NDIVX=10
        IF (NDIVX.LE.0) NDIVX=10
        IF (AKEY.NE.BLANK) WRITE (6,3000)
        IF (AKEY.EQ.BLANK) RETURN
        READ (5,1000) AKEY,NDIVY1,YYMIN1,YYMAX1
        IF (NDIVY1.LE.0) NDIVY1 = 6
        IF (AKEY.NE.BLANK) WRITE (6,3000)
        IF (AKEY.EQ.BLANK) RETURN
        IF (PROB.NE.EIGP)
$  READ (5,1000) AKEY,NDIVY2,YYMIN2,YYMAX2
        IF (NDIVY2.LE.0) NDIVY2 = 6
        IF (PROB.EQ.XIBVP)
$  READ(5,1200)AKEY,NDIVT,TTMIN,TTMAX,(NNODEII(I),I=1,3)
        IF (NNODEII(I).LT.0.OR.NNODEII(I).GT.NUMNP) NNODEI(I)=0
80      CONTINUE
90      CONTINUE
        READ (5,1000) AKEY
100     CONTINUE
```

```
        WRITE (6,3000)
        RETURN
200     CONTINUE
        IF (PROB.EQ.XIBVP)
$    READ (5,1000) AKEY,NDIVY1,YYMIN1,YYMAX1
        IF (NDIVY1.LE.0) NDIVY1 = 6
        IF (PROB.EQ.XIBVP)
$    READ(5,1200)AKEY,NDIVT,TTMIN,TTMAX,(NNODEI(I),I=1,3)
        IF (PROB.EQ.XIBVP)
        DO 220 I=1,3
        IF (NNODEI(I).LT.0.OR.NNODEI(I).GT.NUMNP) NNODEI(I)=0
220     CONTINUE
        IF (NDIVT.LE.0) NDIVT=10
        READ (5,1000) AKEY
        WRITE (6,3000)
        RETURN
C-----------------------------------------------
1000    FORMAT(A4,1X,I5,2F10.5)
1100    FORMAT(5X,I5,2F10.5,3I5)
1200    FORMAT(A4,1X,I5,2F10.5,3I5)
3000    FORMAT(//'-----PREPROCESSING COMPLETE. BEGIN SOLUTION.')
        END
        SUBROUTINE FORMK
C
C       THIS ROUTINE FORMS THE SYSTEM MATRICES BY PERFORMING
C       TWO TASKS FOR EACH ELEMENT IN THE MESH:
C          1.  FORMING THE ELEMENT MATRICES BY CALLING THE
C              APPROPRIATE ELEMENT ROUTINES, AND
C          2   ASSEMBLING THE SYSTEM MATRICES FROM THE
C              ELEMENT MATRICES.
C
        COMMON /GENRL/ NUMNP,NUMEL,NUMPP,NUMIC,NUMEBC,NUMNBC,MBAND
        COMMON /ELMNTS/ NTYPE(200),NGPINT(200),NPHYS(200),ICON(8,200)
        COMMON /SYSEQS/ K(300,50),F(300),A(300)
C       REAL K
        DIMENSION KEFF(300,50)
        REAL KEFF
        EQUIVALENCE (K,KEFF)
        COMMON /EVEQS/ KEIG(61,61),MEIG(61,61)
        REAL KEIG,MEIG
        COMMON /INVAL/ INTRVL,NSTEPS,NTIME,TIME,DT,THETA,AO
        COMMON /IVEQS/ CK(300,50),ANM1(300),FNM1(300),
$    FN(300),FO(300),CLUMP1(300),DELU(300),
$    UO(300),KEFFDU(300),KEFFUO(300)
        REAL KEFFDU, KEFFUO
        COMMON /ELSPEC/ NNODES(30),INTDEF(30),IOPTAU(30)
        COMMON /ELSTF/ KE(8,8),FE(8),ME(8,8),CE(8,8)
        COMMON /PROBLM/ DIMEN,PROB,TITLE(18)
        COMMON /ELVAR/ X1,ETA,JAC,JACINV(2,2),XX(8),YY(8),PHI(8),
$    DPHDX(8),DPHDY(8),DPHDET(8),IELMO,NNO
        REAL JAC,JACINV,JAC
        DATA ELFORM/'FORM'/
        DATA BDVP/'BDVP'/,EIGP/'EIGP'/,XIBVP/'1BVP'/
        DATA ONED/'1DIM'/,TWOD/'2DIM'/
C------ INITIALIZE ARRAYS FOR THE SYSTEM MATRICES
        NLIM = NUMP
        ONEMTH = 1.-THETA
        IF (PROB.EQ.EIGP) NLIM=61
        IF (PROB.EQ.EIGP) MBAND=61
        DO 80 I=1,NLIM
        F(I)=0.
        IF (PROB.EQ.XIBVP) CLUMP(I) = 0.
        DO 60 J=1,MBAND
        IF (PROB.EQ.BDVP.OR.PROB.EQ.XIBVP) K(I,J) = 0.
        IF (PROB.EQ.EIGP) KEIG(I,J) = 0.
```

```fortran
      IF (PROB.EQ.EIGP) MEIG(I,J) = 0.
      IF (PROB.EQ.XIBVP) CK(I,J) = 0.
 60   CONTINUE
 80   CONTINUE
C
C********* BEGIN LOOP OVER ALL ELEMENTS ************
C
      DO 500 IELNO=1,NUMEL
      ITYPE = NTYPE(IELNO)
      NND = NNODES(ITYPE)
C------ FORM ELEMENT MATRICES
      IF (ITYPE.EQ.11) CALL ELEM11(ELFORM)
      IF (ITYPE.EQ.12) CALL ELEM12(ELFORM)
      IF (ITYPE.EQ.13) CALL ELEM13(ELFORM)
      IF (ITYPE.EQ.14) CALL ELEM14(ELFORM)
      IF (ITYPE.EQ.21) CALL ELEM21(ELFORM)
      IF (ITYPE.EQ.22) CALL ELEM22(ELFORM)
      IF (ITYPE.EQ.23) CALL ELEM23(ELFORM)
      IF (ITYPE.EQ.24) CALL ELEM24(ELFORM)
C------ ASSEMBLE THE SYSTEM MATRICES FROM THE ELEMENT MATRICES
      IF (PROB.EQ.BDVP) GO TO 100
      IF (PROB.EQ.EIGP) GO TO 200
      IF (PROB.EQ.XIBVP.AND.THETA.EQ.0.) GO TO 300
      IF (PROB.EQ.XIBVP.AND.THETA.NE.0.) GO TO 400
C---------- ASSEMBLE MATRICES FOR BOUNDARY-VALUE PROBLEM ***********
C
 100  CONTINUE
      DO 140 I=1,NND
      II = ICON(I,IELNO)
      F(II) = F(II)+FE(I)
      DO 120 J=1,NND
      JJ = ICON(J,IELNO)
      JSHIFT = JJ-(II-1)
      IF (JJ.GE.II) K(II,JSHIFT) = K(II,JSHIFT)+KE(I,J)
 120  CONTINUE
 140  CONTINUE
      GO TO 500
C---------- ASSEMBLE MATRICES FOR EIGENPROBLEM ***********
C
 200  CONTINUE
      DO 240 I=1,NND
      II = ICON(I,IELNO)
      DO 220 J=1,NND
      JJ = ICON(J,IELNO)
      JSHIFT = JJ-(II-1)
      KEIG(II,JJ) = KEIG(II,JJ)+KE(I,J)
      MEIG(II,JJ) = MEIG(II,JJ)+ME(I,J)
 220  CONTINUE
 240  CONTINUE
      GO TO 500
C---------- ASSEMBLE MATRICES FOR
C              EXPLICIT INITIAL-BOUNDARY-VALUE PROBLEM ***********
C
C----- COMPUTE SCALE FACTOR
 300  CONTINUE
C          FOR LUMPING 2-D CAPACITY MATRIX
      IF (DIMEN.NE.TWOD) GO TO 330
      CEDIAG = 0.
      CESUM = 0.
      DO 320 I=1,NND
      CEDIAG = CEDIAG+CE(I,I)
      DO 310 J=1,NND
      CESUM = CESUM+CE(I,J)
 310  CONTINUE
 320  CONTINUE
      SCALE = CESUM/CEDIAG
C
 330  CONTINUE
      DO 350 I=1,IELNO
      II = ICON(I,IELNO)
      F(II) = F(II)+FE(I)
      DO 340 J=1,NND
      JJ = ICON(J,IELNO)
      JSHIFT = JJ-(II-1)
      IF (JJ.GE.II) K(II,JSHIFT)=K(II,JSHIFT)+KE(I,J)
C            (LUMP THE CAPACITY MATRIX)
      IF (DIMEN.EQ.ONED) CLUMP(II) = CLUMP(II)+CE(I,J)
      IF (DIMEN.EQ.TWOD) CLUMP(II) = CLUMP(II)+SCALE*CE(I,I)
 340  CONTINUE
 350  CONTINUE
      GO TO 500
C---------- ASSEMBLE MATRICES FOR
C              IMPLICIT INITIAL-BOUNDARY-VALUE PROBLEM ***********
C
 400  CONTINUE
      DO 440 I=1,IELNO
      II = ICON(I,IELNO)
      F(II) = F(II)+FE(I)
      DO 420 J=1,NND
      JJ = ICON(J,IELNO)
      JSHIFT = JJ-(II-1)
      IF (JJ.GE.II) KEFF(II,JSHIFT)=KEFF(II,JSHIFT)+
     $                CE(I,J)/DT+THETA*KE(I,J)
      IF (JJ.GE.II) CK(II,JSHIFT)=CK(II,JSHIFT)+
     $                CE(I,J)/DT-ONEMTH*KE(I,J)
 420  CONTINUE
 440  CONTINUE
 500  CONTINUE
      RETURN
      END
C
      SUBROUTINE JACOB1(NDIM)
C
C     THIS ROUTINE DOES THE JACOBIAN CALCULATIONS
C
      COMMON /ELVAR/ XI,ETA,JAC,JACINV(2,2),XX(8),YY(8),PHI(8),
     $              DPHDX(8),DPHDY(8),DPHDET(8),IELNO,NND
      REAL JACINV,JAC
C
      IF (NDIM.EQ.2) GO TO 200
C-----FOR 1-D CASE
C---------COMPUTE JACOBIAN AT XI
      JAC = 0.
      DO 110 I=1,NND
      JAC = JAC+DPHDXI(I)*XX(I)
 110  CONTINUE
C---------CHECK POSITIVITNESS OF JACOBIAN
      IF (JAC.LT.1.E-8) CALL ERROR(19,IELNO)
C---------COMPUTE INVERSE OF THE JACOBIAN AT XI
      JACINV(1,1) = 1./JAC
      RETURN
C-----FOR 2-D CASE
C---------COMPUTE JACOBIAN AT (XI,ETA)
 200  CONTINUE
      DXDXI = 0.
      DXDET = 0.
      DYDXI = 0.
      DYDET = 0.
      DO 210 I=1,NND
      DXDXI = DXDXI+DPHDXI(I)*XX(I)
      DXDET = DXDET+DPHDET(I)*XX(I)
      DYDXI = DYDXI+DPHDXI(I)*YY(I)
      DYDET = DYDET+DPHDET(I)*YY(I)
 210  CONTINUE
      JAC = DXDXI*DYDET-DXDET*DYDXI
C---------CHECK POSITIVITNESS OF JACOBIAN
      AVG = (ABS(DXDXI)*ABS(DXDET)+ABS(DYDXI)*ABS(DYDET))/4.
      IF (JAC.LT.AVG*1.E-8) CALL ERROR(19,IELNO)
C---------COMPUTE INVERSE OF THE JACOBIAN AT (XI,ETA)
      JACINV(1,1) = DYDET/JAC
      JACINV(1,2) = -DYDXI/JAC
      JACINV(2,1) = -DXDET/JAC
      JACINV(2,2) = DXDXI/JAC
      RETURN
C
      END
      SUBROUTINE PROP(IPHY,X,Y,AX,AY,B,G,M)
C
C     THIS ROUTINE CALCULATES THE PHYSICAL PROPERTIES
C
C-----INPUT
C         IPHY - PHYSICAL PROPERTY NUMBER
C         X    - X COORDINATE
C         Y    - Y COORDINATE
C-----OUTPUT
C         AX - ALPHA-X AT X,Y
C         AY - ALPHA-Y AT X,Y
C         B  - BETA    AT X,Y
C         G  - GAMMA   AT X,Y
C         M  - MU      AT X,Y
C
      COMMON /PHYS/  ALPHAX(6,20),ALPHAY(6,20),BETA(6,20),
     $               GAMMA(6,20),MU(6,20)
      REAL MU
      REAL M
C
      AX = ALPHAX(1,IPHY)  + ALPHAX(2,IPHY)*X   + ALPHAX(3,IPHY)*Y  +
     $     ALPHAX(4,IPHY)*X + ALPHAX(5,IPHY)*Y*Y + ALPHAX(6,IPHY)*X*X +
      AY = ALPHAY(1,IPHY)  + ALPHAY(2,IPHY)*X   + ALPHAY(3,IPHY)*Y  +
     $     ALPHAY(4,IPHY)*X + ALPHAY(5,IPHY)*Y*Y + ALPHAY(6,IPHY)*X*X +
      B  = BETA(1,IPHY)    + BETA(2,IPHY)*X     + BETA(3,IPHY)*Y    +
     $     BETA(4,IPHY)*X + BETA(5,IPHY)*Y*Y + BETA(6,IPHY)*X*X +
      G  = GAMMA(1,IPHY)   + GAMMA(2,IPHY)*X    + GAMMA(3,IPHY)*Y   +
     $     GAMMA(4,IPHY)*X + GAMMA(5,IPHY)*Y*Y + GAMMA(6,IPHY)*X*X +
      M  = MU(1,IPHY)      + MU(2,IPHY)*X       + MU(3,IPHY)*Y      +
     $     MU(4,IPHY)*X + MU(5,IPHY)*Y*Y + MU(6,IPHY)*X*X +
      RETURN
C
      END
      SUBROUTINE ELEM11(OPT)
C
C     THIS SUBROUTINE IS THE MASTER ROUTINE FOR ELEMENT 11, A
C     2-NODE 1-D CO-LINEAR ELEMENT.
C
      IF (OPT.EQ.'FORM'. THE ELEMENT STIFFNESS AND MASS MATRICES AND
C                        LOAD VECTOR (DUE TO INTERIOR LOADS ONLY)
C                        ARE FORMED.
      IF (OPT.EQ.'FLUX'. THE FLUXES ARE CALCULATED AT THE ONE GAUSS
C                        POINT AND THE TWO NODES
C
      COMMON /ELSTF/ KE(8,8),FE(8),ME(8,8),CE(8,8)
      REAL KE,ME
      COMMON /NODES/ X(300),Y(300)
      COMMON /ELMNTS/ NTYPE(200),NGPINT(200),ICON(8,200)
      REAL MU
```

```
      COMMON /INTLD/  FINT(200)
      COMMON /NODSOL/ IU(300),TAUXNP(300),TAUYNP(300)
      COMMON /ELFLUX/ XGP(4,200),YGP(4,200),TAUXGP(4,200),TAUYGP(4,200),
     $                TAUGP(4,200),A(300)
      COMMON /SYSEQS/ K(300,50),F(300),A(300)
      REAL K
      COMMON /ELVAR/  XI,ETA,JAC,JACINV(2,2),XX(8),YY(8),PHI(8),
     $                DPHDX(8),DPHDY(8),DPHDXI(8),DPHDET(8),IELND,NND
      REAL JACINV,JAC
      COMMON /PROBLM/ DIMEN,PROB,TITLE(18)
      DATA FORM/'FORM'/, FLUX/'FLUX'/
C
C-----BRANCH TO FLUX CALCULATION IF OPT = 'FLUX'
C     FORM MATRICES IF OPT = 'FORM'
      IF (OPT .EQ. FLUX) GO TO 200
C
C-----FORM ELEMENT MATRICES AND VECTORS*******************
C
      I = ICON(1,IELND)
      J = ICON(2,IELND)
      IPHY = NPHYS(IELND)
      XCENT = (XX(I)+XX(J))/2.
      YCENT = 0.
      L = ABS(X(J)-X(I))
      KE(1,1) =  ALPHA/L+BETA*L/3.
      KE(1,2) = -ALPHA/L+BETA*L/6.
      KE(2,1) = KE(1,2)
      KE(2,2) = KE(1,1)
      FE(1) = FINT(IELND)*L/2.
      FE(2) = FE(1)
      ME(1,1) = GAMMA*L/3.
      ME(1,2) = GAMMA*L/6.
      ME(2,1) = ME(1,2)
      ME(2,2) = ME(1,1)
      CE(1,1) = MU*L/3.
      CE(1,2) = MU*L/6.
      CE(2,1) = CE(1,2)
      CE(2,2) = CE(1,1)
      RETURN
C
C-----CALCULATE FLUXES********************
C
  200 CONTINUE
      I = ICON(1,IELND)
      J = ICON(2,IELND)
      IPHY = NPHYS(IELND)
C-----CALCULATE FLUXES AT THE GAUSS POINT (CENTER).
      XGP(1,IELND) = (XX(I)+XX(J))/2.
      YGP(1,IELND) = 0.
      XCENT = XGP(1,IELND)
      YCENT = 0.
      CALL PROP(IPHY,XCENT,YCENT,ALPHA,ALPHAY,BETA,GAMMA,MU)
      TAUXGP(1,IELND) = -ALPHA*(A(J)-A(I))/(X(J)-X(I))
      TAUYGP(1,IELND) = 0.
C-----CALCULATE ELEMENT NODAL FLUXES AND SUM FOR NODAL FLUX
C     CALCULATION IN CALFLX (ELEMENT NODAL FLUXES SAME AS
C     GAUSS POINT FLUX)
      TAUXNP(I) = TAUXNP(I) + TAUXGP(1,IELND)
      TAUXNP(J) = TAUXNP(J) + TAUXGP(1,IELND)
      IU(I) = IU(I)+1
      IU(J) = IU(J)+1
      RETURN
C
      END


      SUBROUTINE ELEM12(OPT)
C
C     THIS SUBROUTINE IS THE MASTER ROUTINE FOR ELEMENT 12. A
C     3-NODE 1-D CO-QUADRATIC ELEMENT.
C
C     IF OPT='FORM':  THE ELEMENT STIFFNESS AND MASS MATRICES AND
C                     LOAD VECTOR (DUE TO INTERIOR LOADS ONLY)
C                     ARE FORMED.
C     IF OPT='FLUX':  THE FLUXES ARE CALCULATED AT THE TWO GAUSS
C                     POINTS AND THE THREE NODES.
C
      COMMON /ELSTF/  KE(8,8),FE(8),ME(8,8),CE(8,8)
      REAL KE,ME
      COMMON /NODES/  X(300),Y(300)
      COMMON /ELMNTS/ NTYPE(200),NGPINT(200),NPHYS(200),ICON(8,200)
      REAL MU
      COMMON /INTLD/  FINT(200)
      COMMON /NODSOL/ IU(300),TAUXNP(300),TAUYNP(300)
      COMMON /ELFLUX/ XGP(4,200),YGP(4,200),TAUXGP(4,200),TAUYGP(4,200),
     $                TAUGP(4,200),A(300)
      COMMON /SYSEQS/ K(300,50),F(300),A(300)
      REAL K
      COMMON /ELVAR/  XI,ETA,JAC,JACINV(2,2),XX(8),YY(8),PHI(8),
     $                DPHDX(8),DPHDY(8),DPHDXI(8),DPHDET(8),IELND,NND
      REAL JACINV,JAC
      COMMON /GLINT/  GP(5,5),WT(5,5)
      COMMON /PROBLM/ DIMEN,PROB,TITLE(18)
      DIMENSION XXI(2),TAU(2),SI(3)
      DATA EIGP/'EIGP'/, XIBVP/'IBVP'/
      DATA SI/-1.,0.,1./
      DATA FORM/'FORM'/, FLUX/'FLUX'/
C
C-----BRANCH TO FLUX CALCULATION IF OPT = 'FLUX'
C     FORM MATRICES IF OPT = 'FORM'
      IF (OPT .EQ. FLUX) GO TO 200
C
C-----FORM ELEMENT MATRICES AND VECTORS*******************
C
      YYY = 0.
      IPHY = NPHYS(IELND)
      DO 20 I=1,3
      J = ICON(1,IELND)
      XX(I) = X(J)
      ME(J,I) = 0.
      FE(I) = 0.
   10 DO J=1,3
      KE(J,I) = 0.
      ME(J,I) = 0.
      CE(I,I) = 0.
   20 CONTINUE
      NINT = NGPINT(IELND)
C-----EVALUATE TERMS IN ELEMENT EQUATIONS
      DO 60 INT=1,NINT
      XI = GP(NINT,INTX)
      CALL SHAP12
      XXX = 0.
      DO 30 I=1,3
   30 XXX = XXX+PHI(I)*XX(I)
      CALL PROP(IPHY,XXX,YYY,ALPHA,ALPHAY,BETA,GAMMA,MU)
      DO 50 I=1,3
      FE(I) = FE(I)+FINT(IELND)*PHI(I)*WT(NINT,INTX)*JAC
      DO 40 J=1,3
      KE(I,J) = KE(I,J)+WT(NINT,INTX)*JAC*
     $          (DPHDX(I)*ALPHA*DPHDX(J)+
     $          PHI(I)*BETA*PHI(J))
      IF (PROB.EQ.EIGP)  ME(I,J)=ME(I,J)+WT(NINT,INTX)*JAC*
     $                   PHI(I)*GAMMA*PHI(J)
      IF (PROB.EQ.XIBVP) CE(I,J)=CE(I,J)+WT(NINT,INTX)*JAC*
     $                   PHI(I)*MU*PHI(J)
   40 CONTINUE
   50 CONTINUE
   60 CONTINUE
C-----FOR SYMMETRY
      DO 100 I=1,3
      DO 90 J=1,3
      KE(J,I) = KE(I,J)
      ME(J,I) = ME(I,J)
      CE(J,I) = CE(I,J)
   90 CONTINUE
  100 CONTINUE
      RETURN
C
C-----CALCULATE FLUXES********************
C
  200 CONTINUE
      YYY = 0.
      IPHY = NPHYS(IELND)
      DO 220 I=1,3
      J = ICON(1,IELND)
      XX(I) = X(J)
  220 CONTINUE
C-----CALCULATE FLUXES AT THE 2 GAUSS POINTS
      DO 260 INT=1,2
      XI = GP(2,INT)
      CALL SHAP12
      XXI(INT) = 0.
      DUDX = 0.
      DO 240 I=1,3
      J = ICON(I,IELND)
      XXI(INT) = XXI(INT)+PHI(I)*XX(I)
      DUDX = DUDX+DPHDX(I)*A(J)
  240 CONTINUE
      XXX = XXI(INT)
      CALL PROP(IPHY,XXX,YYY,ALPHA,ALPHAY,BETA,GAMMA,MU)
      TAU(INT) = -ALPHA*DUDX
      XGP(INT,IELND) = XXI(INT)
      TAUXGP(INT,IELND) = TAU(INT)
      TAUYGP(INT,IELND) = 0.
  260 CONTINUE
C-----CALCULATE ELEMENT NODAL FLUXES BY INTERPOLATING THE GAUSS
C     POINT FLUXES AND SUM FOR NODAL FLUX CALCULATION IN CALFLX
      X112 = GP(2,1)-GP(2,2)
      X121 = -X112
      DO 300 I=1,3
      JNDD = ICON(I,IELND)
      SI11 = SI(I)-GP(2,1)
      SI12 = SI(I)-GP(2,2)
      TAUXNP(JNDD) = TAUXNP(JNDD) + SI12/X112*TAU(1)+
     $               SI11/X121*TAU(2)
      IU(JNDD) = IU(JNDD) + 1
  300 CONTINUE
      RETURN
C
      END
      SUBROUTINE SHAP12
C
C     THIS ROUTINE EVALUATES SHAPE FUNCTIONS AT XI FOR THE 3-NODE
C     1-D ELEMENT
C
      COMMON /ELVAR/  XI,ETA,JAC,JACINV(2,2),XX(8),YY(8),PHI(8),
```

(Continued)

```
        REAL JACINV,JAC           DPHDX(8),DPHDY(8),DPHDXI(8),DPHDET(8),IELNO,NND
     $
C*******SHAPE FUNCTIONS**********
C
        PHI(1) = 1./2.*       XI*(XI-1.)
        PHI(2) = -(XI+1.)*    (XI-1.)
        PHI(3) = 1./2.*(XI+1.)*XI
C
C-------DERIVATIVES OF THE SHAPE FUNCTIONS
C
        DPHDXI(1) = XI-1./2.
        DPHDXI(2) = -2.*XI
        DPHDXI(3) = XI+1./2.
C-------DO JACOBIAN CALCULATIONS
        CALL JACOBI1
C-------DERIVATIVES WRT X AT XI
C
        DO 20 I=1,3
        DPHDX(I) = JACINV(1,1)*DPHDXI(I)
   20   CONTINUE
        RETURN
C
        END
        SUBROUTINE ELEM13(OPT)
C
C       THIS SUBROUTINE IS THE MASTER ROUTINE FOR ELEMENT 13, A
C       4-NODE 1-D CO-CUBIC ELEMENT.
C
C       IF OPT='FORM'  THE ELEMENT STIFFNESS AND MASS MATRICES AND
C                      LOAD VECTOR (DUE TO INTERIOR LOADS ONLY)
C                      ARE FORMED.
C       IF OPT='FLUX'  THE FLUXES ARE CALCULATED AT THE THREE GAUSS
C                      POINTS AND THE FOUR NODES.
C
        COMMON /ELSTF/  KE(8,8),FE(8),ME(8,8),CE(8,8)
        REAL KE,ME
        COMMON /NODES/  X(300),Y(300)
        COMMON /ELMNTS/ NTYPE(200),NGPINT(200),NPHYS(200),ICON(8,200)
        REAL MU
        COMMON /ELVAR/  XI,ETA,JAC,JACINV(2,2),XX(8),YY(8),PHI(8),
     $                  DPHDX(8),DPHDY(8),DPHDXI(8),DPHDET(8),IELNO,NND
        REAL JACINV,JAC
        COMMON /GLINT/  GP(5,5),WT(5,5)
        COMMON /PROBLM/ DIMEN PROB,TITLE(18)
        DIMENSION XXI(3),TAU(3),SI(4)
        DATA EIGP/'EIGP'/,XIBVP/'XIBVP'/
        DATA SI/-1.,-0.3333333,0.3333333,1./
        DATA FORM/'FORM'/,FLUX/'FLUX'/
C
C-------BRANCH TO FLUX CALCULATION IF OPT = 'FLUX'
C       FORM MATRICES IF OPT = 'FORM'
        IF (OPT .EQ. FLUX) GO TO 200
C*******FORM ELEMENT MATRICES AND VECTORS*********
C
C
        YYY = 0.
        IPHY = NPHYS(IELNO)
        DO 20 I=1,4
```

```
        J = ICON(I,IELNO)
        XXI(I) = X(J)
        FE(I) = 0.
        DO 10 J=1,4
           KE(J,I) = 0.
           ME(J,I) = 0.
           CE(J,I) = 0.
   10   CONTINUE
   20   CONTINUE
        NINT = NGPINT(IELNO)
C-------EVALUATE TERMS IN ELEMENT EQUATIONS
        DO 60 INTX=1,NINT
           XI = GP(NINT,INTX)
           CALL SHAP13
           XXX = 0.
           DO 30 I=1,4
              XXX = XXX+PHI(I)*XX(I)
   30      CONTINUE
           DO 50 I=1,4
              FE(I) = FE(I)+FINT(IELNO)*PHI(I)*WT(NINT,INTX)*JAC
              DO 40 J=1,4
                 KE(I,J) = KE(I,J)+WT(NINT,INTX)*JAC*
     $                 (DPHDX(I)*ME(I,J)*DPHDX(J)*
     $                  PHI(I)*BETA*PHI(J))
        IF (PROB.EQ.EIGP) KE(I,J)=KE(I,J)+WT(NINT,INTX)*JAC*
     $                  PHI(I)*GAMMA*PHI(J)
        IF (PROB.EQ.XIBVP)CE(I,J)=CE(I,J)+WT(NINT,INTX)*JAC*
     $                  PHI(I)*MU*PHI(J)
   40         CONTINUE
   50      CONTINUE
   60   CONTINUE
C-------FOR SYMMETRY
        DO 100 I=1,4
           DO 90 J=1,4
              KE(J,I) = KE(I,J)
              ME(J,I) = ME(I,J)
              CE(J,I) = CE(I,J)
   90      CONTINUE
  100   CONTINUE
        RETURN
C
C*******CALCULATE FLUXES*********
C
  200   CONTINUE
        YYY = 0.
        IPHY = NPHYS(IELNO)
        DO 220 I=1,4
           J = ICON(I,IELNO)
           XXI(I) = X(J)
  220   CONTINUE
C-------CALCULATE FLUXES AT THE 3 GAUSS POINTS
        DO 260 INT=1,3
           XI = GP(3,INT)
           CALL SHAP13
           XXI(INT) = 0.
           DUDX = 0.
           DO 240 I=1,4
              XXX = XXI(INT)
              XXI(INT) = XXI(INT)+PHI(I)*XX(I)
              DUDX = DUDX+DPHDX(I)*A(J)
           CALL PROP(IPHY,XXX,YYY,ALPHA,ALPHAY,BETA,GAMMA,MU)
           TAU(INT) = -ALPHA*DUDX
           XGP(INT,IELNO) = XXI(INT)
```

```
           YGP(INT,IELNO) = 0.
           TAUXGP(INT,IELNO) = TAU(INT)
           TAUYGP(INT,IELNO) = 0.
  260   CONTINUE
C-------CALCULATE ELEMENT NODAL FLUXES BY INTERPOLATING THE GAUSS
C       POINT FLUXES AND SUM FOR NODAL FLUX CALCULATION IN CALFLX
C
        XI12 = GP(3,1)-GP(3,2)
        XI13 = GP(3,1)-GP(3,3)
        XI21 = -XI12
        XI23 = GP(3,2)-GP(3,3)
        XI31 = -XI13
        XI32 = -XI23
        DO 300 I=1,4
           JNOD = ICON(I,IELNO)
           SI11 = SI(I)-GP(3,1)
           SI12 = SI(I)-GP(3,2)
           SI13 = SI(I)-GP(3,3)
           TAUXNP(JNOD) = TAUXNP(JNOD) + SI12*SI13/XI12*SI13/XI13*TAU(1)+
     $                SI11/XI21*SI13/XI23*TAU(2)+
     $                SI11/XI31*SI12/XI32*TAU(3)
           IU(JNOD) = IU(JNOD) + 1
  300   CONTINUE
        RETURN
C
        END
        SUBROUTINE SHAP13
C
C       THIS ROUTINE EVALUATES SHAPE FUNCTIONS AT XI FOR THE 4-NODE
C       1-D ELEMENT
C
        COMMON /ELVAR/  XI,ETA,JAC,JACINV(2,2),XX(8),YY(8),PHI(8),
     $                  DPHDX(8),DPHDY(8),DPHDXI(8),DPHDET(8),IELNO,NND
        REAL JACINV,JAC
C
C*******SHAPE FUNCTIONS*********
C
        PHI(1) =  -9./16.*        (XI+1./3.)*(XI-1./3.)*(XI-1.)
        PHI(2) =  27./16.*(XI+1.)*          (XI-1./3.)*(XI-1.)
        PHI(3) = -27./16.*(XI+1.)*(XI+1./3.)*          (XI-1.)
        PHI(4) =   9./16.*(XI+1.)*(XI+1./3.)*(XI-1./3.)
C
C-------DERIVATIVES WRT XI AT XI
        DPHDXI(1) =  -9./16.*(3.*XI*XI-2.*XI-1./9.)
        DPHDXI(2) =  27./16.*(3.*XI*XI-2./3.*XI-1.)
        DPHDXI(3) = -27./16.*(3.*XI*XI+2./3.*XI-1.)
        DPHDXI(4) =   9./16.*(3.*XI*XI+2.*XI-1./9.)
C-------DO JACOBIAN CALCULATIONS
        CALL JACOBI1
C-------DERIVATIVES WRT X AT XI
        DO 20 I=1,4
           DPHDX(I) = JACINV(1,1)*DPHDXI(I)
   20   CONTINUE
        RETURN
C
        END
        SUBROUTINE ELEM14(OPT)
C
C       THIS SUBROUTINE IS THE MASTER ROUTINE FOR ELEMENT 14, A
C       5-NODE 1-D CO-QUARTIC ELEMENT.
C
C       IF OPT='FORM'  THE ELEMENT STIFFNESS AND MASS MATRICES AND
C                      LOAD VECTOR (DUE TO INTERIOR LOADS ONLY)
C                      ARE FORMED.
```

(Continued)

```fortran
C
C     IF OPT='FLUX'   THE FLUXES ARE CALCULATED AT THE FOUR GAUSS
C                     POINTS AND THE FIVE NODES.
C
      COMMON /ELSTF/  KE(8,8),FE(8),ME(8,8),CE(8,8)
      REAL KE,ME
      COMMON /NODES/  X(300),Y(300)
      COMMON /ELMNTS/ NTYPE(200),NGPINT(200),NPHYS(200),ICON(8,200)
      REAL MU
      COMMON /ELVAR/  XI,ETA,JAC,JACINV(2,2),XX(8),YY(8),PHI(8),
     $                DPHDX(8),DPHDY(8),DPHDXI(8),DPHDET(8),IELNO,NND
      REAL JACINV,JAC
      COMMON /GLINT/  GP(5,5),WT(5,5)
      REAL JACINV,JAC
      COMMON /NODSOL/ IU(300),TAUXNP(300),TAUYNP(300)
      COMMON /ELFLUX/ XGP(4,200),YGP(4,200),XX(8),YY(8),PHI(8),
     $                DPHDX(8),DPHDY(8),DPHDXI(8),DPHDET(8),IELNO,NND
      COMMON /SYSEQS/ K(300,50),F(300),A(300)
      COMMON /PROBLM/ DIMEN,PROB,TITLE(18)
      DIMENSION XXI(4),TAU(4),SI(5)
      DATA EIGP/'EIGP'/, XIBVP/'IBVP'/
      DATA SI/-1.,-0.5,0.,0.5,1./
      DATA FORM/'FORM'/, FLUX/'FLUX'/
C
C--------BRANCH TO FLUX CALCULATION IF OPT = 'FLUX'
C        FORM MATRICES IF OPT = 'FORM'
C
      IF (OPT .EQ. FLUX) GO TO 200
C
C********FORM ELEMENT MATRICES AND VECTORS***********
C
      YYY = 0.
      IPHY = NPHYS(IELNO)
      DO 20 I=1,5
      J = ICON(I,IELNO)
      XX(I) = X(J)
      KE(I,J) = 0.
      ME(I,J) = 0.
      CE(I,J) = 0.
   10 CONTINUE
   20 CONTINUE
      NINT = NGPINT(IELNO)
C
C--------EVALUATE TERMS IN ELEMENT EQUATIONS
C
      DO 60 INTX=1,NINT
      XI = GP(NINT,INTX)
      CALL SHAP14
      XXX = 0.
      DO 30 I=1,5
      XXX = XXX+PHI(I)*XX(I)
   30 CONTINUE
      CALL PROP(IPHY,XXX,YYY,ALPHA,ALPHAX,BETA,GAMMA,MU)
      DO 40 J=1,5
      KE(I,J) = KE(I,J)+WT(NINT,INTX)*JAC*
     $          (DPHDX(I)*ALPHA*DPHDX(J)+
     $           PHI(I)*BETA*PHI(J))
      ME(I,J) = ME(I,J)+WT(NINT,INTX)*JAC*
     $          PHI(I)*GAMMA*PHI(J)
   40 CONTINUE
      IF (PROB .EQ. EIGP) ME(I,J)=ME(I,J)+WT(NINT,INTX)*JAC*
     $          PHI(I)*MU*PHI(J)
      IF (PROB .EQ. XIBVP)CE(I,J)=CE(I,J)+WT(NINT,INTX)*JAC*
     $          PHI(I)*MU*PHI(J)
   50 CONTINUE
   60 CONTINUE
C
C--------FOR SYMMETRY
      DO 100 I=1,5
      DO 90 J=1,5
      KE(J,I) = KE(I,J)
      ME(J,I) = ME(I,J)
      CE(J,I) = CE(I,J)
   90 CONTINUE
  100 CONTINUE
      RETURN
C
C********CALCULATE FLUXES**************
C
  200 CONTINUE
      IPHY = NPHYS(IELNO)
      DO 220 I=1,5
      J = ICON(I,IELNO)
      XXI(I) = X(J)
  220 CONTINUE
C
C--------CALCULATE FLUXES AT THE 4 GAUSS POINTS
C
      DO 260 INT=1,4
      XI = GP(4,INT)
      CALL SHAP14
      XXI(INT) = 0.
      DUDX = 0.
      DO 240 I=1,5
      J = ICON(I,IELNO)
      XXI(INT) = XXI(INT)+PHI(I)*XX(I)
      DUDX = DUDX+DPHDX(I)*A(J)
  240 CONTINUE
C
C--------CALCULATE ELEMENT NODAL FLUXES BY INTERPOLATING THE GAUSS
C        POINT FLUXES AND SUM FOR NODAL FLUX CALCULATION IN CALFLX
C
      XXX = XXI(INT)
      CALL PROP(IPHY,XXX,YYY,ALPHA,ALPHAX,BETA,GAMMA,MU)
      TAU(INT) = -ALPHA*DUDX
      XGP(INT,IELNO) = XXI(INT)
      TAUXGP(INT,IELNO) = TAU(INT)
  260 CONTINUE
C
C--------CALCULATE ELEMENT NODAL FLUXES
C
      X112 = GP(4,1)-GP(4,2)
      X113 = GP(4,1)-GP(4,3)
      X114 = GP(4,1)-GP(4,4)
      X121 = -X112
      X123 = GP(4,2)-GP(4,3)
      X124 = GP(4,2)-GP(4,4)
      X131 = -X113
      X132 = -X123
      X134 = GP(4,3)-GP(4,4)
      X141 = -X114
      X142 = -X124
      X143 = -X134
      DO 300 I=1,5
      JNOD = ICON(I,IELNO)
      SI11 = SI(I)-GP(4,1)
      SI12 = SI(I)-GP(4,2)
      SI13 = SI(I)-GP(4,3)
      SI14 = SI(I)-GP(4,4)
      TAUXNP(JNOD)=TAUXNP(JNOD)+SI12*SI13*SI14/X112/X113/X114*TAU(1)
     $            +SI11*SI13*SI14/X121/X123/X124*TAU(2)
     $            +SI11*SI12*SI14/X131/X132/X134*TAU(3)
     $            +SI11*SI12*SI13/X141/X142/X143*TAU(4)
      IU(JNOD) = IU(JNOD) + 1
  300 CONTINUE
      RETURN
      END
C
      SUBROUTINE SHAP14
C
C     THIS ROUTINE EVALUATES SHAPE FUNCTIONS AT XI FOR THE 5-NODE
C                                                        1-D ELEMENT
C
      COMMON /ELVAR/  XI,ETA,JAC,JACINV(2,2),XX(8),YY(8),PHI(8),
     $                DPHDX(8),DPHDY(8),DPHDXI(8),DPHDET(8),IELNO,NND
      REAL JACINV,JAC
C
C--------SHAPE FUNCTIONS**************
C
      PHI(1) =  2./3.        *(XI+1./2.)*XI*(XI-1./2.)*(XI-1.)
      PHI(2) = -8./3.*(XI+1.)      *XI*(XI-1./2.)*(XI-1.)
      PHI(3) =    4.*(XI+1.)*(XI+1./2.)  *(XI-1./2.)*(XI-1.)
      PHI(4) = -8./3.*(XI+1.)*(XI+1./2.)*XI*      (XI-1.)
      PHI(5) =  2./3.*(XI+1.)*(XI+1./2.)*XI*(XI-1./2.)
C
C--------DERIVATIVES WRT XI AT XI
C
      DPHDXI(1) =  2./3.*(4.*XI*XI*XI-3.*XI*XI-1./2.*XI+1./4.)
      DPHDXI(2) = -8./3.*(4.*XI*XI*XI-3./2.*XI*XI-2.*XI+1./2.)
      DPHDXI(3) =    4.*(4.*XI*XI*XI-5./2.*XI)
      DPHDXI(4) = -8./3.*(4.*XI*XI*XI+3./2.*XI*XI-2.*XI-1./2.)
      DPHDXI(5) =  2./3.*(4.*XI*XI*XI+3.*XI*XI-1./2.*XI-1./4.)
C
C--------DO JACOBIAN CALCULATIONS
C
      CALL JACOBI(1)
C
C--------DERIVATIVES WRT X AT XI
C
      DO 20 I=1,5
   20 DPHDXI(I) = JACINV(1,1)*DPHDXI(I)
      RETURN
      END
C
      SUBROUTINE ELEM21(OPT)
C
C     THIS SUBROUTINE IS THE MASTER ROUTINE FOR ELEMENT 21.  A
C     3-NODE 2-D CO-LINEAR TRIANGLE.
C
C     IF OPT='FORM'   THE ELEMENT STIFFNESS AND MASS MATRICES AND
C                     LOAD VECTOR (DUE TO INTERIOR LOADS ONLY)
C                     ARE FORMED
C     IF OPT='FLUX'   THE FLUXES ARE CALCULATED AT ONE GAUSS POINT
C                     AND THE TWO NODES.
C
      COMMON /ELSTF/  KE(8,8),FE(8),ME(8,8),CE(8,8)
      REAL KE,ME
      COMMON /NODES/  X(300),Y(300)
      COMMON /ELMNTS/ NTYPE(200),NGPINT(200),NPHYS(200),ICON(8,200)
      REAL MU
      COMMON /INTLD/  FINT(200)
      COMMON /NODSOL/ IU(300),TAUXNP(300),TAUYNP(300)
      COMMON /ELFLUX/ XGP(4,200),YGP(4,200),TAUXGP(4,200),TAUYGP(4,200)
      COMMON /CENTRD/ UC(200),XC(200),YC(200),TAUXC(200),TAUYC(200)
      COMMON /SYSEQS/ K(300,50),F(300),A(300)
      REAL K
      COMMON /ELVAR/  XI,ETA,JAC,JACINV(2,2),XX(8),YY(8),PHI(8),
     $                DPHDX(8),DPHDY(8),DPHDXI(8),DPHDET(8),IELNO,NND
      REAL JAC,JACINV
      COMMON /PROBLM/ DIMEN,PROB,TITLE(18)
      DIMENSION B(3),C(3)
      DATA EIGP/'EIGP'/, XIBVP/'IBVP'/
      DATA FORM/'FORM'/, FLUX/'FLUX'/
C
C--------SET UP VARIABLES COMMON TO BOTH PARTS
```

(Continued)

```
      I = ICON(1,IELNO)
      J = ICON(2,IELNO)
      L = ICON(3,IELNO)
      IPHY = NPHYS(IELNO)
      B(1) = Y(J)-Y(L)
      B(2) = Y(L)-Y(I)
      B(3) = Y(I)-Y(J)
      C(1) = X(L)-X(J)
      C(2) = X(I)-X(L)
      C(3) = X(J)-X(I)
      AREA = (B(2)*C(3)-B(3)*C(2))/2.
      XCENT = (X(I)+X(J)+X(L))/3.
      YCENT = (Y(I)+Y(J)+Y(L))/3.
      CALL PROP(IPHY,XCENT,YCENT,ALPHAX,ALPHAY,BETA,GAMMA,MU)
C---------BRANCH TO FLUX CALCULATION IF OPT = 'FLUX'
      IF (OPT.EQ.FLUX) GO TO 200
C
C*********CALCULATE FLUXES*********
C
      IF (AREA.LE..O) CALL ERROR(20,IELNO)
      DO 20 I=1,3
      FE(I) = FINT(IELNO)*AREA/3.
      DO 10 J=1,3
      IF (I.EQ.J) FACTOR=6.
      IF (I.NE.J) FACTOR=12.
      KE(I,J) = (B(I)*ALPHAX*B(J)+
     $          C(I)*ALPHAY*C(J))/(4.*AREA)+
     $          BETA*AREA/FACTOR
      IF (PROB.EQ.EIGP) ME(I,J) = GAMMA*AREA/FACTOR
      IF (PROB.EQ.XIBVP) CE(I,J) = MU*AREA/FACTOR
   10 CONTINUE
   20 CONTINUE
C---------FOR SYMMETRY
      DO 40 J=1,3
      DO 30 J=1,3
      KE(J,I) = KE(I,J)
      ME(J,I) = ME(I,J)
      CE(J,I) = CE(I,J)
   30 CONTINUE
   40 CONTINUE
      RETURN
C
C*********CALCULATE FLUXES*********
C
  200 CONTINUE
C---------CALCULATE FLUXES AT THE GAUSS POINT (CENTER)
      XGP(1,IELNO) = XCENT
      YGP(1,IELNO) = YCENT
      TAUXGP(1,IELNO) = -ALPHAX*(A(1)*B(1)+A(J)*B(2)+A(L)*B(3))
      TAUYGP(1,IELNO) = -ALPHAY*(A(1)*C(1)+A(J)*C(2)+A(L)*C(3))
C---------CALCULATE ELEMENT NODAL FLUXES AND SUM FOR NODAL FLUX
C         CALCULATION IN CALFLX (ELEMENT NODAL FLUXES SAME AS
C         GAUSS POINT FLUX)
      DO 220 I=1,3
      JNOD = ICON(11,IELNO)
      IU(JNOD) = IU(JNOD)+1
      TAUXNP(JNOD) = TAUXNP(JNOD)+TAUXGP(1,IELNO)
      TAUYNP(JNOD) = TAUYNP(JNOD)+TAUYGP(1,IELNO)
  220 CONTINUE
C---------CALCULATE FUNCTION, FLUXES, AND COORDINATES AT
C         CENTROID FOR CONTOUR PLOTTING
      UC(IELNO) = (A(I)+A(J)+A(L))/3.
      TAUXC(IELNO) = TAUXGP(1,IELNO)
      TAUYC(IELNO) = TAUYGP(1,IELNO)
      XC(IELNO) = XGP(1,IELNO)
      YC(IELNO) = YGP(1,IELNO)
      RETURN
      END
      SUBROUTINE ELEM22(OPT)
C
C     THIS SUBROUTINE IS THE MASTER ROUTINE FOR ELEMENT 22. A
C     6-NODE 2-D CO-QUADRATIC TRIANGLE.
C
C     IF OPT='FORM'  THE ELEMENT STIFFNESS AND MASS MATRICES AND
C                    LOAD VECTOR (DUE TO INTERIOR LOADS ONLY)
C                    ARE FORMED.
C
C     IF OPT='FLUX'  THE FLUXES ARE CALCULATED AT FOUR INTEGRATION
C                    POINTS AND THE SIX NODES.
C
      COMMON /ELSTF/ KE(8,8),FE(8),ME(8,8),CE(8,8)
      REAL KE,ME
      COMMON /NODES/ X(300),Y(300)
      COMMON /ELMNTS/ NTYPE(200),NGPINT(200),NPHYS(200),ICON(8,200)
      REAL MU
      COMMON /INTLD/ FINT(200)
      COMMON /NODSOL/ IU(300),TAUXNP(300),TAUYNP(300)
      COMMON /ELFLUX/ XGP(4,200),YGP(4,200),TAUXGP(4,200),TAUYGP(4,200)
      COMMON /CENTRO/ UC(200),XC(200),YC(200),TAUXC(200),TAUYC(200)
      COMMON /SYSEQS/ K(300,50),F(300),A(300)
      REAL K
      COMMON /ELVAR/ XI,ETA,JAC,JACINV(2,2),XX(8),YY(8),PHI(8),
     $               DPHDX(8),DPHDY(8),DPHDXI(8),DPHDET(8),IELND,NND
      COMMON /TRIINT/ NTIP(7),GPXI(7,7),GPET(7,7),WGHT(7,7)
      REAL JAC,JACINV
      COMMON /PROBLM/ DIMEN,PROB,TITLE(18)
      COMMON /TRANS/ TRI(6,4),TR2(8,4)
      DIMENSION GX(4),GY(4)
      DATA EIGP/'EIGP'/, XIBVP/'IBVP'/
      DATA FORM/'FORM'/, FLUX/'FLUX'/
C---------BRANCH TO FLUX CALCULATION IF OPT = 'FLUX'
C         FORM MATRICES IF OPT = 'FORM'
      IF (OPT.EQ.FLUX) GO TO 200
C
C*********FORM ELEMENT MATRICES AND VECTORS*********
C
      IPHY = NPHYS(IELND)
      DO 20 I=1,6
      J = ICON(I,IELND)
      XX(I) = X(J)
      YY(I) = Y(J)
      FE(I) = 0.
      DO 10 J=1,6
      KE(I,J) = 0.
      ME(I,J) = 0.
      CE(I,J) = 0.
   10 CONTINUE
   20 CONTINUE
      N1 = NGPINT(IELND)
      NINT = NTIP(N1)
C---------INTEGRATE TERMS
      DO 100 INT=1,NINT
      XI = GPXI(N1,INT)
      ETA = GPET(N1,INT)
      WGT = WGHT(N1,INT)/2.
      CALL SHAP22
      XXX = 0.
      YYY = 0.
      DO 30 I=1,6
      XXX = XXX+PHI(I)*XX(I)
      YYY = YYY+PHI(I)*YY(I)
   30 CONTINUE
      CALL PROP(IPHY,XXX,YYY,ALPHAX,ALPHAY,BETA,GAMMA,MU)
      DO 80 I=1,6
      FE(I) = FE(I)+FINT(IELND)*PHI(I)*WGT*JAC
      DO 60 J=1,6
      KE(I,J) = KE(I,J)+WGT*JAC*
     $          (DPHDX(I)*ALPHAX*DPHDX(J)+
     $          DPHDY(I)*ALPHAY*DPHDY(J)+
     $          PHI(I)*BETA*PHI(J))
      IF(PROB.EQ.EIGP) ME(I,J) = ME(I,J)+WGT*JAC*
     $          PHI(I)*GAMMA*PHI(J)
      IF(PROB.EQ.XIBVP) CE(I,J) = CE(I,J)+WGT*JAC*
     $          PHI(I)*MU*PHI(J)
   60 CONTINUE
   80 CONTINUE
C---------FOR SYMMETRY
      DO 160 I=1,6
      DO 140 J=1,6
      KE(J,I) = KE(I,J)
      ME(J,I) = ME(I,J)
      CE(J,I) = CE(I,J)
  140 CONTINUE
  160 CONTINUE
      RETURN
C
C*********CALCULATE FLUXES*********
C
  200 CONTINUE
      IPHY = NPHYS(IELND)
      DO 220 I=1,6
      J = ICON(I,IELND)
      XX(I) = X(J)
      YY(I) = Y(J)
  220 CONTINUE
C---------CALCULATE FLUXES AT THE FOUR INTEGRATION POINTS
      DO 300 INT=1,4
      XI = GPXI(4,INT)
      ETA = GPET(4,INT)
      CALL SHAP22
      XXI = 0.
      YYI = 0.
      DUDX = 0.
      DUDY = 0.
      DO 260 I=1,6
      IK = ICON(I,IELND)
      XXI = XXI+PHI(I)*XX(I)
      YYI = YYI+PHI(I)*YY(I)
      DUDX = DUDX+DPHDX(I)*A(IK)
      DUDY = DUDY+DPHDY(I)*A(IK)
  260 CONTINUE
      CALL PROP(IPHY,XXI,YYI,ALPHAX,ALPHAY,BETA,GAMMA,MU)
      XGP(INT,IELND) = XXI
      YGP(INT,IELND) = YYI
      TAUXGP(INT,IELND) = -ALPHAX*DUDX
      TAUYGP(INT,IELND) = -ALPHAY*DUDY
  300 CONTINUE
C---------CALCULATE ELEMENT NODAL FLUXES USING A LEAST SQUARES FIT
C         THROUGH GAUSS POINTS AND SUM FOR NODAL FLUX
C         CALCULATION IN CALFLX
      DO 320 I=1,6
      JNOD = ICON(I,IELND)
      IU(JNOD) = IU(JNOD)+1
      DO 310 L=1,4
```

```fortran
          TAUXNP(JNOD) = TAUXNP(JNOD)+TR1(1,L)*TAUXGP(L,IELNO)
          TAUYNP(JNOD) = TAUYNP(JNOD)+TR1(1,L)*TAUYGP(L,IELNO)
310     CONTINUE
320   CONTINUE
C-------CALCULATE FUNCTION, FLUXES, AND COORDINATES AT
C             CENTROID FOR CONTOUR PLOTTING
C
      XI = 1./3.
      ETA = 1./3.
      CALL SHAP22
      UC(IELNO) = 0.
      DO 330 I=1,6
        IK = ICON(I,IELNO)
        UC(IELNO) = UC(IELNO)+PHI(I)*A(IK)
330   CONTINUE
      XC(IELNO) = XGP(1,IELNO)
      YC(IELNO) = YGP(1,IELNO)
      TAUXC(IELNO) = TAUXGP(1,IELNO)
      TAUYC(IELNO) = TAUYGP(1,IELNO)
      RETURN
C
      END
      SUBROUTINE SHAP22
C
C     EVALUATE SHAPE FUNCTIONS AND DERIVATIVES AT POINT (XI,ETA)
C     FOR 6-NODE TRIANGLE
C
      COMMON /ELVAR/  XI,ETA,JAC,JACINV(2,2),XX(8),YY(8),PHI(8),
     $                DPHDX(8),DPHDY(8),DPHDXI(8),DPHDET(8),IELNO,NND
      REAL JACINV
C
C*********SHAPE FUNCTIONS*********
      PHI(1) = (1.-2.*XI-2.*ETA)*(-1.-XI-ETA)
      PHI(2) = XI*(-1.+2.-XI)
      PHI(3) = ETA*(-1.-2.-ETA)
      PHI(4) = 4.*XI*(1.-XI-ETA)
      PHI(5) = 4.*XI*ETA
      PHI(6) = 4.*ETA*(1.-XI-ETA)
C
C-------DERIVATIVES WRT XI AT (XI,ETA)
      DPHDXI(1) = -3.+4.*XI+4.*ETA
      DPHDXI(2) = -1.+4.*XI
      DPHDXI(3) = 0.
      DPHDXI(4) = 4.*(1.-2.*XI-ETA)
      DPHDXI(5) = 4.*ETA
      DPHDXI(6) = -4.*ETA
C-------DERIVATIVES WRT ETA AT (XI,ETA)
      DPHDET(1) = -3.+4.*XI+4.*ETA
      DPHDET(2) = 0.
      DPHDET(3) = -1.+4.*ETA
      DPHDET(4) = -4.*XI
      DPHDET(5) = 4.*XI
      DPHDET(6) = 4.*(1.-XI-2.*ETA)
C-------DO JACOBIAN CALCULATIONS
      CALL JACOBI(2)
C-------DERIVATIVES WRT X AND Y AT (XI,ETA)
      DO 50 I=1,6
        DPHDX(I) = JACINV(1,1)*DPHDXI(I)+JACINV(1,2)*DPHDET(I)
        DPHDY(I) = JACINV(2,1)*DPHDXI(I)+JACINV(2,2)*DPHDET(I)
50    CONTINUE
      RETURN
C
      END


      SUBROUTINE ELEM23(OPT)
C
C     THIS SUBROUTINE IS THE MASTER ROUTINE FOR ELEMENT 23.  A
C     4-NODE 2-D CO-LINEAR QUADRILATERAL.
C
C     IF OPT='FORM'  THE ELEMENT STIFFNESS AND MASS MATRICES AND
C                    LOAD VECTOR (DUE TO INTERIOR LOADS ONLY)
C                    ARE FORMED.
C
C     IF OPT='FLUX'  THE FLUXES ARE CALCULATED AT ONE GAUSS POINT
C                    AND THE FOUR NODES.
C
      COMMON /ELSTF/  KE(8,8),FE(8),ME(8,8),CE(8,8)
      REAL KE,ME
      COMMON /NODES/  X(300),Y(300)
      COMMON /ELMNTS/ NTYPE(200),NGPINT(200),NPHYS(200),ICON(8,200)
      REAL MU
      COMMON /INTLD/  FINT(200)
      COMMON /NODSOL/ IU(300),TAUXNP(300),TAUYNP(300)
      COMMON /ELFLUX/ XGP(4,200),YGP(4,200),TAUXGP(4,200),TAUYGP(4,200)
      COMMON /CENTRD/ UC(200),XC(200),YC(200),TAUXC(200),TAUYC(200)
      COMMON /SYSEQS/ K(300,50),F(300),A(300)
      REAL K
      COMMON /ELVAR/  XI,ETA,JAC,JACINV(2,2),XX(8),YY(8),PHI(8),
     $                DPHDX(8),DPHDY(8),DPHDXI(8),DPHDET(8),IELNO,NND
      COMMON /GLINT/  GP(5,5),WT(6,5)
      REAL JAC,JACINV
      COMMON /PROBLM/ DIMEN,PROB,TITLE(18)
      DATA EIGP/'EIGP'/, XIBVP/'IBVP'/
      DATA FORM/'FORM'/, FLUX/'FLUX'/
C
C-------BRANCH TO FLUX CALCULATION IF OPT = 'FLUX'
C       FORM MATRICES IF OPT = 'FORM'
      IF (OPT.EQ.FLUX) GO TO 200
C
C*********FORM ELEMENT MATRICES AND VECTORS*********
C
      IPHY = NPHYS(IELNO)
      DO 20 I=1,4
        XX(I) = X(J)
        YY(I) = Y(J)
        FE(I) = 0.
        DO 10 J=1,4
          KE(I,J) = 0.
          ME(I,J) = 0.
          CE(I,J) = 0.
10      CONTINUE
20    CONTINUE
      NINT = NGPINT(IELNO)
C-------INTEGRATE TERMS
      DO 100 INTX=1,NINT
        XI = GP(NINT,INTX)
        DO 90 INTY=1,NINT
          ETA = GP(NINT,INTY)
          WGT = WT(NINT,INTX)*WT(NINT,INTY)
          CALL SHAP23
          XXX = 0.
          YYY = 0.
          DO 30 I=1,4
            XXX = XXX+PHI(I)*XX(I)
            YYY = YYY+PHI(I)*YY(I)
30        CONTINUE
          CALL PROP(IPHY,XXX,YYY,ALPHAX,ALPHAY,BETA,GAMMA,MU)
          DO 80 I=1,4
            FE(I) = FE(I)+FINT(IELNO)*PHI(I)*WGT*JAC
            DO 70 J=1,4
              KE(I,J) = KE(I,J)+WGT*JAC*
     $                  (DPHDX(I)*ALPHAX*DPHDX(J)+
     $                  DPHDY(I)*ALPHAY*DPHDY(J)+
     $                  PHI(I)*BETA*PHI(J))
              IF(PROB.EQ.EIGP) ME(I,J) = ME(I,J)+WGT*JAC*
     $                  PHI(I)*GAMMA*PHI(J)
              IF(PROB.EQ.XIBVP) CE(I,J) = CE(I,J)+WGT*JAC-
     $                  PHI(I)*MU*PHI(J)
70          CONTINUE
80        CONTINUE
90      CONTINUE
100   CONTINUE
C-------FOR SYMMETRY
      DO 150 I=1,4
        DO 140 J=1,4
          KE(J,I) = KE(I,J)
          ME(J,I) = ME(I,J)
          CE(J,I) = CE(I,J)
140     CONTINUE
150   CONTINUE
      RETURN
C
C*********CALCULATE FLUXES*********
C
200   CONTINUE
      IPHY = NPHYS(IELNO)
      DO 220 I=1,4
        J = ICON(I,IELNO)
        XX(I) = X(J)
        YY(I) = Y(J)
220   CONTINUE
C-------CALCULATE FLUX AT THE 1 GAUSS POINT (CENTER)
      XI = GP(1,1)
      ETA = GP(1,1)
      CALL SHAP23
      XXX = 0.
      YYY = 0.
      DUDX = 0.
      DUDY = 0.
      DO 240 I=1,4
        J = ICON(I,IELNO)
        XXX = XXX+PHI(I)*XX(I)
        YYY = YYY+PHI(I)*YY(I)
        DUDX = DUDX+DPHDX(I)*A(J)
        DUDY = DUDY+DPHDY(I)*A(J)
240   CONTINUE
      CALL PROP(IPHY,XXX,YYY,ALPHAX,ALPHAY,BETA,GAMMA,MU)
      XGP(1,IELNO) = XXX
      YGP(1,IELNO) = YYY
      TAUXGP(1,IELNO) = -ALPHAX*DUDX
      TAUYGP(1,IELNO) = -ALPHAY*DUDY
C-------CALCULATE ELEMENT NODAL FLUXES AND SUM FOR NODAL FLUX
C       CALCULATION IN CALFLX (ELEMENT NODAL FLUXES SAME AS
C       GAUSS POINT FLUX)
      DO 260 I=1,4
        JNOD = ICON(I,IELNO)
        IU(JNOD) = IU(JNOD)+1
        TAUXNP(JNOD) = TAUXNP(JNOD)+TAUXGP(1,IELNO)
        TAUYNP(JNOD) = TAUYNP(JNOD)+TAUYGP(1,IELNO)
260   CONTINUE
C-------CALCULATE FUNCTION, FLUXES, AND COORDINATES AT
C             CENTROID FOR CONTOUR PLOTTING
      XI = 0.
      ETA = 0.
      CALL SHAP23
      UC(IELNO) = 0.
```

(Continued)

817

```fortran
             DO 280 I=1,4
                IK = ICON(I,IELNO)
                UC(IELNO) = UC(IELNO)+PHI(I)*A(IK)
        280  CONTINUE
             XC(IELNO) = XGP(1,IELNO)
             YC(IELNO) = YGP(1,IELNO)
             TAUXC(IELNO) = TAUXGP(1,IELNO)
             TAUYC(IELNO) = TAUYGP(1,IELNO)
             RETURN
C
             END
             SUBROUTINE SHAP23
C
C   EVALUATE SHAPE FUNCTIONS AND DERIVATIVES AT POINT (XI,ETA)
C                    FOR 4-NODE QUADRILATERAL
C
             COMMON /ELVAR/ XI,ETA,JAC,JACINV(2,2),XX(8),YY(8),PHI(8),
            $               DPHDX(8),DPHDY(8),DPHDXI(8),DPHDET(8),IELNO,NND
             COMMON /GLINT/
             REAL JAC,JACINV
C
C-------CONSTANTS
             XIP = 1.+XI
             XIM = 1.-XI
             ETP = 1.+ETA
             ETM = 1.-ETA
             XI2 = XIP*XIM
             ET2 = ETP*ETM
C
C*********SHAPE FUNCTIONS**********
             PHI(1) = XIM*ETM/4.
             PHI(2) = XIP*ETM/4.
             PHI(3) = XIP*ETP/4.
             PHI(4) = XIM*ETP/4.
C
C-------DERIVATIVES WRT XI AT (XI,ETA)
             DPHDXI(1) = -ETM/4.
             DPHDXI(2) =  ETM/4.
             DPHDXI(3) =  ETP/4.
             DPHDXI(4) = -ETP/4.
C
C-------DERIVATIVES WRT ETA AT (XI,ETA)
             DPHDET(1) = -XIM/4.
             DPHDET(2) = -XIP/4.
             DPHDET(3) =  XIP/4.
             DPHDET(4) =  XIM/4.
C
C-------DO JACOBIAN CALCULATIONS
             CALL JACOB1(2)
C
C-------DERIVATIVES WRT X AND Y AT (XI,ETA)
             DO 50 I=1,8
                DPHDX(I) = JACINV(1,1)*DPHDXI(I)+JACINV(1,2)*DPHDET(I)
                DPHDY(I) = JACINV(2,1)*DPHDXI(I)+JACINV(2,2)*DPHDET(I)
         50  CONTINUE
             RETURN
C
             END
             SUBROUTINE ELEM24(OPT)
C
C   THIS SUBROUTINE IS THE MASTER ROUTINE FOR ELEMENT 24, AN
C   8-NODE 2-D CO-QUADRATIC QUADRILATERAL.
C
C   IF OPT='FORM'  THE ELEMENT STIFFNESS AND MASS MATRICES AND
C                  LOAD VECTOR (DUE TO INTERIOR LOADS ONLY)
C                  ARE FORMED.
C
C   IF OPT='FLUX'  THE FLUXES ARE CALCULATED AT FOUR GAUSS POINTS
C                  AND THE EIGHT NODES.
C
             COMMON /ELSTF/ KE(8,8),FE(8),ME(8,8),CE(8,8)
             REAL KE,ME
             COMMON /NODES/ X(300),Y(300)
             COMMON /ELMNTS/ NTYPE(200),NGPINT(200),NPHYS(200),ICON(8,200)
             REAL MU
             COMMON /INTLD/ FINT(200)
             COMMON /NDSOL/ IU(300),TAUXNP(300),TAUYNP(300)
             COMMON /ELFLUX/ XGP(4,200),YGP(4,200),TAUXGP(4,200),TAUYGP(4,200)
             COMMON /CENTRD/ UC(200),XC(200),YC(200),TAUXC(200),TAUYC(200)
             COMMON /SYSEQS/ K(300,50),F(300),A(300)
             REAL K
             COMMON /ELVAR/ XI,ETA,JAC,JACINV(2,2),XX(8),YY(8),PHI(8),
            $               DPHDX(8),DPHDY(8),DPHDXI(8),DPHDET(8),IELNO,NND
             COMMON /GLINT/ GP(5,5),WT(5,5)
             REAL JAC,JACINV
             COMMON /PROBLM/ DIMEN,PROB,TITLE(18)
             COMMON /TRANS/ TR1(6,4),TR2(8,4)
             DIMENSION GX(4),GY(4)
             DATA EIGP/'EIGP'/, XIBVP/'IBVP'/
             DATA FORM/'FORM'/, FLUX/'FLUX'/
C-------BRANCH TO FLUX CALCULATION IF OPT = 'FLUX'.
C       FORM MATRICES IF OPT = 'FORM'.
             IF (OPT.EQ.FLUX) GO TO 200
C
C*********FORM ELEMENT MATRICES AND VECTORS**********
C
             IPHY = NPHYS(IELNO)
             DO 20 I=1,8
                J = ICON(I,IELNO)
                XX(I) = X(J)
                YY(I) = Y(J)
                KE(I,J) = 0.
                ME(I,J) = 0.
                CE(I,J) = 0.
         10  CONTINUE
         20  CONTINUE
C-------INTEGRATE TERMS
             NINT = NGPINT(IELNO)
             DO 100 INTX=1,NINT
                XI = GP(NINT,INTX)
                DO 90 INTY=1,NINT
                   ETA = GP(NINT,INTY)
                   WGT = WT(NINT,INTX)*WT(NINT,INTY)
                   CALL SHAP24
                   XXX = 0.
                   YYY = 0.
                   DO 30 I=1,8
                      XXX = XXX+PHI(I)*XX(I)
                      YYY = YYY+PHI(I)*YY(I)
         30  CONTINUE
                   CALL PROP(IPHY,XXX,YYY,ALPHAX,ALPHAY,BETA,GAMMA,MU)
                   DO 80 I=1,8
                      FE(I) = FE(I)+FINT(IELNO)*PHI(I)*WGT*JAC
                      DO 70 J=1,8
                         KE(I,J) = KE(I,J)+WGT*JAC*
            $             (DPHDX(I)*ALPHAX*DPHDX(J)+
            $              DPHDY(I)*ALPHAY*DPHDY(J)+
            $              PHI(I)*BETA*PHI(J))
                         IF(PROB.EQ.EIGP) ME(I,J) = ME(I,J)+WGT*JAC*
            $             PHI(I)*GAMMA*PHI(J)
                         IF(PROB.EQ.XIBVP) CE(I,J) = CE(I,J)+WGT*JAC*
            $             PHI(I)*MU*PHI(J)
         70  CONTINUE
         80  CONTINUE
        100  CONTINUE
C-------FOR SYMMETRY
             DO 150 I=1,8
                DO 140 J=1,8
                   KE(J,I) = KE(I,J)
                   ME(J,I) = ME(I,J)
                   CE(J,I) = CE(I,J)
        140  CONTINUE
        150  CONTINUE
             RETURN
C
C*********CALCULATE FLUXES**********
        200  CONTINUE
             IPHY = NPHYS(IELNO)
             DO 220 I=1,8
                J = ICON(I,IELNO)
                XX(I) = X(J)
                YY(I) = Y(J)
        220  CONTINUE
C-------CALCULATE FLUXES AT THE 4 GAUSS POINTS
             IGP = 0
             DO 260 IS=1,2
                XI = GP(2,IR)
                DO 280 IR=1,2
                   IGP = IGP+1
                   ETA = GP(2,IS)
                   CALL SHAP24
                   XXX = 0.
                   YYY = 0.
                   DUDX = 0.
                   DUDY = 0.
                   DO 240 I=1,8
                      J = ICON(I,IELNO)
                      XXX = XXX+PHI(I)*XX(I)
                      YYY = YYY+PHI(I)*YY(I)
                      DUDX = DUDX+DPHDX(I)*A(J)
                      DUDY = DUDY+DPHDY(I)*A(J)
        240  CONTINUE
                   CALL PROP(IPHY,XXX,YYY,ALPHAX,ALPHAY,BETA,GAMMA,MU)
                   XGP(IGP,IELNO) = XXX
                   YGP(IGP,IELNO) = YYY
                   TAUXGP(IGP,IELNO) = -ALPHAX*DUDX
                   TAUYGP(IGP,IELNO) = -ALPHAY*DUDY
        260  CONTINUE
        280  CONTINUE
C-------CALCULATE ELEMENT NODAL FLUXES USING A LEAST SQUARES FIT
C       THROUGH GAUSS POINTS AND SUM FOR NODAL FLUX CALCULATION
C       IN CALFLX
             DO 320 I=1,8
                JNOD = ICON(I,IELNO)
                IU(JNOD) = IU(JNOD)+1
                DO 300 L=1,4
                   TAUXNP(JNOD) = TAUXNP(JNOD)+TR2(I,L)*TAUXGP(L,IELNO)
                   TAUYNP(JNOD) = TAUYNP(JNOD)+TR2(I,L)*TAUYGP(L,IELNO)
        300  CONTINUE
        320  CONTINUE
C-------CALCULATE FUNCTION, FLUXES, AND COORDINATES AT
C       CENTROID FOR CONTOUR PLOTTING
             XI = 0.
             ETA = 0.
             CALL SHAP24
```

(Continued)

818

```fortran
      UC(IELNO) = O.
      DO 340 I=1,8
         J = ICON(I,IELNO)
         UC(IELNO)=UC(IELNO)+PHI(I)*A(J)
340   CONTINUE
C
C------CALCULATE FLUXES AND COORDINATES AT CENTROID
C         BY AVERAGING GAUSS PT VALUES (SAME AS FITTING
C         A PLANE THROUGH GAUSS PT VALUES)
C
      TAUXC(IELNO) = O.
      TAUYC(IELNO) = O.
      XC(IELNO) = O.
      YC(IELNO) = O.
      DO 360 I=1,4
         TAUXC(IELNO) = TAUXC(IELNO)+TAUXGP(I,IELNO)/4.
         TAUYC(IELNO) = TAUYC(IELNO)+TAUYGP(I,IELNO)/4.
         XC(IELNO) = XC(IELNO)+XGP(I,IELNO)/4.
         YC(IELNO) = YC(IELNO)+YGP(I,IELNO)/4.
360   CONTINUE
      RETURN
C
      END
      SUBROUTINE SHAP24
C
C       EVALUATE SHAPE FUNCTIONS AND DERIVATIVES AT POINT (XI,ETA)
C          FOR 8-NODE QUADRILATERAL
C
      COMMON /ELVAR/  XI,ETA,JAC,JACINV(2,2),XX(8),YY(8),PHI(8),
     $            DPHDX(8),DPHDY(8),DPHDXI(8),DPHDET(8),IELNO,NND
C
      REAL JACINV
C
C------CONSTANTS
      XIP = 1.+XI
      XIM = 1.-XI
      ETP = 1.+ETA
      ETM = 1.-ETA
      XI2 = XIP*XIM
      ET2 = ETP*ETM
C
C*******SHAPE FUNCTIONS*******
C
      PHI(1) = XIM*ETM*(-XI-ETA-1.)/4.
      PHI(2) = XIP*ETM*( XI-ETA-1.)/4.
      PHI(3) = XIP*ETP*( XI+ETA-1.)/4.
      PHI(4) = XIM*ETP*(-XI+ETA-1.)/4.
      PHI(5) = XI2*ETM/2.
      PHI(6) = XIP*ET2/2.
      PHI(7) = XI2*ETP/2.
      PHI(8) = XIM*ET2/2.
C
C------DERIVATIVES OF THE SHAPE FUNCTIONS
C
C------DERIVATIVES WRT XI AT (XI,ETA)
      DPHDXI(1) = ETM*( 2.*XI+ETA)/4.
      DPHDXI(2) = ETM*( 2.*XI-ETA)/4.
      DPHDXI(3) = ETP*( 2.*XI+ETA)/4.
      DPHDXI(4) = ETP*( 2.*XI-ETA)/4.
      DPHDXI(5) = -XI*ETM
      DPHDXI(6) = ET2/2.
      DPHDXI(7) = -XI*ETP
      DPHDXI(8) = -ET2/2.
C------DERIVATIVES WRT ETA AT (XI,ETA)
      DPHDET(1) = XIM*( XI+2.*ETA)/4.
      DPHDET(2) = XIP*(-XI+2.*ETA)/4.
      DPHDET(3) = XIP*( XI+2.*ETA)/4.
      DPHDET(4) = XIM*(-XI+2.*ETA)/4.
      DPHDET(5) = -XI2/2.
      DPHDET(6) = -XIP*ETA
      DPHDET(7) = X12/2.
      DPHDET(8) = -XIM*ETA
C
C------DO JACOBIAN CALCULATIONS
C
      CALL JACOB(2)
C
C------DERIVATIVES WRT X AND Y AT (XI,ETA)
C
      DO 50 I=1,8
         DPHDX(I) = JACINV(1,1)*DPHDXI(I)+JACINV(1,2)*DPHDET(I)
         DPHDY(I) = JACINV(2,1)*DPHDXI(I)+JACINV(2,2)*DPHDET(I)
50    CONTINUE
      RETURN
C
      END
      SUBROUTINE APLYBC
C
C       THIS ROUTINE APPLIES THE NATURAL AND ESSENTIAL BOUNDARY
C          CONDITIONS TO THE SYSTEM MATRICES.
C
      COMMON /GENRL/  NUMNP,NUMEL,NUMPP,NUMILC,NUMEBC,NUMNBC,MBAND
      COMMON /ESSBC/  IUEBC(100),UEBC(100)
      COMMON /NATBC/  INP(100),JNP(100),KNP(100),TAUNBC(100)
      COMMON /NODES/  X(300),Y(300)
      COMMON /SYSEQS/ K(300,50),F(300),A(300)
      COMMON /ELVAR/  XI,ETA,JAC,JACINV(2,2),XX(8),YY(8),PHI(8),
     $            DPHDX(8),DPHDY(8),DPHDXI(8),DPHDET(8),IELNO,NND
      REAL K
      DIMENSION KEFF(300,50)
      REAL KEFF
      EQUIVALENCE (KEFF,K)
      COMMON /EVEQS/  KEIG(6,6),MEIG(6,6)
      REAL KEIG,MEIG
      COMMON /IVEQS/  CK(300,50),ANM1(300),FNM1(300),
     $            FN(300),FO(300),CLUMP(300),DELU(300),
     $            UO(300),KEFFDU(300),KEFFUO(300)
      REAL KEFFDU,KEFFUO
      COMMON /PROBLM/ DIMEN,PROB,TITLE(18)
      COMMON /INVAL/  INTRVL,NSTEPS,NTIME,TIME,DT,THETA,AO
      DIMENSION BDYTAU(3)
      REAL L
      DATA BDVP/'BDVP'/, EIGP/'EIGP'/, XIBVP/'IBVP'/
C
C********** APPLY THE NONZERO NATURAL BOUNDARY CONDITIONS, IF ANY, *******
C              BY ADDING TERMS TO LOAD VECTOR
C
      IF (NUMNBC.EQ.0) GO TO 80
C
      DO 80 N=1,NUMNBC
         II = INP(N)
         JJ = JNP(N)
         KK = KNP(N)
C-------IF JJ.EQ.0 WE HAVE A CONCENTRATED FLUX AT NODE II
         IF (JJ.NE.0) GO TO 20
         F(II) = F(II)+TAUNBC(N)
         GO TO 80
20       CONTINUE
C-------IF KK.EQ.0 WE HAVE A DISTRIBUTED FLUX ON A SIDE
C           OF A 3-NODE TRIANGLE OR A 4-NODE QUADRILATERAL
         IF (KK.NE.0) GO TO 40
         L = SQRT((X(JJ)-X(II))**2+(Y(JJ)-Y(II))**2)
         F(II) = F(II)+TAUNBC(N)*L/2.
         F(JJ) = F(JJ)+TAUNBC(N)*L/2.
         GO TO 80
40       CONTINUE
C-------WE HAVE A DISTRIBUTED FLUX ON A SIDE OF A 6-NODE TRIANGLE
C           OR AN 8-NODE QUADRILATERAL
         XX(1) = X(II)
         XX(2) = X(KK)
         XX(3) = X(KK)
         YY(1) = Y(II)
         YY(2) = Y(II)
         YY(3) = Y(KK)
         CALL BDYFLX(N,XX,YY,BDYTAU)
         F(II) = F(II) + BDYTAU(1)
         F(JJ) = F(JJ) + BDYTAU(2)
         F(KK) = F(KK) + BDYTAU(3)
80    CONTINUE
90    CONTINUE
C
C********** APPLY THE ESSENTIAL BOUNDARY CONDITIONS, IF ANY, **********
C              BY IMPOSING CONSTRAINT EQUATIONS
C
      IF (NUMEBC.LE.0) RETURN
      IF (PROB.EQ.BDVP) GO TO 100
      IF (PROB.EQ.EIGP) GO TO 200
      IF (PROB.EQ.XIBVP.AND.THETA.EQ.0.) GO TO 300
      IF (PROB.EQ.XIBVP.AND.THETA.NE.0.) GO TO 400
C
C********** MODIFY BOUNDARY-VALUE PROBLEM
C
100   CONTINUE
      DO 140 IEB=1,NUMEBC
         I = IUEBC(IEB)
         DO 120 J=2,MBAND
            IR1 = I+(J-1)
            IR2 = I-(J-1)
            IF (IR1.LE.NUMNP) F(IR1)=F(IR1)-K(I,J)*UEBC(IEB)
            IF (IR1.LE.NUMNP) K(I,J)=0.
            IF (IR2.GT.0) F(IR2)=F(IR2)-K(IR2,J)*UEBC(IEB)
            IF (IR2.GT.0) K(IR2,J)=0.
120      CONTINUE
         F(I) = UEBC(IEB)
         K(I,1) = 1.
140   CONTINUE
      RETURN
C
C********** MODIFY EIGENPROBLEM
C
200   CONTINUE
      DO 240 IEB=1,NUMEBC
         J = IUEBC(IEB)
         DO 220 I=1,NUMNP
            KEIG(J,I) = 0.
            KEIG(I,J) = 0.
            MEIG(J,I) = 0.
            MEIG(I,J) = 0.
220      CONTINUE
         KEIG(J,J) = 1.
         MEIG(J,J) = 1.
240   CONTINUE
      RETURN
C
C********** MODIFY EXPLICIT INITIAL-BOUNDARY-VALUE PROBLEM
C
300   CONTINUE
C------CALCULATE STEP CHANGE IN ESSENTIAL BOUNDARY CONDITIONS
C          (FOR RAMPING IN SUBROUTINE CALFUN)
      DO 310 I=1,NUMNP
         DELU(I) = 0.
310   CONTINUE
      RNS = NSTEPS
      DO 320 IEB=1,NUMEBC
         I = IUEBC(IEB)
```

(Continued)

```fortran
      DELU(I) = (UEBC(IEB)-A(I))/RNS
 320  CONTINUE
      RETURN
C
C********* MODIFY IMPLICIT INITIAL-BOUNDARY-VALUE PROBLEM
C
C------- CALCULATE STEP CHANGE IN ESSENTIAL BOUNDARY CONDITIONS
C------- (FOR RAMPING IN SUBROUTINE CALFUN)
      DO 410 I=1,NUMNP
      UO(I) = 0.
      DELU(I) = 0.
      KEFFDU(I) = 0.
      KEFFUO(I) = 0.
 410  CONTINUE
      RNS = NSTEPS
      DO 420 IEB=1,NUMEBC
      I = IUEBC(IEB)
      UO(I) = A(I)
      DELU(I) = (UEBC(IEB)-A(I))/RNS
 420  CONTINUE
      RETURN
C
C------- MODIFY THE EFFECTIVE STIFFNESS MATRIX (KEFF)
      DO 450 IEB=1,NUMEBC
      I = IUEBC(IEB)
      DO 440 J=2,MBAND
      IR1 = I+(J-1)
      IR2 = I-(J-1)
      IF (IR1.LE.NUMNP) KEFFDU(IR1)=KEFFDU(IR1)+
     $                  KEFF(I,J)*DELU(I)
      IF (IR1.LE.NUMNP) KEFFUO(IR1)=KEFFUO(IR1)+
     $                  KEFF(I,J)*UO(I)
      IF (IR1.LE.NUMNP) KEFF(I,J)=0.
      IF (IR2.GT.0) KEFFDU(IR2)=KEFFDU(IR2)+DELU(IR2)
      IF (IR2.GT.0) KEFFUO(IR2)=KEFFUO(IR2)*UO(IR2)
      IF (IR2.GT.0) KEFF(IR2,J)=0.
 440  CONTINUE
      KEFF(I,1) = 1.
 450  CONTINUE
      RETURN
C
      END
      SUBROUTINE BDYFLX(N,XX,YY,BDYTAU)
C
C------- THIS ROUTINE CALCULATES THE BOUNDARY FLUX INTEGRAL FOR
C------- A DISTRIBUTED FLUX ON THE BOUNDARY OF A 6-NODE
C------- TRIANGLE OR AN 8-NODE QUADRILATERAL, USING A
C------- 3-POINT GAUSSIAN QUADRATURE FORMULA.  THE
C------- FLUX IS ASSUMED CONSTANT.
C
      COMMON /NATBC/ INP(100),JNP(100),KNP(100),TAUNBC(100)
      COMMON /GLINT/ GP(3,5),WT(3,5)
      DIMENSION XX(3),YY(3),BDYTAU(3),PHI(3)
C
      DO 20 I=1,3
      BDYTAU(I) = 0.
 20   CONTINUE
      DO 50 IPT=1,3
      XI = GP(3,IPT)
      PHI(1) = -XI*(1.-XI)/2.
      PHI(2) = 1.-XI*XI
      PHI(3) = XI*(1.+XI)/2.
      DPH1 = XI - 1./2.
      DPH2 = -2*XI
      DPH3 = XI + 1./2.
      BDYJAC = SQRT((DPH1*XX(1)+DPH2*XX(2)+DPH3*XX(3))**2 +
     $              (DPH1*YY(1)+DPH2*YY(2)+DPH3*YY(3))**2)
      DO 40 I=1,3
      BDYTAU(I) = BDYTAU(I) + WT(3,IPT)*(TAUNBC(N)*PHI(I)*BDYJAC)
 40   CONTINUE
 50   CONTINUE
      RETURN
      END
      SUBROUTINE CALFUN
C
C------- THIS ROUTINE SOLVES THE SYSTEM EQUATIONS FOR THE FUNCTION VALUES
C
      COMMON /PROBLM/ DIMEN,PROB,TITLE(18)
      COMMON /SYSEQS/ K(300,50),F(300),A(300)
      REAL K
      DIMENSION KEFF(300,50),FEFF(300),FNS(300),AN(300)
      REAL KEFF
      EQUIVALENCE (KEFF,K),(FNS,F),(AN,A)
      COMMON /ESSBC/ IUEBC(100),UEBC(100)
      COMMON /GENRL/ NUMNP,NUMEL,NUMPP,NUMELC,NUMEBC,NUMBC,MBAND
      COMMON /EVEQS/ KEIG(61),MEIG(61,61)
      REAL KEIG,MEIG
      COMMON /EIGVEC/ XVEC(61,61),EIGV(61)
      COMMON /IVEQS/ CK(300,50),ANM1(300),FNM1(300),
     $               FN(300),FO(300),CLUMP(300),DELU(300),
     $               UO(300),KEFFDU(300),KEFFUO(300)
      REAL KEFFDU, KEFFUO
      COMMON /INVAL/ INTRVL,NSTEPS,NTIME,TIME,DT,THETA,AO
      COMMON /TTMPLT/ TPLOT(501),FTPLOT(3,501),NNODEI(3)
      DIMENSION D(61)
      DATA BDVP/'BDVP'/, EIGP/'EIGP'/, XIBVP/'XIBVP'/
C
C------- BRANCH TO PROPER EQUATION SOLVING METHOD
      IF (PROB.EQ.BDVP) GO TO 100
      IF (PROB.EQ.EIGP) GO TO 200
      IF (PROB.EQ.XIBVP.AND.THETA.EQ.0.) GO TO 300
      IF (PROB.EQ.XIBVP.AND.THETA.NE.0.) GO TO 400
C
C******** BOUNDARY-VALUE PROBLEM
C
 100  CONTINUE
      CALL GAUSEL(NUMNP,MBAND,K,F,A,0)
      RETURN
C
C******** EIGENPROBLEM
C
 200  CONTINUE
C------- FOR THIS EIGEN ROUTINE, WE MUST FILL UP REMAINING
C------- DIAGONALS OF KEIG AND MEIG WITH 1'S.
      NSIZE = NUMNP
      N1 = NSIZE+1
      IF (N1.GT.61) GO TO 260
      DO 250 I=N1,61
      KEIG(I,1) = 1.
      MEIG(I,1) = 1.
 250  CONTINUE
 260  CONTINUE
      NLIM = 61
      RTOL = 1.E-12
      NSMAX = 15
      IFPR = 0
      IOUT = 6
      CALL GENJAC(KEIG,MEIG,XVEC,EIGV,D,NLIM,RTOL,NSMAX,IFPR,IOUT)
      RETURN
C
C******** EXPLICIT INITIAL-BOUNDARY-VALUE PROBLEM
C
 300  CONTINUE
C------- DEFINE SOLUTION VECTOR AT BEGINNING OF INTERVAL
      DO 310 I=1,NUMNP
      ANM1(I) = AN(I)
 310  CONTINUE
C------- SOLVE RECURRENCE RELATION OVER ALL TIME STEPS
      RNS = NSTEPS
      DO 380 N = 1, NSTEPS
      TIME = TIME+DT
      RNM1 = N-1
      DO 350 I=1,NUMNP
      KA = K(I,1)*ANM1(I)
      DO 340 J=2,MBAND
      IR1 = I+(J-1)
      IR2 = I-(J-1)
      IF (IR1.LE.NUMNP) KA=KA+K(I,J)*ANM1(IR1)
      IF (IR2.GT.0) KA=KA+K(IR2,J)*ANM1(IR2)
 340  CONTINUE
C------- CALCULATE NAT +INT LOADS AT STEP N-1 BY RAMPING
      FNM1(I) = FO(I)+(RNM1/RNS)*(FNS(I)-FO(I))
C------- SOLVE FOR FUNCTION AT STEP N (IGNORE ESSENTIAL BC)
      IF (CLUMP(I).EQ.0.) CALL ERROR(33,I)
      AN(I) = ANM1(I)+(DT/CLUMP(I))*(FNM1(I)-KA)
 350  CONTINUE
C------- MODIFY FUNCTION FOR RAMPED ESSENTIAL BC
      DO 360 IEB=1,NUMEBC
      I = IUEBC(IEB)
      AN(I) = ANM1(I)+DELU(I)
 360  CONTINUE
C------- SAVE FUNCTION VALUES FOR NEXT TIME STEP
      DO 370 I=1,NUMNP
      ANM1(I) = AN(I)
      IF (NTIME.GE.501) GO TO 380
      NTIME = NTIME+1
C------- SAVE TIME AND FUNCTION VALUES FOR TIME PLOTS
      TPLOT(NTIME) = TIME
      NNDE = NNODEI(L)
      IF (NNDE.NE.0) FTPLOT(L,NTIME) = AN(NNODE)
 370  CONTINUE
 375  CONTINUE
 380  CONTINUE
C------- SAVE NAT + INT LOAD AT END OF INTERVAL
      DO 390 I=1,NUMNP
      FO(I) = FNS(I)
 390  CONTINUE
      RETURN
C
C******** IMPLICIT INITIAL-BOUNDARY-VALUE PROBLEM
C
 400  CONTINUE
C------- SAVE LOAD VECTOR AND FUNCTION AT START OF INTERVAL
      DO 410 I=1,NUMNP
      FNM1(I) = FO(I)
      ANM1(I) = A(I)
 410  CONTINUE
      ONEMTH = 1.-THETA
      RNS = NSTEPS
C------- FORWARD REDUCTION OF EFFECTIVE STIFFNESS MATRIX (KEFF)
      CALL GAUSEL(NUMNP,MBAND,KEFF,FEFF,AN,1)
C------- SOLVE RECURRENCE RELATIONS OVER ALL TIME STEPS
      DO 480 N=1,NSTEPS
      TIME = TIME+DT
```

```
      RN = N
C------- CALCULATE NAT + INT LOADS AT STEP N BY RAMPING
      DO 420 I=1,NUMNP
      FN(I) = FO(I)+(RN/RNS)*(FNS(I)-FO(I))
  420 CONTINUE
C------- CALCULATE EFFECTIVE LOAD VECTOR (FEFF) AT STEP N
      DO 440 I=1,NUMNP
      CKA = CK(I,1)*ANM1(I)
      DO 430 J=2,MBAND
      IR1 = I+(J-1)
      IR2 = I-(J-1)
      IF (IR1.LE.NUMNP)CKA=CKA+CK(I,J)*ANM1(IR1)
      IF (IR2.GT.O) CKA=CKA+CK(IR2,J)*ANM1(IR2)
  430 CONTINUE
      FEFF(I) = ONEMTH*FNM1(I)+THETA*FN(I)+CKA
  440 CONTINUE
C------- MODIFY EFFECTIVE LOAD VECTOR FOR RAMPED ESSENTIAL BC
      DO 445 I=1,NUMNP
      FEFF(I) = FEFF(I)-KEFFUO(I)*RN+KEFFDU(I)
  445 CONTINUE
      DO 450 IEB=1,NUMEBC
      I = IUEBC(IEB)
      FEFF(I) = ANM1(I)+DELU(I)
  450 CONTINUE
C------- REDUCTION OF LOAD VECTOR: BACKSUBSTITUTION
      CALL GAUSL2(NUMNP,MBAND,KEFF,FEFF,AN,O)
C------- SAVE VALUES FOR NEXT STEP
      DO 460 I=1,NUMNP
      ANM1(I) = AN(I)
      FNM1(I) = FN(I)
  460 CONTINUE
C------- SAVE TIME AND FUNCTION VALUES FOR TIME PLOTS
      IF (NTIME.GE.50)) GO TO 480
      NTIME = NTIME+1
      TPLOT(NTIME) = TIME
      DO 470 L=1,3
      NNDE = NNODEI(L)
      IF (NNDE.NE.O) FTPLOT(L,NTIME) = AN(NNDE)
  470 CONTINUE
  480 CONTINUE
C------- SAVE NAT + INT LOADS AT END OF INTERVAL
      DO 490 I=1,NUMNP
      FO(I) = FNS(I)
  490 CONTINUE
      RETURN
C-------
      END
      SUBROUTINE GAUSEL(NSIZE,MBAND,K,F,A,ICONT)
C
C     THIS ROUTINE SOLVES THE LINEAR ALGEBRAIC SYSTEM OF EQUATIONS
C     K*A=F BY GAUSSIAN ELIMINATION WITHOUT PIVOTING.  THE MATRIX K
C     IS SYMMETRIC, POSITIVE-DEFINITE, AND BANDED; ROWS IN THE UPPER
C     TRIANGLE ARE SHIFTED TO THE LEFT UNTIL DIAGONAL TERMS ARE IN
C     THE FIRST COLUMN. SOLUTION IS RETURNED IN VECTOR A.
C     ADDITIONAL RIGHT HAND SIDES MAY BE SOLVED USING THE SAME
C     REDUCED STIFFNESS MATRIX K BY ENTERING SUBROUTINE AT 'ENTRY GAUSL2'.
C
C-------INPUT VARIABLES
C     NSIZE - ORDER OF MATRIX K
C     MBAND - HALF BANDWIDTH OF K (INCLUDING DIAGONAL)
C     K(NSIZE,MBAND) - EFFECTIVE STIFFNESS MATRIX
C     F(NSIZE) - EFFECTIVE LOAD VECTOR
C     ICONT - FLAG TO CONTROL SOLUTION
C        EQ O - DO COMPLETE SOLUTION
C        EQ 1 - DO FORWARD REDUCTION OF K ONLY
C-------OUTPUT VARIABLES
```

```
C     A(NSIZE) - SOLUTION VECTOR
C
      DIMENSION K(300,50),F(300),A(300)
      REAL K
C-------
      NSIZM1 = NSIZE-1
C-------- FORWARD REDUCTION OF MATRIX
      DO 30 N=1,NSIZM1
      DO 20 L=2,MBAND
      IF(K(N,L).EQ.O..OR.I.GT.NSIZE) GO TO 20
      C = K(N,L)/K(N,1)
      J = N+L-1
      DO 10 M=L,MBAND
      K(I,J) = K(I,J)-C*K(N,M)
   10 CONTINUE
      K(N,L) = C
   20 CONTINUE
   30 CONTINUE
      IF (ICONT.EQ.1) RETURN
C-------- FORWARD REDUCTION OF RIGHT HAND SIDE
      ENTRY GAUSL2(NSIZE,MBAND,K,F,A,ICONT)
      NSIZM1 = NSIZE-1
      DO 60 N=1,NSIZM1
      I = N+L-1
      IF (I.LE.NSIZE) F(I) = F(I)-K(N,L)*F(N)
   50 CONTINUE
      F(N) = F(N)/K(NSIZE,1)
   60 CONTINUE
      F(NSIZE) = F(NSIZE)/K(NSIZE,1)
C-------- BACK SUBSTITUTION
      DO 90 N=1,NSIZM1
      J = NSIZE-N
      A(J) = F(J)
      DO 80 L=2,MBAND
      M=J+L-1
      IF(M.LE.NSIZE) A(J) = A(J)-K(J,L)*A(M)
   80 CONTINUE
   90 CONTINUE
      RETURN
C-------
      END
      SUBROUTINE GENJAC (A,B,X,EIGV,D,N,RTOL,NSMAX,IFPR,IOUT)
C
C     REFERENCE: BATHE AND WILSON, "NUMERICAL METHODS IN FINITE
C                ELEMENT ANALYSIS", PP 458-460.
C                TITLE OF SUBROUTINE-JACOBI
C
C     P R O G R A M
C     TO SOLVE THE GENERALIZED EIGENPROBLEM USING THE
C     GENERALIZED JACOBI ITERATION
C
C------- INPUT VARIABLES
C     A (N,N)  - STIFFNESS MATRIX (ASSUMED POSITIVE DEFINITE)
C     B (N,N)  - MASS MATRIX (ASSUMED POSITIVE DEFINITE)
C     X (N,N)  - MATRIX STORING EIGENVECTORS ON SOLUTION EXIT
C     EIGV (N) - VECTOR STORING EIGENVALUES ON SOLUTION EXIT
C     D (N)    - WORKING VECTOR
C     N        - ORDER OF MATRICES A AND B
C     RTOL     - CONVERGENCE TOLERANCE (USUALLY SET TO 1O **-12)
C     NSMAX    - MAXIMUM NUMBER OF SWEEPS ALLOWED
```

```
C     IFPR   = FLAG FOR PRINTING DURING ITERATION
C                         (USUALLY SET TO 15)
C        EQ O    NO PRINTING
C        EQ 1    INTERMEDIATE RESULTS ARE PRINTED
C     IOUT   = OUTPUT DEVICE NUMBER
C
C-- OUTPUT --
C     A (N,N)  - DIAGONALIZED STIFFNESS MATRIX
C     B (N,N)  - DIAGONALIZED MASS MATRIX
C     X (N,N)  - EIGENVECTORS STORED COLUMNWISE
C     EIGV (N) - EIGENVALUES
C
C-------
      IMPLICIT REAL*8(A-H,O-Z)
      ABS(X)=DABS(X)
      SQRT(X)=DSQRT(X)
C-------
C     THIS PROGRAM IS USED IN SINGLE PRECISION ARITHMETIC ON
C     CDC EQUIPMENT AND DOUBLE PRECISION ARITHMETIC ON IBM
C     OR UNIVAC MACHINES. ACTIVATE, DEACTIVATE, OR ADJUST ABOVE
C     CARDS FOR SINGLE OR DOUBLE PRECISION ARITHMETIC
C-------
      DIMENSION A(N,N),B(N,N),X(N,N),EIGV(N),D(N)
C
C     INITIALIZE EIGENVALUE AND EIGENVECTOR MATRICES
C
      DO 10 I=1,N
      IF( B(I,I).GT.O.) GO TO 4
      II = 1
      WRITE (IOUT,2020) II
      STOP
    4 D(I)=A(I,I)/B(I,I)
   10 EIGV(I)=D(I)
      DO 30 I=1,N
      DO 20 J=1,N
   20 X(I,J)=O.
   30 X(I,I)=1.0
      IF(N.EQ.1) RETURN
C
C     INITIALIZE SWEEP COUNTER AND BEGIN ITERATION
C
      NSWEEP=O
      NR=N-1
   40 NSWEEP=NSWEEP+1
      IF(IFPR.EQ.1) WRITE(IOUT,2000)NSWEEP
C
C     CHECK IF PRESENT OFF-DIAGONAL EL IS LARGE ENOUGH TO REQ ZEROING
C
      EPS=(.O1**NSWEEP)**2
      DO 210 J=1,NR
      JJ=J+1
      DO 210 K=JJ,N
      EPTOLA=(A(J,K)*A(J,K))/(A(J,J)*A(K,K))
      EPTOLB=(B(J,K)*B(J,K))/(B(J,J)*B(K,K))
      IF ( (EPTOLA.LT.EPS*A(J,J)*A(K,K)) .AND.
     1     (EPTOLB.LT.EPS*B(J,J)*B(K,K)) ) GO TO 210
C
C     IF ZEROING IS REQ, CAL THE ROTATION MATRIX ELEMENTS CA AND CG
C
      AKK=A(K,K)*B(J,K)-B(K,K)*A(J,K)
      AJJ=A(J,J)*B(J,K)-B(J,J)*A(J,K)
      AB=A(J,J)*B(K,K)-A(K,K)*B(J,J)
      CHECK=(AB*AB+4.*AKK*AJJ)/4.
      IF(CHECK)50,60,60
      II = 2
   50 WRITE(IOUT,2020) II
```

(Continued)

```
      STOP
   60 SQCH=SQRT(CHECK)
      D1=AB/2.+SQCH
      D2=AB/2.-SQCH
      DEN=D1
      IF(ABS(D2).GT.ABS(D1))DEN=D2
      IF(DEN)80,70,80
   70 CA=0.
      CG=-B(J,K)/B(K,K)
      GO TO 90
   80 CA=AKK/DEN
      CG=-AJJ/DEN
C
C     PERFORM THE GEN ROTATION TO ZERO THE PRESENT OFF-DIAGONAL ELEM
C
   90 IF(N-2)100,190,100
  100 JP1=J+1
      JM1=J-1
      KP1=K+1
      KM1=K-1
      IF(JM1-1)130,110,110
  110 DO 120 I=1,JM1
      AJ=A(I,J)
      BJ=B(I,J)
      AK=A(I,K)
      BK=B(I,K)
      A(I,J)=AJ+CG*AK
      B(I,J)=BJ+CG*BK
      A(I,K)=AK+CA*AJ
  120 B(I,K)=BK+CA*BJ
  130 IF(KP1-N)140,140,160
  140 DO 150 I=KP1,N
      AJ=A(J,I)
      BJ=B(J,I)
      AK=A(K,I)
      BK=B(K,I)
      A(J,I)=AJ+CG*AK
      B(J,I)=BJ+CG*BK
      A(K,I)=AK+CA*AJ
  150 B(K,I)=BK+CA*BJ
  160 IF(JP1-KM1)170,190,190
  170 DO 180 I=JP1,KM1
      AJ=A(J,I)
      BJ=B(J,I)
      AK=A(I,K)
      BK=B(I,K)
      A(J,I)=AJ+CG*AK
      B(J,I)=BJ+CG*BK
      A(I,K)=AK+CA*AJ
  180 B(I,K)=BK+CA*BJ
  190 AK=A(K,K)
      BK=B(K,K)
C
C     UPDATE THE EIGENVECTOR MATRIX AFTER EACH ROTATION
C
  200 DO 200 I=1,N
      XJ=X(I,J)
      XK=X(I,K)
      X(I,J)=XJ+CG*XK
  200 X(I,K)=XK+CA*XJ

  210 CONTINUE
C
C     UPDATE THE EIGENVALUES AFTER EACH SWEEP
C
      DO 220 I=1,N
      IF(B(I,I).GT.0.) GO TO 220
      II = 3
      WRITE(IOUT,2020) II
      STOP
  220 EIG(I)=A(I,I)/B(I,I)
      IF(IFPR.EQ.0)GO TO 290
      WRITE(IOUT,2030)
      WRITE(IOUT,2010)(EIG(I),I=1,N)
C
C     CHECK FOR CONVERGENCE
C
  230 DO 240 I=1,N
      TOL=RTOL*D(I)
      DIF=ABS(EIG(I)-D(I))
      IF(DIF.GT.TOL)GO TO 280
  240 CONTINUE
C
C     CHECK ALL OFF-DIAGONAL ELS TO SEE IF ANOTHER SWEEP IS REQUIRED
C
      EPS=RTOL**2
      DO 250 J=1,NR
      JJ=J+1
      DO 250 K=JJ,N
      EPSA=(A(J,K)*A(J,K))/(A(J,J)*A(K,K))
      EPSB=(B(J,K)*B(J,K))/(B(J,J)*B(K,K))
      IF ( (EPSA.LT.EPS*A(J,J)*A(K,K)) .AND.
     $     (EPSB.LT.EPS*B(J,J)*B(K,K)) ) GO TO 280
      GO TO 280
  280 CONTINUE
C
C     FILL OUT BOTTOM TRI OF RESULTANT MATRICES AND SCALE EIGENVECTORS
C
  255 DO 260 I=1,N
      DO 260 J=1,N
      A(J,I)=A(I,J)
  260 B(J,I)=B(I,J)
      DO 270 J=1,N
      BB=SQRT(B(J,J))
      DO 270 K=1,N
  270 X(K,J)=X(K,J)/BB
      RETURN
C
C     UPDATE  D  MATRIX AND START NEW SWEEP, IF ALLOWED
C
  280 DO 290 I=1,N
  290 D(I)=EIG(I)
      IF(NSWEEP.LT.NSMAX)GO TO 40
      GO TO 255
 2000 FORMAT(27HOSWEEP NUMBER IN *JACOBI* =,I4)
 2010 FORMAT(1H0,6E20.12)
 2020 FORMAT(3H1I*,I2/25HO*** ERROR  SOLUTION STOP  /
     $     31H MATRICES NOT POSITIVE DEFINITE )
 2030 FORMAT(36HOCURRENT EIGENVALUES IN *JACOBI* ARE /)
      END
      SUBROUTINE CALFLX
C
C     THIS ROUTINE CONTROLS THE CALCULATION OF FLUXES IN EACH
C     ELEMENT, AND CALCULATES THE NODAL FLUXES.
C
C     CALCULATE FLUXES AT OPTIMAL FLUX SAMPLING POINTS (GAUSS PNTS)
C     XGP(J,IELNO) - X COORDINATE AT GAUSS PT J IN EL NO. IELNO
C     YGP(J,IELNO) - Y COORDINATE AT GAUSS PT J IN EL NO. IELNO
C     TAUXGP(J,IELNO) - X FLUX AT GAUSS PT J IN EL NO. IELNO
C     TAUYGP(J,IELNO) - Y FLUX AT GAUSS PT J IN EL NO. IELNO
C
C     CALCULATE FLUXES AT NODES
C     TAUXNP(I) - X FLUX AT NODE I
C     TAUYNP(I) - Y FLUX AT NODE I
C
      COMMON /ELMNTS/ NTYPE(200),NGPINT(200),NPHYS(200),ICON(8,200)
      COMMON /SYSEQS/ K(300,50),F(300),A(300)
      REAL K
      COMMON /ELSPEC/ NNODES(30),INTDEF(30),IOPTAU(30)
      COMMON /GENRL/ NUMNP,NUMEL,NUMPP,NUMILC,NUMEBC,NUMNBC,MBAND
      COMMON /LIMITS/ MAXNP,MAXEL,MAXPP,MAXIL,MAXEBC,MAXNBC,MAXBND
      COMMON /NODSOL/ IU(300),TAUXNP(300),TAUYNP(300)
      COMMON /ELFLUX/ XGP(4,200),YGP(4,200),TAUXGP(4,200),TAUYGP(4,200)
      COMMON /ELVAR/  XI,ETA,JAC,JACINV(2,2),XX(8),YY(8),PHI(8),
     $                DPHDX(8),DPHDY(8),DPHDXI(8),DPHDET(8),IELNO,NND
      REAL JACINV,JAC
      COMMON /PROBLM/ DIMEN,PROB,TITLE(18)
      DATA EIGP / EIGP /
      DATA FLUX / FLUX /
C
      IF (PROB.EQ.EIGP) RETURN
C     --INITIALIZE ARRAYS
      DO 40 IELNO=1,NUMEL
      DO 20 J=1,4
      XGP(J,IELNO) = 0.
      YGP(J,IELNO) = 0.
      TAUXGP(J,IELNO) = 0.
      TAUYGP(J,IELNO) = 0.
   20 CONTINUE
      DO 60 I=1,NUMNP
      TAUXNP(I) = 0.
      TAUYNP(I) = 0.
      IU(I) = 0
   60 CONTINUE
C     --CALCULATE FLUXES AT GAUSS POINTS AND NODES IN EACH ELEMENT
      DO 100 IELNO=1,NUMEL
      ITYP = NTYPE(IELNO)
      NND = NNODES(ITYP)
      IF (ITYP.EQ.11) CALL ELEM11(FLUX)
      IF (ITYP.EQ.12) CALL ELEM12(FLUX)
      IF (ITYP.EQ.13) CALL ELEM13(FLUX)
      IF (ITYP.EQ.14) CALL ELEM14(FLUX)
      IF (ITYP.EQ.21) CALL ELEM21(FLUX)
      IF (ITYP.EQ.22) CALL ELEM22(FLUX)
      IF (ITYP.EQ.23) CALL ELEM23(FLUX)
      IF (ITYP.EQ.24) CALL ELEM24(FLUX)
  100 CONTINUE
C     --CALCULATE NODAL FLUXES BY AVERAGING THE SUMMED ELEMENT
C       NODAL FLUXES FROM ELEMENT SUBROUTINES
      DO 150 I=1,NUMNP
      AIU = IU(I)
      IF (IU(I).GT.0) TAUXNP(I) = TAUXNP(I)/AIU
      IF (IU(I).GT.0) TAUYNP(I) = TAUYNP(I)/AIU
  150 CONTINUE
      RETURN
C
      END
      SUBROUTINE PRINTS
C
C     THIS ROUTINE CALLS THE ROUTINES WHICH OUTPUT THE RESULTS
C
      COMMON /PROBLM/ DIMEN,PROB,TITLE(18)
```

(Continued)

```
      DATA BDVP/'BDVP'/,EIGP/'EIGP'/, XIBVP/'IBVP'/
C
      WRITE (6,1900)
C------FOR BOUNDARY-VALUE PROBLEMS
      IF (PROB.EQ.BDVP) CALL BIVPRI
C------FOR EIGENPROBLEMS
      IF (PROB.EQ.EIGP) CALL EIGPRI
C------FOR INITIAL VALUE PROBLEMS
      IF (PROB.EQ.XIBVP) CALL BIVPRI
      RETURN
C
 1900 FORMAT(///,' *****SOLUTION COMPLETE.  BEGIN POSTPROCESSING. ')
      END
      SUBROUTINE BIVPRI
C
C     THIS ROUTINE PRINTS THE OUTPUT IN TABULAR FORM FOR FOR
C        THE BOUNDARY-VALUE PROBLEM AND THE
C        INITIAL-BOUNDARY-VALUE PROBLEM
C
      COMMON /NODES/ X(300),Y(300)
      COMMON /GENRL/ NUMNP,NUMEL,NUMPP,NUMILC,NUMEBC,NUMNBC,MBAND
      COMMON /ELMNTS/ NTYPE(200),NGPINT(200),NPHYS(200),ICDN(8,200)
      COMMON /ELSPEC/ NNODES(30),INTDEF(30),IDPTAU(30)
      COMMON /GENRL/ NUMNP,NUMEL,NUMPP,NUMILC,NUMEBC,NUMNBC,MBAND
      REAL K
      COMMON /SYSEOS/ K(300,50),F(300),A(300)
      COMMON /INVAL/ INTRVL,NSTEPS,NTIME,TIME,DT,THETA,AO
      COMMON /NODSOL/ IU(300),TAUXNP(300),TAUYNP(300)
      COMMON /ELFLUX/ XGP(4,200),YGP(4,200),TAUXGP(4,200),TAUYGP(4,200)
      COMMON /PROBLM/ DIMEN,PROB,TITLE(18)
      DATA BDVP/'BDVP'/,EIGP/'EIGP'/, XIBVP/'IBVP'/
      DATA ONED/'1DIM'/, TWOD/'2DIM'/
C------WRITE ELEMENT OUTPUT
C
      WRITE (6,2000)
      WRITE (6,2100) DIMEN,PROB,TITLE
      IF (PROB.EQ.XIBVP) WRITE (6,2200) INTRVL,TIME
      IF (DIMEN.EQ.ONED) WRITE(6,2310)
      IF (DIMEN.EQ.TWOD) WRITE(6,2320)
      DO 200 IELNO=1,NUMEL
      NND = NNODES(ITYP)
      ITYP = NTYPE(IELNO)
      NGPTAU = IDPTAU(ITYP)
      IF (DIMEN.EQ.ONED) WRITE(6,2311) IELNO,
     $      (I,XGP(I,IELNO),TAUXGP(I,IELNO),I=1,NGPTAU)
      IF (DIMEN.EQ.TWOD) WRITE(6,2321)
     $      IELNO,(I,XGP(I,IELNO),YGP(I,IELNO),
     $      TAUXGP(I,IELNO),TAUYGP(I,IELNO),I=1,NGPTAU)
  200 CONTINUE
C------WRITE NODAL OUTPUT
C
      WRITE (6,2000)
      WRITE (6,2100) DIMEN,PROB,TITLE
      IF (PROB.EQ.XIBVP) WRITE (6,2200) INTRVL,TIME
      IF (DIMEN.EQ.ONED) WRITE (6,2410)
      IF (DIMEN.EQ.TWOD) WRITE (6,2420)
      DO 300 I=1,NUMNP
      IF (DIMEN.EQ.ONED) WRITE(6,2510) I,X(I),A(I),TAUXNP(I)
      IF (DIMEN.EQ.TWOD) WRITE(6,2520) I,X(I),Y(I),A(I),TAUXNP(I),
     $      TAUYNP(I)
  300 CONTINUE
      RETURN
C------UNAFEM-UNDERSTANDING AND APPLYING THE FEM /
C
 2000 FORMAT(1H ,20A4)
     $      ')
 2100 FORMAT(//,'  .20A4)
 2200 FORMAT(//,' OUTPUT FOR INTERVAL NUMBER ',I3,'.  TIME = ',1PE10.3)


 2310 FORMAT(//,' ELEMENT OUTPUT'//
     $   ' ELEM.',3X,'GAUSS PT.',6X,'X',3X,3X,'-ALPHA*DU/DX'/
 2320 FORMAT(//,' ELEMENT OUTPUT'//
     $   ' ELEM.',3X,'GAUSS PT.',6X,'X',3X,6X,'Y',3X,2X,'-ALPHAX*DU/DX'.
     $   2X,'-ALPHAY*DU/DY'/
 2311 FORMAT(/14,5X,14,3X,F10.4,E15.4/(9X,14,3X,F10.4,E15.4))
 2321 FORMAT(/14,5X,14,3X,2F10.4,2E15.4/(9X,14,3X,2F10.4,2E15.4))
 2410 FORMAT(//,' NODAL OUTPUT'//
     $   ' NODE.,7X,'X',12X,'U',3X,'-ALPHA*DU/DX'/)
 2420 FORMAT(//,' NODAL OUTPUT'//
     $   ' NODE.,6X,'X',9X,'Y',8X,'U',8X,'-ALPHAX*DU/DX  -ALPHAY*DU/DY'/)
 2510 FORMAT(14,2X,F10.4,2E13.4)
 2520 FORMAT(14,1X,2F10.4,E13.4,2E14.4)
      END
      SUBROUTINE EIGPRI
C
C     THIS ROUTINE PRINTS THE EIGENPROBLEM RESULTS
C
      COMMON /PROBLM/ DIMEN,PROB,TITLE(18)
      COMMON /GENRL/ NUMNP,NUMEL,NUMPP,NUMILC,NUMEBC,NUMNBC,MBAND
      COMMON /ESSBC/ IUEBC(100),UEBC(100)
      COMMON /EIGVEC/ XVEC(61,61),EIGV(61)
      COMMON /SMLEIG/ EIGSML(5),VECSML(61,5),NUMEIG
      DIMENSION IBCLST(61),LSTSML(5)
C
C     SEARCH TO FIND 5 SMALLEST EIGENVALUES
C
      NLIM = NUMNP
C------CREATE LIST OF ESSENTIAL BC IN NODE ORDER
      DO 20 I=1,NLIM
         IBCLST(I) = 0
   20 CONTINUE
      DO 40 I=1,NUMEBC
         INODE = IUEBC(I)
         IBCLST(INODE) = 1
   40 CONTINUE
C------SEARCH EIGENVALUES FOR 5 SMALLEST
      NUMEIG = NLIM-NUMEBC
      IF (NUMEIG.GT.5) NUMEIG=5
      DO 100 J=1,NUMEIG
         ESM = 1.0E20
         ISML = 0
         DO 80 I=1,NLIM
            IF (EIGV(I).LE.ESM .AND. IBCLST(I).NE.1)ESM=EIGV(I)
            IF (EIGV(I).LE.ESM .AND. IBCLST(I).NE.1)ISML=I
   80    CONTINUE
         LSTSML(J) = ISML
         EIGSML(J) = EIGV(ISML)
         EIGV(ISML) = 1.0E20
   90    CONTINUE
         DO 90 I=1,NLIM
            VECSML(I,J) = XVEC(I,ISML)
   90    CONTINUE
  100 CONTINUE
C------WRITE OUT EIGENVALUES AND EIGENVECTORS
      WRITE (6,2000)
      WRITE (6,2100) DIMEN,PROB,TITLE
      WRITE (6,2200) (J,J=1,NUMEIG)
      WRITE (6,2300) (EIGSML(J),J=1,NUMEIG)
      WRITE (6,2400) (J,J=1,NUMEIG)
      DO 110 I=1,NLIM
         WRITE (6,2500) I,(VECSML(I,J),J=1,NUMEIG)
  110 CONTINUE
      RETURN
C------UNAFEM-UNDERSTANDING AND APPLYING THE FEM /
 2000 FORMAT(1H ,20A4)


 2100 FORMAT(//,'  .20A4)
 2200 FORMAT(//,' EIGENPROBLEM-LOWEST 5 EIGENVALUES & EIGENVECTORS'//
     $   20X,'EIGENVALUES'//14X,I1,11X,I1,11X,I1,11X,I1,11X,I1)
 2300 FORMAT(5X,5E12.4)
 2400 FORMAT(//20X,'EIGENVECTORS'//
     $   ' NODE.,9X,I1,11X,I1,11X,I1,11X,I1,11X,I1)
 2500 FORMAT(14,1X,5E12.4)
      END
      SUBROUTINE PLOTS
C
C     THIS ROUTINE CALLS THE SPECIFIC PLOT ROUTINES
C
      COMMON /KEYWRD/ AKEY,PLTKEY
      COMMON /PROBLM/ DIMEN,PROB,TITLE(18)
      DATA BDVP/'BDVP'/,EIGP/'EIGP'/, XIBVP/'IBVP'/
      DATA XNDPLT/'NPLO'/
C
      WRITE (6,2000)
      IF (PLTKEY.EQ.XNDPLT) RETURN
C------PLOT BOUNDARY-VALUE PROBLEMS
      IF (PROB.EQ.BDVP) CALL BIVPLO
C------PLOT EIGENPROBLEMS
      IF (PROB.EQ.EIGP) CALL EIGPLO
C------PLOT INITIAL BOUNDARY-VALUE PROBLEMS
      IF (PROB.EQ.XIBVP) CALL BIVPLO
C
      RETURN
C
 2000 FORMAT(' UNAFEM-UNDERSTANDING AND APPLYING THE FEM')
      END
      SUBROUTINE EIGPLO
C
C     THIS ROUTINE PLOTS EIGENVECTORS
C
      COMMON /GENRL/ NUMNP,NUMEL,NUMPP,NUMILC,NUMEBC,NUMNBC,MBAND
      COMMON /SMLEIG/ EIGSML(5),VECSML(61,5),NUMEIG
      COMMON /NODES/ X(300),Y(300)
      COMMON /SYSEOS/ K(300,50),F(300),A(300)
      REAL K
      COMMON /NODSOL/ IU(300),TAUXNP(300),TAUYNP(300)
      COMMON /CENTRO/ UC(200),XC(200),YC(200),TAUXC(200),TAUYC(200)
      COMMON /ELMNTS/ NTYPE(200),NGPINT(200),NPHYS(200),ICDN(8,200)
      COMMON /ELSPEC/ NNODES(30),INTDEF(30),IDPTAU(30)
      COMMON /GRAFIC/ XMIN,XXMAX,NDIVX,YMIN,YYMAX,NDIVY1,
     $      YYMIN,YYMAX2,NDIVY2,TTMIN,TTMAX,NDIVT
      COMMON /PROBLM/ DIMEN,PROB,TITLE(18)
      DIMENSION XP(5),YP(5),LABEL1(11),LABEL2(11),LABEL3(11).
     $      XQ(2),YQ(2),CONTR(11),FCENT(8,4)
      DATA FCENT/3*.333333,5*0.,3*-.11111,3*.444444,2*0.,4*.25.
     $      4*0.,4*-.25,4*.5/
      DATA ONED/'1DIM'/, TWOD/'2DIM'/
      DATA EEND/'END '/,EEND2/'END2'/
C
      IF (DIMEN.EQ.TWOD) GO TO 200
C------PLOT 1-D EIGENVECTORS
      NLIM = NUMNP
      CALL MAXMIN(NUMNP,X,XMAX,XMIN,JMAX,JMIN)
      TL = XMAX-XMIN
      XDIV = NDIVX
      IF (XXMIN.LT.XXMAX) XMIN=XXMIN
      IF (XXMIN.LT.XXMAX) XMAX=XXMAX
      XSTP = (XMAX-XMIN)/XDIV
      DO 100 IPL=1,NUMEIG
         DO 10 I=1,NLIM
```

(Continued)

```fortran
10    A(I) = VECSML(I,IPL)
      CONTINUE
      CALL MAXMIN(NUMNP,A,YMAX,YMIN,JMAX,JMIN)
      YMAX = YMAX+0.25*ABS(YMAX)
      YMIN = YMIN-0.25*ABS(YMIN)
      YDIV = 6
      IF (YMIN1.LT.YMAX1) YMAX=YMAX1
      IF (YMIN1.LT.YMAX1) YMIN=YMIN1
      IF (YMIN1.LT.YMAX1) YDIV=NDIVY1
      YSTP = (YMAX-YMIN)/YDIV
      CALL BGNPL(IPL)
      CALL PAGE(8.,6.)
C-------THESE STATEMENTS ARE MACHINE DEPENDENT
C------- FOR THE CRAY-1
      ENCODE (20,4000,LABEL1) IPL
      ENCODE (26,4100,LABEL2) EIGSML(IPL)
C------- FOR THE H6000
      ENCODE (LABEL1,4000) IPL
      ENCODE (LABEL2,4100) EIGSML(IPL)
C------- FOR THE IBM 370
      CALL CORE (LABEL1,20)
      WRITE (99,4000) IPL
      WRITE (99,4100) EIGSML(IPL)
      CALL INTAXS
      CALL TITLE('.','XW.','EIGENVECTOR#.',7.,4.)
      CALL YAXANG(0.)
      CALL XINTIK
      CALL YINTIK
      CALL FRMTIK
      CALL XTICKS(2)
      CALL YTICKS(2)
      CALL HEADIN(LABEL1,0.167,2)
      CALL GRAF(XMIN,XSTP,XMAX,YMIN,YSTP,YMAX)
      CALL THKCUR(0.013,1)
      DO 50 IELNO=1,NUMEL
      ITYPE = NTYPE(IELNO)
      NND = NNODES(ITYPE)
      DO 30 I=1,NND
      J = ICON(I,IELNO)
      XP(I) = X(J)
      XX(I) = A(J)
30    CONTINUE
      INUM = (ABS(XP(NND)-XP(1))/TL)*500.
      ANUM = INUM
      DX = (XP(NND)-XP(1))/ANUM
      XQ(1) = XP(1)
      YQ(1) = YP(1)
      N1 = NND-1
      DO 40 I=2,INUM1
      XX = XX+DX
      IF (I.EQ.INUM1) XX=XP(NND)
      CALL POLYIN(N1,XP,YP,XX,YY)
      XQ(2) = XX
      YQ(2) = YY
      IF (XQ(1).LE.XMAX.AND.XQ(1).GE.XMIN.AND.
     $    XQ(2).LE.XMAX.AND.XQ(2).GE.XMIN.AND.
     $    YQ(1).LE.YMAX.AND.YQ(1).GE.YMIN.AND.
     $    YQ(2).LE.YMAX.AND.YQ(2).GE.YMIN)
     $    CALL CURVE(XQ,YQ,2,0)
      XQ(1) = XX
      YQ(1) = YY
40    CONTINUE
50    CONTINUE
      CALL RESET('THKCUR')
      CALL RESET('XTICKS')
      CALL RESET('YTICKS')
      CALL RESET('YINTIK')
      CALL RESET('XINTIK')
      CALL RESET('FRMTIK')
      CALL RESET('INTAXS')
      CALL ENDGR(0)
C--------- PLOT MESH
      CALL OREL(0.,-1.)
      CALL TITLE('#','#','#.',7.,-1.)
      CALL MESH1D(XMAX,XMIN)
      CALL HEIGHT(0.100)
      CALL MESSAG('MESH#.',3.4,0.010)
      CALL RESET('HEIGHT')
      CALL ENDGR(0)
      CALL ENDPL(0)
100   CONTINUE
      RETURN
C--------- PLOT 2-D EIGENVECTORS
200   CONTINUE
      DO 210 IPL=1,NUMEIG
      A(I) = VECSML(I,IPL)
210   CONTINUE
      CALL MAXMIN(NUMNP,A,VMAX,VMIN,JMAX,JMIN)
      CNTRSP = ABS((VMAX-VMIN)/10.)
      CONTR(1) = VMIN
      DO 220 I=2,11
      CONTR(I) = CONTR(I-1)+CNTRSP
220   CONTINUE
C--------- CALCULATE ELEMENT CENTROIDAL VALUES FOR EIGENVALUE IPL
      DO 300 IELNO=1,NUMEL
      ITYPE = NTYPE(IELNO)
      KN = ITYPE-20
      NND = NNODES(ITYPE)
      XC(IELNO) = 0.
      YC(IELNO) = 0.
      UC(IELNO) = 0.
      DO 250 I=1,NND
      J = ICON(I,IELNO)
      XC(IELNO) = XC(IELNO) + FCENT(I,KN)*X(J)
      YC(IELNO) = YC(IELNO) + FCENT(I,KN)*Y(J)
      UC(IELNO) = UC(IELNO) + FCENT(I,KN)*A(J)
250   CONTINUE
300   CONTINUE
C--------- SET UP PLOT
      CALL SETUP2(XMIN,XMAX,YMIN,YMAX,MAX,DAX,XOR,YOR,XSTEP,YSTEP)
      CALL TITLE('#','#','(#.',(#.',MAX,DAX)
C--------- THESE STATEMENTS ARE MACHINE DEPENDENT
C--------- FOR THE CRAY-1
      ENCODE (20,4000,LABEL1) IPL
      ENCODE (26,4100,LABEL2) EIGSML(IPL)
      ENCODE (29,4200,LABEL3) CNTRSP
C--------- FOR THE H6000
      ENCODE (LABEL1,4000) IPL
      ENCODE (LABEL2,4100) EIGSML(IPL)
      ENCODE (LABEL3,4200) CNTRSP
C--------- FOR THE IBM 370
      CALL CORE (LABEL1,20)
      CALL CORE (LABEL2,26)
      WRITE (99,4000) IPL
      WRITE (99,4100) EIGSML(IPL)
      CALL CORE (LABEL3,29)
      WRITE (99,4200) CNTRSP
      CALL HEADIN(LABEL1,0.165,3)
      CALL HEADIN(LABEL2,0.11,3)
      CALL HEADIN(LABEL3,0.11,3)
      CALL GRAF(XOR,XSTEP,XMAX,YOR,YSTEP,YMAX)
C--------- PLOT CONTOURS
      CALL CNTRPL(A,CONTR)
C--------- PLOT MAX AND MIN CONTOUR VALUES
      CALL MARKER(9)
      XQ(1) = X(JMIN)
      YQ(1) = Y(JMIN)
      XQ(2) = X(JMAX)
      YQ(2) = Y(JMAX)
      CALL CURVE(XQ,YQ,2,-1)
      CALL RLREAL(VMIN,-2,XQ(1),YQ(1))
      CALL RLREAL(VMAX,-2,XQ(2),YQ(2))
      CALL ENDPL(0)
400   CONTINUE
      RETURN
C
4000  FORMAT('EIGENVECTOR NO. ',I2,'#.')
4100  FORMAT('EIGENVALUE = ',E11.4,'#.')
4200  FORMAT('CONTOUR SPACING = ',E9.2,'#.')
      END
      SUBROUTINE BIVPLO
C
C     THIS ROUTINE PLOTS THE BOUNDARY-VALUE AND
C         INITIAL-BOUNDARY-VALUE RESULTS
C
      COMMON /GENRL/ NUMNP,NUMEL,NUMPP,NUMILC,NUMEBC,NUMNBC,MBAND
      COMMON /NODSOL/ IU(300),TAUXNP(300),TAUYNP(300)
      COMMON /SYSEQS/ K(300,50),F(300),A(300)
      REAL K
      COMMON /NODES/ X(300),Y(300)
      COMMON /ELMNTS/ NTYPE(200),NGPINT(200),NPHYS(200),ICON(8,200)
      COMMON /ELSPEC/ NNODES(30),INTDEF(30),IDPTAU(30)
      COMMON /ELFLUX/ XGP(4,200),YGP(4,200),TAUXGP(4,200),TAUYGP(4,200)
      COMMON /CENTRD/ UC(200),XC(200),YC(200),TAUXC(200),TAUYC(200)
      COMMON /GRAFIC/ XXMIN,XXMAX,NDIVX,YYMIN,YYMAX,NDIVY1,
     $               YYMIN2,YYMAX2,NDIVY2,TTMIN,TTMAX,NDIVT
      COMMON /PROBLM/ DIMEN,PROB,TITLE2(18)
      COMMON /INVAL/ INTRVL,NSTEPS,NTIME,TIME,DT,THETA,A0
      DIMENSION XP(8),YP(8),XQ(2),YQ(2),XD(800),FX(800),CONTR(11)
      DIMENSION LABEL1(15),LABEL2(15),LABEL3(15)
      LOGICAL PLOTS
      DATA BDVP/'BDVP'/, EIGP/'EIGP'/, XIBVP/'IBVP'/
      DATA ONED/'1DIM'/, TWOD/'2DIM'/
      DATA PLOTS/.FALSE./
      DATA EEND1/'END '/,EEND2/' END'/
C
      IF (DIMEN.EQ.TWOD) GO TO 400
C--------- PLOT 1-D FUNCTION AND FLUXES
      PLOTS = .TRUE.
      CALL MAXMIN(NUMNP,X,XMAX,XMIN,JMAX,JMIN)
      CALL MAXMIN(NUMNP,A,YMAX,YMIN,JMAX,JMIN)
      TL = XMAX-XMIN
      YMAX = YMAX+0.25*ABS(YMAX)
      YMIN = YMIN-0.25*ABS(YMIN)
      XDIV = NDIVX
      IF (XXMIN.LT.XXMAX) XMIN=XXMIN
      IF (XXMIN.LT.XXMAX) XMAX=XXMAX
      YDIV = NDIVY1
      IF (YYMIN.LT.YYMAX1) YMAX=YYMAX1
      IF (YYMIN.LT.YYMAX1) YMIN=YYMIN1
      XSTP = (XMAX-XMIN)/XDIV
```

(Continued)

```
      YSTP = (YMAX-YMIN)/YDIV
C--------PLOT FUNCTION
      CALL BGNPL(1)
      CALL PAGE(8.,6.)
      CALL PHYSOR(0.5,1.5)
      CALL HEIGHT(0.11)
      CALL INTAXS
      CALL TITLE('#.','#.','U#.',',7.,4.)
      CALL YAXANG(0.)
      CALL XINTIK
      CALL YINTIK
      CALL FRMTIK
      CALL XTICKS(2)
      CALL YTICKS(2)
C------ THESE STATEMENTS ARE MACHINE DEPENDENT
C------ FOR THE CRAY-1
      IF (PROB.EQ.XIBVP) ENCODE(55,4100,LABEL1)TIME
C------ FOR THE H6000
      IF (PROB.EQ.XIBVP) ENCODE(LABEL1,4100)TIME
C------ FOR THE IBM 370
      IF (PROB.EQ.XIBVP) CALL CORE(LABEL1,55)
      IF (PROB.EQ.XIBVP) WRITE (99,4100) TIME
      IF (PROB.EQ.XIBVP) CALL HEADIN(LABEL1,0.167,2)
      IF (PROB.EQ.BDVP)
$ CALL HEADIN('FINITE ELEMENT SOLUTION - FUNCTION#.',0.167,2)
      CALL HEADIN('APPROXIMATE SOLUTION U(X) WITHIN EACH ELEMENT#.',
     1  0.11,2)
      CALL GRAF(XMIN,XSTP,XMAX,YMIN,YSTP,YMAX)
      CALL GRID(1,1)
      CALL THKCUR(0.013,1)
      DO 50 IELND=1,NUMEL
      ITYPE = NTYPE(IELND)
      NND = NNODES(ITYPE)
      DO 30 I=1,NND
      J = ICON(I,IELND)
      XP(I) = X(J)
      YP(I) = A(J)
      CALL CURVE(XQ,YQ,2,0)
      XQ(1) = XX
      YQ(1) = YY
   30 CONTINUE
      INUM = (ABS(XP(NND)-XP(1))/TL)*500.
      ANUM = INUM
      DX = (XP(NND)-XP(1))/ANUM
      XX = XP(1)
      XQ(1) = XP(1)
      YQ(1) = YP(1)
      N1 = NND-1
      INUM1 = INUM-1
      DO 40 I=2,INUM1
      XX = XX+DX
      IF (I.EQ.INUM1) XX=XP(NND)
      CALL POLY(N1,XP,YP,XX,YY)
      XQ(2) = XX
      YQ(2) = YY
      IF(XQ(1).LE.XMAX.AND.XQ(1).GE.XMIN.AND.
     $    XQ(2).LE.XMAX.AND.XQ(2).GE.XMIN.AND.
     $    YQ(1).LE.YMAX.AND.YQ(1).GE.YMIN.AND.
     $    YQ(2).LE.YMAX.AND.YQ(2).GE.YMIN)
     $    CALL CURVE(XQ,YQ,2,0)
      XQ(1) = XX
      YQ(1) = YY
   40 CONTINUE
   50 CONTINUE
      CALL RESET('XTICKS')
      CALL RESET('YTICKS')
      CALL RESET('XINTIK')
      CALL RESET('YINTIK')
      CALL RESET('FRMTIK')

      CALL RESET('INTAXS')
      CALL RESET('HEIGHT')
      CALL RESET('THKCUR')
      CALL ENDGR(0)
      CALL DREL(0.,-1.)
      CALL TITLE('#.','#.','#.',',7.,-1.)
      CALL MESH1D(XMAX,XMIN)
      CALL HEIGHT(0.100)
      CALL MESSAG('MESH#.',3,4,0.010)
      CALL RESET('HEIGHT')
      CALL ENDGR(0)
      CALL ENDPL(0)
C--------ORDER FLUXES
      IC = 0
      DO 100 IELND=1,NUMEL
      ITYPE = NTYPE(IELND)
      NGPTAU = IOPTAU(ITYPE)
      DO 80 I=1,NGPTAU
      IC = IC+1
      XD(IC) = XGP(I,IELND)
      FX(IC) = TAUXGP(I,IELND)
   80 CONTINUE
  100 CONTINUE
      IF (IC.EQ.1) GO TO 210
      IC1 = IC-1
      DO 200 I=1,IC1
      I1 = I+1
      DO 190 J=I1,IC
      IF(XD(I).LE.XD(J)) GO TO 190
      XDUM = XD(J)
      XD(J) = XD(I)
      XD(I) = XDUM
      FDUM = FX(J)
      FX(J) = FX(I)
      FX(I) = FDUM
  190 CONTINUE
  200 CONTINUE
  210 CONTINUE
      CALL MAXMIN(IC,FX,YMAX,YMIN,JMAX,JMIN)
      YMAX = YMAX+0.25*ABS(YMAX)
      YMIN = YMIN-0.25*ABS(YMIN)
      YDIV = NDIV2
      IF (YMIN2.LT.YMAX2) YMAX=YMAX2
      IF (YMIN2.LT.YMAX2) YMIN=YMIN2
      YSTP = (YMAX-YMIN)/YDIV
C--------PLOT FLUX
      CALL BGNPL(2)
      CALL PAGE(8.,6.)
      CALL PHYSOR(0.5,1.5)
      CALL RDUC(0)
      CALL GRLC(2)
      CALL HEIGHT(0.11)
      CALL INTAXS
      CALL TITLE('#.','Xw.','#2a#0*DU/DX#.',7.,4.)
      CALL YAXANG(0.)
      CALL XINTIK
      CALL YINTIK
      CALL FRMTIK
      CALL XTICKS(2)
      CALL YTICKS(2)
C------ THESE STATEMENTS ARE MACHINE DEPENDENT
C------ FOR THE CRAY-1
      IF (PROB.EQ.XIBVP) ENCODE(51,4150,LABEL1)TIME
C------ FOR THE H6000
      IF (PROB.EQ.XIBVP) ENCODE(LABEL1,4150)TIME

C------ FOR THE IBM 370
      IF (PROB.EQ.XIBVP) CALL CORE(LABEL1,51)
      IF (PROB.EQ.XIBVP) WRITE (99,4150) TIME
      IF (PROB.EQ.XIBVP) CALL HEADIN(LABEL1,0.167,3)
      IF (PROB.EQ.BDVP)
$ CALL HEADIN('FINITE ELEMENT SOLUTION - FLUX#.',0.167,3)
      CALL HEADIN('STRAIGHT LINE SEGMENTS CONNECTING SUCCESSIVE GAUSS PO
1INT VALUES#.',0.11,3)
      CALL HEADIN('CIRCLES ARE NODAL VALUES INTERPOLATED FROM GAUSS POIN
1T VALUES#.',0.11,3)
      CALL GRAF(XMIN,XSTP,XMAX,YMIN,YSTP,YMAX)
      CALL GRID(1,1)
      CALL MARKHT(0.035)
      DO 250 IELND=1,NUMEL
      ITYPE = NTYPE(IELND)
      NND = NNODES(ITYPE)
      N1 = NND-1
      DO 230 I=1,N1
      XP(I) = XGP(I,IELND)
      YP(I) = TAUXGP(I,IELND)
  230 CONTINUE
      N2 = N1-1
      I1 = ICON(1,IELND)
      XQ(1) = X(I1)
      XX = XQ(1)
      CALL POLY(N2,XP,YP,XX,YY)
      YQ(1) = YY
      I2 = ICON(NND,IELND)
      XX = X(I2)
      XQ(2) = XX
      CALL POLY(N2,XP,YP,XX,YY)
      YQ(2) = YY
      CALL MARKER(1)
      IF(XQ(1).LE.XMAX.AND.XQ(1).GE.XMIN.AND.
     $    XQ(2).LE.XMAX.AND.XQ(2).GE.XMIN.AND.
     $    YQ(1).LE.YMAX.AND.YQ(1).GE.YMIN.AND.
     $    YQ(2).LE.YMAX.AND.YQ(2).GE.YMIN)
     $    CALL POLY(N2,XP,YP,XX,YY)
      YQ(2) = YY
  250 CONTINUE
      CALL THKCUR(0.013,1)
      CALL CURVE(XD,FX,IC,0)
      CALL RESET('XTICKS')
      CALL RESET('YTICKS')
      CALL RESET('XINTIK')
      CALL RESET('YINTIK')
      CALL RESET('FRMTIK')
      CALL RESET('INTAXS')
      CALL RESET('MARKHT')
      CALL RESET('THKCUR')
      CALL ENDGR(0)
C--------PLOT MESH
      CALL DREL(0.,-1.)
      CALL TITLE('#.','#.','#.',7.,-1.)
      CALL MESH1D(XMAX,XMIN)
      CALL HEIGHT(0.100)
      CALL MESSAG('MESH#.',3,4,0.010)
      CALL RESET('HEIGHT')
      CALL ENDGR(0)
      CALL ENDPL(0)
      RETURN
  400 CONTINUE
C--------PLOT 2-D FUNCTION : U
      CALL MAXMIN(NUMNP,A,UMAX,UMIN,JMAX,JMIN)
      CNTRSP = ABS((UMAX-UMIN)/IC.)
      CONTR(1) = UMIN
      DO 420 I=2,11
```

```fortran
      CONTR(I) = CONTR(I-1)+CNTRSP
  420 CONTINUE
C-----SET UP PLOT
      CALL SETUP2(XMIN,XMAX,YMIN,YMAX,HAX,DAX,XOR,YOR,XSTEP,YSTEP)
      CALL TITLE('#','#','#',.HAX,DAX)
C-----THESE STATEMENTS ARE MACHINE DEPENDENT
C-----FOR THE CRAY-1
      IF (PROB.EQ.XIBVP) ENCODE (43,4210,LABEL1) TIME
      ENCODE (29,4250,LABEL3) CNTRSP
C-----FOR THE H6000
      IF (PROB.EQ.XIBVP) ENCODE (LABEL1,4210) TIME
      ENCODE (LABEL3,4250) CNTRSP
C-----FOR THE IBM 370
      IF (PROB.EQ.XIBVP) CALL CORE (LABEL1,43)
      IF (PROB.EQ.XIBVP) WRITE (99,4210) TIME
      CALL CORE (LABEL3,29)
C-----PLOT CONTOURS
      CALL CNTRPL(A,CONTR)
C-----PLOT MAX AND MIN CONTOUR VALUES
      CALL MARKER(9)
      XQ(1) = X(JMIN)
      YQ(1) = Y(JMIN)
      XQ(2) = X(JMAX)
      YQ(2) = Y(JMAX)
      CALL CURVE(XQ,YQ,2,-1)
      CALL RLREAL(VMIN,-2,XQ(1),YQ(1))
      CALL RLREAL(VMAX,-2,XQ(2),YQ(2))
      CALL ENDPL(0)
C-----PLOT 2-D FLUX - ALPHAX*DU/DX
      CALL MAXMIN(NUMNP,TAUXNP,VMAX,VMIN,JMAX,JMIN)
      CNTRSP = ABS((VMAX-VMIN)/10.)
      CONTR(1) = VMIN
      DO 520 I=2,11
      CONTR(I) = CONTR(I-1)+CNTRSP
  520 CONTINUE
C-----SET UP PLOT
      CALL SETUP2(XMIN,XMAX,YMIN,YMAX,HAX,DAX,XOR,YOR,XSTEP,YSTEP)
      CALL TITLE('#','#','#',.HAX,DAX)
C-----THESE STATEMENTS ARE MACHINE DEPENDENT
C-----FOR THE CRAY-1
      IF (PROB.EQ.XIBVP) ENCODE (43,4210,LABEL1) TIME
      ENCODE (29,4250,LABEL3) CNTRSP
C-----FOR THE H6000
      IF (PROB.EQ.XIBVP) ENCODE (LABEL1,4210) TIME
      ENCODE (LABEL3,4250) CNTRSP
C-----FOR THE IBM 370
      IF (PROB.EQ.XIBVP) CALL CORE (LABEL1,43)
      IF (PROB.EQ.XIBVP) WRITE (99,4210) TIME
      CALL CORE (LABEL3,29)
C-----PLOT CONTOURS
      CALL CNTRPL(TAUXNP,CONTR)
C-----PLOT MAX AND MIN CONTOUR VALUES
      CALL MARKER(9)
      XQ(1) = X(JMIN)

      YQ(1) = Y(JMIN)
      XQ(2) = X(JMAX)
      YQ(2) = Y(JMAX)
      CALL CURVE(XQ,YQ,2,-1)
      CALL RLREAL(VMIN,-2,XQ(1),YQ(1))
      CALL RLREAL(VMAX,-2,XQ(2),YQ(2))
      CALL ENDPL(0)
C-----PLOT 2-D FLUX - ALPHAY*DU/DY
      CALL MAXMIN(NUMNP,TAUYNP,VMAX,VMIN,JMAX,JMIN)
      CNTRSP = ABS((VMAX-VMIN)/10.)
      CONTR(1) = VMIN
      DO 620 I=2,11
      CONTR(I) = CONTR(I-1)+CNTRSP
  620 CONTINUE
C-----SET UP PLOT
      CALL SETUP2(XMIN,XMAX,YMIN,YMAX,HAX,DAX,XOR,YOR,XSTEP,YSTEP)
      CALL TITLE('#','#','#',.HAX,DAX)
C-----THESE STATEMENTS ARE MACHINE DEPENDENT
C-----FOR THE CRAY-1
      IF (PROB.EQ.XIBVP) ENCODE (43,4210,LABEL1) TIME
      ENCODE (29,4250,LABEL3) CNTRSP
C-----FOR THE H6000
      IF (PROB.EQ.XIBVP) ENCODE (LABEL1,4210) TIME
      ENCODE (LABEL3,4250) CNTRSP
C-----FOR THE IBM 370
      IF (PROB.EQ.XIBVP) CALL CORE (LABEL1,43)
      IF (PROB.EQ.XIBVP) WRITE (99,4210) TIME
      CALL CORE (LABEL3,29)
C-----PLOT CONTOURS
      CALL CNTRPL(TAUYNP,CONTR)
C-----PLOT MAX AND MIN CONTOUR VALUES
      CALL MARKER(9)
      XQ(1) = X(JMIN)
      YQ(1) = Y(JMIN)
      XQ(2) = X(JMAX)
      YQ(2) = Y(JMAX)
      CALL CURVE(XQ,YQ,2,-1)
      CALL RLREAL(VMIN,-2,XQ(1),YQ(1))
      CALL RLREAL(VMAX,-2,XQ(2),YQ(2))
      CALL ENDPL(0)
      RETURN
C
 4100 FORMAT('FINITE ELEMENT SOLUTION AT TIME',1PE9.2,'   -   FUNCTION#')
 4150 FORMAT('FINITE ELEMENT SOLUTION AT TIME',1PE9.2,'   -   FLUX#')
 4210 FORMAT('FINITE ELEMENT SOLUTION AT TIME',1E9.2,'#,#')
 4250 FORMAT('CONTOUR SPACING = ',1E9.2,'#,#')
      END
      SUBROUTINE TIMPLT
C
C     THIS ROUTINE PLOTS THE FUNCTION VS TIME FOR UP TO THREE NODES
C
      COMMON /INVAL/ INTRVL,NSTEPS,NTIME,TIME,DT,THETA,AO
      COMMON /TTMPLT/ TPLOT(501),FTPLOT(3,501),NNODEI(3)
      COMMON /GRAFIC/ XXMIN,XXMAX,NDIVX,YYMIN,YYMAX1,NDIVY1,
     $                YYMIN2,YYMAX2,NDIVY2,TTMIN,TTMAX,NDIVT
      COMMON /KEYWRD/ AKEY,PLTKEY
      DATA XNOPLT/NPLO/
      DIMENSION TP(501),FP(501),LABEL1(3)
C
      IF (PLTKEY.EQ.XNOPLT) RETURN

      CALL MAXMIN(NTIME,TPLOT,TMAX,TMIN,JMAX,JMIN)
      TL = TMAX-TMIN
      TDIV = NDIVT
      IF (TTMIN.LT.TTMAX) TMIN=TTMIN
      IF (TTMIN.LT.TTMAX) TMAX=TTMAX
      TSTP = (TMAX-TMIN)/TDIV
      DO 200 IPLOT=1,3
      NNDE = NNODEI(IPLOT)
      IF (NNDE.EQ.0) GO TO 200
      NLAST = 0
      DO 10 I=1,NTIME
      IF (TPLOT(I).LT.TMIN.OR.TPLOT(I).GT.TMAX) GO TO 10
      NLAST = NLAST+1
      TP(NLAST) = TPLOT(I)
      FP(NLAST) = FTPLOT(IPLOT,I)
   10 CONTINUE
      CALL MAXMIN(NTIME,FP,YMAX,YMIN,JMAX,JMIN)
      YMAX = YMAX +0.25*ABS(YMAX)
      YMIN = YMIN -0.25*ABS(YMIN)
      YDIV = NDIVY1
      IF (YYMIN1.LT.YYMAX1) YMAX=YYMAX1
      IF (YYMIN1.LT.YYMAX1) YMIN=YYMIN1
      YSTP = (YMAX-YMIN)/YDIV
      CALL BGNPL(IPLOT)
      CALL PAGE(8.,6.)
      CALL PHYSOR(0.5,1.5)
      CALL HEIGHT(0.11)
      CALL INTAXS
      CALL TITLE('#','   TIME#','U#',.7,.4.)
      CALL YAXANG(0.)
      CALL XINT1K
      CALL YINT1K
      CALL FRMT1K
      CALL XTICKS(2)
      CALL YTICKS(2)
C-----THESE STATEMENTS ARE MACHINE DEPENDENT
C-----FOR THE CRAY-1
      ENCODE(10,4000,LABEL1) NNDE
C-----FOR THE H6000
      ENCODE(LABEL1,4000) NNDE
C-----FOR THE IBM 370
      WRITE (99,4000) NNDE
      CALL CORE(LABEL1,10)
      CALL HEADIN('FINITE ELEMENT SOLUTION#',.0.167,3)
      CALL HEADIN('FUNCTION VS TIME#',.0.11,3)
      CALL HEADIN(LABEL1,0.11,3)
      CALL GRAF(TMIN,TSTP,TMAX,YMIN,YSTP,YMAX)
      CALL GRID(1,1)
      CALL THKCUR(0.013,1)
      CALL CURVE(TP,FP,NLAST,0)
      CALL RESET('XTICKS')
      CALL RESET('YTICKS')
      CALL RESET('XINT1K')
      CALL RESET('YINT1K')
      CALL RESET('FRMT1K')
      CALL RESET('INTAXS')
      CALL RESET('HEIGHT')
      CALL RESET('THKCUR')
      CALL ENDPL(0)
  200 CONTINUE
      WRITE (6,3000)
      RETURN
C
 3000 FORMAT(///,'*****POSTPROCESSING COMPLETE.  RETURN FOR NEXT PROB.')
 4000 FORMAT('NODE',I4,'#',)
      END
```

(Continued)

```fortran
      SUBROUTINE MAXMIN(N,X,XMAX,XMIN,JMAX,JMIN)
C
C     THIS ROUTINE FINDS MAX AND MIN OF THE X VECTOR
C
C
      DIMENSION X(1)
C
      JMAX = 1
      JMIN = 1
      XMAX = -1.E20
      XMIN = 1.E20
      DO 50 I=1,N
      IF (X(I).LT.XMIN) JMIN=I
      IF (X(I).LT.XMIN) XMIN=X(1)
      IF (X(I).GT.XMAX) JMAX=I
      IF (X(I).GT.XMAX) XMAX=X(1)
   50 CONTINUE
      RETURN
C
      END
      SUBROUTINE MESH1D(XMAX,XMIN)
C
C     THIS ROUTINE PLOTS THE 1-D MESH
C
C
      COMMON /ELMNTS/ NTYPE(200),NGPINT(200),NPHYS(200),ICON(8,200)
      COMMON /GENRL/ NUMNP,NUMEL,NUMPP,NUMILC,NUMEBC,NUMBC,MBAND
      COMMON /NODES/ X(300),Y(300)
      COMMON /ELSPEC/ NNODES(30),INTDEF(30),IDPTAU(30)
      DIMENSION XQ(2),YQ(2)
C
      XSTEP = (XMAX-XMIN)/7
      CALL GRAPH(XMIN,XSTEP,0.,1.)
      CALL HEIGHT(0.100)
      DY = 0.04
      DYY = 0.08
      DO 50 IELNO=1,NUMEL
      ITYPE = NTYPE(IELNO)
      NND = NNODES(ITYPE)
      NND1 = NND-1
      DO 30 IJ=1,NND,NND1
      I = ICON(IJ,IELNO)
      J = ICON(N-1,IELNO)
      XQ(1) = X(I)
      XQ(2) = X(J)
      CALL MARKER(10)
      CALL CURVE(XQ,YQ,2,1)
      XQI = XQ(1)
      XQJ = XQ(2)
      Y1 = 0.40
      DO 20 II=1,3
      YQ(1) = Y1
      YQ(2) = Y1+DY
      Y1 = Y1+DYY
      CALL CURVE(XQ,YQ,2,0)
   20 CONTINUE
   30 CONTINUE
      CALL MARKHT(0.035)
      YQ(1) = 0.30
      YQ(2) = 0.30
      DO 40 N=1,NND1
      CALL RLINT(1,XQI,0.15)
      CALL RLINT(J,XQJ,0.15)
   40 CONTINUE
   50 CONTINUE
```

```fortran
      CALL RESET('HEIGHT')
      CALL RESET('MARKER')
      CALL RESET('MARKHT')
      RETURN
C
      END
      SUBROUTINE MESH2D(XOR,YOR,XMAX,YMAX,XSTEP,YSTEP)
C
C     THIS ROUTINE PLOTS THE 2-D MESH
C
C
      COMMON /ELMNTS/ NTYPE(200),NGPINT(200),NPHYS(200),ICON(8,200)
      COMMON /GENRL/ NUMNP,NUMEL,NUMPP,NUMILC,NUMEBC,NUMBC,MBAND
      COMMON /NODES/ X(300),Y(300)
      COMMON /ELSPEC/ NNODES(30),INTDEF(30),IDPTAU(30)
      DIMENSION XQ(3),YQ(3),XS(51),YS(51)
C
C---- DRAW THE OUTLINE (SOLID) AND ELEMENTS(DOTTED)
      CALL GRAF(XOR,XSTEP,XMAX,YOR,YSTEP,YMAX)
C---- SEARCH ALL ELEMENTS FOR EXTERIOR AND INTERIOR SIDES
C         AND PLOT EACH SIDE OF EACH ELEMENT
      DO 400 IELNO1 = 1,NUMEL
      ITYPE1 = NTYPE(IELNO1)
      NND1 = NNODES(ITYPE1)
      NSIDE1 = NND1
      DO 380 ISIDE1 = 1,NSIDE1
      IF (NND1.GT.4) NSIDE1=NSIDE1/2
      IS1 = ICON(ISIDE1,IELNO1)
      ISD = ISIDE1+1
      IF (ISIDE1.EQ.NSIDE1) ISD=1
      JS1 = ICON(ISD,IELNO1)
      DO 300 IELNO2 = 1,NUMEL
      IF (IELNO2.EQ.IELNO1) GO TO 300
      ITYPE2 = NTYPE(IELNO2)
      NND2 = NNODES(ITYPE2)
      NSIDE2 = NND2
      DO 250 ISIDE2 = 1,NSIDE2
      IF (NND2.GT.4) NSIDE2=NSIDE2/2
      IS2 = ICON(ISIDE2,IELNO2)
      ISD2 = ISIDE2+1
      IF (ISIDE2.EQ.NSIDE2) ISD2=1
      JS2 = ICON(ISD2,IELNO2)
      CALL DASH
      ICNT = 1
      IF (((IS1.EQ.IS2.AND.JS1.EQ.JS2).OR.
     $     (IS1.EQ.JS2.AND.JS1.EQ.IS2)).AND.
     $     (IELNO2.LT.IELNO1)) GO TO 380
      IF ((IS1.EQ.IS2.AND.JS1.EQ.JS2).OR.
     $    (IS1.EQ.JS2.AND.JS1.EQ.IS2)) GO TO 310
      CALL RESET('DASH')
      ICNT = 0
  250 CONTINUE
  300 CONTINUE
  310 CONTINUE
      IF (NND1.GT.4) GO TO 340
C---- PLOT FOR ELEMENTS 21 AND 23
      XQ(1) = X(IS1)
      XQ(2) = X(JS1)
      YQ(1) = Y(IS1)
      YQ(2) = Y(JS1)
      CALL MARKER(10)
      CALL CURVE(XQ,YQ,2,1)
      GO TO 380
C---- PLOT FOR ELEMENTS 22 AND 24
  340 CONTINUE
      KS1 = ICON(ISIDE1*NSIDE1,IELNO1)
      XQ(1) = X(IS1)
      XQ(2) = X(KS1)
```

```fortran
      XQ(3) = X(JS1)
      YQ(1) = Y(IS1)
      YQ(2) = Y(KS1)
      YQ(3) = Y(JS1)
      CALL MARKER(10)
      IF (ICNT.EQ.0) GO TO 950
C----      PLOT INTERIOR
      CALL CURVE(XQ,YQ,3,1)
      GO TO 380
C----      PLOT EXTERIOR
  350 CONTINUE
      TH1 = ATAN2((YQ(2)-YQ(1)),(XQ(2)-XQ(1)))
      TH2 = ATAN2((YQ(3)-YQ(2)),(XQ(3)-XQ(2)))
      IF (ABS(TH2-TH1).LE.0.1) GO TO 370
      DO 360 INC=1,51
      AINC = INC-1
      S = -1.0+AINC*0.04
      XS(INC) = -S*(1.-S)/2*XQ(1)+(1.+S)*XQ(2)+
     $          S*(1.+S)/2*XQ(3)
      YS(INC) = -S*(1.-S)/2*YQ(1)+(1.-S)*(1.+S)*YQ(2)+
     $          S*(1.+S)/2*YQ(3)
  360 CONTINUE
      CALL CURVE(XS,YS,51,25)
      GO TO 380
  370 CONTINUE
      CALL CURVE(XQ,YQ,3,1)
  380 CONTINUE
  400 CONTINUE
C---- PLOT NODE NUMBERS
      CALL MARKHT(.055)
      DO 450 N = 1,NUMNP
      XP = X(N)-0.125*XSTEP
      YP = Y(N)-0.125*YSTEP
      CALL RLINT(N,XP,YP)
  450 CONTINUE
      CALL RESET('MARKHT')
      CALL RESET('DASH')
      CALL RESET('MARKER')
  500 CONTINUE
      RETURN
C
      END
      SUBROUTINE POLY(N,X,XX,YY)
C
C     THIS ROUTINE EVALUATES AN NTH DEGREE POLYNOMIAL AT XX.
C     THE VALUE OF THE POLYNOMIAL IS RETURNED IN YY.
C
      DIMENSION X(1),Y(1)
C
      N1 = N+1
      YY = 0.
      DO 20 I=1,N1
      T = 1.
      DO 10 J = 1,N1
      IF (I.NE.J) T=T*(XX-X(J))/(X(I)-X(J))
   10 CONTINUE
      YY = YY+T*Y(I)
   20 CONTINUE
      RETURN
C
      END
      SUBROUTINE CNTRPLA(A,CONTR)
C
C     THIS ROUTINE PLOTS THE EXTERIOR OF THE MODEL AND
C        THE CONTOURS IN EACH ELEMENT
C
C
```

(Continued)

```
      COMMON /ELMNTS/ NTYPE(200),NGPINT(200),NPHYS(200),ICON(8,200)
      COMMON /GENRL/ NUMNP,NUMEL,NUMPP,NUMILC,NUMEBC,NUMBC,MBAND
      COMMON /NODES/ X(300),Y(300)
      COMMON /ELSPEC/ NNODES(30),INTDEF(30),IOPTAU(30)
      DIMENSION XQ(3),YQ(3),XX(4),YY(4),FF(4),A(1),CONTR(11)
      DIMENSION NDS(30),XS(51),YS(51)
      DATA NDS /20*0,3,3,4,4,6*0/
C-----
C-----DRAW THE EXTERIOR
      DO 400 IELNO1 = 1, NUMEL
      ITYPE1 = NTYPE(IELNO1)
      NND1 = NNODES(ITYPE1)
      NSIDE1 = NND1
      IF (NND1.GT.4) NSIDE1=NSIDE1/2
      DO 380 ISIDE1 = 1,NSIDE1
      IS1 = ICON(ISIDE1,IELNO1)
      ISD = ISIDE1+1
      IF (ISIDE1.EQ.NSIDE1) ISD=1
      JS1 = ICON(ISD,IELNO1)
      DO 300 IELNO2 = 1,NUMEL
      IF (IELNO2.EQ.IELNO1) GO TO 300
      ITYPE2 = NTYPE(IELNO2)
      NND2 = NNODES(ITYPE2)
      NSIDE2 = NND2
      IF (NND2.GT.4) NSIDE2=NSIDE2/2
      DO 250 ISIDE2 = 1, NSIDE2
      IS2 = ICON(ISIDE2,IELNO2)
      ISD2 = ISIDE2+1
      IF (ISIDE2.EQ.NSIDE2) ISD2=1
      JS2 = ICON(ISD2,IELNO2)
      IF (IS1.EQ.IS2.AND.JS1.EQ.JS2).OR.
     $ (IS1.EQ.JS2.AND.JS1.EQ.IS2)) GO TO 380
250   CONTINUE
300   CONTINUE
      IF (NND1.GT.4) GO TO 340
C-----PLOT FOR ELEMENTS 21 AND 23
      XQ(1) = X(IS1)
      XQ(2) = X(JS1)
      YQ(1) = Y(IS1)
      YQ(2) = Y(JS1)
      CALL MARKER(10)
      CALL CURVE(XQ,YQ,2,1)
      GO TO 380
340   CONTINUE
C-----PLOT FOR ELEMENTS 22 AND 24
      KS1 = ICON(ISIDE1+NSIDE1,IELNO1)
      XQ(1) = X(IS1)
      XQ(2) = X(KS1)
      XQ(3) = X(JS1)
      YQ(1) = Y(IS1)
      YQ(2) = Y(KS1)
      YQ(3) = Y(JS1)
      CALL MARKER(10)
      TH1 = ATAN2((YQ(2)-YQ(1)),(XQ(2)-XQ(1)))
      TH2 = ATAN2((YQ(3)-YQ(2)),(XQ(3)-XQ(2)))
      IF (ABS(TH2-TH1).LE.0.1) GO TO 370
      DO 360 INC=1,51
      AINC= INC-1
      S = -1.0+AINC*0.04
      XS(INC) = -S*(1.-S)/2.*XQ(1)+(1.-S)*(1.+S)*XQ(2)+
     $          S*(1.+S)/2.*XQ(3)
      YS(INC) = -S*(1.-S)/2.*YQ(1)+(1.-S)*(1.+S)*YQ(2)+
     $          S*(1.+S)/2.*YQ(3)
360   CONTINUE
      CALL CURVE(XS,YS,51,25)
      GO TO 380
370   CONTINUE
      CALL CURVE(XQ,YQ,3,1)
380   CONTINUE
400   CONTINUE
C-----PLOT CONTOURS
C-----CHECK FOR ZERO CONTOURS (DON'T PLOT IF DIFF<1E-7)
      DEL = ABS(CONTR(2)-CONTR(1))
      IF (DEL.LE.1.E-7) RETURN
      DO 600 IELNO=1,NUMEL
      ITYPE = NTYPE(IELNO)
      NSIDE = NDS(ITYPE)
      IF (NSIDE.EQ.0) CALL ERROR(35,IELNO)
C-----STORE NODAL VALUES IN ELEMENT NODAL VECTORS
      DO 420 NOD=1,NSIDE
      I = ICON(NOD,IELNO)
      XX(NOD) = X(I)
      YY(NOD) = Y(I)
      FF(NOD) = A(I)
420   CONTINUE
C-----PLOT CONTOURS IN EACH ELEMENT
C     (TWO CONTOURS OF SAME VALUE CAN OCCUR IN QUAD)
      DO 520 ICNT=2,10
      IF (ICNT.EQ.2.OR.ICNT.EQ.6.OR.ICNT.EQ.10) CALL DASH
      IS = 0
      V = CONTR(ICNT)
      DO 500 ISIDE=1,NSIDE
C-----CHECK SIDE ISIDE FOR CONTOUR VALUE
      NOD1 = ISIDE
      NOD2 = ISIDE+1
      IF (ISIDE.EQ.NSIDE) NOD2=1
      IF ((FF(NOD1).GT.V.AND.FF(NOD2).GT.V).OR.
     $ (FF(NOD1).LT.V.AND.FF(NOD2).LT.V))GO TO 460
      IS = IS+1
      V1 = V-FF(NOD1)
      V2 = V-FF(NOD2)
      V3 = FF(NOD2)-FF(NOD1)
      IF (ABS(V3).LT.1.E-7) GO TO 440
      XQ(IS) = (XX(NOD2)*V1-XX(NOD1)*V2)/V3
      YQ(IS) = (YY(NOD2)*V1-YY(NOD1)*V2)/V3
      GO TO 450
440   CONTINUE
      XQ(IS) = (XX(NOD2)+XX(NOD1))/2.
      YQ(IS) = (YY(NOD2)+YY(NOD1))/2.
450   CONTINUE
C-----PLOT CONTOUR
      IF (IS.EQ.2) CALL CURVE(XQ,YQ,2,0)
      IF (IS.EQ.2) IS=0
460   CONTINUE
      CALL RESET('DASH')
500   CONTINUE
520   CONTINUE
600   CONTINUE
      RETURN
C-----
      END
```

Index

Name Index

Page numbers in boldface indicate references in text; page numbers followed by n refer to material in footnotes.

Subject Index

Page numbers in boldface indicate major discussions; page numbers followed by n refer to material in footnotes.